SECOND EDITION
MICHIGAN FLORA
UPPER PENINSULA

Calypso bulbosa

STEVE W. CHADDE

ABBREVIATIONS

BOTANICAL
(END) endangered in Michigan
(THR) threatened in Michigan
APG III Angiosperm Phylogeny Group III system of 2009
CC coefficient of conservatism (p. 3)
spp. species (plural)
subsp. subspecies
syn synonym
var. variety
× hybrid (or times)
* An asterisk following a species name in the key means that the species is listed under Additional Species and is not fully described (usually introduced or adventive species of limited occurrence in the Upper Peninsula)

GEOGRAPHICAL
UP Upper Peninsula
IR Isle Royale
WUP western Upper Peninsula
CUP central Upper Peninsula
EUP eastern Upper Peninsula
c central
n northern
ne northeastern
nw northwestern
s southern
se southeastern
sw southwestern
w western

MEASUREMENT
cm centimeter
dm decimeter
m meter
mm millimeter

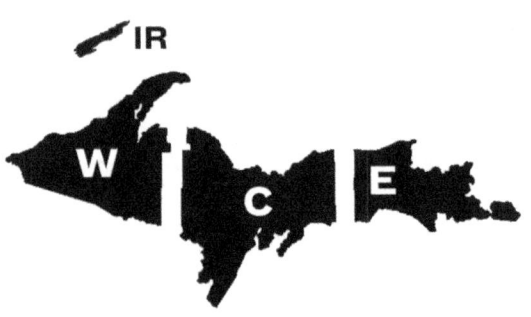

Upper Peninsula geographic regions
IR Isle Royale
W western UP (Baraga, Gogebic, Houghton, Iron, Keweenaw, Ontonagon counties)
C central UP (Alger, Delta, Dickinson, Marquette, Menominee, Schoolcraft counties)
E eastern UP (Chippewa, Luce, Mackinac counties)
see state map, page 4

MICHIGAN FLORA: UPPER PENINSULA
Second Edition

Steve W. Chadde

Copyright © 2016 by Steve W. Chadde
All rights reserved.
Printed in the United States of America.

ISBN: 978-1-951682-06-4

Grateful acknowledgment is given to the Biota of North America Program (*www.bonap.org*) for permission to use their data to determine plant species distribution in the Upper Peninsula.

The author can be reached via email: steve@chadde.net

VERSION 2.2 (04/12/2020)

CONTENTS

INTRODUCTION 1

descriptive flora

FERNS AND FERN RELATIVES

Aspleniaceae | Spleenwort Family 5
Athyriaceae | Lady Fern Family 6
Blechnaceae | Chain Fern Family 7
Cystopteridaceae | Bladder Fern Family 7
Dennstaedtiaceae | Bracken Fern Family.... 9
Dryopteridaceae | Wood Fern Family... 10
Equisetaceae | Horsetail Family 12
Isoetaceae | Quillwort Family 16
Lycopodiaceae | Clubmoss Family...... 16
Onocleaceae | Sensitive Fern Family.... 21
Ophioglossaceae | Adder's-Tongue Fam... 22
Osmundaceae | Royal Fern Family 26
Polypodiaceae | Polypody Fern Family... 27
Pteridaceae | Maidenhair Fern Family... 28
Selaginellaceae | Selaginella Family..... 29
Thelypteridaceae | Marsh Fern Family .. 30
Woodsiaceae | Cliff Fern Family 31

CONIFERS

Cupressaceae | Cypress Family 33
Pinaceae | Pine Family................ 34
Taxaceae | Yew Family 36

DICOTS

Adoxaceae | Muskroot Family 37
Amaranthaceae | Amaranth Family 39
Anacardiaceae | Sumac Family 44
Apiaceae | Carrot Family 46
Apocynaceae | Dogbane Family....... 53
Aquifoliaceae | Holly Family 56
Araliaceae | Ginseng Family........... 56
Aristolochiaceae | Birthwort Family 58
Asteraceae | Aster Family............. 58
Balsaminaceae | Touch-Me-Not Family .. 115
Berberidaceae | Barberry Family 116
Betulaceae | Birch Family 117
Boraginaceae | Borage Family......... 119
Brassicaceae | Mustard Family........ 126
Cabombaceae | Watershield Family ... 143

Cactaceae | Cactus Family 143
Campanulaceae | Bellflower Family ... 144
Cannabaceae | Hempfamily 147
Caprifoliaceae | Honeysuckle Family... 148
Caryophyllaceae | Pink Family........ 152
Celastraceae | Bittersweet Family 164
Ceratophyllaceae | Hornwort Family .. 165
Cistaceae | Rock-Rose Family 166
Cleomaceae | Cleome Family......... 167
Convolvulaceae | Morning-Glory Fam.... 167
Cornaceae | Dogwood Family 168
Crassulaceae | Stonecrop Family...... 170
Cucurbitaceae | Cucumber Family..... 171
Dipsacaceae | Teasel Family 171
Droseraceae | Sundew Family 172
Elaeagnaceae | Oleaster Family 173
Elatinaceae | Waterwort Family....... 174
Ericaceae | Heath Family 175
Euphorbiaceae | Spurge Family 183
Fabaceae | Pea Family 188
Fagaceae | Beech Family............. 196
Gentianaceae | Gentian Family 198
Geraniaceae | Geranium Family 200
Grossulariaceae | Currant Family..... 202
Haloragaceae | Water-Milfoil Family .. 205
Hamamelidaceae | Witch-Hazel Family... 207
Hypericaceae | St. John's-Wort Family... 207
Juglandaceae | Walnut Family 210
Lamiaceae | Mint Family............. 211
Lentibulariaceae | Bladderwort Family ... 221
Limnanthaceae | Meadowfoam Family .. 224
Linaceae | Flax Family................ 225
Linderniaceae | Lindernia Family 226
Lythraceae | Loosestrife Family 226
Malvaceae | Mallow Family 227
Melastomataceae | Melastome Family... 229
Menispermaceae | Moonseed Family .. 230
Menyanthaceae | Buckbean Family ... 230
Molluginaceae | Carpetweed Family .. 230
Montiaceae | Montia Family 230
Moraceae | Mulberry Family 231

Dicots, cont.

Myricaceae | Bayberry Family 231
Nyctaginaceae | Four-O'clock Family . 232
Nymphaeaceae | Water-Lily Family ... 232
Oleaceae | Olive Family 234
Onagraceae | Evening-Primrose Family... 235
Orobanchaceae | Broom-Rape Family.. 240
Oxalidaceae | Wood-Sorrel Family 245
Papaveraceae | Poppy Family......... 246
Penthoraceae | Penthorum Family 248
Phrymaceae | Lopsed Family 248
Plantaginaceae | Plantain Family...... 250
Polemoniaceae | Phlox Family 260
Polygalaceae | Milkwort Family 262
Polygonaceae | Buckwheat Family 264
Portulacaceae | Purslane Family 273
Primulaceae | Primrose Family 273
Ranunculaceae | Buttercup Family 276
Rhamnaceae | Buckthorn Family 286
Rosaceae | Rose Family.............. 287
Rubiaceae | Madder Family.......... 309
Rutaceae | Rue Family............... 313
Salicaceae | Willow Family........... 313
Santalaceae | Sandalwood Family 321
Sapindaceae | Soapberry Family 322
Sarraceniaceae | Pitcherplant Family .. 324
Saxifragaceae | Saxifrage Family...... 325
Scrophulariaceae | Figwort Family 327
Solanaceae | Potato Family........... 328
Thymelaeaceae | Mezereum Family.... 331
Ulmaceae | Elm Family.............. 332
Urticaceae | Nettle Family........... 334
Verbenaceae | Verbena Family 335
Violaceae | Violet Family 337
Vitaceae | Grape Family 343

Monocots

Acoraceae | Calamus Family 344
Alismataceae | Water-Plantain Family ... 344
Araceae | Arum Family.............. 346
Commelinaceae | Spiderwort Family .. 349
Cyperaceae | Sedge Family........... 349
Dioscoreaceae | Yam Family 422
Eriocaulaceae | Pipewort Family...... 422
Hydrocharitaceae | Tape-Grass Family .. 422
Hypoxidaceae | Liliid Monocot Family... 425
Iridaceae | Iris Family 425
Juncaceae | Rush Family............. 427
Juncaginaceae | Arrow-Grass Family .. 434
Liliaceae | Lily Family 434
Orchidaceae | Orchid Family......... 443
Poaceae | Grass Family 457
Pontederiaceae | Pickerelweed Family.... 510
Potamogetonaceae | Pondweed Family ... 511
Scheuchzeriaceae | Scheuchzeria Fam. ... 518
Smilacaceae | Greenbrier Family...... 519
Typhaceae | Cat-Tail Family 520
Xyridaceae | Yellow-Eyed-Grass Family .. 523

FAMILY KEYS 525
REFERENCES 549
GLOSSARY 551
INDEX 559

INTRODUCTION

MICHIGAN FLORA: UPPER PENINSULA is a field-oriented guide to essentially all of the vascular plants ferns and fern relatives, conifers, and flowering plants (dicots, monocots) considered native to the Upper Peninsula (UP), plus nearly all of the vascular plant species considered as introduced or adventive to the peninsula. Excluded are a number of ornamental and cultivated plants that do not typically spread from their planting sites. The focus of the Flora is to provide an up-to-date identification guide to every vascular plant species in the UP that is reproducing without human assistance, using a combination of keys, descriptions, habitat information, illustrations, and distribution data. Included are nearly 1,900 species within ca. 670 genera and 123 plant families.

Michigan's plant life has been the subject of numerous studies for nearly 150 years. Recently, the Field Manual of Michigan Flora (2012) by Anton Reznicek and the late Edward Voss updated Voss's earlier 3-volume Michigan Flora, originally published during the 1970s 90s. However, as in Voss's original work, the Field Manual excluded ferns and fern relatives, so prominent throughout the forests, wetlands, and rocky places of the Upper Peninsula. Most of the information included in the Field Manual (and with the addition of the fern groups) is available at the website maintained by the University of Michigan Herbarium: www.michiganflora.net, and I have adapted a number of their excellent keys in this Flora. My hope is that Michigan Flora: Upper Peninsula will be a useful field reference to the plant life of the Upper Peninsula, and by including illustrations for nearly all the UP's species, may be more easily accessible for a larger number of students of the region's flora.

Arrangement of the Taxa

All plants treated in this flora belong to three informal groups, presented in order: (1) ferns and fern relatives, (2) gymnosperms (conifers), and, by far the largest group, (3) angiosperms. The angiosperms are subdivided into two classes, the dicotyledons or "dicots" (sometimes termed Magnoliopsida) and the monocotyledons or "monocots" (sometimes termed Liliopsida). These subdivision names derive from the observation that the dicots most often have two cotyledons, or embryonic leaves, within each seed. The monocots usually have only one, but the rule is not absolute either way. From a diagnostic point of view, the number of cotyledons is neither a particularly handy nor a reliable character, but provides a simple way to organize plant families into smaller groups. Dicots include many familiar trees, shrubs, and "wildflowers," such as those of the Aster Family; the monocots include the large grass and sedge families, and also smaller familes such as Juncaceae, Orchidaceae, and Typhaceae. The family keys, beginning on page 525, are intended to be used for completely unknown specimens. However, recognition of plant family characters will greatly speed the process of plant identification.

Within each of the divisions, families are listed alphabetically. Under each family, genera and species are also listed alphabetically. If there is more than one genus in a family, a key is provided to the genera. Likewise, if there is more than one species in a genus, a key to the species is provided. For each species treated in the text, the following information is provided: scientific name, common name, synonyms (other formerly accepted scientific names), whether native or introduced in the state or of conservation concern (state-listed as endangered or threatened), a description of the plant's vegetative, floral and fruiting characters, flowering period, habitat, wetland indicator status (page 2), and coefficient of conservatism (page 3). Data from the BONAP database (Biota of North America Program, www.bonap.org), were used to determine each species' occurrence within one or more geographic region's of the UP (based on county boundaries): **IR** - Isle Royale; **W** - western UP; **C** - central UP; **E** - eastern UP (see map, page 4). Known occurrences within a region are indicated by bold type. Finally, most descriptions are accompanied by a line drawing showing the main features of each species.

Placement of monocots and dicots genera within families, with several exceptions, follows that of the Angiosperm Phylogeny Group III system (AGP III) of 2009. The APG was formed in the late 1990s, when researchers from major institutions around the world gathered with the goal of providing a modern, widely accepted classification of angiosperms. Their first attempt at a new system was published in 1998 (the agp system). To date, two revisions have been published, in 2003 (AGP II) and in 2009 (AGP III), each superseding the previous system.

The major exception to APG III is the retention of the traditional Liliaceae (Lily Family) to facilitate field use of the Flora; however the APG III families segregated from the traditional Liliaceae are noted for each genera and presented in tabular form on page 3. Another exception is the retention of the Dipsacaceae (Teasel Family), and not including the

UP's two genera within Caprifoliaceae (Honeysuckle Family). Any other deviations from APG III regarding generic placement are noted in the text.

Similarly, fern families have been updated to reflect recent realignments of families and genera (Christenhusz et al. 2011, Smith et al. 2006). I have used the term "fern relatives" for the clubmosses, spike-mosses and quillworts, even though genetic research suggests that the relationship to true ferns is not close.

Nomenclature of genera and species is not based on a single source, but in general, conforms to that of the published volumes of The Flora of North America series (*www.efloras.org*), the BONAP database (Biota of North America Program, *www.bonap.org*), and The Plant List, a collaboration between the Royal Botanic Gardens (Kew), and the Missouri Botanical Garden (*www.theplantlist.org*). Common names largely reflect those of the BONAP database, or sometimes are names in popular use locally.

Nativity

Native Species

Native plants are those assumed to have been present in some part of Michigan prior to European settlement. Also included are plants believed to have arrived in the state more recently via natural migration from areas adjoining Michigan and in which they are clearly native.

Introduced Species

Several different terms can be used to indicate the status of introduced (or non-native) species in Michigan's flora. Each introduced plant is different in terms of its origin, persistence, rate of spread, etc. Introduced species include those that were deliberately planted and which may escape and reproduce locally. Waifs are species normally found in cultivation, such as tomatoes or watermelons, that are occasionally found sprouted from seeds in waste areas and yards; these usually persist for only a single season. Some ornamental species such as lilacs and day-lilies are planted and usually remain in place but continue to live indefinitely, often growing at old home sites. Adventive species are those that appear here and there, usually from accidental introductions or as escapes from cultivation, but are apparently not firmly established nor spreading in the state. Naturalized species are those non-native species, such as timothy (*Phleum pratense*) that have arrived deliberately or accidentally in Michigan, and have become firmly established as part of the flora; populations may be large or cover extensive areas. Most introduced species in the Flora are of Asian or European origin. These species from other continents are also referred to as exotic species.

Invasive Species

Most invasive species are ecological pioneers and colonizers which, once introduced, are able to quickly establish themselves especially where the native community has been disturbed. In Michigan, examples of invasive species include purple loosestrife (*Lythrum salicaria*), common buckthorn (*Rhamnus cathartica*) and introduced honeysuckles such as *Lonicera morrowii*, *L. tatarica* and *L.* × *bella* which can replace native understory shrubs in woods and thickets. Other invasives are more recent introductions but are already spreading into relatively undisturbed places; notable examples include glossy buckthorn (*Frangula alnus*) and garlic-mustard (*Alliaria petiolata*).

Wetland Status Categories

The National Wetland Plant List (NWPL) is a list of wetland plants and their assigned indicator statuses. An indicator status reflects the likelihood that a particular plant occurs in a wetland or upland. The five indicator statuses are:

- **OBL:** Obligate Wetland Plants that almost always occur in wetlands (i.e. almost always in standing water or seasonally saturated soils.
- **FACW:** Facultative Wetland Plants that usually occur in wetlands, but may occur in non-wetlands. These plants predominately occur with hydric soils, often in geomorphic settings where water saturates the soils or floods the soil surface at least seasonally.
- **FAC:** Facultative Plants that occur in wetlands and non-wetland habitats. These plants can grow in hydric, mesic, or xeric habitats. The occurrence of these plants in different habitats represents responses to a variety of environmental variables other than just hydrology, such as shade tolerance, soil pH, and elevation, and they have a wide tolerance of soil moisture conditions.
- **FACU:** Facultative Upland Plants that usually occur in non-wetlands but may occur in wetlands. These plants predominately occur on drier or more mesic sites in geomorphic settings where water rarely saturates the soils or floods the soil surface seasonally.
- **UPL:** Obligate Upland Plants that almost never occur in wetlands (or in standing water or saturated soils). Typical growth forms include herbaceous, shrubs, woody vines, and trees.

Plants in the Flora with no indicator rating are considered UPL (Obligate Upland) for wetland delineation purposes. The indicator status ratings are based on the 2013 definitions prepared for the

National Wetland Plant List (Lichvar 2013). The indicators are routinely used in wetlands research and delineation studies, and also provide insight into each species' habitat preferences.

A species' indicator status may vary based on the region of its occurrence. Michigan lies within two delineation regions:
- **NC:** Northcentral-Northeast Region (Northcentral Subregion)
- **MW:** Midwest Region

The Upper Peninsula lies within the Northcentral-Northeast Region. For more information, visit the website of the National Wetland Plant List: *https://wetland_plants.usace.army.mil*.

Conservation Status

Because of their rarity, primarily as a result of habitat loss, 83 plant species are listed as endangered by the state of Michigan, and a further 204 species are listed as threatened. Those species which are reported from the Upper Peninsula are noted in the Flora by either (END), endangered, or (THR), threatened.
- **Endangered:** a species is considered endangered if the species is threatened with extinction throughout all or a significant portion of its range within Michigan.
- **Threatened:** species is considered threatened if the species is likely to become endangered within the foreseeable future throughout all or a significant portion of its range within Michigan.

Coefficients of Conservatism and Floristic Quality Assessment

The Flora includes a *coefficient of conservatism* for the majority of native species (abbreviated CC in the species descriptions). The coefficient (or C value) is used to evaluate the floristic quality of a natural area, and is based on that species' tolerance for disturbance and fidelity to a particular pre-settlement plant community type.

C values range from 0 to 10 and represent an estimated probability that a plant is likely to occur in a landscape relatively unaltered from what is believed to be a pre-settlement condition. For example, a C value of 0 is given to plants such as box elder (Acer negundo), that have demonstrated little fidelity to any remnant natural community (i.e. it may be found almost anywhere). Conversely, a C value of 10 is applied to plants like sage willow (*Salix candida*) that are almost always restricted to a high quality natural area. Introduced plants, by definition, were not part of the pre-settlement flora, so no C value is given. C values were obtained from the Michigan Flora website (*www.michiganflora.net*).

The aggregate conservatism of all the plants inhabiting a site determine its floristic quality. See Swink and Wilhelm (1994) for a thorough discussion of the method and how to calculate a Floristic Quality Index.

APG III LILIACEAE REORGANIZATION

Alliaceae
　Allium
Asparagaceae
　Asparagus
　Convallaria
　Maianthemum
　Muscari
　Ornithogalum
　Polygonatum
Colchicaceae
　Uvularia
Liliaceae
　Clintonia
　Erythronium
　Lilium
　Medeola
　Prosartes
　Streptopus
Melanthiaceae
　Anticlea
　Trillium
Tofieldiaceae
　Triantha
Xanthorrhoeaceae
　Hemerocallis

4 MICHIGAN MAP

Upper Peninsula geographic regions
- **IR** Isle Royale
- **W** western UP (Baraga, Gogebic, Houghton, Iron, Keweenaw, Ontonagon counties)
- **C** central UP (Alger, Delta, Dickinson, Marquette, Menominee, Schoolcraft counties)
- **E** eastern UP (Chippewa, Luce, Mackinac counties)

FERNS AND FERN RELATIVES

Aspleniaceae
SPLEENWORT FAMILY

Asplenium Spleenwort

Mostly small ferns, with short rootstocks covered by old petiole-bases, and a tuft of small to medium-sized leaves (fronds); the blades firm, simple, pinnate, or 2-pinnate, and often evergreen. Veins free or forking. Sori (clusters of spore containers) elongate, occurring along the veinlets. Indusium (covering over sori) usually membranous, attached lengthwise along one side of the sorus. Species usually of rock crevices and where shaded and mossy.

ADDITIONAL SPECIES
Asplenium ruta-muraria L. (Wall-rue), small native fern; in Michigan known only from dolomite cliffs along Drummond Island shoreline.

1	Blades simple	2
1	Blades once-pinnate	3
2	Leaf tip long tapering to a gradual point .... **A. rhizophyllum**	
2	Leaf tip abruptly pointed or rounded; rare in Chippewa and Mackinac counties .... **A. scolopendrium**	
3	Blades ovate in outline; pinnae only 3–5 per side; Drummond Island .... **A. ruta-muraria***	
3	Blades linear in outline, pinnae ca. 6–40 per side	4
4	Leaves of two types, the fertile long and upright, the sterile shorter and spreading, pinnae with conspicuous basal lobes overlapping rachis; Chippewa and Mackinac counties .... **A. platyneuron**	
4	Leaves all alike; pinnae bases not overlapping rachis	5
5	Rachis purple-brown throughout entire length .... **A. trichomanes**	
5	Rachis green, dark only at base .... **A. viride**	

Asplenium platyneuron (L.) B.S.P.
EBONY-SPLEENWORT native
IR | WUP | CUP | **EUP** FACU CC 2

Tufted fern with 2 types of leaves; the fertile leaves stiff and upright, 20–40 cm long and 2.5–4 cm wide, gradually tapered to the base; pinnae linear-oblong or basal pinnae triangular, widely separated; rachis satiny chestnut purple; sterile leaves shorter and widely spreading; sori linear-oblong and on the veins.

Open woods and fields, especially on sandy and loamy soils; less often on rock; Chippewa and Mackinac counties.

The stiff, upright fertile blades are distinctive.

Asplenium rhizophyllum L.
WALKING FERN (THR) native
IR | **WUP** | **CUP** | **EUP** CC 10

Camptosorus rhizophyllus (L.) Link

Tufted fern with fronds 5–30 cm long or longer, clustered at the end of an erect or ascending scaly rhizome. Leaves evergreen, entire, 1–3 cm wide at the cordate or auriculate base, usually tapering to a long caudate tip; veins reticulate; sori elongate, scattered along the veins; indusium attached on one side of the sorus.

Shaded rocks, usually of limestone, less commonly on sandstone or rarely quartzite.

The tips of the arching blades often root to form new plants, hence the name walking fern.

Asplenium scolopendrium L.
HART'S-TONGUE FERN (END) native
IR | WUP | CUP | **EUP** CC 10

Phyllitis scolopendrium var. *americanum* Fern.

Fronds 15–40 cm long or longer, from a short caudex; blade simple, strap-like, cordate-auriculate at base, usually tapered to a point at the tip; petiole (stipe) short, covered with narrow, curling, long-tapering scales. Sori oblong, nearly at right angles to the midrib, and located on either side of adjacent veinlets so that the sori appear to have double indusia opening along the middle.

Mossy dolomite boulders in moist deciduous forests, usually on talus below escarpments; rare in Chippewa and Mackinac counties.

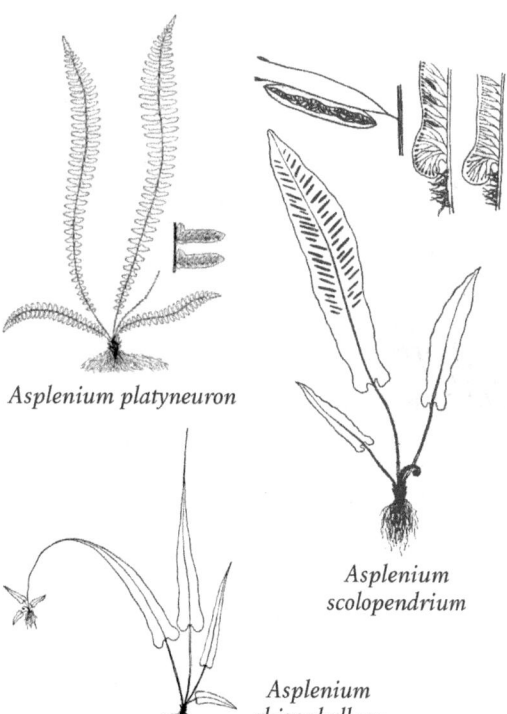

Asplenium platyneuron

Asplenium scolopendrium

Asplenium rhizophyllum

ASPLENIACEAE | Spleenwort Family

Asplenium trichomanes L.
MAIDENHAIR-SPLEENWORT — native
IR | WUP | CUP | EUP — CC 10

Fern with leaves 6–20 cm long or longer, forming a dense tuft from a compact rhizome. Petiole and rachis purple brown; old rachises persistent; blades linear, pinnate; pinnae usually opposite or subopposite, oval, rounded to cuneate at the inequilateral base and slightly toothed on the sides and at the blunt apex; sori linear, situated on the veins between the midrib and the margin.

Sheltered rock crevices on sandstone, limestone or quartzite, where moist or dry.

Asplenium viride Huds.
GREEN SPLEENWORT — native
IR | WUP | CUP | EUP — CC 10
Asplenium trichomanes-ramosum L.

Fern with leaves 2–14 cm long, tufted from a short rhizome. Petioles darkened below, bright green above; rachis green; blades linear to linear-lance-shaped, pinnate; pinnae rounded or rhomboid-ovate, crenate; sori elongate, borne near the indistinct midrib, becoming confluent at maturity.

In crevices of shaded, wet limestone cliffs and talus.

A distinctive species less common than *Asplenium trichomanes*, with which it sometimes grows. The green rachis (stem of the blade) distinguishes it from our other spleenworts.

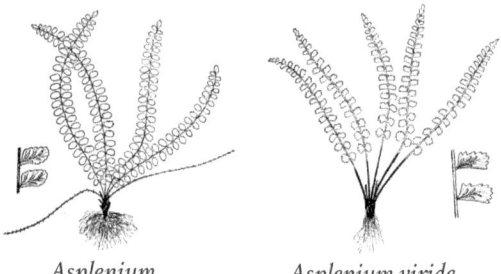

Asplenium trichomanes   *Asplenium viride*

Athyriaceae
LADY FERN FAMILY

Medium to large ferns. *Deparia* and *Diplazium*, formerly considered as species of *Athyrium*, are now treated as separate species. The degree of blade division helps separate our two species (see key), and in contrast to *Deparia*, *Athyrium* has a more deeply grooved rachis, which is continuous from rachis to costa (vs. discontinuous in *Deparia*).

1 Blades 2-pinnate (with the pinnae again divided), the pinnules also sometimes deeply lobed .... **Athyrium filix-femina**

1 Blades with deeply lobed pinnae (1-pinnate-pinnatifid) .... **Deparia acrostichoides**

Athyrium Lady Fern

Athyrium filix-femina L.
LADY FERN — native
IR | WUP | CUP | EUP — FAC CC 4
Athyrium angustum (Willd.) K. Presl

Clumped fern, rhizomes short and ascending. Leaves deciduous, sterile and fertile leaves similar; petioles with brown, linear scales; blades elliptic, 2-pinnate, broadest at middle or slightly below middle; pinnae short-stalked or stalkless; sori generally somewhat curved to hook-shaped, less often straight; indusia elongate, laterally attached.

Common; moist deciduous woods, thickets, streambanks, wetland margins, shaded rock outcrops.

Deparia Silvery Glade Fern

Deparia acrostichoides (Swartz) M. Kato
SILVERY SPLEENWORT — native
IR | WUP | CUP | EUP — FAC CC 6
Athyrium thelypterioides (Michx.) Desv.

Large fern from creeping rhizomes. Leaves deciduous, 50–100 cm long, sterile and fertile leaves alike; petioles straw-colored (but dark red-brown at base), with brown lance-shaped scales; blades lance-shaped to oblong in outline, tapered at tip and distinctly narrowed toward base; deeply lobed, the segments blunt to somewhat tapered at their tip, margins entire to slightly lobed. Sori crowded, elongate, straight or sometimes curved, the indusia silvery and shiny when young.

Moist, rich deciduous or mixed woods, especially in swales, ravines, depressions; streambanks.

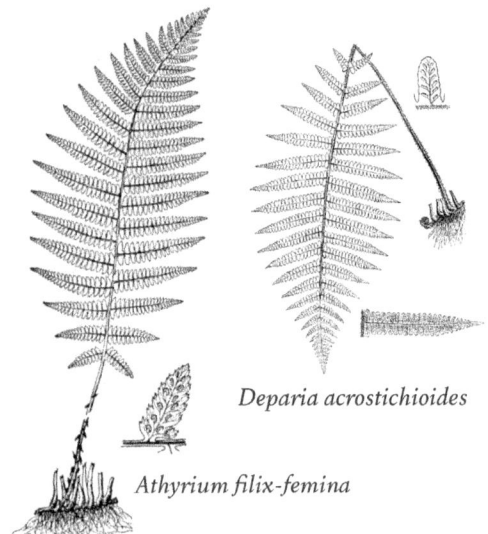

Deparia acrostichoides

Athyrium filix-femina

Blechnaceae
CHAIN FERN FAMILY

Woodwardia *Chain Fern*

Woodwardia virginica (L.) Sm.
VIRGINIA CHAIN FERN *native*
IR | WUP | CUP | **EUP** CC 10

Fern with leaves 6-10 dm long, scattered along the creeping rhizome and sometimes forming large colonies with the blades all facing in one direction; petioles (stipes) long, satiny; blades oblong-lanceolate, 10-30 cm wide, pinnate-pinnatifid; pinnules obtuse, with finely serrulate margins. Veins united to form a single series of areolae next to the midrib of both the pinnae and the pinnules, then free to the margin. Sori oblong, usually becoming confluent at maturity, one to each areole.

Sphagnum bogs, openings in acidic swamps, sandy or peaty lakeshores where strongly acidic; Chippewa County.

The chains of sori on the areolae adjacent to midrib are distinctive (right).

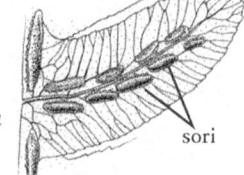

Woodwardia virginica — sori

Cystopteridaceae
BLADDER FERN FAMILY

Small to medium ferns; two genera in the UP: *Cystopteris* and *Gymnocarpium*.

1 Blades ternate (divided into 3 more or less equal parts); indusium absent................... **Gymnocarpium**
1 Blades 1-pinnate; indusium present ...... **Cystopteris**

Cystopteris *Bladder Fern*

Delicate, medium-sized ferns, with 2-3-pinnate blades arising from short creeping rhizomes. Veins free. Indusium hood-shaped, thin, and withering, attached at one side and arching over the rounded sori.

1 Blades elliptic to lance-shaped, typically widest at or slightly below middle of blade; rachis and pinnule midribs without glandular hairs 2
1 Blades elliptic to triangle-shaped, usually widest at base; rachis and pinnule midribs sparsely to densely covered with glandular hairs 4
2 Stems covered with yellow hairs; leaves clustered 1-4 cm below protruding apex of stem .... **C. protrusa**
2 Stems without hairs; leaves clustered at apex of stem 3
3 Pinnae usually at acute angle to rachis and often curving toward apex of blade; pinnae margins rounded-toothed .... **C. tenuis**
3 Pinnae usually perpendicular to rachis and not curving toward apex of blade; pinnae margins sharp-toothed.. .... **C. fragilis**
4 Rachis often with bulblets, rachis and midribs usually densely glandular-hairy; blades narrowly to broadly triangle-shaped, apex of blade long-tapered. **C. bulbifera**
4 Rachis occasionally with bulblets, rachis and midribs usually only sparsely glandular-hairy; ovate to lance-shaped, widest above base........... **C. laurentiana**

Cystopteris bulbifera (L.) Bernh.
BULBLET FERN; BLADDER FERN *native*
IR | **WUP** | **CUP** | **EUP** FACW CC 5

Clumped fern, rhizomes short and thick. Leaves deciduous, 30–100 cm long, sterile and fertile leaves similar but sterile blades usually shorter than fertile; petioles much shorter than blades; blades lance-shaped, 6–15 cm wide at base, long tapered to tip, with 20–30 pairs of pinnae; the veins ending in a notch (sinus). Sori round, on a small vein; indusia hoodlike and attached at its base, covered with scattered, short-stalked glands. Green bulblets, 4–5 mm wide, are produced on lower side of rachis (main stem of leaf) toward upper end of blade, these falling and forming new plants.

Rocky streambanks, ravines, seepy slopes, cedar swamps, and moist, shaded, often calcium-rich rocks and cliffs.

Distinguished from *Cystopteris fragilis*, a common fern of moist woods, by the blade broadest at base, most veins ending in a notch, and the small bulblets on underside of rachis. In fragile fern, blade broadest above its base, most veins end in a tooth, and bulblets are absent.

Cystopteris fragilis (L.) Bernh.
BRITTLE BLADDER FERN *native*
IR | WUP | CUP | **EUP** FACU CC 4

Fern with leaves 10-35 cm long or longer, tufted from short creeping rhizomes. Blades lance-shaped, 3-8 cm wide or wider near the base, bipinnate; pinnae pinnatifid to lobed, and at least the basal pinnules varying from orbicular to triangular and rounded to the base; veins mostly ending in a tooth or on the unnotched margin. Indusium up to 1 mm long and more or less cleft at the apex.

Sheltered crevices in cliffs, moist banks, and wooded talus slopes.

Petioles translucent (when held to a light), with veins of the blade extending to the very tips of the teeth, and a smooth rachis.

8 CYSTOPTERIDACEAE | Bladder Fern Family
ferns and fern relatives

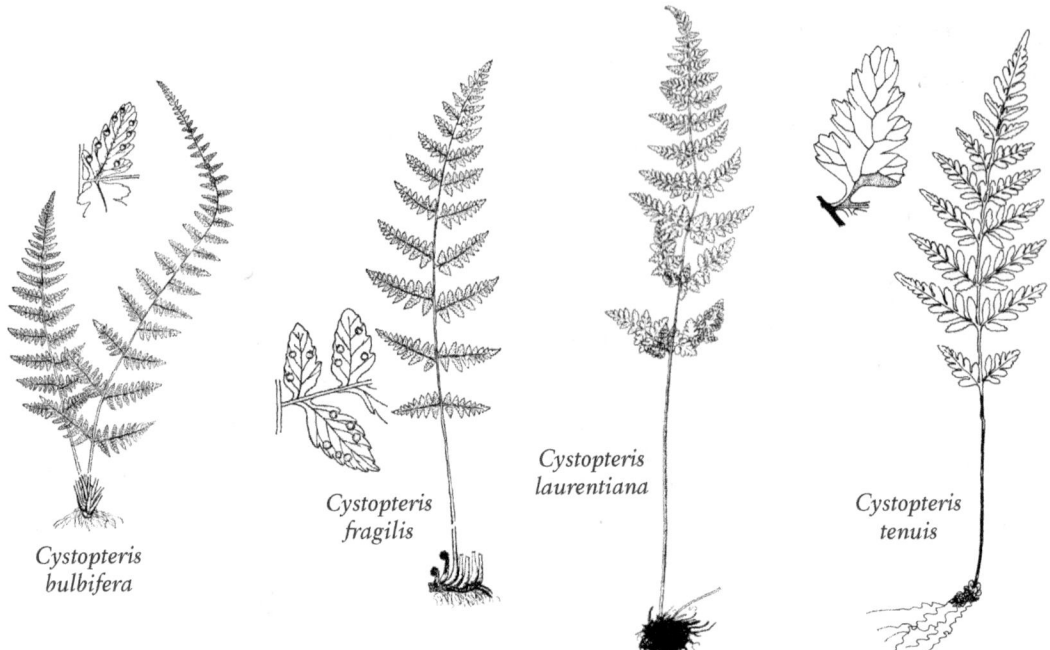

Cystopteris bulbifera

Cystopteris fragilis

Cystopteris laurentiana

Cystopteris tenuis

Cystopteris laurentiana (Weatherby) Blasdell
ST. LAWRENCE BLADDER FERN native
IR | WUP | CUP | EUP CC 9
Cystopteris fragilis var. *laurentiana* Weatherby
Fern with tufted leaves from a short creeping rhizome, to about 60 cm long. Petioles light brown to red-tinged. Blades ovate, to about 30 cm long and 12 cm wide; sterile blades usually shorter than the fertile. Indusium to 1 mm wide, very finely glandular.
  Calcareous rock or rocky slopes.
  This species combines the attributes of its presumed parents, *C. fragilis* var. *fragilis* and *C. bulbifera*. It is usually an upright, vigorous plant larger than typical *C. fragilis*. The veins extend both to the teeth-tips and to the sinuses.

Cystopteris protrusa (Weatherby) Blasdell
LOWLAND BLADDER FERN native
IR | WUP | **CUP** | EUP FACU CC 5
Cystopteris fragilis var. *protrusa* Weatherby
Fern with leaves 20–50 cm long, scattered along a creeping rhizome. Petioles greenish, straw-colored, or pale brown. Blades lance-shaped, to 25 cm long, 5–10 cm wide; sterile blades usually shorter, bipinnate; pinnules sharply toothed, ovate-lance-shaped; lower pinnules tapered to a stalk-like base; veins mostly ending in a tooth or on the unnotched margin. Indusium to 0.5 mm long, shallowly toothed or entire at its tip.
  Under deciduous trees on riverbottom benches.
  Distinguished from *C. fragilis* by the long internodes on the rhizome, the greenish or straw-colored petioles, the softer, larger blades, and the lower pinnules, which taper to a stalk-like base.

Cystopteris tenuis (Michx.) Desv.
UPLAND BRITTLE BLADDER FERN native
IR | WUP | CUP | EUP CC 5
Cystopteris fragilis var. *mackayi* Lawson
Similar to *C. fragilis*, but the pinnules oblong to lance-shaped and evenly wedge-shaped at the base; the indusium about 0.5 mm long and shallowly toothed or entire at its tip.
  In habitats similar to *Cystopteris fragilis*, but more often on streambanks, rotted logs, and moist openings.

Gymnocarpium *Oak Fern*
Small ferns with 3-parted delicate blades, the blades glabrous or glandular, arising singly from slender rootstocks. Sori round. Indusium absent. Veins free, simple, or forking. Northern oak fern (*G. dryopteris*) is common; our other two species are much less frequent.

1 The two lower divisions of the blade nearly as long as the terminal division; blades membranous and thin; rachis glabrous . . . . **G. dryopteris**
1 The two lower divisions of the blade about half the length of terminal division; blades firm and somewhat stiff; rachis in part densely glandular 2
2 Innermost pinnules of lowest pair of pinnae only slightly longer than opposite upper pinnules; upper blade surface glabrous; acidic or neutral rock . . . . . **G. jessoense**
2 Innermost pinnules of lowest pair of pinnae much longer than opposite upper pinnules; upper blade surface moderately glandular; limestone and calcareous rock . . . . **G. robertianum**

Gymnocarpium dryopteris (L.) Newman
NORTHERN OAK FERN native
IR | WUP | CUP | EUP FACU CC 5

Small, delicate fern with leaves to 30 cm long or longer, arising singly from a slender blackish rhizome. Blades glabrous or nearly so, triangular in outline, 3-parted; the pinnae pinnate-pinnatifid. Sori small, located near the margin.

Cool, moist coniferous and mixed woods, base of talus slopes, swamp margins.

The small, delicate, triangular blades oriented parallel to the ground and yellow-green in color are distinctive.

Gymnocarpium jessoense (Koidzumi) Koidzumi
NAHANNI OAK FERN (END) native
IR | WUP | CUP | EUP CC 10

Small fern with leaves to 30 cm long, arising singly from a slender blackish rhizome. Blades narrowly triangular in outline, 3-parted, bipinnate-pinnatifid, glandular; the innermost lower pinnules usually only slightly longer than the corresponding upper pinnules. Sori small, located near the margin.

Shaded cliffs and talus.

The blade and rachis are glandular, as in limestone oak fern (*G. robertianum*); Nahanni oak fern, however, is smaller and more slender, the pinnae usually curve upwards, and the pinnules curve outwards.

Gymnocarpium robertianum (Hoffmann) Newman
LIMESTONE OAK FERN (THR) native
IR | WUP | CUP | EUP FACU CC 10

Small to medium fern with leaves to 40 cm long, arising singly from a slender blackish rhizome. Blades triangular in outline, 3-parted, bipinnate-pinnatifid, glandular, the innermost basal pinnules of the lowermost pair of pinnae usually much longer than the corresponding upper pinnules. Sori small, located near the margin.

Limestone cliffs, outcrops and pavements (alvars).

The long-triangular, glandular blades (including glands on the upper surface), with pinnules at right angles are distinctive.

Dennstaedtiaceae
BRACKEN FERN FAMILY

Ferns with leaves all alike, arising singly from creeping rhizomes. Blade once-pinnate to compound, glabrous or hairy. Sori near or at blade margin on vein tips or on a marginal vein; indusia present, free, or fused with portion of blade margin to form a cup, or hidden by the revolute blade margin.

Pteridium *Bracken Fern*
Pteridium aquilinum (L.) Kuhn
BRACKEN FERN native
IR | WUP | CUP | EUP FACU CC 0

Coarse fern with leaves 30–70 cm or more long, often forming large colonies from the creeping rhizomes. Blades triangular in outline, usually 3-parted, 30–50 cm wide; lower pinnules more or less pinnatifid; upper pinnules entire, glabrous or slightly hairy on underside, with revolute margins. Sporangia borne in marginal sori on the underside of the pinnules; sporangia covered by a nearly continuous false outer indusium formed by the revolute pinnae margin.

Ubiquitous in open drier woods, pine plantations, old fields, and sandy clearings.

Plants growing in shade tend to have more or less horizontal blades; blades of plants growing in sun tend to be upright and stiff.

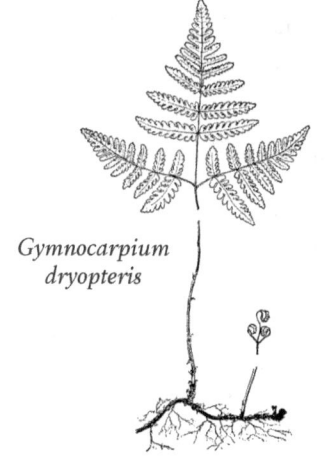
Gymnocarpium dryopteris

Gymnocarpium robertianum

Pteridium aquilinum

pinnule with marginal sori

Dryopteridaceae
WOOD FERN FAMILY

Medium to large ferns; rhizomes short, stout and scaly. Leaves dark green, sometimes evergreen; petioles shorter than blades, straw-colored or green, with chaffy scales near base. Sterile and fertile leaves alike or slightly different; sterile leaves sometimes persisting over winter; blades 1–3 pinnate, the smallest segments commonly toothed or lobed, veins simple to 1- or 2-branched. Sori round, on underside veins of pinnae; indusia round to kidney-shaped.

1 Fronds 1-pinnate-pinnatifid to more divided, the pinnae pinnatifid or themselves fully divided, lacking a prominent basal lobe, light green to dark green, herbaceous to nearly leathery; indusia kidney-shaped . **Dryopteris**
1 Fronds 1-pinnate, the pinnae toothed and each with a slight to prominent lobe near the base on the side towards the leaf tip, dark green, leathery or nearly so; indusia peltate (umbrella-like) . . . . **Polystichum**

Dryopteris *Wood-Fern*

Medium to large ferns; rhizomes short, stout and scaly, often covered with old petiole bases. Leaves dark green, sometimes evergreen; petioles shorter than blades, straw-colored or green, with chaffy scales near base. Sterile and fertile leaves alike or slightly different; sterile leaves sometimes persisting over winter; blades 1–3 pinnate, the smallest segments commonly toothed or lobed, veins simple to 1- or 2-branched. Sori round, on underside veins of pinnae; indusia round to kidney-shaped.

HYBRIDS
Hybrids may be recognized by an appearance intermediate between the parent species, and the presence of abortive spores. Four *Dryopteris* hybrids are reported from the UP:
Dryopteris × boottii (Tuckerman) Underwood: *D. cristata × D. intermedia;* leaves more dissected than *D. cristata*.
Dryopteris × triploidea Wherry: *D. carthusiana × D. intermedia;* leaves similar to parents but often somewhat larger.
Dryopteris × montgomeryi: *D. filix-mas × D. marginalis*.
Dryopteris × slossoniae: *D. cristata × D. marginalis*.

1 Blades small, very scaly on underside; old leaves forming conspicuous, persistent curled tufts at base of plant . . . . **D. fragrans**
1 Blades large, scales few or absent 2
2 Sori on margins of blade segments; blades leathery gray-green and paler on underside . . . . . **D. marginalis**
2 Sori near middle of smallest blade segments 3
3 Lowest pinnules on lowest pinnae stalkless 4
3 Lowest pinnules on lowest pinnae stalked 6
4 Leaves of 2 types; the sterile leaves shorter than fertile; the pinnae of fertile leaves usually turned to a nearly horizontal position . . . . **D. cristata**
4 Sterile and fertile leaves similar; pinnae of fertile leaves in same plane as blade 5
5 Blade widest near middle; petiole much shorter than length of blade . . . . **D. filix-mas**
5 Blade widest at or near base; petiole longer than blade . . . . **D. goldiana**
6 Lowermost inner pinnule shorter than adjacent lower pinnule . . . . **D. intermedia**
6 Lowermost inner pinnule longer than next outer one 7
7 Lower basal pinnule on basal pinna closer to the second upper pinnule than to the inner or first upper pinnule . . . . **D. expansa**
7 Lower basal pinnule on basal pinna closer to the inner upper pinnule than to the second upper pinnule . . . . **D. carthusiana**

Dryopteris carthusiana (Villars) H. P. Fuchs
SPINULOSE WOOD-FERN native
IR | WUP | CUP | EUP FACW CC 5
Dryopteris austriaca var. spinulosa (O.F. Müll.) Fisch., *Dryopteris spinulosa* (O.F. Müll.) Watt
Clumped fern, rhizomes short-creeping. Leaves all alike, deciduous, smooth except for chaffy, pale brown scales near base of petioles; blades 2- to nearly 3-pinnate, 2–6 dm long and 1–4 dm wide, tapered to tip, slightly narrowed at base; pinnae usually 10–15 pairs, alternate to nearly opposite, narrowly lance-shaped; pinnules toothed to deeply lobed, mostly 5–40 mm long and 3–10 mm wide, the teeth tipped with a small spine; innermost lower pinnule longer than next outer one and 2–3x longer than opposite upper pinnule. Sori halfway between midvein and margin; indusia 1 mm wide, without stalked glands.
Moist to wet woods, hummocks in swamps, thickets; also drier sand dunes and ridges.

Dryopteris cristata (L.) A. Gray
CRESTED WOOD-FERN native
IR | WUP | CUP | EUP OBL CC 6
Clumped fern, rhizomes short-creeping with ascending tips. Sterile and fertile leaves somewhat different, the outer sterile leaves waxy, persistent and smaller than inner fertile leaves; fertile leaves deciduous, 3–8 dm long. Blades 1-pinnate to nearly 2-pinnate, narrowly lance-shaped, 2–6 dm long and 7–15 cm wide, tapered to tip, narrowed at base; pinnae 5–9 cm long and to 4 cm wide, typically twisted to a nearly horizontal position, giving a "venetian blind" appearance to blades; pinnae segments to 20 mm long and 8 mm wide, with small spine-tipped teeth;

petioles with sparse, pale brown, long-tapered scales. Sori round, midway between midvein and margin; indusia smooth, 1 mm wide.

Swamps, thickets, open bogs, fens and seeps.

Dryopteris expansa (K. Presl) Fraser-Jenkins & Jermy
SPREADING WOOD FERN native
IR | WUP | CUP | EUP FAC CC 9
Dryopteris assimilis S. Walker

Fern with leaves to 1 m long, forming a large, more or less upright crown at the end of the upright, chaffy rhizome. Petioles usually shorter than the blades, with brown-tinged, often dark-centered, ovate scales. Blades broadly triangular to ovate, abruptly tapering to the tip, twice pinnate to tripinnate; pinnae short-stalked; basal pinnae triangular, inequilateral; the inner, lowermost pinnule on each lowermost pinna closer to the second upper pinnule than to the first upper pinnule. Indusia glabrous, or rarely finely glandular.

Cool moist woods and thickets.

Dryopteris filix-mas (L.) Schott
MALE FERN native
IR | WUP | CUP | EUP CC 10

Large fern with leaves to 1 m long or longer, forming a crown from an upright, scaly rhizome. Petioles usually short, densely covered with long-tapered pale brown scales. Blades lance-shaped, narrowed towards the base, dark green on upper surface and lighter below; pinnae lance-linear. Sori usually only on the upper half of the frond on the lower three-quarters of the pinnules; indusia glabrous.

Rich woods, rocky slopes, and at the base of shaded rock outcrops often on limestone.

The taper of the blades at their base and tip and the presence of sori only on the upper half of fertile fronds are characteristic; male fern is somewhat similar to marginal wood fern (*D. marginalis*), but the blades are much less leathery.

Dryopteris fragrans (L.) Schott
FRAGRANT WOOD FERN native
IR | WUP | CUP | EUP CC 10

Fern of rocky habitats with leaves to 30–40 cm long, forming a spreading or ascending crown from the rhizome; old leaves curled, shriveled, and persistent at the plant base. Petioles to 15 cm long, glandular and chaffy. Blades leathery, tapered from the middle to the base and tip; pinnae overlapping and often inrolled, densely chaffy with brownish scales; pinnae lance-shaped, pinnately cleft or crenate; rachises and pinnae glandular. Indusia large and often overlapping, whitish, becoming tan, often ragged on their margins.

Cliffs and talus slopes (often somewhat calcareous).

Dryopteris fragrans is somewhat similar to the smaller rusty cliff fern (*Woodsia ilvensis*) of similar habitats, but the curled, persistent old leaves drooping below fragrant wood fern are distinctive.

Dryopteris goldiana (Hook.) A. Gray
GOLDIE'S WOOD-FERN native
IR | WUP | CUP | EUP FAC CC 10

Clumped fern; rhizomes short-creeping, to 1 cm thick, densely scaly. Leaves to 1 m long; blades 30–60 cm long and 20–40 cm wide, deciduous late in season, the upper part abruptly narrowed to a small, tapered tip, the tip often mottled with white; pinnae with small, often rounded teeth; petioles brown, slightly shorter than blades, with narrow, pale brown scales 1–2 cm long, lower scales with a dark midstripe. Sori close to midveins, with a smooth indusia 1–2 mm across.

Moist hardwood forests, shaded streambanks, talus slopes; soils rich in humus and usually neutral.

Dryopteris intermedia (Muhl.) A. Gray
FANCY WOOD-FERN native
IR | WUP | CUP | EUP FAC CC 5
Dryopteris spinulosa var. *intermedia* (Muhl. ex Willd.) Underw.

Clumped fern, rhizomes ascending. Leaves in an open vaselike cluster of evergreen leaves; blades broadest just above base and abruptly tapered near tip, 2–5 dm long and 1–2 dm wide, 2-pinnate; pinnae at right angles to stem, lowermost inner pinnule usually shorter than next outer pinnule, pinnules toothed and tipped with small spines; petioles 1/3 as long as blade, with pale brown scales with a darker center, petioles and stems with small, gland-tipped hairs. Sori midway between midvein and margin, the indusia 1 mm wide, covered with stalked glands.

Moist hardwood and mixed hardwood-conifer forests, hummocks in swamps; soils rich in humus, slightly acid to neutral.

Dryopteris marginalis (L.) A. Gray
MARGINAL WOOD FERN native
IR | WUP | CUP | EUP FACU CC 5

Fern with leaves mostly 25-60 cm long, crowded to form a crown; lower part of the petiole covered with light brown lance-shaped scales. Blades 10-20 cm wide, dark green above, gray-green below, leathery, lance-shaped to ovate, 2-pinnate; pinnae lance-shaped; pinnules oblong, entire to deeply lobed. Sori located near the margins; indusia smooth, whitish, becoming light brown.

Rocky woods and ravines.

The leathery or spongy character of the nearly evergreen blades and the nearly marginal sori are characteristic.

EQUISETACEAE | Horsetail Family

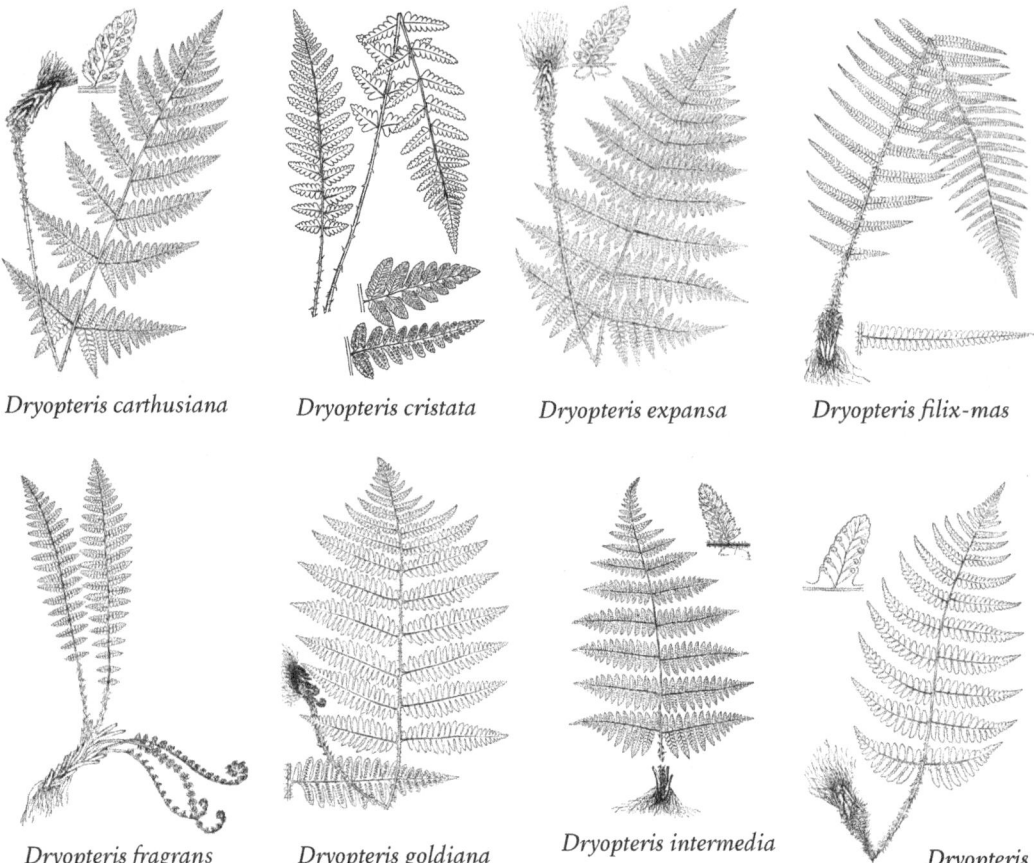

Dryopteris carthusiana   *Dryopteris cristata*   *Dryopteris expansa*   *Dryopteris filix-mas*

Dryopteris fragrans   *Dryopteris goldiana*   *Dryopteris intermedia*   *Dryopteris marginalis*

Polystichum *Holly Fern*

Large, tufted ferns with mostly evergreen, leathery blades; the petioles usually scaly, arising from short, stout, chaffy rhizomes. Sori round; indusia round, attached at the center.

1 Blades once-pinnate .... **P. lonchitis**
1 Blades twice-pinnate .... **P. braunii**

Polystichum braunii (Spenner) Fée
BRAUN'S HOLLY FERN   native
IR | **WUP** | **CUP** | **EUP**   CC 8

Medium to large fern with leaves to 1 m long, forming a crown at the end of a short ascending rhizome. Petiole about one-sixth the length of the blade, chaffy. Blades dark green, broadly lance-shaped, narrowed at the base; rachis with persistent chaff; pinnae lance-shaped; pinnae generally once-pinnate, the margins with incurved bristle-tipped teeth. Sori in two rows near the midrib; indusia often erose.

Rocky woods, along rocky streams, and on shaded cliffs within moist northern forests.

Polystichum lonchitis (L.) Roth
HOLLY FERN   native
IR | **WUP** | **CUP** | **EUP**   CC 10

Small to medium fern; leaves shiny, 10–60 cm long. Petioles very short, chaffy. Blades linear or nearly so, tapered to the tip and base. Middle and upper pinnae sickle-shaped in outline; upper side of pinnae base auriculate; teeth of margin tipped by a small spine; the pinnae near the base of the blade reduced to small triangle-shaped auricles. Sori round, in two rows, occurring midway between the midvein and the margin; indusium entire.

Limestone cliffs, moist rocky slopes, talus slopes, and occasionally coniferous woods.

Equisetaceae
HORSETAIL FAMILY

Equisetum *Horsetail, Scouring-Rush*

Rushlike herbs with dark rhizomes. Stems annual or perennial, grooved, usually with large central cavity and smaller outer cavities, unbranched or with whorls of branches at nodes. Leaves reduced

EQUISETACEAE | Horsetail Family

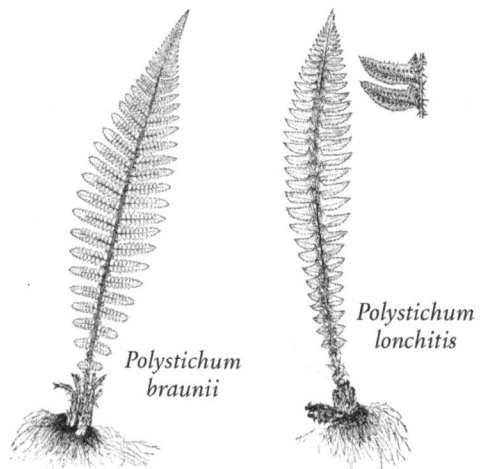

Polystichum braunii

Polystichum lonchitis

to scales, united into a sheath at each node; top of sheath divided into dark-colored teeth. Spores in cones at tips of green or brown fertile stems.

ADDITIONAL SPECIES & HYBRIDS

Equisetum × ferrissii Clute: *E. hyemale × E. laevigatum*.

Equisetum × mackayi (Newm.) Brichan: *E. hyemale × E. variegatum*.

Equisetum telmateia Ehrh. (Giant horsetail), collected in a alder thicket in Keweenaw County by O. A. Farwell in 1890 and 1895; disjunct from western North America, and known in the east only from this locality. However, the species is likely now no longer present in Michigan.

1. Stems evergreen (annual in *E. laevigatum*); unbranched or with a few scattered branches, branches not in regular whorls (scouring rushes) 2
1. Stems annual; usually with regular whorls of branches, sometimes unbranched (horsetails) 5
2. Stems solid (central cavity absent); stems small, slender and sprawling .... **E. scirpoides**
2. Stems hollow (central cavity present); stems larger, usually upright 3
3. Stems 1–3 dm tall, with 5–12 ridges, central cavity to 1/3 diameter of stem .... **E. variegatum**
3. Stems usually taller, with 16–50 ridges, central cavity more than half diameter of stem 4
4. Cones with a distinct, small sharp tip; stem sheaths with a black band at tip and base .... **E. hyemale**
4. Cones blunt-tipped, sheaths with black band at tip only .... **E. laevigatum**
5. Stems unbranched 6
5. Stems with regular whorls of branches 10
6. Stems green 7
6. Stems brown or flesh-colored 8
7. Stems with 9–25 shallow ridges; central cavity more than half diameter of stem; sheath teeth entirely black or with narrow white margins............ **E. fluviatile**
7. Stems with 5–10 strongly angled ridges; central cavity less than 1/3 diameter of stem; sheath teeth with white margins and dark centers .... **E. palustre**
8. Sheath teeth papery and red-brown, teeth joined and forming several broad lobes .... **E. sylvaticum**
8. Sheath teeth black or brown, not papery, separate or joined in more than 4 small groups 9
9. Stems withering after spores mature, remaining unbranched. .... **E. arvense**
9. Stems persistent, becoming branched and green .... **E. pratense**
10. First internode of each branch shorter than the subtending sheath of the main stem 11
10. First internode of each branch equal or longer than the subtending sheath of the main stem 12
11. Stems with 9–25 shallow ridges; central cavity more than half diameter of stem; sheath teeth more than 12, entirely black or with narrow white margins........... .... **E. fluviatile**
11. Stems with 5–10 strongly angled ridges; central cavity about same size as outer cavities; sheath teeth 5–6, with white margins and dark centers .... **E. palustre**
12. Stem branches themselves branched; sheath teeth papery and red-brown, teeth joined and forming several broad lobes .... **E. sylvaticum**
12. Stem branches unbranched; sheath teeth black or brown, not papery, separate or joined in more than 4 small groups 13
13. Stem branches ascending; teeth of branch sheaths gradually tapering to a slender tip............. **E. arvense**
13. Stem branches spreading; teeth of branch sheaths broadly triangular .... **E. pratense**

Equisetum arvense L.
COMMON OR FIELD HORSETAIL *native*
IR | WUP | CUP | EUP FAC CC 0

Stems annual, upright from creeping, branched, tuber-bearing rhizomes covered with dark hairs. Sterile and fertile stems unalike; sterile stems appearing in spring as fertile wither, green, regularly branched, 1–6 dm tall and 2–5 mm wide, with 10–14 shallow ridges, the ridges usually rough-to-touch; central cavity 1/3–2/3 stem diameter; sheaths with 6–14 persistent, black-brown teeth 1–2 mm long; branches numerous in dense whorls, usually without branchlets, upright or spreading, 3–5-angled, solid. Fertile stems flesh-colored, shorter than sterile stems and with larger sheaths, maturing in early spring and soon withering, unbranched, to 3 dm tall and 8 mm wide; sheaths with 8–12 dark brown teeth. Cones blunt-tipped, long-stalked at end of stem, 0.5–3 cm long.

Common; streambanks, meadows, moist woods, ditches, roadsides and along railroads; calcareous fens.

EQUISETACEAE | Horsetail Family

Equisetum fluviatile L.
WATER-HORSETAIL — native
IR | WUP | CUP | EUP — OBL CC 7

Stems annual, fertile and sterile stems alike, to 1 m or more tall, from smooth, shiny, light brown, creeping rhizomes; stems with 9–25 shallow, smooth ridges; central cavity large, about 4/5 stem diameter; stem sheaths green, 6–10 mm long; teeth 12–24, persistent, 2–3 mm long, dark brown to black, sometimes with narrow white margins; branches none or few, to many and regularly whorled from middle nodes, spreading, without branchlets, 4–6-angled, hollow. Cones 1–2 cm long at tips of stems, long-stalked, blunt-tipped, deciduous, maturing in summer.

In standing water of marshes, ponds, peatlands, ditches and swales.

Equisetum hyemale L.
COMMON SCOURING-RUSH — native
IR | WUP | CUP | EUP — FAC CC 2

Equisetum affine Engelm.

Stems evergreen, persisting for more than 1 year, fertile and sterile stems alike, from black, slender rhizomes; stems mostly unbranched or with few, short, upright branches from upper nodes, to 15 dm tall but usually shorter, 4–14 mm wide, with 14–50 rounded, very rough ridges; central cavity at least 3/4 stem diameter; stem sheaths 5–15 mm long, with a dark band at tip and usually also at base, the teeth dark brown to black with chaffy margin, 2–4 mm long, deciduous or persistent. Cones stalkless or short-stalked at tips of stems, sharp-pointed, eventually deciduous, 1–2.5 cm long, maturing in summer, or old stems sometimes developing branches with cones in the following spring.

Often forming dense colonies in seeps, wet to moist meadows, shores and streambanks, ditches, roadsides and along railroads; usually where sandy or gravelly.

Equisetum laevigatum A. Braun
SMOOTH SCOURING-RUSH — native
IR | WUP | CUP | EUP — FACW CC 2

Stems mostly annual, fertile and sterile stems alike, from brown or black rhizomes; stems mostly unbranched or with a few upright branches, 3–10 dm tall and 3–8 mm wide, smooth and rather soft, with 10–32 ridges; central cavity 2/3–3/4 stem diameter; stem sheaths with a single dark band at tip, or rarely lowest sheaths with a dark band at base or entirely black; teeth dark brown or black with chaffy margins, free or partly joined in pairs, 1–4 mm long, soon deciduous. Cones short-stalked at tips of stems, rounded with a small sharp point, maturing in early summer and eventually deciduous.

Wet meadows, low prairie, streambanks, floodplains, seeps, and ditches, often where sandy or gravelly.

Equisetum palustre L.
MARSH-HORSETAIL — native
IR | WUP | CUP | EUP — FACW CC 8

Stems annual, erect, fertile and sterile stems alike, from creeping, branched, shiny black rhizomes; stems 2–8 dm tall, with 5–10 pronounced ridges, the ridges mostly smooth; central cavity small, 1/6–1/3 stem diameter; sheaths green, loose and flared upward; teeth 5–6, free or partly joined, persistent, 3–7 mm long, brown to black, with pale, translucent margins; branches few and irregular, to many and whorled at upper nodes, upright, without branchlets, 5–6-angled, hollow. Cones long-stalked at tips of stems, 1–3 cm long, blunt-tipped, maturing in summer, deciduous.

Wetland margins, streambanks, alder thickets, fens; often in shallow water.

Equisetum pratense Ehrh.
MEADOW-HORSETAIL — native
IR | WUP | CUP | EUP — FACW CC 10

Stems annual and erect, sterile and fertile stems unalike, from creeping, dull black rhizomes. Sterile

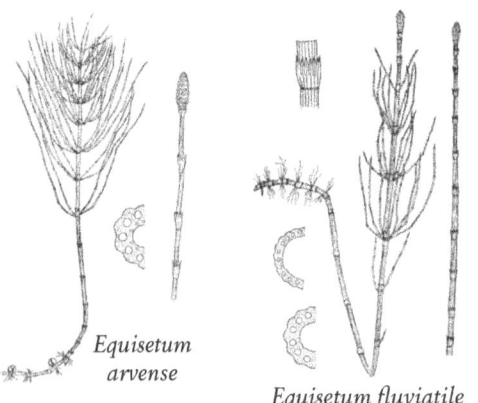

Equisetum arvense

Equisetum fluviatile

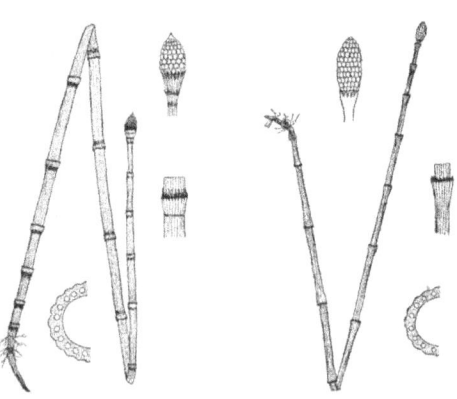

Equisetum hyemale

Equisetum laevigatum

stems regularly branched, 2–5 dm tall and 1–3 mm wide; 8–18-ridged, the ridges roughened by silica on middle and upper stem; central cavity 1/3–1/2 stem diameter; main stem sheaths 2–6 mm long, the teeth persistent, 1–2 mm long, free or partly joined in pairs, brown with white margins and a dark midstripe; branches slender, many in regular whorls from middle and upper nodes, without branchlets, horizontal or drooping, mostly 3-angled, solid. Fertile stems uncommon, appearing in early spring before sterile stems and persisting, at first unbranched, fleshy and brown (without chlorophyll), later becoming green at nodes and producing many small green branches, mostly 1–3 dm tall; sheaths and teeth about twice as long as on sterile stems. Cones long-stalked at tips of stems, to 2.5 cm long, blunt-tipped, deciduous.

Moist woods, streambanks, and meadows.

Equisetum scirpoides Michx.
DWARF SCOURING-RUSH *native*
IR | WUP | CUP | EUP FAC CC 7

Stems evergreen, very slender, fertile and sterile stems alike, from widely branching rhizomes; stems 5–30 cm long and only 0.5–1 mm wide, in dense clusters, usually unbranched and zigzagged, upright or trailing; central cavity absent, 3 small outer cavities present; sheaths green with broad black band at tip, loose and flared above, with 3–4 teeth; teeth with white, chaffy margin, ± persistent, but tips usually soon deciduous. Cones black, small, 3–5 mm long, sharp-tipped.

Mossy places and moist, shaded woods, the stems often partly buried in humus.

Equisetum sylvaticum L.
WOODLAND-HORSETAIL *native*
IR | WUP | CUP | EUP FACW CC 5

Stems annual, erect, sterile and fertile stems unalike, from creeping, shiny light brown rhizomes, tubers occasionally present. Sterile stems green, 3–7 dm tall and 1.5–3 mm wide, with 10–18 ridges, rough-to-touch with sharp, hooked silica spines; central cavity 1/2–2/3 stem diameter; sheaths green at base, red-brown and flaring at tip; teeth brown, 3–5 mm long, joined in 3–5 broad lobes. Stems densely branched in regular whorls from the nodes, the branches themselves branched, often curving downward, 4–5-angled, solid. Fertile stems at first pink-brown (without chlorophyll), fleshy, unbranched, becoming green and branched as in sterile stems; sheaths and teeth larger than in sterile stems. Cones 1.5–3 cm long, stalked, blunt-tipped, deciduous.

Wet or swampy woods, thickets, usually in partial shade.

Equisetum variegatum Schleicher
VARIEGATED SCOURING-RUSH *native*
IR | WUP | CUP | EUP FACW CC 6

Stems evergreen, fertile and sterile stems alike, from creeping, much-branched, smooth rhizomes; may form thick colonies; stems 1–3 dm tall and 1–2.5 mm wide, with 5–12 shallow, rough ridges, branched near base and otherwise usually unbranched; central cavity 1/4–1/3 stem diameter, smaller outer cavities present; sheaths green at base with a broad black band above; teeth persistent, with a dark brown or black midstripe and wide white margins, abruptly narrowed to a hairlike, deciduous tip 0.5–1 mm long. Cones to 1 cm long, strongly sharp-tipped, maturing in summer or persisting unopened until following spring.

Wet calcareous open areas such as shores, low places in dunes, borrow pits and ditches.

Equisetum variegatum commonly forms hybrids with *E. hyemale* and *E. laevigatum*, sometimes making identification of this species difficult.

Equisetum variegatum

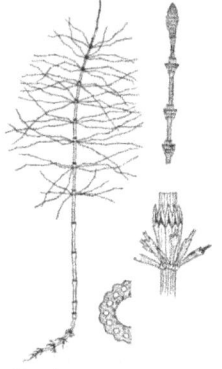
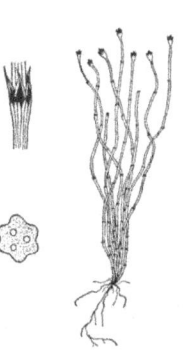

Equisetum palustre *Equisetum pratense* *Equisetum scirpoides* *Equisetum sylvaticum*

Isoetaceae
QUILLWORT FAMILY
Isoetes Quillwort

Perennial aquatic or emergent herbs. Leaves simple, entire, linear, from a 2–3 lobed rhizome (corm). Outermost and innermost leaves typically sterile. Outer fertile leaves have a pocketlike structure (sporangia) bearing whitish spores (megaspores; about 0.5 mm in diameter, magnification needed to see features); inner fertile leaves have numerous small microspores.

1 Megaspores conspicuously covered with small spines . . . . **I. echinospora**
1 Megaspores not spiny. . . . . **I. lacustris**

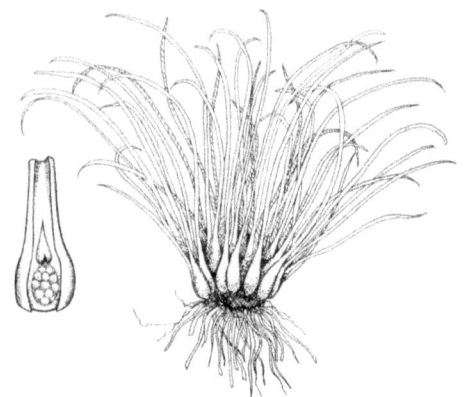

Isoetes echinospora

Isoetes echinospora Durieu
SPINY-SPORED QUILLWORT native
IR | WUP | CUP | EUP OBL CC 9

Isoetes braunii Durieu, *Isoetes muricata* Durieu

Leaves linear, 7–25 or more, 5–15 cm long and 0.5–1.5 mm wide, usually erect, soft, bright green to yellow-green, tapered from base to a very long, slender tip, without peripheral strands from base; corm 2-lobed. Sporangium 4–8 mm long, usually brown-spotted when mature, half or more covered by a membranous flap (velum). Megaspores round, white, 0.3–0.6 mm wide, covered with short, sharp to blunt spines.

Shallow water (to 1 m deep) of lakes, ponds and slow-moving rivers; plants rooted in mud, sand, or gravel.

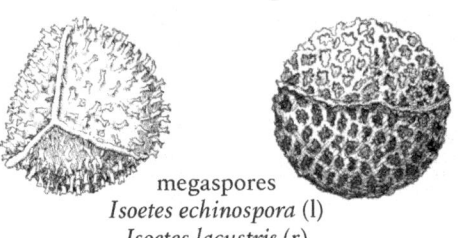

megaspores
Isoetes echinospora (l)
Isoetes lacustris (r)

Isoetes lacustris L.
LAKE QUILLWORT native
IR | WUP | CUP | EUP OBL CC 9

Isoetes hieroglyphica A.A. Eaton
Isoetes macrospora Durieu

Leaves several to many, 5–20 cm long and 1–2 mm wide, stiff and erect or with leaf tips curved downward, dark green, fleshy and twisted, peripheral strands from base usually absent; corm 2-lobed. Sporangium to 5 mm long, usually not spotted; membranous flap (velum) covering up to half of sporangium. Megaspores round, white, 0.6–0.8 mm wide, with ridges forming an irregular netlike pattern.

Underwater in shallow to deep water of cold lakes, ponds and streams.

Lycopodiaceae
CLUBMOSS FAMILY

Low, trailing, evergreen herbs resembling large mosses. Leaves needlelike or scalelike, alternate or opposite on stem. Spore-bearing leaves (sporophylls) similar to vegetative leaves or in conelike clusters at tips of upright stems.

1 Horizontal stems absent; sporangia in axils of unmodified leaves . . . . **Huperzia**
1 Horizontal stems present; sporangia in axils of modified, reduced sporophylls, the sporophylls grouped into upright or nodding strobili. 2
2 Strobili upright on leafy peduncles, the peduncle leaves not reduced in size; wetland species . . . . **Lycopodiella inundata**
2 Strobili sessile or on peduncles, the peduncles if present with scattered, small leaves; mostly upland species . . 3
3 Ultimate shoots and their leaves 5–12 mm wide, rounded in cross-section; leaves not strongly overlapping . . . . **Lycopodium**
3 Ultimate shoots and their leaves to 6 mm wide, 4-angled or flattened in cross-section; leaves overlapping . . . . **Diphasiastrum**

Diphasiastrum Ground-Pine

Small plants of drier habitats resembling miniature trees; branches flattened or 4-angled in cross-section; leaves 4-ranked, neither spine- nor hair-tipped. Strobili (cones) stalked, the stalks branched into segments of equal length.

HYBRIDS

Diphasiastrum × habereri (House) Holub, hybrid between *D. digitatum × D. tristachyum*.

Diphasiastrum × zeilleri (Rouy) Holub, hybrid between *D. complanatum* × *D. tristachyum*.

1. Strobili 1 (–2) on short to elongated, rarely forked stalks (peduncles); base of strobilus with a few sporophylls scattered along peduncle; stomata on both leaf surfaces. .... **D. sabinifolium**
1. Strobili 2–4 on forked stalks; base of strobilus compact and abruptly distinct from peduncle; stomata on lower leaf surfaces only. 2
2. Stem branchlets cordlike, nearly square in cross-section, usually waxy blue-green color .... **D. tristachyum**
2. Stem branchlets flat in cross-section, usually green 3
3. Branchlets regularly fan-shaped and arching, without conspicuous constrictions between seasonal growth; most strobili with sterile tips .... **D. digitatum**
3. Branchlets irregular, with conspicuous constrictions; most strobili without sterile tips .... **D. complanatum**

Diphasiastrum complanatum (L.) Rothm.
NORTHERN RUNNING-PINE   native
IR | WUP | CUP | EUP   CC 5
Lycopodium complanatum L.
Horizontal stems mostly below the surface of the ground; leaves scale-like. Upright stems to about 30 cm tall, with forking branchlets. Branchlets flattened, often strongly constricted between yearly growths, 2–4 mm wide. Leaves 4-ranked. Strobili 1 or 2 on peduncles.

Woods and clearings.

Conspicuous annual constrictions present, giving plants a somewhat irregular appearance, in contrast to the regularity of fan ground-pine (*D. digitatum*). The strobili are also irregular in number per peduncle (varying from 1–4), and the naked peduncles are very slender.

Diphasiastrum digitatum (Dill.) Holub
FAN GROUND-PINE   native
IR | WUP | CUP | EUP   CC 3
Lycopodium digitatum Dill.
Horizontal stems mostly on or near the surface of the ground; leaves distant, scale-like. Upright stems to about 30 cm high, the branchlets of the branches arched and fan-like; constrictions between yearly growth absent or only slightly evident. Branchlets 2–3 mm wide. Leaves 4-ranked. Strobili mostly 3 or 4 on peduncles; peduncle branched at one point.

Dry woods and clearings.

The branchlets are very regular and fan-like, annual constrictions are lacking, and the strobili are usually in groups of 4 on long, naked peduncles.

Diphasiastrum sabinifolium (Willd.) Holub
SAVIN-LEAF CREEPING-CEDAR   native
IR | WUP | CUP | EUP   CC 10
Lycopodium sabinifolium Willd.
Horizontal stems mostly creeping on ground surface. Erect stems dichotomously forked, to 20 cm long; sterile branchlets flattened. Leaves 4-ranked, awl-like, nearly similar lengths; leaves of upper and lower side appressed; lateral leaves slightly larger and appressed to stem for about half their length, the free portion spreading and incurved at tip. Strobili stalked, the stalks sometimes forked; sporophylls near base of strobili widely spaced.

Sandy woods and meadows, often where disturbed as in blowout areas.

Diphasiastrum tristachyum (Pursh) Holub
DEEP-ROOT GROUND-PINE   native
IR | WUP | CUP | EUP   CC 7
Lycopodium tristachyum Pursh
Horizontal stems usually deeply buried; leaves scale-like. Upright stems to 30 cm tall. Sterile branches ascending to loosely divergent, flattened, 1–1.5 cm wide. Leaves 4-ranked, blue-green, lance-shaped. Strobili 2–6 on leafy-bracted peduncles.

Dry, sometimes sandy woods and clearings.

The branches are vase-shaped and crowded, bluish green and white waxy on their underside; annual constrictions are present along the branches; the peduncles often branch and then branch again, resulting in 4 strobili.

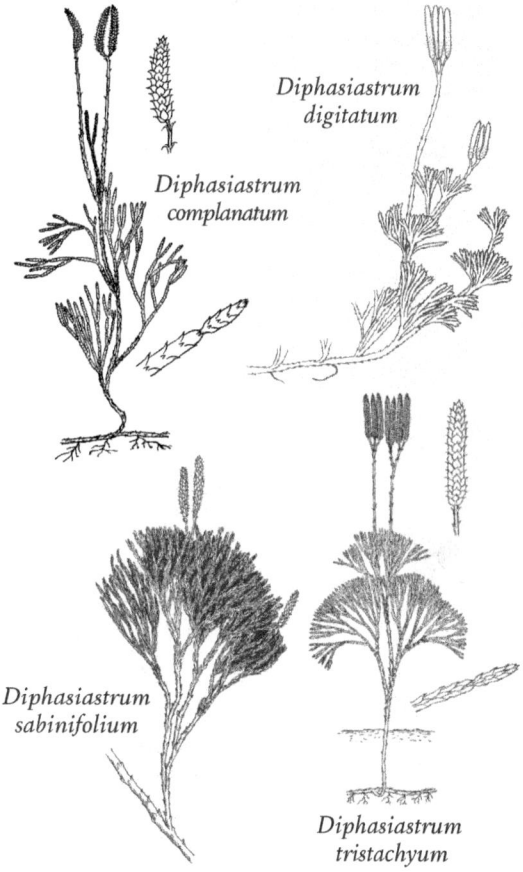

Diphasiastrum digitatum
Diphasiastrum complanatum
Diphasiastrum sabinifolium
Diphasiastrum tristachyum

LYCOPODIACEAE | Clubmoss Family

Huperzia Fir-Moss

Low evergreen perennials with erect shoots; leaves spreading or appressed and upright. Spores borne at base of upper leaves.

1 Leaves obovate, widest above middle, spreading to reflexed, upper portion of at least the larger leaves with distinct teeth; shoots shaggy with conspicuous annual constrictions; usually growing on soil . . . . **Huperzia lucidula**
1 Leaves lance-shaped, widest below the middle, leaves (at least those on the upper stem) often ascending, entire or with a few small teeth; annual constrictions absent or faint **2**
2 Leaves lance-shaped with sides nearly parallel; stomates on upper surface of each leaf number 2-50 (view fresh leaves under 20x lens to see the light-colored, dot-like stomates) . . . . **Huperzia porophila**
2 Leaves lance-shaped (as above) or ovate or triangular; if leaf shape is inconclusive, then number of stomates on upper leaf surface is greater than 60 . . . . **3**
3 Leaves near base of plant essentially same size as those on upper portion; gemmae formed in a single whorl at end of the annual growth . . . . **Huperzia selago**
3 Leaves near base of plant conspicuously longer than those on upper portion; gemmae formed throughout upper portions of shoot . . . . **Huperzia appressa**

Huperzia appressa (Desv.) Á. & D. Löve
MOUNTAIN FIR-MOSS native
IR | **WUP** | **CUP** | **EUP** CC 10

Huperzia appalachiana Beitel & Mickel

Stems short, to only 10 cm long; clustered; annual constrictions absent. Leaves ascending, narrowly lance-shaped or with the sides parallel; stomates on both surfaces; margins entire; upper stem leaves smaller than those of lower stem. Sporangia in distinct zones on upper stems.

Cliffs, talus slopes, where open and exposed, on moss or thin soil.

Huperzia lucidula (Michaux) Trev.
SHINING FIR-MOSS native
IR | **WUP** | **CUP** | **EUP** FAC CC 5

Lycopodium lucidulum Michx.

Stems light green, creeping and rooting, upcurving stems forked several times, to 25 cm high, crowded with shiny dark green leaves which persist for more than one season. Leaves in mostly 6 rows, spreading or curved downward, in alternating groups of longer sterile and shorter fertile leaves, giving shoots a ragged look. Sterile leaves 6–12 mm long, toothed and broadest above middle; sporophylls barely widened and with small teeth or entire at tip. Small two-lobed buds (gemmae) produced in some upper leaf-axils; these may sprout into new plants after falling onto moist humus.

Moist to wet conifer and hardwood forests.

Huperzia porophila (Lloyd & Underwood) Holub
ROCK CLUBMOSS native
IR | **WUP** | **CUP** | EUP FACU CC -

Lycopodium porophilum Lloyd & Underwood
Lycopodium selago var. *porophilum* (Lloyd & Underwood)

Shoots erect, 12–15 cm long, leaves of mature portion slightly smaller than leaves of juvenile portion; annual constrictions distinct to indistinct. Leaves reflexed at base, ascending at stem apex (forming cluster) and spreading for most of stem length,

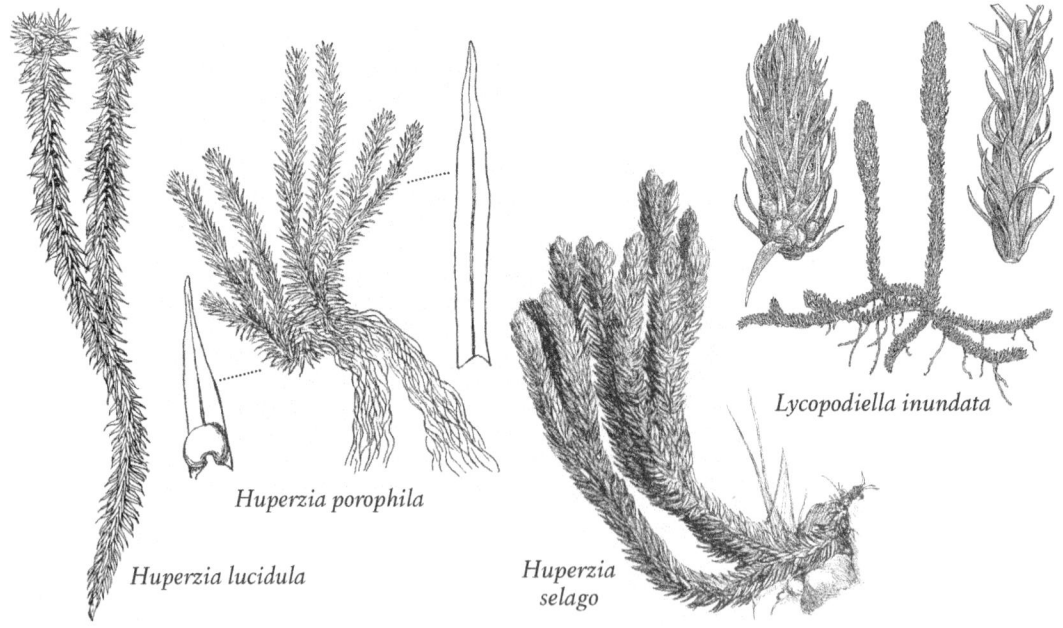

Huperzia porophila

Huperzia lucidula

Huperzia selago

Lycopodiella inundata

sparse, yellow-green to green, lustrous; largest leaves lance-shaped with roughly parallel sides, 5–8 mm long; smallest leaves triangular, widest at base, 3–6 mm long; margins almost entire or a few large teeth. Gemmiferous branchlets produced in 1–3 pseudowhorls at end of annual growth; gemmae 4–5 mm long; lateral leaves 1–1.5 mm wide, acute, widest above middle.

Shaded sandstone ledges.

Huperzia selago L.
NORTHERN FIR-MOSS *native*
IR | WUP | CUP | EUP   FACU CC 10
Lycopodium selago L.

Horizontal stems short; upright stems forked from base, 6–20 cm long and 2–3 mm wide (stem only). Leaves persistent, yellow-green, in 8–10 rows, 3–6 mm long and to 1 mm wide, swollen and concave at base, gradually tapered to tip, mostly without teeth, uniform in length; leaves appressed to stem, giving stems a smooth, cylindric outline. Sporophylls similar to vegetative leaves; sporangia produced early in season in leaf axils, followed later by sterile leaves. Upper axils with small, 2-lobed reproductive buds (gemmae).

Cedar swamps, streambanks, sandy lake shores, usually where mossy; also in wet, sandy borrow pits.

Lycopodiella *Bog Clubmoss*

Small plants of wet places with horizontal stems creeping on ground surface, and fertile, upright stems. Strobili (cones) formed on upper part of upright stems, stalkless and covered with leaves.

Lycopodiella inundata (L.) Holub
NORTHERN BOG CLUBMOSS *native*
IR | WUP | CUP | EUP   OBL CC 7
Lycopodium inundatum L.

Low, creeping perennial of wet habitats. Stems elongate and trailing, deciduous but with evergreen buds at tips, rooting throughout. Upright shoots unbranched, leafy, scattered along horizontal stems. Leaves in 8–10 rows, those on underside of trailing stems twisted upward, narrowly lance-shaped, margins ± entire. Fertile branches few, erect, to 1 dm high, with spreading leaves. Spores borne in terminal, leafy cones; the cones 1.5–5 cm long and 6–12 mm wide; sporophylls green, base widened and with a pair of teeth.

Acidic, open sphagnum bogs, wet sandy shores and streambanks; disturbed wetlands, sandy borrow pits.

ADDITIONAL SPECIES
Lycopodiella subappressa J.G. Bruce, W.H. Wagner, & Beitel, native; Mackinac County.

Lycopodium *Ground-Pine*

Plants mainly trailing on ground. Roots emerging from point of origin on underside of main stems. Horizontal stems on substrate surface or subterranean, long-creeping. Upright shoots scattered along horizontal stem, 5–16 mm diameter, round or flat in cross section, unbranched or with 1–4 lateral branchlets. Leaves not imbricate, linear to linear lance-shaped; leaves on horizontal stems scattered, appressed; leaves on lateral branchlets mostly 6-ranked or more, monomorphic with few exceptions, appressed, ascending to spreading, margins entire to dentate. Gemmae absent. Strobili single and sessile, or multiple and pedunculate; peduncle, when present, conspicuously leafy; sporophylls extremely reduced, much shorter than peduncle or stem leaves. Sporangia kidney-shaped.

1 Stroboli stalked; upright stems with 2–5 branches, not forming tree-like shapes; leaves tipped with hairs ... 2
1 Stroboli not stalked 3
2 Stroboli mostly single on stalk (rarely in pairs and then nearly stalkless); leaves apppressed or ascending on stem .... **L. lagopus**
2 Stroboli 2–5; leaves spreading or somewhat ascending .... **L. clavatum**
3 Stems creeping and horizontal; stroboli single at end of upright, mostly unbranched shoot 4
3 Stems upright, much-branched and tree-like; stroboli 1–7 at end of shoot 5
4 Longest leaves 6–9 mm long, spreading or reflexed, usually at least obscurely toothed near tip .. **L. annotinum**
4 Longest leaves less than 6 mm long, all but lowest leaves ascending, usually entire .... **L. canadense**
5 Branches flat in cross-section; leaves of unequal sizes . ... **L. obscurum**
5 Branches round in cross-section; leaves of equal sizes 6
6 Leaves on main stem below branches dark green and appressed to stem, soft to touch .... **L. hickeyi**
6 Leaves on stem below branches pale green and spreading, prickly .... **L. dendroideum**

Lycopodium annotinum L.
STIFF GROUND-PINE *native*
IR | WUP | CUP | EUP   FAC CC 5
Spinulum annotinum (L.) A. Haines

Stems elongated, prostrate, mostly unbranched, rooting at intervals; leaves uniform but the lower leaves turned upward. Erect stems simple to forked several times, increasing annually to 20 cm or more in height. Leaves 8-ranked, more or less stiff and hard, linear-subulate to linear-oblance-shaped, with a sharp spinule. Strobili sessile at the ends of leafy stems.

Moist woods and clearings, subalpine forests, and exposed rocky and peaty habitats.

Lycopodium canadense Nessel
NORTHERN CLUB-MOSS　native
IR | WUP | CUP | EUP　CC 6

Spinulum canadense (Nessel) A. Haines

Leaves of erect stems 2.5–6.0 mm long, lance-shaped to lance-oblong, flat, obscurely toothed, strongly ascending to tightly appressed against the stem; this species reported to have usually more than 25 stomata on upper leaf surface, in *L. annotinum*, stomata usually absent. Perhaps best considered a minor variant of *L. annotinum*.

Local in dry, often open sandy or rocky woods and shores, mostly near Lake Superior.

Lycopodium clavatum L.
RUNNING GROUND-PINE　native
IR | WUP | CUP | EUP　FAC CC 4

Stems elongated, horizontal on the surface of the ground, forking, rooting at intervals; leaves uniform, but lower leaves turned upward. Erect branches at first simple, becoming dichotomous; fertile branches with a leafy-bracted peduncle bearing 2 to several sessile or short-stalked strobiles. Leaves linear-subulate, incurved-spreading, usually tipped with a soft white hair-like bristle. Bracts of strobili yellow, fimbriate-erose, at least the lower with white filiform tips.

Dry woods and clearings.

Mature fruiting plants present no problems in identification; young or sterile plants sometimes confused with *L. annotinum*. The extended, soft, hair-like bristles on the leaf tips are distinctive.

Lycopodium dendroideum Michx.
TREE GROUND-PINE　native
IR | WUP | CUP | EUP　FACU CC 5

Dendrolycopodium dendroideum (Michx.) Haines, *Lycopodium obscurum* var. *dendroideum* (Michx.) D.C. Eat.

Subterranean stems creeping, branching, and rhizome-like, with broad scale-like leaves; aerial stems upright, 10–30 cm high, simple below, forking above, constricted between the seasonal growth. Lower leaves strongly divergent; leaves of lateral branchlets in 2 dorsal, 2 ventral, and 2 lateral ranks; leaves strongly decurrent, the free part linear-attenuate. Strobili sessile and terminal on the main axis, or dominant branches and produced in the second, third, or fourth growing season.

Woods and clearings.

Quickly identified by grasping the base of an aerial stem; this will feel distinctly prickly because of the stiff divergent leaves.

Lycopodium hickeyi W.H. Wagner, Beitel & Moran
PENNSYLVANIA GROUND-PINE　native
IR | WUP | CUP | EUP　CC 5

Dendrolycopodium hickeyi (W.H. Wagner, Beitel & R.C. Moran) A. Haines, *Lycopodium obscurum* var. *isophyllum* Hickey

Similar to *L. obscurum* in that leaves of lower portion of stem are strongly appressed to slightly divergent; leaves of branchlets are all of equal size and linear-attenuate.

Woods.

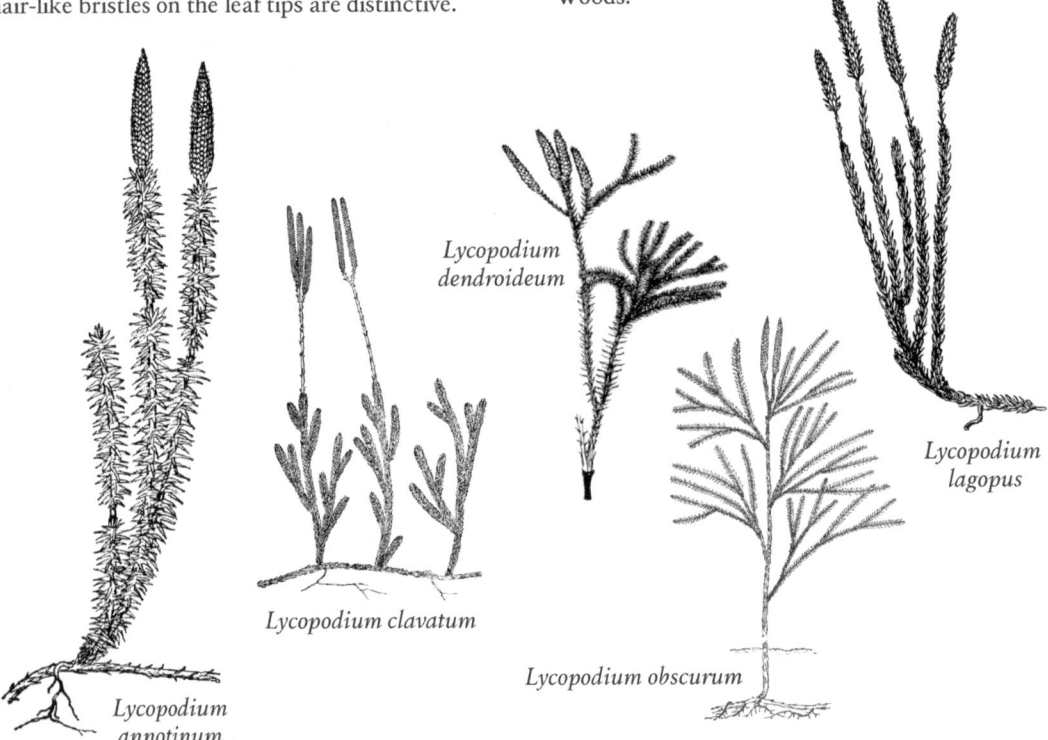

Lycopodium dendroideum

Lycopodium clavatum

Lycopodium annotinum

Lycopodium obscurum

Lycopodium lagopus

Lycopodium lagopus (Laestad.) Zinserl.
ONE-CONE GROUND-PINE native
IR | WUP | CUP | EUP FACU CC 5
Lycopodium clavatum L. var. *lagopus* Laestad.
Horizontal stems on substrate surface. Upright shoots clustered, dominant main shoot branches 2–3, mostly in lower half; lateral branchlets few and like the upright shoots; annual bud constrictions abrupt and conspicuous, branches mostly erect. Leaves ascending to appressed, medium green, 3–5 mm long, margins entire; apex with narrow hairlike tip 1-3 mm long. Peduncles 3.5–12.5 cm long, unbranched, with remote pseudo-whorls of appressed leaves. Strobili solitary (if double, usually nearly sessile).
Fields and openings in woods.

Lycopodium obscurum L.
PRINCESS-PINE native
IR | WUP | CUP | EUP FACU CC 5
Dendrolycopodium obscurum (L.) Haines
Similar to *L. dendroideum*, from which it may be distinguished by the strongly appressed to slightly divergent leaves on the lower portion of the aerial shoot. Leaves of the lateral branchlets arranged in 1 dorsal, 1 ventral, and 4 lateral ranks; leaves of ventral rank linear-attenuate to long triangular, smaller than leaves of other ranks; leaves of other ranks linear-acuminate to linear-acute.
Dry to moist woods.

Onocleaceae
SENSITIVE FERN FAMILY

Large coarse ferns with creeping hairy rhizomes (*Onoclea*) or with stolons on ground surface (*Matteuccia*); sterile and fertile fronds strongly different, the sterile fronds deciduous, pinnatifid to 1-pinnate-pinnatifid; fertile fronds persistent. Sori enclosed under recurved margin of pinna segment (outer false indusium) and a tiny true inner indusium (membranous or of hairs).

1   Sterile blades solitary from creeping rhizomes, deeply divided into lobes (or the lowermost divisions pinnae)
    .... **Onoclea sensibilis**
1   Sterile blades in a circle from a thick crown; pinnate with lobed pinnules....... **Matteuccia struthiopteris**

Matteuccia *Ostrich Fern*
Matteuccia struthiopteris (L.) Todaro
OSTRICH FERN native
IR | WUP | CUP | EUP FAC CC 3
Large, colony-forming fern; rhizomes deep and long-creeping, black, scaly, producing erect leafy crowns. Sterile leaves upright, 1-pinnate, to 2 m tall and 15–50 cm wide; blades much longer than petioles, abruptly narrowed to tip, gradually tapered to base, stems ± hairy; each pinnae deeply divided into 20 or more pairs of pinnules, these 3–6 mm wide at base and rounded at tip; veins not netlike. Fertile leaves stiff and erect within a circle of sterile leaves, green at first, turning brown or black, much shorter than sterile leaves (to 6 dm tall), produced in mid to late summer and often persisting into following year; fertile blades 1-pinnate, pinnae upright or appressed, 2–6 cm long and 2–4 mm wide, the margins inrolled and covering the sori; indusia with a jagged margin.

Wet and swampy woods, streambanks, seeps, and ditches.

Onoclea *Sensitive Fern*
Onoclea sensibilis L.
SENSITIVE FERN native
IR | WUP | CUP | EUP FACW CC 2
Medium fern, in clumps of several leaves, spreading by branching rhizomes and forming large patches. Leaves upright, with petioles about as long as blades. Sterile leaves deciduous, 1-pinnate at base, deeply cleft upward; the stem broader-winged toward the tip; blades 15–40 cm long and 15–35 cm wide, with 8–12 pairs of opposite pinnae, these deeply wavy-margined or coarsely toothed, 1–5 cm wide, with scattered white hairs on underside veins, the veins joined and netlike. Fertile leaves produced in late summer and persisting over winter, shorter than sterile leaves; fertile blades 1-pinnate, pinnae upright, divided into beadlike pinnules with inrolled margins covering the sori; veins not joined. Sori round and covered by a hoodlike indusia, becoming dry and hard.

Swampy woods and low places in forests, wet meadows, calcareous fens, roadside ditches, wet or moist wheel ruts; sometimes weedy.

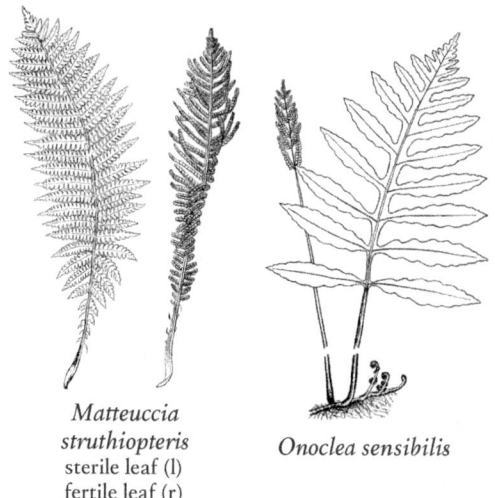

Matteuccia struthiopteris sterile leaf (l) fertile leaf (r)

Onoclea sensibilis

Leaves of *Onoclea* susceptible to damage from even light frosts, hence the common name of 'sensitive fern.'

Ophioglossaceae
ADDER'S-TONGUE FAMILY

Perennial herbs from short, erect rhizomes having several fleshy roots. Plants produce one leaf each year on a single stalk (stipe), with bud for next year's leaf at base of stipe. Leaves divided into a fertile segment (sporophyll) and a sterile expanded blade. Sterile blades entire (*Ophioglossum*), or lobed or 1–3x pinnately divided (*Botrychium*, *Botrypus*, *Sceptridium*). Spores in numerous round sporangia borne on simple or branched fertile blades.

1 Sterile blades simple, entire; veins netlike; sporangia imbedded in rachis of spike .. **Ophioglossum pusillum**
1 Sterile blades pinnately lobed or dissected; veins forked; sporangia exposed, often on a branched structure ... 2
2 Sterile blades somewhat leathery, persisting over winter; blades with distinct stalk, usually 5-25 cm long in fertile plants; fertile portion of frond joining sterile portion at or near ground level .... **Sceptridium**
2 Plants deciduous, withering in fall; blades usually ± unstalked, large (more than 5 cm long) or much smaller (in some species of *Botrychium*), herbaceous or sometimes fleshy; fertile portion of frond joining sterile portion well above ground level.................... 3
3 Sterile blades triangular, 3–4x pinnate, stalkless and mostly 5–25 cm wide; fertile portion erect, appearing to be a continuation of stipe ...... **Botrypus virginianus**
3 Sterile blades short-triangular, oblong, or linear; lobed (simple) to 3-pinnate (usually 1-pinnate to 2-pinnate-pinnatifid), mostly 1-5 cm wide; fertile portion of blade upright or spreading................... **Botrychium**

Botrychium *Grape-Fern; Moonwort*

Mostly small plants with one leaf, the blade divided into sterile and fetile segments. Sterile portion of blade pinnately divided or lobed, fertile portion branched to form a panicle bearing the sporangia.

ADDITIONAL SPECIES

Botrychium acuminatum W.H. Wagner (Tailed moonwort), sand dunes, sandy woods and meadows; Alger and Ontonagon counties; endangered.
Botrychium hesperium (Maxon & R. T. Clausen) W. H. Wagner & Lellinger, sand dunes sandy fields and sandy dry woods; Alger, Chippewa, and Keweenaw counties; threatened.

1 Sterile blade (trophore) simple to lobed, lobes rounded to square and angular, stalks usually 1/2 to 2/3 length of sterile blade; rare plant of shaded woods (*Botrychium mormo*), or more common and often in open grassy fields (*Botrychium simplex*)............. 2
1 Sterile blade pinnately lobed (either if actual pinnae or simply lobed), lobes of varying shapes, stalk usually less than 1/4 length of sterile blade; plants often in open sunny places; dunes, streambanks, roadsides, on trails and openings in forests, etc....................... 3
2 Segments of sterile leaves rounded, margins mostly entire; plants herbaceous in texture, green; habitats various, in forests or open ground .... **B. simplex**
2 Segments of sterile leaves angular, outer margins often coarsely dentate; plants ± succulent; shiny yellow-green; rare in shady forest understories, often concealed by leaf litter .... **B. mormo**
3 Basal pinnae or segments of sterile blade with venation like the ribs of a fan; midrib absent................. 4
3 Basal pinnae or segments of sterile blade with pinnate venation; midrib present.......................... 8
4 Basal pinnae broadly fan-shaped (almost perfect half moons) with narrow stalks .... **B. lunaria**
4 Basal pinnae narrowly fan- or wedge-shaped to nearly linear 5
5 Sterile blade at least partially folded longitudinally when alive (conduplicate), usually not more than 4 cm long by 1 cm wide; pinnae up to 5 pairs; basal pinnae usually 2-parted....................................... 6
5 Sterile blade flat or folded only at base when alive, usually up to 10 long by 2.5 cm wide; pinnae up to 10 pairs; basal pinnae unlobed, or if lobed, not usually 2-parted 7
6 Sterile blade very fleshy; fertile portion of blade usually less than 1.5 times length of sterile blade; pinnae mostly linear; basal pinna lobes ± equal; plants appearing in late spring .... **B. campestre**
6 Sterile blade herbaceous; fertile portion usually 1.5–4 times the length of vegetative blades; pinnae asymmetrically fan-shaped; basal pinna lobes unequal; plants appearing in summer .... **B. pallidum**
7 Sterile blade narrowly oblong (sterile blade widest above base), firm to herbaceous; pinnae fan-shaped, margins shallowly crenate............ **B. minganense**
7 Sterile blade narrowly deltate (sterile blade widest at lowest pinna pair); pinnae spatulate to linear spatulate, margins entire to very coarsely and irregularly dentate .... **B. spathulatum**
8 Fertile portion of blade 3-parted, with 3 major branches from near base of stalk at sterile blade **B. lanceolatum**
8 Fertile portion of blade unbranched or with loosely pinnate branches smaller than the single main stem .... **B. matricariifolium**

Botrychium campestre W.H. Wagner & Farrar
IOWA MOONWORT (THR) native
IR | **WUP** | **CUP** | **EUP** CC 9
Small plants, sometimes less than 5 cm tall and hard to see if surrounded by taller vegetation. Plants succulent, developing early in growing season, with

aboveground portion drying by mid-summer. Sterile blade fleshy, once pinnate, sessile to the common stalk, oblong in outline, longitudinally folded; to about 4 cm long and 1.3 cm wide; usually divided into 5 pairs of linear or linear-spatulate segments; margins crenate or dentate and also usually notched into 2 or several smaller segments. Common stalk short, less than 3 cm long. Sporophore often large relative to size of sterile blade.

Inconspicuous in prairies, dunes, grassy railroad sidings, and fields over limestone. Leaves appear in early spring and wither in late spring and early summer, long before those of other moonworts.

Botrychium lanceolatum (Gmel.) Angstr.
TRIANGLE MOONWORT native
IR | WUP | CUP | EUP   FACW CC 7

Plants 6–30 cm tall, dark green, smooth, appearing in early summer and persisting to fall. Stems about 5x longer than blades; blades triangular in outline, 1–8 cm long and 1–5 cm wide, stalkless or on a short stalk to 6 mm long, divided into 2–5 pairs of sharp-pointed, toothed pinnae, lower-most pair the largest. Fertile segment 2–9 cm long, mostly twice pinnate, on a stalk 1–3 cm long. Ours are var. *angustisegmentum*, found from Nfld and Ontario to Minn, becoming increasingly rare southward to NJ and Ohio.

Moist humus-rich woods, hummocks in swamps, streambanks.

Botrychium lunaria (L.) Sw.
COMMON MOONWORT native
IR | WUP | CUP | EUP   FACW CC 7

Botrychium neolunaria Stensvold & Farrar

Plants 3–20 cm tall, rubbery-textured, appearing in late spring and withering in summer. Leaf blades 1.5–7 cm long and 1–3 cm wide, stalkless or on a short stalk to 5 mm long; pinnately divided into 3–6 pairs of stalkless pinnae, the pinnae fan-shaped, wider than long and without a midrib; petioles 1.5–3 cm long. Fertile segments 0.5–7 cm long, on stalks about as long as the segments.

Cool, moist sandy soils in woods.

Moonwort has a long and illustrious history in early herbals; the seeds could reputedly make one invisible or could be used to unlock doors. Fully mature plants are distinctive, but plants somewhat similar to *B. minganense*.

Botrychium matricariifolium (A. Braun ex Dowell) A. Braun ex Koch
DAISY-LEAF MOONWORT native
IR | WUP | CUP | EUP   FACU CC 5

Plants to about 30 cm tall, membranous to fleshy. Leaf blades narrowly deltoid to ovate, short-stalked, inserted above the middle, pinnatifid to bipinnate-pinnatifid; segments of blade blunt and usually toothed. Fertile segment paniculate. Spores mature in June and July.

Acidic soil in old sandy and sterile fields, dry wooded slopes, rocky woods, moist cedar woods, and rich swamps.

This species is somewhat larger than *B. simplex*. The shape of the blade is variable (deltoid to ovate) but it is stalked, and the toothed segments are distinctive (compare *B. lanceolatum*).

Botrychium minganense Victorin
MINGAN MOONWORT native
IR | WUP | CUP | EUP   CC 7

Plants to 30 cm long, somewhat membranous. Leaf blades narrowly oblong, sessile or nearly so, inserted below the middle, pinnate or occasionally pinnate-pinnatifid at the base; segments of blades opposite, obovate, rhomboidal or oblong, frequently incised, remote. Fertile segment paniculate. Spores mature in July and August. *Botrychium minganense* can be distinguished from *B. lunaria* by its yellowish green hue and by its trough-shaped sterile segments, which are ascending rather than at right angles to the stalk and which rarely overlap with each other.

Moist hardwood forests, aspen-balsam-fir woods, old clearings; soils mostly circumneutral.

Botrychium mormo W.H. Wagner
LITTLE GOBLIN MOONWORT (THR) native
IR | WUP | CUP | EUP   CC 10

Our tiniest moonwort, uncommon; leaves appearing in late spring to fall or sometimes not appearing above the leaf litter. Plants to about 8-10 cm high but often smaller; yellow-green, somewhat shiny. Sterile blade variable; blade of well-developed plants with 2-3 pairs of small blunt lobes; blade in smaller-plants may be nearly absent. Sporophore to about 3 cm long, with several sporangia embedded in the fleshy stalk.

Extremely sporadic in mature deciduous forests of sugar maple or basswood, and sometimes with eastern hemlock or northern white cedar. Sites shaded and moist; soils are loams, with a rich litter layer.

Botrychium pallidum W.H. Wagner
PALE MOONWORT native
IR | WUP | CUP | EUP   CC 10

Plants 2.5-7 cm tall, waxy pale green to whitish. Sterile blade to 4 cm long by 1 cm wide, 1-pinnate, with up to 5 pairs of fan-shaped pinnae, each pair of pinnae often folded towards each other; basal pinnae usually divided into 2 unequal lobes, the upper lobe larger; margins entire to irregularly toothed. Sporangia sometimes present on lobes of lower pinnae. Sporophore longer than sterile blade, the sporangia on short branches from main stalk.

OPHIOGLOSSACEAE | Adder's-Tongue Family

Fields, dry sand and gravel ridges, roadsides, wet depressions, marshy lakeshores, tailings basins, second-growth forests; soils sandy.

Botrychium simplex E. Hitchc.
LEAST MOONWORT native
IR | WUP | CUP | EUP   FAC CC 5

Plants to ca. 15 cm tall, rather fleshy. Blades simple, lobed or pinnately divided, inserted at the base or towards the middle; segments of blade oblong, rhomboid or kidney-shaped, and usually overlapping, with the basal segments occasionally pinnatifid. Fertile segment simple or compound. Spores mature in late May and June.

Pastures, meadows, lakeshores, and gravelly slopes; easily overlooked in the field due to its small size and grassy habitat.

Botrychium spathulatum W.H. Wagner
SPOON-LEAF MOONWORT (THR) native
IR | WUP | CUP | EUP   CC 10

Leaf single, erect, to 12 cm long; shiny yellowish-green, leathery. Sterile blade sessile or short-stalked (less than 1 mm long); pinna pairs mostly 4-5 (7), spoon- or fan-shaped, widest at tip; lowest pinnae largest, commonly folded over rachis; pinnae mostly widely spaced and not overlapping, outer pinna margins entire or lobed. Sporophore 1-2x length of the trophophore; 1-2 times pinnately divided into segments bearing the sporangia. Leaves appearing late spring through summer.

Sand dunes, old fields, grassy railways, often where underlain by limestone.

Botrychium spathulatum has long been confused with the more common *B. minganense*, with which it often grows in the Lake Superior region; leaves appear later in *B. spathulatum* than in *B. minganense*.

Botrypus *Rattlesnake Fern*

Botrypus virginianus (L.) Holub
RATTLESNAKE FERN native
IR | WUP | CUP | EUP   FACU CC 5

Botrychium virginianum (L.) Sw.

Plants 40–75 cm tall, appearing in spring, withering in autumn, not overwintering. Blade (trophophore) broadly triangular, sessile, to 25 cm long and to 1.5x as wide, 3–4x pinnate, thin and herbaceous. Pinnae to 12 pairs, usually somewhat overlapping and slightly ascending; pinnules lance-shaped and deeply lobed, the lobes linear, sharply toothed and pointed at tip. Spore-bearing portion (sporophore) 2-pinnate, 0.5–1.5x length of trophophore.

Moist to fairly dry deciduous woods; occasional in swamps of cedar and black spruce. Rattlesnake fern is widespread in North America, occurring across Canada and most of the USA.

Ophioglossum *Adder's-Tongue*

Ophioglossum pusillum Raf.
NORTHERN ADDER'S-TONGUE native
IR | WUP | CUP | EUP   FACW CC 6

Plants erect, 7–30 cm tall, from slender rhizomes. Leaves 1, entire, on a stalk 3–15 cm long; blades upright, oval to ovate, rounded to acute at tip, 3–8 cm long and 1–4 cm wide, conspicuously net-veined. Sporangia in 2 rows in a terminal, unbranched fertile segment, 1–5 cm long and 2–4 mm wide, on a stalk 6–15 cm long.

Wet sandy meadows and beaches, moist depressions, wetland margins.

Sceptridium *Grape Fern*

Small to medium leathery ferns found in a variety of moist to dry, open to shaded habitats, often where sandy. Sterile blades dissected, winter-green; sporophore short-lived, withering by late summer. Sterile blade and sporophore joined near or below ground level.

1 Sterile blade segments deeply cut more than half way to the midvein, the entire blade lacerate . **S. dissectum**

1 Sterile blade segments finely to coarsely toothed.... 2

2 Ultimate segments of blade ± uniform in size; sterile blade segments finely toothed to ± entire; dissection of blade into segments extending to within 1 cm of apex at tips of blades 3

2 Ultimate segments of blades variable in size, the apical segments much longer than the laterals; sterile blade segments coarsely and ± irregularly toothed or cut; dissection of blade into segments stopping at ca. 1–2.5 cm from apex at tips of blades 4

3 Segments of sterile blade rounded at base; symmetrically tapered to an often ± blunt or even rounded apex; larger segments mostly 9–17 mm long; margins nearly entire or finely and inconspicuously toothed .... **S. multifidum**

3 Segments of sterile blade usually (obliquely) asymmetrical and angular, cuneate to the apex; larger segments mostly 4–9 mm long; margins clearly finely dentate, especially visible in immature leaves.... **S. rugulosum**

4 Overwintering leaves green, not bronze; larger (terminal) segments of vegetative blades narrowly to broadly ovate, obtuse to rounded at apex, ± symmetrical at base; margins toothed but never lacerate **S. oneidense**

4 Overwintering leaves bronze-colored (or green if covered by leaves); larger (terminal) segments of sterile blades lance-shaped, acute, and strongly asymmetric at base; margins toothed to irregularly cut . **S. dissectum**

OPHIOGLOSSACEAE | Adder's-Tongue Family

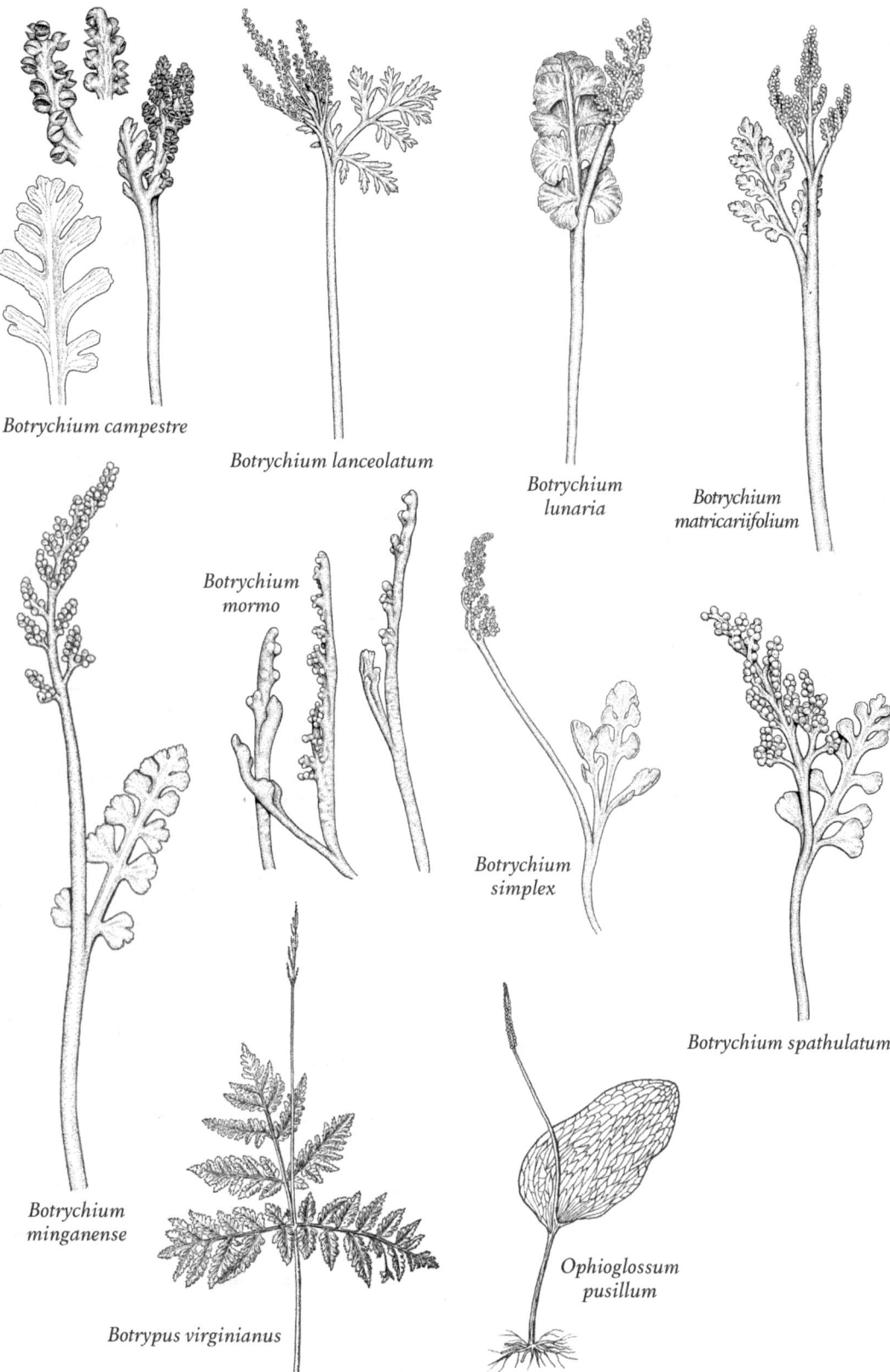

Sceptridium dissectum (Spreng.) Lyon
CUT-LEAF GRAPE FERN native
IR | **WUP** | **CUP** | **EUP** FAC CC 5
Botrychium dissectum Spreng.
Botrychium obliquum Muhl.
Leaves to 30 cm long; stem and blade less leathery than *S. multifidum;* blades long-petioled, triangular, 3-parted, attached at or near the base; ultimate divisions of blade cut in linear segments; segments more or less notched at the apex. Fertile segment paniculate. Spores mature Sept–Nov.

Sterile hilltops, dry pastures, dry woodlands, and grassy banks. Blades are often bronze or turn reddish in late fall.

Sceptridium multifidum
(Gmel.) Nishida ex Tagawa
LEATHERY GRAPE FERN native
IR | **WUP** | **CUP** | **EUP** FACU CC 5
Botrychium multifidum (Gmel.) Trev.
Leaves to 20 cm long; stem and blade leathery; blades evergreen, long-petioled, 3-parted, attached near the base of the plant; ultimate segments of blade crowded, sometimes imbricate, ovate, more or less the same size, obtuse or somewhat acute. Fertile segment paniculate. Spores mature in Aug and Sept.

Grassy hillsides, sterile fields, exposed meadows, and sandy open places.

Sceptridium oneidense (Gilbert) Holub
BLUNT-LOBE GRAPE FERN native
IR | **WUP** | CUP | EUP CC 7
Botrychium oneidense (Gilbert) House
Leaves 40 cm long or longer; stem and blade somewhat leathery; blades triangular, 3-partedly decompound, little divided, attached at or near the base; chief terminal segments of blade broadly ovate and obtuse. Fertile segment paniculate. Spores mature in Sept–Oct.

Rich moist woods. The broad, rounded divisions and the shaded habitat are characteristic.

Sceptridium rugulosum (W..H. Wagner) Skoda & Holub
TERNATE GRAPE FERN native
IR | **WUP** | **CUP** | **EUP** CC 6
Botrychium rugulosum W.H. Wagner
Leaves 25 cm long or longer, thin and membranous; blades inserted at the base, 3-parted, with the three major divisions stalked; ultimate segments of blade all about the same size, ovate to oblong, acutish, serrate or entire, and concave. Fertile segment paniculate. Spores mature Aug–Oct.

Swampy woods, brushy fields, wooded streambanks.

Osmundaceae
ROYAL FERN FAMILY

Osmunda *Royal Fern*
Perennial ferns with large rootstocks and exposed crowns covered with old roots and stalks, sending up tufts of coarse leaves. Leaves 1–2-pinnate, differentiated into sterile and fertile segments. Sporangia in round clusters, spores green. The fibrous roots (osmunda fibre) were formerly used as a medium for growing orchids and bromeliads.

1   Leaves 2-pinnate, pinnae ± entire; sporangia on upper half of fertile leaves .... **O. regalis**
1   Leaves 1-pinnate, sterile pinnae deeply cleft; sporangia only near middle of fertile leaves, or fertile and sterile leaves separate 2
2   Fertile and sterile leaves separate, fertile leaves cinnamon-colored, sterile leaves with a tuft of wool in axil of pinnae .... **O. cinnamomea**
2   Fertile pinnae near middle of vegetative leaves, with sterile pinnae above and below fertile portion, fertile portion green-black, pinnae mostly without tuft of wool in axil .... **O. claytoniana**

Osmunda cinnamomea L.
CINNAMON-FERN native
IR | **WUP** | **CUP** | **EUP** FACW CC 5
Osmundastrum cinnamomeum (L.) K. Presl
Large clumped fern, to 1 m or more tall. Blades of

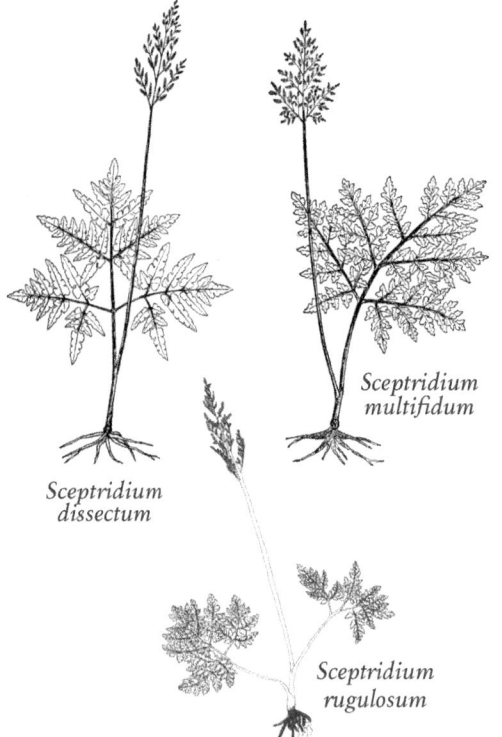

Sceptridium multifidum

Sceptridium dissectum

Sceptridium rugulosum

sterile leaves to 30 cm wide, gradually tapered to tip, 1-pinnate, with conspicuous tuft of white or brown woolly hairs at base of each pinna, pinnae stalkless and deeply cleft into segments, with fringe of short hairs on margins; petioles densely hairy when young. Fertile leaves at center of crown, surrounded by taller sterile leaves, without leafy tissue, arising in spring or early summer and turning cinnamon brown, withering and inconspicuous by midsummer.

Swamps, bog-margins, wooded stream-banks, and low wet places; soils acid.

Osmunda claytoniana L.
INTERRUPTED FERN   native
IR | WUP | CUP | EUP   FAC CC 6

Clumped fern to 1 m or more tall; often forming large colonies. Outer leaves usually sterile, inner leaves larger and with 2–5 pairs of fertile pinnae in middle of blade; fertile segments to 6 cm long and 2 cm wide and much smaller than vegetative segments above and below them; sporangia clusters at first green-black, turning dark brown and withering. Blades 4–10 dm long and 15–30 cm wide; pinnae stalkless and deeply cut into segments, with smooth or slightly hairy margins. Petioles covered with tufts of woolly hairs when young, becoming smooth or sparsely hairy with age, the hairs not forming tufts at pinna-bases (as in *O. cinnamomea*).

Moist or seasonally wet depressions in forests, hummocks in swamps, low prairie, wet roadsides; often in drier places than *Osmunda cinnamomea* or *O. regalis*.

Osmunda regalis L.
ROYAL FERN   native
IR | WUP | CUP | EUP   OBL CC 5

Large fern to 1 m or more tall. Blades broadly ovate in outline, 4–8 dm long and to 3–5 dm wide, 2-pinnate into ± opposite divisions (pinnules), these well-spaced, oblong, rounded at tips, with entire or finely toothed margins. Fertile leaves with uppermost several pinnae replaced by sporangia clusters. Petioles smooth, green or red-green, to 3/4 length of blade.

Bogs, swamps, alder thickets and shallow pools; soils usually acidic.

Polypodiaceae
POLYPODY FERN FAMILY

Polypodium *Polypody*

Polypodium virginianum L.
ROCK POLYPODY   native
IR | WUP | CUP | EUP   CC 8

Polypodium vulgare auct. non L. p.p.

Evergreen, colony-forming fern. Leaf blades to 50 cm long from a creeping rhizome; oblong lance-shaped; pinnatifid, leathery; veins free. Sori round, midway between the midvein and margin, and occurring on the upper segments; indusia absent.

In shallow humus on rocks, in crevices, on woodland banks, and rarely on mossy stumps and in crotches of trees.

Easily identified by the small evergreen blades and its colony-forming habit on rocky slopes, talus, boulders, and ledges.

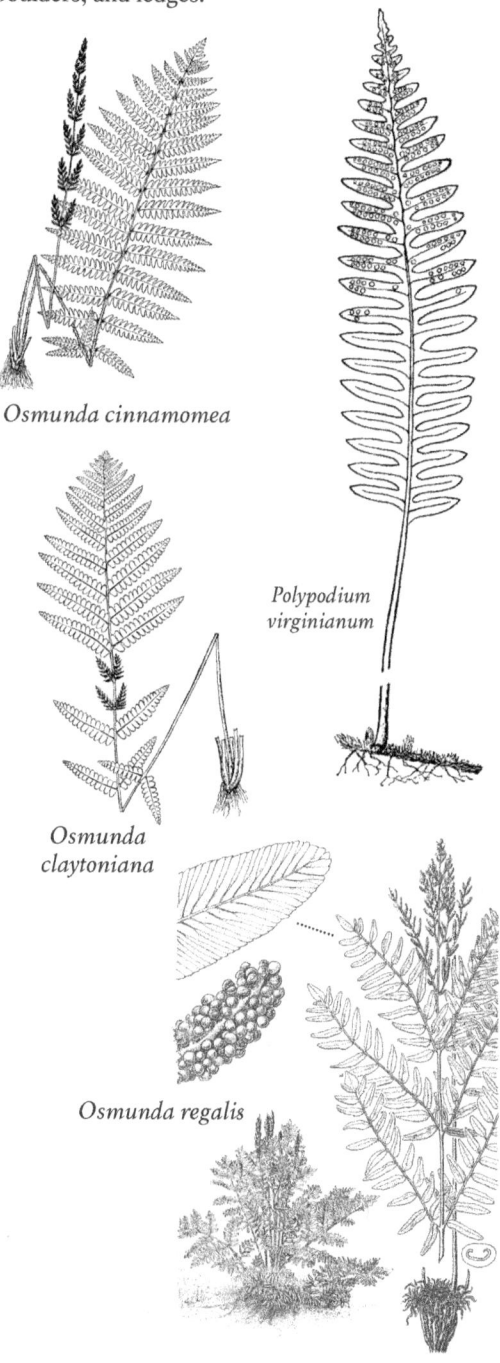

Osmunda cinnamomea

Polypodium virginianum

Osmunda claytoniana

Osmunda regalis

Pteridaceae
MAIDENHAIR FERN FAMILY

Delicate to coarse ferns, deciduous, or evergreen. Blades pinnate to decompound. Sori marginal, protected by the indusium, which opens toward the margin, or by the reflexed margins of the pinnae, or borne along the veins and lacking an indusium.

1. Blade segments separate from one another .... **Adiantum pedatum**
1. Blade segments not separate and distinct........... 2
2. Leaves of two types, the fertile much longer than the sterile; petioles dark brown near base, green above ... . **Cryptogramma stelleri**
2. Fertile and sterile leaves mostly similar; petioles dark brown to black .... **Pellaea**

Adiantum *Maidenhair Fern*

Adiantum pedatum L.
NORTHERN MAIDENHAIR FERN native
IR | WUP | CUP | EUP FACU CC 6

Leaves 30–60 cm tall, in colonies arising from horizontal rhizomes. Petioles lustrous purple-brown, forking at the summit into two arching rachises, each of which is divided several times, thus forming a semicircular blade 15–35 cm wide or wider. Pinnules short-stalked, obliquely triangular oblong; terminal pinnule fan-shaped; main vein along the lower margin; upper margin cleft, with lobes thus formed blunt. Sori elongate, borne on the upper margins of the lobes of the pinnules; indusium formed by the inrolled margin.

Wooded, sometimes rocky slopes, in humus-rich soil.

The usually arching and palmately divided lustrous purple brown rachises and fan-shaped pinnules with the main vein along the lower margin are distinctive.

Cryptogramma *Rockbrake*

Cryptogramma stelleri (Gmel.) Prantl
FRAGILE ROCKBRAKE native
IR | WUP | CUP | EUP FACU CC 10

Small rock fern with dimorphic leaves, from short much branched or slender elongate rhizomes. Leaves glabrous, deciduous, dimorphic, scattered along the horizontal rhizome. Sterile leaves almost flaccid, 3-10 cm long; blades ovate to ovate-deltoid, bipinnate; pinnules oblong or ovate; petioles purplish. Fertile leaves stiffer than sterile blades, 9-21 cm long; pinnules lance-shaped to oblong. Sori marginal, covered by a continuous indusium formed by the reflexed margin.

Moist, shaded, usually calcareous crevices and cliffs. Plants may be easily overlooked as they turn brown later in the season.

ADDITIONAL SPECIES
Cryptogramma acrostichoides R. Br. (American rockbrake), small, tufted fern, in Michigan restricted to crevices of dry, open acidic rock on Isle Royale, where locally common. Leaves to 25 cm long by 4 cm wide, dimorphic, the sterile leaves evergreen and much shorter than the deciduous fertile leaves; petiole brown at base, turning straw-colored, then green upwards; grooved; leaf blade 2-pinnate, with 5-6 pairs of pinnae. Sori elongate, near margins of narrowed segments; margins strongly inrolled and covering the sori.

Pellaea *Cliffbrake*

Small tufted plants from compact rootstocks. Blades firm; petioles and rachises wiry; pinnae gray green; veins free. Sori marginal and confluent under the inrolled and altered margin of the fertile pinnules. Plants gray-green in color that blends with the limestone rock crevices and ledges where they are found.

1. Sterile and fertile leaves different; petiole and rachis scurfy and with incurved hairs .... **P. atropurpurea**
1. Sterile and fertile leaves similar; petiole and rachis glabrous or with a few spreading hairs ..... **P. glabella**

Pellaea atropurpurea (L.) Link
PURPLE-STEM CLIFFBRAKE (THR) native
IR | WUP | CUP | EUP CC 10

Leaves dimorphic; fertile frond 10-35 cm long, 3.5-8 cm wide, longer than the sterile frond. Petioles and rachis dark purple brown; dull, pubescent, with more or less appressed, incurved hairs. Pinnae rigid, evergreen, bluish green, simple above, bipinnate below; fertile pinnae linear to oblong or narrowly ovate, with the lower pinnules stalked; sterile pinnules ovate-oblong. Sori situated around the margins of the fertile pinnules; indusium formed by the inrolled margin of pinnule.

Dry, steep, exposed, limestone rock slopes or cliffs, limestone paving, and tops of large talus boulders.

This species looks somewhat like *P. glabella* but may be distinguished from it by the usually taller, more upright habit, with fertile blades that are more divided, that are darker blue green to olive green, and that have markedly hairy petioles and rachis.

Pellaea glabella Mett.
SMOOTH CLIFFBRAKE native
IR | WUP | CUP | EUP CC 10

Leaves all similar, 10-25 cm long or longer, usually shorter than those of *P. atropurpurea*, open and spreading out beyond the rock face. Petioles and

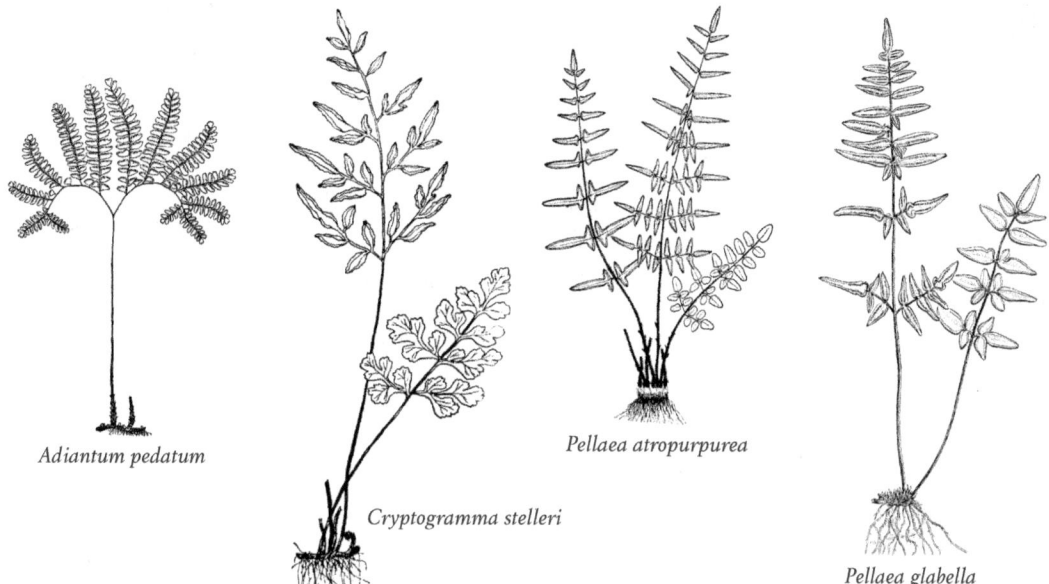

Adiantum pedatum

Cryptogramma stelleri

Pellaea atropurpurea

Pellaea glabella

rachis dark reddish brown, smooth, and lustrous. Pinnae rigid, evergreen, bluish green, simple above, pinnate below; basal pinnae persistent; pinnules sessile or nearly so, oblong lance-shaped. Sori situated around the margins of the fertile pinnules; indusium formed by the inrolled margin of pinnule.

Crevices of dry, sometimes partly shaded, limestone cliffs. Smooth cliffbrake is a distinctive species of high, steep limestone cliffs; it grows from small, tight, crevices and blends well with the background; often there are no other plants associated with it.

Selaginellaceae
SELAGINELLA FAMILY

Selaginella *Spikemoss*
Trailing, evergreen herbs with branched, leafy stems, rooting at branching points. Leaves small and overlapping. Spore-bearing leaves similar to vegetative leaves and clustered in cones at ends of branches. Megaspores 4 in each sporangium, yellow or white; microspores numerous and very small, red or yellow, covered with small spines.

1 Leaves and stems firm and evergreen, the plants forming small tufts 2.5–5 cm high; leaves crowded, very narrow, tipped by a bristle; cones four-angled . . . . **S. rupestris**

1 Leaves and stems lax and subevergreen or deciduous; plants forming small mats; cones nearly round in cross-section 2

2 Leaves in four rows, of two kinds: large and spreading, and small and appressed to stem; cones 0.5–1 cm long . . . . **S. eclipes**

2 Leaves in many rows, all alike, with hairs on margins; fertile branches upright, cones 2–4 cm long . . . . **S. selaginoides**

Selaginella eclipes W. R. Buck.
HIDDEN SPIKEMOSS   native
IR | WUP | **CUP** | **EUP**   CC 5

Selaginella apoda subsp. *eclipes* (W.R.Buck) Skoda
Plants forming large, yellow-green mats of branching, trailing stems with upright tips. Stems slender, to 0.4 mm wide. Leaves scalelike, in 4 rows, of 2 types, the larger leaves spreading, 1–2 mm long and 1 mm wide; the smaller leaves appressed to stem, up to 1 mm long and 0.5 mm wide. Cones 0.5–2 cm long, cylindric in 4 rows, the sporophylls similar to lateral leaves and slightly larger. Megaspores white, with a netlike surface.

Open fens, wet meadows, sandy or marly lakeshores and riverbanks; especially where calcium-rich.

Selaginella rupestris (L.) Spring
LEDGE SPIKE-MOSS   native
IR | WUP | **CUP** | **EUP**   CC 8

Lycopodium rupestre L.
Stems prostrate, forming open mats. Leaves decurrent on the sides of the stem; leaves linear-lance-shaped, about 2.8 mm long (including the ca. 0.7 mm long scabrous seta), grooved on the back, ciliate. Sporophylls narrowly ovate, apiculate, ciliate, about as long as the leaves.

Sand dunes and open or shaded, dry, often igneous rocky bluffs. In eastern UP, known only from Drummond Island.

Selaginella selaginoides (L.) Link
NORTHERN SPIKEMOSS native
IR | WUP | CUP | EUP   FACW CC 10
Trailing, evergreen plants forming small mats. Stems branched, leafy, rooting at branching points. Sterile stems prostrate, 2–5 cm long; fertile stems upright, deciduous, 5–10 cm high and 0.5 mm wide (stem only), changing upward into broader sporophylls. Leaves overlapping in multiple spiral rows, all alike, 2–4 mm long and 1 mm wide, with sharp tips and sparsely hairy margins; spore-bearing leaves similar to vegetative leaves and clustered in cones at ends of branches; cones ± cylindric but with 4 rounded angles, 1.5–3 cm long and to 5 mm wide. Megaspores 4 in each sporangium, yellow-white, with low rounded projections on the 3 flat surfaces.

Calcium-rich fens, mossy hummocks in cedar swamps.

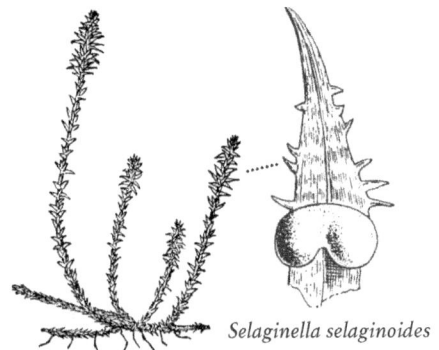
Selaginella selaginoides

Thelypteridaceae
MARSH FERN FAMILY

Medium-sized deciduous ferns, spreading by rhizomes to form colonies; sterile and fertile fronds usually alike, 1-pinnate to pinnate-pinnatifid, with transparent needle-like hairs; sori usually on veins (but not marginal) on pinna underside; indusia present and often soon withering, or absent (*Phegopteris*).

1 Leaf blades 7-25 (-30) cm long, triangular, not more than 2x longer than wide; rachis with wings between the pinnae; sori without indusia; midribs of pinnae lacking a groove on upper surface.............. **Phegopteris**
1 Leaf blades (15-) 20-100 cm long, lance-shaped, oblong-lance-shaped, or triangular, more than 2x longer than wide; rachis without wings between the pinnae; sori with kidney-shaped indusia; midribs of pinnae with a groove on upper surface 2
2 Leaf blade broadest near middle, gradually reduced to base; stipe less than 1/3 length of blade; plants of upland and wetland habitats **Parathelypteris noveboracensis**
2 Leaf blade broadest near base, the pinnae stopping abruptly; stipe 2/3 to fully as long as blade; plants of wetlands..................... **Thelypteris palustris**

Parathelypteris *New York Fern*

Parathelypteris noveboracensis (L.) Ching
NEW YORK FERN native
IR | WUP | CUP | EUP   CC 5
Thelypteris noveboracensis (L.) Nieuwl.
Fronds 25-55 cm long, arising from a slender rhizome and may form dense colonies. Blades elliptic to elliptic-lanceolate, 9-15 cm wide; lower pinnae very reduced, pinnate-pinnatifid; pinnae somewhat hairy on the rachis and veins; pinnules oblong, somewhat blunt. Sori round, situated near the margin; indusia glandular-ciliate.

Moist to wet deciduous or mixed conifer-hardwood forests; often on low hummocks in wettest habitats.

Phegopteris *Beech-Fern*

Deciduous ferns with creeping rhizomes. Leaf blades triangular in outline, 1-pinnate-pinnatifid (the pinnae deeply lobed); rachis winged, with spreading, ovate-lance-shaped scales; veins free, simple or often forked, veins of segment reaching margin or nearly so; underside hairs unbranched, unicellular. Sori round to oblong, indusia absent.

1 Rachis not winged between lowest two pinnae pairs; blades usually longer than wide .... **P. connectilis**
1 Rachis winged between all pinnae pairs; blades usually wider than long .... **P. hexagonoptera**

Phegopteris connectilis (Michaux) Watt
NORTHERN BEECH-FERN native
IR | WUP | CUP | EUP   FACU CC 5
Dryopteris phegopteris (L.) C.Chr.
Thelypteris phegopteris (L.) Sloss
Fern with long, slender, scaly and densely hairy rhizomes. Leaves triangular, 15–25 cm long and 6–15 cm wide; blades 1-pinnate, the pinnalike divisions joined by a wing along rachis, except for lowermost pair which are free and angled downward; pinnules oblong, rounded at tip, and usually hairy; petioles longer than blades, hairy, with narrow, brown scales.

Cool moist woods, thickets, streambanks, sphagnum moss hummocks, shaded rock crevices.

Phegopteris hexagonoptera (Michx.) Fée
BROAD BEECH-FERN native
IR | WUP | CUP | EUP   FACU CC 8
Thelypteris hexagonoptera (Michx.) Weatherby
Leaves 30-60 cm long or longer. Leaf blades broadly triangular, 15-30 cm wide or wider, about as broad as long, tapering to the top, pinnate-pinnatifid; middle and upper pinna-like divisions lance-shaped; lower pinna-like divisions unequally ovate to lance-shaped, not projected forward; all divisions connected by a wing; the segments, especially of the lower pinnae, often deeply pinnatifid. Petiole naked except

at the base; rachis not chaffy or with almost colorless scales; rachis and veins minutely glandular puberulent. Sori small, near the margin.

Rich, often rocky, woods and wooded slopes.

P. hexagonoptera has all the divisions of the blade connected to the rachis, including the basal pair. In shape, the blade is more broadly triangular in broad beech fern than in northern beech fern. The shape of the basal segments is unlike that in *P. connectilis*, being widest in the middle and lobed again rather than entire.

Thelypteris *Marsh-Fern*

Thelypteris palustris Schott
MARSH-FERN   native
IR | **WUP** | **CUP** | **EUP**   FACW CC 2

Small to medium ferns from slender rhizomes; rhizomes slender, spreading and branching. Leaves deciduous, ± hairy, erect, 20–60 cm long and to 15 cm wide; blades broadly lance-shaped, short-hairy on rachis and midveins, tapered to tip and only slightly narrowed at base; 1-pinnate, pinnae in 10–25 pairs, mostly alternate, narrowly lance-shaped, to 2 cm wide. Sterile and fertile leaves only slightly different; sterile leaves thin and delicate, pinnules blunt-tipped, 3–5 mm wide, veins once-forked. Fertile leaves longer than sterile leaves; pinnules oblong, 2-4 mm wide, the margins rolled under, veins mostly 1-forked; petioles longer than blades, black at base, hairless and without scales. Sori round, located halfway between midvein and margin, sometimes partly covered by the rolled under margin; indusia irregular in shape, usually with a fringe of hairs.

Swamps, low areas in forests, sedge meadows, open bogs, calcareous fens, marshes.

Woodsiaceae
CLIFF FERN FAMILY

Woodsia *Cliff Fern*

Small tufted ferns arising from compact rootstocks. Indusium of thread-like or plate-like segments, more or less arched over the round sori. Sometimes confused with *Cystopteris*, and to distinguish *Woodsia* from *Cystopteris*, check if the indusium is attached below the sorus (*Woodsia*) or is hooded (attached at one side and arched over the sorus, *Cystopteris*); petioles are opaque in *Woodsia* and translucent in *Cystopteris* (visible when held up to the light in the field); the veins in *Woodsia* are less distinct and appear to stop short of the margin; in *Cystopteris*, the veins clearly extend to the margin; in *Woodsia*, old petiole bases persist as either an even or uneven stubble (see key).

HYBRIDS

Woodsia × abbeae Butters is a hybrid between *W. ilvensis* and *W. oregana*; reported for Marquette and Ontonagon counties.

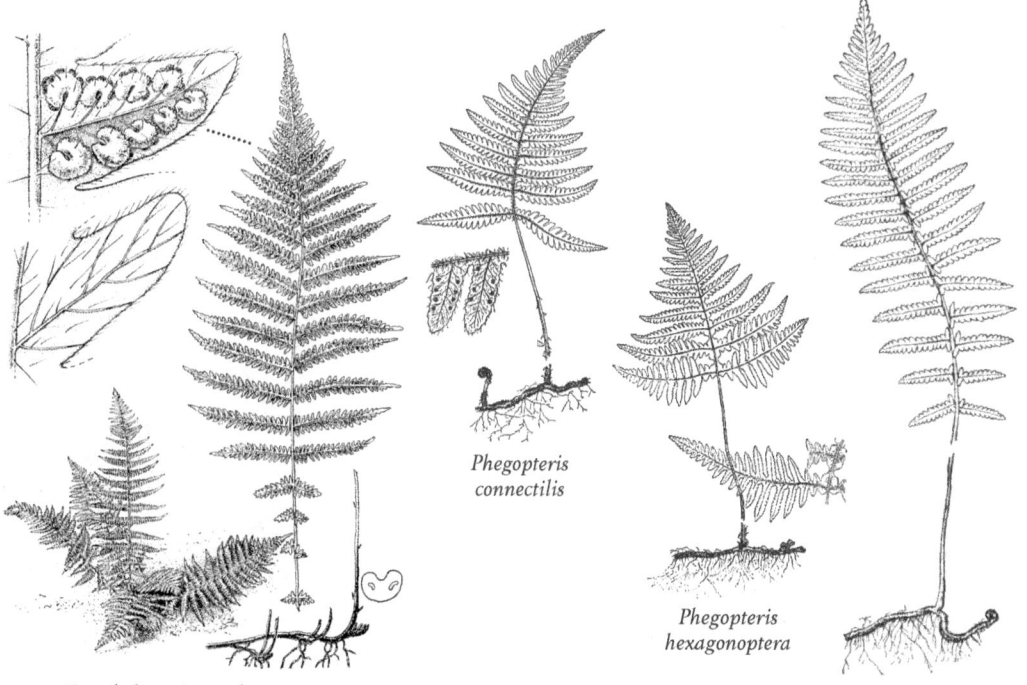

Parathelypteris noveboracensis

Phegopteris connectilis

Phegopteris hexagonoptera

Thelypteris palustris

WOODSIACEAE | Cliff Fern Family

1. Petioles jointed at base, the persistent bases (stubble) all about same length 2
1. Petioles not jointed at base; the persistent bases of the petioles of differing lengths 3
2. Blade narrow, nearly smooth; rare .... **W. alpina**
2. Blade wider, densely hairy; widespread species .... **W. ilvensis**
3. Indusia of narrow, thread-like segments; rachis smooth or with only fine glandular hairs; fern of limestone cliffs and ledges .... **W. oregana**
3. Indusia of several wide segments; rachis glandular-hairy; fern of various types of mostly loose rock .... **W. obtusa**

Woodsia alpina (Bolton) S.F. Gray
ALPINE WOODSIA (END) native
IR | WUP | CUP | EUP CC 10

Small deciduous fern; leaves in dense clusters; sterile and fertile leaves alike; 10-20 cm long. Blades lance-shaped, broadest below middle; 1-pinnate-pinnatifid (barely so), bright green; pinnae 8 to 15 pairs, largest pinnae with 1-3 pairs of pinnules, the shorter ones merely fan-shaped; margins nearly entire, merely lobed; veins free, simple or forked. Petioles red-brown or dark purple when mature, jointed above base at swollen node halfway up petiole, red-brown lance-shaped scales at base, fewer upwards. Rootstock erect, with numerous persistent stipe bases of nearly equal length. Sori round, near the margin; indusium dissected into hairlike segments enveloping sorus, unravelling with maturity; sporangia brownish.

Crevices and ledges of moist, partially shaded cliffs of acidic rock; rare in Keweenaw (including Isle Royale) and Marquette counties.

Woodsia ilvensis (L.) R. Br.
RUSTY CLIFF FERN native
IR | WUP | CUP | EUP CC 10

Small fern with leaves 10–30 cm long, 2–3 cm wide, oblong lance-shaped, pinnate to bipinnate; pinnae oblong lance-shaped; margins of the segments crenate and usually somewhat inrolled. Petioles jointed, with the old petiole-bases persistent and about the same length; rachis and undersurface of the blade usually brown-chaffy. Sori round and close together on the underside; indusia of many long ciliate segments.

Dry, often exposed, rocks and crevices of cliff faces and talus, the rock usually acidic.

Plants are both scaly and glandular.

Woodsia obtusa (Spreng.) Torr.
BLUNT-LOBED WOODSIA (THR) native
IR | WUP | CUP | EUP CC 10

Small fern with leaves 10–30 cm long, 2–10 cm wide. Blades broadly lance-shaped, pinnate; pinnae mostly separate from one another; lower pinnae triangular; middle and upper pinnae ovate lance-shaped, pinnatifid, or pinnate at the base. Petioles not jointed; rachis straw-colored, glandular-hairy. Sori round, located near the margins; indusia covering the sori, splitting into several jagged lobes later in the season.

Usually on talus, occasionally on shaded ledges and rocky slopes.

Woodsia obtusa is an erect, rather robust species. In aspect it resembles *Cystopteris fragilis*, with which it often grows. *Woodsia obtusa* is stiffer, with glands and scales on the axes and veins.

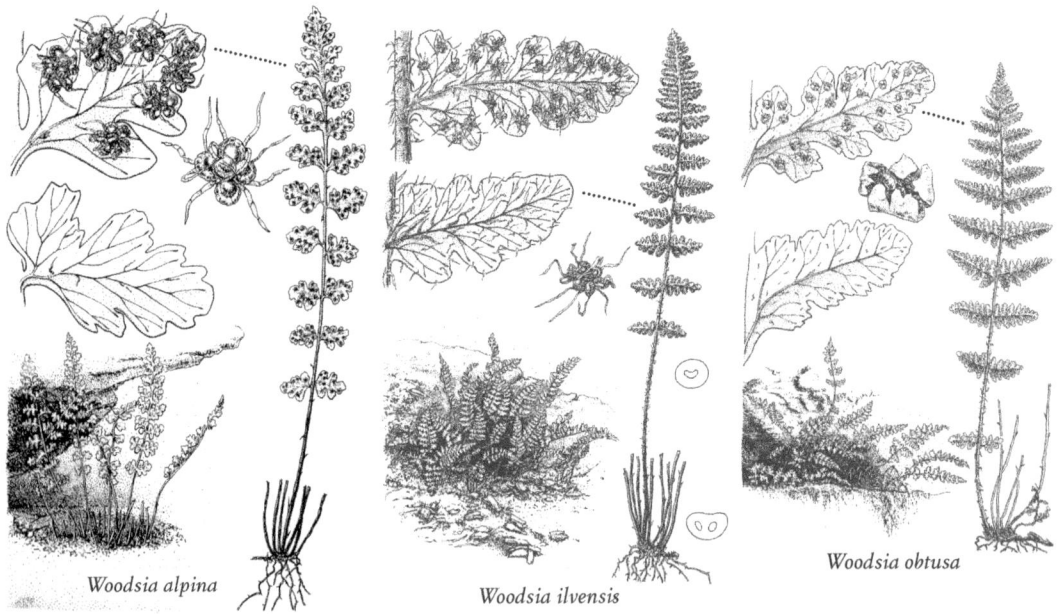

Woodsia alpina

Woodsia ilvensis

Woodsia obtusa

Woodsia oregana D.C. Eat.
OREGON WOODSIA native
IR | WUP | CUP | EUP CC 10

Small fern with leaves mostly 10–30 cm long, 1–3 cm wide. Blades linear-lance-shaped; pinnae opposite, mostly separate from one another, triangular-oblong; pinnules oblong, blunt, with marginal crenulate-serrate teeth often inrolled. Petioles not jointed; rachis dark brown at base, becoming straw-colored above, glabrous or finely glandular, usually without scales. Sori round; indusia of narrow and threadlike segments.

Crevices of calcareous ledges and cliffs. Somewhat similar to *Woodsia ilvensis* but usually without scales and on calcareous rather than acid rock; the stubble is uneven rather than even as in *W. ilvensis*.

CONIFERS

Cupressaceae
CYPRESS FAMILY

Trees or shrubs; leaves opposite or whorled, sometimes dimorphic, separated by short internodes or overlapping. Flowers monoecious or dioecious, solitary, axillary or terminal. Staminate flowers of several stamens with short filaments. Pistillate flowers with opposite or whorled carpels or with 1–3 terminal ovules. Fruit a cone, or becoming fleshy and berry-like.

1 Trees or shrubs; cones berry-like and fleshy . **Juniperus**
1 Trees; cones woody or leathery . . . . **Thuja**

Juniperus *Juniper*

Trees or shrubs; leaves evergreen, scale-like or subulate, opposite or in whorls of 3. Flowers monoecious or dioecious, axillary and short-stalked, or terminal. Staminate flowers catkin-like; stamens numerous, opposite or whorled. Pistillate flowers of several scales, the lower sterile, the terminal sometimes fertile, or sometimes sterile and the ovules terminal. Cone-scales at maturity becoming fleshy and coalescent, forming an indehiscent, usually colored, berry-like fruit with 1–10 seeds; seeds plump, wingless.

1 Leaves in clusters of three, linear and sharp-pointed, jointed at base; cones in leaf axils . . . . **J. communis**
1 Leaves scale-like, not jointed; cones at end of branches
. . . . **J. horizontalis**

Juniperus communis L.
COMMON JUNIPER native
IR | WUP | CUP | EUP FACU CC 4

Evergreen shrub. Leaves in whorls of 3, crowded, linear, sharp-pointed, 6–15 mm long, marked with a median white stripe above. Fruit bluish or black, 6–13 mm in diameter, normally 3-seeded. Ours var. *depressa* Pursh., with branches soon becoming decumbent and forming large circular patches eventually several meters in diameter, usually flat-topped and 0.5–2 m tall; leaves spreading or ascending.

Dry woods, old fields, dried bogs, rocky bluffs.

Juniperus horizontalis Moench
CREEPING JUNIPER native
IR | WUP | CUP | EUP FACU CC 10

Evergreen shrub; branches prostrate, often greatly elongate, bearing numerous erect branchlets 1–3 dm tall. Leaves mostly scale-like and appressed, varying from ovate, 1–2 mm long, to oblong and to 4 mm long. Fruit blue, 5–8 mm in diameter, on short recurved pedicels; seeds 3–5 mm long, not pitted.

On rocks, sandy openings, sandy or gravelly shores, sand dunes and beach ridges along Lake Michigan.

Thuja *Arbor-Vitae*
Thuja occidentalis L.
NORTHERN WHITE CEDAR native
IR | WUP | CUP | EUP FACW CC 4

Shade-tolerant tree to 20 m tall, cone-shaped with widely spreading branches, sometimes layered at base, trunk to 1 m wide or more, bark reddish or gray-brown, in long shreddy strips; twigs flattened, in fanlike sprays. Leaves scale-like and overlapping, 3–6 mm long and 1–2 mm wide, yellow-green, aromatic, persisting for 1–2 years. Seed cones small, brown, 1 cm long, maturing in fall and persisting over winter.

Thuja occidentalis

Juniperus communis   *Juniperus horizontalis*

Cold, poorly drained swamps where *Thuja* may form dense stands; soils neutral or basic, usually highly organic, water not stagnant; also along streams, on gravelly and sandy shores of Great Lakes, and dry soils over limestone.

Pinaceae

PINE FAMILY

Resinous trees with evergreen or deciduous, needlelike leaves. Male and female cones separate but borne on same tree. Male cones small and soft, falling after pollen is shed. Female cones larger, with woody scales arranged in a spiral. Seeds on upper surface of scales.

1	Leaves grouped into clusters	2
1	Leaves not in clusters, alternate on branches	3
2	Leaves evergreen, in clusters of 2–5 needles	**Pinus**
2	Leaves deciduous, with many leaves in each cluster	**Larix laricina**
3	Cones upright; leaves attached directly to branch, not leaving a bump when shed	**Abies balsamea**
3	Cones drooping; leaves attached to a persistent short stalk	4
4	Leaves flat in cross-section, soft	**Tsuga canadensis**
4	Leaves four-sided in cross-section, stiff	**Picea**

Abies *Fir*

Abies balsamea (L.) Miller
BALSAM FIR *native*
IR | WUP | CUP | EUP FAC CC 3

Shade-tolerant tree to 25 m tall, crown spirelike, trunk to 6 dm wide; bark thin, smooth and gray, becoming brown and scaly with age; lower branches often drooping; twigs sparsely short-hairy. Leaves evergreen, linear, 12–25 mm long and 1–2 mm wide, blunt or with a small notch at tip, flat in cross-section, twisted at base and arranged in 1 plane (especially on lower branches), or spiraled on twigs.

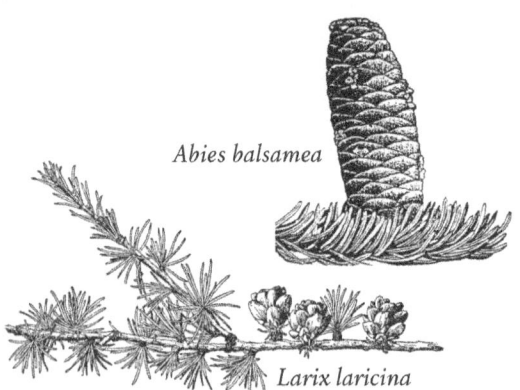

Abies balsamea

Larix laricina

Seed cones 5–10 cm long and 1.5–3 cm wide, with broadly rounded scales.

Cold boreal forests, swamps, and moist forests.

Larix *Larch*

Larix laricina (Duroi) K. Koch
TAMARACK; EASTERN LARCH *native*
IR | WUP | CUP | EUP FACW CC 5

Shade-intolerant tree to 20 m tall, crown narrow, trunk to 6 dm wide; bark smooth and gray when young, becoming scaly and red-brown; twigs yellow-brown, ± horizontal or with upright tips. Leaves deciduous, in clusters of 10–20, linear, 1–2.5 cm long and less than 1 mm wide, soft, blunt-tipped, bright green, turning yellow in fall. Seed cones 1–2 cm long and 0.5–1 cm wide, ripening in fall and persisting on trees for 1 year.

Cold, poorly drained swamps, bogs and wet lakeshores; reaching southern range limit in northern Indiana.

ADDITIONAL SPECIES
Larix decidua P. Mill. (European larch), sometimes planted and occasionally escaping; leaves longer than the native *L. laricina* (to 2.5 cm or more long); Houghton County.

Picea *Spruce*

Evergreen trees; bark thin and scaly, resin blisters common in white spruce (*Picea glauca*). Leaves linear, square in cross-section, stiff, spreading in all directions around twig, jointed at the base to a short projecting sterigma which persists on the leafless branches. Cones borne on last year's branches, drooping, with persistent scales much exceeding the bracts. Seeds wing-margined. Several species are commonly cultivated, especially Norway spruce (*Picea abies*).

1	Leaves mostly 0.5–1.2 cm long, blunt-tipped; twigs with rust-colored hairs; cones ovoid, less than twice as long as wide when open	**P. mariana**
1	Leaves mostly 1–2 cm long, sharp-tipped; twigs glabrous; cones more than twice as long as wide	2
2	Cones 2.5–5 cm long; native species	**P. glauca**
2	Cones larger, 12–15 cm long; introduced species	**P. abies**

Picea abies (L.) Karst.
NORWAY SPRUCE *introduced*
IR | WUP | **CUP** | EUP

Trees to 30 m; crown cone-shaped; bark gray-brown, scaly; branches short and stout, the upper ascending, the lower drooping; twigs stout, reddish brown, usually glabrous. Leaves 1–2.5 cm, 4-angled in cross section, rigid, light to dark green, bearing stomates on all surfaces, apex blunt-tipped. Seed cones (10–

)12–16 cm long; scales diamond-shaped, widest near middle, 18–30 x 15–20 mm, thin and flexuous, margin at apex erose to toothed, apex extending 6–10 mm beyond seed-wing impression.

Introduced from Europe, *P. abies* is the most widely cultivated spruce in North America, widely planted as a shade tree, sometimes used for reforestation, and now locally naturalized in woods; reported for Alger County.

Picea abies

Picea glauca (Moench) Voss
WHITE SPRUCE native
IR | WUP | CUP | EUP FACU CC 3
Moderately shade-tolerant tree to 30 m tall (often smaller), crown conelike, trunk to 60 cm or more wide; bark thin, gray-brown; branches slightly drooping, hairless. Leaves evergreen, linear, 1.5–2 cm long, 4-angled in cross-section, stiff, waxy blue-green, sharp-tipped. Cones 2.5–6 cm long, scales fan-shaped, rounded at tip, the tip entire.

Moist to sometimes wet forests; absent from wetlands where water is stagnant.

Picea mariana (Miller) BSP.
BLACK SPRUCE native
IR | WUP | CUP | EUP FACW CC 6
Moderately shade-tolerant tree to 25 m tall (often smaller), crown narrow, often clublike at top, trunk to 25 cm wide; bark thin, scaly, gray-brown; branches short and drooping, often layered at base. Leaves evergreen, linear, 6–18 mm long, 4-angled in cross-section, stiff, waxy blue-green, mostly blunt-tipped. Seed cones 1.5–3 cm long, scales irregularly toothed, persisting for many years.

Cold, acid, sphagnum bogs, swamps, and lakeshores; often where water is slow-moving and low in oxygen; less common in calcium-rich, well-aerated swamps dominated by northern white cedar (*Thuja occidentalis*).

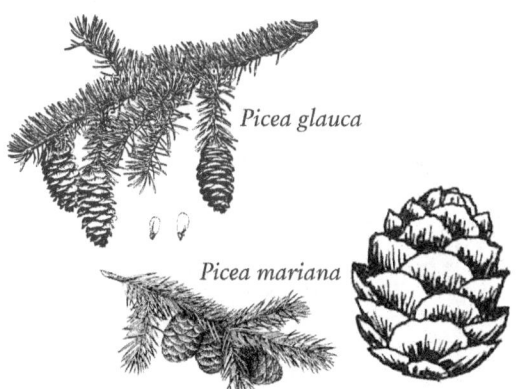

Picea glauca

Picea mariana

Black spruce (*Picea mariana*) can be distinguished from white spruce (*Picea glauca*) by its shorter needles, the branches with fine, white to red-brown hairs, the smaller, rounded seed cones with toothed scale margins, and its occurrence in generally wetter (and sometimes stagnant) habitats.

Pinus *Pine*

Trees (ours) with dimorphic branches and leaves, the foliage leaves borne on dwarf branches only, solitary or in clusters of 2–5. Staminate flowers catkin-like, in fascicles at the base of the current year's growth, each composed of numerous spirally imbricate stamens. Pistillate flowers forming a cone consisting of numerous spirally imbricate cone-scales, each subtended by a bract and bearing 2 inverted ovules at the base. Fruit a hard woody cone, maturing at the end of the second or third season and often long persistent on the tree; seeds winged.

1 Leaves in clusters of 5; cones 10–25 cm long **P. strobus**
1 Leaves in clusters of 2; cones less than 10 cm long ... 2
2 Leaves 10–15 cm long .... **P. resinosa**
2 Leaves 2.5–10 cm long 3
3 Cones persistent on branches; bark dark gray ....
 .... **P. banksiana**
3 Cones not persistent on branches; bark of upper trunk becoming orange-brown .... **P. sylvestris**

Pinus banksiana Lamb.
JACK PINE native
IR | WUP | CUP | EUP FACU CC 5
Usually a small tree, but occasionally to 20 m or more tall, with spreading branches. Leaves in pairs, usually somewhat curved, 2–3.5 cm long, 1–1.5 mm wide. Cones erect or strongly ascending, usually somewhat curved or unsym- metrical, conic, yellowish-brown, 3–5 cm long; seeds about 1.5 cm long.

Dry or sterile, sandy or rocky soil.

Pinus resinosa Soland.
RED PINE native
IR | WUP | CUP | EUP FACU CC 6
Tree, occasionally 40 m tall. Leaves in clusters of 2, slender, soft and flexible, 10–15 cm long, dark green. Cones spreading, conic-ovoid, 4–8 cm long; seeds 1.5–2 cm long.

Dry sandy or rocky soil.

Pinus strobus L.
EASTERN WHITE PINE native
IR | WUP | CUP | EUP FACU CC 3
Tall tree, occasionally as much as 70 m tall, with thick furrowed bark. Leaves in clusters of 5, very slender, pale green and glaucous, 8–13 cm long.

Cones cylindric, 10–15 cm long; seeds, including the wing, 2–3 cm long.

In many different habitats, but preferring fertile or well-drained, sandy soil.

Pinus sylvestris L.
SCOTCH PINE *introduced*
IR | **WUP** | CUP | EUP

Tree to 30 m tall, the larger branches conspicuously orange-brown. Leaves in pairs, bluish-green, stiff, usually twisted, 3–7 cm long, about 1.5 mm wide. Cones soon reflexed, short-ovoid to oblong, often bent, 3–6 cm long.

Native of Europe, where it is an important source of lumber; planted and persisting, and occasionally escaping; Houghton County.

Tsuga *Hemlock*

Tsuga canadensis (L.) Carr.
EASTERN HEMLOCK *native*
IR | **WUP** | **CUP** | **EUP** FACU

Tree to 30 m tall, the bark often purple-brown; twigs pubescent. Leaves linear, 8–15 mm long, blunt, marked beneath with two white strips of stomata, usually minutely spinulose on the margin (best detected by touch), on short (about 1 mm) petioles, spirally disposed on the twigs but forming a flat spray by twisting of the petioles. Cones at maturity pendulous on short peduncles, their scales much larger than the minute bracts; thickly ellipsoid, 12–20 mm long; seeds winged.

Moist soil, rocky ridges and hillsides.

An important component of our northern forests, useful for lumber and tanbark; often cultivated in a number of horticultural forms.

Taxaceae

YEW FAMILY

Taxus *Yew*

Taxus canadensis Marsh.
CANADA YEW; GROUND HEMLOCK *native*
IR | **WUP** | **CUP** | **EUP** FACU CC 5

Straggling evergreen shrub; stems ascending, to 2 m tall. Leaves spirally arranged on the ste, linear, 1–2 cm long, 1–2 mm wide, abruptly narrowed to a sharp point, tapering at base to a poorly defined petiole. Staminate flowers solitary in the axils; pistillate flowers in pairs, each subtended by its pair of scales. Fruit a fleshy red aril, about 5 mm long, open at the top.

Coniferous and mixed woods. A favored winter browse for deer.

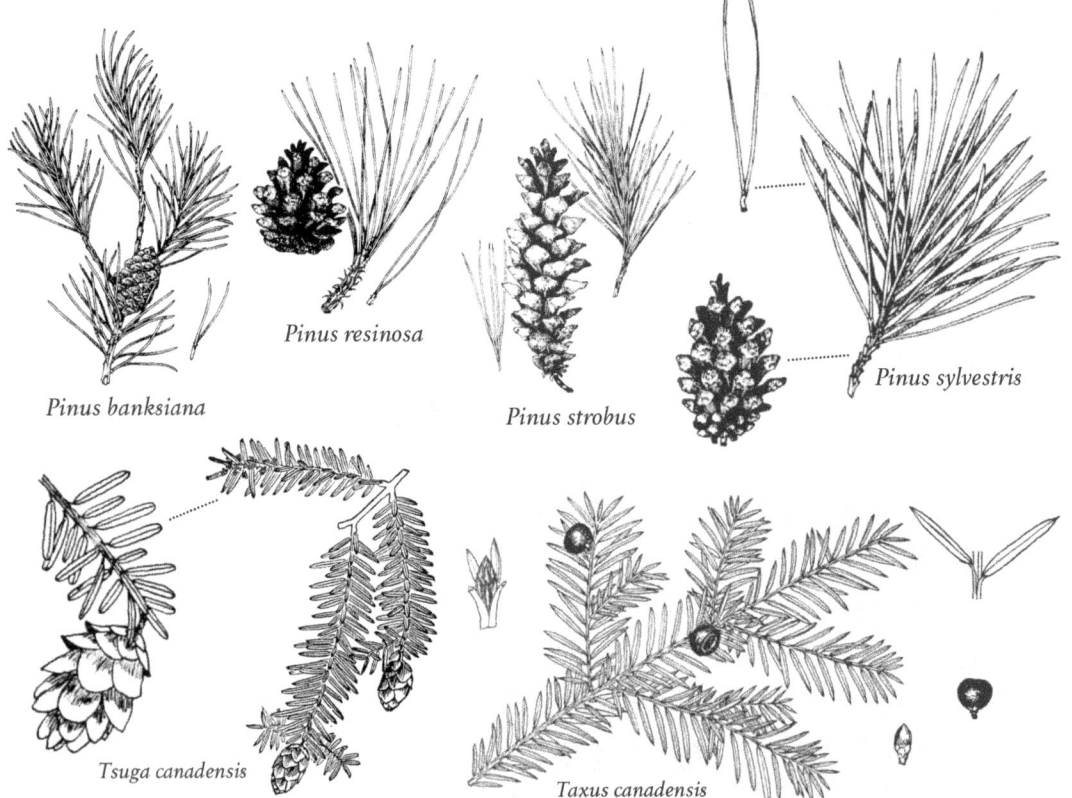

Pinus banksiana

Pinus resinosa

Pinus strobus

Pinus sylvestris

Tsuga canadensis

Taxus canadensis

DICOTS

Adoxaceae
MUSKROOT FAMILY

Our 2 woody genera, *Sambucus* and *Viburnum*, were previously included in Caprifoliaceae.

1 Leaves pinnately compound; fruit with 3 (or more) seed-like pits.................................. **Sambucus**
1 Leaves simple; fruit with only 1 pit......... **Viburnum**

Sambucus *Elder*

Shrubs or small trees. Stems pithy, the bark with wartlike lenticels. Leaves pinnately divided. Flowers in large, rounded terminal clusters; small, perfect; calyx lobes tiny or none; corolla spreading, united at base, 5-lobed, white; stamens 5; stigmas 3. Fruit a red or dark purple, berrylike drupe with 3 nutlets.

1 Flowers opening in summer after leaves developed, in broad, nearly flat clusters; fruit purple-black, edible; leaflets usually 7 .... **S. canadensis**
1 Flowers opening in late spring with unfolding leaves, in pyramid-shaped or rounded clusters; fruit red, inedible; leaflets usually 5 .... **S. racemosa**

Sambucus canadensis L.
COMMON ELDER   *native*
IR | **WUP** | **CUP** | **EUP**   FACW CC 3

Sambucus nigra subsp. *canadensis* (L.) R. Bolli

Shrub to 3 m tall; spreading underground and forming thickets. Young stems soft or barely woody, smooth; older stems with warty gray-brown bark; inner pith white. Leaves large, opposite, pinnately divided into 5–11 (usually 7) leaflets, the lower pair of leaflets sometimes divided into 2–3 segments; leaflets lance-shaped to oval, tapered to a long sharp tip, base often asymmetrical, smooth or hairy on underside, especially along veins; margins with sharp, forward-pointing teeth. Flowers small, white, 5-parted, 3–5 mm wide, numerous, in flat or slightly rounded clusters 10–15 cm wide at ends of stems. Fruit a round, purple-black, berrylike drupe, edible. July–Aug (blooming when fruit of *Sambucus racemosa* is about ripe).

Floodplain forests, swamps, wet forest depressions, thickets, shores, meadows, roadsides, fence-rows.

Sambucus racemosa L.
RED-BERRIED ELDER   *native*
IR | **WUP** | **CUP** | **EUP**   FACU CC 3

Sambucus pubens Michx.

Shrub to 3 m tall. Stems soft or barely woody; twigs yellow-brown and hairy, branches with warty gray-brown bark; inner pith red-brown. Leaves large, opposite, pinnately divided into 5–7 (usually 5) leaflets, the leaflets lance-shaped to ovate, tapered to a long sharp tip, smooth or hairy on underside; margins with small, sharp, forward-pointing teeth. Flowers small, white, 5-parted, 3–4 mm wide, many, in elongate, pyramidal or rounded clusters at ends of stems, the clusters 5–12 cm long and usually longer than wide. Fruit a round, red, berrylike drupe, inedible. May–June (flowers opening with developing leaves).

Occasional in swamps and thickets; more common in moist deciduous forests, roadsides and fencerows.

Sambucus canadensis

Sambucus racemosa

Viburnum *Squashberry; Arrow-Wood*

Shrubs or small trees. Leaves simple, entire, toothed, often palmately lobed. Flowers white or pink, in rounded clusters at ends of stems, sometimes outer florets larger and sterile. Calyx lobes and corolla lobes each 5; stamens 5; style 1; stigmas 1-3. Fruit a fleshy drupe with a single large seed; white, yellow, pink, or orange at first, maturing to orange, red, or blue-black.

ADDITIONAL SPECIES

Viburnum recognitum Fern. (Smooth arrow-wood), native of e USA; leaves coarsely toothed, with prominent veins on leaf underside; Marquette and Schoolcraft counties.

1 Leaves not lobed; pinnately veined................. 2
1 Leaves 3-lobed; palmately veined from base of leaf .. 5
2 Leaves entire, wavy-margined or finely sharp-toothed; lateral veins not terminating in the teeth 3
2 Leaves with large spreading teeth; lateral veins terminating in the teeth .... **V. rafinesquianum**
3 Leaf underside with branched hairs .... **V. lantana**
3 Leaf underside glabrous or scurfy, without branched hairs... 4
4 Inflorescence stalked; leaf margins entire, wavy or with fine rounded teeth .... **V. nudum**
4 Inflorescence sessile or nearly so; leaf margins sharply toothed................................... **V. lentago**

5 Outer flowers large and sterile, much larger than inner flowers .... **V. opulus**
5 Flowers all similar 6
6 Leaf underside densely hairy with branched hairs, margins coarsely toothed; fruit blue-black. . **V. acerifolium**
6 Leaf underside glabrous or hairy on veins only, margins with fine, sharp teeth; fruit yellow, orange or red; Keweenaw County (incl. Isle Royale) .... **V. edule**

Viburnum acerifolium L.
MAPLE-LEAF ARROW-WOOD *native*
IR | **WUP** | **CUP** | EUP   CC 6

Shrub 1–2 m tall. Younger stems, petioles, lower leaf surface and inflorescence finely stellate-pubescent. Leaf blades 6–12 cm long and about as wide, 3-lobed above the middle, coarsely toothed except in the sinuses and toward the broadly rounded or subcordate base; petioles slender, 1–2 cm long. Flowers in cymes on terminal peduncles, with usually 7 rays; corollas 4–5 mm wide; stamens long-exsert, on filaments 3–4 mm long. Fruit a drupe, purple-black when ripe, ellipsoid or subglobose, about 8 mm long; seed a lenticular stone, with 3 shallow grooves on one side and 2 on the other. May–June.

Moist or dry woods.

Viburnum edule (Michx.) Raf.
SQUASHBERRY   (THR) *native*
IR | **WUP** | CUP | EUP   FACW CC 10

Viburnum pauciflorum Bach. Pyl. ex Torr. & A. Gray

Shrub 1–2 m tall. Stems upright or spreading; twigs brown-purple, smooth, often angled or ridged. Leaves opposite, mostly shallowly 3-lobed and palmately veined (leaves at ends of stems often unlobed), 5–12 cm long and 3–12 cm wide, tapered to a sharp tip; underside veins hairy; margins coarsely toothed; petioles 1–3 cm long, Flowers creamy-white, small, in few-flowered, stalked clusters 1–3 cm wide, on short, 2-leaved branches from lateral buds on last year's shoots. Fruit a round, berrylike drupe 6–10 mm long, yellow at first, becoming orange or red. June–July; fruit ripening in late-summer.

Moist, shaded talus slopes; Keweenaw County (including Isle Royale).

Viburnum lantana L.
WAYFARING-TREE   *introduced*
IR | **WUP** | **CUP** | **EUP**

Tall shrub. Young stems, naked winter buds, petioles, and lower leaf-surface gray-pubescent with stellate hairs. Leaf blades oblong to ovate, 5–10 cm long, acute or obtuse, finely serrate, rounded or cordate at base, pinnately veined; petioles 1–3 cm long. Flowers in short-stalked cymes, about 7-rayed; flowers all alike, about 4 mm wide. Fruit a red drupe, 8–10 mm long; seed a stone, furrowed on both sides. June.

Native of Eurasia, occasionally escaped.

Viburnum lentago L.
NANNY-BERRY   *native*
IR | **WUP** | **CUP** | **EUP**   FAC CC 4

Tall shrub or small tree, glabrous throughout or minutely scurfy on the inflorescence or petiole. Leaf blades ovate, varying to oblong or orbicular, 5–8 cm long, all or the uppermost abruptly and sharply acuminate, sharply and finely serrate, the teeth often incurved and callous-tipped. Flowers in sessile cymes, 5–10 cm wide, with 3–5 (rarely 7) rays; flowers 4–8 mm wide. Fruit a drupe blue-black with a whitish bloom, ellipsoid to subglobose, 8–14 mm long; seed a flat, oval stone, scarcely grooved. May–June.

Woods and roadsides.

Viburnum nudum L.
WITHE-ROD; WILD RAISIN   *native*
IR | **WUP** | **CUP** | **EUP**   FACW CC 6

Viburnum cassinoides L.

Shrub to 4 m tall. Young stems brown-scurfy at first, becoming smooth; winter buds gold-brown. Leaves opposite, oval to oblong, 5–10 cm long and 3–5 cm wide, tapered to an abrupt, blunt tip, main vein on underside brown-hairy; margins entire or shallowly toothed; petioles grooved, 5–15 mm long. Flowers creamy-white, unpleasantly scented, all-alike, in more or less flat-topped clusters 5–10 cm wide at ends of stems. Fruit a round to oval drupe, 6–10 mm long, yellow-white at first, then pink, ripening to blue or blue-black and covered with a waxy bloom. May–July.

Cedar swamps, open bogs, fens, floodplain forests, wetland margins; occasional in drier woods.

Viburnum opulus L.
HIGH-BUSH CRANBERRY   *native-introduced*
IR | **WUP** | **CUP** | **EUP**   FACW CC –

Viburnum trilobum Marshall

Shrub, 3–4 m tall. Young stems smooth. Leaves opposite, maple-like, sharply 3-lobed and palmately veined, 5–10 cm long and about as wide, the lobes tapered to sharp tips; smooth or hairy beneath, especially on the veins; margins entire or coarsely toothed, petioles grooved, 1–3 cm long, with several club-shaped glands present near base of blade. Flowers white, in large, flat-topped clusters 5–15 cm wide at ends of stems; outer flowers sterile with large petals, surrounding the inner, smaller fertile flowers. Fruit an orange to red, round or oval drupe, 10–15 mm long. June.

Swamps, fens, streambanks, shores, ditches.

Our native plants are var. *americanum*:

1 Larger petiolar glands less than 1 mm long (rarely absent), usually stalked and flat topped ....
... var. *americanum* (syn: *Viburnum. trilobum* Marshall)
1 Larger petiolar glands (0.8–) 0.9–1.5 (–2) mm long, usually sessile and with the apex indented .... var. *opulus*

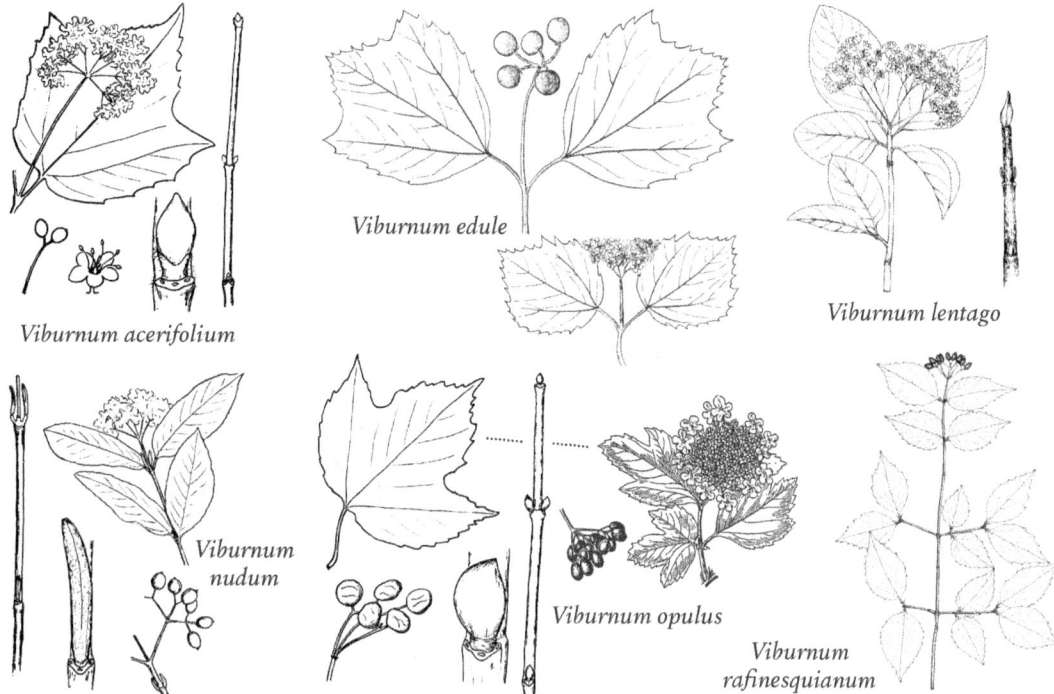

Viburnum rafinesquianum J.A. Schultes
DOWNY ARROW-WOOD native
IR | **WUP** | **CUP** | EUP CC 5
Viburnum affine Bush ex C.K. Schneid.

Shrub to 1.5 m tall, the younger stems glabrous or very sparsely stellate. Leaf blades ovate lance-shaped to ovate, or even subrotund, 3–7 cm long, acuminate or acute, at base obtuse, truncate or subcordate, usually downy hairy on underside; petioles short, pubescent, most with a pair of linear stipules near their base. Flowers in cymes, these sessile or on peduncles to 6 cm long, 4–7-rayed; hypanthium glandular. Fruit a blue-black, flat-ellipsoid drupe, 6–8 mm long; seed a flattened stone, shallowly grooved on both sides. May–June.

Dry, sandy or rocky forests, usually with oak, hickory, or pine; sometimes in wetlands and thickets along rivers.

Amaranthaceae
AMARANTH FAMILY

Our species annual or perennial herbs. Leaves simple, alternate, or occasionally opposite (*Salicornia*). Flowers small, often aggregated into large spikes, panicles, or heads, in some species with conspicuous colored bracts. Flowers perfect or unisexual; sepals usually 5; petals absent; ovary superior, 1-chambered. Fruit a 1-seeded utricle; seeds lenticular. The Amaranthaceae now includes former members of the Chenopodiaceae.

1. Leaves opposite; either much reduced and scale like or with white silky-woolly hairs on both surfaces....... 2
1. Leaves alternate (or the lower sometimes opposite), well developed but without white silky-woolly pubescence 3
2. Leaves linear-lanceolate; stem and leaves (both surfaces) with white silky-woolly hairs .... **Froelichia**
2. Leaves much reduced and scale-like, scarious, glabrous, connate; stem branches succulent, glabrous, appearing jointed, the flowers entirely sunk in the fleshy internodes................................... **Salicornia**
3. Leaf tips with a sharp spine over 0.5 mm (usually ca. 1 mm, even longer on bracts subtending flowers); leaves filiform, ± terete; fruit horizontal, 1–1.3 mm long, slightly broader, covered by the perianth; tepals with transverse keel or wing sometimes longer than body of tepal .... **Salsola**
3. Leaf tips at most with mucro less than 0.5 mm long; leaves various in width, flat; fruit and perianth various 4
4. Flowers unisexual (plants monoecious or dioecious); tepals and bracts acute, scarious or fruit in most if not all flowers enveloped by a pair of bracteoles (perianth absent)... 5
4. Flowers mostly bisexual; fruit not enveloped by bracts but perianth may cover it; bracts herbaceous or firm and hardened, not scarious........................... 6
5. Bracts and tepals all acute, scarious ..... **Amaranthus**
5. Bracts beneath pistillate flowers broad and usually tuberculate and toothed with margins partly fused, obtuse to acute but herbaceous in texture (or even hardened in one species), tepals herbaceous . **Atriplex**

6   Leaves linear to narrowly lanceolate, less than 4 (–6) mm broad, entire, 1 (–3)-nerved 7
6   Leaves usually at least 4 mm broad, toothed to sinuate or crenulate on the margin (if entire, then pinnate- or 3-nerved and not linear) 9
7   Inflorescence and leaves beneath farinose; flowers crowded on short branches that exceed their subtending bracts . . . . **Chenopodium pratericola**
7   Inflorescence and leaves not farinose; flowers 1–3 in the axils of longer bracts. 8
8   Leaves and bracts green to the tips, not mucronate; bracts long-ciliate, especially basally; fruit horizontal, round, less than 1 mm long, enclosed by the perianth (each tepal with a transverse wing) . . **Bassia scoparia**
8   Leaves tipped with a non-green sharp mucro less than 0.5 mm long; bracts glabrous to pubescent but not long-ciliate; fruit various . . . . **Corispermum**
9   Fruit horizontal, completely encircled by the connate wing of the perianth; styles 3 . . . . **Cycloloma**
9   Fruit horizontal or vertical, but the perianth without connate wing; styles usually 2 10
10 Tepals with transverse (but separate) wings; leaves entire, not over 5 mm wide; fruit horizontal; bracts long-ciliate, especially basally . . . . **Bassia scoparia**
10 Tepals not transversely winged (may be keeled); leaves and fruit various; bracts not ciliate 11
11 Leaves (and rest of plant) neither glandular nor pubescent, but farinose in some species . . . . . **Chenopodium**
11 Leaves with yellow to orange resinous glands or gland-tipped hairs at least beneath, not farinose; bruised plant strongly aromatic . . . . **Dysphania**

Amaranthus *Amaranth*

Annual herbs; stems erect, ascending, or prostrate, usually much branched. Leaves alternate, petiolate, entire or sinuate, stipules absent. Flowers in small clusters in the axils, or aggregated into axillary or terminal, simple or panicled spikes; flowers small, each subtended by bracts, the bracts sometimes colored and showy; stamens and pistils in different flowers on the same or different plants; calyx of 3–5 scarious or membranous sepals separate to the base; stamens 2–5; ovary short and broad, compressed; style short or none; stigmas 2 or commonly 3, pubescent. Fruit a thin-walled or leathery utricle, indehiscent or commonly opening at the middle, crowned by the persistent stigmas; seed flattened or lenticular.

1   Flowers all or nearly all in small clusters from the leaf axils (a small terminal panicle may also be present) . . 2
1   Flowers mainly in elongate, spike-like, terminal clusters (small axillary clusters may be present) 3
2   Plants bushy tumbleweeds . . . . **A. albus**
2   Plants prostrate . . . . **A. blitoides**
3   Sepals obtuse, upper portion curved outward . . . . **A. retroflexus**
3   Sepals acute, straight or nearly so. . . . . **A. powellii**

Amaranthus albus L.
TUMBLEWEED *native*
IR | **WUP** | **CUP** | **EUP** FACU CC 0

Plants bushy-branched, to 1 m high and wide; stems whitish. Leaves of the flowering branches elliptic to oblong or obovate, 5–30 mm long, pale green, obtuse or rounded, attenuate at base to a long petiole; early leaves often up to 8 cm long. Flowers in short dense axillary clusters; bracts rigid, subulate, about twice as long as the flowers; sepals of the pistillate flowers commonly 3, uneven, the longest about equaling the utricle. Fruit a lenticular utricle, 1–2 mm long, opening at the middle, wrinkled when dry; seeds lenticular, to 1 mm wide.

Disturbed areas such as roadsides and railways; also sandy lakeshores and streambanks.

Amaranthus blitoides S. Wats.
MAT AMARANTH *introduced*
IR | **WUP** | **CUP** | EUP FACU

Stems prostrate, much branched, 2–6 dm long. Leaves numerous, often crowded, pale green, oblong to obovate, 1–4 cm long, obtuse or rounded, attenuate into a long petiole. Flowers in short dense axillary clusters; bracts about equaling the sepals, acuminate, scarcely aristate; sepals of the pistillate flowers normally 5, occasionally 4, ovate to oblong, unequal in length. Utricle thick-lenticular, 2–2.5 mm long, about equaling the longest sepal, smooth or nearly so, circumscissile at the middle; seed nearly circular, 1.4–1.7 mm wide.

Disturbed areas such as yards and along roads and railways. Native of the Great Plains, weedy throughout much of Michigan.

Amaranthus powellii S. Wats.
GREEN AMARANTH *introduced*
IR | WUP | **CUP** | EUP

Stems to 2 m tall, freely branched, glabrous or finely hairy. Leaves long-petioled, lance-ovate, mostly to 10 cm long in well developed plants. Inflorescence terminal, stiff, dense and spike-like, unbranched or with a few widely spaced long branches, dull greenish, not showy; bracts about 5 mm long, much longer than the sepals and fruits, with a very thick, excurrent midrib; sepals 3–5, with simple midvein, those of the pistillate flowers sharply acute, unequal, 2–3 mm long, the longer (outer) ones generally surpassing the fruit; stamens as many as the sepals; Fruit slightly rugose; seeds dark brown, 1–1.3 mm wide.

Weedy in cultivated fields and on roadsides; Dickinson County.

Amaranthus retroflexus L.
RED-ROOT AMARANTH *introduced*
IR | **WUP** | **CUP** | EUP FACU

Stems stout, erect, usually branched, finely villous, up to 2 in. tall. Leaves long-petioled, ovate or rhombic-ovate, up to 1 dm long. Terminal panicle of several or many, short, densely crowded, ovoid, obtuse spikes, the whole 5–20 cm long; similar but smaller panicles produced from the upper axils; bracts rigid, subulate, much longer than the calyx, 4–8 mm long; sepals of the pistillate flowers 5, oblong lance-shaped, rounded or truncate, mucronate, much exceeding the utricle, 3–4 mm long. Utricle compressed, 1.5–2 mm long, circuinscissile at the middle, the upper part rugulose; seeds round-obovate, dark red-brown, 1–1.2 mm long.

Weedy along roadsides and in fields and gardens, rarely along sandy lakeshores.

Amaranthus powellii resembles *A. retroflexus* in general habit but is nearly glabrous, with sharply acute sepals.

Atriplex *Spearscale; Orache*

Atriplex patula L.
HALBERD-LEAF ORACHE *introduced*
IR | **WUP** | **CUP** | EUP

Taprooted annual herb. Stems erect to sprawling, usually branched, 2–10 dm long. Leaves alternate (or the lowest opposite), lance-shaped or triangular, 2–8 cm long and 1–6 cm wide, with outward pointing basal lobes, gray and mealy when young, becoming dull green and smooth with age; petioles present, or absent on upper leaves. Flowers tiny, green; either staminate or pistillate but on same plant, usually intermixed in crowded spikes from leaf axils and at ends of stems, the spikes without bracts or with a few small bracts near base of spikes; staminate flowers with a 5-lobed calyx and 5, stamens; pistillate flowers without sepals or petals, surrounded by 2 sepal-like, small bracts, these expanding and enclosing fruit when mature. Fruit a lens-shaped utricle, dark brown to black, 1–3 mm wide. Aug–Sept.

Shores, streambanks and mud flats, usually where brackish; disturbed places.

Bassia *Smotherweed*

Bassia scoparia (L.) A.J. Scott
MEXICAN-FIREWEED *introduced*
IR | **WUP** | **CUP** | EUP FACU

Kochia scoparia (L.) Schrad.

Annual hairy herb. Stems erect, bushy-branched, to 1 m tall, usually villous above. Leaves linear to narrowly lance-shaped, sessile. Flowers solitary or paired, perfect or pistillate, sessile in the axils of bracts, forming dense, axillary or terminal spikes; calyx 5-lobed, at maturity 2.5 mm wide, each sepal incurved over the fruit and bearing a short dorsal wing; stamens 5; styles 2 or 3, exserted, villous, to 2 mm long; pericarp thin, free; seed 1.5 mm wide. Late summer.

Native of Asia; occasionally escaped from cultivation, especially along railroads and highways where salt applied in winter; plants turn bright red in fall.

Chenopodium *Goosefoot*

Taprooted annual herbs. Stems erect to spreading. Leaves alternate, mostly lance-shaped to broadly triangular, somewhat fleshy and often mealy on lower surface. Flowers perfect, small and numerous,

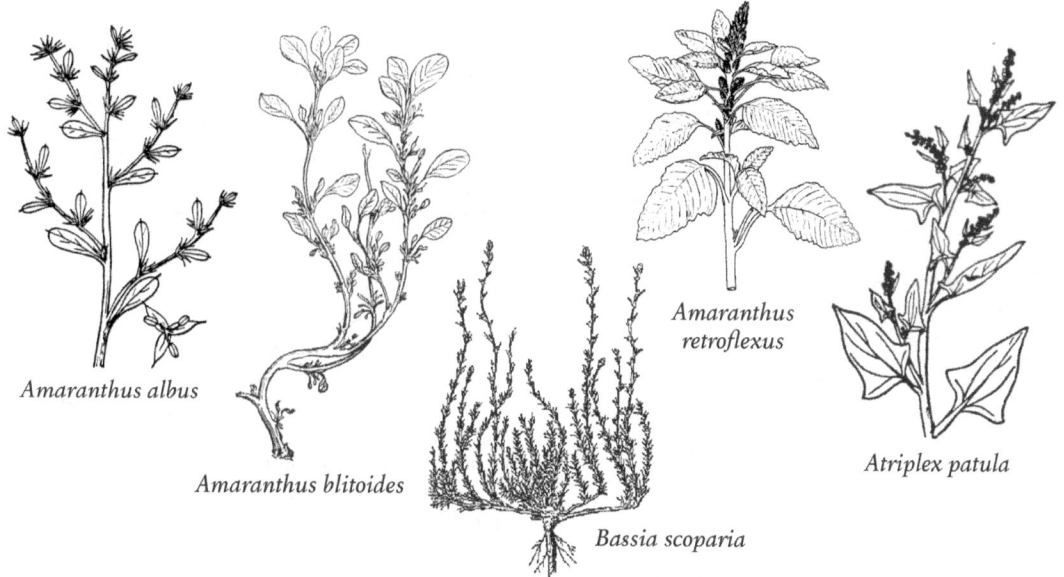

Amaranthus albus

Amaranthus blitoides

Bassia scoparia

Amaranthus retroflexus

Atriplex patula

green or red-tinged, in dense spike-like clusters from leaf axils or at ends of stems, the spikes with small leafy bracts; sepals often curved over the fruit; petals absent; stamens 1–5; styles 2–3. Fruit a 1-seeded utricle; seeds with edge vertical or horizontal.

ADDITIONAL SPECIES
Chenopodium bonus-henricus L. (Good King Henry), introduced; Houghton County.

1 Seeds erect; sepals mostly 3 2
1 Seeds horizontal; sepals 5 3
2 Leaves white-mealy on underside, dull green above . . . . **C. glaucum**
2 Leaves not white-mealy when mature, green on upper and lower sides . . . . **C. capitatum**
3 Mature sepals rounded to conform with fruit, the midvein not much raised . . . . **C. simplex**
3 Mature sepals raised, folded, or hood-like, the calyx appearing somewhat star-shaped 4
4 Leaves narrow, linear to lance-shaped, mostly entire . . . . **C. pratericola**
4 Leaves wider, lance-shaped to ovate or triangular 5
5 Seed loosely enclosed by dry, brittle pericarp . . . . **C. standleyanum**
5 Seed tightly enclosed by thin, membranous pericarp . . . . **C. album**

Chenopodium album L.
LAMB'S QUARTERS; PIGWEED *native*
IR | **WUP** | **CUP** | EUP   FACU CC 0

Annual; leaves and inflorescence often red or reddish late in the season. Stems stout, erect, usually much branched, to 1 m or more tall. Leaves green or more or less white-mealy, broadly rhombic-ovate to lance-shaped, 3–10 cm long, broadly cuneate at base, the larger almost always toothed. Flowers in dense glomerules, these forming interrupted or continuous spikes grouped into a terminal panicle; calyx more or less white-mealy, its segments covering the fruit. Pericarp thin and delicate, when dry minutely rugulose-reticulate; seeds black, shining, usually 1–1.5 mm wide, smooth or sculptured. Highly variable.

Fields, gardens, roadsides, waste ground, dry woods, and barrens.

Chenopodium capitatum (L.) Ambrosi
STRAWBERRY-BLITE *native*
IR | **WUP** | **CUP** | EUP   CC 5

Annual herb. Stems erect or ascending, 2–6 dm tall, branched from the base. Lower petioles often exceeding the blades, the upper much shorter. Leaves triangular or triangular-hastate, up to 10 cm long, acute, broadly truncate to an acute base, above the lateral angles entire to coarsely sinuate-dentate. Flowers in globose clusters 5–10 mm wide at anthesis, in the upper axils or in a terminal leafless spike; calyx deeply 5-parted, the segments oblong to obovate, concave, at maturity commonly enlarged, fleshy, confluent, bright red; seeds vertical, dull black, about 1.5 mm wide, flattened, narrowly margined.

Woodland clearings, often following a fire, roadsides, and waste places.

Chenopodium glaucum L.
OAK-LEAF GOOSEFOOT INTRODUCED
IR | **WUP** | **CUP** | EUP   FACW

Annual herb. Stems upright to sprawling, 1–6 dm long, usually branched from base, sometimes red-tinged. Leaves lance-shaped to ovate, 1–4 cm long and to 2 cm wide, dull green above, densely white-mealy on underside (especially when young); margins entire, wavy, or with few rounded teeth; petioles slender, shorter on upper leaves. Flowers in small, often branched, spike-like clusters from leaf axils, the spikes often shorter than leaves; sepals mostly 3; petals absent; seeds dark brown, shiny, 1 mm wide. Aug–Oct.

Shores, streambanks, and disturbed areas such as railroad ballast and barnyards, soils often brackish. Introduced from Eurasia.

Chenopodium pratericola Rydb.
DESERT GOOSEFOOT *introduced*
IR | **WUP** | **CUP** | EUP

Annual herb. Stems erect, to 8 dm tall, commonly with many ascending branches. Leaves erect or ascending, linear to narrowly lance-shaped, cuspidate, entire, 1–3-nerved without apparent secondary veins, cuneate to a short petiole, glabrous or glabrescent above, densely and usually completely white-mealy beneath, the principal ones commonly 2–4 cm long, 2–5 mm wide. Inflorescence white-mealy throughout, of numerous small glomerules disposed in short, terminal or subterminal, erect or ascending spikes, forming a slender panicle; sepals carinate at maturity. Pericarp easily separable; seeds black, shining, about 1 mm wide.

Occasional on lakeshores, prairies, barrens, and waste ground. Native of w USA, considered adventive in Mich; Keweenaw and Menominee counties.

Chenopodium simplex (Torr.) Raf.
MAPLE-LEAF GOOSEFOOT *native*
IR | **WUP** | **CUP** | EUP   CC 1

Annual herb. Stems erect, bright green, to 1.5 m tall. Leaves long-petioled, broadly ovate to deltoid, 5–20 cm long, truncate to rounded or cordate at base, bearing on each side 1–4 large teeth separated by broadly rounded sinuses. Inflorescence a loose, sparsely flowered, terminal panicle of short, interrupted spikes, the branches often white-mealy; calyx sparsely or not at all mealy; seeds horizontal, loosely

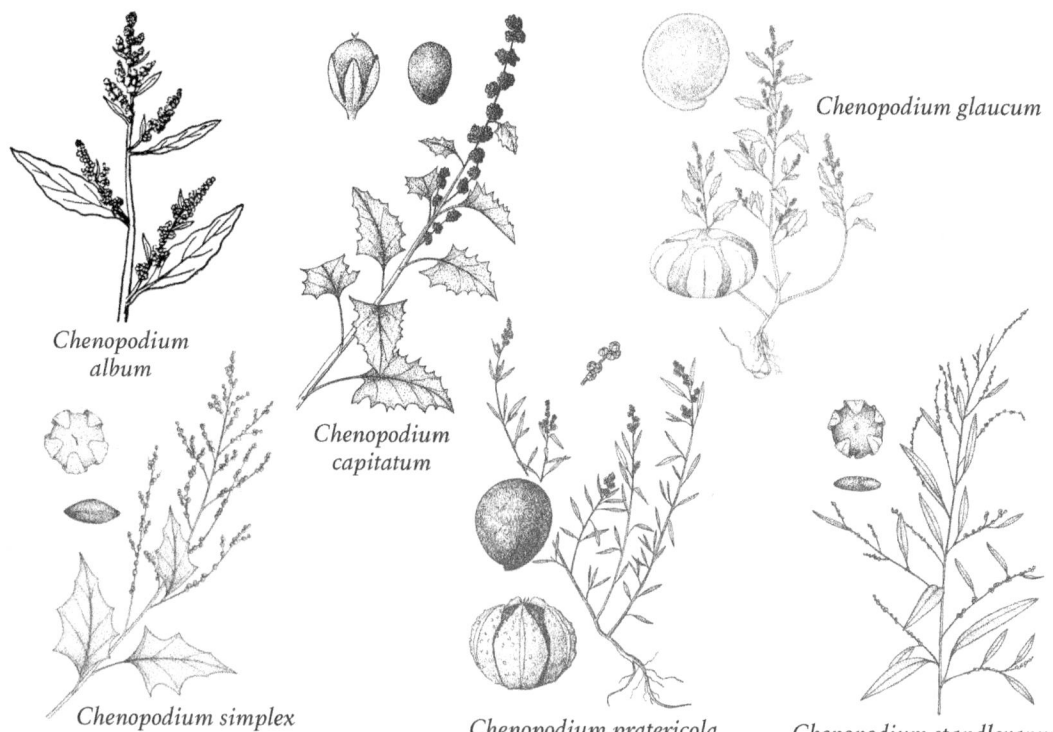

Chenopodium album
Chenopodium glaucum
Chenopodium capitatum
Chenopodium simplex
Chenopodium pratericola
Chenopodium standleyanum

or tightly enclosed in the readily separable pericarp, shiny-black, 1.5–2.5 mm wide, with a bluntly keeled margin.

Disturbed ground and moist woods.

Chenopodium standleyanum Aellen
WOODLAND GOOSEFOOT   *native*
IR | **WUP** | **CUP** | EUP   CC 5

Annual herb. Stems slender, erect or arched, to 1 m tall. Leaves green or sparsely white-mealy, lance-shaped or rarely ovate, to 8 cm long, entire or the larger with a few low teeth, acute or cuneate at base. Flowers single to few in small glomerules, these forming short interrupted spikes, the latter grouped into a loose, open, slender, often nodding, terminal panicle; calyx more or less white-mealy, scarcely covering the fruit. Pericarp smooth, papery, easily separable; seed black, shining, about 1 mm wide, smooth to faintly striate.

Dry open woods.

Corispermum *Bugseed*

Corispermum americanum (Nutt.) Nutt.
BUGSEED   *native*
IR | WUP | **CUP** | EUP   FACU CC 3
  Corispermum orientale Lam.

Annual herb. Stems slender, much branched, 16 dm tall, often pubescent when young, especially about the inflorescence. Leaves often deciduous early, linear, 1–6 cm long, 1–3 mm wide, glabrous or sparsely pubescent. Spikes densely to loosely flowered, 2–10 cm long, 3–8 mm wide; bracts ovate, 4–10 mm long, long-acuminate, concealing the fruits, the lowest often approximating the leaves in shape and size. Fruit obovate, 2–4 mm long, with a pale firm wing to 0.5 mm wide.

Sandy shores and soils, occasionally adventive along railways and waste places.

ADDITIONAL SPECIES
Corispermum pallasii Steven, (Siberian bugseed), introduced; Alger and Mackinac counties.
Corispermum villosum Rydb. (Hairy bugseed), adventive in Marquette County.

Cycloloma *Winged-Pigweed*

Cycloloma atriplicifolium (Spreng.) Coult.
WINGED-PIGWEED   *native*
IR | WUP | **CUP** | EUP   FACU CC 3

Annual branched herb. Stems 1–8 dm tall, pubescent when young, soon glabrescent. Leaves pale green, early deciduous, lance-shaped in outline, coarsely and irregularly sinuate-toothed, the lower up to 8 cm long, the upper progressively reduced. All terminal branchlets bearing flowers, forming spikes 2–6 cm long. Flowers closely sessile, subtended by tiny bracts, perfect or pistillate; calyx persistent, 5-lobed to about the middle, the segments usually keeled, incurved over the ovary; stamens 5, flattened; styles 2 or commonly 3. Fruit plano-convex, purple-black,

Dysphania Wormseed
Previously included in genus *Chenopodium*.

1. Leaf blades ± copiously covered with short spreading gland-tipped hairs; stem with abundant stalked glands; flowers in branched axillary cymes . . . . **D. botrys**
1. Leaf blades with sessile glands on the underside; stem glabrous or pubescent, but not or only sparsely glandular; flowers in axillary and terminal spike-like inflorescences . . . . **D. ambrosioides**

Dysphania ambrosioides (L.) Mosyakin & Clemants
MEXICAN TEA; WORMSEED *introduced*
IR | **WUP** | CUP | EUP   FACU

Chenopodium ambrosioides L.

Annual or perennial, unpleasantly aromatic. Stems erect, to 1 m tall, with numerous ascending branches. Leaves beset with minute yellow glands, the lower leaf blades lance-shaped to ovate, to 12 cm long, deeply sinuate-pinnatifid to merely serrate, cuneate at base, the upper progressively reduced, less toothed or entire. Flowers sessile in small glomerules which are disposed in slender, elongate, bracted or bractless spikes, these in turn forming large or small, open or compact, terminal panicles; calyx glabrous to minutely pubescent, not obviously glandular. Seeds horizontal or vertical, thick-lenticular, 0.7–1 mm wide, dark brown, shining.

Native of tropical America; naturalized in gardens, roadsides, and waste places; Houghton County.

Dysphania botrys (L.) Mosyakin & Clemants
JERUSALEM-OAK *introduced*
IR | **WUP** | CUP | EUP   FACU

Chenopodium botrys L.

Annual, pubescent with short glandular hairs, strongly but not unpleasantly aromatic. Stems 2–6 dm tall, simple or branched from near the base. Leaves oblong to ovate in outline, the lower up to 8 cm long, sinuate-pinnatifid, the upper much smaller, pinnately lobed to coarsely toothed, or those of the flowering branches entire. Panicles numerous, axillary, forming large, more or less cylindric, terminal inflorescences; calyx glandular-pubescent, its segments rounded on the back. Seeds all or mostly horizontal, thick-lenticular, dull, dark brown, to 0.8 mm wide.

Native of Europe; a weed in waste places.

Froelichia Cottonweed
Annual herbs (ours). Leaves narrow, opposite. Flowers perfect, each subtended by a scarious bract and 2 bractlets, in elongate, woolly, terminal spikes. Calyx tubular, becoming conic and hard in fruit, 5-lobed, densely woolly. Stamens 5, the filaments united into a tube equaling the calyx. Ovary ovoid; style slender; stigma capitate. Fruit a membranous utricle.

Froelichia gracilis (Hook.) Moq.
SLENDER COTTONWEED *introduced*
IR | WUP | **CUP** | EUP

Stems branched from near the base, rather spreading, 3–7 dm long. Leaves mostly below the middle of the stem, linear to narrowly lance-shaped, commonly 26 cm long, 3–8 mm wide. Spikes 1–3 cm long; calyx at maturity conic, 4 mm long, densely woolly, bearing two lateral rows of short spines, and three basal, facial, slightly deflexed, spine-like tubercles.

Dry soil. Great Plains native, considered adventive in Michigan; Dickinson County.

Salsola Russian-Thistle

Salsola tragus L.
PRICKLY RUSSIAN-THISTLE *introduced*
IR | **WUP** | **CUP** | EUP   FACU

Salsola kali L. subsp. *tenuifolia* Moq.

Annual herb. Stems much branched, 3–8 dm tall, glabrous or pubescent. Lower leaves cylindric; upper leaves shorter, stiff, dilated at base, long-spined at tip, each subtending a solitary flower or a short spike of 2 or 3 flowers. Flowers perfect, single or few in axils of the shorter and spinier upper leaves, each subtended by a pair of bractlets; calyx deeply 5-lobed, the segments at maturity incurved over the fruit; stamens usually 5; styles 2. Variable. Late summer.

Anacardiaceae
SUMAC FAMILY

Woody plants, juice often milky. Leaves alternate, chiefly compound. Flowers small, regular, perfect or unisexual, 5-merous. Stamens 5, inserted beneath a disk surrounding the ovary. Pistil 1, 3-carpellary. Ovary 1-celled, sessile on the disk; styles 3. Fruit a 1-seeded, dry or fleshy drupe.

1. Flowers in dense inflorescences, these terminal or lateral on previous years twigs; fruit red, glandular-hairy. . . . . **Rhus**
1. Flowers in loose clusters from leaf axils; fruit whitish, nearly smooth . . . . **Toxicodendron**

Rhus Sumac
Trees or shrubs. Leaves pinnately compound, of 3 to many leaflets. Flowers lateral or terminal, polygamo-dioecious. Calyx 5-lobed. Petals 5, white or greenish. Stamens 5. Ovary 1-celled. Fruit a drupe.

ANACARDIACEAE | Sumac Family

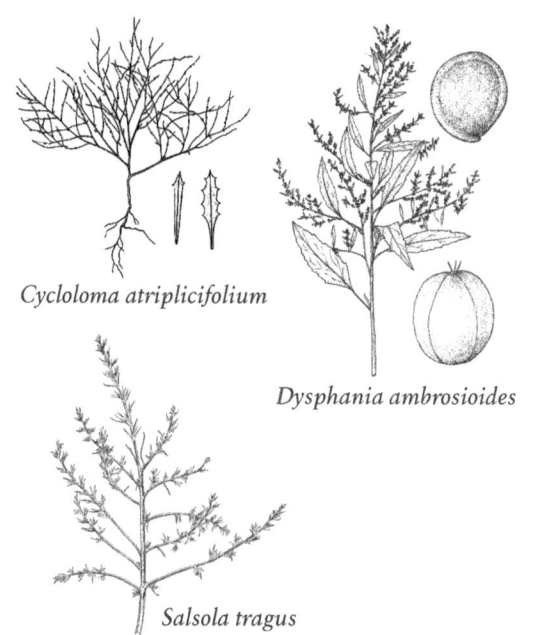

Cycloloma atriplicifolium

Dysphania ambrosioides

Salsola tragus

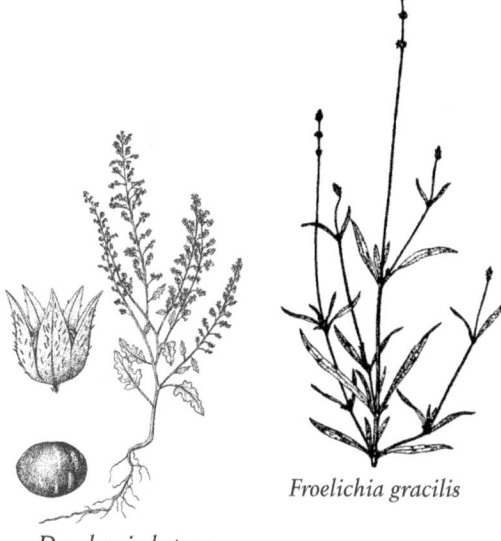

Dysphania botrys

Froelichia gracilis

HYBRIDS

Rhus × borealis Greene, *R. glabra* and *R. typhina*, reported from eastern and western UP.

1 Bushy shrubs with 3 sessile leaflets; Drummond Island only . . . . **R. aromatica**
1 Sparsely branched shrubs or small trees; leaflets several to many; widespread in the UP 2
2 Twigs and leaf petioles glabrous . . . . **R. glabra**
2 Twigs and petioles densely hairy . . . . **R. typhina**

Rhus aromatica Ait.
SQUAW-BUSH   native
IR | WUP | CUP | **EUP**   CC 7

Bushy shrub, often forming thickets. Leaflets 3, all sessile or nearly so, the terminal elliptic to rhombic-ovate, 4–8 cm long, with usually 3–6 coarse rounded teeth on each side in the distal half, the lateral smaller, elliptic to ovate lance-shaped, with similar teeth, at least the outer margin rounded to the petiole. Flowers in several short (1–2 cm) spike-like clusters, forming a panicle about 1 dm long, opening before or with the leaves, sessile or on pedicels no longer than the calyx. Bracts glabrescent just below the strongly ciliate apex. Drupes bright red, densely pubescent. April–May.

Dry woods, hills, sand dunes, and rocky soil; Chippewa County (Drummond Island only).

Rhus glabra L.
SMOOTH SUMAC   native
IR | **WUP** | **CUP** | **EUP**   CC 2

Usually a sparsely branched shrub, but sometimes to 6 m tall, the younger branches and petioles glabrous and somewhat glaucous. Leaflets 11–31, lance-shaped to narrowly oblong, 5–10 cm long, commonly serrate, much paler beneath. Panicle dense, often 2 dm long. Fruit bright red, densely beset with minute obovoid hairs about 0.2 mm long. June–July.

Dry soil, old fields, roadsides, and margins of woods.

Rhus typhina L.
STAGHORN SUMAC   native
IR | **WUP** | **CUP** | **EUP**   CC 2

Rhus hirta (L.) Sudw.

Tall shrub or small tree to 10 m tall. Younger branches, petioles, and leaf-rachis densely and softly hirsute. Leaflets 9–29, lance-shaped to narrowly oblong, 5–12 cm long, finely or coarsely serrate, paler beneath. Fruit red, densely covered with slender hairs 1–2 mm long. June–July.

Dry soil.

Toxicodendron *Poison-Ivy*

Toxicodendron rydbergii (Small) Greene
WESTERN POISON-IVY   native
IR | **WUP** | **CUP** | **EUP**   FAC CC 3

Rhus radicans var. *rydbergii* (Small ex Rydb.) Rehder

Strongly rhizomatous shrub, forming colonies. Stems to 1(–3) m tall, nearly erect, simple or sparingly branched. Leaflets broadly ovate, tending to be openly folded along the midrib rather than flat, glabrous on both sides or strigose beneath and often with a line of minute, curly hairs along the midrib above. Inflorescence unbranched or sparingly branched, usually with fewer than 25 flowers. Fruit a white or yellowish drupe, 4-7 mm thick, smooth, sessile or subsessile and crowded in an erect inflorescence.

Shores, dunes, and open slopes; forest openings and along margins; roadsides and railroads; soil usually sandy, gravelly, or rocky.

All parts of these plants may cause an allergic skin reaction if handled.

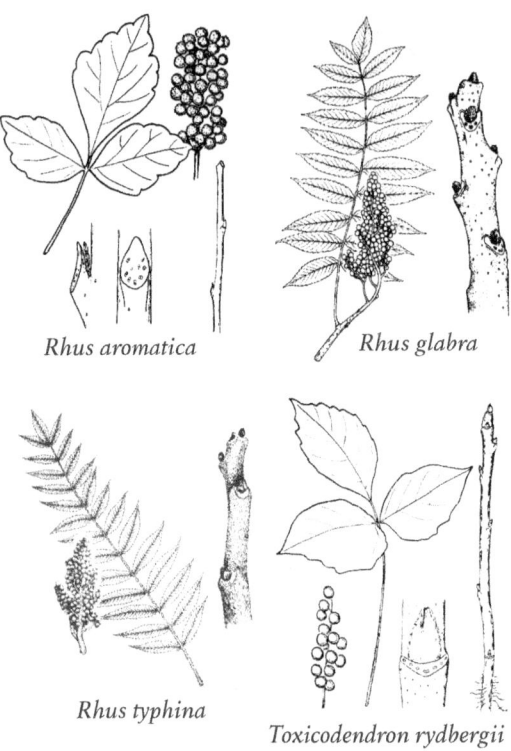

Rhus aromatica

Rhus glabra

Rhus typhina

Toxicodendron rydbergii

Apiaceae

CARROT FAMILY

Biennial or perennial aromatic herbs with hollow stems, some very toxic. Leaves alternate and sometimes also from base of plant, mostly compound; petioles sheathing stems. Flowers small, perfect (with both staminate and pistillate parts), regular, in flat-topped or rounded umbrella-like clusters (umbels); sepals 5 or absent; petals 5, white or greenish. Fruit 2-chambered, separating into 2, 1-seeded fruit when mature.

ADDITIONAL SPECIES

Levisticum officinale W.D.J. Koch (Garden lovage), introduced; Delta County.

Myrrhis odorata (L.) Scop. (Anise), introduced; Mackinac County.

Petroselinum crispum (P. Mill.) Nyman ex A.W. Hill (Parsley), introduced; Baraga and Houghton counties.

KEY TO APIACEAE GROUPS

1 Inflorescence neither a true umbel nor a compound umbel................................... **Sanicula**
1 Inflorescence a true umbel or a compound umbel.... 2
2 Ovary and fruit pubescent, tuberculate, bristly, or prickly................................... **Group A**
2 Ovary and fruit glabrous......................... 3
3 Leaves divided into distinct and separate leaflets of about uniform shape, these often more than 2 cm wide
.... **Group B**
3 Leaves much dissected or 2 or more times compound, the segments ovate, oblong, linear, or thread-like and less than 1 cm wide............................... 4
4 Plants flowering........................... **Group C**
4 Plants fruiting............................ **Group D**

GROUP A

Inflorescence a true umbel or compound umbel; most leaves compound, dissected, or deeply divided; fruit and ovary pubescent, covered with small bumps, or bristly or prickly.

1 Principal leaves palmately or once-pinnately compound or divided, the leaflets sometimes again divided..... 2
1 Principal leaves twice or more compound........... 4
2 Leaflets large, mostly 1 dm wide or more; fruit pubescent........................ **Heracleum maximum**
2 Leaflets less than 1 dm wide; fruit bristly or spiny.....
.... **Sanicula**
3 Leaves highly dissected into segments less than 1 cm wide............................... **Daucus carota**
3 Leaves with sharply toothed leaflets, the leaflets 1 cm wide or more..................................... 4
4 Umbel branches 2–8; fruit not winged..... **Osmorhiza**
4 Umbel branches 18–35; fruit winged................
.... **Angelica atropurpurea**

GROUP B

Inflorescence a true umbel or a compound umbel; fruit and ovary glabrous; leaves divided into distinct leaflets of uniform shape, these often more than 2 cm wide.

1 Principal leaves 1-compound (or sometimes simple in *Zizia*).. 2
1 Main leaves 2- or 3-times compound............... 7
2 Upper leaf-sheaths expanded, 1 cm or more wide when flattened; flowers white; fruit flattened and wing-margined...................... **Heracleum maximum**
2 Upper leaf sheaths not expanded, less than 1 cm wide; flowers and fruit various......................... 3
3 Taprooted introduced weeds of waste places and disturbed areas...................................... 4

dicots APIACEAE | Carrot Family

3	Native species with fibrous or tuberous-thickened roots, most common in woods or wetlands. 5
4	Flowers yellow; fruit wing-margined . **Pastinaca sativa**
4	Flowers white; fruit not winged . **Pimpinella saxifraga**
5	Leaves with 3 leaflets (or the basal ones simple and toothed) 6
5	Leaves with 5 or more leaflets; flowers white . . . . **Sium suave**
6	Flowers white or greenish-white . . . . **Cryptotaenia canadensis**
6	Flowers yellow . . . . **Zizia**
7	Leaflets entire; flowers yellow . . **Taenidia integerrima**
7	Leaflets toothed or lobed; flowers yellow or white . . . 8
8	Plants flowering 9
8	Plants fruiting 13
9	Flowers yellow or cream-colored 10
9	Flowers white 11
10	Central flower of each umbellet sessile . . . . **Zizia**
10	Flowers all on pedicels . . . . **Pastinaca sativa**
11	Sheaths of the upper leaves expanded, 1 cm or more wide when flattened . . . . **Angelica atropurpurea**
11	Sheaths of the upper leaves less than 1 cm wide 12
12	Main lateral veins of the leaflets oriented toward the sinuses between the teeth; some of the roots tuberous-thickened; base of stem thickened, hollow and cross-partitioned . . . . **Cicuta**
12	Main lateral veins mostly oriented toward the teeth on the margin; stem and roots not modified as in *Cicuta* . . . . **Aegopodium podagraria**
13	Fruit evidently winged 14
13	Fruit not winged, or only slightly so 15
14	Dorsal ribs prominent; plants taprooted . . . . **Angelica atropurpurea**
14	Dorsal ribs very slender, only slightly raised above the fruit surface . . . . **Pastinaca sativa**
15	Main lateral veins of the leaflets oriented toward the sinuses between the teeth; base of the stem thickened, hollow and cross-partitioned . . . . **Cicuta**
15	Main lateral veins mostly oriented toward the teeth on the margin, or more net-like and oriented toward neither the teeth or sinuses; base of stem not modified as in *Cicuta* 16
16	Stylopodium (disk-like swelling at base of style) absent; stem from fibrous or fleshy roots . . . . **Zizia**
16	Stylopodium well developed **Aegopodium podagraria**

GROUP C

Inflorescence a true umbel or a compound umbel; fruit and ovary glabrous; leaves dissected or 2 or more times compound, the segments ovate, oblong, linear, or thread-like and less than 1 cm wide; plants flowering.

1	Flowers yellow 2
1	Flowers white (or rarely pink) 3
2	Introduced, often weedy plants; sepals absent . . . . **Anethum graveolens**
2	Native, non-weedy perennials; sepals evident (check with a hand lens) . . . . **Thaspium**
3	Stem purple-spotted; plants coarse, well-branched, biennial herbs to 3 m high . . . . **Conium maculatum**
3	Stem not purple-spotted 4
4	Plants annual or biennial 5
4	Plants perennial 6
5	Bractlets fringed with hairs . . . . . **Anthriscus sylvestris**
5	Bractlets absent . . . . **Carum carvi**
6	Plants with bulblets in axils of some of the upper leaves . . . . **Cicuta**
6	Plants not with bulblets in upper leaf axils . . . . **Anthriscus sylvestris**

GROUP D

Inflorescence a true umbel or a compound umbel; fruit and ovary glabrous; leaves dissected or 2 or more times compound, the leaf segments ovate, oblong, linear, or thread-like and less than 1 cm wide; plants fruiting.

1	Fruit dorsally flattened . . . . **Anethum graveolens**
1	Fruit nearly round in cross-section or somewhat compressed laterally 2
2	Stems purple-spotted; coarse, branched, biennial herb to 3 m high . . . . **Conium maculatum**
2	Stems not purple-spotted 3
3	Fruit lance-shaped or linear and with a beak 1–3 mm long, ribs absent; bractlets fringed with hairs . . . . **Anthriscus sylvestris**
3	Fruit not beaked, the ribs evident; bractlets entire or absent 4
4	Plants perennial . . . . **Cicuta**
4	Plants annual or biennial . . . . **Carum carvi**

Aegopodium *Goutweed*

Aegopodium podagraria L.
BISHOP'S GOUTWEED *introduced (invasive)*
IR | **WUP** | **CUP** | **EUP** FAC

Perennial herb from a creeping rhizome. Stems erect, branched, 4–9 dm tall. Lower leaves long-petioled, mostly 1- or 2-times parted with 9 leaflets but often irregular; leaflets oblong to ovate, 3–8 cm long, margins sharply serrate; upper leaves reduced, short-petioled, chiefly once-parted. Flowers in dense umbels, these terminal and lateral, 6–12 cm wide, long-peduncled, rising above the leaves; primary rays 15–25, nearly equal; petals white. Fruit oblong-ovoid, 3–4 mm long, flattened laterally, tipped by the 2 conspicuous stylopodia.

Native of Eurasia; cultivated in gardens and sometimes escaped, especially where moist and partially shaded.

48 APIACEAE | Carrot Family

Anethum *Dill*

Anethum graveolens L.
DILL *introduced*
IR | WUP | **CUP** | **EUP**

Strongly scented annual herb. Stems to 15 dm tall, branched above, glabrous and more or less glaucous throughout. Leaves pinnately dissected into numerous filiform segments, the lower long-petioled, the upper shorter-petioled and smaller. Flowers in terminal and lateral compound umbels, overtopping the leaves; primary rays usually 30–40, widely spreading, about equal; sepals absent; petals yellow. Fruit oblong or elliptic, flattened; ribs prominent, the lateral conspicuously winged. July–Aug.

Native of s Europe; cultivated commercially and in kitchen gardens and escaped into waste ground.

Angelica *Angelica*

Angelica atropurpurea L.
PURPLE-STEM ANGELICA *native*
IR | **WUP** | **CUP** | EUP OBL CC 6

Perennial herb. Stems stout, 2–3 m tall, more or less smooth, often streaked with purple and green. Leaves alternate, lower leaves 3-parted, 1–3 dm long, on long petioles; upper leaves smaller, less compound, on shorter petioles, or reduced to bladeless sheaths; leaflets ovate to lance-shaped, smooth, 4–10 cm long; margins sharp toothed. Flowers in rounded small clusters (umbelets), these grouped into large rounded umbels 1–2 dm wide; petals white to green-white. Fruit oval, 4–6 mm long, winged. May–July.

Springs, seeps, calcareous fens, streambanks, shores, marshes, sedge meadows, wet depressions in forests; often where calcium-rich.

Anthriscus *Chervil*

Anthriscus sylvestris (L.) Hoffmann
WILD CHERVIL *introduced*
IR | **WUP** | **CUP** | EUP

Annual or biennial herb. Stems freely branched, to 1 m tall. Leaves 2–3x compound; leaflets dentate to incised. Flowers in large compound umbels, terminal and from the upper axils, with 6–10 primary rays up to 4 cm long; umbellets few-flowered; bractlets ovate lance-shaped, 3–6 mm long; sepals absent; petals white. Fruit lance-shaped, about 6 mm long, the beak about 1 mm long. May–July.

Native of Europe; rarely established in waste places.

Carum *Caraway*

Carum carvi L.
CARAWAY *introduced*
IR | **WUP** | **CUP** | **EUP** UPL

Glabrous biennial herb, from a taproot. Stems to 1 m tall. Leaflets pinnately dissected into linear segments 5–15 mm long. Flowers in terminal and lateral compound umbels; primary rays several to many; peduncles 5–13 cm long; primary rays 7–14, commonly 2–4 cm long; umbellets small; involucel of a few minute bracts or none; pedicels very unequal; sepals absent; petals white or rarely pink. Fruit elliptic to oblong, 3–4 mm long, about half as wide, prominently ribbed. June–Aug.

Native of Eurasia; sometimes cultivated and often weedy in waste places.

Cicuta *Water-Hemlock*

Biennial or perennial toxic herbs. The tuberous roots, chambered stem base and young shoots of common water-hemlock (*Cicuta maculata*) are especially toxic. Leaves alternate, 2–3-pinnate; leaflets narrow or lance-shaped, entire or toothed; leaf veins ending in the lobes (sinuses) and not at teeth as in other mem-

Aegopodium podograria

Anethum graveolens

Angelica atropurpurea

Anthriscus sylvestris

bers of this family. Flowers white or green, in few to many umbels; umbels usually without bracts, umbellets bracted. Fruit oval or round, flattened, ribbed.

1   Upper leaflet axils usually with bulblets; leaflets to 5 mm wide .... **C. bulbifera**
1   Bulblets absent; leaflets usually much more than 5 mm wide .... **C. maculata**

Cicuta bulbifera L.
BULBLET-BEARING WATER-HEMLOCK *native*
IR | **WUP** | **CUP** | **EUP**   OBL CC 5

Biennial or perennial herb, toxic; fibrous-rooted or with a few thickened, tuberlike roots. Stems slender, upright, 3–10 dm tall, not thickened at base. Leaves alternate along stem, to 15 cm long and 10 cm wide, pinnately divided; leaflets mostly linear, 1–5 mm wide, margins sparsely toothed to entire; upper leaves reduced in size, undivided or with few segments, with 1 to several bulblets 1–3 mm long, in axils. Flowers white, in umbels 2–4 cm wide. Fruit round, 1–2 mm wide, but rarely maturing. Aug–Sept.

Streambanks, lake and pond shores, marshes, swamps, open bogs, thickets, springs, ditches.

Cicuta maculata L.
COMMON WATER-HEMLOCK *native*
IR | **WUP** | **CUP** | **EUP**   OBL CC 5

Biennial or perennial herb. Stems single or several together, often branched, 1–2 m long, distinctly hollow above the chambered and tuberous-thickened base. Leaves from base of plant and alternate on stem, mostly 10–30 cm long and 5–20 cm wide; basal leaves larger and longer stalked than stem leaves; leaflets linear to lance-shaped, 3–10 cm long and 5–35 mm wide; margins toothed. Flowers white, in several to many umbels, these 6–12 cm wide in fruit, on stout stalks 5–15 cm long. Fruit round to ovate, 2–4 mm long, with prominent ribs. June–Sept.

Wet meadows, marshes, swamps, moist to wet forests, thickets, shores, streambanks, springs.

Considered most toxic plant in North America.

Conium *Poison-Hemlock*
Conium maculatum L.
POISON-HEMLOCK *introduced (invasive)*
IR | **WUP** | **CUP** | **EUP**   FACW CC 4

Biennial herb with a strong, unpleasant odor. Stems stout, branched, purple-spotted, 1–2 m long. Leaves alternate, 2–4 dm long, 3–4x pinnately divided, the leaflets toothed or sharply lobed. Flowers white, in many umbelets, these grouped in umbels to 6 cm wide. Fruit ovate, ribbed, 3 mm long. June–July. Weed of shores, streambanks, waste ground and roadsides, especially on moist, fertile soil.

Very toxic, fatal if eaten.

Cryptotaenia *Honewort*
Cryptotaenia canadensis (L.) DC.
CANADIAN HONEWORT *native*
IR | **WUP** | **CUP** | **EUP**   FAC CC 2

Perennial glabrous herb. Stems branched, 3–8 dm tall. Leaves 3-foliate, lower leaves long-petioled, the upper on short petioles dilated as far as the leaflets. Leaflets lance-shaped to obovate, 4–15 cm long, irregularly often doubly serrate or sometimes lobed. Flowers in numerous loose, irregular, compound umbels arising terminally and from the upper axils; primary rays 2–7, ascending, 1–5 cm long, somewhat unequal; umbellets few-flowered; involucel none or of 1–3 minute bractlets; pedicels very unequal; sepals low or obsolete; corolla white. Fruit dark, slightly flattened, tipped by the slender stylopodium, 5–8 mm long; the ribs evident but low and obtuse. June–July.

Moist rich woods, swamps.

Daucus *Carrot*
Daucus carota L.
QUEEN ANNE'S-LACE *introduced (invasive)*
IR | **WUP** | **CUP** | **EUP**

Biennial herb, with a stout taproot. Stems 5–10 dm tall, glabrous, scabrous, or commonly rough-hairy.

Carum carvi

Cicuta bulbifera

Cicuta maculata

Conium maculatum

Leaves pinnately compound, the ultimate divisions linear or lance-shaped. Umbels compound, terminal and from the upper axils, long-peduncled, usually many-rayed; terminal umbel erect, commonly 7–15 cm wide, the lateral ones usually smaller; the outer primary rays curve inward after anthesis, producing a congested cluster. Flowers white or rarely pinkish, the central one of each umbellet often purple. Fruit 3–4 mm long, flattened dorsally, the primary ribs low and inconspicuous, bearing a row of short bristles, the four secondary ribs prominently winged, divided into a row of hooked or straight spines. June–Sept.

Native of Eurasia; established as a weed in fields, roadsides, waste ground, and open woods.

The cultivated carrot is a race of this species.

Heracleum *Cow-Parsnip*

Heracleum maximum Bartr.
AMERICAN COW-PARSNIP   native
IR | WUP | CUP | EUP   FACW CC 3
Heracleum lanatum Michx.
Heracleum sphondylium L. subsp. *montanum*

Large perennial herb. Stems stout, hairy, 1–2 m long. Leaves alternate, nearly round in outline, divided into 3 leaflets; leaflets 1–4 dm long and as wide, margins coarsely toothed. Flowers white, in large umbels, the terminal umbel 1–2 dm wide. Fruit obovate, 8–12 mm long and nearly as wide, often hairy. May–July.

Streambanks, thickets, wet meadows, moist forest openings and disturbed areas.

ADDITIONAL SPECIES

Heracleum mantegazzianum Sommier & Levier (Giant hogweed), introduced; Gogebic County.
Heracleum sphondylium ssp. **sibiricum** (L.) Simonkai, introduced; Houghton County.

Osmorhiza *Sweet-Cicely*

Erect perennial herbs from thickened roots and glabrous to pubescent stems 4–8 dm tall; our 3 species similar in general appearance and foliage. Leaves ternate; leaflets several, the lower petioled, the upper subsessile, the ultimate segments ovate to lance-shaped, serrate or lobed. Flowers in terminal and lateral umbels, these usually surpassing the leaves; primary rays of the umbel mostly 3–6, widely ascending; involucre present or absent. Umbellets few-flowered; involucel present or lacking. Sepals none; petals white or greenish white. Fruit elongate, slightly flattened, ribbed, the base prolonged into bristly tails.

ADDITIONAL SPECIES

Osmorhiza depauperata Phil. (Blunt-fruit sweet-cicely), reported from mixed woods in Keweenaw County (including Isle Royale); threatened.

1 Umbels without bracts at base of umbel branches .... **O. berteroi**
1 Umbels with bracts at base of umbel branches 2
2 Plants anise-scented; styles 2 mm long, becoming 3–4 mm long in fruit .... **O. longistylis**
2 Plants unscented; styles less than 1.5 mm long (even in fruit) .... **O. claytonii**

Osmorhiza berteroi DC.
MOUNTAIN SWEET-CICELY   native
IR | WUP | CUP | EUP   FACU CC 5
Osmorhiza chilensis Hook. & Arn.

Involucel lacking. Fruit concavely narrowed to the acute summit, 14–18 mm long. Stylopodium ovoid-conic, commonly longer than thick; styles at maturity outwardly curved, 0.4–0.7 mm long including the stylopodium. June.

Moist woods.

Osmorhiza claytonii (Michx.) C.B. Clarke
HAIRY SWEET-CICELY   native
IR | WUP | CUP | EUP   FACU CC 4

Stems commonly sparsely villous, but vary from densely villous to nearly glabrous. Styles at anthesis distinctly shorter than the petals, in fruit nearly straight and parallel, 1.2–1.5 mm long including the

Cryptotaenia canadensis

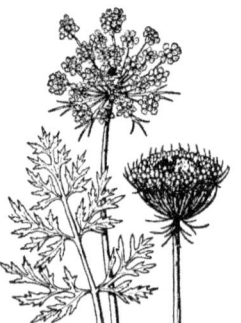

Daucus carota

Heracleum maximum

Osmorhiza claytonii

stylopodium. Mericarps 2–2.5 cm long. May–June. Moist woods.

Osmorhiza longistylis (Torr.) DC.
ANISEROOT native
IR | **WUP** | **CUP** | **EUP** FACU CC 3
Styles at anthesis about 2 mm long, much exceeding the petals, in fruit nearly straight and parallel, 3–3.5 mm long including the stylopodium. Mericarps about 2 cm long. May–June.
Moist woods.

Pastinaca *Parsnip*
Pastinaca sativa L.
WILD PARSNIP *introduced (invasive)*
IR | **WUP** | **CUP** | **EUP**
Stout biennial herb, to 1.5 m tall. Lower leaves long-petioled, the upper on shorter, wholly sheathing petioles, all typically 1-pinnate; leaflets 5–15, usually oblong to ovate, 5–10 cm long, variously serrate or lobed, or in vigorous plants sometimes completely divided into 2–5 segments. Umbels large, 1–2 dm wide, compound, the terminal soon overtopped by the lateral ones; primary rays unequal, 15–25; involucre and involucel usually lacking; sepals minute or none; petals yellow. Fruit broadly elliptic or obovate, strongly flattened, 5–7 mm long, with low ribs, the lateral ribs broadly and thinly winged.

Native of Eurasia; long in cultivation and thoroughly established as a weed in waste places, fields, and roadsides.

Skin irritant if handled.

Pimpinella *Burnet Saxifrage*
Pimpinella saxifraga L.
BURNET SAXIFRAGE *introduced*
IR | WUP | **CUP** | **EUP**
Perennial herb. Stems 3–6 dm tall, filled with pith. Lower stem leaves 1-pinnate, the leaflets varying from ovate or subrotund and merely serrate to deeply pinnately dissected. Upper leaves much reduced, the uppermost consisting of sheaths only or of sheaths with a few small linear leaflets at the summit. Umbels peduncled, terminal and lateral, compound; involucre none or rarely of 1–few bracts; primary rays 8–20; sepals minute or lacking; petals white. Fruit glabrous, ovoid, 2–2.5 mm long, the ribs 5, narrow.

Native of Eurasia; escaped or adventive in waste places.

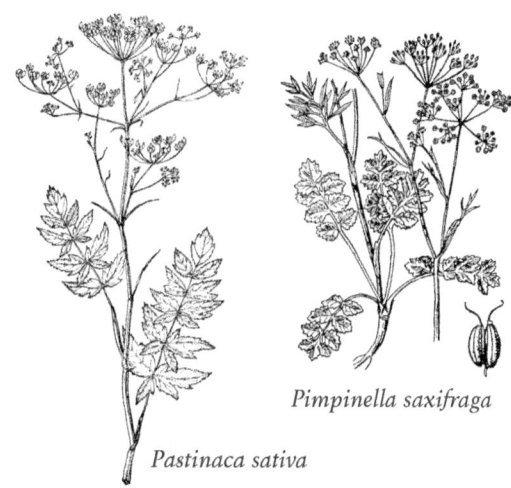

Pimpinella saxifraga

Pastinaca sativa

Sanicula *Black-Snakeroot*
Biennial or perennial herbs; stems arising from a cluster of fibrous or sometimes tuberous roots. Leaves palmately divided into 3–5 segments, the basal long-petioled, the cauline progressively reduced, and the uppermost short-petioled to nearly sessile. Umbels irregular, with spreading primary branches of unequal length, the umbellets dense or almost capitate, commonly with 3 sessile or short-pediceled perfect flowers, their hypanthium bristly, and several staminate flowers with smooth hypanthium, all or mostly on much longer pedicels; sepals narrow, joined at base, persistent; petals greenish white to greenish yellow. Fruit ovoid to subglobose, slightly flattened laterally, ribs absent, densely covered with hooked bristles.

1 Styles shorter than bristles of the fruit; staminate flowers 2–7; Delta County .... **S. trifoliata**
1 Styles longer than bristles of the fruit; staminate flowers 12–25 in each umbellet 2
2 Staminate flowers longer than the fruit; sepals awl-shaped, 1–2 mm long .... **S. marilandica**
2 Staminate flowers shorter than the fruit; sepals lance-shaped or ovate, 0.5 mm long .... **S. odorata**

Sanicula marilandica L.
MARYLAND BLACK-SNAKEROOT *native*
IR | **WUP** | **CUP** | **EUP** FACU CC 4
Leaves 5-parted, often appearing 7-parted, serrate, doubly serrate, or toward the apex incised. Bractlets of the involucel resembling the bracts but smaller. Flowers greenish white, the fertile ones sessile; calyx lobes lance-subulate, 1–1.5 mm long, equaling or slightly shorter than the petals. Anthers greenish white. Fruit nearly sessile, 4–6 mm long, narrowed and with shorter bristles toward the base. Styles recurved, exceeding the bristles. June–Aug.

Moist or dry woods.

Sanicula odorata (Raf.) Pryer & Phillippe
CLUSTERED BLACK-SNAKEROOT *native*
IR | **WUP** | **CUP** | **EUP** FAC CC 2
Sanicula gregaria Bickn.
Leaves 3–5-parted, the segments sharply serrate to

incised. Bractlets of the involucel small, subscarious. Flowers greenish yellow, the fertile on pedicels 0.5–1 mm long; calyx lobes ovate-lance-shaped to ovate, obtuse or subacute, much shorter than the petals. Anthers bright yellow. Fruit subglobose, about 3 mm long. Styles conspicuous, recurved, exceeding the bristles. June–Aug.

Moist or dry woods.

Sanicula trifoliata Bickn.
LARGE-FRUIT BLACK-SNAKEROOT native
IR | WUP | **CUP** | EUP CC 6

Leaves 3-parted, the leaflets coarsely and doubly serrate to incised, the lateral often deeply lobed. Bractlets ovate, subscarious. Flowers white, the fertile sessile, the staminate few, on slender pedicels up to 8 mm long and much exceeding the fertile ones; sepals lance-subulate, exceeding the petals. Fruit ovoid to oblong, 6–8 mm long including the comparatively few bristles, the sepals lance-shaped, connivent, exceeding the bristles, forming a conspicuous beak 2–2.5 mm long. June–Aug.

Moist or dry woods; Delta County.

Sium *Water-Parsnip*

Sium suave Walt.
HEMLOCK WATER-PARSNIP native
IR | **WUP** | **CUP** | EUP OBL CC 5

Perennial emergent herb. Stems single, smooth, 5–20 dm long, strongly ribbed upward; stem base thickened and hollow with cross-partitions. Leaves once-pinnate, on long, hollow stalks (shorter stalked above); leaflets 7–17 per leaf, linear to lance-shaped, 5–10 cm long and 3–15 mm wide; margins with fine, sharp, forward-pointing teeth; finely dissected underwater leaves often present from spring to midsummer. Flowers white or green-white, 1–2 mm wide, in stalked umbels 4–12 cm wide at ends of stems and from side branches. Fruit oval, 2–3 mm long, with prominent ribs. July–Sept.

Wet forest depressions, marshes, swamps, streambanks, lakeshores, ditches; usually in shallow water.

Taenidia *Pimpernel*

Taenidia integerrima (L.) Drude
YELLOW-PIMPERNEL native
IR | **WUP** | **CUP** | EUP CC 8

Perennial herb. Stems branched, 4–11 dm tall, glabrous and somewhat glaucous. Lower leaves long-petioled, commonly 3x compound, the upper 1–2x compound, with short, wholly sheathing petioles; leaflets normally entire, ovate to oblong or elliptic. Umbels terminal and lateral, loose and irregular, primary rays numerous, the outer elongate, to 9 cm long, the inner often much shorter; involucre none; umbellets many-flowered, involucel none. Inner flowers of each umbellet staminate and short-pediceled, the marginal long-pediceled and fertile; calyx teeth tiny or none; petals yellow. Fruit elliptic to broadly ovate-oblong, 3–4 mm long, flattened; ribs faint. May–June.

Dry woods and rocky hillsides.

Zizia *Alexanders*

Branched perennial herbs 3–8 dm tall, glabrous or nearly so, from a cluster of thickened roots. Leaves mostly 1–3x compound. Umbels compound; primary rays several to many, the inner often much shorter; involucre none. Umbellets many-flowered; involucel of a few short linear lance-shaped bractlets; pedicels very unequal, the central flower commonly sessile. Sepals short, triangular. Petals bright yellow. Fruit ovate to oblong, flattened laterally; ribs 5 on each mericarp, varying from low and narrow to narrowly winged.

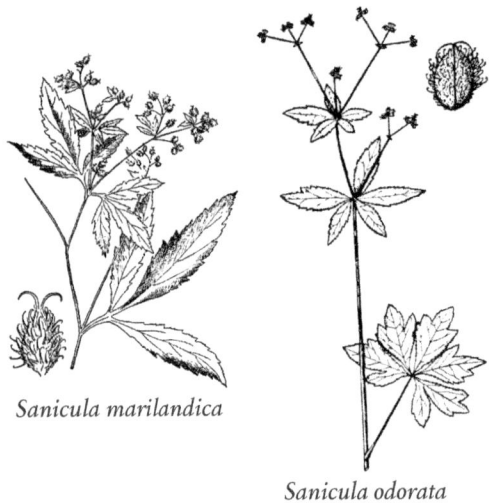

Sanicula marilandica

Sanicula odorata

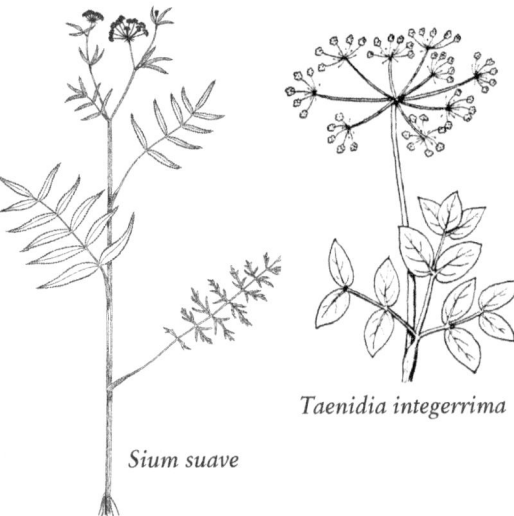

Taenidia integerrima

Sium suave

1 Basal leaves simple, with heart-shaped blades; stem leaves with 3 leaflets; rare in Gogebic and Mackinac counties .... **Z. aptera**
1 Basal leaves compound and similar to the stem leaves, with 5–11 leaflets .... **Z. aurea**

Zizia aptera (Gray) Fern.
HEART-LEAF ALEXANDERS (THR) *native*
IR | **WUP** | CUP | **EUP** FACU CC 9

Perennial herb. Basal leaves and occasionally the lower stem leaves simple, long-petioled, deltoid-ovate or round-ovate to oblong-ovate, 4–12 cm long, cordate at base. Stem leaves once or twice 3-parted, the leaflets ovate lance-shaped to obovate-oblong. Primary rays at anthesis 1–3 cm long, at maturity ascending, up to 5 cm long. Fruit oblong-ovate, 3–4 mm long. May–June.

Moist meadows and open woods; Gogebic and Mackinac counties.

Zizia aurea (L.) W.D.J. Koch
GOLDEN ALEXANDERS *native*
IR | **WUP** | **CUP** | EUP FAC CC 6

Perennial herb. Lower leaves twice 3-parted, the upper leaves once 3-parted or irregularly compound; leaflets ovate to lance-shaped, finely serrate with ascending teeth averaging 5–10 per centimeter of margin. Primary rays commonly 10–18, the outer ones of the terminal umbel becoming 3–5 cm long and stiffly ascending at maturity of the fruit. Fruit oblong-ovoid, 3–4 mm long and about half as wide. May–June.

Moist fields and meadows.

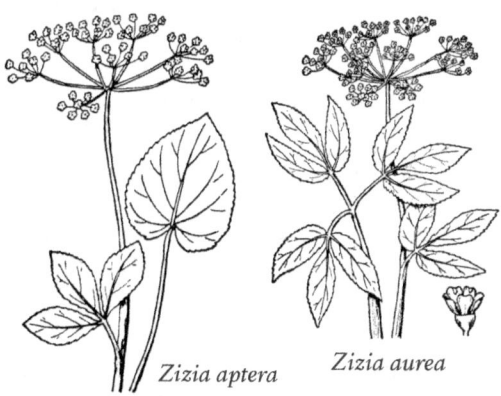

Zizia aptera   *Zizia aurea*

Apocynaceae
DOGBANE FAMILY

Our species herbs or twining woody vines; most species have milky juice. Leaves opposite, alternate, or sometimes whorled. Flowers 5-merous, regular, perfect. Fruit a capsule or follicle; seeds often bearing long hairs. Family now includes former members of Asclepiadaceae; *Apocynum* differs by having corolla lobes overlapping and twisted in bud, and stamens without a crown.

1 Plants trailing, subwoody, evergreen; flowers solitary in leaf axils; corolla blue; seeds glabrous .... **Vinca**
1 Plants erect or twining, herbaceous and not evergreen; flowers in terminal or axillary cymes or umbels; corolla various colors, not blue; seeds with tuft of silky hairs 2
2 Corolla lobes erect to spreading; flowers in small terminal (and sometimes axillary) cymes; mature fruits 3–5 mm wide.............................. **Apocynum**
2 Corolla lobes strongly reflexed at maturity; flowers in umbels; mature fruits 6–35 mm in diameter . **Asclepias**

Apocynum *Dogbane*

Perennial herbs with tough fibrous stems. Leaves opposite, mucronate. Flowers small, white or pink, in branched terminal cymes. Calyx deeply divided into triangular or lance-shaped lobes. Corolla white or pinkish, campanulate or short-cylindric, with 5 short lobes, bearing within a tooth or scale near the base of the tube opposite each lobe. Anthers lance-shaped, joined, adherent to the stigma and prolonged into a cone beyond it. Ovaries 2, subtended by 5 nectaries; style none; stigma large, 2-lobed. Fruit a cylindric follicle, pendulous; seeds numerous, bearing long soft hairs (coma).

1 Corolla pink, 5–8 mm long; leaves widely spreading or drooping .... **A. androsaemifolium**
1 Corolla white, 3–4 mm long; leaves ascending .... **A. cannabinum**

Apocynum androsaemifolium L.
SPREADING DOGBANE *native*
IR | **WUP** | **CUP** | **EUP** CC 3

Perennial herb. Stems more or less inclined from tbe vertical, the branches chiefly alternate. Leaves petiolate, more or less drooping, oblong lance-shaped to ovate, commonly 3–8 cm long, pilose beneath. Principal cymes terminal; secondary cymes of smaller size in the upper axils; calyx lobes triangular, a third to half as long as the corolla tube; corolla campanulate, 6–10 mm long, pink to nearly white, marked with red within, the lobes spreading or recurved. May–Aug.

Upland woods, occasionally in fields and roadsides.

Apocynum cannabinum L.
INDIAN-HEMP *native*
IR | **WUP** | **CUP** | **EUP** FAC CC 3

Perennial herb. Stems erect, branched above, 1–1.5 m tall. Leaves varying from oblong lance-shaped to ovate or broadly elliptic, acute to rounded at the mucronate apex and base, glabrous or pubescent beneath, regularly on conspicuous petioles commonly 5–10 mm long. Flowers white or greenish white. Follicles usually 10–15 cm long; seed-coma 2–2.5 cm long. June–Sept.

Dry or moist open places.

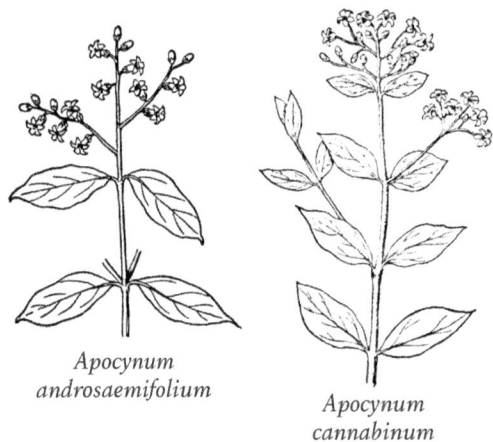

Apocynum androsaemifolium

Apocynum cannabinum

Asclepias Milkweed

Perennial herbs from a thick root or deep rhizome and with milky juice (except in *A. tuberosa*). Stems usually simple. Leaves opposite (in some species whorled or rarely alternate), entire. Flowers small or medium-sized, in peduncled, terminal or axillary umbels; calyx lobes, corolla lobes, and stamens each 5; corolla lobes usually meeting in bud without overlapping; corolla deeply divided, at anthesis reflexed and concealing the calyx; anthers united with stigma forming an organ known as the gynostegium; pollen of each anther-sac united into a waxy mass known as a pollinium; ovaries 2; styles 2; stigma 1. Fruit a pod-like follicle, normally produced in pairs, commonly erect, lance-shaped or linear lance-shaped, acuminate; seeds with long silky hairs (coma).

1	Leaves linear, less than 4 mm wide, mostly whorled .. .. **A. verticillata**
1	Leaves more than 5 mm wide 2
2	Flowers orange; juice not milky; leaves mostly alternate .... **A. tuberosa**
2	Flowers not orange; juice milky; leaves opposite 3
3	Reflexed corolla lobes red-purple, mostly less than 5 mm long .... **A. incarnata**
3	Reflexed corolla lobes various colors, 5 mm or more long 4
4	Umbels nodding; flowers white or greenish **A. exaltata**
4	Umbels mostly erect; flowers cream-colored or tinged with red or purple 5
5	Upper stems, peduncles and pedicels white-woolly; hoods 1 cm long or more.................. **A. speciosa**
5	Pubescence varies but not white-woolly; hoods less than 1 cm long 6
6	Plants small, slender, to 60 cm high; umbels usually single; rare in Menominee County .... **A. ovalifolia**
6	Plants large and coarse, more than 60 cm high; umbels usually 2 or more .... **A. syriaca**

Asclepias exaltata L.
POKE MILKWEED native
IR | **WUP** | **CUP** | EUP CC 6

Stems 8–15 dm tall, glabrous or puberulent in narrow lines. Leaves thin, broadly elliptic, 1–2 dm long, acuminate at both ends, glabrous, or puberulent beneath, on petioles 1–2 cm long. Umbels several from the upper axils, loosely few-flowered, the slender pedicels spreading or often drooping; corolla white to pale dull purple, 7–10 mm long. Hoods white or pink, about 4 mm long, about equaling the gynostegium, the lateral margins adjacent, each terminating in an erect tooth 1–1.5 mm long, the rest of the hood truncate. Horns subulate, nearly erect, conspicuously exsert. Pods erect on deflexed pedicels, puberulent, about 15 cm long. June–July.

Moist upland woods.

Asclepias incarnata L.
SWAMP MILKWEED native
IR | **WUP** | **CUP** | EUP OBL CC 6

Perennial herb, from thick rhizomes; plants with milky juice. Stems stout, to 1.5 m long, branched above, smooth except for short, appressed hairs on upper stem. Leaves opposite, simple, mostly lance-shaped, 6–15 cm long and 1–5 cm wide, tapered to a sharp tip, margins entire, petioles short. Flowers pink to purple-red, numerous in umbels at ends of stems and from upper leaf axils, perfect, regular; sepals 5, spreading; petals 5, 4–6 mm long and curved downward; stamens 5; flowers with 5 petal-like hoods, each with an awl-shaped horn projecting from the opening. Fruit a follicle (1-chambered and opening on 1 side only) with many seeds, the seeds having tufts of white hairs. June–Aug.

Openings in conifer swamps, marshes, streambanks, ditches, open bogs and fens; often in shallow water.

Asclepias ovalifolia Dcne.
DWARF MILKWEED (END) native
IR | WUP | **CUP** | EUP CC 10

Stems slender, 2–5 dm tall. Leaves firm in texture, ovate lance-shaped to oblong or elliptic, commonly 3–6 cm., rarely to 10 cm long, cuneate to rounded at base, finely pubescent beneath; petioles 2–8 mm long. Umbels solitary and terminal, or a few in the upper axils, loosely few-flowered; corolla greenish white to greenish purple, 6–7 mm long. Hoods as in *A. syriaca* but smaller, 4.5–5 mm long. Pods merely thinly pubescent. June–July.

Dry prairies; Menominee County.

APOCYNACEAE | Dogbane Family

Asclepias speciosa Torr.
SHOWY MILKWEED *introduced*
IR | WUP | CUP | **EUP** FAC

Stems stout, to 1 m tall, pubescent. Leaves ovate, ovate lance-shaped, or ovate-oblong, 10–15 cm long, acute, broadly rounded or subcordate at base, pubescent beneath; petioles 3–6 mm long. Umbels usually few, terminal and subterminal but also in the upper axils, 5–7 cm wide; peduncles stout, 3–7 cm long; corolla greenish purple, 10–12 mm long. Hoods 11–15 mm long, abruptly narrowed below the middle into an oblong-linear tip. Horns short, inflexed. Pods densely tomentose and beset with soft filiform processes. July–Aug.

Moist prairies; common in w USA, considered introduced in Michigan; Mackinac County.

Asclepias syriaca L.
COMMON MILKWEED *native*
IR | **WUP** | **CUP** | **EUP** UPL CC 1

Stems tall and stout, mostly simple, pubescent. Leaves thick, narrowly or broadly elliptic to ovate or oblong, 10–15 cm long, acute or apiculate, softly pubescent beneath, on distinct petioles 5–15 mm long. Umbels often numerous, terminal and in the upper axils, compactly many-flowered; peduncles stout, 3–10 cm long; corolla green suffused with purple, varying from almost purple to almost green, 8–10 mm long. Hoods pale purple, somewhat divergent, 6–8 mm long, surpassing the gynostegium, the lateral margins bearing a prominent, sharp, triangular lobe at or near the middle. Horns short, inflexed. Pods erect on deflexed pedicels about 1 dm long, tomentose and beset with soft filiform to conic processes.

Fields, meadows, and roadsides; often weedy.

Asclepias tuberosa L.
BUTTERFLY WEED *native*
IR | WUP | **CUP** | EUP CC 5

Stems ascending or erect, 3–7 dm tall, villous or hirsute, simple to much branched above, the branches ascending or widely spreading, often flexuous. Leaves alternate or on the branches opposite, linear to lance-shaped or oblong lance-shaped, 5–10 cm long, pubescent, cuneate to truncate or subcordate at base. Umbels varying from solitary and terminal to numerous, often from most of the axils of divergent branches; corolla yellow to orange-red, 7–10 mm long. Hoods yellow to orange, 5–7 mm long, greatly exceeding the gynostegium, nearly straight, erect, the lateral margins bearing an obscure tooth below the middle. Pods 8–12 cm long, erect. June–Aug.

Dry or moist prairies and upland woods, especially in sandy soil; Dickinson County.

Our only *Asclepias* without milky juice. Plants variable in habit and shades of flower color.

Asclepias verticillata L.
WHORLED MILKWEED *native*
IR | WUP | **CUP** | EUP UPL CC 1

Stems slender, erect, 2–5 dm tall, simple to the inflorescence, pubescent in lines. Leaves very numerous in whorls of 3–6, narrowly linear, 2–5 cm long, 1–2 mm wide, revolute. Umbels several from the upper nodes; peduncles 1–3 cm long; petals white or greenish, 4–5 mm long. Hoods white or greenish white, somewhat divergent, 1.5–2 mm long, about equaling the gynostegium, their margins entire. Horns subulate, much surpassing the hoods, slightly narrowed over the stamens. Pods slender, erect on erect pedicels, 4–5 cm long. June–Aug.

Dry or moist fields, roadsides, upland woods, and prairies; Menominee and Schoolcraft counties.

Vinca *Periwinkle*

Vinca minor L.
LESSER PERIWINKLE *introduced (invasive)*
IR | **WUP** | **CUP** | **EUP**

Perennial trailing herb. Stems trailing or scrambling, to 1 m long, forming mats. Leaves leathery, opposite, lance-elliptic, 3–5 cm long, entire, petiolate. Flowers blue or rarely white, solitary in 1 axil only of a pair

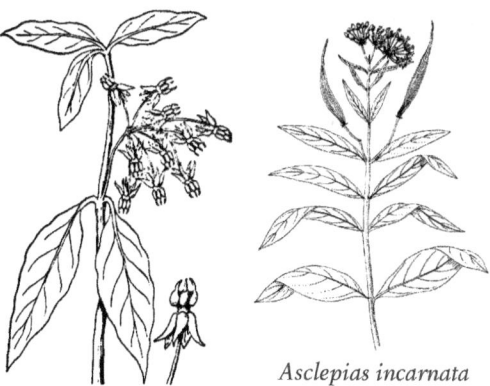

Asclepias exaltata

Asclepias incarnata

Asclepias speciosa

Asclepias syriaca

56 AQUIFOLIACEAE | Holly Family

of leaves; calyx deeply 5-parted; corolla large, salverform, corolla tube 8–12 mm long, the limb 2–3 cm wide; ovaries 2, accompanied by 2 nectaries. Fruit a linear, few-seeded follicle; seeds naked. April–May. Native of s Europe, planted as a groundcover and escaped to roadsides and open woods.

Aquifoliaceae
HOLLY FAMILY

Ilex *Holly*
Shrubs. Leaves usually alternate, toothed or entire, not lobed. Flowers from leaf axils, 4–8-parted, usually either staminate or pistillate, sometimes perfect, on same or different plants. Fruit a fleshy berrylike drupe with 4–9 stones.

1 Leaves tipped with a short, sharp point, margins mostly entire or with a few scattered teeth; petals linear; sepals tiny or absent........................ **I. mucronata**
1 Leaves not tipped with a short, sharp point, margins toothed; petals oblong; sepals evident .. **I. verticillata**

Ilex mucronata (L.) Powell, Savolainen & Andrews
MOUNTAIN HOLLY; CATBERRY native
IR | **WUP** | **CUP** | **EUP** OBL CC 7
Nemopanthus mucronatus (L.) Loes.
Much-branched shrub to 3 m tall; young twigs purple-tinged. Leaves deciduous, alternate, oval or ovate, 3–6 cm long and 2–3 cm wide, bright green above, dull and paler below, tip of leaf with a small, sharp point; margins entire or with small scattered teeth, on purple-red stalks 1 cm long. Flowers very small, yellow-white, on threadlike stalks from leaf axils; staminate flowers usually in small groups, pistillate flowers single. Fruit a purple-red berrylike drupe, 5–6 mm wide. May–June.

Open bogs (especially along outer moat), swamps, thickets, wet depressions in forests, lakeshores.

Ilex verticillata (L.) Gray
WINTERBERRY native
IR | **WUP** | **CUP** | **EUP** FACW CC 5
Shrub to 5 m tall; twigs smooth, finely ridged. Leaves deciduous, alternate, obovate to oval, tapered to a tip, dull green above, paler below; margins with incurved teeth. Flowers small, green-white, on short stalks from leaf axils, opening before leaves fully expanded in spring; staminate flowers in crowded clusters, pistillate flowers 1 or several in a group. Fruit a berrylike drupe, orange or red, 5–6 mm wide and persisting into winter. June.

Swamps, open bogs, thickets, shores and streambanks.

Araliaceae
GINSENG FAMILY

Shrubs or herbs, rarely trees. Leaves usually alternate, compound or rarely simple, the petiole not sheathing at base and usually adnate to the stipules; flowers small, umbellate. Flowers regular, epigynous, perfect or unisexual, 5–10-merous. Calyx small, its limb truncate to denticulate. Petals valvate or scarcely imbricate, usually distinct, deciduous at maturity.

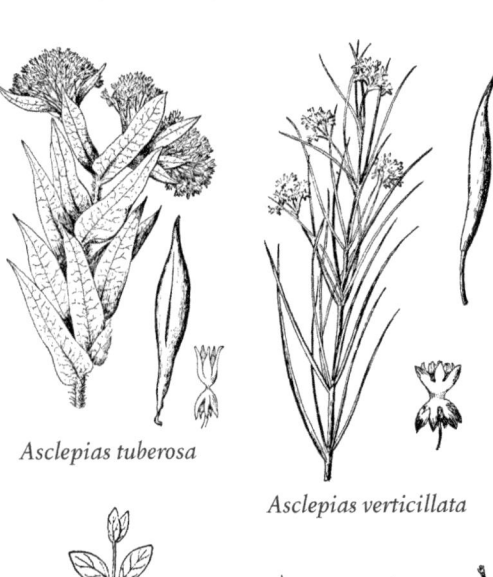

Asclepias tuberosa

Asclepias verticillata

Vinca minor

Ilex mucronata

Ilex verticillata

Stamens usually as many as the petals, rarely more. inserted on a tusk within the calyx; anthers short, longitudinally dehiscent. Ovary inferior, 2–12-celled, with one ovule in each cell. Styles as many as the cells of the ovary, distinct or more or less connate. Fruit a berry or a leathery drupe.

ADDITIONAL SPECIES

Oplopanax horridus (Small) Miq. (Devil's-club); primarily a species of northwestern North America, disjunct on islands in northern Lake Superior, including Porphyry Island off the Sibley Peninsula and the Slate Islands of Ontario. In Michigan, Devil's-club occurs only at northeast end of Isle Royale, several of the nearby offshore islands, and on Passage Island. Habitats are mostly wooded ravines with balsam-fir and paper birch. Stems (to 2 m tall) and leaf veins have sharp spines; flowers in umbels clustered in a large panicle; fruit bright red; threatened.

1 Leaves simple, palmately lobed . . . . **Hydrocotyle**
1 Leaves compound 2
2 Leaves alternate or basal, mostly 2–3 times compound; carpels 5 . . . . **Aralia**
2 Leaves in 1 whorl, once-palmately compound; carpels 2 or 3 . . . . **Panax**

Aralia *Sarsaparilla*

Herbs or shrubs. Stems herbaceous or slightly woody at the base only, rarely thorny (*A. hispida* bristly at the base). Leaves pinnately or 3-partedly compound. Flowers white or greenish, in 2–many umbels in each inflorescence. Petals and stamens each 5. Cells of the ovary 4–6, usually 5. Styles 4–6, usually 5, free or somewhat connate at base. Fruit a berry, tipped by the persistent styles; seeds usually 5.

1 Plants with flowers on a leafless scape . . **A. nudicaulis**
1 Plants with leafy stems 2
2 Lower stems bristly; umbels several (3–13) in a loose cluster . . . . **A. hispida**
2 Stems smooth; umbels very many, in a large terminal panicle . . . . **A. racemosa**

Aralia hispida Vent.
BRISTLY SARSAPARILLA native
IR | WUP | CUP | EUP CC 3

Perennial herb from a stout rhizome. Stems to 1 m tall, bristly near the base with sharp slender spines and often decreasingly so above. Leaves few, on petioles usually shorter than the blade, bipinnate; leaflets oblong to ovate or lance-shaped, up to 10 cm long but usually much smaller, acute or short-acuminate, sharply serrate. Umbels several, in a loose, open, terminal inflorescence; styles connate about half their length. Berry globose, nearly black. June–July.

Dry woods, especially in sandy or sterile soil.

Aralia nudicaulis L.
WILD SARSAPARILLA native
IR | WUP | CUP | EUP FACU CC 5

Acaulescent perennial herb, the leaves and peduncle arising from a long rhizome. Petiole erect, to 5 dm tall. Leaves 3-parted, each division pinnately 3–5-foliolate; leaflets lance-elliptic to obovate, up to 15 cm long and 8 cm wide, acuminate, finely serrate, the lateral ones asymmetric at base. Peduncles usually much shorter than the petioles, bearing 2–7, commonly 3 umbels; styles distinct to the base. Fruit nearly black. May–June.

Moist or dry woods.

Aralia racemosa L.
SPIKENARD native
IR | WUP | CUP | EUP FACU CC 8

Stout perennial herb to 2 m tall, lacking thorns or bristles. Leaves few, widely spreading, up to 8 dm long, the three primary divisions pinnately compound; leaflets ovate, variable in size in the same leaf, the larger up to 15 cm long, sharply and often doubly serrate, acuminate, obliquely cordate at base. Inflorescence a large panicle with numerous umbels; styles connate at base only. Fruit dark purple. July.

Rich woods.

Aralia hispida

Aralia nudicaulis

Aralia racemosa

Hydrocotyle *Pennywort*

Hydrocotyle americana L.
AMERICAN MARSH-PENNYWORT native
IR | **WUP** | **CUP** | **EUP** OBL CC 6

Small perennial herb. Stems slender and creeping, 10–20 cm long, often rooting at nodes. Leaves round to kidney-shaped, 1–5 cm wide; petioles long; margins with 7–12 shallow lobes. Flowers white, in nearly stalkless umbels from nodes; umbels 2–7-flowered. Fruit of 2 compressed carpels, more or less round in outline, 1–2 mm wide, ribbed. June–Sept.

Conifer swamps, streambanks, shores, wet forest depressions. Formerly included in Apiaceae.

Panax *Ginseng*

Perennial herbs, the unbranched stems rising from a deep-seated, thickened or tuber-like root, bearing a single whorl of once palmately compound leaves, usually 3 in number. Flowers in usually a single long-peduncled terminal umbel. Petals and stamens each 5; petals white or greenish; ovary 2–3 celled; styles 2 or 3. Fruit a small, 3-angled or flattened, 2–3-seeded berry.

1 Leaflets long-stalked; rare in Gogebic County . . . . **P. quinquefolius**
1 Leaflets sessile; common . . . . **P. trifolius**

Panax quinquefolius L.
AMERICAN GINSENG (THR) native
IR | **WUP** | CUP | EUP C C 10

Perennial herb. Stems 2–6 dm tall, root fusiform. Leaflets 3–7, usually 5, obovate to obovate, 6–15 cm long, conspicuously serrate, on long petiolules. Flowers greenish white, all or mostly perfect; peduncle 1–12 cm long; styles usually 2. Fruit a bright red berry, about 1 cm wide. June–July.

Rich deciduous woods, now rare due to heavy collecting; in the UP, known only from Gogebic County, where at north edge of species' range.

Panax trifolius L.
DWARF GINSENG native
IR | **WUP** | **CUP** | **EUP** CC 8

Perennial herb. Stems 1–2 dm tall; root globose. Leaflets 3–5, sessile or nearly so, lance-shaped to elliptic or oblong lance-shaped, 4–8 cm long, finely serrate. Flowers white or tinged with pink, often unisexual; peduncle 2–8 cm long; tyles usually 3. Fruit a yellow berry, about 5 mm wide. April–May.

Rich woods.

Aristolochiaceae
BIRTHWORT FAMILY

Asarum *Wild Ginger*

Asarum canadense L.
CANADIAN WILD GINGER native
IR | **WUP** | **CUP** | EUP UPL CC 5

Perennial herb; rhizome slender, branched, pubescent; producing annually a pair of leaves, between which arises the solitary, short-peduncled flower. Leaves 2, cordate, entire, at anthesis commonly 8–12 cm wide, larger at maturity, pubescent, especially on the long petiole. Flowers axillary, red-brown, 2–4 cm long, on a stout, pubescent pedicel 2–5 cm long; calyx tubular at base, deeply 3-lobed, the lobes spreading to reflexed, purple inside; petals absent or tiny and awl-shaped; ovary inferior, 6-celled. Fruit a capsule, bursting irregularly; seeds large, ovoid, wrinkled. April–May.

Rich woods, usually in small colonies.

Asteraceae
ASTER FAMILY

Annual, biennial or perennial herbs. Leaves simple or compound, opposite, alternate, or whorled. Flowers perfect (with both staminate and pistillate parts) or single-sexed (sometimes sterile) and of 2 types: ray

Hydrocotyle americana   *Panax quinquefolius*

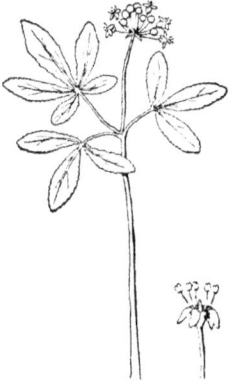

Panax trifolius

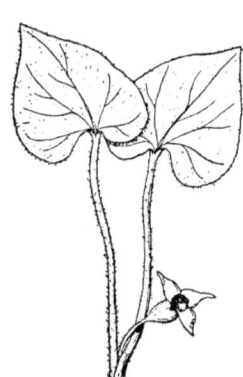

Asarum canadense

dicots

(or ligulate) and disk (or tubular). Ray flowers joined at base and have a long, flat, segment above (the ray); disk flowers tube-shaped with 5 lobes or teeth at tip.

Flowers are clustered in 1 of 3 types of heads resembling a single flower and attached to a common surface (receptacle): ray flowers only (as in dandelion, *Taraxacum*); disk flowers only (discoid, as in tansy, *Tanacetum*); and heads with both ray and disk flowers (radiate), the ray flowers surrounding the disk flowers (as in sunflower, *Helianthus*).

In addition to flowers, the receptacle may also have scales called chaff; if no scales present, the receptacle is termed naked. Each head is surrounded by involucral bracts (sometimes called phyllaries); collectively, the bracts are termed the involucre, comparable to the group of sepals (calyx) subtending an individual flower. Fertile flowers have 1 pistil tipped by a 2-cleft style (undivided in sterile flowers); stamens 5; ovary (and achene) often topped by several to many scales, awns or hairs (the pappus). Fruit a seedlike achene (sometimes termed cypsela in Asteraceae).

ADDITIONAL SPECIES

Several other members of the Aster Family occur in the UP, and, if not included in further described genera, are listed below. Many are introduced garden escapes, and most are adventive and not truly established in our flora.

Calendula officinalis L. (Pot-marigold), reported for Houghton County but likely not persistent.

Canadanthus modestus (Lindl.) Nesom (Canada-aster), native; Keweenaw County.

Cosmos bipinnatus Cav. (Garden cosmos), annual, leaves opposite, rays about 8, rose or lilac; cultivated and occasionally escaped; Houghton County.

Jacobea vulgaris Gaertn. (Stinking Willie); reported from Gogebic and Ontonagon cos.

Leucanthemella serotina (L.) Tzvelev (Giant daisy), reported for Iron County.

Logfia arvensis (L.) Holub (Field false cotton-rose), introduced; reported from Gogebic and Mackinac cos.

Mycelis muralis (L.) Dumort. (Wall-lettuce), introduced, Mackinac County.

Scorzoneroides autumnalis (L.) Moench (August-flower), introduced in several UP locations.

Tetraneuris herbacea Greene (Lakeside daisy); native; known from a single Michigan site, at the edge of a *Thuja* forest in marly soil over limestone in Mackinac County; a common Great Plains species, endangered in Michigan.

KEY TO GROUPS A, B, C

1. Heads entirely of flat ligulate flowers (these all bisexual); sap milky; leaves alternate or basal (never opposite) . . . . GROUP A
1. Heads all or partly of regular disk flowers (sometimes thread-like, sometimes very deeply lobed); ligulate (ray) flowers sterile or pistillate and only around the margin of the head; sap watery; leaves various (opposite in some species). 2
2. Pappus none or entirely of scales, teeth, flattened spines, or a few (to 8) stiff awns. . . . . GROUP B
2. Pappus all or primarily of numerous long soft hairs or bristles. . . . . GROUP C

KEY TO GENERA OF GROUP A

1. Pappus none or a crown or entirely of scales (without separate bristles). 2
1. Pappus at least partly of bristles or soft hairs 4
2. Leaves all basal . . . . **Leontodon**
2. Leaves all or partly along the stem 3
3. Flowers blue (rarely white or pink), in heads 3–4.5 cm wide; pappus of numerous tiny scales; stem leaves sessile; phyllaries (at least the outer ones at maturity) nearly always with few to many gland-tipped hairs. . . . . **Cichorium**
3. Flowers yellow, in heads ca. 1 cm or less broad; pappus none; stem leaves (below inflorescence) tapered to a petiolar base; phyllaries glabrous or nearly so **Lapsana**
4. Pappus composed of both scales and bristles (scales may be short and inconspicuous); involucre a single series of phyllaries . . . . **Krigia**
4. Pappus composed entirely of hairs or bristles; involucre various 5
5. Pappus at least partly of plumose bristles. 6
5. Pappus entirely of simple bristles or hairs (at most barbed or scabrous) 9
6. Plants with leaves of fertile stems all or partly on the stem. . . . . **Tragopogon**
6. Plants with leaves in a basal rosette, the stems (simple or branched) at most with very reduced bracts 7
7. Receptacle with a ± membranous scale subtending each flower and achene and bearing an elongate bristle-like tip (evident even among the flowers, and equaling the pappus); achene with a slender beak longer than the body; involucre 11–16mm long (or to 25 mm in fruit), the phyllaries glabrous or (usually) with few to many bristles on the midrib only . . . . **Hypochaeris**
7. Receptacle without scales or chaff; achene beakless; involucre ca. 7–14 mm long, even in fruit, the phyllaries glabrous or ± hairy all across 8
8. Stem usually branching, with a few heads on long pedicels; hairs all simple, not forked; outermost achenes in head with pappus of one row of plumose hairs, as in the other achenes . . . . **Scorzoneroides***
8. Stem simple, the heads on naked scapes; hairs (especially those on the leaves) minutely 2 (–3)-forked at the apex [under magnification]; outermost achenes in head with pappus solely of scales, the other achenes with pappus of plumose hairs surrounded by short simple hairs. . . . . **Leontodon**

9 Heads solitary on completely naked scapes; leaves all in a basal rosette; scapes glabrous or with a little loose tomentum............................**Taraxacum**

9 Heads (1–) 2–several on naked to leafy stems; leaves basal or on the stem; scapes on 1-headed plants rather densely pubescent with stellate and/or long straight hairs.. 10

10 Leaves all or mostly in a basal rosette at flowering time (or on stoloniferous basal shoots), entire and unlobed.
....**Hieracium**

10 Leaves all or mostly along the stem at flowering time, entire or toothed or lobed, or if largely basal, then toothed..11

11 Involucres cup-shaped or bell-shaped, at least two-thirds as broad as long at anthesis; achenes scarcely if at all beaked....................................... 12

11 Involucres (at least at anthesis) cylindrical to urn-shaped, at least twice as long as broad; achenes long-beaked, short-beaked, or beakless............ 14

12 Stem leaves tapered to base or sessile but not auriculate-clasping; pappus light to darkish brown**Hieracium**

12 Stem leaves (at least the upper ones) ± auriculate or clasping at the base; pappus pure white, copious ... 13

13 Leaf margin not spiny; involucre 5–10 mm long; achenes ± terete (but ribbed)....................... **Crepis**

13 Leaf margin prickly with prolonged ± spiny-tipped teeth; involucre ca. 8–18 mm long at maturity; achenes at least somewhat flattened................ **Sonchus**

14 Blades of stem leaves (in outline) less than twice as long as broad, entire or with lobes 3 (–5) and somewhat palmate in aspect (or involucre hairy); achenes neither flattened nor beaked..................... **Prenanthes**

14 Blades of stem leaves (in outline) over three times as long as broad, entire or with lobes 5–7 and clearly pinnate; involucres glabrous; achenes strongly flattened, beaked (usually) or not........................... 15

15 Florets 5 per head; longer phyllaries 5 or fewer .... **Mycelis***

15 Florets and phyllaries more numerous....... **Lactuca**

KEY TO GENERA OF GROUP B

1 Involucre of just 2 series of phyllaries distinctly different in size, texture, and/or orientation; heads at least 1 cm broad; stem leaves opposite or whorled 2

1 Involucre imbricate, with phyllaries of more than 2 lengths, or with a single series of phyllaries, or heads less than 1 cm broad; stem leaves various 6

2 Leaves all or mostly aquatic (submersed), apparently whorled, fully dissected with filiform segments (usually at least 1 pair of simple, opposite, toothed leaves on emersed stem beneath head); heads radiate, 1–2 at the end of long leafy (submersed) branches; pappus awns elongating to 2 (–4) cm in fruit .... **Bidens**

2 Leaves not aquatic (if finely divided, nevertheless terrestrial); heads various; pappus awns, if any, less than 1 cm long 3

3 Pappus of distinct barbed awns.................... 4

3 Pappus of 2 teeth or scales or absent 5

4 Ray flowers none or clear yellow; leaves various (but narrowly lobed in only one species of wet places); native, common .... **Bidens**

4 Ray flowers red to white or orange; leaves deeply pinnately lobed into narrow or even filiform segments; rarely escaped into disturbed places .... **Cosmos***

5 Outer phyllaries various (often less than half as long as inner), at most barely connate at the very base; leaves various: simple, or palmately lobed or compound, or if pinnately lobed at least the central lobe usually much larger than the lateral ones; achenes strongly flattened, not beaked, in most species winged........ **Coreopsis**

5 Outer phyllaries more than half as long as inner, clearly connate at the base (for ca. 0.5–1 mm above the receptacle); leaves deeply pinnatisect with uniformly narrowly lance-shaped to thread-like lobes; achenes little if at all flattened, beaked, without wings . **Cosmos***

6 Margins of phyllaries with a well defined hyaline or scarious border at least around the tip; leaves alternate.. 7

6 Margins of all or most phyllaries not hyaline or scarious (or if so, the leaves opposite); leaves opposite or alternate or all basal.................................. 20

7 Receptacle, and hence the disk, conical or high-hemispherical, the disk (including florets) at least 6 mm broad at the base; leaves all deeply dissected 8

7 Receptacle flat or slightly convex, the disk in some species less than 5 mm wide; leaves dissected or not 12

8 Heads rayless; bruised plant with fruity (pineapple) aroma .... **Matricaria discoidea**

8 Heads with ray flowers; bruised plant with or without odor 9

9 Ray flowers yellow .... **Anthemis tinctoria**

9 Ray flowers white 10

10 Receptacle with chaff (except around the marginal flowers in *A. cotula*); leaf segments and axis pubescent, often all or mostly 0.5–2 mm wide, flat ..... **Anthemis**

10 Receptacle without chaff; leaf segments mostly revolute-filiform, essentially glabrous, less than 0.5 mm wide 11

11 Involucres ca. 2–3 mm long; pappus none; achene ribs 5, ± weak; plant aromatic when bruised. .... **Matricaria chamomilla**

11 Involucres ca. 4–5.5 mm long; pappus present (though minute); achene ribs 3 strongly thickened or wing-like; plant nearly or quite odorless..... **Tripleurospermum**

12 Leaves unlobed (but closely and regularly toothed), glabrous or only slightly silky when mature......... 13

12 Leaves lobed or dissected (or at least deeply toothed near base) or if unlobed then densely tomentose at least beneath 16

13 Rays ca. 4–6 mm long; leaves not over 8 mm wide .... **Achillea ptarmica**

13 Rays over 1 mm long or absent; leaves over 1 mm wide 14

ASTERACEAE | Aster Family

14 Disk less than 1 cm wide; rays none; foliage sweet-smelling; heads numerous on very short pedicels ..... **Tanacetum balsamita**

14 Disk and conspicuous white rays each over 1 cm; foliage not aromatic; heads 1 to several on each stem or branch 15

15 Leaves not glandular-punctate, with low rounded teeth; heads 1–2 on long peduncles .... **Leucanthemum**

15 Leaves glandular-punctate, with large acute teeth; heads several on each stem or branch .... **Leucanthemella***

16 Rays present, white 17

16 Rays absent or yellow 19

17 Rays 1–3 mm long; receptacle chaffy with an elongate scale at the base of each floret .. **Achillea millefolium**

17 Rays mostly 4–5 mm long; receptacle without chaff . 18

18 Heads solitary on long stems, 3–6 cm wide; foliage scentless, the leaves shallowly once-pinnately lobed .. **Leucanthemum**

18 Heads several to many, less than 2 cm wide, ± corymbose; foliage strong-scented, the leaves bipinnatifid .. **Tanacetum parthenium**

19 Disk conspicuous (yellow), 5–22 mm wide (if less than 10 mm, then the leaves nearly glabrous); inflorescence ± corymbiform .... **Tanacetum**

19 Disk inconspicuous, usually less than 5 mm wide (if 5–10 mm, then the leaves densely tomentose beneath); heads in an elongate panicle-like inflorescence .... **Artemisia**

20 Heads without ray flowers 21

20 Heads with small to conspicuous ray flowers 24

21 Lower surface of leaves densely white-woolly (stem never spiny); corollas present ... **Adenocaulon bicolor**

21 Lower surface of leaves glabrate or, if densely pubescent, the hairs neither white nor woolly; corollas often absent on pistillate flowers 22

22 Leaves all alternate; involucre of pistillate heads covered with hook-tipped spines, forming a beaked bur in fruit; staminate heads crowded immediately above the pistillate heads, with separate phyllaries ... **Xanthium**

22 Leaves (at least the middle and lower) opposite; involucre of all heads without hooked spines (at most with horn-like protuberances); staminate (or all) heads in a branched cymose inflorescence or in elongate raceme-like or spike-like inflorescences and the phyllaries united into a cup 23

23 Staminate and pistillate heads separate, the former on elongate branches, the latter at the bases of these branches (rare plants unisexual); larger leaves lobed except on depauperate plants .... **Ambrosia**

23 Staminate and pistillate flowers in the same heads (pistillate at the margin); leaves simple (but coarsely toothed) .... **Cyclachaena**

24 Rays white to pink or rose-purple 25

24 Rays yellow to orange, sometimes mostly red-purple with yellow bands or tips 28

25 Heads solitary (one per stem or at end of a branch); rays showy, conspicuously longer than the involucre 27

25 Heads several per stem or branch; rays inconspicuous, shorter than the involucre (but may protrude beyond it) 27

26 Plants small (less than 2 dm tall), scapose; rays ca. 1 cm or shorter .... **Bellis perennis**

26 Plants tall (well over 2 dm), leafy-stemmed; rays ca. 3–6 cm long .... **Echinacea**

27 Leaves alternate .... **Parthenium**

27 Leaves opposite .... **Galinsoga**

28 Disk flowers sterile, the ovary smaller than on fertile ray flowers, and the style undivided; achenes flattened parallel to the phyllaries, wing-margined .... **Silphium**

28 Disk flowers fertile, the ovary at least as large as on any fertile ray flowers, and the mature style forked; achenes thick and angled or ± flattened at right angles to the phyllaries, wingless (except in *Verbesina*) 29

29 Leaves in a basal rosette; flowers solitary .... **Tetraneuris herbacea***

29 Leaves (at least some) from the stem; flowers normally more than one per stem 30

30 Leaves opposite, at least toward base of plant (fragmentary specimens lacking lower part of the stem may be inadequate for identification) 31

30 Leaves all alternate 32

31 Ray flowers sterile, without style; achenes ± compressed but not angled; leaves of some plants alternate on upper part of stem; phyllaries in most species acute or acuminate .... **Helianthus**

31 Ray flowers fertile, with style; achenes 3–4-angled; leaves all opposite; phyllaries rounded or blunt at apex .... **Heliopsis**

32 Phyllaries sticky or gummy (appearing strongly varnished when dry), the outer ones with prolonged narrow recurved tip; leaves sessile, clasping, toothed, and with shiny glandular dots on both surfaces; receptacle without chaffy bracts (but long pappus bristles evident) .... **Grindelia squarrosa**

32 Phyllaries not gummy, without recurved tip; leaves various but without above combination; receptacle with or without chaff 33

33 Stem (at least upper half) with prominent green wings decurrent from leaf bases the full length of the internode .... **Helenium**

33 Stem not at all winged 34

34 Receptacle and disk flat or nearly so 35

34 Receptacle (and hence disk) conspicuously dome-shaped, conical, or even cylindrical 36

35 Ray flowers sterile; achenes wingless; chaff of receptacle consisting of elongate flattish bracts easily seen among the disk flowers; plants common (native or introduced), annual or perennial .... **Helianthus**

35 Ray flowers pistillate; achenes winged or chaff none; rare escape, annual .... **Calendula***

36 Rays strongly drooping; achenes ± flattened; receptacle with chaff subtending both ray and disk flowers, with the bracts curved (like a hood) and densely hairy at the apex, concealing the florets in bud; stem leaves very deeply pinnately lobed or compound, with narrow segments (or leaflets)........................ **Ratibida**

36 Rays mostly spreading; achenes plump, angled; receptacle with chaff none or only on the disk, of various nature but not hood-like; stem leaves variously lobed or usually unlobed (in some species of *Rudbeckia* with few broad palmate or pinnate lobes, in *Gaillardia* sometimes pinnately lobed and toothed)..................... 37

37 Receptacle with irregular bristle-like hairs not subtending individual florets; pappus of several awn-tipped scales; achenes long-hairy, especially near base; rays mostly 3–5-lobed or -toothed at apex ...... **Gaillardia**

37 Receptacle with a chaffy scale subtending each disk floret (and partly surrounding the achene); pappus none or a tiny crown; achenes glabrous; rays 2-lobed or -toothed at apex .... **Rudbeckia**

KEY TO GENERA OF GROUP C

1 Leaves all or partly opposite or whorled 2
1 Leaves all alternate or basal 5
2 Rays bright yellow, showy, greatly exceeding the involucre (ca. 14–25 mm long); leaf blades ± cordate, the middle and lower on winged petioles.......... **Arnica**
2 Rays none; leaf blades tapered to petiole or sessile (or even connate around stem)....................... 3
3 Leaves whorled; heads pale pink to reddish or .. purple .... **Eutrochium**
3 Leaves opposite (sometimes a few upper ones alternate); heads white or blue........................ 4
4 Leaf blades petiolate, ovate, less than twice as long as wide (rarely 2.5 times)..................... **Ageratina**
4 Leaf blades sessile, or if petiolate, lanceolate and ca. 2–4 times as long as wide .... **Eupatorium**
5 Ray flowers ± bright yellow or orange (disk flowers also yellow)... 6
5 Ray flowers none or white, blue, pink, or purple (disk flowers may be yellow, and marginal ones may also be enlarged and deeply lobed) 11
6 Phyllaries in a single series (all or essentially all ± equal in length and overlapping at most a little toward the base, but sometimes a few short bracteoles at base of involucre)................................ **Packera**
6 Phyllaries imbricate (of various lengths, overlapping) 7
7 Heads small, the disk (receptacle) less than 5 mm wide 8
7 Heads large, the disk (receptacle) 7 mm or more wide 9
8 Leaves narrowly linear-lanceolate, entire, usually punctate with shiny dots, best developed on the middle of the stem; heads all or mostly sessile or subsessile in little clusters of 2 or more, in a ± corymbiform inflorescence .... **Euthamia**

8 Leaves various but if entire and linear-lanceolate, then not punctate and/or best developed at the base and/or the inflorescence not corymbiform and/or the heads on distinct pedicels .... **Solidago**
9 Leaves entire (or very nearly so) .... **Heterotheca**
9 Leaves toothed...................................... 10
10 Teeth of leaves with prolonged spiny tips .... **Grindelia ciliata**
10 Teeth of leaves blunt or rounded .... **Inula**
11 Rays present and showy, extending beyond the involucre at least as great a distance as the length of the involucre 12
11 Rays none or very inconspicuous (extending beyond the involucre, if at all, less than its length), (do not be misled by some species where the marginal disk flowers may be much larger than the central ones)............. 24
12 Phyllaries all or almost all ± equal in length 13
12 Phyllaries of several lengths, the involucre clearly imbricate 17
13 Stem leaves consisting of modified broad multi-veined petiolar bracts with at most a rudimentary blade, on a thick (often over 5 mm diameter) stem...... **Petasites**
13 Stem leaves not so modified, on a stem not over 5 (–6) mm thick 14
14 Pedicels and phyllaries with dense stalked glands ... 15
14 Pedicels and phyllaries without glands or at most with obscure glands 16
15 Leaf blades only slightly auriculate; phyllaries usually green throughout; achenes sparsely strigose; stems arising singly from an elongate rhizome .... **Canadanthus modestus***
15 Leaf blades strongly auriculate; phyllaries usually flushed with purple, especially on the acuminate tip; achenes densely strigose; stems arising in clumps from short, thick rhizomes **Symphyotrichum novae-angliae**
16 Phyllaries at least sparsely pubescent on back, with no distinct expanded apex on central stripe; rays usually white to pink (showy on the few-headed *E. pulchellus*). .... **Erigeron**
16 Phyllaries glabrous on back, with central narrow green stripe expanded into a ± diamond-shaped green area below the tip; rays bright blue or purple .... **Symphyotrichum**
17 Pappus hairs thickened (slenderly clavate) toward apex; rays white, 12–18; achenes glabrous; leaves ± oblanceolate......................... **Solidago ptarmicoides**
17 Pappus hairs not thickened (if obscurely so, achenes pubescent); leaves various (but not oblanceolate if rays white and 12–18)................................... 18
18 Blades of at least the lower and/or basal leaves with distinct petioles and cordate to broadly rounded or truncate bases 19
18 Blades of leaves not cordate or if somewhat so, the leaves sessile, not distinctly petioled 20
19 Involucres mostly 7–10 mm long, with outer phyllaries 1–2.2 mm wide; inflorescence corymbiform ... **Eurybia**

19 Involucres less than 6.5 (–7.5) mm long, with no phyllaries more than 1 mm wide; inflorescence elongate (racemose or paniculate) .... **Symphyotrichum**
20 Phyllaries with ± dense stalked glands 21
20 Phyllaries without glands 22
21 Leaf blades only slightly auriculate; phyllaries usually green throughout; achenes sparsely strigose; stems arising singly from an elongate rhizome .... **Canadanthus modestus***
21 Leaf blades strongly auriculate; phyllaries usually flushed with purple, especially on the acuminate tip; achenes densely strigose; stems arising in clusters from short, thick rhizomes .... **Symphyotrichum novae-angliae**
22 Middle phyllaries with the green zone ± abruptly expanded at or above the middle into a distinct short or elongate broad diamond-shaped area, the lower portion of the phyllaries mostly whitish with a very narrow darker midvein; heads numerous in an elongate to broadly triangular paniculate inflorescence (not flat-topped)...................... **Symphyotrichum**
22 Middle phyllaries with broad dark (green or reddish) zone bordering midvein mostly ± uniform in width (at mid-length, nearly or quite as wide as the pale area on each side); heads 1–7 or numerous in a flat-topped inflorescence 23
23 Leaves elliptic, mostly (7–) 10–30 mm wide, glabrous or pubescent but eglandular and reticulate with strongly contrasting dark veinlets and tiny pale areoles ca. 0.6 mm or less across, with flat margins; heads ca. 2 cm or less across, numerous in flat-topped corymbs, with white rays .... **Doellingeria**
23 Leaves linear to very narrowly elliptic (less than 7 mm wide), glandular beneath but not strongly reticulate-veined, with revolute margins; heads ca. 2.5–4 cm across, solitary to few on long pedicels, with pink rays .... **Oclemena**
24 Phyllaries (at least the outer ones) spine-tipped (hooked in *Arctium*) or strongly fringed or lacerate-margined, or the lobes of the leaves spine-tipped, or both conditions (phyllaries and leaves) present; disk corollas usually very deeply lobed 25
24 Phyllaries and leaves all without spines; disk corolla deeply lobed or not............................ 28
25 Spine at tip of phyllary hooked; leaves very large and ovate, not spiny .... **Arctium**
25 Spine (if any) at tip of phyllary straight; leaves various 26
26 Phyllaries lacerate or fringed at the tip (including a terminal spine or not); leaves without spines.. **Centaurea**
26 Phyllaries with a terminal spine or none; leaves with spine-tipped lobes or spiny margins 27
27 Pappus bristles simple (at most slightly barbed); internodes of stem continuously or intermittently winged its entire length............................. **Carduus**
27 Pappus bristles plumose (feathery with very long fine lateral branches); internodes of most species not (or only slightly) winged .... **Cirsium**

28 Middle phyllaries wholly scarious or with prominent pale to brownish scarious tips and/or margins totaling 1/4 or more of the length or width of the phyllary; corollas (except in staminate *Antennaria* with pilose phyllaries) thread-like (i.e., extremely slender, only 0.1–0.3 mm wide)................................. 29
28 Middle phyllaries with scarious margin none or very narrow (1/6 the length of phyllary or less, rarely approaching 1/4 in *Liatris*); corollas in most species more than 0.3 mm wide.................................. 34
29 Plants mostly stoloniferous with leafy rosettes; stem leaves much reduced, ± remote; pappus bristles (at least in pistillate or bisexual flowers) united in a ring at the base................................... **Antennaria**
29 Plants annual or rhizomatous, with neither stolons nor basal rosettes; stem leaves numerous, much overlapping; pappus bristles often separate 30
30 Phyllaries tomentose nearly or quite to the tip; receptacle chaffy except at the middle (the chaff grading into the sparse phyllaries, but glabrate) .. **Logfia arvensis***
30 Phyllaries glabrous on at least the apical half; receptacle without chaff.. 31
31 Phyllaries (except at base) pure pearly white, appearing distinctly longitudinally striate (from tiny creases); leaves smooth and glabrous above or with loose white tomentum (rarely with a few tiny gland-tipped hairs hidden in the tomentum); plant rhizomatous and (fresh or dry) without sweetish odor .... **Anaphalis**
31 Phyllaries off-white to brownish, not appearing striate from tiny creases; leaves at least in common species with short-gland-tipped hairs or at least roughened above; plants tap-rooted (rhizome only in *Gnaphalium sylvatica*) and the common species (fresh or dry) with sweetish (maple syrup-like) odor especially when crushed.... 32
32 Inflorescence elongate (spicate or racemose); pappus bristles united in a ring at the base; plants rhizomatous .... **Gnaphalium**
32 Inflorescence corymbiform (or heads crowded at ends of branches, or sometimes spike-like in *Gnaphalium uliginosum* with involucres only 2–3 mm long); pappus bristles separate; plants tap-rooted 33
33 Involucre ca. 2—3 mm long; plants (except the most depauperate) bushy-branched; heads in clusters overtopped by subtending leaves............. **Gnaphalium**
33 Involucre 4.5–7 mm long; plants rarely branched (except at the top or in the inflorescence); heads not overtopped by subtending leaves..... **Pseudognaphalium**
34 Flowers purple or pink 35
34 Flowers white, cream, or yellow.................. 38
35 Rays present, tiny 36
35 Rays none 37
36 Involucre to 4 mm long .... **Conyza canadensis**
36 Involucre at least 5 mm long.......... **Erigeron acris**
37 Phyllaries in a single series, of ± equal length (tiny bracteoles may be at base), smooth; stem leaves reduced to wide bracts (at most a tiny blade at apex); basal leaves ca. 1–3 dm wide ...... **Petasites hybridus**

37 Phyllaries imbricate, ± glandular-punctate; stem leaves normal; basal leaves much less than 1 dm wide . **Liatris**

38 Phyllaries in 2 or more series of ± equal length or imbricate... 39

38 Phyllaries in a single series of ± equal length, in some species with bracteoles very much shorter and usually narrower at base of involucre..................... 42

39 Involucre 2.7–4 mm long......................... 40

39 Involucre at least 5 mm long 41

40 Pubescence mostly of spreading, longer hairs; pappus of a single series .... **Conyza**

40 Pubescence mostly of appressed antrorse hairs 0.5 mm or shorter; pappus double, with an outer series of minute scale-like bristles .... **Erigeron strigosus**

41 Pedicels and often phyllaries with minute gland-tipped hairs; leaves lanceolate to oblanceolate; achenes glabrous............................ **Erigeron acris**

41 Pedicels and phyllaries eglandular; leaves narrowly linear (less than 4 mm wide); achenes pubescent......... .... **Symphyotrichum ciliatum**

42 Flowers yellow; stem leaves deeply pinnatifid, at least toward the base............................... 43

42 Flowers white to cream; stem leaves at most toothed 44

43 Upper half of unbranched portion of stem less leafy than the lower half (i.e., leaves sparser and distinctly smaller); biennial or perennial, often with crowded basal leaves .... **Packera**

43 Upper half of unbranched portion of stem nearly or quite as leafy as the lower half (i.e., leaves ± equally numerous and of similar size and shape, or only gradually reduced upwards); annual, without basal rosette of persistent leaves........................... **Senecio**

44 Stem leaves reduced to wide petiolar (parallel-veined) bracts with at most a rudimentary blade at the apex .. .. **Petasites**

44 Stem leaves with well developed blades 45

45 Leaves from the stem (remote and few) and basal, long-petioled, coarsely and broadly toothed, shallowly lobed, or entire, with several main palmate or longitudinal veins; pappus bristles scabrous or minutely antrorse-barbed; perennial..................... **Arnoglossum**

45 Leaves all from the stem (crowded and overlapping), sharply toothed, pinnately veined with 1 midrib; pappus bristles smooth; annual .... **Erechtites**

Achillea *Yarrow*

Perennial herbs. Leaves alternate, subentire to pinnately dissected. Inflorescence more or less corymbiform, of several to many relatively small heads. Heads radiate or rarely discoid, the rays mostly 5–12, pistillate and fertile, rarely neutral, white, sometimes pink or rarely yellow. Involucral bracts imbricate in 3–4 series, dry, with scarious or hyaline margins and often greenish midrib. Receptacle conic or convex, chaffy throughout. Disk flowers about 10–75, perfect and fertile. Fruit a compressed achene; pappus none.

1 Leaves finely dissected; plants tomentose; ubiquitous. .... **A. millefolium**

1 Leaves nearly entire; plants nearly hairless; uncommon adventive species .... **A. ptarmica**

Achillea millefolium L.
COMMON YARROW *native*
IR | WUP | CUP | EUP   FACU CC –

Aromatic rhizomatous perennial, sparsely to rather densely villous throughout. Stems about 2–10 dm tall. Leaves pinnately dissected, the blade about 3–15 cm long and to 2.5 cm wide, the basal petiolate, all but the lowermost stem leaves sessile. Heads numerous in a flat or round-topped, short and wide, paniculate-corymbiform inflorescence, the disk about 2–4 mm wide; involucre in ours mostly 4–5 mm high; rays about 5, white or occasionally pink, 2–3 mm long; disk flowers about 10–30. June–Oct.

Common in fields, prairies, lawns, beaches, and waste places.

Achillea ptarmica L.
SNEEZEWEED *introduced*
IR | WUP | CUP | EUP

Rhizomatous perennial. Stems 3–6 dm tall, villous above, often nearly glabrous below. Leaves glabrous or nearly so, linear or lance-linear, about 3–10 cm long and 2–6 mm wide, sessile, closely and rather shallowly serrate to subentire. Heads several or numerous in an open corymbiform inflorescence, the disk about 4–8 mm wide; involucre about 4–5 mm high; rays commonly 8–10, white, 3–5 mm long; disk flowers about 50–75. July–Sept.

Beaches, roadsides, and waste places; native to n Europe and Asia. Forms escaped from cultivation are often double, with more than the usual number of ray flowers.

Adenocaulon *Trailplant*

Adenocaulon bicolor Hook.
AMERICAN TRAILPLANT *native*
IR | WUP | CUP | EUP   CC 8

Fibrous-rooted perennial herb. Stems 3–7 dm tall. Leaves mostly basal, long-petioled, 3–15 cm wide, deltoid-ovate to cordate, nearly glabrous above, white woolly below. Inflorescence an open panicle; heads discoid; involucral bracts about 2 mm long, reflexed in fruit.

Deciduous and mixed forests, especially hemlock-hardwoods and in moist ravines and along trails. A species of the Pacific Northwest, disjunct in the Black Hills of South Dakota and in the upper Great Lakes region; in Michigan, known only from the UP in the counties bordering Lake Superior.

Ageratina *Snakeroot*

Ageratina altissima (L.) R. M. King & H. Rob.
WHITE SNAKEROOT   *native*
IR | **WUP** | **CUP** | EUP   FACU CC 4
Eupatorium rugosum Houtt.

Perennial herb. Stems mostly 5–8 dm tall, finely pubescent. Leaves opposite, ovate to broadly ovate, base subcordate, 4–11 cm long and 3–8 cm wide, smaller upward; petioles 1–3 cm long; margins coarsely serrate. Inflorescence flat-topped or flat dome-shaped; involucre 4–5 mm long, involucral bracts acuminate to obtuse, all about the same length; florets 12–24, corolla white; pappus of bristles. Aug–Oct.

Floodplain forests, cedar swamps, thickets, streambanks, wooded ravines, sometimes where disturbed.

A toxic substance in this plant can cause 'trembles,' a fatal disease of cattle which have browsed on it and transmittable to humans by their milk, in whom the consequent milk sickness caused many deaths in the 19th century.

Ambrosia *Ragweed*

Coarse annual or perennial herbs. Leaves opposite or alternate, mostly lobed or dissected. Heads unisexual, small. Staminate heads in a spike-like or raceme-like bractless inflorescence; involucre 5–12-lobed; receptacle flat, its bracts slender. Pistillate heads borne below the staminate ones, in the axils of leaves or bracts; involucre closed, nut-like, usually with a single series of tubercles or short erect spines near the apex; pistil solitary, without corolla; pappus absent.

The pollen is wind-borne, and some species are among the most important causes of hay-fever in the USA.

1 Leaves palmately 3–5 lobed or unlobed; large annual plant to 2 m or more tall . . . . **A. trifida**
1 Leaves 1–2 times pinnately lobed or divided; plants usually less than 1 m tall 2
2 Perennials, forming colonies from creeping underground roots; leaves usually coarsely lobed. . . . **A. psilostachya**
2 Taprooted annuals; leaves finely divided . . . . **A. artemisiifolia**

Ambrosia artemisiifolia L.
COMMON RAGWEED   *native*
IR | **WUP** | **CUP** | EUP   CC 0

Annual weed. Stems branching at least above, variously hairy or subglabrous, mostly 3–10 dm tall. Leaves opposite below, alternate above, petiolate, 1–2x pinnatifid, ovate or elliptic in outline, commonly 4–10 cm long. Sterile heads short-pedunculate. Fruiting involucre short-beaked, about 3–5 mm long, with several short sharp spines. Aug–Oct.

Waste places.

Ambrosia psilostachya DC.
PERENNIAL RAGWEED   *native*
IR | **WUP** | **CUP** | EUP   FAC

Similar to *A. artemisiifolia*. Perennial from a creeping rhizome. Leaves thicker, short-petiolate or subsessile, usually only once pinnatifid, averaging narrower in outline, sometimes 10 cm long. Fruiting involucre tuberculate above, sometimes obscurely so. July–Oct.

Waste places, usually in dry or sandy soil.

Ambrosia trifida L.
GIANT RAGWEED   *native (invasive)*
IR | **WUP** | **CUP** | EUP   FAC CC 0

Annual weed, to sometimes as much as 5 m tall.

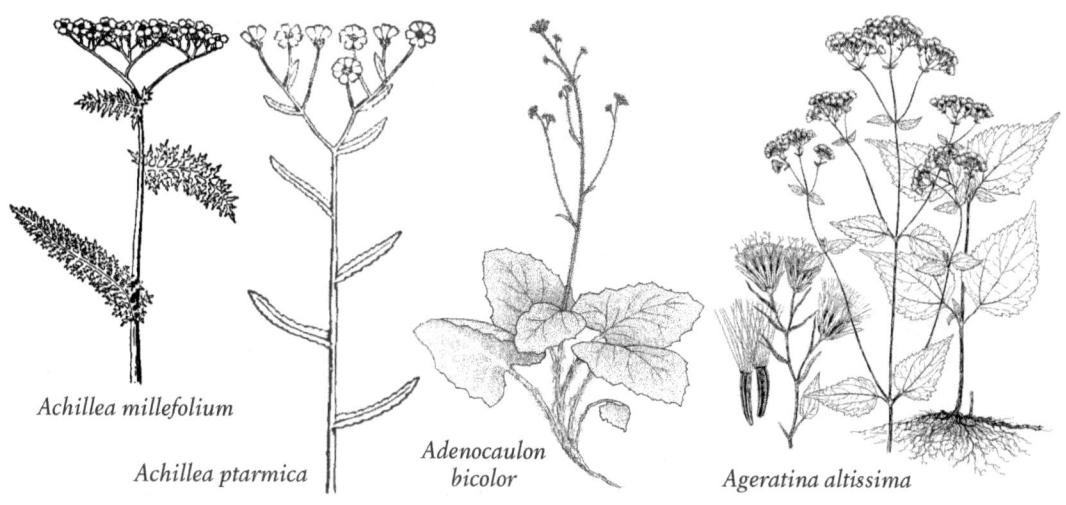

Achillea millefolium

Achillea ptarmica

Adenocaulon bicolor

Ageratina altissima

Stems spreading-hirsute or hispid above, often glabrous or glabrate below. Leaves opposite, petiolate, broadly elliptic to more commonly ovate or suhorbicular, serrate, palmately 3–5-lobed, or, especially in depauperate specimens, lobeless, often 2 dm long or more, more or less scabrous on both sides. Sterile involucres unilaterally 3-nerved. Fertile involucres about 5–10 mm long in fruit, several-ribbed, each rib bearing a short spine at the tip. July–Oct.

Moist soil and waste places.

Anaphalis *Pearly-Everlasting*

Anaphalis margaritacea (L.) Benth.
PEARLY-EVERLASTING   native
IR | WUP | CUP | EUP   CC 3

White-woolly perennial herbs. Stems erect, simple or branched, commonly 3–9 dm tall, leafy, loosely white-woolly. Leaves alternate, lance-shaped or linear, to about 12 cm long and 1.5 cm wide, sessile, commonly less pubescent above than beneath, or green and glabrous above, the margins entire and often revolute, basal leaves soon deciduous. Heads 1 cm wide or less, numerous and crowded in a short wide inflorescence; some flowers bearing both stamens and pistils. Involucre about 5–7 mm high, the bracts pearly white. Achenes papillate; pappus, in both staminate and pistillate flowers, of distinct capillary bristles. Variable. July–Aug.

Chiefly in dry woods and clearings.

Antennaria *Pussytoes*

Perennial woolly herbs. Leaves basal and alternate on the stem. Flowers dioecious, rarely incompletely so. Heads many-flowered, disciform or discoid, solitary to many in a crowded inflorescence. Involucral bracts imbricate in several series, scarious at least at the tip, often colored. Receptacle naked, flat or convex. Staminate flowers with scanty pappus, the bristles commonly barbellate or clavate. Pistillate flowers with filiform-tubular corolla, bifid style, and copious bristles slightly united at the base. Achenes terete or slightly compressed. Most of our species are partly or wholly apomictic, producing seeds without fertilization.

1 Rosette leaves large, 3- or 5-nerved . . . . **A. parlinii**
1 Rosette leaves small, 1-nerved or obscurely 3-nerved (best seen on old basal leaves from the previous year) 2
2 New basal leaves of the season essentially glabrous above or very soon becoming so (may appear hairy along the margin from tomentum of underside) . . . . **A. howellii**
2 New basal leaves pubescent above when young (becoming glabrous only in age); Keweenaw Co. . . **A. neglecta**

Antennaria howellii Greene
SMALL PUSSYTOES   native
IR | WUP | CUP | EUP   CC 2

Antennaria neglecta var. *howellii* (Greene) Cronquist
Plants mostly with pistillate flowers only, staminate plants rare; spreading by short stolons. Stems to 35 cm tall, sometimes with gland-tipped hairs. Basal leaves oblanceolate to ovate, 2–5 cm long, tips mucronate, upper surface tomentose, underside green-glabrous or gray-pubescent. Stem leaves linear, 1–4 cm long. Heads 3–15 in corymbiform clusters. Involucral bracts white, cream, or light brown, sometimes rose at base.

Many types of dry, open places, including rock ledges and outcrops, openings in sandy or rocky woods; sometimes on moist shores, roadsides, and in fields and lawns.

Antennaria neglecta Greene
FIELD PUSSYTOES   native
IR | **WUP** | CUP | EUP   CC 3

Antennaria neodioica Greene
Plants 1–4 dm tall, with short and leafy or longer and merely bracteate stolons. Basal leaves and those at the ends of the stolons densely and persistently tomentose beneath, only sparsely so (and eventually glabrate) above, or the upper side glabrous from the first, relatively small, mostly under 1.5 cm wide, 1-nerved or obscurely 3-nerved. Pistillate involucres 7–10 mm long; variously sexual or apomictic. April–June.

Dry woods and open places; Keweenaw County.

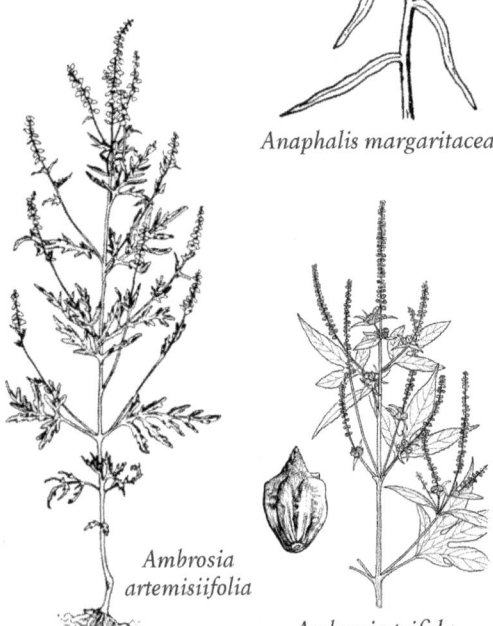

Anaphalis margaritacea

Ambrosia artemisiifolia

Ambrosia trifida

dicots **ASTERACEAE** | Aster Family

Antennaria parlinii Fernald
PARLIN'S PUSSYTOES *native*
IR | WUP | CUP | EUP CC 2

Antennaria plantaginifolia var. *parlinii* Cronq. Similar to *A. plantaginifolia* and sometimes included in it. Stems usually with purple glandular hairs (especially near tips of young flowering stems).

Dry open places, including rock outcrops, banks, grassy roadsides, hillsides, and open woods; sometimes in shaded forests.

Anthemis *Chamomile*

Annual or perennial, usually aromatic herbs. Leaves alternate, dissected. Flowers in campanulate or nearly hemispheric heads terminating the branches. Heads radiate or rarely discoid, the rays elongate, white or yellow, pistillate or neutral; involucral bracts subequal or more commonly imbricate in several series, the margins more or less scarious or hyaline; receptacle convex to conic or hemispheric, chaffy at least toward the middle. Disk flowers numerous, perfect, yellow. Fruit a terete, angled, or somewhat compressed achene; pappus a short crown, or more commonly none. *Anthemis* and *Matricaria* are very similar, separated on the receptacle chaffy in *Anthemis* and naked in *Matricaria*.

1 Ray flowers yellow....................**A. tinctoria**
1 Ray flowers white 2
2 Ray flowers pistillate and fertile; receptacle chaffy throughout.........................**A. arvensis**
2 Ray flowers usually neutral, sterile; receptacle chaffy only near middle........................**A. cotula**

Anthemis arvensis L.
CORN CHAMOMILE *introduced*
IR | WUP | **CUP** | EUP

More or less branching annual, 1–5 dm tall, often with several stems from the base, more or less finely hairy. Leaves about 2–5 cm long, bipinnatifid, with narrowly winged rachis. Heads several or numerous, pedunculate at the ends of the branches; involucre thinly villous or villous-tomentose; disk about 6–12 mm wide, becoming ovoid in age; rays about 15–20, pistillate, white, about 6–13 mm long; receptacle chaffy throughout, its bracts thin, the pale cuspidate awn–tips equaling or shorter than the disk-flowers. Achenes quadrangular, about 10-nerved. Pappus a minute border, or obsolete. May–Aug.

Fields and waste places; native of Europe; Marquette County.

Anthemis cotula L.
STINKING CHAMOMILE *introduced*
IR | **WUP** | **CUP** | **EUP** FACU

More or less branched, usually subglabrous, ill-smelling annual 1–6 dm tall. Leaves about 2–6 cm long, 2 or 3x pinnatifid, with very narrow segments. Heads more or less numerous, short-pedunculate at the ends of the branches, the disk about 5–10 mm wide, becoming ovoid or short-cylindric at maturity; involucre sparsely villous; rays about 10–20, white, neutral, 5–11 mm long; receptacle chaffy only toward the middle, its bracts narrow, tapering to the apex, scarcely awned. Achenes subterete, about 10-ribbed, glandular-tuberculate. Pappus none. May–Oct.

Native of Europe; fields and waste places.

Anthemis tinctoria L.
GOLDEN-CHAMOMILE *introduced*
IR | **WUP** | **CUP** | EUP

Cota tinctoria (L.) J. Gay ex Guss.

Short-lived perennial. Stems 3–7 dm tall, sparingly branched above or simple, finely hairy at least above. Leaves pinnatifid, about 2–5 cm long, with winged rachis and deeply toothed or pinnatifid segments, villous or almost floccose beneath. Heads solitary and long-pedunculate at the ends of the branches, the disk about 12–18 mm wide; involucre thinly tomentose; rays about 20–30, pistillate, yellow, about 7–15 mm long receptacle chaffy throughout, its bracts narrow, with firm yellow awn-tips equaling the disk-flowers. Achenes compressed-quadrangular, more or less striate-nerved. Pappus a very short crown. June–July.

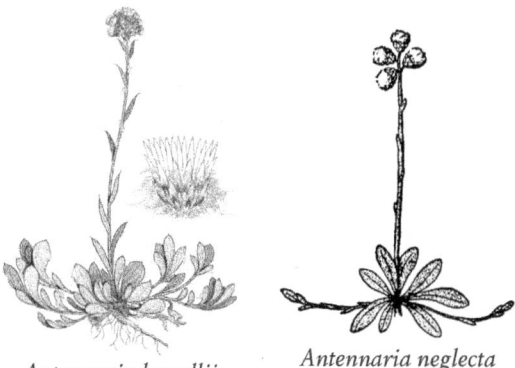

Antennaria howellii

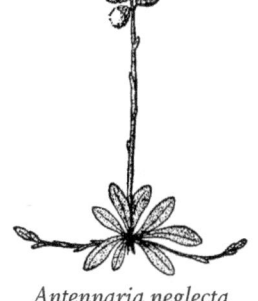

Antennaria neglecta

Anthemis arvensis

Anthemis cotula

Fields and waste places; native of Europe.

Arctium *Burdock*

Arctium minus Bernh.
LESSER BURDOCK *introduced (invasive)*
IR | WUP | CUP | EUP FACU

Coarse biennial herb, to 1.5 m tall or rarely taller. Leaves large, alternate, petiolate, the lower petioles mostly hollow; blade narrowly to very broadly ovate, to about 5 dm long and 4 dm wide, thinly tomentose and often eventually glabrate beneath, nearly glabrous above. Branches of the inflorescence ascending to widely spreading, raceme-like, the heads mostly short-pedunculate or subsessile, 1.5–3 cm thick, glabrous or slightly glandular to sometimes tomentose, usually a little shorter than the flowers, stramineous or purplish, the inner bracts often more flattened than the others and scarcely hooked. Flowers all tubular and perfect, the corolla pink or purplish, with long slender lobes. Achenes oblong, slightly compressed, few-angled, many-nerved, truncate at the apex, glabrous. Receptacle flat, densely bristly; pappus of numerous short, separately deciduous bristles.

Roadsides, railroads, fields, fencerows, farmyards, around old buildings, disturbed places.

Arnica *Arnica*

Two species of *Arnica* are present in Michigan, both restricted to the Keweenaw Peninsula or to Isle Royale.

1 Lower leaves cordate; Keweenaw County (including Isle Royale)............................ **A. cordifolia**
1 Lower leaves lanceolate to narrowly ovate; Isle Royale only............................ **A. lonchophylla**

Arnica cordifolia Hook.
HEART-LEAF ARNICA (END) *native*
IR | WUP | CUP | EUP CC 9

Perennial herb. Stems 15–45 cm tall, spreading by rhizomes and forming colonies, with mostly single stems borne along the rhizome. Plants are long hairy, and often covered with glands. Leaves opposite on the stem, with the lowest leaves heart-shaped, shallowly toothed, and 5–10 cm long and 5–8 cm wide, on long petioles; upper leaves much smaller, usually ovate, with the petioles shorter or nearly absent. Flowers in large, daisy-like heads, the heads 1–3 at the ends of stems; ray flowers yellow and 3–7 cm wide, and number 10–15; disk flowers yellow, on a receptacle 1.5–2.5 cm wide. June-July.

Along roadsides and borders of rocky or open, mixed or deciduous woods. A species of western North America, disjunct in the Black Hills of South Dakota, the Keweenaw Peninsula, and on Isle Royale.

Arnica lonchophylla Greene
LONG-LEAVED ARNICA (END) *native*
IR | WUP | CUP | EUP CC 10

Perennial herb. Stems to 4 dm tall, erect to somewhat lax, single or loosely clustered from a scaly, branching rhizome; stems stipitate-glandular and also covered with spreading hairs. Lowermost leaves narrowly elliptic to ovate, prominently several-veined, petiolate, 5–15 cm long and 1–3 cm wide; upper leaves strongly reduced, sessile. Heads 1–7, erect, campanulate; involucral bracts 1–2 cm long; ray florets ca. 8(6–10), ligule yellow-orange, 1–2 cm long; disk corollas 6–9 mm long, the lower part slender and tubular, 3–4 mm long. Achenes 4.5–6 mm long, pubescent; pappus of many white capillary bristles. June-July.

Anthemis tinctoria

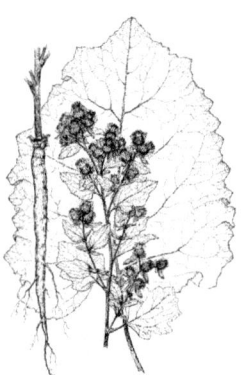

Arctium minus

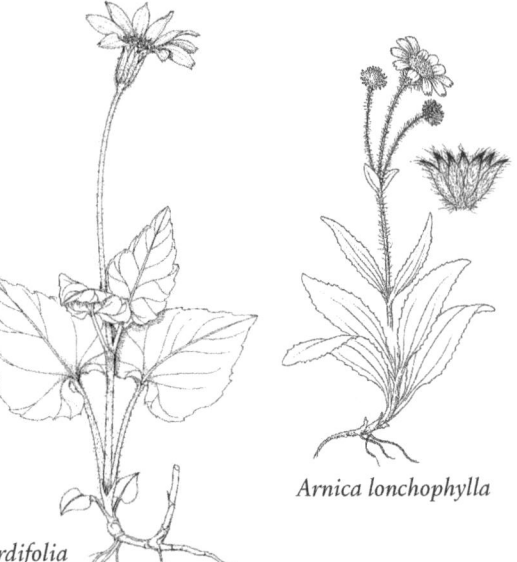

Arnica cordifolia

Arnica lonchophylla

Isle Royale, on shaded mossy ledges and crevices of rock cliffs near Lake Superior.

Arnoglossum *Indian Plantain*

Large perennial herbs with basal or alternate leaves. Flower heads with white disk flowers only, the ray flowers absent. Fruit an achene, tipped by a pappus of numerous, slender bristles.

Arnoglossum plantagineum Raf.
GROOVE-STEM INDIAN-PLANTAIN *native*
IR | WUP | CUP | **EUP**   FAC CC 10

Cacalia plantaginea (Raf.) Shinners
Cacalia tuberosa Nutt.

Glabrous perennial herb with a short tuberous-thickened base and fleshy-fibrous roots. Stems stout, erect, 6–18 dm tall, striate-angled. Leaves thick and firm, entire or slightly toothed, with several prominent longitudinal nerves converging toward the summit; basal and lowermost stem leaves conspicuously long-petioled, the blade 6–20 cm long and 2–10 cm wide, commonly elliptic and tapering to the base, sometimes ovate and subtruncate at the base in robust plants; stem leaves few, conspicuously reduced upwards, becoming sessile or subsessile. Otherwise similar to our other species. June–July.

Wet to sometimes dry prairies; marshy or boggy places; Mackinac County.

A. plantagineum

Artemisia *Wormwood; Sage*

Annual, biennial, or perennial herbs, or shrubs, usually aromatic, with alternate, entire to dissected leaves and few to numerous small, ovoid to campanulate or hemispheric heads in a spiciform, raceme-like, or panicle-like inflorescence. Heads discoid, sometimes with only perfect flowers, sometimes the outer pistillate, the central ones then sometimes sterile. Involucral bracts dry, imbricate, at least the inner scarious or with scarious margins. Receptacle flat to convex or hemispheric, naked or densely beset with long hairs. Achenes ellipsoid or obovoid to nearly prismatic, scarcely compressed, usually glabrous. Pappus none.

ADDITIONAL SPECIES
Artemisia abrotanum L. (Southern wormwood), introduced; Houghton County.

1 Plants perennial and somewhat woody at base; leaves covered with silky hairs; receptacle hairy ....
....**A. absinthium**
1 Plants annual, biennial, or perennial; leaves hairy to glabrous; receptacle naked 2
2 Disk flowers sterile; mature plants usually glabrous...
....**A. campestris**
2 Disk flowers fertile 3
3 Leaves glabrous or nearly so, pinnately divided or dissected ....**A. biennis**
3 Leaves densely hairy at least on one surface, simple or dissected... 4
4 Leaves finely divided into thread-like segments; uncommon garden escape ....**A. pontica**
4 Leaves entire or the leaf segments broader 5
5 Leaves green and nearly glabrous above, white hairy below; uncommon weed ....**A. vulgaris**
5 Leaves hairy on upper and lower sides.............. 6
6 Leaves entire or irregularly toothed; common ....
....**A. ludoviciana**
6 Leaves obtusely lobed; uncommon garden escape ....
....**A. stelleriana**

Artemisia absinthium L.
COMMON WORMWOOD *introduced*
IR | **WUP** | **CUP** | **EUP**

Fragrant perennial herb or shrub. Stems 4–10 dm tall, finely sericeous or eventually glabrate. Leaves silvery-sericeous on both sides, or eventually nearly glabrous above, the lower long-petiolate and 2–3 times pinnatifid, with mostly oblong segments about 1.5–4 mm wide, the blade rounded-ovate in outline, about 3–8 cm long; upper leaves progressively less divided and shorter-petiolate, the divisions often more acute. Inflorescence ample, leafy; involucre about 2–3 mm high, finely and densely sericeous; flowers all fertile, the marginal pistillate; receptacle beset with numerous long white hairs between the flowers. Achenes glabrous, nearly cylindric, but narrowed to the base and rounded at the summit. July–Sept.

Native of Europe; fields and waste places.

Artemisia biennis Willd.
BIENNIAL WORMWOOD *introduced*
IR | **WUP** | **CUP** | **EUP**   FACW

Taprooted, annual or biennial herb. Stems erect, to 1 m or more long, often branched, smooth, only faintly scented. Leaves alternate, pinnately dissected nearly to middle, 5–12 cm long and 2–5 cm wide, the segments linear and toothed. Flowers in stalkless heads from upper leaf axils; the heads composed of many small green disk flowers, grouped into spike-like inflorescences, with leafy bracts much longer than the clusters of heads; pappus none. Fruit a small oblong achene. Aug–Sept.

Sandy lakeshores, streambanks, ditches, mud flats, disturbed areas; often where seasonally flooded. Native to nw USA, throughout Michigan as a weed.

Artemisia campestris L.
FIELD SAGEWORT *native*
IR | WUP | CUP | EUP CC 5

Scarcely odorous perennial with a taproot and generally several glabrous to villous stems 1–10 dm tall from a branching caudex. Basal leaves crowded, about 2–10 cm long including the petiole, 0.7–4 cm wide, 2x or 3x pinnatifid or 3-parted, with mostly linear-filiform divisions seldom more than 2 mm wide, glabrous to sericeous, persistent, or, especially in the larger forms, sometimes deciduous; cauline leaves similar but smaller and less divided, the uppermost often 3-parted or simple. Inflorescence small and spike-like to diffuse and panicle-like; involucre glabrous to densely villous-tomentose, 2–4.5 mm high; outer flowers pistillate and fertile; disk-flowers sterile, with abortive ovary; receptacle glabrous. Achenes subcylindric, glabrous, those of the disk flowers abortive. July–Sept.

Open places, often in sandy soil.

Artemisia ludoviciana Nutt.
WHITE SAGE (THR) *native*
IR | WUP | CUP | EUP CC 8

Aromatic rhizomatous perennial. Stems about 3–10 dm tall, simple to the inflorescence, more or less white-tomentose, at least above. Leaves narrowly to broadly lance-shaped, 3–11 cm long and 4–15 mm wide, entire or irregularly toothed, or sometimes deeply lobed, but the lobes generally at least 2 mm wide or more, densely and persistently white-tomentose beneath and often also above, or more thinly hairy and soon glabrate above. Inflorescence ample or narrow; involucre tomentose, 2.5–4 mm high; flowers all fertile, the outer pistillate; disk corollas 2–3 mm long; receptacle glabrous. Achenes ellipsoid, not nerved or angled, essentially glabrous. July–Oct.

Prairies, dry ground, and waste places.

Artemisia pontica L.
ROMAN WORMWOOD *introduced*
IR | WUP | CUP | **EUP**

Perennial with a creeping rhizome. Stems 4–10 dm tall, fragrant, simple or nearly so, the twigs puberulent or eventually glabrate. Leaves 1–3 cm long, white-tomentose on both sides, more thinly so and sometimes eventually glabrate above, twice or thrice pinnatifid, with short divergent segments scarcely 1 mm wide, ordinarily with a pair of stipule-like lobes or auricles at the base. Inflorescence relatively narrow, elongate; involucre 2–3 mm high, tomentulose; receptacle glabrous; flowers all fertile, the outer pistillate. Achenes glabrous, 4–5-angled, broadest at or near the truncate summit. Aug–Sept.

Native of Europe, escaped from cultivation and sparingly established in dry open places; Chippewa and Houghton counties.

Artemisia stelleriana Bess.
DUSTY MILLER *introduced*
IR | **WUP** | CUP | EUP FACU

Perennial from a creeping rhizome, inodorous. Stems 3–7 dm tall, simple to the inflorescence, densely white-tomentose. Leaves white-tomentose on both sides more densely so beneath, obovate, 3–10 cm long, including the petiole, and 1–5 cm wide, with a few rounded relatively broad lobes, which may be again slightly lobed. Inflorescence narrow and often dense, elongate; heads relatively large, the involucre 6–7.5 mm high, the disk corollas 3–4 mm long; receptacle glabrous; flowers all fertile, the outer pistillate. Achenes glabrous, subterete, but narrowed to the base and rounded at the summit. May–Sept.

Native of Asia, escaped from cultivation to sandy places; Keweenaw County.

Artemisia vulgaris L.
MUGWORT *introduced*
IR | **WUP** | **CUP** | **EUP**

Aromatic perennial herb with a stout rhizome. Stems 0.5–1.5 m tall, simple or branched above, glabrous or nearly so below the inflorescence. Leaves green and glabrous or nearly so above, densely white-tomentose beneath, chiefly obovate or ovate in outline, about 5–10 cm long and 3–7 cm wide, the principal ones cleft nearly to the midrib into ascending, unequal segments which are again toothed or more deeply cleft, and ordinarily with one or two pairs of stipule-like lobes at the base. Inflorescence generally ample and leafy; involucre 3.5–4.5 mm high, more or less tomentose; receptacle glabrous; flowers all fertile, the outer pistillate; disk corollas about 2.0–2.8 mm long. Achenes ellipsoid, not nerved or angled, essentially glabrous. July–Oct.

Fields, roadsides, and waste places; Old World native, now established throughout most of e North America.

Bellis *English Daisy*

Bellis perennis L.
LAWN DAISY; ENGLISH DAISY *introduced*
IR | WUP | CUP | EUP

Perennial, more or less spreading-hairy. Leaves basal, elliptic or obovate to orbicular, the blades dentate or denticulate, to 3.5 cm long and 2 cm wide, narrowed to margined petioles of equal or greater length. Heads solitary atop a scape 5–15 cm high; involucral bracts herbaceous, equal; receptacle conic,

ASTERACEAE | Aster Family

Artemisia absinthium

Artemisia biennis

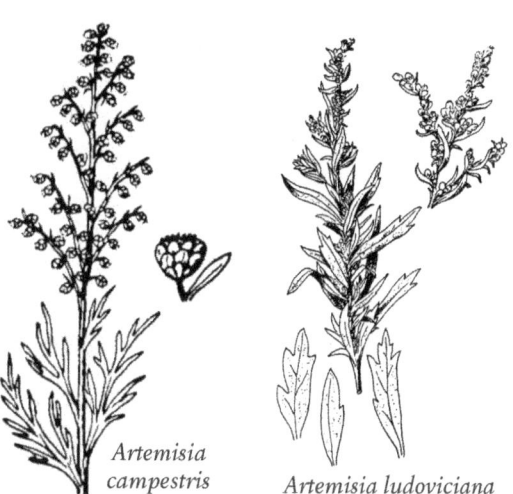
Artemisia campestris   *Artemisia ludoviciana*

naked. Rays many, pistillate, white to pink or purple; disk flowers yellow. Achenes compressed, mostly 2-nerved; pappus absent. April–Nov.

Weedy in lawns or waste places.

Bellis perennis

Bidens *Beggarticks*

Weedy annual herbs; perennial in the aquatic *B. beckii*. Leaves opposite (or whorled in *B. beckii*), simple, lobed, or pinnately divided. Flower heads with both disk and ray flowers, or with disk flowers only; ray flowers often about 8, yellow; involucral bracts in 2 series, the outer row leaflike and spreading, the inner row much shorter and erect; receptacle more or less flat and chaffy. Fruit a flattened achene; pappus of 2–5 barbed awns which persist atop the achene; the body of achene barbed or with stiff hairs (at least on the angles), the "stick-tights" facilitating dispersal of seed by animals.

1 Plants aquatic; underwater leaves whorled, dissected into narrow segments . . . . **B. beckii**
1 Plants not aquatic (sometimes emergent); leaves not as above 2
2 Leaves simple and toothed, or sometimes lobed; achenes 3–4-awned. 3
2 Leaves all (or mostly) pinnately divided or compound; achenes 2-awned. 5
3 Leaves with a petiole 1–4 cm long. . . . . **B. tripartita**
3 Leaves mostly sessile 4
4 Heads nodding when mature; outer involucral bracts widely spreading . . . . **B. cernua**
4 Heads mostly upright; outer involucral bracts erect or nearly so . . . . **B. tripartita**
5 Heads with both disk and ray flowers, the rays over 1 cm long . . . . **B. trichosperma**
5 Heads with disk flowers only, or with short rays less than 5 mm long 6
6 Outer involucral bracts 2–5 (usually 4), not fringed with hairs. . . . . **B. discoidea**
6 Outer involucral bracts 6 or more, fringed with hairs (at least near base) 7
7 Disk flowers orange; outer involucral bracts mostly 6–8 . . . . **B. frondosa**
7 Disk flowers yellow; outer involucral bracts 10 or more . . . . **B. vulgata**

Bidens beckii Torr. ex Spreng.
BECK'S WATER-MARIGOLD   *native*
IR | WUP | CUP | EUP   OBL CC 10
Megalodonta beckii (Torr.) Greene

Perennial aquatic herb. Stems 0.4–2 m long, little-branched. Underwater leaves opposite or whorled, dissected into threadlike segments; emersed leaves simple, opposite, lance-shaped to ovate, margins with forward-pointing teeth, petioles absent. Flower heads single or few at ends of stems; rays 6–10, gold-yellow, 1–1.5 cm long, notched at tip; involucral bracts smooth. Fruit an achene, more or less round in section, 10–15 mm long; pappus of 3–6 slender awns, longer than achenes, the upper portion of awn with downward-pointing barbs. June–Sept.

Quiet, shallow to deep water of lakes, ponds, rivers and streams.

Bidens cernua L.
NODDING BUR-MARIGOLD   *native*
IR | WUP | CUP | EUP   OBL CC 3

Annual herb. Stems often branched, to 1 m long, smooth or with spreading hairs. Leaves opposite, smooth, lance-shaped to oblong lance-shaped, 3–

16 cm long and 0.5–5 cm wide; margins with sharp, forward-pointing teeth and often rough-to-touch; petioles absent, the leaves usually clasping at base. Flower heads many, globe-shaped, 1.5–3 cm wide, usually nodding after flowering; rays yellow, 6–8, to 1.5 cm long, or absent; outer involucral bracts 4–8, unequal in length, the margins often fringed with hairs. Fruit a more or less straight-sided achene, 5–7 mm long, with downward-pointing barbs on margins; pappus with 4 (sometimes 2) awns, the awns with downward-pointing barbs. July–Oct.

Exposed, sandy or muddy shores, streambanks, marshes, forest depressions, wet meadows, ditches and other wet places.

Bidens discoidea (Torr. & Gray) Britt.
SMALL BEGGARTICKS native
IR | **WUP** | CUP | EUP   FACW CC 7

Annual herb. Stems smooth, 3–10 dm long. Leaves opposite, smooth, divided into 3-leaflets, the leaflets lance-shaped, the terminal leaflet largest, to 10 cm long and 4 cm wide; margins with coarse, forward-pointing teeth; petioles slender, 1–6 cm long. Flower heads many on slender stalks, the disk to 1 cm wide; rays absent; outer involucral bracts usually 4, leaflike, much longer than disk. Fruit a flattened achene, 3–6 mm long; pappus of 2 awns to 2 mm long, with short, upward pointing bristles. Aug–Sept.

Hummocks or logs in swamps, exposed muddy shores; usually where shaded; Baraga and Houghton counties.

Bidens frondosa L.
DEVIL'S-PITCHFORK native
IR | **WUP** | **CUP** | **EUP**   FACW CC 1

Annual herb. Stems erect, 2–10 dm tall, branched, purple-tinged, more or less smooth. Leaves pinnately divided into 3–5 segments, the segments lance-shaped, to 10 cm long and 3 cm wide, underside sometimes with short hairs; margins with coarse, forward-pointing teeth; petioles slender, 1–6 cm long. Flower heads many on long, leafless stalks; disk flowers orange, the disk to 1 cm wide; rays absent or very small; the outer involucral bracts usually 8, green and leaflike, longer than disk, fringed with hairs on margins. Fruit a flattened, nearly black achene, 5–10 mm long; pappus of 2 slender awns with downward-pointing barbs. July–Oct.

Wet, sandy or gravelly shores, forest depressions, streambanks, pond margins; weedy in wet disturbed areas.

Bidens trichosperma (Michx.) Britton
CROWNED BEGGARTICKS native
IR | WUP | **CUP** | EUP   OBL CC 7

Bidens coronata (L.) Britt.

Annual or biennial herb. Stems branched, 3–15 dm tall, smooth, often purple. Leaves opposite, smooth, to 15 cm long, pinnately divided into 3–7 narrow leaflets; margins coarsely toothed or deeply lobed to sometimes entire; petioles 3–15 mm long. Flower heads with both disk and ray flowers, large and numerous on slender stalks; rays about 8, gold-yellow, 1–2.5 cm long; outer involucral bracts 6–10, to 1 cm long, short-hairy on margins, inner bracts shorter. Fruit a flattened achene, 5–9 mm long, with long, stiff hairs on margins; pappus of 2 short, scale-like awns, 1–2 mm long. July–Oct.

Bogs, fens, tamarack swamps, shores, streambanks, marshes, sand bars; Delta County.

Bidens tripartita L.
THREE-LOBE BEGGARTICKS native
IR | **WUP** | **CUP** | **EUP**   FACW CC 5

Bidens acuta (Wieg.) Britt.
Bidens comosa (Gray) Wieg.
Bidens connata Muhl.

Annual herb. Stems yellow, 1–12 dm tall, branched, smooth. Leaves opposite, lance-shaped to oval, 3–15 cm long and 0.5–5 cm wide, margins with coarse, forward-pointing teeth, rough-to-touch; petioles absent, or leaves tapered to a short, winged petiole. Flower heads 1–2.5 cm wide, several to many, remaining erect after flowering; disk flowers yellow-green; rays absent; outer involucral bracts leaflike, 5–10 or more, 2–4x longer than head. Fruit an achene, 3–7 mm long, downwardly barbed on the margins; pappus of 3 downward-pointing barbed awns, the awns shorter than the achenes. Aug–Oct.

Exposed shores, streambanks, mudflats, forest depressions, pond, wet meadows, ditches and other wet places.

Bidens vulgata Greene
TALL BEGGARTICKS native
IR | **WUP** | **CUP** | EUP   FAC CC 0

Bidens puberula Wieg.

Annual herb. Stems to 2 m tall, smooth or upper stem and leaves short-hairy. Leaves opposite, pinnately divided into 3–5 segments, the segments lance-shaped, to 15 cm long and 5 cm wide, with prominent veins; margins with sharp, forward-pointing teeth; petioles present. Flower heads on stout, leafless stalks, disk flowers yellow; ray flowers usually present, small, yellow; outer involucral bracts about 13, leaflike. Fruit a flattened, olive-green or brown achene, 10–12 mm long; pappus of 2 awns with downward-pointing barbs. Aug–Oct.

Streambanks, wet meadows, wet forests; weedy in moist disturbed areas.

Similar to devil's pitchfork (*Bidens frondosa*), but usually larger.

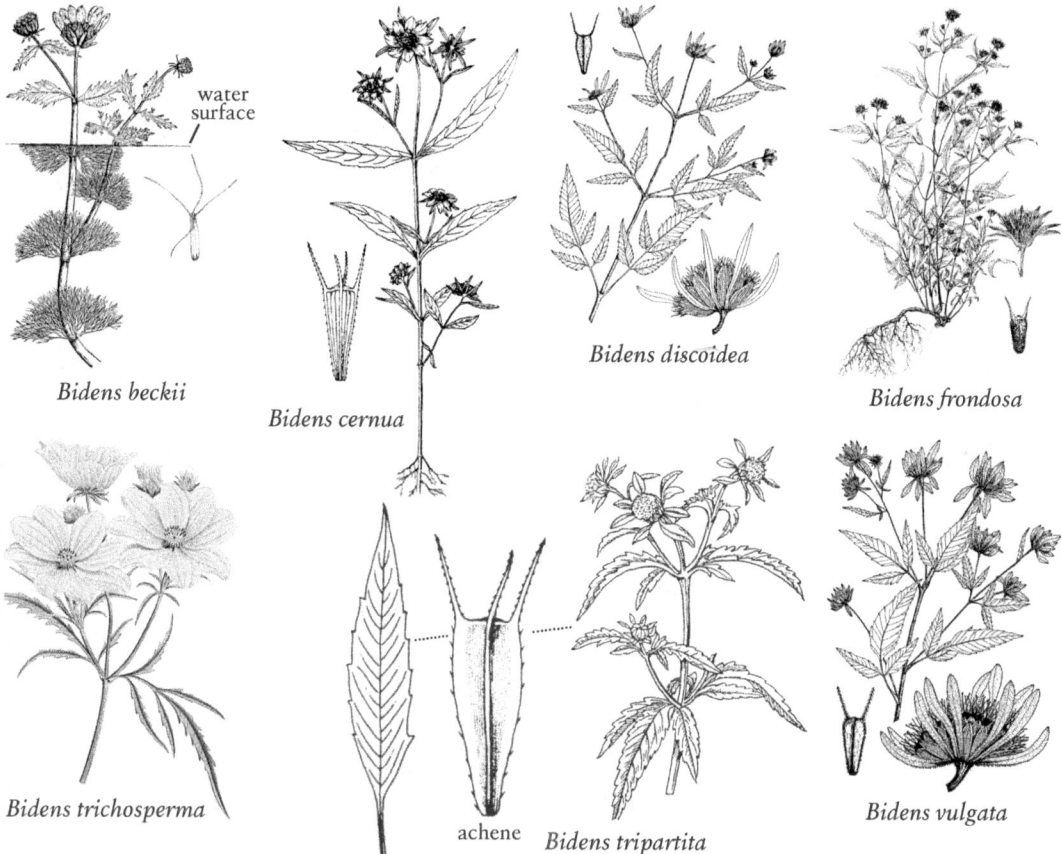

Bidens beckii
Bidens cernua
Bidens discoidea
Bidens frondosa
Bidens trichosperma
achene Bidens tripartita
Bidens vulgata

Carduus Thistle

Annual, biennial, or perennial spiny herbs. Stems generally winged by the decurrent leaf-bases. Leaves alternate, serrate to more often pinnately lobed or pinnatifid. Heads discoid, the flowers all tubular and perfect, or the plants sometimes dioecious by abortion, clustered or solitary at the ends of the branches. Involucral bracts imbricate in several series, mostly spine-tipped. Receptacle flat or convex, densely bristly. Corollas purple or reddish to white or rarely yellow. Achenes glabrous, obovate, quadrangular or somewhat flattened; 5–10-nerved or nerveless. Pappus of numerous capillary bristles, deciduous in a ring.

Closely related to *Cirsium*, from which it is distinguished primarily by the non-plumose pappus.

ADDITIONAL SPECIES
Carduus crsipus L. (Curly plumeless-thistle), introduced; Gogebic County.

1 Heads nodding; involucral bracts 2 mm wide or more; Schoolcraft County . . . . **C. nutans**
1 Heads not nodding; involucral bracts narrow, less than 2 mm wide; Mackinac County . . . . **C. acanthoides**

Carduus acanthoides L.
SPINY PLUMELESS-THISTLE *intro. (invasive)*
IR | WUP | CUP | **EUP**

Biennial herb 3–10 dm tall, very strongly spiny, the stem tough. Leaves deeply lobed or pinnatifid, to 25 cm long and 8 cm wide, loosely villous beneath, chiefly along the midrib and main veins, with long multicellular hairs, or glabrous, glabrous or similarly hairy over the surface above. Heads clustered or solitary at the ends of the branches, erect, small, the disk mostly 1.5–2.5 cm wide when pressed; involucral bracts narrow, rarely any of them as much as 2 mm wide, erect or loosely spreading, the usually middle and outer spine-tipped, the inner softer and flatter, large, scarcely spiny. July-Oct.

Roadsides, pastures, and waste places; native of Europe; Mackinac County.

Carduus nutans L.
MUSK THISTLE *introduced (invasive)*
IR | WUP | **CUP** | EUP   FACU

Biennial or rarely annual herb 3–10 dm or rarely 20 dm tall. Leaves glabrous, or long-villous chiefly along the midrib and main veins beneath, deeply lobed, to 25 cm long and 10 cm wide. Heads mostly solitary and nodding at the ends of the branches, usually large, the disk 4–8 cm wide when pressed, sometimes smaller and scarcely more than 1.5 cm; middle and outer involucral bracts conspicuously broad (2–8 mm), with long, flat, spreading or reflexed, spine-pointed tip; inner bracts narrower and softer, scarcely spiny, often purplish. June–Oct. Native of Europe and w Asia, established on roadsides and waste places; Schoolcraft County.

Carduus acanthoides

Carduus nutans

Centaurea *Knapweed; Star-Thistle*

Annual, biennial, or perennial herbs with alternate or all basal, entire to pinnatifid leaves, and solitary to numerous, small to large heads. Heads discoid, the flowers sometimes all tubular and perfect, or more commonly the marginal ones sterile, with enlarged, irregular, falsely radiate corolla. Involucral bracts imbricate in several series, either spine-tipped or more often some of them with enlarged appendages. Receptacle nearly flat, densely bristly. Corollas purple or blue to yellow or white, with slender tube and long narrow lobes. Achenes obliquely or laterally attached to the receptacle. Pappus of several series of graduated bristles or narrow scales, often much reduced, or wanting.

ADDITIONAL SPECIES

Centaurea macrocephala Puschk. ex Willd. (Globe knapweed), introduced, reported from Gogebic and Houghton counties.
Centaurea montana L. (Mountain cornflower), introduced and known from several UP counties and Isle Royale.
Centaurea nigrescens Willd. (Tyrol knapweed).

1   Involucral bracts tipped by short to long spines. . . . . . .
    . . . . **C. diffusa**
1   Involucral bracts not tipped by long spines; leaf bases not decurrent. 2
2   Leaves pinnately divided into linear-elliptic lobes; common weed . . . . **C. stoebe**
2   Leaves entire or toothed, sometimes few-lobed 3
3   Plants annual (or winter-annuals); leaves linear, less than 1 cm wide . . . . **C. cyanus**
3   Plant perennial; leaves wider, at least some lower leaves more than 1 cm wide 4
4   Chaffy tips of involucral bracts tan to dark brown. . . . . .
    . . . . **C. jacea**
4   Chaffy tips of involucral bracts black . . . . **C. nigra**

Centaurea cyanus L.
GARDEN CORNFLOWER *introduced*
IR | **WUP** | **CUP** | EUP   UPL

Annual or winter-annual, 2–12 dm tall. Herbage loosely white-tomentose when young, the leaf undersides often persistently so. Leaves linear, entire, or the lower ones slightly toothed or with a few lobes, to 13 cm long and 8 mm wide (excluding the lobes). Heads terminating the branches; involucre mostly 11–16 mm high, its bracts striate, with a pectinate or lacerate fringe near the tip; flowers mostly blue, sometimes purple or white, the marginal ones with enlarged irregular corolla; pappus mostly 2–3 mm long. May–Oct.

Native to the Mediterranean region, cultivated as an ornamental; weedy in fields, roadsides, and waste places.

Centaurea diffusa Lam.
WHITE KNAPWEED *introduced*
IR | **WUP** | **CUP** | EUP

Diffusely branched annual 1–6 dm tall. Herbage sparsely rough-puberulent under the thin and deciduous tomentum. Leaves small, more or less pinnatifid, the lower ones deciduous, the reduced ones of the inflorescence mostly entire. Heads numerous; involucre 8–10 mm high, the middle and outer bracts tipped with a slender spine 1.5–4 mm long; flowers few, ochroleucous or sometimes red, the marginal ones not enlarged; pappus wanting. July–Sept.

Native of Europe, adventive weed in waste places.

Centaurea jacea L.
BROWN-RAY KNAPWEED *introduced*
IR | **WUP** | **CUP** | **EUP**

Perennial, to 12 dm tall, glabrous or somewhat arachnoid. Leaves toothed or shallowly lobed to entire, the basal ovate or lance-shaped to elliptic, long-petiolate, the cauline reduced upwards and becoming sessile. Heads terminating the often numerous branches; involucre mostly 12–18 mm high, a little narrower to a little broader than high; appendages of the involucral bracts well developed, broad, tan to dark brown, the middle and outer ones rather irregularly lacerate, the inner less so and often deeply bifid; marginal flowers almost always enlarged. Pappus none. June–Sept.

dicots   **ASTERACEAE** | Aster Family

Fields, roadsides, and waste places; native of Europe.

Centaurea nigra L.
LESSER KNAPWEED *introduced*
IR | **WUP** | **CUP** | EUP

Perennial, 2–8 dm tall; herbage rough-puberulent, and sometimes arachnoid when young. Leaves entire or toothed, the basal broadly lance-shaped, petiolate, the cauline reduced upwards and becoming sessile. Heads terminating the often numerous branches; involucre mostly 12–19 mm high, broader than high; appendages of the involucral bracts well developed, conspicuously blackish at least in part, the middle and outer deeply and fairly regularly pectinate, seldom any of them markedly bifid; ray-like marginal flowers mostly wanting. Pappus about 1 mm long or less. July–Oct.

Fields, roadsides, and waste places; native of Europe, now widely established in ne USA and se Canada; Delta and Keweenaw counties.

Centaurea stoebe L.
SPOTTED KNAPWEED   *introduced (invasive)*
IR | **WUP** | **CUP** | **EUP**

Centaurea maculosa Lam.

Biennial or short-lived perennial, mostly 3–12 dm tall. Herbage with a thin and loose, soon deciduous tomenturn, also sparsely scabrous-puberulent. Leaves obscurely to evidently glandular-punctate, pinnatifid with narrow lobes, or the reduced ones of the inflorescence entire. Heads terminating the numerous branches, constricted upwards; involucre mostly 10–13 mm high, its bracts striate, the middle and outer ones with short, dark, pectinate tips; flowers pink-purple, the marginal ones enlarged. Pappus to 2 mm long, or rarely wanting. June–Oct.

Native of Europe; aggressive weed of fields, roadsides, waste places.

Cichorium *Chicory*

Cichorium intybus L.
CHICORY   *introduced (invasive)*
IR | **WUP** | **CUP** | **EUP**   FACU

Perennial with milky juice, from a long deep taproot. Stems branching, hirsute or glabrous, 3–17 dm tall. Leaves alternate, lower leaves oblong lance-shaped, petiolate, toothed or pinnatifid, 8–25 cm long and 1–7 cm wide, becoming reduced, sessile, and entire or merely toothed upwards. Heads sessile or short-pedunculate, borne 1–3 together in the axils of the much reduced upper leaves. Flowers all ligulate and perfect, blue or occasionally white; involucral bracts biseriate, the outer shorter. Achenes glabrous, 2–3 mm long, 5-angled, or the outer slightly compressed; pappus of 2–3 series of scales, sometimes minute. July–Oct.

Roadsides, fields, and waste places; native of Eurasia, now a cosmopolitan weed.

The roasted root is used as an adulterant or substitute for coffee.

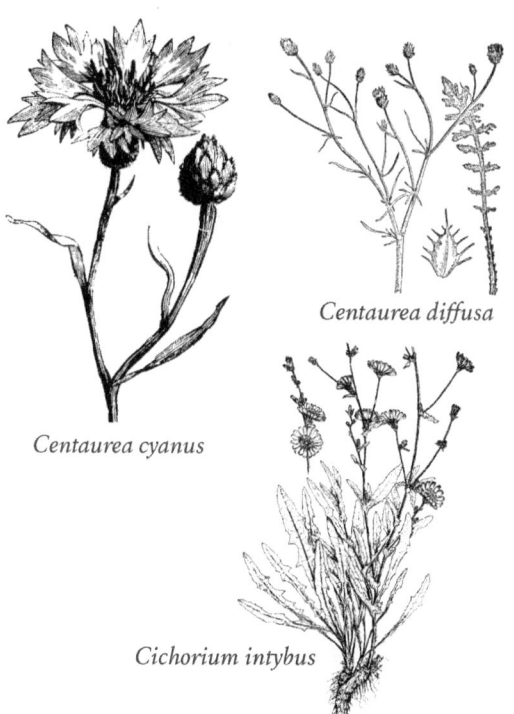

Centaurea diffusa

Centaurea cyanus

Cichorium intybus

Centaurea jacea

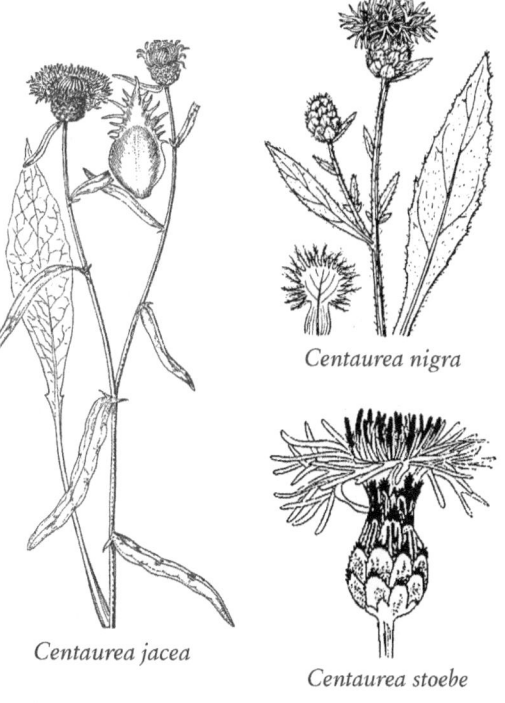

Centaurea nigra

Centaurea stoebe

ASTERACEAE | Aster Family

Cirsium Thistle

Biennial or perennial herbs. Stems and leaves often spiny. Leaves from base of plant or alternate on stem. Flower heads of pink to purple disk flowers only; involucral bracts tipped with spines. Fruit a smooth achene; pappus of many slender bristles.

1. Involucral bracts tipped by spines mostly more than 2 mm long. 2
1. Involucral bracts tipped by short spines to only 1 mm long. 5
2. Leaves densely white-hairy on both sides, especially on underside rare native species along Lake Michigan dunes . . . . **C. pitcheri**
2. Leaves coarsely hairy, with cobwebby hairs on underside 3
3. Stem leaves decurrent at base, forming conspicuous spiny wings on stem . . . . **C. vulgare**
3. Stem leaves not decurrent; stems not appearing winged 4
4. Leaves green, with coarse hairs; stems 3–5 dm tall from persistent basal rosettes; dry to mesic prairies; rare in dry or moist prairies . . . . **C. hillii**
4. Leaves densely white-hairy on underside; stems 6 dm tall or more, the basal rosettes not persistent . . . . **C. discolor**
5. Colony-forming perennial herb from deep, creeping rhizomes; common weed of dry to moist places . . . . **C. arvense**
5. Biennial herbs; moist to wet habitats 6
6. Leaf bases not decurrent; stem not winged; involucral bracts usually with cobwebby hairs; flowers deep rose-purple; native and not weedy . . . . **C. muticum**
6. Leaf bases decurrent, forming spiny wings on stem; involucral bracts usually without cobwebby hairs; flowers pale pink-purple; introduced and weedy. . . . **C. palustre**

Cirsium arvense (L.) Scop.
CANADIAN THISTLE *introduced (invasive)*
IR | **WUP** | **CUP** | **EUP** FACU

Perennial with deep-seated rhizomes, mostly 3–15 (or 20) dm tall, subglabrous, or the leaves more or less white-tomentose beneath. Heads more or less numerous, polygamo-dioecious, the pappus of the pistillate heads surpassing the corollas, that of the starminate heads surpassed by the corollas; involucre 1–2 cm high, its bracts all innocuous, or the outer with weak spine-tips about 1 mm long; flowers pink-purple or occasionally white. Achenes about 4 mm long. July–Aug.

A noxious weed of fields and waste places; native of Eurasia, now statewide.

Cirsium discolor (Muhl.) Spreng.
FIELD THISTLE *native*
IR | WUP | **CUP** | EUP UPL CC 4

Similar to *C. altissimum* and intergrading with it; perhaps not specifically distinct. Differs chiefly in its deeply pinnatifid, generally firmer and spinier leaves; plants averaging smaller (mostly 1–2 m) than *C. altissimum*, the peduncles tending to be a little leafier, and the heads often a little broader-based. July–Oct.

Fields, open woods, river bottoms, and waste places.

Cirsium hillii (Canby) Fern.
HILL'S THISTLE *native*
IR | WUP | **CUP** | **EUP** CC 8

Perennial, producing basal rosettes of leaves the first year, followed by flowers one or two years later. Very similar to *Cirsium pumilum* in general form (a species of ne USA), and sometimes treated as a subspecies of it. Perennial with elongate, thickened roots, traversed longitudinally by a ring of several small cavities. Achenes mostly 4–5 mm long. June–Aug. Prairies and other open places; Chippewa and Menominee counties.

Cirsium muticum Michx.
SWAMP THISTLE *native*
IR | **WUP** | **CUP** | **EUP** OBL CC 6

Stout biennial herb. Stems 0.5–2 m long, branched in head, with long, soft hairs when young, becoming more or less smooth. Leaves deeply lobed into pinnate segments, 1–2 dm long, underside often with matted, cobwebby hairs, becoming more or less smooth with age; margins toothed and often tipped with spines; petioles present on lower leaves, stem leaves sessile. Flower heads of purple or pink disk flowers only, single on leafless stalks over 1 cm long at ends of stems; involucre 2–3.5 cm high; the involucral bracts overlapping, densely hairy with cottony hairs (especially on margins), sometimes tipped with a short spine 0.5 mm long. Achenes 5 mm long; pappus of long, slender bristles. Aug–Oct.

Swamps, thickets, calcareous fens, sedge meadows, streambanks, shores.

Cirsium palustre (L.) Scop.
EUROPEAN SWAMP THISTLE *introduced*
IR | **WUP** | **CUP** | **EUP**

Biennial herb. Stems 0.5–2 m tall, spiny. Leaves to 20 cm long, deeply lobed into pinnate segments, covered with loosely matted hairs or more or less smooth, tapered at base and continued downward on stem as spiny wings; margin teeth spine-tipped. Flower heads of purple disk flowers only, on short stalks mostly less than 1 cm long; involucre 1–2 cm high; the involucral bracts overlapping, not spine-tipped. Achenes 3 mm long; pappus of slender bristles to 1 cm long. June–Aug.

Introduced and spreading into roadside ditches and adjacent wetlands, including swamps, thickets and fens, especially where disturbed; resembling the native *C. muticum* in these habitats.

Cirsium pitcheri (Torr.) Torr. & Gray
DUNE THISTLE (THR) *native*
IR | WUP | **CUP** | EUP   CC 10

Plants biennial or short-lived perennials, commonly 5–10 dm tall; stems and lower surfaces of the leaves densely and persistently white-tomentose, the upper surfaces of the leaves thinly so. Leaves deeply pinnatifid, with narrow rachis and long, remote, linear, entire or few-toothed, weakly spine-tipped lobes to about 8 cm long and 7 mm wide. Heads several; involucre mostly 2.5–3 cm high, its bracts slightly tomentose especially marginally, well imbricate, the inner long-acuminate, the others with weak spine-tips mostly 1–2 mm long; flowers commonly ochroleucous. Achenes mostly 5–7.5 mm long. June–Aug.

Uncommon plants of sand dunes along Lakes Michigan, Huron, and Superior. In their first year plants form a rosette of leaves flattened against the dune sands; the following year or after several years, plants bloom and then die.

Cirsium vulgare (Savi) Ten.
BULL THISTLE *introduced (invasive)*
IR | WUP | **CUP** | EUP   FACU

Biennial weed mostly 5–15 dm tall. Stems conspicuously spiny-winged by the decurrent leaf-bases, copiously spreading-hirsute to sometimes arachnoid. Leaves pinnatifid, the larger ones with the lobes again toothed or lobed, scabrous-hispid above, thinly white-tomentose to sometimes green and merely hirsute beneath. Heads several, purple; involucre 2.5–4 cm high, its bracts all spine-tipped, without any well developed glutinous dorsal ridge. Achenes less than 4 mm long. June–Oct.

Pastures, fields, roadsides, and waste places; native of Eurasia, now widely established as a weed in North America.

Conyza *Horseweed*

Conyza canadensis (L.) Cronq.
HORSEWEED *introduced*
IR | WUP | **CUP** | EUP   FACU

Erigeron canadensis L.
Leptilon canadense (L.) Britt.

Coarse annual herb, 1–15 dm tall, simple or nearly so to the inflorescence. Stems coarsely spreading-hirsute to nearly coarsely ciliate near the base, numerous. Leaves alternate, oblong lance-shaped, more or less pubescent, gradually reduced upwards, the stem leaves to about 8 cm long and 8 mm wide, the basal leaves larger and relatively broader, but generally deciduous before flowering time. Heads, except in depauperate plants, numerous in a long and open inflorescence. Involucre about 3–4 mm high, glabrous or nearly so, the bracts strongly imbricate, brown or with distinct brown midvein. Pistillate flowers very numerous and slender, with very short, narrow, and in-

Conyza canadensis

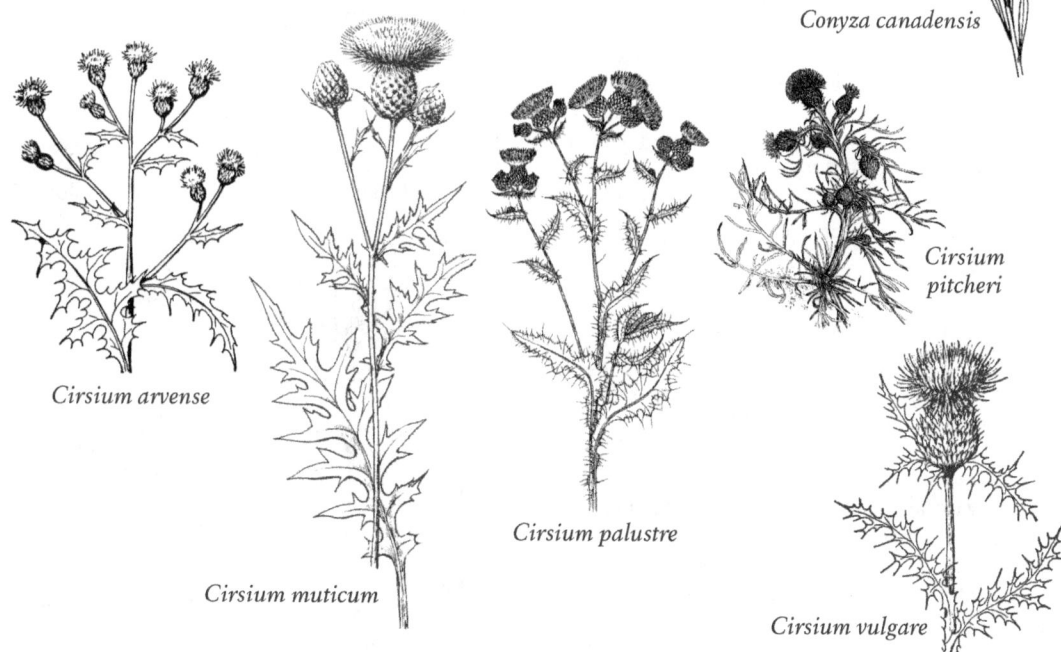

Cirsium arvense

Cirsium muticum

Cirsium palustre

Cirsium pitcheri

Cirsium vulgare

conspicuous white or sometimes pinkish, about equaling the pappus. Receptacle flat or nearly so; pappus of capillary bristles, sometimes with a short outer series. Late summer and fall. Achenes 1–2-nerved, or nerveless.

A weed in waste places.

Closely related to *Erigeron*.

Coreopsis *Tickseed*

Annual or perennial herbs or subshrubs. Leaves opposite or rarely alternate, entire to pinnatifid or 3-parted. Heads radiate, the rays conspicuous, usually neutral, yellow or rarely pink or white, sometimes marked with reddish brown at the base. Involucral bracts biseriate and dimorphic, all joined at the base, the outer narrower, usually shorter than the inner. Receptacle flat or slightly convex, its bracts thin and flat. Disk flowers tubular and perfect. Achenes flattened parallel to the bracts of the involucre, usually winged, not beaked. Pappus of 2 smooth or upwardly barbed, short awns or teeth, or a minute crown, or obsolete.

1 Stems with less than 5 pairs of leaves, the leaves clustered on the lower two-thirds of the stem, simple or sometimes with 1 or 2 lobes............ **C. lanceolata**
1 Stems with 5 or more pairs of leaves, the main leaves with 3 or more lobes or divisions....... **C. grandiflora**

Coreopsis grandiflora Hogg
BIGFLOWER TICKSEED *introduced*
IR | WUP | **CUP** | **EUP**

Similar to *C. lanceolata*, occasionally annual. Leafy nearly to the summit, the peduncles not greatly elongate. Leaves pinnatifid into linear-filiform to narrowly lance-shaped divisions, or the lowermost ones entire, the lateral lobes rarely more than 5 mm wide, the terminal often a little wider. Outer involucral bracts lance-shaped. Rays averaging a little shorter and narrower than in *C. lanceolata*. Achenes often with a large callous ventral excrescence at top and bottom. The typical form is essentially glabrous. May–June.

Rather dry, often sandy places.

Coreopsis lanceolata L.
LANCE-LEAF TICKSEED *native*
IR | **WUP** | **CUP** | **EUP** FACU CC 8

Perennial with a short woody caudex. Stems usually several, 2–6 dm tall, glabrous, or, especially near the base, spreading-villous, leafy below, subnaked and elongate above. Leaves spatulate to linear or lance-linear, simple or with 1 or 2 pairs of small lateral lobes, glabrous to villous or hirsute, the lower long-petiolate, to 20 cm long (including petioles) and 17 mm wide, the others reduced and sessile or nearly so. Heads few or solitary on long naked peduncles, the disk about 1–2 cm wide; outer involucral bracts about 8–10, lance-shaped to oblong-ovate, glabrous except sometimes near the tip, more or less scarious-margined, about 5–10 mm long; inner involucral bracts longer and broader than the outer; rays about 1.5–3 cm long, often over 1 cm broad, yellow; receptacle bracts flat and chaffy below, somewhat awn-like above. Achenes with thin flat wings, orbicular, about 2–3 mm long, black. Pappus of 2 short chaffy teeth. May–July.

Dry, often sandy places.

Crepis *Hawk's-Beard*

Crepis tectorum L.
NARROW-LEAF HAWK'S-BEARD *introduced*
IR | **WUP** | **CUP** | **EUP**

Annual taprooted herb with milky juice, glabrous or pubescent. Stems mostly 1–10 dm tall. Basal leaves well-developed, petiolate, the blade lance-shaped or oblong lance-shaped, denticulate to pinnately parted, to 15 cm long and 4 cm wide; stem leaves sessile, auriculate, linear or nearly so. Heads several, 30–70-flowered; involucre cylindric or campanulate, the principal bracts in 1 or 2 series, the outer bracts 12–15, subulate, about 1/3 as long as the inner bracts, finely tomentose and sometimes also glandular. Flowers all ligulate and perfect, yellow. Fruit an achene 2.5–4.5 mm long, dark purplish brown when mature, 10-ribbed; pappus of many whitish bristles. June–July.

Native of Eurasia; naturalized in waste places.

Cyclachaena *Marsh-Elder*

Cyclachaena xanthiifolia (Nutt.) Fresen.
CARELESSWEED *native*
IR | **WUP** | **CUP** | EUP FAC CC 0

Iva xanthifolia Nutt.

Coarse, branching annual weed. Stems 4–20 dm tall, the glabrous below, becoming viscid-villous in

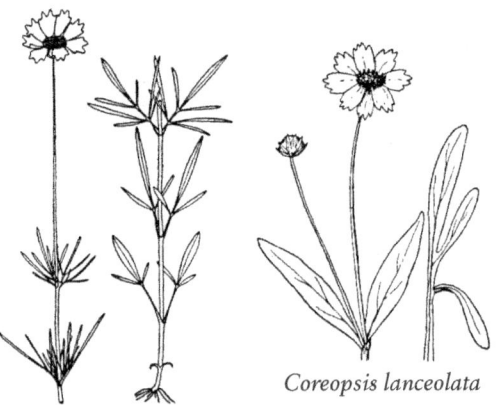

Coreopsis lanceolata

Coreopsis grandiflora

the inflorescence. Leaves opposite, except the upper, ovate, scaberulous above, paler and often finely hairy beneath, commonly 5–20 cm long and 2.5–15 cm wide, coarsely and often doubly serrate, long-petiolate. Inflorescence large, panicle-like, the heads small, numerous, greenish-white, subsessile, not subtended by leaves; involucre of 1–3 equal or imbricate series of bracts, viscid-hairy or subglabrous, about 1.5–3 mm high, the 5 herbaceous outer bracts larger than the 5 more membranous inner; corolla of the pistillate flowers nearly obsolete. Achenes glabrous or glandular, obovate, compressed. Pappus none. Aug–Oct.

Bottomlands and moist waste places.

Doellingeria
Flat-Topped White Aster

Doellingeria umbellata (P. Mill.) Nees
TALL FLAT-TOPPED WHITE ASTER   *native*
IR | **WUP** | **CUP** | **EUP**   FACW CC 5
Aster pubentior Cronq., *Aster umbellatus* P. Mill.
Perennial herb, from thick rhizomes. Stems 0.5–2 m long, upper stem with appressed, short hairs. Leaves alternate, lance-shaped to oblong lance-shaped, 4–15 cm long and 1–4 cm wide, rough-to-touch above, densely short-hairy below; margins entire; petioles short, or absent on upper leaves. Flower heads usually many, 1–1.5 cm wide, in a ± flat-topped inflorescence; involucre 3–5 mm high, the involucral bracts short-hairy and overlapping; rays 5–10, white, 5–8 mm long. Fruit a nerved achene; pappus whitish. July–Sept.

Openings in swamps and moist forests, thickets, streambanks, sedge meadows, calcareous fens, roadside ditches.

Echinacea *Coneflower*

Echinacea pallida (Nutt.) Nutt.
PRAIRIE CONEFLOWER   *introduced*
IR | **WUP** | CUP | **EUP**
Perennial from a strong woody taproot. Stems 1–10 dm tall, sparsely to densely spreading-hirsute. Leaves simple, alternate, evidently hirsute with loose or spreading hairs that may be tuberculate-strumose at the base, linear to lance-shaped or lance-elliptic, gradually tapering to the base, the blade mostly 5–20 times as long as wide, or the basal a little wider, to about 20 cm long and 4 cm wide, entire, the basal long-petiolate, the upper stem leaves less so. Heads solitary, conspicuously long- pedunculate, the disk 1.5–3 cm wide; involucral bracts hirsute, in 2–4 subequal or slightly imbricate series, with spreading or reflexed green tip; rays purple, rarely pale, about 2–8 cm long, more or less drooping, especially when longer. Achenes 4-angled, glabrous or sparsely pubescent on the angles; pappus a short toothed crown. May–Aug.

Dry or dry-mesic prairies, roadsides, and along railroads; native of central USA south of Michigan, considered adventive in the UP.

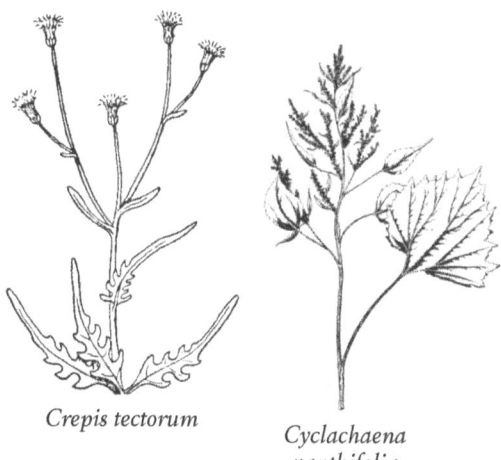

Crepis tectorum

Cyclachaena xanthifolia

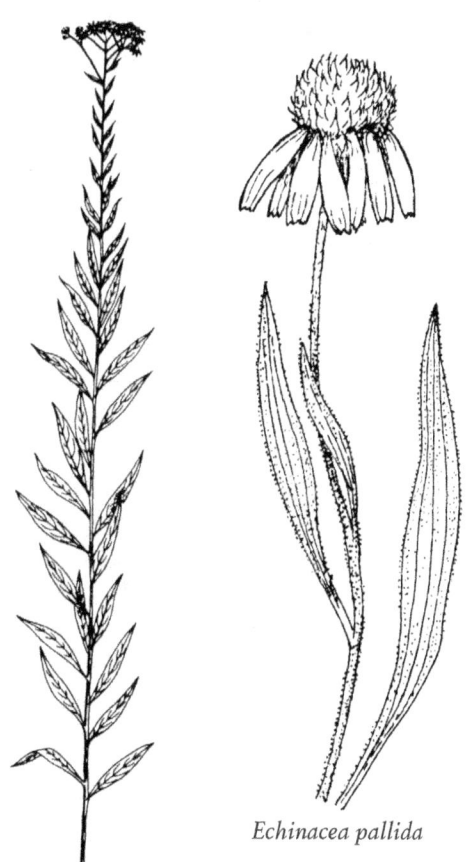

Doellingeria umbellata

Echinacea pallida

Erechtites *Fireweed*

Erechtites hieraciifolia (L.) Raf.
FIREWEED *native*
IR | **WUP** | **CUP** | **EUP** FACU CC 2
Senecio hieraciifolius L.

Fibrous-rooted annual herb. Stems erect, 0.1–2.5 m tall, slightly succulent. Leaves alternate, of various sizes to sometimes 20 cm long and 8 cm wide, sharply serrate with callous-tipped teeth, sometimes also irregularly lobed. Heads cylindric to ovoid, several to many in a flat-topped or elongate inflorescence, or in depauperate plants often solitary; discoid, whitish; involucre about 1–1.5 cm high, the bracts glabrous or finely strigose, green with pale margins, 0.5–2 mm wide. Achenes about 2–3 mm long, finely strigose between the mostly 10–12 ribs, with a white annular ring at the tip; pappus of numerous bright white bristles, eventually deciduous. Aug, Sept.

Various habitats, including dry woods, marshes, and waste places, often abundant after fires.

Erigeron *Daisy; Fleabane*

Biennial to perennial herbs with simple, alternate leaves. Flower heads with both disk and ray flowers; disk flowers yellow; rays white to pink, very narrow, only to about 0.5 mm wide; involucral bracts in 1–2 series, linear, about equal in length, green in middle and at base, translucent at tip and on upper margins. Fruit a flattened achene; pappus of 20–30 slender, rough bristles.

ADDITIONAL SPECIES
Erigeron acris L. (Bitter fleabane), a circumpolar species; in Michigan known from Houghton, Keweenaw (including Isle Royale), and Marquette counties; threatened.
Erigeron hyssopifolius Michx. (Hyssop-leaf fleabane), Keweenaw County (1890 collection only) and fens and swamps in Mackinac County; threatened.

1	Pappus of the ray flowers short, less than 1 mm long; weedy annual herbs	2
1	Pappus of long bristles; biennial or perennial herbs	3
2	Plants 6 dm or more tall; stems leafy; pubescence on middle of stem long and spreading	**E. annuus**
2	Plants to 7 dm tall; stem leaves few; pubescence mostly short and appressed	**E. strigosus**
3	Disk flowers less than 4 mm long; rays very narrow, less than 1 mm wide	**E. philadelphicus**
3	Disk flowers 4–6 mm long; rays about 1 mm wide; Menominee County	**E. pulchellus**

Erigeron annuus (L.) Pers.
EASTERN DAISY FLEABANE *native*
IR | **WUP** | **CUP** | **EUP** FACU CC 0

Annual or rarely biennial. Stems 6–15 dm tall, amply leafy, more or less hirsute, the hairs spreading except near the top. Basal leaves elliptic to suborbicular, coarsely toothed, to 10 cm long and 7 cm wide, more or less abruptly long-petiolate; stem leaves numerous, broadly lance-shaped, all except sometimes the uppermost sharply toothed, or rarely nearly entire. Heads several to very numerous; involucre 3–5 mm high, finely glandular, and sparsely hairy with long, flattened, transparent hairs; disk 6–10 mm broad. Rays about 80–125, white or rarely pinkish or bluish, about 4–10 mm long and 0.5–1.0 mm wide; disk corollas 2.0–2.8 mm long. Achenes 2-nerved. Pappus of the disk-flowers double, of 10–15 fragile bristles and several very short slender scales less than 1 mm long; pappus of the ray flowers of short scales only, lacking the longer bristles. Early and middle summer.

A weed over most of n USA and s Canada.

Erigeron philadelphicus L.
PHILADELPHIA DAISY *native*
IR | **WUP** | **CUP** | **EUP** FAC CC 2

Biennial or short-lived perennial herb. Stems 1 to several, branched in head, 2–7 dm long, usually long-hairy. Leaves alternate, lower leaves spatula-shaped, 5–15 cm long and 1–4 cm wide, tapered to a short petiole; upper leaves smaller, lance-shaped, clasping at base, hairy to nearly smooth, rounded at tip; margins entire or with rounded teeth. Flower heads few to many, with both disk and ray flowers, 1.5–2.5 cm wide; involucre 3–6 mm high, the involucral bracts hairy; rays many, white to deep pink, 5–10 mm long and to 0.5 mm wide. Fruit a short-hairy achene; pappus of long rough bristles. May–Aug.

Wet meadows, shores, wet woods, floodplains, springs; also weedy in open disturbed areas, lawns.

Erigeron pulchellus Michx.
ROBIN'S PLANTAIN *native*
IR | WUP | **CUP** | EUP FACU CC 5

Biennial or short-lived perennial, with a simple caudex and slender stoloniform rhizomes. Stems 1.5–6 dm tall. Basal leaves oblong lance-shaped to suborbicular, mostly 2–12 cm long and 6–50 mm wide, commonly more or less toothed; stem leaves ovate to lance-shaped or oblong, reduced upwards. Heads solitary or few; involucre 5–7 mm high; disk 10–20 mm wide. Rays 50–100, 6–10 mm long, about 1 mm wide or a little more, blue, or sometimes pink or white; disk corollas 4.0–6.0 mm long. Achenes 2–4-nerved; pappus simple. Spring.

Woods and streambanks; Menominee County.

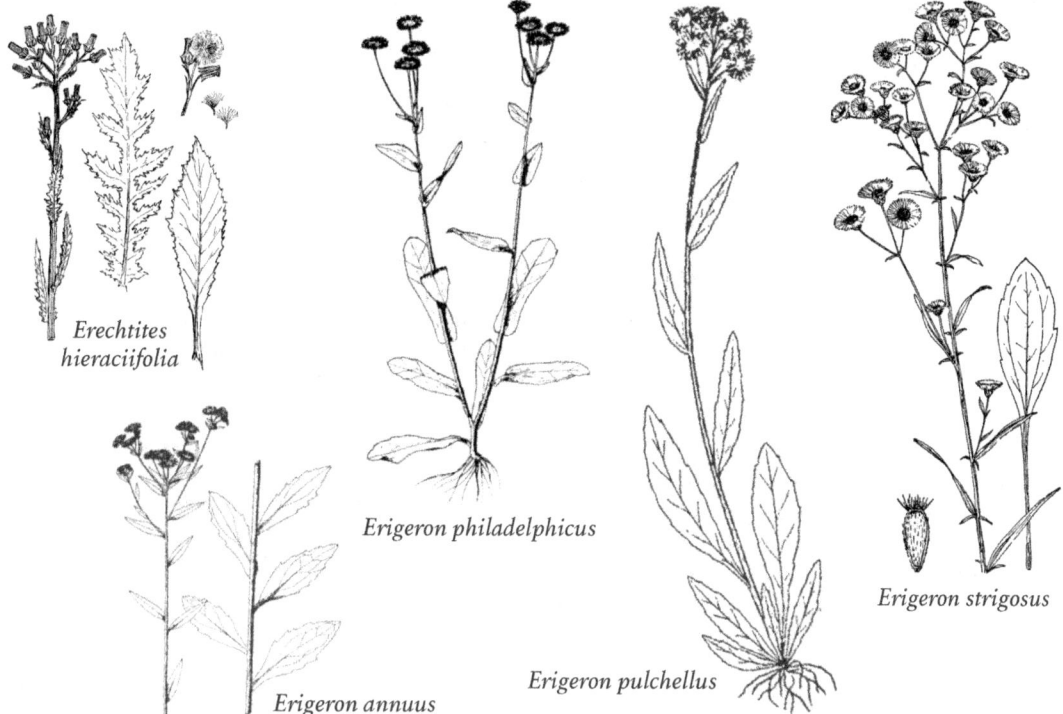

Erigeron strigosus Muhl.
ROUGH FLEABANE native
IR | **WUP** | **CUP** | **EUP** FACU CC 4

Erigeron ramosus (Walt.) B.S.P.
Annual or rarely biennial. Stems 3–7 dm tall, sparsely leafy, more or less hairy, the hairs spreading or usually appressed. Basal leaves mostly oblong lance-shaped to elliptic, entire or toothed, the blade and petiole together not more than 15 cm long and 2.5 cm wide; stem leaves linear to lance-shaped, entire, or the lower ones slightly toothed, rarely the middle ones slightly toothed also. Heads several to very numerous, involucre 2–5 mm high, obscurely glandular and more or less hairy, the hairs long or short; disk about 5–12 mm broad. Rays about 50–100, white, or sometimes pinkish or bluish, to 6 mm long, 0.4–1.0 mm wide; disk corollas 1.5–2.6 mm long. Achenes 2-nerved; pappus as in *E. annuus*. Early and midsummer.

A weed in much of the USA and s Canada.

Eupatorium *Joe-Pye-Weed; Boneset*
Eupatorium perfoliatum L.
BONESET native
IR | **WUP** | **CUP** | **EUP** FACW CC 4

Perennial herb from a thick rhizome. Stems stout, erect, 3–15 dm tall, with long, spreading hairs. Leaves opposite, mostly joined at the broad base and perforated by the stem (upper leaves sometimes separate), lance-shaped, 6–20 cm long and 1.5–5 cm wide, both sides dotted with yellow glands; margins finely toothed and rough-to-touch; petioles absent. Flower heads of dull white disk flowers only, in a flat-topped inflorescence; involucre 3–6 mm high, the involucral bracts green with white margins, hairy, overlapping in 3 series. Fruit a black achene, 1–2 mm long; pappus of long slender bristles. July–Sept.

Marshes, wet meadows, low prairie, shores, streambanks, ditches, cedar swamps, thickets, calcareous fens.

Often growing with spotted joe-pye-weed (*Eutrochium maculatum*).

Eurybia *Wood-Aster*
Eurybia macrophylla (L.) Cass.
LARGE-LEAF WOOD-ASTER native
IR | **WUP** | **CUP** | **EUP** UPL CC 4

Aster macrophyllus L.
Perennial with creeping rhizomes, sometimes also with a short branched caudex, producing abundant clusters of basal leaves on short sterile shoots. Stems 2–12 dm tall, glandular in the inflorescence or sometimes throughout, often also spreading-hairy. Leaves thick and firm, varying from essentially glabrous on both sides to scabrous above and hairy beneath, and sometimes glandular; margins crenate or serrate; basal and lower stem leaves cordate, 4–20 cm long and 3–15 cm wide, long-petiolate, the middle and upper leaves gradually or abruptly reduced, becoming

sessile. Inflorescence corymbiform, flat or round-topped, its bracts commonly few and broad; involucre 7–11 mm high, usually glandular and sometimes also short-hairy, its bracts firm, imbricated, the green tips sometimes obscure; ray florets pistillate and fertile; rays commonly 9–20, lilac- or purple-tinged, 7–15 mm long; disk florets bisexual and fertile, yellow, becoming purple at maturity. Pappus persistent and bristly. July–Oct.

Woodlands.

Eurybia macrophylla

Euthamia *Flat-Topped Goldenrod*

Perennial herbs, spreading by rhizomes. Stems leafy. Leaves alternate, covered with resinous dots; margins entire; petioles absent or very short. Flower heads small, of yellow disk and ray flowers, in a more or less flat-topped cluster at ends of stems; involucre somewhat sticky. Fruit an achene; pappus of slender white bristles.

1 Main stem leaves with 1 strong longitudinal vein (midrib) and sometimes with a pair of weak veins, the widest blades less than 3 mm wide; leaves and upper stem glabrous; uppermost leaves and leafy bracts satiny and with conspicuous dark glandular dots............ .... **E. caroliniana**
1 Main stem leaves with 3 strong longitudinal veins, and often with an additional pair of weaker veins; the widest blades 3–10 mm wide; leaves and upper stem usually short-hairy to scabrous; uppermost leaves and leafy bracts dull, the glandular dots often obscure .... **E. graminifolia**

Euthamia caroliniana (L.) Greene ex Porter & Britt.
SLENDER GOLDENTOP   native
IR | **WUP** | **CUP** | **EUP**   FAC CC 10
Euthamia remota Greene

Perennial herb, spreading by rhizomes. Stems smooth, 3–8 dm long, branched in head. Leaves alternate, linear, 2–8 cm long and 2–3 mm wide, often with clusters of smaller leaves in axils of main leaves, 1-veined or with another pair of fainter veins, with glandular dots; margins entire, sometimes rough-to-touch; petioles absent. Flower heads of yellow disk and ray flowers, in flat-topped clusters at ends of stems; involucre sticky, 3–5 mm long; the involucral bracts overlapping. Fruit a short-hairy achene; pappus of slender white bristles.

Sandy or mucky shores (especially on recently exposed lakeshores), interdunal swales; occasionally in dry habitats such as jack pine woods; main range Atlantic coast.

Euthamia graminifolia (L.) Greene
COMMON FLAT-TOPPED GOLDENROD   native
IR | **WUP** | **CUP** | **EUP**   FAC CC 3
Solidago graminifolia (L.) Salisb.

Perennial herb, spreading by rhizomes. Stems erect, 5–15 dm tall, smooth to hairy, usually branched in head. Leaves alternate, linear to narrowly lance-shaped or oval, 3–15 cm long and 3–10 mm wide, 3-veined, with small glandular dots; margins entire, smooth or rough-to-touch; petioles absent or very short. Flower heads small, in flat-topped clusters at ends of stems; with yellow disk and ray flowers, the rays small, to 1 mm long; involucre 3–5 mm high, somewhat sticky, the involucral bracts overlapping in several series, yellow or green-tipped. Fruit a finely hairy achene, 1 mm long; pappus of many white, slender bristles. Aug–Sept.

Shores, wet meadows, low prairie, springs, fens, swamps, interdunal wetlands, often where sandy or gravelly; also weedy in abandoned fields.

Eutrochium *Joe-Pye-Weed*

Eutrochium maculatum (L.) E. Lamont
SPOTTED JOE-PYE-WEED   native
IR | **WUP** | **CUP** | **EUP**   OBL CC 4
Eupatorium maculatum L.

Perennial herb from a thick rhizome. Stems stout, erect, 5–20 dm long, spotted or tinged with purple, short-hairy above, especially on branches of head.

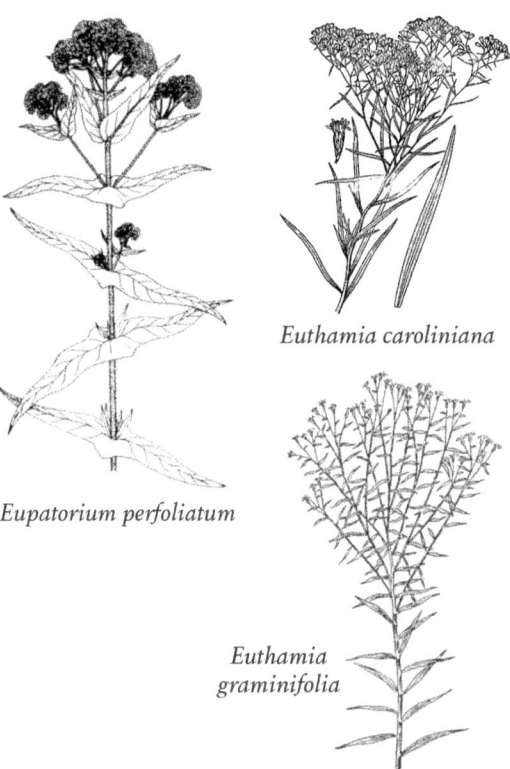

Euthamia caroliniana

Eupatorium perfoliatum

Euthamia graminifolia

dicots   **ASTERACEAE** | Aster Family

Leaves in whorls of mostly 4–5, lance-shaped to ovate, 5–20 cm long and 2–7 cm wide, upper surface with sparse short hairs, underside often densely short-hairy; margins with sharp, forward-pointing teeth; petioles to 2 cm long. Flower heads of light pink to purple disk flowers only, the inflorescence more or less flat-topped; involucres 6–9 mm high, purple-tinged, the involucral bracts overlapping. Fruit a black, angled achene, 2–4 mm long; pappus of long, slender bristles. July–Sept.

Wet meadows, marshes, shores, streambanks, ditches, cedar swamps, bogs, calcareous fens.

Gaillardia *Blanket-Flower*

Gaillardia aristata Pursh
COMMON BLANKET-FLOWER   *introduced*
IR | **WUP** | **CUP** | EUP

Usually perennial. Stems 1 or several from the base, simple or not much branched, pubescent, about 2–7 dm tall. Leaves narrow, rarely as much as 15 cm long and 2.5 cm wide, sometimes all basal in depauperate plants, entire to coarsely toothed, hairy. Heads solitary or few, long-pedunculate, the disk commonly 1.5–3 cm wide, purple or brownish purple, rarely yellow; involucral bracts mostly acuminate or attenuate-acuminate, usually loosely hairy; rays yellow, or more or less suffused with purple at the base, about 1–3 cm long; disk corollas densely woolly toward the summit. Receptacle convex to subglobose, with chaffy or spinelike setae. Achenes partly or wholly covered by a basal tuft of long ascending hairs; pappus of 6–10 awned scales. May–Sept.

Plains, meadows, and other open places; native of western states, considered adventive in Michigan.

Galinsoga *Quickweed*

Annual herbs. Leaves opposite. Flower heads small, campanulate or hemispheric; radiate, the rays few, short, broad, only slightly surpassing the disk, white or pink, pistillate and fertile; involucral bracts few, broad, membranous but greenish in part, each subtending a ray, and sometimes joined at the base; receptacle conic, chaffy throughout, its bracts membranous, nearly flat. Disk-flowers perfect; style-branches flattened, with short, minutely hairy appen- dages. Achenes 4-angled; pappus of several to many scales, often fimbriate or awn-tipped, that of the rays often reduced or absent.

1 Leaves nearly entire to shallowly toothed; pappus of ray flowers absent . . . . **G. parviflora**
1 Leaves all sharply toothed; ray flowers with well developed pappus; Mackinac County . . . . . **G. quadriradiata**

Galinsoga parviflora Cav.
GALLANT-SOLDIER   *introduced*
IR | WUP | CUP | **EUP**   UPL

Freely branching annual. Stems 2–7 dm tall, glabrous or sparsely pubescent with appressed hairs. Leaves ovate or lance-ovate, petiolate, 2–7 cm long and 1–4 cm wide, serrulate or crenulate, nearly glabrous or sparsely appressed-hairy. Peduncles appressed-hairy, or with spreading gland-tipped hairs. Heads numerous in leafy cymes, small, the disk about 3–6 mm wide; rays white; pappus nearly absent; pappus-scales of the disk-flowers conspicuously fimbriate, nearly as long as the corolla. Achenes of the disk sparsely hairy or glabrous. June–Nov.

Waste places; Mackinac County.

Galinsoga quadriradiata Cav.
SHAGGY-SOLDIER   *introduced*
IR | **WUP** | **CUP** | EUP   FACU

Similar to and more common than *G. parviflora*. Stems more pubescent, the hairs coarser and spreading. Leaves more coarsely toothed. Peduncles with gland-tipped hairs. Rays with well developed pappus-scales about equaling the tube; pappus-scales of

Eutrochium maculatum

Gaillardia aristata

Galinsoga parviflora

Galinsoga quadriradiata

the disk-flowers less fimbriate than in *G. parviflora*, sometimes evidently shorter than the corolla. Achenes of the disk finely hairy, the hairs apressed or spreading.

Native of South and Central America; now found as a weed.

Gnaphalium *Cudweed*

Annual or perennial woolly herbs. Leaves alternate, entire. Involucre ovoid or campanulate, the bracts more or less imbricate, scarious at the tip or nearly throughout. Heads disciform; flowers yellow or whitish, the numerous outer ones slender and pistillate, the few inner ones coarser and perfect. Pappus of capillary bristles. Achenes small, terete or slightly compressed.

1 Plants perennial; inflorescence narrow; achenes sparsely hairy........................ **G. sylvaticum**
1 Plants annual; inflorescence various; achenes smooth or covered with small bumps .... **G. uliginosum**

Gnaphalium sylvaticum L.
WOODLAND CUDWEED (THR) *native*
IR | **WUP** | **CUP** | EUP   CC 5

Omalotheca sylvatica (L.) Schultz-Bip. & F.W. Schultz

Perennial. Stems erect, simple, thinly woolly, about 1–6 dm tall. Leaves nearly glabrous on upper surface, linear, the larger basal and lower stem leaves mostly 3–8 mm wide, upper stem leaves mostly 2–3 mm wide. Inflorescence narrow, somewhat leafy-bracteate, with 10–many heads; involucre scarcely woolly, or woolly only at base, 5–7 mm high, light straw-colored, some or all of them with a conspicuous, dark brown spot above the middle, the tips lighter-colored and transparent. Achenes sparsely stiff hairy; pappus bristles united at base, falling in a ring. July–Sept.

Open woods; Alger and Gogebic counties.

Gnaphalium uliginosum L.
MARSH CUDWEED *introduced*
IR | **WUP** | **CUP** | **EUP**   FAC CC 3

Annual. Stems 5–25 cm tall, branching, densely and often loosely white-woolly, the leaves sparsely so. Leaves numerous, linear, up to 4 cm long and 5 mm wide. Heads glomerate in numerous small clusters in axils and at ends of the branches, overtopped by their subtending leaves; involucre about 2–3 mm high, woolly at the base; bracts greenish or brown, often with lighter-colored tips. Achenes papillate or smooth; pappus bristles distinct, falling separately. July–Oct. Streambanks and waste places, where wet or dry.

Grindelia *Gumweed*

Grindelia squarrosa (Pursh) Dunal
CURLY-TOP GUMWEED *introduced*
IR | **WUP** | **CUP** | EUP   FACU

Biennial or sometimes perennial. Stems branched above and sometimes also at base, 1–10 dm tall, sometimes woody at base. Leaves alternate, punctate and resinous, finely serrulate to entire, or, sometimes coarsely toothed; middle and upper leaves ovate or oblong, 3–7 cm long, 4–20 mm wide. Heads several to many, radiate or occasionally discoid, the rays mostly 15–45, yellow, pistillate and fertile; the disk about 1–2 cm wide; disk-flowers yellow, the inner and often also the outer sterile. Receptacle flat or convex, naked. Involucral bracts strongly sticky-resinous, imbricate in several series, the green tips reflexed, especially the outer. Rays 20–35, 7–15 mm long, or absent. Fruit a compressed to 4-angled achene, scarcely nerved, 2–3 mm long; pappus awns 2–8. July–Sept.

Open or waste places.

ADDITIONAL SPECIES

Grindelia ciliata (Nutt.) Spreng. (Spanishgold), Teeth of leaf margins with long, spine-like tips, and the phyllaries not resinous; adventive in Schoolcraft County.

Helenium *Sneezeweed*

Annual or perennial herbs. Leaves alternate, glandular-dotted, usually decurrent on stem. Flower heads solitary to numerous, radiate or rarely discoid, the rays pistillate or neutral, yellow or sometimes partly purple, 3-lobed, not very numerous; involucral bracts in 2–3 series, subequal or the inner shorter, soon deflexed, the outer sometimes joined at the base; receptacle convex to ovoid or conic, naked. Disk flowers numerous, perfect, the corolla lobes glandular-puberulent. Achenes 4–5-angled, generally pubescent on the angles and ribs; pappus of several scarious or hyaline, often awn-tipped scales.

1 Disk flowers yellow, 5-lobed at tip; stem leaves more than 1 cm wide; common .... **H. autumnale**
1 Disk flowers dark brown, 4-lobed; stem leaves to 1 cm wide; uncommon .... **H. flexuosum**

Helenium autumnale L.
COMMON SNEEZEWEED *native*
IR | **WUP** | **CUP** | **EUP**   FACW CC 5

Perennial herb. Stems single or clustered, erect, 3–13 dm tall, smooth or finely hairy, branched in head. Leaves alternate, bright green, lance-shaped to oval, 4–12 cm long and 0.5–3.5 cm wide, glandular-dotted, usually short-hairy, ; margins entire to shallowly toothed; petioles absent, the blades tapered to a narrow base extending downward as wings on stem.

dicots ASTERACEAE | Aster Family 85

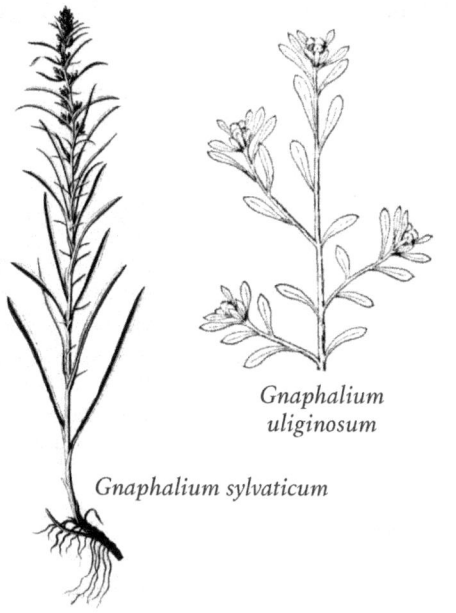

Gnaphalium uliginosum

Gnaphalium sylvaticum

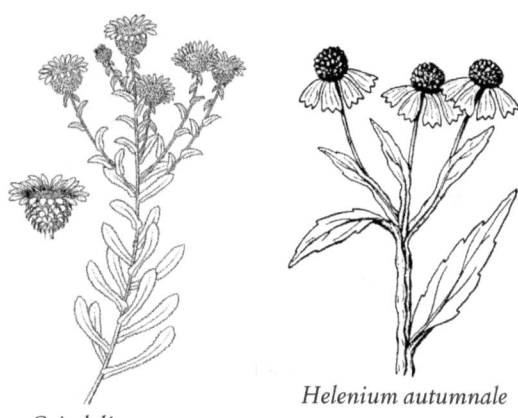

Grindelia squarrosa

Helenium autumnale

Flower heads more or less round, 1.5–4 cm wide; few to many on slender stalks in a leafy inflorescence, with both disk and ray flowers, the disk flowers yellow to brown, the rays yellow and drooping, 1.5–2.5 cm long; involucral bracts in 2–3 series, linear, short-hairy, bent downward with age. Fruit a finely hairy, 4–5-angled achene, 1–2 mm long; pappus of several translucent, awn-tipped scales. July–Sept.

Wet meadows, shores, streambanks, marshes, fens, tamarack swamps.

Helenium flexuosum Raf.
PURPLE-HEAD SNEEZEWEED *introduced*
IR | **WUP** | **CUP** | EUP   FAC
Fibrous-rooted perennial. Stems 2–10 dm tall, finely hairy, winged by the decurrent leaf-bases. Leaves smaller, less numerous, and more erect than in H. autumnale, entire or nearly so, densely glandular-dotted and also finely hairy, the lowermost oblong lance-shaped, commonly deciduous by flowering time, the others lance-shaped, sessile, not much reduced upwards except in the inflorescence, 3–12 cm long and 5–20 mm wide. Heads numerous, occasionally few or even solitary, in an open, leafy-bracteate inflorescence; involucral bracts lance-shaped, finely hairy, soon deflexed; rays yellow, sometimes tinged with purple at the base, 5–20 mm long, 3-lobed; disk purple

Helenium flexuosum

or brownish purple, 6–14 mm wide, subglobose, a little more elongate than in *H. autumnale*. Pappus scales ovate or lance-shaped, awn-tipped. June–Oct.

Moist ground and waste places; Baraga and Delta counties.

Helianthus *Sunflower*
Large perennial herbs (annual in several species), with fibrous or fleshy roots and short to long rhizomes. Stems unbranched or branched above. Leaves usually opposite on lower part of stem and alternate above, lance-shaped, margins entire or with forward-pointing teeth; petioles present. Flower heads large, mostly 1 to several (rarely many), at ends of stems and branches, with yellow disk and ray flowers, the rays large and showy; involucre of several series of narrow, overlapping bracts; receptacle chaffy. Fruit a flattened achene; pappus of 2 deciduous, awn-tipped scales.
HYBRIDS
Helianthus × laetiflorus Pers. (pro sp.), hybrid between *H. pauciflorus* and *H. tuberosus*; reported from Marquette County.

1 Disk flowers red-purple to brown; receptacle flat or nearly so; leaves mostly alternate; plants annual. 2
1 Disk flowers yellow, or rarely red-brown or purple; receptacle convex to conical; leaves opposite or alternate; plants perennial. 3
2 Involucral bracts lance-shaped, gradually tapered to a tip; chaff of receptacle bearded at tip with white hairs.
. . . . **H. petiolaris**
2 Involucral bracts ovate, abruptly narrowed to a slender tip; chaff not with white hairs . . . . **H. annuus**
3 Disk flowers reddish-brown or yellow; stem leaves few to several, reduced in size upward on the stem 4
3 Disk flowers yellow; stem leaves numerous, well developed 5
4 Disk flowers reddish brown; stems with more than 6 pairs of leaves, the leaves only gradually reduced in size upward on the stem . . . . **H. pauciflorus**

4 Disk flowers yellow; stems with less than 6 pairs of leaves, greatly reduced in size upward on stem . . . . **H. occidentalis**

5 Stems glabrous or nearly so, sometimes glaucous; fine hairs may be present within the inflorescence 6

5 Stems pubescent 7

6 Leaves sessile or with short petioles less than 5 mm long; the blades less than 5 cm wide; heads solitary or few, the disk less than 1.5 cm wide . . . . . **H. divaricatus**

6 Leaves abruptly narrowed at base to a petiole 5 mm or more long; the blades often more than 5 cm wide; heads few to many; disks more than 1.5 cm wide **H. strumosus**

7 Leaves ovate, sessile, heart-shaped to clasping at base, the blades densely gray-hairy; rare in Ontonagon County . . . . **H. mollis**

7 Leaves lance-shaped, contracted or tapered to a short petiole, the blades pubescent but not densely covered with gray hairs 8

8 Leaves lance-shaped, less than 3.5 cm wide, mostly alternate 9

8 Leaves ovate, often more than 3.5 cm wide, the upper leaves opposite or alternate 10

9 Leaves often somewhat folded along their midrib; stems pubescent, the hairs fine, white, and appressed . . . . **H. maximiliani**

9 Leaves not folded; stems pubescent, the hairs coarse and spreading . . . . **H. giganteus**

10 Upper leaves alternate, tapered at base to a winged petiole more than 1.5 cm long; involucral bracts becoming dark with age, especially near their base . . . . **H. tuberosus**

10 Leaves opposite, tapered at base to an unwinged petiole less than 1.5 cm long; involucral bracts remaining green; Menominee County . . . . **H. hirsutus**

Helianthus annuus L.
COMMON SUNFLOWER *introduced*
IR | **WUP** | **CUP** | EUP FACU

Coarse annual herb. Stems branching, more or less hirsute or hispid, 1–3 m tall, or much smaller and simpler in depauperate forms. Leaves, except the lowermost, chiefly alternate, ovate or broader, at least the lower ones cordate except in depauperate plants, acute to acuminate, mostly toothed, petiolate, scabrous on both sides. Heads mostly several or numerous, large, the disk 2 cm wide or more, smaller when depauperate; involucral bracts chiefly ovate or ovate-oblong and abruptly narrowed above the middle to the acuminate tip, occasionally narrower and more tapering, commonly more or less hispid or hirsute, and ciliate on the margins, sometimes merely shortly scabrous-hispid; disk red-purple (yellow in some cultivated forms); receptacle flat or nearly so, its bracts merely pubescent at the tip. July–Sept.

Prairies and dry places.

Helianthus divaricatus L.
WOODLAND SUNFLOWER *native*
IR | **WUP** | **CUP** | EUP CC 5

Perennial with woody or fibrous roots and long rhizomes. Stems mostly 0.5–1.5 m tall, commonly hispidulous in the inflorescence, otherwise essentially glabrous and often glaucous. Leaves all opposite, sessile or borne on short petioles to 5 mm long, narrowly lance-shaped to broadly ovate lance-shaped, broadest near the truncate or broadly rounded base, tapering thence to the slender acuminate tip, commonly 6–15 cm long and 1–8 cm wide, shallowly toothed or subentire, 3-nerved near the base, pinnately veined above, scabrous on the upper surface, loosely hirsute or hispidulous on the lower, at least on the midrib and main veins. Heads solitary or few to occasionally fairly numerous, the disk yellow, 1–1.5 cm wide; involucral bracts rather loose, or the outer subappressed, in 2 or 3 subequal series, lance-shaped, sometimes a little surpassing the disk, ciliolate toward the base, scaberulous on the back; rays about 8–15, 1.5–3 cm long. July–Sept.

Dry woodlands; Dickinson and Gogebic counties.

Helianthus giganteus L.
GIANT SUNFLOWER *native*
IR | **WUP** | **CUP** | EUP FACW CC 5

Perennial herb, with short rhizomes and thick, fleshy roots. Stems 1–3 m long, often purple, with coarse hairs or sometimes nearly smooth, often branched in head. Upper leaves generally alternate, lower leaves opposite; lance-shaped, 6–20 cm long and 1–4 cm wide, base with 3 main veins, upper surface very rough-to-touch, underside with short, stiff hairs; margins toothed to more or less entire; petiole short or absent. Flower heads 3–6 cm wide, several to many, on long stalks in an open inflorescence; with yellow disk and ray flowers, the rays 1.5–3 cm long; involucral bracts narrow, awl-shaped, green or dark near base, hairy or margins fringed with hairs. Fruit a smooth achene; pappus of 2 awl-shaped scales. July–Sept.

Wet meadows, low prairie, sedge meadows, fens, floodplain forests, streambanks.

Helianthus hirsutus Raf.
HAIRY SUNFLOWER *native*
IR | WUP | **CUP** | EUP CC 10

Similar to *H. divaricatus*. Stems with coarse spreading hairs that are enlarged at the base, and often also with some shorter appressed hairs. Leaves ascending with short petioles mostly 5–15 mm long, occasionally subcordate at the base in robust plants, often more densely hairy beneath than in *H. divaricatus*. Heads averaging larger than in *H. divaricatus* and the involucral bracts often a little wider, the disk often 2 cm wide, the rays 10–15, 1.5–3.5 cm long. July–Oct.

Dry wooded or open places; uncommon in Menominee County.

Helianthus maximiliani Schrad.
MAXIMILIAN SUNFLOWER *introduced*
IR | WUP | CUP | EUP

Perennial from short rhizomes and thickened, often fleshy roots. Stems 0.5–3 m tall, conspicuously pubescent, especially upwards, with mostly short, white, appressed hairs. Leaves lance-shaped, gradually narrowed to the short winged petiole, commonly 7–15 cm long and 1–3 cm wide, subentire or occasionally evidently toothed, pinnately veined, not 3-nerved, strongly scabrous on both sides, usually some of them falcate, the upper mostly alternate (all opposite in depauperate forms). Heads several or occasionally solitary, the disk 1.5–2.5 cm wide, yellow; involucral bracts narrow, often much exceeding the disk, canescent with short white hairs; rays 10–25, 1.5–4 cm long. June–Sept.

Prairies and waste ground, often in sandy soil.

Helianthus mollis Lam.
ASHY SUNFLOWER (THR) *native*
IR | **WUP** | CUP | EUP CC 9

Perennial with crown-buds, fibrous roots, and well developed rhizomes. Stems mostly 0.5–1 m tall, copiously spreading-villous with rather long white hairs, often also finely hairy. Leaves opposite or the uppermost alternate, sessile, subcordate, ascending, broadly lance-shaped, mostly 6–15 cm long and 3–7 cm wide, inconspicuously toothed or subentire, more or less 3-nerved near the base, pinnately veined above, densely hairy with short spreading hairs on both sides. Heads solitary or few, the disk yellow, mostly 2–3 cm wide; involucral bracts slightly imbricate, lance-shaped, densely white-hairy and often also finely glandular, the tips commonly loose; rays mostly 15–30, 1.5–3.5 cm long. July–Sept.

Mostly in dry, often sandy places and roadsides; rare in Ontonagon County.

Helianthus occidentalis Riddell
NAKED-STEMMED SUNFLOWER *native*
IR | WUP | **CUP** | EUP FACU CC 8

Rhizomatous perennial. Stems 0.5–1.5 m tall, sparsely hairy, often becoming densely villous near the base. Leaves mostly opposite, 3–8 pairs below the inflorescence, the lower ones much the largest, long-petiolate; blades 3-nerved, ovate, to 15 cm long and 7 cm wide, entire or minutely and sparsely toothed, short-hairy on both sides or eventually nearly glabrous; middle and upper leaves reduced and distant, the stem often appearing almost naked. Heads solitary to numerous, generally few, the disk 1–1.5 cm wide, yellow; involucral bracts imbricate, lance-shaped, at least the inner with loose acuminate tips; rays about 10–20, 1–3 cm long. Aug–Oct.

Dry soil; Menominee County.

Helianthus pauciflorus Nutt.
STIFF SUNFLOWER *native*
IR | **WUP** | **CUP** | EUP CC 5

Perennial with well developed stout creeping rhizomes. Stems stout, 0.5–2 m tall, more or less scabrous or hispid to subglabrous. Leaves nearly all opposite, 9–15 pairs below the inflorescence (fewer in depauperate plants), lance-shaped, broadest below the middle, mostly 3-nerved, tapering to the short petioles, toothed or nearly entire, mostly 5–15 cm long and 1.5–6 cm wide, the middle ones seldom much smaller than the lowermost ones present at flowering time, scabrous on both sides. Heads several or solitary, the disk purple, rarely yellow, 1–2.5 cm wide; involucral bracts imbricate, appressed, ovate, conspicuously ciliolate, generally otherwise glabrous; rays commonly 15–20, about 1.5–3 cm long; pappus nearly always with some short bristles in addition to the 2 longer awns. Aug–Sept.

Dry prairies.

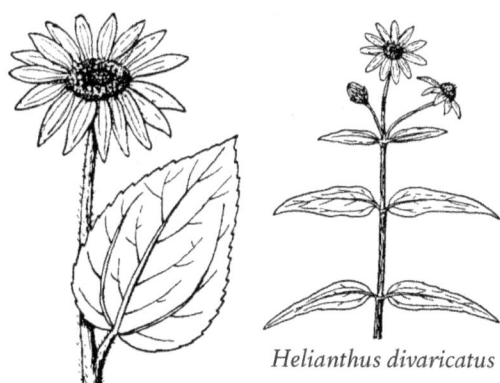

Helianthus annuus

Helianthus divaricatus

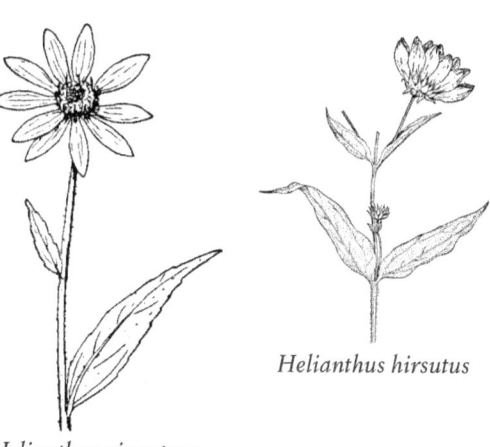

Helianthus giganteus

Helianthus hirsutus

ASTERACEAE | Aster Family

Helianthus petiolaris Nutt.
PLAINS SUNFLOWER *introduced*
IR | **WUP** | **CUP** | EUP

Annual herb, similar to *H. annuus* but smaller, seldom over 1 m tall, with narrower, more often entire, rarely cordate leaves, which may be more densely hairy beneath, and with smaller heads, the disk 1–2.5 cm wide. Involucral bracts lance-shaped, tapering gradually to the tip, shortly scabrous-hispid, seldom at all ciliate or with any long hairs. Central receptacle bracts conspicuously white-bearded at the tip. June–Sept.

Fields and waste places; adventive in Houghton and Schoolcraft counties.

Helianthus strumosus L.
PALE-LEAF WOODLAND SUNFLOWER *native*
IR | **WUP** | **CUP** | EUP   CC 4

Perennial with rather woody roots and well developed rhizomes. Stems 1–2 m tall, short-hairy in the inflorescence, otherwise glabrous or with only a few scattered long hairs, often glaucous. Leaves opposite, or the uppermost alternate, broadly lance-shaped, 8–20 cm long and 2.5–9 cm wide, shallowly toothed or subentire, more or less abruptly contracted or sometimes broadly rounded at base, commonly with a short decurrence on the 6–30 mm petiole, scabroushispidulous above, some of the hairs with broad, white, slightly raised base (strumose), lower surface green and moderately short-hairy to nearly glabrous and glaucous, 3-nerved near the base, pinnately veined above. Heads several or solitary; disk yellow, 1–2 cm wide; involucral bracts lance-shaped, somewhat loose, especially the long acuminate tips, ciliolate on the margins; rays 8–15, 1.5–4 cm long. July–Sept.

Chiefly in woods.

Helianthus tuberosus L.
JERUSALEM-ARTICHOKE *native*
IR | **WUP** | **CUP** | EUP   CC 6

Perennial with tuber-bearing rhizomes. Stems stout, 1–3 m tall, pubescent with mostly spreading hairs. Leaves alternate, or sometimes all but the uppermost opposite, broadly lance-shaped, 10–25 cm long and 4–12 cm wide, on winged petioles 2–8 cm long; margins serrate; densely and coarsely rough-hairy on the upper surface, sparsely to densely velvety-hairy on the lower surface, 3-nerved near the base, pinnately veined above. Heads usually several or numerous; disk 1.5–2.5 cm wide, yellow; involucral bracts usually rather dark, especially near the base, narrowly lance-shaped, loose especially above the middle, ciliate on the margins; rays 10–20, 2–4 cm long. Aug–Oct.

Moist soil and waste places; escaped from cultivation and also native.

Cultivated since pre-Columbian times for its edible tubers.

Heliopsis *Sunflower-Everlasting*

Heliopsis helianthoides (L.) Sweet
SUNFLOWER-EVERLASTING *native*
IR | **WUP** | **CUP** | EUP   FACU CC 5

Short-lived perennial, generally with a short caudex and fibrous roots. Stems 5–15 dm tall, glabrous or more or less scabrous. Leaves opposite, ovate, serrate, often subtruncate at base, 5–15 cm long and 2.5–8 cm wide, borne on petioles about 5–35 mm long. Heads solitary or several, sometimes numerous,

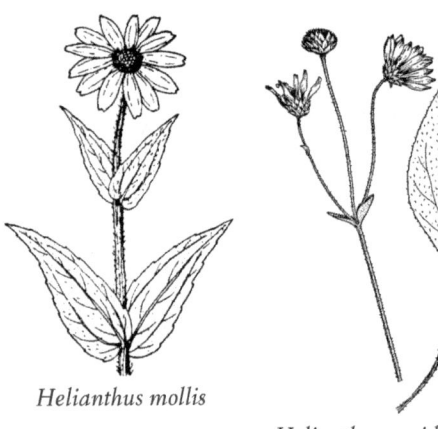

Helianthus mollis

Helianthus occidentalis

Helianthus pauciflorus

Helianthus strumosus

Helianthus tuberosus

naked-pedunculate; radiate, the rays yellow, pistillate, fertile or rarely sterile, persistent on the achenes and becoming papery; receptacle conic, chaffy throughout, its bracts concave and clasping, subtending the rays as well as the disk-flowers; disk-flowers perfect and fertile; the disk about 1–2.5 cm wide; rays 8–15, pale yellow, 1.5–4 cm long. Achenes quadrangular, glabrous; pappus none, or of a short irregular crown or a few teeth. June–Oct.

Dry woodlands, prairies, and waste places.

Heterotheca *Golden Aster*

Heterotheca villosa (Pursh) Shinners
HAIRY GOLDEN ASTER *introduced*
IR | WUP | **CUP** | EUP   CC –

Chrysopsis villosa (Pursh) Nutt.

Perennial herb, pubescent throughout with long or short, appressed or spreading hairs. Stems several, 2–10 dm tall. Leaves numerous, nearly alike, oblong-elliptic to oblong lance-shaped, 2–7 cm long and 3–15 mm wide, entire or denticulate, the lower short-petiolate. Heads several, radiate, the rays yellow, pistillate and fertile; disk flowers yellow disk 0.8–2.5 cm wide; receptacle flat or a little convex, naked; involucre strigose or hirsute and sometimes also glandular, the bracts commonly purple-tipped. Achenes narrowly obovate, 3–5-nerved; outer pappus of coarse bristles. July–Oct.

Dry, open, often sandy places; native of western USA, adventive in Schoolcraft County.

Hieracium *Hawkweed*

Fibrous-rooted perennial herbs with milky juice, with a rhizome which may be elongate or shortened into a caudex. Leaves alternate or all basal, entire or more or less toothed. Heads solitary to numerous, small or large, in a corymbiform or panicle-like inflorescence. Flowers all ligulate and perfect, yellow to red-orange. Involucre cylindric to hemispheric, its bracts imbricate. Achenes terete, mostly narrowed toward the base, truncate or occasionally narrowed toward the summit, strongly ribbed. Pappus of numerous whitish to brownish capillary bristles.

ADDITIONAL SPECIES

Several other introduced *Hieracium* are reported for the UP:

Hieracium lachenalii K.C. Gmel., known from scattered locations across the UP.

Hieracium maculatum Sm. (Spotted Hawkweed), Mackinac County.

Hieracium murorum L., scattered locations across the UP.

Hieracium venosum L. (Rattlesnake-weed), Chippewa and Mackinac counties.

1 Plants with flowers on a naked stalk (scape); leaves clustered at base. 2
1 Flowers not with on a scape; leaves not clustered at base 4
2 Flowers bright orange-red . . . . **H. aurantiacum**
2 Flowers yellow 3
3 Leaves glabrous, narrowly oblong lance-shaped; stolons absent. . . . . **H. piloselloides**
3 Leaves with tan or white hairs, oblong lance-shaped; stolons present, arching. . . . . **H. caespitosum**
4 Leaves broadly elliptic, coarsely toothed, tapered to long, hairy petioles; stems glabrous . . . . **H. lachenalii***
4 Leaves various, petioles short or absent; stems glabrous or hairy. 5
5 Leaves spatula-shaped, lower leaves with petioles, upper leaves sessile; involucres and peduncles with black glands . . . . **H. scabrum**
5 Leaves lance-shaped to oblong lance-shaped, sessile, toothed; involucres and peduncles without glands. . . . . **H. umbellatum**

Hieracium aurantiacum L.
DEVIL'S-PAINTBRUSH *introduced (invasive)*
IR | WUP | **CUP** | EUP

Perennial with slender stolons and normally with a slender elongate rhizome, commonly 1–6 dm tall, the stem naked or with a single (rarely 2) more or less reduced leaf, conspicuously long-setose, also becoming stellate-tomentose and hispid with gland-tipped hairs above. Basal leaves oblong lance-shaped or narrowly elliptic, blunt, 4–20 cm long (including the petiole), 1–3.5 cm wide, long-setose on both sides, or nearly glabrous above; leaves of the stolons few, similar but much smaller. Heads 5–25 in a compact corymbiform inflorescence; flowers red-orange, becoming deeper red in drying; involucre 5–8 mm high, long-setose, hispid with blackish gland-tipped hairs. Achenes about 2 mm long; pappus slightly sordid. June–Sept.

Native of Europe; fields, roadsides, and meadows; common.

Heliopsis helianthoides

Heterotheca villosa

Hieracium caespitosum Dumort.
YELLOW KING-DEVIL *introduced*
IR | WUP | CUP | EUP

Hieracium pratense Tausch

Perennial with a short or more often elongate rhizome and commonly with short stout stolons, the stem 2.5–9 dm tall, sparsely to rather densely long-setose, becoming stellate–tomentose and hispid with blackish gland-tipped hairs above, naked or with only one or two (rarely 3) reduced leaves. Basal leaves oblong lance-shaped or narrowly elliptic, 4–25 cm long (including the petiole), 1–3 cm wide, long-setose on both sides, sometimes sparsely so above, commonly slightly stellate beneath. Heads several or rather numerous in a compact corymbiform inflorescence, the rays yellow; the involucre 6–8 mm high, hispid with blackish, gland-tipped hairs, commonly also sparsely long-setose and slightly stellate. Achenes 1.5–2 mm long; pappus slightly sordid. May–Sept.

A weed in fields, pastures, and along roadsides, occasionally in dry woods; native of Europe; Alger and Keweenaw (including Isle Royale) counties.

Hieracium piloselloides Vill.
TALL HAWKWEED *introduced (invasive)*
IR | WUP | CUP | EUP

Hieracium florentinum All.

Perennial from a usually rather short praemorse rhizome. Stems 2–10 dm tall, naked or with 1 or 2, rarely as many as 5, reduced leaves. Herbage glaucous, sparsely long-setose or subglabrous, the peduncles becoming stipitate- glandular and somewhat stellate. Basal leaves oblong lance-shaped, mostly 3–18 cm long (including the petiole) and 5–18 mm wide, 5–12 times as long as wide. Heads mostly 3–75 in a corymbiform inflorescence, the rays yellow; the involucre 6–8 mm high, hispid with blackish, mostly gland-tipped hairs and somewhat stellate. Achenes 1.5–2 mm long; pappus slightly sordid. June–Sept.

Mostly in fields, meadows, pastures, roadsides, and waste places; native of Europe; common.

Hieracium scabrum Michx.
ROUGH HAWKWEED *native*
IR | WUP | CUP | EUP CC 3

Perennial from a short mostly simple caudex. Stems 2–14 dm tall, setose with spreading hairs seldom as much as 5 mm long, at least near the base, becoming stellate and glandular upwards, densely so in the inflorescence. Leaves sparsely or moderately setose on both sides, more densely so on the petiole and midrib beneath; basal and often also the lowermost stem leaves ordinarily deciduous, the lower leaves oblong lance-shaped to elliptic, 5–20 cm long (including the usually short petiole), 1–4.5 cm wide, the others progressively reduced upwards, soon becoming sessile. Inflorescence open-corymbiform (especially in smaller specimens) to more elongate and cylindric; heads mostly 40–100-flowered; the rays yellow; involucre 6–9 mm high, hispid with blackish mostly gland-tipped hairs, especially toward the base. Achenes 2–3 mm long; pappus tawny. July–Sept.

Open ground and dry woods, especially in sandy soil.

Hieracium umbellatum L.
NARROW-LEAF HAWKWEED *native*
IR | WUP | CUP | EUP CC 7

Hieracium canadense Michx., *Hieracium kalmii* L., *Hieracium scabriusculum* Schwein.

Perennial from a short caudex. Stems 1.5–15 dm tall, often spreading-hairy below, sometimes stellate-puberulent above. Leaves stellate-puberulent to subglabrous, and often long-hairy beneath, the basal and lowermost stem ones small and soon deciduous, the others, except for the strongly reduced upper ones, mostly rather numerous, nearly alike in size and shape, sessile and tending to be broadly rounded and somewhat clasping at the base, elliptic to ovate, mostly 3–12 cm long and 7–40 mm wide, usually with a few irregularly spaced sharp teeth. Inflorescence loosely corymbiform to often umbel-like, or the heads occasionally solitary, the peduncles stellate-puberulent, occasionally with some longer spreading hairs as well; heads mostly 40–110-flowered, the rays yellow; involucre 6–13 mm high, its bracts imbricate in several series, glabrous or obscurely puberulent, occasionally with a few longer hairs. Achenes 2.5–3.5 mm long; pappus tawny or yellowish. July–Sept.

Woods, beaches, and fields, especially in sandy soil; common.

Hypochaeris *Cat's-Ear*

Hypochaeris radicata L.
HAIRY CAT'S-EAR *introduced*
IR | WUP | CUP | EUP FACU

Perennial from a caudex, fibrous-rooted, or more often several of the roots en– larged. Stems 1.5–6 dm tall, striate-angled, branched above or in small plants simple, often spreading-hispid below. Basal leaves hispid, oblong lance-shaped, toothed or pinnatifid, 3–3.5 cm long and 0.5–7 cm wide. Heads usually several, terminating the branches; involucre 1–1.5 cm high at anthesis, sometimes nearly 2.5 cm in fruit, its bracts imbricate, glabrous or hispid. Body of the achene 4–5 mm long, from a little longer to more often much shorter than the slender beak, the prominent nerves and the lower part of the beak scabrous. Some of the outer pappus-bristles often shorter than the inner. May–Sept.

Roadsides, pastures, fields, and waste places; native of Eurasia; now widely established.

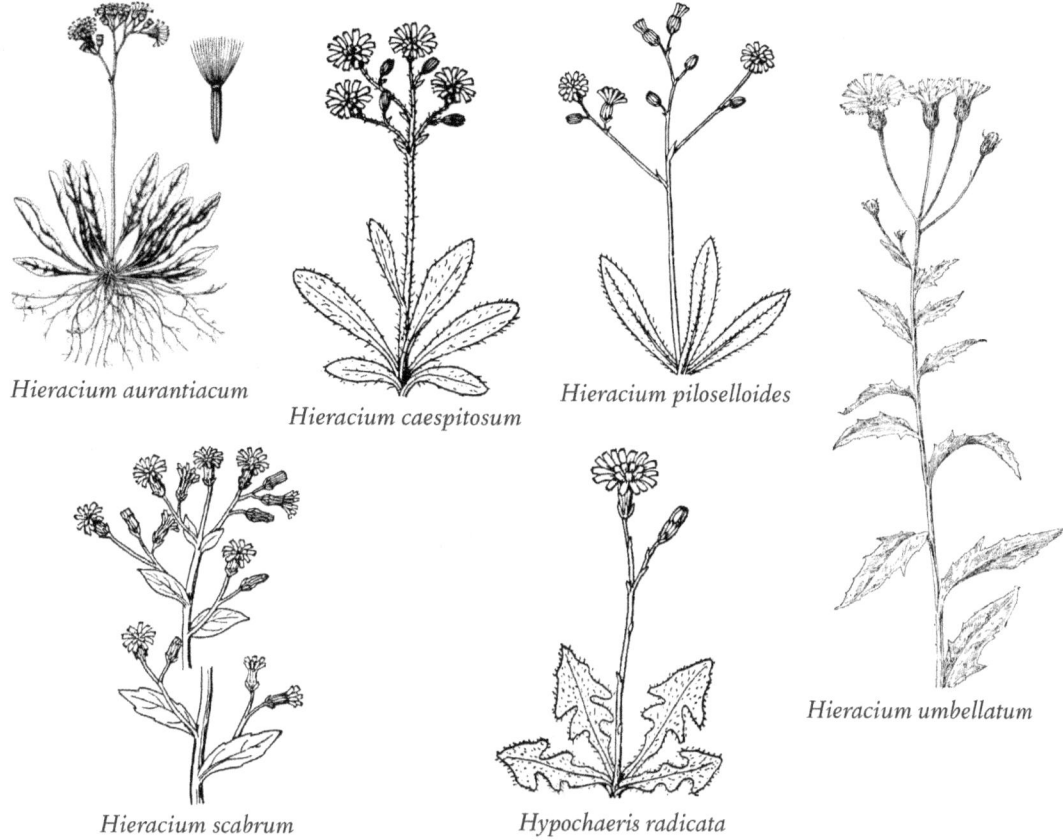

Hieracium aurantiacum
Hieracium caespitosum
Hieracium piloselloides
Hieracium scabrum
Hypochaeris radicata
Hieracium umbellatum

Similar to *Leontodon*, from which it is distinguished primarily by its chaffy-bracted receptacle.

Inula *Elecampane*

Inula helenium L.
ELECAMPANE *introduced*
IR | WUP | **CUP** | **EUP** FACU

Coarse perennial herb. Stems to 2 m tall, finely spreading-hairy. Leaves alternate, irregularly and shallowly dentate, densely velvety beneath, sparsely spreading-hairy or subglabrous above, the lower long-petioled and elliptic; blades sometimes 5 dm long and 2 dm wide, the upper leaves becoming ovate, sessile, and cordate-clasping. Heads few, pedunculate, large, the disk 3–5 cm wide, the involucre 2–2.5 cm high, the bracts imbricate in several series; outer bracts broad, herbaceous, densely short-hairy; inner bracts narrow, subscarious, glabrous; receptacle flat or convex, naked. Outer flowers pistillate, with yellow rays over 1 cm long; inner flowers perfect, tubular, yellow. Achenes slender, glabrous, 4-angled; pappus bristles few to many. July–Aug.

Introduced from Europe; cultivated and escaped to fields and waste places where sometimes forming large colonies; Chippewa and Menominee counties.

Krigia *Dwarf-Dandelion*

Krigia biflora (Walt.) Blake
ORANGE DWARF-DANDELION *native*
IR | **WUP** | **CUP** | EUP FACU CC 5

Fibrous-rooted perennial with milky juice. Stems 2–8 dm tall. Herbage glabrous, except often for some spreading glandular hairs under the heads, somewhat glaucous. Basal leaves oblong lance-shaped to elliptic, 3–25 cm long, including the petiole, 7–50 mm wide, entire or toothed to sometimes lobed or pirmatifid; stem leaves few, sessile and clasping, often much reduced, the uppermost often subopposite and with several long peduncles in their common axil. Heads several; flowers all ligulate and perfect, orange; involucre mostly 7–14 mm high, its bracts 9–18, narrow, becoming reflexed in age. Achenes nerved or ribbed, transversely rugulose; pappus of 20–35 fragile bristles and some short inconspicuous scales. May–Oct.

Woods, roadsides, and fields.

Lactuca *Lettuce*

Annual, biennial, or perennial herbs with milky juice. Leaves alternate, entire to pinnatifid. Heads usually numerous in a panicle-like inflorescence. Flowers all ligulate and perfect, yellow, blue, or white, the corolla tube generally more than half as long as the ligule. Involucre cylindric, often broadening at the base in fruit, generally imbricate. Achenes compressed, winged or strongly nerved marginally, with 1–several lesser nerves on each face, expanded at the summit where the pappus is attached. Pappus of capillary bristles.

ADDITIONAL SPECIES

Lactuca hirsuta Muhl. ex Nutt. (Hairy lettuce), native biennial, our plants glabrous or nearly so; dry, sandy places; Menominee County.

1 Achenes with one central nerve on each side......... .... **L. canadensis**
1 Achenes with several prominent nerves on each face .2
2 Achene tipped by a slender beak as long as or longer than achene body...................... **L. serriola**
2 Achene with a short beak, shorter than body, or the beak absent .... **L. biennis**

Lactuca biennis (Moench) Fern.
TALL BLUE LETTUCE *native*
IR | **WUP** | **CUP** | **EUP** FAC CC 2

Robust annual or biennial. Stems glabrous, 6–20 dm tall. Leaves glabrous, or hairy on the main veins beneath, sometimes sagittate at the base, commonly 10–40 cm long and 4–20 cm wide. Heads numerous in an elongate, rather narrow, panicle-like inflorescence, often crowded, mostly 15–34-flowered (rarely to 54); fruiting involucre 10–14 mm high. Achenes 4–5.5 mm long, thin-edged, prominently several-nerved on each face, tapering to the beakless or shortly beaked tip; pappus light brown. July–Sept.

Moist places.

Lactuca canadensis L.
TALL LETTUCE *native*
IR | **WUP** | **CUP** | **EUP** FACU CC 2

Annual, or usually biennial. Stems 3–25 dm tall, the herbage glabrous or occasionally coarsely hirsute, often more or less glaucous. Leaves entire or toothed to pinnately lobed, sagittate or sometimes narrowed to the base, mostly 10–35 cm long and 1.5–12 cm wide. Heads numerous, relatively small, mostly 13–22-flowered, the flowers yellow; fruiting involucre mostly 10–15 mm long. Achenes blackish, flat, with a median nerve on each face, rugulose, the body 3–4 mm long, beaked; pappus 5–7 mm long. July–Sept.

Fields, waste places, woods.

Lactuca serriola L.
PRICKLY LETTUCE *introduced*
IR | **WUP** | **CUP** | **EUP** FACU
Lactuca scariola L.

Biennial or winter annual. Stems 3–15 dm tall, often prickly below, otherwise glabrous. Leaves prickly on the midrib beneath, and more finely prickly toothed on the margins, otherwise generally glabrous, pinnately lobed, commonly twisted at base to lie in a vertical position, sagittate-clasping, oblong or oblong lance-shaped in outline, mostly 5–30 cm long and 1–10 cm wide, the upper much reduced. Heads numerous in a long, often diffuse inflorescence, commonly 18–24-flowered (rarely 13–27), the flowers yellow, often drying blue; involucre 10–15 mm high in fruit. Achenes gray or yellowish gray, the body compressed, 3–4 mm long and a third as wide, prominently several-nerved on each face, spinulose above, at least marginally, the slender beak about equaling the body. July–Sept.

A weed in fields and waste places; native of Europe, now naturalized throughout most of USA.

Lapsana *Nipplewort*

Lapsana communis L.
COMMON NIPPLEWORT *introduced*
IR | **WUP** | **CUP** | **EUP**

Branching annual herb with milky juice, hirsute to nearly glabrous. Stems 1.5–15 dm tall. Leaves alternate, thin, petiolate; blade ovate, toothed or occasionally basally lobed, 2.5–10 cm long and 2–7 cm wide, progressively less petiolate and narrowed upwards. Heads several or numerous in a corymbiform or panicle-like inflorescence, naked-pedunculate, mostly 8–15-flowered. Flowers all ligulate and perfect, yellow; involucre cylindric-campanulate, 5–8 mm high; bracts nearly equal, uniseriate, keeled. Achenes curved, narrowed to both ends, 3–5 mm long, glabrous; pappus none. June–Sept.

Native of Eurasia; now established in woods, fields, and waste ground.

Leontodon *Hawkbit*

Leontodon saxatilis Lam.
LITTLE HAWKBIT *introduced*
IR | **WUP** | **CUP** | **EUP** UPL
Leontodon taraxacoides (Vill.) Mérat

Fibrous-rooted perennial herb with milky juice. Stems 1–3.5 dm tall, the scapes simple and ordinarily naked. Leaves basal, oblong lance-shaped, hispid-hirsute, 4–15 cm long, 6–25 mm wide, usually shallowly lobed. Heads solitary; involucre 6–11 mm high, glabrous or hairy; receptacle not chaffy-bracted Flowers all ligulate and perfect, yellow. Achenes fusiform, nerved, short-beaked, 3–6 mm long, scabrous; pappus of the inner flowers partly of plumose bristles, partly of shorter outer scales which may be tipped with a scabrous bristle. June–Sept.

Native of Europe; a weed in lawns and waste places.

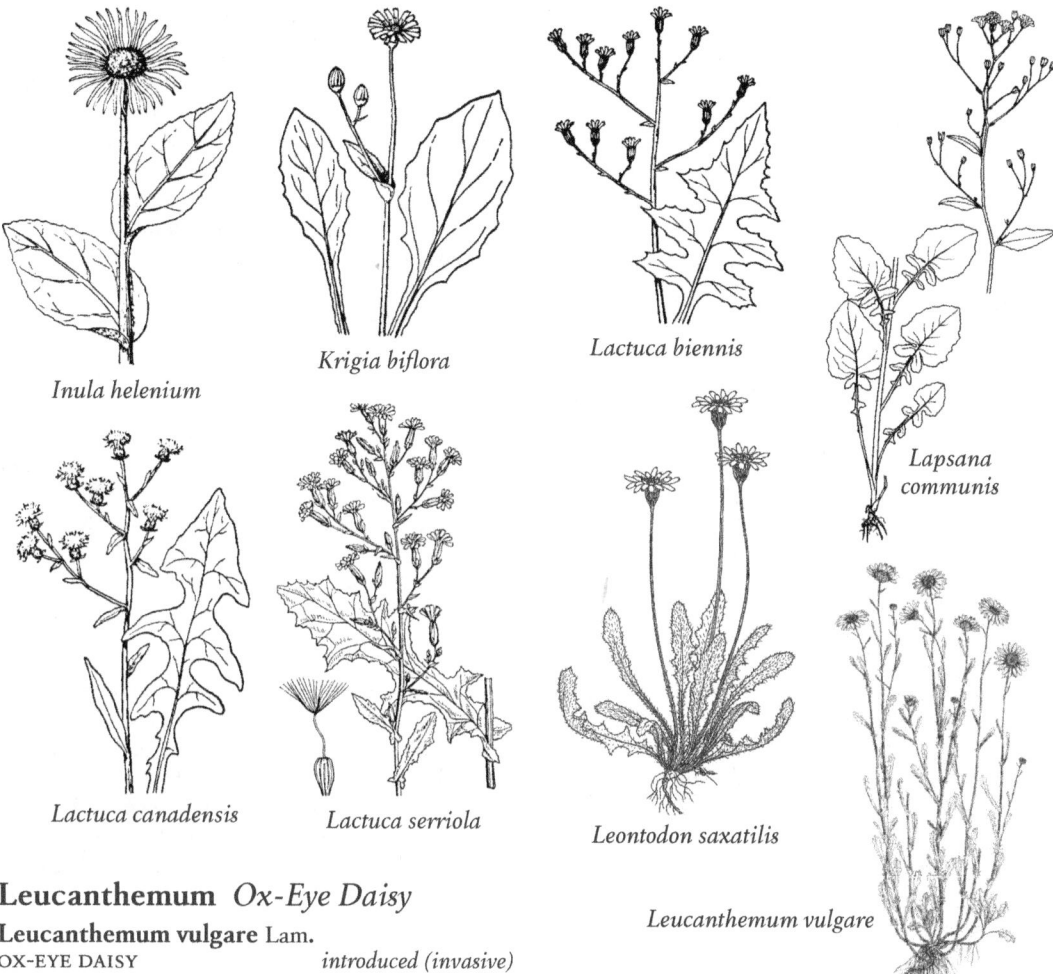

Inula helenium

Krigia biflora

Lactuca biennis

Lapsana communis

Lactuca canadensis

Lactuca serriola

Leontodon saxatilis

Leucanthemum vulgare

Leucanthemum *Ox-Eye Daisy*

Leucanthemum vulgare Lam.
OX-EYE DAISY   introduced (invasive)
IR | WUP | CUP | EUP

Chrysanthemum leucanthemum L.
Rhizomatous perennial. Stems 2–8 dm tall, simple or nearly so, glabrous or sparsely hairy. Leaves alternate, glabrous or hairy; basal leaves oblong lance-shaped or spatulate, petiolate, 4–15 cm long, crenate and often also lobed or cleft; stem leaves reduced and becoming sessile, pinnatifid or nearly entire. Heads hemispheric or flattened, solitary at the ends of the branches, naked-pedunculate, radiate; involucral bracts imbricate in 2–4 series, dry, scarious or hyaline at least at the margins and tips, the midrib sometimes greenish; receptacle flat or convex, naked; disk about 1–2 cm wide; rays about 15–30, white, 10–20 mm long; disk flowers tubular and perfect, the corolla with 4 or more commonly 5 lobes. Achenes terete, about 10-ribbed; pappus none. May–Oct.

Fields, roadsides, waste places; native of Europe and Asia, naturalized throughout most of temperate North America; common in the UP.

Liatris *Blazing Star; Gay Feather*

Perennial herbs, mostly with an evident corm, rarely from a more elongate crown or stout rhizome. Leaves alternate, entire, more or less punctate, the basal ones usually the largest. Inflorescence spike-like. Heads discoid, the flowers all tubular and perfect, 3–100 or more in each head. Involucral bracts imbricate in several series. Receptacle naked. Corollas pink-purple or occasionally white. Achenes about 10-ribbed, pubescent. Pappus of 1 or 2 series of barbellate or plumose capillary bristles.

1 Inflorescence a dense spike, heads with mostly 5–10 flowers; leaves mostly less than 1 cm wide; Schoolcraft County.......................... **L. pycnostachya**
1 Inflorescence an open spike or raceme; heads with 14 or more flowers; larger leaves 1–4 cm wide 2
2 Middle involucral bracts green and herbaceous throughout, or with entire or slightly fringed chaffy margins; Keweenaw County...................... **L. scariosa**

2 Middle involucral bracts with distinctly chaffy, torn margins.................................. **L. aspera**

Liatris aspera Michx.
TALL GAYFEATHER *native*
IR | **WUP** | **CUP** | EUP CC 4

Plants 4–12 dm tall, the herbage short-hairy, or glabrous throughout. Leaves 25-90 below the inflorescence, the lowermost ones 5–40 cm long (including the long petiole) and 7–45 mm wide, the middle and upper ones gradually reduced and becoming sessile. Heads generally numerous in an elongate spiciform inflorescence, or the peduncles occasionally more elongate and to 5 cm long; terminal head not evidently enlarged; involucre 8–15 mm high, campanulate or subhemispheric, glabrous, its bracts loosely spreading, often purplish upwards, with conspicuous, lacerate, often crisped margins; flowers 16-35 in each head, the corolla hairy within toward the base; pappus barbellate. Aug–Oct.

Dry open places and thin woods, especially in sandy soil.

Liatris pycnostachya Michx.
THICK-SPIKE BLAZING STAR *introduced*
IR | WUP | **CUP** | EUP FAC

Perennial from a woody corm or rootstock, 6–15 dm tall, more or less hirsute in the inflorescence or throughout. Leaves numerous, linear or nearly so, the lowermost ones 10-50 cm long and 3–13 mm wide, reduced upwards. Heads sessile, crowded in an elongate, densely spiciform inflorescence; involucre subcylindric or narrowly turbinate, 8–11 mm high, its bracts tapering to an acuminate, conspicuously squarrose tip, or the inner ones sometimes merely loosely erect; flowers mostly 5–7, or reputedly to 12, the corolla glabrous or nearly so within; pappus strongly barbellate. July–Sept.

Moist or dry prairies and open woods; native west of Michigan, adventive in Schoolcraft County.

Liatris scariosa (L.) Willd.
DEVIL'S-BITE *native*
IR | **WUP** | CUP | EUP CC 5

Plants 3–8 dm tall, the herbage glabrous or pubescent. Leaves 8–20 below the inflorescence, the lowermost 8–25 cm long, including the long petiole, and 2–5 cm wide, the others much reduced and becoming sessile. Heads about 4–30, their peduncles mostly 1–5 cm long; involucre 9–15 mm high, subhemispheric, glabrous or puberulent, its bracts loose, broadly rounded, often purplish above, the middle and inner with narrow scarious margins; flowers mostly 25-40, the corolla hairy toward the base within; pappus barbellate. Aug–Sept.

Dry open places; reported for Keweenaw County.

Matricaria *Mayweed*

Annual or perennial herbs. Leaves alternate, pinnatifid or pinnately dissected. Inflorescence corymbiform, terminating the branches. Heads radiate or discoid, the rays white, pistillate and usually fertile, or sometimes absent. Involucral bracts dry, 2–3-seriate, not much imbricate, with scarious or hyaline margins. Receptacle naked, hemispheric to more commonly conic, or elongate. Disk-corollas yellow, 4–5-toothed. Achenes generally nerved on the margins and ventrally, nerveless dorsally. Pappus a short crown or none.

1 Heads with greenish disk flowers only... **M. discoidea**
1 Heads with white rays; disk flowers yellow.......... 2
2 Receptacle hemispheric; achenes with wing-like ribs.. .... **M. maritima**
2 Receptacle cone-shaped at maturity; achenes ribbed but these not enlarged and wing-like.. **M. chamomilla**

Matricaria chamomilla L.
WILD CHAMOMILE *introduced*
IR | **WUP** | **CUP** | EUP

Matricaria recutita L.

Glabrous branching aromatic annual about 2–8 dm tall. Leaves about 2–6 cm long, bipinnatifid, the ultimate segments linear or filiform. Heads numerous, the disk 6–10 mm wide; rays 10–20, white, 4–10 mm long; disk corollas 5-toothed; receptacle conic. Achenes with 2 nearly marginal and 3 ventral, raised but not at all wing-like ribs, smooth on the back and between the ribs; pappus a short crown or more commonly none. May–Sept.

Native of Europe and Asia, now on roadsides and in waste places.

Similar to *Anthemis cotula*.

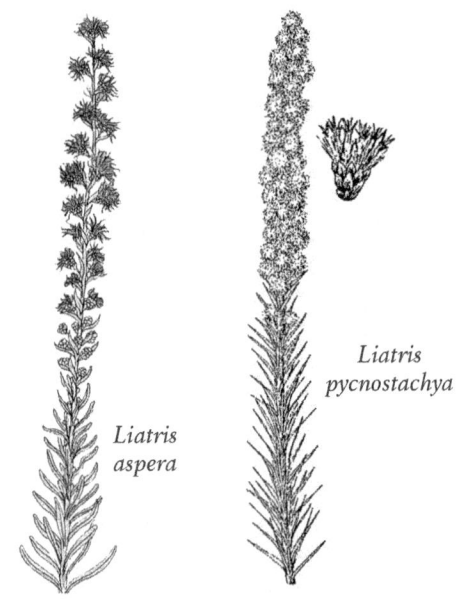

Liatris aspera

Liatris pycnostachya

Matricaria discoidea DC.
PINEAPPLE-WEED *introduced*
IR | **WUP** | **CUP** | **EUP** FACU

Matricaria matricarioides (Less.) Porter

Pineapple-scented glabrous annual. Stems leafy, branching, 5–40 cm tall. Leaves 1–5 cm long, 1–3x pinnatifid, the ultimate segments short, linear or filiform. Heads several or numerous, rayless, the disk about 5–9 mm wide; involucral bracts with broad hyaline margins; disk corollas 4-toothed; receptacle conic, pointed. Achenes with 2 marginal and 1 or several weak nerves; pappus a short crown. May–Sept.

Roadsides and waste places; a common weed.

Matricaria maritima L.
SCENTLESS CHAMOMILE *introduced*
IR | **WUP** | CUP | EUP FAC

Chamomilla inodora (L.) Gilib.
Tripleurospermum inodorum (L.) Sch. Bip., *T. maritimum* subsp. *inodorum* (L.) Appleq., *T. perforata* (Merat) M. Lainz

Annual, biennial, or occasionally perennial, nearly scentless herb 1–6 dm tall, glabrous or nearly so. Leaves 2–8 cm long, bipinnatifid, the ultimate segments mostly elongate, linear or linear-filiform. Heads several or numerous, the disk 8–15 mm wide; rays 12–25, white, 6–13 mm long; disk corollas 5-toothed; receptacle hemispheric, rounded. Achenes with 2 marginal and 1 ventral, thickened, almost wing-like ribs, minutely rugose on the back and between the ribs; pappus a short crown. July–Sept.

Native of Europe, waste places; reported from Baraga County and Isle Royale.

Oclemena *Bog Aster*

Oclemena nemoralis (Ait.) Greene
LEAFY BOG ASTER *native*
IR | **WUP** | **CUP** | **EUP** OBL CC 10

Aster nemoralis Ait.

Rhizomatous perennial. Stems 1–6 dm tall, covered with harsh fine hairs. Leaves all cauline, firm, upper surface scabrous, underside ± puberulent, at least along main veins (and sometimes also glandular), linear or lance-linear to elliptic or oblong, acute or obtuse, entire or nearly so, the margins commonly revolute, sessile, 12–50 mm long and 2–12 mm wide. Heads conspicuously slender-pedunculate, solitary or several in a minutely bracteate corymb-like inflorescence; involucre 5–7.5 mm long, the bracts imbricate, sharply pointed, purplish-tinged or faintly greenish, viscidulous-puberulent to subglabrous; rays 13–27, pink or lilac-purple, 9–15 mm long. Fruit a glandular achene.

Peatlands, especially where underlain by limestone and sphagnum moss is uncommon.

Packera *Groundsel*

Erect perennial herbs (ours). Leaves alternate or from base of plant, stalked near base, stalkless and usually smaller upward. Flower heads with both disk and ray flowers, few to many in clusters at ends of stems; disk flowers perfect and yellow, the rays yellow; involucral bracts in 1 series and not overlapping, of equal lengths; receptacle flat or convex, not chaffy. Fruit an achene, nearly round in section; pappus of slender bristles.

ADDITIONAL SPECIES

Packera insulae-regalis R. R. Kowal (ined.); previously included in *Senecio pauperculus*. Apparently endemic to Isle Royale in rocky openings; believed to be of hybrid origin between *P. paupercula* and octoploid *P. indecora*. Flowering stems similar to those of *P. paupercula*; characters of the basal rosettes are the best way to separate the two species (see key).

1. Blades of basal rosette leaves (except sometimes the smallest) tapering at the base.......... **P. paupercula**
1. Blades of basal rosette leaves cordate or ± truncate (abruptly contracted to the petiole) 2
2. Blades of major basal rosette leaves strongly cordate; colonial from elongated, branching, horizontal rhizomes at surface of ground; rays present and golden yellow. . . .. **P. aurea**
2. Blades of rosette leaves abruptly contracted to the petiole; elongate rhizomes absent; rays absent or present and medium yellow................................ 3
3. Rays absent or less than 6.5 mm long; basal rosette leaf blades glabrous........................ **P. indecora**
3. Rays present, 5–9 mm long; basal rosette leaf blades sparsely arachnoid pubescent; Isle Royale only .... **P. insulae-regalis***

Packera aurea (L.) Á. & D. Löve
HEART-LEAVED GROUNDSEL *native*
IR | **WUP** | **CUP** | **EUP** FACW CC 5

Senecio aureus L.

Perennial herb, from a spreading crown or rhizome. Stems single or clumped, 3–8 dm long, slightly hairy

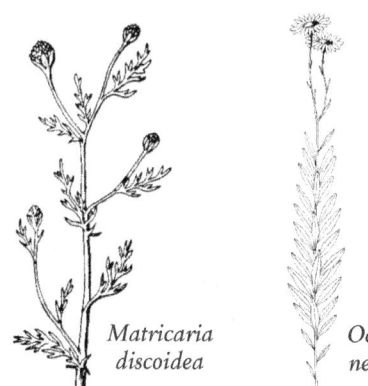

Matricaria discoidea

Oclemena nemoralis

when young, soon becoming smooth. Basal leaves heart-shaped, 5–10 cm long and to as wide, often purple-tinged, on long petioles, the margins with rounded teeth; stem leaves much smaller and more or less pinnately lobed, becoming sessile. Flower heads several to many, the disk 5–10 mm wide, rays gold-yellow, 6–13 mm long; involucre 5–8 mm high, the involucral bracts often purple-tipped. Fruit a smooth achene; pappus of slender white bristles. May–July.

Floodplain forests, wet forest depressions, swamp openings and hummocks, sedge meadows, thickets, fens, ditches.

Packera indecora (Greene) Á. & D. Löve
RAYLESS MOUNTAIN GROUNDSEL   (THR)   native
IR | **WUP** | **CUP** | EUP   FACW CC 10
Senecio indecorus Greene

Fibrous-rooted perennial with a simple or slightly branched crown. Herbage glabrous or soon glabrate except sometimes for sparse tomentum in the axils. Stems 3–8 dm tall. Leaves relatively thin, the basal ones elliptic or broadly ovate, tapering or subtruncate at the base, serrate, evidently petiolate, the blade to 6 cm long and 4 cm wide; stem leaves sharply incised, the lobes irregularly again few-toothed, reduced and becoming sessile upwards. Heads mostly 6–40, rarely fewer, yellow, discoid or rarely with short rays; involucre mostly 7–10 mm high, its bracts often purple-tipped. Achenes glabrous. July–Aug.

Uncommon in moist woods, streambanks, swales, and bogs.

Packera paupercula (Pursh) Á. & D. Löve
RAYLESS ALPINE GROUNDSEL   native
IR | **WUP** | **CUP** | **EUP**   FAC CC 3
Senecio pauperculus Michx.

Fibrous-rooted perennial with a short, simple or slightly branched crown, occasionally also with very short slender stolons. Stems 1–5 dm tall, the herbage lightly floccose-tomentose when young, generally soon glabrate, except frequently at the very base and in the leaf-axils. Basal leaves mostly oblong lance-shaped to elliptic, occasionally suborbicular, generally tapering to the petiolar base, crenate or serrate to subentire, seldom over 12 cm long and 2 cm wide, generally much smaller; stem leaves more or less pinnatifid, the lower sometimes larger than the basal, the others reduced and becoming sessile. Heads relatively few, seldom more than 20, the disk 5–12 mm wide; involucre 4–7 mm high, its bracts carinate-thickened or thin and flat, often purple-tipped; rays 5–10 mm long, yellow, rarely wanting. Achenes glabrous or hispidulous. May–July.

Meadows, prairies, streambanks, beaches, and cliffs; common.

Parthenium *Feverfew*

Parthenium integrifolium L.
WILD QUININE   native
IR | **WUP** | **CUP** | EUP   CC –

Perennial from a tuberous-thickened, usually short root. Stems simple or branched above, finely hairy, 5–10 dm tall. Leaves alternate, large and sometimes few, crenate-serrate, scabrous to nearly glabrous above, or sometimes hirsute along the midrib and main veins, sparsely scaberulous beneath, the basal long-petiolate, with ovate blades 7–20 cm long and 4–10 cm wide; stem leaves with progressively shorter petioles and more or less reduced, the upper leaves often sessile and clasping. Heads radiate, several or numerous in a flat-topped or slightly rounded inflorescence; involucre of 2–4 series of imbricate dry bracts, the disk 4–7 mm wide; rays scarcely 2 mm long, white or whitish. Achenes obovate, black, 3 mm long, their subtending bracts joined at base to the 2 or 3 adjacent receptacular bracts and partly enclosing the achene, the whole falling as a unit. June–Sept.

Prairies, dry woods, roadsides, and along railroads, where likely escapes from cultivation; Alger and Keweenaw counties.

Petasites *Sweet Colt's-Foot*

Perennial herbs, spreading by rhizomes. Leaves mostly from base of plant on long petioles, arrowhead-shaped or palmately lobed, with white woolly hairs on underside; stem leaves alternate, reduced to bracts. Flowering before or as leaves expand in spring, the heads white, the flowers mostly staminate and pistillate and on different plants, the heads sometimes with both staminate and pistillate flowers, the staminate heads usually with disk flowers only, the pistillate heads with all disk flowers or sometimes with short rays; several to many in the inflorescence; involucral bracts in a single series; receptacle not chaffy. Fruit a linear, ribbed achene; pappus of many white, slender bristles.

ADDITIONAL SPECIES
Petasites hybridus (L.) G. Gaertn., B. Mey. & Scherb. European native with large round or kidney-shaped leaves to 4 dm or more wide; locally established in moist places in city of Marquette, Marquette County.

1 Flowers pink-purple, all without rays; introduced species, locally established in Marquette County . . . . **P. hybridus***
1 Flowers creamy white, marginal pistillate ones often with rays; native 2
2 Leaf blades palmately lobed . . . . **P. frigidus**

2   Leaf blades arrowhead-shaped, toothed and not lobed
 .... **P. sagittatus**

Petasites frigidus (L.) Fries
NORTHERN SWEET COLT'S-FOOT　　　native
IR | WUP | CUP | EUP　　　　　　　FACW　CC 10
Petasites palmatus (Ait.) Gray
Perennial herb, spreading by rhizomes. Stems 1–6 dm long, smooth or short-hairy in the head. Leaves mostly from base of plant, triangular to nearly round in outline, palmately lobed, 5–30 cm wide, upper surface green and smooth, underside densely white-hairy, sometimes becoming smooth with age; margins coarsely toothed; petioles of basal leaves 1–3 dm long; stem leaves small and bractlike, 2–6 cm long. Flower heads nearly white, staminate and pistillate flowers mostly on separate plants; rays of pistillate heads to 7 mm long, involucre 4–9 mm high. Fruit a narrow achene; pappus of many slender bristles. May–June.

Wet conifer forests and swamps, wet trails and clearings, aspen woods.

Petasites sagittatus (Banks) Gray
ARROW-LEAF S. COLT'S-FOOT　　(THR) native
IR | WUP | CUP | EUP　　　　　　　FACW　CC 10
Petasites frigidus var. *sagittatus* (Banks ex Pursh) Cherniawsky
Perennial herb. Stems 3–6 dm tall, sparsely covered with woolly white hairs. Leaves mostly from base of plant, arrowhead-shaped, 10–30 cm long and 3–30 cm wide, upper surface smooth to sparsely hairy, densely white hairy below; margins wavy with outward-pointing teeth; petioles 1–3 dm long; the stem leaves reduced in size. Flower heads more or less white; rays of pistillate heads 8–9 mm long. Fruit a linear achene; pappus of slender bristles. May–June.

Uncommon in wet meadows, marshes, sedge meadows, open swamps. Sometimes treated as a var. of *P. frigidus*.

Prenanthes *Rattlesnake-Root*
Perennial herbs with milky juice and tuberous-thickened roots. Leaves alternate. Inflorescence corymbiform or panicle-like to thyrsoid, the heads in most species nodding. Flowers all ligulate and perfect, pink or purple to white, cream-color, or pale yellow. Involucre of 4–15 principal bracts and several much reduced outer ones, or the outer occasionally better developed and almost passing into the inner. Achenes elongate, cylindric or slightly tapering to the summit, glabrous, mostly reddish brown, in our species ribbed-striate. Pappus of numerous deciduous capillary bristles. Our species sometimes placed in genus *Nabalus*.

1   Inflorescence an open panicle; lower leaves on long petioles, broadly ovate to triangular, toothed to deeply palmately lobed........................... **P. alba**
1   Inflorescence a narrow raceme-like panicle; lower leaves spatula-shaped, gradually narrowed to petiole ....
 .... **P. racemosa**

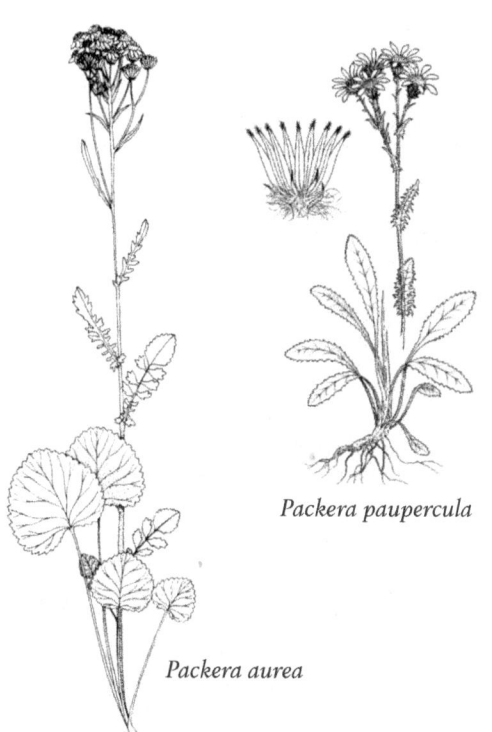

Packera paupercula

Packera aurea

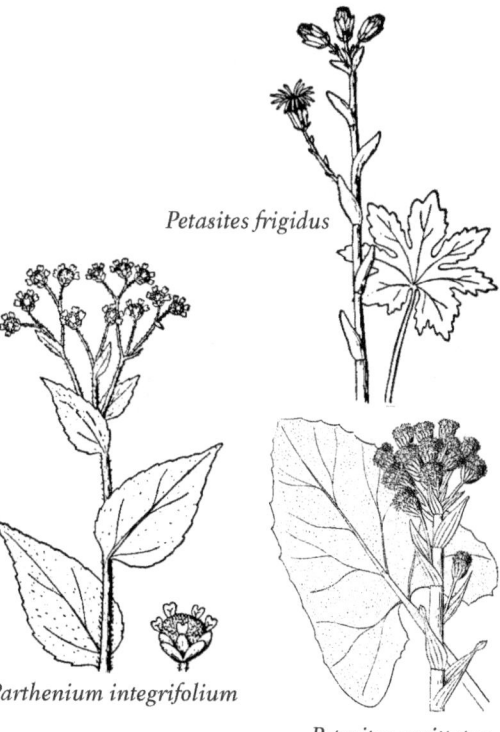

Petasites frigidus

Parthenium integrifolium

Petasites sagittatus

Prenanthes alba L.
WHITE RATTLESNAKE-ROOT native
IR | WUP | CUP | EUP   FACU CC 5
Nabalus albus (L.) Hook.

Perennial herb. Stems stout, commonly 4–15 dm tall, the herbage more or less glaucous. Leaves glabrous above, paler and often hairy beneath, very variable in size and shape, the lower ones long-petioled, palmately few-lobed to sagittate and merely coarsely toothed, becoming smaller, less cut, and less petiolate upwards, the upper leaves often entire. Inflorescence elongate panicle-like, the heads nodding, 10–15-flowered, the flowers fragrant, greenish or yellowish white; involucre 11–14 mm long, generally somewhat purplish, its principal bracts 8, glabrous, but more or less densely papillate with white, waxy-appearing cells; pappus cinnamon-brown. Aug–Sept.

Moist deciduous woods, rock outcrops.

Prenanthes racemosa Michx.
GLAUCOUS WHITE LETTUCE native
IR | WUP | CUP | EUP   FACW CC 8
Nabalus racemosus (Michx.) DC.

Perennial herb. Stems slender, erect, ridged, 4–18 dm tall, smooth and somewhat waxy, hairy in the head. Leaves thick, smooth and waxy; lower leaves oval to obovate, 10–20 cm long and 2–10 cm wide; margins shallowly toothed; petioles long and winged; stem leaves becoming smaller upwards, stalkless and partly clasping the stem. Flower heads many in a narrow, elongate inflorescence, of ray flowers only, pink or purplish; involucre 9–14 mm high, purple-black, long-hairy. Achenes linear; pappus of straw-colored bristles. Aug–Sept.

Sandy or gravelly shores, streambanks, wet meadows, low prairie, fens.

Pseudognaphalium
Rabbit-Tobacco

Biennial glandular herbs (ours), sometimes aromatic. Stems usually erect, woolly-tomentose, sometimes glandular. Leaves basal and along the stem or mostly cauline, alternate, usually sessile; blades mostly narrowly lance-shaped, bases often clasping the stem, margins entire. Inflorescence corymbiform or panicle-like, sometimes a terminal cluster; heads disciform. Corollas yellowish. Fruit a glabrous achene. Pappus of 10–12 barbellate bristles.

1 Leaves decurrent at base; stems glandular-hairy, sometimes also woolly hairy .... **P. macounii**
1 Leaves not decurrent at base .... **P. obtusifolium**

Prenanthes alba

Prenanthes racemosa

Pseudognaphalium macounii (Greene) Kartesz
CLAMMY RABBIT-TOBACCO native
IR | WUP | CUP | EUP   CC 2
Gnaphalium macounii Greene
Pseudognaphalium helleri (Britt.) A. Anderb.
Pseudognaphal. micradenium (Britt.) A. Anderb.

Similar to *P. obtusifolium*. Stems glandular-hairy, becoming woolly in the inflorescence, rarely somewhat woolly to near the base, as well as glandular. Leaves distinctly decurrent at the base, the upper surface glandular-hairy, the lower surface usually woolly, or sometimes glandular-hairy. July-Sept.

Common in dry, open places.

Pseudognaphalium obtusifolium (L.) Hilliard & Burtt
FRAGRANT RABBIT-TOBACCO native
IR | WUP | CUP | EUP   CC 2
Gnaphalium obtusifolium L.

Annual or perhaps sometimes biennial, fragrant, 1–8 dm tall, erect. Stems thinly white-woolly, commonly becoming subglabrous or sometimes a little glandular toward the base. Leaves numerous, linear lance-shaped, up to about 10 cm long and 1 cm wide, sessile, white-woolly beneath, green and from glabrous to slightly glandular or slightly woolly above. Inflorescence branched and many-headed except in depauperate plants, flat or round-topped and often elongate, the final clusters with the heads somewhat glomerate. Involucre yellowish white or somewhat dingy, campanulate, woolly only near the base, 5–7 mm high; pappus bristles distinct, falling separately. Achenes glabrous. July–Oct.

Open, often sandy places.

When crushed, plants have a characteristic maple syrup scent.

Ratibida *Coneflower*

Ratibida pinnata (Vent.) Barnh.
GLOBULAR CONEFLOWER native
IR | **WUP** | **CUP** | **EUP**   CC 4

Perennial from a stout woody rhizome or sometimes a short caudex. Stems mostly 4–12 dm tall, strigose above, strigose or more commonly spreading-hirsute below. Leaves alternate, pinnatifid, loosely hirsute, the segments lance-shaped, coarsely toothed to entire. Heads several or occasionally solitary, naked-pedunculate; involucre a single series of green, sub-herbaceous, linear or lance-linear bracts; receptacle columnar, about 12–20 mm high, its chaffy bracts subtending the rays as well as the disk-flowers; rays about 5–10, pale yellow, mostly 2.5–6 cm long, spreading or reflexed; style-appendages elongate, acuminate. Achenes smooth, compressed at right angles to the involucral bracts; pappus none. June–Aug.

Prairies and dry woods.

Rudbeckia *Coneflower*

Perennial herbs. Stems and leaves rough-hairy. Leaves alternate. Flower heads with both disk and ray flowers, the rays yellow to orange; involucral bracts green, overlapping; receptacle rounded, chaffy. Fruit a smooth, 4-angled achene; pappus none or a short crown. The genus includes the well-known black-eyed Susan (*Rudbeckia hirta*), widespread in dry places.

1 Main leaves deeply lobed 2
1 Leaves unlobed 3
2 Disk yellow; stems glabrous or nearly so; largest leaves 5–7 lobed . . . . **R. laciniata**
2 Disk dark purple-red; stems pubescent; largest leaves 3-lobed . . . . **R. triloba**
3 Leaves ovate, margins toothed; chaff of receptacle glabrous, tapered to a sharp prolonged point; Houghton County . . . . **R. triloba**
3 Leaves and margins various; chaff of receptacle not tapered to a sharp point; widespread . . . . **R. hirta**

Rudbeckia hirta L.
BLACK-EYED SUSAN native
IR | **WUP** | **CUP** | **EUP**   FACU CC 1

Biennial or short-lived perennial, sometimes flowering the first year. Stems 3–10 dm tall, more or less hirsute throughout. Leaves variable in size and shape,

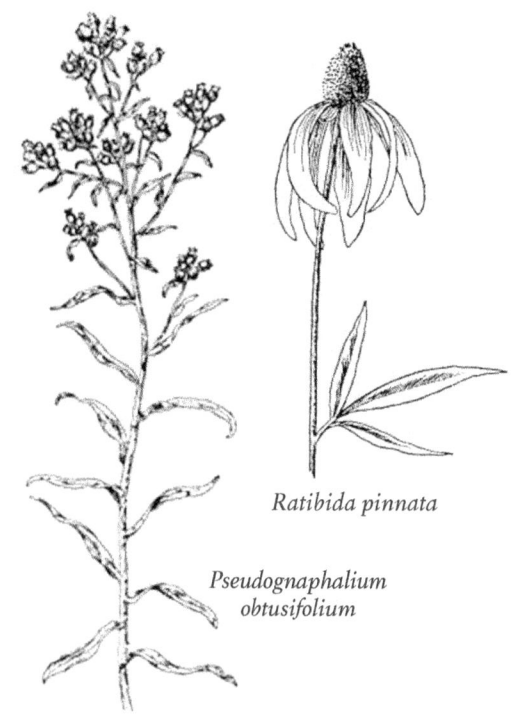

Ratibida pinnata

Pseudognaphalium obtusifolium

toothed or subentire, the basal and lower stem leaves mostly oblong lance-shaped to elliptic and long-petiolate, the others lance-linear to oblong or ovate, mostly sessile. Heads several or solitary, mostly long-pedunculate, the disk hemispheric or ovoid, 12–20 mm wide, dark purple or brown, rarely yellow; involucral bracts copiously hirsute, nearly equal, green, spreading, sometimes elongate and equaling the rays; rays about 8–20, orange or orange-yellow, sometimes darker or marked with purple near the base, commonly 2–4 cm long; receptacular bracts more or less hispid near the tip, often also ciliate on the margins. Achenes quadrangular; pappus none. June–Oct.

Various habitats, common in disturbed or waste places, meadows, and roadsides.

Rudbeckia laciniata L.
CUTLEAF CONEFLOWER native
IR | **WUP** | **CUP** | **EUP**   FACW CC 6

Perennial herb, from a woody base. Stems branched, 5–30 cm tall, smooth and often waxy. Leaves alternate, to 30 cm wide, deeply lobed, nearly smooth to hairy on underside; margins coarsely toothed as well as lobed, or entire on upper leaves; petioles long on lower leaves, becoming short above. Flower heads several to many at ends of stems, with both disk and ray flowers, disk flowers green-yellow, rays lemon-yellow, drooping, 3–6 cm long; involucral bracts of unequal lengths; receptacle round at first, becoming cylindric. Fruit a 4-angled achene; pappus a short toothed crown. July–Sept.

Floodplain forests, swamps, streambanks, thickets, ditches; usually in partial or full shade.

Rudbeckia triloba L.
THREE-LOBED CONEFLOWER *native*
IR | **WUP** | CUP | EUP   FACU CC 5

Short-lived perennial. Stems mostly 5–15 dm tall, moderately spreading-hirsute to subglabrous. Leaves thin, sharply toothed to subentire, moderately appressed hairy or nearly glabrous, the basal leaves broadly ovate or subcordate and long-petiolate, the stem leaves mostly narrower and short-petiolate or sessile, usually some of the larger leaves deeply 3-lobed. Heads several or numerous, the disk dark purple, hemispheric or ovoid, 8–15 mm wide; involucral bracts narrow, nearly equal, green and more or less leafy, spreading or reflexed, mostly strigose and ciliate-margined; rays 6–12, yellow, or partly or wholly orange, 1–3 cm long; receptacular bracts glabrous, equaling or usually a little exceeding the disk corollas, abruptly narrowed to a distinct awn-point. Achenes equably quadrangular; pappus a minute crown. July–Oct.

Woods and moist soil; Houghton County. Our form is var. *triloba*, with the larger leaves merely 3-lobed.

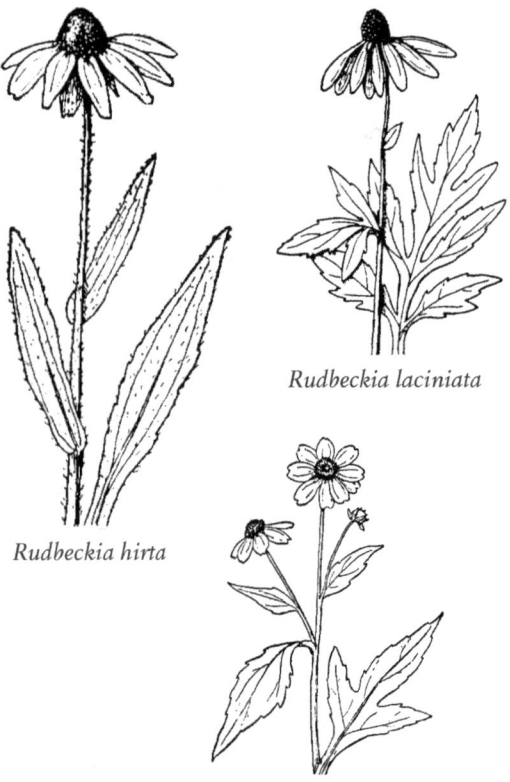

Rudbeckia laciniata

Rudbeckia hirta

Rudbeckia triloba

Senecio *Groundsel; Ragwort*

Erect annual herbs (ours). Leaves alternate or from base of plant, stalked near base, stalkless and usually smaller upward. Flower heads with both disk and ray flowers, few to many in clusters at ends of stems; disk flowers perfect and yellow, the rays yellow; involucral bracts in 1 series and not overlapping, of equal lengths; receptacle flat or convex, not chaffy. Fruit an achene, nearly round in section; pappus of slender bristles. Several former members of this genus now placed in *Packera*.

1. Rays absent; bracts below involucral bracts with distinct black tips .... **S. vulgaris**
1. Rays present but tiny; bracts below involucral bracts not black-tipped; Gogebic County........... **S. sylvaticus**

Senecio sylvaticus L.
WOODLAND RAGWORT *introduced*
IR | **WUP** | CUP | EUP

Annual taprooted herb. Stems 1.5–8 dm tall, generally simple to the inflorescence, leafy throughout. Herbage pubescent with crisp loose hairs, or subglabrous, scarcely or not at all glandular. Leaves all more or less pinnatifid and irregularly toothed, 2–12 cm long and 4–40 mm wide. Heads several or numerous, the disk 3–7 mm wide involucre 5–7 mm high; bracteoles inconspicuous or wanting, not black-tipped; rays much reduced, less than 2 mm long; pappus copious, equaling or surpassing the disk corollas. Achenes canescent. July–Sept.

Native of Europe, established in dry soil and waste places; Gogebic County.

Senecio vulgaris L.
OLD-MAN-IN-THE-SPRING *introduced*
IR | **WUP** | **CUP** | **EUP**   FACU

Simple or strongly branched annual with a more or less evident taproot. Stems 1–4 dm tall, leafy throughout; herbage sparsely crisp-hairy or subglabrous. Leaves coarsely and irregularly toothed or more often pinnatifid, 2–10 cm long and 5–45 mm wide, the lower tapering to the petiole, the upper sessile and clasping. Heads several or numerous, strictly discoid, the flowers all tubular and perfect; disk usually 5–10 mm wide; involucre about 5–8 mm high; bracteoles well developed, black-tipped; pappus very copious, equaling or generally surpassing the corollas. Achenes strigillose, chiefly along the angles. May–Oct.

Native of the Old World; waste places.

Silphium *Rosinweed*

Tall perennial herbs, with resinous juice. Leaves opposite or all from base of plant, broadly ovate. Flower heads with yellow disk and ray flowers, in clusters at ends of stems; involucral bracts overlapping, receptacle more or less flat, chaffy. Fruit an achene; pappus none or of 2 small scales from top of achene.

1 Leaves alternate, deeply lobed .... **S. laciniatum**
1 Leaves basal, unlobed .... **S. terebinthinaceum**

Silphium laciniatum L.
COMPASS-PLANT (THR) native
IR | WUP | **CUP** | EUP CC 9

Coarse perennial with a woody taproot. Stems 1.5–3 m tall, hispid or hirsute with spreading hairs, sometimes slightly glandular as well. Leaves alternate, deeply pinnatifid, hirsute chiefly along the midrib and main veins beneath, the lower very large, sometimes 4 dm long, progressively reduced upwards, the uppermost entire and well under 1 dm long. Heads several or numerous in a narrow sometimes raceme-like inflorescence, large, the disk commonly 2–3 cm wide; involucre scabrous-hispid, commonly 2–4 cm long, exceeding the disk, its bracts ovate, squarrose, not much imbricate; rays about 15–30, 2–5 cm long. July–Sept.

Prairies; rare in Delta and Menominee counties.

The large basal leaves tend to align themselves facing in an east-west direction.

Silphium terebinthinaceum Jacq.
BASAL-LEAVED ROSINWEED native
IR | WUP | **CUP** | EUP FAC CC 6
Silphium rumicifolium Small

Taprooted perennial herb. Stems 1–3 m long, branched above, more or less leafless, smooth. Main leaves all from base of plant, ovate, leathery, 1–5 dm long and 1–3 dm wide, usually heart-shaped at base, usually rough-to-touch, margins sharply toothed, petioles long; stem leaves few and reduced to large bracts. Flower heads many in an open inflorescence, with yellow disk and ray flowers, the disk 1.5–3 cm wide, the rays 2–3 cm long; involucre 1–2.5 cm high, the involucral bracts smooth, loose and overlapping. Achenes obovate, narrowly winged; pappus reduced to 2 small teeth at top of achene. July–Sept.

Low prairie, fens; especially where calcium-rich; Delta and Menominee counties.

Solidago *Goldenrod*

Erect perennials, spreading by rhizomes or from a crown. Leaves alternate, margins entire or toothed. Flower heads small, many, in flat-topped (corymb-like), rounded (panicle-like) or spike-like clusters at ends of stems; the flowers sometimes mostly on 1 side of inflorescence branches (secund) in species with panicle-like heads; the heads with yellow disk and ray flowers; involucral bracts in several overlapping series, papery at base and tipped with green;

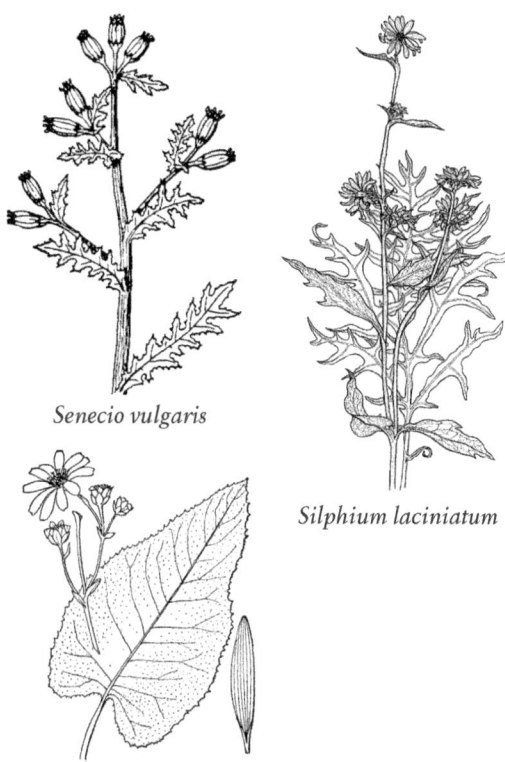

Senecio vulgaris

Silphium laciniatum

Silphium terebinthinaceum

receptacle flat or convex, not chaffy. Fruit an achene, angled or nearly round in cross-section; pappus of many slender white bristles. *Solidago houghtonii*, *S. ohioensis*, *S. ptarmicoides*, and *S. rigida* are the flat-topped goldenrods, sometimes separated into genus *Oligoneuron*.

1 Heads in a more or less flat-topped cluster at end of stem 2
1 Heads in an elongate or pyramid-shaped cluster 5
2 Leaf blades of middle and upper stem ovate to elliptic; stems and leaves densely hairy; common species of dry to mesic habitats; Delta and Menominee counties .... **S. rigida**
2 Leaf blades linear to lance-shaped or oblong lance-shaped, glabrous apart from rough leaf margins; stems glabrous or nearly so, or slightly hairy below inflorescence 3
3 Rays white, 12 or more; upper stem leaves broadest above their middle .... **S. ptarmicoides**
3 Rays yellow, 10 or less; upper stem leaves broadest at or below middle 4
4 Rays 1.5–3 mm long and involucre 3.5–6 mm long; pedicels smooth and glabrous or rough-hairy......... .... **S. ohioensis**
4 Rays mostly 3–5 mm long and involucre ca. 5–8 mm long; pedicels rough-hairy; rare species near shores of Lakes Huron and Michigan .... **S. houghtonii**

5 Inflorescence terminal, usually more or less pyramid-shaped and slightly nodding at top; inflorescence branches curving; the heads mostly on upper side of the branches 6
5 Flower heads spiraled around branches of inflorescence and not all on one side of branch 15
6 Stem leaves with 3 prominent veins (midrib plus 2 distinct lateral veins) 7
6 Stem leaves with prominent midrib and weaker lateral veins 11
7 Stem pubescent for all or most of its length 9
7 Stem glabrous (or nearly so) below inflorescence ... 10
8 Involucres 2-3 mm long. .... **S. canadensis**
8 Involucres 3-6 mm long 9
9 Mid to upper leaves serrate, glabrous or scabrous above, pubescent on the veins beneath; stem pilose chiefly above the middle .... **S. canadensis**
9 Mid to upper leaves minutely serrate to entire, scabrous above, densely pubescent beneath; stem grayish with close puberulence throughout, except sometimes near the base .... **S. altissima**
10 Basal leaves absent; stem leaves elliptic, withering by flowering time, numerous, not reduced in size; distinctly 3-nerved; inflorescence branches densely hairy; flowering in Aug–Sept .... **S. gigantea**
10 Basal leaves present; basal and lower stem leaves oblong lance-shaped to elliptic, with long petioles, persistent, middle and upper stem leaves few, smaller than basal leaves; leaves obscurely 3-nerved; inflorescence branches glabrous or nearly so; plants begin flowering in July .... **S. juncea**
11 Stems pubescent, at least on upper half. 12
11 Stems glabrous (or sometimes with fine hairs in inflorescence) 13
12 Margins of stem leaves entire or with rounded teeth; stems and leaves finely and densely short-hairy; lower leaves oblong lance-shaped, tapered to a winged petioles; upper stem leaves reduced in size; plants of dry habitats .... **S. nemoralis**
12 Margins sharply toothed; lower leaves elliptic to lance-shaped, same size as upper stem leaves, but withering during season; plants of moist or shaded places .... **S. rugosa**
13 Bases of lowest stem leaves clasping stem; wet habitats .... **S. uliginosa**
13 Lower leaf bases not clasping stem 14
14 Stems strongly ridged; leaf upperside very rough-to-touch; wet habitats; Menominee County .... **S. patula**
14 Stems round with only small ridges; leaf upperside smooth or only slightly roughened; dry to mesic meadows and woods .... **S. juncea**
15 Basal and lower stem leaves smaller than mid-stem leaves; middle and upper stem leaves with sharp teeth, longer than the axillary inflorescences. 16
15 Basal and lower stem leaves larger than mid-stem leaves; middle and upper stem leaves entire or with rounded teeth, not longer than inflorescences. 17
16 Stems ribbed, zigzagged from node to node; leaves ovate, narrowed at base to a winged petiole; common. .... **S. flexicaulis**
16 Stems not ribbed or zigzagged; leaves narrowly elliptical, sessile; Mackinac County .... **S. caesia**
17 Stem and leaves pubescent 18
17 Stems glabrous or nearly so, sparse fine hairs may be present in inflorescence 19
18 Achenes glabrous .... **S. hispida**
18 Achenes densely covered with upward-pointing appressed hairs. .... **S. simplex**
19 Lower stem leaves (including petioles) mostly 7–16 times longer than wide; petioles clasping stem; wet habitats .... **S. uliginosa**
19 Lower stem leaves usually 3–8 times longer than wide; petioles not clasping stem; dry, often sandy habitats .. .. **S. speciosa**

Solidago altissima L.
TALL GOLDENROD native
IR | WUP | CUP | EUP FACU CC 1
Solidago canadensis var. *scabra* Torr. & Gray
Perennial from creeping rhizomes. Stems 5–20 dm tall, usually short-hairy throughout. Lower stem leaves usually withered by flowering; sessile or subpetiolate; blades oblong lance-shaped, 5–15 cm long and 7–20 mm wide, relatively thick and firm, entire to finely serrate, strongly 3-nerved, upper surface finely strigose, underside scabrous; much reduced upwards. Heads many, in a secund, pyramidal, panicle-like inflorescence, branches divergent and recurved; involucres narrowly campanulate, 2.5–4.5 mm long, the bracts in 3 series, unequal; ray florets 8–13; disk florets 3–6; corollas 2–4 mm long. Achenes sparsely to moderately hairy.

Many types of wet to dry habitats.

S. altissima is sometimes treated as *S. canadensis* var. *scabra*. The short hairs on the leaves give fresh plants a gray-green color not seen in *S. canadensis*. Subject to insect galls on the stems.

Solidago caesia L.
WREATH GOLDENROD native
IR | WUP | CUP | **EUP** FACU CC 6
Perennial from a short stout caudex-like rhizome with abundant fibrous roots, sometimes with long-creeping rhizomes as well. Stems 3–10 dm tall, glaucous. Leaves serrate, glabrous or slightly hairy above and along the midrib beneath; basal and lowermost stem leaves deciduous by flowering time, smaller than those above, which are lance-shaped, 6–12 cm long and 1–3 cm wide. Inflorescence an series of axillary clusters, all but the uppermost of which are exceeded by their narrow subtending leaves, or in robust forms many of the clusters elongate and

themselves leafy-bracteate; involucre glabrous, about 3–4.5 mm high, its bracts narrow, obtuse or rounded, obscurely several-nerved; rays mostly 3–4, occasionally 5. Achenes hairy. Sept–Oct.

Deciduous forests mostly near Lakes Huron and Michigan in Mackinac County.

Solidago canadensis L.
COMMON GOLDENROD   native
IR | WUP | CUP | EUP   FACU CC 1

Perennial from creeping rhizomes, without a well-developed caudex. Stems 3–13 dm tall, more or less puberulent at least above the middle. Leaves thin, sharply serrate to subentire, glabrous or slightly scabrous above, commonly finely hairy on the midrib and main veins beneath; basal leaves absent, or, like the lower stem leaves, reduced and soon deciduous; mid- and upper stem leaves numerous and crowded, only gradually reduced upwards, lance-linear, tapering to the sessile base, 3-nerved, 5–13 cm long and 5–18 mm wide. Inflorescence terminal, panicle-like, with conspicuously recurved-secund branches; involucre about 2–3 mm high, its bracts imbricate in several series, yellowish, without well defined green tips; rays mostly 10–17, only about 1–1.5 mm long. Achenes short-hairy. July–Sept.

Open, moist or dry places; common.

Solidago flexicaulis L.
ZIGZAG GOLDENROD   native
IR | WUP | CUP | EUP   FACU CC 6

Perennial with creeping rhizomes. Stems 3-12 dm tall, grooved, glabrous below the inflorescence. Leaves sharply and often coarsely serrate or dentate, hirsute beneath, at least on the midrib and main veins, or rarely glabrous, glabrous or sparsely hairy above; basal and lowermost stem leaves deciduous by flowering time; upper leaves ovate to elliptic, 7–15 cm long and 3–10 cm wide, abruptly contracted to the winged petiole. Inflorescence a series of short raceme-like clusters, the lower in the axils of ordinary scarcely reduced foliage leaves, these progressively reduced upwards, the uppermost becoming inconspicuous and shorter than their axillary clusters; involucre 4–6 mm. high, its bracts strongly imbricate, glabrous; rays 3–4. Achenes short-hairy. Aug–Oct.

Woods.

Solidago gigantea Ait.
SMOOTH GOLDENROD   native
IR | WUP | CUP | EUP   FACW CC 3

Solidago serotina Ait. non Retz.

Perennial herb, from stout rhizomes, often forming colonies. Stems 0.5–2 m tall, mostly smooth, sometimes waxy, short-hairy on upper branches. Leaves lance-shaped to oval, 6–15 cm long and 1–4 cm wide, prominently 3-veined, tapered to a stalkless or short petiolelike base, glabrous or sparsely hairy on underside veins; margins with sharp, forward-pointing teeth. Flower heads many, in large panicle-like clusters, on 1 side of the spreading branches (secund), with yellow disk and ray flowers, the rays 2–3 mm long; involucre 2–5 mm high, the involucral bracts linear. Achenes 1–2 mm long. July–Sept.

Wet meadows, streambanks, swamps, floodplain forests, thickets, marshes, calcareous fens, ditches; also in moist to dry open woods and roadsides.

Canada goldenrod (*Solidago canadensis*) similar but generally smaller and densely short-hairy on leaf undersides and upper stem.

Solidago hispida Muhl.
HAIRY GOLDENROD   native
IR | WUP | CUP | EUP   CC 3

Perennial with a stout branched caudex and fibrous roots. Stems 1–10 dm tall, the herbage generally spreading-hirsute throughout. Basal and lowermost stem leaves well developed and generally persistent, broadly oblong lance-shaped, crenate or serrate to entire, petiolate, the blade and petiole 3–20 cm long and 1–5 cm wide; stem leaves reduced upwards and becoming sessile. Inflorescence terminal, elongate and narrow, generally more or less leafy-bracteate toward the base, the lower clusters often elongate and stiffly ascending, but not secund; involucre 4–6 mm high, the bracts imbricate in several series, yellowish; rays about 7–14, usually deep yellow. Achenes glabrous, at least when mature. July–Oct.

Dry woodlands and rocky shores.

Solidago houghtonii Torr. & Gray
HOUGHTON'S GOLDENROD   (THR) native
IR | WUP | CUP | EUP   CC 10

Oligoneuron houghtonii (Torr. & Gray) Nesom

Native perennial herb, from a crown. Stems slender, 2–5 dm long, smooth below, short-hairy in the inflorescence. Leaves alternate, largest at base of plant, becoming smaller upward, linear, to 20 cm long and 1–2 cm wide; margins entire but rough-to-touch; lower leaves tapered to a petiolelike base, petioles absent on upper leaves. Flower heads often many, crowded in flat-topped to rounded clusters at ends of stems, with yellow disk and ray flowers; involucre 5–8 mm high, the involucral bracts rounded at tip. Fruit a smooth, angled achene; pappus of slender, sometimes feathery hairs. July–Sept.

Moist flats and depressions between dunes near northern shores of Lakes Huron and Michigan, also in nearby fens, limestone pavements on Drummond Island; usually where calcium-rich. Rare in UP (Chippewa, Mackinac, Schoolcraft counties) and northern LP of Mich; also known from western NY and southern Ontario (total range for this species); federally listed as threatened.

Solidago juncea Ait.
EARLY GOLDENROD — native
IR | **WUP** | **CUP** | EUP — CC 3

Perennial with a stout branched caudex and fibrous roots, frequently with long creeping rhizomes as well. Stems 3–12 dm tall, essentially glabrous throughout except for the scabrous or ciliate leaf-margins, but sometimes short-hairy on one or both surfaces or in the inflorescence. Basal leaves tufted and persistent, 15–40 cm long and 2–7.5 cm wide, with narrowly elliptic, serrate blades tapering to the long petiole; stem leaves progressively reduced, becoming sessile. Inflorescence terminal, panicle-like, dense, generally about as wide as long or even wider, with recurved-secund branches; involucre glabrous, 3–5 mm high, its bracts imbricate; rays minute, usually 7–12. Achenes persistently short-hairy. June–Oct.

Dry open places and open woods, especially in sandy soil.

One of the earliest goldenrods to flower.

Solidago nemoralis Ait.
GRAY GOLDENROD — native
IR | **WUP** | **CUP** | EUP — CC 2

Perennial with a branching caudex and fibrous roots. Stems 1–10 dm tall, the herbage densely puberulent with loosely spreading hairs. Leaves weakly 3-nerved; basal leaves well-developed, tufted and persistent, oblong lance-shaped, long-petiolate, 5–25 cm long and 8–40 mm wide, toothed; stem leaves progressively reduced, less petiolate, and less toothed upwards, the lowermost similar to the basal, but often deciduous. Inflorescence terminal, panicle-like, sometimes elongate and nodding at the apex; sometimes larger with long, recurved, secund branches; involucre 3–6 mm high, its bracts imbricate in several series, glabrous except for the ciliolate margins; rays short, 5–9. Achenes pubescent. Aug–Oct.

Dry woods and open places, especially in sandy soil.

Solidago ohioensis Frank
OHIO FLAT-TOPPED GOLDENROD — native
IR | **WUP** | **CUP** | **EUP** — CC 8

Oligoneuron ohioense (Frank) G.N. Jones

Perennial herb, from a crown. Stems 5–10 dm tall, smooth. Leaves largest at base of plant and becoming smaller upward, lance-shaped to oblong lance-shaped, to 2 dm long and 1–5 cm wide, pinnately-veined, margins entire or slightly toothed near tip, rough-to-touch; tapered to a long petiole on lower leaves, upper leaves sessile. Flower heads many in a branched, flat-topped to rounded inflorescence at ends of stems, with yellow disk and ray flowers; involucre smooth, 4–5 mm high, the involucral bracts rounded at tip. Achenes glabrous. July–Sept.

Wet, sandy or gravelly shores, streambanks, sedge meadows, calcareous fens, low prairie; soils often calcium-rich.

Solidago patula Muhl.
ROUGH-LEAVED GOLDENROD — native
IR | WUP | **CUP** | EUP — OBL CC 6

Perennial herb, from a crown. Stems 5–20 dm tall, lower stem smooth, strongly angled, upper stem short-hairy. Leaves largest at base of plant, 1–3 dm long and 4–10 cm wide, becoming smaller upward, oval to ovate, pinnately veined, upper surface rough-to-touch, underside smooth; margins with large, forward-pointing teeth; petioles long and winged on lower leaves, upper leaves sessile. Flower heads in a panicle-like head, the branches spreading and curved downward at tip, flower heads mostly on 1 side of branches, with yellow disk and ray flowers; involucre 3–4 mm high, the involucral bracts tapered to a sharp or rounded tip. Achenes sparsely hairy. Aug–Sept.

Swamps, thickets, calcareous fens, sedge meadows; Menominee County.

Solidago ptarmicoides (Nees) Boivin
PRAIRIE FLAT-TOPPED GOLDENROD — native
IR | **WUP** | **CUP** | **EUP** — CC 6

Aster ptarmicoides (Nees) Torr. & Gray
Oligoneuron album (Nutt.) Nesom

Perennial with a branched caudex and fibrous roots, the old leaf-bases persisting and becoming chaffy-fibrous. Stems 1–7 dm tall, scabrous at least above. Leaves firm, glabrous or scabrous, entire or with a few remote teeth, commonly 3-nerved, 3–20 cm long and 1.5–10 mm wide, the lower narrowly lance-shaped and petioled, sometimes tufted, persistent, and larger than those above; upper leaves becoming sessile and linear. Heads 3–60 in an open, minutely bracteate, corymbiform inflorescence; involucre 4–7 mm high, glabrous, its bracts imbricate, often with strongly thickened; rays 10–25, white, 5–9 mm long. Achenes several-nerved, glabrous. July–Sept.

Prairies and other open, usually dry places.

Solidago rigida L.
STIFF GOLDENROD — native
IR | WUP | **CUP** | EUP — CC 5

Oligoneuron rigidum (L.) Small

Perennial from a stout branched caudex. Stems 2.5–15 dm tall. Herbage densely pubescent with short spreading hairs, but sometimes nearly glabrous. Leaves firm, slightly toothed or entire, the basal and lowermost cauline leaves well developed and usually persistent, with elliptic, or broadly lance-shaped blades 6–25 cm long and 2–10 cm wide, often exceeded by the long petiole; stem leaves progressively reduced and less petiolate upwards, the middle ones sessile or nearly so. Inflorescence terminal, dense, corymbiform; heads large and many-flowered; involucre 5–9 mm high, its bracts conspicuously striate; rays 8–14. Achenes 10–15 nerved, glabrous, or hairy at the tip. July–Oct.

Prairies and other dry open places, especially in sandy soil; Delta and Menominee counties.

Solidago rugosa P. Mill.
WRINKLE-LEAVED GOLDENROD *native*
IR | WUP | **CUP** | **EUP** FAC CC 3

Perennial from long creeping rhizomes, without a well developed caudex. Stems 3–15 (rarely 25) dm tall, spreading-hairy. Basal and lowermost stem leaves reduced and soon deciduous; upper leaves numerous, crowded, and only gradually reduced upwards, rugose-veiny, not 3-nerved, glabrous or scabrous above, hirsute on the midrib and main veins beneath, and usually over the surface as well, lance-ovate, strongly serrate, subsessile, not clasping, the larger leaves 3.5–13 cm long and 13–40 mm wide. Inflorescence terminal, panicle-like, usually ample and spreading, with recurved-secund branches much exceeding their reduced subtending leaves; involucre 2.5–4 mm high, its bracts imbricate; rays 6–11. Achenes short-hairy. Aug–Oct.

Moist woods; swamps and peatlands (especially on margins); fields, fencerows, ditches; often in disturbed areas in forests and swamps as along trails and in clearings.

Solidago simplex Kunth
MOUNT ALBERT GOLDENROD *native*
IR | **WUP**| **CUP** | **EUP** FACU CC 10

Solidago spathulata DC.

Perennial with a short usually branched caudex and fibrous roots. Stems 1–9 dm tall, glabrous or nearly so except for fine hairs in the inflorescence (or sometimes more hairy throughout in var. gillmanii). Basal and lower stem leaves narrowly oblong lance-shaped, 2–30 cm long (including the petiole), 4–40 mm wide, toothed to nearly entire, often irregularly ciliolate-margined, usually persistent; stem leaves progressively reduced, becoming sessile upwards. Inflorescence terminal, varying from few-headed to dense, but not secund, its leaves reduced and inconspicuous. Involucre 3–9 mm high, often glutinous, its bracts imbricate in several series; rays mostly 9–10. Achenes short-hairy. Aug–Oct.

Sand dunes and beaches.

Plants of the rare var. *gillmanii* are robust, mostly 3–9 dm tall, with large heads (involucre 6–9 mm high) in a long, often branched inflorescence, and reported from Schoolcraft County.

Solidago speciosa Nutt.
SHOWY GOLDENROD *native*
IR | **WUP**| **CUP** | EUP CC 5

Perennial with a stout woody caudex and fibrous roots. Stems 3–15 dm tall, coarsely puberulent in the inflorescence, otherwise glabrous or slightly scabrous. Leaves thick and firm, numerous, entire or the lower slightly toothed, sometimes gradually increasing in size toward the base, the persistent lower ones abruptly petiolate, to as much as 30 cm long and 10 cm wide, sometimes all smaller and nearly uniform in size, the lower then generally deciduous. Inflorescence terminal, not secund, usually with crowded, stiffly ascending branches, sometimes becoming looser and more open; the heads conspicuously pedicellate; involucre 3–5 mm high, its bracts imbricate, glutinous, yellowish; rays 6–8, 3–5 mm long. Achenes glabrous, seldom over 2 mm long. Aug–Oct.

Open woods, fields, prairies, and plains.

Solidago uliginosa Nutt.
NORTHERN BOG-GOLDENROD *native*
IR | **WUP**| **CUP** | **EUP** OBL CC 4

Perennial herb, from a branched crown. Stems stout, 5–15 dm long, glabrous but finely hairy in the inflorescence. Leaves largest at base of plant, 5–35 cm long and 1–5 cm wide, becoming smaller upwards, lance-shaped to oblong lance-shaped, glabrous; lower leaves tapered to long petioles, somewhat clasping stem, upper leaves stalkless; margins finely toothed, or entire on upper leaves, rough-to-touch. Inflorescence long, crowded, and spike-like, the branches ascending, straight or curved downward at tip, the heads sometimes secund; involucre 3–5 mm high. Achenes more or less glabrous. Aug–Sept.

Common in conifer swamps, fens, open bogs, low prairie, wet meadows, interdunal wetlands, Lake Superior rocky shore.

Sonchus *Sow-Thistle*

Annual or perennial herbs with milky juice. Leaves alternate or all basal, entire to pinnatifid or dissected, mostly auriculate, often prickly-margined. Heads solitary to usually several or many in an irregular corymb-like inflorescence. Flowers all ligulate and perfect, yellow, few to more often numerous (our species with usually ca. 120–160 flowers in each head). Involucre ovoid or campanulate, its bracts generally imbricate in several series. Achenes flattened, about 6–20-ribbed, beakless, glabrous. Pappus of numerous white capillary bristles which tend to fall connected, sometimes with a few stouter ones which fall separately. The perennial species are troublesome farm weeds.

1 Perennial with creeping rhizomes; leaf bases auriculate and clasping stem, the auricles small ...... **S. arvensis**
1 Taprooted annuals; leaf bases auriculate and clasping stem, the auricles large and conspicuous 2

106 ASTERACEAE | Aster Family

dicots

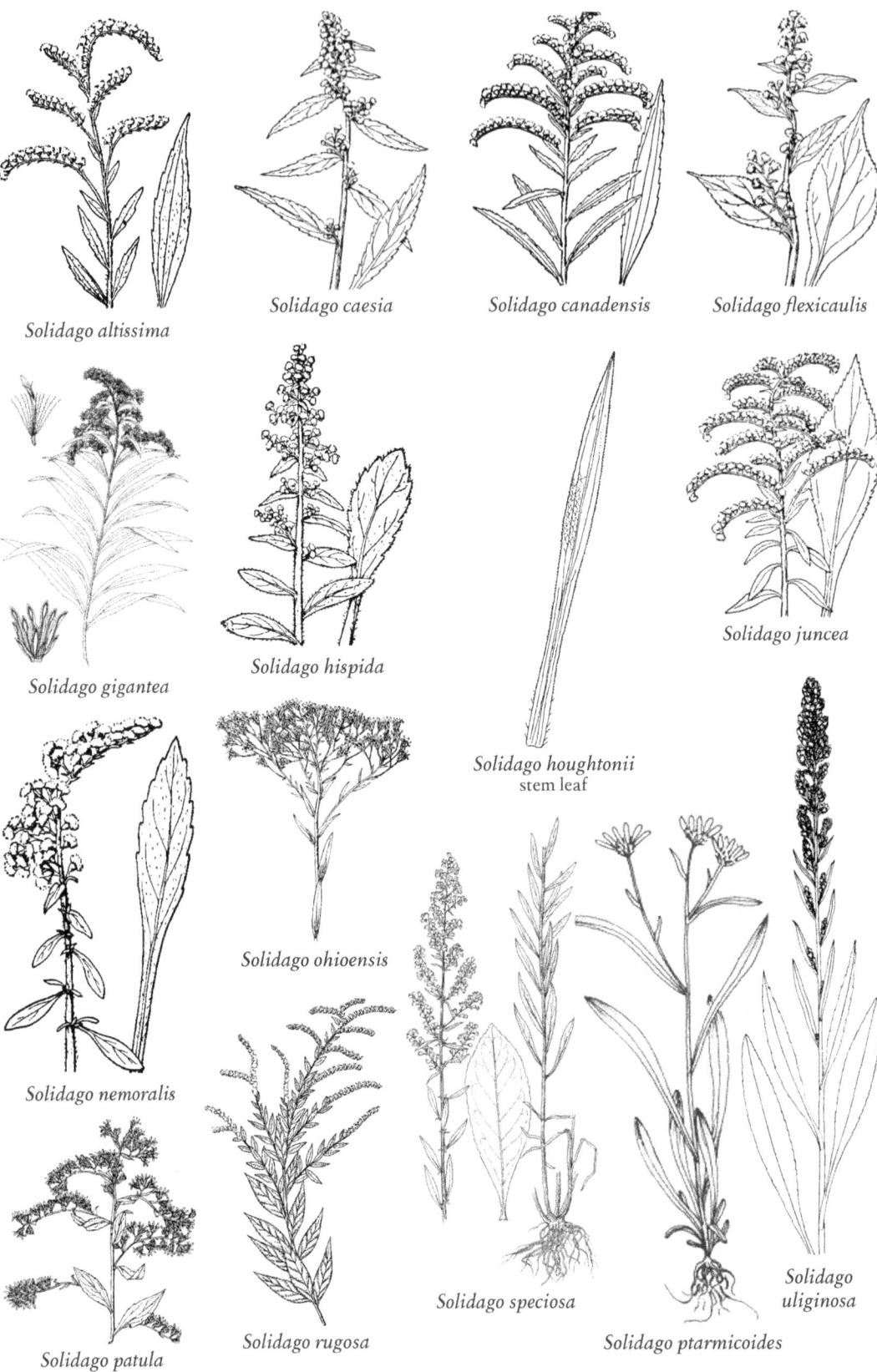

2 Leaf margins sparsely prickly............ **S. oleraceus**
2 Leaf margins with numerous spine-tipped teeth ......
.... **S. asper**

Sonchus arvensis L.
PERENNIAL SOW-THISTLE *introduced*
IR | WUP | CUP | EUP FACU

Perennial with long vertical roots, and extensively spreading by horizontal, rhizome-like, often deep-seated roots. Stems 4–20 dm tall, glabrous below the inflorescence and often somewhat glaucous. Leaves prickly-margined, the lower and middle ones usually pinnately lobed or pinnatifid, commonly 6–40 cm long and 2–15 cm wide, becoming less lobed and often more strongly auriculate upwards, the upper reduced and distant. Heads several in an open corymbiform inflorescence, relatively large, commonly 3–5 cm wide in flower, the fruiting involucre 15–22 mm long; involucre and peduncles more or less copiously provided with coarse, spreading, gland-tipped hairs, the involucre with some small and obscure tufts of tomenturn as well. Achenes about 2.5–3.5 mm long, with 5 or more prominent longitudinal ribs on each face, strongly rugulose. July–Oct.

A cosmopolitan weed of European origin.

Sonchus asper (L.) Hill
SPINY-LEAF SOW-THISTLE *introduced*
IR | WUP | CUP | EUP FACU

Similar to *S. oleraceus*, but usually more prickly. Leaves pinnatifid, or frequently obovate and lobeless, with rounded, not acute auricles. Achenes with 3 or rarely 4–5 evident longitudinal ribs on each face, not rugulose, although there may be minute projections from the marginal ribs. July–Oct.

A cosmopolitan weed; native of Europe.

Sonchus oleraceus L.
COMMON SOW-THISTLE *introduced*
IR | WUP | CUP | EUP FACU

Annual with a short taproot. Stems 1–10 dm tall, glabrous except sometimes for a few spreading gland-tipped hairs on the involucre and peduncle. Leaves pinnatifid to occasionally merely toothed, the margins rather weakly or scarcely prickly, 6–30 cm long and 1–15 cm wide, all but the lowermost prominently auriculate, the auricles with well rounded margins but eventually sharply acute; leaves progressively less divided upwards, and more or less reduced. Heads several in a corymbiform inflorescence, relatively small, only about 1.5–2.5 cm wide in flower; receptacle expanding and becoming conspicuously pale and indurate in fruit; fruiting involucre mostly 9–13 mm high. Achenes 2.5–3 mm long, rugulose and 3–5-ribbed on each face. July–Oct.

A cosmopolitan weed; native of Europe.

Sonchus arvensis

Sonchus asper

Sonchus oleraceus

Symphyotrichum *Wild Aster*

Mostly perennial herbs (annual in S. ciliatum). Leaves simple, alternate. Flower heads with both ray and disk flowers (disk flowers only in S. ciliatum); ray flowers white, pink, blue or purple, usually more than 0.5 mm wide (in contrast to the very narrow rays in Erigeron); disk flowers red, purple or yellow; involucral bracts in 2 or more series, usually overlapping; receptacle naked (not chaffy), flat or nearly so; pappus of numerous hairlike bristles. The traditional genus *Aster* has been split into several segregate genera to reflect differences with European species. Most species native to e USA are now placed within the genus *Symphyotrichum*, with the following exceptions for Michigan species: *A. umbellatus* in the genus *Doellingeria*; *A. furcatus*, *A. macrophyllus*, and *A. schreberi* in the genus *Eurybia*. Initially controversial, this classification is now widely accepted and is followed here.

1 Leaves, at least the lower ones, heart-shaped at base and with petioles................................ 2
1 Leaves not both heart-shaped and petioled 4
2 Leaves entire or nearly so; involucral bracts with a short, diamond-shaped green tip; Menominee County ......
.... **S. oolentangiense**
2 Leaves toothed; involucral bracts various 3

3   Inflorescence with relatively few heads, often less than 50; peduncles and inflorescence branches with only a few bracts .... **S. ciliolatum**
3   Inflorescence with many heads, often over 100; peduncles and inflorescence branches with many bracts .... **S. urophyllum**
4   Involucre strongly glandular .... **S. novae-angliae**
4   Plants not glandular 5
5   Base of leaf strongly clasping stem 6
5   Base of leaf not clasping stem (or only slightly so) ... 7
6   Involucral bracts (at least the inner), long-tapered to a slender tip .... **S. puniceum**
6   Involucral bracts rounded or short-tapered to the tip . ... **S. laeve**
7   Leaves densely silvery-hairy; margins entire; Keweenaw County .... **S. sericeum**
7   Leaves glabrous or pubescent, not silvery-hairy; margins entire to toothed 8
8   Taprooted annual herb; rays absent .... **S. ciliatum**
8   Perennial herbs with fibrous roots and from rhizomes or crowns; rays well developed 9
9   Most involucral bracts with a slender green tip, the margins inrolled .... **S. pilosum**
9   Involucral bracts flat, the margins not inrolled 10
10 Tips of outer involucral bracts loose or recurved, tapered to a very small spine-tip; leaves entire; Menominee County .... **S. ericoides**
10 Involucral bracts appressed or only slightly loose, the tips not spine-tipped; leaves various 11
11 Leaves hairy on underside, at least along the midvein 12
11 Leaf underside glabrous 13
12 Leaf underside hairy; plants with creeping rhizomes; Dickinson and Menominee counties ..... **S. ontarionis**
12 Leaf underside glabrous except for hairs on the midvein; plants without creeping rhizomes ..... **S. lateriflorum**
13 Heads very small and numerous, the rays 3–6 mm long; heads often arranged on 1-side of the inflorescence branches .... **S. lanceolatum**
13 Heads larger or few in number, not arranged on 1 side of the branches 14
14 Peduncles long, with many large bracts along its length, the bracts often more than 2 cm long; Mackinac County .... **S. dumosum**
14 Peduncles either short or with only a few bracts along its length 15
15 Rays bright blue-violet; stem leaves ± clasping (most bases, especially lower on the stem, circling more than half the circumference of the stem); involucral bracts nearly equal in length (or the outer ones over half as long as the inner ones) .... **S. robynsianum**
15 Rays usually white; stem leaves not clasping (bases circling half or less the stem circumference); involucral bracts of different lengths, imbricate 16
16 Slender plants of bogs and other wetlands; inflorescence short-stalked and wide in outline .... **S. boreale**
16 Plants stouter, not in bogs; inflorescence elongate ... . **S. lanceolatum**

Symphyotrichum boreale (Torr. & Gray) A. & D. Löve
NORTHERN BOG-ASTER   native
IR | WUP | CUP | EUP   OBL CC 9
Aster borealis (Torr. & Gray) Prov.
Aster junciformis Rydb.

Perennial herb, from rhizomes 1–2 mm wide. Stems erect, slender, 3–8 dm tall and to 2 mm wide, unbranched below, usually branched in the head; smooth except for lines of short, appressed hairs below base of upper leaves. Leaves alternate, linear, 4–12 cm long and 2–6 mm wide, sometimes slightly clasping at base, margins rough-to-touch, petioles absent. Flower heads usually few to rarely many, in an open, broad inflorescence; the heads 1.5–2 cm wide; involucre 5–7 mm high, the involucral bracts overlapping, often purple at tips and on margins; ray flowers 20–50, white to light blue or lavender, 1–1.5 cm long. Fruit an achene; pappus of pale hairs. Aug–Sept.

Conifer swamps, calcareous fens, open bogs, wet meadows, shores and seeps.

Symphyotrichum ciliatum (Ledeb.) Nesom
WESTERN ANNUAL ASTER   *introduced*
IR | WUP | **CUP** | **EUP**   FAC
Aster brachyactis Blake

Taprooted annual herb. Stems unbranched and erect, to branched and spreading, 2–6 dm long, smooth. Leaves alternate, linear, 2–10 cm long and mostly 2–5 mm wide, margins fringed with scattered hairs, petioles absent. Flower heads several to many, in an open inflorescence which forms much of plant; flower heads bell-shaped, 1–2 cm wide, involucre 5–10 mm high, the involucral bracts mostly green, linear, of equal length or slightly overlapping; ray flowers absent. Achenes flattened, 1–2 mm long; pappus of many long, soft hairs. Aug–Sept. Shores (including along Great Lakes), streambanks, wet meadows, roadside ditches, usually where brackish.

Native of northern Great Plains; considered adventive in Michigan.

Symphyotrichum ciliolatum (Lindl.) A. & D. Löve
NORTHERN HEART-LEAVED ASTER   native
IR | WUP | CUP | EUP   CC 4
Aster ciliolatus Lindl.
Aster lindleyanus Torr. & Gray

Perennial with long creeping rhizomes, sometimes also with a short branched caudex. Stems 2–12 dm tall, hirsute, especially in the inflorescence and on the lower leaf surfaces, sometimes glabrous throughout. Basal and lower stem leaves petiolate and cordate or subcordate, 4–12 cm long and 2–6 cm wide, sharply serrate, often deciduous, those above abruptly narrowed to the broadly winged petiole, and often less toothed, or the upper leaves sessile and entire. Inflorescence open, relatively few-headed, the heads often less than 50, rarely more than 100, the branches and peduncles sparsely or scarcely bracteate, the bracts narrow, the peduncles of very unequal length, generally some of them over 1 cm long; involucre 5–8 mm high, its slender bracts slightly or moderately imbricate, glabrous except for the sometimes ciliolate margins, their green tips relatively narrow and elongate; rays mostly 12–25, blue, 8–15 mm long. Achenes glabrous or nearly so, gray or stramineous, 3–6-nerved. July–Oct.

Woods and clearings.

Symphyotrichum dumosum (L.) Nesom
BUSHY ASTER   *native*
IR | WUP | CUP | **EUP**   FAC   CC 7
Aster dumosus L.

Perennial herb, spreading by long or short rhizomes. Stems branched above, 3–10 dm long, smooth or with fine hairs on upper stem. Leaves alternate, all from stem, linear to narrowly lance-shaped, 3–10 cm long and to 1 cm wide, much smaller and bractlike on branches, upper surface usually rough-to-touch, underside smooth, margins rough-to-touch, petioles absent. Flower heads many in an open, branched inflorescence, on stalks 2 cm or more long; involucre 4–6 mm long, smooth, the involucral bracts broad and green at tip, overlapping; ray flowers 15–30, usually white, rarely blue to pale violet, 5–10 mm long. Achenes hairy; pappus white. Aug–Oct.

Moist to wet sandy or mucky shores, interdunal swales, sedge meadows, sometimes where calcium-rich; also in drier oak and jack pine woods; Mackinac County.

Symphyotrichum ericoides (L.) Nesom
WHITE HEATH ASTER   *native*
IR | WUP | **CUP** | EUP   FACU CC 3
Aster ericoides L.

Perennial from well developed creeping rhizomes, pubescent with appressed or spreading hairs, or the leaves sometimes subglabrous. Stems 3–10 dm tall, occasionally more. Leaves numerous, linear, sessile, rarely as much as 6 cm long and 7 mm wide, the lower and often also the middle ones soon deciduous, those of the branches reduced and divaricate, often becoming mere bracts. Heads numerous, small, commonly somewhat secund on the divergent or recurved branches; involucre about 3–5 mm high, its bracts more or less strongly imbricate in several series, the outer spinulose and more or less squarrose, some or all of the bracts coarsely ciliolate-margined; rays 8–20, white, rarely blue or pink, 3–5 mm long. Achenes hairy.

Dry, open places; Menominee County.

Symphyotrichum laeve (L.) A. & D. Löve
SMOOTH BLUE ASTER   *native*
IR | **WUP** | **CUP** | EUP   FACU CC 5
Aster laevis L.

Perennial from a short stout rhizome or branched caudex, occasionally with short creeping red rhizomes as well. Stems 3–10 dm tall. Herbage glabrous throughout, except occasionally for some puberulent lines in the inflorescence, commonly somewhat glaucous. Leaves thick and firm, variable in size and shape but the larger ones over 1 cm wide, entire or sometimes toothed, sessile and more or less strongly clasping, or the lower tapering to winged petioles and scarcely clasping; leaves of the inflorescence reduced and often bractlike, clasping at their base. Heads several or numerous in an open inflorescence; involucre 5–9 mm high, its appressed bracts imbricate in several series, with short, commonly with rhombic

Symphyotrichum boreale

Symphyotrichum ciliatum

Symphyotrichum ciliolatum   *Symphyotrichum dumosum*

green tips; rays mostly 15–25, blue or purple, 8–15 mm long. Achenes nearly glabrous; pappus reddish or sometimes white. Aug–Oct.

Open, usually dry places.

Symphyotrichum lanceolatum (Willd.) Nesom
EASTERN LINED ASTER   native
IR | WUP | CUP | EUP   FACW CC 2
Aster lanceolatus Willd., *Aster hesperius* Gray, *Aster interior* Wieg.
Symphyotrichum simplex (Willd.) A.& D. Löve

Perennial herb, forming colonies from long rhizomes. Stems 0.5–1.5 m long, upper stems with lines of hairs. Leaves alternate, all on stem, lance-shaped to linear, 8–15 cm long and 3–30 mm wide, upper surface smooth or slightly rough-to-touch, margins toothed or sometimes entire; petioles absent or blades tapered to petiolelike base, sometimes slightly clasping stem. Flower heads many in an elongate leafy inflorescence; the involucre 3–6 mm high, the involucral bracts tapered to a green tip, smooth or margins fringed with hairs, strongly overlapping; ray flowers 20–40, usually white, sometimes lavender or blue, 4–12 mm long. Fruit an achene; pappus white. Aug–Oct.

Marshes, wet meadows, fens, swamp openings, low prairie, streambanks and shores; common.

Symphyotrichum lateriflorum (L.) A. & D. Löve
GOBLET-ASTER; FAREWELL-SUMMER   native
IR | WUP | CUP | EUP   CC 2
Aster lateriflorus (L.) Britt.
Aster hirsuticaulis Lindl.

Perennial from a branching caudex or short stout rhizome, with numerous fibrous roots. Stems several, 3–12 dm tall, curly-villous to glabrous. Leaves scabrous or nearly glabrous above, glabrous beneath except for the usually puberulent midrib; basal and lower stem leaves soon deciduous, or the basal occasionally persistent, obovate to lance-shaped, tending to taper from the middle to both ends, entire or serrate, mostly 5–15 cm long and 5–30 mm wide, petiolate; upper leaves sessile or nearly so. Heads numerous in a widely branched or sometimes more simple inflorescence; involucre glabrous, mostly 4–5.5 mm high, its bracts imbricate in few series, with broad green tips, often suffused with purple upwards; rays 9–14, white or slightly purple-tinged, 4–6.5 mm long; lobes of the disk corollas recurved. Achenes few-nerved, somewhat hairy. Aug–Oct.

Various habitats, most commonly in open woods, dry open places, and on beaches.

Symphyotrichum novae-angliae (L.) Nesom
NEW ENGLAND ASTER   native
IR | WUP | CUP | EUP   CC 3
Aster novae-angliae L.

Perennial herb, from a short rhizome or crown. Stems stout, erect, 4–10 dm long, with stiff, spreading, sometimes gland-tipped hairs. Leaves alternate, lance-shaped, 3–7 cm long and 1–2.5 cm wide, upper surface rough-to-touch or with short hairs, underside soft hairy, base of leaf strongly clasping stem, margins entire, petioles absent. Flower heads several to many, in clusters at ends of branches, 1.5–3 cm wide; involucre 7–12 mm high, the involucral bracts awl-shaped, glandular-hairy, sometimes purple; ray flowers 40 or more, blue-violet to less often red or pink, 1–2 cm long. Achenes hairy; pappus red-tinged. Aug–Oct.

Wet meadows, low prairie, shores, thickets, calcareous fens, roadsides; usually in moist or wet open areas.

Symphyotrichum ericoides

Symphyotrichum laeve

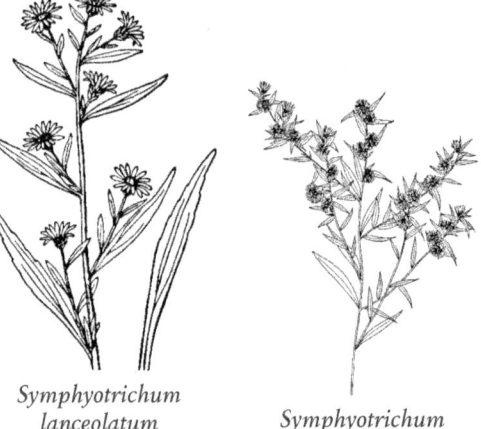

Symphyotrichum lanceolatum

Symphyotrichum lateriflorum

Symphyotrichum ontarionis (Wieg.) Nesom
ONTARIO ASTER　　　　　　　　　　　　　native
IR | WUP | **CUP** | EUP　　　　　　　　　FAC CC 6
　Aster ontarionis Wieg.
Perennial herb, from long creeping rhizomes. Stems branched, 3–8 dm long, upper stems with short spreading hairs. Leaves alternate, thin, oblong lance-shaped, 5–10 cm long and 1–3 cm wide (upper leaves smaller), upper surface rough-hairy to nearly smooth, underside finely to densely hairy; margins with sharp, forward-pointing teeth above middle of blade; petioles absent. Flower heads 1–2 cm wide, on short stalks from short leafy branches; involucre smooth to finely hairy, 5–7 mm high, the involucral bracts overlapping; ray flowers white, 9 or more. Fruit an achene. Sept–Oct.

　Floodplain forests, river terraces, thickets; Dickinson and Menominee counties. Similar to *S. lateriflorum*, but with long rhizomes rather than a crown or short rhizomes.

Symphyotrichum oolentangiense (Riddell) Nesom
PRAIRIE HEART-LEAVED ASTER　　　　　native
IR | WUP | **CUP** | EUP　　　　　　　　　　　CC 4
　Aster oolentangiensis Riddell
Perennial from a branched caudex or short rhizomes, with numerous fibrous roots. Stems 2–15 dm tall, scabrous-puberulent to occasionally nearly glabrous. Leaves thick and firm, entire or occasionally shallowly serrate, scabrous-hispid above, the hairs on the lower surface softer, and usually longer and looser than those on the upper; basal and usually also the lower stem leaves long-petiolate, cordate, lance-shaped or ovate, 4–13 cm long and 1.2–6 cm wide, those above abruptly smaller, narrower, less petiolate, and generally not at all cordate, the upper leaves sessile and lance-shaped or linear. Inflorescence open, panicle-like, with narrow and usually numerous bracts, the peduncles often very long; involucre 4.5–8 mm high, its bracts imbricate in several series, with a diamond-shaped green tip, glabrous except for the often ciliolate margins; rays commonly 10–25, blue, or rarely pink, 5–12 mm long. Achenes glabrous or nearly so, 3–5-nerved, usually pale. Aug–Oct.

　Prairies and dry open woods; Menominee County.

Symphyotrichum pilosum (Willd.) Nesom
WHITE OLDFIELD ASTER　　　　　　　　native
IR | **WUP** | **CUP** | EUP　　　　　　　FACU CC 1
　Aster pilosus Willd.
Perennial herb, from a large crown. Stems to 1.5 m long, more or less smooth (var. pringlei) or stems and leaves with spreading hairs (var. pilosum). Lower leaves oblong lance-shaped, 5–10 cm long and 1–2 cm wide, petioled; upper leaves smaller, linear, stalkless; margins entire or slightly toothed; petioles fringed with hairs; basal leaves and lower stem leaves soon deciduous (or basal leaves persistent). Flower heads at ends of small branches, forming an open inflorescence; involucre urn-shaped, narrowed near middle and flared upward, 3–5 mm high, smooth, involucral bracts overlapping to nearly equal in length, green-tipped; ray flowers 15–35, white. Fruit an achene.

　Sandy and gravelly shores, interdunal swales, wet meadows; often where calcium-rich; sometimes weedy in disturbed fields and roadsides.

Symphyotrichum puniceum (L.) A. & D. Löve
PURPLE-STEM ASTER　　　　　　　　　　native
IR | WUP | CUP | EUP　　　　　　　　　　OBL CC 5
　Aster firmus Nees, *Aster puniceus* L.
Large perennial herb, from a short rhizome or crown, sometimes also with short stolons. Stems stout, red-purple, 0.5–2 m long, unbranched, or branched in head, with long stiff hairs or sometimes nearly smooth. Leaves alternate, lance-shaped to oblong lance-shaped, 6–18 cm long and 1–4 cm wide, rough-to-touch to nearly smooth above, underside smooth or with long hairs on midvein; margins with scattered sharp teeth or sometimes entire; petioles absent, base of leaf clasping. Flower heads numerous, 1.5–2.5 cm wide; involucre 6–10 mm high, involucral bracts about equal in length, smooth or fringed with

Symphyotrichum novae-angliae

Symphyotrichum ontarionis

Symphyotrichum oolentangiense

Symphyotrichum pilosum

hairs, green and spreading; ray flowers 20–50, blue (rarely white). Fruit a smooth achene; pappus more or less white. Aug–Sept.

Swamps, sedge meadows, thickets, calcareous fens, streambanks, shores, springs, roadside ditches; common.

Symphyotrichum robynsianum (J.Rousseau) Brouillet & Labrecque
LONG-LEAVED ASTER native
IR | WUP | CUP | EUP FACW CC 9

Aster longifolius sensu Semple & Heard, non Lam.
Perennial, forming colonies from long-creeping rhizomes. Stems 10–80 cm tall, single, erect, often red-tinged, glabrous, upper portion with lines of hairs. Leaves stiff, glabrous to scabrous, lower leaves withering by flowering, long-petiolate, petioles narrowly winged, blades lance-shaped, 20 cm long and 3–5 mm wide, bases clasping the stem; margins sparsely serrulate or entire, somewhat revolute; upper stem leaves sessile, smaller, slightly clasping or not, margins entire. Inflorescence elongate, open, panicle-like or raceme-like; branches ascending, with 1–3 heads per branch; branch leaves small; peduncles glabrous or densely pilose in lines. Involucre 5–8.5 mm long, its bracts in 3–4 series, their margins scarious, erose, and hyaline. Rays 20–35; corollas dark blue-violet, seldom white; disk florets 23–40; corollas yellow. Achene tan, obovoid, 5–6-nerved; pappus pinkish.

Moist open sandy, gravelly, or rocky places, including shores, limestone alvars, seasonally wet swales, rocky shores of Lake Superior.

The long leaves usually overtop most or all of the inflorescence.

Symphyotrichum sericeum (Vent.) Nesom
WESTERN SILVERY ASTER (THR) native
IR | WUP | CUP | EUP CC 10

Aster sericeus Vent.
Perennial from a short branched caudex, with numerous fibrous roots. Stems 2–7 dm tall, several, wiry, branched upwards, thinly sericeous, or glabrate below. Leaves sericeous on both sides, entire, the basal leaves oblong lance-shaped and petiolate, but these and the leaves on the lower half of the stem soon deciduous, the others sessile, lance-shaped to elliptic, to 4 cm long and 1 cm wide. Heads several or numerous in a widely branched inflorescence, often clustered at the ends of the branches; involucre sericeous, 6–10 mm high, its broad bracts in several series but seldom much imbricate, their leafy tips often loose or spreading; rays 15–25, deep violet to rose purple or rarely white, 8–15 mm. long. Achenes glabrous, closely 8–12-nerved. Aug–Oct.

Dry prairies and other open places; southern Lower Peninsula and a single old collection from Keweenaw County.

The very silky-hairy leaves are distinctive.

Symphyotrichum urophyllum (Lindl.) Nesom
ARROW-LEAVED ASTER native
IR | WUP | CUP | EUP CC 2

Aster sagittifolius Willd.
Perennial with a branched caudex or short rhizome and numerous fibrous roots. Stems 4–12 dm tall, glabrous or nearly so below the inflorescence, or the upper part occasionally puberulent in lines. Leaves rather thick, shallowly toothed, glabrous or scabrous above, glabrous or hirsute beneath; lowermost leaves lance-ovate, cordate, 6–15 cm long and 2–6 cm wide, long-petiolate; upper leaves narrowed to the often broadly winged petiole, or sessile. Inflorescence panicle-like, elongate, with ascending bracteate branches; the heads often very numerous, borne on branches rarely more than 1 cm long, thus appearing crowded; involucre 4–6 mm high, its imbricate bracts glabrous except for the sometimes ciliolate margins, slender with elongate green tips; rays 8–20, usually pale blue or lilac, sometimes white, 4–8 mm long. Achenes pale, glabrous, 4–5-nerved. Aug–Oct.

Streambanks, woodlands; less often in open places.

Tanacetum *Tansy*

Annual or perennial herbs, sometimes somewhat woody at the base. Leaves alternate, pinnately dissected. Heads small or medium-sized, corymbiform or solitary, hemispheric to campanulate; discoid or nearly so, the outer flowers pistillate, with a short tubular corolla in some species expanded into a short yellow ray, or the pistillate flowers rarely wanting. Involucral bracts imbricate, dry, the margins and tips commonly scarious. Receptacle flat or convex, naked. Disk-flowers perfect, with 5-toothed tubular yellow corolla. Achenes mostly 5-ribbed, commonly glandular. Pappus a short crown, or none.

ADDITIONAL SPECIES
Tanacetum coccineum (Willd.) Grierson; Eurasian native; known from Michigan by a 2014 roadside collection in Keweenaw County; flowers bright pink, fading to pale pink or whitish; leaves finely dissected.

1 Leaves undivided (though regularly toothed)..**T. balsamita**
1 Leaves pinnatifid or bipinnatifid 2
2 Rays present, white; Houghton and Keweenaw counties ....**T. parthenium**
2 Rays absent or yellow............................. 3
3 Heads 13–20 mm wide; leaves ± hairy; rare along Lake Michigan beaches................... **T. bipinnatum**
3 Heads 5–10 mm wide; leaves glabrous or nearly so; common ....**T. vulgare**

dicots   **ASTERACEAE** | Aster Family 113

Tanacetum balsamita L.
COSTMARY   *introduced*
IR | **WUP** | **CUP** | EUP
  Balsamita major Desf.
  Chrysanthemum balsamita L.
Coarse fragrant perennial. Stems 5–12 dm tall, strigose above, glabrous below. Leaves silvery-strigose when young, more or less glabrate in age, crenate, sometimes with a few reduced basal pinnae, the basal with elliptic blades 10–25 cm long and 2.5–8 cm wide, on a petiole of about equal length; stem leaves smaller, sessile or nearly so, seldom over 10 cm long, numerous. Heads numerous in a corymbiform inflorescence, the disk 4–7 mm wide; rays usually wanting, if present white and less than 1 cm long; involucral bracts narrow, with conspicuous, expanded, hyaline tip. Achenes subterete, about 10-ribbed; pappus a minute border or crown. Aug–Oct.

  Eurasian native, escaped from cultivation to roadsides and other waste places.

Tanacetum bipinnatum (L.) Sch.Bip.
EASTERN TANSY   (THR) *native*
IR | WUP | **CUP** | **EUP**   CC 10
  Tanacetum huronense Nutt.
Rhizomatous perennial, villous throughout. Stems 1–8 dm tall. Leaves few or fairly numerous, sessile or short-petiolate, scarcely punctate; stem leaves 5–20 cm long and 1.5–8 cm wide, the basal often longer and persistent, all dissected, with narrow ultimate segments, the main rachis scarcely or not at all winged, that of each pinna only narrowly so. Heads 1–15, rarely more, the disk about 10–18 mm wide; rays inconspicuous or more commonly evident, sometimes as much as 4 mm long; pappus a short toothed crown. July–Aug.

  Sandy beaches, dunes, and cracks in limestone pavement (alvars).

Tanacetum parthenium (L.) Schultz-Bip.
FEVERFEW   *introduced*
IR | **WUP** | CUP | EUP
  Chrysanthemum parthenium (L.) Bernh.
  Matricaria parthenium L.
Perennial with a taproot or stout caudex. Stems 3–8 dm tall, generally glabrous below, puberulent above. Leaves finely puberulent at least beneath, pinnatifid, with rounded, incised or again pinnate segments, evidently petiolate, the blade to 8 cm long and 6 cm wide. Heads several or numerous in a corymbiform inflorescence, the disk 5–9 mm wide; involucral bracts narrow, the inner with sharply marked hyaline tips; rays 10–20, or more numerous in double forms, 4–8 mm long, white. Achenes subterete, about 10-ribbed; pappus a minute crown, or obsolete. June–Sept.

  Native of Europe, sometimes escaping from cultivation to waste places; Houghton and Keweenaw counties.

Tanacetum vulgare L.
COMMON TANSY   *introduced (invasive)*
IR | **WUP** | **CUP** | **EUP**   FACU
Coarse aromatic perennial with a stout rhizome, glabrous or nearly so throughout. Stems about 4–15 dm tall. Leaves numerous, 1–2 dm long and nearly half as wide, sessile or short-petiolate, punctate, pinnatifid, with evidently winged rachis, the pinnae

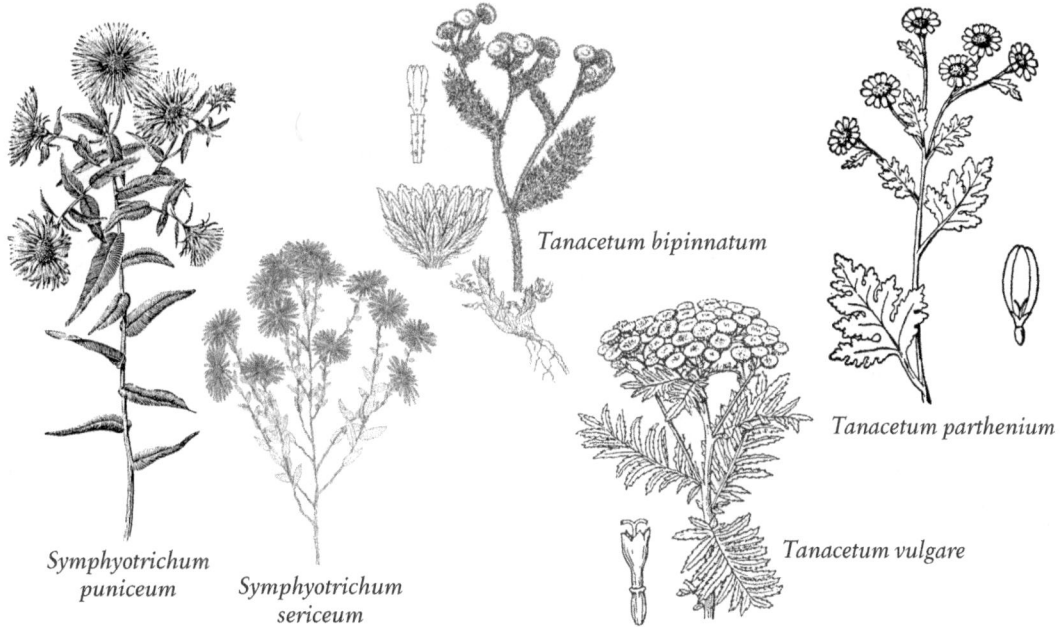

Tanacetum bipinnatum

Tanacetum parthenium

Tanacetum vulgare

Symphyotrichum puniceum

Symphyotrichum sericeum

again pinnatifid or deeply lobed, with broadly winged rachis, the pinnules often again toothed. Heads discoid, numerous, commonly about 20–200, the disk about 5–10 mm wide; pappus a minute crown, almost obsolete. Aug–Oct.

Native of the Old World, escaped from cultivation to roadsides, fields and waste places.

Taraxacum *Dandelion*

Perennial, scapose, taprooted herbs with milky juice. Leaves all basal, forming a rosette, entire to pinnatifid. Flowers all ligulate and perfect, mostly numerous. Heads solitary, erect; involucral bracts biseriate, the outer usually shorter than the inner and often reflexed. Fruit a columnar achene, longitudinally ribbed. Pappus of numerous capillary bristles.

ADDITIONAL SPECIES
Taraxacum palustre (Lyons) Symons (Marsh dandelion), introduced; Chippewa County.

1. Outer phyllaries straight, dark gray to nearly black; leaves toothed or shallowly lobed, usually less than 2 cm wide; Chippewa County.................**T. palustre***
1. Outer phyllaries recurved, usually green to purple or brown tinged; leaves deeply lobed to ± entire, often more than 2 cm broad, especially if ± entire; widespread species.......................................2
2. Leaves generally deeply lobed or cut to midrib; mature achenes red-brown.............**T. erythrospermum**
2. Leaves various, deeply lobed to entire; mature achenes tan or olive-green ....**T. officinale**

Taraxacum erythrospermum Andrz. ex Besser
RED-SEED DANDELION   *introduced*
IR | **WUP** | **CUP** | **EUP**

Taraxacum laevigatum (Willd.) DC.
Similar to *T. officinale*, often more slender. Leaves generally very deeply cut for their whole length, the lobes narrow, the terminal one seldom much larger than the lateral ones. Heads a little smaller, the involucre mostly 1–2 cm high, its inner bracts mostly 11–13, often somewhat corniculate, the outer bracts appressed to reflexed, a third to a little more than half as long as the inner. Body of the achene becoming bright red or reddish purple at maturity, commonly somewhat rugulose below as well as muricate above; beak usually stramineous, from more than twice as long to occasionally only half as long as the body. April–June.

Native of Eurasia, now established throughout Michigan in fields, pastures, lawns, and other disturbed places, but less common than *T. officinale*.

Taraxacum officinale G.H. Weber
COMMON DANDELION   *introduced*
IR | **WUP** | **CUP** | **EUP**   FACU

Leaves commonly sparsely hairy beneath and on the midrib, otherwise generally glabrous, or sometimes completely so, oblong lance-shaped, mostly 6–40 cm long and 0.7–15 cm wide, pinnatifid or lobed, the terminal lobe tending to be larger than the others, tapering to a narrow, scarcely or obscurely winged petiolar base. Scape 5–50 cm tall, glabrous or more or less villous, especially upwards. Heads usually large, the involucre mostly 1.5–2.5 cm high, the inner bracts mostly 13–20, these at first erect, finally reflexed, the mature achenes and pappus then forming a conspicuous ball easily disintegrated by the wind; outer bracts a little shorter and scarcely wider than the inner, reflexed. Body of the achene 3–4 mm long, pale gray-brown to olive-brown, muriculate above or sometimes to near the base, about half or a third as long as the slender beak; pappus white. March–Dec.

Native of Europe and adjacent Asia, now a cosmopolitan weed of lawns and disturbed sites.

Tragopogon *Goat's-Beard*

Biennial or perennial lactiferous herbs with a taproot. Leaves alternate, linear, entire, clasping, commonly grass-like. Heads solitary at the ends of the branches. Flowers all ligulate and perfect, yellow or purple. Involucre cylindric or campanulate, the bracts uniseriate, equal. Achenes linear, terete or angled, 5–10-nerved, narrowed at the base, slender-beaked, or the outer occasionally beakless. Pappus of a single series of plumose bristles, united at the base, the plume-branches interwebbed, several of the bristles commonly longer than the others and naked at the apex. Our species may hybridize with the others where they grow together.

1. Flowers purple; uncommon garden escape...................................**T. porrifolius**
1. Flowers yellow; common weeds....................2
2. Peduncle enlarged or inflated below the head; leaf tips not recurved ....**T. dubius**
2. Peduncle not enlarged; leaf tips recurved . **T. pratensis**

Tragopogon dubius Scop.
MEADOW GOAT'S-BEARD   *introduced*
IR | **WUP** | **CUP** | **EUP**

Tragopogon major Jacq.
Similar to *T. porrifolius* but with yellow flowers and often smaller and less robust. Leaves averaging narrower. Involucral bracts sometimes more numerous. Achenes (including beak) seldom over 3.5 cm long. May–July.

Native of Europe; roadsides and waste places.

Tragopogon porrifolius L.
SALSIFY *introduced*
IR | **WUP** | CUP | EUP

Glabrous perennial 4–10 dm tall. Leaves to 30 cm long and nearly 2 cm wide. Peduncles more or less enlarged and fistulous under the heads. Involucral bracts most commonly 8, sometimes 5–11, mostly 2.5–4 cm long in flower and longer than the purple rays, elongating to 4–7 cm in fruit. Achenes mostly 2.5–4 cm long, long-beaked, the body only 10–16 mm long, the marginal ones strongly scabrous-muricate, the central ones nearly smooth. April–Aug.

Native of Europe; escaped from cultivation to roadsides and waste places, mostly in moist soil.

Tragopogon pratensis L.
JACK-GO-TO-BED-AT-NOON *introduced*
IR | **WUP** | CUP | EUP

Glabrous perennial 1.5–8 dm tall. Leaves to 30 cm long and nearly 2 cm wide, often much narrower. Peduncles not at all enlarged in flower, scarcely so in fruit. Involucral bracts most commonly 8, mostly 12–24 mm long in flower, equaling or shorter than the yellow rays, elongating to 18–38 mm in fruit. Achenes 12–24 mm long, relatively shorter beaked, the body nearly or quite as long as in the other two species. May–Aug.

Native of Europe; roadsides, fields, and waste places.

Xanthium *Cocklebur*

Xanthium strumarium L.
COMMON COCKLEBUR *native*
IR | WUP | **CUP** | EUP   FAC CC –

Weedy taprooted annual herb; plants variable in size and habit, rough-to-touch or sometimes nearly smooth. Stems 2–15 dm long, often brown-spotted. Leaves alternate, ovate to nearly round, sometimes with 3–5 shallow lobes, 3–15 cm long and 2–20 cm wide, margins with blunt teeth; petioles 3–10 cm long. Flower heads either staminate or pistillate, the staminate flowers brown, in clusters of small round heads at ends of stems above the larger pistillate heads; pistillate heads in several to many clusters from leaf axils, each head with 2 flowers, with a spiny involucre enclosing the head; petals absent. Fruit a brown bur formed by the involucre, 1.5–3 cm long, covered with hooked prickles; achenes thick, 1 in each of the 2 chambers of the bur. Aug–Sept.

Shores, streambanks, wet meadows, sand bars, dried depressions, often where disturbed; also in cultivated and abandoned fields, roadsides and waste places; Marquette and Menominee counties.

Balsaminaceae
TOUCH-ME-NOT FAMILY

Impatiens *Touch-Me-Not*

Impatiens capensis Meerb.
JEWELWEED *native*
IR | **WUP** | CUP | EUP   FACW CC 2

Impatiens biflora Walt.

Glabrous annual herb. Stems hollow, succulent, 3–10 dm long, usually branched above. Leaves simple, alternate, ovate to oval, 3–9 cm long and 1.5–4 cm wide, tapered to tip or rounded and tipped with a short slender point, margins shallowly and irregularly toothed; petioles longest on lower leaves, shorter upward, 0.5–5 cm long. Flowers orange-yellow, irregular, pouchlike and spurred, hanging from the petioles in few-flowered racemes from upper leaf axils; 1.5–3 cm long, usually mottled with red-brown spots, the spur recurved parallel to the sac and to

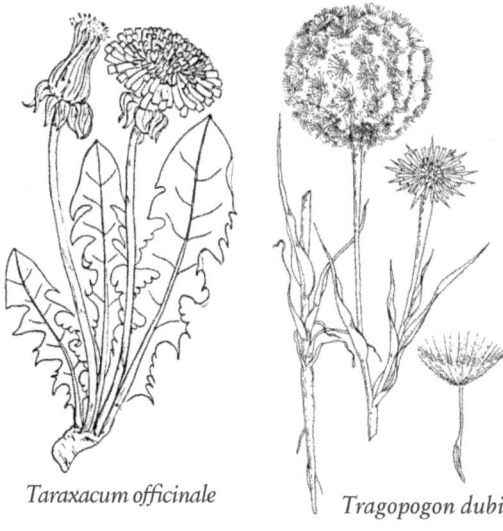

Taraxacum officinale   *Tragopogon dubius*

Tragopogon pratensis

Xanthium strumarium

half its length; sepals 3, petal-like; petals 3; stamens 5. Fruit a 5-valved capsule about 2 cm long, splitting when mature if jarred or touched, and scattering the seeds away from parent plants. July–Sept.

Swamps, low areas in woods, floodplain forests, thickets, streambanks, shores, marshes, fens, springs; often where disturbed.

Small, cleistogamous (self-fertile) flowers lacking petals are sometimes produced in summer and are often the only flowers on plants growing in shaded situations.

Berberidaceae
BARBERRY FAMILY

Herbs or shrubs. Leaves alternate or basal, simple, lobed, or compound. Flowers solitary, racemose or cymose; perfect, all parts free and distinct. Sepals 4 or 6, sometimes early deciduous, in some genera petal-like. Petals as many as or more than the sepals, petaloid or reduced to nectaries. Stamens as many as the petals. Ovary 1-celled. Fruit a berry or capsule.

1 Plants spiny shrubs . . . . **Berberis**
1 Plants smooth perennial herbs 2
2 Flowers in a small panicle-like cyme . . . .
. . . . **Caulophyllum thalictroides**
2 Flowers single . . . . **Podophyllum peltatum**

Berberis *Barberry*

Spiny shrubs. Leaves of the shoots reduced to alternate, simple or 3-branched spines, with clusters of small foliage leaves in their axils. Flowers yellow, in elongate racemes, umbel-like clusters, or sometimes solitary. Sepals 6, petal-like, subtended by 2 or 3 small bracts. Petals 6, usually smaller than the sepals and with 2 glands at their base. Stamens 6, appressed to the sepals until irritated, when they rapidly bend toward the center. Fruit a red, one to few-seeded berry.

ADDITIONAL SPECIES

Berberis aquifolium Pursh (Oregon grape); native of Pacific Northwest, a popular ornamental evergreen shrub sometimes escaped from cultivation to nearby woods; Alger, Baraga, and Houghton counties.

1 Leaves compound, evergreen; stem without spines (but leaflets spiny-toothed); ripe fruits dark blue . . . .
. . . . **B. aquifolium***
1 Leaves simple, deciduous; stem spiny at nodes; ripe fruits red 2
2 Leaves entire; flowers single or in clusters of 2–4 . . . . .
. . . . **B. thunbergii**

2 Leaves tipped by a small spine; flowers in racemes of 10–20 flowers; Mackinac County . . . . **B. vulgaris**

Berberis thunbergii DC.
JAPANESE BARBERRY *introduced (invasive)*
IR | **WUP** | **CUP** | **EUP** FACU

Densely and divaricately branched shrub to 2 m tall; spines usually simple. Leaves obovate to spatulate, usually obtuse, entire, narrowed at base to a short petiole. Flowers solitary or in small clusters of 2–4, about 8 mm wide. Fruit about 1 cm long. May.

Native of Japan; commonly planted for low hedges and escaped along roadsides and in thickets.

Berberis vulgaris L.
EUROPEAN BARBERRY *introduced (invasive)*
IR | WUP | CUP | **EUP** FACU

Freely branched shrub to 3 m tall. Leaves obovate to obovate-oblong, 2–5 cm long, obtuse or acute, finely spinulose-denticulate, the veinlets prominently reticulate beneath. Racemes usually 3–6 cm long, with 10–20 flowers on pedicels 5–10 mm long; petals entire. Fruit about 1 cm long.

Native of Europe; formerly widely planted and frequently escaped along roadsides and fences and in open woods; now largely purposefully exterminated as the alternate host of black rust of wheat; Mackinac County.

Caulophyllum *Blue Cohosh*

Caulophyllum thalictroides (L.) Michx.
BLUE COHOSH *native*
IR | **WUP** | **CUP** | **EUP** CC 5

Smooth perennial herb. Stems erect, 3–8 dm tall, glaucous when young, bearing above the middle a single large, sessile, 3-parted leaf, and another smaller leaf just below the panicle. Leaflets obovate-oblong, 2–5-lobed above the middle, 5–8 cm long when fully grown. Flowers yellowish green or greenish purple, nearly 1 cm wide, in a panicle 3–6 cm long; sepals 6, petal-like, subtended by 3 or 4 sepal-like bracts; petals 6, reduced to small gland-like bodies much shorter than the sepals and opposite them; stamens 6; ovary soon ruptured by the enlarging seeds, which ripen exposed on short stout stalks and resembling drupes; seeds dark blue, 5–8 mm long, on stalks of nearly the same length. April–May.

Rich moist woods.

Podophyllum *May-Apple*

Podophyllum peltatum L.
MAY-APPLE *native*
IR | **WUP** | **CUP** | EUP FACU CC 3

Herb, from a perennial rhizome, usually colony-forming. Flowering stem 3–5 dm tall, bearing a pair

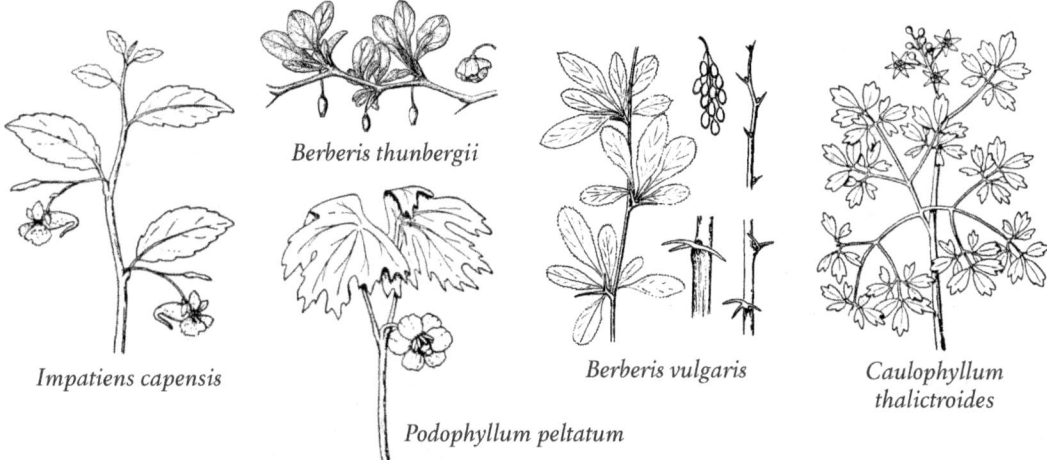

Impatiens capensis
Berberis thunbergii
Podophyllum peltatum
Berberis vulgaris
Caulophyllum thalictroides

of leaves and a short-peduncled, solitary, terminal flower. Sterile plants bearing a single, large, peltate, deeply radially lobed, terminal leaf; fertile plants bearing 2 half-round, similarly lobed leaves. Flowers 3–5 cm wide, on a short nodding peduncle; sepals 6, falling early; petals 6–9, white; stamens 2x as many as the petals; ovary ovoid, with a large sessile stigma. Fruit a yellow, ovoid, fleshy, many-seeded berry 4–5 cm long. May; fruit ripe in Aug.

Moist, preferably open woods; Houghton and Menominee counties.

The ripened fruit is edible in small amounts, toxic if consumed in large quantities; rhizome, leaves and seeds toxic.

Betulaceae
BIRCH FAMILY

Medium to large trees, or shrubs. Leaves deciduous, simple, alternate, with toothed margins and pinnate veins. Flowers small, staminate and pistillate flowers separate on same plant, crowded into catkins (aments) that open in spring before leaves fully open; staminate catkins hang downward; conelike pistillate catkins erect or drooping. Fruit a small, 1-seeded, winged nutlet.

1 Plants in flower 2
1 Plants in fruit 6
2 Pistillate flowers 1 or several in a cluster ..... **Corylus**
2 Pistillate flowers in catkins 3
3 Each bract of staminate catkin with 1 flower, this without sepals 4
3 Each bract of staminate catkin with 3–6 flowers, each with sepals 5
4 Staminate catkins in groups of 1. **Carpinus caroliniana**
4 Staminate catkins usually in clusters of 3 .... **Ostrya virginiana**
5 Pistillate bracts 3-lobed; stamens 2............ **Betula**
5 Pistillate bracts 5-lobed; stamens 3–5 .... **Alnus**
6 Each fruit (nut) subtended by leaf-like bracts........ 7
6 Fruit without leafy bracts, in the axil of a small scaly bract.. 9
7 Shrubs; nut 1 cm long or more .... **Corylus**
7 Trees; nut to 6 mm long........................... 8
8 Bark furrowed and shredding, gray-brown, bracts saclike, enclosing the nut.......... **Ostrya virginiana**
8 Bark smooth and gray; bracts not enclosing the nut... .... **Carpinus caroliniana**
9 Bracts woody, widely spreading from rachis of cone... .... **Alnus**
9 Bracts papery, ascending..................... **Betula**

Alnus *Alder*

Thicket-forming shrubs, or an introduced tree. Leaves deciduous, ovate, toothed on margins. Staminate and pistillate flowers separate on same plant, staminate flowers in long, drooping catkins which fall after shedding pollen; pistillate flowers in short, persistent conelike clusters. Fruit a flattened achene with winged or thin margins.

1 Twigs and young leaves sticky, leaves with small, sharp teeth; catkins on long stalks; fruit broadly winged .... **A. viridis**
1 Twigs and young leaves not sticky, leaves unevenly double-toothed; catkins stalkless or on short stalks; fruit narrowly winged .... **A. incana**

Alnus incana (L.) Moench
SPECKLED ALDER; TAG ALDER native
IR | WUP | CUP | EUP FACW CC 5
Alnus rugosa (Du Roi) Spreng.

Thicket-forming shrub to 5 m tall; twigs red-brown, waxy, with conspicuous pale lenticels. Leaves ovate to oval, broadest near or below middle, 6–14 cm long and 4–7 cm wide, dark green and smooth above, paler and hairy below; margins sharply toothed and shallowly lobed; petioles 1–2.5 cm long. Flowers in catkins clustered at ends of branches; staminate catkins developing in late summer, short-stalked, elongate, 4–9 cm long; pistillate catkins appear in late summer, stalkless, rounded, 1–2 cm long and to 1 cm wide, the scales unlobed, becoming conelike, persistent. Fruit a flat nutlet, narrowly winged on margin, 2–4 mm long. April–June.

Swamps, thickets, bog margins, shores and streambanks.

Alnus viridis (Vill.) Lam. & DC.
GREEN ALDER *native*
IR | WUP | CUP | EUP   FAC CC 8
Alnus crispa (Ait.) Pursh

Thicket-forming shrub to 4 m tall; bark red-brown to gray; twigs brown, sticky, somewhat hairy, lenticels pale and scattered. Leaves round-oval, bright green above, slightly paler and shiny below, sticky when young, margins wavy with small, sharp teeth; petioles 6–12 mm long. Flowers in catkins; staminate catkins stalked, slender, developing in late summer and expanding in spring; pistillate catkins appear in spring, becoming long-stalked, blunt and conelike, persistent, 1–2 cm long. Fruit a nutlet, 2–3 mm long, with a pale, thin wing.

Lakeshores, wet depressions in woods, rock outcrops, beaches along Lake Superior.

Betula *Birch*

Trees or shrubs, often with multiple stems from base; bark sometimes peeling in thin layers. Leaves deciduous, alternate, sharply toothed. Staminate and pistillate flowers separate on same plant, catkins appearing in fall, opening the following spring, staminate flowers in drooping slender catkins; pistillate flowers in erect conelike catkins. Fruit a wing-margined achene (samara).

1 Shrub to 2 m tall; bark not shredding; leaves to 5 cm long. . . . . **B. pumila**
1 Small to large trees; bark shredding with age 2
2 Bark white; samara wings as wide or wider than body . . . . **B. papyrifera**
2 Bark yellow-gray; samara wings narrower than body . . . . **B. alleghaniensis**

Betula alleghaniensis Britt.
YELLOW BIRCH *native*
IR | WUP | CUP | EUP   FAC CC 7
Betula lutea Michx. f.

Medium to large tree to 25 m tall; bark on young trees thin and smooth with conspicuous horizontal lenticels, becoming yellow-gray and shredding into thin, shaggy horizontal strips; bark of old trees breaking into large plates; twigs hairy when young, becoming smooth and shiny, wintergreen-scented when crushed. Leaves alternate, simple, ovate, tapered to a short, sharp tip, dark green above, paler yellow-green below, 6–12 cm long, margins coarsely double-toothed, petioles grooved and hairy. Staminate and pistillate flowers in catkins, separate on same tree, appearing before leaves in spring; staminate catkins drooping, yellow-purple, 7–10 cm long; pistillate catkins erect, green, 2–4 cm long, more or less stalkless. Fruit a winged nutlet, 3–5 mm wide. April–May.

Moist forests with sugar maple; also occasional in swamps, thickets, and forest depressions with red maple, black ash, black spruce, eastern hemlock and *Alnus incana*.

Betula papyrifera Marsh.
WHITE BIRCH; PAPER BIRCH *native*
IR | WUP | CUP | EUP   FACU CC 2

Trees, usually 20 m or shorter; trunks single or sometimes 2 or more. Bark of young trunks and branches dark reddish brown, smooth; in maturity creamy to chalky white, peeling in paper-thin sheets; lenticels pale, horizontal, in maturity dark, much expanded. Twigs without strong odor and taste of wintergreen, slightly to moderately pubescent, infrequently with small, scattered, resinous glands. Leaf blade ovate, with 9 or fewer pairs of lateral veins, 5–9 cm long and 4–7 cm wide, base rounded or truncate; lower surface pubescent, often velvety-hairy along major veins and in vein axils; margins coarsely or irregularly doubly serrate. Flowers in pendulous, cylindric catkins, 2.5–5 cm long, readily shattering with fruits in late fall; scales pubescent to glabrous. Samaras with wings as wide as or slightly wider than body. Late spring.

Moist, open, upland forest, especially where rocky; also on sand dunes swamps and sometimes in swampy woods; especially characteristic after fire or timber harvests, when seedlings are often abundant.

Includes *B. papyrifera* var. *cordifolia* (Regel) Fern., sometimes considered a separate species (*B. cordifolia* Regel).

The bark, which has a high oil content making it waterproof, was used for a wide variety of building and clothing purposes by Native Americans.

Betula pumila L.
BOG BIRCH *native*
IR | WUP | CUP | EUP   OBL CC 8
Betula glandulosa var. *glandulifera* (Regel) Gleason

Shrub 1–3 m tall; bark dull gray or brown; twigs gray, short-hairy and dotted with resin glands, becoming red-brown and waxy with age. Leaves leathery, rounded to obovate, 2–4 cm long and 1–3 cm wide, dark green above, paler and often waxy below; margins coarsely toothed, the teeth blunt or sharp; petioles 3–6 mm long. Flowers in catkins; staminate catkins stalkless, cylindric, 15–20 mm long and 2–3 mm wide; pistillate catkins stalked, cylindric, 1–2 cm long and 5 mm wide; scales 3-lobed. Fruit a flat, winged, rounded nutlet, 2–3 mm long and 2–4 mm wide. May.

Swamps, bogs, fens, seeps; often where calcium-rich.

Carpinus *Hornbeam*

Carpinus caroliniana Walt.
HORNBEAM; IRONWOOD native
IR | **WUP** | **CUP** | EUP   FAC CC 6

Tall shrub or small tree up to 10 m tall, with fluted trunk and smooth, blue-gray or ashy gray bark. Leaves oblong to oblong-ovate, 5–12 cm long; margins sharply and often doubly serrate. Staminate catkins slender, pendulous; scales ovate, each subtending a single naked flower composed of several stamens. Pistillate catkins slender, 2–5 cm long; scales ovate, deciduous; pistillate flowers in pairs, each subtended by a minute bract adnate at base to 2 minute bractlets; calyx minute. Fruit a small ribbed nutlet.

Moist woods.

Corylus *Hazelnut*

Shrubs or small trees. Leaves doubly serrate. Staminate catkins elongate, cylindric, emerging in autumn, reaching anthesis in early spring. Stamens 4, the filaments deeply bipartite, each division bearing a half-anther. Pistillate catkins small, ovoid, resembling a leaf-bud, the few closely imbricate scales concealing the flowers, except the elongate protruding stigmas. Flower subtended by a minute bract and 2 bractlets, becoming greatly enlarged at maturity and enclosing the hard-shelled edible nut. Ovary inferior, surmounted by the minute calyx.

1 Twigs and leaf petioles with glandular bristles . . . . **C. americana**
1 Twigs and petioles not glandular-bristly. . . . **C. cornuta**

Corylus americana Walt.
AMERICAN HAZELNUT native
IR | **WUP** | **CUP** | EUP   FACU CC 5

Shrub 1–3 m tall, the young twigs and petioles more or less pubescent (hairs red when young) and normally beset with stout stipitate glands. Leaves broadly ovate to obovate, finely doubly serrate, broadly rounded to cordate at base, paler and more or less pubescent beneath. Involucral bracts pubescent but not bristly, closely surrounding the nut and prolonged beyond it into a broadly dilated, flattened beak, cut at the summit into broadly triangular lobes, the whole 1.5–3 cm long. Nut compressed, 1–1.5 cm long.

Dry or moist woods and thickets; in the UP, known only from the counties bordering Wisconsin.

Corylus cornuta Marsh.
BEAKED HAZELNUT native
IR | **WUP** | **CUP** | EUP   FACU CC 5

Shrub 1–3 m tall, the young twigs villous, later nearly glabrous. Leaves oblong or obovate, broadly rounded to subcordate at base, pale green beneath, pubescent, especially on the veins and in the vein-axils beneath; margins coarsely doubly serrate. Involucre usually densely bristly toward the base, closely surrounding the nut and prolonged beyond it into a long slender beak cut at the summit into narrowly triangular lobes, the whole 4–7 cm long. Nut short-ovoid, scarcely compressed, 1–1.5 cm long.

Moist woods and thickets.

Ostrya *Hop-Hornbeam*

Ostrya virginiana (P. Mill.) K. Koch
HOP-HORNBEAM native
IR | **WUP** | **CUP** | EUP   FACU CC 5

Tree or tall shrubs to 20 m tall, with light brown scaly bark; twigs and petioles at first pilose, eventually nearly glabrous, occasionally also stipitate-glandular. Leaves alternate, narrowly to broadly oblong or ovate; margins sharply and often doubly serrate. Catkins opening with the leaves in spring; staminate catkins elongate, densely flowered, composed of spirally arranged scales, tipped with a sharp point and each subtending a cluster of several stamens; filaments shortdivided at the summit, each branch bearing a half-anther; pistillate catkins short-cylindric, 3–5 cm long, loosely flowered, the ovate, hairy bracts early deciduous; calyx minute. Fruit a flattened-ovoid nutlet about 5 mm long.

Moist or dry woods and banks.

Boraginaceae
BORAGE FAMILY

Annual or perennial herbs with usually bristly stems and alternate, bristly leaves; plants glabrous in eastern bluebells (*Mertensia virginica*). Flowers typically in a spirally coiled, spike-like head that uncurls as flowers mature; flowers perfect (with both staminate and pistillate parts), with 5 petals, 4–5 sepals, and 5 stamens. Fruit a dry capsule with 4 nutlets.

120 BORAGINACEAE | Borage Family

dicots

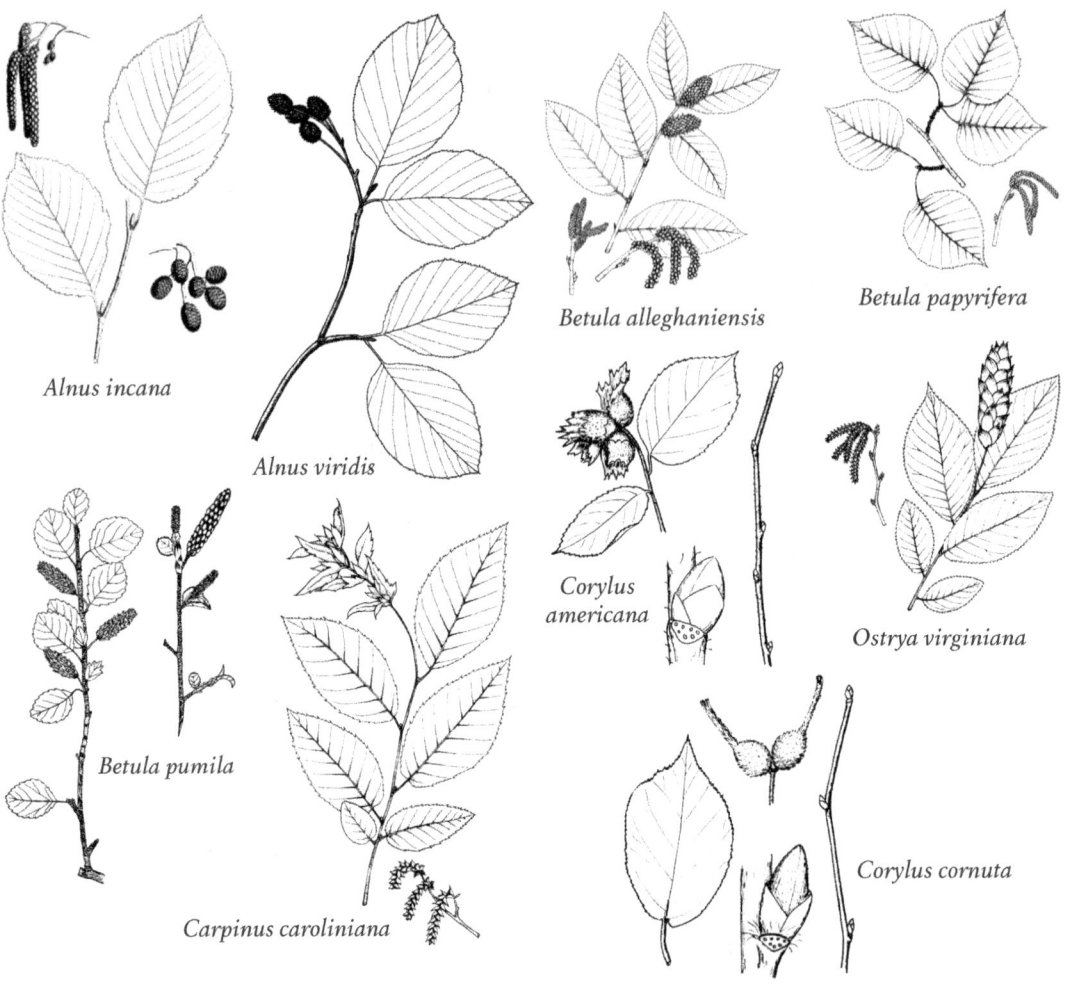

Hydrophyllum previously included in Hydrophyllaceae and lack the deeply 4-lobed ovary of other Boraginaceae.

ADDITIONAL SPECIES

Anchusa azurea P. Mill. (Italian bugloss), introduced; reported from Houghton County.

Borago officinalis L. (Borage), annual herb, 2–6 dm tall, with hirsute stem and bright blue flowers in large terminal cymes; sometimes cultivated as a salad herb; escaped in waste places and roadsides; reported from Keweenaw and Ontonagon counties.

Phacelia franklinii (R. Br.) A. Gray (Franklin's phacelia), native; main range mostly northwest of the Great Lakes; found here only along the north shore of Lake Superior, in Michigan only known from Isle Royale in open rocky and gravelly places; threatened.

1 Leaves shallowly palmately lobed to deeply pinnately divided . . . . **Hydrophyllum**
1 Leaves simple, entire 2
2 Plants glabrous . . . . **Mertensia virginica**
2 Plants hairy 3
3 Plants in flower 4
3 Plants in fruit 14
4 Corolla rotate, the upper portion of the corolla lobes reflexed and disk-like . . . . **Borago officinalis*
4 Corolla not rotate 5
5 Flowers irregular; stamens conspicuously exserted . . . . **Echium vulgare**
5 Flowers regular; stamens not longer than corolla 6
6 Corolla blue or purple, or leaf base extending downward along stem (decurrent) 7
6 Corolla not blue or purple; the leaf bases not decurrent along the stem 11
7 Flowers more than 1 cm long 9
7 Flowers less than 1 cm long; leaf bases not decurrent 10
8 Leaf bases decurrent along stem . . . . **Symphytum**
8 Leaf bases not decurrent along stem . . . . **Mertensia paniculata**

9	Leaves 2 cm or more wide, or calyx lobes 5 mm long or more .... **Cynoglossum**
9	Leaves less than 2 cm wide and calyx lobes less than 5 mm long.. 10
10	Flowers all subtended by bracts .... **Lappula**
10	Only lowest flowers with bracts............ **Myosotis**
11	Style 2-lobed, stigmas 2 .... **Lithospermum**
11	Style not lobed, stigma 1 12
12	Leaves 2 cm or more wide.................. **Hackelia**
12	Leaves less than 2 cm wide 13
13	Flowers with subtending bracts .... **Lappula**
13	Most flowers without subtending bracts ..... **Myostis**
14	Nutlets covered with bristly hairs, the hairs hooked at tip 15
14	Nutlets not covered with bristly hairs............... 17
15	Leaves less than 1 cm wide................. **Lappula**
15	Leaves more than 1 cm wide 16
16	Sepals when mature more than 5 mm long .... **Cynoglossum**
16	Sepals when mature less than 5 mm long .... **Hackelia**
17	All flowers subtended by bracts.................... 18
17	All, or at least upper, flowers without subtending bracts 20
18	Lateral veins on leaves conspicuous ... **Lithospermum**
18	Lateral veins absent or very faint.................. 19
19	Plants covered with coarse, stiff hairs, the hairs 2–3 mm long.................................. **Echium vulgare**
19	Pubescence various, the hairs less than 2 mm long.... .... **Lithospermum**
20	Calyx lobes to 6 mm long 21
20	Calyx lobes more than 6 mm long 22
21	Nutlets smooth and shiny .... **Myosotis**
21	Nutlets wrinkled and dull ...... **Mertensia paniculata**
22	Leaves less than 1 cm wide, the petioles not winged... .... **Amsinckia lycopsoides**
22	Leaves mostly more than 1 cm wide, the petioles usually winged.. 23
23	Plants usually more than 60 cm high; flowers in scorpioid (coiled) cymes .... **Symphytum**
23	Plants shorter, less than 60 cm high; flowers not in scorpioid cymes .... **Borago officinalis***

Cynoglossum *Hound's-Tongue*

Biennial or perennial herbs. Leaves large, usually pubescent. Flowers pediceled in elongating, bractless, axillary and terminal racemes. Calyx deeply parted, in fruit reflexed by the growth of the nutlets. Corolla broadly funnelform, the short tube closed by 5 appendages at the throat. Stamens included in the corolla tube. Style slender; stigma 1. Nutlets with conspicuous, stout, hooked bristles.

1	Flowers red-purple; leaves many, continuing upward on stem into inflorescence, not clasping stem **C. officinale**
1	Flowers blue; leaves few, not in inflorescence, the upper leaves clasping at base .... **C. boreale**

Cynoglossum boreale Fern.
WILD COMFREY *native*
IR | WUP | CUP | EUP CC 7

Cynoglossum virginianum subsp. *boreale* (Fern). A. Haines

Perennial herb. Stems erect, unbranched, 4–8 dm tall. Basal leaves elliptic-oblong, the blades 1–2 dm long, tapering at base and decurrent upon the long petiole; stem leaves sessile, progressively smaller, some broadly clasping at base, some often narrowed below and more or less expanded at the very base. Racemes 1–4, usually 3, at maturity 1–2 dm long, terminating a long, erect, terminal peduncle; calyx at anthesis 3–4 mm long; corolla blue, 8–12 mm wide, its broadly rounded lobes more or less overlapping; fruiting pedicels 5–15 mm long, recurved. Nutlets 6–8 mm long, uniformly bristly over the exterior surface. May–June.

Upland woods.

Cynoglossum officinale L.
HOUND'S-TONGUE *introduced*
IR | WUP | CUP | EUP UPL

Biennial herb, the stem and foliage finely and usually softly pubescent. Stems erect, branched above, 6–10 dm tall. Basal and lower leaves oblong or oblong lance-shaped, to 3 dm long, tapering to a long petiole-like base; upper leaves progressively shorter and proportionately narrower, the uppermost sessile. Racemes numerous, divaricate, eventually 1–2 dm long; mature pedicels spreading; corolla dull red or red-purple, rarely white, about 8 mm wide. Nutlets 5–8 mm long, uniformly bristly, surrounded by a low ridge with crowded bristles. May–Aug.

Native of Eurasia; established in fields, meadows, and open woods.

Echium *Viper's-Bugloss*

Echium vulgare L.
COMMON VIPER'S-BUGLOSS *introduced*
IR | WUP | CUP | EUP

Biennial, very hispid herb. Stems erect, simple or branched, 4–8 dm tall. Leaves linear-oblong to oblong lance-shaped, the basal to 15 cm long, the upper progressively smaller, the uppermost bractlike. Cymes numerous, in the axils of the upper foliage leaves, at first coiled, straightening with age. Flowers sessile, crowded, subtended by linear bracts; calyx deeply parted; corolla blue, rarely white, pubescent, narrowly campanulate, somewhat curved, conspicuously longer on the upper side, 12–20 mm long; stamens inserted near the middle of the corolla

BORAGINACEAE | Borage Family

tube, the slender filaments unequal in length, long-exsert; style elongate, usually pubescent, shortly 2-cleft at the summit. Nutlets ovoid, 3-angled, rough.

Native of s Europe, weedy in waste places, roadsides, and meadows, usually where sandy or gravelly.

Hackelia *Stickseed; Beggar's-Lice*

Perennial herbs with numerous, usually paired racemes terminating the axillary branches; racemes bracteate for at least part of their length; flowers small, blue or white. Corolla deeply 5-cleft into narrow lobes. Corolla salverform or broadly funnelform, the tube shorter than to scarcely surpassing the calyx, the throat nearly closed by the small appendages. Stamens and style short, included in the corolla tube. Fruiting pedicels short, recurved or reflexed. Nutlets attached by a lance-shaped to ovate area occupying the middle third only, the terminal third free, the basal third free and often with 2 low divergent keels; dorsal area lance-shaped, bordered by a row of hooked bristles and in some species bearing similar bristles on the surface.

1 Corolla blue; widest stem leaves to 2.5 cm broad...... .... **H. deflexa**
1 Corolla white; widest stem leaves to 3–5 cm broad.... .... **H. virginiana**

Hackelia deflexa (Wahlenb.) Opiz
NODDING STICKSEED   *native*
IR | **WUP** | **CUP** | EUP   CC 2

Stems to 1 m tall, freely branched above, the branches all terminated by racemes (usually paired). Lower leaves long-tapering to a petiole-like base; middle and upper leaves oblong-elliptic, to 10 cm long, sharply pointed at both ends. Racemes eventually 5–10 cm long, spreading; bracts linear or lance-shaped, reduced above and often absent beyond the middle of the raceme, opposite the flowers or alternate with them; fruiting pedicels 2–4 mm long, abruptly deflexed at base; corolla white or pale blue, about 2 mm wide. Nutlets with an ovate dorsal area 2–3 mm long, with a few short bristles on the back, bearing a marginal row of flat hooked bristles. May–Aug.

Moist woods, thickets, and hillsides.

Hackelia virginiana (L.) I. M. Johnston
BEGGAR'S-LICE   *native*
IR | **WUP** | **CUP** | EUP   FACU CC 1

Stems to 1 m tall, freely branched above and bearing numerous racemes. Lower leaves narrowed to a petiole, to 2 dm long; middle and upper leaves oblong-elliptic, 5–10 cm long, about 1/4 to 1/3 as wide, sharply narrowed to both ends, sessile; uppermost leaves, above the lowest flowering branch, progressively reduced and passing into the small, lance-shaped to linear bracts. Racemes eventually 5–15 cm long, spreading; bracts often alternate with the flowers, those beyond the middle of the raceme minute or lacking; fruiting pedicels 2–5 mm long, reflexed or recurved; corolla white or pale blue, about 2 mm wide. Nutlets forming a globose cluster of 4, bearing about 10–15 erect bristles as long as the marginal ones. July–Sept.

Dry or moist upland woods.

Hydrophyllum *Waterleaf*

Perennial herbs from horizontal rhizomes. Leaves large, lobed or divided. Flowers several to many in a repeatedly forked cyme. Sepals separate to below the middle or nearly to the base. Corolla campanulate to tubular, lobed to or below the middle, the lobes erect or somewhat spreading, white to purple. Stamens equaling the corolla or exsert, the slender filaments usually villous. Style 1, shortly bifid at the summit. Ovary 1-celled. Capsule globose, hispid or pubescent.

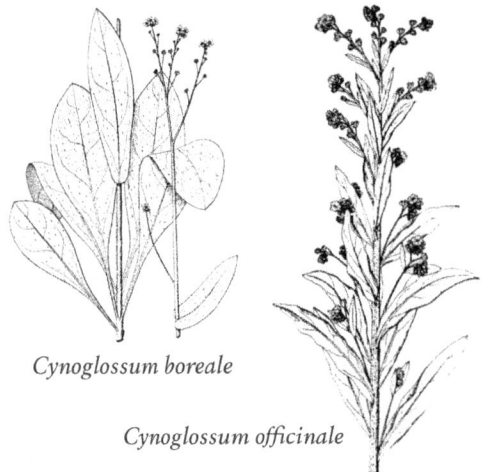

Cynoglossum boreale

Cynoglossum officinale

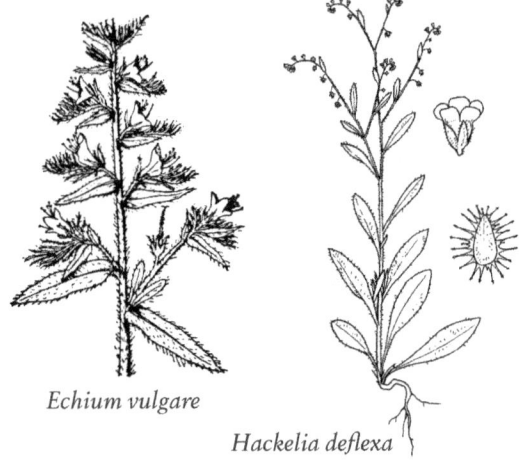

Echium vulgare

Hackelia deflexa

1. Leaves pinnately compound, divided into 5–7 leaflets or lobes; Alger and Mackinac counties . . . **H. virginianum**
1. Leaves palmately lobed; Delta County . . . . **H. appendiculatum**

Hydrophyllum appendiculatum Michx.
GREAT WATERLEAF *native*
IR | WUP | **CUP** | EUP CC 7

Stems at anthesis 3–6 dm tall, the upper portion and inflorescence densely pubescent with short hairs 0.3–0.5 mm long and also conspicuously hirsute with longer hairs 2–3 mm long. Stem leaves mostly overtopped by the cymes, orbicular in outline, 6–15 cm wide at anthesis, shallowly 5–7 lobed; sepals separate nearly to the base, lance-shaped, densely hirsute, alternating with small but conspicuous reflexed appendages; corolla lavender or pink-purple, 9–13 mm long, the lobes about equaling the tube; stamens as long as or slightly longer than the corolla. May–June.

Rich moist woods; Delta County.

Hydrophyllum virginianum L.
EASTERN WATERLEAF *native*
IR | WUP | **CUP** | EUP FAC CC 4

Stems 3–8 dm tall at anthesis, the upper portion, cymes, pedicels, and back of the sepals strigose with short hairs rarely to 0.5 mm long. Stem leaves broadly ovate triangular in outline, 1–2 dm long and usually somewhat wider, pinnately lobed almost to the midvein, the segments usually 5, occasionally 7 or 9, the terminal one and the basal pair often 2–3-lobed, all with sharply acute apex and similar, strongly ascending teeth. Cymes very dense at anthesis; sepals sparsely hirsute; corolla white to pale pink-purple, 7–10 mm long; stamens long-exsert. May–June.

Moist or wet woods, or open wet places; Alger and Mackinac counties.

Lappula *Stickseed; Beggar's-Lice*

Lappula squarrosa (Retz.) Dumort.
TWO-ROW STICKSEED *introduced*
IR | WUP | **CUP** | EUP

Roughly pubescent annual herb. Stems 2–8 dm tall, usually simple to above the middle, then freely branched, each branch terminating in an elongate bracteate raceme of small blue flowers. Leaves linear, linear-oblong, or linear-oblong lance-shaped, 2–5 cm long, usually ascending, acute or obtuse, narrowed to a sessile base, roughly hirsute. Racemes numerous, eventually 5–10 cm long; pedicels at maturity erect or ascending, 5–10 mm apart, 1–2 mm long; bracts linear or lance-shaped, 3–10 mm long; corolla blue, 2–3 mm wide, salverform or broadly funnelform, the tube about as long as the calyx, the throat closed by 5 scales; calyx deeply 5-cleft into narrow lobes; stamens and style short, included in the calyx. Nutlets 3–4 mm long, the lance-shaped face surrounded by 2 rows of bristles, those of the inner row usually the longer (a single nutlet viewed from the end will show whether the bristles are in a single or double row on each margin). May–Sept.

Native of Asia and the Mediterranean region; established as a weed in waste places.

Fruit is needed to separate *Lappula* from the similar *Hackelia*.

Lithospermum
Gromwell; Puccoon; Stoneseed

Perennial herbs (ours) with pubescent stem and foliage. Leaves narrow. Flowers solitary in the axils or crowded into a terminal leafy-bracted cyme. Calyx lobes narrow, separate nearly to the base. Corolla funnelform or salverform; the tube slender or wide, appendaged at the summit. Stamens inserted in the corolla tube, the anthers included or partly exsert. Style shortly 2-lobed. Nutlets bony, ovoid to nearly globose, smooth or pitted, usually only 1 or 2 ripening in each flower.

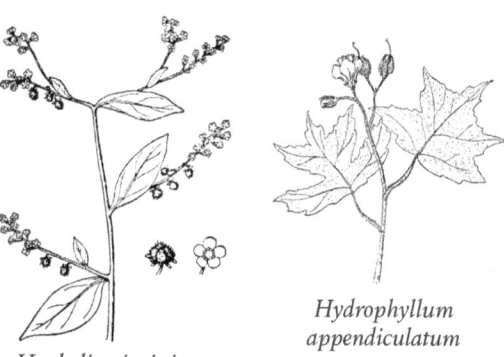

Hackelia virginiana

Hydrophyllum appendiculatum

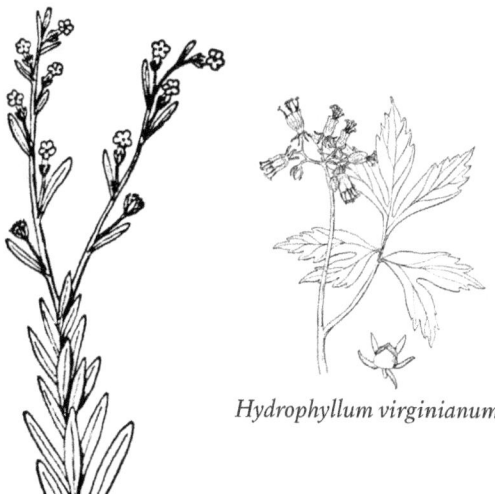

Hydrophyllum virginianum

Lappula squarrosa

BORAGINACEAE | Borage Family

1. Flowers white to pale yellow from leaf axils .... **L. officinale**
1. Flowers light to deep yellow or yellow-orange in a terminal inflorescence............................. 2
2. Plants soft-hairy; calyx lobes less than 5 mm long; Menominee County .... **L. canescens**
2. Plants stiffly hairy; corolla lobes more than 5 mm long .... **L. caroliniense**

Lithospermum canescens (Michx.) Lehm.
HOARY PUCCOON *native*
IR | WUP | **CUP** | EUP   CC 10

Perennial herb. Stems often several from a thickened vertical root, at anthesis 1–4 dm tall, usually simple. Leaves ascending, lance-shaped, 2–6 cm long (or the lowest greatly reduced), softly and densely canescent, obtuse. Inflorescence of 1–3 densely flowered, terminal or subterminal, leafy-bracted cymes; calyx lobes linear, 3–6 mm long, densely villous; corolla bright yellow, the tube about 8 mm long, the limb 10–15 mm wide. Nutlets yellowish white, smooth and shining. April–May.

Moist or dry prairies and dry open woods; Menominee County.

Lithospermum caroliniense (Walt.) MacM.
PLAINS PUCCOON *native*
IR | WUP | **CUP** | **EUP**   CC 10

Perennial herb. Stems erect from a stout woody root, at anthesis 3–6 dm tall, very leafy, simple or branched above, villous or hirsute. Leaves linear to lance-shaped, 3–6 cm long, roughly hirsute, the hairs often papillate at base. Cymes at first dense, leafy-bracted, becoming elongate and racemiform after anthesis; calyx lobes linear, 9–11 mm long, hirsute; corolla tube 10–14 mm long, pubescent at the base within; corolla-limb bright orange-yellow, 15–25 mm wide. Nutlets ivory-white, smooth and shining. May–July.

In dry, moist or preferably sandy soil, upland woods, shores, and prairies.

Lithospermum officinale L.
EUROPEAN GROMWELL *introduced*
IR | WUP | **CUP** | **EUP**

Perennial from a thick root. Stems erect, usually much branched above, to 1 m tall, strigose, the principal internodes usually less than 2 cm long, often only 5 mm. Leaves nearly sessile, lance-shaped to oblong, mostly 6–15 mm wide, scabrous above, with 2 or 3 conspicuous veins on each side of the midvein. Flowers solitary in the axils of the crowded upper leaves, 3–15 mm apart at maturity, white or nearly so, 4–5 mm long; calyx lobes nearly as long as the corolla. Nutlets ovoid, 3–3.5 mm long, white to pale brown, shining, smooth or sparsely pitted. May–Aug.

Native of Eurasia; introduced as a weed of waste places.

Mertensia *Bluebells*

Perennial herbs; plants smooth or hairy. Leaves alternate and entire. Flowers usually blue (pink in bud), tube-, funnel- or bell-shaped, petals widened and shallowly lobed at tip; in small clusters at ends of stems and branches. Fruit a smooth or wrinkled nutlet.

1. Leaves and sepals hairy .... **M. paniculata**
1. Leaves and sepals without hairs; Schoolcraft County .. .. **M. virginica**

Mertensia paniculata (Ait.) G. Don
NORTHERN BLUEBELLS *native*
IR | **WUP** | **CUP** | EUP   FAC CC 8

Perennial herb. Stems erect, 3–10 dm long, branched above, smooth or with sparse hairs. Basal leaves ovate, rounded at base; stem leaves lance-shaped to ovate, 5–15 cm long, tapered to a tip, hairy, entire; petioles short on lower leaves, upper leaves more or less stalkless. Flowers blue-purple, narrowly bell-shaped, 10–15 mm long, on slender stalks, in few-flowered racemes at ends of stems and branches; sepal lobes lance-shaped, 3–6 mm long, with dense, short hairs. Fruit a nutlet. June–July.

Conifer swamps, streambanks, seeps.

Mertensia virginica (L.) Pers.
VIRGINIA BLUEBELLS (END) *native*
IR | WUP | **CUP** | EUP   FAC CC 10

Perennial herb; plants smooth. Stems upright, 3–7 dm long. Leaves oval to obovate, entire, 5–15 cm long, rounded or blunt at the tip; upper leaves stalkless, lower leaves with winged petioles. Flowers showy, blue-purple, trumpet-shaped, 5-lobed at tip, 2–3 cm long, stalked, in a cluster at end of stem; sepals rounded at tip, 3 mm long. Fruit a nutlet. April–May.

Floodplain forests, moist deciduous forests, streambanks; rare in Schoolcraft County.

Myosotis
Forget-Me-Not; Scorpion Grass

Perennial (sometimes annual) herbs; plants with short, appressed hairs. Leaves alternate and entire. Flowers blue, tube-shaped and abruptly flared outward at tip, in a 1-sided raceme. Fruit a nutlet.

1. Calyx hairs all straight-tipped, appressed 2
1. Calyx hairs mostly hooked at tip, spreading 3
2. Plants without stolons; lobes of sepals as long or longer than corolla tube; flowers up to 6 mm wide; nutlets longer than style .... **M. laxa**

BORAGINACEAE | Borage Family

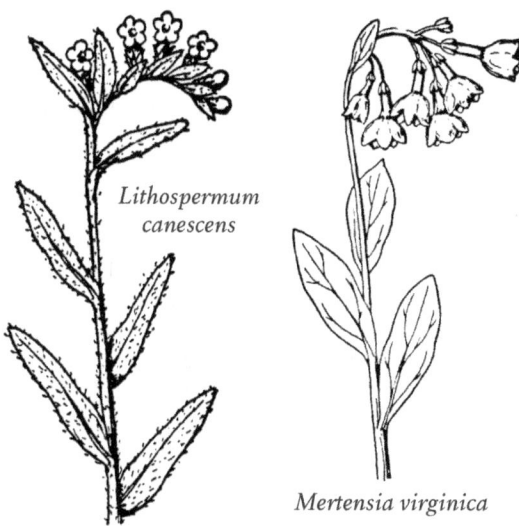
Lithospermum canescens
Mertensia virginica

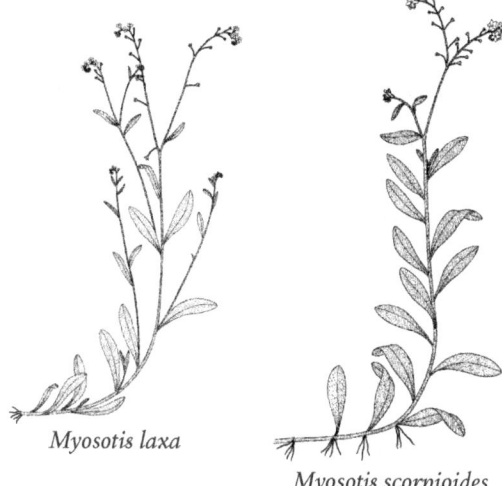
Myosotis laxa
Myosotis scorpioides

2   Plants creeping and spreading by stolons; lobes of sepals shorter than corolla tube; flowers mostly 6 mm or more wide; nutlets shorter than style .... **M. scorpioides**
3   Pedicels equal to length of calyx (when in full-flower or fruit); corolla 3 mm or more wide 4
3   Pedicels shorter than calyx; corolla 1–2 mm wide 5
4   Expanded part of petal cupped, less than 4 mm wide.. .... **M. arvensis**
4   Expanded part of petal flat, 5–10 mm wide .... **M. sylvatica**
5   Corolla white; calyx appearing 2-lipped, the 3 upper lobes shorter than the 2 lower; Keweenaw County .... **M. verna**
5   Corolla blue, calyx lobes nearly equal ...... **M. stricta**

Myosotis arvensis (L.) Hill
ROUGH FORGET-ME-NOT   *introduced*
IR | WUP | CUP | EUP   FACU
Annual or biennial. Stems simple or branched, eventually 3–5 dm long. Leaves firm, oblong, varying to lance-shaped or oblong lance-shaped, the larger 2–5 cm long. Racemes becoming 1–2 dm long, usually completely bractless; fruiting pedicels 5–15 mm apart, divergent, 5–9 mm long; mature calyx 3–4.5 mm long, pubescent with both hooked and appressed hairs, the lobes slightly longer than the tube; corolla blue or white, broadly funnelform, the limb 2–3 mm wide. Nutlets 1.3–1.7 mm long. Summer.

Native of Eurasia; established in fields and roadsides.

Myosotis laxa Lehm.
SMALLER FORGET-ME-NOT   *native*
IR | WUP | CUP | EUP   OBL   CC 6
Short-lived perennial (sometimes annual) herb. Stems slender, 1–4 dm long, often lying on ground at base, but not creeping, with fine, short, appressed hairs. Leaves oblong or spatula-shaped, 2–6 cm long. Flowers blue, on stalks usually much longer than the flower, in 1-sided clusters at ends of stems, the clusters becoming open; sepals covered with short hairs, sepal lobes shorter than the tube; petal lobes shorter or slightly longer than the tube. Fruit a nutlet distinctly longer than the style. June–Sept.

Cedar swamps, wet shores and streambanks.

Myosotis scorpioides L.
TRUE FORGET-ME-NOT   *introduced (invasive)*
IR | WUP | CUP | EUP   OBL
Myosotis palustris (L.) Hill
Perennial herb. Stems 2–6 dm long, with short, appressed hairs, often creeping at base and producing stolons. Leaves 3–8 cm long and 0.5–2 cm wide, lower leaves oblong lance-shaped, upper leaves oblong or oval; stalkless or the lower leaves on short petioles. Flowers blue with a yellow center, tube-shaped, abruptly flared at tip, in a 1-sided raceme at ends of stems, becoming open; flower stalks spreading in fruit; sepals with short, appressed hairs, sepal lobes equal or shorter than the tube. Fruit a nutlet shorter than the style. May–Sept.

Streambanks, shores, ditches, swamps, wet depressions in forests.

Myosotis stricta Link
BLUE SCORPION-GRASS   *introduced*
IR | WUP | **CUP** | **EUP**
Annual. Stems slender, branched from the base, 1–2 dm tall, forking at the second or third node into the primary racemes. Leaves oblong lance-shaped, 8–20 mm long, all except the very lowest bearing flowers in their axils. Racemes constituting three-fourths of the height of the plant, or even more, the lower flowers subtended by foliaceous bracts; fruiting pedicels 1–1.5 mm long, more or less hirsute with obtuse hooked hairs; mature calyx about 4 mm long,

narrowed at the base, densely hirsute with hooked hairs, the lobes about as long as the tube; corolla blue, funnelform, about 1.5 mm wide, its tube not exceeding the calyx. Nutlets 1–1.2 mm long. April–July.

Native of Eurasia; locally introduced in dry waste places.

Myosotis sylvatica Ehrh.
GARDEN FORGET-ME-NOT *introduced (invasive)*
IR | WUP | CUP | EUP

Perennial. Stems eventually to 5 dm long, bearing several racemes. Leaves oblong to lance-shaped or spatulate, thin and soft, the larger 3–7 cm long. Racemes bractless, seldom more than 1 drn long; fruiting pedicels 5–15 mm apart, ascending or spreading, to 9 mm long; mature calyx 4–5 mm long, much shorter than the pedicel, densely pubescent with hooked hairs, the narrowly triangular lobes much longer than the tube; corolla blue or rarely white, salverform, the limb 5–8 mm wide. Nutlets 1.5–2 mm long. April–Sept.

Native of Eurasia; commonly cultivated for ornament and sometimes escaped near gardens.

Myosotis verna Nutt.
SPRING FORGET-ME-NOT *native*
IR | WUP | CUP | EUP   FACU CC 6

Myosotis virginica auct. non (L.) B.S.P.

Stems erect, often branched from the base, sometimes beginning to flower when only 1 dm tall, eventually to 4 dm tall. Leaves linear to oblong, oblong lance-shaped, 1–6 cm long. Racemes commonly bracteate at base or as far as the middle; pedicels at maturity shorter than the fruiting calyx, nearly straight, erect or appressed, usually less than 1 cm apart, the calyx itself ascending or spreading from the axis, at maturity 4–6 mm long, hirsute, 3 of its lobes distinctly shorter than the others; corolla white, 1–2 mm wide. Nutlets, 1.2–1.5 mm long. April–July.

Dry soil of upland woods and fields; Keweenaw County.

Symphytum *Comfrey*

Symphytum officinale L.
COMMON COMFREY *introduced*
IR | WUP | CUP | EUP

Perennial pubescent herb. Stems erect, branched, 5–10 dm tall, sparsely to densely hirsute. Leaves decurrent as two wings on the stem, lance-shaped or ovate lance-shaped, acute or acuminate, the lower to 2 dm long, narrowed at base to a winged petiole, the upper progressively smaller, on shorter petioles or sessile. Flowers numerous in dense scorpioid cymes; calyx deeply parted into linear lobes; corolla 12–18 mm long, dull blue or dull yellow; stamens inserted at the summit of the corolla tube. Nutlets brown, shining, smooth or slightly wrinkled dilated below to a finely toothed ring (visible under a 10x lens). June–Aug.

Disturbed places, sometimes escaping from gardens.

Brassicaceae
MUSTARD FAMILY

Annual, biennial or perennial herbs. Leaves simple or compound, alternate on stems or basal, smooth or hairy, some species with branched or star-shaped hairs. Flowers in terminal or lateral clusters (racemes), the lower portion often fruiting while tip in flower, the stalks elongating in fruit. Flowers perfect (with both staminate and pistillate parts), cross-shaped, with 4 sepals and 4 yellow, white, pink or purple petals; stamens 6, the outer 2 stamens shorter than the inner 4; pistil 1, style 1, ovary superior. Fruit a cylindrical or round pod (silique or silicle) with 2 chambers and 1 to many seeds in 1 or 2 rows in each chamber.

ADDITIONAL SPECIES
Braya humilis (C. A. Mey.) B. L. Rob. (Low northern rock cress), native; *Braya* is a boreal and arctic

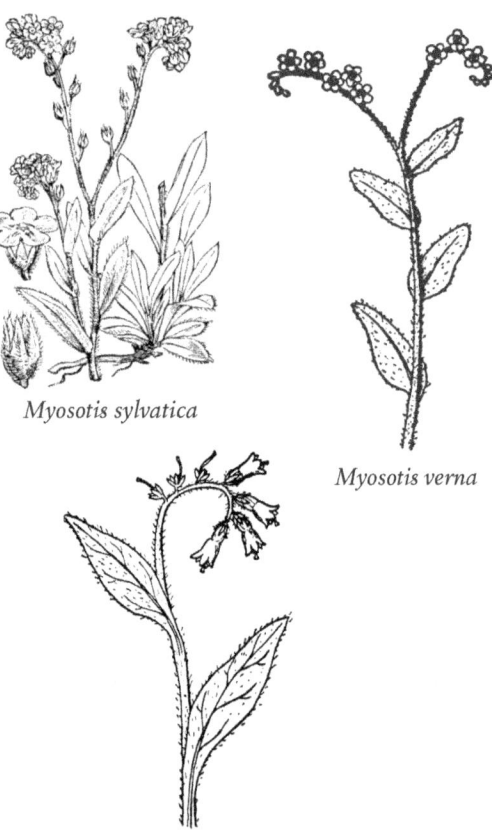

Myosotis sylvatica

Myosotis verna

Symphytum officinale

genus, collected in Keweenaw Co. at Copper Harbor in 1895 and on rocks at Isle Royale in 1933. The fruit is pubescent as well as somewhat constricted between the seeds. Threatened.

Chorispora tenella (Pall.) DC. (Purple rocket); introduced from Asia in dry, open, disturbed places; reported from Dickinson County.

Iberis sempervirens L. (Evergreen candytuft); European subshrubby perennial forming small evergreen mounds; cultivated and rarely escaping to disturbed open places. The flowers are white and the inflorescences elongate in fruit.

Iberis umbellata L. (Globe candytuft); introduced Mediterranean annual, occasionally spreading from cultivation to shores and disturbed places.

Lobularia maritima (L.) Desv. (Sweet alyssum), introduced ornamental, sometimes escaping to roadsides and disturbed places.

Lunaria annua L. (Money-plant, Honesty), introduced biennial ornamental, sometimes escaping to roadsides and weedy places; reported for Schoolcraft County.

1. Petals yellow, yellow-tinged, or orange 2
1. Petals white, greenish, pink, purple, or absent 20
2. Leaves simple, not deeply lobed 3
2. At least the lower leaves lobed, pinnately lobed, or pinnately compound. 10
3. Plants glabrous throughout 4
3. Plants pubescent, at least near the base 7
4. Stem leaves not clasping at their base. . . . . **Rorippa**
4. Stem leaves lobed at base and clasping the stem. 5
5. Leaves entire . . . . **Conringia orientalis**
5. At least some leaves wavy-margined or finely toothed 6
6. Uppermost leaves entire or nearly so, narrow (5–10 times as long as wide); fruit glabrous; petals 7–10 mm long. . . . . **Brassica**
6. Uppermost leaves (e.g., at base of main branches of inflorescence) coarsely toothed, 2–4 times as long as wide; fruit glabrous or densely hairy; petals mostly 10–15 mm long . . . . **Sinapis**
7. At least some of the stem leaves clasping the stem . . . . **Camelina**
7. None of the leaves clasping stem 8
8. Ovary and fruit soon becoming much longer than wide . . . . **Erysimum**
8. Ovary and fruit rounded in outline and usually to 1–2 times longer than wide 9
9. Fruit about as long as wide; pubescence of well-branched hairs. . . . . **Alyssum alyssoides**
9. Fruit much longer than wide; leaves loosely pubescent, the hairs with few branches. . . . . **Draba**
10. Pedicels in the lower portion of the raceme subtended by leafy bracts. . . . . **Erucastrum gallicum**
10. Pedicels not subtended by bracts 11
11. Petals up to 5 mm long (including the slender or tapered claw), or if 6 mm long then plant a creeping or rhizomatous perennial. 12
11. Petals more than 5 mm long; plants never creeping perennials 15
12. Lobes of leaves rounded, nearly round in outline to broadly oval or obovate, the terminal lobe much larger than the lateral lobes; plants glabrous throughout . . . . **Barbarea**
12. Lobes of leaves pointed, mostly distinctly longer than wide, the terminal lobe similar in size to the lateral lobes; plants glabrous or pubescent 15
13. Leaves 2–3 times pinnately dissected; plants with tiny stalked glands, or pubescent with branched hairs, or both . . . . **Descurainia**
13. Leaves 1–2 times pinnately dissected; plants without glands, glabrous or pubescent, the hairs unbranched 14
14. Fruits linear, more than 5x longer than wide; style very short; plants taprooted annuals or winter annuals . . . . **Sisymbrium**
14. Fruits spherical to oblong, less than 5x longer than wide (if longer, then the plant a rhizomatous perennial); style stout to elongate; plants annuals or perennials **Rorippa**
15. Leaves all on lower half of stem . . . **Diplotaxis muralis**
15. Leaves present above the middle of stem 16
16. Basal rosette leaves often present at flowering time; plants glabrous throughout or nearly so; uppermost leaves usually with at least one pair of lobes. **Barbarea**
16. Basal rosette leaves usually absent or mostly withered at flowering time; plants glabrous or pubescent; uppermost leaves variously toothed or wavy-margined, but not distinctly lobed 17
17. Fruit bristly hairy, or the leaves clasping; plants glabrous throughout . . . . **Brassica**
17. Fruit not bristly hairy; leaves not clasping; plants not glabrous throughout 18
18. Petals pale yellow with dark veins; fruit strongly twisted and long-beaked . . . . **Raphanus raphanistrum**
18. Petals pale to deep yellow but without dark veins; fruit not twisted (or only slightly so), beak present or absent 19
19. Fruit widely spreading, beakless or nearly so, up to 1 mm wide; sepals to 4 mm long. . . . . **Sisymbrium**
19. Plants not as above . . . . **Brassica**
20. Leaves 3-parted or deeply palmately divided . . . . **Cardamine**
20. Leaves neither 3-parted nor palmately divided. 21
21. Petals pink or purple. 22
21. Petals white, greenish, or absent. 25
22. At least the lower and middle stem leaves pinnately lobed or dissected . . . . **Raphanus raphanistrum**
22. None of the stem leaves pinnately lobed or dissected 23

23 Plants succulent, glabrous or nearly so .... **Cakile edentula**
23 Plants pubescent, at least near the base or on the leaves 24
24 Stem leaves numbering less than 10, ovate or lance-ovate, entire to wavy-margined or wavy-toothed; petals less than 1.5 cm long .... **Cardamine**
24 Stem leaves more than 10, lance-shaped to ovate or oblong lance-shaped, entire to finely toothed; petals mostly 1.5 cm or more long ...... **Hesperis matronalis**
25 At least the upper leaves sessile and clasping at their base 26
25 Leaves not clasping the stem 36
26 Stem leaves pinnately compound; plants aquatic or subaquatic .... **Nasturtium officinale**
26 Leaves not all pinnately compound; plants never aquatic 27
27 Ovaries and fruit becoming linear, more than 5 times longer than wide 28
27 Ovaries and fruit up to 2 times longer than wide.... 33
28 Fruiting pedicels ± spreading, divaricate, or reflexed; fruit straight or somewhat curved and clearly spreading from the axis or even pendent .... **Boechera**
28 Fruiting pedicels strongly ascending to appressed; fruit straight, erect and closely appressed to the stem ... 29
29 Stem and leaves entirely glabrous or with a very few scattered hairs at the very base of the plant (especially on leaf margins and petioles); sepals ca. half as long as the petals; mature fruit 1.4–2.5 (–3.3) mm wide, with seeds in 2 rows in each locule .... **Boechera stricta**
29 Stem and leaves pubescent, at least at the base, with spreading simple or stellate hairs; sepals ca. 2/3 as long as the petals; mature fruit less than 1.3 mm wide, with seeds crowded into 1 row in each locule 30
30 Fruit rather strongly flattened; style-beak clearly narrower than mature fruit; stem pubescent with simple and/or forked (or stellate) hairs on at least the lower half or third, and leaves on the same portion ± pubescent (often stellate).............. **Arabis pycnocarpa**
30 Fruit ± terete or 4-angled, slightly if at all flattened at maturity; style-beak nearly or quite as wide as the fruit; stem pubescence only on the lowermost 1–3 full-grown internodes, and only lowermost leaves pubescent .... **Turritis glabra**
31 Plants glabrous throughout 32
31 Plants pubescent, at least below middle 34
32 Upper leaves nearly round in outline, less than 1.5 times longer than wide, clasping, the stem appearing to perforate the leaf.......................... **Lepidium**
32 Upper leaves lance-shaped to narrowly ovate, more than 2 times longer than wide, clasping the stem ... 33
33 Inflorescence of corymb-like racemes; fruit widest at their base, less than 5 mm long; style present**Lepidium**
33 Inflorescence of unbranched racemes, or branched 1–2 times but not corymb-like; fruit oval or obovoid, more than 5 mm long; style absent .... **Thlaspi arvense**

34 Pedicels densely hairy; stem leaves unlobed. **Lepidium**
34 Pedicels glabrous or nearly so; leaves lobed or unlobed 35
35 Basal leaves deeply incised to pinnate, present at flowering time; stem leaves strongly reduced in size, pinnate (or if entire then less than 1 cm wide); fruit triangle-shaped, truncate at the tip ... **Capsella bursa-pastoris**
35 Basal leaves absent or shriveled at flowering time; stem leaves well developed, entire to finely toothed, mostly more than 1 cm wide; fruit nearly round in outline .... **Lepidium**
36 Plants aquatic, the submersed leaves simple, all basal and awl-shaped, or dissected into numerous linear, thread-like segments 37
36 Plants not aquatic, or if aquatic then the leaves not as above 38
37 Plants small with all leaves basal and awl-shaped, usually flowering and fruiting under water .... **Subularia aquatica**
37 Plants larger, leaves dissected into numerous linear, thread-like segments .... **Rorippa aquatica**
38 Nearly all stem leaves deeply lobed, or pinnately divided or compound, or stem leaves absent 39
38 All or nearly all of the stem leaves simple, not lobed or pinnately divided 44
39 Lowest leaves 2–3 times pinnately divided or pinnately compound 40
39 Lowest leaves only once-pinnate (the segments sometimes with a few teeth).......................... 41
40 Upper leaves simple to lobed or once-pinnate; fruit nearly round in outline.................... **Lepidium**
40 All leaves 2–3 times pinnately divided or pinnately compound; fruit linear...................... **Descurainia**
41 Petals more than 6 mm long...................... 42
41 Petals to 5 mm long 43
42 Leaves pinnately compound, the lobes entire and all similar, the lobes of the upper leaves linear, the lobes of the lower leaves round in outline or nearly so .... **Cardamine**
42 Leaves pinnately lobed, the terminal lobe much larger than the smaller lateral lobes; upper leaves similar to the lower leaves, though usually smaller .... **Raphanus raphanistrum**
43 Plants aquatic, rooting at the nodes; stems succulent; petals about 5 mm long........ **Nasturtium officinale**
43 Plants not aquatic; stems slender and firm; petals to 4 mm long .... **Cardamine**
44 Basal and lower leaves with distinct petioles, rounded to heart-shaped at their base 45
44 Leaves narrowed to sessile or nearly sessile base ... 48
45 Stem leaves deeply toothed to pinnately divided; fruit to 6 mm long, but usually soon falling .... **Armoracia rusticana**
45 Stem leaves shallowly toothed or wavy-margined, or nearly entire; fruit more than 6 mm long, persistent 46

46 Stem leaves truncate to heart-shaped at base, petioles 5 mm or more long; petals less than 7 mm long; fruiting pedicels stout, about 5 mm long .... **Alliaria petiolata**

46 Stem leaves not both heart-shaped at base and on petioles as long as 5 mm; petals more than 7 mm long; fruiting pedicels slender, or if stout then much longer than 5 mm.. 47

47 Stem leaves fewer than 10, ovate, the margins entire to wavy or wavy-toothed; petals less than 1.5 cm long ... . **Cardamine**

47 Stem leaves more than 10, lance-shaped to ovate or oblong lance-shaped, the margins finely toothed; petals usually 1.5 cm or more long...... **Hesperis matronalis**

48 Ovaries and fruit less than 2 times longer than wide 49

48 Ovaries and fruit more than 2 times longer than wide 51

49 Fruit ascending or erect; the fruit and stems densely woolly hairy .... **Berteroa incana**

49 Fruit widely spreading; plants glabrous to pubescent.. 50

50 Plants glabrous or with tiny hairs, the hairs unbranched .... **Lepidium**

50 Plants pubescent, the hairs branched................ .... **Alyssum alyssoides**

51 Fruit linear, more or less round in cross-section, to 1 mm wide; plants blooming and then withering by late spring .... **Arabidopsis**

51 Fruit linear, flat, 1 mm or more wide; plants withering or persisting 52

52 Fruit to 18 mm long; plants usually not leafy above the middle; early blooming and withering by late spring .. .. **Draba**

52 Fruit longer than 18 mm; plants usually leafy throughout; plants persisting to late summer or early fall .... **Boechera**

Alliaria *Garlic-Mustard*

Alliaria petiolata (Bieb.) Cavara & Grande
GARLIC-MUSTARD *introduced (invasive)*
IR | **WUP** | **CUP** | **EUP** FACU
   Alliaria alliaria (L.) Britt.
   Alliaria officinalis Andrz.
Biennial, garlic-scented herb, glabrous or nearly so, the hairs simple. Stems erect, to 1 m tall, simple or little branched. Basal leaves in more or less evergreen rosettes; lower leaves kidney-shaped; stem leaves deltoid, 3–6 cm long and wide, coarsely toothed. Petals white, spatulate, 5–6 mm long; pedicels at maturity stout, about 5 mm long. Fruit widely divergent, 4–6 cm long; seeds black, nearly cylindric, about 3 mm long. May–June.

Native of Europe; invasive in rich, moist, shaded soil; also on roadsides or rarely in swamps; at scattered locales across the UP.

Alyssum *Madwort*

Alyssum alyssoides (L.) L.
PALE MADWORT *introduced*
IR | **WUP** | **CUP** | **EUP**
Annual herb. Stem, leaves, inflorescence, and fruits stellate-pubescent. Stems 5–25 cm tall, simple and erect, or branched from the base only. Leaves oblong lance-shaped, 6–15 mm long, entire, obtuse. Flowers pale yellow or nearly white, about 2 mm wide; petals narrowly oblong. Fruit on widely divergent pedicels, circular, 3–4 mm long, flat at the margin, convex toward the center; seeds 2 in each cell. May–June.

Native of Europe; weedy in waste places.

ADDITIONAL SPECIES

Alyssum murale Waldst. & Kit. (Yellowtuft), European species rarely escaped from cultivation to roadsides and waste places; Marquette Co. The flowers are a much deeper and brighter yellow than the pale petals of *A. alyssoides*.

Arabidopsis *Thalecress*

Annual or biennial herbs, more or less pubescent with branched hairs. Leaves mostly in a basal rosette. Sepals oblong, obtuse. Petals white, spatulate. Fruit linear, nearly terete, many-seeded.

1 Petals mostly 5–8 mm long, seeds ca. 1 mm long .... **A. lyrata**

1 Petals mostly 2–4 mm long, seeds to 0.5 mm long; Luce County .... **A. thaliana**

Arabidopsis lyrata (L.) O'Kane & Al-Shehbaz
LYRE-LEAF ROCKCRESS *native*
IR | **WUP** | **CUP** | **EUP** FACU CC 7
   Arabis lyrata L.
Biennial herb. Stems erect or ascending, branched from base, 1–4 dm tall, hirsute below (very rarely glabrous), glabrous or glabrescent above. Basal leaves

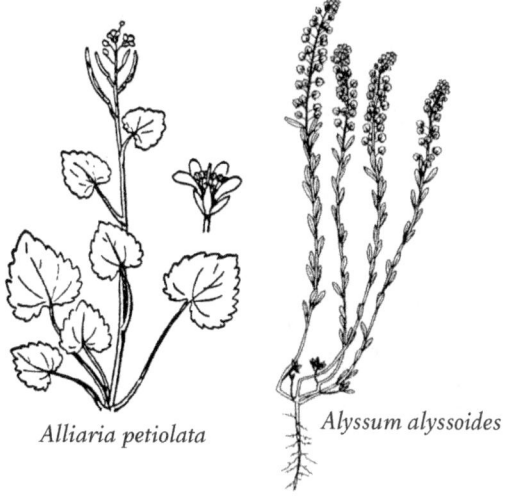

Alliaria petiolata   *Alyssum alyssoides*

spatulate, 2–4 cm long, entire to pinnately lobed; stem leaves linear to narrowly spatulate, the lowest sometimes with a few teeth or shallow lobes. Petals white, 3–8 mm long; pedicels at maturity widely ascending, 6–15 mm long. Fruit maintaining about the same direction as the pedicel, 2–4.5 cm long, occasionally shorter, about 1 mm wide; seeds wingless, about 1 mm long. April–June.

Dry woods and fields, especially in sandy soil; sand dunes.

Arabidopsis thaliana (L.) Heynh.
MOUSE-EAR CRESS *introduced*
IR | WUP | CUP | **EUP**

Annual or winter-annual. Stems branched from base and sparsely above, 1–4 dm tall. Leaves chiefly in a basal rosette, oblong, 1–5 cm long, stellate-pubescent. Stem leaves smaller, linear to narrowly oblong. Sepals pilose; petals white, 2–3 mm long. April–June.

Widespread in e USA; introduced in Michigan on disturbed, usually sandy sites, including cultivated land, fields, and oak forests; Luce County.

Arabis *Rockcress*

Most of our native species formerly in genus *Arabis* are now placed in genus *Boechera*, with either broader fruit (if tightly appressed) or the fruit spreading, reflexed or pendent; *Arabis glabra* (L.) Bernh. now placed in genus *Turritis*.

ADDITIONAL SPECIES
Arabis caucasica Willd. (Gray rockcress), garden escape, reported from Houghton and Mackinac counties.

Arabis pycnocarpa M. Hopkins
HAIRY ROCK CRESS *native*
IR | WUP | CUP | EUP   FACU CC 6

Arabis hirsuta (L.) Scop.
Biennial herb. Stems erect, 2–8 dm tall, pubescent at least at base with simple or branched hairs. Stem leaves oblong to linear lance-shaped, 1–3 cm long, sessile and more or less clasping, at least the lower pubescent. Mature pedicels erect, slender, 7–10 mm long; petals white. Fruit erect, flat, linear, 3–5 cm long, about 1 mm wide; seeds in one row, flattened and narrowly winged, 1–1.2 mm long. May–June.

Woods, often where calcareous.

Armoracia *Horse-Radish*

Armoracia rusticana P.G. Gaertn. B. Mey. & Scherb.
HORSE-RADISH *introduced*
IR | WUP | CUP | EUP

Glabrous perennial herb from thick roots. Stems erect, to 1 m tall. Lower leaves long-petioled, the blade oblong, 1–3 dm long, cordate at base; upper leaves smaller, short-petioled to sessile, lance-shaped. Racemes several, terminal and from the upper axils; petals white, obovate, 6–8 mm long; pedicels after anthesis ascending, 8–12 mm long. Fruit obovoid, inflated, 2-celled, eventually up to 6 mm long but usually falling early; seeds apparently never maturing. May–July.

Native of se Europe and w Asia; commonly cultivated and escaped into moist soil of ditches, shores, roadsides, and disturbed places.

Barbarea *Yellow-Rocket*

Biennial herbs, smooth or with a few simple hairs. Basal leaves pinnatifid with a large terminal lobe and 2 to several small lateral lobes; stem leaves smaller, entire to pinnatifid. Petals yellow, spatulate to obovate. Short stamens partly surrounded at base by a semicircular gland; long stamens separated by a short erect gland. Ovary cylindric, narrowed to a slender style. Fruit linear, terete or obscurely 4-angled, several-seeded, tipped by the persistent style.

1 Petals 6–8 mm long; beak of fruit 2–3 mm long .... **B. vulgaris**
1 Petals to 5 mm long; beak of fruit less than 2 mm long. .... **B. orthoceras**

Barbarea orthoceras Ledeb.
AMERICAN YELLOW-ROCKET *native*
IR | WUP | CUP | EUP   OBL CC 10

Biennial herb; plants smooth or with sparse covering of unbranched hairs. Stems 3–8 dm long, unbranched, or branched above. Leaves simple or with 1–4 pairs of lateral lobes, the middle and upper leaves deeply lobed. Flowers in racemes; on short stalks to 1 mm long, the stalks clublike at tip; petals yellow, 3–5 mm long. Fruit upright, 2–4 cm long, with a beak 0.5–2 mm long. June–July.

Rocky shores, swamps and wet woods.

Barbarea vulgaris Ait. f.
GARDEN YELLOW-ROCKET *introduced*
IR | WUP | CUP | EUP   FAC

Stems erect, branched above, 2–8 dm tall. Basal leaves petioled, with 1–4 pairs of small, elliptic to ovate, lateral lobes and a large ovate to rotund terminal lobe; stem leaves progressively shorter petioled and with fewer lobes, the upper sessile or clasping, angulately toothed, repand, or entire. Flowers yellow, about 8 mm wide, crowded at anthesis; pedicels at maturity 3–6 mm long, about 0.5 mm thick. Fruit 1.5–3 cm long, the beak 2–3 mm long. April–June.

Native of Europe; naturalized as a common weed in damp soil of fields, roadsides, and gardens.

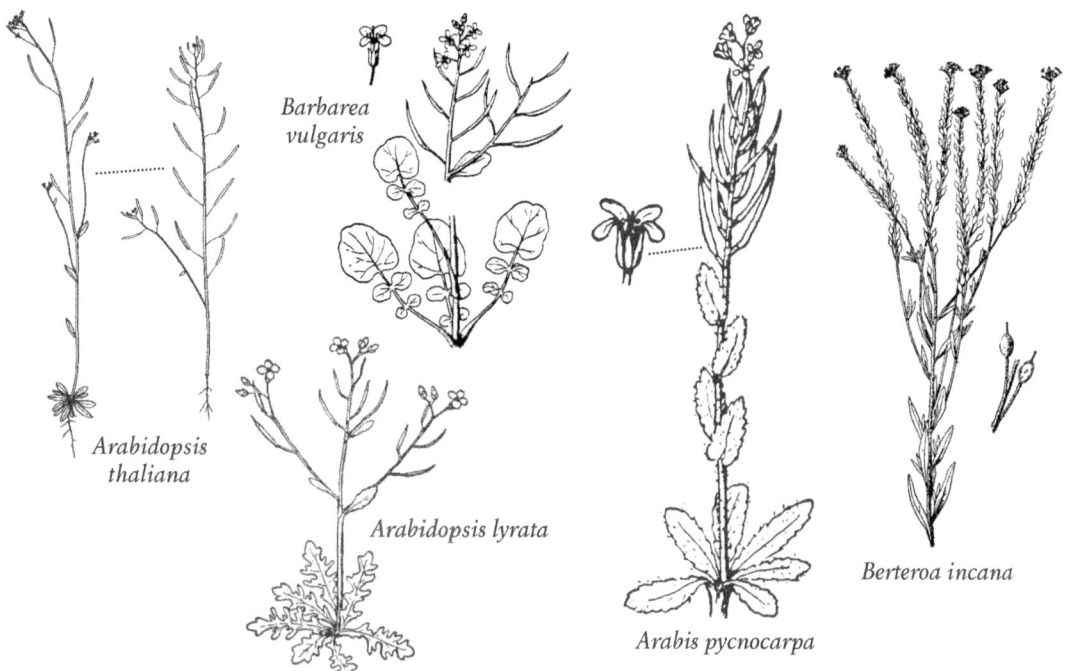

Berteroa *Hoary Alyssum*

Berteroa incana (L.) DC.
HOARY ALYSSUM *introduced*
IR | WUP | CUP | EUP

Annual herb. Stem, foliage, and inflorescence finely canescent, the hairs stellate with radiating branches. Stems stiffly erect, usually branched above, to 7 dm tall. Leaves oblong lance-shaped, 2–5 cm long, acute, entire. Sepals ascending; petals white, 2-lobed, about 3 mm wide. Fruit elliptic, thinly pubescent, 5–8 mm long, 3–4 mm wide; seeds 3–6 in each cell. May–Sept.

Native of Europe; now established as a weed.

Boechera *Rockcress*

Biennial or perennial herbs. Basal leaves petioled, the stem leaves smaller and usually sessile; pubescence usually present, of simple, forked, or stellate hairs. Sepals erect or spreading, the outer pair sometimes saccate at base. Petals white, yellowish, or pink, spatulate to oblong or obovate. Ovary cylindric. Fruit linear, elongate, flat or subterete, many-seeded. Seeds flattened, often with a marginal wing. Includes former native members of genus *Arabis*.

1 Pedicels ascending to strongly appressed, even after anthesis; the fruit erect, closely appressed to the stem; sepals various . . . . **B. stricta**
1 Pedicels becoming distinctly reflexed before the petals wither, the fruit pendent 2
2 Stems with stellate pubescence only at the very base, otherwise glabrous; sepals glabrous; mature fruit spreading to loosely pendent, the pedicels more arched than reflexed . . . . **B. grahamii**
2 Stems with stellate pubescence at least on lower half; sepals at least sparsely stellate-pubescent; mature fruit strongly pendent, the pedicels sharply reflexed . . . .
 . . . . **B. retrofracta**

Boechera grahamii (Lehmann) Windham & Al-Shehbaz
SPREADING ROCKCRESS *native*
IR | WUP | CUP | EUP   FACU CC 6
Arabis divaricarpa A. Nels.

Biennial herb. Stems erect, to 1 m tall, glabrous except at the very base. Basal leaves oblong lance-shaped, finely stellate-pubescent on both sides; stem leaves linear lance-shaped, erect or nearly so, 2–5 cm long, sessile, auriculate at base, entire, glabrous on both sides. Petals pinkish or white, 5–8 mm long; pedicels at maturity widely spreading, 6–12 mm long. Fruit at first erect, soon widely spreading, linear, straight or nearly so, 3–9 cm long, 1.2–2.2 mm wide, the valves 1-nerved to or beyond the middle; seeds in 1 row and broadly quadrate to orbicular, or in 2 rows and oblong. June–July.

Sandy or rocky soil.

Boechera retrofracta (Graham) Á. & D. Löve
ROCK CRESS *native*
IR | WUP | CUP | EUP   CC 10
Arabis holboelii Hornem.

Biennial or perennial herb. Stems erect, 1–9 dm tall, 1 to several from a simple or branching crown, simple or branched above, pubescent throughout with appressed or spreading hairs, sometimes glabrous above. Basal leaves linear-oblanceolate to spatulate, entire to somewhat dentate, densely pubescent with stellate hairs, 1–5 cm long and to 6 mm wide; stem leaves auriculate and clasping to nonauriculate, oblong to lanceolate, 1–4 cm long and to 6 mm wide, lower surface densely pubescent, upper surface pubescent to glabrous, entire. Flowers in a loose raceme; sepals oblong, 2–4 mm long; petals spatulate with a narrow claw, purplish- pink to whitish, 5–10 mm long. Fruit a glabrous silique, straight to slightly curved, reflexed to pendulous, 5–7 cm long and 1– 2.5 mm wide; fruiting pedicels straight to somewhat curved, reflexed to loosely descending, pubescent or glabrous; seeds narrowly winged all around, about 1 mm wide, mostly in a single row. May– July.

Sand dune ridges, rock ledges and summits.

On sand dunes, plants sometimes have a distinctive habit in that the rosettes are above the ground level.

Boechera stricta (Graham) Al-Shehbaz
DRUMMOND'S ROCKCRESS *native*
IR | WUP | CUP | EUP   FACU CC 6
Arabis drummondii A. Gray
Biennial herb. Stems erect, 3–9 dm tall, glabrous, or at base very thinly pubescent with 2–pronged hairs, often glaucous. Stem leaves sessile, lance-shaped to narrowly oblong, 2–8 cm long, acute, entire or with a few teeth, auriculate at base, usually wholly glabrous. Petals pink to purple (often drying whitish)5–9 mm long; fruiting pedicels erect, 10–15 mm long. Fruit straight, erect, flat, 4–7 cm long or rarely longer, 1.5–2.5 mm wide or occasionally wider; seeds in 2 rows. May–Aug.

Moist or dry places, often where calcareous.

Brassica *Mustard*

Coarse annual or biennial herbs. Leaves (at least the lower) pinnatifid. Sepals erect or spreading, often sac-like at base. Petals yellow (ours), varying to nearly white in some cultivated species, obovate, clawed. Ovary nearly cylindric, scarcely narrowed to the short style. Fruit nearly terete to angled, more or less elongate, few–several-seeded, terminated by a conspicuous beak sometimes containing a seed at its base. Seeds nearly globose, in one row.

Many species have been long-cultivated, and plants may persist in gardens over winter, blooming the second year.

The oilseeds known as canola are sometimes varieties of *Brassica rapa* but are mostly *B. napus* and *B. juncea*.

ADDITIONAL SPECIES
Brassica napus L. (Turnip), cultivated and sometimes escaping but not persistent.

1 Middle and upper leaves clasping stem .... **B. rapa**
1 Leaves not clasping stem.......................... 2
2 Fruit to 2 cm long, strongly appressed........ **B. nigra**
2 Fruit becoming more than 2 cm long, not strongly appressed................................ **B. juncea**

Brassica juncea (L.) Czern.
CHINESE MUSTARD   *introduced*
IR | **WUP** | CUP | EUP
Annual herb, glabrous, often somewhat glaucous. Stems branched, 3–10 dm tall. Lower leaves to 2 dm long, pinnatifid and dentate, the upper progressively reduced, short-petioled to sessile. Flowers 12–15 mm wide. Pedicels at maturity ascending, 10–15 mm long. Fruit ascending, nearly terete, 1.5–4 cm long; beak slender, subulate, 6–9 mm long; seeds brown, about 2 mm long, conspicuously and evenly reticulate. June–Oct.

Established as a weed in waste places and fields; Houghton and Keweenaw counties.

Brassica nigra (L.) W.D.J. Koch
BLACK MUSTARD   *introduced*
IR | **WUP** | CUP | EUP
Annual herb. Stems simple or branched, to 15 dm tall, usually bristly below, glabrate or glabrous above. Leaves all petioled, ovate to obovate, the lower commonly lobed, the upper merely dentate. Flowers 8–10 mm wide. Pedicels at maturity erect, 3–4 mm long. Fruit erect, quadrangular, 1–2 cm long, smooth; beak slender, 2.5–4 mm long; seeds brown, 1.5–2 mm long, minutely roughly reticulate. Summer and fall.

Naturalized in fields and waste places; Delta and Keweenaw counties.

Brassica rapa L.
FIELD MUSTARD; TURNIP   *introduced*
IR | **WUP** | CUP | EUP
Brassica campestris L.
Annual or biennial herb, green or glaucous, glabrous or nearly so. Stems branched, to 8 dm tall. Lower leaves petioled, more or less pinnately lobed; upper leaves oblong to lance-shaped, dentate or entire, sessile and clasping. Flowers about 10 mm wide; pedicels at maturity widely ascending, 1–2 cm long. Fruit ascending to nearly erect, terete or nearly so, 3–5 cm long; ; beak slender, 8–15 mm long; seeds dark brown, 1.5–2 mm long, minutely roughened. May–Oct.

Naturalized as a weed of fields and waste ground.

Cakile *Sea-Rocket*

Cakile edentula (Bigelow) Hook.
SEA-ROCKET   *native*
IR | **WUP** | **CUP** | **EUP**   FACU CC 5
Cakile lacustris (Fern.) Pobed.
Succulent annual. Stems much branched and bushy,

Boechera retrofracta
Boechera stricta

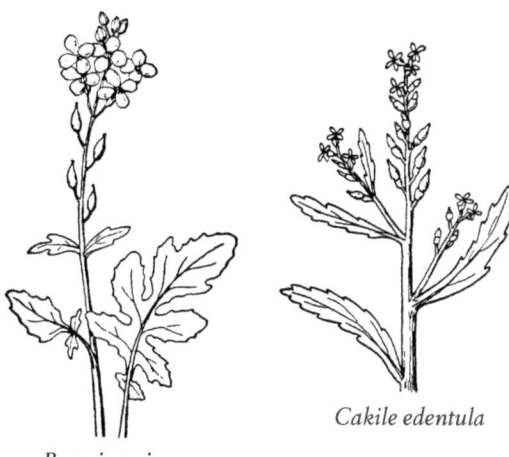
Brassica nigra
Cakile edentula

2–3 dm tall. Leaves oblong lance-shaped or spatulate, sinuately toothed, varying to pinnately lobed or nearly entire. Petals obovate, pale purple, about 5 mm wide; pedicels at maturity 2–5 mm long; short stamens subtended by a minute gland; long stamens separated by a larger gland. Fruit corky at maturity when dry, divided into 2 joints of different shape, the lower persistent, 1-seeded or seedless, the upper always conspicuously longer, eventually deciduous, fertile, usually 1-seeded, ovoid or lance-ovoid; seed suspended in the lower joint, erect in the upper.

Atlantic and Great Lakes coastal species; in UP most common along sandy beaches and low dunes near Lakes Huron and Michigan, often spreading by pieces of the floating fruit.

Camelina *False Flax*

Annual or winter-annual herbs, bearing both simple and branched hairs; stems branched above. Basal leaves narrowly spatulate; stem leaves linear to lance-shaped, clasping by a sagittate-auriculate base. Sepals erect, obtuse, the outer slightly saccate at base. Petals yellow, spatulate. Short stamens flanked at base by a pair of semicircular glands. Style slender, persistent. Fruit obovoid or pyriform, somewhat keeled, narrowed to the base and short-stipitate.

1. Fruit to 7 mm long and 5 mm wide; lower stem pubescent with both spreading and appressed hairs . . . . **C. microcarpa**
1. Fruit more than 7 mm long and more than 5 mm wide; stem glabrous or with tiny hairs . . . . **C. sativa**

Camelina microcarpa DC.
LITTLE-POD FALSE FLAX   *introduced*
IR | WUP | **CUP** | EUP   UPL

Stems erect, 3–7 dm tall, rough–pubescent, as is also the foliage, with both simple and branched hairs, the former 1–2 mm long. Fruit erect, 5.5–8 mm long, 3–4.5 mm wide, obscurely rugulose. April–June. Fields and waste places, usually in sandy soil; Marquette County.

Camelina sativa (L.) Crantz
LARGE-SEED FALSE FLAX   *introduced*
IR | **WUP** | **CUP** | EUP   FACU

Similar to *C. microcarpa* in foliage and habit. Stems and leaves glabrous to sparsely pubescent, the simple hairs not projecting beyond the stellate. Fruit commonly 7–10 mm long, 5–7 mm wide, inconspicuously veiny. May–June. Fields and waste places, usually where sandy; Keweenaw and Marquette counties.

Capsella *Shepherd's-Purse*

Capsella bursa-pastoris (L.) Medik.
SHEPHERD'S-PURSE   *introduced*
IR | **WUP** | **CUP** | **EUP**   FACU

Annual herb, pubescent with stellate hairs. Stems 1–6 dm tall, sparingly branched. Basal leaves oblong, 5–10 cm long, pinnately lobed; stem leaves much smaller, lance-shaped to linear, entire or denticulate, auriculate at base. Racemes at anthesis congested, at maturity greatly elongate, often forming half the total height of the plant; pedicels at maturity widely spreading, 1–2 cm long; sepals short-oblong, ascending; petals white, obovate, about 2 mm wide and about 2x as long as the sepals. Fruit oblong cordate, flattened, 5–8 mm long, truncate to notched at the tip. Spring.

Weedy in lawns, gardens, and waste places. Where sheltered, this is one of the first plants to bloom in spring.

134 BRASSICACEAE | Mustard Family

dicots

Cardamine *Bittercress; Toothwort*

Annual, biennial or perennial herbs, smooth or with short hairs near base of stem. Leaves simple to pinnately divided, the basal leaves often different in shape than stem leaves. Flowers in racemes or umbel-like clusters; sepals green to yellow, early deciduous; petals usually white. Fruit a 2-chambered, linear pod (silique), the seeds in a single row in each chamber.

ADDITIONAL SPECIES

Cardamine flexuousa With. (Woodland bittercress), European annual of moist roadsides and shores; at scattered UP locations.

1	Leaves simple to pinnately compound	2
1	Leaves palmately 3–5 parted or compound	4
2	Petals 8 mm or more long	**C. pratensis**
2	Petals to 4 mm long	3
3	Leaflets of stem leaves linear	**C. parviflora**
3	Leaflets ovate	**C. pensylvanica**
4	Leaves divided into 4–7 linear segments	**C. concatenata**
4	Leaves divided into 3 ovate segments	5
5	Leaves usually 2	**C. diphylla**
5	Leaves usually 3	**C. maxima**

Cardamine concatenata (Michx.) Sw.
CUT-LEAF TOOTHWORT *native*
IR | **WUP** | CUP | EUP   FACU CC 5
Dentaria laciniata Muhl.

Perennial herb; rhizome constricted at intervals, the segments 2–3 cm long. Stems 2–4 dm tall, pubescent above and on the rachis. Basal and stem leaves similar, basal leaves usually absent at anthesis, stem leaves typically in a whorl of 3 above the middle of the stem, deeply 3-parted, the segments linear or lance-shaped, nearly entire to laciniately toothed, the segments often deeply bifid, the whole leaf appearing 5-parted. Sepals 5–8 mm long; petals 12–19 mm long. April–May.

Moist rich woods.

Cardamine diphylla (Michx.) Wood
BROAD-LEAF TOOTHWORT *native*
IR | **WUP** | CUP | EUP   FACU CC 5
Dentaria diphylla Michx.

Perennial herb, rhizome continuous. Stems glabrous, 2–4 dm tall. Stem leaves commonly 2, opposite or nearly so; leaflets 3, coarsely crenately toothed, about half as long as wide; basal leaves similar. Sepals 5–8 mm long; petals 11–17 mm long. April–May.

Rich woods.

Cardamine maxima (Nutt.) Wood
THREE-LEAF TOOTHWORT (THR) *native*
IR | **WUP** | CUP | EUP   CC 10
Dentaria maxima Nutt.

Perennial herb, rhizome constricted at intervals. Stems erect or ascending, 1.5–4 dm tall, glabrous. Basal leaves usually present at anthesis, resembling those of the stem. Stem leaves normally 3, alternate, 3-parted, the segments petiolulate, ovate lance-shaped, crenately toothed, the lateral segments unsymmetrical, or with a lateral lobe, or bifid nearly to the base. Sepals 6–8 mm long; petals 12–15 mm long. April–May.

Rare in rich deciduous woods, often along streams.

Cardamine parviflora L.
SMALL-FLOWERED BITTERCRESS *native*
IR | **WUP** | CUP | EUP   FAC CC 10
Cardamine arenicola Britt.

Annual herb. Stems usually solitary, simple to much branched, 1–3 dm tall. Terminal leaflet of the basal leaves oblong to broadly cuneate-obovate. Stem leaves 2–4 cm long, with 3–6 pairs of segments, the terminal linear to cuneate-oblong, entire or toothed, the lateral not much smaller, linear or oblong, usually 1–3 mm wide and entire, not decurrent. Pedicels and fruit as in *C. pensylvanica*.

Usually in dry soil.

Cardamine pensylvanica Muhl.
PENNSYLVANIA BITTERCRESS *native*
IR | **WUP** | CUP | EUP   FACW CC 1

Biennial herb. Stems erect or spreading, to 6 dm long, usually hairy on lower stem. Leaves pinnately divided into 2–5 pairs of lateral leaflets and a single terminal segment, 4–8 cm long and 1–4 cm wide, the leaflets entire or with a few teeth or lobes; the terminal leaflet largest, 1–4 cm long and 1–2 cm wide; petioles shorter than blades, becoming shorter upward. Flowers in a raceme; sepals 1–2 mm long; petals white, 2–4 mm long. Fruit an upright silique, 2–3 cm long and to 1 mm wide, with a style-beak to 2 mm long, on stalks 5–15 mm long. May–Sept.

Streambanks, swamps, and wet forests (often where seasonally flooded); wet, disturbed areas.

Cardamine pratensis L.
CUCKOO-FLOWER *native*
IR | **WUP** | CUP | EUP   CC 10

Perennial upright herb. Stems 2–5 dm long. Basal leaves on long petioles, divided into 3–8 broad leaflets, 5–20 mm long, the terminal segment largest and more or less entire; lower stem leaves similar to basal ones, becoming shorter and with shorter petioles upward on stem; stem leaves with 7–17 oval to linear leaflets. Flowers in a crowded raceme; petals white, 8–15 mm long. Fruit an upright silique, 2.5–4 cm long, with a style-beak 1–2 mm long, on

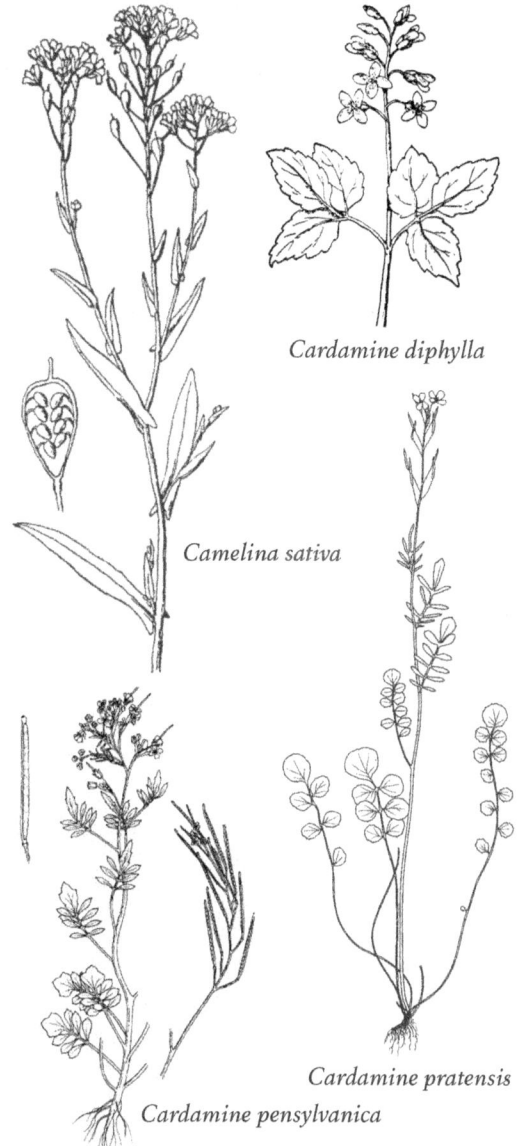

stalks 8–15 mm long. May–June.

Peatlands, tamarack and cedar swamps, wet depressions in forests.

Conringia *Hare's-Ear-Mustard*

Conringia orientalis (L.) Dumort.
HARE'S-EAR-MUSTARD   *introduced*
IR | **WUP** | **CUP** | **EUP**

Annual glabrous herb, often glaucous. Stems erect, branched above or simple, up to 8 dm tall. Leaves entire, pale green, the lower narrowed to the base, the upper oval or oblong, cordate-clasping. Sepals erect, saccate at base; petals yellowish white, narrowly obovate, 10–12 mm long, long-clawed; ovary cylindric, gradually tapering to the short style; pedicels and fruit widely divergent, the fruit elongate, slender,

8–12 cm long, 2–3 mm thick, 4-angled; seeds in 1 row, oblong, 2–3 mm long, granular-roughened. May–Aug.

Native of Eurasia; naturalized or adventive in waste places.

Descurainia *Tansy-Mustard*

Annual or biennial herbs, more or less pubescent or canescent with wholly or partly branched hairs; leaves 1–3-pinnate with very numerous small segments. Sepals ovate, obtuse. Petals yellow or pale yellow, small, sometimes barely surpassing the calyx, obovate or spatulate. Filaments slender; anthers ovate or oblong. Staminal glands minute or none. Ovary cylindric; style very short, as thick as the ovary; stigma capitate; ovules numerous. Fruits linear or clavate, terete or slightly 4-angled, tipped with the very short persistent style; valves with a prominent midnerve. Seeds elliptic or oblong, in one or two rows.

1   Plants green, with glandular hairs; fruit less than 13 mm long. . . . . D. pinnata
1   Plants gray-green, with stellate hairs, the hairs not glandular; fruit 13 mm or more long . . . . D. sophia

Descurainia pinnata (Walt.) Britt.
TANSY-MUSTARD   native
IR | WUP | CUP | EUP   CC 0

Annual herb. Stems erect, simple, or abundantly branched below, or branched at the inflorescence, 2–7 dm tall. Leaves oblong in outline, the lower leaves largest, bipinnate, or pinnate with deeply pinnatifid segments; the upper leaves progressively reduced, less divided, the uppermost 1-pinnate. Flowers 2–4 mm wide.; raceme after anthesis elongate, up to 3 dm long; pedicels at maturity widely divergent, 5–20 mm long. Fruit narrowly clavate, 5–13 mm long, 1–2 mm wide; seeds in 2 rows.

Usually in disturbed places; roadsides, railroads, fields, gravel pits, shores.

Descurainia sophia (L.) Webb
HERB-SOPHIA   introduced
IR | WUP | CUP | EUP

Annual herb, plants canescent throughout. Stems erect, usually much branched, 3–8 dm tall. Leaves ovate to obovate in outline, or the upper narrower, 2 to 3x pinnate into linear segments. Flowers about 3 mm wide; raceme after anthesis loose and open; pedicels widely ascending, 8–14 mm long. Fruit narrowly linear, 15–25 mm. long, 0.5–1 mm wide; seeds in 1 row.

Disturbed places, roadsides, railroads.

Diplotaxis *Wallrocket*
Diplotaxis muralis (L.) DC.
ANNUAL WALLROCKET   introduced
IR | WUP | CUP | EUP

Annual herb, branched from the base, glabrous or sparingly pubescent. Flowering stems erect or decumbent, 2–5 dm tall. Leaves chiefly basal or near the base, oblong lance-shaped, long-attenuate to the base, shallowly or deeply toothed or pinnatifid. Racemes at maturity elongate, the lower pedicels remote, usually 1–1.5 cm long. Sepals erect, not saccate; petals yellow, gradually narrowed to the claw; short stamens subtended by a kidney-shaped gland; each pair of long stamens subtended by a short prismatic gland. Fruit ascending, linear, 2–3 cm long, about 2 mm wide, tipped by a short but distinct beak, not stipitate; seeds numerous, smooth, in 2 distinct rows. May–Sept.

Native of Europe; naturalized in waste places, especially in sandy soil; Houghton County.

Draba *Whitlow-Grass*

Annual, biennial, or perennial herbs, in some species woody at base. Leaves entire or dentate, more or less pubescent with simple, branched, or stellate hairs, or with 2 types of hairs together. Sepals ascending or erect, blunt. Petals yellow or white, rounded, sometimes bifid, narrowed below to a claw, or in certain species sometimes reduced or absent. Fruit a 2-valved silicle, rarely as much as 5x longer than wide.

ADDITIONAL SPECIES
Draba incana L. (Twisted whitlow-grass), native; a northern species of eastern North America, isolated in Michigan on Passage Island and Gull Islands of Isle Royale National Park, where it grows in exposed rock crevices above Lake Superior; threatened.

1   Plants annuals or winter-annuals. 2
1   Plants perennial. 3
2   Petals cleft to near middle; leaves all basal . . . D. verna
2   Petals not cleft; at least some leaves along stem; Delta County. . . . . D. nemorosa
3   Stem very leafy (at least 20 cauline leaves between basal rosette and lowest pedicel); axis of inflorescence densely pubescent; fruit not twisted. . . . . D. incana*
3   Stem with few cauline leaves but many basal leaves (often with short leafy basal shoots); axis of inflorescence sparsely pubescent or glabrate. 4
4   Fruit densely covered with stellate hairs . . . . . . D. cana
4   Fruit glabrous or only sparsely hairy 5
5   Fruit usually twisted . . . . D. arabisans
5   Fruit straight; Isle Royale. . . . . D. glabella

Draba arabisans Michx.
ROCK WHITLOW-GRASS   native
IR | WUP | CUP | EUP   CC 10

Perennial. Stems 1–4 dm tall, simple to branched above, glabrous to stellate-pubescent. Basal leaves narrowly oblong lance-shaped to spatulate, up to 6 cm long, often sharply toothed, uniformly stellate-pubescent; stem leaves few to several, oblong to obovate, narrowed or acute at base, often dentate. Mature racemes loose, up to 10 cm long, commonly glabrous or nearly so; petals white, 4–6 mm long. Fruit glabrous, lance-shaped to narrowly oblong, 7–12 mm long, about one-fourth as wide, soon twisted, in poorly grown examples shorter, ovate, and straight. May–June.

Uncommon on rocks and cliffs.

Draba cana Rydb.
HOARY WHITLOW-GRASS (THR) *native*
IR | WUP | **CUP** | **EUP** CC 10

Perennial. Stems 1–3 dm tall. Basal rosettes very dense, the numerous spatulate, densely stellate-pubescent leaves commonly less than 3 cm long; stem leaves lance-shaped to ovate, usually 5–20 mm long. Raceme elongate, at maturity constituting 1/2 to 3/4 of the total plant height; lower flowers remote, often axillary, only the upper crowded; petals white, about 4 mm long. Fruit oblong, 5–12 mm long, usually 1.5–2 mm wide, stellate-pubescent. May–July.

Rocky limestone ledges, cliffs, and gravelly or rocky soil; rare in UP; more common in Rocky Mountains.

Draba glabella Pursh
SMOOTH WHITLOW-GRASS (END) *native*
IR | WUP | **CUP** | EUP CC 10

Perennial. Stems 1–4 dm tall, stellate-pubescent at least below, glabrous or glabrate in the raceme. Basal leaves oblong lance-shaped, to 3 cm long, entire or denticulate, finely and evenly stellate-pubescent; stem leaves oblong to ovate, usually dentate. Racemes loose, commonly 5–8 cm, occasionally to 12 cm long; petals white, 3–5 mm long. Fruit oblong lance-shaped or oblong, 7–11 mm long, 2.4–3.5 mm wide, straight, glabrous or sparsely hirtellous. May–July.

In Michigan, known only from Isle Royale.

Draba nemorosa L.
WOODLAND WHITLOW-GRASS *introduced*
IR | WUP | **CUP** | EUP

Annual or winter-annual, to 3 dm tall. Basal leaves ovate lance-shaped to oval, elliptic, or obovate, 1–2.5 cm long; stem leaves similar, few to several, all below the middle of the stem, often dentate, pubescent with both simple and branched hairs. Raceme at maturity loose and elongate, up to 2 dm long, glabrous; petals pale yellow, becoming white in age, about 2 mm long; pedicels at maturity widely spreading or ascending, to 3 cm long, glabrous. Fruit ascending to erect, oblong-linear, 5–10 mm long.

Dry soil, prairies and hillsides; Delta County.

Draba verna L.
SPRING WHITLOW-GRASS *introduced*
IR | **WUP** | **CUP** | **EUP**

Annual or winter-annual. Leaves in a crowded basal rosette, oblong lance-shaped, 1–2 cm long, pubescent with simple and branched hairs. Scapes several, leafless, very slender, usually 10–15 cm tall; pedicels ascending, elongating in fruit and the lowest to 3 cm long. Flowers white, 2–3 mm wide; petals bifid nearly to the middle. Fruit narrowly to broadly elliptic, usually 4–10 mm long, 2–3.5 mm wide, glabrous. April–May.

Native of Europe; naturalized in fields and on roadsides.

Erucastrum *Dog-Mustard*

Erucastrum gallicum (Willd.) O.E. Schulz
COMMON DOG-MUSTARD *introduced*
IR | **WUP** | **CUP** | **EUP**

Annual or biennial, with the general aspect of a *Brassica*; pubescence of simple hairs or none. Stems

Diplotaxis muralis

Draba cana

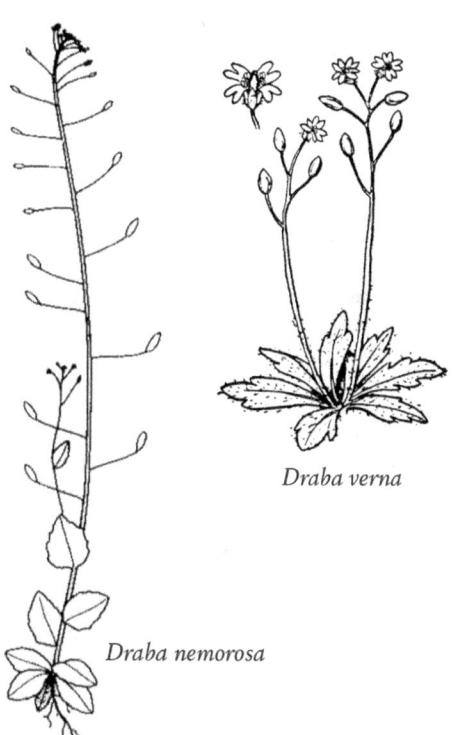

Draba verna

Draba nemorosa

erect or ascending, 3–6 dm tall, branched from the lower nodes. Basal and lower leaves oblong lance-shaped in outline, to 15 cm long, sparsely pubescent, deeply pinnatifid, the segments dentate, the terminal segment largest; stem leaves progressively reduced, the uppermost only 1–2 cm long; Mature racemes greatly elongate; pedicels slender, ascending; sepals ascending; petals yellow, spatulate, about 7 mm long. Fruit 4-angled, usually upwardly curved, the body 2–2.5 cm long, the beak about 3 mm long; seeds numerous, in 1 row. May–Sept.

Waste places.

Erysimum *Wallflower*

Annual to perennial herbs, with narrow, entire, dentate, or pinnatifid leaves, more or less pubescent with appressed, 2–4-pronged hairs. Petals yellow to orange (in our species), obovate or spatulate, abruptly or gradually narrowed to a long claw. Ovary linear-cylindric, pubescent; style very short; stigma capitate, 2-lobed. Fruits elongate, more or less 4-angled, thinly to densely pubescent; valves with a prominent midnerve. Seeds numerous, in one row. All our species are more or less densely pubescent on the stem, leaves, sepals, and fruit, and more or less so on the back of the petals, especially at the base of the blade. The 2-pronged hairs lie lengthwise; V-shaped hairs have the single prong directed backward.

1 Petals 15–25 mm long; fruit 5–10 cm long; Mackinac County . . . . **E. capitatum**
1 Petals 10 mm long or less 2
2 Petals less than 6 mm long; fruit to 3 cm long . . . . **E. cheiranthoides**
2 Petals 6 mm or more long 3
3 Leaves very finely toothed . . . . **E. hieraciifolium**
3 Leaves entire . . . . **E. inconspicuum**

Erysimum capitatum (Dougl.) Greene
WESTERN WALLFLOWER *introduced*
IR | WUP | CUP | **EUP**

Erysimum asperum (Nutt.) DC.

Biennial herb. Stems erect, simple or branched above, 2–10 dm tall. Leaves linear to oblong lance-shaped, entire or with a few low teeth, thinly to densely pubescent. Racemes at maturity greatly elongate, the stout pedicels divergent, 7–15 mm long; sepals 10 mm long; petals bright yellow to orange-yellow, 15–25 mm long, the blade about half as long as the very slender exserted claw. Fruit ascending, 4–10 cm long, 4-angled. May–June.

Prairies, sandy open woods; Mackinac County.

Erysimum cheiranthoides L.
WORM-SEED WALLFLOWER *introduced*
IR | WUP | CUP | EUP   FACU

Annual herb. Stems erect, simple or sparingly branched, 2–10 dm tall. Leaves linear to oblong lance-shaped, entire or barely sinuate, thinly pubescent hut bright green, tapering to the base. Mature racemes elongate, the rachis straight, the pedicels very slender, widely divergent, 6–14 mm, commonly 8–12 mm long; sepals 2–3.5 mm long; petals bright yellow, 3.5–5.5 mm long. Fruit ascending to erect, 12–25 mm, commonly 15–20 mm long. June–Aug.

Usually in wet soil, but also appearing as a weed in fields and roadsides.

Erysimum hieraciifolium L.
EUROPEAN WALLFLOWER *introduced*
IR | **WUP** | **CUP** | EUP

Biennial or perennial herb, much like *E. inconspicuum*, usually but not always with somewhat broader leaves, these to 1 or 1.5 cm wide and sometimes evidently toothed; hairs of the upper leaf surface mainly 3-pronged, with an admixture of 4-pronged ones. June–July.

European weed; Delta and Keweenaw counties.

Erysimum inconspicuum (S. Wats.) MacM.
SHY WALLFLOWER *introduced*
IR | **WUP** | **CUP** | **EUP**

Perennial herb. Stems erect, commonly simple, occasionally sparingly branched, 3–8 dm tall. Leaves mostly erect or ascending, linear to oblong lance-shaped, entire or obscurely and remotely sinuate-dentate, the stem leaves rarely more than 5 mm wide, canescent. Mature racemes elongate and wand-like, the stout ascending pedicels 3–9 mm long; sepals densely stellate, 5–7 mm long; petals pale yellow, 6–10 mm long. Fruit erect or nearly so, 1.5–4 cm long. May–Aug.

Dry soil of prairies, plains, and upland woods.

Hesperis *Dame's Rocket*

Hesperis matronalis L.
DAME'S ROCKET *introduced (invasive)*
IR | **WUP** | **CUP** | **EUP**   FACU

Perennial herb. Stems erect, 5–10 dm tall, simple or branched above. Leaves lance-shaped, short-petioled or sessile, remotely and sharply denticulate, pubescent above with simple hairs, below chiefly with branched hairs. Flowers fragrant; sepals erect, the outer narrow; the inner broad, saccate at base; petals purple, varying to pink or white, 2–2.5 cm long, the blade obovate; ovary cylindric; stigma 2-lobed. Fruit widely spreading on stout pedicels, linear, terete or nearly so, 5–10 cm long, somewhat constricted between the seeds; seeds numerous, large, 3–4 mm long, angularly fusiform, in 1 row. May–June.

Formerly cultivated for ornament; frequently escaped along roads and fencerows and in open woods.

Lepidium *Pepperwort*

Annual, biennial, or perennial herbs. Leaves linear to elliptic, entire, toothed, or pinnatifid. Sepals blunt. Petals small, white (rarely yellowish), linear to spatulate, sometimes notched at tip. Stamens 6, or by abortion 4 or 2. Ovary flat. Fruit a flattened silicle, thin or somewhat distended over the seeds, ovate to circular or obovate, often winged, commonly notched at tip, tipped by the persistent style or stigma.

ADDITIONAL SPECIES
Lepidium appelianum Al-Shehbaz [syn: *Cardaria pubescens* (C.A. Mey.) Jarmolenko] introduced from Asia; Alger County.

1 At least upper stem leaves sessile and auriculate, sagittate, or clasping at base........................ 2
1 Stem leaves petiolate or subsessile, never auriculate, sagittate, or clasping at base 3
2 Plants annual, not rhizomatous; fruit broadly winged at apex; racemes elongated in fruit........ **L. campestre**
2 Plants perennial, rhizomatous; fruit wingless at apex; racemes not elongated in fruit; Alger County.........
.... **L. appelianum***
3 Fruit obovate, widest above middle; petals absent or often rudimentary; rachis of raceme puberulent with cylindrical or clavate hairs............ **L. densiflorum**
3 Fruit orbicular, widest at middle; petals present or rarely rudimentary; rachis of raceme puberulent with curved hairs, rarely glabrous................. **L. virginicum**

Lepidium campestre (L.) Ait. f.
FIELD-CRESS   *introduced*
IR | WUP | CUP | EUP
Biennial herb; densely short-pubescent. Stems erect, 2–5 dm tall, simple to much branched. Basal leaves elongate, oblong lance-shaped, entire to shallowly lobed; stem leaves erect or ascending, lance-shaped to narrowly oblong, 2–4 cm long, entire or denticulate, sessile, clasping by acute auricles. Racemes dense, up to 15 cm long at maturity; fruiting pedicels 4–8 mm long, widely divergent. Fruit oblong-ovate, 5–6 mm long, about three–fourths as wide, convex below, concave above, broadly winged distally, the short style barely exsert. May–June.

Native of Europe; a weed of sandy waste ground, fields, and roadsides.

Lepidium densiflorum Schrad.
PRAIRIE PEPPERWORT *introduced*
IR | WUP | CUP | EUP   FACU
Annual herb. Stems 2–5 dm tall, thinly short-pubescent. Basal leaves 4–7 cm long or rarely longer, coarsely dentate or pinnatifid; stem leaves shorter, linear or narrowly ohlance-shaped, mostly entire, sharply acute. Racemes at maturity erect, 5–10 cm long; petals none, or shorter than the sepals, linear to narrowly spatulate. Fruit broadly oval to obovate, 2–3.3 mm long, narrowly winged distally; stigma included in the notch. May–June.

Dry, sandy or gravelly disturbed places; native to Great Plains and western USA, considered adentive in Michigan and eastern USA.

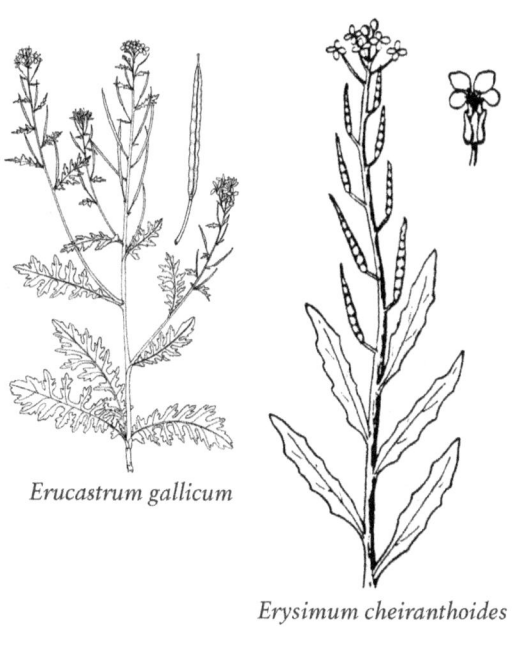

Erucastrum gallicum

Erysimum cheiranthoides

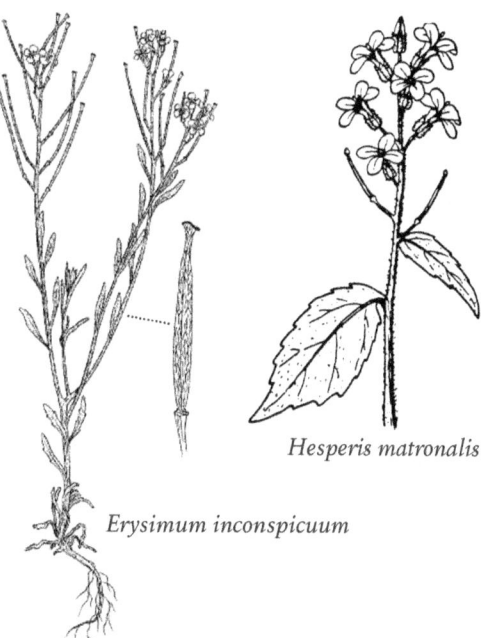

Hesperis matronalis

Erysimum inconspicuum

Lepidium virginicum L.
POOR-MAN'S PEPPER *native*
IR | **WUP** | **CUP** | **EUP** FACU CC 0

Annual herb. Stems erect, 1–5 dm tall, simple to much branched. Basal leaves oblong lance-shaped in outline, sharply toothed to pinnatifid, occasionally bipinnatifid. Upper leaves smaller, oblong lance-shaped to linear, dentate to entire, acute, narrowed to the base. Racemes numerous, many-flowered, up to 1 dm long; petals present, equaling to twice as long as the sepals. Fruit broadly elliptic to circular, 2.5–4 mm long, narrowly winged across the top; style included in the notch. May–June.

Dry fields, gardens, roadsides, and waste places.

Nasturtium *Watercress*

Nasturtium officinale R.Br.
WATERCRESS *introduced (invasive)*
IR | **WUP** | CUP | **EUP** OBL

Rorippa nasturtium-aquaticum (L.) Hayek

Perennial herb; plants smooth, often forming la. Stems underwater, floating, or trailing on mud; rooting from lower nodes. Leaves 4–12 cm long and 2–5 cm wide, pinnately divided into 3–9 segments, the lateral segments round to ovate in outline, the terminal segment largest; margins entire or with a few shallow rounded teeth; petioles present. Flowers in 1 to several racemes per stem, flat-topped and elongating in fruit; flowers 5 mm wide, sepals green-white, oblong, 1–3 mm long; petals white, sometimes purple-tinged, obovate, 4–5 mm long. Fruit a linear, often curved pod (silique), 1–2.5 cm long and 2 mm wide, tipped with a short style beak to 1 mm long. May–Sept.

Seeps, slow-moving streams, ditches, cedar swamps, especially in cold spring-fed waters; Mackinac County; naturalized throughout most of USA and s Canada.

ADDITIONAL SPECIES

Nasturtium microphyllum (Boenn. ex Rchb.) Rchb., introduced perennial known from Alger and Schoolcraft counties; very similar to *N. officinale* (sometimes treated as a variety) and found in same aquatic habitats; distinguished by narrower mature fruit and 2 rows of seed under each valve rather than 1 row.

Raphanus *Radish*

Raphanus raphanistrum L.
WILD RADISH *introduced*
IR | **WUP** | **CUP** | EUP

Coarse annual herb from a stout taproot, pubescence of simple hairs. Stems stout, 3–8 dm tall, usually sparsely hispid. Lower leaves obovate in outline, 1–2 dm long, pinnatifid into 5–15 oblong segments, the lower very small, the upper progressively larger; upper leaves much smaller, oblong to lance-shaped, entire, dentate, or few-lobed. Sepals obtuse, somewhat saccate at base; petals yellow, becoming white in age, 10–15 mm long; mature pedicels ascending, 8–15 mm long. Fruit nearly cylindrical when fresh, when dry becoming prominently several-ribbed and constricted between the 4–10 seeds, the body 2–4 cm long, the beak 1–3 cm long. June–Aug.

Native of Eurasia; weedy on roadsides, fields, waste places.

ADDITIONAL SPECIES

Raphanus sativus L. (Radish), cultivated and rarely escaping from gardens.

Rorippa *Yellowcress*

Annual, biennial or perennial herbs; plants smooth or with unbranched hairs. Leaves sometimes in a basal rosette in young plants, toothed to pinnately divided, petioles short or absent. Flowers small, in racemes at ends of stems or from lateral branches; sepals green to yellow, deciduous by fruiting time;

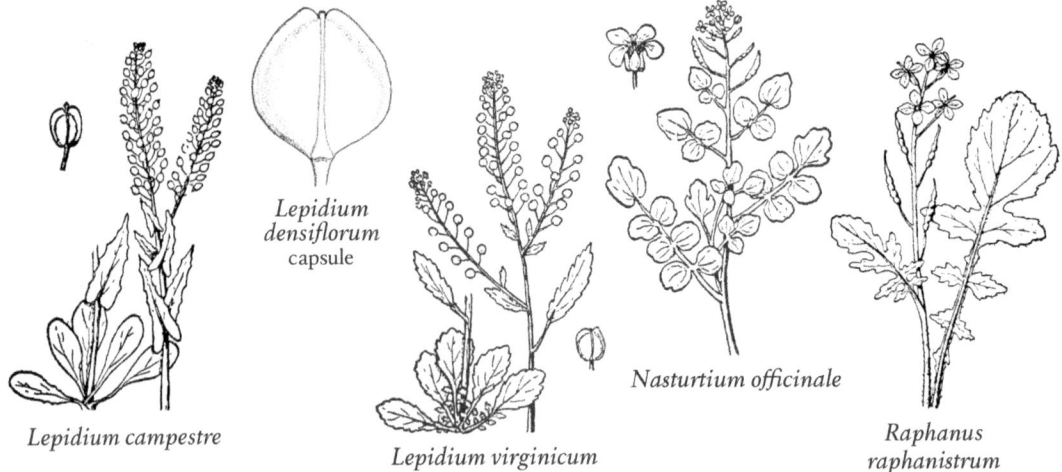

Lepidium densiflorum capsule

Lepidium campestre

Lepidium virginicum

Nasturtium officinale

Raphanus raphanistrum

petals yellow or white, shorter to longer than sepals. Fruit a short-cylindric to linear pod (silique), mostly 2-chambered, the seeds in 2 rows.

1 Plant truly aquatic, the submersed leaves dissected in a bipinnate pattern into filiform segments (midvein present, the lateral segments again dissected), frequently detaching readily from the stem; petals white. . . . . **R. aquatica**
1 Plant terrestrial or aquatic but even if in water the leaves with definite flat lobes (not bipinnately dissected) and not falling from the stem; petals yellow . 2
2 Plants annual or biennial, taprooted; petals shorter than or equal to sepals. . . . . **R. palustris**
2 Plants perennial, roots creeping; petals longer than sepals . . . . **R. sylvestris**

Rorippa aquatica (Eaton) E.J. Palmer & Steyerm.
LAKECRESS (THR) native
IR | **WUP** | **CUP** | EUP   OBL CC 8
 Armoracia aquatica (Eaton) Wieg.
 Armoracia lacustris (Gray) Al-Shehbaz & Bates
 Neobeckia aquatica (Eaton) Greene
Perennial, fibrous-rooted herb. Stems and leaves smooth, usually underwater. Underwater leaves pinnately dissected into many threadlike segments; emersed leaves, if present, lance-shaped, 3–7 cm long, coarsely toothed. Flowers on spreading stalks to 1 cm long; sepals turning upright; petals white, 6–8 mm long. Fruit oval, 5–8 mm long, 1-chambered, tipped by a persistent slender style 2–4 mm long, but apparently rarely maturing. June–Aug.

Uncommon in Lake Superior estuaries, quiet water in lakes, rivers and streams; muddy shores.

Rorippa palustris (L.) Bess.
COMMON YELLOWCRESS   native
IR | **WUP** | **CUP** | EUP   OBL CC 1
Annual or biennial herb. Stems erect, usually 1, to 1 m long, unbranched or branched upward. Leaves lance-shaped to obovate, mostly pinnately divided; the blades oblong to oblong lance-shaped, 5–30 cm long and 2–6 cm wide, middle stem leaves usually with basal lobes and clasping stem, smooth to densely hairy on lower surface; margins deeply lobed and slightly wavy; petioles short or absent. Flowers in racemes at ends of stems and from leaf axils, the terminal raceme flowering and fruiting first, the oldest siliques on lowest portions of raceme; sepals green, 1–3 mm long, early deciduous; petals yellow, drying white, 2–3 mm long. Fruit a round to short-cylindric pod, 3–10 mm long and 1–3 mm wide, straight-sided or slightly tapered to tip, on stalks 3–10 mm long.

June–Sept. Marshes, wet meadows, shores, streambanks, ditches and other wet places.

Rorippa sylvestris (L.) Bess.
CREEPING YELLOWCRESS   *introduced*
IR | **WUP** | **CUP** | EUP   OBL
Perennial herb, spreading by rhizomes and sometimes stolons. Stems erect, branched above, 2–6 dm long; smooth, or sparsely hairy on lower stem; basal rosettes present on young plants. Stem leaves pinnately divided, oblong in outline, 3–15 cm long and 2.5 cm wide, gradually reduced in size upward on stem, margins usually toothed; petioles present on lower leaves, absent on upper leaves. Flowers in racemes at ends of stems and from upper leaf axils, all flowering at about same time or the oldest siliques on lower portion of terminal racemes; sepals yellow-green, 2–3 mm long; petals yellow, 3–5 mm long. Fruit a linear pod (silique), 4–10 mm long and to 1 mm wide, usually upright on spreading stalks 5–10 mm long. June–Aug.

Introduced from Europe; sometimes weedy in wet forests, lakeshores, muddy streambanks and ditches.

Sinapis *White-Mustard*
Annual introduced herbs. Similar to *Brassica* (and previously included in that genus) but the flowers tend to be larger.

Rorippa palustris

Rorippa aquatica

Rorippa sylvestris

BRASSICACEAE | Mustard Family

1. Fruit glabrous, 2 mm wide.................. **S. arvensis**
1. Fruit bristly, 4 mm wide..................... **S. alba**

Sinapis alba L.
WHITE-MUSTARD *introduced*
IR | **WUP** | CUP | EUP

Brassica alba Rabenh. non L., *Brassica hirta* Moench

Annual, more or less stiffly pubescent. Stems 3–7 dm tall. Leaves obovate in outline, the lower up to 2 dm long, lyrately lobed; upper leaves progressively reduced, less lobed or merely dentate. Flowers about 15 mm wide; pedicels at maturity widely divergent, about 1 cm long. Fruit divergent or ascending, commonly bristly, at least when young, the body 4–8-seeded, 10–15 mm long, the valves much distended over the seeds; beak 1–2 cm long, flat, often curved; seeds pale brown, smooth, about 2 mm wide. May–July.

Occasionally cultivated for its seeds which are used to make the condiment mustard; established as a weed in fields and waste places but seldom abundant; Keweenaw County.

Sinapis arvensis L.
CORN-MUSTARD *introduced*
IR | **WUP** | **CUP** | EUP

Brassica arvensis Rabenh. non L.
Brassica kaber (DC.) L.C. Wheeler

Annual. Stems 2–8 dm tall, usually sparsely hirsute. Leaves obovate in outline, the lower sometimes lobed but more often merely coarsely toothed; upper leaves progressively smaller, coarsely toothed, roughly pubescent to nearly glabrous. Flowers about 15 mm wide; pedicels at maturity ascending, about 5 mm long. Fruit ascending, linear, nearly terete, the body 1–2 cm long, 1.5–2.5 mm thick, smooth or somewhat pubescent; beak commonly about half as long as the body, flattened-quadrangular; seeds dark brown, smooth, 1–1.5 mm wide. May–July.

Common weed of fields, gardens, and waste ground.

Sisymbrium Hedge-Mustard

Our species annual or winter-annual herbs, with simple hairs. Leaves (at least the lower) deeply pinnatifid. Sepals obtuse, ascending. Petals small, yellow, obovate. Ovary cylindric; style short, scarcely differentiated. Fruits elongate, linear, terete or slightly quadrangular, tipped with the minute persistent style. Seeds in one row, oblong, nearly or quite smooth.

1. Fruit erect, appressed; pedicels erect, 2–3 mm long ... . **S. officinale**
1. Fruit widely spreading; pedicels spreading, 5 mm long or more.............................. **S. altissimum**

Sisymbrium altissimum L.
TUMBLING MUSTARD *introduced*
IR | **WUP** | **CUP** | EUP FACU

Stems erect, usually much branched, to 1 m tall, glabrous or sparsely pilose. Leaves petioled, deeply pinnately parted, the lower into 5–8 pairs, the upper into 2–5 pairs of segments; segments varying from linear and entire to lance-shaped and serrate. Racemes greatly elongate after anthesis; pedicels nearly or quite as thick as the fruit, ascending, 5–10 mm long; petals pale yellow, 6–8 mm long. Fruit ascending or spreading, slender, linear, 5–10 cm long, 1–1.5 mm wide. June–July.

Native of Eurasia; established as a weed in fields and waste ground.

Sisymbrium officinale (L.) Scop.
HEDGE-MUSTARD *introduced*
IR | **WUP** | **CUP** | EUP

Stems erect, 3–8 dm tall, branched above or simple. Lower leaves petioled, deeply pinnatifid; segments oblong to ovate or the terminal sometimes rotund, angularly toothed; upper leaves sessile or nearly so, few-lobed or 3-lobed or entire, the lateral lobes widely divergent. Flowers bright yellow, about 3 mm wide. Racemes erect, simple or with straight, widely divergent branches; pedicels at maturity 2–3 mm long, closely appressed, distally thickened and as wide as the fruit at the summit. Fruit closely appressed, subulate, 10–15 mm long, 1–1.5 mm wide at base. May–Oct.

Native of Europe; established as a weed in gardens, roadsides, and waste ground.

Subularia Water-Awlwort
Subularia aquatica L.
AMERICAN WATER-AWLWORT (END) *native*
IR | **WUP** | CUP | EUP OBL CC 10

Small, native annual aquatic herb; plants underwater or sometimes on muddy shores. Stems 3–10 cm long. Leaves all basal, awl-shaped or linear, 1–5 cm long. Flowers small, 2–10, widely separated in a raceme; sepals persistent, petals white. Fruit a short, oval or oblong pod (silicle), 2–4 mm long. June–Aug. Cold lakes in shallow water to 1 m deep.

In Michigan, known from Chippewa and Keweenaw counties (including Isle Royale); a circumboreal species, south in North America to New England, Mich, Minn, Colo and Calif.

Thlaspi Pennycress
Thlaspi arvense L.
FIELD PENNYCRESS *introduced*
IR | **WUP** | **CUP** | EUP UPL

Glabrous annual herb. Stems 1–5 dm tall, simple to much branched. Stem leaves sessile, oblong to lance-

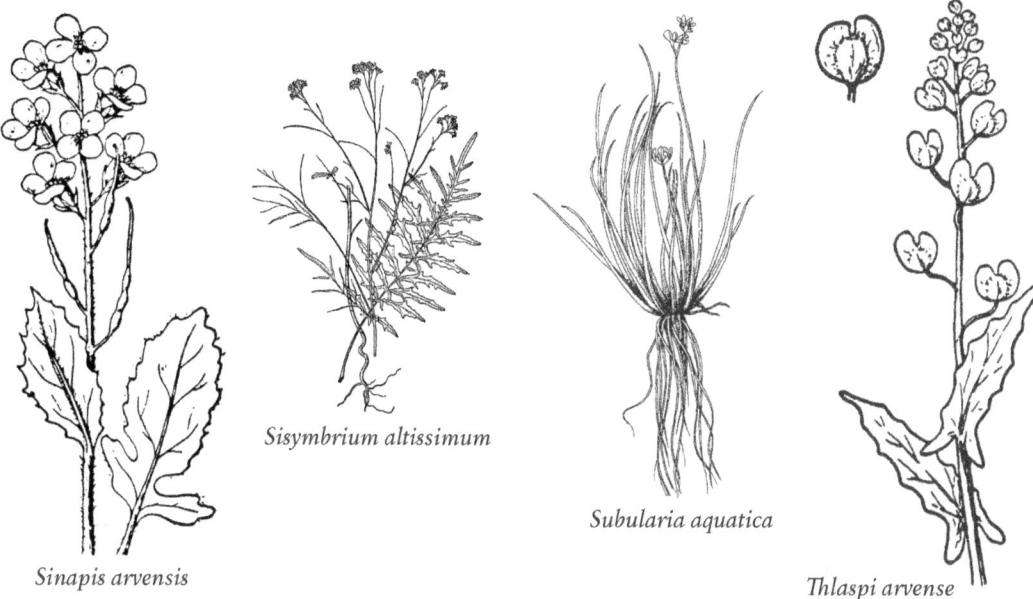

Sisymbrium altissimum

Subularia aquatica

Sinapis arvensis

Thlaspi arvense

shaped, entire or few-toothed, with 2 narrow auricles at base 1–5 mm long. Petals white, spatulate to obovate, about 3 mm wide, 2x as long as the sepals; pedicels at maturity ascending. Fruit circular to broadly elliptic, 10–14 mm long, strongly flattened and distended over the seeds, keeled or winged at the margin, tip deeply (2–3 mm) notched. April–June.

Native of Europe; roadsides and waste ground.

Turritis *Tower-Mustard*

Turritis glabra L.
TOWER-MUSTARD native
IR | WUP | CUP | EUP CC 3
Arabis glabra (L.) Bernh.

Biennial herb. Stems commonly hirsute near the base with mostly simple hairs, occasionally with some or many branched hairs, glabrous and glaucous above. Lower leaves more or less pubescent, usually with Y-shaped hairs; stem leaves overlapping in the lower part of the stem, more remote above, all lance-shaped or lance-oblong, sessile and auriculate-clasping, usually glabrous and glaucous. Petals 3–6 mm long; mature pedicels erect, 7–16 mm long. Fruit erect, nearly terete, overlapping, 5–9 cm long, 0.8–1.3 mm wide; seeds angular, very narrowly winged. May–June.

Dry sandy fields, gravel pits, roadsides, gravelly shores.

Cabombaceae
WATERSHIELD FAMILY

Brasenia *Watershield*

Brasenia schreberi J.F. Gmel.
WATERSHIELD native
IR | WUP | CUP | EUP OBL CC 6

Perennial aquatic herb; underwater portions of plant with a slippery, jellylike coating. Stems to 2 m long. Leaf blades floating, oval, 4–12 cm long and half as wide; petiole attached to center of blade underside. Flowers perfect (with both staminate and pistillate parts), dull-purple, on emergent stalks to 15 cm long from leaf axils; sepals 3, petals 3, 12–15 mm long. Fruit an oblong capsule, 3–5 mm long. July.

Quiet ponds and lakes; water usually acid.

Cactaceae
CACTUS FAMILY

Opuntia *Prickly Pear*

Opuntia fragilis (Nutt.) Haw.
LITTLE PRICKLY PEAR END native
IR | WUP | CUP | EUP CC 10

Branched and jointed perennial plants, the joints ("pads") greatly flattened, with conspicuous yellow flowers (ours), and with short prickles from the areoles (glochids). Stems spreading, forming dense mats to 5 dm wide; joints orbicular to obovate, very turgid, 2–5 cm long, easily detached; areoles crowded, nearly all armed with 3–7 spines. Flowers regular,

144 CAMPANULACEAE | Bellflower Family

perfect, about 5 cm wide, sepals and petals numerous in several series Stamens numerous, shorter than the petals. Fruit dry, inedible; seeds wingless. May–July.

Dry sand prairies and shallow, dry soil over rock outcrops; rare in Marquette County.

Campanulaceae
BELLFLOWER FAMILY

Perennial herbs. Stems usually with milky juice. Leaves simple, alternate. Flowers in racemes at ends of stems or single from upper leaf axils, perfect (with both staminate and pistillate parts), 5-parted, regular and funnel-shaped (*Campanula*) or irregular (*Lobelia*); petals blue, white or scarlet; stamens separate or joined into a tube around style. Fruit a many-seeded capsule. *Lobelia* is sometimes placed in the Lobeliaceae, but that family discontinued under APG III.

1. Flowers irregular; stamens joined to form a tube around the style. . . . . **Lobelia**
1. Flowers regular; stamens separate 2
2. Plants annual; flowers sessile in leaf axils; leaves clasping stem . . . . **Triodanis perfoliata**
2. Plants perennial; flowers on slender pedicels; leaves not clasping stem . . . . **Campanula**

Campanula *Bellflower; Harebell*

Perennial herbs (ours). Leaves alternate. Flowers conspicuous, solitary or in various types of inflorescence. Sepals 5, triangular to linear. Corolla rotate, campanulate, or funnelform, deeply or shallowly 5-lobed, in our species blue or violet to white. Stamens attached at the very base of the corolla; filaments widened at the base; anthers distinct. Ovary 3–5-celled with elongate style. Capsule short, usually conspicuously ribbed, opening by 3 or 5 lateral pores; seeds numerous. A number of introduced species are cultivated for their handsome flowers and are occasionally reported as escaped (e.g., *C. carpatica* Jacq., *C. glomerata* L., *C. latifolia* L., *C. medium* L., and *C. persicifolia* L.).

1. Stems weak, reclining on other plants . **C. aparinoides**
1. Stems upright 2
2. Flowers on short pedicels in an erect, 1-sided raceme . . . . **C. rapunculoides**
2. Flowers solitary or in loose, open clusters on slender pedicels 3
3. Plants glabrous or nearly so; stem leaves linear. . . . . **C. rotundifolia**
3. Stems bristly; stem leaves lance-shaped; Mackinac County . . . . **C. trachelium**

Campanula aparinoides Pursh
MARSH BELLFLOWER   native
IR | WUP | CUP | EUP   OBL CC 7

Perennial herb, spreading by slender rhizomes. Stems slender, weak, usually reclining on other plants, 2–6 dm long, 3-angled, rough-to-touch. Leaves linear or narrowly lance-shaped, larger below and smaller upward on stem, 2–8 cm long and 2–8 mm wide, tapered to a sharp tip; margins and midvein on leaf underside often rough; petioles absent. Flowers single on long slender stalks from upper leaf axils, funnel-shaped, sepals triangular to lance-shaped, 2–5 mm long; petals pale blue to white, 5–12 mm long. Fruit a capsule, opening near its base to release seeds. July–Sept.

Sedge meadows, marshes, calcareous fens, conifer swamps (cedar, tamarack), thickets, open bogs; soils often calcium-rich.

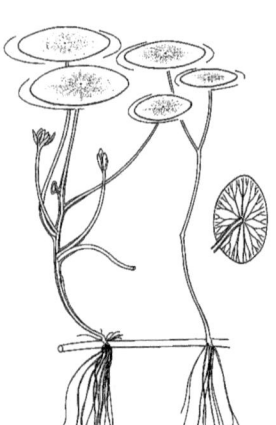

Brasenia schreberi

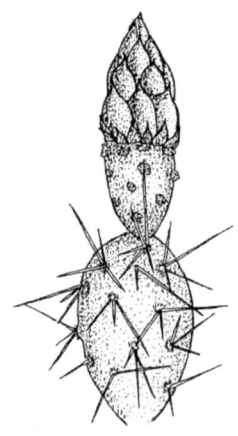

Opuntia fragilis

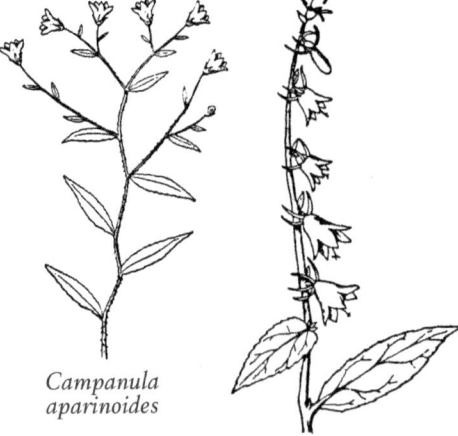

Campanula aparinoides

Campanula rapunculoides

CAMPANULACEAE | Bellflower Family

Campanula rapunculoides L.
CREEPING BELLFLOWER *intro. (invasive)*
IR | WUP | CUP | EUP

Erect perennial herb. Stems 4–10 dm tall from a creeping rhizome, usually unbranched, smooth or sparsely pubescent. Leaves coarse, irregularly serrate, usually sparsely pubescent beneath, the lower long-petioled, ovate, rounded to subcordate at base, the upper progressively narrower, short-petioled or subsessile. Inflorescence strict, unbranched, secund, forming a 1-sided raceme with much reduced bracts; lower pedicels sometimes elongate to 5 cm long, the upper 1–5 mm long; calyx lobes reflexed; corolla blue, somewhat nodding, 2–3 cm long. Capsule opening from pores near base. July–Aug.

Native of Eurasia, forming persistent weedy colonies on roadsides, railroads, and disturbed places.

Campanula rotundifolia L.
BLUEBELL-OF-SCOTLAND *native*
IR | WUP | CUP | EUP FACU CC 6

Perennial herb. Stems simple or branched from the base, 1–5 dm tall, smooth or nearly so, bearing 1 to several spreading or drooping flowers. Basal leaves broadly ovate or subcordate, seldom persistent to the flowering season; stem leaves linear, 2–10 cm long, decreasing above; pedicels slender. Sepals linear, 4–10 mm long; corolla blue (rarely white), bell-shaped, 12–20 mm long or occasionally somewhat larger, its lobes much shorter than the tube. Capsule turbinate or short-cylindric, opening by pores near the base. June–Sept.

Dry woods, meadows, cliffs and beaches. Variable in habit, stature, and number and size of flowers.

Campanula trachelium L.
THROATWORT *introduced*
IR | WUP | CUP | **EUP**

Erect perennial herb. Stems 3–10 dm tall, angular, often bristly hairy above. Leaves bristly with short hairs, coarsely and irregularly serrate, the lower on long petioles, cordate or triangular, the upper narrower but scarcely shorter, lance-shaped, narrowed to a very short petiole. Flowers in loose terminal or axillary clusters; hypanthium and lance-shaped sepals (10–13 mm long) bristly with pale hairs; corolla blue or violet, bell-shaped, 25–40 mm long. Aug.

Native of Eurasia and n Africa; roadsides and waste places; Mackinac County.

Lobelia *Lobelia*

Mostly perennial herbs. Stems single, usually with milky juice. Leaves alternate. Flowers irregular, in racemes at ends of stems; white, bright red, or pale to dark blue, often with white or yellow markings; 2-lipped, the 3 lobes of lower lip spreading, the 2 lobes of upper lip erect or pointing forward, divided to base, the anthers projecting through the split; stamens 5, joined to form a tube around style, the lower 2 anthers hairy at tip and shorter than other 3. Fruit a capsule. Most species toxic if taken internally.

1. Stem leaves narrow, to 4 mm wide, margins entire or with a few small teeth; or leaves all from base of plant 2
1. Stem leaves broader, 1–5 cm wide, margins toothed . . 3
2. Leaves all from base of plant, hollow, round in cross-section; plants usually underwater ..... **L. dortmanna**
2. Leaves all from stem, flat and linear; wetland habitats. .... **L. kalmii**
3. Flowers small, to 1.5 cm long 4
3. Flowers larger, 2–4 cm long 5
4. Inflorescence usually branched; hypanthium equalling the corolla or nearly so, inflated in fruit ..... **L. inflata**
4. Inflorescence unbranched; hypanthium shorter than corolla, not much inflated in fruit .... **L. spicata**
5. Flowers bright red (rarely white), 3 cm or more long .. .. **L. cardinalis**
5. Flowers blue with white stripes on lower lip, less than 2.5 cm long .... **L. siphilitica**

Lobelia cardinalis L.
CARDINAL-FLOWER *native*
IR | WUP | **CUP** | **EUP** OBL CC 7

Perennial herb. Stems erect, usually unbranched, 5–15 dm long, hairy to smooth. Leaves lance-shaped to oblong, 10–15 cm long and 3–5 cm wide, margins toothed; lower leaves on short petioles, upper leaves more or less stalkless. Flowers bright scarlet (rarely white), in racemes 1–4 dm long, the racemes with small, leafy, linear bracts; flowers 2–4 cm long, on hairy pedicels 5–15 mm long. July–Sept.

Floodplain forests, swamps, thickets, streambanks, shores and ditches; sometimes in shallow water.

Campanula rotundifolia

CAMPANULACEAE | Bellflower Family

Lobelia dortmanna L.
WATER LOBELIA *native*
IR | WUP | CUP | EUP   OBL CC 10

Perennial herb; plants usually underwater or sometimes on exposed shores. Stems upright, hollow, smooth, with milky juice. Leaves in dense rosettes at base of plants, fleshy, hollow, linear, 3–8 cm long, rounded at tip; stem leaves tiny. Flowers pale blue or white, 1–2 cm long, in a few-flowered raceme; sepals 2 mm long. July–Sept.

Shallow water of acid lakes and ponds; wet, sandy shores.

Lobelia inflata L.
INDIAN-TOBACCO *native*
IR | **WUP** | **CUP** | **EUP**   FACU CC 0

Annual or biennial herb. Stems erect, usually branched, villous, to 1 m tall. Leaves sessile or subsessile, obovate, 5–8 cm long, 1.5–3.5 cm wide, more or less serrate, usually pubescent. Racemes terminating the branches, 1–2 dm long; lower bracts foliaceous, the upper gradually reduced; pedicels 3–8 mm long, glabrous or puberulent, bracteolate at the base. Flowers 7–10 mm long; sepals linear, 3–5 mm long; corolla blue or white, the lower lip pubescent; hypanthium much inflated in fruit. July–Oct.

Open woods in moist or dry soil, disturbed places such as roadsides, ditches, borrow pits, trails, utility line clearings.

The long, irregular hairs at base of stem are distinctive. Long used in herbal medicine.

Lobelia kalmii L.
BROOK LOBELIA *native*
IR | WUP | CUP | EUP   OBL CC 10

Small perennial herb. Stems erect, smooth, 1–4 dm long, unbranched or with a few branches above, sometimes with a rosette of small, obovate leaves at base of plant. Stem leaves linear, 1–5 cm long and 1–5 mm wide, blunt to sharp-tipped, margins with a few small teeth. Flowers blue with a white center, 6–10 mm long, in an open raceme, the flowers on pedicels 4–10 mm long. July–Oct.

Wet, sandy or gravelly shores, wet meadows, interdunal wetlands, conifer swamps (cedar, tamarack), rock ledges and crevices; usually where calcium-rich.

Lobelia siphilitica L.
GREAT BLUE LOBELIA *native*
IR | **WUP** | **CUP** | EUP   FACW CC 4

Perennial herb. Stems stout, erect, 3–12 dm long. Leaves oblong or oval, smaller upward, 6–12 cm long and 1–3 cm wide, tip sharp or blunt, margins irregularly toothed, petioles absent. Flowers dark blue, in crowded racemes 1–3 dm long; the lower lip blue and white-striped, 1.5–2.5 cm long, on ascending stalks 4–10 mm long; sepals triangular to lance-shaped, 5–20 mm long, usually with narrow lobes near base. Aug–Sept.

Swamps, streambanks, wet meadows, calcareous fens.

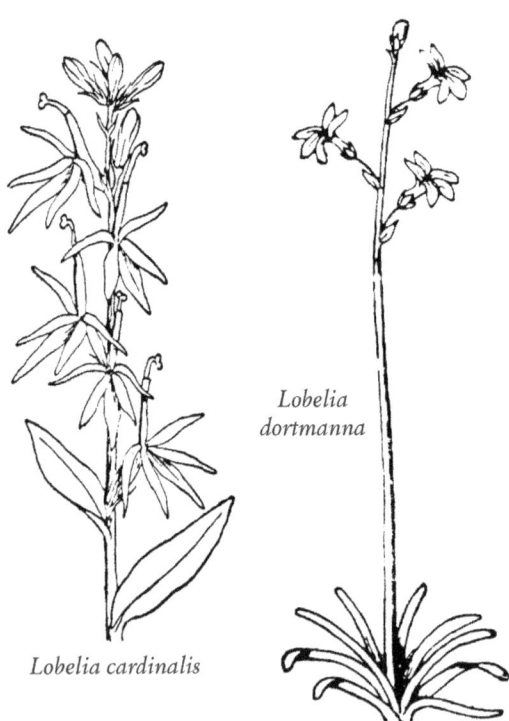

Lobelia cardinalis

Lobelia dortmanna

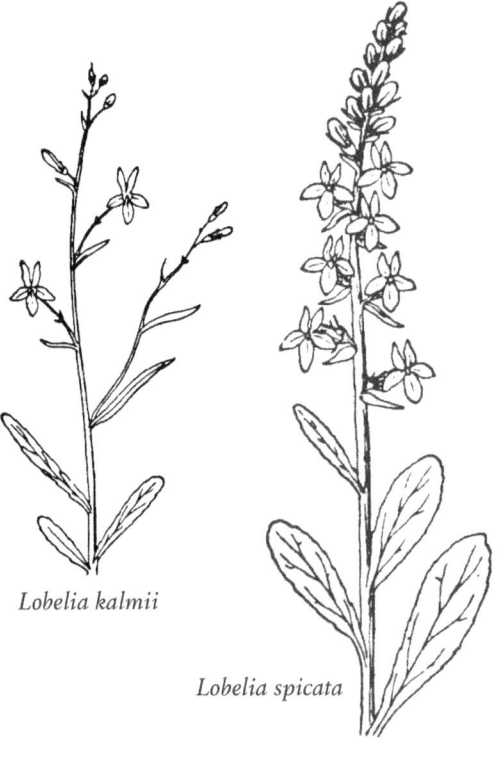

Lobelia kalmii

Lobelia spicata

dicots

Lobelia spicata Lam.
SPIKED LOBELIA   *native*
IR | **WUP** | **CUP** | **EUP**   FAC CC 4

Perennial herb. Stems unbranched, 3–10 dm long, hairy toward base. Leaves obovate to lance-shaped, 5–10 cm long, hairy, becoming smaller above. Flowers pale blue to white, 6–10 mm long, on stalks 2–4 mm long, in a slender, crowded raceme; base of sepals often with distinct, curved lobes (auricles), 1–2 mm long. May–Aug.

Moist to wet prairies (sometimes where disturbed), swamp margins.

Triodanis Venus'-Looking-Glass

Triodanis perfoliata (L.) Nieuwl.
CLASPING-LEAF V.'-LOOKING-GLASS   *native*
IR | **WUP** | CUP | EUP   FACU CC 6

Annual herb. Stems erect, simple or sparingly branched, 1–10 dm tall, the angular stems glabrous to finely rough-hairy. Leaves numerous, orbicular or broadly ovate, 1–3 cm long, cordate-clasping at the sessile base, usually toothed, often pubescent, palmately veined. Flowers in sessile axillary cymes, the lower flowers usually cleistogamous with minute corolla; the upper with a blue, deeply 5-lobed subrotate corolla about 15 mm wide; sepals 5; stamens 5; ovary 3-celled; stigmas 3, pubescent. Capsule ellipsoid, 5–8 mm long, opening near the middle to expose 1–3 linear pores; seeds lenticular. May–June.

Open woods, old fields, roadsides; Houghton County.

Cannabaceae

HEMP FAMILY

Trees (*Celtis*), erect herbs (*Cannabis*), or twining herbs (*Humulus*). Leaves alternate or opposite, simple to palmately lobed or compound. Inflorescences axillary to the upper (often reduced) leaves, the staminate relatively loose, branched, and many-flowered, the pistillate more compact and few-flowered. Flowers unisexual, small and inconspicuous, the staminate with 5 erect stamens opposite the 5 sepals, the pistillate with a short, entire, membranous calyx enclosing the ovary (*Humulus*), or the calyx often much reduced in *Cannabis*; petals absent; style short, with 2 elongate, filiform stigmas. Fruit a drupe (*Celtis*) or an achene, in *Humulus* invested by the persistent calyx.

1 Trees; leaves simple; fruits pediceled drupes; Menominee County. . . . . **Celtis occidentalis**
1 Herbs or herbaceous vines; larger leaves compound or lobed; fruits sessile achenes subtended by bracts 2

2 Plants erect; leaves compound . . . . **Cannabis sativa**
2 Plants twining; leaves not compound . . . . **Humulus**

Cannabis Hemp

Cannabis sativa L.
HEMP; MARIJUANA   *introduced*
IR | **WUP** | CUP | EUP   FACU

Erect annual herb. Stems slender, 1–2 m tall. Leaves parted to the base into 5–9 serrate leaflets, the lower commonly opposite, the upper smaller, usually alternate, with fewer leaflets or undivided; leaflets linear to narrowly lance-shaped, pubescent, 5–15 cm long, petioled. Flowers green, ordinarily dioecious; staminate flowers numerous in small clusters from the upper axils, forming a leafy panicle; sepals 5, imbricate; stamens 5; pistillate flowers in small clusters on short lateral leafy branches from the upper axils, each flower closely surrounded by an abruptly acuminate bract; calyx barely lobed, surrounding only the base of the ovary; stigmas 2, elongate, filiform. Fruit a thick-lenticular achene, 3–4 mm long, enclosed within the expanding bract.

Native of Asia; cultivated for its valuable fiber in many parts of the world and sometimes escaped; collected from Keweenaw County.

The pistillate inflorescence is the source of the narcotic marijuana.

Celtis Hackberry

Celtis occidentalis L.
COMMON HACKBERRY   *native*
IR | WUP | **CUP** | EUP   FAC CC 5

Medium tree; bark cork-like with wart-like protuberances. Leaves alternate, simple, ovate lance-shaped to broadly ovate, distinctly asymmetrical at base, 3-nerved; those of fertile branches 6–12 cm long, conspicuously serrate, abruptly acuminate to very long-acuminate. Plants monoecious or polygamous; staminate flowers in small clusters near the base of the twigs of the season; calyx deeply 5-parted; stamens 5, opposite the calyx lobes, exsert; fertile flowers solitary or sometimes paired from the upper axils of the same twigs; stamens present or none; ovary ovoid, 1-celled; style very short; bearing 2 elongate, recurved stigmas. Fruit an ellipsoid drupe 7–13 mm long, dark red to nearly black at maturity, with thin sweet pulp and a hard stone. Variable. May (appearing after the leaves).

Usually in rich moist forests, on riverbanks, and in ravines; Menominee County.

Trees sometimes with witch's brooms when disease causes a proliferation of branch tips.

148 CAPRIFOLIACEAE | Honeysuckle Family

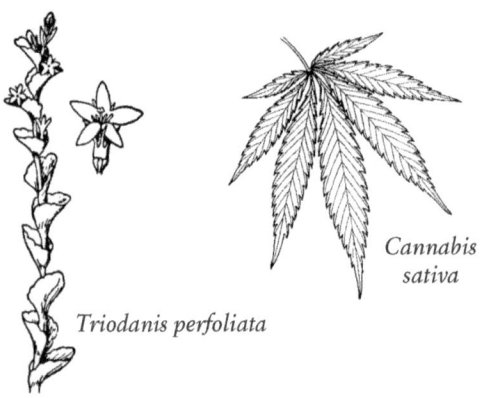

Triodanis perfoliata

Cannabis sativa

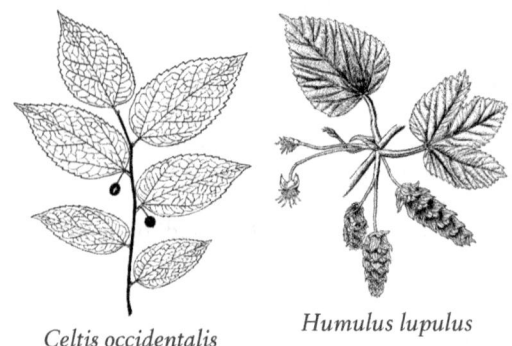

Celtis occidentalis

Humulus lupulus

Humulus *Hops*
Humulus lupulus L.
COMMON HOPS native
IR | WUP | CUP | EUP FACU CC 3

Twining, herbaceous, perennial vine to 10 m long. Stems rough. Leaves opposite, usually lobed; principal leaves as wide as long, cordate at base, 3-lobed to below the middle, the lateral lobes obliquely ovate-oblong, the terminal lance-shaped, constricted at base; upper leaves commonly broadly ovate, unlobed. Dioecious. Flowers small, in axillary clusters; staminate panicles 5–15 cm long; sepals 5, distinct; stamens 5; pistillate spikes about 1 cm long at anthesis, in pairs, each pair subtended by a foliaceous bract; with conspicuous slender stigmas; calyx membranaceous, unlobed. Fruit (hops) an achene enclosed within the persistent calyx and covered by the expanded bracts, straw-colored, cylindric, 3–6 cm long, the bracts entire and mostly blunt, very glandular at base. July–Aug.

Moist soil.

Yellow glands secreting a bitter substance, lupulin, occur on many parts of the plant but are most numerous on the fruit, which is important in beer-making.

Caprifoliaceae
HONEYSUCKLE FAMILY

Shrubs or vines, with opposite, mostly simple leaves. Flowers perfect (with both staminate and pistillate parts), mostly 5-parted. Fruit a fleshy berry or dry capsule. Family now includes members of the former Valerianaceae (*Valeriana, Valerianella*), herbs with opposite, simple or divided leaves, and numerous small flowers in terminal, panicled or capitate cymes.

1 Flowers axillary or on paired pedicels on a peduncle .. .. **Valeriana**
1 Flowers numerous, in rather dense terminal inflorescences (at ends of stem and branches)............. 2
2 Plants small, creeping, evergreen; flowers paired and nodding at tips of slender stalks .... **Linnaea borealis**
2 Plants larger shrubs or coarse herbs, upright, deciduous 3
3 Plants herbaceous; flowers from leaf axils .. **Triosteum**
3 Plants woody vines or shrubs; flowers various 4
4 Leaf margins toothed or lobed ..... **Diervilla lonicera**
4 Leaf margins entire................................ 5
5 Corolla bell-shaped, less than 1 cm long .... **Symphoricarpos**
5 Corolla tube-shaped, mostly more than 1 cm long .... **Lonicera**

Diervilla *Bush-Honeysuckle*
Diervilla lonicera P. Mill.
NORTHERN BUSH-HONEYSUCKLE native
IR | WUP | CUP | EUP CC 4

Shrub to 12 dm tall, at first simple, branched in age. Leaves opposite, oblong lance-shaped, 8–15 cm long, finely serrate, usually ciliate, otherwise nearly glabrous; petioles 3–10 mm long. Peduncles 3–7-flowered, the terminal flower usually sessile, the lateral pediceled; sepals narrowly linear, 2–7 mm long, persistent and accrescent in fruit; corolla funnelform, nearly regular, 5-lobed, 12–20 mm long, at first yellow, becoming reddish in age, hairy within; stamens 5, about equaling the corolla; ovary 2-celled; style exsert. capsule slender, beaked, 10–15 mm long, tardily dehiscent. June–July.

Dry woods, usually where sandy or rocky, forest borders, old dunes, sandy bluffs, railways, fencerows; increasing after disturbance such as fire.

Sometimes segregated into family Diervillaceae.

Linnaea *Twinflower*
Linnaea borealis L.
TWINFLOWER native
IR | WUP | CUP | EUP FAC CC 6

Low, evergreen, trailing vine. Stems slightly woody, to 1–2 m long, with numerous short, erect, leafy branches to 10 cm long; branches green to red-brown, finely hairy; older stems woody, 2–4 mm wide. Leaves opposite, simple, evergreen, oval to round, 1–2 cm long, blunt at tip, upper surface and margins with short, straight hairs; margins rolled under, with a few rounded teeth near tip; petiole short, short-hairy. Flowers small, pink to white, bell-shaped, shallowly 5-lobed, slightly fragrant, in nodding pairs atop a Y-shaped stalk to 10 cm long, the stalk with gland-tipped hairs and 2 small bracts at the fork and a pair of smaller bracts at base of each flower. Fruit a small, dry, 1-seeded capsule. June–Aug.

Hummocks in cedar swamps and thickets, moist conifer woods, on rotten logs and mossy boulders.

Placed in family Linnaeaceae by some authors.

Lonicera *Honeysuckle*

Shrubs or woody vines. Leaves opposite, simple, entire. Flowers long and tubular or funnel-shaped, in pairs from leaf axils. Fruit a few-seeded, blue or red berry.

ADDITIONAL SPECIES AND HYBRIDS

Lonicera × bella Zabel, a hybrid between *L. morrowii* and *L. tatarica*, is common (and invasive) across the UP.

1 Plants woody, climbing vines; flowers in opposite, sessile, 3-flowered clusters, producing a whorl of 6 flowers 2
1 Shrubs; flowers paired at ends of peduncles from leaf axils 4
2 Leaves hairy on upper surface . . . . **L. hirsuta**
2 Leaves glabrous on upper surface 3
3 Uppermost joined leaves waxy on upper surface, rounded or notched at tip; flowers yellowish . . . . **L. reticulata**
3 Uppermost joined leaves green on upper surface, pointed at tip; flowers reddish . . . . **L. dioica**
4 Style glabrous; fruit black, red, or blue; native species. 5
4 Style with coarse, stiff hairs; fruit red; introduced species (except *L. oblongifolia*) 7
5 Fruit black; bracts below flowers oval-shaped; Keweenaw County (incl. Isle Royale) . . . **L. involucrata**
5 Fruit red or blue; bracts below flowers awl-shaped to spatula-shaped. 6
6 Fruit red; ovaries appearing separate and divergent . . . . **L. canadensis**
6 Fruit blue; ovaries appearing united . . . . **L. villosa**
7 Corolla strongly 2-lipped; the upper lip lobed to half its length . . . . **L. oblongifolia**
7 Corolla only weakly 2-lipped; the upper lip lobed to base or nearly so 9
8 Leaves glabrous on underside; flower peduncles 1.5 cm long or more . . . . **L. tatarica**
8 Leaves hairy on underside; flower peduncles to 1.5 cm long. . . . . **L. morrowii**

Lonicera canadensis Bartr.
FLY-HONEYSUCKLE native
IR | WUP | CUP | EUP FACU CC 5

Shrub to 2 m tall, with straggling branches. Leaves ovate to oblong, 3–12 cm long, acute or obtuse, ciliate, glabrous to sparsely pubescent beneath. Peduncles 2–3 cm long; bracts linear or subulate, from much shorter than to slightly exceeding the ovaries; bractlets to 0.5 mm long, or obsolete; corolla yellowish, 12–22 mm long, distinctly saccate at base, glabrous, its lobes a third to a half as long as the tube; ovaries glabrous, strongly divergent at anthesis; style glabrous. Berries red. May–June.

Common in dry or moist woods, occasionally swamps.

Lonicera dioica L.
LIMBER HONEYSUCKLE native
IR | WUP | CUP | EUP FACU CC 5

Climbing woody vine with glabrous branches. Leaves ovate or obovate, 5–12 cm long, rounded or narrowed at base, glaucous beneath; uppermost one or two pairs united into a rhombic or doubly ovate disk narrowed to an obtuse or acute tip or rounded and mucronate. Spike short-peduncled (usually 5–20

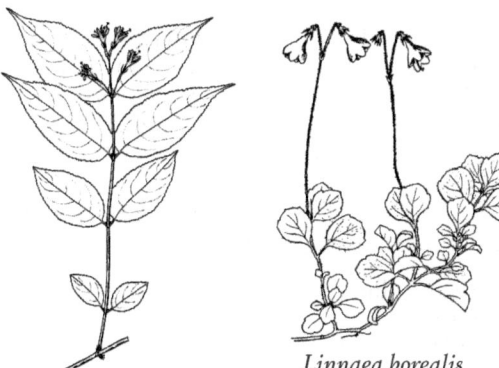

Diervilla lonicera

Linnaea borealis

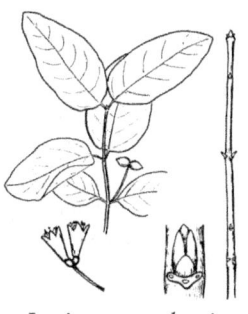

Lonicera canadensis

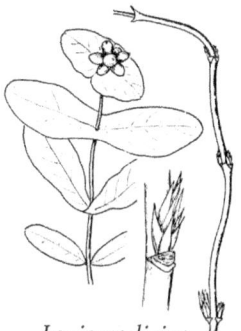

Lonicera dioica

mm); corolla yellow to deep maroon, 1.5–2.5 cm long, abruptly bulging just above the base, hairy inside. Berries red, 8–12 mm long, in clusters surrounded by leafy cup-like bracts, inedible and extremely bitter. May–June.

Moist woods and thickets, occasionally on dunes or in swamps.

Lonicera hirsuta Eat.
HAIRY HONEYSUCKLE   native
IR | **WUP** | **CUP** | EUP   FAC CC 6

Twining woody vine, the younger stems pubescent and glandular. Leaves dull green, sessile or short-petioled, broadly oval, 6–12 cm long, more or less pubescent on both sides, underside also glaucous; upper 1 or 2 pairs acuminate, connate into a rhombic-elliptic to subrotund disk. Spikes bearing 1–4 crowded whorls of flowers; corolla usually yellow, corolla tube 10–18 mm long, the pubescent tube slightly bulging near the base. Berries orange to red, sessile, subtended by a pair of saucer-shaped bracts. June–July.

Moist woods, particularly on margins and in clearings, often where sandy or rocky; occasionally in white cedar swamps.

The pubescent upper surface of the leaves distinguish this species from *L. dioica* and *L. reticulata* which also have connate terminal leaves.

Lonicera involucrata Banks
BLACK TWINBERRY   (THR) native
IR | **WUP** | CUP | EUP   FACU CC 10

Shrub 1–3 m tall. Leaves ovate to obovate, 6–12 cm long, short-acuminate, tapering at base to a short petiole, villosulous beneath. Peduncles 2–4 cm long; bracts oval, foliaceous, 1–2 cm long, persistent; corolla yellow, 10–13 mm long, saccate at base, pubescent, its lobes nearly equal, about half as long as the tube; ovary 3-celled; style glabrous. Berries purple-black.

Rare in cold moist woods; Keweenaw County (including Isle Royale).

Lonicera morrowii Gray
MORROW'S HONEYSUCKLE   intro. (invasive)
IR | **WUP** | CUP | **EUP**   FACU

Tall shrub, the older branches hollow. Leaves ovate to oblong, 3–6 cm long, rounded at base to a short petiole, softly pubescent beneath. Peduncles 5–15 mm long; bracts linear lance-shaped, 1–3 times as long as the ovary; bractlets half as long as to equaling the ovary; sepals ciliate; corolla white, fading yellow, pubescent, saccate at base, the lips equaling or longer than the tube; upper lip 4-lobed to its base; style hirsute. Fruit a dark red berry 7–8 mm wide; seeds numerous. May–June.

Native of Asia; escaping and invasive; Gogebic and Mackinac counties.

Lonicera oblongifolia (Goldie) Hook.
SWAMP FLY-HONEYSUCKLE   native
IR | **WUP** | **CUP** | EUP   OBL CC 8

Thicket-forming shrub 1–1.5 m tall; branches upright, with shredding bark and solid pith; twigs green to purple, smooth. Leaves opposite, oblong or oval, 3–8 cm long and 1–4 cm wide, rounded or blunt at tip, underside hairy when young, becoming smooth; margins entire, not fringed with hairs; petioles absent or to 1–2 mm long. Flowers yellow-white, tube-shaped with 2 spreading lips, 10–15 mm long, in pairs at ends of slender stalks up to 4 cm long from leaf axils. Fruit an orange-red to red (or sometimes purple), few-seeded berry composed of the 2 joined ovaries. May–June.

Cedar and tamarack swamps, fens, open bogs, wet streambanks and shores; often over limestone.

Lonicera reticulata Raf.
GRAPE HONEYSUCKLE   native
IR | **WUP** | CUP | EUP   CC –

Lonicera prolifera (Kirchn.) Booth

Woody climber with glabrous stems to 5 m long. Leaves thinly hairy beneath, the lower broadly oval, sessile or nearly so, 4–8 cm long; upper 2–4 pairs more or less connate, the uppermost completely so into a suborbicular disk rounded or retuse at the ends and glaucous above. Spike short-peduncled, the whorls of flowers usually crowded; hypanthium glabrous and glaucous; corolla 2–3 cm long, glabrous outside, hairy inside, gradually enlarged above the base. Berries orange-red to red, seeds about 3 mm long. May–June.

Moist woods and thickets; Houghton and Keweenaw counties, where likely escapes from cultivation.

Lonicera tatarica L.
TARTARIAN HONEYSUCKLE   intro. (invasive)
IR | **WUP** | **CUP** | EUP   FACU

Tall shrub, the older branches hollow. Leaves ovate to oblong, 3–6 cm long, rounded to subcordate at base, glabrous. Peduncles 15–25 mm long; bracts subulate, shorter than to exceeding the ovary; bractlets broadly ovate, about a third as long as the ovary; sepals entire; corolla white to pink, glabrous, barely gibbous at base, the lips equaling or exceeding the tube; upper lip 4-lobed to its base; style hirsute. Fruit a shiny orange or red berry to 1 cm wide. May–June.

Native of e Europe and Asia, an old favorite in cultivation, escaping and invasive.

Lonicera villosa (Michx.) J.A. Schultes
WATERBERRY   native
IR | **WUP** | **CUP** | **EUP**   FACW CC 8

Lonicera caerulea L.

Shrub to 1 m tall; branches upright, red-brown to

gray, outer thin layers soon peeling to expose red-brown inner layers; twigs purple-red, with long, soft hairs. Leaves opposite, oval to oblong, 2–6 cm long and 1–3 cm wide, blunt or rounded at tip, upper surface dark green, underside paler and hairy, especially on veins; margins fringed with hairs and often rolled under; petioles absent or to 1–2 mm long. Flowers yellow, tubular to funnel-shaped, 10–15 mm long, in pairs on short hairy stalks from axils of lower leaves. Fruit an edible dark blue berry consisting of the 2 joined ovaries. May–July.

Cedar and tamarack swamps, thickets, fens, shores.

Symphoricarpos *Snowberry*

Low bushy shrubs. Leaves ovate-oblong to rotund, short-petioled, entire or sometimes crenate on the rapidly growing branches. Flowers small, white or pink, terminating the stem or also in the upper axils. Corolla funnelform or campanulate, regular or nearly so, 4- or 5-lobed, usually bearded within. Calyx short, 4–5-toothed. Stamens (4 or) 5. Ovary 4-celled. Fruit a 2-seeded white or red berry.

1 Fruit red; corolla to 4 mm long . . . . **S. orbiculatus**
1 Fruit white; corolla 5 mm or more long 2
2 Style exserted, 4 mm or more long . . . .
. . . . **S. occidentalis**
2 Style included, to 3 mm long . . . . **S. albus**

Symphoricarpos albus (L.) Blake
SNOWBERRY *native*
IR | WUP | CUP | EUP   FACU CC 5

Shrub to 1 m tall, or sometimes dwarf, the younger branches pubescent or glabrous. Leaves ovate or oval, usually 2–3 cm long. Flowers in pairs on short pedicels or in short, few-flowered, interrupted spikes; corolla campanulate, slightly ventricose, 6–8 mm long, the lobes from half as long as to sometimes equaling the tube; style glabrous, 2–3 mm long. Fruit white; seeds 5 mm long.

Dry or rocky soil.

Symphoricarpos occidentalis Hook.
WESTERN SNOWBERRY *native*
IR | WUP | CUP | EUP   FACU CC -

Shrub to 1 m tall, the younger parts finely pubescent. Leaves ovate to ovate-oblong, usually 3–6 cm long, often coarsely crenate. Flowers several to many, sessile in short dense spikes, terminal or from the upper axils; corolla funnelform, 6–8 mm long, its lobes equaling the tube; stamens exserted; style mostly 4–8 mm long, typically pilose near the middle. Fruit white; seeds narrowly elliptic, 5 mm long. June–Aug.

Dry or rocky soil.

Symphoricarpos orbiculatus Moench
CORALBERRY *introduced*
IR | WUP | CUP | EUP   FACU

Freely branched shrub, the slender purplish stems to 15 dm tall, pubescent toward the summit. Leaves oval or ovate, usually 2–4 cm long, obtuse or rounded

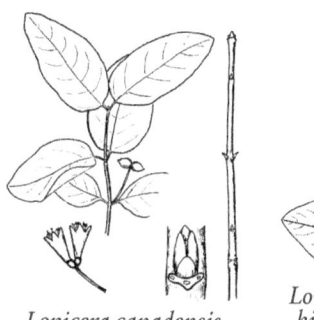
Lonicera canadensis

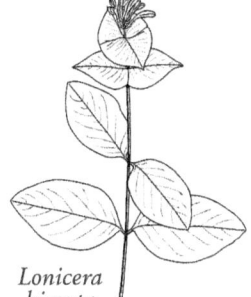
Lonicera hirsuta

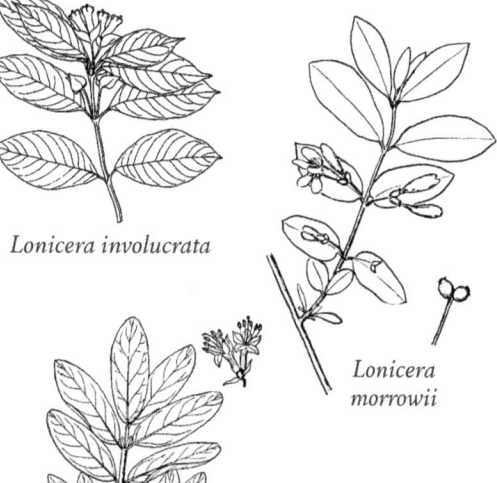
Lonicera involucrata

Lonicera morrowii

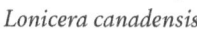
Lonicera oblongifolia

Lonicera tatarica

Lonicera villosa

at both ends, pubescent beneath. Flowers in dense clusters from most upper axils, sessile or nearly so; corolla 3–4 mm long, its lobes half as long as the tube. Fruit red, persistent through the winter; seeds 3 mm long. June–Aug.

Dry or rocky soil and margin of woods; southern USA native, considered adventive in Michigan; Keweenaw County.

Triosteum *Feverwort; Horse-Gentian*

Triosteum aurantiacum Bickn.
HORSE-GENTIAN *native*
IR | **WUP** | **CUP** | **EUP** CC 5

Coarse, erect, pubescent, perennial herb. Leaves distinct, tapering to a narrow base (or seldom 1-3 pairs with connate base 1-2 cm wide), the hairs of the stem mostly over 0.5 mm long. Flowers solitary or in small clusters in their axils. Corolla narrowly campanulate, gibbous at the base, unequally 5-lobed, purplish-red, the style about equaling the corolla or shortly included. Sepals 5, linear, elongate. Stamens 5. Fruit a bright orange-red berry crowned by the persistent sepals, enclosing a few hard oblong seeds. May–July.

Rich woods and thickets.

Valeriana *Valerian*

Perennial, strongly scented herbs. Leaves from base of plant and opposite along stem, simple to pinnately divided. Flowers somewhat irregular, in branched heads at ends of stems; calyx inrolled when young, later expanding and spreading; petals joined into a tube-shaped, 5-lobed corolla; stamens 3. Fruit a 1-chambered achene. *Valeriana* sometimes placed in family Valerianaceae.

1 Basal and stem leaves mostly with 6 or more pairs of leaflets . . . . **V. officinalis**
1 Basal leaves entire or with several lobes; stem leaves sparse. . . . . **V. uliginosa**

Valeriana officinalis L.
ALLHEAL *introduced*
IR | **WUP** | CUP | **EUP**

Perennial herb. Stems stout, erect, 6–15 dm tall, usually pubescent at the nodes. Basal and stem leaves similar, pinnately divided into 11–21 lance-shaped dentate segments; petioles of the upper leaves progressively shorter. Panicle large and open, its lower branches often remote (to 10 cm) from the upper, to 10 cm long; corolla obconic, the tube about 4 mm, the lobes about 1 mm long. Fruit lance-oblong, 4.5–5 mm long by about half as wide, glabrous. May–

Aug. Native of Europe and Asia; escaped from gardens, where it is commonly cultivated, to roadsides, ditches, fields, shores, and forest margins.

Valeriana uliginosa (Torr. & Gray) Rydb.
BOG VALERIAN *native*
IR | WUP | **CUP** | **EUP** CC 10
Valeriana sitchensis subsp. *uliginosa* (Torr. & Gray) F.G. Mey.

Perennial herb, from a stout rhizome or crown. Stems 3–8 dm long, more or less smooth. Leaves thin, net-veined; basal leaves obovate, 6–14 cm long and 1–3 cm wide, tapered to a long petiole; margins entire or with several lobes; stem leaves 2–6 pairs, pinnately divided into 3–15 ovate segments, margins often fringed with fine hairs; petioles present. Flowers all perfect, in clusters at ends of stems; corolla 5-lobed, pale pink, 5–7 mm long. Fruit a smooth, ovate achene, 3–4 mm long. May–July.

Openings in conifer swamps (especially white cedar and tamarack), marshes, calcareous fens, wet meadows; soils often alkaline.

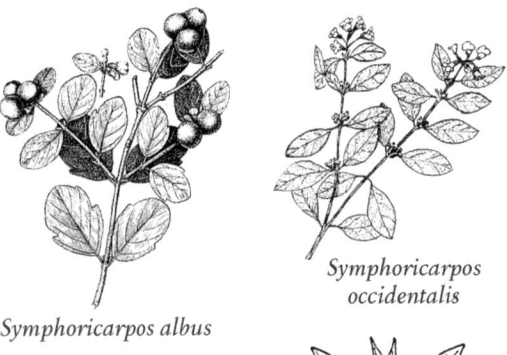

Symphoricarpos albus

Symphoricarpos occidentalis

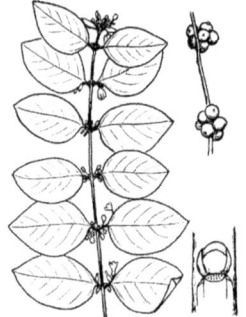

Symphoricarpos orbiculatus

Triosteum aurantiacum

Caryophyllaceae

PINK FAMILY

Annual or perennial herbs. Leaves simple, entire, mostly opposite but sometimes alternate or whorled. Stems often swollen at nodes. Flowers perfect (with both staminate and pistillate parts) or imperfect, in open or compact heads at ends of stems or from leaf axils; sepals usually 5, separate or joined into a tube;

CARYOPHYLLACEAE | Pink Family

Valeriana officinalis

Valeriana uliginosa

petals 5 (sometimes 4), separate, often lobed or toothed, sometimes absent; stamens 3–10, anthers often distinctly colored. Fruit a few- to many-seeded capsule.

ADDITIONAL SPECIES
Herniaria hirsuta L. (Hairy rupturewort), introduced in Alger and Marquette counties.

1 Leaves with chaffy or membranous stipules 2
1 Leaves without stipules 3
2 Leaves whorled; styles 5 . . . . **Spergula arvensis**
2 Leaves opposite; styles 3 . . . . **Spergularia**
3 Sepals joined to form a lobed tube 4
3 Sepals not joined, distinct from one another 11
4 Styles 2. 5
4 Styles 3–5 9
5 Calyx subtended by 1–3 pairs of bracts. 6
5 Calyx not subtended by bracts 7
6 Calyx with 20 or more nerves . . . . **Dianthus**
6 Calyx 5-nerved; Delta County . . **Petrorhagia saxifraga**
7 Flowers to 1 cm long . . . . **Gypsophila**
7 Flowers 2 cm long or more 8
8 Calyx tube-shaped, nerved . . . . . . **Saponaria officinalis**
8 Calyx ovoid, angled and wing . . . . . **Vaccaria hispanica**
9 Styles mostly 3 . . . . **Silene**
9 Styles mostly 5 10
10 Lobes of calyx much longer than tube formed by sepals . . . . **Agrostemma githago**
10 Lobes of calyx much shorter than tube. . . . . **Silene**
11 Petals absent, or entire or somewhat fringed at tip. . 12
11 Petals deeply cleft at tip into 2 segments 16
12 Petals absent; fruit a 1-seeded utricle. . . . . . **Scleranthus**
12 Petals usually present; fruit a several- to many-seeded capsule. 13
13 Styles 4 or 5, equaling the number of sepals. . . . **Sagina**
13 Styles (or style-branches) 3, fewer than the sepals . . 14

14 Leaves linear-subulate, the principal stem leaves subtending dense axillary clusters; plant entirely glabrous; capsule dehiscing into 3 valves . . . . **Minuartia**
14 Leaves ovate to elliptic or lanceolate, mostly without axillary tufts; plants puberulent at least on the stem; capsule with the 3 valves again split, resulting in a total of 6 teeth 15
15 Ripe seeds minutely and regularly roughened (tuberculate) and unappendaged; leaves ovate-elliptic, acute to acuminate, but not over 7 (–9) mm long; petals shorter than sepals; annuals . . . . **Arenaria serpyllifolia**
15 Ripe seeds smooth and shiny, with a pale appendage at the point of attachment; leaves lance-shaped to elliptic, mostly over 10 mm long; petals exceeding sepals; perennials . . . . **Moehringia**
16 Fruit cylindric; styles usually 5 (3 in *Cerastium nutans*) . . . . **Cerastium**
16 Fruit ovoid or oblong; styles usually 3 (5 in *Stellaria aquatica*) . . . . **Stellaria**

Agrostemma *Corncockle*

Agrostemma githago L.
COMMON CORNCOCKLE   *introduced*
IR | WUP | CUP | EUP

Annual herb. Stems to 1 m tall, thinly hairy. Leaves opposite, entire, linear or lance-shaped, 8–12 cm long and 5–10 mm wide. Flowers reddish. solitary at the ends of the branches, on pedicels to 2 dm long; calyx tube 12–18 mm long, 10-ribbed, calyx lobes 5, much longer than the tube; petals 5, 2–3 cm long, oblong lance-shaped stamens 10; styles (4)5. Capsule 14–18 mm long, dehiscent by (4)5 ascending teeth. July–Sept.

Native of Europe, originally a weed in grainfields and waste places, but now uncommon.

Arenaria *Sandwort*

Arenaria serpyllifolia L.
THYME-LEAF SANDWORT   *introduced*
IR | WUP | CUP | EUP   FAC

Finely hairy annual herb. Stems diffuse, 5–20 cm long, the internodes usually much longer than the leaves. Leaves usually 8–10 pairs, 3–5 mm long, 3–4 mm wide, ovate, sparsely rough hairy, often pustulate, 3–5-nerved. Inflorescence a terminal cyme, short or extending to the middle of the stem; bracts leaf-like; pedicels subcapillary, 4–8 mm long; sepals 5, 2.5–3.5 mm long, ovate lance-shaped, frequently glandular; petals 5, obovate, usually shorter than the sepals; stamens normally 10. Capsule ovoid, exceeding the sepals, dehiscent by 6 nearly equal teeth; seeds numerous, about 0.5 mm wide, kidney-shaped, gray-black.

Native of Eurasia; sandy or rocky places.

CARYOPHYLLACEAE | Pink Family

Cerastium *Mouse-Ear Chickweed*

Low annual or perennial herbs. Leaves opposite. Flowers solitary or more commonly in terminal cymes. Sepals 5. Petals 5, notched at tip to bifid, seldom entire. Stamens normally 10, sometimes 5. Styles normally 5, sometimes 4 or 3. Capsule usually exceeding the sepals, cylindric, membranous, dehiscent by 10 short teeth, frequently curved. Ovary 1-celled. Seeds numerous, kidney-shaped, papillate.

ADDITIONAL SPECIES

Cerastium brachypodum (Engelm. ex Gray) B.L. Robins (Short-stalk chickweed), native annual of rock outcrops and limestone pavement (alvar); Baraga County, Chippewa County (Drummond Island only); plants similar to *C. nutans* and sometimes treated as *C. nutans* var. *brachypodum* Engelm. ex Gray; threatened.

Cerastium tomentosum L. (Snow-in-summer), introduced ornamental, rarely escaping.

1. Leaves, stems, and sepals densely white-tomentose... **C. tomentosum***
1. Leaves, stems, and sepals with mostly straight or irregular hairs, not white, sometimes sticky 2
2. Petals large and showy, longer than the sepals........ **C. arvense**
2. Petals about equal to the sepals.................... 3
3. Bracts of upper inflorescence with papery, non-green margins and tips 4
3. Bracts of inflorescence green, herbaceous 5
4. Matted perennial; stamens 10 .... **C. fontanum**
4. Small annual; stamens 4–5........ **C. semidecandrum**
5. Longest pedicels distinctly less than twice as long as the capsules, mostly equaling or shorter than capsules ... . **C. brachypodum***
5. Longest pedicels twice as long as the capsules or longer .... **C. nutans**

Cerastium arvense L.
FIELD CHICKWEED — *native*
IR | WUP | CUP | EUP — FACU CC

Matted perennial. Stems ascending or erect, glabrous to densely villous, glandular or nonglandular, mostly 1.5–4 dm tall. Leaves linear to lance-shaped, 2–7 cm long, 1–15 mm wide, glabrous or pubescent, glandular or nonglandular; primary leaves usually subtending conspicuous axillary fascicles or short sterile shoots. Inflorescence with few to many flowers, the pedicels very slender; sepals lance-shaped, mostly 5–8 mm long, 2–3x exceeded by the conspicuous white petals. Capsule cylindric, somewhat to much exceeding the sepals; seeds papillate-tuberculate, reddish brown. April–Aug.

Rocky, gravelly, or sandy areas, chiefly in calcium- or magnesium-rich soils, weedy in abandoned fields and meadows.

Cerastium fontanum Baumg.
MOUSE-EAR CHICKWEED — *introduced*
IR | WUP | CUP | EUP — FACU

Cerastium vulgatum L.

Perennial. Stems tufted, spreading or erect, viscid-pubescent, 1.5–5 dm long. Leaves oblong to lance-shaped, 1–2 cm long, 3–12 mm wide, 1-nerved, sparingly pubescent, the lower leaves oblong lance-shaped. Inflorescence at length rather open, the mature pedicels 5–12 mm long; sepals 4.5–6 mm long, oblong lance-shaped, pubescent, rather strongly 1-nerved toward the base, scarious-margined; petals about as long as the sepals. Capsule 8–10 mm long, 2–3 mm wide, cylindric, sometimes curved; seeds rugose-papillate, reddish brown. April–Oct.

Native of Eurasia; widely naturalized in fields, woods, and waste places and frequently a troublesome weed, especially in lawns.

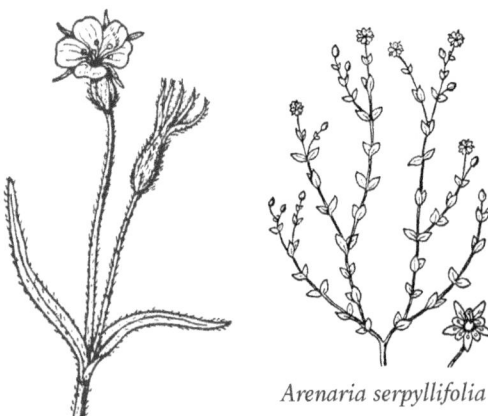

Agrostemma githago

Arenaria serpyllifolia

Cerastium arvense

Cerastium fontanum

dicots **CARYOPHYLLACEAE** | Pink Family

Cerastium nutans Raf.
NODDING MOUSE-EAR CHICKWEED *native*
IR | WUP | **CUP** | EUP FACU CC 4

Cerastium brachypodum (Engelm.) B.L. Robins. Annual. Stems weak or ascending, usually branched, viscid-pubescent, 1–4.5 dm long. Leaves narrowly oblong lance-shaped or oblong lance-shaped, commonly 1.5–5 cm long, 5–10 mm wide. Inflorescence open and loosely cymose; sepals lance-shaped, 4–5 mm long, acute or obtusish, scarious-margined; petals more or less equaling, frequently conspicuously exceeding the sepals; occasionally lacking. Capsule 8–15 mm long, cylindric, straight or curved; seeds coarsely papillose, pale reddish brown. April–June.

Moist or dry woods or open places; Delta and Dickinson counties.

Cerastium semidecandrum L.
FIVE-STAMEN MOUSE-EAR CHICKWEED *intro.*
IR | WUP | **CUP** | EUP

Viscid-pubescent annual. Stems 0.5-2 dm tall. Leaves relatively small, mostly 0.5–1.5 cm long; basal leaves oblong lance-shaped; stem leaves ovate. Iinflorescence often compact, varying to moderately open, the pedicels usually deflexed in fruit: bracts scarious-margined; sepals 3–5 mm long, lance-shaped, stipitate-glandular, scarious-margined; petals shorter than the sepals, only shallowly notched; stamens 5 (sometimes 10). Capsule 4.5-7 mm long; seeds smooth.

Native of Eurasia; dry, sandy disturbed places.

Dianthus *Pink*

Biennial or perennial, usually glaucous herbs. Leaves narrow. Flowers solitary or in paniculate or capitate cymes. Calyx subtended by 1–3 pairs of bracts, cylindric, with 30 or more nerves. Petals 5, without auricles or appendages. Stamens 10. Styles 2. Capsule dehiscent by 4 or 5 teeth. Seeds disc-shaped, apiculate.

Many species are well known in cultivation; the carnation is *D. caryophyltus* L.

ADDITIONAL SPECIES
Dianthus carthusianorum L. (Clusterhead), introduced; Houghton County.
Dianthus sylvestris Wulfen L. (Woodland pink), introduced; Houghton County.

1	Leaves not linear, more than 9 mm wide .. **D. barbatus**
1	Leaves linear, less than 9 mm wide 2
2	Annual; calyx and bracts hairy .... **D. armeria**
2	Perennial; calyx and bracts glabrous or only very finely hairy 3
3	Leaves hard and stiff; flowers clove-scented .... **D. plumarius**
3	Leaves soft and lax; flowers not clove-scented .... **D. deltoides**

Dianthus armeria L.
DEPTFORD PINK *introduced*
IR | **WUP** | **CUP** | EUP

Biennial. Stems 2–6 dm tall, usually strigose below the nodes, otherwise glabrous. Basal leaves numerous; stem leaves 5–10 pairs, linear to linear lance-shaped, 3–8 cm long, puberulent, ciliate. Cymes congested, 3–9-flowered, the lower, lance-subulate, erect bracts frequently surpassing the flowers; calyx 12–18 mm long, 20–25-nerved; petals 2–2.5 cm long, the blades 4–5 mm long, pink or rose, dentate. Capsule equaling the calyx, dehiscent by 4 recurved teeth. Summer.

Native of Europe, established as a weed.

Dianthus barbatus L.
SWEETWILLIAM *introduced*
IR | **WUP** | **CUP** | EUP

Glabrous perennial. Stems stout, 3–6 dm tall. Stem leaves 5–19 pairs, lance-shaped to oblong lance-shaped, 6–10 cm long, 1–1.8 cm wide; basal leaves somewhat wider. Cymes densely corymbose; primary bracts leaflike; secondary bractlets awn–tipped, ciliolate; calyx 15–18 mm long, about 40-nerved; petals 15–25 mm long, the blades dark red, pink, or whitish, crenate-denticulate. Capsule about 1 cm long. Summer.

Native of the Old World; escaped from cultivation and locally established.

Dianthus deltoides L.
MAIDEN PINK *introduced*
IR | **WUP** | **CUP** | EUP UPL

Perennial with a very slender creeping rootstalk. Stems slender, branched, 1–4 dm tall, glabrous or hispidulous-puberulent. Basal leaves oblong lance-shaped, 1.5–3 cm long, 1.5–3 mm wide; stem leaves 5–10 pairs, linear lance-shaped, acute, 2–4 cm long. Flowers solitary on pedicels 1–4 cm long; calyx 12–15 mm long, with 30–40 nerves, subtended by 1 or 2 pairs of obovate, abruptly acuminate or awned bracts; petals purple-red, lavender, or white, 1.5–2 cm long, the blades 4–8 mm long, sharply denticulate. Capsule narrowly ellipsoid–lance-shaped, about equaling the calyx. Summer.

Native of Europe; often cultivated and locally escaped into waste places.

Dianthus plumarius L.
FEATHERED PINK *introduced*
IR | **WUP** | **CUP** | EUP

Perennial. Stems simple, 1–3 dm tall, glabrous, frequently glaucous. Leaves linear, 2–8 cm long, 1.5–3 mm wide, 3-nerved, the 3–8 pairs of stem leaves somewhat shorter than the basal ones. Inflorescence 1–5-flowered, the pedicels 1–3 cm long; calyx 15–20 mm long, 40-nerved, the lobes 4–5 mm long. Bracts 1 or 2 pairs, 5–10 mm long, abruptly mucronate or short-awned; petals red to white, 1.5–2 cm long, fringed. Capsule somewhat exceeding the calyx, dehiscent by 4 subobtuse teeth. Summer.

Native of Europe; escaped from cultivation and locally established.

Gypsophila Baby's-Breath

Annual or perennial Eurasian herbs. Cymes paniculately much branched or the flowers solitary. Calyx campanulate to turbinate, short, 5-nerved, ebracteate. Petals white to pinkish, scarcely differentiated into claw and blade, without auricles or appendages. Styles 2. Capsule globose to ovoid-oblong, dehiscent by 4–6 ascending teeth. Seeds obovoid, ovoid, or kidney-shaped.

ADDITIONAL SPECIES
Gypsophila elegans Bieb. (Showy baby's-breath), Mackinac and Menominee counties.

1	Plants annual	**G. muralis**
1	Plants perennial	2
2	Larger leaves to 1 cm wide, narrowed and not clasping at base	**G. paniculata**
2	Leaves 1–2 cm wide, broad and clasping at base	**G. scorzonerifolia**

Gypsophila muralis L.
LOW BABY'S-BREATH   *introduced*
IR | **WUP** | CUP | EUP

Annual herb. Stems slender, diffusely branched 5–15 cm tall, glabrous or puberulent. Leaves 5–15 mm long, 1–2 mm wide, linear. Pedicels capillary, 1–2 cm long, spreading or ascending from the axils of all but the lower leaves; calyx 5-nerved, 3–4 mm long; petals oblong lance-shaped, 6–10 mm long, emarginate, pink or purplish. Capsule ellipsoid-ovoid, 3–4 mm long; seed about 0.6 mm long, low-tuberculate. June–Sept.

Native of Eurasia; established locally as a weed; Gogebic and Iron counties.

Gypsophila paniculata L.
TALL BABY'S-BREATH   *introduced (invasive)*
IR | WUP | CUP | EUP

Perennial herb; plants large and tumbleweed-like, much-branched and covered with tiny flowers. Inflorescence paniculate, diffusely branched; petals petals white or rarely pinkish, about 3 mm long. Capsule globose.

Sandy roadsides, fields, ditches and railroad embankments, a troublesome weed of sand dunes; Houghton and Schoolcraft counties.

Gypsophila scorzonerifolia Ser.
GARDEN BABY'S-BREATH   *introduced*
IR | WUP | CUP | EUP

Perennial herb. Stems 5–20 dm tall, several from a stout base. Leaves basal and cauline, bases clasping, oblong lance-shaped, glaucous. Petals white with pink tinge. Capsule globose.

Roadsides, shores, dunes, quarries, sometimes where very calcareous; Houghton County.

Minuartia Stitchwort

Previously included in genus *Arenaria*.

1	Petals clearly longer than the sepals; seeds 0.7–0.8 mm wide	**M. michauxii**
1	Petals shorter than to equaling the sepals; seeds 0.6–0.7 mm wide	**M. dawsonensis**

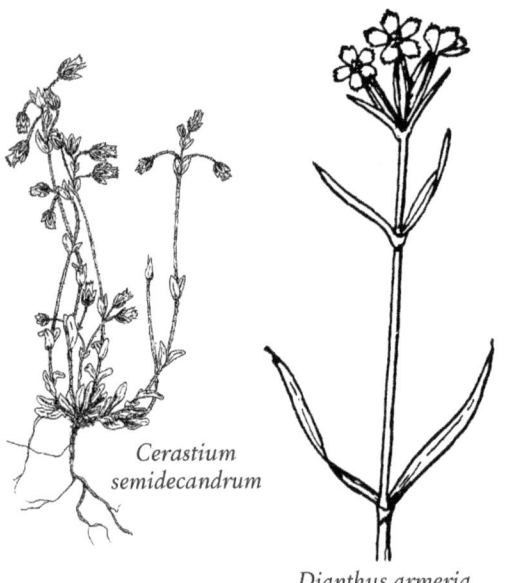

Cerastium semidecandrum

Dianthus armeria

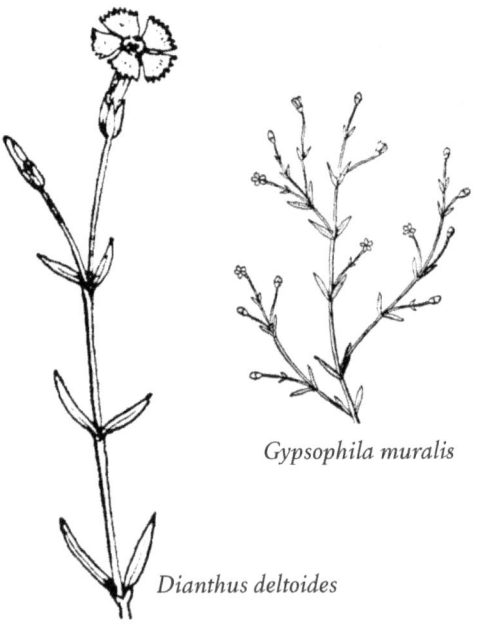

Gypsophila muralis

Dianthus deltoides

CARYOPHYLLACEAE | Pink Family

Minuartia dawsonensis (Britt.) House
ROCK STITCHWORT native
IR | WUP | **CUP** | **EUP** CC 10
Perennial herb, sometimes mat-forming, similar to *M. michauxii* but typically smaller and less common. Stems erect to ascending, green, to 30 cm long, glabrous. Leaves overlapping or crowded near base, with a small sheath around the stem; blades straight to slightly curved, 1-veined, linear, 4–15 mm long and to 2 mm wide, shiny; axillary leaves present among lower stem leaves. Inflorescence 7–15-flowered in open cymes; bracts subulate; sepals 3-veined, ovate, 2.5-4 mm long, tips green to purple; petals lanceolate to spatulate, or petals absent. Capsules ovoid, 3.5–4.5 mm long, longer than the sepals; seeds dark brown to black, suborbiculate, tuberculate. Late spring–summer.

Moist to dry gravelly, rocky, or sometimes calcareous places; Isle Royale, and Mackinac and Schoolcraft counties.

Minuartia michauxii

Minuartia michauxii

Minuartia michauxii (Fenzl) Farw.
MICHAUX'S STITCHWORT native
IR | WUP | CUP | **EUP** CC 10
Arenaria michauxii (Fenzl) Hook.f.
Perennial or annual herb, diffuse, completely glabrous or sometimes pubescent, the prostrate branches with numerous short sterile shoots or the plants merely cespitose. Stems 1–4 dm long, leafy for 1/3 to 2/3 of their length, the primary leaves with short, leafy, fascicle-like sterile shoots in their axils. Primary leaves 8–30 mm, usually 10–20 mm long, subulate, somewhat involute, 3-nerved. Inflorescence open, forked; pedicels slender, 5–20 mm long; sepals 3.5–6.5 mm long, broadly lance-shaped, scarious-margined, 3-nerved or rarely 1-nerved; petals oblong lance-shaped, 5–8 mm long, entire. Capsule equaling or somewhat exceeding the sepals, the 3 valves dehiscent from near the middle to the base; seed kidney-shaped, brown-black, low-tuberculate. July–Sept.

Dry woods of oak and jack pine, sand dunes, limestone pavements ("alvars," Drummond Island); Chippewa and Schoolcraft counties.

Moehringia Grove-Sandwort

Moehringia macrophylla (Hook.) Fenzl
LARGE-LEAF GROVE-SANDWORT (THR) native
IR | WUP | CUP | EUP CC 10
Arenaria macrophylla Hook.
Perennial herbs. Stems prostrate or ascending to erect. Leaves not congested at or near base of flowering stem; lance-shaped to elliptic, 2–5 cm long, 3–8 mm wide. Inflorescence an open cyme. Sepals mostly 3–5 mm long, broadly lance-shaped. Capsule shorter than the sepals, globose-oblong, the 6 segments dehiscing halfway or nearly to the base, the teeth recurved; seed obliquely kidney-shaped, smooth, glossy, reddish brown. Seeds have strophioles, spongy seed appendages that attract ants; foraging ants gather the seeds, eat only the strophiole, and plant the seeds in their nests. May–Aug.

Open to partly shaded rocks and cliffs.

Petrorhagia *Saxifrage-Pink*

Petrorhagia saxifraga (L.) Link
SAXIFRAGE-PINK introduced
IR | WUP | **CUP** | EUP
Tufted perennial. Stems decumbent at base, 1–4 dm tall, glabrous or sparsely hispidulous. Leaves linear-subulate, 5–10 mm long, 1 mm wide. Flowers solitary or in cymes; calyx 4–5 mm long, 5-ribbed, closely subtended by 2 or 3 pairs of lance-shaped scarious bracts; petals purple to pink, 5–6 mm long, the blades broadly notched; stamens 10; styles 2. Capsule globose-ovoid, dehiscent by 4 teeth; seeds disciform. Summer.

Native of Europe; rarely escaped from cultivation and established mostly as a roadside weed; Delta County.

Sagina *Pearlwort*

Small, tufted perennial herbs. Leaves filiform, subulate. Flowers small, inconspicuous, terminal or axillary; petals 4 or 5 white, entire or notched; styles as many as the sepals; capsules many-seeded.

1 Petals distinctly longer than the sepals; styles 1–1.5 mm long; Keweenaw County (incl. Isle Royale) .. **S. nodosa**
1 Petals much shorter than the sepals, inconspicuous; styles less than 0.5 mm long.......... **S. procumbens**

CARYOPHYLLACEAE | Pink Family

Sagina nodosa (L.) Fenzl
KNOTTY PEARLWORT (THR) *native*
IR | **WUP** | CUP | EUP   CC 10

Tufted, mat-forming, perennial herb, 5-15 cm high. Stems erect, ascending, or decumbent. Lower stem leaves triangular in cross-section, 5-20 mm long; upper leaves paired, scale-like; some leaf-pairs there may have a small, sterile shoot at their base. Flowers small, white, delicate, borne singly at ends of slender branches; sepals to 2 mm long; petals to 4 mm long. capsules with several black, rough-surfaced seeds.

Crevices of rocks along shore of Lake Superior, where the spray from waves keeps the plants moist; Keweenaw County (including Isle Royale); disjunct from arctic Canada.

Sagina procumbens L.
BIRD-EYE PEARLWORT   *introduced*
IR | **WUP** | CUP | EUP   FAC

Branched perennial or perhaps sometimes annual herb. Stems prostrate, spreading, or ascending, 2–10 cm long, glabrous or minutely puberulent; short shoots or leaf-fascicles often conspicuous at the nodes. Leaves opposite, linear, 3–15 mm long, mucronate, often minutely ciliolate. Flowers lateral or in terminal cymes, on filiform pedicels, often nodding after anthesis, at length becoming erect; sepals 4, sometimes 5, broadly ovate, 2–2.5 mm long, spreading at maturity; petals white, shorter than the sepals or lacking; stamens 5 or 4; styles 5 or 4, alternate with the sepals. Capsule about equaling the sepals, the valves 4, sometimes 5; seeds many, 0.5 mm wide, grooved, dark reddish brown. May–Sept.

Moist soil and rocky places, weedy in paths and pavements.

Saponaria *Soapwort; Bouncing-Bet*

Saponaria officinalis L.
BOUNCING-BET   *introduced*
IR | **WUP** | CUP | EUP   FACU

Perennial herb from a horizontal rhizome and forming colonies. Stems coarse, erect, 4–8 dm tall, simple or branched, glabrous. Leaves 7–10 cm long, 2–4 cm wide, elliptic to elliptic-ovate, glabrous, rarely puberulent. Inflorescence congested and subcapitate to open and oblong-pyramidal, to 15 cm long; primary bracts foliaceous, the ultimate ones scarious. Flowers fragrant, frequently double; calyx 1.5–2.5 cm long, 20-nerved, the lobes long triangular, the tube often deeply bilobed; petals white or pinkish, appendages conspicuous, awl-shaped; stamens exsert; styles 2 (rarely 3). Capsule elliptic-oblong, dehiscent by 4 (rarely 6) teeth; seeds uniformly reticulate. Summer.

Native of the Old World; formerly in cultivation and now commonly weedy on roadsides, railways and waste places.

Name from the Latin, *sapo*, soap, alluding to the mucilaginous juice which forms a lather with water.

Scleranthus *Knawel*

Herbs with forked stems. Leaves opposite, subulate, joined at the base. Flowers perfect. Calyx 5-lobed, the tube becoming thick and indurate, enclosing the membranous utricle. Stamens 1–10, inserted on a disc in the throat of the calyx tube. Ovary ovoid; styles 2, free to the base. Seed ovoid, beaked.

1 Perennial; sepals blunt-tipped, with a broad white margin...................**S. perennis**
1 Annual; sepals pointed, with only a tiny white margin ....**S. annuus**

Scleranthus annuus L.
ANNUAL KNAWEL   *introduced*
IR | **WUP** | CUP | EUP   FACU

Annual herb; plants to 15 cm tall, diffuse, spreading, glabrous or puberulent. Flowers sessile or subsessile; calyx lobes equal or exceeding the calyx tube; stamens usually 5–10. Seeds 1–1.3 mm long, straw-colored. Summer.

Native of Eurasia; a weed of dry, usually sandy fields, roadsides, and waste places.

Petrorhagia saxifraga   *Sagina nodosa*   *Sagina procumbens*   *Saponaria officinalis*

Scleranthus perennis L.
PERENNIAL KNAWEL *introduced*
IR | **WUP** | **CUP** | EUP

Perennial herb. Differs from *S. annuus* by its perennial habit, its obtuse or rounded sepals with a conspicuous white-scarious border to 0.5 mm wide near the tip.

Introduced species of dry open places; Houghton and Marquette counties.

Silene *Catchfly; Campion*

Annual or perennial herbs. Leaves opposite, entire. Inflorescence simple or branched, sometimes reduced to a few-flowered or 1-flowered cyme. Flowers perfect or sometimes unisexual. Calyx sometimes inflated. Petals 5, the claw narrow, expanded distally into more or less prominent auricles, provided ventrally with a pair of appendages; blade usually exsert, entire, lobed or dissected. Stamens normally 10. Styles normally 3, sometimes 4, or occasionally 5. Ovary usually stipitate. Capsule 3-celled (or 1-celled), dehiscent normally by 6 teeth. Seeds kidney-shaped to globose, sometimes covered with small bumps.

1	Calyx not glandular or inflated	2
1	Calyx glandular, often inflated	3
2	Plants densely covered with white woolly hairs	**S. coronaria**
2	Plants hairy but not white-woolly	**S. viscaria**
3	Styles 5	**S. latifolia**
3	Styles 3	4
4	Calyx glabrous	5
4	Calyx hairy or glandular	9
5	Corolla pink	**S. armeria**
5	Corolla white	6
6	Annual; calyx tight around capsule	**S. antirrhina**
6	Perennial; calyx somewhat inflated	7
7	Flowers 5 or fewer, single in leaf axils; rare in Ontonagon County	**S. nivea**
7	Flowers in clusters of more than 5	8
8	Perennial; calyx very inflated	**S. vulgaris**
8	Biennial; calyx only slightly inflated	**S. csereii**
9	Flowers in long, upright, often one-sided racemes	**S. dichotoma**
9	Flowers in an open cyme, often nodding	**S. noctiflora**

Silene antirrhina L.
SLEEPY CATCHFLY *native*
IR | **WUP** | **CUP** | **EUP** CC 2

Annual, glabrous or more or less puberulent. Stems 2–8 dm tall, simple or branched, erect or sometimes decumbent, usually with glutinous zones below the upper nodes. Basal leaves oblong lance-shaped; stem leaves oblong lance-shaped to linear, usually 3–6 cm long, 2–12 mm wide, glabrous or puberulent, the margins ciliate near the base. Inflorescence open, strict or with divaricate branches. Flowers numerous, rarely few or solitary; calyx 4–10 mm long, 10-nerved; petals white or pink, equaling or exceeding the calyx, frequently obsolete, 2-lobed; appendages minute or lacking. Capsule ovoid, 4–10 mm long, 3-celled; seeds with 3 or 4 rows of dorsal papillae. Summer.

In waste places or sandy soil.

Silene armeria L.
NONE-SO-PRETTY *introduced*
IR | **WUP** | **CUP** | EUP CC 2
Atocion armeria (L.) Raf.

Annual with glabrous or rarely sparsely puberulent stems about 3 dm tall. Leaves ovate lance-shaped; basal leaves 2–5 cm long, 5–15 mm wide, sessile; upper leaves clasping. Inflorescence simple or open and compound, the ultimate cymes congested; calyx tubular, 13–17 mm long, 10-nerved; corolla pink or lavender; auricles lacking; appendages linear, 2–3 mm long; blades 4–7 mm long, obovate. Capsule nearly completely 3-celled; seeds 0.6 mm wide, rugose; carpophore (the stalk supporting the fruit) 7–8 mm long. June–July.

Native of Europe; once popular in cultivation; escaped as a weed in waste places.

Silene chalcedonica (L.) E.H.L. Krause
MALTESE CROSS *introduced*
IR | **WUP** | **CUP** | EUP
Lychnis chalcedonica L.

Perennial with hirsute stems 5–10 dm tall. Basal leaves spatulate or lance-shaped; stem leaves 10–20 pairs, lance-shaped to ovate lance-shaped, 5–12 cm long, 2–5 cm wide, sparingly pubescent or glabrate, serrulate-ciliate. Inflorescence terminal, congested; flowers numerous, red, rose, or white; calyx tubular, 12–17 mm long at maturity, coarsely 10-ribbed, the ribs strigose- hirsute, the lobes lance-shaped, 2.5–3.5 mm long; petals 14–18 mm long, the claws ciliate; appendages tubular, 2–3 mm long; blades 7–9 mm long, deeply bilobed. Capsule about 1 cm long. June–Sept.

Native of Asia; occasionally escaped from cultivation and persisting.

Silene coronaria (Desr.) Clairv. ex Rchb.
ROSE CAMPION *introduced*
IR | **WUP** | CUP | EUP
Lychnis coronaria (L.) Desr.

Perennial, gray-tomentose, the stout stems rarely branched, 4–8 dm tall. Basal leaves 5–10 cm long, 1–3 cm wide; stem leaves 5–10 pairs, lance-shaped to oblong lance-shaped, usually smaller than the basal. Inflorescence few-flowered, the pedicels 5–10 mm long; calyx 12–15 mm long, 10-ribbed, the lobes narrowly lance-shaped, 4–7 mm long, connivent, twisted; petals crimson, 2–3 cm long, without auricles; appendages narrowly lance-shaped, 1.5–2.5 mm long, the blades broadly obovate, 10–15 mm long. Capsules 12–16 mm long, ovoid, dehiscent by 5 teeth. June–Aug.

Native of Europe; escaped from cultivation; Keweenaw County.

Silene csereii Baumg.
BALKAN CATCHFLY *introduced*
IR | **WUP** | **CUP** | **EUP**

Robust glaucous perennial; leaves 2–4 cm wide. Inflorescence narrow, raceme-like; calyx 7–9 mm long, in fruit becoming 10–12 mm long, but little inflated and the veins not conspicuously reticulate; petals as in S. vulgaris but root much stouter and inflorescence more elongate. Seeds conspicuously papillate. Summer.

Native of Europe; disturbed places. Often confused with *S. vulgaris*.

Silene dichotoma Ehrh.
FORKED CATCHFLY *introduced*
IR | WUP | CUP | **EUP**

Annual, with simple or sparingly branched, strongly hirsute stems 3–8 dm tall. Leaves lance-shaped to oblong lance-shaped, 3–8 cm long, 3–35 mm wide, the lower leaves usually with ciliate petioles, the upper sessile. Inflorescence usually once or several times dichotomous, the ultimate branches racemose. Flowers mostly perfect; calyx narrowly tubular, 10–15 mm long, the 10 green nerves stiffly hirsute; corolla white to reddish; auricles lacking; appendages truncate, about 0.2 mm long; blades cuneate, 5–9 mm long, deeply 2-lobed; stamens usually exsert but sometimes vestigial. Capsule 3-celled; seeds 1–1.3 mm wide, finely and regularly rugose; carpophore 2–4 mm long. Summer.

Native of Eurasia; disturbed places such as fields, roadsides, and railways; Mackinac County.

Silene latifolia Poir.
WHITE CAMPION *introduced*
IR | **WUP** | **CUP** | **EUP**
 Lychnis alba Mill.

Dioecious annual or perennial from a stout root. Stems 4–12 dm tall, coarsely pubescent, glandular above. Stem leaves as many as 10 pairs, lance-shaped to broadly elliptic, 3–10 cm long, 1–4 cm wide, 3–5-nerved, puberulent to hirsute, the lower petiolate, the upper sessile. Inflorescence usually much branched, the primary bracts foliar; flowers white, characteristically unisexual, opening in the evening; calyx 15–20 mm long, tubular in anthesis, becoming distended at maturity, 10-nerved in the staminate flower, 20-nerved in the pistillate, the lobes lance-shaped, 3–5 mm long; petals 2–4 cm long, the claw exsert, auriculate, the appendages 1–1.5 mm long, erose, the blade deeply bilobed. Capsule ovoid, 10–15 mm long, dehiscent by 10 erect or spreading teeth. Summer.

Native of Europe; a very common weed.

Silene nivea (Nutt.) Muhl.
SNOWY CATCHFLY (THR) *native*
IR | **WUP** | CUP | EUP FACW CC 10

Perennial herb, spreading by rhizomes; plants smooth or with a few short hairs. Stems 2–3 dm long. Leaves mostly on stem, opposite, lance-shaped or oblong, 5–10 cm long and 1–3 cm wide, sessile or on short petioles. Flowers few, mostly in leaf axils; sepals joined to form a tube 1.5 cm long; petals white, stamens 10, styles 3. Fruit a 1-chambered capsule. June–July.

Streambanks, wooded ravines; rare in Ontonagon County (species at northern range limit).

Silene noctiflora L.
NIGHT-FLOWERING CATCHFLY *introduced*
IR | **WUP** | **CUP** | **EUP**

Annual, 2–8 dm tall, with simple or branched, coarsely hirsute stems and leaves. Leaves ovate lance-shaped, 5–12 cm long, 2–4 cm wide; basal leaves somewhat narrowed to a petiole; stem leaves narrower, sessile. Inflorescence loosely branched; flowers often unisexual; calyx about 15 mm long in anthesis, in fruit 2.5–5 cm long, the 10 nerves glandular, the lobes linear lance-shaped, 5–9 mm long; corolla white or pink; auricles 1–1.5 mm long; appendages broad, 0.5–1.5 mm long, entire or erose; blades 7–10 mm long, deeply 2-lobed. Capsule 3-celled; seeds 0.8–1 mm wide, uniformly rugose-papillate; carpophore 1–3 mm long. July–Sept.

Native of Europe; disturbed places such as roadsides, railways, fields. Plants superficially resemble *S. latifolia* which normally has 5 styles.

Silene vulgaris (Moench) Garcke
BLADDER-CAMPION *introduced*
IR | **WUP** | **CUP** | **EUP**

Robust perennial. Stems 2–8 dm tall, from a creeping rhizome, usually glabrous and glaucous, often decumbent. Leaves ovate lance-shaped, 3–8 cm long, 1–3 cm wide, abruptly acuminate, sometimes ciliolate; stem leaves often clasping. Inflorescence open, 5–30-flowered; calyx campanulate, 1 cm long, papery in texture with reticulate veins, in fruit becoming much inflated; corolla white; auricles lacking; appendages inconspicuous or lacking; blades 3.5–6 mm long, deeply bilobed. Capsule 3-celled; seeds 1–1.5 mm, warty; carpophore 2–3 mm long. Summer.

Native of Europe, weedy in waste places.

Spergula *Spurry*

Spergula arvensis L.
CORN SPURRY *introduced*
IR | **WUP** | **CUP** | **EUP**

Fleshy annual herb. Stems simple or much-branched, to 40 cm long, sparingly glandular-puberulent. Leaves whorled, 2–5 cm long, narrowly linear, clustered at the nodes in two opposite sets of 6–8; stipules small, connate. Inflorescence terminal, dichotomously branched, the pedicels reflexed; sepals 5, ovate, 2–3 mm long, obtuse, glandular-puberulent; petals 5, white, shorter than or somewhat surpassing the sepals; stamens 10 or sometimes 5; styles normally 5. Capsule broadly ovoid, somewhat longer than the sepals; seeds 1–1.5 mm wide, blackish, minutely roughened. May–Aug.

Native of Europe; a weed of cultivated ground and waste places.

Spergularia *Sandspurry*

Spergularia rubra (L.) J.& K. Presl
ROADSIDE SANDSPURRY *introduced*
IR | WUP | CUP | EUP FACU

Low-growing, succulent, annual or short-lived perennial herb. Stems simple or much branched with slender, prostrate or ascending stems 5–30 cm long; glabrous, or sparsely glandular-pubescent below the inflorescence. Leaves opposite, linear-filiform, mucronate, scarcely fleshy, 4–25 mm long, to 1 mm wide; stipules conspicuous, triangular-acuminate, 2.5–5 mm long. Flowers in branched terminal cymes; sepals and petals each 5; sepals 3.5–5 mm long, lance-shaped; petals pink, shorter than the sepals; stamens 6–10, more often 10. Fruit a 3-valved capsule 3.5–5 mm long; seeds tiny, dark brown, papillate, not winged. May–Sept.

Native of Europe; sandy or gravelly soil.

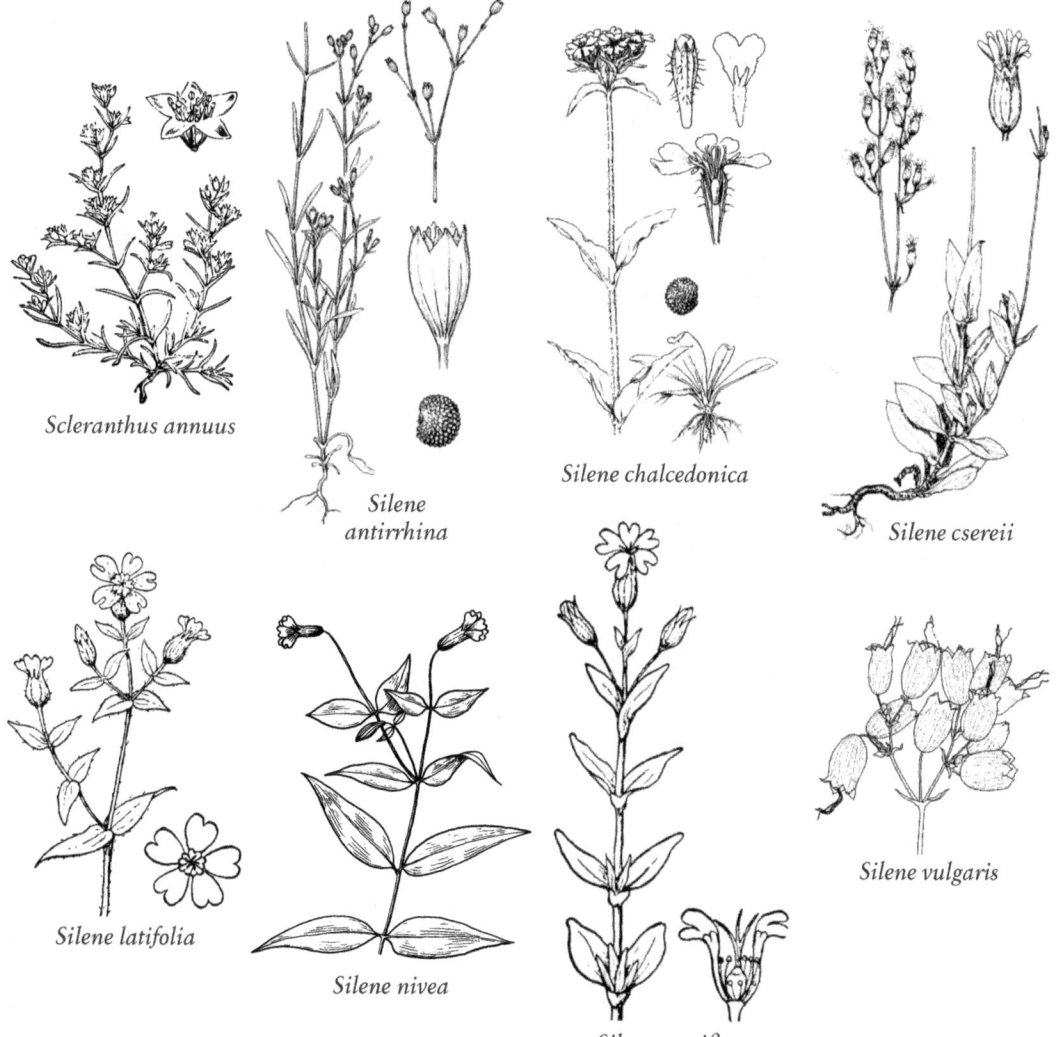

Scleranthus annuus

Silene antirrhina

Silene chalcedonica

Silene csereii

Silene latifolia

Silene nivea

Silene noctiflora

Silene vulgaris

CARYOPHYLLACEAE | Pink Family

Stellaria *Chickweed*

Low, spreading or erect perennials (ours), mostly without hairs. Stems slender, 4-angled. Flowers single in forks of stems or in few-flowered clusters at ends of stems; sepals green with translucent margins; petals white, lobed or deeply cleft (sometimes absent in *S. borealis*); stamens 10 or less; styles 3. Fruit an ovate or oblong capsule.

1	Plants large, the stems to 8 dm long; styles 5 . . . . **S. aquatica**	
1	Plants smaller; styles 3–4 2	
2	Leaves wider, not linear, mostly more than 1 cm wide . . . . **S. media**	
2	Leaves narrow, linear or lance-shaped, less than 1 cm wide 3	
3	Flowers in branched cymes 4	
3	Flowers single in forks of stems 7	
4	Petals usually absent, or shorter than the sepals . . . . **S. borealis**	
4	Petals much longer than sepals 5	
5	Inflorescence less open; pedicels erect or ascending . . . . **S. longipes**	
5	Inflorescence open and branched; pedicels spreading 6	
6	Flowers numerous; sepals 4.5–5.5 mm long, with 3 prominent nerves; seeds bumpy . . . . **S. graminea**	
6	Flowers few; sepals to 4.5 mm long, only weakly 3-nerved; seeds smooth . . . . **S. longifolia**	
7	Stems 25 cm or more long; seeds smooth . . **S. borealis**	
7	Stems to 20 cm long; seeds rough; rare in Luce County . . . . **S. crassifolia**	

Stellaria aquatica (L.) Scop.
GIANT CHICKWEED *introduced*
IR | WUP | CUP | EUP FAC

Myosoton aquaticum (L.) Moench

Perennial herb, spreading by rhizomes. Stems sprawling and matted, to 8 dm long, rooting at nodes, covered with gland-tipped hairs. Leaves ovate to lance-shaped, 2–8 cm long and 1–4 cm wide, petioles short or absent. Flowers in open, leafy clusters at ends of stems; sepals 5–9 mm long; petals white, much longer than sepals. Fruit a capsule; seeds 0.8 mm long, covered with small bumps. June–Oct.

Streambanks, ponds, wet or moist disturbed areas, often in partial shade.

Stellaria borealis Bigelow
NORTHERN STITCHWORT *native*
IR | WUP | CUP | EUP FACW CC 6

Perennial herb, spreading by rhizomes. Stems sprawling, to 5 dm long, branched, angled. Leaves lance-shaped, narrowed at base, 1–5 cm long and 2–8 mm wide, margins hairy. Flowers in clusters at ends of stems; sepals 2–4 mm long; petals usually absent. Fruit a dark capsule, longer than sepals; seeds 0.8 mm long, with very small bumps. June–Aug.

Openings and hollows in conifer forests, margins of ponds and marshes.

Stellaria crassifolia Ehrh.
FLESHY STITCHWORT (END) *native*
IR | WUP | CUP | **EUP** FACW CC 10

Perennial herb. Stems sprawling and matted to erect, freely branched, 8–30 cm long, fleshy, glabrous. Leaves soft, oval to lance-shaped, narrowed at base, 1–3 cm long and 1–3 mm wide. Flowers single in forks of stem, nodding on stalks 1–3 cm long; sepals 2–4 mm long; petals longer than sepals. Fruit an ovate capsule, to 5 mm long and longer than the sepals; seeds red-brown, to 1 mm long. June–July.

Streambanks and wet shores; Luce County.

Stellaria graminea L.
GRASS-LEAF STITCHWORT *introduced*
IR | WUP | CUP | EUP

Perennial herb. Stems 3–5 dm long, weak, 4-angled, the angles prominent and sometimes scabrous. Leaves 1.5–5 cm long, 1.5–7 mm wide, linear to linear lance-shaped, the base often obtuse and ciliate. Inflorescence terminal, many-flowered, diffuse, frequently extending to the middle of the stem; bracts scarious, ciliolate; pedicels slender, spreading or reflexed; sepals lance-shaped, in fruit 4.5–5.5 mm long, strongly 3-nerved, the margins scarious, commonly ciliolate, at least at the base; petals exceeding the sepals. Capsule straw-colored, more or less equaling or somewhat surpassing the sepals; seeds 0.8–1.2 mm long, kidney-shaped, dark reddish brown, covered with small bumps. May–July.

Native of Europe; introduced in grassy places, fields, roadsides, and waste land.

Stellaria longifolia Muhl.
LONG-LEAVED STITCHWORT *native*
IR | WUP | CUP | EUP FACW CC 5

Perennial herb. Stems sprawling, prominently 4-angled, usually freely branched, 1–5 dm long. Leaves spreading to ascending, linear to lance-shaped, 2–5 cm long and 1–6 mm wide, widest at or above middle, tapered at both ends. Flowers in branched clusters at ends of stems; sepals 3–5 mm long; petals longer than sepals. Fruit a green-yellow to brown capsule, usually longer than the sepals; seeds light brown, about 1 mm long. May–July.

Wet meadows and marshes, shrub thickets, swamps, streambanks, pond margins.

Stellaria longipes Goldie
LONG-STALK STARWORT *native*
IR | WUP | **CUP** | EUP FACU CC 10

Perennial herb, spreading by rhizomes. Stems erect, or sprawling and matted, 5–30 cm long. Leaves upright, stiff and shiny-waxy, linear or lance-shaped,

1–4 cm long and 1–4 mm wide, widest near base, tapered to tip. Flowers in branched clusters at ends of stems or appearing lateral from stem; sepals 4–5 mm long; petals slightly longer than sepals. Fruit a straw-colored to shiny purple capsule, longer than the sepals; seeds reddish brown, about 1 mm long. May–July.

Wet meadows, ditches and thickets.

Stellaria media (L.) Vill.
COMMON CHICKWEED *introduced*
IR | **WUP** | **CUP** | **EUP**   FACU

Weakly tufted annual. Stems to 4 dm long, with ascending branches, puberulent in lines. Leaves usually 1–3 cm long, ovate or obovate, glabrous, frequently pustulate; upper leaves sessile; lower leaves with petioles that may exceed the length of the blade, often ciliate toward the base or on the petioles. Flowers solitary or in few-flowered, terminal, leafy cymes; pedicels ascending, reflexed, frequently pubescent; sepals oblong, 3.5–6 mm long, pubescent and pustulate; petals shorter than the sepals; stamens 3–5. Capsule ovoid, somewhat surpassing the sepals; seeds 1–1.2 mm long, suborbicular, reddish brown, conspicuously covered with small bumps.

Introduced from the Old World but often appearing to be native; now a cosmopolitan weed of waste places, cultivated areas, meadows, and woodlands.

Vaccaria *Cowcockle*

Vaccaria hispanica (P. Mill.) Rauschert
COWCOCKLE *introduced*
IR | **WUP** | **CUP** | **EUP**

Annual herb from a slender taproot. Stems branching above, 2–6 dm tall, glabrous and glaucous. Stem leaves 5–10 cm long, 24 cm wide, lance-shaped to ovate lance-shaped, clasping or the lower connate. Inflorescence a loose, open, paniculate cyme; calyx strongly 5-ribbed, 12–17 mm long, ovoid; petals 18–22 mm long, pink, without auricles or appendages,

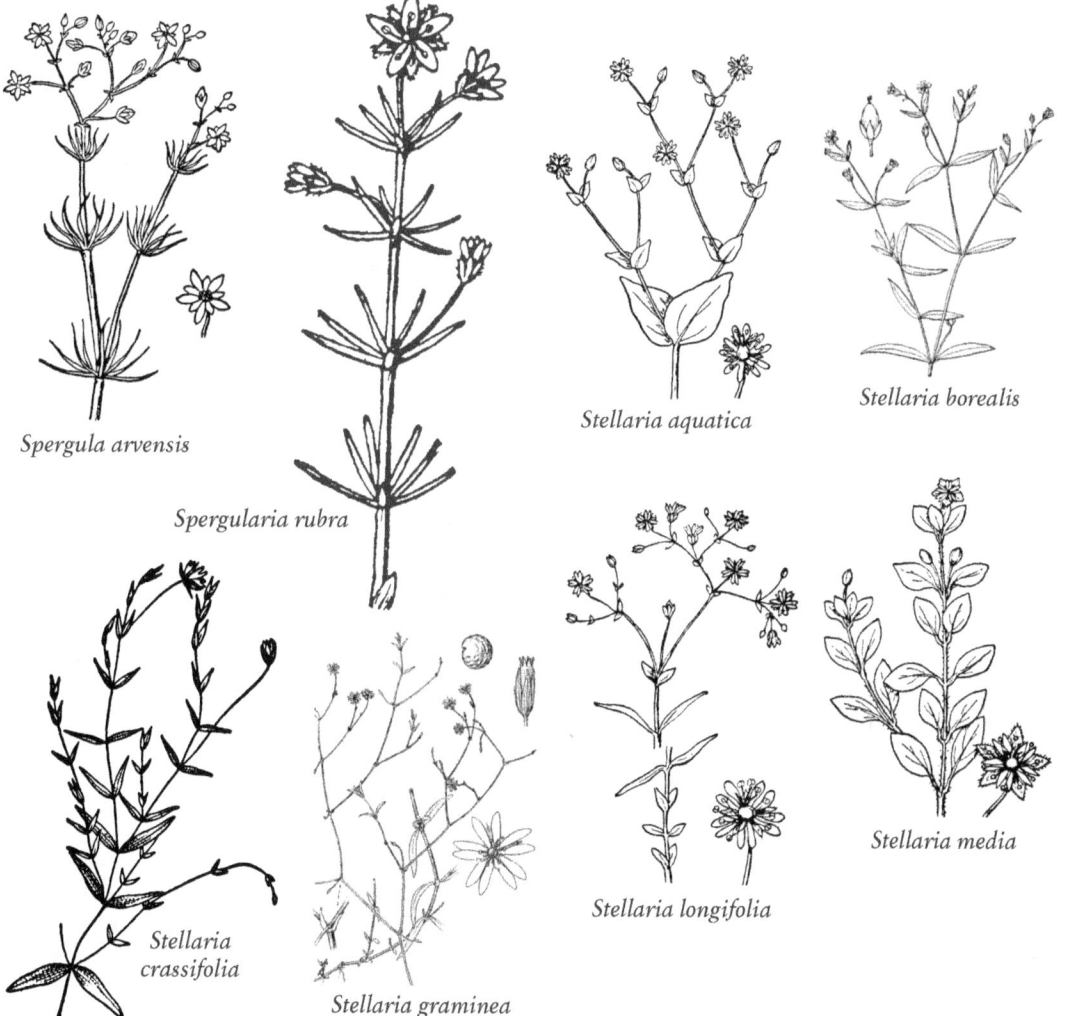

Spergula arvensis

Spergularia rubra

Stellaria aquatica

Stellaria borealis

Stellaria crassifolia

Stellaria graminea

Stellaria longifolia

Stellaria media

the blades 6–8 mm long, obovate, retuse; stamens exsert. Capsule 6–8 mm long, dehiscent by 4 teeth; seeds 2–2.5 mm long, minutely tuberculate, reddish brown to black.

Native of Europe; Keweenaw County.

Celastraceae
BITTERSWEET FAMILY

Shrubs (*Euonymus*), vines (*Celastrus*), or glabrous perennial herbs (*Parnassia*). Leaves with simple, evergreen or deciduous, opposite or alternate leaves and small, axillary or terminal, solitary or clustered flowers. Flowers perfect or unisexual, regular, polypetalous, usually 4–5-merous. Stamens as many as the petals and alternate with them; in *Parnassia*, staminodes (infertile stamens) attached to base of petals and divided into threadlike segments tipped with glandular knobs. Pistil 1, inserted on or surrounded by the disk; ovary 2–5-celled. Fruit a capsule. Celastraceae now includes members of genus *Parnassia*.

1 Perennial herbs; leaves basal with a single stem leaf .. .. **Parnassia**
1 Twining vines; leaves alternate............ **Celastrus**

Celastrus *Bittersweet*

Celastrus scandens L.
AMERICAN BITTERSWEET native
IR | WUP | CUP | EUP   FACU CC 3

Woody twiners, climbing several meters high. Leaves deciduous, alternate, elliptic or oblong to ovate, acuminate, serrulate, 5–10 cm long. Panicles terminal, 3–8 cm long. Flowers dioecious or polygamo-dioecious, small, whitish or greenish, 5-merous. Staminate flowers with 5 stamens about as long as the petals, inserted on the margin of the cup-shaped disk. Pistillate flowers with rudimentary stamens and a well developed ovary, stout columnar style, and 3-lobed stigma. Fruit subglobose, nearly 1 cm long, several in a cluster, 3-valved, each valve covering 1 or 2 seeds enclosed in a fleshy bright red-orange aril; seeds ellipsoid, 6 mm long. May–June.

Roadsides and thickets, usually in rich soil; occasionally cultivated and now established in open woods and thickets.

Parnassia *Grass-of-Parnassus*

Glabrous perennial herbs. Leaves all from base of plant but often with 1 stalkless leaf near middle of stalk, margins entire; petioles present. Flowers large, white, single at ends of stalks; calyx 5-lobed; petals white, veined, spreading; fertile stamens 5, alternating with petals; staminodes (infertile stamens) attached to base of petals and divided into threadlike segments tipped with glandular knobs; stigmas 4. Fruit a 4-chambered capsule with numerous seeds.

1 Sepals with narrow translucent margins; staminodes (sterile stamens) 3-parted, not widened at base; petals 12–16 mm long; leaves leathery and somewhat succulent .... **P. glauca**
1 Sepal margins green; staminodes 5 to many-parted; petals 5–13 mm long; leaves thin and membranous... 2
2 Leaves broadly rounded or heart-shaped at base; petals 8–13 mm long....................... **P. palustris**
2 Leaves narrowed to base; petals 5–9 mm long .... **P. parviflora**

Parnassia glauca Raf.
FEN GRASS-OF-PARNASSUS   native
IR | WUP | **CUP** | **EUP**   OBL CC 8

Leaves from base of plant and usually with 1, more or less sessile stem leaf; broadly ovate to nearly round, 2–7 cm long and 1–5 cm wide; margins entire; petioles long. Flowers single atop a stalk 1–4 dm long; sepals ovate, 2–5 mm long, with a narrow, translucent margin; petals white with green veins, 1–2 cm long; staminodes 3-parted from near base, shorter than to equal to stamens. Fruit a capsule about 1 cm long. Aug–Sept.

Calcareous fens and wet meadows.

Parnassia palustris L.
ARCTIC GRASS-OF-PARNASSUS   (THR) native
IR | WUP | CUP | EUP   OBL CC 10

Leaves from base of plant and usually with 1, clasping, heart-shaped leaf below middle of stalk; ovate to nearly round, 1–3 cm long; margins entire; petioles long and slender. Flowers single atop a stalk 1.5–4 dm long; sepals lance-shaped, to 1 cm long, green throughout; petals white with green veins, ovate, 1–1.5 cm long, longer than sepals; staminodes many-parted from the widened tip, 5–9 mm long. Fruit a capsule. July–Sept.

Calcareous fens, shores, streambanks and wet, seepy meadows.

Parnassia parviflora DC.
SMALL-FLOWERED GRASS-OF-PARNASSUS   native
IR | **WUP** | **CUP** | **EUP**   CC 10

Parnassia palustris var. *parviflora* (DC.) Boivin

Leaves from base of plant, with a sessile, non-clasping leaf near middle of stalk; oval to ovate, 1–3 cm long, narrowed to base; margins entire; petioles present. Flowers single atop slender stalks 0.5–3 dm long; sepals ovate, 3–7 mm long, nearly as long as petals, green throughout; petals white with green veins, oval, 5–10 mm long; staminodes 5–7-parted near the slightly widened tip, shorter than the stamens. Fruit a capsule. July–Aug.

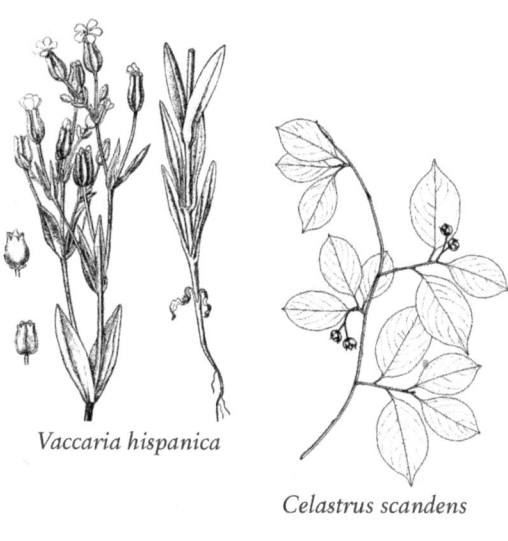

Vaccaria hispanica

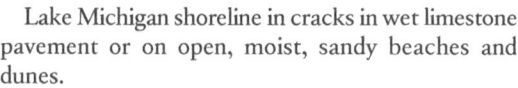

Celastrus scandens

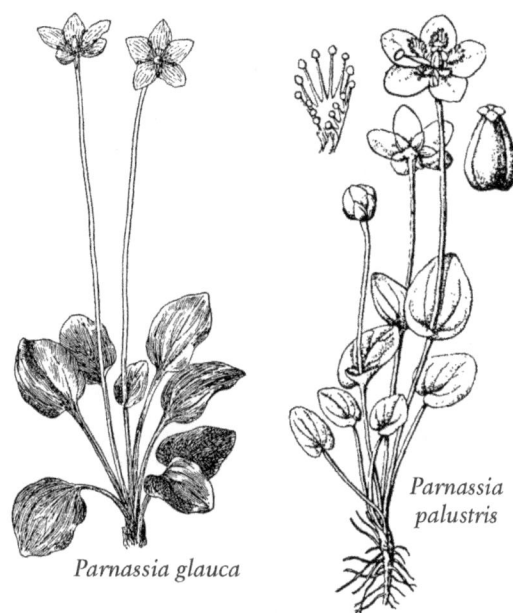

Parnassia glauca

Parnassia palustris

Lake Michigan shoreline in cracks in wet limestone pavement or on open, moist, sandy beaches and dunes.

Ceratophyllaceae
HORNWORT FAMILY

Ceratophyllum
Coon's-Tail; Hornwort

Aquatic perennial herbs, often forming large patches; roots absent, but plants usually anchored to substrate by pale, modified leaves. Stems slender, branched. Leaves in whorls, with more than 4 leaves per node, whorls crowded at ends of stems, dissected 2–3 times into narrow segments. Flowers small, inconspicuous in leaf axils, staminate and pistillate flowers separate on same plant, staminate usually above pistillate on stems. Our only genus of aquatic vascular plants with whorled, forked leaves.

1 Leaves usually stiff, forked 1–2 times, margins coarsely toothed; achenes with 2 spines near base **C. demersum**
1 Leaves limp, some larger leaves forked 3–4 times, margins not toothed; achenes with 2 spines near base and several spines on margin . . . . **C. echinatum**

Ceratophyllum demersum L.
COON'S-TAIL   *native*
IR | WUP | CUP | EUP   OBL CC 1

Aquatic perennial herb. Stems long, branched. Leaves in whorls of 5–12 at each node, stiff, 1–3 cm long, 1–2-forked; leaf segments linear, 0.5–1 mm wide, coarsely toothed. Fruit an oval achene, 4–6 mm long, with 2 spines at base.

Shallow to deep water of lakes, ponds, backwater areas, ditches; water typically neutral or alkaline.

Ceratophyllum echinatum Gray
SPINELESS HORNWORT   *native*
IR | WUP | CUP | EUP   OBL CC 10

Ceratophyllum muricatum Cham.

Aquatic perennial herb. Similar to *C. demersum*, but leaves usually limp, larger leaves usually 3- or sometimes 4-forked, the segments narrower and mostly without teeth. Fruit an achene with 2 spines at base and several unequal spines on achene body.

Lakes, ponds and quiet water of rivers and streams; water typically acid.

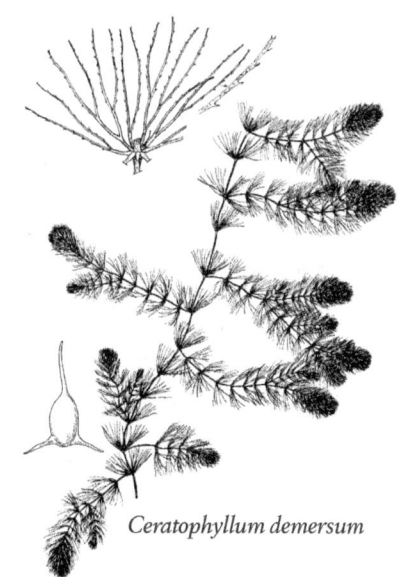

Ceratophyllum demersum

Cistaceae

ROCK-ROSE FAMILY

Herbs or shrubs. Leaves simple, alternate, opposite, or appearing whorled. Flowers cymose, perfect, regular except the calyx, 3–5-merous. Sepals 5, the outer 2 much smaller than the inner 3. Petals small to large, soon deciduous, or lacking in some flowers. Stamens irregular in number, often very numerous, the filaments distinct. Ovary 1-celled; style 1, short or elongate. Fruit a capsule, usually separating completely to the base and enclosed by the persistent calyx.

1 Plants shrubby; leaves small and scale-like .... **Hudsonia tomentosa**
1 Plants herbaceous; leaves linear to ovate 2
2 Leaves densely white-hairy, the hairs branched; petals 5, yellow, conspicuous .... **Crocanthemum**
2 Leaves nearly glabrous to densely hairy, the hairs unbranched; petals 3, dark red, tiny........... **Lechea**

Crocanthemum *Frostweed*

Perennial herbaceous or suffrutescent plants from short or elongate rhizomes. Stems and leaves with stellate-pubescence. Leaves narrow. Sepals 3 or 5; when 5, the outer two much narrower than the inner. Petals yellow, 5 in the first flowers of the season, soon deciduous; petals mostly absent in the later flowers. Stamens numerous. Ovary 1-celled; style short; stigmas large, capitate. Capsule short. At the beginning of the blooming season, plants have one or a few flowers with petals 1–2 cm long. Later, plants produce numerous smaller apetalous flowers. Both types are normally followed by mature capsules, those from apetalous flowers usually much smaller, containing similar seeds. Our species formerly included in genus *Helianthemum*.

1 Petal-bearing flowers 1 .... **C. canadense**
1 Petal-bearing flowers 3 or more.......... **C. bicknellii**

Crocanthemum bicknellii (Fernald) Janch.
HOARY FROSTWEED native
IR | WUP | **CUP** | EUP   CC 10
Helianthemum bicknellii Fern.
Perennial herb. Stems solitary or few together, erect or nearly so, simple or sparsely branched at first anthesis, later producing numerous floriferous branches from the upper axils. Leaves linear-oblong to oblong lance-shaped, 2–3 cm long on the main axis, much smaller on the branches. Petaliferous flowers 5–12 in a loose terminal raceme, 15–25 mm wide; sepals densely stellate, the outer nearly or quite as long as the inner. Capsules of the petaliferous flowers 4–5 mm wide. Apetalous flowers numerous and crowded on short axillary branches, their capsules about 2 mm wide, strongly triquetrous; seeds minutely netveined. June–July; flowering 2–3 weeks later than *C. canadense*.

Dry, usually sandy soil; Dickinson County.

The terminal flowers are not much surpassed by the lateral branches, and their capsules, if persistent in late summer, are near the top of the plant.

Crocanthemum canadense (L.) Britton
LONG-BRANCH FROSTWEED native
IR | **WUP** | **CUP** | EUP   CC 8
Helianthemum canadense (L.) Michx.
Perennial herb. Resembling *C. bicknellii* in habit and foliage. Petaliferous flowers solitary or occasionally 2, 2–4 cm wide; sepals densely stellate and also villous with apparently simple hairs 1–1.5 mm long, the outer about two-thirds as long as the inner. Capsules of the petaliferous flowers 4–6 mm wide. Apetalous flowers fewer and less crowded than in *C. bicknellii*, their capsules commonly 3–3.5 mm wide with convex sides; seeds minutely papillose. Late May–June.

Dry sandy soil, open upland woods.

The terminal flowers are soon surpassed by the lateral branches and their capsules, usually long persistent, are then far below the top of the plant.

Hudsonia *Golden-Heather*

Hudsonia tomentosa Nutt.
SAND GOLDEN-HEATHER native
IR | **WUP** | **CUP** | EUP   CC 10
Perennial. Stems prostrate or bushy, much branched, woody at base, forming dense mats or bushes 2–6 dm across. Leaves scale-like, lance-shaped, 1–4 mm long, closely appressed and imbricate, the entire surface nearly concealed by the pubescence. Flowers 6–10 mm wide, each solitary at the end of a short, leafy, lateral branch; petals 5, yellow, narrowly elliptic, early deciduous; stamens 10–30; ovary 1-celled, glabrous. Capsule ovoid, glabrous, much shorter than the mature calyx, few-seeded. May–July.

Beaches, sand dunes, and dry sandy woods of oak and jack pine.

Lechea *Pinweed*

Perennial herbs, with solitary or few erect stems. Leaves small, alternate (occasionally appearing opposite or whorled), entire, sessile or short-petioled, 1-nerved. Flowers many, tiny, red, in leafy panicles. Sepals 5. Petals 3, smaller than the sepals. Stamens commonly 5–15. Ovary 1-celled; style none; stigmas 3, plumose. Capsule 3-valved, maturing 1–6 seeds, enclosed wholly or largely by the persistent calyx. Flowering in mid–late summer but rarely seen with expanded petals; at anthesis the sepals spread widely but soon return to an erect position. Late in the season basal shoots with numerous crowded leaves are produced.

ADDITIONAL SPECIES
Lechea pulchella Raf. (Leggett's pinweed), native; dry to moist sandy habitats; Dickinson County; threatened.

1   Leaves with a hard shiny ± conical brownish tip ca. 0.2–0.3 mm long; seeds 2–3; plant ± diffusely branched above with the branches spreading; rare in Dickinson County . . . . **L. pulchella***
1   Leaves pointed but without such a differentiated tip (flat and green to the margin); fruiting calyx mostly at least 1.5 mm broad, nearly spherical; seeds mostly 4–6; plant compact above, the branches ascending . . . . **L. intermedia**

Lechea intermedia Leggett
SAVANNA PINWEED native
IR | **WUP** | **CUP** | EUP CC 6

Stems 2–6 dm tall, thinly appressed-pubescent. Leaves of the basal shoots oblong lance-shaped to narrowly elliptic, 3–7 mm long, sparsely pilose on the midrib and margin beneath or essentially glabrous; stem leaves linear-oblong, very sparsely pubescent on the midrib beneath or glabrous. Panicle occupying 1/3 to 1/2 of the plant, slenderly cylindric, the lateral branches seldom more than 5 cm long; pedicels equaling or surpassing the sepals; calyx and capsule together subglobose. Capsule subglobose, barely exceeding the sepals; seed shaped like a section of orange, pale brown, partly and irregularly invested with a gray membrane.

Dry sterile or sandy soil.

Cleomaceae
CLEOME FAMILY

Polanisia *Clammyweed*
Polanisia dodecandra (L.) DC.
LARGE CLAMMYWEED native
IR | WUP | **CUP** | EUP CC 5

Viscid-pubescent, branched, annual herb. Stems 2–5 dm tall. Leaves alternate trifoliolate; leaflets oval, elliptic, or oblong, 2–5 cm long, acute; petiole about as long as blade. Flowers small, white or pinkish, in terminal racemes; sepals 4, separate or slightly united at base; petals 4, unequal, long-clawed at base, notched at tip, 4–8 mm long, the blade usually exceeding the claw; stamens 6 or more, little exsert, the longest rarely 2 mm longer than the petals; receptacle bearing a large, fleshy, 2-lobed gland on the upper side; ovary glandular-viscid. Capsule linear, 2–4 cm long, dehiscent by 2 valves; seeds numerous, kidney-shaped, net-veined. July–Sept.

Dry sandy or gravelly soil, especially along streams, also in waste places and along railroads; Dickinson County.

Convolvulaceae
MORNING-GLORY FAMILY

Herbs (ours), often twining, with alternate simple leaves and small to very large flowers. Flowers regular, perfect, mostly 5-merous. Sepals usually distinct to the base, imbricate, often of unequal size. Corolla rotate, funnelform, salverform, or tubular, entire or deeply to shallowly lobed. Stamens as many as the corolla lobes and alternate with them, inserted near the base of the corolla. Ovary usually 2–3-celled; styles 1 or 2; stigmas linear to capitate. Fruit in most genera a capsule.

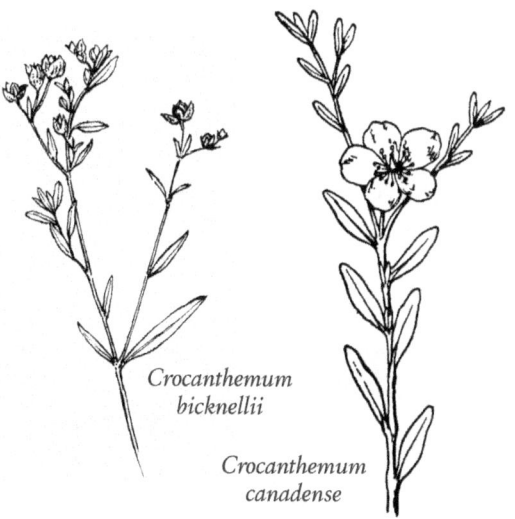

Crocanthemum bicknellii

Crocanthemum canadense

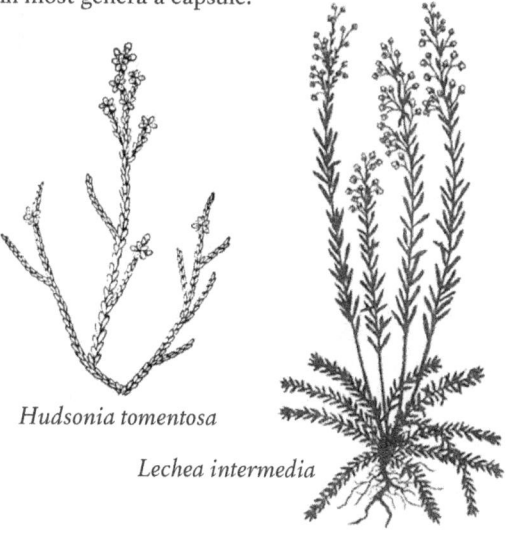

Hudsonia tomentosa

Lechea intermedia

168 CONVOLVULACEAE | Morning-Glory Family
dicots

1. Plants leafless, non-green, annual parasitic vines ..... **Cuscuta**
1. Plants leafy, green, not parasitic 2
2. Bracts leaf-like, attached just below the calyx and nearly concealing it .... **Calystegia**
2. Bracts small, attached much below the calyx .... **Convolvulus arvensis**

Calystegia *Bindweed*
Much like *Convolvulus*, but the bracts usually large, inserted just beneath the calyx, and more or less concealing it; ovary more or less 1-chambered, the partition incomplete; stigmas oblong, cylindric, blunt; flowers usually solitary; ours rhizomatous perennials.

1. Petioles of leaves subtending flowers with petiole more than half length of blade midvein .... **C. sepium**
1. Petioles of leaves subtending flowers with petiole much less than half length of midvein .... **C. spithamaea**

Calystegia sepium (L.) R. Br.
HEDGE-BINDWEED *native*
IR | **WUP** | **CUP** | EUP FAC CC 2

Stems twining or occasionally trailing, to 3 m long. Leaves long-petioled, triangular to oblong in outline, hastate or sagittate, 5–10 cm long, 1/4 to 3/4 as wide. Peduncles 5–15 cm long; bracts ovate or oblong, 1–2 cm long, nearly cordate at the base; corolla pink or white, 4–7 cm long.

Open streambanks, ditches, marshes, wet meadows; often forming tangled masses of stems.

Calystegia spithamaea (L.) Pursh
LOW BINDWEED *native*
IR | **WUP** | **CUP** | EUP CC 8

Stems erect, at least to and including the flowering portion, the remainder often elongating and eventually declined. Leaves obovate-oblong, 3–8 cm long, acute to obtuse or rounded, at base rounded, truncate, or cordate, always more or less pubescent. Peduncles few, 2–8 cm long, produced 5–20 cm above the base of the plant; bracts oblong or ovate, very rarely cordate at base; corolla white or pink, 4–7 cm long. May–July.

Dry rocky or sandy soil, fields, open woods, often with bracken.

Convolvulus *Bindweed*
Convolvulus arvensis L.
FIELD BINDWEED *introduced (invasive)*
IR | **WUP** | **CUP** | EUP

Perennial, deeply rooted herb. Stems trailing or climbing, to 1 m long, often forming dense tangled mats. Leaves variable, triangular to oblong in outline, 2–5 cm long, cordate-ovate or hastate, glabrous or finely pubescent, the basal lobes spreading or descending. Flowers borne mostly 1–2 together on axillary peduncles exceeding the subtending leaves; bracts borne 5–20 mm below the flower; sepals elliptic, 3–5 mm long; corolla funnelform, usually white, sometimes pink, 15–20 mm long; stamens included, inserted near the base of the corolla. Capsule globose, 2–4-celled. May–Sept.

Native of Europe; naturalized in fields, roadsides, and waste places; often a troublesome weed.

Cuscuta *Dodder*
Our single known species an annual, yellow or brown, parasitic, twining vine. Leaves reduced to minute scales. Flowers small, yellow or whitish, in cymose clusters, mostly 4–5-merous. Corolla campanulate to cylindric, lobed. Stamens inserted at the sinuses of the corolla, usually shorter than its lobes. Opposite each stamen near the base of the corolla tube is (in almost all species) a usually toothed or fringed scale. Ovary 2-celled. Fruit a capsule. Only a few species of the genus cause serious damage to crop plants, but all members of the genus are considered noxious weeds. Dodder seeds germinate in soil like autonomous plants. If young plants contact a suitable species, they soon attach themselves to a host-plant by means of numerous haustoria borne along the stem. Connection with the soil is then soon lost. Our species bloom in late summer and most of them live on a wide variety of host-plants.

Cuscuta gronovii Willd.
COMMON DODDER *native*
IR | WUP | **CUP** | EUP CC 3

Flowers 2.5–4 mm long, sessile or subsessile in dense clusters; calyx short, its lobes broadly round-ovate to subrotund, overlapping, scarcely reaching the middle of the corolla tube; corolla lobes broadly ovate, obtuse, spreading, shorter than the tube. Scales commonly copiously fringed, reaching to the sinuses of the corolla. Styles nearly as long as the ovary. Capsule commonly globose-ovoid, about 3 mm wide; seeds about 1.5 mm long.

Reported for Menominee County.

Cornaceae
DOGWOOD FAMILY

Cornus *Dogwood*
Shrubs, or herbaceous shoots from a woody rhizome in bunchberry (*Cornus canadensis*). Leaves mostly opposite, sometimes alternate or whorled, simple, entire. Flowers in a rounded or flat-topped cluster, 4-parted, sepals and petals small. Fruit a berrylike drupe with 1–2 hard seeds.

CORNACEAE | Dogwood Family

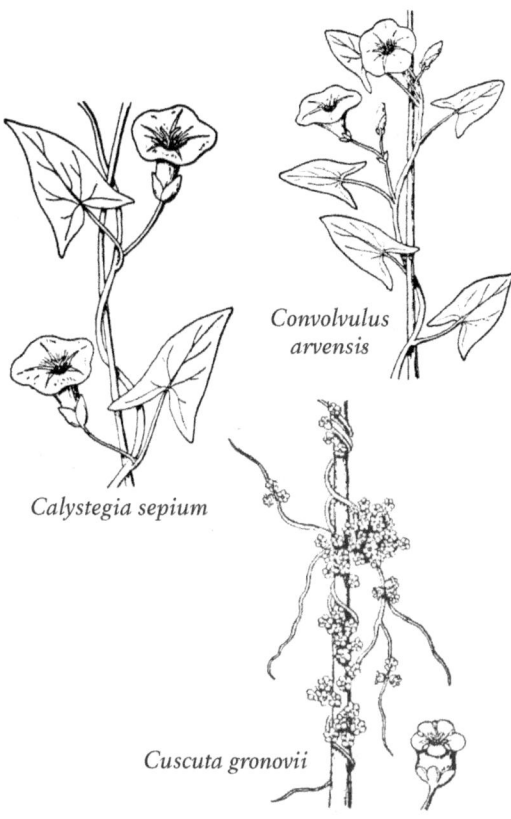

Calystegia sepium

Convolvulus arvensis

Cuscuta gronovii

1 Plants herbaceous from a woody rhizome, less than 3 dm tall; leaves whorled. . . . . **C. canadensis**
1 Taller shrubs, 5 dm or more tall; leaves opposite or alternate 2
2 Leaves alternate on stems. . . . . **C. alternifolia**
2 Leaves opposite 3
3 Twigs yellow or yellow-green with purple spots; leaves round in outline or nearly so. . . . . **C. rugosa**
3 Twigs not yellow, or if yellow not spotted; leaves longer than wide 4
4 Fruit white; young twigs densely short-hairy . . . . **C. obliqua**
4 Fruit blue; young twigs smooth or nearly so 5
5 Twigs gray; leaves with fewer than 5 pairs of lateral veins . . . . **C. racemosa**
5 Twigs red; leaves with 5 or more pairs of lateral veins. . . . . **C. alba**

Cornus alba L.
RED OSIER-DOGWOOD　　　　　　　　　　　native
IR | WUP | CUP | EUP　　　　　　　　　　　FACW CC 2
Cornus sericea L., *Cornus stolonifera* Michx.
Many-stemmed shrub, 1–3 m tall, forming thickets; branches upright or prostrate and rooting; twigs and young branches red; pith white. Leaves opposite, green, ovate to oval, mostly 5–15 cm long and 2–7 cm wide, tapered to a tip, soft hairy on underside;

margins entire; petioles to 2.5 cm long. Flowers small, white, many in flat-topped or slightly rounded clusters. Fruit a round, white or blue-tinged, berrylike drupe, 6–9 mm wide. May–Aug.

Swamps, marshes, shores, streambanks, floodplain forests, shrub thickets, calcareous fens; also on sand dunes.

Cornus alternifolia L. f.
PAGODA D.; ALTERNATE-LEAF DOGWOOD　　native
IR | WUP | CUP | EUP　　　　　　　　　　　FACU CC 5
Swida alternifolia (L. f.) Small
Shrub, to 5 m tall; twigs red-green or brown, somewhat shiny, alternate on stems, pith white. Leaves alternate, sometimes crowded and appearing whorled near ends of stems, oval to ovate, 5–12 cm long and 3–7 cm wide, tapered to a sharp tip, underside finely hairy; lateral veins 4–5 pairs, these curving toward tip of blade; margins entire; petioles to 5 cm long. Flowers small, creamy-white, in crowded, flat-topped or rounded clusters at ends of stems. Fruit a round, blue, berrylike drupe, 6 mm wide, atop a red stalk. May–July.

Swamps, thickets, streambanks and springs; also in drier deciduous and mixed forests.

Cornus canadensis L.
BUNCHBERRY; DWARF CORNEL　　　　　　native
IR | WUP | CUP | EUP　　　　　　　　　　　FAC CC 6
Perennial from horizontal, woody rhizomes, often forming large colonies. Stems erect, green, 1–2 dm tall, with a pair of small bracts on lower stem, topped with a whorl-like cluster of 4–6 leaves. Leaves oval to obovate, 4–7 cm long, tapered at both ends; lateral veins 2–3 pairs, arising from midvein below middle of blade; margins entire; petioles short or absent. Flowers small, yellow-green or creamy-white in a single cluster at end of a stalk 1–3 cm long; flowers surrounded by 4 white or pinkish, petal-like showy bracts, 1–2 cm long, these soon deciduous. Fruit a cluster of round, bright red berrylike drupes, the drupes 6–8 mm wide. June–July.

Cedar swamps, thickets and moist conifer forests, often on hummocks or rotting logs; also in drier, mixed conifer-deciduous forests.

Cornus obliqua Raf.
SILKY DOGWOOD　　　　　　　　　　　　　native
IR | WUP | CUP | EUP　　　　　　　　　　　FACW CC 2
Cornus amomum subsp. *obliqua* (Raf.) J.S. Wilson
Shrub, 1–3 m tall; older branches red and gray-streaked, young twigs gray, finely hairy; pith brown. Leaves opposite, oval to ovate, 5–12 cm long and 2–5 cm wide, usually less than half as wide as long, tapered to a sharp tip, lateral veins 4–6 on each side, underside finely hairy; margins entire; petioles 1–2 cm long, often curved and causing the leaves to

droop. Flowers small, creamy-white, in flat-topped or slightly rounded, hairy clusters. Fruit a round, blue or blue-white, berrylike drupe, 8 mm wide, atop a long stalk. June–July (our latest flowering dogwood).

Conifer swamps, marshes, open bogs, calcareous fens, lakeshores, streambanks, wet dunes.

Cornus racemosa Lam.
GRAY DOGWOOD native
IR | WUP | **CUP** | **EUP** FAC CC 1

Cornus foemina ssp. *racemosa* (Lam.) J.S. Wilson
Shrub, 1–3 m tall, often forming dense thickets; twigs red, becoming gray or light brown; pith usually brown. Leaves opposite, lance-shaped to oval, 4–9 cm long and 2–4 cm wide, abruptly tapered to a rounded tip, underside with short hairs; lateral veins 3 or 4 on each side of midvein; margins entire; petioles to 1 cm long. Flowers small, creamy-white, ill-scented, in numerous, open, elongated clusters. Fruit a round, berrylike drupe, at first lead-colored, becoming white, 5 mm wide, on red stalks. June–July.

Lakeshores, streambanks, swamps, thickets, marshes, moist woods, low prairie.

Cornus rugosa Lam.
ROUND-LEAF DOGWOOD native
IR | WUP | CUP | EUP CC 6

Shrub 1–3 m tall, the younger branches yellowish green, often shaded or mottled with red; pith white. Leaves ovate to rotund, usually 7–12 cm long, abruptly acuminate, broadly cuneate or usually rounded at base, minutely scaberulous above, softly white-pubescent beneath with erect, curled or curved hairs usually 0.5–1 mm long; lateral veins 7 or 8 on each side. Inflorescence flat or slightly convex. Drupes light blue, about 6 mm wide. May–July.

Moist or dry, sandy or rocky soil.

Crassulaceae
STONECROP FAMILY

Usually succulent plants of diverse habit. Leaves simple. Flowers usually cymose, regular, 4–5-merous or occasionally more, usually perfect.

ADDITIONAL SPECIES
Phedimus spurius (M. Bieb.) 't Hart (Caucasian-stonecrop), introduced; cultivated and sometimes escaping to shores, rock outcrops, and disturbed places. Leaves opposite, ciliate-margined; flowers whitish to pink.

Sedum *Stonecrop*

Succulent perennial herbs. Leaves thick or terete, alternate, opposite, or whorled. Flowers yellow or red-purple, 4–5-merous. Sepals united at base. Petals separate, often lance-shaped. Stamens 8 or 10, the epipetalous ones usually adnate to the petals at base. Nectarial scales short, at the base of the ovaries. Pistils distinct or nearly so, tapering into a stout style. Fruit a group of follicles; seeds numerous, small.

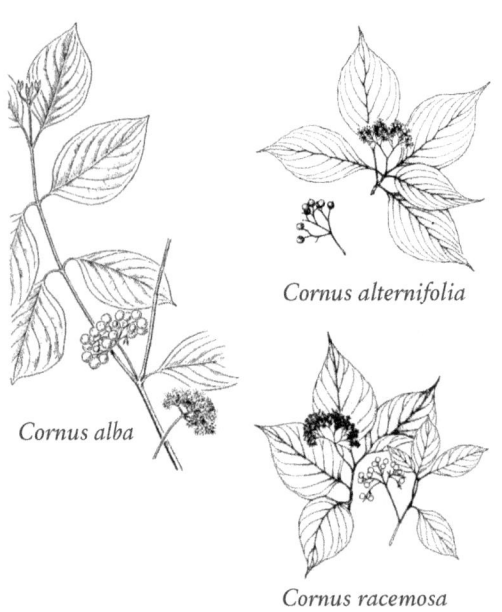

Cornus alba

Cornus alternifolia

Cornus racemosa

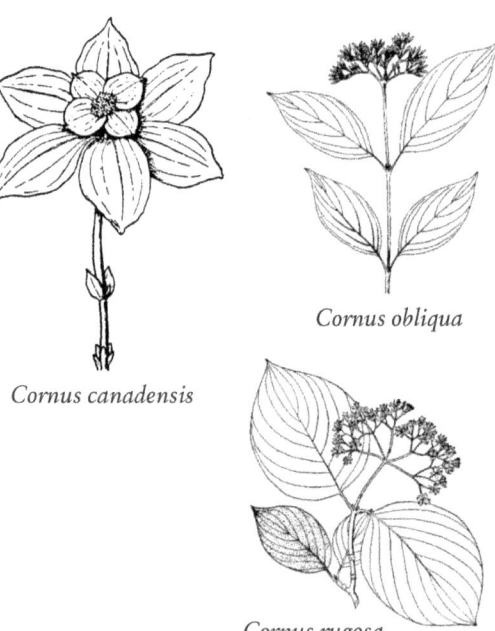

Cornus canadensis

Cornus obliqua

Cornus rugosa

1 Petals deep pink...................... **S. purpureum**
1 Petals yellow .... **S. acre**

Sedum acre L.
MOSSY STONECROP *introduced*
IR | **WUP** | **CUP** | **EUP**
Perennial from creeping stems, forming dense mats. Flowering stems 5–10 cm long. Leaves crowded, imbricate, terete, ovoid, 2–5 mm long, blunt. Inflorescence of a few short branched cymes. Flowers yellow, 8–10 mm wide; petals lance-shaped, spreading, about twice as long as the sepals. June–July.

Native of Eurasia; cultivated and escaped in dry, sandy soil.

Sedum purpureum (L.) J.A. Schultes
LIVE FOREVER *introduced*
IR | **WUP** | **CUP** | **EUP**
Hylotelephium telephium (L.) H. Ohba.
Sedum telephium L. subsp. *purpureum* (L.) Schinz & Keller

Perennial from a thick caudex. Stems stout, erect, 3–6 dm tall. Leaves fleshy, green or bluish green, not glaucous, oblong to obovate, 3–6 cm long, commonly with several to many conspicuous irregular teeth, varying to entire. Inflorescence repeatedly branched, convex to hemispheric, the branchlets narrowly winged. Flowers densely crowded, red-purple, 6–8 mm wide; sepals triangular, about a third as long as the ovate lance-shaped petals. Late summer.

A highly variable Eurasian species, long cultivated for ornament and occasionally escaped.

Cucurbitaceae
CUCUMBER FAMILY

Annual or perennial vines, trailing or climbing by tendrils, with small to large, mostly white or yellow or greenish flowers, and simple, alternate, often lobed leaves. Flowers monoecious or dioecious, regular. Calyx 4–6 (usually 5)-lobed, sometimes to the very base of the tube. Stamens 1–5, distinct or wholly or partly united, usually 3. Ovary 1- or 3-celled; styles united, with a thick stigma. Fruit a dry or fleshy pepo, few–many-seeded.

ADDITIONAL SPECIES
The Eurasian **Citrullus lanatus** (Thunb.) Matsum. & Nakai (watermelon) may occur in the UP near gardens. Native of America are **Cucurbita pepo** L., (pumpkin), and **Cucurbita maxima** Duchesne (squashes). These species may grow in waste ground, especially where their seeds have been discarded, but none are persistent in the flora.

Echinocystis *Wild Cucumber*
Echinocystis lobata (Michx.) Torr. & Gray
WILD CUCUMBER *native*
IR | **WUP** | **CUP** | **EUP** FACW CC 2

Annual vining herb, to 5 m or more long. Leaves round in outline, with 3–7 (usually 5) sharp, triangular lobes; petioles 3–8 cm long. Flowers white; staminate flowers 8–10 mm wide, with lance-shaped lobes, in long, upright racemes; pistillate flowers 1 to several on short stalks from leaf axils. Fruit green, ovate, inflated, 3–5 cm long, with soft prickles. Aug–Sept.

Floodplain forests, wet deciduous forests, streambanks, thickets, and waste ground.

Dipsacaceae
TEASEL FAMILY

Herbs. Leaves opposite, simple or divided. Flowers in dense heads subtended by a many-leaved involucre; each flower also often subtended by a receptacular bract. Flowers perfect or polygamo-monoecious, more or less zygomorphic, 4–5-merous. Sepals minute, but sometimes with conspicuous appendages. Corolla tubular to narrowly campanulate. Stamens 4 or 2. Ovary 1-celled; style 1, slender. Included in Caprifoliaceae in the 2009 Angiosperm Phylogeny Group III system.

1 Stems prickly............................ **Dipsacus**
1 Stems not prickly.................. **Knautia arvensis**

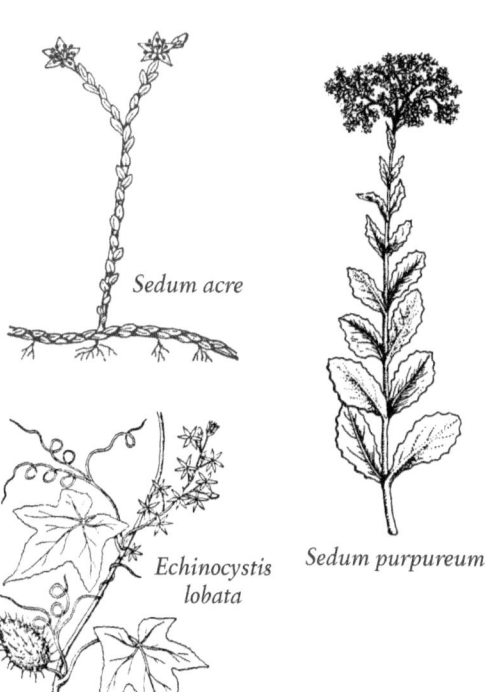

Sedum acre

Echinocystis lobata

Sedum purpureum

DROSERACEAE | Sundew Family

Dipsacus *Teasel*

Coarse, tall, biennial or perennial herbs, little branched, with prickly stems. Leaves large, sessile or connate. Flowers small, in dense ovoid to cylindric heads. Calyx short, 4-angled or 4-lobed. Corollas 4-lobed, the marginal ones not enlarged. Bracts of the involucre linear, often elongate. Receptacular bracts ovate or lance-shaped, acuminate into an awn surpassing the flowers. Involucel 4-angled, truncate or 4-toothed at the summit. True calyx very short, hairy, without appendages.

1 Flowers white; leaves deeply pinnately lobed......... .... **D. laciniatus**
1 Flowers purple; leaves not divided .... **D. fullonum**

Dipsacus fullonum L.
FULLER'S TEASEL *introduced*
IR | **WUP** | **CUP** | EUP   FACU

Dipsacus sylvestris Huds.

Stems stout, erect, 0.5–2 m. tall, increasingly prickly above. Basal leaves oblong lance-shaped, crenate; stem leaves lance-shaped, entire, sessile or connate, prickly on the midvein beneath. Heads ovoid to cylindric, 3–10 cm long, terminating long naked peduncles; involucral bracts curved-ascending, linear, prickly, some of them surpassing the head; bracts of the head exceeding the flowers, ending in a straight awn. Calyx silky, 1 mm long; corolla slender, pubescent, 10–15 mm long, the tube white, the short (1 mm) lobes pale purple. July–Sept.

Naturalized from Europe; roadsides and waste ground.

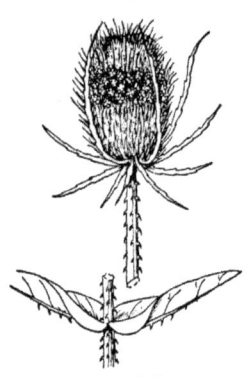
Dipsacus fullonum

Dipsacus laciniatus L.
CUT-LEAF TEASEL *introduced*
IR | **WUP** | **CUP** | EUP   FACU

Resembling *D. fullonum* in habit and flowers. Stem leaves irregularly pinnately lobed or divided, more or less bristly. July–Aug. Native of Europe; Dickinson County.

Knautia *Bluebuttons*

Knautia arvensis (L.) Coult.
BLUEBUTTONS *introduced*
IR | **WUP** | **CUP** | EUP

Perennial herb. Stems erect, pubescent, 4–8 dm tall, the upper internodes progressively longer. Leaves pinnately divided into 5–15 narrowly lance-shaped segments, the upper pairs progressively reduced. Flowers in dense hemispheric heads terminating elongate peduncles; heads 2–4 cm wide; involucral bracts ovate lance-shaped, 10–15 mm long; calyx with 8–12 setaceous teeth; corolla lilac-purple, 4-lobed, more or less irregular, the marginal ones distinctly so, 9–12 mm long. Achenes hairy, 4-ribbed; calyx-limb about 1 mm long, bearing 8–12 setaceous appendages. June–Sept.

Native of Europe; fields, roadsides, waste places.

Knautia arvensis

Droseraceae
SUNDEW FAMILY

Drosera *Sundew*

Perennial herbs. Leaves all from base of plant, covered with stalked, sticky glands that trap and digest insects. Flowers white, several, on 1 side of erect, leafless stalks, the stalks nodding at tip; with 5 petals and 5 sepals; stamens mostly 5, styles 3. Fruit a dry, many-seeded capsule.

1 Leaves widely spreading, the blades round, wider than long................................. **D. rotundifolia**
1 Leaves upright, blades linear or broad at tip and tapered to base, longer than wide 2
2 Leaf blades linear, 10–20 times longer than wide; young petals pink............................. **D. linearis**
2 Leaf blades broad near tip and narrowed to base, 2–7 times longer than wide; young petals white 3
3 Blades 2–3 times longer than wide, petioles without hairs; flower stalks from side of plant base and curving upward.............................. **D. intermedia**
3 Blades 5–7 times longer than wide, petioles with some hairs; flower stalks erect from center of plant base.... .... **D. anglica**

Drosera anglica Huds.
ENGLISH SUNDEW *native*
IR | **WUP** | **CUP** | EUP   OBL CC 10

Perennial insectivorous herb. Leaf blades obovate to spatula-shaped, 15–35 mm long and 3–4 mm wide, upper surface covered with gland-tipped hairs; petioles 3–6 cm long, smooth or with few glandular hairs. Flowers 1–9 in a racemelike cluster atop a stalk 6–25 cm tall; flowers 6–7 mm wide; sepals 5–

6 mm long; petals white, 6 mm long, spatula-shaped. Seeds black, 1 mm long, with fine lines. June–Aug.

Floating sphagnum mats, calcareous fens, wet areas between dunes.

Similar to *D. intermedia* but rarely occurring together; plants of *D. anglica* are generally larger, with shorter petioles (1–3x as long as blades vs. 2.5–3.5x as long in *D. intermedia*).

Drosera intermedia Hayne
SPOON-LEAF SUNDEW   *native*
IR | WUP | CUP | EUP   OBL CC 8

Perennial insectivorous herb. Leaves in a basal rosette and also usually along lower stem; spatula-shaped, 2–4 mm wide, upper surface covered with long, gland-tipped hairs; petioles smooth, 2–5 cm long. Flowers on stalks to 20 cm tall; sepals 3–4 mm long; petals white, 4–5 mm long. Seeds red-brown, to 1 mm long, covered with small bumps. July–Sept.

Low spots in open bogs, sandy shores, often in shallow water.

Drosera linearis Goldie
SLENDER-LEAF SUNDEW   *native*
IR | WUP | CUP | EUP   OBL CC 10

Perennial insectivorous herb. Leaf blades linear, 2–5 cm long and 2 mm wide; petioles smooth, 3–7 cm long. Flowers 1–4 atop stalks 6–15 cm tall; flowers 6–8 mm wide; sepals 4–5 mm long; petals obovate, 6 mm long, white. Seeds black, less than 1 mm long, with small craterlike pits on surface. June–Aug.

Calcareous fens, wet areas between dunes near Great Lakes; rarely in sphagnum moss.

Drosera rotundifolia L.
ROUND-LEAF SUNDEW   *native*
IR | WUP | CUP | EUP   OBL CC 6

Small, perennial insectivorous herb. Leaf blades more or less round, wider than long, 2–10 mm long and as wide or wider, covered with long, red, gland-tipped hairs; abruptly tapered to a petiole longer than blade; petioles 2–5 cm long covered with gland-tipped hairs. Flowers 2–15 in a more or less 1-sided, racemelike cluster, on a leafless stalk 10-30 cm tall; flowers 4–7 mm wide, sepals 5, 4–5 mm long; petals white to pink, longer than sepals; stamens 5, shorter than petals. Seeds light brown, shiny and with fine lines, 1–1.5 mm long. July–Aug.

Swamps and open bogs, usually in sphagnum; wet sandy shores and openings; our most common sundew.

Elaeagnaceae
OLEASTER FAMILY

Shrubs or trees. Leaves opposite or alternate, covered with small scales (lepidote). Flowers small, solitary or clustered, perfect or unisexual, regular, usually 4-merous. Hypanthium in the staminate flowers saucer-shaped to cup-shaped, in perfect or pistillate flowers prolonged into a short or elongate tube and persistently enclosing the ovary. Sepals present, often colored. Petals none. Stamens inserted at or just below the summit of the hypanthium, as many or twice as many as the sepals. Ovary 1-celled. Fruit drupe-like, the dry mature ovary enclosed by the base of the hypanthium.

1   Leaves alternate; stamens 4 .... **Elaeagnus umbellata**
1   Leaves opposite; stamens 8 .... **Shepherdia**

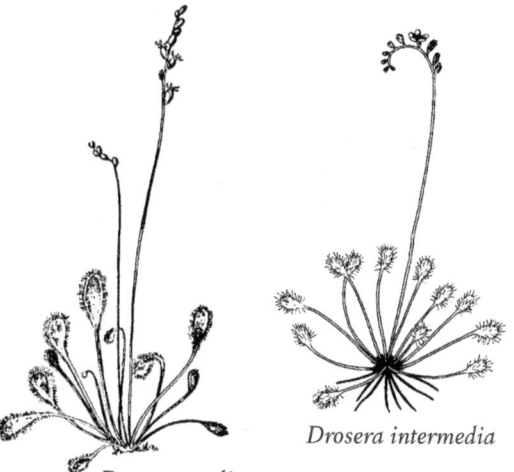

Drosera anglica

Drosera intermedia

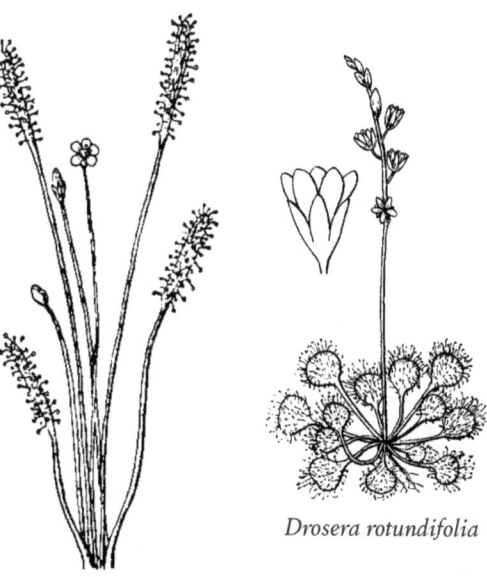

Drosera linearis

Drosera rotundifolia

Elaeagnus
Russian-Olive; Silver-Berry

Elaeagnus umbellata Thunb.
AUTUMN OLIVE *introduced (invasive)*
IR | **WUP** | CUP | EUP

Shrubby tree to 5 m. Leaves alternate, soon green and glabrescent above; hypanthium tube about twice as long as the sepals. Flowers perfect or unisexual in small lateral clusters on twigs of the current year; stamens 4, inserted near the summit of the hypanthium, scarcely exsert. Fruit red, finely dotted with pale scales, juicy and edible, 6–8 mm wide, on pedicels about 1 cm long. May–June.

Native of e Asia, originally introduced as an ornamental shrub and for wildlife habitat, now spreading to many dry and wet habitats; in the UP, reported from Ontonagon County.

Shepherdia *Buffalo-Berry*

Shrubs or small trees. Leaves opposite, usually silvery and scaly (lepidote). Flowers dioecious, in small clusters on twigs of the previous season, the pistillate flowers usually few. Sepals 4, greenish yellow within. Stamens 8, alternating with the lobes of the disk. Fruit a red or yellow-red berry.

1 Upper surface of leaf green and nearly glabrous . . . . **S. canadensis**
1 Upper and lower leaf surfaces covered with silvery scales; Delta County . . . . **S. argentea**

Shepherdia argentea (Pursh) Nutt.
SILVER BUFFALO-BERRY *introduced*
IR | WUP | **CUP** | EUP FACU

Widely branched shrub or small tree to 6 m tall, the stiff branches often ending in short stout spines. Leaves silvery-lepidote on both sides, oblong lance-shaped to oblong, 2–5 cm long, usually 7–12 mm wide, obtuse at the tip, narrowed to the base. Flowers similar to those of the preceding species, usually somewhat smaller with erect sepals. Fruit scarlet, about 5 mm long, edible but sour. April–May. Riverbanks.

Native of Great Plains, considered adventive in Delta County.

Shepherdia canadensis (L.) Nutt.
RUSSET BUFFALO-BERRY *native*
IR | **WUP** | CUP | EUP CC 7

Unarmed shrub 1–3 m tall. Leaves ovate lance-shaped to ovate, varying to narrowly lance-shaped or elliptic, 3–5 cm long, obtuse, obtuse to rounded or subcordate at base, green and nearly glabrous above, densely lepidote beneath. Staminate flowers 4–6 mm wide; sepals ovate, spreading, much exceeding the erect stamens; pistillate flowers similar, the mouth of the hypanthium closed by a dense tomentum. Fruit yellowish red, inedible, 5–7 mm long. April–May.

Dry, sandy or stony, often calcareous soil.

Elatinaceae
WATERWORT FAMILY

Elatine *Waterwort*

Elatine minima (Nutt.) Fisch. & C.A. Mey.
SMALL WATERWORT *native*
IR | **WUP** | **CUP** | EUP OBL CC 10

Small, branched, annual herb, forming small mosslike mats; plants smooth, with branches to 5 cm long. Leaves opposite, oblong to obovate, rounded at tip, to 4 mm long, with small membranous stipules; petioles absent. Flowers small, perfect, single and stalkless in leaf axils, sepals 2, petals 2. Fruit a round capsule; seeds with rows of small, rounded pits.

Shallow water and wet shores along lakes and ponds, usually where sandy or mucky.

Elaeagnus umbellata

Shepherdia argentea

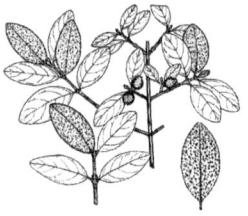

Shepherdia canadensis

Elatine minima

Ericaceae

HEATH FAMILY

Ericaceae now includes former members of Monotropaceae and Pyrolaceae. The traditional Ericaceae are shrubs or scarcely woody shrubs. Leaves evergreen or deciduous, mostly alternate, simple, with entire or toothed margins. Flowers usually perfect (with both staminate and pistillate parts), urn- or vase-shaped, mostly white, pink, or cream-colored; stamens as many (or twice as many) as petals. Fruit a berry or dry capsule.

Former Monotropaceae (*Monotropa, Pterospora*) are mycotropic perennial herbs without chlorophyll, variously white to pink, red, purple, yellow or brown in color. Leaves with much-reduced, scale-like, alternate. Flowers solitary or in a bracteate raceme, regular, perfect, mostly 4–5-merous; petals distinct or connate into a lobed tube, commonly about the same color as the stem; stamens mostly 6–10, mostly twice as many as the sepals or petals, distinct or shortly connate at base. fruit a capsule or berry; seeds numerous and tiny.

Former Pyrolaceae (*Moneses, Orthilia, Pyrola*) are perennial herbs or half-shrubs, most dependent on wood-rotting fungi (mycotrophic). Leaves alternate to sometimes opposite or nearly whorled, often shiny, evergreen or deciduous. Flowers perfect, 5-parted, waxy and nodding. Fruit a capsule.

ADDITIONAL SPECIES
Calluna vulgaris (L.) Hull (Heather), introduced shrub, reported from Mackinac County.

1. Leaves reduced to non-green scales; plants entirely white, yellow, reddish, orange, or maroon. 2
1. Leaves green (these sometimes small and needle-like or scale-like); plants normal green color. 3
2. Stems to 2 dm tall; petals free; fruit erect . **Monotropa**
2. Stems 3–10 dm tall; petals joined and urn-like; fruit nodding on curved pedicels; rare **Pterospora andromedea**
3. Leaves scale-like or needle-like, less than 1.5 mm wide. 4
3. Leaves with expanded flat blades, more than 1.5 mm wide; plants herbaceous or woody 5
4. Leaves scale-like; fruit a capsule; introduced small shrub . . . . **Calluna vulgaris***
4. Leaves needle-like; fruit berry-like; rare trailing shrub . . . . **Empetrum nigrum**
5. Leaves in a basal rosette; plants herbaceous 6
5. Leaves opposite, alternate, or whorled; plants woody (sometimes small subshrubs woody only at the base or prostrate creepers). 9
6. Style ± strongly bent downward, at least 4 mm long; inflorescence a ± symmetrical raceme. . . . . **Pyrola**
6. Style straight, short or long; inflorescence various (usually 1-flowered or a 1-sided raceme). 7
7. Inflorescence 1-flowered, the corolla 15–20 mm wide, flat (petals widely spreading); anthers prolonged into a short cylindrical tube below the pore; valves of capsule glabrous; style (not including prominent stigma lobes) 3–5 mm long . . . . **Moneses uniflora**
7. Inflorescence racemose, the corolla 3–7 mm wide, ± bell-shaped (petals close about reproductive parts); anthers not prolonged into tubes; valves of capsule with cobwebby fibers on the margins when dehiscing; style various (but stigma only very shallowly lobed) 8
8. Raceme 1-sided; style 2.5–6 (–6.5) mm long, protruding at maturity from the corolla; sepal margins finely toothed or erose . . . . **Orthilia secunda**
8. Raceme symmetrical; style 1.5 mm or less long, scarcely if at all protruding beyond the corolla; sepal margins entire. . . . . **Pyrola minor**
9. Leaves opposite or whorled; flowers 8–20 mm wide. 10
9. Leaves all alternate; flowers in most species less than 8 mm wide. 11
10. Leaves coarsely few-toothed; woody only at base, forming colonies by rhizomes . . . . . . **Chimaphila umbellata**
10. Leaves entire; true woody, clump-forming shrub. . . . . **Kalmia**
11. Leaves narrow, linear to linear lance-shaped, more than 7 times longer than wide; margins revolute. . . . . **Andromeda glaucophylla**
11. Leaves ovate to oblong; margins various 12
12. Leaves less than 7 times longer than wide, dark green and leathery on upper surface, densely covered with woolly rust-red hairs on underside; margins revolute. . . . . **Rhododendron groenlandicum**
12. Leaves never both revolute and with rust colored hairs on underside. 13
13. Leaves scurfy, densely scale-covered (especially on upper surface). . . . . **Chamaedaphne calyculata**
13. Leaves various but not scurfy 14
14. Fruit fleshy; leaves evergreen or deciduous. 15
14. Fruit dry or mealy; leaves evergreen 19
15. Plants trailing; leaves evergreen 16
15. Upright shrubs; leaves deciduous 18
16. Leaves with small bristles; fruit white. . . . . **Gaultheria hispidula**
16. Leaves glabrous; fruit red when ripe 17
17. Wet habitats; leaves small, to 15 mm long (cranberries) . . . . **Vaccinium**
17. Drier habitats; leaves 2 cm long or more . . . . **Gaultheria procumbens**
18. Leaves with shiny, orange-yellow resinous glands (especially on underside) . . . . **Gaylussacia baccata**
18. Leaves without glands (blueberries) . . . . . . **Vaccinium**
19. Plants with stiff hairs; leaves ovate to broadly elliptic . . . . **Epigaea repens**

19 Plants nearly glabrous or only finely hairy; leaves spatula-shaped, mostly widest above middle . . . .
. . . . **Arctostaphylos uva-ursi**

Andromeda *Bog-Rosemary*

Andromeda glaucophylla Link
BOG-ROSEMARY native
IR | WUP | CUP | EUP   OBL CC 10

Andromeda polifolia L. var. *glaucophylla* (Link) DC. Low upright or trailing shrub, 3–6 dm tall. Stems gray to blackish; twigs brown, with hairs in lines running down stems, or sometimes smooth. Leaves evergreen and leathery, often blue-green, linear or narrowly oval, 2–5 cm long and 3–10 mm wide, the tip sharp-pointed and tipped with a small spine, the base tapered to the stem or a short petiole, dark green above and whitened below by short stiff hairs; margins entire and distinctly rolled under. Flowers in drooping clusters at ends of branches, white or often pink, urn-shaped, 5-parted, 5–6 mm long, on curved stalks to 8 mm long. Fruit a rounded capsule to 5 mm wide, the style persistent from indented top of capsule; fruit drooping at first, but erect when mature. May–June.

Sphagnum bogs, black spruce and tamarack swamps.

Arctostaphylos *Bearberry*

Arctostaphylos uva-ursi (L.) Spreng.
RED BEARBERRY native
IR | WUP | CUP | EUP   CC 8

Shrub, forming low mats to 1 m wide. Stems prostrate, freely branched. Leaves leathery, evergreen, alternate, entire, obovate, 1–3 cm long, obtuse or rounded at tip, tapering to the base. Flowers 5-merous, in short terminal racemes; calyx saucer-shaped, the sepals imbricate, distinct to the base, about 1.5 mm long; corolla white or tinged with pink, 4–6 mm long, the 5 rounded lobes spreading or recurved; stamens 10, much shorter than the corolla; ovary 5-celled, subtended by a 10-lobed disk. Fruit a bright red drupe, dry or mealy, 6–10 mm wide, with 5 bony nutlets. May–June.

Sandy or rocky soil.

Bearberry has a long history of medicinal uses, especially amongst American Indians; it is also the main component of a smoking mix known as kinnikinnick.

Chamaedaphne *Leatherleaf*

Chamaedaphne calyculata (L.) Moench
LEATHERLEAF native
IR | WUP | CUP | EUP   OBL CC 8

Upright shrub to 1 m tall. Older stems gray, the outer bark shredding to expose the smooth, red inner bark; twigs brown, with fine hairs and covered with small, round scales. Leaves evergreen and leathery, becoming smaller toward ends of flowering branches, oval to ovate, 1–5 cm long and 3–15 mm wide, the tip rounded or pointed, brown-green and smooth above, pale brown with a covering of small, round scales below; margins entire or with small rounded teeth; petioles short. Flowers white, urn-shaped or cylindric, in 1-sided, leafy racemes, hanging from axils of reduced leaves near ends of branches; 5-parted, 5–7 mm long, on stalks 2–5 mm long. Fruit a brown, rounded capsule to 6 mm wide, the hairlike style persistent from indented top of capsule; capsules persisting on branches for several years. May–June.

Open bogs, lakeshores and streambanks, often forming low, dense thickets.

Chimaphila *Prince's Pine*

Chimaphila umbellata (L.) W. Bart.
PRINCE'S PINE; PIPSISSEWA native
IR | WUP | CUP | EUP   CC 8

Low, perennial, evergreen half-shrubs, from a creeping rhizome. Stems spreading, the flowering branches erect, 1–3 dm tall. Leaves thick, oblong lance-shaped, 3–6 cm long, acute or mucronate, sharply dentate especially toward the tip, nearly entire below the middle, tapering to a short petiole. Flowers 4–8, 10–15 mm wide, white or pink, corymbose on long peduncles; petals 5, distinct, widely spreading; stamens 10; ovary 5-celled, depressed-globose. Capsule erect, globose, opening from the top downward. June–Aug.

Dry woods, especially in sandy soil.

Empetrum *Crowberry*

Empetrum nigrum L.
BLACK CROWBERRY (THR) native
IR | WUP | CUP | EUP   CC 10

Much-branched low shrub to 3 dm tall; sometimes forming mats 1–2 m wide; branchlets minutely glandular. Leaves evergreen and needlelike, dark green and leathery, linear-oblong, only 4–8 mm long, rounded or blunt at tip, narrowed at base to a short stalk, margins rolled under. Flowers usually unisexual, 3-parted, small, pink to purple, single in axils of upper leaves. Fruit black, berrylike, 4–6 mm wide, somewhat juicy, with 6–9 hard nutlets. July–Aug.

Bare rock outcrops, cedar or black spruce swamps, sandy bluffs and old dune ridges under pines; rare in Alger, Keweenaw (including Isle Royale), Luce and Mackinac counties.

Previously placed in its own family (Empetraceae), but now considered a wind-pollinated member of the Ericaceae.

dicots   **ERICACEAE** | Heath Family

Epigaea *Trailing Arbutus*

Epigaea repens L.
TRAILING ARBUTUS   *native*
IR | **WUP** | **CUP** | EUP   CC 7

Prostrate, creeping, evergreen shrub, often dioecious. Stems branched, 2–4 dm long, hirsute. Leaves leathery, alternate, entire, ovate or oblong, 2–10 cm long, apiculate, rounded to cordate at base, more or less pilose, especially when young; on pubescent petioles about half as long as the blade. Flowers pink to white, fragrant, perfect or unisexual, 5-merous, each closely subtended by 2 ovate bracts nearly or quite as long as the calyx, in short, crowded, terminal and axillary spikes 2–5 cm long; sepals distinct to the base, strongly imbricate; corolla salverform, the tube 8–15 mm long, densely pubescent within, the lobes 6–8 mm long; stamens included; ovary 5-celled. Capsule depressed-globose, hirsute, subtended by the persistent calyx and bracts, white-pulpy within. April–May.

Sandy or rocky acid soil.

Gaultheria *Teaberry*

Shrubs of diverse aspect, with alternate persistent leaves and usually white flowers in racemes or particles or (in our species) solitary in or just above the axils and closely subtended by 2 bracteoles. Flowers 4–5-merous. Calyx campanulate to saucer-shaped, deeply divided. Corolla tubular to campanulate, shallowly lobed. Stamens included. Ovary 4–5-celled. Capsule thin-walled, completely enclosed in the fleshy, white or colored, expanded calyx, forming a dry or mealy berry.

1 Leafy stems upright; flowers 5-parted; berries red . . . . **G. procumbens**
1 Leafy stems prostrate; flowers 4-parted; berries white . . . . **G. hispidula**

Gaultheria hispidula (L.) Muhl.
CREEPING SNOWBERRY   *native*
IR | **WUP** | **CUP** | EUP   FACW CC 8

Low, creeping, matted shrub. Stems 2–4 dm long, covered with brown hairs. Leaves crowded, evergreen, oval to nearly round, 4–10 mm long and to 5 mm wide, abruptly tapered to tip, green above, underside paler, with brown, bristly hairs; margins rolled under; petioles short. Flowers few, single in leaf axils, white, bell-shaped, 4-parted, 2–4 mm long, on curved stalks 1 mm long. Fruit a translucent, juicy, white berry 5–10 mm wide, slightly wintergreen-flavored. May–June.

Open bogs, swamps, wet conifer woods, often in moss on hummocks or downed logs.

Andromeda glaucophylla

Arctostaphylos uva-ursi

Chamaedaphne calyculata

Chimaphila umbellata

Empetrum nigrum

Epigaea repens

Gaultheria hispidula

Gaultheria procumbens

ERICACEAE | Heath Family

Gaultheria procumbens L.
WINTERGREEN native
IR | WUP | CUP | EUP FACU CC 5

Low, creeping shrub. Leafy stems erect from a horizontal rhizome, 1–2 dm tall, bearing a few leaves crowded near the summit. Leaf blades elliptic or oblong, 2–5 cm long, entire or crenulate, glabrous; petioles 2–5 mm long. Flowers 5-merous, on nodding pedicels 5–10 mm long; calyx saucer-shaped; corolla barrel-shaped, 7–10 mm long, the rounded lobes about 1 mm long. Berry bright red, 7–10 mm wide, wintergreen-flavored. July–Aug.

Dry or moist woods in acid soil.

Gaylussacia *Huckleberry*

Gaylussacia baccata (Wangenh.) K. Koch
BLACK HUCKLEBERRY native
IR | WUP | CUP | EUP FACU CC 7

Medium shrub. Stems upright, much-branched, 3–10 dm long; branches brown, finely hairy when young, dark and smooth with age. Leaves alternate, deciduous, leathery, oval, 2–5 cm long and 1–2.5 cm wide; dark green above, paler below, both sides with shiny, orange-yellow resinous dots; margins entire, often fringed with small hairs; petioles 2–4 mm long. Flowers yellow-orange or red-tinged, cylindric, 5-lobed, 4–6 mm long, in more or less 1-sided racemes from lateral branches, the flowers on short, gland-dotted stalks 4–5 mm long. Fruit a red-purple to black, berrylike drupe, 6–8 mm long, with 10 nutlets; edible but seedy. May–June.

Open bogs, usually with tamarack and leatherleaf (*Chamaedaphne calyculata*); more common in dry, acid, sandy or rocky habitats.

Kalmia *Laurel*

Evergreen shrubs. Leaves opposite, entire and leathery (our species). Flowers showy, 5-parted, in lateral or terminal clusters. Fruit a rounded capsule.

1 Stems round; leaves mostly in whorls of 3, distinctly petioled, green beneath; inflorescences axillary, the branches terminating in leafy shoots; calyx and pedicels glandular; Chippewa County .... **K. angustifolia**
1 Stems flattened; leaves opposite, sessile or nearly so, strongly whitened beneath; inflorescences terminal; calyx and pedicels not glandular; widespread in the UP .... **K. polifolia**

Kalmia angustifolia L.
SHEEP-LAUREL native
IR | WUP | CUP | **EUP** FAC? XXX CC 7

Medium shrub, 6–10 dm tall. Older stems smooth and gray; twigs round in section, brown and finely hairy when young. Leaves opposite or in whorls of 3, evergreen and leathery, oval to oval, 2–5 cm long and 5–20 mm wide, tip blunt or rounded; dark green above, paler and smooth or with scattered stalked glands below; margins entire and somewhat rolled under; petioles 3–10 mm long. Flowers showy, deep pink, several to many in lateral clusters from axils of previous year's leaves, saucer-shaped, 5-parted, 9–12 mm wide, on stalks to 2 cm long. Fruit a round capsule to 5 mm wide, the style persistent. Clusters of capsules may persist for several years. June–July.

Bogs and borders of peatlands; moist coniferous forests or swamps, moist depressions in jack pine, oak, or aspen forests. Chippewa County; more common in the northern Lower Peninsula.

Kalmia polifolia Wangenh.
BOG-LAUREL native
IR | WUP | CUP | EUP OBL CC 10

Low evergreen shrub to 6 dm tall. Older stems dark; twigs swollen at nodes, flattened and 2-edged in section, smooth, pale brown when young. Leaves opposite, evergreen and leathery, linear to narrowly oval, 1–4 cm long and 6–12 mm wide, tip blunt or narrowed to an abrupt point; dark green and smooth above, white below with a covering of short, white hairs, midrib on underside with large purple, stalked glands; margins entire and rolled under; petioles absent. Flowers showy, pale to rose-pink, in terminal clusters at ends of current year's branches, saucer-shaped, 5-parted, 8–11 mm wide, on stalks to 3 cm long. Fruit a rounded capsule to 6 mm wide, tipped by the persistent style, the capsules in upright clusters. May–June.

Sphagnum peatlands, black spruce and tamarack swamps.

Moneses *Single-Delight*

Moneses uniflora (L.) Gray
ONE-FLOWERED SHINLEAF native
IR | WUP | CUP | EUP FAC CC 8

Pyrola uniflora L.

Low perennial herb from a very slender creeping rhizome. Stems to 10 cm long. Leaves deciduous, mostly at base of plant, opposite or in whorls of 3, nearly round, margins entire or finely toothed. Flowers white, single at end of long stalk, nodding, 1–2 cm wide; petals 5 distinct, widely spreading; stamens 10; ovary subglobose, concave at the summit, 5-celled. Capsule subglobose, opening from top downward. July–Aug.

Cedar swamps, wet conifer or mixed conifer and deciduous forests.

Monotropa *Indian-Pipe*

White, yellow, pink, or red plants, turning black in drying, parasitic on soil-fungi. Stems erect. Leaves small, scale-like. Flowers nodding, of the same color as the stem. Corolla urn-shaped or broadly tubular; petals 4 or 5, distinct, all or some saccate at base. Sepals none, or 2–5. Stamens 8 or 10; filaments slender, pubescent. Ovary 4–5-celled; style short, thick. Capsules erect, ovoid to subglobose.

1 Flowers single . . . . **M. uniflora**
1 Flowers few to many in a raceme . . . . . **M. hypopithys**

Monotropa hypopithys L.
PINESAP *native*
IR | **WUP** | **CUP** | **EUP** CC 6

Hypopitys americana (DC.) Small
Hypopitys monotropa Crantz

Stems 1–3 dm tall, often gregarious, yellow, tawny, pink, or red, never pure white, more or less pubescent. Raceme dense, at first nodding, erect at anthesis. Flowers 8–12 mm long, the lower usually 4-merous, the terminal often 5-merous and larger; sepals lance-shaped, erect; style shorter than the ovary; stigma more or less villous at the margin.

Moist or dry woods, usually in acid soil.

Monotropa uniflora L.
ONE-FLOWER INDIAN-PIPE *native*
IR | **WUP** | **CUP** | **EUP** FACU CC 5

Stems 1–2 dm tall, usually solitary, commonly pure waxy white, rarely pink or red. Flower solitary, nodding, odorless, 10–17 mm long; sepals often absent; petals broadly oblong, slightly widened distally; style longer than the ovary; stigma glabrous. June–Aug.

Rich moist woods; occasionally in sphagnum moss in bogs.

Orthilia *Sidebells*

Orthilia secunda (L.) House
ONE-SIDED SHINLEAF *native*
IR | **WUP** | **CUP** | **EUP** FAC CC 7

Pyrola secunda L.

Perennial herb. Leaves elliptic to subrotund, 1.5–4 cm long, entire to crenulate, often separated by conspicuous internodes. Scape 8–20 cm tall, bearing a crowded secund raceme; sepals semicircular to ovate, 0.5–1 mm long; petals white or greenish, about 5 mm long; style elongate, exsert at anthesis, 5-lobed. June–July. Moist woods and mossy bogs. Separated from other members of family by its ovary subtended by a 10-lobed hypogynous disk, and petals with 2 rounded projections at base.

Moist coniferous and mixed forests, mossy bogs; sometimes in deciduous forests.

Pterospora *Pinedrops*

Pterospora andromedea Nutt.
WOODLAND PINEDROPS (THR) *native*
IR | **WUP** | **CUP** | **EUP** CC 10

Parasitic on soil fungi and forming a rounded subterranean mass of rhizoids. Stems erect, simple, brown or purplish, 3–10 dm tall, glandular-pubescent,

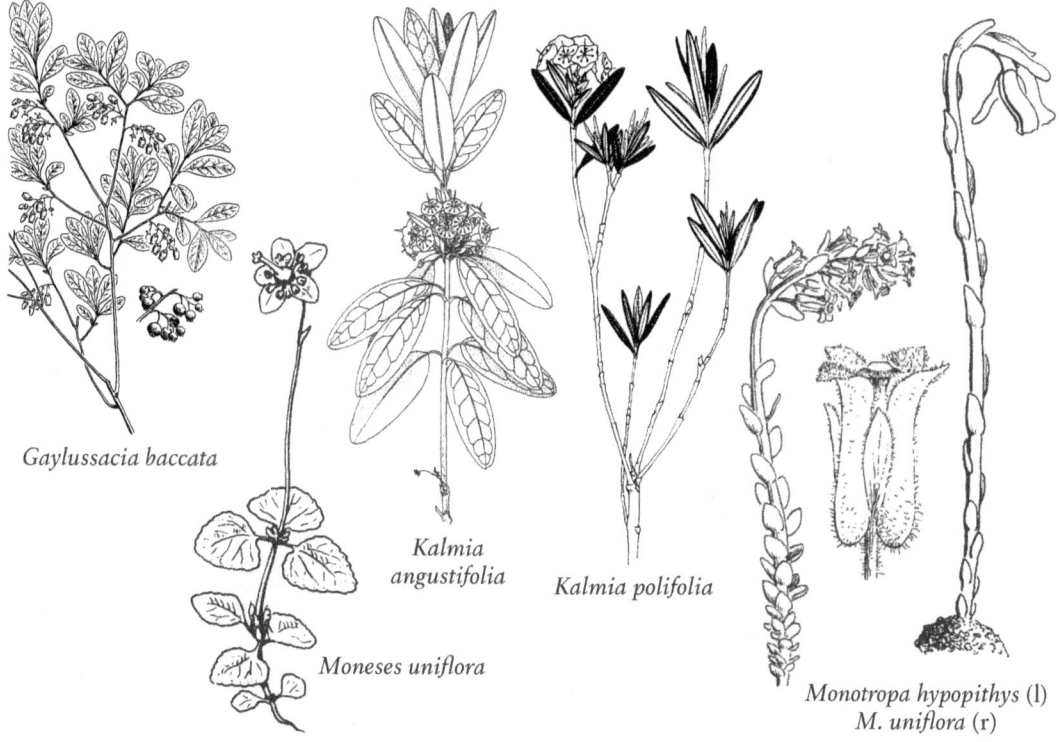

Gaylussacia baccata

Kalmia angustifolia

Moneses uniflora

Kalmia polifolia

Monotropa hypopithys (l)
M. uniflora (r)

180 ERICACEAE | Heath Family

bearing numerous scale-like leaves especially toward the base. Raceme many-flowered, 1–3 dm long; sepals densely glandular-pubescent; corolla nodding, urn-shaped, white, 5-lobed, persistent on the fruit 6–7 mm long, its spreading-recurved lobes 1–2 mm long; stamens 10; ovary 5-celled. Capsule depressed-globose, nearly 1 cm wide; seeds 0.3 mm long, with a terminal round wing, 0.7–1 mm wide. June–Aug.

Under conifer trees (especially white pine, as saprophytic on pines), usually in drier, clayey soils; rare.

Pyrola *Wintergreen; Shinleaf*

Perennial herbs from creeping rhizomes. Leaves few, broad, petiolate, nearly basal. Flowers in an erect, terminal, long-peduncled raceme; regular, 5-merous. Stamens 10. Ovary 5-celled; style short or elongate; stigma 5-lobed. Capsule dehiscent from the base upward.

1	Style straight, the stamens closely surrounding style	**P. minor**
1	Style curved downward; anthers of stamen not surrounding style	2
2	Sepal lobes longer than wide	3
2	Sepal lobes shorter than wide	4
3	Leaves to 3 cm long; sepals ovate	**P. chlorantha**
3	Leaves 3–7 cm long; sepals triangular	**P. elliptica**
4	Petals white; sepals oblong	**P. americana**
4	Petals pink; sepals triangular	**P. asarifolia**

Pyrola americana Sweet
AMERICAN WINTERGREEN native
IR | WUP | CUP | EUP FACU CC 7

Pyrola rotundifolia L.

Perennial herb. Leaves firm in texture, broadly elliptic to subrotund, rarely somewhat ovate or obovate, 2.5–7 cm long, broadly rounded above, rounded, truncate, or short-cuneate at base; always somewhat decurrent on the petioles. Scapes 1.5–3 dm tall, usually with 1 or 2 scale-leaves below the raceme; sepals oblong or ovate-oblong, nearly twice as long as wide, erose or undulate, not overlapping at base; petals white. July–Aug.

Dry or moist woods, rarely bogs.

Pyrola asarifolia Michx.
PINK SHINLEAF native
IR | WUP | CUP | EUP FACW CC 8

Perennial herb. Stems to 3 dm long. Leaves persisting over winter, all near base of plant, kidney-shaped, 3–4 cm long and 3–5 cm wide, margins shallowly rounded-toothed; flower stalk with 1–3 small, scale-like leaves. Flowers nodding in a raceme; sepals triangular, 2–3 mm long; petals 5, 5–7 mm long, pink to pale purple. Fruit a capsule opening from base upward. June–Aug.

Cedar swamps, peatlands, marly wetlands, and interdunal wetlands.

Pyrola chlorantha Sw.
GREEN-FLOWER WINTERGREEN native
IR | WUP | CUP | EUP FACU CC 8

Perennial herb. Leaves obovate or broadly elliptic to subrotund, 1–3 cm long, often shorter than the petiole, rounded to truncate at the summit, rounded to broadly cuneate at base but scarcely decurrent. Scapes 1–2.5 dm tall; petals white, more or less veined with green; sepals broadly ovate-triangular, broader than long, obtuse or subacute. June–Aug.

Dry woods.

Pyrola elliptica Nutt.
ELLIPTIC SHINLEAF native
IR | WUP | CUP | EUP FACU CC 6

Perennial herb. Leaves broadly elliptic, broadly oblong, or somewhat obovate, 3–7 cm long, commonly longer than the petiole, subacute to rounded at the summit, acute to rounded at base but always decurrent partway down the petiole. Scapes 1.5–3 dm tall; petals white, more or less veined with green; sepals triangular, about as broad as long, very shortly acuminate. June–Aug.

Dry upland woods.

Pyrola minor L.
LITTLE SHINLEAF native
IR | WUP | CUP | EUP FAC CC 10

Perennial herb. Leaves elliptic or round-oblong, 2–4 cm long, rounded or truncate at both ends. Scape 5–15 cm tall, bearing a loose, non-secund raceme of 5–15 nodding flowers; petals white, about 5 mm long; style at anthesis about 2 mm long, about equaling the petals; stigma deeply 5-lobed. June–Aug.

Moist, northern boreal forests near Lake Superior; UP at southern edge of range in eastern North America; species more common in western USA.

Rhododendron *Rhododendron*

Rhododendron groenlandicum (Oeder) Kron & Judd
RUSTY LABRADOR-TEA native
IR | WUP | CUP | EUP OBL CC 8

Ledum groenlandicum Oeder

Medium shrub, to 1 m tall. Older stems gray or red-brown; twigs covered with woolly, curly brown hairs. Leaves alternate, evergreen and leathery, fragrant when rubbed, narrowly oval to oblong, 2.5–5 cm long and 5–20 mm wide, rounded at tip; dark green and smooth above, the midvein sunken; underside covered with tan to rust-colored curly hairs; margins entire and rolled under; petioles short. Flowers creamy-white, in rounded clusters at ends of branches, 5-parted, to 1 cm wide, on finely hairy stalks 1–2 cm long. Fruit a lance-shaped, 5-celled capsule 5–6 mm long, the style persistent and hairlike; capsules splitting at base to release numerous small seeds, the empty capsules persistent on stems for several years. May–June.

Sphagnum bogs, swamps, wet conifer forests.

Many species in this genus have been introduced and cultivated for their attractive flowers.

Rhododendron groenlandicum

Vaccinium *Blueberry*

Deciduous or evergreen shrubs. Leaves alternate, simple. Flowers 4- or 5-parted, single in leaf axils or in clusters in axils or at ends of branches; ovary inferior. Fruit a many-seeded, red, blue, or black berry. The genus may be divided into 3 subgroups: Blueberries (*V. angustifolium*, *V. corymbosum*, *V. membranaceum*, *V. myrtilloides*, *V. pallidum*), cranberries (*V. macrocarpon*, *V. oxycoccos*, *V. vitis-idaea*), and bilberries (*V. caespitosum*, *V. uliginosum*).

ADDITIONAL SPECIES

Vaccinium uliginosum L. (Alpine bilberry), native small shrub; leaves similar to those of bearberry (*Arctostaphylos uva-ursi*), a common species of similar habitats; flowers white or pinkish, with 5 lobes on the urn-shaped corolla; berries blue or blackish and somewhat waxy. A circumpolar arctic-alpine species, at its southern border in the Great Lakes region at Isle Royale (especially Passage Island), where it grows in crevices and margins of rock pools; threatened. The fruit of both this and *V. cespitosum* is blue-glaucous, as in *V. ovalifolium* and *V. myrtilloides*.

1 Leaves deciduous; berries blue to blue-black 2
1 Leaves evergreen; berries red. 9
2 Flowers and fruit 1 (–2) in axils of leaves; calyx lobes rounded or virtually absent, deciduous, leaving at most a ring atop the fruit; twigs smooth or wrinkled 3
2 Flowers and fruit in terminal or lateral racemes or crowded clusters; calyx lobes acute (to obtuse), persistent as a small toothed crown on the fruit; twigs warty or wrinkled 6
3 Leaves (at least the larger ones) over 1.5 cm wide; plant over 5 dm tall; new branchlets ± sharply ridged, angled, or 2-edged; pedicels 4–12 mm long; berries 10–18 mm wide 4
3 Leaves all less than 1.5 cm wide (mostly ca. 1 cm); plants less than 5 dm tall; new branchlets terete or obscurely angled; pedicels 1–6 mm long; berries less than 10 mm wide 5
4 Leaf blades toothed their whole length, clearly acute; mature fruit purple-black......... **V. membranaceum**
4 Leaf blades entire (or obscurely toothed on lower half), nearly obtuse to rounded at apex; mature fruit glaucous blue .... **V. ovalifolium**
5 Leaf blades toothed, membranous, bright green; flowers solitary in axils of foliage leaves, on current year's growth; bud scales 2 .... **V. caespitosum**
5 Leaf blades entire, leathery, blue-green; flowers 1–2 from buds on previous year's stem; bud scales more than 2; Isle Royale.................. **V. uliginosum***
6 Leaf margins with small bristle-tipped teeth......... 7
6 Leaves entire; margins sometimes finely hairy....... 8

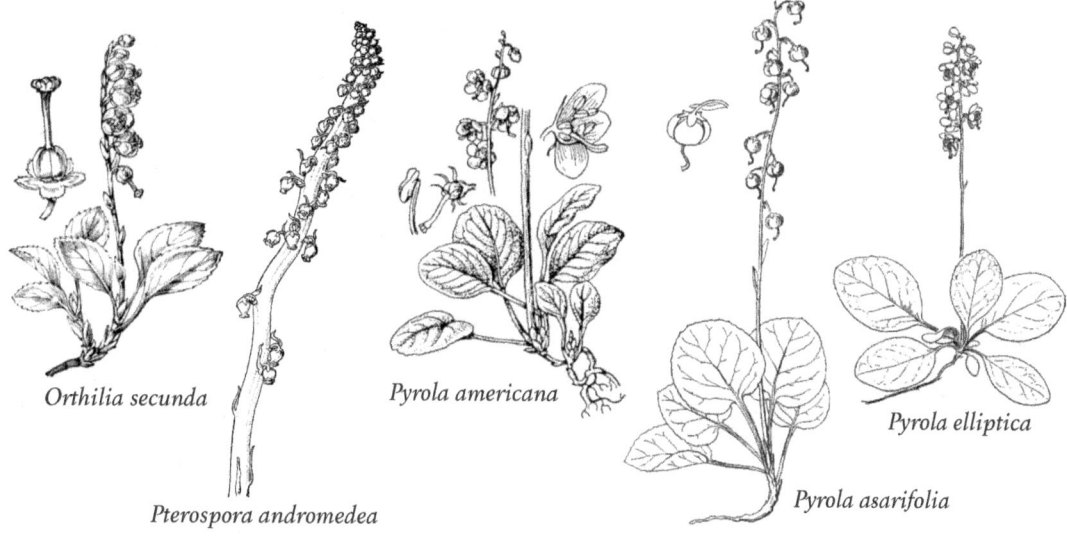

Orthilia secunda

Pyrola americana

Pyrola elliptica

Pterospora andromedea

Pyrola asarifolia

7 Margin teeth many, closely spaced; branches spreading .... **V. angustifolium**
7 Margin teeth few, margins finely hairy; branches stiffly erect .... **V. pallidum**
8 Stems and leaves velvety-hairy .... **V. myrtilloides**
8 Stems and leaves glabrous or with sparse hairs .... **V. pallidum**
9 Leaf underside with black bristly glands; Isle Royale... .... **V. vitis-idaea**
9 Leaf underside without black glands............... 10
10 Leaves blunt or rounded at tip (and sometimes notched), pale below; bracts on flower stalk green and leaflike (more than 1 mm wide) ...... **V. macrocarpon**
10 Leaves tapered to pointed tip, white below; bracts on flower stalk red and narrow (less than 1 mm wide) .... **V. oxycoccos**

Vaccinium angustifolium Ait.
LOWBUSH BLUEBERRY   native
IR | WUP | CUP | EUP   FACU CC 4

Low shrub 1–6 dm tall, forming colonies from surface runners. Older stems red-brown to black; twigs green-brown, with hairs in lines down stems, or sometimes smooth. Leaves deciduous, bright green oval, 2–5 cm long and 5–15 mm wide, smooth on both sides or sparsely hairy on veins; margins finely toothed with bristle-tipped teeth; petioles very short. Flowers in clusters, opening before or with leaves, white or pale pink, narrowly bell-shaped, 5-parted, 4–6 mm long. Fruit blue and wax-covered, 5–12 mm wide, edible and sweet. Flowering April–June, fruit ripening July–Aug.

Sphagnum peatlands and wetland margins; also in dry, sandy openings and forests.

Vaccinium caespitosum Michx.
DWARF BILBERRY   (THR) native
IR | WUP | CUP | EUP   FACU CC 9

Shrub. Stems branched, 1–2 dm tall. Leaves deciduous, thin, subsessile, oblong lance-shaped or cuneate to obovate, 1–3 cm long, obtuse or rounded at the summit, finely aristate-serrulate, tapering to the base. Flowers solitary in the axils of the lower leaves of the current season's branches, on decurved pedicels about 3 mm long; sepals very short and broad; corolla about 5 mm long, usually pink. Berries blue, 6–8 mm wide. May–June, fruit ripens in Aug.

Openings in pine barrens, often with bracken fern.

Vaccinium macrocarpon Ait.
LARGE CRANBERRY   native
IR | WUP | CUP | EUP   OBL CC 8

Evergreen trailing shrub. Stems slender, to 1 m or more long, with branches to 2 dm tall. Leaves leathery, oblong-oval, 5–15 mm long and 2–5 mm wide, rounded or blunt at tip, pale on underside; margins flat or slightly rolled under; petioles absent or very short. Flowers white to pink, 1 cm wide, 4-lobed, the lobes turned back at tips, single or in clusters of 2–6, on stalks 1–3 cm long, the stalks with 2 bracts above middle of stalk, the bracts green, 2–4 mm long and 1–2 mm wide. Fruit red, 1–1.5 cm wide, edible but tart, often over-wintering. Flowering June–July, fruit ripening Aug–Sept.

Sphagnum bogs, swamps and peaty pond margins. *V. macrocarpon* is the cultivated cranberry.

Vaccinium membranaceum Torr.
TALL BILBERRY   native
IR | WUP | CUP | EUP   UPL CC 8

Erect, branching, nearly glabrous shrub. Stems to 15 dm long, twigs somewhat angled. Leaves deciduous, thin, dull, often yellow-green, oval, oblong or ovate, green on both sides and nearly smooth, margins finely serrate, 2-7 cm long and 1.5-3 cm wide when mature. Flowers noddmg, solitary on short axillary peduncles; corolla depressed-globular, greenish or purplish, usually 5-toothed; stamens 10, included; Fruit a large, dark-purple to black, edible, juicy, somewhat acidic berry. Flowering June–July; fruit ripe July-August.

Forested dunes, rocky woods, under hardwoods and/or conifers. A species of western North America, in the east, known only from the UP and Ontario.

Vaccinium myrtilloides Michx.
VELVET-LEAF BLUEBERRY   native
IR | WUP | CUP | EUP   FACW CC 4

Low shrub, often forming colonies. Stems 3–6 dm long, red-brown to black with numerous wartlike lenticels; young twigs green-brown, densely velvety white-hairy. Leaves deciduous, thin and soft, oval, 2–5 cm long and 1–2.5 cm wide, dark green above, paler and soft hairy below, not waxy; margins entire and finely hairy; petioles very short. Flowers in clusters at ends of short, leafy branches, opening with leaves, creamy or green-white, tinged with pink, bell-shaped or short-cylindric, 5-parted, 4–5 mm long. Fruit blue, wax-covered, 6–9 mm wide; edible but tart. Flowering May–July, fruit ripening July–Sept.

Sphagnum bogs and swamps; also in dry to moist woods and clearings.

Vaccinium ovalifolium Smith
OVAL-LEAF BLUEBERRY   native
IR | WUP | CUP | EUP   WET?XXX CC 9

Straggling shrub. Stems to 8-12 dm; young twigs brownish and glabrous, with sharp longitudinal ridges or conspicuously angled; older branches purplish-gray to blackish with flaking outer bark. Leaves alternate, simple, and deciduous; blade thin, broadly oval to nearly round, 1-3 cm long and to 1.8 cm wide, blunt or rounded at the tip and often with a small abrupt point (mucro); rounded or broadly wedge-shaped at the base, dull green above, paler beneath, glabrous on both surfaces; margins entire or with a few small, widely spaced, gland-tipped teeth near the base: petioles 1-2 mm long. Flowers solitary on short stalks from the axils of the lower leaves on the current year's growth; corolla about 6 mm long, 5-parted, broadly bell-shaped and pinkish, opening when the leaves are only half grown or sometimes before the leaves appear; June. Fruit a dark blue to blue-black berry with a bloom. 6-9 mm wide.

A Pacific Northwest species, disjunct in the Great Lakes region (Keweenaw Peninsula east to Algoma District of Ontario) and also near the mouth of the St. Lawrence River. While often found with *V. membranaceum* on forested dunes and rocks, this species seems more often to be in deep shade in forests of hemlock, beech and maple, or mixed conifers.

Vaccinium oxycoccos L.
SMALL CRANBERRY   *native*
IR | WUP | CUP | EUP   OBL   CC 8

Small, evergreen trailing shrub. Stems slender, 0.5 m or more long, with upright branches 1–2 dm tall. Leaves leathery, ovate to oval or narrowly triangular, 2–10 mm long and 1–3 mm wide, pointed or rounded at tip, strongly whitened on underside; margins flat or strongly rolled under; petioles absent or very short. Flowers pale pink, 1 cm wide, 4-lobed, the lobes turned back at tips, single or in clusters of 2–4, on stalks 1–3 cm long, the stalks with 2 bracts at or below middle of stalk, the bracts red, scale-like, to 2 mm long and less than 1 mm wide. Fruit pale and red-speckled when young, becoming red, 6–12 mm wide, edible but tart. Flowering June–July, fruit ripening Aug–Sept.

Wet, acid, sphagnum bogs.

Vaccinium pallidum Ait.
EARLY LOWBUSH BLUEBERRY   *native*
IR | WUP | CUP | EUP   CC 7

Shrub, 3–10 dm tall, plants often colonial. Leaves firm in texture, elliptic, 3–5 cm long, about half as wide, acute, finely serrulate, pale green or somewhat glaucous beneath, glabrous or pubescent along the veins. Corolla 4–6 mm long, greenish white or tinged with pink. Fruit dark blue to black, sometimes glaucous, 5–7 mm wide.

Dry upland woods; Alger and Iron counties.

Vaccinium vitis-idaea L.
MOUNTAIN CRANBERRY   (END) *native*
IR | WUP | CUP | EUP   FAC CC 10

Low evergreen, trailing shrub. Older stems brown-black with peeling bark, branching, the branches upright, slender, 1–2 dm long, often forming mats;

twigs green-brown to red, more or less smooth. Leaves alternate, leathery, oval to oval, 0.5–2 cm long and 4–15 mm wide, rounded or slightly indented at tip; upper surface dark green, shiny and smooth, paler and with dark bristly glands below; margins entire and rolled under; petioles hairy, 1–2 mm long. Flowers white to pink, bell-shaped and 4-lobed, style longer than petals, several in 1-sided clusters at ends of branches, the flowers on short glandular stalks, the stalks with 2 small bracts at base. Fruit a dark red berry, to 1 cm wide, persisting over winter, tart but edible, especially the following spring. June–July.

Sphagnum bogs; also in drier, sandy or rocky places; Isle Royale. Michigan UP, northern Minnesota and northern Wisconsin at south edge of species' range.

Mountain cranberry can be distinguished from the more common cranberries (*V. macrocarpon* and *V. oxycoccos*) by the black, bristly, glandular dots on leaf underside. Gathered in Europe (where known as lingen or red whortleberry) and North America (where available) and cooked and eaten like commercial cranberries.

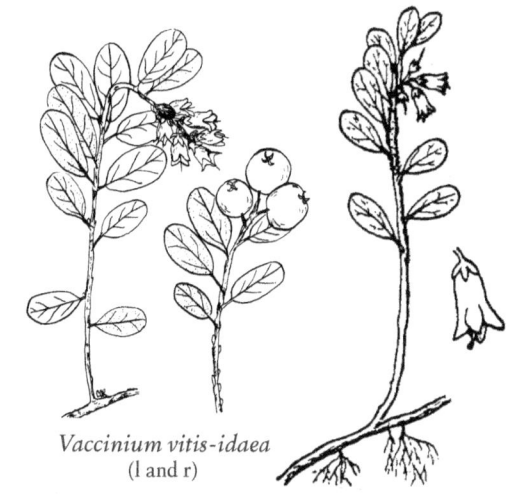

Vaccinium vitis-idaea
(l and r)

Euphorbiaceae
SPURGE FAMILY

Herbs (ours). Leaves usually alternate and simple. Flowers mostly tiny but in some species subtended by conspicuous bracts or involucral appendages. Plants monoecious or dioecious. Flowers commonly unisexual, very rarely perfect, regular. Calyx present or absent; petals absent in most genera. Stamens 1 to many, the filaments sometimes branched. Pistil 1; ovary usually 3-celled. Fruit usually a dehiscent capsule. Seeds with copious endosperm.

EUPHORBIACEAE | Spurge Family

dicots

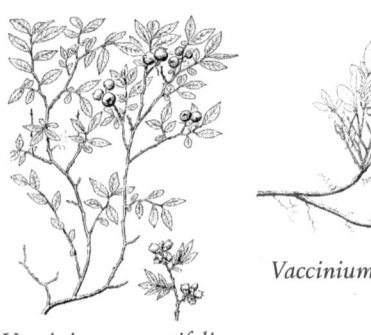
Vaccinium angustifolium

Vaccinium caespitosum

Vaccinium macrocarpon   *Vaccinium membranaceum*

Vaccinium ovalifolium

Vaccinium pallidum

Vaccinium myrtilloides

Vaccinium oxycoccos

A large, chiefly tropical family, with about 6,500 species, some of which are of economic importance. Species of the Amazonian genus *Hevea* are the chief source of rubber; *Manihot esculenta* Crantz, native of Brazil, yields cassava (tapioca), a staple food in the tropics. Many species are toxic and a few are used in medicine.

1. Plant with watery juice; stem pubescent with incurved hairs . . . . **Acalypha rhomboidea**
1. Plant with milky juice; stems glabrous or variously pubescent . . . . **Euphorbia**

Acalypha *Copperleaf*

Acalypha rhomboidea Raf.
RHOMBIC COPPERLEAF   native
IR | **WUP** | **CUP** | EUP   FACU CC 0

Acalypha virginica L. var. *rhomboidea* (Raf.) Cooperrider

Annual herb. Stems erect, simple or branched, 2–6 dm tall, glabrous, puberulent in lines, or puberulent throughout with incurved hairs, often with a few spreading hairs also. Leaves alternate, leaf blades ovate lance-shaped to ovate, commonly with a distinct tendency to rhombic; petioles slender, divaricate, those of the larger leaves regularly more than half as long as the blades. Plants monoecious; pistillate bracts 5–7-lobed, usually stipitate-glandular (visible under 10x lens); staminate spikes little if any exceeding the bracts. Petals none; sepals of the staminate flowers 4, stamens 8 or more with slender filaments and linear, frequently coiled or curved anthers. Sepals of the pistillate flowers 3–5; styles slender, elongate, irregularly branched. Capsule 3-celled; seeds solitary in each cell.

Dry or moist soil of open woods, roadsides, waste places; Baraga and Menominee counties.

Euphorbia *Spurge*

Annual or perennial herbs of diverse form; with milky, often highly acrid juice. Flowers greatly reduced, the staminate flowers consisting of a single stamen only, the pistillate flowers of a single pistil only. Several staminate flowers surround one pistillate flower inserted at the base of a cup-shaped involucre to form an inflorescence termed a cyathium. Around the margin of the cyathium are 4–5 glands, and in some species these have petal-like appendages, so that the whole cyathium mimics a single flower. Ovary 3-celled; styles 3, each bifid. Capsule 3-lobed, 3-seeded. The milky juice is poisonous and for some people produces a dermatitis similar to that caused by poison-ivy.

1. Annual herbs; leaf bases symmetrical; glands of cyathium 1, without appendages; stems erect, with ascending branches . . . . **E. davidii**

EUPHORBIACEAE | Spurge Family

1. Annual or perennial herbs; leaf bases various; glands of cyathium 4 or 5, not appendaged or with petal-like appendages; stems erect to prostrate 2
2. Annual herbs, the stems prostrate or ascending; leaves all opposite; leaf bases typically unequal............ 3
2. Annual or perennial herbs, the stems erect, leaves mostly alternate; leaf bases equal or nearly so....... 9
3. Stems with long soft hairs; capsules with stiff appressed hairs .... **E. maculata**
3. Stems glabrous or pubescent; capsules glabrous..... 4
4. Leaves entire; seeds round, smooth, with a white seed coat... 5
4. Leaves finely toothed, at least along upper portion of blade; seeds angular, smooth, bumpy, or ridged, the seed coat brown or blackish 6
5. Capsule 3–4 mm long .... **E. polygonifolia**
5. Capsule less than 2 mm long .... **E. geyeri**
6. Stems pubescent, at least on the upper portions, leaves mostly more than 1 cm long, the margins toothed.... 7
6. Stems glabrous; leaves mostly less than 1 cm long, the margins toothed only near leaf tip and along one side near leaf base...................................... 8
7. Stems erect or ascending, nearly glabrous or with lines of hairs; mature leaves more than 15 mm long **E. nutans**
7. Stems prostrate or nearly so, sparsely hairy; mature leaves less than 15 mm long............ **E. vermiculata**
8. Leaves linear, finely sharp-toothed, rounded at tip; seeds with small ridges............. **E. glyptosperma**
8. Leaves oblong to ovate, the lower two-thirds of the blade entire, finely toothed at the blunt tip; seeds smooth, or faintly wrinkled or pitted... **E. serpyllifolia**
9. Inflorescence glands with conspicuous white, petal-like appendages..................................... 10
9. Inflorescence glands without white, petal-like appendages 11
10. Upper leaves and bracts green with conspicuous white patches; capsules pubescent; plants annual .... **E. marginata**
10. Leaves and bracts green; capsules glabrous; plants perennial .... **E. corollata**
11. Leaves finely sharp-toothed .... **E. helioscopia**
11. Leaves entire.................................... 12
12. Plants annual or short-lived perennials; stem leaves broadly ovate to obovate; inflorescence usually with 3 main branches; seeds pitted .... **E. peplus**
12. Plants perennial; stem leaves linear to lance-shaped; inflorescence with 5 or more branches; seeds smooth 13
13. Main stem leaves slender and linear, 1–3 cm long and 1–3 mm wide, crowded................ **E. cyparissias**
13. Main stem leaves 3–7 cm long, mostly 3–10 mm wide, less crowded......................... **E. virgata**

Euphorbia corollata L.
FLOWERING SPURGE *native*
IR | WUP | **CUP** | **EUP** CC 4

Perennial from a deep root, glabrous to villous on stem and leaves. Stems 3–10 dm tall, usually simple below, umbellately or paniculately branched above. Stem leaves alternate, linear to elliptic, usually 3–6 cm long; leaves subtending the primary branches similar, whorled; leaves of the inflorescence smaller, opposite or alternate. Involucres numerous, forming a panicle-like cyme often 3 dm wide; pedicels, except a few lower ones, less than 1 cm long; petal-like appendages white (very rarely green), conspicuous, ovate to obovate, 1.5–4 mm long. June–Sept.

Dry woods and old fields. Highly variable.

Euphorbia cyparissias L.
CYPRESS SPURGE *introduced (invasive)*
IR | WUP | **CUP** | **EUP**

Perennial by horizontal rhizomes, gregarious. Stems 2–4 dm tall. Stem leaves very numerous, crowded, linear, 1–3 cm long, 1-nerved; leaves subtending the umbel similar; leaves of the umbel broadly cordate. Rays of the umbel usually 10 or more. Capsule about 3 mm long, slightly granular-roughened, seldom produced.; seeds plump, smooth, 1.5–2 mm long. April–July.

Native of Eurasia; established on roadsides and waste ground.

Euphorbia davidii Subils
DAVID'S SPURGE *introduced*
IR | WUP | **CUP** | EUP

Erect annual, 2–6 dm tall, often branched, with hairy herbage. Leaves all or mostly opposite, petiolate, linear to ovate, coarsely toothed to subentire. Inflorescence congested, mingled with reduced green leaves; involucres 2–3 mm long, with fimbriate lobes and a conspicuous, fleshy, flattened, bilabiate gland; styles bifid half their length or deeper. Fruit smooth, 5 mm thick; seeds ovoid, rough-tuberculate, 2.5–3 mm long, usually carunculate. July–Sept.

Dry soil; established as a weed on roadsides and waste places, especially in cindery soil; Delta County.

Euphorbia geyeri Engelm.
DUNE SPURGE *introduced*
IR | WUP | **CUP** | EUP

Chamaesyce geyeri (Engelm.) Small

Annual, glabrous throughout. Stems branched from the base, usually prostrate, 1–3 dm long. Leaves oblong to broadly elliptic, 5–10 mm long, entire, broadly rounded to shallowly retuse at the apex. Petal-like appendages inconspicuous, short-ovate, rarely more than 2x as long as the gland; stamens in each involucre 5–17. Capsule about 2 mm long; seeds not compressed, smooth, roundly 3-angled, about 1.5 mm long.

Sandy prairies and dunes; native west of Michigan, considered adventive here; Schoolcraft County.

Euphorbia glyptosperma Engelm.
RIB-SEED SANDMAT *native*
IR | **WUP** | **CUP** | **EUP** CC –

Chamaesyce glyptosperma (Engelm.) Small
Annual, glabrous throughout. Stems mostly prostrate, freely branched, 1–3 dm long, often forming mats. Leaves oblong or ovate, 4–15 mm long, strongly inequilateral, minutely serrulate (often visible only under a lens), especially on the rounded summit. Appendages rather conspicuous. Capsule depressed-ovoid, sharply 3-angled, about 1.5 mm long; seeds about 1 mm long, sharply 4-angled, marked with 3 or 4 conspicuous transverse ridges.

Dry sandy soil.

Euphorbia helioscopia L.
SUN SPURGE *introduced*
IR | WUP | **CUP** | **EUP**

Annual, smooth or nearly so. Stems 2–5 dm tall, the upper internodes usually progressively longer. Stem leaves spatulate, 1.5–5 cm long, very blunt or retuse, finely and sharply serrulate; leaves subtending the primary umbel similar but proportionately wider; leaves of the umbel somewhat oblique, broadly elliptic to obovate. Rays of the primary umbel 5, on well grown plants repeatedly branched. Involucres about 2 mm long. Capsule smooth, about 3 mm long; seeds ovoid, 2–2.5 mm long, conspicuously areolate.

Native of Europe; Mackinac (especially on Mackinac Island) and Marquette counties, waste places.

Euphorbia maculata L.
SPOTTED SANDMAT *native*
IR | **WUP** | **CUP** | EUP FACU CC 0

Chamaesyce maculata (L.) Small
Stems prostrate or nearly so, to 4 dm long, often forming circular mats, sparsely to densely villous. Leaves dark green, often with a red spot, oblong or ovate-oblong, varying to linear-oblong, usually 5–15 mm long, almost always widest below the middle. Involucre cleft on one side; ovary and capsule strigose. Capsule about 1.5 mm long; seeds 4-angled, about 1 mm long, with a few inconspicuous transverse ridges.

Common as a weed in lawns, gardens, and waste places, also in meadows and open woods.

Euphorbia marginata Pursh
SNOW-ON-THE-MOUNTAIN *introduced*
IR | **WUP** | CUP | EUP

Annual. Stems erect, 3–8 dm tall, softly villous, especially above. Stem leaves alternate, sessile, broadly ovate to elliptic or obovate-oblong, 4–10 cm long; leaves subtending the inflorescence whorled; those of the usually 3-rayed inflorescence smaller, margined with white or entirely white. Cymes crowded; involucres pubescent, the 5 lobes deeply fimbriate, the 5 petal-like appendages white, conspicuous, kidney-shaped to broadly ovate. Capsule 3-lobed, 6–7 mm wide; seeds ovoid, about 4 mm long, tuberculate. Summer.

Native of Great Plains; cultivated for ornament and often escaped farther east; considered adventive in Michigan; Houghton County.

The milky juice is extremely acrid.

Euphorbia nutans Lag.
EYEBANE *native*
IR | WUP | **CUP** | EUP FACU CC 0

Chamaesyce nutans (Lag.) Small
Annual. Stems to 8 dm tall, obliquely ascending at least in the upper half; the lower half often erect; the younger parts puberulent, often in a single longitudinal strip, with usually incurved hairs to 0.3 mm. the older parts glabrous or nearly so. Leaves opposite, oblong or oblong-ovate, 1–3.5 crn long and about a third as wide, serrulate, usually conspicuously inequilateral. Fruit 2–2.5 mm long, strongly 3-lobed, glabrous; seeds gray or pale brown, 1–1.5 mm long. June–Oct.

Dry or moist soil; weedy in lawns and gardens; Delta County.

Euphorbia peplus L.
PETTY SPURGE *introduced*
IR | **WUP** | CUP | EUP

Annual. Stems 1–3 dm tall, usually with many lateral branches. Stem leaves obovate, obovate, or nearly rotund, usually petiolate, 1–2 cm long, rounded to retuse at the summit; leaves subtending the umbel broadly ovate to obovate; leaves of the umbel smaller, ovate. Rays of the umbel 3, repeatedly dichotomous. Capsule with 2 low longitudinal keels on each valve; seeds about 1.5 mm long, marked on the outer face with 4 rows of 3 or 4 large pits, on the inner face with 2 longitudinal furrows. Summer.

Native of Eurasia; widely introduced in North America; Houghton County.

Euphorbia polygonifolia L.
SEASIDE SANDMAT *native*
IR | WUP | CUP | EUP CC 10

Chamaesyce polygonifolia (L.) Small
Annual, glabrous throughout. Stems usually prostrate, divergently branched or forming mats. Leaves linear-oblong to oblong lance-shaped, usually 8–15 mm long, entire, slightly inequilateral at base. Petal-like appendages of the involucre very small or none. Capsule 3–3.5 mm long; seeds smooth, compressed-ovoid, gray, 2–2.6 mm long.

Sand dunes and sandy beaches along Lake Michigan; Menominee County.

dicots **EUPHORBIACEAE** | Spurge Family 187

Euphorbia serpyllifolia Pers.
THYME-LEAVED SPURGE *introduced*
IR | **WUP** | **CUP** | EUP
Chamaesyce serpyllifolia (Pers.) Small
Annual, glabrous throughout. Stems usually prostrate and much branched, 1–3 dm long. Leaves linear-oblong to obovate, 6–12 mm long, very inequilateral, minutely serrulate on the rounded summit. Petal-like appendages of the involucre minute. Capsule sharply 3-angled, 1.5–2 mm long; seeds quadrangular, 1–1.4 mm long, smooth or very minutely rugulose.
Dry rocky soil.

Euphorbia vermiculata Raf.
WORM-SEED SANDMAT *introduced*
IR | WUP | **CUP** | EUP
Chamaesyce vermiculata (Raf.) House
Annual. Stems prostrate to ascending, to 4 dm long, hirsute more or less uniformly from base to tip with spreading hairs 0.5–1.5 mm long. Leaves obliquely ovate-oblong to ovate, 0.5–2 cm long, serrulate. Capsule 1.5–2 mm long, strongly 3-lobed, glabrous; seeds gray or pale brown, 1–1.3 mm long. Fields, roadsides, and waste ground.
Native in ne USA; considered adventive in Michigan.

Euphorbia virgata Waldst. & Kit.
LEAFY SPURGE *introduced (invasive)*
IR | **WUP** | **CUP** | **EUP**
Euphorbia esula L.
Perennial from a deep root. Stems erect, 3–7 dm tall, glabrous, usually with numerous alternate flowering branches below the umbel. Stem leaves linear to narrowly oblong, 3–7 cm long, obtuse to mucronate; leaves subtending the umbel shorter and broader, lance-shaped to ovate; leaves of the umbel broadly cordate or kidney-shaped. Rays of

Euphorbia virgata

Acalypha rhomboidea

Euphorbia corollata

Euphorbia cyparissias

Euphorbia glyptosperma

Euphorbia polygonifolia

Euphorbia helioscopia

Euphorbia maculata

Euphorbia peplus

Euphorbia serpyllifolia

the primary umbel 7 or more. Capsule 2.5–3 mm long, finely granular; seeds brown, globose-ovoid, about 2 mm long. Summer.

Native of Eurasia; widely established in dry, disturbed places in North America; a troublesome noxious weed, first collected in Michigan in 1885.

Fabaceae
PEA FAMILY

Perennial herbs, shrubs and trees. Leaves alternate, pinnately divided, the terminal leaflet sometimes modified as a tendril (*Lathyrus*, *Vicia*). Flowers in simple or branched racemes, perfect (with both staminate and pistillate parts), irregular, 5-lobed (only 1 lobe in *Amorpha*), the upper lobe (banner) larger than the other lobes, with 2 outer, lateral petals (wings), and 2 inner petals which are partly joined (the keel), and enclosing the 10 stamens and style; pistil 1, ovary 1-chambered, maturing into a pod.

ADDITIONAL SPECIES

Anthyllis vulneraria L. (Common kidney-vetch), introduced perennial herb, reported from Mackinac County. Stems usually branched from the base, 2–5 dm tall; leaflets mostly 5–11, the terminal leaflet often much larger than the lateral ones; flowers yellow.

Caragana arborescens (Siberian peashrub), introduced and sometimes escaped from hedgerow plantings in the western UP.

Hedysarum alpinum L. (Alpine sainfoin), native perennial, a circumpolar boreal species long known as far south as the north shore of Lake Superior; in Michigan, known only from from a meadow overlying limestone along the Escanaba River in Delta County.

1	Trees or shrubs............................2
1	Herbs....................................4
2	Trees....................**Robinia pseudoacacia**
2	Shrubs...................................3
3	Twigs and petioles covered with stiff hairs; petals 5; pods linear, stiffly hairy, with several seeds............................**Robinia hispida**
3	Twigs and petioles not covered with stiff hairs; corolla of a single purple petal; pods short, with 1 or 2 seeds; Baraga County................**Amorpha fruticosa**
4	Leaves even-pinnate, ending in a tendril............5
4	Leaves not even-pinnate.......................6
5	Style round in cross-section, pubescent near the tip............................**Vicia**
5	Style flattened, pubescent along the inner side............................**Lathyrus**
6	Leaves divided into 3 or rarely 5 leaflets............7
6	Leaves with 5 or more leaflets..................10
7	Leaflet margins entire...................**Lespedeza**
7	Leaflet margins toothed.........................8
8	Flowers reflexed in long slender racemes, white or yellow; pods small, straight, reflexed..........**Melilotus**
8	Flowers in rounded clusters.......................9
9	Pods straight; stamens joined to the corolla **Trifolium**
9	Pods curved or coiled; stamens free from the corolla......................**Medicago**
10	Leaflets 5–11....................................11
10	Leaflets 11–31..................................14
11	Flowers yellow, borne in umbels..................12
11	Flowers not yellow, borne in axillary or terminal racemes.....................................13
12	Bracts subtending flower heads entire.........**Lotus**
12	Bracts subtending flower heads 3-parted............................**Anthyllis vulneraria***
13	Stems twining or climbing; leaflets usually 5–7; flowers in axillary racemes.................**Apios americana**
13	Stems erect; leaflets 7–11; flowers blue (or pink or white), in terminal racemes.................**Lupinus**
14	Flowers in umbel-like clusters; pods linear, 4-angled, jointed.........................**Securigera varia**
14	Flowers in racemes; pods neither 4-angled nor jointed............................**Astragalus**

Amorpha *Indigo-Bush*

Amorpha fruticosa L.
FALSE INDIGO-BUSH *introduced*
IR | **WUP** | CUP | EUP FACW

Much-branched shrub. Stems mostly 1–3 m tall; twigs tan to gray. Leaves pinnately divided, 5–15 cm long; leaflets 9–27, oval to obovate, 1–4 cm long and 0.5–3 cm wide, upper surface smooth, underside short-hairy, margins entire, petioles 2–5 cm long, stipules absent. Flowers dark purple, in dense spike-like racemes 2–15 cm long at ends of stems; petals 1-lobed, only the banner present, 3–5 mm long, folded to enclose the 10 stamens. Pod oblong, curved near tip, 5–7 mm long, spotted with glands, with 1–2 seeds. June–July.

Native south and west of Michigan, considered an escape from cultivation here; Baraga County.

Amphicarpaea *Hog-Peanut*

Amphicarpaea bracteata (L.) Fern.
AMERICAN HOG-PEANUT *native*
IR | **WUP** | **CUP** | EUP FAC CC 5

Twining perennial herb. Stems to 1 m long. Leaves pinnately compound into 3 petioled leaflets. Flowers in racemes or panicles peduncled from many of the axils, each pedicel subtended at base by a striate-veined bractlet, bearing several to many pale purple to whitish flowers 12–18 mm long; calyx slightly irregular, the tube short-cylindric, the lobes apparently

4 through the fusion of the upper two; standard obovate, narrowed to the base, sometimes auricled below the middle; wings and keel slightly shorter, with elongate slender claws exceeding the blades. Pods (of petaliferous flowers) flat, oblong, pointed at both ends, usually 3-seeded, coiled after dehiscence. Besides the pod-producing petaliferous flowers, the plants bear nearly or completely apetalous flowers near the base of the stem, producing, often under the ground, small 1-seeded pods. Aug–Oct.

Woods and thickets.

Apios *Groundnut*

Apios americana Medik.
GROUNDNUT native
IR | WUP | **CUP** | **EUP** FACW CC 3

Perennial herbaceous vine, rhizomes with a necklace-like series of 2 or more edible tubers; plants with milky juice. Stems to 1 m long, climbing over other plants. Leaves pinnately divided; main leaves with 5–7 leaflets; leaflets ovate, 4–6 cm long, tapered to a point, smooth to short-hairy beneath, margins entire. Flowers brown-purple, 10–13 mm long, single or paired, in crowded racemes from leaf axils. Pods linear, 5–10 cm long. July–Aug.

Floodplain forests, thickets, shores, wet meadows, low prairie.

Astragalus *Milk-Vetch*

Ours perennial herbs from a stout taproot, caudex, or rhizome. Leaflets numerous. Flowers white, yellowish white, or purple, in long or short axillary racemes. Calyx tube campanulate to cylindric; calyx lobes short, triangular or subulate. Standard obovate to rotund, usually exceeding the wings. Stamens 10. Fruit a pod of various forms. Seeds few to many. *Astragalus* are most common in arid or semiarid regions of Asia and w North America.

1 Pod one-chambered .... **A. neglectus**
1 Pod divided to form 2 chambers .... **A. canadensis**

Astragalus canadensis L.
CANADIAN MILK-VETCH (THR) native
IR | **WUP** | **CUP** | EUP FAC CC 9

Stems erect, to 15 dm tall, glabrous to thinly strigose. Stipules connate, lance-shaped to deltoid, 3–6 mm long. Leaflets 13–29, oblong or elliptic, 1–3 cm long, 5–15 mm wide, glabrous or rarely strigose above, more or less strigose beneath with T-shaped hairs. Racemes long-peduncled, 5–12 cm long. Flowers white or yellowish white, spreading or somewhat reflexed, 12–15 mm long. Pods numerous in a crowded raceme, ovoid or oblong, 10–18 mm long, 2-celled, glabrous, nearly terete in cross-section. May–Aug.

Open woodlands, river banks and shores, usually in moist soil.

Astragalus neglectus (Torr. & Gray) Sheldon
COOPER'S MILK-VETCH native
IR | **WUP** | **CUP** | **EUP** FACU CC 9

Stems erect, 4–7 dm tall, nearly or quite glabrous. Leaflets 11–17, oblong to elliptic or obovate, 1–3 cm long, a fourth to a third as wide, glabrous above, strigose beneath. Racemes several, scarcely surpassing the subtending leaf, many-flowered; calyx tube cylindric, 4–6 mm long, strigose, the lobes about a third as long; corolla commonly white, 11–14 mm long. Pods erect, sessile, straight, ovoid, inflated, 1–2 cm long, glabrous. June.

Riverbanks and lakeshores, especially on limestone; disturbed forests and fields.

Desmodium *Tick-Trefoil*

Desmodium canadense (L.) DC.
SHOWY TICK-TREFOIL native
IR | **WUP** | **CUP** | EUP FACU CC 3

Perennial herb. Stems erect, branched above, to 2 m tall, pubescent. Leaves petioled, 3-parted; stipules linear-subulate, to 8 mm long, ciliate, otherwise glabrous; petioles 2–20 mm, usually about 10 mm long, the petiole and leaf-rachis together up to 1/2 as long as the terminal leaflet. Leaflets oblong or lance-oblong, appressed-pubescent beneath, the terminal 5–9 cm long. Flowers small, white to purple or violet, sometimes marked with yellow, often greenish in age, in densely flowered elongate racemes with conspicuous ovate lance-shaped bracts. Flowers 10–13 mm long; calyx 5–7 mm long. Fruit an indehiscent pod, elevated on a stalk above the persistent calyx and stamens, more or less beset with hooked hairs. petiole of the fruit 2–4 mm long; joints commonly 3–5. July–Aug.

Moist soil, thickets, and riverbanks.

Hylodesmum *Tick-Clover*

Hylodesmum glutinosum (Willd.) H. Ohashi & R. R. Mill
POINTED-LEAF TICK-CLOVER native
IR | **WUP** | **CUP** | EUP UPL CC 5

Desmodium glutinosum (Muhl.) Wood

Stems erect, 1–4 dm tall, bearing near the summit several long-petioled leaves and prolonged into a terminal panicle 3–8 dm long. Lateral leaflets asymmetrically ovate, acuminate; terminal leaflet round-ovate, 7–15 cm long and nearly as wide, long-acuminate. Flowers 6–8 mm long; calyx 2.5–3 mm long; petiole of the fruit 6–12 mm long, glabrous; joints seldom more than 3. July.

Rich deciduous woods.

FABACEAE | Pea Family

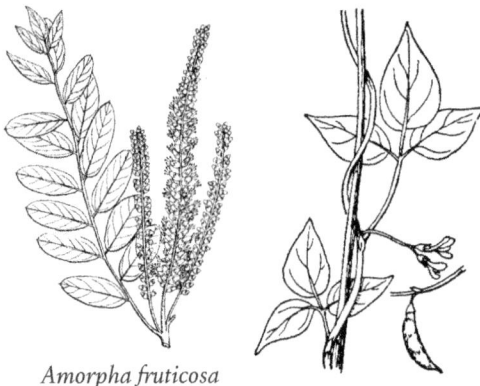

Amorpha fruticosa

Amphicarpaea bracteata

Apios americana

Astragalus canadensis

Desmodium canadense   *Hylodesmum glutinosum*

Lathyrus *Vetchling; Wild Pea*

Perennial herbs (ours). Leaves terminated by a tendril. Flowers few to many in a raceme; corolla small or medium-sized, red-purple to white or yellow; standard broadly obovate; wings obovate; keel upwardly curved. Pods flat to terete, 2–many-seeded. Most of our species superficially resemble *Vicia*.

ADDITIONAL SPECIES

Lathyrus pratensis L. (Yellow vetchling), Eurasian, known from several UP locations on roadsides and in fields.

Lathyrus sylvestris L. (Narrow-leaf vetchling), introduced and occasional in western and central UP.

1	Leaflets 1-pair; introduced species	2
1	Leaflets 2 or more pairs; native species	3
2	Stem winged	**L. latifolius**
2	Stem not winged	**L. tuberosus**
3	Flowers yellow-white	**L. ochroleucus**
3	Flowers purple, rarely white	4
4	Stipules leafy, nearly as large as the adjacent leaflets	**L. japonicus**
4	Stipules much smaller than leaflets	5
5	Stems usually winged; racemes with usually 10–20 flowers; moist habitats	**L. palustris**
5	Stems not winged; racemes with 2–6 flowers; dry woods	**L. venosus**

Lathyrus japonicus Willd.
BEACH-PEA   *native*
IR | WUP | CUP | EUP   FACU CC 10
Lathyrus maritimus Bigelow

Perennial, plants typically glabrous. Stems stout, decumbent to nearly erect, up to 1 m long. Stipules foliaceous, broadly ovate, 1.5–4 cm long, 1–2.5 cm wide, essentially symmetrical and therefore attached at the middle of the broadly truncate base. Leaflets 3–6 pairs, oblong to obovate, 3–5 cm long, about half as wide, the lowest pair near the base of the petiole. Peduncles equaling or shorter than the subtending leaves, bearing usually 5–10 purple flowers about 2 cm long; calyx irregular, the lowest lobe linear lance-shaped, nearly 2x as long as the triangular upper ones. June–Aug.

Beaches and lakeshores.

Lathyrus latifolius L.
EVERLASTING-PEA   *introduced*
IR | WUP | CUP | EUP

Perennial. Stems climbing or trailing, to 2 m long, broadly winged, 5–10 mm wide. Stipules lance-shaped, with a basal lobe, foliaceous, 1.5–4 cm long, usually wider than the stem. Petiole broadly winged, about as wide as the stem. Leaflets 2, lance-shaped to elliptic, 4–8 cm long, 1–3 cm wide. Peduncles 10–20 cm long, bearing a raceme of 4–10 handsome flowers 1.5–2.5 cm long; corolla purple, varying to pink or white; calyx lobes very unequal. June–Aug.

Native of s Europe; cultivated and escaping to roadsides and vacant land.

Lathyrus ochroleucus Hook.
CREAM VETCHLING *native*
IR | **WUP** | **CUP** | **EUP** CC 8

Glabrous perennial. Stems to 8 dm long. Stipules semi-ovate, 1.5–3 cm long, the larger ones toothed, usually shorter than the petiole. Leaflets 3–5 pairs, thin, elliptic to ovate lance-shaped, 2.5–5 cm long. Racemes shorter than the subtending leaf, bearing 5–10 yellowish white flowers 12–18 mm long; calyx irregular, the upper lobes triangular, to half as long as the lower. May–July.

Dry upland woods and thickets.

Lathyrus palustris L.
MARSH VETCHLING *native*
IR | **WUP** | **CUP** | **EUP** FACW CC 7

Perennial vining herb, spreading by rhizomes. Stems to 1 m long, strongly winged, climbing and clinging to surrounding plants by tendrils. Leaves pinnately divided, with 4–8 leaflets and a terminal leaflet modified into a tendril; leaflets linear to lance-shaped, 2–7 cm long and 3–20 mm wide; stipules prominent, more or less arrowhead-shaped, 1–3 cm long; margins entire; petioles absent. Flowers in racemes from leaf axils, 2–6 flowers per raceme, red-purple, drying blue to blue-violet; sepals irregular, 7–10 mm long; petals 12–20 mm long. Fruit a flat, many-seeded pod, 3–5 cm long. June–Aug.

Conifer swamps, wet meadows, marshes, streambanks, calcareous fens, low prairie.

Lathyrus tuberosus L.
EARTH-NUT VETCHLING *introduced*
IR | **WUP** | **CUP** | **EUP**

Perennial, with tuberously thickened roots. Stems wingless, 4-angled, to 8 dm long, branched above. Stipules lance-shaped, 5–12 mm long, with a single short basal lobe. Leaflets 2, oblong lance-shaped to elliptic, 1.5–3 cm long. Peduncles 5–8 cm long, bearing a crowded raceme of 2–5 red-purple flowers about 15 mm long, the standard broader than long; calyx lobes broadly triangular, shorter than the tube. June–Aug.

Native of Europe and w Asia; occasional on roadsides, fields, railways.

Lathyrus venosus Muhl.
VEINY VETCHLING *native*
IR | **WUP** | **CUP** | EUP FAC CC 8

Perennial. Stems stout, to 1 m long, glabrous or sparsely pubescent. Stipules narrowly lance-shaped to semi-ovate in outline. Leaflets commonly 8, 10, or 12, narrowly to broadly elliptic, 3–6 cm long, rounded to the mucronate tip. Peduncles stout, shorter than to equaling the subtending leaves, bearing a dense raceme of 10–30 purple flowers about 15 mm long; calyx oblique, the lower teeth much exceeding the upper but shorter than the tube. June–July.

Moist woods and thickets; Dickinson, Keweenaw and Menominee counties.

Lespedeza *Bush-Clover*

Lespedeza capitata Michx.
ROUND-HEAD BUSH-CLOVER *native*
IR | WUP | **CUP** | EUP FACU CC 5

Perennial herb. Stems usually erect, 6–15 dm tall, simple or branched above, sparsely to densely villous. Leaves small, trifoliolate; leaflets oblong, to 4.5 cm long and 1.8 cm wide, glabrous to sericeous above, thinly to densely sericeous or velutinous beneath. Petioles 2–5 mm long. Spikes numerous, forming a thyrsoid inflorescence, subglobose to short-ovoid, 12–25 mm long, with densely crowded flowers; peduncles rarely longer than the spikes and usually shorter than the subtending leaves. Flowers purple to yellowish white, 8–12 mm long, in sessile or peduncled axillary clusters, each flower subtended at base by 2 (sometimes 4) small

Lespedeza capitata

Lathyrus japonicus

Lathyrus ochroleucus

Lathyrus palustris

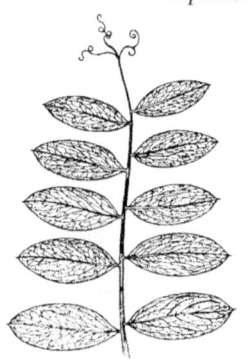

Lathyrus venosus

bractlets; standard short-clawed; wings straight, clawed and auriculate; calyx lobes villous, 6–10 mm long. Fruit oval to elliptic, 1-seeded, pubescent, conspicuously shorter (about 2 mm long) than the calyx. July–Sept.

Open dry woods, sand dunes, and prairies; Dickinson and Menominee counties. Plants growing on sand dunes are occasionally prostrate.

The genus differs from *Desmodium* by its 1-seeded fruits and in the absence of leaflet petioles.

Lotus *Trefoil*

Lotus corniculatus L.
BIRD'S-FOOT-TREFOIL *introduced (invasive)*
IR | WUP | CUP | EUP FACU

Lotus corniculatus var. *arvensis* (Schkuhr) Ser. Perennial herb. Stems prostrate, ascending, or erect, to 6 dm long. Leaves nearly sessile, without stipules; pinnately compound; leaflets 5, elliptic to oblong lance-shaped, 5–15 mm long, the terminal one sessile. Flowers in long-peduncled head-like umbels from the upper axils; pedicels 1–3 mm long; calyx lobes linear-triangular, about equaling the tube; corolla yellow to brick-red, about 14 mm long. Pods 2–4 cm long. June–Aug.

Native of Europe; established in fields, meadows, and roadsides.

Lupinus *Lupine*

Lupinus polyphyllus Lindl.
BLUE-POD LUPINE *introduced*
IR | WUP | CUP | EUP FACU

Perennial herb. Plants taller and coarser than *L. perennis* (found just south of the UP), to 1 m or more tall; leaves palmately compound, leaflets 11–17. Flowers white, yellow, pink, or blue, in terminal racemes or spikes 2–4 dm long; standard suborbicular with strongly reflexed sides; wings united toward the summit; keel petals strongly convex on the lower side, prolonged into a beak-like tip; calyx deeply bilabiate, the upper lip 2-toothed, the lower entire or 3-lobed; stamens 10, forming a closed tube for about half their length. Pods oblong, flattened, with 2–several seeds.

Escape from cultivation, especially along roadsides.

Medicago *Alfalfa; Medick*

Herbs with 3-foliolate serrulate leaves, the terminal leaflet stalked. Flowers in axillary heads or short head-like racemes of small yellow or blue flowers. Calyx tube campanulate, the 5 lobes nearly equal. Standard obovate or oblong, longer than the oblong erect wings; keel blunt, shorter than the wings. Stamens all free from the corolla. Pod straight or coiled, glabrous or spiny, usually indehiscent, 1–several-seeded.

1 Plants perennial from a long taproot; flowers blue-violet or sometimes yellow . . . . **M. sativa**
1 Plants annual or biennial; flowers yellow . . **M. lupulina**

Medicago lupulina L.
BLACK MEDICK *introduced*
IR | WUP | CUP | EUP FACU

Annual herb. Stems prostrate, widely spreading, or ascending, to 8 dm long; 4-angled. Leaflets elliptic to obovate, 1–2 cm long. Stipules lance-shaped, entire or toothed. Peduncles slender, much exceeding the subtending leaves, bearing a globose to short-cylindric head up to 1 cm long; peduncles and calyx glandular. Flowers yellow, 2–4 mm long. Pods nearly black, 2–3 mm long, kidney-shaped, 1-seeded, the conspicuous veins tending to be longitudinal. May–Sept.

Native of Europe and w Asia; common as a troublesome weed of roadsides, lawns, fields, railroads, and disturbed places.

Medicago sativa L.
ALFALFA *introduced*
IR | WUP | CUP | EUP UPL

Perennial herb. Stems erect or decumbent, to 1 m tall. Leaflets oblong lance-shaped, 1.5–3 cm long, toothed at the tip. Stipules ovate lance-shaped, toothed. Peduncles erect, about equaling the subtending leaves, with a subglobose to short-cylindric head 1–3 cm long. Flowers blue, nearly 1 cm long, on pedicels 2–3 mm long; sepals linear lance-shaped, 2–3 mm long, about equaling the tube. Pod coiled into a loose spiral of 1–3 complete turns, finely pubescent. June–Sept.

Native probably of c and w Asia; long in cultivation and valued for hay and forage; commonly escaped or introduced on roadsides, fields, and disturbed places.

Melilotus *Sweet-Clover*

Annual or biennial herbs. Leaves 3-foliolate, serrulate, the terminal leaflet stalked. Flowers white or yellow, in elongate peduncled racemes from the upper axils; stipules partially adnate to the petiole. Calyx eventually deciduous, the tube campanulate, the lobes nearly equal. Petals separate; standard oblong to obovate, usually longer than the others; wings and keel coherent. Ovary short, sessile or somewhat stipitate. Pod ovate to rotund, slightly compressed to nearly globose, 1–4-seeded.

Melilotus officinalis (L.) Lam.
YELLOW SWEET-CLOVER *intro. (invasive)*
IR | **WUP** | **CUP** | **EUP** FACU
Melilotus albus Medik.

Stems erect or ascending, 5–15 dm tall. Leaflets oblong lance-shaped to obovate, 1–2.5 cm long. Racemes 5–15 cm long including the peduncle. Pedicels 1.5–2 mm long, decurved. Flowers yellow, 5–7 mm long. Summer.

Native of Eurasia; established as a weed of waste places.

White sweet-clover (*M. albus*) now treated as a variant of *M. officinalis*.

Robinia *Locust*

Trees or shrubs. Leaves odd-pinnate, stipules setaceous or modified into spines. Flowers white, pink, or purple, in axillary racemes. Calyx tube broadly campanulate, bilabiate; lower 3 calyx lobes about equal; upper 2 lobes connate for a third or more of their length. Corolla large; standard more or less reflexed; wings and keel long-clawed with a rounded lobe at base of the blade; the keel petals strongly upwardly curved. Pods elongate, flat, many-seeded.

1 Tree to 25 m tall; twigs and petioles glabrous; flowers white . . . . **R. pseudoacacia**
1 Shrubs to 3 m tall; stems bristly with short stiff hairs; flowers pink or rose-colored. . . . . **R. hispida**

Robinia hispida L.
BRISTLY LOCUST *introduced*
IR | **WUP** | CUP | EUP

Stoloniferous shrub 1–2 m tall. Stems, peduncles, and calyx densely or sparsely hispid with glandular hairs up to 5 mm long, those of the stem persisting and becoming indurate. Leaflets 7–13, ovate-oblong to nearly rotund, commonly 3–6 cm long. Racemes usually 3–6-flowered. Flowers rose or pink-purple, 2.5–3 cm long; ovary densely glandular. Pods densely hispid, rarely developed. June–July.

Sometimes planted and occasionally escaped to roadsides and open woods; Ontonagon County.

Robinia pseudoacacia L.
BLACK LOCUST *introduced (invasive)*
IR | WUP | CUP | EUP FACU

Tree up to 25 m tall, the younger stems and peduncles finely pubescent. Stipules frequently modified into stout woody thorns. Leaflets 7–19, oval or elliptic, 2–4 cm long, mucronate at the truncate or rounded apex. Racemes drooping, many-flowered, 1–2 dm long. Flowers white, very fragrant, 2–2.5 cm long; calyx finely pubescent, the upper lip truncate or broadly notched; ovary glabrous. Pods very flat, smooth, 5–10 cm long. June.

Planted and sometimes escaped to roadsides, open woods, and waste land.

Lotus corniculatus

Lupinus polyphyllus

Medicago lupulina

Medicago sativa

Robinia hispida

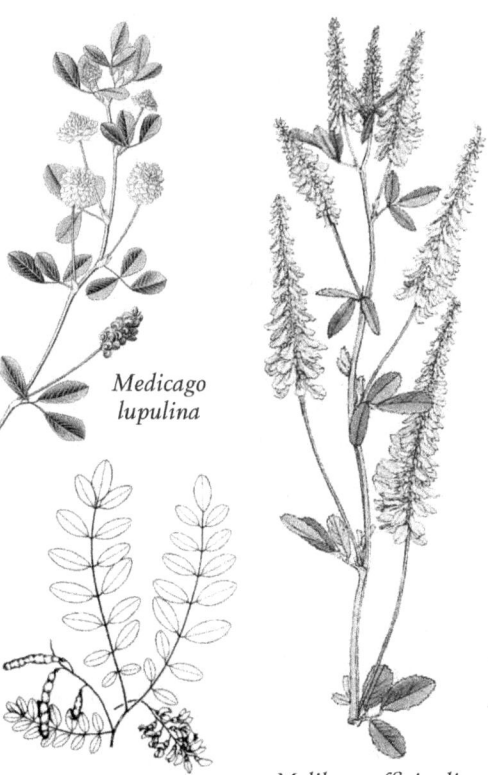

Robinia pseudoacacia

Melilotus officinalis

Securigera Crown-Vetch

Securigera varia (L.) Lassen
PURPLE CROWN-VETCH   *introduced*
IR | WUP | **CUP** | **EUP**

Coronilla varia L.

Perennial herb. Stems ascending, 3–5 dm long. Leaves sessile, 6–15 cm long; leaflets 11–21, oblong to obovate, 1–2 cm long, acute to rounded or retuse at the summit. Peduncles stout, equaling or surpassing the subtending leaves. Flowers in long-peduncled axillary umbels of 10–15 flowers, pink, the keel tipped with purple; calyx tube campanulate, bilabiate, the broad lower lip with 3 short triangular teeth, the upper lip narrow, triangular, shallowly cleft; petals about equal in length, clawed; standard orbicular; wings ovate-oblong; keel-petals upwardly curved. Pods linear, 4-angled, 2–4 cm long, with 3–7 joints. May–Sept.

Native of Eurasia and n Africa; introduced or escaped, mostly along roadsides.

Trifolium Clover

Annual, biennial, or perennial herbs. Leaves 3-foliolate, serrulate. Flowers in heads, spikes, or headlike racemes or umbels. Calyx tube campanulate to tubular. Petals all separate or more or less united into a tube, usually withering and persistent after anthesis; standard ovate to obovate, often folded about the wings or with only the tip outwardly curved. Fruit short, straight, often included in the persistent calyx, 1–6-seeded. Valuable for forage and several species extensively cultivated. In the absence of fruit, our yellow-flowered species may be distinguished from *Medicago* by their strongly bilabiate calyx.

ADDITIONAL SPECIES
Trifolium incarnatum L. (Crimson clover), introduced; Keweenaw County.

1 Flowers white, pink, or purple 2
1 Flowers yellow, turning brown with age 6
2 Flowers on short pedicels, these becoming reflexed with age 3
2 Flowers sessile or nearly so. 4
3 Flowers white; very common. . . . . **T. repens**
3 Flowers pink or purple-tinged . . . . **T. hybridum**
4 Heads cylindric-shaped 5
4 Heads nearly globose to ovoid . . . . **T. pratense**
5 Flowers white, shorter than the calyx . . . . . . **T. arvense**
5 Flowers crimson, longer than the calyx **T. incarnatum***
6 Leaflets sessile; stipules linear. . . . . **T. aureum**
6 Terminal leaflet on a short petiole; stipules ovate to lance-shaped . . . . **T. campestre**

Trifolium arvense L.
RABBIT-FOOT CLOVER   *introduced*
IR | **WUP** | **CUP** | EUP

Annual. Stems erect, freely branched, 1–4 dm tall, softly pubescent. Petioles 4–10 mm long, exceeded by the subulate tips of the stipules. Leaflets narrowly oblong lance-shaped, 1–2 cm long, serrulate only at the tip. Heads gray to pale brown, densely flowered, ovoid to cylindric, 1–3 cm long, on peduncles 1–3 cm long; calyx densely villous with hairs up to 2 mm long, the tube to 2 mm long, the lobes 3–5 mm long, exceeding and partly concealing the small whitish or pinkish corolla. May–Sept.

Native of Eurasia and n Africa; a weed of sterile soil, roadsides, old fields, and waste places.

Trifolium aureum Pollich
GREATER HOP CLOVER   *introduced*
IR | **WUP** | **CUP** | **EUP**

Annual or biennial. Stems much-branched, mostly erect, 2–5 dm tall, appressed-hairy. Leaflets all sessile or nearly so, oblong lance-shaped, 1–2 cm long; petioles 5–12 mm long, about equaling the lance-oblong stipules. Heads short-cylindric, 1–2 cm long, on peduncles 1–4 cm from the upper axils; pedicels 0.5 mm. Flowers 5–7 mm long; calyx strongly 2-lipped, glabrous, the tube 5-nerved, 1 mm, the lobes lance-linear; corolla yellow; the standard obovate, conspicuously striate in age, usually serrulate; the wings somewhat spreading at the tip. May–Sept.

Native of Eurasia, weedy on roadsides and in waste places.

Trifolium campestre Schreb.
LESSER HOP CLOVER   *introduced*
IR | **WUP** | **CUP** | EUP

Annual. Stems much-branched, 1–4 dm tall, pubescent. Leaflets obovate, 8–15 mm long, the terminal one on a stalk 1–3 mm long; heads globose to short-cylindric, 8–15 mm long, compact, with usually 20–30 flowers; petioles 8–12 mm long, 2x as long as the stipules. Flowers 3.5–5 mm long; calyx as in T. aureum; corolla yellow, the standard obovate, with 5 conspicuous diagonal veins on each side, much exceeding the spoon-shaped, slightly divergent wings. May–Sept.

Native of Eurasia and n Africa; weedy on roadsides and in waste places.

Trifolium hybridum L.
ALSIKE CLOVER   *introduced*
IR | **WUP** | **CUP** | **EUP**   FACU

Perennial. Stems erect or ascending, 3–8 dm tall. Stipules ovate lance-shaped, tapering to a long slender point. Leaflets oval to elliptic, broadly rounded to retuse at the summit. Heads numerous, not involucrate, globose, on peduncles 2–8 cm long.

Flowers distinctly pediceled, 7–10 mm long; calyx glabrous, the linear lobes somewhat unequal, 1.7–2.5 mm long, slightly exceeding the tube; corolla white and pink, turning brown after anthesis; the standard obovate, about 2 mm longer than the obtuse wings. Summer.

Native of Eurasia, commonly escaped.

Trifolium pratense L.
RED CLOVER *introduced*
IR | WUP | CUP | EUP   FACU

Perennial. Stems erect, decumbent, or ascending, to 8 dm tall, sparsely to densely appressed-pubescent. Stipules oblong, the free portion abruptly narrowed to a short awn. Lower leaves long-petioled; upper leaves short-petioled to sessile. Heads sessile or on peduncles up to 2 cm long, globose or round-ovoid. Flowers 13–20 mm long; calyx glabrous to sparsely pilose, the tube 3–4 mm long, the lobes setaceous, one 4–7 mm long, four 2–5 mm long; corolla magenta, varying to nearly white; standard obovate, equaling or slightly exceeding the oblong obtuse wings. May–Aug.

Native of Europe, commonly planted for forage and escaped to fields and on roadsides.

Trifolium repens L.
WHITE CLOVER *introduced*
IR | WUP | CUP | EUP   FACU

Perennial. Stems creeping, sending up long-petioled leaves and long-peduncled heads without involucres. Leaflets broadly elliptic to obovate, rounded or notched at the tip, 1–2 cm long. Flowers distinctly pediceled, 7–11 mm long; calyx glabrous, the tube to 3 mm long, its lobes narrowly triangular, unequal, the longest about equaling the tube; corolla white or tinged with pink; the standard elliptic-obovate, rounded at the summit, exceeding the obtuse wings. Summer.

Native of Eurasia, commonly planted and escaped to lawns and roadsides.

Vicia *Vetch*

Annual or perennial herbs. Leaves 1-pinnate, with small stipules, the terminal leaflets in most species metamorphosed into tendrils. Flowers in short or elongate racemes from the axils, or in sessile or subsessile, few-flowered, axillary clusters. Calyx regular or irregular, often swollen at base. Standard with a broad claw overlapping the wings, its blade obovate to subrotund; wings oblong or narrowly obovate, adherent to and usually exceeding the keel. Ovary sessile or short-stipitate. Pods flat to terete. Seeds 2 to many.

ADDITIONAL SPECIES
Vicia sepium L. (Hedge vetch), introduced from Europe; collected trailside on Isle Royale in 1959.

1. Flowers single or in pairs, sessile from leaf axils .... V. sativa
1. Flowers in racemes on stalks from leaf axils 2
2. Flowers white or white tinged with purple, 3–7 mm long .... V. tetrasperma
2. Flowers blue or white, 8 mm long or more........... 3
3. Calyx with a large swollen bump on one side of base .. .. V. villosa
3. Calyx only slightly swollen on one side of base 4
4. Margins of stipules sharply toothed; flowers 15–30 mm long .... V. americana
4. Margins of stipules entire; flowers to 13 mm long .... V. cracca

Vicia americana Muhl.
AMERICAN VETCH *native*
IR | WUP | CUP | EUP   FACU CC 5

Perennial. Stems trailing or climbing, to 1 m long. Leaflets usually 4–7 pairs, elliptic to oblong, 1.5–3 cm long, 5–12 mm wide, obtuse to broadly rounded at the mucronate tip. Stipules all or mostly sharply serrate. Racemes shorter than the subtending leaves, loose, bearing 2–9 blue-purple flowers 17–27 mm long; calyx tube 3.5–5.5 mm long. May–July.

Moist woods.

Vicia cracca L.
BIRD-VETCH *introduced*
IR | WUP | CUP | EUP

Perennial. Stems climbing or trailing, to 1 m long. Leaflets usually 5–10 pairs, linear, 1.5–3 cm long, mucronate. Stipules entire. Racemes long-peduncled, dense, secund, equaling or exceeding the subtending leaf, bearing numerous crowded blue flowers 9–13 mm long; calyx tube swollen at base, 2–3 mm long. Blade of the standard about as long as the claw. June–Aug.

Fields, roadsides, meadows.

Vicia sativa L.
COMMON VETCH *introduced*
IR | WUP | CUP | EUP   FACU

Annual. Stems slender, to 1 m long, ascending, erect, or tending to climb. Leaflets commonly 4–8 pairs, oblong to elliptic or obovate, 3–5 cm long, mucronate. Stipules often sharply serrate, bearing a glandular spot beneath. Flowers commonly paired in the axils, nearly sessile, violet or purple, rarely white, 2–3 cm long; calyx tube campanulate, 5–7 mm long; lobes nearly equal, the upper 3–7 mm, the lower 4–9 mm long. Pod flattened, brown; seeds flattened.

Native of Europe; in cultivation since antiquity; may persist after cultivation or escape into fields and roadsides.

196 FABACEAE | Pea Family

Vicia tetrasperma (L.) Schreb.
LENTIL VETCH *introduced*
IR | WUP | CUP | **EUP**

Annual. Stems slender, branched, decumbent or climbing, 3–6 dm long. Leaflets commonly 3 or 4 pairs, occasionally more, linear-oblong, 1–2 cm long. Peduncles very slender, 1–3 cm long, bearing usually 1 or 2, sometimes 3 or 4, light purple to white flowers 4–6.5 mm long. Lowest calyx-lobe linear, about equaling the tube, the upper much shorter and triangular. Pod flat, glabrous, 1–1.5 cm long, usually 4-seeded. May–Aug.

Eurasian introduction; Chippewa County.

Vicia villosa Roth
HAIRY VETCH *introduced*
IR | **WUP** | **CUP** | **EUP**

Annual. Stems to 1 m long, more or less villous throughout, always softly villous above and in the racemes with hairs 1–2 mm long. Leaflets usually 5–10 pairs, narrowly oblong to linear lance-shaped, obtuse and mucronate to acute, 1–2.5 cm long. Racemes long-peduncled, dense, secund, bearing usually 10–30 flowers; calyx irregular, villous, the tube 2–4 mm long, swollen at base; upper lobes linear-triangular, 0.8–1.5 mm long; lateral and lower lobes linear above a triangular base, the lowest 3–5 mm long, long-villous; corolla slender, 12–20 mm long, the spreading blade of the standard less than 1/2 as long as the claw. June–Aug.

Native of Europe; introduced in fields, roadsides, and waste places.

Fagaceae
BEECH FAMILY

Trees or shrubs. Leaves alternate, simple, entire to lobed; the stipules early deciduous. Plants monoecious. Staminate flowers in catkins or heads; corolla none; calyx small, deeply 4–8-parted; stamens 3–20. Pistillate flowers solitary, or in small clusters or short spikes, more or less enclosed by an involucre of numerous bracts; ovary commonly 3-celled or 6-celled, with 2 ovules in each cell, of which only one matures. Fruit a 1-seeded nut, wholly or partly surrounded by the expanded involucre.

1 Leaves entire, or toothed or lobed with fewer than 9 pairs of lateral veins . . . . **Quercus**

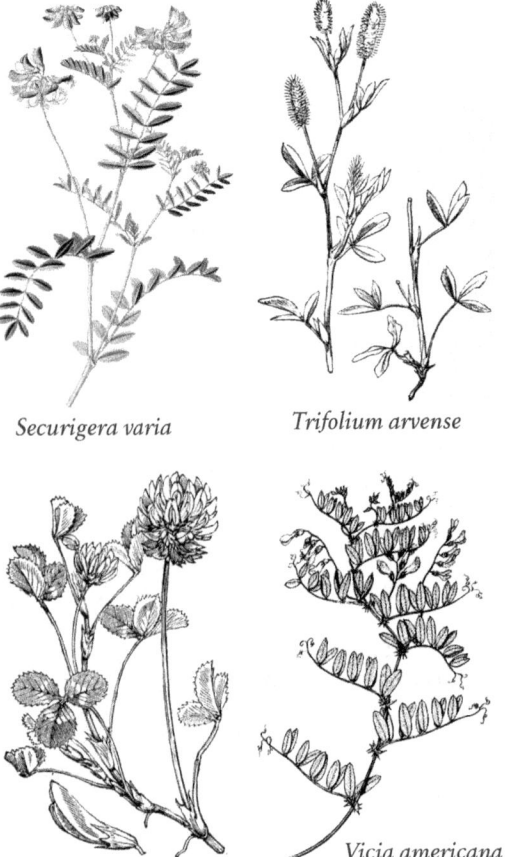

Securigera varia

Trifolium arvense

Trifolium repens

Vicia americana

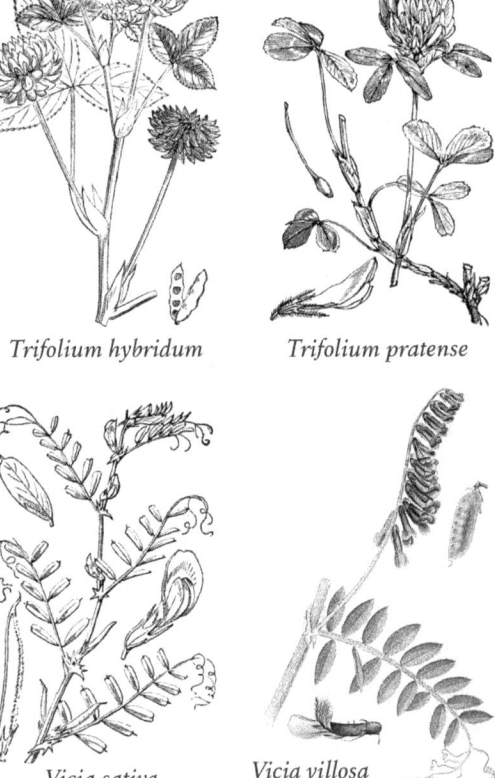

Trifolium hybridum

Trifolium pratense

Vicia sativa

Vicia villosa

1. Leaf margins various; lateral veins more than 9 pairs . 2
2. Bark smooth, silvery-gray; terminal buds single, narrow and more than 1 cm long; fruit bristly, the nut 3-angled .... **Fagus grandifolia**
2. Bark becoming furrowed or scaly; terminal buds 2 or more, less than 1 cm long; fruit not bristly, the nuts rounded................................... **Quercus**

1. Lobes rounded, not tipped by bristles 2
1. Lobes acute, bristle-tipped 3
2. Acorn 1.5–3 cm long within a deep, fringed cup; branches often with corky ridges...... **Q. macrocarpa**
2. Acorn 1.3–2 cm long, only about one-quarter covered by warty cup; branches not corky-ridged......... **Q. alba**
3. Acorn cup covering lower one-third to lower one-half of acorn .... **Q. ellipsoidalis**
3. Acorn cup saucer-shaped; covering only base of acorn .... **Q. rubra**

Fagus *Beech*

Fagus grandifolia Ehrh.
AMERICAN BEECH *native*
IR | WUP | **CUP** | **EUP**   FACU CC 6

Tree, to 30 m tall, with smooth gray bark. Leaves simple, alternate, straight-veined, a vein running to each tooth, short-petioled, ovate to obovate, serrate to denticulate, densely silky when young, at maturity glabrous above, beneath usually silky on the midvein and sometimes more or less pubescent on the surface. Flowers appearing with the leaves, the staminate flowers from the lower axils in small heads on drooping peduncles, subtended by deciduous bracts, stamens 8–16; pistillate flowers from the upper axils, usually in pairs at the end of a short peduncle, subtended and largely concealed by numerous subulate bracts; calyx adnate to the ovary, with 6 short acuminate lobes; ovary 3-celled; styles 3. Fruit a sharply 3-angled, 1-seeded nut, borne in pairs enclosed within the expanded 4-valved involucre.

Beech-maple forests; the UP at northern and western edge of species' range.

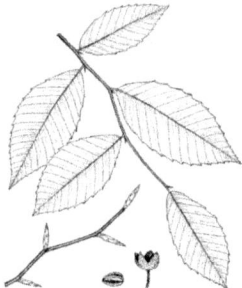

Fagus grandifolia

Quercus *Oak*

Deciduous trees (or rarely shrubby). Leaves alternate, simple, lobed, pinnately veined. Plants monoecious (staminate and pistillate flowers separate but on same tree). Staminate flowers in slender naked catkins, catkins appearing with the leaves. Calyx divided to the base into 3–7 (usually 6) segments. Stamens 3–12. Pistillate flowers solitary or in small spikes, each subtended by a bract and surrounded by an involucre of many scales. Ovary 3-celled. Fruit a nut (acorn) partially enclosed by a cuplike structure (cupule). Acorns are important food for many mammals and birds but are usually avoided by humans because of their tannin content.

HYBRIDS

Quercus × schuettei Trel., *Q. bicolor* and *Q. macrocarpa*; Menominee County.

Quercus alba L.
WHITE OAK *native*
IR | WUP | **CUP** | **EUP**   FACU CC 5

Tall tree with light gray bark and widely spreading branches; twigs soon glabrescent. Leaves obovate, glabrous or very sparsely and obscurely pubescent beneath at maturity; lobes variable, 3, 4, or rarely 5 pairs, ascending, oblong to ovate, rounded or rarely acute. Acorns sessile or on peduncles to 4 cm long; cup deeply saucer-shaped, pubescent within, covering a fourth to a third of the nut; nut ovoid to cylindric-ovoid, 1.5–2.5 cm long.

Upland woods of oak-hickory and beech-maple, sandy plains with other oaks and jack pine.

Quercus ellipsoidalis E.J. Hill
NORTHERN PIN OAK *native*
IR | **WUP** | **CUP** | **EUP**   CC 4

Medium-sized tree, the twigs soon glabrescent. Leaves smooth on both sides except for small tufts of stellate hairs in the vein-axils beneath; lateral lobes 2–3 pairs, separated by rounded sinuses, usually extending more than half-way to the midvein, sometimes broadest at the base, but usually widened and several-toothed distally, the sinuses then elliptic. Acorn cup turbinate, 9–14 mm wide, with closely appressed puberulent scales, covering about a third of the nut; nut ovoid to ellipsoid, 12–20 mm long.

Dry upland soil.

Quercus macrocarpa Michx.
BUR-OAK *native*
IR | **WUP** | **CUP** | **EUP**   FACU CC 5

Low shrub to tall tree, the latter with rough, deeply furrowed bark. Leaves obovate to obovate, cuneate at base, pale beneath with a close, fine, stellate pubescence, with 4–7 pairs of blunt or acute lateral lobes, a pair of sinuses near the middle usually deeper than the others. Acorns sessile or on a stout peduncle; cup deeply saucer-shaped to sub-globose, covering 1/3 to nearly all the nut, pubescent within, the marginal scales acuminate into slender awns forming a terminal fringe; nut depressed-ovoid, broadly rounded to almost retuse at the summit, to narrowly ovoid, 1–4 cm wide.

Moist woods and alluvial floodplains.

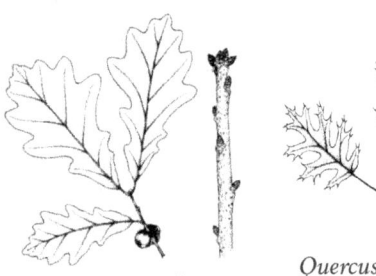
Quercus alba

Quercus ellipsoidalis

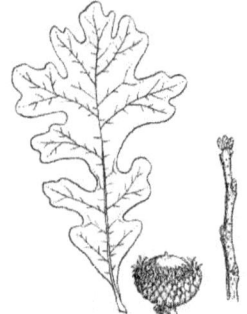
Quercus macrocarpa

Quercus rubra

Quercus rubra L.
NORTHERN RED OAK native
IR | WUP | CUP | EUP FACU CC 5
Medium to large tree. Young twigs glabrous, dark reddish brown. Leaves dull green, 10–20 cm long, soon glabrous throughout or with small tufts of pubescence persistent in the leaf-axils, 7–11-lobed, lobes roughly triangular in outline, broadest at the base, bristle-tipped and usually with a few lateral teeth, little if any longer than the central body of the blade. Acorn cup saucer-shaped or almost turbinate, 1.5–2 cm wide, enclosing about 1/4 to 1/3 of the nut; nut ovoid, 2–2.5 cm long.

Mesic forests, ridges, sandy plains with jack pine.

Gentianaceae
GENTIAN FAMILY

Annual, biennial or perennial herbs; plants usually glabrous. Leaves simple, entire, opposite or whorled, stem leaves without petioles. Flowers often showy, perfect (with both staminate and pistillate parts), regular, single at end of stems or in clusters; petals 4-5, blue, purple, white or green, joined for at least part of their length; stamens 4 or 5. Fruit a 2-chambered, many-seeded capsule enclosed by the withered, persistent petals.

1 Leaves reduced to small, narrow scales less than 3 mm long . . . . **Bartonia**
1 Leaves not scale-like, well developed 2
2 Flowers pink . . . . **Centaurium pulchellum**
2 Flowers blue, green tinged with purple, or white 3
3 Petals 4, spurred at base; flowers green, tinged with purple . . . . **Halenia deflexa**
3 Petals 4, with fringed lobes; or petals 5 and not spurred; blue, purple or white 4
4 Petals 4, fringed; flowers on stalks longer than the flowers; seeds covered with small bumps. . . . **Gentianopsis**
4 Petals 5, not fringed; flower stalks short or absent; seeds smooth . . . . **Gentiana**

Bartonia *Screwstem*

Slender annual or biennial herbs. Stems pale green to yellow or purple. Leaves reduced to small opposite or alternate scales. Flowers small, 4-parted, green-white to green-yellow, bell-shaped, in slender panicles or racemes at ends of stems.

1 Mid-stem leaves alternate; anthers 0.3–0.5 mm long . . . . **B. paniculata**
1 Mid-stem leaves opposite or subopposite; anthers 0.5–1 mm long . . . . **B. virginica**

Bartonia paniculata (Michx.) Muhl.
TWINING SCREWSTEM (THR) native
IR | WUP | CUP | **EUP** OBL CC 10
Annual or biennial herb. Stems slender, 2–4 dm long, upright or lax. Leaves small and scale-like, 1–2 mm long, mostly alternate, or the lower leaves opposite. Flowers yellow-white or greenish, 2–4 mm long, in panicles 5–20 cm long; the flowers on slender, arched and spreading stalks; sepals awl-shaped, 2 mm long; petals lance-shaped; anthers yellow. Fruit a capsule. Aug–Sept. Tamarack swamps, fens, sphagnum bogs, open wetlands; Chippewa and Luce counties; more common in se USA.

Bartonia virginica (L.) B.S.P.
YELLOW SCREWSTEM native
IR | WUP | **CUP** | **EUP** FACW CC 7
Annual or biennial herb. Stems slender, erect, yellow-green, 1–4 dm long. Leaves mostly opposite, small and scale-like, 1–2 mm long. Flowers green-yellow or green-white, 3–4 mm long, in a slender raceme or panicle, the branches and flower stalks opposite and upright; sepals awl-shaped; petals oblong, tapered to a rounded tip. Fruit a capsule 2–3 cm long. Aug–Sept.

Swamps (often in sphagnum moss), open bogs, wet woods and depressions, sandy shores and ditches; Chippewa and Schoolcraft counties.

Centaurium *Centaury*

Centaurium pulchellum (Sw.) Druce
BRANCHED CENTAURY *introduced*
IR | WUP | CUP | **EUP** FACU

Annual herb. Stems much branched, often from the base, 1–2 dm tall. Leaves sessile, lance-shaped or ovate lance-shaped, 1–2 cm long. Inflorescence a many-flowered terminal cyme; bracteal leaves linear. Flowers 4-merous; calyx tubular, about 9 mm long, deeply cleft into narrow segments; corolla tube slightly exceeding the calyx; corolla lobes pink, about 4 mm long; stamens inserted in the throat of the corolla; filaments slender, exsert; ovary elongate, 1-celled. Capsule oblong, thin-walled, invested by the persistent calyx and withered corolla. June–Sept.

Native of Europe; local in fields and waste places, often where salted in winter; Mackinac County.

ADDITIONAL SPECIES
Centaurium erythraea Rafn (Forking Centaury), adventive biennial of moist, sandy, disturbed places such as fields, meadows, shores, and roadside ditches; Delta County.

Gentiana *Gentian*

Perennial herbs, with thick, fibrous roots. Leaves opposite or whorled, simple, margins entire, petioles absent. Flowers large, blue, green-white or yellow, 5-parted, in clusters near ends of stems; petals forming a tubelike, shallowly lobed flower, the lobes alternating with a folded membrane as long or longer than petal lobes; stamens 5. Fruit a 2-chambered capsule.

1 Flowers blue (rarely white), remaining closed, the corolla lobes absent or reduced to small points . . . . **G. andrewsii**
1 Flowers blue, white or yellowish, opening, the corolla lobes prominent 2
2 Leaves deep green, linear or nearly so, not over 1 cm wide; calyx lobes green throughout, exposed by the spreading, narrow involucral leaves . . . . **G. linearis**
2 Leaves ± pale green, lanceolate to ovate, at least the larger ones over 1 cm wide; calyx lobes green only near tip, usually hidden by the wide, enveloping involucral leaves . . . . **G. rubricaulis**

Gentiana andrewsii Griseb.
BOTTLE-GENTIAN *native*
IR | **WUP** | **CUP** | **EUP** FACW CC 5

Perennial herb. Stems erect, single or few together, 2–8 dm long, unbranched, smooth. Leaves opposite, lance-shaped, 4–12 cm long and 1–3 cm wide, margins entire. Flowers 1 to many, stalkless in upper leaf axils, 3–5 cm long; sepals forming a tube around petals, the sepal lobes unequal, fringed with hairs; petals forming a tubelike flower, usually remaining closed, the folds between petal lobes finely fringed (use hand lens to see this) and longer than the petal lobes. Fruit a capsule; seeds winged. Aug–Sept.

Wet meadows, swamps and wet woods, thickets, low prairie, shores, ditches.

Gentiana linearis Froel.
NARROW-LEAF GENTIAN (THR) *native*
IR | **WUP** | **CUP** | EUP CC 10

Perennial herb. Stems smooth, 2–8 dm long. Leaves dark green, linear to narrowly lance-shaped, 4–9 cm long and less than 1 cm wide, margins entire. Flowers opening slightly, several in a cluster at end of stem and upper leaf axils; sepal lobes green, linear, 4–10 mm long; petals blue (sometimes white), 3–5 cm long, the lobes ovate, rounded, not fringed. Fruit a capsule; seeds winged. Aug–Sept.

Wet meadows, shores, streambanks, thickets; rare in Baraga, Keweenaw and Marquette counties; disjunct from main range of northeastern USA.

Gentiana rubricaulis Schwein.
GREAT LAKES GENTIAN *native*
IR | **WUP** | **CUP** | **EUP** OBL CC 7

Gentiana linearis var. *lanceolata* A. Gray
Perennial herb. Stems smooth, 3–7 dm long. Leaves pale green, lance-shaped, 4–8 cm long and 2–3 cm wide, margins entire. Flowers 3–5 cm long, green-blue below, blue above, narrowly open, in a cluster at end of stem; sepal lobes oblong, 4–12 mm long, chaffy and translucent near base. Fruit a capsule; seeds winged.

Wet meadows, peatlands, streambanks, thickets, conifer swamps, Lake Superior rocky shores; soils usually calcium-rich.

Gentianopsis *Fringed-Gentian*

Gentianopsis virgata (Raf.) Holub
LESSER FRINGED-GENTIAN *native*
IR | WUP | **CUP** | **EUP** OBL CC 8

Gentianopsis procera (Holm) Ma
Glabrous annual herb. Stems simple or few-branched, 1–5 dm long. Basal leaves spatula-shaped; stem leaves opposite, sessile, linear to linear lance-shaped, 2–5 cm long and 2–7 mm wide, tapered to a blunt tip, the base not clasping stem; margins entire. Flowers bright blue, 2–5 cm long, mostly 4-parted, single on stalks at ends of stems; sepal tube 6–15 mm long; petals forming a tubelike flower, flared toward tip, petal lobes ragged toothed across tips, often fringed on sides, without a folded membrane between the lobes (present in *Gentiana*). Fruit a capsule. Sept–Oct.

Sandy and gravelly shores, wet meadows, fens, intradunal wetlands near Lake Michigan; soils usually calcium-rich.

200 GENTIANACEAE | Gentian Family   *dicots*

Halenia *Spurred Gentian*

Halenia deflexa (Sm.) Griseb.
SPURRED GENTIAN   *native*
IR | WUP | CUP | EUP   FAC CC 7

Annual herb. Stems erect, simple or few-branched, rounded 4-angled, 15–40 cm long. Leaves opposite, lower leaves spatula-shaped, narrowed to a petiole; stem leaves lance-shaped to ovate, 2–5 cm long and 1–2.5 cm wide, sessile; margins entire. Flowers green, tinged with purple, 10–12 mm long, 4-parted, on stalks to 4 cm long, in loose clusters of 5–9 flowers at ends of stems; petals lance-shaped, usually with downward-pointing spurs at base, the spurs to 5 mm long. Fruit an oblong capsule. July–Aug.

Cedar swamps, moist conifer woods (especially along shores), old logging roads.

Geraniaceae
GERANIUM FAMILY

Annual or perennial herbs. Leaves usually opposite, simple or compound, palmately toothed, lobed, or divided. Flowers 5-merous, regular or somewhat zygomorphic, all perfect or part of them sterile. Sepals 5. Petals 5, pink to purple. Fruit a carpel prolonged at maturity into beaks and eventually separating; seed 1 in each carpel.

1 Leaves simple, palmately lobed, or 3-parted; anthers usually 10 . . . . **Geranium**
1 Leaves pinnately compound; anthers 5 . . . . **Erodium cicutarium**

Erodium *Stork's Bill; Filaree*

Erodium cicutarium (L.) L'Hér.
REDSTEM-FILAREE   *introduced*
IR | **WUP** | **CUP** | EUP

Winter-annual or biennial herb. Stems at first anthesis very short, with the leaves mostly basal, later diffusely

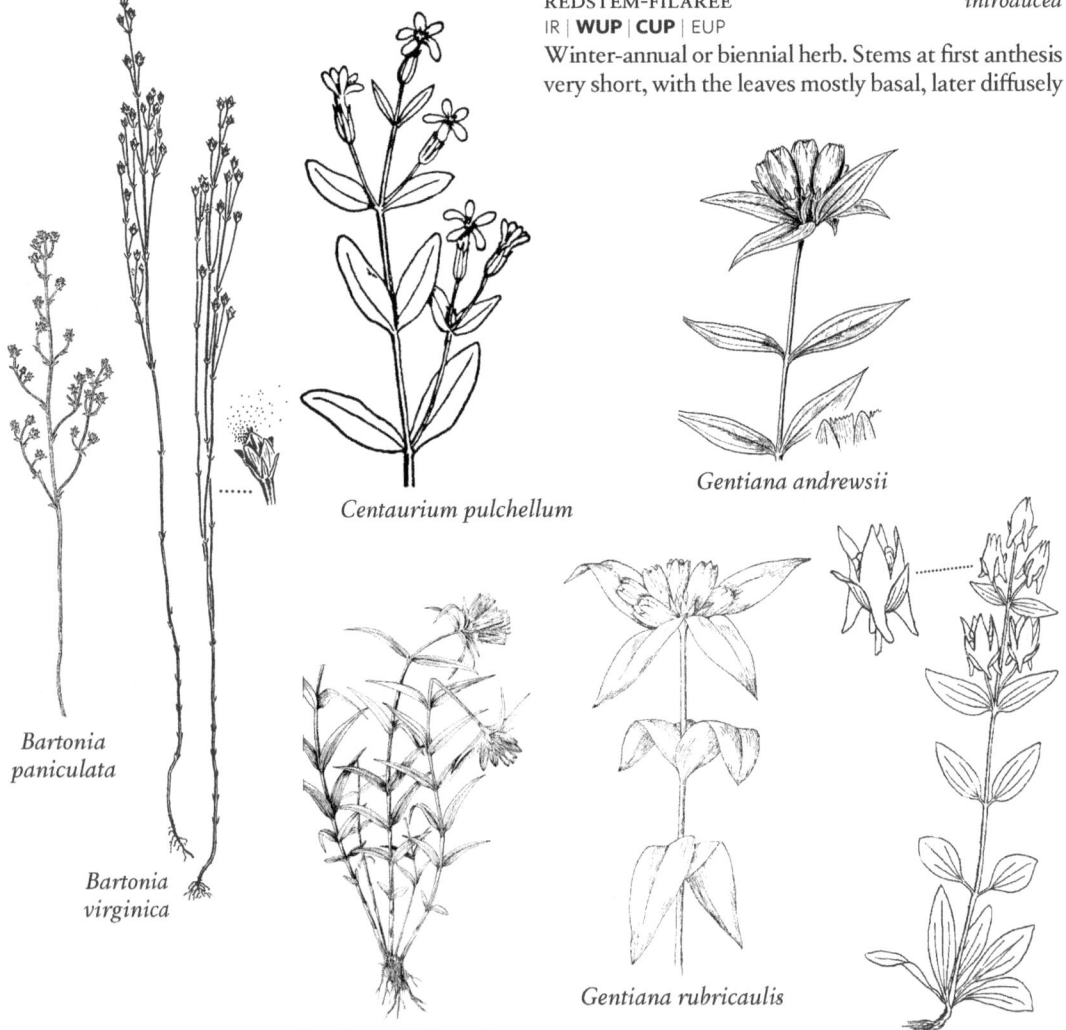

Centaurium pulchellum

Bartonia paniculata

Bartonia virginica

Gentianopsis linearis

Gentiana andrewsii

Gentiana rubricaulis

Halenia deflexa

branched, to 4 dm long. Leaves alternate, oblong lance-shaped in outline, pinnately compound, with several sessile, deeply and irregularly cleft pinnae each 1–2.5 cm long. Flowers in long-peduncled, 2–8 flowered cymes; pedicels 1–2 cm long; sepals 5; petals 5, obovate, the 2 upper often differing in size from the lower 3; glands 5, minute, alternating with the petals; ovary 5-celled, each carpel containing a single seed and prolonged into a long beak which separates from the other carpels and becomes spirally twisted when dry. April–Sept.

Native of the Mediterranean region; weedy, especially in fallow fields.

Geranium
Wild Geranium; Crane's-Bill

Annual or perennial herbs. Leaves palmately lobed, cleft, or divided, the stem leaves chiefly opposite. Flowers small or medium-sized, usually pink to purple, usually pedicellate in pairs at the ends of axillary peduncles. Sepals 5, imbricate. Petals 5, imbricate, alternating at base with 5 glands. Stamens 10. Axis of the ovary prolonged at maturity into a long beak.

ADDITIONAL SPECIES
Geranium sanguineum L. (Bloody crane's-bill); Eurasian perennial of dry, disturbed places; Marquette and Schoolcraft counties.

1	Perennial, spreading by rhizomes; petals more than 11 mm long; leaves few, large . . . . **G. maculatum**	
1	Annuals or short-lived perennials; petals less than 11 mm long; stem leaves several to many 2	
2	Sepals rounded or acute, not awn-tipped . **G. pusillum**	
2	Sepals narrowed to long, awn-like tips 3	
3	Leaves divided to their base . . . . **G. robertianum**	
3	Leaves deeply divided but not to their base 4	
4	Flower pedicels covered with glandular, spreading hairs . . . . **G. bicknellii**	
4	Pedicel hairs without glands . . . . **G. carolinianum**	

Geranium bicknellii Britt.
NORTHERN CRANE'S-BILL   *native*
IR | **WUP** | **CUP** | **EUP**   CC 4

Annual. Stems erect, usually with many ascending branches, eventually to 5 dm long. Leaves pentagonal in outline, the principal ones cleft nearly to the base with usually 5 segments, these deeply incised with several parrowly oblong lobes. Peduncles 2-flowered, the elongate pedicels glandular-villous; sepals at anthesis 7–9 mm long, including the conspicuous subulate tips; petals pink-purple, about equaling the sepals. Mature fruit, including the calyx, 20–25 mm long, the beak 4–5 mm long, the body 3 mm long, sparsely hirsute. May–Sept.

Open woods and fields, usually where sandy or gravelly.

Geranium carolinianum L.
CAROLINA CRANE'S-BILL   *native*
IR | **WUP** | CUP | **EUP**   CC 4

Annual. Stems several from the base, freely branched, eventually to 6 dm long, villous with spreading or somewhat retrorse hairs, becoming glandular in the inflorescence. Principal leaves kidney-shaped, 3–7 cm wide, deeply cleft into 5–9 oblong to obovate, deeply toothed or lobed segments. Flowers in compact, many-flowered, umbel-like, terminal clusters Peduncles mostly 2-flowered, the pedicels to twice as long as the calyx; sepals to 1 cm long, about equaling the pink petals. Fruit about 2.5 cm long, the stylar beak 1–2 mm long, the body hirsute with long antrorse hairs; seeds very obscurely reticulate. May–Aug.

Dry, barren, rocky or sandy soil, limestone pavement, disturbed places. Chippewa (Drummond Island only) and Keweenaw (including Isle Royale) counties.

Geranium maculatum L.
SPOTTED CRANE'S-BILL   *native*
IR | **WUP** | **CUP** | EUP   FACU CC 4

Perennial from a thick rhizome. Stems erect, 3–7 dm long. Basal leaves long-petioled, pedately 5–7-cleft into wedge-like segments; stem leaves a single pair, resembling the basal but short-petioled. Flowers few to several, rose-purple, 2.5–4 cm wide; calyx and pedicels pubescent but not glandular. Fruit erect, the beak 2–3 mm long; seed minutely reticulate. April–June.

Dry or moist woods.

Geranium pusillum L.
SMALL-FLOWER CRANE'S-BILL   *introduced*
IR | **WUP** | CUP | **E**UP

Annual or biennial. Stems to 5 dm long, diffusely branched, spreading or ascending. Basal leaves long-petioled, the blades rotund, 3–6 cm wide, deeply 7–9 cleft, the divisions cuneate, palmately lobed at the summit; upper leaves progressively smaller, shorter-petioled, 7–3-cleft. Flowers numerous on densely but minutely glandular pedicels; sepals at anthesis 2.5–4 mm long, more or less hirsute, especially at the margin, nearly as long as the red–violet corolla. Fruit, including the calyx, 9–12 mm long, the body closely but minutely strigose. Summer.

Native of Europe; established as a weed in fields and waste land.

Geranium robertianum L.
HERB-ROBERT   *native*
IR | **WUP** | **CUP** | **EUP**   CC 3

Annual or winter-annual. Stems to 6 dm long, branched, spreading, villous. Leaves triangular in

202 GROSSULARIACEAE | Currant Family

outline, 3-divided, the lateral segments often again divided and all segments pinnately lobed or cleft. Peduncles from most of the upper nodes, usually 2-flowered. Flowers pink to red-purple, 10–15 mm wide. Carpels detached from the cauline beak, each terminating in 2 slender filaments. May–Sept.

Damp rich woods.

Grossulariaceae
CURRANT FAMILY

Ribes *Currant; Gooseberry*

Medium shrubs. Stems upright to spreading, smooth, or with spines at nodes and sometimes also with bristles between nodes. Leaves alternate, palmately veined and palmately 3–5-lobed, margins toothed. Flowers 1 to several in short clusters, or few to many in racemes; green to white or yellow, perfect, regular, ovary inferior; sepals 5; petals 5, shorter than sepals; stamens 5, alternate with petals, styles 2. Fruit a many-seeded berry, usually topped by persistent, dry flower parts. *Ribes* are of two types: currants and gooseberries. Currants lack spines and bristles (except in *R. lacustre*) and the stalk of berry is jointed at its tip so that berries detach from stalks. Gooseberries have spines and bristles and the berry stalk is not jointed so that stalks remain attached to berries when picked.

1 Stems with spines or bristles, at least at the nodes; flowers single or in corymb-like clusters of 2–3 (gooseberries) 2
1 Flowers in racemes of 5 or more; stems without spines or bristles (except in *R. lacustre*; currants) 4
2 Ovary and fruit usually bristly; calyx lobes shorter than corolla tube .... **R. cynosbati**
2 Ovaries and fruit smooth (or bristly in the rare *R. oxyacanthoides*); calyx lobes longer than corolla tube 3
3 Leaves with glands, at least on underside veins; fruit bristly or with gland-tipped hairs to sometimes smooth .... **R. oxyacanthoides**
3 Leaves without glands; fruit smooth ..... **R. hirtellum**
4 Ovary and fruit bristly with gland-tipped hairs 5
4 Ovary and fruit neither bristly nor with gland-tipped hairs 6
5 Stems densely bristly .... **R. lacustre**
5 Stems unarmed .... **R. glandulosum**
6 Leaf underside dotted with shiny resinous glands; fruit black 7
6 Leaf underside without resinous glands 8
7 Flowers yellow to greenish; calyx glabrous or sparsely hairy; inflorescence bracts longer than pedicels .... **R. americanum**
7 Flowers white to greenish-white; calyx hairy; inflorescence bracts much shorter than pedicels 9
8 Flowers and fruit in upright racemes; native species .. .. **R. hudsonianum**
8 Flowers and fruit in drooping racemes; occasional garden escape .... **R. nigrum**
9 Flowers golden-yellow; fruit black .... **R. odoratum**
9 Flowers yellow-green; fruit red 10
10 Pedicels with scattered hairs and short-stalked glands .... **R. triste**
10 Pedicels glabrous .... **R. rubrum**

Ribes americanum P. Mill.
WILD BLACK CURRANT native
IR | **WUP** | **CUP** | **EUP** FACW CC 6

Shrub, 1–1.2 m tall. Stems without spines or bristles, young stems finely hairy; branches upright to spreading; twigs gray-brown and smooth, black with age. Leaves 3–8 cm long and 3–10 cm wide, 3-lobed and usually with 2 additional shallow lobes at base, dotted with shiny, yellow to brown resinous glands, especially on underside, smooth or short-hairy above, hairy below; margins coarsely toothed; petioles hairy and resin-dotted, 3–6 cm long. Flowers creamy-

Geranium bicknellii

Geranium carolinianum

Geranium maculatum *Geranium robertianum*

white to yellow, bell-shaped, 8–12 mm long; 6–15 in drooping racemes 3–8 cm long; each flower with a linear bract longer than the flower stalk, the stalks 2–3 mm long; sepals 4–5 mm long, rounded; petals blunt, 2–3 mm long; stamens about equaling petals. Fruit an edible, smooth, black berry. April–June.

Moist to wet forests, swamps, marsh and lake borders, streambanks.

Ribes cynosbati L.
EASTERN PRICKLY GOOSEBERRY   *native*
IR | WUP | CUP | EUP   FACU CC 4

Shrub to 6–9 dm tall, branches upright to spreading. Stems and branches with 1–3 spines at nodes, outer bark peeling off, inner bark brown-purple to black; young stems brown-gray, finely hairy. Leaves 3–8 cm long and 3–7 cm wide, 3–5-lobed, the lobes rounded at tips; upper surface dark green, sparsely hairy, underside paler, finely hairy and with gland-tipped hairs along veins; margins with coarse, round teeth; petioles 2.5–4 cm long, finely hairy and with scattered gland-tipped hairs. Flowers green-yellow, bell-shaped, 6–9 mm long, in clusters of 2–3 from spurs on old wood, on stalks with gland-tipped hairs. Fruit a red-purple berry, covered with stiff, brown spines. May–June.

Moist hardwood forests (where one of our most common gooseberries); occasional in wet woods, swamps, thickets and streambanks.

Ribes glandulosum Grauer
SKUNK CURRANT   *native*
IR | WUP | CUP | EUP   FACW CC 5

Shrub to 8 dm tall. Stems sprawling, spines and bristles absent. Stems and leaves with skunklike odor when crushed; older stems smooth and dark as outer bark peels off; young stems smooth to finely hairy, brown-gray. Leaves 2–8 cm long and 4–8 cm wide, 3–5-lobed, smooth above, paler and finely glandular hairy below (at least along veins); margins toothed or double-toothed; petioles 3–6 cm long, finely hairy. Flowers yellow-green to purple, saucer-shaped, in loose upright clusters 3–6 cm long, on slender stalks; bracts very small, the stalks and bracts with gland-tipped hairs; sepals 2 mm long; petals 1–2 mm long. Fruit a dark red berry with bristles and gland-tipped hairs, 6 mm wide. June.

Cedar and tamarack swamps, cool wet woods, thickets and streambanks.

Ribes hirtellum Michx.
HAIRY-STEM GOOSEBERRY   *native*
IR | WUP | CUP | EUP   FACW CC 6

Shrub to 9 dm tall. Stems upright, outer bark pale, soon peeling to expose dark inner layer; young stems gray and smooth, or with 1–3 slender spines at nodes and scattered bristles between nodes. Leaves 2.5–5 cm long and 2–5 cm wide, with 3 or 5 pointed lobes, upper surface dark green, smooth to sparsely hairy, lower surface paler, hairy at least along veins, without glands; margins coarsely toothed and fringed with hairs; petioles 1–3 cm long, hairy, some of which are gland-tipped. Flowers green-yellow to purple, bell-shaped, 6–9 mm long, in clusters of 2–3 on short, smooth stalks; stamens as long or longer than sepals, the bracts fringed with long hairs. Fruit an edible, smooth, dark blue-black berry. June.

Cedar and tamarack swamps, thickets, shores, rocky openings.

Ribes hudsonianum Richards.
HUDSON BAY CURRANT   *native*
IR | WUP | CUP | EUP   OBL CC 10

Shrub, 6–9 dm tall. Stems upright, spines and bristles absent; bark gray, with scattered yellow resin dots, peeling to expose inner purple-black bark. Leaves 5–9 cm long and 6–13 cm wide, 3–5-lobed, with unpleasant odor when rubbed, upper surface dark green and mostly hairless, underside paler, smooth to hairy and with yellow resin dots; margins coarsely toothed, the teeth with a hard tip; petioles 2.5–8 cm long, with fine hairs and resin dots. Flowers white, bell-shaped, 4–5 mm long, in small clusters on threadlike stalks. Fruit a smooth, blue-black berry, barely edible. June.

Cedar swamps, wet conifer woods and streambanks.

Ribes lacustre (Pers.) Poir.
BRISTLY BLACK GOOSEBERRY   *native*
IR | WUP | CUP | EUP   FACW CC 6

Shrub to 1 m tall. Stems upright or spreading, densely bristly, long-spiny at nodes; older bark gray, peeling to expose dark inner bark. Leaves 4–8 cm long and 4–7 cm wide, with 3–5 deeply parted, pointed lobes, upper surface dark green and mostly smooth, underside paler with scattered gland-tipped hairs; margins cleft into rounded teeth; petioles 2.5–4 cm long, with gland-tipped hairs. Flowers yellow-green to pinkish, saucer-shaped, 4–5 mm wide, on stalks with dark, gland-tipped hairs, in arching or drooping clusters. Fruit palatable but insipid, red, becoming black or dark purple, covered with gland-tipped hairs. May–June.

Moist conifer woods, swamps, thickets, and rock outcrops.

Ribes nigrum L.
GARDEN BLACK CURRANT   *introduced*
IR | **WUP** | CUP | EUP

Shrub. Leaves 3–5–lobed about to the middle, dotted with resinous glands beneath. Racemes drooping, the pedicels (2–8 mm long) much exceeding the minute ovate bracts; hypanthium above the ovary

204 GROSSULARIACEAE | Currant Family

short-campanulate; sepals greenish purple within; ovary commonly with sessile resinous glands. Fruit black, glabrous, sweet-tasting.

Native of Eurasia; occasionally planted and rarely escaped; Houghton County.

Ribes odoratum H. Wendl.
BUFFALO-CURRANT *introduced*
IR | **WUP** | CUP | EUP FACU

Ribes aureum Pursh var. *villosum* DC.

Erect shrub. Leaves orbicular to cuneate-obovate in outline, broadly cuneate to truncate at base, deeply 3-lobed or rarely 5-lobed, ciliate, finely puberulent to glabrate beneath, the lobes entire in their lower half, entire or few-toothed across the summit. Flowers golden yellow; hypanthium above the ovary tubular, 11–15 mm long; sepals obovate, 5–6.5 mm long, broadly rounded above; petals 2.5–3.5 mm long, erose at the summit. Fruit yellow or black, edible. Apr–June.

Cliffs and rocky hillsides; cultivated and sometimes escaped; Houghton County.

Ribes oxyacanthoides L.
NORTHERN GOOSEBERRY *native*
IR | **WUP** | **CUP** | EUP FACU CC 9

Shrub to 1 m tall. Stems upright with 1–3 spines to 1 cm long at nodes and smaller spines scattered between nodes; young stems gray-brown and finely hairy. Leaves 2.5–5 cm long and 2–5 cm wide, with 3–5 blunt or rounded lobes, upper surface sparsely hairy, some hairs tipped with glands, underside resin-dotted, hairy, some gland-tipped, especially along veins; margins coarsely toothed and hairy; petioles 0.5–3 cm long, with short hairs and scattered glands. Flowers green-yellow, bell-shaped, 6–9 mm long, in clusters of 2–3 on short stalks; stamens shorter than petals. Fruit a smooth, edible, blue-black berry. June.

Uncommon on rocky and sandy shores, rocky openings, cold moist woods.

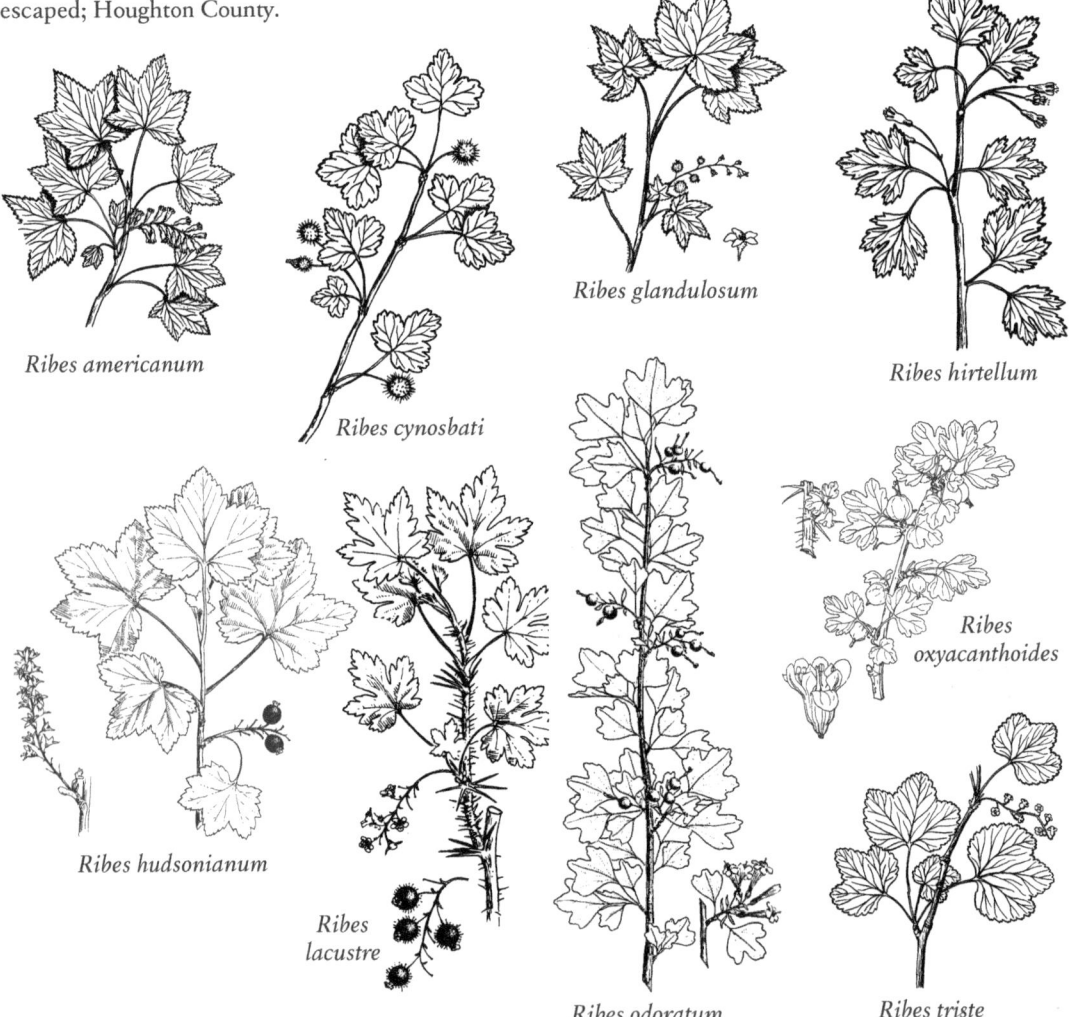

Ribes americanum

Ribes cynosbati

Ribes glandulosum

Ribes hirtellum

Ribes hudsonianum

Ribes lacustre

Ribes oxyacanthoides

Ribes odoratum

Ribes triste

Ribes rubrum L.
GARDEN RED CURRANT *introduced*
IR | **WUP** | **CUP** | **EUP**
Ribes sativum Syme

Erect shrub. Stems erect, nearly glabrous, crisped-puberulent; spines at nodes absent; prickles on internodes absent. Leaves commonly 5-lobed, the lateral lobes spreading. Flowers cream to pinkish. Fruit a bright red berry, glabrous, sour.

Native of the Old World; long in cultivation and occasionally escaped.

Ribes triste Pallas
SWAMP RED CURRANT *native*
IR | **WUP** | **CUP** | **EUP** OBL CC 6

Low shrub, 0.4–1 m tall. Stems spreading or lying on ground and rooting at nodes, spines and bristles absent; older stems smooth, purple-black, young stems short-hairy. Leaves 4–10 cm long and 4–10 cm wide, with 3–5 broad lobes, dark green and mostly smooth above, paler and usually finely hairy below; margins with both rounded and sharp teeth, the teeth with a hard tip; petioles 2.5–6 cm long, with scattered gland-tipped hairs. Flowers green-purple, 4–5 mm wide, on stalks 1–4 mm long, in drooping clusters of 5–12. Fruit a red berry, glabrous, sour-tasting. May–June.

Wet woods swamps, alder thickets, seeps.

Haloragaceae
WATER-MILFOIL FAMILY

Perennial aquatic herbs. Leaves alternate or whorled, finely dissected. Flowers small, stalkless in axils of leaves or bracts, 3- or 4-parted, regular, perfect (with both staminate and pistillate parts), or imperfect, petals small or absent. Fruit small and nutlike, dividing into 3 or 4 segments (mericarps).

1. Flowers 4-parted; leaves mostly whorled, emersed leaves reduced to small bracts .... **Myriophyllum**
1. Flowers 3-parted; leaves alternate, emersed leaves not bract-like. .... **Proserpinaca palustris**

Myriophyllum Water-Milfoil

Perennial aquatic herbs. Stems submerged, sparsely branched, freely rooting at lower nodes. Leaves mostly whorled (alternate in *M. farwellii*), pinnately divided into threadlike segments, upper leaves often reduced to bracts. Flowers small, mostly imperfect, stalkless in axils of upper emersed leaves (the floral bracts) or axils of underwater leaves; staminate flowers above pistillate flowers; perfect flowers (if present) in middle portion of spike; sepals inconspicuous; petals 4 or absent; stamens 4 or 8; pistil 4-chambered. Fruit nutlike, 4-lobed, each lobe (mericarp) with 1 seed, rounded on back or with a ridge or row of small bumps.

1. Leaves simple, reduced to small, blunt-tipped scales; stems erect and crowded from creeping rhizomes .... **M. tenellum**
1. Leaves dissected into narrow segments. 2
2. Leaves alternate, opposite, or scattered on stem .... **M. farwellii**
2. Foliage leaves all whorled (or appearing so in *M. heterophyllum*) 3
3. Flowers and bracts below flowers alternate on stem .. .. **M. alterniflorum**
3. Flowers and bracts below flowers whorled 4
4. Bracts surrounding staminate flowers deeply cleft.... .... **M. verticillatum**
4. Bracts surrounding staminate flowers sharply toothed or entire. 5
5. Bracts sharply toothed and much longer than flowers . ... **M. heterophyllum**
5. Bracts surrounding staminate flowers entire and not longer than flowers. 6
6. Leaf segments mostly 5–12 on each side of midrib; small bulbs (turions) produced at ends of stems and in upper leaf axils. .... **M. sibiricum**
6. Leaf segments many, 12–20 on each side of midrib; turions absent. .... **M. spicatum**

Myriophyllum alterniflorum DC.
ALTERNATE-FLOWER WATER-MILFOIL *native*
IR | **WUP** | **CUP** | **EUP** OBL CC 10

Perennial herb. Stems very slender. Leaves in whorls of 3–5, usually less than 1 cm long and shorter than the stem internodes, pinnately divided. Flower spikes raised above water surface, 2–5 cm long; bracts mostly alternate, linear, shorter than the flowers; staminate flowers with 4 pink petals; stamens 8. Fruit segments 1–2 mm long, rounded on back and base.

Acidic lakes, Lake Superior coastline.

Myriophyllum farwellii Morong
FARWELL'S WATER-MILFOIL (THR) *native*
IR | **WUP** | **CUP** | **EUP** OBL CC 10

Perennial herb; plants entirely underwater, turions present at ends of stems. Leaves 1–3 cm long, dissected into threadlike segments, all or most leaves alternate, or more or less opposite, or irregularly scattered on stems. Flowers underwater, single in axils of foliage leaves; pistillate flowers with 4 purple petals; stamens 4, tiny. Fruit 2 mm long, each fruit segment with 2 small, bumpy, longitudinal ridges.

Rare in ponds and small lakes.

Myriophyllum heterophyllum Michx.
TWO-LEAF WATER-MILFOIL native
IR | WUP | CUP | EUP OBL CC 6
Perennial herb. Stems stout, to 3 mm wide, often red-tinged, to 1 m or more long. Leaves appearing whorled due to the very short internodes, 1.5–4 cm long, divided into threadlike segments. Flowers in spikes raised above water surface, 5–30 cm long; floral bracts whorled, smaller than foliage leaves, ovate, sharply toothed, spreading or curved downward. Flowers both perfect and imperfect; petals of staminate and perfect flowers 1–3 mm long; stamens 4. Fruit olive, more or less round, 2 mm long; fruit segments rounded or with 2 small ridges, beaked by the curved stigma. June–Aug.

Lakes, ponds and pools in streams; sometimes where calcium-rich.

Myriophyllum sibiricum Komarov
COMMON WATER-MILFOIL native
IR | WUP | CUP | EUP OBL CC 10
Myriophyllum exalbescens Fern.
Perennial herb; plants often whitish when dried. Stems to 1 m or more long. Leaves in whorls of 3–4, 1–4 cm long, with mostly 5–10 threadlike segments on each side of midrib; internodes between whorls about 1 cm long. Flowers in spikes with whorled flowers and bracts, raised above water surface, red, clearly different than underwater stems, 4–10 cm long; flowers imperfect, the upper staminate, the lower pistillate; floral bracts much smaller than the leaves, oblong to obovate; staminate flowers with pinkish petals (absent in pistillate flowers), 2–3 mm long; stamens 8, the yellow-green anthers conspicuous when flowering. Fruit more or less round, 2–3 mm long, the segments rounded on back. June–Sept.

Shallow to deep water of lakes, ponds, marshes, ditches and slow-moving streams; sometimes where calcium-rich.

When flowering, the numerous red spikes of this species are conspicuous on water surface. *M. spicatum*, introduced from Eurasia, is similar but has more finely divided leaves (12–24 threadlike segments on each side of midrib) and larger floral bracts.

Myriophyllum spicatum L.
EURASIAN WATER-MILFOIL *intro. (invasive)*
IR | WUP | CUP | EUP OBL
Perennial herb, similar to *M. sibiricum*. Stems widening below head and curved to a horizontal position, usually many-branched near water surface, internodes between leaves mostly 1–3 cm long, turions absent. Leaves with more leaf segments per side (mostly 12–20) than in *M. sibiricum;* lower flower bracts often divided into comblike segments and often longer than the flowers. Fruit segments 2–3 mm long. Aug–Sept.

Lakes and ponds. Introduced from Europe and spreading in lakes throughout e USA; known from scattered locations in the UP.

Myriophyllum tenellum Bigelow
SLENDER WATER-MILFOIL native
IR | WUP | CUP | EUP OBL CC 10
Perennial herb. Stems slender, 10–30 cm long, mostly upright and unbranched. Leaves absent or reduced to a few spaced scales. Flowers in spikes raised above water surface, 2–5 cm long; flower bracts mostly alternate, oblong to obovate, entire, shorter to slightly longer than the flowers. Fruit segments rounded on back and at base, 1 mm long.

Acidic lakes; often forming large colonies, especially in deep water.

Myriophyllum verticillatum L.
WHORLED WATER-MILFOIL native
IR | WUP | CUP | EUP OBL CC 6
Perennial herb, similar to *M. sibiricum*, but plants often larger. Stems 5–25 dm long. Leaves in whorls of 4–5, with 9–17 threadlike segments along each side of midrib, 1–5 cm long; lower and middle internodes between whorls mostly less than 1 cm long. Flowers perfect, or the lower pistillate and upper staminate; in spikes 4–12 cm long, the floral bracts much smaller than the leaves, with comblike segments, mostly longer than the flowers; petals blunt-tipped, 2–3 mm long, smaller in pistillate flowers; stamens 8. Fruit more or less round, 2–3 mm long, the segments rounded on back. July–Sept.

Lakes, ponds, quiet rivers.

Proserpinaca Mermaid-Weed

Proserpinaca palustris L.
COMMON MERMAID-WEED native
IR | WUP | CUP | EUP OBL CC 6
Perennial aquatic herb, often forming large colonies. Stems horizontal at base and often rooting; the flower-bearing branches erect, 1–4 dm tall. Leaves alternate; underwater leaves, if present, ovate in outline, 2–4 cm long, deeply divided into linear segments; emersed leaves narrowly lance-shaped, 2–6 cm long, margins with sharp, forward-pointing teeth. Flowers small, perfect, green or purple-tinged, 1–3 in axils of emersed leaves, stalkless; sepals triangle-shaped, persistent; petals absent, stamens 3, stigmas 3. Fruit nutlike, 3-angled, 2–5 mm long and as wide, with 3 seeds. June–Aug.

Shallow water of ponds, streambanks and ditches, muddy shores, sedge meadows; usually where seasonally flooded.

Myriophyllum alterniflorum

Myriophyllum sibiricum

Myriophyllum farwellii

Myriophyllum spicatum

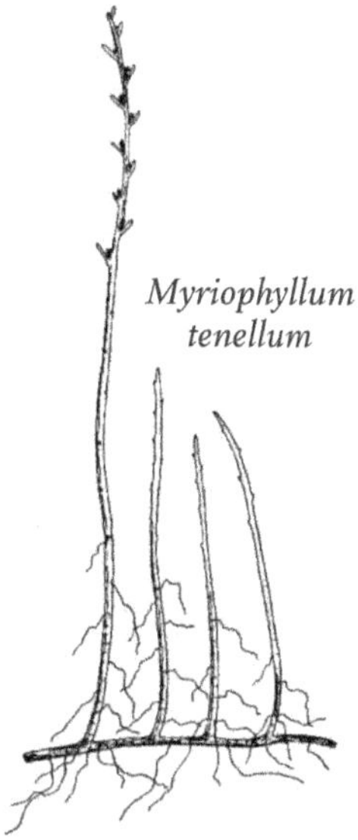

Myriophyllum tenellum

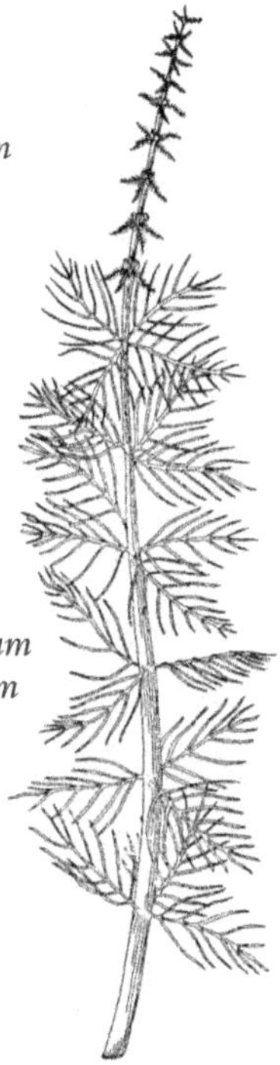

Myriophyllum verticillatum

Proserpinaca palustris

Hamamelidaceae
WITCH-HAZEL FAMILY

Hamamelis Witch-Hazel

Hamamelis virginiana L.
AMERICAN WITCH-HAZEL *native*
IR | **WUP** | **CUP** | EUP   FACU CC 5

Tall shrub to 5 m tall, with scurfy or glabrous twigs. Leaves broadly obovate, with several to many rounded teeth, base broadly rounded or subcordate, green on both sides, glabrous or stellate-pubescent beneath. Flowers in short-pediceled axillary clusters; 4-merous; sepals small, triangular, dull yellowish brown within; petals bright yellow or suffused with red, spreading, 1.5–2 cm long; stamens 4, opposite the sepals and much shorter than them, alternating with 4 small scale-like staminodia; styles 2. Fruit ovoid before dehiscence, 1–1.5 cm long, the hypanthium often bearing the persistent sepals; seeds black, eventually discharged explosively from the capsule. Oct–Nov; fruit ripe a year later.

Moist woods.

Witch hazel extract is derived from the bark.

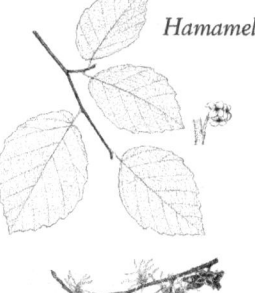

Hamamelis virginiana

Hypericaceae
ST. JOHN'S-WORT FAMILY

Glabrous annual or perennial herbs (shrubby in *Hypericum kalmianum*). Stems usually unbranched below, branched in head. Leaves simple, opposite, dotted with dark or translucent glands (visible when held to light), especially on underside; margins entire; petioles absent. Flowers few to many in clusters at ends of stems or from upper leaf axils, perfect, regular, sepals 5, petals 5, yellow or pink to green or purple; stamens 9–35, separate or joined near base into 3 or more groups; styles 3, ovary superior. Fruit a 3-chambered, many-seeded capsule.

1　Petals yellow; stamens 15–many .... **Hypericum**
1　Petals pink or purple; stamens 9 .... **Triadenum**

Hypericum St. John's-Wort

Shrubs or herbs. Leaves opposite, sometimes dotted with black and/or small transparent glands; margins entire. Flowers in clusters at ends of stems and upper leaf axils, yellow, perfect, regular, sepals 5, petals 5, stamens 5–many, separate or joined into 3 or 5 bundles. Fruit a capsule.

1　Styles joined at base, persisting on capsule as a straight beak; stamens many, distinct 2
1　Styles free to base, the capsules not beaked; stamens few to many, joined at base into 3 or 5 bundles 3
2　Small shrubs to 1 m tall .... **H. kalmianum**
2　Perennial herbs, slightly woody at base.. **H. ellipticum**
3　Plants 1–2 m tall; leaves 5 cm long or more; flowers 4 cm or more wide; styles 5. .... **H. ascyron**
3　Plants usually less than 1 m tall; leaves less than 5 cm long; flowers to 3 cm wide; styles 3 4
4　Petals spotted with black dots; stamens in 3 weak groups 5
4　Petals not spotted with black dots; stamens in 5 weak groups 6
5　Flowers 15 mm or more wide, petals black-dotted only on margins; capsules oblong cone-shaped; common introduced weed .... **H. perforatum**
5　Flowers 6–10 mm wide, petals and sepals with black dots and lines; capsules nearly round to ovate; native species. .... **H. punctatum**
6　Sepals broadest near or above middle; capsule rounded at tip .... **H. boreale**
6　Sepals lance-shaped, broadest below middle; capsule tapered to tip. 7
7　Leaves 1-nerved (sometimes 3-nerved), tapered to base; sepals 2–4 mm long .... **H. canadense**
7　Leaves 5–7-nerved, rounded at base and broadest below middle; sepals 5–6 mm long .... **H. majus**

Hypericum ascyron L.
GREAT ST. JOHN'S-WORT　　　　　　　　　native
IR | WUP | CUP | EUP　　　　　　　　　FAC CC 8

Hypericum pyramidatum Ait.

Perennial herb. Stems upright, branched, 6–20 dm long. Leaves lance-shaped to oval, 4–10 cm long and 1–4 cm wide, base often clasping stem; petioles absent. Flowers few, 4–6 cm wide, mostly single on stalks from upper leaf axils; stamens numerous, joined at base into 5 bundles; petals bright yellow; styles 5, not persisting. Fruit an ovate, 5-chambered capsule, 15–30 mm long. July–Aug.

Streambanks, ditches, fen and marsh margins.

Hypericum boreale (Britt.) Bickn.
NORTHERN ST. JOHN'S-WORT　　　　　　native
IR | WUP | CUP | EUP　　　　　　　　　OBL CC 5

Perennial herb, from slender rhizomes. Stems 1–4 dm long, round or slightly 4-angled, branched above. Leaves oval or oblong, rounded at ends and nearly clasping stem, 3–5-nerved, larger leaves 1–2 cm long and 0.5–1 cm wide; petioles absent. Flowers in clusters at ends of stems and from upper leaf axils; sepals blunt-tipped; petals yellow, 3 mm long; stamens 8–15; styles 3 (sometimes 4), less than 1 mm long. Fruit a 1-chambered purple capsule, 3–5 mm long. July–Sept.

Pond and marsh margins, low areas between dunes, open bogs.

Hypericum canadense L.
LESSER CANADIAN ST. JOHN'S-WORT　　　native
IR | WUP | CUP | EUP　　　　　　　　　FACW CC 6

Annual or perennial herb, with short leafy stolons from base of plant. Stems upright, branched, 1–6 dm long. Leaves linear, 1–4 cm long and 1–4 mm wide, blunt-tipped, mostly 1-nerved, bracts much smaller; petioles absent. Flowers in open clusters at ends of stems and from upper leaf axils; sepals lance-shaped, 3–5 mm long; petals yellow, 2–3 mm long; stamens 12–22; styles 3 (sometimes 4), less than 1 mm long. Fruit a purple capsule 4–6 mm long. July–Sept.

Sandy shores, wetland margins, ditches.

Hypericum ellipticum Hook.
PALE ST. JOHN'S-WORT　　　　　　　　native
IR | WUP | CUP | EUP　　　　　　　　　OBL CC 9

Perennial herb, spreading by rhizomes. Stems 2–5 dm long, branched only in head. Leaves oval, 1–4 cm long and 1–1.5 cm wide, rounded at tip, narrowed at base and sometimes clasping stem; petioles absent. Flowers few to many, in clusters at ends of stems; sepals to 6 mm long; petals pale yellow, 6–7 mm long; stigmas 3 (sometimes 4), small. Fruit a 1-chambered capsule, 5–6 mm long, rounded to a short beak formed by the persistent styles. July–Aug.

Streambanks, sandy shores and flats, thickets, bogs.

Hypericum kalmianum L.
KALM'S ST. JOHN'S-WORT　　　　　　　native
IR | WUP | CUP | EUP　　　　　　　　　FACW CC 10

Branched shrub to 1 m tall; branches 4-angled, twigs flattened. Leaves linear, 2–4 cm long and 3–8 mm wide, often waxy on underside; margins sometimes rolled under; petioles absent. Flowers in clusters of 3–7 at ends of stems, yellow, 2–3.5 cm wide; stamens many, not joined; styles 5. Fruit a 5-chambered, ovate capsule, 7–10 mm long, beaked by the persistent style base. June–Sept.

Dunes (especially wet areas between dunes) and rocky lakeshores, mostly near Great Lakes, often on limestone or where calcium-rich.

Hypericum majus (Gray) Britt.
GREATER CANADIAN ST. JOHN'S-WORT   *native*
IR | WUP | CUP | EUP   FACW CC 4

Perennial herb, spreading from rhizomes or stolons. Stems upright, unbranched or branched above, 1–6 dm long. Leaves lance-shaped, 2–4 cm long and 3–10 mm wide, dotted with brown sunken glands, 5–7-nerved from base; leaf tip rounded, leaf base rounded or heart-shaped and weakly clasping; petioles absent. Flowers few to many in clusters at ends of stems and from upper leaf axils; sepals lance-shaped, 4–6 mm long; petals yellow, equal to sepals but then shriveling to half the length of sepals; stamens 14–21, not joined; styles to 1 mm long. Fruit a red-purple ovate capsule, 5–7 mm long. July–Sept.

Streambanks, sandy, mucky or calcareous shores, low areas between dunes, marshes, wetland margins.

Hypericum perforatum L.
COMMON ST. JOHN'S-WORT   *introduced (invasive)*
IR | WUP | CUP | EUP   UPL

Perennial herb. Stems 4–6 dm tall, with numerous very leafy decussate branches. Leaves sessile, linear-oblong, commonly 2–4 cm long, on the main axis, about half as large on the branches. Flowers numerous, forming a large rounded or flattened compound cyme; sepals narrowly lance-shaped, acuminate, 4–6 mm long, with few or no black glands; petals oblong, 8–10 mm long, black-dotted near the margin. Seeds 1–1.3 mm long. June–Sept.

Native of Europe; a common weed in the UP in fields, meadows, and roadsides.

Hypericum punctatum Lam.
SPOTTED ST. JOHN'S-WORT   *native*
IR | WUP | CUP | EUP   FAC CC 4

Perennial herb. Stems erect, 5–10 dm tall, with few branches below the inflorescence. Leaves oblong-elliptic, the larger ones commonly 4–6 cm long, and more than 1 cm wide, blunt or even retuse. Inflorescence usually small, compact and crowded. Flowers short-pediceled, 8–15 mm wide; sepals heavily dotted and lined with black, ovate or oblong, obtuse or broadly acute, 2.5–4 mm long; petals copiously dotted with black; styles 2–4 mm long, often persistent. Capsule ovoid, 4–6 mm long; seeds less than 1 mm long. June–Aug.

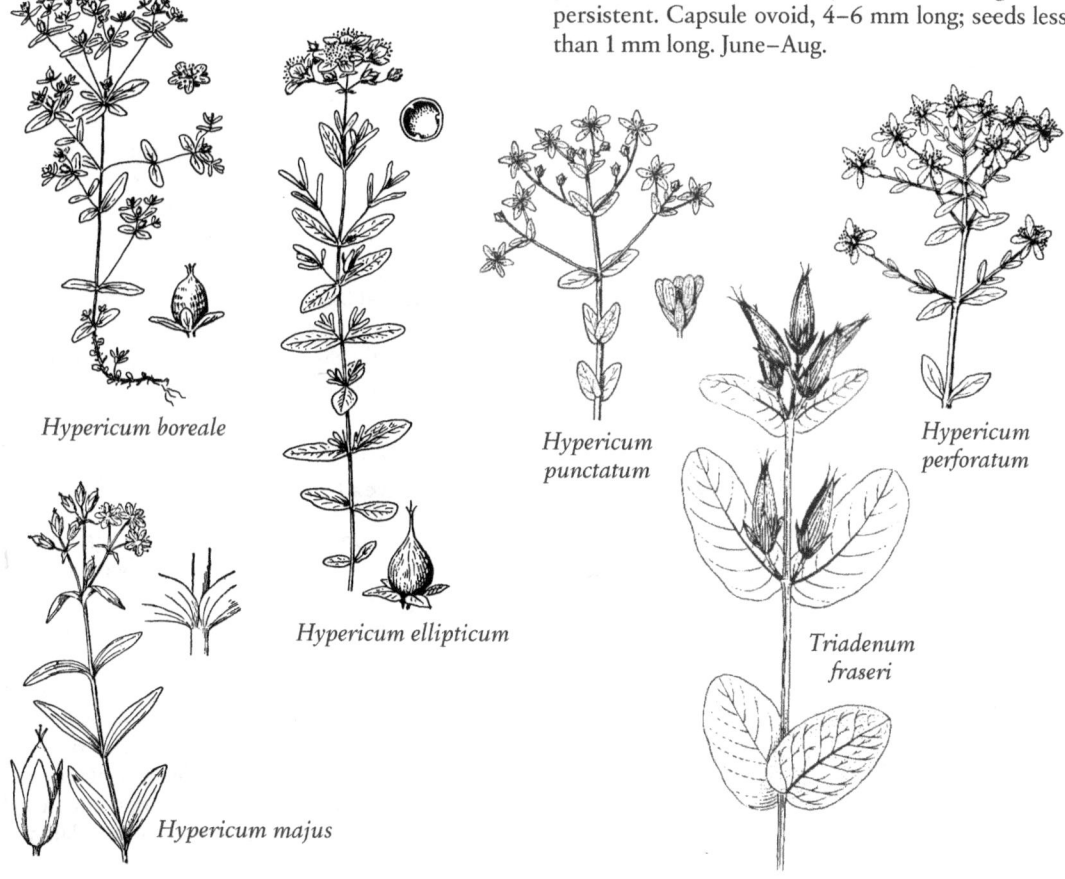

Hypericum boreale

Hypericum ellipticum

Hypericum majus

Hypericum punctatum

Hypericum perforatum

Triadenum fraseri

Moist or dry soil, fields and open woods; Delta and Menominee counties.

Triadenum *Marsh St. John's-Wort*

Triadenum fraseri (Spach) Gleason
FRASER'S MARSH-ST. JOHN'S-WORT *native*
IR | WUP | **CUP** | EUP OBL CC 6

Hypericum virginicum L. var. *fraseri* (Spach) Fern.

Glabrous Perennial herb, with creeping rhizomes. Stems upright, mostly unbranched, red, smooth, 3–6 dm long. Leaves opposite, oval or ovate, 3–6 cm long and 1–3 cm wide, pinnately veined, rounded at tip, rounded or heart-shaped and clasping at the base, with dark dots and transparent glands on underside; margins entire. Flowers in clusters at ends of stems and from leaf axils; sepals 3–5 mm long, rounded at tip; petals pink to green-purple, 5–8 mm long; stamens 9, joined at base into 3 bundles, the bundles alternating with orange glands; styles 1–2 mm long. Fruit a purple, cylindric capsule, 7–12 mm long, abruptly narrowed to the 1 mm long persistent style beak. July–Aug.

Marshes, sedge meadows, open bogs, fens, sandy and calcium-rich shores.

Juglandaceae
WALNUT FAMILY

Trees. Leaves alternate, odd-pinnate. Flowers monoecious. Staminate flowers in elongate catkins, each composed of a 2–6-lobed calyx, subtended by a narrow bract, and bearing few to many stamens on its upper side. Pistillate flowers terminating the young branches, each subtended by a perianth-like, cup-shaped involucre formed of connate bracts. Ovary 1-celled, tipped with 2 plumose stigmas. Fruit large, consisting of a fleshy or woody exocarp enclosing a nut; embryo large and oily, without endosperm.

1 Leaflets mostly 5–9, the terminal leaflet largest . **Carya**
1 Leaflets 11–23, the lateral leaflets largest . . . . . . **Juglans**

Carya *Hickory*

Trees with hard heavy wood. All species are more or less stellate-pubescent, at least when young, and leaves, buds, and fruit also copiously covered with resin when young. Leaves odd-pinnate, the 3 terminal leaflets the largest. Flowers appear in spring as the leaves open. Staminate catkins slender, elongate, borne in peduncled groups of 3 at the summit of the previous year's growth or the base of that of the current year. Staminate calyx 2- or 3-lobed. Stamens 3–10, commonly 4. Pistillate flowers solitary or in spikes of 2–10, terminating the branches, each subtended by a cup-shaped, 4-lobed, perianth-like involucre. Fruit a hard-shelled nut, enclosed within the expanded, 4-valved involucre.

1 Leaflets 7–9; bud scales sulfur-yellow; bark smooth or with shallow ridges; Menominee County. . . . . **C. cordiformis**
1 Leaflets 5; bud scales not yellow; bark shaggy; Delta and Mackinac counties. . . . . **C. ovata**

Carya cordiformis (Wangenh.) K. Koch
BITTERNUT HICKORY *native*
IR | WUP | **CUP** | EUP FACU CC 5

Tree; bark scaly. Winter-buds bright orange-yellow. Leaflets commonly 7 or 9, occasionally 5, rarely 11, the lateral lance-shaped to ovate lance-shaped, the terminal commonly long-cuneate at base and nearly or quite sessile. Involucre obovoid to subglobose, often somewhat flattened, 2.5–3.5 cm long, winged chiefly above the middle, splitting about to the middle; nut subglobose to obovoid, 1.5–3 cm long, at least two-thirds as thick, obscurely angled, otherwise smooth, tipped with a slender persistent point. Kernel bitter.

Dry or moist forests; Menominee County.

Carya ovata (P. Mill.) K. Koch
SHAGBARK HICKORY *native*
IR | WUP | **CUP** | EUP FACU CC 5

Tree; bark light gray, soon separating into long plates. Leaflets 5, or 7 on sprouts, pubescent beneath when young, soon becoming nearly or wholly glabrous, the terminal obovate, much larger and proportionately wider than the lateral. Involucre subglobose to broadly obovoid, 3.5–5 cm long, thick-walled, eventually splitting to the base; nut compressed, 2–3 cm long, rounded at base, usually sharp-pointed.

Rich moist soil; Delta and Mackinac counties.

Variable in the size and shape of the nuts and in the pubescence of the leaves.

Juglans *Walnut*

Juglans cinerea L.
BUTTERNUT; WHITE WALNUT *native*
IR | WUP | **CUP** | EUP FACU CC 5

Tree to 30 m tall; bark grayish brown, with smooth ridges; twig pith dark brown. Leaves glandular-pubescent, odd-pinnate, the median lateral leaflets the largest; leaflets commonly 11–17, oblong lance-shaped, somewhat pointed. Staminate catkins protruding from the buds in autumn, elongating in spring, densely flowered, pendulous. Calyx spreading, 3–6-lobed, with 8–40 stamens on its upper side. Pistillate flowers in short spikes terminating the branches, composed of a 3-lobed, cup-shaped involucre. Fruit clammy-

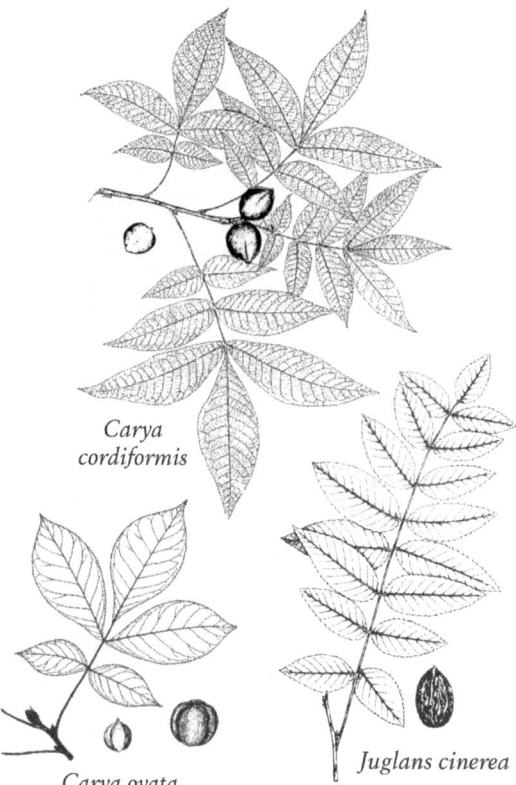

Carya cordiformis

Carya ovata

Juglans cinerea

glandular. Nut ovoid, ovoid-oblong, or short-cylindric, very rough, marked with 2 or 4 obscure longitudinal ridges, indehiscent but 2-valved.

Rich moist soil.

Lamiaceae
MINT FAMILY

Perennial, often aromatic, herbs. Stems usually 4-angled. Leaves simple, opposite, sharply toothed or deeply lobed. Flowers in leaf axils or in heads or spikes at ends of stems, perfect (with both staminate and pistillate parts), nearly regular to irregular; sepals 5-toothed or sometimes 2-lipped; petals white, pink, blue or purple, often 2-lipped; stamens 2 or 4; ovary 4-lobed, splitting into 4, 1-seeded nutlets when mature.

ADDITIONAL SPECIES

Origanum vulgare L. (Wild marjoram), introduced and rarely escaping from gardens; Marquette and Ontonagon counties.

Satureja hortensis L. (Summer savory), corolla pale pink-purple to white, 5–7 mm long; native of Mediterranean region and sw Asia, and long cultivated as a culinary herb, reported from Gogebic and Houghton counties.

Thymus pulegioides L. (Lemon thyme), introduced and cultivated in gardens; reported as an escape in several UP locations.

1. Calyx with a distinct cap or protuberance on the upper side of the tube .... **Scutellaria**
1. Calyx without a cap or protuberance on the tube 2
2. Upper lip of the corolla very short, or its lobes adjacent to the margins of the lower lip, the corolla thus appearing to be 1-lipped 3
2. Upper lip of the corolla well developed, entire or 2-lobed, or the corolla regular or nearly so 4
3. Lower lip 5-lobed, the 2 lobes nearest its base representing the upper lip .... **Teucrium canadense**
3. Lower lip 3-lobed, or appearing 4-lobed if the center lip is notched .... **Ajuga genevensis**
4. Stamens included and hidden within the corolla tube . ... **Glechoma hederacea**
4. Stamens exserted beyond the throat of the corolla .. 5
5. Stamens 2 .... **Group A**
5. Stamens 4 6
6. Inflorescence appearing axillary, the verticils (whorls of flowers around the stem) several to many, subtended by normal leaves and separated from one another by normal interodes, or the uppermost subtending leaves smaller and internodes shorter (not including plants with axillary spikes or racemes) .... **Group B**
6. Inflorescence appearing terminal, the verticils 1 to many, all or mostly subtended by bract-like leaves different from the main leaves, or separated by much shorter internodes (and including plants with lateral or axillary spikes) 7
7. Flowers single in the axils of each bract-like leaf, the verticils with 1 or 2 flowers ...... **Physostegia virginiana**
7. Flowers 2–many in the axil of each bract-like leaf, the verticils with 4 or more flowers .... **Group C**

GROUP A

1. Calyx distinctly 2-lipped 2
1. Calyx regular or nearly so, the lobes alike in size and shape 3
2. Flowers in loose, few-flowered verticils in the axils of foliage leaves, blue, 3–4 mm long .... **Hedeoma**
2. Flowers in terminal inflorescences .... **Salvia**
4. Corolla very irregular, 15–50 mm long .... **Monarda**
4. Corolla regular or nearly so, to 5 mm long .... **Lycopus**

GROUP B

1. Calyx regular or nearly so, the lobes of the upper and lower lips similar in shape and size 2
1. Calyx distinctly 2-lipped, the lobes of the upper and lower lips of different size and shape 7
2. Corolla about equally 4- or 5-lobed 3
2. Corolla strongly 2-lipped, the upper lip concave and arched over the stamens 4

3	Flowers 1–3 in each axil, and 2–6 in each verticil; Drummond Island . . . . **Trichostema brachiatum**
3	Flowers many in each axil . . . . **Mentha**
4	Flowers distinctly pediceled, forming loosely flowered cymules (the clusters making up a cyme) . . . . **Glechoma hederacea**
4	Flowers sessile in the cymules 5
5	Calyx lobes tapered to a slender tip but not spiny . . . . **Lamium**
5	Calyx lobes prolonged into short stiff spines 6
6	Lower corolla lip with 2 yellow or white protuberances at its base . . . . **Galeopsis**
6	Lower corolla lip without protuberances . . . . **Leonurus**
7	Stamens projecting beyond the corolla . . . . **Mentha**
7	Stamens ascending under upper lip of the corolla but not longer than the lip . . . . **Clinopodium**

GROUP C

1	Stamens ascending under the upper corolla lip but not longer than the lip. 2
1	At least some of the stamens protruding from the corolla 8
2	Calyx distinctly 2-lipped and irregular 3
2	Calyx regular or nearly so, the lobes all alike or differing in size only. 5
3	One calyx lobe (the upper center lobe) longer and wider than the other 4 . . . . **Dracocephalum**
3	Three calyx lobes (which form the upper lip) differing from the other 2 lobes 4
4	Bracts broadly rounded, abruptly tapered at the tip to a short sharp point. . . . . **Prunella vulgaris**
4	Bracts awl-shaped, coarsely hairy . . . . **Clinopodium**
5	Leaves linear, entire, sessile 6
5	Leaves wider than linear, or the margins toothed, or petioled 7
6	Stems finely pubescent . . . . **Clinopodium**
6	Stems glabrous, or with small hairs on the angles only . . . . **Stachys**
7	Calyx 15-nerved; lower verticils often with distinct peduncles . . . . **Nepeta cataria**
7	Calyx 5-10-nerved; lower verticils sessile . . . . . **Stachys**
8	Inflorescence a dense or loose raceme in which the component verticils are plainly visible; flowers on distinctly short pedicels . . . . **Mentha**
8	Inflorescence otherwise 9
9	Inflorescence a group of terminal heads or crowded cymes, often with secondary heads or cymes in some of the upper axils, never a spike or raceme . . . . **Pycnanthemum**
9	Inflorescence a dense spike, or with one or 2 lower verticils sometimes separate from the others; flowers sessile or nearly so . . . . **Agastache**

Agastache *Giant-Hyssop*

Agastache foeniculum (Pursh) Kuntze
BLUE GIANT-HYSSOP   *introduced*
IR | **WUP** | **CUP** | EUP

Perennial herb. Stems erect, to 1 m tall, simple or branched above. Leaves ovate, the larger to 9 cm long, rounded or truncate at base, coarsely serrate, glabrous above, beneath whitened with a very fine pubescence, the hairs scarcely visible under a 10x lens; petioles less than 1.5 cm long. Flowers small, numerous in dense verticils subtended by inconspicuous bracteal leaves; spikes solitary and terminal, or with additional spikes from short axillary branches; bracteal leaves broadly ovate. Calyx nearly regular, the tube cylindric, slightly longer on the upper side, the lobes 3-nerved, similar in size and shape, puberulent at anthesis, 5–7 mm long, its lobes blue, 1.5–2 mm long. Corolla blue, almost 1 cm long, surpassing the calyx, the upper lip shallowly 2-lobed; the lower lip 3-lobed. Stamens 4, exsert beyond the corolla lobes, the 2 lower stamens curved upward under the upper corolla-lip, the 2 upper stamens curved downward. Nutlets minutely pubescent at their tip. July–Aug.

Dry upland woods and prairies; native west of Michigan, adventive in the UP.

Ajuga *Bugle*

Ajuga genevensis L.
BLUE BUGLE   *introduced*
IR | **WUP** | **CUP** | EUP

Perennial herb, without basal stolons. Flowering stems erect, 1–3 dm tall, villous. Leaves ovate to spatulate, 2–5 crn long, the larger leaves usually sinuate-dentate; lower leaves tapering to a petiole; upper leaves narrowed or rounded to a sessile base. Flowers in whorls of 4–6 in the axils of bracteal leaves, forming a terminal leafy spike; calyx villous, 6–8 mm long, its lobes linear lance-shaped, somewhat longer than the tube, one lobe slightly shorter than the other four; corolla blue, 10–15 mm long; upper lip very short, 2–lobed; lower lip dilated immediately beyond the upper lip, the lateral lobes reaching to the middle of the median lobe; stamens 4, unequal in length, reaching to the end of the lateral lobes; ovary shallowly 4-lobed. April–June.

Native of Europe and n Asia; cultivated for ornament and escaped in lawns, gardens, and roadsides; Houghton and Marquette counties.

ADDITIONAL SPECIES
Ajuga reptans L. (Carpet bugle), introduced in Alger and Houghton counties.

Clinopodium *Wild Basil*

Annual or perennial herbs. Calyx tubular to campanulate, conspicuously 10–13-nerved, often hairy in the throat, 2-lipped or regular, the lobes subulate to triangular. Corolla tube widened toward the summit; upper lip flat or slightly concave, straight to somewhat spreading, entire; lower lip deflexed, 3-lobed. Stamens 4, ascending under the upper lip of the corolla, the upper pair distinctly shorter than the lower. Nutlets smooth.

1 Stem leaves broadly elliptic-ovate, mostly 1 cm or more wide; flowers subtended by narrow, awl-shaped bracts, their margins fringed with long hairs. . . . . . . **C. vulgare**
1 Stem leaves linear to lance-shaped, usually less than 1 cm wide; flowers not with narrow bracts. 2
2 Plants glabrous . . . . **C. arkansanum**
2 Plants pubescent . . . . **C. acinos**

Clinopodium acinos (L.) Kuntze
BASIL-THYME   *introduced*
IR | WUP | **CUP** | EUP

Acinos arvensis (Lam.) Dandy
Satureja acinos (L.) Scheele

Annual. Stems erect, 1–2 dm tall, usually branched from the base and occasionally above, finely pubescent. Leaves obovate to elliptic, 6–12 mm long, scabrous, entire or with a few low teeth, on petioles 1–2 mm long. Flowers 1–3 in each of the upper axils; pedicels 2–4 mm long, apparently arising directly from the axil, not from the end of a peduncle; calyx 5–6 mm long, swollen on the lower side, constricted at the throat, the lips equal and about 1/2 as long as the tube; lower lip cleft to the base into 2 narrow lobes; upper lip widened to the tip and almost truncate, the lobes 3 small teeth about 0.5 mm long; corolla pale purple, 7–10 mm long. June–Sept.

Native of Europe; weedy on roadsides and waste places.

Clinopodium arkansanum (Nutt.) House
LIMESTONE WILD BASIL   *native*
IR | WUP | **CUP** | EUP   FACW CC 10

Clinopodium glabrum (Nutt.) Kuntze
Satureja arkansana (Nutt.) Briq.
Satureja glabella var. *angustifolia* (Torr.) Sven.

Perennial. Stems glabrous except with a minute pubescent area at each node, from short stolons; stolons for the following year usually developed while the plant is in bloom and bearing ovate petioled leaves 3–10 mm long. Leaves linear, entire, 1–2 cm long. Flowers 2–8 at each node in the upper half of the plant, subtended by progressively reduced leaves and each flower by 2 linear bracts; pedicels 3–10 mm long; calyx glabrous, 4–6 mm long, the lips about 1/2 as long as the tube; corolla pale purple, 8–15 mm long. May–Aug.

Sandy beaches, calcareous soil.

Clinopodium vulgare L.
WILD BASIL   *native*
IR | **WUP** | **CUP** | EUP   CC 3

Satureja vulgaris (L.) Fritsch

Perennial from short stolons. Stems erect, simple or occasionally branched above, 2–6 dm tall. Leaves ovate, 2–4 cm long, entire or with a few low teeth, on petioles to 1 cm long or the upper nearly sessile. Flowers numerous in a dense, subglobose, terminal, head-like glomerule, or in vigorous plants with 1 or 2 similar glomerules in the uppermost axils, mingled with numerous hirsute bracts about as long as the calyx; calyx tubular, hirsute throughout, 9–10 mm long, the lips nearly as long as the tube; upper lip cleft about half its length, its lobes subulate above a triangular base; lower lip cleft to its base into subulate lobes; corolla pale purple, rose-purple, or pink, varying to white, 12–15 mm long.

Dry or moist upland woods.

Dracocephalum *Dragonhead*

Dracocephalum parviflorum Nutt.
AMERICAN DRAGONHEAD   *native*
IR | WUP | **CUP** | EUP   FACU CC 3

Erect perennial herb; stems simple or branched, 2–8 dm tall, finely pubescent to nearly glabrous. Leaves lance-shaped, 3–8 cm long, several-nerved from the base, sharply serrate with a few teeth. Flowers in dense glomerules aggregated into a terminal, globose or short-cylindric spike 2–10 cm long and 2–3 cm wide, sometimes also with a separate lower glomerule. Bracts lance-shaped, about equaling the calyx, the few teeth ending Calyx tubular, 2-lipped, the 4 lower lobes similar, the uppermost lobe much wider; calyx lobes nearly as long as the tube, sharply tipped, strongly 3-nerved. Corolla blue, barely longer than the calyx, weakly bilabiate, the tube elongate, gradually widened upwards, the limb much shorter; upper lip straight, 2-lobed; lower lip deflexed, 3-lobed, the median lobe notched. Stamens 4, ascending under the upper corolla-lip. Nutlets oblong, smooth. in short awns.

Native of western USA, considered adventive in Michigan.

Galeopsis *Hemp-Nettle*

Galeopsis tetrahit L.
BRITTLE-STEM HEMP-NETTLE   *introduced*
IR | WUP | **CUP** | EUP

Annual herb; stems simple or branched, 3–8 dm tall, swollen at the nodes, hispid, often densely so, with long, straight, somewhat reflexed hairs. Leaves

LAMIACEAE | Mint Family

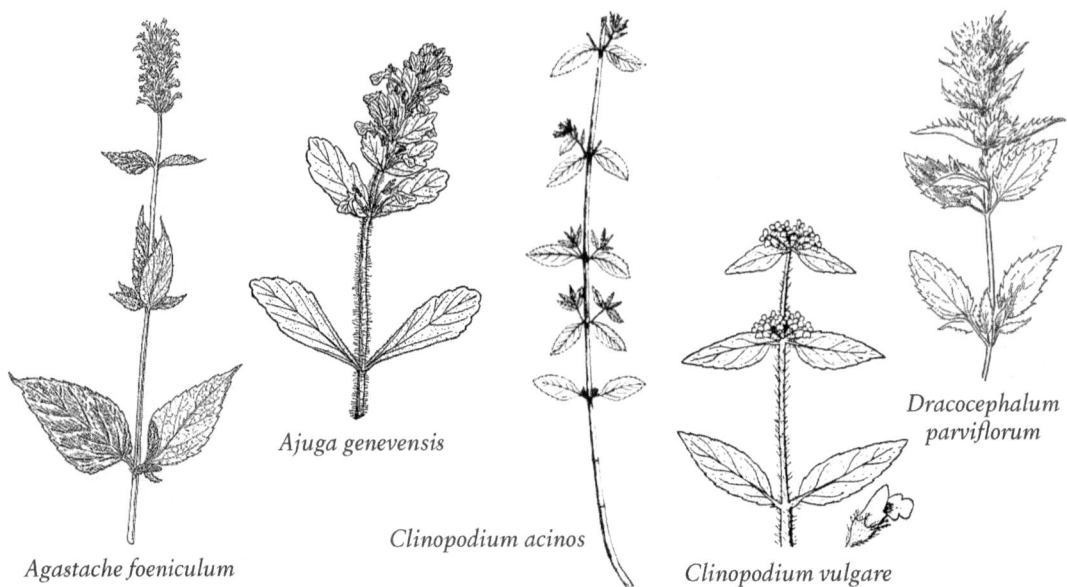

Agastache foeniculum
Ajuga genevensis
Clinopodium acinos
Clinopodium vulgare
Dracocephalum parviflorum

lance-shaped to ovate, 5–10 cm long, acuminate, crenate-serrate, pubescent on both sides, on petioles 1–3 cm long. Flowers white or pink or variegated, commonly with two yellow spots, borne in 2–6 dense verticils in the axils of the upper foliage leaves. Calyx tube broadly tubular to campanulate, with 10 conspicuous ribs and usually 10 intermediate ones; calyx lobes all equal, narrowly triangular, the strong midnerve excurrent as a prominent spine; calyx enlarged in fruit, its lobes eventually 5–10 mm long. Corolla strongly 2-lipped, the tube exceeding the calyx, the upper lip entire, erect, concave, the lower lip 3-lobed, bearing 2 protuberances at its base. Stamens 4, ascending under the upper corolla lip, the lower pair slightly the longer. Nutlets obovate, smooth, 3–4 mm long. June–Sept.

Native of Eurasia; introduced as a weed of gardens, roadsides, waste places, and forests.

Most of our plants are var. *bifida* (Boenn.) Lej. & Courtois, sometimes separated as *G. bifida* Boenn., with the middle lobe of the lower lip ± notched rather than entire and squarish.

Glechoma *Ground-Ivy*

Glechoma hederacea L.
GROUND-IVY   introduced (invasive)
IR | WUP | **CUP** | EUP   FACU
Perennial herb. Stems slender, creeping, eventually to 1 m long, villous to nearly glabrous. Leaves rotund to kidney-shaped, 1.5–4 cm wide, conspicuously crenate, long-petioled. Flowers blue, usually 3 in each axil; bractlets subulate, shorter than the calyx; calyx tubular, 5.5–9 mm long, 15-nerved, the 5 lobes triangular, about equal, about a 1/3 as long as the tube, with 3 nerves subtending each lobe, the middle one excurrent into a short awn; corolla much-surpassing the calyx, 2-lipped, upper lip shallowly 2-lobed; lower lip much larger, the lateral lobes short and rounded, the median lobe dilated; stamens 4, ascending under the upper corolla lip, and about equaling it. April–June.

Native of Eurasia; widely naturalized in yards, roadsides, cemeteries, and moist woods.

Hedeoma *False Pennyroyal*

Hedeoma hispida Pursh
ROUGH FALSE PENNYROYAL   native
IR | WUP | **CUP** | EUP   CC 3
Small, strongly scented, annual herb; stems simple or branched from the base, occasionally branched above, 5–20 cm tall, pubescent with recurved hairs. Leaves linear, 1–2 cm long, sessile, entire. Flowers small, blue, pediceled, in axillary few-flowered verticils. Calyx tubular in anthesis, strongly 13-ribbed, 2-lipped, villous in the throat, upper lip of calyx cleft to or below the middle into narrow ciliate teeth. Corolla tubular, weakly 2-lipped, the upper lip erect, the lower spreading, 3-lobed. Stamens 2, ascending under the upper corolla-lip and about equaling it. Nutlets ovoid, smooth. May–Aug.

Dry soil, sand dunes and barrens.

Lamium *Dead Nettle*

Annual or perennial herbs, commonly spreading or decumbent. Leaves broad, crenate. Flowers white to red or purple, in verticils of 6–12, subtended by scarcely reduced leaves, forming a short, crowded or somewhat interrupted, terminal spike. Calyx campanulate, almost regular, 5-nerved, the lobes nearly equal, nearly as long as to longer than the tube, triangular at base, tapering to a long, slender, but not spine-like tip. Corolla tube very slender at base, near the summit abruptly dilated, the upper lip erect, constricted at base; lateral lobes of the lower lip essentially obsolete; lowest median lobe constricted at base, as broad as long or broader, emarginate to deeply 2-lobed. Stamens 4, ascending under the upper corolla-lip, the lower pair the longer.

ADDITIONAL SPECIES

Lamium purpureum L. (Red henbit), annual introduced herb of waste places, known from Gogebic County, resembling *L. amplexicaule*.

1   Upper leaves sessile and clasping stem, lower leaves on long petioles . . . . **L. amplexicaule**
1   All leaves with petioles . . . . **L. maculatum**

Lamium amplexicaule L.
HENBIT   *introduced*
IR | WUP | CUP | **EUP**

Annual. Stems branched from the base, weak, ascending or decumbent, 1–4 dm long, bearing a few small long-petioled leaves at the base, usually with 1 or 2, rarely 3, greatly elongate internodes, above which the internodes are much shorter and subtended by sessile leaves. Leaves subrotund, 1–3 cm wide, deeply crenate. Flowers in dense verticils in the axils of the upper leaves, the uppermost verticils often adjacent; calyx 5–7 mm long, densely villous, its setaceous lobes directed forward, about equaling the tube; corolla pink to purplish, 12–18 mm long, the upper lip a fourth to a third as long as the tube. March–Nov. Native of Eurasia and n Africa; introduced as a weed in fields, gardens, and waste places, especially in moist fertile soil; Mackinac County.

Lamium maculatum L.
SPOTTED DEAD NETTLE   *introduced*
IR | WUP | **CUP** | **EUP**

Perennial. Stems erect or ascending from a decumbent base, 2–6 dm tall. Leaves all petioled, usually with a white stripe along the midvein, ovate or deltoid. Calyx 8–10 mm long, the lobes mostly shorter than to equaling the tube; corolla red-purple, sometimes white, 2–2.5 cm long, the upper lip more than 1/2 as long as the tube, densely short-pubescent, the tube with a transverse constriction near the base. April–Sept. Native of Eurasia, escaped from cultivation on roadsides and waste places; Mackinac and Schoolcraft counties.

Leonurus *Motherwort*

Leonurus cardiaca L.
MOTHERWORT   *introduced*
IR | **WUP** | **CUP** | EUP

Strongly-scented perennial herb. Stems stout, erect, to 1.5 m. tall, finely pubescent on the angles and nodes. Leaves long-petioled, the larger broadly ovate, palmately lobed and sharply toothed, the upper progressively smaller and narrower, those subtending verticils commonly oblong and merely 3-toothed. Flowers crowded in dense verticils subtended by bracteal leaves and by linear bracts, forming long, interrupted, terminal spikes. bracts subulate, rarely 1/2 as long as the calyx; calyx tube 5-angled, 5-ribbed, nearly glabrous, 3–4 mm long; calyx lobes nearly as long as the tube, the lower two somewhat the larger and strongly deflexed; corolla pale pink, the upper lip white-villous. Stamens 4, about equal, ascending under the upper lip of the corolla. Nutlets obpyramidal, 3–4-angled, truncate and pubescent at their tips. June–Aug.

Native of c Asia; formerly cultivated as a home remedy and now established in waste places, roadsides, and gardens.

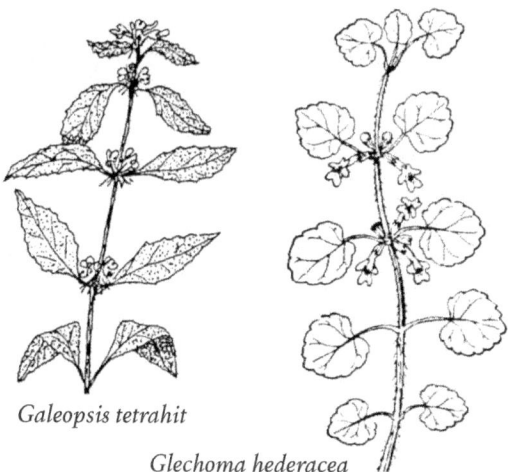

Galeopsis tetrahit

Glechoma hederacea

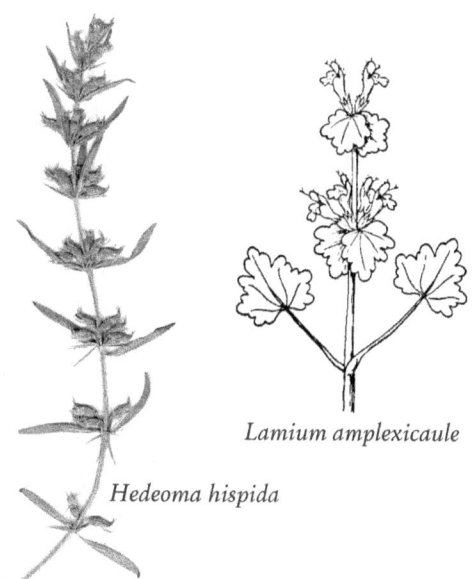

Lamium amplexicaule

Hedeoma hispida

Lycopus Water-Horehound

Perennial, unscented herbs. Stems erect, 4-angled. Leaves opposite, coarsely toothed or deeply lobed, smaller on upper stems; petioles short or absent. Flowers small, in clusters in middle and upper leaf axils, often appearing whorled; white to pink, the sepals and petals often dotted on outer surface, 4-lobed, stamens 2. Fruit a nutlet.

1 Sepal lobes slender, 1–3 mm long, longer than nutlets, the midvein prominent . . . . **L. americanus**
1 Sepal lobes broad, triangular to ovate, to 1 mm long, shorter than to about as long as nutlets, the midvein not prominent 2
2 Leaves mostly less than 3 cm wide; stamens and styles visible, longer than petals; outer rim of nutlets taller than the inner rim . . . . **L. uniflorus**
2 Larger leaves 3 cm or more wide; stamens and styles hidden by petals; inner and outer rim of nutlets same height, the 4 nutlets appearing flat-topped across tops; Ontonagon County . . . . **L. virginicus**

Lycopus americanus Muhl.
CUT-LEAF WATER-HOREHOUND native
IR | WUP | CUP | EUP OBL CC 2

Perennial herb, spreading by rhizomes, tubers absent. Stems erect, often branched, 2–8 dm long, upper stems smooth or short-hairy. Leaves opposite, lance-shaped, 3–8 cm long and 1–4 cm wide, with glandular dots, smooth or rough on upper surface, underside veins short-hairy; margins coarsely and irregularly deeply toothed or lobed, the lowest teeth largest; nearly stalkless or on short petioles. Flowers in dense, whorled clusters in leaf axils; sepal lobes narrow, sharp-tipped, 1–3 mm long, longer than fruit; petals white, sometimes pink to purple-dotted, 4-lobed, the upper lobe wider and notched. Fruit a nutlet, 1–2 mm long. July–Sept.

Marshes, wet meadows, shores, streambanks, ditches, calcareous fens, wetland margins.

Our most common water-horehound.

Lycopus uniflorus Michx.
NORTHERN WATER-HOREHOUND native
IR | WUP | CUP | EUP OBL CC 2

Perennial herb. Stems smooth or short-hairy, 1–5 dm long. Leaves opposite, lance-shaped to oblong, 3–6 cm long and 1–3 cm wide, margins with a few outward-pointing teeth, petioles short or nearly absent. Flowers in dense, whorled clusters in leaf axils; sepal lobes broad, triangular to ovate, soft, rounded at tip, to 1 mm long, shorter to as long as nutlets; petals white or pink, 2–3 mm long, 5-lobed, longer than sepals. Fruit a nutlet 1–1.5 mm long. Aug–Sept.

Swamps, streambanks, thickets, open bogs, calcareous fens, ditches; often growing with *L. americanus*.

Lycopus virginicus L.
VIRGINIA WATER-HOREHOUND THR native
IR | WUP | CUP | EUP OBL CC 8

Perennial herb, spreading by stolons (tubers usually absent). Stems 2–6 dm long, with dense covering of appressed hairs. Leaves opposite, lance-shaped to oval, 5–10 cm long and 1.5–5 cm wide, long-hairy, lower surface usually also with short, feltlike hairs; margins coarsely toothed, the lowest tooth just below middle of blade, the margin below tooth concave and petiolelike. Flowers in whorled clusters from leaf axils; sepals shorter than nutlets; petals white, 4-lobed, (upper lobe often notched). Fruit a nutlet, 1–2 mm long, the group of 4 nutlets more or less flat across tips. July–Sept.

Floodplain forests; rare in Ontonagon County.

Mentha Mint

Perennial herbs, spreading by rhizomes or stolons, with erect stems, serrate leaves, and small, blue to lavender flowers borne in the axils of the leaves or in terminal spikes or heads. Calyx regular or weakly 2-lipped, tubular to campanulate, 10–13-nerved, the lobes broadly triangular to subulate. Corolla tube slightly widened to the summit; corolla limb nearly regular and apparently 4-lobed; upper lobe, corresponding to the upper lip, usually somewhat wider than the others. Stamens 4, essentially uniform in length, straight, somewhat divergent, exsert from the corolla. Nutlets ovoid, smooth or roughened. All species bloom in summer.

ADDITIONAL SPECIES

Mentha aquatica L. (Water mint), introduced lemon-scented perennial, reported from Chippewa County.
Mentha × gracilis Sole (pro sp.), introduced hybrid between *M. arvensis* and *M. spicata*; Gogebic and Marquette counties.
Mentha × villosa Huds. (pro sp.), introduced; Marquette and Schoolcraft counties.

1 Flowers in axillary whorls sepearated by internodes of normal length . . . . **M. arvensis**
1 Flowers in terminal spikes or heads, the internodes short 2
2 Main leaves with petioles; peppermint-scented . . . . **M. × piperita**
2 Main leaves sessile or nearly so; spearmint-scented . . . . **M. spicata**

LAMIACEAE | Mint Family

Mentha arvensis L.
AMERICAN WILD MINT *native*
IR | WUP | CUP | EUP   FACW CC 3
Mentha canadensis L.

Perennial herb, strongly mint-scented, spreading by rhizomes and often also by stolons. Stems 2–8 dm long, 4-angled, hairy at least on stem angles. Leaves opposite, ovate to lance-shaped, 2–7 cm long and 0.5–3 cm wide, smooth or hairy; margins with sharp, forward-pointing teeth; petioles short. Flowers small, white or light pink to lavender, hairy, crowded in whorled clusters in middle and upper leaf axils; sepals 2–3 mm long, hairy and glandular; petals more or less regular to slightly 2-lipped, 4–6 mm long, glandular on outside, 4- or 5-lobed; stamens and style longer than petals. Fruit a smooth nutlet to 1 mm long, enclosed by the persistent sepals. July–Sept.

Wet meadows, marshes, swamps, thickets, streambanks, ditches, springs and other wet places; our only native *Mentha*.

Mentha × piperita L.
PEPPERMINT *introduced*
IR | **WUP** | **CUP** | **EUP**   OBL

Perennial herb. Stems erect, glabrous or very nearly so, to 1 m. tall. Leaves lance-shaped or oblong lance-shaped, 4–8 cm long, sharply serrate, obtuse or rounded at base, glabrous; petioles of the principal leaves 4–15 mm long, those on the branches much shorter. Spikes 1 to several, terminating the stem and the upper lateral branches, 2–8 cm long, continuous, about 1 cm wide (excluding the corollas); calyx tubular, 3–4 mm long, the tube glabrous, the lance-subulate lobes glabrous or sparsely pilose.

Considered to have originated by hybridization between *M. aquatica* and *M. spicata*; resembling the former in its large tubular calyx and blunt spikes, the latter in its elongate spikes and narrow leaves.

Of European origin; cultivated as an herb and commercially for its oil; escaped in wet soil.

Mentha spicata L.
SPEARMINT *introduced*
IR | **WUP** | CUP | **EUP**   FACW

Perennial herb. Stems erect, to 5 dm tall, nearly or quite glabrous. Leaves sessile or subsessile, oblong lance-shaped, 2–6 cm long, sharply serrate, rounded or obtuse at base, nearly or quite glabrous. Spikes several, terminating the stem and short branches from the upper axils, 3–12 cm long, about 6 mm wide (excluding the corollas), continuous or somewhat interrupted, often tapering; calyx campanulate, 1.5–2 mm long, the tube glabrous, the lobes more or less pilose.

Native of Europe; commonly cultivated as a flavoring herb and an occasional escape in the UP.

Monarda *Beebalm*

Erect perennial herbs (ours). Leaves lance-shaped to ovate, sessile or petiolate. Flowers conspicuous, densely aggregated into head-like clusters terminating the branches or also borne in the upper axils, subtended by foliaceous bracts and with linear bractlets. Calyx tubular, 13–15-nerved, regular, the 5 lobes alike or nearly so, much shorter than the tube. Corolla strongly bilabiate, the upper lip narrow, entire, straight or curved, the lower somewhat broader, spreading or deflexed, 3-lobed or with a central projecting tooth. Stamens 2, ascending under the upper corolla lip. Nutlets oblong, smooth.

1   Flowers yellowish, dotted with purple; stamens and style not exserted. . . . . **M. punctata**
1   Flowers lavender, white or scarlet; stamens and style strongly exserted beyond corolla; heads 2 or more and forming an interrupted spike 2
2   Corolla lavender (rarely white). . . . . **M. fistulosa**
2   Corolla bright scarlet; Marquette County . . **M. didyma**

Monarda didyma L.
SCARLET BEEBALM; OSWEGO TEA *native*
IR | WUP | **CUP** | EUP   FACU CC 9

Perennial. Stems 7–15 dm tall, simple or branched, glabrous or sparsely pilose, especially at the nodes. Leaves thin, ovate, deltoid-ovate, or nearly lance-shaped, 7–15 cm long, 2.5–6 cm wide, acuminate, serrate, broadly acute to commonly rounded at base, on petioles 1–4 cm long. Heads 2–4 cm wide (excluding the corollas); bracteal leaves lance-shaped, longer than the calyx, usually tinged with red; calyx 10–14 mm, long, glabrous to minutely puberulent, nearly or quite glabrous in the throat; calyx lobes 1–2 mm long, subulate above a triangular base; corolla bright crimson, 3–4.5 cm long, the upper lip about half as long as the tube. July–Sept.

Moist woods and thickets; uncommon in Marquette County; sometimes cultivated for ornament.

Monarda fistulosa L.
WILD BERGAMOT *native*
IR | **WUP** | **CUP** | **EUP**   FACU CC 2

Perennial. Stems erect, often branched, 5–12 dm tall, usually pubescent, at least above, rarely glabrous. Leaves 6–10 cm long, commonly deltoid-lance-shaped, varying to lance-shaped or rarely ovate, more or less serrate; rounded, truncate, or broadly acute at base, pubescent or essentially glabrous above or canescent beneath, on petioles 1–1.5 cm long. Heads 1.5–3 cm wide (excluding the corollas); bracteal leaves lance-shaped or ovate; calyx 7–10 mm long, puberulent, its throat densely hirsute within; calyx lobes subulate, 1–2 mm long; corolla pale lavender, 2–3 cm long, the upper lip densely

218 LAMIACEAE | Mint Family

dicots

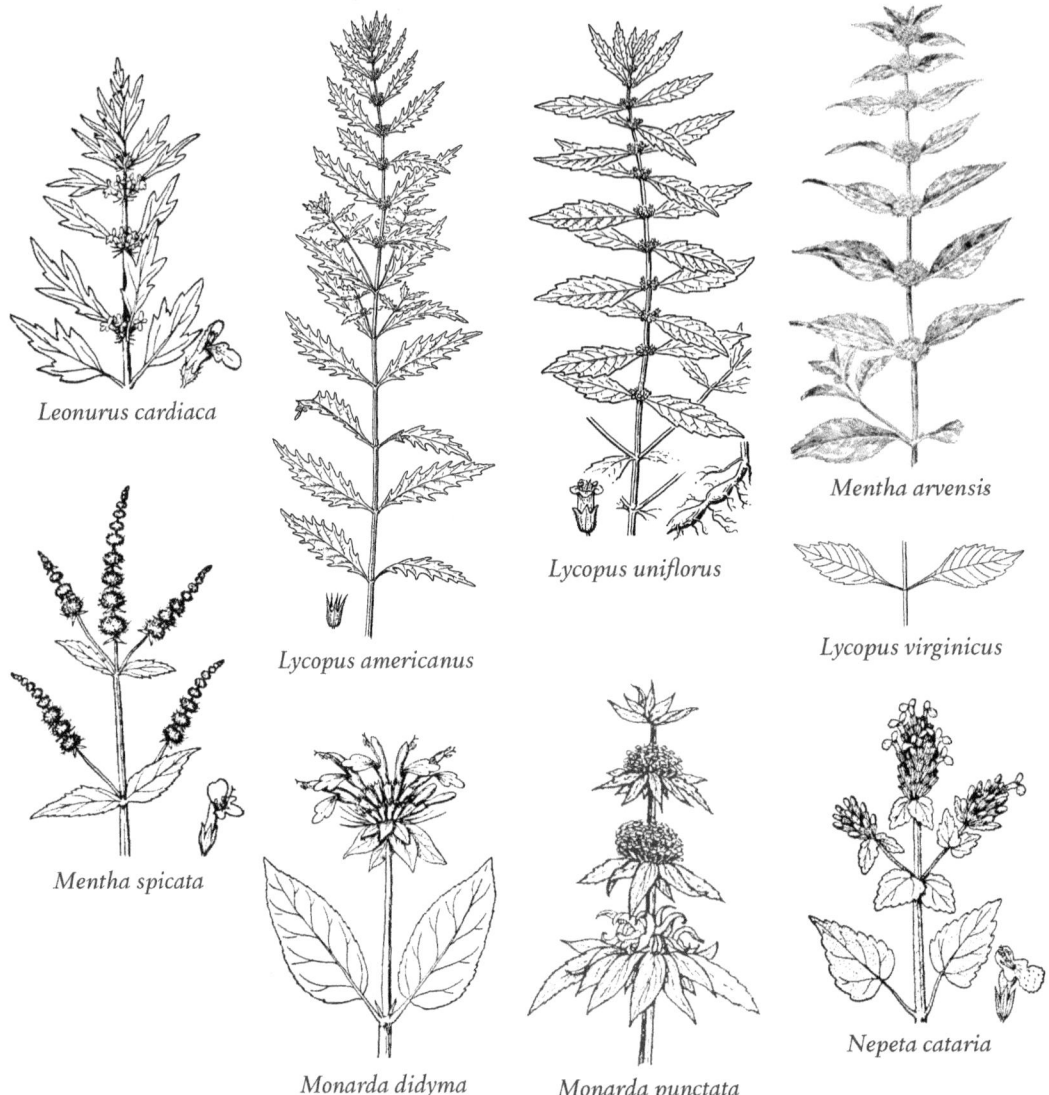

villous at the summit. June–Sept.
Upland woods, thickets, and prairies.

Monarda punctata L.
HORSE-MINT *native*
IR | **WUP** | **CUP** | **EUP** CC 4
Perennial. Stems simple or branched, 3–10 dm tall, thinly canescent. Leaves lance-shaped or narrowly oblong, 2–8 cm long, more or less pubescent. Glomerules 2–5, or solitary on depauperate plants, the bracteal leaves lance-shaped to ovate, much exceeding the calyx, spreading or reflexed, often pale green to nearly white, or tinged with purple; calyx 5–9 mm long, densely villous in the throat, more or less villous externally at the summit, its lobes 1–1.5 mm long; corolla pale yellow, spotted with purple, the upper lip arched. June–Sept.

Sand dunes, sandy fields and prairies, open oak and pine woods; roadsides, railroads, and disturbed places.

Nepeta *Catnip*

Nepeta cataria L.
CATNIP *introduced*
IR | **WUP** | **CUP** | **EUP** FACU
Perennial herb; stems, undersides of leaves, and inflorescences covered with grayish hairs. Stems erect, much branched, to 1 m tall. Leaves narrowly to broadly deltoid, 3–8 cm long, truncate or subcordate at base, coarsely crenate-dentate, on petioles about 1/2 as long as the blade. Flower clusters continuous or interrupted, 2–6 cm long, rather loosely many-flowered; calyx tubular, weakly 2-lipped, at anthesis

about 7 mm long, its lobes about 1/2 as long as the tube; upper 3 lobes each 2–3-nerved, somewhat longer and wider than the 1-nerved lower lobes; corolla 10–12 mm long, dull white, the lower lobe dotted with pink or purple; stamens 4, ascending under the upper corolla lip and nearly equaling it, the upper pair slightly longer. July–Oct.

Native of se Europe and sw Asia, formerly cultivated for reputed medicinal properties; now established in waste places, fencerows, and roadsides.

Physostegia *False Dragonhead*

Physostegia virginiana (L.) Benth.
OBEDIENCE *native*
IR | **WUP** | **CUP** | **EUP**   FACW CC 8

Perennial herb, spreading by rhizomes. Stems erect, 5–15 dm long, often branched near top, 4-angled. Leaves opposite, oval to oblong lance-shaped, 2–15 cm long and 1–4 cm wide, sometimes smaller upward; margins with sharp teeth; sessile, not clasping. Flowers in several racemes 5–20 cm long, the stalks short-hairy; sepals 4–8 mm long, often with some gland-tipped hairs; petals pink-purple or white with purple spots, 1.5–3 cm long, short-hairy to smooth. Fruit a nutlet, 2–3 mm long. July–Sept.

Sedge meadows, low prairie, shores, swamps, floodplain forests, thickets and ditches. Sometimes cultivated for its attractive flowers.

Prunella *Self-Heal*

Prunella vulgaris L.
SELF-HEAL *native-introduced*
IR | **WUP** | **CUP** | **EUP**   FAC CC 0

Perennial herb. Stems upright or sometimes spreading, 1–5 dm long, 4-angled. Leaves opposite, lance-shaped to oval or ovate, 2–8 cm long and 1–4 cm wide; lower leaves wider than upper; margins entire or with a few small teeth; petioles present. Flowers in dense spikes 2–5 cm long and 1–2 cm wide, with obvious bracts; sepals to 1 cm long, green or purple, with spine-tipped teeth; corolla 2-lipped, the upper lip hoodlike and entire, lower lip shorter and 3-lobed; petals blue-violet (rarely pink or white), 1–2 cm long; stamens 4, about as long as petals. Fruit a smooth nutlet. Subsp. vulgaris, introduced from Europe and found in mostly disturbed places, has broad leaves half as wide as long. The native subsp. *lanceolata* has narrower leaves, 1/3 as wide as long. June–Oct.

Common in many types of wetlands (especially where disturbed): swamps, wet forest depressions, wet trails, streambanks; also in drier forests, fields and lawns.

Pycnanthemum *Mountain-Mint*

Erect herbs, perennial from rhizomes, simple to the inflorescence or branched. Leaves linear to ovate, sessile or petioled, entire to serrate. Flowers small, in crowded or head-like cymes terminating the stem and its branches, or also sessile or peduncled in the axils of the upper leaves. Calyx tubular, 10–13-nerved, regular or more or less 2-lipped; calyx lobes erect, triangular, commonly shorter than the tube. Corolla 2-lipped, only slightly irregular, purple to white, the lower lip commonly spotted with purple; upper lip entire, lower lip 3-lobed. Stamens 4, exsert. Nutlets smooth, or pubescent at the summit. Our species all bloom in summer.

1. Stems and leaves glabrous; Gogebic County.......... .... **P. tenuifolium**
1. Stems pubescent on the angles, leaves rough-to-touch; Menominee County.................. **P. virginianum**

Pycnanthemum tenuifolium Schrad.
NARROW-LEAF MOUNTAIN-MINT *native*
IR | **WUP** | CUP | EUP   FAC CC 6

Perennial barely scented herb. Stems 5–8 dm tall, glabrous, very leafy by the production of numerous short axillary branches. Leaves linear, entire, glabrous, those of the central axis 2–5 cm long, 2–4 mm wide, those of the axillary branches smaller; lateral veins 1 or 2, rarely 3, on each side of the midvein, all arising in the basal 1/4 of the leaf. Heads numerous, dense, hemispheric, 3–8 mm wide, on peduncles 3–15 mm long. Outer bracts lance-shaped, glabrous, sometimes exceeding the heads; inner bracts closely appressed, about as long as the calyx, the conspicuous midvein prolonged into a awl-like point. Calyx lobes narrowly triangular, puberulent, usually 1–1.5 mm long. Dry soil of upland woods and prairies; Gogebic County.

Pycnanthemum virginianum
(L.) T. Dur. & B.D. Jackson
VIRGINIA MOUNTAIN-MINT *native*
IR | WUP | **CUP** | EUP   FACW CC 5

Perennial, strongly scented herb. Stems to 1 m long, branched above, 4-angled, angles short-hairy. Leaves numerous, opposite, narrowly lance-shaped, 3–4 cm long and to 1 cm wide (leaves in heads much smaller), upper surface smooth, with 3–4 pairs of lateral veins, undersides often finely hairy on midvein; margins entire but fringed with short, rough hairs; more or less sessile. Flowers small, 2-lipped, in branched, crowded clusters at ends of stems and branches from upper leaf axils; sepals short woolly hairy; petals white, purple-spotted. Fruit a 4-parted nutlet. July–Sept.

Wet meadows, marshes, tamarack swamps, calcareous fens, low prairie; Menominee County.

Salvia *Sage*

Salvia pratensis L.
PRAIRIE-MEADOW SAGE *introduced*
IR | WUP | **CUP** | EUP

Perennial herb. Stems 3–6 dm tall. Leaves mostly basal, ovate-oblong, 7–12 cm long, irregularly serrate or crenate, long-petioled; stem leaves few, smaller. Flowers in verticils subtended by usually much reduced bracteal leaves, forming a terminal, interrupted, spike-like raceme 1–2 dm long; flowers 4–8 at each node, subtended by broadly ovate bracteal leaves; calyx tubular to campanulate, 2-lipped, villous, the upper lip broadly ovate, minutely 3-toothed, about 2/3 as long as the tube and much shorter than the lobes of the lower lip; corolla blue, 1.5–2 cm long, the upper lip arched into a half-circle; stamens 2, ascending under the upper lip. June–Aug.

Native of s and c Europe; occasionally weedy in fields; Delta County.

Scutellaria *Skullcap*

Perennial herbs, spreading by rhizomes. Stems erect or spreading, 4-angled. Leaves opposite, ovate to lance-shaped, margins toothed, petioled or nearly sessile. Flowers blue or blue with white markings, single on short stalks in axils of middle and upper leaves, or in racemes from leaf axils; calyx 2-lipped, with a rounded bump on upper side; corolla 2-lipped, pubescent on outer surface, upper lip hoodlike, lower lip more or less flat, 3-lobed; stamens 4, ascending into the upper corolla lip. Fruit a 4-parted nutlet.

ADDITIONAL SPECIES
Scutellaria x churchilliana Fern. (pro sp.), native; hybrid between *S. galericulata* and *S. lateriflora*; Marquette and Gogebic counties.

1	Flowers in racemes from leaf axils. . . . . . . **S. lateriflora**	
1	Flowers single in leaf axils. 2	
2	Corolla 15–20 mm long; leaves 2 or more times longer than wide; common. . . . . **S. galericulata**	
2	Corolla to 10 mm long; leaves less than 2 times longer than wide; Chippewa and Menominee counties. . . . . **S. parvula**	

Scutellaria galericulata L.
HOODED SKULLCAP *native*
IR | WUP | **CUP** | EUP OBL CC 5

Scutellaria epilobiifolia A. Hamilton
Perennial. Stems erect or spreading, 2–8 dm long, unbranched or branched, 4-angled, short-hairy at least on angles of upper stem. Leaves opposite, lance-shaped to narrowly ovate, 2–6 cm long and 0.5–2.5 cm wide, upper surface smooth, underside short-hairy; margins with low, rounded, forward-pointing teeth; petioles very short. Flowers 2-lipped, single in leaf axils (and paired at nodes), on stalks 1–3 mm long; sepals 3–6 mm long; petals blue, marked with white, 15–25 mm long. Fruit a nutlet. June–Sept.

Common on shores, streambanks, marshes, wet meadows, swamps, thickets, bogs, ditches.

Scutellaria lateriflora L.
BLUE SKULLCAP *native*
IR | WUP | **CUP** | EUP OBL CC 5

Perennial. Stems 2–6 dm long, usually branched, 4-angled, short-hairy on upper stem angles or smooth. Leaves opposite, ovate to lance-shaped, 3–8 cm long and 1.5–5 cm wide, smooth; margins coarsely toothed; petioles 0.5–2 cm long. Flowers 2-lipped, in elongate racemes from leaf axils; sepals 2–4 mm long; petals blue (rarely pink or white), 5–8 mm long. Fruit a nutlet. July–Sept.

Common on shores, streambanks, wet meadows, marshes, swamps, shaded wet areas.

Scutellaria parvula Michx.
LITTLE SKULLCAP (THR) *native*
IR | WUP | **CUP** | EUP FACU CC 9

Small perennial. Stems erect, often several from the end of a rhizome, 1–2 dm tall, pubescent, usually densely so, with spreading glandular hairs, puberulent on the angles also with minute recurved eglandular hairs. Principal stem leaves sessile, ovate, usually 10–15 mm long, distinctly hirsute on the whole upper surface; lateral veins 3–5 on each side of the midvein, not anastomosing or scarcely so. Flowers axillary, blue, 7–9 mm long, the short pedicels pubescent, the hairs of the pedicels spreading; calyx glandular-pubescent. May–June.

Upland woods, dry prairies, sandstone bluffs; rare in Chippewa and Menominee counties.

Stachys *Hedge-Nettle*

Erect perennial herbs, spreading by rhizomes; plants usually hairy. Stems 4-angled. Leaves opposite, margins entire or toothed, stalkless or with short petioles. Flowers in interrupted spikes at ends of stems, appearing whorled in more or less evenly spaced clusters; sepals more or less regular, with 5 equal teeth; corolla 2-lipped, petals pink, often with purple spots or mottles, upper lip concave, entire, lower lip spreading, 3-lobed; stamens 4, ascending under the upper lip. Fruit a dark brown, 4-lobed nutlet, loosely enclosed by the persistent sepals.

1	Plants glabrous; leaf petioles 8–25 mm long. . . . . **S. tenuifolia**
1	Plants pubescent, at least on the stem angles; leaves sessile or with petioles to 10 mm long. 2

2 Leaves ± linear, entire to lightly toothed, glabrous, sessile or subsessile . . . . **S. hyssopifolia**
2 Leaves lanceolate to ovate, regularly serrate, usually at least sparsely pubescent (at least on underside), petioles to 10 mm long . . . . **S. pilosa**

Stachys hyssopifolia Michx.
HYSSOP HEDGE-NETTLE   native
IR | WUP | **CUP** | EUP   FACW CC 10

Perennial herb. Stems 3–5 dm long, often branched from base, 4-angled, smooth, or sometimes with fine hairs at nodes and on stem angles. Leaves opposite, smooth, linear, 2–6 cm long and 3–10 mm wide, uppermost leaves reduced to short bracts; margins entire or with a few low teeth; ± stalkless. Flowers in spaced, several-flowered clusters at ends of stems and on branches from leaf axils; sepals smooth or with a few hairs; petals light purple, smooth. Fruit a nutlet. July–Sept.

Sandy shores, wet depressions, meadows, sometimes where marly or peaty; Delta and Schoolcraft counties.

Stachys pilosa Nutt.
HEDGE-NETTLE   native
IR | **WUP** | **CUP** | EUP   FACW CC 5

Perennial herb Stems erect, rarely branched, 5–10 dm tall, villous on the sides and angles; hairs of the stem widely spreading. Leaves lance-shaped to ovate, 5–10 cm long, 2–4 cm wide, softly pubescent on both sides, usually with short fine hairs and longer bristles mingled, sharply serrate, sessile or nearly so. Verticils usually 6-flowered, the subtending leaves narrowly lance-shaped; calyx tube densely glandular-pubescent and also hirsute, the tube 3.5–5 mm long; calyx lobes glandular-hirsute, narrowly triangular, nearly as long as the tube, tapering to a stiff subulate tip. July–Aug.

Damp ground, ditch banks, beaches, and wet prairies.

Stachys tenuifolia Willd.
SMOOTH HEDGE-NETTLE   native
IR | **WUP** | **CUP** | EUP   FACW CC -

Perennial herb. Stems 4–10 dm long, 4-angled, smooth, or with downward-pointing, bristly hairs on stem angles. Leaves opposite, lance-shaped to ovate, 6–14 cm long and 2–6 cm wide, more or less smooth; margins with sharp, forward-pointing teeth; petioles slender, 1–2 cm long or absent. Flowers in interrupted spikes at ends of stems or also in upper leaf axils; sepals 5–7 mm long, glabrous; petals pale red to purple, 1.5–2.5 cm long. Fruit a nutlet. July–Sept.

Floodplain forests, shores, streambanks, thickets, wet meadows; Gogebic, Marquette, and Onotnagon counties.

Teucrium *Germander*
Teucrium canadense L.
AMERICAN GERMANDER   native
IR | WUP | **CUP** | **EUP**   FACW CC 4

Perennial herb, spreading by rhizomes. Stems 3–10 dm long, mostly unbranched, 4-angled, long-hairy. Leaves opposite, lance-shaped or oblong, 4–12 cm long and 1.5–5 cm wide, upper surface smooth or sparsely hairy, underside with dense, matted hairs, margins irregularly finely toothed, petioles 5–15 mm long. Flowers in a dense spike-like raceme, 5–20 cm long; bracts present and narrowly lance-shaped; pedicels 1–3 mm long; sepals more or less regular, purple or green, 4–7 mm long, covered with long silky hairs and very short glandular hairs; corolla irregular, 10–16 mm long, with short gland-tipped hairs, upper lip absent, lower lip large; petals pink to purple; stamens 4, arched over the corolla. Fruit a golden nutlet. July–Sept.

Marshes, wet meadows, shores, streambanks, thickets, floodplain forests, ditches.

Trichostema *False Pennyroyal*
Trichostema brachiatum L.
FALSE PENNYROYAL   (THR) native
IR | WUP | CUP | **EUP**   CC 8

Isanthus brachiatus (L.) B.S.P.

Annual herb with lemon-like scent. Stems erect, much branched, 2–4 dm tall, finely puberulent, becoming glandular in the inflorescence. Leaves acute, narrowed to the base, short-petioled. Flowers 1–3 from the axils of the scarcely reduced nearly linear bracteal leaves, forming a leafy panicle; calyx campanulate, nearly regular, the lobes narrowly triangular, exceeding the tube, at anthesis 2–3 mm, in fruit 3–5 mm long; corolla blue, the tube included in the calyx, upper 4 corolla lobes spreading or ascending, the lower lobe deflexed, usually slightly longer; stamens 4, somewhat exsert from the corolla tube. Nutlets obovoid, 2.5–3 mm long, prominently reticulate. Aug–Sept.

Limestone habitats, rarely in dry sandy fields; Chippewa County (Drummond Island only).

Lentibulariaceae
BLADDERWORT FAMILY

Insectivorous herbs. Leaves in a basal rosette (*Pinguicula*), or floating, or in peat, muck, or wet soil (*Utricularia*). Flowers perfect (with both staminate and pistillate parts), irregular, 2-lipped, sometimes with a spur, 1 to several on an erect stem; stamens 2. Fruit a capsule.

222 LAMIACEAE | Mint Family

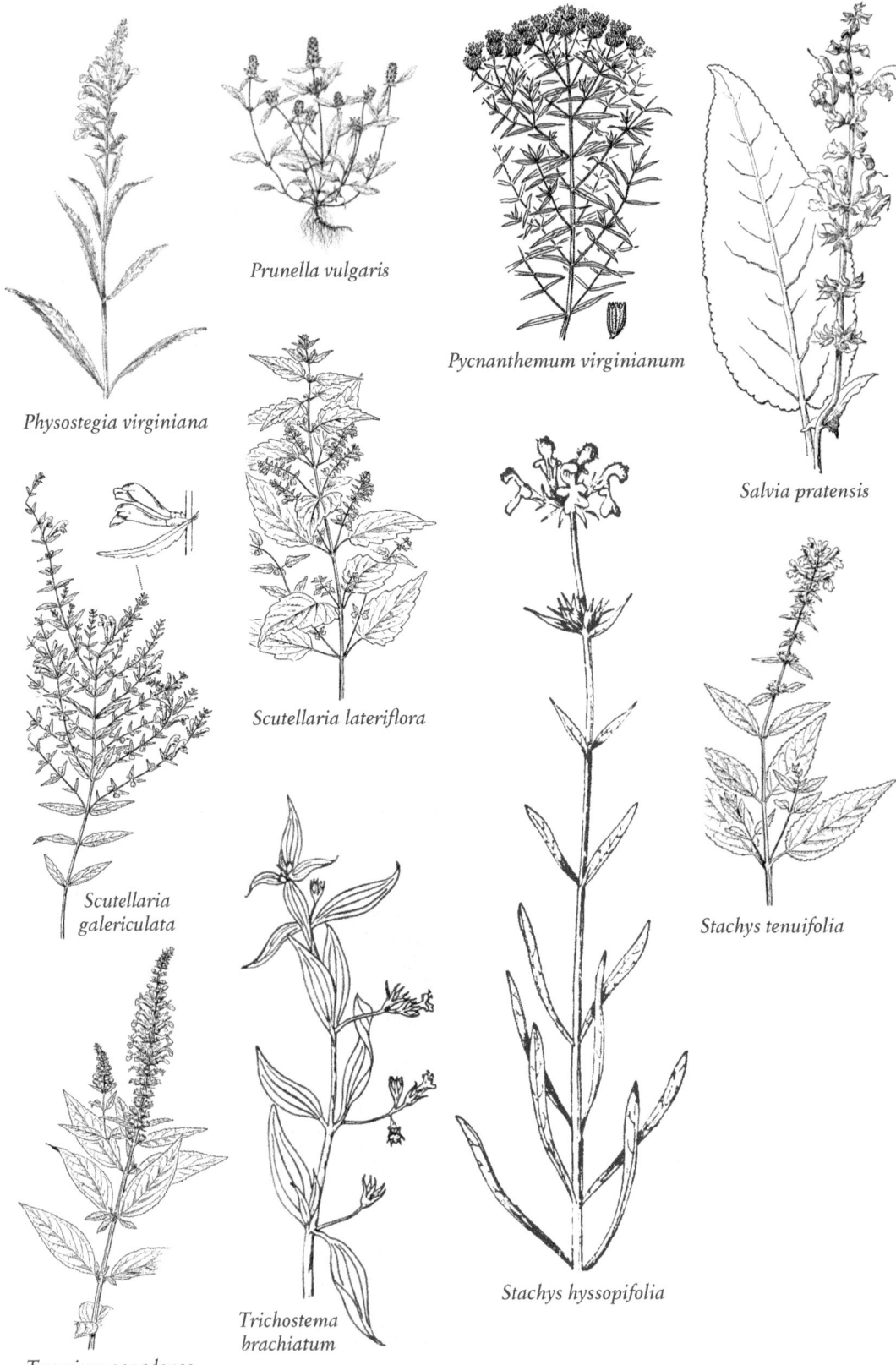

1. Leaves ovate or oval, in a basal rosette; flowers single on a bractless stalk . . . . **Pinguicula vulgaris**
1. Leaves linear or dissected into narrow segments; flowers 1, or several in a raceme, each flower subtended by a bract . . . . **Utricularia**

Pinguicula *Butterwort*

Pinguicula vulgaris L.
COMMON BUTTERWORT   native
IR | WUP | CUP | EUP   OBL CC 10

Perennial herb. Leaves 3–6 in a basal rosette, ovate or oval, 2–5 cm long, blunt-tipped, narrowed to base, upper surface sticky; margins inrolled. Flowers single atop a leafless stalk (scape) 5–15 cm long; corolla violet-purple, spurred, 2-lipped, the upper lip 3-lobed, the lower lip 2-lobed, 1.5–2 cm long (including spur). Fruit a 2-chambered capsule. June–July.

Mostly in rock crevices in cool sandstone cliffs along Lake Superior; usually occurring with Mistassini primrose (*Primula mistassinica*).

Small insects are trapped by the sticky leaf surface.

Utricularia *Bladderwort*

Mostly aquatic, annual or perennial herbs. Leaves underwater, alternate, entire or dissected into many linear segments, some with bladders which trap tiny aquatic invertebrates; or leaves in wet soil and rootlike or absent. Flowers perfect, irregular, 1 to several in a raceme atop stalks raised above water or soil surface, each flower subtended by a small bract; corolla yellow or purple, similar to a snapdragon flower, 2-lipped, the upper lip erect, entire or slightly 2-lobed, lower lip entire or 3-lobed, the corolla tube extended backward into a sac or spur, stamens 2. Fruit a many-seeded capsule.

1. Flowers purple or pink 2
1. Flowers yellow 3
2. Flowers 2–5 atop a stout stalk; plants floating in water, masses of leaves present . . . . **U. purpurea**
2. Flowers one atop a slender stalk; plants not free-floating, rooted in muck, appearing leafless . **U. resupinata**
3. Scapes appearing leafless; leaves simple or absent; plants of peat or moist sand or marl . . . . **U. cornuta**
3. Scapes with leaves at base, the leaves dissected and with bladderlike traps; plants mostly floating in water. 4
4. Leaf divisions flat in cross-section 5
4. Leaf divisions round in cross-section or threadlike . . . 6
5. Bladders borne on leaves; smallest leaf divisions entire (visible with a 10x hand lens); flower with a sac or spur much shorter than lower lip . . . . **U. minor**
5. Bladders on branches separate from leaves; smallest leaf divisions finely toothed, the teeth spine-tipped; flower with a spur as long as lower lip . . **U. intermedia**
6. Plants large; leaves floating; scapes 1 mm or more wide; flowers 13 mm or more long, 5 or more per head; larger bladders more than 2 mm wide . . . . **U. macrorhiza**
6. Plants smaller; leaves floating or creeping; scapes threadlike; flowers to 12 mm long, 1–3 per head; larger bladders mostly less than 2 mm wide 7
7. Plants forming tangled masses, creeping on bottom in shallow water, or on muck or drying pond edges; often with emergent scapes with at least 1 normal flower; cleistogamous flowers absent . . . . **U. gibba**
7. Plants forming a delicate mass of floating leaves; emergent scapes with normal flowers rare; cleistogamous flowers common, on stalks 4–8 mm long . . . . **U. geminiscapa**

Utricularia cornuta Michx.
HORNED BLADDERWORT   native
IR | WUP | CUP | EUP   OBL CC 10

Annual or perennial herb. Stems and leaves underground, roots with tiny bladders. Flowers yellow, with a downward-pointing spur 6–15 mm long, on stalks 1–2 mm long, 1–6 atop an erect stalk 10–25 cm long; bracts ovate, 1–2 mm long. Fruit a rounded capsule. June–Sept.

Acid lakes, shores, peatlands, calcareous pools between dunes, borrow pits.

Utricularia geminiscapa Benj.
HIDDEN-FRUIT BLADDERWORT   native
IR | WUP | CUP | EUP   OBL CC 8

Annual or perennial herb, similar to *U. macrorhiza* but smaller. Stems floating below water surface, sparsely branched. Leaves alternate, 1–2 cm long, branched into 4–7 segments and without bladders, or unbranched with bladders. Flowers yellow, 2–5 atop a slender stalk, 5–15 cm long, bracts below flowers 2–3 mm long; individual flower stalks 4–8 mm long, these arched when plants fruiting; cleistogamous flowers without petals more commonly produced, these single on leafless stalks 5–15 mm long along stems and often 1 at base of scape. July–Aug.

Acid lakes, pools in open bogs.

Utricularia gibba L.
CREEPING BLADDERWORT   native
IR | WUP | CUP | EUP   OBL CC 8

Annual or perennial herb. Stems creeping on bottom in shallow water, mostly less than 10 cm long, radiating from base of flower stalk (scape) and forming mats. Leaves alternate, scattered, to 5 mm long, 1–2-forked into threadlike segments; bladders present. Flowers 1–3, yellow, 5–6 mm long, with a thick, blunt spur shorter than lower lip, atop a single stalk 5–10 cm long. Fruit a rounded capsule. July–Sept.

Exposed shores, lakes, ponds, marshes, fens.

Utricularia intermedia Hayne
FLAT-LEAF BLADDERWORT   native
IR | WUP | CUP | EUP   OBL CC 10

Annual herb. Stems very slender, creeping along bottom in shallow water. Leaves alternate, 0.5–2 cm long, mostly 3-parted near base, then again divided 1–3x, the segments linear and flat, margins with small, bristly teeth; bladders 2–4 mm wide, borne on branches separate from leaves. Flowers yellow, 2–4 atop an emergent stalk 5–20 cm long; individual flower stalks to 15 mm long, remaining erect in fruit; spur nearly as long as lower lip. Fruit a capsule. June–Aug.

Shallow water (usually alkaline), marly pools between dunes, calcareous fens, marshes, ponds and rivers. bogs and swamps.

Utricularia macrorhiza Le Conte
GREATER BLADDERWORT   native
IR | WUP | CUP | EUP   OBL CC 6

Utricularia vulgaris L.

Perennial herb. Stems floating below water surface, sparsely branched, often forming large mats. Leaves alternate, 1–5 cm long, 2-forked at base and repeatedly 2-forked into segments of unequal length, the segments more or less round in section, becoming smaller with each branching, the final segments threadlike; bladders 1–4 mm wide, borne on leaf segments. Flowers yellow, 6–20 atop a stout stalk 6–25 cm long; lower flower lip 1–2 cm long, sometimes much smaller on late-season flowers, upper lip more or less equal to lower lip; spur about 2/3 as long as lower lip; stalks bearing individual flowers curved downward in fruit. Fruit a capsule. June–Aug.

Shallow water of lakes, ponds, peatlands, marshes and rivers. Our most common bladderwort.

Utricularia minor L.
LESSER BLADDERWORT   native
IR | WUP | CUP | EUP   OBL CC 10

Perennial herb. Stems few-branched, 10–30 cm long, creeping on bottom in shallow water or on wet soil. Leaves alternate, to 1 cm long, with few divisions, the segments slender, flat, the smallest segments strongly tapered to tip, margins entire; bladders 1–2 mm wide, 1–5 on leaves. Flowers pale yellow, 2–8 atop a threadlike stalk 4–15 cm long; individual flower stalks to 1 cm long, curved downward in fruit; lower lip of flower 4–8 mm long, 2x longer than upper lip; spur small, to half length of lower lip. Fruit a capsule. June–Aug.

Fens, open bogs, sedge meadows and marshes; often in shallow water and where calcium-rich.

Utricularia purpurea Walt.
PURPLE BLADDERWORT   native
IR | WUP | CUP | EUP   OBL CC 10

Vesiculina purpurea (Walt.) Raf.

Annual or perennial herb. Stems underwater, to 1 m long. Leaves in whorls of 5–7, branched into threadlike segments, many segments tipped by a bladder. Flowers red-purple, 1–4 atop a stalk 3–15 cm long; corolla 1 cm long, lower lip 3-lobed, with a yellow spot near base; spur short and appressed to lower lip. Fruit a capsule. July–Sept.

Acid lakes and ponds in water to 1 m deep, peatlands, marshes.

Utricularia resupinata B.D. Greene
LAVENDER BLADDERWORT   native
IR | WUP | CUP | EUP   OBL CC 10

Annual or perennial herb. Stems delicate, on water surface in shallow water or creeping just below soil surface. Leaves alternate, 3-parted from base, the middle segment erect and linear, to 3 cm long; the 2 lateral segments slender, rootlike, with bladders. Flowers purple, 1 cm long, single atop an erect stalk 2–10 cm long; bract tubelike, surrounding the stem, its margin notched; flower tipped backward on stalk and facing upward; lower lip 3-lobed; spur more or less horizontal. Fruit a rounded capsule. July–Aug.

Shallow to deep water, wet lake and pond shores where sandy or mucky.

Limnanthaceae
MEADOWFOAM FAMILY

Floerkea *False Mermaidweed*

Floerkea proserpinacoides Willd.
FALSE MERMAIDWEED   native
IR | WUP | CUP | EUP   FACU CC 7

Annual herb. Stems weak, diffuse or decumbent, 1–3 dm long. Leaves deeply divided into 3–7 linear, oblong lance-shaped, or narrowly elliptic lobes each 1–2 cm long. Peduncles from the upper axils, at first about equaling the petiole, becoming much longer in fruit; sepals ovate lance-shaped, about 3 mm long at anthesis, up to 7 mm at maturity; petals white, oblong lance-shaped, about 2 mm long; stamens (3 or) 6. Carpels (2–) 3, tuberculate. April–May.

Moist woods in rich soil. Distinct among our dicots in its completely 3-merous flowers, and distinguished from our monocots by the very deeply pinnately lobed leaves.

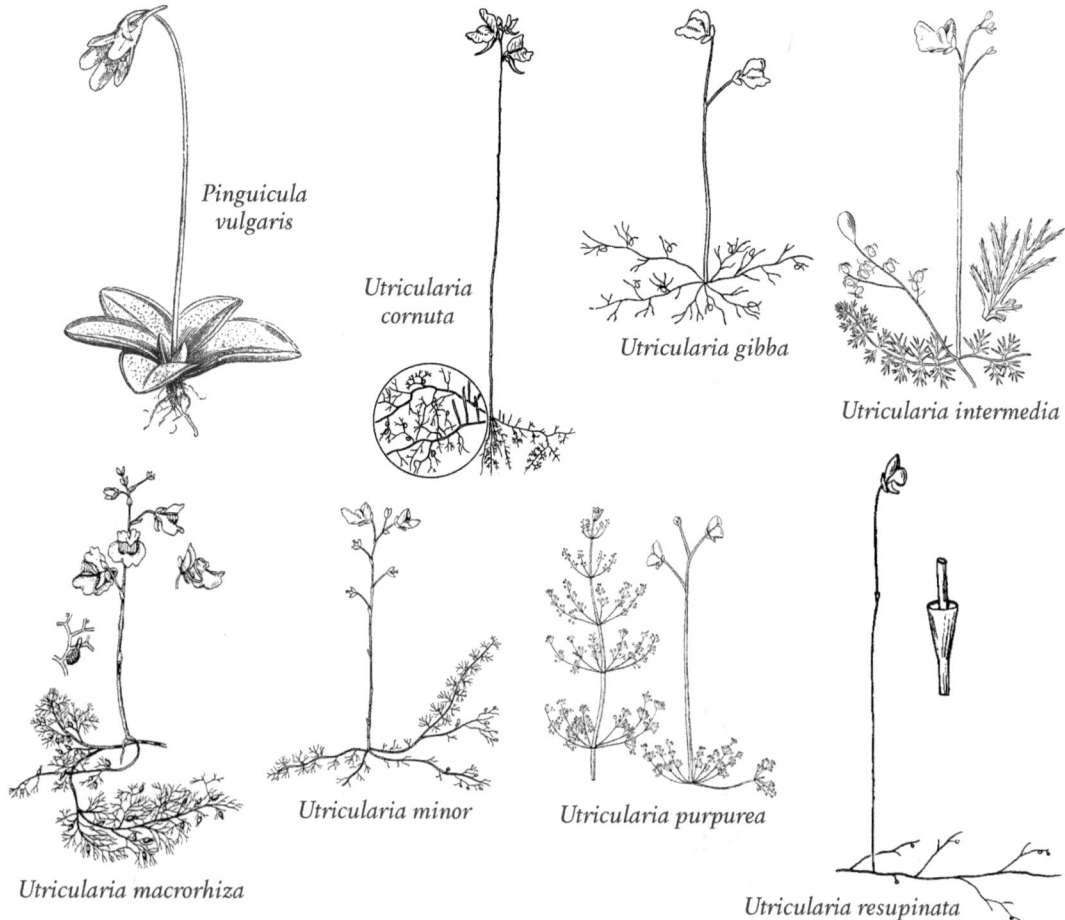

Linaceae
FLAX FAMILY

Linum *Flax*
Annual or perennial herbs. Leaves simple, alternate or opposite, narrow, margins entire, petioles absent. Flowers regular, perfect, 5-parted. Sepals separate, imbricate. Petals yellow or blue. Stamens as many as the petals. Fruit a 10-chambered capsule.

ADDITIONAL SPECIES
Linum catharticum (Fairy flax); introduced, petals white; occasional in the UP.

1   Petals blue; pedicels becoming more than 1 cm long. . . . . **L. usitatissimum**
1   Petals yellow; pedicels to 1 cm . . . . **L. sulcatum**

Linum sulcatum Riddell
GROOVED YELLOW FLAX *native*
IR | WUP | CUP | EUP   CC 8

Annual. Stems erect, branched above, 2–8 dm tall. Leaves narrow, all or chiefly alternate, 1–2 cm long, with a pair of minute dark glands at base. Branches of the inflorescence slender and raceme-like; pedicels to 8 mm long but averaging much shorter. Sepals persistent, 5–6 mm long, long-acuminate, conspicuously glandular-ciliate; petals yellow, 5–8 mm long. Capsule subglobose, about 3 mm long; false septum incomplete, ciliate. June–July.

Dry sandy soil, prairies; uncommon in Menominee County.

Linum usitatissimum L.
CULTIVATED FLAX   *introduced*
IR | **WUP** | **CUP** | EUP

Annual. Stems erect, usually solitary from a slender root, to 1 m tall. Leaves linear lance-shaped, 3-nerved. Sepals 7–9 mm long at maturity, the inner ciliate on the scarious margin; petals blue, 10–15 mm long. Capsule globose, 6–10 mm wide; false septa very incomplete. Summer.

Of unknown origin; cultivated since prehistoric times for its fiber (linen) and more recently for its oil (linseed); sometimes escaped or adventive in fields and roadsides.

Linderniaceae
LINDERNIA FAMILY
Lindernia *False Pimpernel*
Lindernia dubia (L.) Pennell
YELLOW-SEED FALSE PIMPERNEL *native*
IR | **WUP** | **CUP** | **EUP** OBL CC 4

Annual herb. Stems smooth, 1–2 dm long, widely branched. Leaves opposite, ovate to obovate, 5–30 mm long and 3–10 mm wide, the upper leaves smaller; margins entire or with small, widely spaced teeth; petioles absent. Flowers single, on slender stalks 0.5–2.5 cm long from leaf axils; sepals 5, linear; corolla pale blue-purple, 5–10 mm long, 2-lipped, the upper lip 2-lobed, the lower lip 3-lobed and wider than upper lip; fertile stamens 2, staminodes (sterile stamens) 2. Fruit an ovate capsule, 4–6 mm long. June–Sept.

Mud flats, sandbars, shores of temporary ponds and marshes, streambanks.

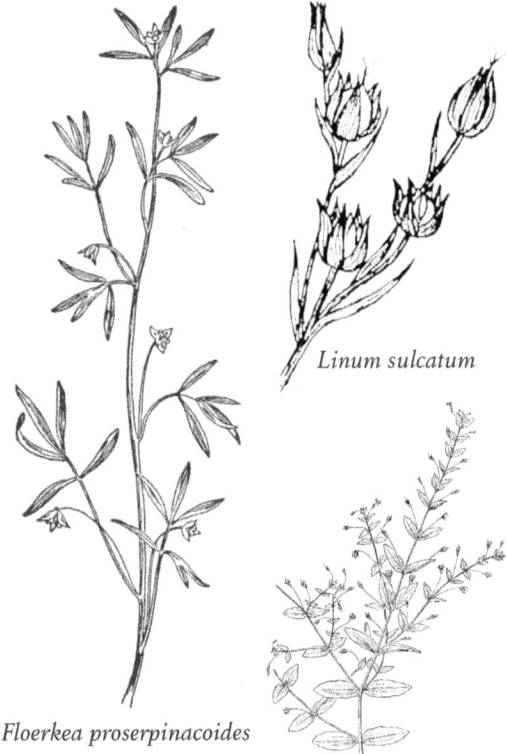

Linum sulcatum

Floerkea proserpinacoides

Lindernia dubia

Lythraceae
LOOSESTRIFE FAMILY

Annual or perennial herbs, sometimes woody at base (*Decodon*). Leaves simple, opposite, or both opposite and alternate, or whorled, margins entire, more or less stalkless. Flowers 1 or several in leaf axils or in spike-like heads at ends of stems; perfect (with both staminate and pistillate parts), regular or irregular; sepal lobes 4 or 6; petals 4 or 6, separate, pink or purple, deciduous; stamens usually 2 times number of petals. Fruit a dry, many-seeded capsule.

1 Plants arching, woody near base; leaves with petioles and mostly whorled . . . . **Decodon verticillatus**
1 Plants perennial herbs; leaves opposite to alternate, or if whorled, leaves without petioles . . . . **Lythrum**

Decodon *Water-Willow*
Decodon verticillatus (L.) Ell.
SWAMP-LOOSESTRIFE *native*
IR | **WUP** | **CUP** | **EUP** OBL CC 7

Perennial herb, woody near base. Stems slender, angled, smooth or slightly hairy, 1–3 m long, arching downward and rooting at tip when in contact with water or mud. Leaves in whorls of 3–4 or opposite, lance-shaped, 5–15 cm long and 1–3 cm wide, smooth above, sparsely hairy below; margins entire; petioles short. Flowers in dense clusters in upper leaf axils; sepals 5–7, short, triangular; petals pink-purple, tapered to base, 10–15 mm long; stamens 10 (rarely 8), alternately longer and shorter than petals. Fruit a more or less round capsule, 5 mm wide. July–Sept.

Shallow water and margins of lakes, ponds, bogs, swamps and marshes; soils mucky; local in Gogebic, Mackinac, and Menominee counties.

Lythrum *Loosestrife*

Perennial herbs. Stems erect, sometimes rather woody at base, usually with ascending branches above, upper stems 4-angled. Leaves opposite, alternate, or rarely whorled, entire, lance-shaped, stalkless, reduced to bracts in the head. Flowers in showy, spike-like heads, 1 to several in axils of upper leaves, regular or somewhat irregular, the stamens and styles of 2 or 3 different lengths. Sepals joined into a tube, the calyx tube cylinder-shaped, green-striped with 8–12 nerves; petals 6, purple, not joined; stamens 6 or 12; ovary 2-chambered. Fruit an ovate capsule, enclosed by the calyx tube.

1 Flowers single in upper leaf axils; stamens usually 6. . . . . **L. alatum**
1 Flowers many in spike-like heads at ends of stems; stamens usually 12 (6 long and 6 short) . . . . **L. salicaria**

Lythrum alatum Pursh
WINGED LOOSESTRIFE *native*
IR | WUP | **CUP** | **EUP** OBL CC 9

Smooth perennial herb, spreading by rhizomes.

Stems usually branched above, 2–8 dm long, somewhat woody at base. Lower leaves usually opposite, upper leaves alternate; lance-shaped, 1–4 cm long and 3–10 mm wide, rounded at base; margins entire; petioles absent. Flowers single in axils of upper, reduced leaves (bracts), short-stalked; calyx tube 4–6 mm long, smooth; petals 6, deep purple, 3–7 mm long; stamens usually 6. Fruit a capsule enclosed by the sepals. June–Aug.

Lakeshores, wet meadows, marshes, low prairie, calcareous fens, ditches; especially where sandy.

Lythrum salicaria L.
PURPLE LOOSESTRIFE *introduced (invasive)*
IR | **WUP** | **CUP** | **EUP** OBL

Perennial herb, spreading and forming colonies by thick, fleshy roots which send up new shoots. Stems erect, 6–15 dm long, 4-angled, with many ascending branches. Leaves opposite or sometimes in whorls of 3, becoming alternate and reduced to bracts in the head; lance-shaped, 3–10 cm long and 0.5–2 cm wide, mostly heart-shaped and clasping at base; margins entire; petioles absent. Flowers large and showy, 2 or more in axils of reduced upper leaves (bracts), in spikes 1–4 dm long at ends of branches; sepals joined, the calyx tube 4–6 mm long, hairy; petals 6, purple-magenta, 7–10 mm long; stamens usually 12, the stamens and styles of 3 different lengths. Fruit a capsule enclosed by the sepals. June–Sept.

Introduced from Europe and sometimes planted as an ornamental, escaping to marshes, wet ditches, streambanks, cranberry bogs and shores, where a serious threat to our native flora and of little value to wildlife. In addition to spreading vegetatively, a single plant may produce thousands of seeds each year. To limit the spread of this species, plants should be pulled (including roots), bagged, and removed from infested sites.

Malvaceae
MALLOW FAMILY

Annual or perennial herbs with upright stems; trees in *Tilia*. Leaves alternate, entire to lobed or dissected, often round or kidney-shaped, palmately veined. Flowers single or in small, narrow clusters from leaf axils, with 5 united sepals (separate in Tilia) and 5 petals; stamens many and joined near base, forming a tube around the style. Fruit a capsule.

1 Trees; with inflorescence apparently borne at the middle of a tongue-shaped bract .... **Tilia**
1 Herbs or shrubs; inflorescences various, but never with a large, tongue-shaped bract 2
2 Calyx without involucral bracts . **Abutilon theophrasti**
2 Calyx subtended by a series of 2 or more bracts 3
3 Involucral bracts 6 or more 4............ **Alcea rosea**
3 Involucral bracts 5 or less 4
4 Flower petals straight across at tip, the tip finely fringed .... **Callirhoe**
4 Petals obovate, rounded at tip, not fringed ..... **Malva**

Abutilon *Velvetleaf*

Abutilon theophrasti Medik.
VELVETLEAF *introduced*
IR | **WUP** | **CUP** | EUP FACU

Annual herb, softly pubescent throughout with stellate hairs. Stems stout, branched, 1–1.5 m tall. Leaves cordate, 10–15 cm long and as wide, toothed, on petioles of about the same length. Peduncles jointed above the middle, at first short, at maturity 2–3 cm long. Flowers yellow, 15–25 mm wide. Head of fruit 2–3 cm wide; carpels commonly 10–15, densely pubescent, with conspicuous, horizontally spreading beaks. July–Oct.

Native of Asia; a weed in fields and waste places; reported from Gogebic and Schoolcraft counties, more abundant south of the UP.

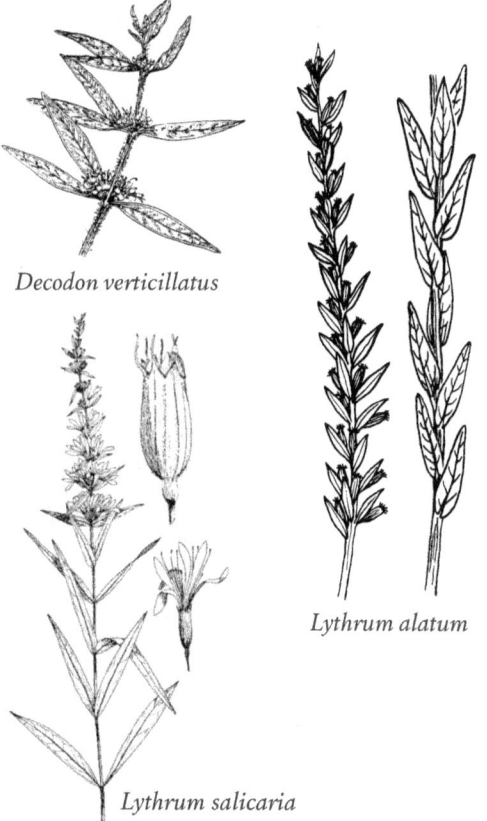

Decodon verticillatus

Lythrum alatum

Lythrum salicaria

MALVACEAE | Mallow Family

Alcea *Hollyhock*

Alcea rosea L.
HOLLYHOCK *introduced*
IR | **WUP** | CUP | EUP
Althaea rosea (L.) Cav.

Perennial herb. Stems erect, mostly unbranched, 1.5–3 m tall. Leaves orbicular, 1–3 dm wide. Flowers vary from white to pink or purplish, about 10 cm wide, on short peduncles from the upper axils, forming an elongate raceme-like inflorescence; bractlets broadly triangular. June–Aug.

Native perhaps to se Europe; cultivated and appearing near gardens, railroads, waste places; Baraga and Houghton counties.

Malva *Mallow*

Annual, biennial, or perennial herbs. Leaves broad, serrate, crenate, lobed, or parted. Flowers solitary or fascicled in the axils. Bractlets of the involucel 3, linear to obovate. Calyx 5-lobed. Petals truncate, notched, or obcordate at tip. Carpels 10–20, beakless, 1-seeded, glabrous, pubescent, or rugose.

ADDITIONAL SPECIES
Malva alcea L., introduced in central and western UP.

1 Leaves deeply divided into 3–7 segments 2
1 Leaf margins entire or only shallowly lobed 3
2 Pedicels and calyx with many short stellate hairs; bractlets subtending calyx to 3x longer than wide. . . . .
 . . . . **M. alcea***
2 Pedicels and calyx with only simple hairs; bractlets subtending calyx usually 3–5x longer than wide. . . . .
 . . . . **M. moschata**
3 Petals purple or pink, more than 1.5 cm long; stems erect
 . . . . **M. sylvestris**
3 Petals white or purple-tinged, less than 1.5 cm long; stems ascending or prostrate 4
4 Stems glabrous or nearly so; petals only slightly longer than calyx lobes. . . . . **M. pusilla**
4 Stems pubescent; petals about 2 times longer than calyx lobes . . . . **M. neglecta**

Malva moschata L.
MUSK MALLOW *introduced*
IR | **WUP** | CUP | **EUP**

Perennial herb. Stems erect, 4–10 dm tall, roughly pubescent. Leaves orbicular in outline, 5–7-parted, the segments of the upper leaves again deeply pinnatifid. Flowers partly solitary on long pedicels from the upper axils but chiefly crowded in terminal clusters; bractlets linear to narrowly lance-shaped or oblong lance-shaped, ciliate, glabrous or nearly so on the back; petals white to pale purple, triangular, 2.4–3 cm long. Mature carpels rounded on the back, not rugose, densely pubescent. June–Sept.

Native of Europe; escaped from cultivation along roadsides and in waste places.

Malva neglecta Wallr.
COMMON MALLOW *introduced*
IR | **WUP** | **CUP** | **EUP**

Biennial herb. Stems prostrate, procumbent, or ascending, to 1 m long, usually branched from the base. Leaves long-petioled, orbicular or kidney-shaped, 3–6 cm wide, shallowly 5–9-lobed, crenate, cordate at base. Flowers fascicled in the axils, on pedicels to 3 cm long; bractlets narrow; petals obcordate, 6–12 mm long, white or slightly tinged with pink or purple. Mature carpels usually 12–15, rounded on the back, not rugose or reticulate, usually finely pubescent, the whole ring of carpels presenting a crenate outline, the depressed central portion of the head about a 1/3 as wide as the head. May–Oct.

Native of Eurasia and n Africa; common as a weed in gardens and waste places.

Malva pusilla Sm.
DWARF MALLOW *introduced*
IR | WUP | **CUP** | EUP
Malva rotundifolia L.

Biennial herb; resembling *M. neglecta* in habit and foliage. Pedicels often little longer than the calyx. Mature carpels 5–11, commonly 10, glabrous or pubescent, conspicuously rugose-reticulate on the back, the margins sharply angled, the whole head of carpels circular in outline; central depressed area about 1/5 the diameter of the whole head.

Native of Europe; occasional as a weed.

Malva sylvestris L.
HIGH MALLOW *introduced*
IR | **WUP** | CUP | EUP

Biennial herb. Stems erect, 4–10 dm tall, sparsely hirsute to glabrate. Leaves orbicular or kidney-shaped in outline, shallowly 3–7-lobed, the lobes broadly rounded, serrate; petioles pubescent only or chiefly in a single line on the upper side; bractlets oblong to ovate or obovate. Flowers fascicled in the upper axils, on peduncles to 5 cm long; petals red-purple, 2–2.5 cm long. Mature carpels rugose-reticulate on the back, glabrous or sparsely pubescent. June–Aug.

Native of Eurasia; occasionally escaped from cultivation; Houghton County.

Tilia *Basswood; Linden*

Tilia americana L.
BASSWOOD; LINDEN *native*
IR | **WUP** | **CUP** | **EUP** FACU CC 5

Tree to 35 m tall; bark gray to light brown, with narrow, well-defined fissures; twigs smooth, red-

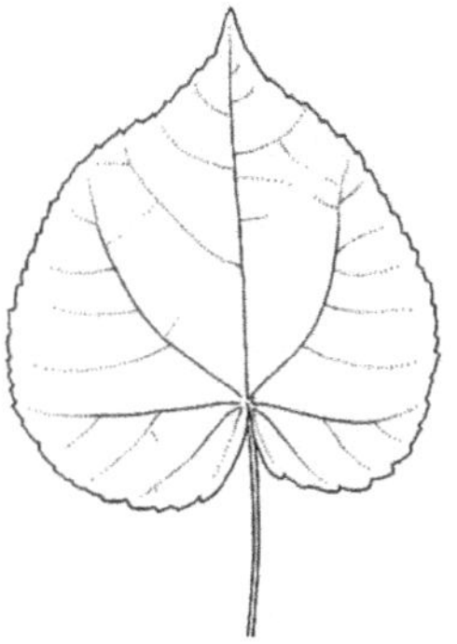

Abutilon theophrasti

Alcea rosea

Malva neglecta

Malva moschata

Malva pusilla

Malva sylvestris

Rhexia virginica

Tilia americana

dish-green, becoming light to dark gray, marked with dark wart-like bumps. Leaves broadly ovate to subrotund, palmately veined, cordate or truncate at the oblique base, sharply serrate, green beneath and glabrous to sparsely stellate-pubescent on the surface, with conspicuous tufts of hairs in the vein axils. Flowers fragrant, perfect, 5-merous, white or cream-colored in axillary cyme-like clusters, the long peduncle adnate about to the middle of a narrow, elongate, short-petioled, foliaceous bract; bracts glabrous or nearly so on both sides; sepals separate to the base petals narrowly oblong, 7–12 mm long, tapering to the base; stamens numerous, either all distinct or united into 5 bundles, one in front of each petal; ovary tomentose, 5-celled. Fruit nutlike, tomentose, 1–2-seeded. July.

Moist fertile soil.

The dried flowers have a long history of medicinal uses, and bees produce a fragrant honey from the blossoms.

Melastomataceae

MELASTOME FAMILY

Rhexia *Meadow-Pitchers*

Rhexia virginica L.

WING-STEM MEADOW-PITCHERS *native*
IR | WUP | CUP | **EUP** OBL CC 9

Perennial herb, roots often with tubers. Stems simple or branched above, 2–6 dm tall, 4-angled and 4-winged, with bristly hairs at nodes, otherwise glabrous to sparsely glandular-hirsute. Leaves ovate, 2–6 cm long and 1–3 cm wide, obtuse or rounded at the sessile base, glabrous or with short, stiff hairs on either side; margins finely toothed. Flowers perfect, regular, 4-merous, in cymes from ends of stems and upper leaf axils; hypanthium tubular at anthesis, in fruit becoming urn-shaped, the basal portion distended by the developing capsule, the terminal portion persisting as a tubular or flaring neck crowned by the persistent sepals; sepals narrow, 2–4 mm long; petals purple, 12–20 mm long; stamens 2x as many as the petals; ovary 2–several-celled. Fruit a 4-chambered capsule; seeds coiled, rough or papillate. July–Sept.

Open shores, moist meadows, thickets (often of *Aronia*); soils acidic, sandy or peaty; Chippewa County.

When growing in water, the stems are often covered toward the base by a spongy aerenchyma.

Menispermaceae
MOONSEED FAMILY

Menispermum *Moonseed*

Menispermum canadense L.
CANADIAN MOONSEED — *native*
IR | WUP | **CUP** | EUP — FAC CC 5

Dioecious woody twiners climbing 2–4 m high. Leaves simple, alternate, broadly ovate to nearly orbicular, 10–15 cm wide and long, palmately veined, shallowly 3–7-lobed to entire; slender-petioled. Flowers small, unisexual, usually 3-merous, regular, in racemes or panicles that arise just above the leaf-axils; perianth segments scarcely differentiated into calyx and corolla, normally in 4 alternating whorls, the 2 outer (calyx) exceeding the 2 inner (corolla); sepals 4–8, longer than the 4–8 petals; stamens 12–24. Drupe bluish-black, 6–10 mm long; stone flattened, thickened into 3 rough ridges over most of its margin. June–July.

Moist woods and thickets; Menominee County.

The drupes, suspected to be toxic, resemble wild grapes.

Menispermum canadense

Menyanthaceae
BUCKBEAN FAMILY

Menyanthes *Buckbean*

Menyanthes trifoliata L.
BUCKBEAN — *native*
IR | WUP | CUP | EUP — OBL CC 8

Perennial glabrous herb, with thick rhizomes covered with old leaf bases. Leaves alternate along rhizomes, palmately divided into 3 leaflets, the leaflets oval to ovate, 3–10 cm long and 1–5 cm wide, entire or sometimes wavy-margined; petioles 5–30 cm long, the base of petiole expanded and sheathing stem. Flowers in racemes on leafless stalks 2–4 dm long and longer than the leaves; bracts mostly 3–5 mm long; individual flowers on stalks 5–20 mm long; flowers perfect, regular, 5-parted, often of 2 types, some with flowers with long stamens and a shorter style, others with a long style and shorter stamens; sepal lobes 2–3 mm long; corolla funnel-shaped, 8–12 mm long, petals white, often purple-tinged, bearded with white hairs on inner surface; stamens 5. Fruit a rounded capsule, 6–10 mm wide; seeds shiny, yellow-brown. May–July.

Open bogs and fens (especially in pools and outer moat), cedar swamps, wet thickets.

Molluginaceae
CARPETWEED FAMILY

Mollugo *Carpetweed*

Mollugo verticillata L.
GREEN CARPETWEED — *introduced*
IR | **WUP** | **CUP** | EUP — FAC

Annual herb. Stems prostrate or ascending, repeatedly forked, forming mats to 4 dm wide. Leaves in whorls of 3–8, narrowly to broadly oblong lance-shaped, 1–3 cm long, long-tapering to a short, scarcely differentiated petiole. Flowers perfect, 2–5 from each node, on pedicels 5–15 mm long, sepals 5; petals 5, pale green to white, 4–5 mm wide; stamens 3 or 4. Capsules ovoid, 3 mm long, many-seeded. June–Sept.

Apparently native of tropical America; now a common weed in moist soil.

Montiaceae
MONTIA FAMILY

Claytonia *Springbeauty*

Glabrous perennial herbs from rounded tubers (ours). Leaves one or few from the base and a single opposite pair on the stem below the loose terminal raceme; the raceme with 5–15 long-pediceled flowers and sometimes a small bract below the lowest flower. Sepals ovate, persistent in fruit. Petals 5, white or pale pink with pink veins, oval or elliptic, spreading. Stamens 5, opposite the petals. Capsule ovoid, opening by inrolling valves. Flowers open from March to May, with stems and leaves withered by early summer.

Menyanthes trifoliata

Mollugo verticillata

In Michigan, the two native species of *Claytonia* are rarely found growing together. In those situations, *C. virginica* reaches the peak of its flowering several days later than *C. caroliniana*.

ADDITIONAL SPECIES
Claytonia sibirica L. (Siberian spring-beauty); plants tap-rooted; uncommon escape from cultivation; reported from woods in Houghton County.

1 Leaves with distinct petiole; blades less than 5 times longer than wide .... **C. caroliniana**
1 Petiole not distinct; blades 5 times or more longer than wide .... **C. virginica**

Claytonia caroliniana Michx.
CAROLINA SPRINGBEAUTY   native
IR | **WUP** | **CUP** | **EUP**   FACU CC 6
Perennial herb. Stem leaves, including the petiole, commonly 3–6 cm long, rarely to 9 cm long, the blade usually 10–15 mm wide, acute at base, clearly distinguished from the petiole.
Cool woods.

Claytonia virginica L.
VIRGINIA SPRINGBEAUTY   native
IR | **WUP** | **CUP** | **EUP**   FACU CC 4
Perennial herb. Stem leaves long-tapering to base, the blade sessile or merging gradually into the short, poorly differentiated petiole, commonly 4–10 mm wide, rarely less than 7 cm long, including the petiole.
Damp woods and fields; much less common than *C. caroliniana*.

Moraceae
MULBERRY FAMILY

Morus *Mulberry*
Morus alba L.
WHITE MULBERRY   introduced
IR | **WUP** | CUP | EUP   FACU
Monoecious or dioecious tree to 15 m tall. Leaves alternate, rotund in outline, palmately veined, serrate, often irregularly several-lobed, often cordate at base, glabrous or nearly so on both sides, or sparsely pubescent with white spreading hairs along the veins beneath. Flowers in cylindric catkins, the staminate longer and more loosely flowered than the pistillate. Calyx deeply 4-parted. Stamens 4. Fruit white, pink, or pale purple to nearly black, edible, resembling a blackberry, composed of the juicy calyx, each enclosing a small seed-like achene, with the remains of the styles protruding.

Apparently native of Asia; long cultivated in Europe and America for its fruit or fiber or as food for the silkworm, and now used for ornament in several horticultural forms; the fruit eaten by birds and thence spreading the seeds to roadsides, vacant land, and open woods; Baraga County.

Myricaceae
BAYBERRY FAMILY
Monoecious or dioecious shrubs, with alternate simple leaves; leaves resinous-dotted and fragrant. Flowers unisexual, without perianth, solitary in the axils of small bracts, aggregated into globose to cylindric catkins. Stamens 2–many, usually 4–8, the short filaments free or connate. Ovary 1-celled, subtended by 2–8 usually minute bractlets; ovule 1, basal; style very short; stigmas 2, linear, elongate.

1 Leaves entire or nearly so; wet habitats... **Myrica gale**
1 Leaves pinnately lobed; dry sandy habitats.......... .... **Comptonia peregrina**

Comptonia *Sweet-Fern*
Comptonia peregrina (L.) Coult.
SWEET-FERN   native
IR | **WUP** | **CUP** | **EUP**   CC 6
Shrub. Stems much branched, to 1 m tall. Leaves linear-oblong, 6–12 cm long, about 1 cm wide, deeply pinnately lobed, resinous-dotted, more or less pubescent. Staminate catkins clustered, cylindric, 1–3 cm long, nodding; bracts quadrangular, resinous, villous, acuminate; pistillate catkins subglobose; bracts similar but concealed by the linear-subulate bractlets. Nutlets ellipsoid, blunt, 3–5 mm long, subtended by the elongate bractlets, the whole fruit bur-like, 1–2 cm wide. April–May.
Dry, especially sandy soil.

Myrica *Bayberry*
Myrica gale L.
SWEET GALE   native
IR | **WUP** | **CUP** | **EUP**   OBL CC 6
Gale palustris Chev.
Much-branched shrub, 6–15 dm tall; bark dark gray to red-brown with small pale lenticels; twigs hairy, dotted with glands. Leaves alternate, deciduous,

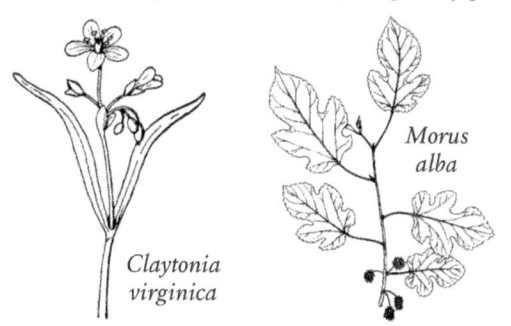

Claytonia virginica

Morus alba

NYCTAGINACEAE | Four-O'Clock Family

wedge-shaped, tapered to base, broadest above middle, 3–6 cm long and 1–2 cm wide, tip rounded and toothed, dark green on upper surface, paler below, dotted with shiny yellow glands, fragrant when rubbed; petioles short, 1–3 mm long. Staminate and pistillate flowers separate and on different plants, appearing before or with unfolding leaves; staminate flowers in catkins 1–2 cm long, with dark brown, shiny triangular scales; pistillate flowers in conelike, brown clusters 10–12 mm long. Fruit a flattened, ovate achene, resin-dotted, 2–3 mm long. April–May.

Lakeshores (where often in shallow water), marshes, swamps, bogs.

Nyctaginaceae
FOUR-O'CLOCK FAMILY

Mirabilis *Four-O'clock*

Perennial herbs, or woody at base. Leaves opposite. Flowers many in terminal panicles. Calyx funnelform, 5-lobed, the short tube closely surrounding the ovary and constricted above it, the ovary apparently inferior. Stamens 3–5. Flowers rose to pink-purple, open in the morning, solitary or in clusters of 2–4, nearly sessile, subtended by a 5-lobed, saucer-shaped or cup-shaped involucre. Fruit prismatic to obovoid, 5-ribbed, mucilaginous when wet. Early in the season plants commonly produce a few solitary peduncled flowers from the forks of the stem; the large panicles appear later.

1　Leaves ovate, heart-shaped or truncate at base; inflorescence not glandular . . . . **M. nyctaginea**
1　Leaves linear to lance-shaped, tapered to the base; inflorescence glandular hairy . . . . **M. albida**

Mirabilis albida (Walter) Heimerl
HAIRY FOUR-O'CLOCK　　　　　　　　　　　　*introduced*
IR | WUP | **CUP** | EUP
　　Mirabilis hirsuta (Pursh) MacM.

Stems erect or decumbent, to 1 m tall, more or less hirsute, especially about the nodes, with spreading hairs 1–2 mm long, becoming glandular-pubescent in the inflorescence. Leaves linear lance-shaped to ovate lance-shaped, the larger commonly 1–2 cm wide. Involucre about 5 mm long, glandular-pubescent, becoming 1–2 cm long at maturity; calyx pink, about 1 cm long. Fruit narrowly obovoid, 4–5 mm long, pubescent, rugose on the sides and ridges. Summer.

Dry prairies, hills, and barrens; native of Great Plains, considered adventive in UP (Marquette and Schoolcraft counties).

Mirabilis nyctaginea (Michx.) MacM.
HEART-LEAF FOUR-O'CLOCK　　　　　　　　　*introduced*
IR | **WUP** | **CUP** | EUP

Stems nearly smooth, branched above, to 1 m tall. Leaves ovate to deltoid-ovate, cordate or truncate at base, glabrous or nearly so, petioles 1–3 cm long. Involucre saucer-shaped, about 1 cm wide, densely ciliate, accrescent in fruit; calyx pinkish-purple, about 10 mm long. Fruit narrowly obovoid, densely pubescent, rough on the sides and on the 5 prominent ribs. May–Aug.

Dry soil, waste places; native west of Michigan, considered adventive in UP.

Nymphaeaceae
WATER-LILY FAMILY

Aquatic, perennial herbs. Stems long and fleshy, from horizontal rhizomes rooted in bottom mud. Leaves large, leathery, mostly floating or emergent above water surface, heart-shaped to shield-shaped, notched at base, margins entire. Flowers showy, single on long stalks and borne at or above water surface, perfect, white or yellow, sepals 4–6, green or yellow; petals numerous, small to large and showy. Fruit a many-seeded, berrylike capsule, opening underwater when mature.

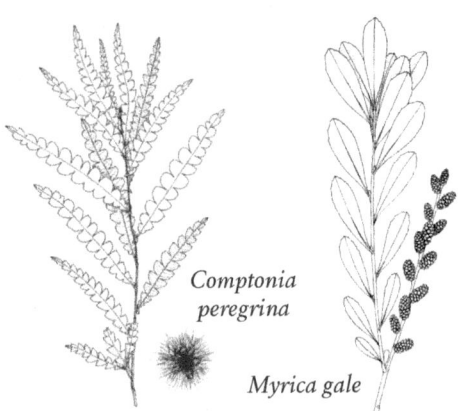

Comptonia peregrina

Myrica gale

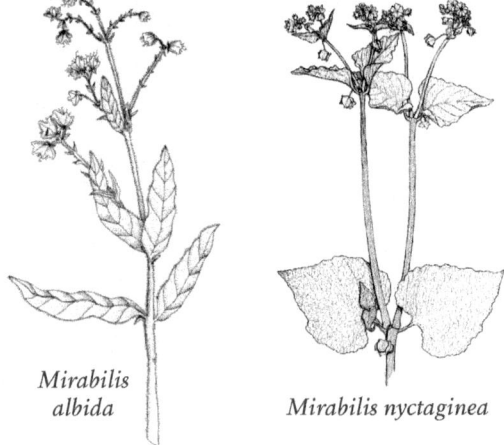

Mirabilis albida

Mirabilis nyctaginea

NYMPHAEACEAE | Water-Lily Family

1 Flowers yellow, often red-tinged, sepals petal-like, true petals small; leaf blades oblong to oval or heart-shaped .... **Nuphar**
1 Flowers white (rarely pink), sepals green, true petals large and showy; leaf blades nearly round............ .... **Nymphaea odorata**

Nuphar *Yellow Water-Lily*

Aquatic herbs. Leaves mostly large and floating or emergent. Sepals 5–6, yellow and petal-like, forming a saucer-shaped flower; petals small and numerous.

1 Disk at base of stigma green or yellow; anthers longer than filaments........................**N. variegata**
1 Disk at base of stigma red; anthers shorter than the filaments..2
2 Leaf sinus 2/3 or more length of the midrib; sepals 5; anthers 1–3 mm long.................**N. microphylla**
2 Leaf sinus about 1/2 length of midrib; sepals 5 or 6; anthers 3–6 mm long.................**N. × rubrodisca**

Nuphar microphylla (Pers.) Fern.
YELLOW POND-LILY (END) *native*
IR | **WUP** | **CUP** | **EUP** OBL CC 10
Nuphar pumila (Timm) DC.

Leaves both underwater and floating; floating leaves 5–10 cm long and 3–8 cm wide, notch at base usually more than half as long as midvein; petioles flattened on upper side; underwater leaves membranous, somewhat larger. Flowers 1.5–2 cm wide, sepals 5, yellow on inner surface; petals small and many; disk at base of stigma red, 3–6 mm wide, with 6–10 rays. Fruit ovate, 15 mm long. July–Aug.
Lakes, ponds and slow-moving streams.

Nuphar × rubrodisca Morong
YELLOW POND-LILY *native*
IR | **WUP** | CUP | **EUP** OBL CC –
Nuphar lutea (L.) Sm. subsp. *rubrodisca* (Morong) Hellquist & Wiersema

Leaves both submersed and floating, the latter commonly 10–15 cm long, the sinus averaging half as long as the midvein; petioles flattened on the under side. Flowers 3–4 cm wide, yellow within or suffused with red; anthers 3–6 mm long; stigmatic disk red, 7–10 mm wide at anthesis, somewhat wider in fruit, 8–13–rayed. Fruit 20–25 mm long, scarcely constricted below the disk. Summer.
Rare in Chippewa and Keweenaw counties.
Considered a hybrid between *Nuphar microphylla* and *N. variegata*.

Nuphar variegata Dur.
YELLOW POND-LILY *native*
IR | **WUP** | **CUP** | **EUP** OBL CC 7

Leaves mostly floating, 10–25 cm wide, notch usually less than half as long as midvein, petioles flattened on upper side and narrowly winged; underwater leaves absent or few. Flowers 2.5–5 cm wide; sepals usually 6, yellow, red-tinged on inner surface; petals small and numerous; anthers 4–7 mm long, longer than filaments; disk at base of stigma green, 1 cm wide, with 10–15 rays. Fruit ovate, 2–4 cm long. June–Aug. Ponds, lakes, quiet streams.

Nymphaea *Water-Lily*

Nymphaea odorata Ait.
WHITE WATER-LILY *native*
IR | **WUP** | **CUP** | **EUP** OBL CC 6
Nymphaea tuberosa Paine

Aquatic perennial herb, rhizomes sometimes with knotty tubers. Leaves floating, round, 1–3 dm wide, with a narrow notch, green and shiny on upper surface, usually purple or red below. Flowers large and showy, white (rarely pink), usually fragrant, 7–20 cm wide, often opening in morning and closing in late afternoon (or remaining open on cool, cloudy days); sepals 4, green, 3–10 cm long; petals 17–25, about as long as sepals, oval, tapered to a rounded tip; stamens 40–100. Fruit round, mostly covered by sepals; seeds 2–4 mm long. June–Aug. Shallow water of ponds and lakes, quiet water of rivers.

ADDITIONAL SPECIES

Nymphaea leibergii Morong (Pygmy pond-lily); a circumpolar species in Michigan known only from a 1963 collection from a stream at the head of Duncan Bay, Isle Royale; endangered.

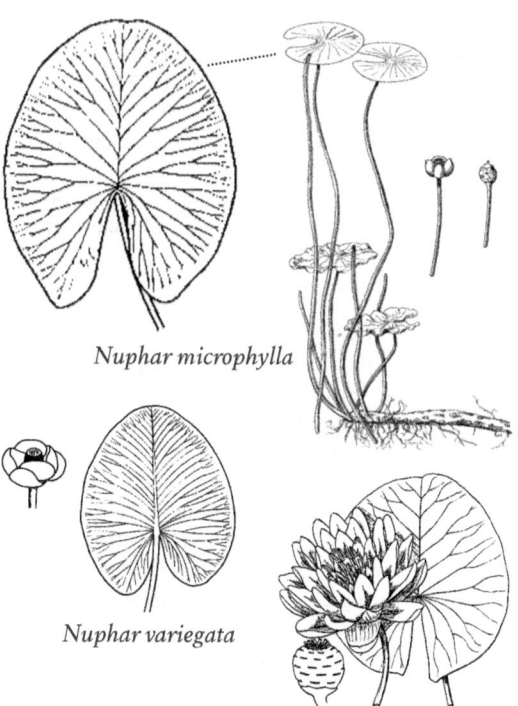

Nuphar microphylla

Nuphar variegata

Nymphaea odorata

Oleaceae

OLIVE FAMILY

Trees or shrubs with opposite, simple or compound leaves. Flowers perfect or unisexual, regular, usually 4-merous. Calyx small or in some genera lacking. Corolla in our genera partially or wholly fused, or lacking (*Fraxinus*). Stamens 2–4, usually 2, inserted on the corolla tube if the corolla is present. Ovary 2-celled; seeds 2–several or rarely 1 in each cell. Fruit a drupe, capsule, or samara.

1	Trees; leaves pinnately compound	**Fraxinus**
1	Shrubs; leaves simple	**Syringa vulgaris**

Fraxinus *Ash*

Medium trees. Leaves deciduous, opposite, pinnately divided into leaflets. Flowers in clusters from axils of previous year's twigs, mostly single-sexed, staminate and pistillate flowers on different trees, rarely perfect, petals absent. Fruit a 1-seeded, winged samara.

1	Twigs densely hairy	2
1	Twigs glabrous	3
2	Lateral leaflets tapered at base to a short winged petiole, or the leaflets sessile	**F. pennsylvanica**
2	Lateral leaflets rounded at base, short petioles present	**F. americana**
3	Leaflets pale or waxy on underside, margins often entire; lateral leaflets on short petioles 5 mm or more long	**F. americana**
3	Leaflets not waxy on underside, margins usually finely toothed; lateral leaflets sessile or on short petioles to 3 mm long	4
4	Lateral leaflets usually number 8, sessile; body of fruit flat in cross-section	**F. nigra**
4	Lateral leaflets usually 4–6, on short petioles; body of fruit round in section	**F. pennsylvanica**

Fraxinus americana L.
WHITE ASH — *native*
IR | **WUP** | **CUP** | **EUP** — FACU CC 5

Tree to 40 m. tall. Leaflets 5–9, usually 7, oblong to ovate or obovate, usually abruptly acuminate, entire or serrulate toward the summit, rounded to broadly acute at base, paler beneath, on wingless leaflet petioles. Samaras linear or oblong, 3–5 cm long, obtuse to retuse at the tip, the wing extending about 1/3 of the length of the terete body; the free portion, above the apex of the body, longer than the body itself; subtending calyx 1–1.5 or rarely 2 mm long, seldom cleft on one side only.

Rich moist woods; a valuable timber tree.

Fraxinus nigra Marsh.
BLACK ASH — *native*
IR | **WUP** | **CUP** | **EUP** — FACW CC 6

Tree to 15 m tall, crown open and narrow; bark gray, thin, flaky; twigs smooth, round in section, dark green, becoming gray. Leaves opposite, pinnately divided into 7–11 stalkless (except for terminal) leaflets; leaflets lance-shaped to oblong, 7–13 cm long and 2.5–5 cm wide, long-tapered to a tip; margins with sharp, forward-pointing teeth. Flowers appear in spring before leaves, in open clusters on twigs of previous year; some perfect, some single-sexed, staminate and pistillate flowers on different trees. Fruit a 1-seeded samara, 2.5–4 cm long and 6–10 mm wide, the wing broad and rounded at tip, deciduous or persisting until following spring. April–May.

Floodplain forests, cedar swamps, wet depressions in forests.

Fraxinus pennsylvanica Marsh.
GREEN ASH — *native*
IR | **WUP** | **CUP** | **EUP** — FACW CC 2

Tree to 15 m tall; bark dark gray or brown, thick, with shallow furrows and netlike ridges; twigs usually hairy for 1–3 years, becoming light gray or red-brown. Leaves opposite, pinnately divided into 7–9 leaflets; leaflets oblong lance-shaped to ovate, 7–13 cm long and 2.5–4 cm wide, upper surface smooth, underside smooth or hairy; margins entire or with few forward-pointing teeth; leaflet petioles short, smooth or hairy. Flowers appear in spring before or with leaves, in compact, hairy clusters on twigs of previous year; single-sexed, staminate and pistillate flowers on different trees. Fruit a 1-seeded, slender samara, 2.5–5 cm long, in open clusters persisting until following spring. April–May.

Floodplain forests, swamps, shores, streambanks.

Syringa *Lilac*

Syringa vulgaris L.
COMMON LILAC — *introduced*
IR | **WUP** | **CUP** | **EUP**

Much-branched deciduous shrub to 6 m tall, spreading and forming thickets. Leaves opposite, simple, entire, ovate, 5–10 cm long, short-acuminate, truncate to cordate at base. Flowers fragrant, in dense panicles 1–2 dm long; 4-merous; calyx small, campanulate, truncate to 4-toothed; corolla salverform, usually lilac, about 1 cm wide; stamens 2, included in the corolla tube. Fruit a 2-celled capsule with 2 seeds in each cell. May.

Native of se Europe, found near abandoned farms. Lilac apparently does not spread by seed but is very long-lived and persists indefinitely after planting.

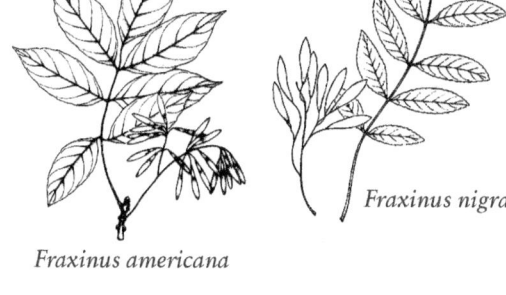
Fraxinus americana
Fraxinus nigra

Syringa vulgaris
Fraxinus pennsylvanica

ADDITIONAL SPECIES
Syringa reticulata (Blume) H. Hara (Japanese tree lilac), introduced small tree from Asia; flowers creamy white, in large panicles, blooming in early summer; reported as an escape from planting in Chippewa and Schoolcraft counties.

Onagraceae
EVENING-PRIMROSE FAMILY

Annual or perennial herbs. Leaves opposite to alternate, simple to pinnately divided, stalkless or short-petioled. Flowers usually large and showy, perfect (with both staminate and pistillate parts), regular, borne in leaf axils or in heads at ends of stems; sepals 8 or 4; petals 4, white, yellow, or pink to rose-purple. Fruit a 4-chambered capsule; seeds many, with or without a tuft of hairs (coma).

1 Petals 2, small, white; leaves opposite; fruit with bristly hairs . . . . **Circaea**
1 Petals 4 (rarely absent), white, pink, or yellow; leaves alternate or opposite; fruit without bristly hairs 2
2 Hypanthium prolonged beyond ovary into a tube below the petals; leaves alternate . . . . **Oenothera**
2 Hypanthium scarcely if at all prolonged beyond ovary, not tube-like; leaves alternate or opposite 3
3 Petals yellow (or absent); seeds without a tuft of hairs . . . . **Ludwigia**
3 Petals pink or rose-purple; seeds with a tuft of hairs (coma) 4
4 Leaves all alternate; flowers in long, terminal, showy racemes . . . . **Chamaenerion**
4 Leaves, except sometimes the upper ones, opposite; flowers all or mostly in the axils of leaves or leafy bracts . . . . **Epilobium**

Chamaenerion *Fireweed*

Chamaenerion angustifolium (L.) Scop.
FIREWEED native
IR | WUP | CUP | EUP CC 3
Epilobium angustifolium L.
Perennial herb. Stems erect, to 2 m or more tall. Leaves lance-shaped or linear lance-shaped, to 2 dm long, narrowed to a sessile or obscurely petioled base, glabrous beneath; veinlets conspicuous. Racemes elongate, many-flowered, the lower flowers often exceeded by the subtending leaves, the upper with short bracts or none; petals purple, or rarely white, 10–15 mm long, clawed at base; style pubescent at base; stigmas soon revolute. Capsules 2.5–7 cm long. June–Sept.

In a variety of habitats, preferring moist soils rich in humus; often abundant after fires.

Circaea *Enchanter's Nightshade*

Perennial herbs with opposite petioled leaves and small white flowers in one to few terminal racemes. Flowers 2-merous. Hypanthium shortly prolonged above the ovary, tubular. Petals obcordate or deeply notched. Stamens 2. Ovary 1–2-celled. Fruit reflexed, obovoid or pear-shaped, usually slightly compressed, beset with soft or stiff hooked bristles.

1 Plants to 6 dm tall; flowers and fruit well-spaced on stalk; calyx lobes more than 1.5 mm long; leaves rounded at base, the margins very shallowly toothed. . . . . **C. canadensis**
1 Plants smaller, to 3 dm tall; open flowers clustered near top of stem; calyx lobes less than 1.5 mm long; leaves usually heart-shaped at base, margins sharply toothed . . . . **C. alpina**

Circaea alpina L.
ALPINE ENCHANTER'S NIGHTSHADE native
IR | WUP | CUP | EUP FACW CC 4
Perennial herb, spreading from rhizomes thickened and tuberlike at ends. Stems weak, 1–3 dm long, mostly smooth. Leaves opposite, ovate, 2–5 cm long

and 1–3 cm wide; margins coarsely toothed; petioles flat on upper side, underside thin-winged along center. Flowers white, in short racemes of 10–15 flowers, becoming 1 dm long in fruit; sepals 1–2 mm long; petals to 2 mm long. Fruit a 1-seeded capsule, 2–3 mm long, covered with soft hooked bristles. June–Aug.

Cedar swamps (where often on rotting logs), low spots in forests.

Circaea canadensis (L.) Hill
COMMON ENCHANTER'S NIGHTSHADE native
IR | **WUP** | **CUP** | **EUP** CC 2
Circaea lutetiana L.
Stems erect, to 1 m tall, glabrous below, becoming minutely villosulous in the inflorescence. Leaves oblong-ovate, commonly 6–12 cm long, acuminate, very shallowly sinuate-denticulate, rounded or barely subcordate at base; petioles rounded or angled on the lower side. Racemes commonly many-flowered, to 2 dm long; petals 2.5–3.5 mm long. Fruit 3.5–5 mm long, beset with stiff bristles, equally 2-celled, 2-seeded, each half bearing normally 3 large and 2 small rounded ridges separated by narrow furrows. June–Aug.

Moist woods.

Epilobium *Willowherb; Fireweed*

Perennial herbs, often producing leafy rosettes or bulblike offsets (turions) at base of stem late in growing season. Leaves simple, opposite, alternate, or opposite below and becoming alternate above; stalkless or short-petioled. Flowers white to pink, single in axils of upper reduced leaves, or in spike or racemes at ends of stems; sepals 4; petals 4; stamens 8, the inner 4 stamens shorter than outer 4; ovary 4-chambered, maturing into a linear, 4-parted capsule, splitting from tip to release numerous brown seeds which are tipped with a tuft of fine hairs (coma).

ADDITIONAL SPECIES
Epilobium parviflorum Schreb. (Small-flower willowherb), introduced perennial of swamps, shores, ditches; Delta and Mackinac counties.

1 Stigma 4-parted............................2
1 Stigma entire, not 4-parted 4
2 Taprooted annual herb; leaves mostly alternate, linear .... **E. brachycarpum**
2 Perennial rhizomatous herbs; leaves mostly opposite, lance-shaped 3
3 Leaves sessile and (except sometimes the smallest) the blades clasping up to halfway around the stem or the largest leaves slightly decurrent; petals 10–17 mm long .... **E. hirsutum**
3 Leaves sessile or subsessile but the blades not clasping or decurrent; petals to 10 mm long ... **E. parviflorum***
4 Leaves entire or nearly so, the margins often revolute; stems round in cross-section, without lines of hairs on stem below base of each leaf 5
4 Leaf margins conspicuously toothed; stems 4-angled in section, with lines of hairs on stems below leaf bases 7
5 Stems with soft, straight hairs............ **E. strictum**
5 Stems finely hairy, the hairs appressed to stem 6
6 Upperside of leaves finely hairy...... **E. leptophyllum**
6 Upper surface of leaves glabrous or nearly so .... **E. palustre**
7 Tuft of hairs attached to tip of seeds (coma) white or nearly so, seeds with a broad, short beak; margins of stem leaves with mostly 10–30 teeth on a side........ .... **E. ciliatum**
7 Coma brown, seeds beakless; leaf margins with more than 30 teeth on a side .... **E. coloratum**

Epilobium brachycarpum K. Presl
TALL ANNUAL WILLOWHERB *introduced*
IR | **WUP** | CUP | EUP
Annual from a slender taproot. Stems 1–6 dm tall, terete, glabrous, the bark exfoliating near the base. Leaves linear or nearly so, rarely more than 3 mm wide, entire or nearly so, not revolute; lateral veins obsolete or very obscure. Petals about 5 mm long. Capsules about 2 cm long, much longer than their pedicels.

Native of western USA, adventive in gravel pit in Houghton County.

Epilobium ciliatum Raf.
AMERICAN WILLOWHERB native
IR | **WUP** | **CUP** | **EUP** FACW CC 3
Epilobium glandulosum Lehm.
Perennial herb, with over-wintering leafy rosettes. Stems often branched, 3–10 dm long, smooth below, short-hairy above, especially in the head (where often with gland-tipped hairs). Leaves opposite, usually alternate near top; lance-shaped to ovate, 3–10 cm long and to 3 cm wide; margins with few, small, forward-pointing teeth; sessile or with short, winged petioles to 6 mm long. Flowers usually nodding when young, on stalks 3–10 mm long, on branches from upper leaf axils; sepals ovate, 2–5 mm long; petals white (or pink), notched at tip, 2–8 mm long. Fruit a linear capsule, 4–8 cm long, with gland-tipped hairs; seeds 1 mm long, the coma white. July–Sept.

Shores, streambanks, marshes, wet meadows, seeps, ditches and other wet places.

Epilobium coloratum Biehler
PURPLE-LEAF WILLOWHERB native
IR | **WUP** | **CUP** | **EUP** OBL CC 3
Perennial herb, producing basal, leafy rosettes in fall; similar to American willowherb (*E. ciliatum*) but larger. Stems 5–10 dm long, much-branched in

the head, smooth below, short-hairy above with hairs often in lines; stems and leaves often purple-tinged. Leaves mostly opposite, becoming alternate and smaller above, lance-shaped, 5–15 cm long and 0.5–3 cm wide, long-tapered to a pointed tip; margins finely toothed, with irregular sharp teeth; short-petioled to sessile. Flowers many on branches from upper leaf axils; sepals lance-shaped, 2–3 mm long; petals pink or white, 3–5 mm long, notched at tip; individual flowers on stalks to 10 mm long. Fruit a linear capsule, 3–5 cm long; seeds 1.5 mm long, the coma brown when mature. July–Sept.

Shores, seeps, swamps and wet woods, wet meadows, fens, ditches.

Epilobium hirsutum L.
HAIRY WILLOWHERB *introduced (invasive)*
IR | WUP | CUP | **EUP**   FACW

Perennial herb, spreading by rhizomes. Stems much-branched, 5–15 dm long, upper stems with a dense covering of soft, straight hairs. Leaves opposite (but bracts alternate), lance-shaped or oblong, 5–10 cm long and 1–3 cm wide, hairy on both sides, somewhat clasping at base; margins with sharp, forward-pointing teeth; petioles absent. Flowers upright on stalks from upper leaf axils, petals red-purple, 10–15 mm long, shallowly notched; stigma 4-lobed. Fruit a hairy, linear capsule, 5–8 cm long; coma nearly white. June–Sept.

Introduced from Eurasia, known from Chippewa and Mackinac counties in marshes, shores, wet meadows, and ditches.

Epilobium leptophyllum Raf.
BOG WILLOWHERB *native*
IR | WUP | CUP | EUP   OBL CC 6

Perennial herb, similar to marsh willowherb (*E. palustre*) but somewhat larger and more hairy. Stems simple or branched, 2–10 dm long, with short, incurved hairs. Leaves opposite or alternate, linear or linear lance-shaped, 2–7 cm long and 1–6 mm wide, upper surface hairy, underside hairy, at least on midvein, lateral veins indistinct; margins entire and rolled under; petioles short or more or less absent. Flowers erect in upper leaf axils on short, slender stalks to 1 cm long; petals light pink, 3–5 mm long, entire or slightly notched at tip. Fruit a linear, finely hairy capsule, 4–5 cm long; the coma yellow-white. July–Sept.

Swamps, marshes, open bogs, sedge meadows, shores, streambanks and springs.

Epilobium palustre L.
MARSH WILLOWHERB *native*
IR | WUP | CUP | EUP   OBL CC 10

Perennial herb, from slender rhizomes or stolons. Stems simple or with a few branches above, 1–6 dm long, upper stem hairy with small incurved hairs. Leaves mostly opposite, lance-shaped, erect or ascending, 2–6 cm long and 3–15 mm wide, tapered to a rounded tip, upper surface smooth or with sparse hairs along midvein, underside smooth or finely hairy along midvein, lateral veins distinct; margins entire and often rolled under; sessile. Flowers few in upper leaf axils, on short stalks; petals white to pink, 3–5 mm long, notched at tip. Fruit a linear, finely hairy capsule; coma pale. July–Aug.

Open bogs and swamps.

Epilobium strictum Muhl.
DOWNY WILLOWHERB *native*
IR | WUP | CUP | EUP   OBL CC 8

Perennial herb, spreading by slender rhizomes; plants densely soft white-hairy. Stems erect, simple or branched above, 3–6 dm long. Lower leaves opposite, upper leaves alternate; lance-shaped, ascending, 2–4 cm long and 3–8 mm wide, tapered to a rounded tip; margins mostly entire, rolled under; sessile. Flowers on slender stalks from upper leaf axils; petals pink, 5–8 mm long, notched at tip. Fruit a linear, densely hairy capsule; coma pale brown. July–Aug.

Conifer swamps, sedge meadows, calcareous fens, marshes; Keweenaw and Luce counties.

Ludwigia *Primrose-Willow*

Perennial herbs. Stems floating, creeping, or upright. Leaves simple, opposite or alternate, entire. Flowers single in leaf axils; sepals 4; petals 4 (or absent), yellow or green, large or very small; stamens 4; stigma unlobed. Fruit a 4-chambered, many-seeded capsule; seeds without a tuft of hairs at tip (coma).

1 Leaves opposite; stems floating, or creeping and rooting at nodes . . . . **L. palustris**
1 Leaves alternate; stems erect or ascending . . . . **L. polycarpa**

Ludwigia palustris (L.) Ell.
MARSH PRIMROSE-WILLOW *native*
IR | WUP | CUP | EUP   OBL CC 4

Perennial herb. Stems weak, creeping and rooting at nodes or partly floating, simple to branched, 1–5 dm long, succulent, smooth or with sparse scattered hairs. Leaves opposite, lance-shaped to ovate, 0.5–3 cm long and 0.5–2 cm wide, shiny green or red, margins entire, tapered at base to a winged petiole to 2 cm long. Flowers single in leaf axils, stalkless; sepals broadly triangular, 1–2 mm long; petals usually absent, or small and red. Fruit a capsule, 2–5 mm long and 2–3 mm wide, somewhat 4-angled, with a green stripe on each angle. July–Sept.

Shallow water or exposed mud of pond margins, lakeshores, streambanks, ditches, springs.

238 ONAGRACEAE | Evening-Primrose Family

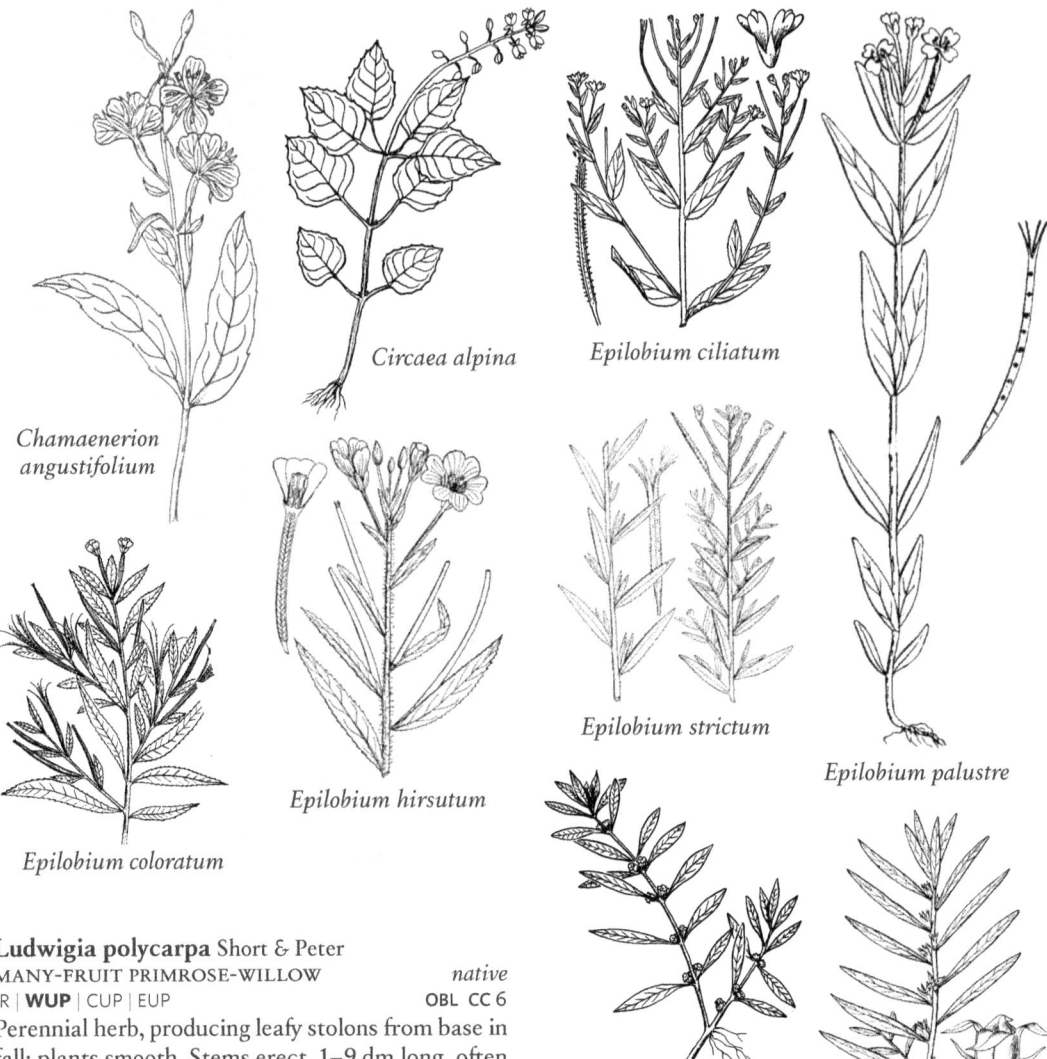

Ludwigia polycarpa Short & Peter
MANY-FRUIT PRIMROSE-WILLOW   native
IR | **WUP** | CUP | EUP   OBL CC 6

Perennial herb, producing leafy stolons from base in fall; plants smooth. Stems erect, 1–9 dm long, often branched, usually 4-angled. Leaves alternate, lance-shaped to oblong lance-shaped, 3–12 cm long and 5–15 mm wide; margins entire; more or less stalkless. Flowers single in leaf axils, stalkless; sepals triangular, 2–4 mm long, usually persistent; petals green and very small or absent. Fruit a short-cylindric, rounded 4-angled capsule, 4–7 mm long and 3–5 mm wide. July–Sept.

Borders of swamps and marshes, muddy shores, wet depressions; Houghton County.

Oenothera *Evening-Primrose*

Annual, biennial, or perennial herbs with alternate, mostly narrow leaves, and small to large, yellow, white, or pink flowers solitary in the axils or forming a terminal raceme. Flowers 4-merous. Hypanthium slenderly tubular (except one species), much prolonged above the ovary, deciduous from the fruit. Sepals at first connate, often with terminal or subterminal appendages, splitting at anthesis and reflexed. Petals commonly yellow, or in a few species white to pink, usually large and conspicuous but ephemeral. Stamens 8, equal or alternately unequal. Ovary 4-celled; stigma entire, 4-lobed, or deeply 4-cleft into elongate segments. Fruit a capsule; seeds numerous.

Four of our species can be considered part of an *Oenothera biennis* complex, including closely related (and sometimes difficult to distinguish) *O. biennis*, *O. oakesiana*, *O. parviflora*, and *O. villosa*.

1 Flowers white, becoming pink . . . . **O. nuttallii**
1 Flowers yellow 2
2 Ovary round in cross-section or nearly so; fruit round or rounded 4-angled in cross-section, abruptly rounded at base 3

dicots ONAGRACEAE | Evening-Primrose Family

2 Ovary 4-angled; fruit sharply 4-angled or 4-winged, tapered to the base. 6
3 Bases of sepals contiguous 4
3 Bases of awl-shaped sepals separate 5
4 Plant green in aspect, with mostly spreading long hairs and often shorter glandular ones . . . . **O. biennis**
4 Plant, especially the upper portion and inflorescence, gray in aspect, with dense appressed non-glandular hairs . . . . **O. villosa**
5 Calyx, ovary, capsule, and upper leaves or bracts densely pubescent with appressed whitish non-glandular hairs; largest leaves typically less than 15 mm wide, finely toothed . . . . **O. oakesiana**
5 Calyx, ovary, capsule, and other parts glabrate to sparsely pubescent, often with some long spreading hairs as well as shorter glandular hairs; largest leaves various, usually at least 15 mm wide, nearly entire . . . . **O. parviflora**
6 Petals to 8 mm long; tip of stem nodding when in bud. . . . . **O. perennis**
6 Flower petals 15–30 mm long; tip of stem erect 7
7 Calyx glabrous or nearly so, the free tips at most 2 mm long (often less than 1 mm); capsules 5–10 mm long . . . . **O. fruticosa**
7 Calyx densely long-hairy, at least on the free tips, which are 2–3.5 mm long; capsules 12–13 mm long . . . . **O. pilosella**

Oenothera biennis L.
COMMON EVENING-PRIMROSE   native
IR | WUP | CUP | EUP   FACU CC 2

Biennial herb. Stems 1–2 m tall, often suffused with red, terete or becoming somewhat angled in the inflorescence. Leaves lance-shaped to oblong, 1–2 dm long, acute or acuminate, entire to repand–dentate, often crisped on the margin, sessile or short–petioled, glabrous to sparsely pubescent but always green. Flowers several to many in a terminal raceme, the bracts resembling the leaves but much smaller; hypanthium tube 3–6 cm long; petals yellow, 15–25 mm long. July–Oct.

Common in fields, roadsides, and waste places.

Oenothera fruticosa L.
SUNDROPS   native
IR | WUP | CUP | EUP   CC 7

Perennial herb. Stems to 1 m tall. Leaves narrow, alternate. Flowers in dense terminal clusters; petals 4, bright yellow, to ca. 2.5 cm long.

Dry fields and open places; sometimes in marshy habitats.

Oenothera nuttallii Sweet
WHITE-STEM EVENING-PRIMR   introduced
IR | WUP | **CUP** | EUP

Perennial herb. Stems erect, 3–8 dm tall, with white bark exfoliating toward the base, glabrous below, minutely glandular-puberulent in the inflorescence.

Leaves linear, 2–8 cm long, entire or nearly so, tapering to the base, glabrous above, puberulent with incurved hairs beneath. Flowers sessile in the upper axils, nodding in the bud; hypanthium tubular, prolonged 2–2.5 cm beyond the ovary; petals white, turning pink, about 2 cm long. Capsule linear-oblong, 1–3 cm long, minutely glandular; seeds in one row in each cell. June–July.

Prairies; adventive in Dickinson County.

Oenothera oakesiana (Gray) J.W. Robbins ex. S. Wats. & Coult.
OAKES' EVENING-PRIMROSE   native
IR | WUP | CUP | EUP   CC 7

Oenothera biennis var. *oakesiana* A. Gray
O. parviflora var. *oakesiana* (Robbins) Fern.

Biennial herb with a taproot. Stems erect to procumbent, 1–6 dm tall. Leaves narrowly oblanceolate, 4–20 cm long, 0.5–3 cm wide. Flower petals yellow. July–Sept. Very similar to *O. parviflora*, *O. biennis* and *O. clelandii*; differs from *O. parviflora* by its inflorescence lacking gland-tipped hairs; differs from *O. biennis* in having sepals separate at base (in *O. biennis* sepals close together at base); differs from *O. clelandii* by its seeds not being pitted and the fruit being broad at the base.

Sandy or rocky shores, dunes, and clearings along the Great Lakes; occasionally inland along railroads, sandy shores, or disturbed places; the common evening-primrose of Great Lakes sand dunes and beaches.

Oenothera parviflora L.
SMALL-FLOWER EVENING-PRIMROSE   native
IR | WUP | CUP | EUP   FACU CC 2

Biennial herb, similar to *O. laciniata* and *O. nuttallii* but plants usually smaller. Leaves commonly narrowly oblong lance-shaped to almost linear, green, glabrous or sparsely pubescent. Raceme crowded, the bracteal leaves commonly as long as the hypanthium; calyx in anthesis densely strigose to nearly glabrous, the actual end of the sepal represented by a prominent lobe or transverse ridge; petals yellow, 10–15 mm long. June–Oct.

Sandy shores, river banks, fields; along railways and roadsides.

Oenothera perennis L.
LITTLE SUNDROPS   native
IR | WUP | CUP | EUP   FAC CC 5

Perennial herb. Stems erect, 2–6 dm tall, usually simple. Principal leaves oblong lance-shaped to elliptic, 3–6 cm long, obtuse, narrowed to a petiole-like base; bracteal leaves shorter and proportionately narrower. Inflorescence nodding, the axis straightening during anthesis, the flowers becoming erect and opening singly; hypanthium prolonged 3–10 mm, averaging 6.5 mm, beyond the ovary; petals 5–9 mm long.

Capsule 5–10 mm long, obovoid, at first minutely glandular-puberulent, often glabrescent in age. June–Aug.

Moist or dry soil, fields, meadows, and open woods.

Oenothera pilosella Raf.
MIDWESTERN SUNDROPS *introduced*
IR | WUP | CUP | **EUP**   FAC

Perennial herb. Stems erect, 2–6 dm tall, hirsute, especially above, with spreading hairs 1–2 mm long. Leaves lance-shaped or lance-elliptic, to 10 cm long, long-hairy on both sides, mostly sessile, acute or acuminate, very minutely denticulate. Inflorescence compact, many-flowered, the buds erect; hypanthium prolonged 15–25 mm beyond the ovary, hirsute; sepals to 2 cm long, their free tips spreading, distinctly hirsute; petals 15–25 mm long, anthers 5–9 mm long. Capsules sessile, 10–15 mm long, hirsute when young, often glabrescent in age. May–July.

Moist soil, meadows, fields, and open woods; Chippewa (Drummond Island only) and Mackinac counties; our populations likely escapes from cultivation.

Oenothera villosa Thunb.
HAIRY EVENING-PRIMROSE *native*
IR | WUP | **CUP** | EUP   FAC CC 4

Biennial herb, appearing grayish due to pubescence. Stems 1–2 m tall, branching above. Leaves oval, tapering to both ends, to 12 cm long and 2.5 cm wide, margins weakly toothed; sessile; pubescent on both surfaces. Flowers in crowded terminal and axillary spikes; sepals recurved; petals yellow. Fruit an elongate capsule to 2.5 cm long, weakly angled, opening by 4 sections separating at the tip and recurving. July–August.

Fields, shores, roadsides, railroads.

Orobanchaceae
BROOM-RAPE FAMILY

Annual, biennial, or perennial herbs; some genera fleshy, without green color, parasitic on the roots of other plants. Leaves opposite, alternate, or the leaves reduced to scales. Flowers mostly perfect, single or few from leaf axils, or numerous in clusters at ends of stems or leaf axils, usually with a distinct upper and lower lip; calyx 2–5-lobed or toothed, persistent in fruit; ppetals 4–5 (petals sometimes absent); stamens usually 4, inserted on the corolla tube. Fruit a several- to many-seeded 2-valved capsule.

Now includes many former members of Scrophulariaceae.

1 Plants non-green, leaves converted to small, non-photosynthetic bracts 2
1 Plants green, leaves present and photosynthetic, green or sometimes strongly tinged with purple (often blackening upon drying) 4
2 Stems well-branched; the flowers nearly sessile in spike-like racemes . . . . **Epifagus virginiana**
2 Stems unbranched or with only a few branches, the flowers single or few on long pedicels, or crowded into dense, spike-like racemes 3

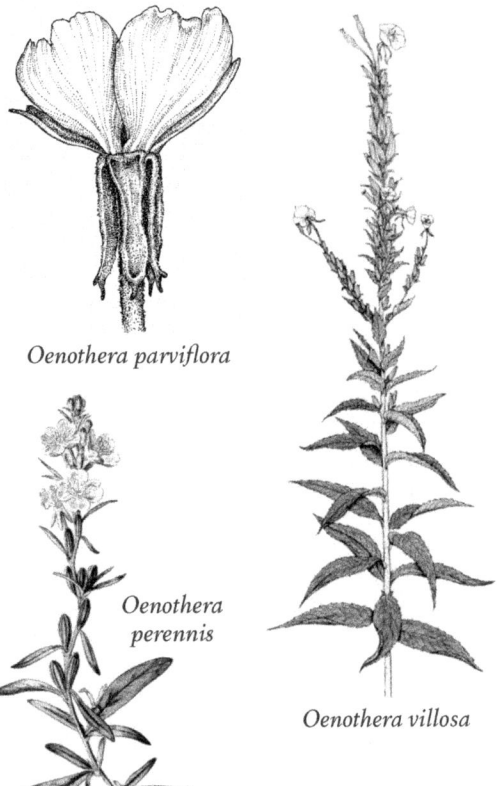

Oenothera parviflora

Oenothera perennis

Oenothera villosa

Oenothera nuttallii

Oenothera biennis

3 Stems thickened and pinecone-like; calyx deeply parted on both upper and lower sides. **Conopholis americana**
3 Stems not pinecone-like; calyx deeply parted on lower side .... **Orobanche**
4 Stem leaves of fertile stems alternate .... **Castilleja**
4 Stem leaves of fertile stems all opposite or nearly so, sometimes alternate below the flowers 5
5 Most leaves deeply pinnately divided; corolla cream-colored or yellow, 1.5–5 cm long 6
5 Stem leaves toothed or entire, not deeply pinnately lobed; corolla various colors, in most species less than 1.5 cm long 7
6 Flowers in dense racemes at ends of stems and branches; corolla strongly 2-lipped .... **Pedicularis**
6 Flowers from leaf axils; corolla 5-lobed and only weakly 2-lipped .... **Aureolaria**
7 Calyx somewhat inflated in flower; conspicuously inflated and laterally compressed in fruit; flowers in a one-sided, leafy spike; leaf margins with large, forward-pointing teeth .... **Rhinanthus minor**
7 Calyx not inflated 8
8 Flowers in the axils of alternate bracts, forming a terminal inflorescence 9
8 Flowers solitary in the axils of opposite or whorled leaves or bracts 10
9 Leaves 3–6 times longer than wide, pinnately veined, the midvein prominent .... **Odontites vulgaris**
9 Leaves less than 2 times longer than wide, palmately veined, with 3–5 prominent veins ... **Euphrasia stricta**
10 Corolla nearly regular, the 5 lobes similar ..... **Agalinis**
10 Corolla irregular, 2-lipped, the upper lip 2-lobed, the lower lip 3-lobed 11
11 Leaves much longer than wide, with short petioles .... **Melampyrum lineare**
11 Leaves less than 2 times longer than wide, sessile .... **Euphrasia**

Agalinis *False Foxglove*

Annual hemiparasitic herbs. Stems slender, erect, branched, usually 4-angled. Leaves opposite, linear, stalkless, smooth, or rough-to-touch on upper surface. Flowers showy, in clusters at ends of branches; sepals joined, the calyx 5-lobed, bell-shaped; petals united, corolla 5-lobed, bell-shaped, and nearly regular, only slightly 2-lipped, pink to purple; stamens 4, of 2 different lengths. Fruit a nearly round, many-seeded capsule.

1 Longer pedicels more than 6 mm long, equaling or longer than the calyx .... **A. tenuifolia**
1 Pedicels less than 6 mm long, shorter than or equaling the calyx 2
2 Corolla 2–3 cm long; calyx lobes less than half length of calyx tube .... **A. purpurea**
2 Corolla less than 2 cm long; calyx lobes nearly as long as calyx tube .... **A. paupercula**

Agalinis paupercula (A.Gray) Britton
SMALL-FLOWER FALSE FOXGLOVE *native*
IR | WUP | **CUP** | **EUP**   OBL CC –
Gerardia purpurea var. *paupercula* Gray

Annual herb. Stems erect, 30–70 cm long, 4-angles, simple or branching. Stem leaves opposite, sessile, linear 2–4 mm wide; branch leaves may be alternate. Flowers in racemes from upper leaf axils; calyx 5-lobed, joined to form a tube at base, separating into lobes above, the lobes nearly as long as the tube; corolla 1.5–2 cm long. Aug–Sept.

Open, sandy wet places.

Very similar to *A. purpurea* and perhaps best considered a variety of that species; *A. paupercula* has smaller flowers and longer calyx lobes than those of *A. purpurea*.

Agalinis purpurea (L.) Pennell
PURPLE FALSE FOXGLOVE *native*
IR | **WUP** | **CUP** | **EUP**   FACW CC 7
Gerardia purpurea L.

Annual herb. Stems slender, 2–8 dm long, 4-angled, smooth to slightly rough, branched and spreading above. Leaves opposite, spreading, linear, 1–5 cm long and 1–3 mm wide; margins entire; petioles absent. Flowers on spreading stalks 2–5 mm long, in racemes on the branches; calyx 4–6 mm long, the lobes less than half the length of the tubular base; corolla purple, 2–3 cm long, the lobes spreading, 5–10 mm long. Fruit a round capsule, 4–6 mm wide. Aug–Sept.

Wet meadows, fens, shores of Great Lakes and along inland lakes and ponds, moist areas between dunes, ditches; usually where sandy, often where calcium-rich.

Agalinis tenuifolia (Vahl) Raf.
COMMON FALSE FOXGLOVE *native*
IR | **WUP** | **CUP** | **EUP**   FACW CC 5
Agalinis besseyana (Britt.) Britt.
Gerardia tenuifolia var. *parviflora* Nutt.

Annual herb. Stems slender, erect, 2–6 dm tall, smooth, usually with many branches. Leaves opposite, spreading, linear, 1–5 cm long and 1–3 mm wide, upper surface slightly rough; margins entire; petioles absent. Flowers on slender, ascending stalks, 1–2 cm long; calyx 3–5 mm long, with short teeth; corolla purple (rarely white), often spotted, 10–15 mm long, the lobes 3–5 mm long. Fruit a round capsule, 4–6 mm wide. Aug–Sept.

Wet meadows, low prairie, fens, shores, streambanks and ditches, usually where sandy.

242 OROBANCHACEAE | Broom-Rape Family

dicots

Aureolaria *False Foxglove*
Aureolaria pedicularia (L.) Raf.
ANNUAL FALSE FOXGLOVE *native*
IR | WUP | **CUP** | EUP CC 7

Annual, partly parasitic on oak roots. Stems much branched, to 1 m tall, finely pubescent and usually densely stipitate-glandular. Principal stem leaves opposite, 3–6 cm long, sessile or subsessile, lance-shaped or ovate in outline, pinnatifid, the 5–8 pairs of pinnae irregularly serrate; uppermost leaves often irregularly alternate, subtending large, yellow, solitary, pediceled flowers. Pedicels curved upward, usually 1–2 cm long at anthesis, stipitate-glandular; calyx campanulate, the 5 lobes spreading, 7–10 mm long, shorter than the tube, the tube stipitate-glandular; corolla usually yellow, 2.5–4 cm long; stamens 4; style slender; stigma 1. Capsule ellipsoid or ovoid, 10–15 mm long; seeds few to several. Aug–Sept.

Dry upland woods; Menominee County.

Castilleja *Indian-Paintbrush*

Annual or perennial hemiparasitic herbs. Leaves alternate, often pinnatifid. Flowers in dense terminal spikes, each subtended by a large, entire or pinnatifid, sometimes brightly colored, bracteal leaf. Calyx tubular, divided into two lateral halves. Corolla tube very slender, usually surpassing the lips, scarcely dilated; upper corolla-lip triangular-acuminate; lower corolla lip much shorter than the upper. Stamens 4, ascending under the upper corolla lip and not exsert. Capsule ovoid or oblong, many-seeded.

1. Floral bracts tipped with bright red, orange, or yellow; mid-stem leaves deeply lobed (rarely simple), hairy . . . . **C. coccinea**
1. Floral bracts cream-colored or yellowish (lower ones often purplish); mid-stem leaves entire, unlobed, glabrous . . . . **C. septentrionalis**

Castilleja coccinea (L.) Spreng.
SCARLET INDIAN-PAINTBRUSH *native*
IR | **WUP** | **CUP** | **EUP** FAC CC 8

Annual, more or less pubescent. Stems usually simple, 2–6 dm tall. Principal stem leaves diverse, varying from rarely entire to commonly 3–5-cleft, the segments linear, the lateral segments almost always shorter or narrower than the terminal; bracteal leaves wholly or mostly scarlet (rarely pale), commonly deeply 3-lobed, occasionally 5-lobed. Spike at first dense, 4–6 cm long, elongating to as much as 2 dm in fruit; calyx 2–3 cm long, thin and membranous, often more or less scarlet, deeply divided into two lateral halves; each half gradually widened distally and at the summit broadly rounded or truncate; corolla greenish yellow, little surpassing the calyx, the minute lower lip less than 1/3 as long as the upper. May–Aug.

Meadows, moist prairies, calcareous sandy or gravelly shores, swamps.

Castilleja septentrionalis Lindl.
LABRADOR INDIAN-PAINTBRUSH (THR) *native*
IR | **WUP** | CUP | EUP CC 10

Pubescent perennial. Stems to 50 cm tall. Leaves linear-lanceolate, sessile; bracteal leaves white, often tinged with purple (especially the lower bracts). June–July.

Rock crevices, ledges, open forests, and sandy places near Lake Superior; Houghton and Keweenaw (including Isle Royale) counties. A species of subarctic eastern North America, ranging from Newfoundland to Great Bear Lake, south into New England and the Lake Superior area in Ontario, Michigan, and Cook Co. Minnesota.

Conopholis *Squawroot*
Conopholis americana (L.) Wallr.
AMERICAN SQUAWROOT *native*
IR | **WUP** | **CUP** | **EUP** CC 10

Unbranched perennial herb; parasitic on the roots of oaks (especially *Quercus rubra*). Stems stout, 5–20 cm tall, entirely or mostly concealed by the numerous, fleshy, overlapping leaf-scales, pale brown or yellowish throughout. Leaf-scales ovate, to 2 cm long. Spike usually constituting half of the plant or more, 1.5–2 cm thick; calyx 8–13 mm long, subtended at base by 1 or 2 minute bractlets; corolla tubular, 10–15 mm long, curved downward, very irregular, the upper lip straight, entire or nearly so; the lower lip decurved, lobed. Capsule ovoid, tipped with the persistent style and capitate stigma. May–June.

Deciduous or mixed forests with oak trees.

Epifagus *Beechdrops*
Epifagus virginiana (L.) W. Bart.
BEECHDROPS *native*
IR | WUP | **CUP** | **EUP** CC 10

Freely branched annual herb; parasitic on the roots of beech trees, with small, scattered, alternate leaf scales and numerous, nearly sessile, solitary, axillary flowers, forming a large panicle. Stems 1–5 dm tall, pale brown, usually marked with fine brown-purple lines, with numerous, elongate, ascending branches, the dead dry stems of the previous season persist through the winter and into the next summer. Leaf-scales triangular-ovate, 2–4 mm long. Flowers dimorphic, the lower small, pistillate, fertile, the upper perfect but sterile. Lower flowers about 5 mm long, calyx cup-shaped, strongly 5-ribbed, the 5 short lobes triangular-subulate; corolla not opening, promptly forced off by the developing ovary and persisting for a time on its tip; stamens none; style short.

OROBANCHACEAE | Broom-Rape Family

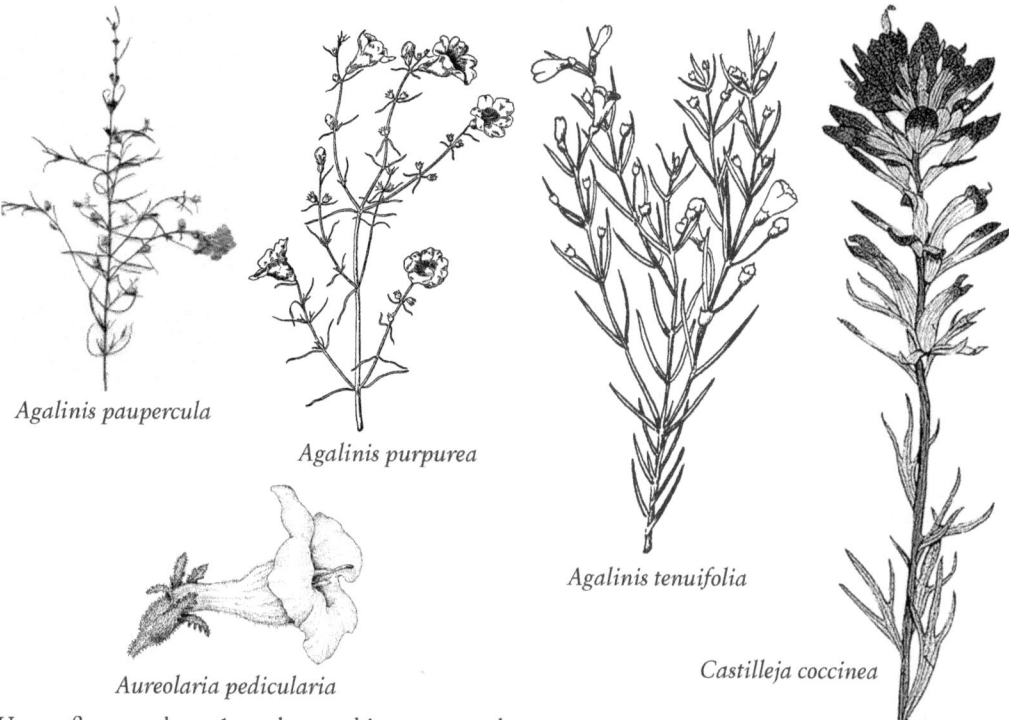

Agalinis paupercula

Agalinis purpurea

Aureolaria pedicularia

Agalinis tenuifolia

Castilleja coccinea

Upper flowers about 1 cm long, white, commonly with two stripes of brown-purple; calyx similar to lower flowers but larger; corolla tubular, shortly 4-lobed, about equaled by the stamens and style. Capsule about 5 mm long, dehiscent across the top. Aug–Sept.

In forests with beech and usually other trees, such as sugar maple, hemlock, oak.

Euphrasia *Eyebright*

Low annual herbs. Leaves opposite, lobed or toothed. Flowers in axils of leaf-like bracts. Calyx 4-lobed; corolla 2-lipped, white or pale blue-purple, with darker purple lines and a bright yellow spot at the base of the lower lip. Fruit a flattened capsule.

ADDITIONAL SPECIES
Euphrasia nemorosa (Pers.) Wallr. (Common eyebright), native annual herb of arctic regions, extending southward in Michigan to of rock crevices and ledges at Scoville Point on Isle Royale and nearby Edwards Island; threatened.

1 Calyx, bracts, and leaves glabrous or at most scabrous on some nerves; calyx lobes and leaf teeth tapered to long bristle tips.......................... **E. stricta**
1 Calyx, bracts, and leaves pubescent; calyx lobes acute to somewhat bristle-tipped....................... 2
2 Stem leaves (not subtending flowers) strongly cuneate at base, the teeth acute to blunt; calyx pubescent principally on the nerves and lobes; plants sometimes flowering as low as the second node ... **E. hudsoniana**
2 Stem leaves broadly cuneate to rounded at base, the teeth rounded to blunt; calyx rather densely hairy throughout; plants flowering only as low as the fifth node; rare on Isle Royale .... **E. nemorosa***

Euphrasia hudsoniana Fernald & Wiegand
HUDSON BAY EYEBRIGHT   (THR) native
IR | WUP | CUP | EUP   CC 8
Low annual herb. Stems erect, to 30 cm tall, simple or branching, minutely villous. Leaves opposite, ovate or elliptic, 1-2 cm long, puberulent, sessile. Flowers from the upper axils of slightly smaller bracts, 7-9 mm long; calyx about equally 4-lobed with acute, ciliate teeth; corolla white to pale lilac, with veins of deeper purple, the upper lip 2-lobed, the lower lip 3-lobed, each lobe notched; stamens included. capsule compressed, pubescent, widened toward the tip; style base short- persistent; seeds wing-edged, ribbed, finely cross-striate.

Rock crevices and ledges at several Isle Royale sites. Often not easily distinguished from *E. nemorosa*.

Euphrasia stricta D. Wolff ex J.F. Lehm.
DRUG EYEBRIGHT   *introduced*
IR | **WUP** | **CUP** | **EUP**
Euphrasia officinalis L.
Low annual herb. Stems 1–3 dm tall, usually freely branched below the middle. Leaves glabrous, opposite, sessile, ovate, conspicuously longer than wide, sharply 3–5-toothed on each margin, the teeth tipped with a hair-like bristle; upper bracteal leaves tending to

become alternate. Flowers small, sessile or nearly so; calyx 4-lobed (the upper median lobe lacking); corolla bilabiate, 6–8 mm long; lower lip 3-lobed, the lateral lobes diverging at an angle of about 60 degrees, white with violet lines; upper lip shallowly 2-lobed or merely notched, suffused with purple; stamens 4, ascending under the upper corolla-lip and not exsert. Capsule usually laterally compressed, truncate or retuse at the tip; seeds several, with about 10 narrow longitudinal wings.

Dry fields, lawns, clearings, trails, roadsides, old railroad grades.

Melampyrum *Cow-Wheat*

Melampyrum lineare Desr.
AMERICAN COW-WHEAT   *native*
IR | WUP | **CUP** | EUP   FACU CC 6

Annual herb, partially parasitic on other plants, often red-tinged when in open habitats. Stems usually branched, 1–4 dm long. Leaves opposite, lower leaves oblong lance-shaped, upper leaves linear or lance-shaped, often toothed near base; petioles short or absent. Flowers from upper leaf axils; calyx tube cup-shaped, calyx lobes 4 or 5, longer than the tube; corolla about 1 cm long, 2-lipped, the upper lip white, the lower pale yellow; stamens 4, ascending under the upper lip and not exsert. Fruit a capsule to 1 cm long. June–Aug.

Common in a wide variety of habitats, ranging from wet to dry forests and openings; in wetlands occasional in swamps and on hummocks in open fens.

Odontites *Eyebright*

Odontites vulgaris Moench
EYEBRIGHT   *introduced*
IR | WUP | **CUP** | EUP

Odontites serotinus Dumort.

Annual herb, usually parasitic on the roots of other plants. Stems 1–4 dm tall, simple or more commonly branched, finely pubescent. Leaves opposite, lance-shaped, sessile with a broad base, 1–3 cm long, roughly pubescent, with 2 or 3 blunt teeth on each margin. Flowers nearly sessile from the upper axils, forming a terminal, often secund spike or raceme; calyx regular, campanulate, about 5 mm long, 4-lobed, shorter than the tube; corolla tubular, about 1 cm long, pubescent, light red; upper lip nearly straight, entire; lower lip spreading, shallowly 3-lobed; stamens 4, slightly unequal, ascending under the upper corolla lip and included by it. Capsule elliptic, pubescent, about 7 mm long, few-seeded, included in the mature calyx. July–Sept.

Native of Europe; established as a weed in fields and waste places; Delta and Menominee counties.

Orobanche *Broom-Rape*

Orobanche uniflora L.
NAKED BROOM-RAPE   *native*
IR | **WUP** | **CUP** | EUP   CC 8

Perennial, parasitic on numerous plants. Aboveground stem to rarely 5 cm long (stem mostly underground), bearing a few overlapping, obovate, glabrous scales. Pedicels 1–4, usually 2, 6–20 cm long, erect, finely glandular-pubescent, without bractlets, each bearing a single white to violet perfect flower about 2 cm long; calyx campanulate, the lobes 5, about equal, triangular-acuminate, slightly longer than the calyx tube; corolla tubular, bilabiate, the tube usually curved downward, much longer than the rounded or acute lobes; stamens 4, about as long as the corolla tube and inserted below its middle. Capsule 2-valved, many-seeded, enclosed by the withering corolla.

Moist woods and streambanks.

Pedicularis *Lousewort*

Perennial herbs (ours). Leaves either opposite, alternate, or scattered, sharply toothed to 2-pinnatifid. Flowers yellow or purple in terminal spikes or racemes, each subtended by a bracteal leaf. Calyx campanulate to tubular, entire or variously lobed, but usually longer on the upper side. Corolla tube gradually enlarged distally; upper lip as long as or longer than the lower, very concave or arched, often laterally compressed; lower lip more or less expanded, with two longitudinal folds below the sinuses. Stamens 4, ascending under the upper corolla-lip and not exsert. Capsule compressed, ovate to oblong, pointed. Seeds several, not winged.

1   Stems glabrous or nealy so; leaves opposite; flowering in late summer . . . . **P. lanceolata**
1   Stems usually long-hairy; leaves alternate; flowering in early summer . . . . **P. canadensis**

Pedicularis canadensis L.
WOOD BETONY   *native*
IR | **WUP** | **CUP** | EUP   FACU CC 10

Perennial herb. Stems several, erect, 1.5–4 dm tall, sparsely villous. Leaves chiefly basal, lance-shaped to oblong lance-shaped, pinnately lobed usually more than halfway to the midvein; lower leaves on petioles often longer than the blade; stem leaves progressively reduced, short-petioled to nearly sessile; bracteal leaves usually toothed only at the tip. Spikes commonly solitary, at anthesis 3–5 cm, in fruit to 20 cm long; corolla yellow, maroon, or yellow on the lower lip and maroon on the upper; the upper lip bearing 2 slender teeth just below the rounded tip; lower lip shorter than the upper. Capsule oblong, about 15 mm long, 2x as long as the mature calyx, opening along the upper side. April–June.

Dry forests and savannas, moist hardwood forests, especially in openings; less often in conifer swamps, meadows, and grasslands.

Pedicularis lanceolata Michx.
SWAMP LOUSEWORT *native*
IR | WUP | **CUP** | **EUP** FACW CC 8

Perennial herb; plants at least partially parasitic on other plants. Stems 3–8 dm long, more or less smooth, unbranched or few-branched. Leaves opposite, or in part alternate, mostly lance-shaped, 4–9 cm long and 1–2 cm wide, pinnately lobed; margins with small rounded teeth; lower leaves short-petioled, upper leaves sessile. Flowers more or less sessile, in spikes at ends of stems and from upper leaf axils; the spikes 2–10 cm long; calyx 2-lobed; corolla yellow, about 2 cm long, the upper lip entire and arched, lower lip upright. Capsule unequally ovate, mostly shorter than the calyx. July–Sept.

Wet meadows, calcareous fens, wetland margins, springs, streambanks.

Rhinanthus *Yellow Rattle*
Rhinanthus minor L.
LITTLE YELLOW RATTLE *introduced*
IR | **WUP** | **CUP** | **EUP** FAC
Rhinanthus crista-galli L.

Annual herb. Stems 2–6 dm tall, simple or branched, retrorsely pubescent on two sides, glabrous on the other two. Leaves narrowly oblong to lanceolate, 1.5–5 cm long, conspicuously serrate; bracteal leaves equaling or exceeding the calyx, sharply laciniate; calyx about 1 cm long at anthesis; corolla 1–2 cm long, mostly yellow, often with pale or colored teeth on the upper lip or with dark markings on the lower. Capsule 10–15 mm long and wide; seeds pale yellowish brown, 4–6 mm long, the wing to 1 mm wide.

Roadsides and waste places; Houghton, Luce, and Schoolcraft counties.

Oxalidaceae
WOOD-SORREL FAMILY

Oxalis *Wood-Sorrel*

Perennial herbs (ours). Leaves basal or alternate on the stem, 3-foliolate, leaflets obcordate. Flowers solitary on axillary peduncles or in cymose or

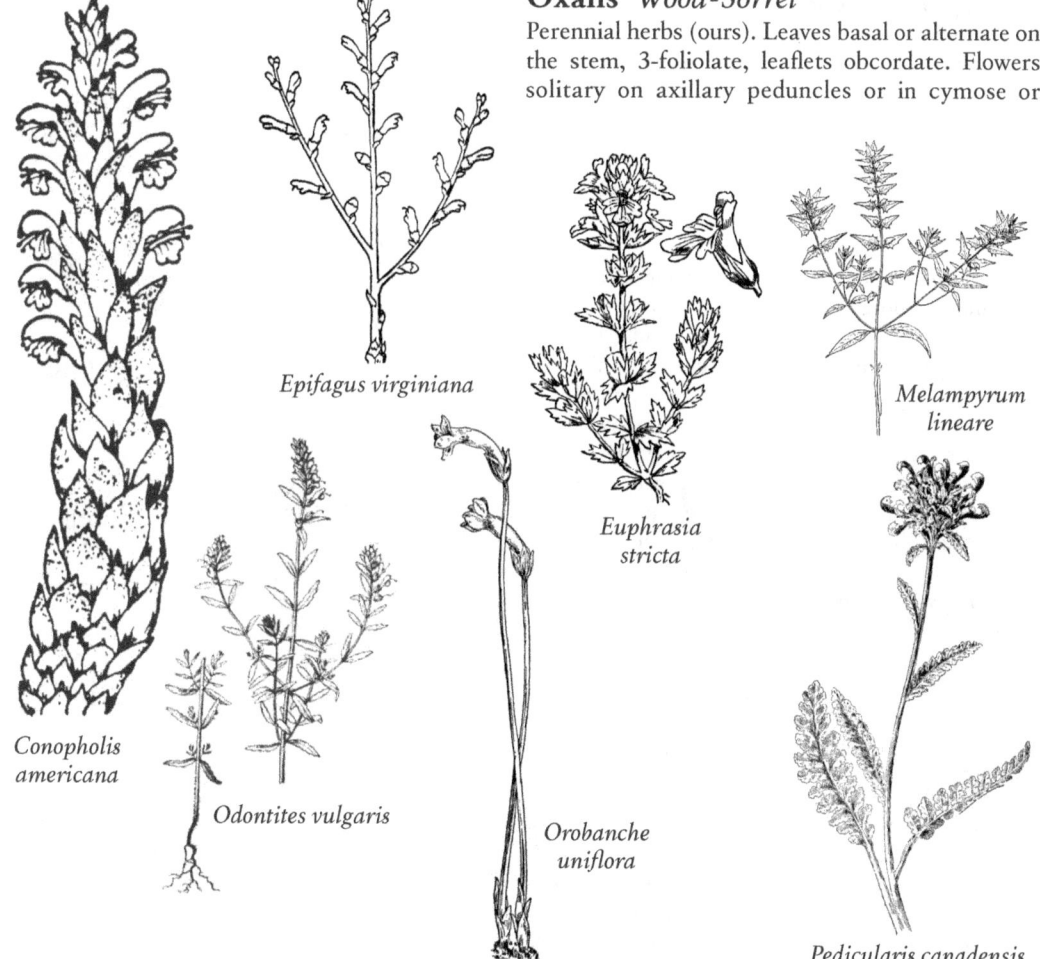

Conopholis americana

Epifagus virginiana

Odontites vulgaris

Euphrasia stricta

Orobanche uniflora

Melampyrum lineare

Pedicularis canadensis

246 PAPAVERACEAE | Poppy Family

umbel-like clusters; perfect, regular, 5-merous, white, yellow, pink, or purple. Sepals usually imbricate. Stamens 10, alternately long and short. Ovary 5-celled; styles 5. Fruit a capsule.

1. Flowers yellow; plants with leaves from stems... **O. stricta**
1. Flowers white to pink; plants without stems, the leaves and scapes all from plant base.......... **O. acetosella**

Oxalis acetosella L.
NORTHERN WOOD-SORREL   *native*
IR | WUP | CUP | EUP   FACU CC 7

Oxalis montana Raf.

Perennial herb, from slender, scaly rhizomes. Leaves single or 3–6 together, all from base of plant, on stalks 4–15 cm long, these joined at base; palmately divided into 3 leaflets, the leaflets notched at tips, sparsely hairy. Flowers perfect, broadly bell-shaped, single atop stalks 6–15 cm long (usually slightly taller than leaves), with a pair of small bracts above middle of stalk; sepals 5, much shorter than petals; petals 5, white or pink and with pink veins, 10–15 mm long. Fruit a smooth, nearly round capsule. May–July.

Hummocks in swamps, wet depressions in forests, moist wetland margins.

Oxalis stricta L.
COMMON YELLOW WOOD-SORREL   *native*
IR | WUP | CUP | EUP   FACU CC 0

Oxalis dillenii Jacq.
Oxalis fontana Bunge

Perennial herb. Stems erect or eventually decumbent, to 5 dm tall, more or less pubescent with ascending or incurved hairs. Leaflets 1–2 cm wide. Pedicels similarly pubescent, at maturity commonly abruptly divaricate or deflexed. Inflorescence umbel-like. Flowers yellow, 5–10 mm long, rarely more than 3 to a peduncle. Capsules 1–3 cm long, densely to thinly gray-pubescent with mostly retrorse hairs 0.1–0.3 mm long, or with longer ones intermingled, or wholly or partly glabrous. Here, includes plants soetimes separated as *Oxalis dillenii* Jacq.

In many different habitats, mostly a common weed of roadsides, railroads, gardens, lawns, fields, and disturbed places; also in trailside in forests.

Papaveraceae
POPPY FAMILY

Herbs, or vines (*Adlumia*), with watery, milky, or colored juice. Leaves alternate or rarely opposite. Flowers regular, perfect. Sepals 2 or 3, early deciduous. Petals 4 or more (rarely absent), separate, conspicuous. Stamens 6, or 12 or more. Ovary 1-celled or falsely 2-celled, of 2 or more carpels. Fruit a capsule, dehiscent by terminal valves or longitudinally (rarely otherwise). Fumariaceae is now included within Papaveraceae, differing in bilateral symmetry of the flowers and watery juice. All of our members of Papaveraceae in the strict sense have colored juice (yellow to red-orange or milky).

ADDITIONAL SPECIES
Glaucium flavum Crantz (Horned poppy), introduced biennial, occasionally escaping from cultivation; Mackinac County; plants very glaucous, lower leaves deeply pinnatifid, upper leaves shallowly lobed and toothed with strongly clasping base.

1. Corolla bilaterally symmetrical; juice watery, clear ... 2
1. Corolla regular, juice colored whitish to yellow or red-orange 4
2. Delicate vine .... **Adlumia fungosa**
2. Upright herbaceous plants 3
3. Corolla white, with 2 spurs; leaves basal ..... **Dicentra**
3. Corolla yellow, 1-spurred; leave alternate .... **Corydalis aurea**
4. Leaf 1 from base of plant; petals 8 or more .... **Sanguinaria canadensis**
4. Plants with leafy stems; petals 4 5
5. Flowers yellow................. **Chelidonium majus**
5. Flowers red, purple, or white................ **Papaver**

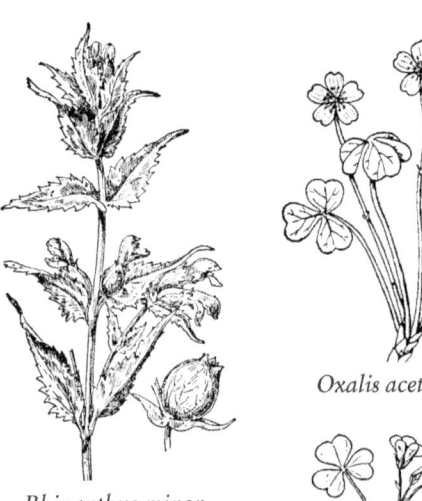

Rhinanthus minor

Oxalis acetosella

Oxalis stricta

Adlumia *Allegheny-Vine*

Adlumia fungosa (Ait.) Greene
ALLEGHENY-VINE   *native*
IR | **WUP** | **CUP** | **EUP**   CC 4

Biennial vine, climbing by the upper part of the rachis of the 3-parted leaves, During the first year acaulescent, with several ascending, non-prehensile, decompound leaves; climbing to 3 m high the second year, with slender elongate stems and large, delicate, prehensile leaves, their rachis elongate and the uppermost leaflets greatly reduced. Flowers pearly pink, in drooping axillary panicles; sepals 2, scale-like, quickly deciduous; corolla bilateral, narrowly flattened-ovoid, the outer petals constricted near the summit to form an ovate appendage, the inner narrow, dilated at the summit into an oval appendage; corolla after anthesis persistent with little change of color, becoming spongy, enclosing the slender 2-valved capsule. June–Sept.

Uncommon in woods, rocky shores, thickets; sometimes where soil has been disturbed.

Capnoides *Rock-Harlequin*

Capnoides sempervirens (L.) Borkh.
ROCK-HARLEQUIN   *native*
IR | **WUP** | **CUP** | **EUP**   CC 5

Corydalis sempervirens (L.) Pers.
Biennial herb, glaucous. Stems slender, erect, 3–6 dm tall, divaricately branched above; principal internodes 4–8 cm long. Lower leaves petioled, upper leaves nearly sessile. Flowers in small panicles at the end of the branches; bracts minute, lance-shaped; corolla pink, tipped with yellow, 12–17 mm long, the tube 8.5–12 mm, the spur 2.5–5 mm long; sepals broadly ovate, 2–4 mm long. Capsule erect or nearly so, 2–4 cm long; seed 1.3–1.5 mm long.

Dry or rocky woods, gravelly shores, especially where disturbed.

Chelidonium *Celandine*

Chelidonium majus L.
CELANDINE   *introduced (invasive)*
IR | WUP | **CUP** | EUP

Biennial herb with saffron-colored juice. Stems branched, 3–8 dm tall. Stem leaves several, alternate, deeply pinnately parted (usually to the midrib) into 5–9 segments; lateral segments often alternate, variously toothed or lobed, the lowest often with a secondary basal division; terminal segments broadly obovate, 3-lobed. Umbel peduncled, several-flowered; sepals 2, deciduous, glabrous; petals 4, yellow, about 1 cm long; stamens numerous, with long slender filaments. Capsule cylindric, glabrous, 3–5 cm long. April–Sept.

Native of Eurasia; established in moist soil; Delta and Schoolcraft counties.

Corydalis *Fumewort*

Corydalis aurea Willd.
GOLDEN CORYDALIS   *native*
IR | **WUP** | **CUP** | **EUP**   CC 5

Biennial herb. Stems 2–6 dm tall, erect or ascending, branched above. Leaves cauline, alternate, 2-pinnately dissected. Flowers short-pediceled, in bracted racemes often surpassed by the upper leaves; bracts lance-shaped; sepals broadly ovate, 1.5–2 mm long, erose; corolla yellow, 12–15 mm long, of which the spur constitutes less than 1/3; outer petals folded distally along the median line into a conspicuous but wingless keel. Capsules slender, 2-valved, terminated by the slender persistent styles, mooth, spreading or drooping, 1–2 cm long; seeds black, shining, 2–2.5 mm wide.

Rocky banks or sandy soil.

Dicentra *Bleedinghearts*

Perennial herbs from rhizomes or a cluster of small tubers. Leaves basal or alternate, compound. Flowers white to red-purple, scapose or in axillary racemes or panicles. Sepals 2, minute. Corolla ovate or cordate, bilaterally symmetrical; petals weakly united, the outer two large, saccate or spurred at base, spreading or ascending at the summit, the inner much narrower and more or less dilated at the summit. Ovary slender, gradually tapering into the long style. Fruit a capsule.

ADDITIONAL SPECIES
Dicentra eximia (Ker Gawl.) Torr. (Wild bleedingheart), introduced ornamental; escaped to rocky woods near city of Marquette, Marquette County.

1 Inflorescence compound; flowers pink to purple; uncommon escape from cultivation . . . . **D. eximia***
1 Inflorescence a simple raceme; flowers white; common native forest species. 2
2 Corolla sac-like, with small rounded spurs about as long as wide; leaves waxy on underside . . . . . **D. canadensis**
2 Corolla with widely spreading spurs, these longer than wide; leaves green or only slightly waxy . **D. cucullaria**

Dicentra canadensis (Goldie) Walp.
SQUIRREL-CORN   *native*
IR | **WUP** | **CUP** | **EUP**   CC 7

Very similar to *D. cucullaria* in foliage, size, and habit. Tubers fewer and about 2x larger; corolla narrowly ovate, the spurs short, broadly rounded, scarcely divergent.

Rich deciduous forests, rarely in swampy or dry forests.

Dicentra cucullaria (L.) Bernh.
DUTCHMAN'S-BREECHES native
IR | WUP | CUP | EUP CC 7

Scapes and leaves from a dense cluster of small, white, grain-like tubers. Leaves broadly triangular in outline, compound, the ultimate segments linear, long-petioled. Scapes 1–3 dm tall, bearing a terminal raceme of 3–12 nodding white flowers, suffused with yellow at the summit; corolla 15–20 mm long; spurs subacute, divergent.

Rich deciduous forests, occasionally in swampy or relatively dry woods.

Papaver *Poppy*

Annual herbs (ours), with colored juice and large, usually long-peduncled flowers terminating the stem and branches. Sepals 2, deciduous. Petals normally 4, white or colored, thin in texture. Stamens numerous. Ovary of 4 to many carpels. Capsule opening by small valves near the margin of the disk.

ADDITIONAL SPECIES
Papaver orientale L. (Oriental poppy), introduced ornamental, reported from Houghton and Schoolcraft counties.

1 Stem leaves clasping at base . . . . **P. somniferum**
1 Stem leaves not clasping . . . . **P. rhoeas**

Papaver rhoeas L.
CORN POPPY introduced
IR | WUP | CUP | EUP

Stems sparsely branched, to 1 m tall, hirsute. Leaves pinnately divided, the pinnae usually lobed or incised. Flowers usually scarlet, varying to purple or white. Capsule subglobose to broadly obovoid, glabrous; stigmatic rays 8–14, usually 10.

Native of Eurasia and n Africa; introduced or escaped but not common.

Papaver somniferum L.
OPIUM POPPY introduced
IR | WUP | CUP | EUP

Stems rather stout, to 1 m tall. Leaves sessile, cordate-clasping, oblong in outline, coarsely toothed or shallowly lobed. Flowers purple or red to white. Fruit subglobose or broadly ovoid, glabrous; stigmatic rays 8–12. June–Aug.

Native of Eurasia, cultivated for ornament and sometimes escaped.

The seeds are commonly used in baking; opium is derived from the milky juice of the capsule.

Sanguinaria *Bloodroot*

Sanguinaria canadensis L.
BLOODROOT native
IR | WUP | CUP | EUP FACU CC 5

Perennial herb with red juice, from a stout rhizome which sends up a single lobed leaf and a large white scapose flower. Leaves orbicular in outline, 3–9-lobed, the lobes undulate to coarsely toothed. Scape 5–15 cm tall at anthesis. Flowers white, varying rarely to pink, 2–5 cm wide; sepals 2; petals typically 8, often more and as many as 16; 4 petals usually longer than the others and the flower quadrangular in outline; stamens numerous. Capsule 3–5 cm long, crowned by the persistent style, dehiscent longitudinally. March–April.

Rich deciduous and floodplain forests.

The leaves continue to expand after anthesis and may grow to 2 dm wide.

Penthoraceae
PENTHORUM FAMILY

Penthorum *Ditch-Stonecrop*

Penthorum sedoides L.
DITCH-STONECROP native
IR | WUP | CUP | EUP OBL CC 3

Perennial herb, spreading by rhizomes; plants often red-tinged. Stems 1–6 dm long, smooth and round in section below, upper stem often angled and with gland-tipped hairs. Leaves alternate, lance-shaped, 2–10 cm long and 0.5–3 cm wide, tapered to tip and base; margins with small, forward-pointing teeth; sessile or on petioles to 1 cm long. Flowers star-shaped, perfect, 3–6 mm wide, on short stalks, in branched racemes at ends of stems; sepals 5, green, triangular, 1–2 mm long; petals usually absent; stamens 10; pistils 5, joined at base and sides to form a ring. Fruit a many-seeded capsule, the seeds about 0.5 mm long. July–Sept.

Streambanks, muddy shores and ditches.

Phrymaceae
LOPSEED FAMILY

Perennial herbs. Calyx tubular, 5-lobed. Fruit a dehiscent capsule. Previously, this family was monotypic with only genus *Phryma*; now includes *Mimulus*.

1 Flowers nearly sessile in pairs in terminal spike-like racemes, subtended by tiny bracts; upper calyx teeth bristle-like; fruit an achene, strongly reflexed . **Phryma**
1 Flowers peduncled, borne singly in the axils of opposite leaves or leafy bracts or in leafy racemes at ends of

PHRYMACEAE | Lopseed Family

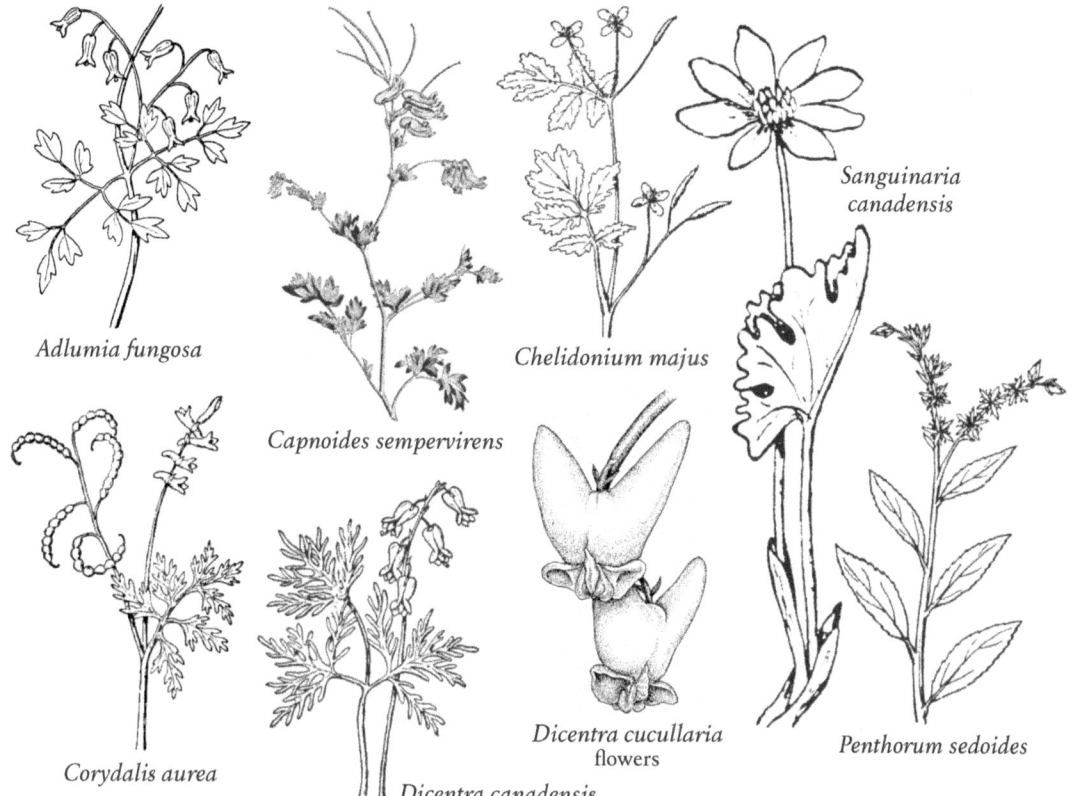

stems; calyx teeth not bristle-like; fruit a capsule; not strongly reflexed......................... **Mimulus**

Mimulus *Monkey-Flower*

Perennial herbs (ours). Leaves opposite, margins shallowly toothed. Flowers often large and showy, single on stalks from leaf axils or in leafy racemes at ends of stems; sepals joined, the calyx tube-shaped; corolla 2-lipped, the upper lip 2-lobed, the lower lip 3-lobed, yellow or blue-violet; stamens 4, of 2 different lengths; stigmas 2. Fruit a cylindric capsule.

1 Flowers blue to violet; stems 4-angled; leaves pinnately veined (with lateral veins arising all along the midrib); stems and pedicels glabrous............. **M. ringens**
1 Corolla yellow; stems terete or many-ridged but not square in cross-section; leaves nearly palmately veined (with lateral veins all near base of blade); stems and pedicels glabrous or pubescent.................... 2
2 Plants ± densely viscid-pubescent (with many hairs longer than 0.5 mm); calyx lobes slightly unequal, half to fully as long as the tube; corolla nearly regular..... .... **M. moschatus**
2 Plants glabrous or with minute, often gland-tipped hairs to 0.5 mm long; calyx lobes very unequal, only the upper large lobe as much as half as long as the tube; corolla 2-lipped.. 3

3 Corolla 3–4.5 cm long, the throat nearly closed by the up-arching lower lip; style 20–25 mm long; plant pubescent with tiny hairs on calyx, pedicels, and stems..... .... **M. guttatus**
3 Corolla 1–2.7 cm long, the throat open; style less than 14 mm long; plants glabrous or with minute glandular hairs 4
4 Style mostly 3–5 mm long............. **M. glabratus**
4 Style mostly 8–11 (–14) mm long.... **M. michiganensis**

Mimulus glabratus Kunth
ROUND-LEAF MONKEY-FLOWER   *native*
IR | **WUP** | **CUP** | **EUP**   OBL CC 10

Perennial herb, spreading by stolons and often forming large mats. Stems succulent, smooth, 0.5–5 dm long, creeping and rooting at nodes, the stem ends angled upward. Leaves opposite, nearly round to broadly ovate, 1–2.5 cm wide, palmately veined, hairy when young, becoming glabrous; margins shallowly toothed or entire; petioles short and winged, or the upper leaves sessile. Flowers yellow, on stalks from leaf axils and at ends of stems; calyx 5–9 mm long, barely toothed, irregular, the upper lobe large, the other lobes smaller; corolla 2-lipped, 9–15 mm long, the throat open and bearded on inner surface. Fruit an ovate capsule, 5–6 mm long. June–Aug.

Cold springs, seeps, and banks of spring-fed streams; usually where calcium-rich.

PLANTAGINACEAE | Plantain Family

Mimulus guttatus DC.
WESTERN MONKEY-FLOWER *native*
IR | **WUP** | CUP | EUP　　　　　　　　　　　　CC 8
Perennial herb. Native annual or perennial herb, spreading by stolons or rhizomes; plants smooth or short-hairy. Stems 0.5–6 dm long, unbranched or branched. Leaves opposite, ovate to obovate, 2–8 cm long and 1–4 cm wide, reduced to bracts in the head; margins with irregular, coarse teeth; petioles short on lower leaves, upper leaves stalkless or clasping. Flowers yellow, showy, in loose racemes at ends of stems; flowers on stalks 1–2.5 cm long; calyx irregular, 10–15 mm long, the upper lobe largest; corolla yellow, often spotted with red or purple, 2-lipped, 2.5–5 cm long, lower lip bearded at base. Fruit an ovate capsule, about as long as calyx tube. July–Sept.

Springs and spring-fed streams; wet, seepy woods; Ontonagon County, where perhaps introduced; native and common in the western USA.

Mimulus michiganensis (Pennell) Posto & Prather
MICHIGAN MONKEY-FLOWER　(END) *native*
IR | WUP | CUP | **EUP**　　　　　　　　　　　　CC 10
Perennial mat-forming herb. Stems to about 40 cm long, reclining at base, rooting at lower leaf nodes and producing additional shoots via stolons. Leaves opposite, broadly ovate, margins nearly entire to coarsely sharp-toothed; petioles shorter than blades. Flowers borne on slender pedicels from the upper leaf axils, bright yellow, tubular, two-lipped, 16–27 mm long; lower lip irregularly red-spotted. Fruit (seldom produced) an oblong capsule 8–10 mm long, seeds many, with longitudinal striations. July–Aug.

Marly springs, in streams in cedar swamps, calcareous shores and ditches, usually near shores of Great Lakes.

Michigan monkey-flower has been on the federal and state list of endangered species since 1990. Its total known range to date, is in the Straits of Mackinac and Grand Traverse regions of Michigan (less than 20 known sites altogether in Mackinac County in the UP, and Benzie, Cheboygan, Emmet, and Leelanau counties (and Beaver Island) in the Lower Peninsula; endemic to Michigan.

Mimulus moschatus Dougl. ex Lindl.
MUSKY MONKEY-FLOWER *native*
IR | **WUP** | CUP | EUP　　　　　　　　　　　　CC 10
Native perennial herb; plants long-hairy, sticky, with a musky odor. Stems 2–4 dm long, lower stems creeping, the tips ascending. Leaves opposite, ovate, 3–6 cm long and 1–2 cm wide, pinnately veined; margins entire or with spaced, coarse teeth; petioles short. Flowers single from upper leaf axils, on slender stalks 1–2 cm long; calyx about 1 cm long, the lobes about equal; corolla yellow, open in the throat, 1.5–2.5 cm long and 2–3x longer than sepals. Fruit a capsule. July–Aug.

Shores, swamp margins, streambanks, springs, ditches, wet forest trails.

Mimulus ringens L.
ALLEGHENY MONKEY-FLOWER *native*
IR | **WUP** | CUP | EUP　　　　　　　　　OBL　CC 5
Smooth perennial herb, from stout rhizomes. Stems usually erect, 3–8 dm long, 4-angled and sometimes winged. Leaves opposite, oblong to lance-shaped, 4–12 cm long and 1–3.5 cm wide, upper leaves smaller; margins with forward-pointing teeth; petioles absent, the base of leaf clasping stem. Flowers single from upper leaf axils, on slender stalks 1–5 cm long and longer than the sepals; calyx regular, angled, 1–2 cm long, the lobes awl-shaped, 3–5 mm long; corolla blue-violet, 2-lipped, 2–3 cm long, the throat nearly closed, the upper lip erect and bent upward, lower lip longer and bent backward. Fruit a capsule, about as long as calyx tube. July–Aug.

Streambanks, oxbow marshes, swamp openings, floodplain forests, muddy shores, ditches; sometimes where disturbed.

Phryma *Lopseed*

Phryma leptostachya L.
AMERICAN LOPSEED *native*
IR | **WUP** | CUP | EUP　　　　　　　　　FACU　CC 4
Perennial herb. Stems erect, 5–10 dm tall, simple or with a few divergent branches. Leaves opposite, ovate, 6–15 cm long; lower petioles to 5 cm long, the upper shorter or the uppermost sessile. Flowers pale purple to white, in elongate, long-peduncled, interrupted spike-like racemes terminating the stem and also from a few upper axils, opposite and horizontal; calyx 2-lipped, the upper 3 lobes bristle-like, with hooked tips when mature, about equaling the tube, the lower 2 very short, broadly triangular; corolla tube scarcely widened upward, the upper lip straight, the lower much longer, spreading, 3-lobed; stamens 4, included. Fruit an achene, contained in the persistent calyx. June–Aug.

Rich deciduous forests, especially moist areas in beech-maple woods, but also in drier forests with oak and sometimes with conifers. Recognized by the distant paired flowers, reflexed fruit, and broad, opposite, petioled leaves.

Plantaginaceae
PLANTAIN FAMILY

Annual or perennial herbs. Leaves simple, entire, all from base of plant. Flowers perfect in a narrow

PLANTAGINACEAE | Plantain Family

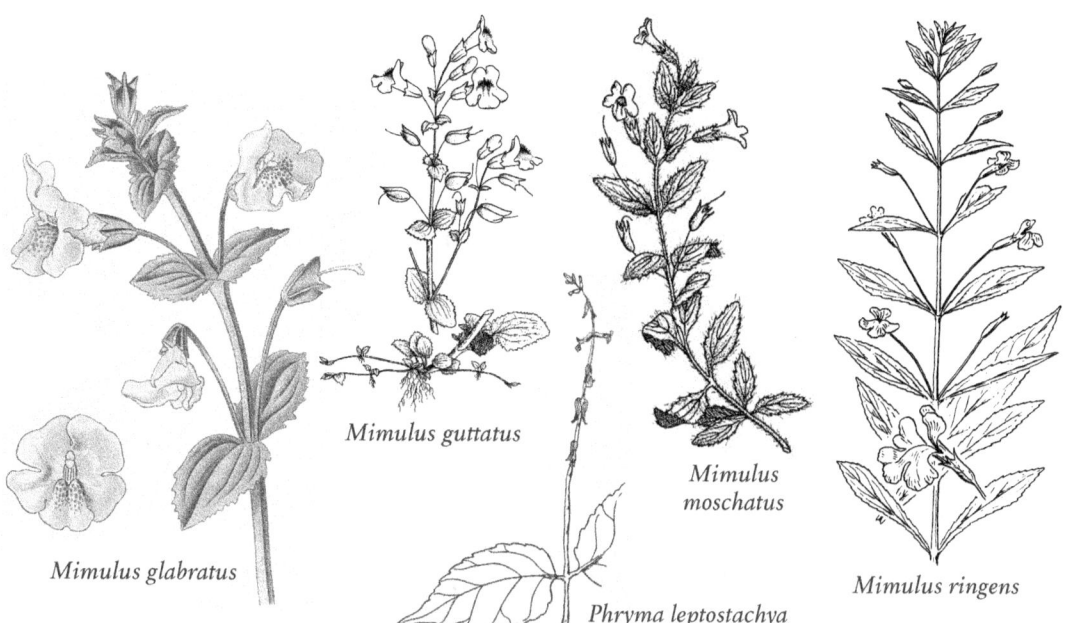

Mimulus glabratus
Mimulus guttatus
Mimulus moschatus
Phryma leptostachya
Mimulus ringens

spike (*Plantago*), each flower subtended by bracts, or single-sexed, the staminate and pistillate flowers on same plant (*Littorella*); flower parts mostly in 4s. Fruit a capsule opening at tip. Plantaginaceae is the correct name for the family that encompasses not only the plantains with their reduced flowers, but also the related larger-flowered genera formerly placed in the Scrophulariaceae as well as highly reduced aquatics, such as *Hippuris* (Hippuridaceae) and *Callitriche* (Callitrichaceae).

1 Flowers tiny, lacking a corolla, or corolla regular and scarious 2
1 Flowers usually conspicuous, with both calyx and corolla present, the corolla petaloid, usually conspicuously bilaterally symmetrical 6
2 Leaves in a basal rosette 3
2 Leaves opposite or whorled on an elongate stem 4
3 Leaves terete (ca. 3 mm or less thick at the middle, thence tapering to tip), 1-veined, glabrous; flowers unisexual (the staminate long-stalked, the pistillate basal); fruit indehiscent; submersed or on moist shores . . . . **Littorella**
3 Leaves flat, in most species with at least 3 prominent veins and/or pubescent; flowers bisexual (in heads or spikes); capsule circumscissile; usually in dry habitats . . . . **Plantago**
4 Leaves in whorls of 6–12 (usually 9) . . . . **Hippuris**
4 Leaves opposite 5
5 Aquatic plants; flowers solitary in the leaf axils, corolla absent . . . . **Callitriche**
5 Introduced plant of dry places; flowers many in short spikes; corolla present . . . . **Plantago arenaria**

6 Stem leaves all or mostly alternate on fertile stems (lowermost leaves sometimes opposite and rosette of larger basal leaves sometimes present) 7
6 Stem leaves all or mostly opposite (rarely whorled) on fertile stems (may be alternate beneath flowers) 12
7 Corolla nearly regular, the lobes equaling or exceeding the tube . . . . **Veronica**
7 Corolla bilaterally symmetrical, ± 2-lipped, the lobes distinctly shorter than the tube (including spur, if any) . . 8
8 Stem trailing or sprawling; leaf blades not over 1.5 times as long as broad; corolla with basal spur; capsules ± spherical, 3–4.5 mm wide . . . . **Cymbalaria**
8 Stem erect; leaf blades (or their principal lobe) over 1.5 times as long as broad; corolla spurred (*Chaenorrhinum, Linaria*) or not; capsules various, mostly longer than broad 9
9 Corolla without spur (at most swollen or saccate at base) . . . . **Digitalis**
9 Corolla with a slender basal spur projecting back between the lower calyx lobes 10
10 Flowers all solitary in axils of leaves (nearly to base of plant); corolla pale purple and white; leaves, calyx, and stem with ± dense gland-tipped hairs **Chaenorrhinum**
10 Flowers in compact or elongate terminal inflorescences (half or less the height of the plant); corolla yellow, red, or blue; leaves, calyx, and usually stem glabrous and eglandular or nearly so 11
11 Corolla 1.3–4 cm long (including spur), yellow (red-pink in a rare weedy annual); seeds strongly wrinkled, tuberculate, ridged, or winged . . . . **Linaria**
11 Corolla 0.6–1.1 cm long, blue; seeds smooth or weakly pebbled . . . . **Nuttallanthus**

12 Inflorescence terminal and branched (± paniculate); stamens 4 fertile plus 1 staminodium . . . . **Penstemon**
12 Inflorescence a spike or raceme (no branched stalks), or flowers all axillary; stamens 2 or 4 fertile, in most genera with no staminodium (or only a very rudimentary one) 13
13 Sepals (at least at anthesis) fused one-third or more the length of calyx; fertile stamens 4 . . **Collinsia parviflora**
13 Sepals separate nearly or quite to the base; fertile stamens 2 or 4 14
14 Corolla 2.3–3.5 cm long; sepals broadly ovate-orbicular, overlapping; stamens 4 fertile plus a filamentous elongate staminodium . . . . **Chelone**
14 Corolla less than 1.5 cm long; sepals linear-lanceolate to somewhat ovate, not conspicuously overlapping; stamens 2 or 4 (including any staminodia) 15
15 Corolla with a spur projecting back at the base; plant with ± dense gland-tipped hairs; leaves linear; fertile stamens 4 . . . . **Chaenorrhinum**
15 Corolla not spurred; plant glabrous or with eglandular hairs (or if with gland-tipped hairs, the leaves not linear); fertile stamens 2 (staminodia filamentous, reduced, or none) 16
16 Leaves in whorls of 3–6, sharply toothed; inflorescence of 1–several dense elongate slenderly tapering spikes or spike-like racemes; corolla tube much longer than the lobes . . . . **Veronicastrum**
16 Leaves opposite, entire or toothed; inflorescence racemose or flowers solitary in axils of alternate or opposite bracts or leaves; corolla tube various 17
17 Corolla 2-lipped, the tube much longer than the lobes; flowers solitary in axils of opposite leaves; sepals 5 . . . . **Gratiola**
17 Corolla often nearly regular, the tube shorter than the lobes (usually a flat limb); flowers in axillary racemes or solitary in axils of bracts or leaves; sepals 4 . . **Veronica**

Callitriche *Water-Starwort*

Small, perennial aquatic herbs with weak, slender stems and fibrous roots. Leaves simple, opposite, all underwater or upper leaves floating; underwater leaves linear, 1-nerved, entire except for shallowly notched tip; floating leaves mostly in clusters at ends of stems, obovate to spatula-shaped, 3–5-nerved, rounded at tip. Flowers tiny, staminate and pistillate flowers usually separate on same plant, each flower with 1 stamen or 1 pistil; single and stalkless in middle and upper leaf axils, or 1 staminate and 1 pistillate flower in each axil, subtended by a pair of thin, translucent, deciduous bracts, or the bracts absent; ovary flattened, oval to round, 4-chambered, separating when mature into 4 nutlets.

1 Leaves all underwater, 1-veined, linear . . . . **C. hermaphroditica**
1 Leaves both underwater and floating; floating leaves 3-veined, spatula-shaped or obovate . . . . **C. palustris**

Callitriche hermaphroditica L.
AUTUMN WATER-STARWORT   *native*
IR | WUP | CUP | EUP   OBL CC 9

Stems 10–30 cm long. Leaves all underwater, alike, linear, 1-nerved, 3–12 mm long and to 1.5 mm wide, shallowly notched at tip, clasping at base, the opposite leaf bases not connected; darker green than our other species. Flowers either staminate or pistillate; single in leaf axils, not subtended by translucent bracts. Fruit flattened, rounded, 1–2 mm long, deeply divided into 4 segments. June–Sept.

Uncommon in shallow to deep water of lakes, ponds, marshes, ditches and streams.

Callitriche palustris L.
VERNAL WATER-STARWORT   *native*
IR | WUP | CUP | EUP   OBL CC 6

Stems 10–20 cm long. Leaves of 2 types; underwater leaves mostly linear, 1–2 cm long and to 1 mm wide, shallowly notched at tip, the leaf pairs connected at base by a narrow wing; floating leaves in clusters at ends of stems or opposite along upper stems, 3–5-nerved, obovate to spatula-shaped, rounded at tip, 5–15 mm long and 2–5 mm wide; leaves intermediate between underwater and floating leaves usually present. Flowers either staminate or pistillate; usually 1 staminate and 1 pistillate flower together in leaf axils, subtended by a pair of translucent bracts, these soon deciduous. Fruit 1–1.5 mm long and about 0.2 mm longer than wide, broadest above middle, narrowly winged near tip, pitted in vertical rows. June–Sept.

Shallow water of lakes, ponds, streams; exposed mudflats.

Chaenorhinum *Dwarf-Snapdragon*

Chaenorhinum minus (L.) Lange
DWARF-SNAPDRAGON   *introduced*
IR | WUP | CUP | EUP

Annual herb. Stems erect, branched, 1–3 dm tall, glandular-pubescent. Leaves linear, 1–2 cm long, obtuse, narrowed to the base but scarcely petiolate. Pedicels 10–15 mm long, arising from many or most leaf-axils; sepals linear-spatulate, unequal, about 3 mm long; corolla 5–6 mm long, blue-purple, with yellow on the palate; spur 1.5–2 mm long. Capsule subglobose, about 5 mm long. June–Sept.

Native of Europe; established in waste places, especially on railway ballast.

PLANTAGINACEAE | Plantain Family

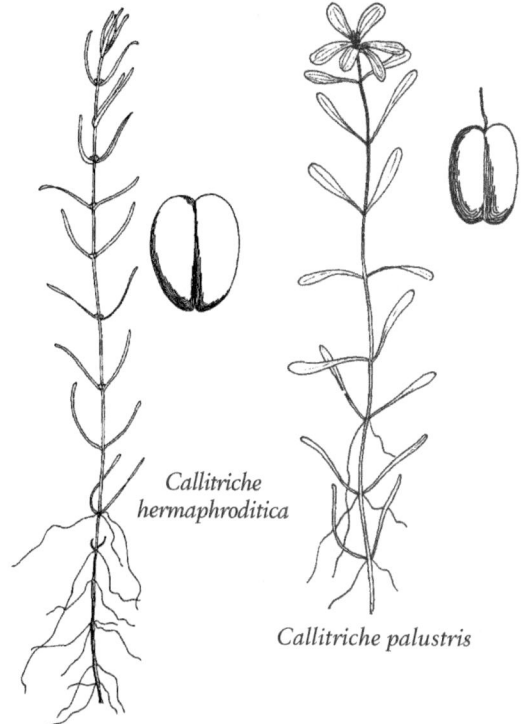

Callitriche hermaphroditica

Callitriche palustris

Chelone *Turtlehead*

Chelone glabra L.
WHITE TURTLEHEAD native
IR | **WUP** | CUP | EUP   OBL CC 7

Perennial herb. Stems erect, 5–10 dm long, rounded 4-angled, unbranched or sometimes branched above. Leaves opposite, lance-shaped, to 15 cm long and 1–3 cm wide, tapered to a sharp tip; margins with sharp, forward-pointing teeth; petioles very short or absent. Flowers in dense spikes at ends of stems, 3–8 cm long; sepals 5; corolla white or light pink, 2.5–3.5 cm long. Fruit an ovate capsule. Aug–Sept.

Swamp openings, thickets, streambanks, shores, wet meadows, marshes, calcareous fens.

Collinsia *Blue-Eyed Mary*

Collinsia parviflora Lindl.
SMALL-FLOWER BLUE-EYED MARY native
IR | **WUP** | CUP | EUP   THR CC 8

Annual herb. Stems to 30 cm tall. Lower leaves opposite, ovate or oblong, entire or few-toothed, glabrous or short-hairy; upper leaves whorled, narrowly oblong or linear, entire, glabrous or short-hairy. Flowers white to blue, long-pediceled, solitary in the leaf axils; calyx nearly regular; corolla irregular, 2-lipped; stamens 4.

Uncommon in crevices and gravelly soil on exposed rock outcrops; disjunct in the UP from the western USA.

Cymbalaria *Kenilworth-Ivy*

Cymbalaria muralis Gaertn., Mey. & Scherb.
KENILWORTH-IVY *introduced*
IR | **WUP** | CUP | EUP

Annual glabrous herb. Stems trailing, rooting at the nodes, 1–4 dm long. Leaves alternate, nearly orbicular in outline with 3–7 shallow palmate lobes, long-petioled. Flowers solitary in the axils, on long slender pedicels; calyx regular, deeply 5-parted; corolla bilabiate, distinctly spurred at base, blue with yellow palate, 7–10 mm long; the throat closed by the prominent palate, the 2 upper lobes erect, the 3 lower spreading; stamens 4. Capsule globose, 3–4 mm wide, rupturing at the tip into 2 pores, the clefts later extending nearly to the base. Summer.

Native of Eurasia; commonly cultivated, escaped near gardens; Houghton County.

Digitalis *Foxglove*

Introduced biennial or perennial herbs (ours). Leaves alternate. Flowers showy, in terminal racemes. Calyx 5-lobed. Corolla short lobed. Stamens 4. in pairs. Some species are a source of cardiac medicines, but plants can be highly toxic or even fatal if consumed.

1 Flowers yellow; leaves glabrous on upperside ....
   .... **D. grandiflora**
1 Flowers purple; leaves pubescent on upperside ....
   .... **D. purpurea**

Digitalis grandiflora P. Mill.
YELLOW FOXGLOVE *introduced*
IR | **WUP** | CUP | EUP

Perennial herb. Stems to 12 dm tall. Leaves glossy. Flowers bell-shaped, pale yellow; inflorescence with dense covering of short, gland-tipped hairs.

Native to Europe and western Asia; in the UP, escaped from cultivation and reported from several roadside locations.

Toxic, potentially fatal if eaten.

Digitalis purpurea L.
COMMON FOXGLOVE *introduced*
IR | **WUP** | CUP | EUP

Biennial herb. Stems to 1 m tall. Leaves ovate, oblong or lanceloate, entire or somewhat toothed, petioled, hairy. Flowers purple or sometimes white, dotted within, large and showy; pediceled; many in long, terminal racemes; calyx nearly regular; corolla irregular, slightly 2-lipped; stamens 4.

A native of western Europe, escaped and persisting in forests, clearings, and forest margins.

Gratiola *Hedge-Hyssop*

Low annual or perennial herbs of shallow water and shores. Leaves opposite. Flowers on stalks from leaf axils; sepals 5; corolla white or yellow, 2-lipped, the upper lip entire or 2-lobed, the lower lip 3-lobed; fertile stamens 2. Fruit a 4-chambered capsule.

1. Plants perennial, spreading by rhizomes; flowers bright yellow; leaves entire, widest at base .... **G. aurea**
1. Plants annual, rhizomes absent; flowers white; leaves toothed, widest near middle of blade ..... **G. neglecta**

Gratiola aurea Pursh
GOLDEN HEDGE-HYSSOP (THR) native
IR | **WUP** | **CUP** | **EUP** OBL CC 10
Gratiola lutea Raf.

Perennial herb. Stems ascending or creeping, 1–3 dm long, somewhat 4-angled, smooth or glandular hairy. Leaves opposite, lance-shaped to ovate, 1–2.5 cm long, with dark, glandular dots; margins entire or with a few small teeth; petioles absent. Flowers on slender stalks 5–15 mm long from leaf axils; sepals lance-shaped, 4–5 mm long; corolla bright yellow, 10–15 mm long. Fruit a round capsule, 2–3 mm long, about as long as sepals. July–Sept.

Shallow water of lakes, wet sandy or gravelly shores; rare in Chippewa and Gogebic counties.

Patches of small, sterile plants may occur with larger plants, in water to 1 m or more deep.

Gratiola neglecta Torr.
CLAMMY HEDGE-HYSSOP native
IR | **WUP** | **CUP** | **EUP** OBL CC 5

Annual herb. Stems erect to horizontal, 5–25 cm long, usually branched, glandular-hairy above. Leaves opposite, linear to lance-shaped, 5–25 mm long and 1–10 mm wide, clasping at base; margins entire to wavy-toothed; petioles absent. Flowers single in the leaf axils, on slender stalks 1–2 cm long, subtended by a pair of small narrow bracts; sepals 5, unequal, 3–6 mm long, enlarging after flowering; corolla white, tube-shaped, slightly 2-lipped, 6–10 mm long; stamens 2. Fruit an ovate capsule, 3–5 mm long. June–Sept.

Mud flats, shores of ponds and marshes.

Hippuris *Mare's-Tail*

Hippuris vulgaris L.
COMMON MARE'S-TAIL native
IR | **WUP** | **CUP** | **EUP** OBL CC 10

Perennial herb, from large, spongy rhizomes. Stems 2–6 dm long, unbranched, underwater and lax, or emersed and upright, densely covered by the closely spaced whorls of leaves. Leaves numerous, in whorls of 6–12, linear, 1–2.5 cm long and 1–3 mm wide, stalkless. Flowers very small, perfect, stalkless and single in upper leaf axils, or often absent; sepals and petals lacking; stamen 1, style 1, ovary 1-chambered. Fruit nutlike, oval, 2 mm long. June–Aug.

Shallow water or mud of marshes, lakes, streams and ditches.

Linaria *Toadflax*

Perennial herbs (ours), almost always glabrous, with erect flowering stems. Leaves numerous, narrow. Flowers several to many in terminal racemes. Calyx deeply 5-parted. Corolla irregular, strongly bilabiate, spurred at base, the upper lip erect, 2-lobed, the lower 3-lobed. Stamens 4. Capsule ovoid to globose.

ADDITIONAL SPECIES
Linaria spartea (L.) Chaz. (Ballast toadflax), introduced, rare escape from cultivation; Menominee County. Plants normally with yellow corolla but corolla of Michigan collection reddish pink.

1. Leaves 6 mm wide or more, ovate, the upper leaves clasping stem .... **L. dalmatica**
1. Leaves less than 5 mm wide, linear, sessile or with petioles .... **L. vulgaris**

Linaria dalmatica (L.) P. Mill.
DALMATIAN TOADFLAX introduced
IR | **WUP** | **CUP** | **EUP**

Stout, glaucous perennial, spreading by rhizomes and forming colonies. Stems 4–12 dm tall, branched above. Leaves numerous, ovate or lance-ovate, 2–5 cm long, palmately veined, sessile and clasping. Flowers short-pedicellate or nearly sessile in elongate racemes, bright yellow, with well developed, orange-bearded palate, the spur about as long as the rest of the corolla. Capsule broadly ovoid-cylindric, 6–8 mm long; seeds irregularly wing-angled. July–Aug.

Native of Europe; roadsides and other disturbed sites.

Linaria vulgaris P. Mill.
BUTTER-AND-EGGS introduced (invasive)
IR | **WUP** | **CUP** | **EUP**

Perennial; spreading by rhizomes and forming colonies. Stems erect, 3–8 dm tall. Leaves very numerous, pale green, 2–5 cm long, 2–4 mm wide, narrowed below to a petiolelike base. Flowers numerous in a compact spike, yellow with orange palate, 2–3 cm long, including the spur. Capsule round-ovoid, 8–12 mm long, the seeds winged. May–Sept.

Native of Europe; fields, roadsides, and waste places.

Littorella *Shoreweed*

Littorella uniflora (L.) Aschers.
AMERICAN SHOREWEED native
IR | **WUP** | **CUP** | **EUP** OBL CC 10
Littorella americana Fern.

Low perennial herb; plants clumped, often forming

mats. Leaves bright green, linear, to 5 cm long and 2–3 mm wide, succulent; margins entire. Flowers only from emersed plants, single-sexed, staminate and pistillate flowers on same plant; staminate flowers 1–2 on stalks to 4 cm long; pistillate flowers stalkless among the leaves; sepals 4 (sometimes 3 in pistillate flowers), lance-shaped, 2–4 mm long, with a dark green midrib and lighter margins; petals joined, 4-lobed; stamens 4, longer than the petals. Fruit a 1-seeded nutlet, 2 mm long and 1 mm wide. July–Aug.

Sandy or mucky lakeshores, or in water 1 m or more deep.

Nuttallanthus *Oldfield-Toadflax*

Nuttallanthus canadensis (L.) D.A. Sutton
OLDFIELD-TOADFLAX   native
IR | WUP | **CUP** | EUP   CC 8

Linaria canadensis (L.) Chaz.
Annual herb. Stems erect, 2–6 dm tall, glabrous, with several procumbent or widely spreading sterile shoots from the base. Leaves narrowly linear, 1–3 cm long, those of the erect stems widely scattered, alternate, those of the sterile shoots smaller, crowded, often opposite or whorled. Racemes congested at anthesis, later elongate; pedicels 2–4 mm long; corolla blue, the lips much longer than the tube; lower lip with 2 short white ridges. Capsule 3–4 mm wide. May–Aug.

Dry, open, sandy or rocky sterile ground; oak savanna, jack pine plains, dried lake beds; Delta and Dickinson counties.

Sometimes misidentified as the less common *Lobelia kalmii*, with which it shares narrow leaves and blue bilaterally symmetrical flowers, but *L. kalmii* is found in wetlands and has milky juice.

Penstemon *Beardtongue*

Perennial herbs (ours), the erect stems rising from a rosette of petioled basal leaves, the stem leaves sessile and often clasping, the upper progressively reduced in size (the leaves of all our species are about alike and are of little value in distinguishing species). Flowers white to blue-violet or red-violet, in terminal clusters. Calyx herbaceous, deeply 5-parted, the lobes usually unequal. Corolla tubular or trumpet-shaped, the tube much longer than the lobes, bilabiate, the upper lip erect, 2-lobed, the lower equaling or longer than the upper, 3-lobed. Fertile stamens 4; sterile stamen present, about as long as the fertile stamens. Capsule ovoid or conic, many-seeded.

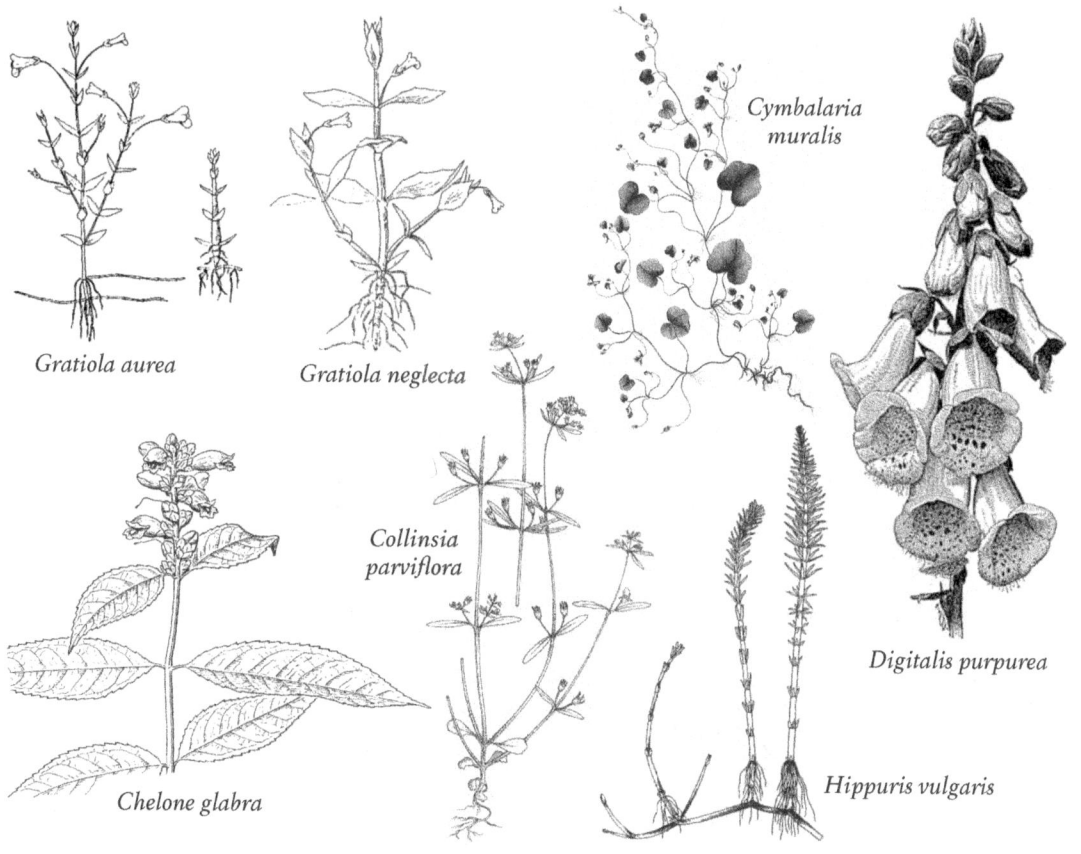

Gratiola aurea

Gratiola neglecta

Cymbalaria muralis

Collinsia parviflora

Digitalis purpurea

Chelone glabra

Hippuris vulgaris

256 PLANTAGINACEAE | Plantain Family

dicots

ADDITIONAL SPECIES
Penstemon calycosus Small (Long-sepal beardtongue), native perennial herb of moist fields, corolla pink-violet uniting to form an open tube; reported from Menominee County; becoming common south of Michigan.

1. Throat of corolla inflated and broader than the tube; larger leaves often more than 1.5 cm wide . . **P. digitalis**
1. Throat of corolla not broader than tube; leaves mostly less than 1.5 cm wide 2
2. Corolla pale to dark violet, the lower lip arched upward and nearly closing the throat; stems long-hairy, the hairs spreading or tangled . . . . **P. hirsutus**
2. Corolla pale violet, the lower lip not arched upward; stems finely hairy, the hairs somewhat appressed to stem . . . . **P. gracilis**

Penstemon digitalis Nutt.
FOXGLOVE BEARDTONGUE   *introduced*
IR | **WUP** | **CUP** | EUP   FAC

Stems to 1.5 m tall, typically glabrous and shining, often suffused with purple or somewhat glaucous. Stem leaves narrowly oblong to narrowly triangular, strictly glabrous beneath, the larger commonly 10–15 cm long. Panicle 1–3 dm long, with erect or strongly ascending branches; calyx at anthesis 6–7 mm long; corolla white or very faintly suffused with violet, usually marked with purple lines within, 23–30 mm long, the tube abruptly dilated near the middle into a wide throat. May–July.

Moist open woods and prairies.

Penstemon gracilis Nutt.
SLENDER BEARDTONGUE   (END) *native*
IR | WUP | **CUP** | EUP   CC 10

Stems 3–5 dm tall, finely puberulent with minute reflexed hairs, often in two longitudinal strips, varying to glabrous. Stem leaves lance-shaped, 5–10 cm long, glabrous or finely pubescent. Inflorescence slender, the short lateral branches erect; calyx 6–9 mm long at anthesis; corolla 15–20 mm long, pale violet.

The sole Michigan collection is from rocky ledges and oak savanna along the Menominee River in Dickinson County.

Penstemon hirsutus (L.) Willd.
HAIRY BEARDTONGUE   *native*
IR | WUP | **CUP** | **EUP**   CC 5

Stems erect, usually several from the same rhizome, 4–8 dm tall, sometimes glabrous below, more often villous, above and in the inflorescence always densely glandular-puberulent, the hairs about 0.5 mm long. Stem leaves lance-shaped to oblong, 5–12 cm long, subentire to dentate, at base rounded or truncate. Calyx at anthesis commonly 5–6 mm long, at maturity more than half as long as the capsule; corolla very slender, not widened toward the summit, about 2.5 cm long, pale violet externally, its mouth nearly or quite closed by the arched base of the lower lip, the tube pubescent internally with pale hairs but lacking purple lines. May–July.

Dry woods and fields; Alger and Mackinac counties.

Plantago *Plantain*

Perennial or annual herbs. Leaves all from base of plant, simple. Flowers small, perfect or single-sexed, green, more or less stalkless in axils of small bracts, grouped into crowded spikes; sepals and petals 4. Fruit a capsule.

Linaria dalmatica

Littorella uniflora

Linaria vulgaris   *Nuttallanthus canadensis*

1 Leaves linear to lance-shaped, more than 5 times longer than wide 2
1 Leaves narrowly ovate to ovate, less than 5 times longer than wide 4
2 Leaves more than 5 mm wide .... **P. lanceolata**
2 Leaves less than 5 mm wide 3
3 Leaves opposite along stem or sometimes whorled; flowers on peduncles from upper leaf axils. **P. arenaria**
3 Leaves all from base of plant; flowers on a naked stalk. .... **P. patagonica**
4 Petioles green; bracts broadly ovate .... **P. major**
4 Petioles red-tinged at base; bracts narrowly lance-shaped. .... **P. rugelii**

Plantago arenaria Waldst. & Kit.
SAND PLANTAIN *introduced*
IR | **WUP** | CUP | EUP

Pubescent annual herb. Stems erect, 1–5 dm tall. Leaves opposite, linear, 3–8 cm long. Peduncles several, opposite from the upper axils. Heads 1–1.5 cm long; lowest bracts rotund at base, abruptly narrowed to a slender point 2–5 mm long; median and upper bracts obovate, rounded at the summit, broadly scarious-margined. July–Aug.

Native of Eurasia; a weed of waste places, roadsides, and railways; Houghton County.

Plantago lanceolata L.
ENGLISH PLANTAIN *introduced*
IR | **WUP** | CUP | EUP   FACU

Perennial herb. Scapes to 6 dm tall, strigose above. Leaves narrowly lance-shaped to oblong lance-shaped, to 3 dm long, including the petiole, gradually tapering to both ends. Spikes very dense, at maturity 1–10 cm long; bracts broadly ovate, with narrow herbaceous center and broad scarious margins, often caudate-tipped. Outer two sepals united into one, broadly obovate, truncate, with two midveins; inner sepals ovate; corolla lobes 2–3 mm long. Capsule ellipsoid, 3–4 mm long, circumscissile near the base; seeds 2, black, 2–3 mm long, deeply concave on the inner face. Variable.

Native of the Old World; a common weed of lawns, roadsides, and waste places.

Plantago major L.
COMMON PLANTAIN *introduced*
IR | **WUP** | CUP | EUP   FACU

Perennial herb; closely resembling P. rugelii in form and often confused with it. Petioles commonly green and pubescent at base. Flowers sessile; bracts shorter than the sepals, broadly ovate or elliptic, with prominent scarious margin and rounded elevated keel; sepals ovate, obtuse, the rounded keel about as wide as the scarious margins. Capsule ellipsoid, 2–4 mm long, circumscissile near the middle; seeds several, commonly 10 or more, about 1 mm long.

Native of Eurasia; naturalized in lawns, roadsides, and waste places.

Plantago patagonica Jacq.
WOOLLY PLANTAIN *introduced*
IR | **WUP** | CUP | EUP

Annual herb, gray-villous throughout. Leaves linear, to 15 cm long, 3–8 mm wide. Scapes 5–15 or rarely 20 cm long. Spikes very dense cylindric, obtuse, 3–10 cm long, 4–6 mm wide; bracts linear, even the lowest scarcely exceeding the flowers.

Dry prairies; adventive in Delta and Gogebic counties.

Plantago rugelii Dcne.
AMERICAN PLANTAIN *native*
IR | **WUP** | CUP | EUP   FAC CC 0

Perennial herb. Leaves broadly elliptic to oval, 5–20 cm long, many-nerved, narrowed at base; petiole margined, at base usually glabrous and tinged with purple. Spikes to 3 dm long, about 5 mm wide, comparatively loose, the axis frequently exposed; pedicels about 0.5 mm long. Bracts lance-shaped, 1/2 to 3/4 as long as the calyx; sepals ovate or oblong, the sharp keel much wider than the scarious margin; corolla lobes less than 1 mm long, reflexed after anthesis. Capsule narrowly ovoid, 4–6 mm long, dehiscent well below the middle; seeds 4–10, black, angular, about 2 mm long.

Lawns, gardens, roadsides, and waste places.

Veronica *Speedwell*

Annual or perennial herbs. Leaves opposite, or becoming alternate in the head. Flowers single or in racemes from leaf axils or at ends of stems; sepals deeply 4-parted, enlarging after flowering; corolla blue or white, 4-lobed, somewhat 2-lipped, the tube shorter than the lobes; stamens 2. Fruit a flattened capsule, lobed or notched at tip; styles usually persistent on fruit.

ADDITIONAL SPECIES
Veronica filiformis Sm. (Thread-stalk speedwell), introduced; Menominee County.

1 Flowers in racemes from leaf axils, or leaves more than 4 cm long, or both; plants perennial 2
1 Flowers single in axils of leafy bracts, or in terminal spikes; leaves less than 3 cm long; plants annual or perennial 8
2 Stems glabrous or nearly so; leaves toothed or entire 3
2 At least the upper stem pubescent; leaves toothed .. 6
3 Leaves with short petioles. 4
3 Leaves sessile 5
4 Leaves mostly widest near leaf base; styles 2.5–3.5 mm long .... **V. americana**
4 Leaves widest above leaf middle; styles to 2.2 mm long .... **V. beccabunga**

258 PLANTAGINACEAE | Plantain Family

dicots

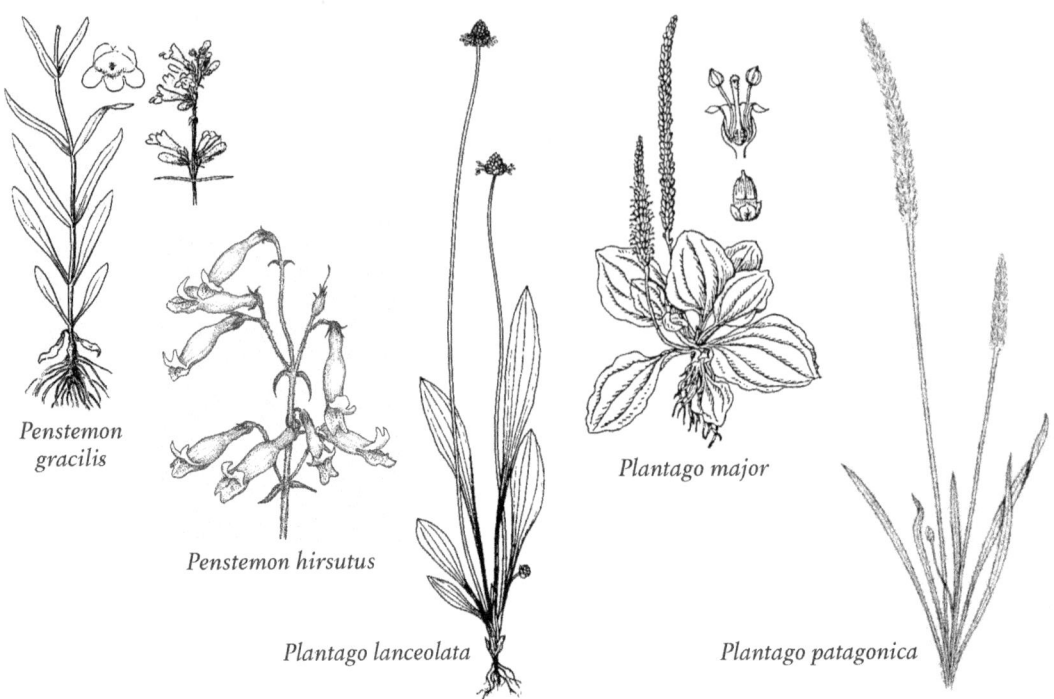

Penstemon gracilis

Penstemon hirsutus

Plantago lanceolata

Plantago major

Plantago patagonica

5   Upper leaves lance-shaped, with wide, clasping bases; rachis of raceme stout and straight; capsules swollen . . . . **V. anagallis–aquatica**

5   Leaves mostly linear, narrowed to a sessile base; rachis of raceme slender and zigzagged; capsules strongly flattened. . . . . **V. scutellata**

6   Flowers on long pedicels, the pedicels longer than the bracts . . . . **V. chamaedrys**

6   Flowers sessile or on short pedicels shorter than the bracts 7

7   Leaves more than 5 cm long, widest near base, on petioles 1 cm or more long; flowers in dense terminal spikes; styles persistent on capsule. . . . . **V. longifolia**

7   Leaves less than 5 cm long, widest near middle, sessile or nearly so; spikes from leaf axils, loosely flowered; styles deciduous. . . . . **V. officinalis**

8   Flowers and capsules on pedicels more than 4 mm long . . . . **V. persica**

8   Flowers and capsules sessile or on short pedicels to 4 mm long 9

9   Middle stem leaves pinnately divided . . . . **V. verna**

9   Leaves entire or toothed, not divided. 10

10 Plants matted perennial herbs; stems finely hairy; leaves glabrous or nearly so (except when young) . . . . **V. serpyllifolia**

10 Plants erect annuals; stems and leaves glabrous or pubescent with mostly spreading hairs 11

11 Flowers blue; the capsules only shallowly notched at tip; stems fleshy; plants glabrous . . . . **V. peregrina**

11 Flowers white; the capsules deeply notched at tip; stems not fleshy; plants pubescent. . . . . **V. arvensis**

Veronica americana Schwein.
AMERICAN-BROOKLIME *native*
IR | WUP | CUP | EUP   OBL CC -

Veronica beccabunga var. *americana* Raf.

Perennial, spreading by rhizomes; plants glabrous and succulent. Stems erect to creeping, 1–6 dm long. Leaves opposite, ovate to lance-shaped (or lower leaves oval), 2–8 cm long and 0.5–3 cm wide, upper leaves tapered to a tip, lower leaves often rounded; margins with forward-pointing teeth; petioles short. Flowers in stalked racemes from leaf axils; the racemes with 10–25 flowers and to 15 cm long; corolla 4-lobed, blue (sometimes white), often with purple stripes. Fruit a more or less round, compressed capsule, 3–4 mm long, slightly notched at tip, the styles persistent, 2–4 mm long. July–Sept.

Streambanks and wet shores, hummocks in swamps, springs.

Veronica anagallis-aquatica L.
BLUE WATER SPEEDWELL *native*
IR | WUP | CUP | EUP   OBL CC 4

Biennial or short-lived perennial, spreading by stolons or leafy shoots produced in fall; plants more or less glabrous. Stems erect to spreading, 1–6 dm long, often rooting at lower nodes. Leaves opposite, lance-shaped to ovate, 2–10 cm long and 0.5–5 cm wide, tapered to a blunt or rounded tip; margins entire or with fine, forward-pointing teeth; petioles absent, the leaves often clasping. Flowers in many-flowered racemes from leaf axils, the racemes 5–12 cm long; corolla 4-lobed, blue or striped with purple, about 5 mm wide. Fruit a round, compressed capsule, 2–4 mm long, notched at tip, the styles persistent, 1–2 mm long. June–Sept.

Wet, sandy or muddy streambanks and ditches; often in shallow water.

Veronica arvensis L.
CORN SPEEDWELL *introduced*
IR | **WUP** | **CUP** | **EUP**   FACU
Annual, villous throughout. Stems erect or nearly so, simple or branched, 1–2 dm tall. Foliage leaves ovate, 6–12 mm long, obtuse, with 2–4 blunt teeth on each side, palmately veined, the lower short-petioled, the upper sessile. Inflorescence often constituting 2/3 of the plant, the bracteal leaves progressively reduced in length and width and mostly entire. Pedicels to 1.5 mm long; calyx lobes oblong, the lower pair 4–5 mm long, the upper pair 3/4 as long; corolla blue, about 2 mm wide. Capsule 3–4 mm wide, nearly as long, deeply notched; style extending about as far as the summit of the capsule lobes. April–June.

Native of Eurasia; established as an inconspicuous weed in gardens, lawns, and fields, and occasionally in open woods.

Veronica beccabunga L.
BROOKLIME *introduced*
IR | **WUP** | **CUP** | **EUP**   OBL
Perennial, closely related to *V. americana* and resembling it in habit. Leaves elliptic or obovate, 2–6 cm long, broadest near or above the middle, usually broadly rounded at the summit, obtuse to rounded at base, crenate. Flowers blue, 5–7 mm wide. Capsules turgid, 3–4 mm wide, nearly as long, not notched, the persistent style about 2 mm long. Summer.

Native of Eurasia, muddy shores and streambanks; Mackinac County.

Veronica chamaedrys L.
GERMANDER SPEEDWELL *introduced*
IR | **WUP** | **CUP** | EUP
Perennial, sparsely pubescent throughout, or in 2 strips only along the stem. Stems prostrate, or ascending at the tip, 2–4 dm long. Leaves sessile or nearly so, ovate, 2–3 cm long, obtuse, serrate, broadly rounded or truncate at base. Racemes few, erect from the upper axils, eventually 8–15 cm long including the peduncle, loosely 10–20-flowered; bracts lance-shaped, half to nearly as long as the slender pedicels; calyx lobes 3–5 mm long; corolla blue with dark blue lines and white orifice, about 1 cm wide; style (after corolla has fallen) 4–5 mm long. Capsules (rarely produced) 4–5 mm wide, very broadly and shallowly notched. May–June.

Native of Europe; introduced in moist gardens, roadsides, and fields.

Veronica longifolia L.
LONG-LEAF SPEEDWELL *introduced*
IR | **WUP** | **CUP** | **EUP**
Perennial. Stems erect, to 1 m tall. Leaves opposite or in whorls of 3, lance-shaped, 4–10 cm long, very sharply serrate; short-petioled. Racemes 1 or few, erect, spike-like, the axis pubescent but not glandular; corolla blue, pubescent in the throat, its lobes 4–5 mm long. Capsule little flattened, 3 mm long, smooth or puberulent, about half as long as the persistent style. June–Aug.

Native of Europe; introduced in fields, roadsides, and waste places.

Veronica officinalis L.
COMMON SPEEDWELL *introduced*
IR | **WUP** | **CUP** | **EUP**   FACU
Perennial, pubescent throughout. Stems prostrate or reclining, the flowering branches erect or ascending, 2–3 dm long. Leaves elliptic, 2.5–5 cm long, obtuse or rounded, uniformly serrate except toward the base, narrowed below to a short petiole. Racemes few, solitary or opposite, spike-like, commonly 3–6 cm long, often interrupted below, on a peduncle of about their own length; bracts linear lance-shaped, about equaling the flowers; calyx lobes oblong, 2–3 mm long; corolla pale violet with darker lines, 5–7 mm wide. Capsule reverse-triangular, glandular-puberulent, 4.5–5 mm wide, truncate or with a broad and very shallow notch; style at maturity to 3 mm long. May–July.

Dry fields and upland woods.

At anthesis the main axis is often developed only slightly beyond the racemes, so that the inflorescence may appear terminal.

Veronica peregrina L.
PURSLANE SPEEDWELL *native*
IR | **WUP** | **CUP** | **EUP**   FAC CC 0
Small annual. Stems upright, 0.5–3 dm long, unbranched or with spreading branches, usually glandular-hairy. Lower leaves opposite, becoming alternate and smaller in the head, oval to linear, 5–25 mm long and 1–5 mm wide, rounded at tip; margins of lower leaves sparsely toothed, upper leaves entire; petioles short or absent. Flowers small, on short stalks from upper leaf axils; corolla 4-lobed, more or less white, about 2 mm wide. Fruit an oblong heart-shaped capsule, 2–4 mm long, notched at tip, the styles not persistent. May–July.

Mud flats, shores, ditches, temporary ponds, swales; also weedy in cultivated fields, lawns and moist disturbed areas.

Veronica persica Poir.
BIRDSEYE SPEEDWELL *introduced*
IR | WUP | **CUP** | EUP

Annual. Stems prostrate or ascending, much branched, 1–3 dm long. Leaves broadly ovate, 8–15 or rarely 25 mm long, obtuse, bearing 3–5 sharp and usually coarse teeth on each side, broadly obtuse to truncate at base; petioles 1–5 mm long. Pedicels solitary in the axils, in fruit more or less recurved and to 2 cm long; calyx lobes ovate lance-shaped, 4–5 mm long at anthesis, to 8 mm long in fruit, conspicuously 3-nerved; corolla blue with deeper blue lines and pale orifice, 8–11 mm wide. Capsule 6.5–10 mm wide, reticulately veined, with a broad triangular notch; style surpassing the capsule. April–Aug.

Native of sw Asia; introduced in gardens, lawns, roadsides, and waste places; Delta County.

Veronica scutellata L.
GRASS-LEAF SPEEDWELL *native*
IR | **WUP** | CUP | EUP   OBL CC 6

Perennial, spreading by rhizomes or leafy shoots produced in fall; plants smooth (or sometimes with sparse hairs). Stems slender, erect to reclining, 1–4 dm long, often rooting at lower nodes. Leaves opposite, linear to narrowly lance-shaped, 3–8 cm long and 2–10 mm wide, tapered to a sharp tip; margins entire or with small, irregularly spaced teeth; petioles absent. Flowers in racemes from leaf axils, the racemes with 5–20 flowers, as long or longer than the leaves; corolla 4-lobed, blue, 6–10 mm wide. Fruit a strongly flattened capsule, 3–4 mm long, notched at tip, the style persistent, 3–5 mm long. June–Sept.

Marshes, pond margins, hardwood swamps, thickets, springs, streambanks, wet depressions.

Veronica serpyllifolia L.
THYME-LEAF SPEEDWELL *introduced*
IR | WUP | **CUP** | EUP   FAC

Perennial. Stems creeping and forming mats, the flowering ones upwardly curved and erect, 1–2 dm tall. Leaves ovate, elliptic, or nearly rotund, mostly 1–1.5 cm, occasionally 2 cm long, obtuse, entire or obscurely crenate, the lower often somewhat narrowed to a short petiole, the upper sessile. Racemes at first short and compact, soon elongating, in fruit to 10 cm long; bracteal leaves elliptic, to 3 cm long; pedicels 2–4 mm long; calyx lobes obtuse. Capsule notched at the apex for about 1/4 of its length, 3–5 mm wide, distinctly wider than long, but much longer than the persistent style. Variable. May–July. Ours subsp. *serpyllifolia* introduced from Europe, with usually usually pale blue flowers with darker blue lines.

Established in fields, meadows, lawns, trails, and old logging roads; extending into moist open woods where it may appear native.

Veronica verna L.
SPRING SPEEDWELL *introduced*
IR | WUP | **CUP** | EUP

Taprooted annual herb. Stems 5-15 cm tall, usually branched from the base. Lower leaves broadly lance-shaped, margins toothed, petioles short; upper leaves sessile and pinnately divided into narrow lobes. Flowers short-stalked, borne singly in the axils of bracts at ends of stems; corolla blue, about 3 mm wide.

A small weed of dry sandy, gravelly, or rocky bare places such as lawns, parking areas, and roadsides.

Veronicastrum *Culver's-Root*

Veronicastrum virginicum (L.) Farw.
CULVER'S-ROOT *native*
IR | **WUP** | CUP | EUP   FAC CC 8

Erect perennial herb. Stems 1–2 m tall, usually with several upright branches. Leaves in whorls of 3–6, lance-shaped; margins with fine, forward-pointing teeth; petioles to 1 cm long. Flowers in erect, spike-like racemes to 15 cm long, the flowers crowded and spreading; corolla white, nearly regular, 4–5 parted, the lobes shorter than the tube; stamens 2, long-exserted from the corolla mouth. Fruit a capsule, 4–5 mm long. June–Aug.

Moist to wet prairies, fens and streambanks; also in drier deciduous woods and sandy grasslands.

Polemoniaceae
PHLOX FAMILY

Perennial herbs (ours). Leaves opposite (*Phlox*) or pinnately divided (*Polemonium*). Flowers perfect (with both staminate and pistillate parts), single or in clusters at ends of stems and from leaf axils; sepals and petals 5-parted and joined for part of length. Fruit a 3-chambered capsule, with usually 1 seed per chamber.

1 Leaves opposite or mostly so.................. **Phlox**
1 Leaves alternate .... **Collomia linearis**

Collomia *Mountain-Trumpet*

Collomia linearis Nutt.
NARROW-LEAF MTN-TRUMPET *introduced*
IR | **WUP** | CUP | EUP   FACU

Annual herb. Stems simple or branched above, 1–4 dm tall, minutely pubescent below, increasingly so above. Leaves alternate, linear to narrowly lance-shaped, entire or pinnatifid, 2–6 cm long, very finely puberulent. Flowers in sessile cymes, several in the

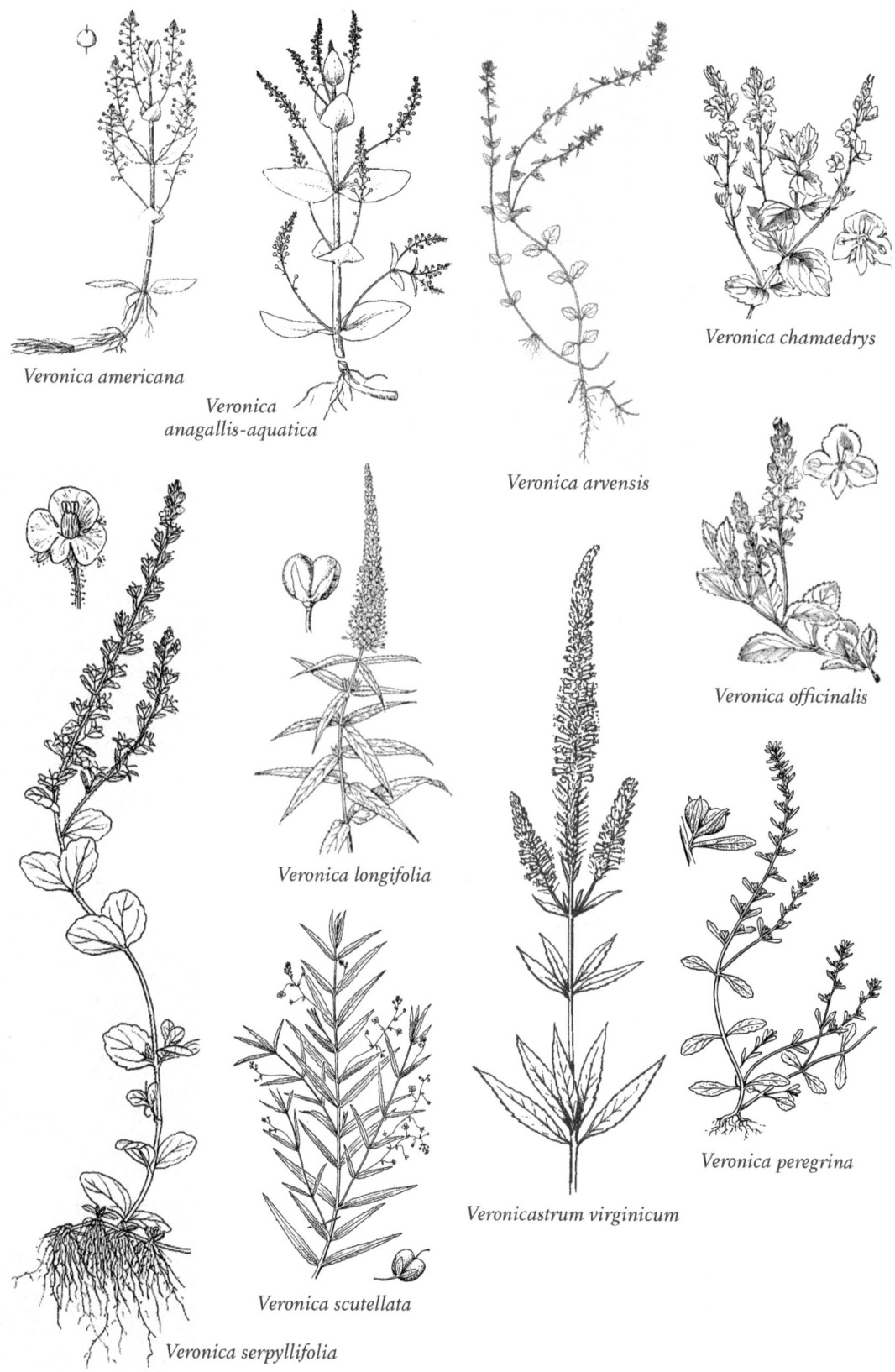

axils of the crowded upper leaves, forming a dense headlike cluster; calyx at anthesis about 6 mm long, at maturity about 10 mm long; corolla narrowly trumpet-shaped, blue-purple to white, 8–12 mm long, its limb 3–4 mm wide; stamens unequal lengths in the corolla tube and not protruding from it. Capsule obovoid, 3-lobed, surrounded by the persistent calyx; seeds 2.5–3 mm long, mucilaginous when wet. May–Aug.

Dry sandy or gravelly grasslands, shores, and disturbed places. Native to western USA, adventive in Menominee and Ontonagon counties.

Phlox *Phlox*

Erect perennial herbs. Leaves opposite, margins entire. Flowers pink, purple or rarely white, in stalked clusters at ends of stems and from upper leaf axils; sepals joined and tubelike; corolla 5-lobed, tubelike but flared outward at tip; stamens 5. Fruit a 3-chambered capsule.

1 Stems somewhat woody, trailing on ground . . . . **P. subulata**
1 Stems herbaceous, upright 2
2 Stems with long soft hairs; flowering ends in mid- to late-June . . . . **P. divaricata**
2 Stems glabrous below inflorescence; flowering begins in July . . . . **P. paniculata**

Phlox divaricata L.
FOREST PHLOX *native*
IR | WUP | **CUP** | EUP FACU CC 5

Perennial herb. Stems erect or decumbent at base, 3–5 dm tall, emitting decumbent basal stolons. Leaves ovate lance-shaped to oblong, 3–5 cm long at anthesis, broadest usually below the middle. Inflorescence a loosely branched, glandular-pubescent cyme, the branches on distinct peduncles, the pedicels often 5–10 mm long; corolla usually pale blue-purple, varying to red-purple or white, 2–3 cm wide, the glabrous tube 1–2 cm long. April–June.

Moist woods; Delta County.

Phlox paniculata L.
FALL PHLOX *introduced*
IR | **WUP** | **CUP** | EUP FACU

Perennial herb. Stems erect, to 2 m tall. Leaves narrowly oblong or lance-shaped, 8–15 cm long, minutely ciliate, narrowed to an acute or obtuse base, usually glabrous above, the conspicuous lateral veins confluent to form a submarginal connecting vein. Inflorescence often large, of several panicled cymes, densely but minutely pubescent; calyx tube minutely puberulent or more commonly glabrous; calyx lobes glabrous; corolla red to purple, varying to white, 1.5–2 cm wide, its tube usually sparsely pubescent; one or more anthers at least partly exsert from the corolla tube. July–Sept.

Rich moist soil; cultivated in numerous horticultural varieties and escaped into roadsides and waste places.

Phlox subulata L.
MOSS-PINK *introduced*
IR | **WUP** | **CUP** | **EUP**

Perennial by a prostrate suffruticose stem, freely branched and producing numerous flowering branches 5–20 cm long. Leaves numerous and crowded, subulate, 5–20 mm long, usually ciliate, often with fascicles of smaller leaves in their axils. Cymes few-flowered; corolla rose-purple to pink or white, 12–20 mm wide, its lobes notched for 1/8 to 1/4 of their length. April–May.

Sandy or gravelly soil and rock-ledges; frequently cultivated and sometimes escaped.

Polygalaceae
MILKWORT FAMILY

Polygala *Milkwort*

Annual, biennial, or perennial herbs. Leaves alternate or verticillate. Flowers perfect, in racemes. Sepals 5, the three outer small, the two inner (termed wings) much larger and often colored like the petals. Petals 3, all more or less united with each other and with the stamen-tube, the two upper ones similar, the lower one keel-shaped or boat-shaped with a fringe-like crest (in our species). Stamens 8 (or 6).

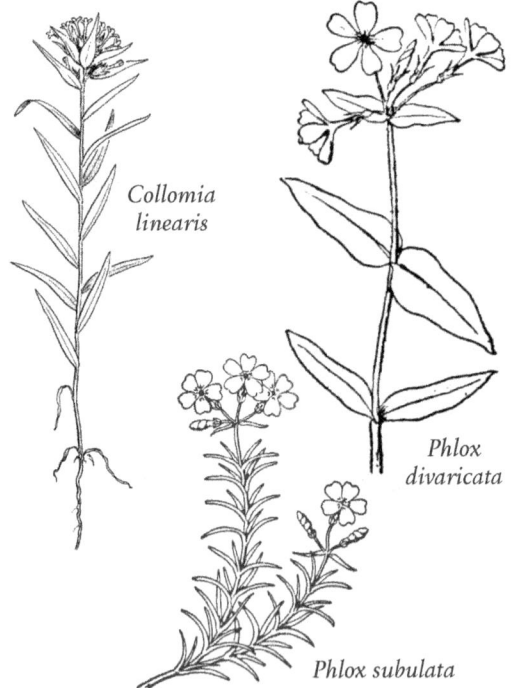

Collomia linearis

Phlox divaricata

Phlox subulata

Ovary 2-celled. Fruit a small capsule.

ADDITIONAL SPECIES

Polygala vulgaris L. (Common milkwort), uncommon Euasian escape from cultivation; flowers blue; reported from Alger County and former Quincy Mine site in Houghton County.

1. Flowers large, 13 mm long or more; stamens 6; leaves few, mostly near top of stem . . . . **P. paucifolia**
1. Flowers smaller, mostly less than 10 mm long; stamens 8; leaves distributed along stem or mostly near base . 2
2. Plants annual, the stems solitary from a small taproot. . . . . **P. sanguinea**
2. Plants biennial or perennial 3
3. Flowers white, in densely flowered spike-like racemes. . . . . **P. senega**
3. Flowers rose-purple to white, in loose racemes . . . . **P. polygama**

Polygala paucifolia Willd.
FRINGED POLYGALA; GAYWINGS   *native*
IR | **WUP** | **CUP** | **EUP**   FACU CC 7
Polygaloides paucifolia (Willd.) J.R. Abbott
Perennial from a slender rhizome. Stems 8–15 cm tall, bearing below several scattered scale-like leaves 2–8 mm long and near the summit 3–6 elliptic to oval leaves 1.5–4 cm long. Flowers 1–4, rose-purple varying to white, the obovate wings about 15 mm long; corolla about equaling the wings; stamens 6. Capsule suborbicular, about 6 mm long and wide, notched at the summit. May–June.

Moist rich woods. Leaves usually pubescent only on the midrib and margin.

Polygala polygama Walt.
RACEMED MILKWORT   *native*
IR | **WUP** | **CUP** | EUP   FACU CC 9
Biennial. Stems several from base, decumbent, glabrous, 1–2.5 dm tall, simple at anthesis, later sparingly branched. Lowest leaves spatulate to obovate, about 1 cm long; stem leaves oblong lance-shaped, 1–3 cm long, 2–7 mm wide, obtuse to subacute. Raceme loose and open, 2–10 cm long. Flowers rose-purple, varying to white; wings obovate, 4–6 mm long, exceeding the corolla.

Dry, usually sandy soil.

Polygala sanguinea L.
PURPLE MILKWORT   *native*
IR | **WUP** | **CUP** | EUP   FACU CC 4
Annual. Stems 1–4 dm tall, erect, simple or branched above. Leaves linear or narrowly elliptic, 1–4 cm long, 1–5 mm wide. Racemes very dense, headlike, rounded or short-cylindric, about 1 cm thick, the floriferous portion 1–2 cm long, the whole axis to 4 cm long, sessile or short-peduncled. Flowers rose-purple, white, or greenish; wings oval, 3–5 mm long, or longer in fruit, blunt, with conspicuous midvein; corolla about half as long as the wings; seed pear-shaped, the two linear lobes of the aril extending beyond the middle. July–Sept.

Fields, meadows, and open woods; Delta and Keweenaw counties.

Polygala senega L.
SENECA-SNAKEROOT   *native*
IR | **WUP** | **CUP** | **EUP**   FACU CC 8
Perennial. Stems commonly several from one base, 1–5 dm tall, usually unbranched, minutely puberulent. Leaves alternate, the lowest reduced or scale-like; stem leaves linear lance-shaped or wider. Racemes dense, 1.5–4 cm long, 6–8 mm thick, on a peduncle 1–3 cm long. Flowers white; wings broadly elliptic, 3–3.5 mm long, exceeding the corolla. Capsule suborbicular; seeds pubescent, 2–3 mm long; aril nearly or quite as long. May–June.

Dry or moist woods and prairies.

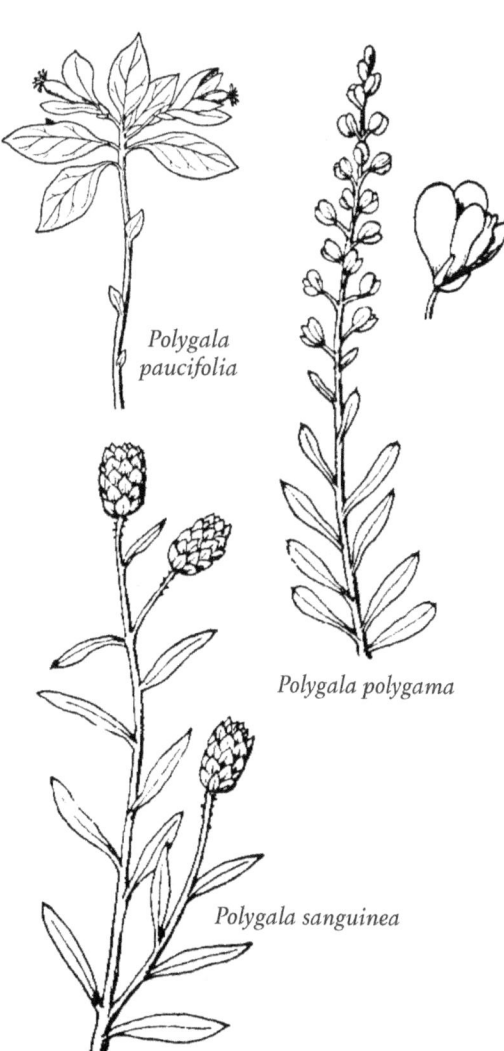

Polygala paucifolia

Polygala polygama

Polygala sanguinea

Polygonaceae
BUCKWHEAT FAMILY

Annual or perennial herbs, plants sometimes vining. Leaves alternate, simple, sometimes wavy-margined, otherwise entire; the nodes usually enlarged. Stipules joined to form a membranous or papery sheath (ocrea) around stem at each node. Flowers in spike-like racemes or small clusters from leaf axils (*Persicaria, Polygonum*), or in crowded panicles at ends of stems (*Rumex*). Flowers small, perfect (with both staminate and pistillate parts), regular, petals absent. In *Rumex* the sepals herbaceous, green to brown, in inner and outer groups, each group with 3 sepals, the 3 inner enlarging after flowering, becoming broadly winged, persisting to enclose the achene; stamens 4–8; ovary 1-chambered, styles 2–3; in other genera of family, sepals more or less petal-like, white to pink or yellow, mostly 5 (sometimes 4). Fruit a 3-angled or lens-shaped achene. Polygonaceae recognized by presence of a stipular sheath (ocrea), which surrounds the stem above the attachment of each leaf. The similar reduced structure in the inflorescence is called an ocreola.

ADDITIONAL SPECIES

Fagopyrum esculentum Moench (Garden buckwheat) occasional escape from cultivation or where seed spilled, but not persisting in our flora.

Rheum x hybridum Murray (Rhubarb), native of Asia, hybrid between *R. rhabarbarum* and *R. rhaponticum*; commonly cultivated and occasionally escaping near gardens. The stalks are edible when cooked; however, the leaf blades are highly toxic.

1 Tepals 6, greenish or reddish, scarcely petaloid, the 3 inner (but not the outer) ones enlarging in fruit and concealing the achene; stigmas a feathery tuft; plants in some species dioecious or polygamous and hence some flowers entirely staminate . . . . **Rumex**
1 Tepals 4–5, white to red and ± petaloid at least along the margins, uniform in size or the outer ones larger; stigmas usually not feathery and plants mostly with bisexual flowers 2
2 Pedicels with a swollen joint near the middle (but not far above the sheathing ocreolae), solitary in each ocreola, the inflorescence thus composed of slender racemes, appearing jointed because of the overlapping ocreolae; leaves not over 1 mm wide; delicate-looking annual . . . . **Polygonella articulata**
2 Pedicels usually jointed near the summit (if at all), often crowded, the inflorescence various; leaves at least 2 mm wide; annual or perennial, not delicate 3
3 Stem and petioles with retrorse prickles; leaves hastate or sagittate (with acute basal lobes) . . . . **Persicaria**
3 Stem and petioles without prickles; leaves various . . . 4
4 Outer tepals winged or keeled in fruit, or plant somewhat twining or vine-like, or both; leaves ovate-cordate to broadly sagittate . . . . **Fallopia**
4 Outer tepals not winged or keeled; plant not twining; leaves various 5
5 Flowers 1–4 at a node, sessile or pediceled in the axils of foliage leaves or bracts; leaf blades jointed at the base, less than 2 (–2.4) cm broad; summit of ocrea silvery white, becoming lacerate-shredded; annuals . . . . **Polygonum**
5 Flowers (or bulblets in *Bistorta vivipara*) numerous in peduncled terminal or axillary spikes, racemes, or panicles, often densely crowded; leaves not jointed at base of blade, in some species over 2.5 cm broad; summit of ocrea tinged with brown, shattering at maturity but not shredding; annuals or perennials 6
6 Leaves mostly basal, reduced in size up the stem and not more than 3 stem leaves present; many flowers in the single spike converted to bulblets; stems simple . . . . **Bistorta vivipara**
6 Leaves cauline, more than 3, basal leaves absent; flowers in the often several to many spikes not converted to bulblets; stems usually branched . . . . **Persicaria**

Bistorta *Bistort*

Bistorta vivipara (L.) Delarbre
ALPINE BISTORT (THR) native
IR | WUP | CUP | EUP CC 10
Polygonum viviparum L.

Glabrous perennial herb from knotty, elongate rhizome; stems to 3.5 dm tall, simple, terminating in a spike-like raceme; basal leaves with long slender petioles and oblong blades; stem leaves linear lanceolate, pale green beneath, with prominent midvein, margins revolute; ocreae buff sparsely puberulent; racemes bulbiliferous except for 1–3 staminate flowers toward the tip; bulbils maroon, pointed, rounded at base, 3–5 mm long; sepals pinkish, stamens 8 with reddish anthers; perfect flowers and fruits not present. Leaves of young plants ovate with very slender petioles. June–July.

Wet cobble beaches and bedrock crevices along the rocky shore of Lake Superior; often growing in the shade of shrubs such as alder (*Alnus*); Keweenaw County (including Isle Royale).

Fallopia *Black-Bindweed*

Annual or perennial, twining, or stout and erect and forming large colonies. In the past, our species typically included in genus *Polygonum*.

1 Stems twining and slender 2
1 Stems erect and stout, 1–3 m tall 3
2 Base of sheathing stipules with stiff, downward-pointing hairs . . . . **F. cilinodis**
2 Base of stipules not with stiff, downward-pointing hairs . . . . **F. convolvulus**

3 Leaf blades heart-shaped at base with rounded basal lobes, the blades often 20 cm or more long from leaf tip to lobe tip; flowers perfect . . . . **F. sachalinense**

3 Leaf blades cut nearly straight across at base, the blades less than 20 cm long; flowers functionally either staminate or pistillate . . . . **F. japonica**

Fallopia cilinodis (Michx.) Holub
FRINGED BLACK BINDWEED *native*
IR | WUP | CUP | EUP CC 3
Polygonum cilinode Michx.

Perennial, pubescent, varying to nearly glabrous. Stems twining, trailing, or occasionally erect, to 2 m long, nearly terete. Leaves ovate, deeply cordate at base; ocreae very oblique, reflexed-bristly at base. Racemes long-peduncled, mostly branched, 4–10 cm long, the small flower clusters remote; perianth white, 1.5–2 mm long; styles separate, divergent. Achenes very glossy, black, scarcely surpassed by the calyx. July–Aug.

Dry woods and thickets.

Plants in open sun are often erect, with stouter red stems, the red color extending into the leaf veins.

Fallopia convolvulus (L.) Á. Löve
BLACK BINDWEED *introduced*
IR | WUP | CUP | EUP FACU
Polygonum convolvulus L.

Annual. Stems trailing or twining, to 1 m long, angled, minutely scabrous in lines, as are also the petioles and often the leaf veins. Leaves hastate to triangular-cordate, broadly V-shaped to cordate at base; ocreae smooth. Racemes interrupted, 2–6 cm long, naked or with a few small leaves at base; flowers in clusters of 3–6; pedicels 1–2 mm long, jointed near the summit; perianth 1.5–2 mm long, green without, white within; outer 3 sepals often narrowly winged on the midrib; styles united. Achenes dull black, 3–4 mm long, not exceeded by the calyx. June–Sept.

Native of Europe; roadsides, railway tracks, and waste ground.

Fallopia japonica (Houtt.) Ronse Decr.
JAPANESE KNOTWEED *introduced (invasive)*
IR | WUP | CUP | EUP FACU
Polygonum cuspidatum Sieb. & Zucc.
Reynoutria japonica (Houtt.)

Perennial and spreading by long rhizomes. Stems stout, 1–3 m tall. Leaves broadly ovate, 8–15 cm long, 5–12 cm wide, abruptly acuminate, broadly truncate at base, the basal angles prominent. Racemes numerous from most of the upper axils, often branched, forming a series of panicles 8–15 cm long; perianth white or greenish white; outer sepals narrowly winged along the midrib; styles 3; stigmas minute. Achenes triangular, about 3 mm long, enclosed by the enlarged calyx. Aug–Sept.

Native of Japan; sometimes planted but often escaping to form large colonies.

Fallopia sachalinensis (F. Schmidt) Ronse Decr.
GIANT KNOTWEED *introduced (invasive)*
IR | WUP | CUP | EUP
Polygonum sachalinense F. Schmidt
Reynoutria sachalinensis (F. Schmidt) Nakai

Closely resembling *F. japonica* in habit, flower, and fruit. Stems sometimes more than 4 m tall. Leaf blades ovate, cordate at base, the basal lobes broadly rounded.

Native of e Asia; occasionally planted and sometimes escaped and forming colonies; Baraga and Houghton counties.

Persicaria
Lady's-Thumb; Smartweed

Annual and perennial herbs. Flowers pink or sometimes white, in terminal spikes. The genus was formerly included in *Polygonum*.

1 Tepals 4; styles elongate, persistent and becoming hard and stiff . . . . **P. virginiana**

1 Tepals usually 5; styles short, not persistent nor becoming hard and stiff 2

2 Stems with downward-pointing prickles on the stem angles . . . . **P. sagittata**

2 Stems smooth to hairy, but not prickly 3

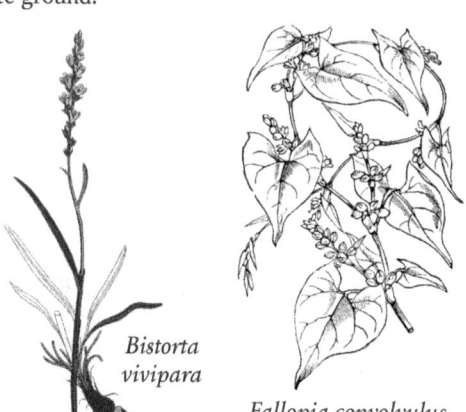

Bistorta vivipara

Fallopia convolvulus

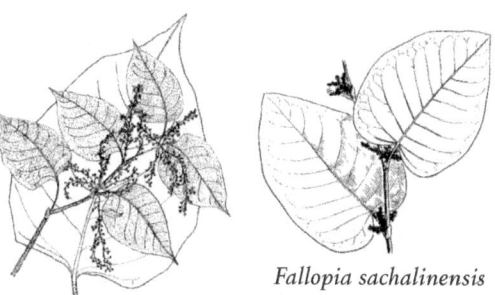

Fallopia japonica

Fallopia sachalinensis

POLYGONACEAE | Buckwheat Family

3 Perennial herbs from rhizomes or stolons 4
3 Taprooted annual herbs 6
4 Flowers in 1 or 2 terminal racemes . . . . **P. amphibia**
4 Flowers in several to many terminal and axillary racemes 5
5 Perianth dotted with glands . . . . **P. punctata**
5 Perianth not dotted with glands . . . **P. hydropiperoides**
6 Sheathing stipules (ocreae) fringed with bristles at tip 7
6 Ocreae entire or irregularly cut, not fringed with bristles 11
7 Perianth dotted with glands 8
7 Perianth not dotted with glands 9
8 Tepals usually 4; achenes dull . . . . **P. hydropiper**
8 Tepals 5; achenes shiny . . . . **P. punctata**
9 Upper stem and peduncles with gland-tipped hairs . . . . **P. careyi**
9 Upper stem and peduncles not with gland-tipped hairs 10
10 Small stipules at base of each inflorescence (ocreolae) fringed with long hairs 2–3 mm long . . . . . . **P. longiseta**
10 Small stipules at base of each inflorescence entire, or with a few short hairs to 1 mm long . . . . . . **P. maculosa**
11 Outer sepals strongly 3-nerved, each nerve ending in an anchor shaped fork; racemes nodding to erect . . . . **P. lapathifolia**
11 Outer sepals with faint, irregularly forked nerves; racemes erect . . . . **P. pensylvanica**

Persicaria amphibia (L.) Delarbre
WATER SMARTWEED native
IR | WUP | CUP | EUP OBL CC 6
Polygonum amphibium L.

Perennial floating or emergent herb, from spreading rhizomes. Stems to 1 m or more long, leaves and habit variable. Submerged plants smooth, usually branched, the branches floating, branch tips often upright and raised above water surface; leaves floating, leathery, oval, 4–20 cm long and 1–4 cm wide, rounded at tip; stipules (ocreae) membranous; petioles 1–8 cm long. Exposed plants hairy; leaves stalkless or with short petioles. Flowers pink to red, in 1–2 spike-like racemes from branch tips, the racemes 2–15 cm long and 1–2 cm wide; sepals 5-lobed to below middle, 4–5 mm long; stamens 5. Achenes lens-shaped, 2–4 mm long, shiny dark brown. June–Sept.

Common in ponds, lakes, marshes, bog pools, backwater areas, quiet streams.

Persicaria careyi (Olney) Greene
CAREY'S SMARTWEED (THR) native
IR | WUP | CUP | EUP FACW CC 9
Polygonum careyi Olney

Annual herb. Stems upright, branched, to 1 m long, with gland-tipped hairs. Leaves lance-shaped; stipules (ocreae) fringed with bristles and covered with stiff, spreading hairs. Flowers in cylindric, drooping racemes 3–6 cm long; sepals pink or rose, 3 mm long; stamens 5 (sometimes to 8). Achenes black, smooth, shiny, 2 mm wide. July–Aug.

Sandy lakeshores and streambanks, marshes, recently burned wetlands; reported from Iron County.

Persicaria hydropiper (L.) Delarbre
MILD WATER-PEPPER *introduced*
IR | WUP | CUP | EUP OBL
Polygonum hydropiper L.

Annual herb. Stems red, erect to sprawling, 2–6 dm long, sometimes rooting at lower nodes, branched or unbranched, peppery-tasting. Leaves lance-shaped, 3–8 cm long and to 2 cm wide, hairless except for short hairs on veins and margins, nearly stalkless or with a short petiole; stipules (ocreae) membranous, 5–15 mm long, swollen and fringed with bristles. Flowers green and usually white-margined, continuous in slender racemes, often nodding at tip; sepals 5, 3–4 mm long, with glandular dots; stamens 4 or 6. Achenes dull, dark brown, 3-angled or lens-shaped, 2–3 mm long. July–Oct.

Muddy shores, streambanks, floodplains, marshes, ditches and roadsides.

Persicaria hydropiperoides (Michx.) Small
SWAMP SMARTWEED native
IR | WUP | CUP | EUP OBL CC 5
Polygonum hydropiperoides Michx.

Perennial herb, spreading by rhizomes. Stems erect to sprawling with upright tips, to 1 m long, usually branched, nearly smooth or with short hairs. Leaves linear to lance-shaped, 4–12 cm long and to 2.5 cm wide, petioles short; stipules (ocreae) membranous, 5–15 mm long, with stiff hairs and fringed with bristles. Flowers green, white or pink, in 2 to several slender racemes, 1–6 cm long, often interrupted near base; sepals 2–3 mm long, 5-lobed to just below middle, without glandular dots or only the inner sepals slightly glandular; stamens 8. Achenes black, shiny, 3-angled with concave sides, 2–3 mm long. July–Sept.

Shallow water or wet soil; ponds, marshes, swamps, bogs and fens, streambanks, lakeshores and ditches.

Persicaria lapathifolia (L.) Delarbre
DOCK-LEAF SMARTWEED native
IR | WUP | CUP | EUP FACW CC 0
Polygonum lapathifolium L.

Annual herb. Stems erect to sprawling, unbranched or few-branched, 2–15 dm long. Leaves lance-shaped, 4–20 cm long and 0.5–5 cm wide, smooth above, often densely short-hairy on leaf undersides; petioles to 2 cm long, smooth to glandular; stipules (ocreae) 5–20 mm long, entire or with irregular, jagged margins. Flowers deep pink, white or green, crowded in erect or nodding racemes 1–5 cm long; sepals 3–4 mm long, 4- or 5-lobed to below middle, the outer 2 sepals strongly 3-nerved; stamens usually 6. Achenes brown, lens-shaped, 2–3 mm long. July–Sept.

Common and weedy in marshes, wet meadows, shores, streambanks, ditches and cultivated fields.

Persicaria longiseta (Bruijn) Kitag.
BRISTLY LADY'S-THUMB   *introduced*
IR | **WUP** | CUP | EUP
Polygonum caespitosum Blume

Annual. Stems glabrous or nearly so, freely branched, soon decumbent, to 1 m long. Leaves thin, dark green, lance-shaped to elliptic or oblong lance-shaped; ocreae minutely strigose or glabrous, ciliate with bristles 5–10 mm long. Racemes dense 2–4 cm long, about 5 mm thick; ocreolae overlapping, their cilia 2–3.5 mm long, often equaling or surpassing the flowers. Achenes black, smooth and shining, trigonous, about 2 mm long.

Native of e Asia; waste places, preferably in moist soil; Iron County.

Persicaria maculosa Gray
LADY'S-THUMB   *introduced*
IR | **WUP** | CUP | EUP   FACU
Polygonum persicaria L.

Annual herb. Stems upright to spreading, 2–8 dm long, unbranched to branched, often red. Leaves lance-shaped, 3–15 cm long and 0.5–3 cm wide, smooth or with few hairs, underside usually dotted with small glands, leaves stalkless or on petioles to 1 cm long; ocreae 5–15 mm long, fringed with bristles, with short hairs. Flowers pink to rose, crowded in straight, cylindric racemes 1–4 cm long and 0.5–1 cm wide; sepals 2–4 mm long, 5-lobed to near middle; stamens 6. Achenes black, shiny achene, lens-shaped or sometimes 3-angled, 2–3 mm long. July–Sept.

Muddy shores, streambanks, ditches and cultivated fields, often weedy.

Persicaria pensylvanica (L.) M. Gómez
PINKWEED   *native*
IR | **WUP** | **CUP** | EUP   FACW CC 0
Polygonum pensylvanicum L.

Annual herb. Stems erect, 3–20 dm long, unbranched to widely branching. Leaves lance-shaped, 3–15 cm long and 1–4 cm wide, smooth except for short hairs on margins; petioles to 2.5 cm long; stipules (ocreae) 0.5–1.5 cm long, entire or with an irregular, jagged margin, hairless, not fringed with bristles. Flowers pink to white, in dense racemes 2–3 cm long, the flower stalks with gland-tipped hairs; sepals 3–5 mm long, 5-parted to below middle, the outer sepals faintly nerved; stamens 8 or less. Achenes dark brown to black, shiny, lens-shaped, to 3 mm long. June–Sept.

Streambanks, exposed shores, marshes, fens, ditches and cultivated fields.

Persicaria punctata (Elliott) Small
DOTTED SMARTWEED   *native*
IR | **WUP** | **CUP** | **EUP**   OBL CC 5
Polygonum punctatum Ell.

Annual or perennial herb. Stems erect to spreading, 4–10 dm long, unbranched to branched. Leaves narrowly lance-shaped or oval, 4–15 cm long and 1–2 cm wide, smooth except for small short hairs on margins, underside usually dotted with small glands; petioles short; stipules (ocreae) 5–15 mm long, smooth or with stiff hairs and fringed with bristles. Flowers green-white; in numerous slender, loosely flowered racemes, interrupted in lower portion, to 10 cm long; sepals 3–4 mm long, with glandular dots, 5-parted to about middle; stamens 6–8. Achenes dark, shiny, lens-shaped or 3-angled, 2–3 mm long. Aug–Sept.

Floodplain forests, marshes, shores, streambanks and cultivated fields.

Persicaria sagittata (L.) H.Gross
ARROW-LEAF TEARTHUMB   *native*
IR | **WUP** | **CUP** | **EUP**   OBL CC 5
Polygonum sagittatum L.

Slender annual herb. Stems 4-angled, weak, usually supported by other plants, 1–2 m long, with downward pointing prickles on stem angles, petioles, leaf midribs and flower stalks. Leaves lance-shaped to oval, arrowhead-shaped at base, 3–10 cm long and to 2.5 cm wide, the basal lobes pointing downward; petioles long on lower leaves, shorter above; stipules (ocreae) 5–10 mm long, with a few hairs on margins. Flowers white or pink; in round racemes to 1 cm long, on long slender stalks at ends of stems or from leaf axils; sepals 3 mm long, 5-parted to below middle. Achenes brown to black, shiny, 3-angled, 2–3 mm long. July–Sept.

Swamps, marshes, wet meadows and burned wetlands.

Persicaria virginiana (L.) Gaertn.
JUMPSEED   *native*
IR | WUP | **CUP** | EUP   FAC CC 4
Antenoron virginianum (L.) Roberty & Vautier
Polygonum virginianum L.

Perennial herb. Stems erect from a rhizome, 5–10 dm tall. Leaves lance-shaped to ovate, to 15 cm long, acute to rounded at base, varying from roughly pubescent to glabrous on either or both sides; petioles to 2 cm long; ocreae pubescent and long-ciliate. Racemes very slender, terminal, 1–4 dm long, the ocreolae much separate toward the base, becoming contiguous or overlapping toward the summit, 1–3-flowered; pedicels divergent, jointed at the tip; sepals 4, greenish white, or suffused with pink, about 2.5

268 POLYGONACEAE | Buckwheat Family

dicots

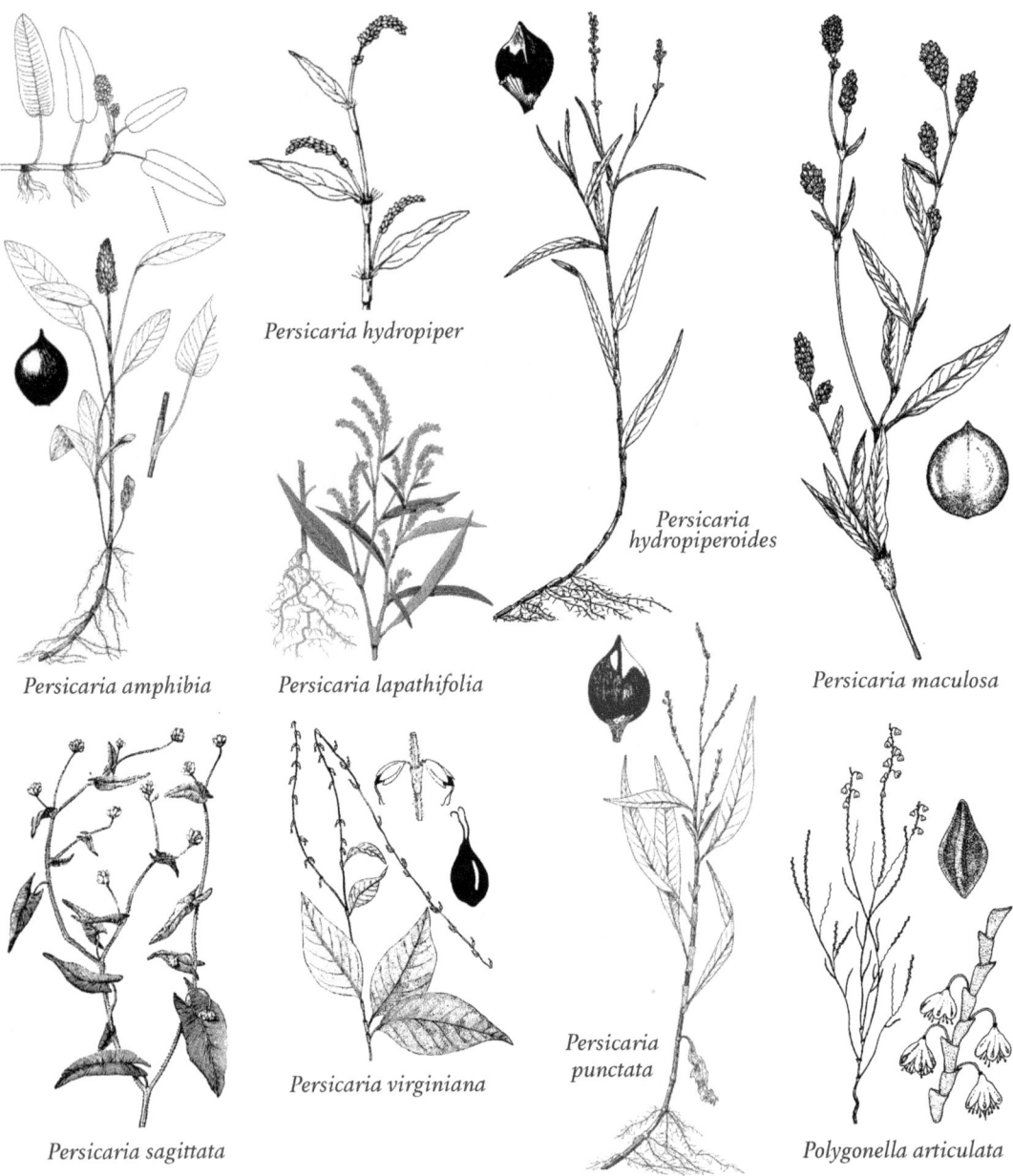

Persicaria hydropiper
Persicaria amphibia
Persicaria lapathifolia
Persicaria hydropiperoides
Persicaria maculosa
Persicaria sagittata
Persicaria virginiana
Persicaria punctata
Polygonella articulata

mm long, the 2 lateral exterior and somewhat smaller than the median, scarcely changed in fruit. Achenes lens-shaped, ovate, about 4 mm long; styles persistent, hooked at the tip. Aug–Sept.

Moist woods; Delta County.

Polygonella *Jointweed*

Polygonella articulata (L.) Meisn.
COASTAL JOINTWEED
IR | **WUP** | **CUP** | **EUP**
native
CC 8

Polygonum articulatum L.

Annual herb. Stems slender, wiry, branched above, 1–4 dm tall. Leaves linear, revolute, 1 mm or less wide, 5–20 mm long; leaves jointed with the summit of the ocrea. Flowers perfect, or a few unisexual by abortion, each flower solitary from the axils of a sheathing bract, in several racemes; pedicels slender, decurved, jointed just above the base; sepals 5, persistent, petal-like, white or greenish to pink or red, 1.5–2 mm long, the outer 2 obovate, keeled toward the summit, the inner 3 elliptic; stamens 8; ovary 3-angled. Fruit a smooth, sharply 3-angled achene, subtended and loosely enclosed by the calyx. July–Aug.

Great Lakes sandy shores and dunes.

Polygonum
Smartweed; Knotweed; Tearthumb

Annual herbs (ours). Stems erect to sprawling, often swollen at nodes. Leaves arrowhead-shaped to lance-shaped or oval; stipules joined to form a tubular sheath (ocrea) around the stem above each node; the ocreae (plural) membranous or papery, entire or with an irregular, jagged margin or fringed with bristles. Flowers small, from leaf axils; sepals usually 5, petal-like, green-white to pink; stamens 8 or less; styles 2–3. Fruit a brown to black achene, lens-shaped or 3-angled.

1	Flowers in the axils of short bracts, all but the lowermost bracts less than 2x longer than the flowers, the inflorescence thus appearing to be a remotely flowered slender spike; plants stiffly erect with leaves mostly linear. . . . . **P. douglasii**
1	Flowers in the axils of foliage leaves, these mostly at least 2x longer than the flowers; plants erect to ascending or prostrate, with leaves acute to blunt. 2
2	Perianth abruptly narrowed above achene ("bottle-shaped"). 3
2	Perianth not narrowed above achene. 4
3	Leaves yellow-green; fruiting perianth divided for about three-fourths of its length. . . . . **P. erectum**
3	Leaves blue-green; fruiting perianth divided for about one-third its length . . . . **P. achoreum**
4	Outer 3 tepals flat, shorter than or equaling inner 2 tepals . . . . **P. aviculare**
4	Outer 3 tepals hood-like, longer than inner 2 tepals. . 5
5	Plants prostrate; leaves 2–4 times longer than wide. . . . . **P. aviculare**
5	Plants upright; leaves 4–12 times longer than wide. . . . . **P. ramosissimum**

Polygonum achoreum Blake
LEATHERY KNOTWEED *native*
IR | **WUP** | **CUP** | **EUP** FACU CC 0

Annual, closely resembling *P. erectum*. Stems erect or ascending, freely branched, 1–5 dm tall. Leaves elliptic to obovate, thin, bright- or bluish green, 1–3 cm long, broadly rounded at the tip; ocreae to 1 cm long, mostly scarious, 3-nerved. Pedicels about equaling the calyx; calyx about 3 mm long, green, unequally 5-lobed to above the middle; outer lobes at maturity distinctly exceeding the inner, cucullate at the tip, exceeding the achene. Achenes dull yellow-brown, finely and uniformly granular.

Sandy and gravelly roadsides, barnyards, gardens, railroads.

Polygonum aviculare L.
YARD KNOTWEED *introduced*
IR | **WUP** | **CUP** | **EUP** FACU CC -

Annual. Stems erect to prostrate, much branched, the branches commonly equaling the central axis, or the latter suppressed. Leaves linear to elliptic or oblong, 1–3 cm long, 1–8 mm wide, narrowed to the base, veinless or faintly veined. Flowers short-pediceled, included in the ocrea or barely exsert; calyx 2–3 mm long, lobed to below the middle; sepals oblong or ovate, green with white or pink margins, appressed at maturity. Achenes ovoid, dark brown, 2–2.5 mm long, finely puncticulate.

Common weed of waste ground, streets, and lawns; also common on beaches and around salt marshes.

Polygonum douglasii Greene
DOUGLAS' KNOTWEED *native*
IR | **WUP** | **CUP** | EUP CC 5

Slender taprooted annual. Stems to 4 dm tall, the branches few, ascending and sometimes crowded. Leaves not crowded, inclined, the lower ones to 4 cm long and 8 mm wide, linear to lanceolate, the petiole very short; leaves smaller above. inflorescences terminal on the branches, of spike-like racemes, or the racemes in a small panicle, the flowers 2–4 at the nodes. Flowers pedicellate, the pedicels becoming reflexed in fruit; perianth 5-lobed, connate only at the base, green with white (pinkish) margins. Achene black, 3-angled, the sides concave, sometimes flattened on 1 or 2 sides. July–Sept.

Sandy soil of open places (sometimes with *Polygonella*) in the counties bordering Lake Michigan in the Upper Peninsula, but also on dry rock outcrops in the northwestern Upper Peninsula and Isle Royale; disjunct from western North America.

Polygonum erectum L.
ERECT KNOTWEED *native*
IR | **WUP** | CUP | EUP FACU CC 0

Annual. Stems erect or ascending, 1–5 dm tall, with numerous branches. Leaves oval to obovate, 1–4 cm long, thin, bright- or bluish green, broadly rounded, acute at base; ocreae to 1 cm long, 3–5-nerved. Pedicels shorter than to equaling the calyx; calyx about 3 mm long, green, unequally 5-lobed to below the middle; outer lobes at maturity distinctly exceeding the inner, narrowly keeled. Achenes dimorphic; either shining, punctate, dark brown, broadly ovoid, included, about 2.5 mm long; or dull brown, ovoid, exsert, 3–3.5 mm long.

Weedy in waste ground; Keweenaw County.

Polygonum ramosissimum Michx.
YELLOW-FLOWER KNOTWEED *native*
IR | WUP | **CUP** | **EUP** FAC CC 7

Annual. Stems erect, 3–10 dm tall, freely branched; lower internodes to 5 cm long, the upper progressively shorter. Leaves linear, mostly 1–6 cm long, 2–5 mm wide, flat; lateral veins inconspicuous or obsolete. Flowers from the upper ocreae only, 1–3 together,

270 POLYGONACEAE | Buckwheat Family

dicots

exsert on pedicels longer than the calyx, forming racemes to 15 cm long; calyx about 3 mm long, 5-parted nearly to the base, the outer 3 sepals notably exceeding the inner, especially in fruit. Achenes black, ovoid, 3 mm long, smooth and shining.

Sandy fields and meadows, sandy or gravelly shores of the Great Lakes.

Rumex *Dock; Sorrel*

Perennial, sometimes weedy herbs (annual in *R. fueginus*). Leaves large and clustered at base of plants, or leafy-stemmed; mostly oblong to lance-shaped, flat to wavy-crisped along margins, usually with petioles. Membranous sheaths around stems present at nodes (ocreae). Flowers in crowded whorls in panicles at ends of stems; flowers small and numerous, green but turning brown; sepals in 2 series of 3, the inner 3 sepals (valves) enlarging, becoming winged and loosely enclosing the achene, giving the appearance of a 3-winged fruit, the midvein of the valve often swollen to produce a grainlike tubercle on the back; stamens 6; styles 3. Fruit a brown, 3-angled achene, tipped with a short slender beak.

ADDITIONAL SPECIES

Rumex occidentalis S. Watson (Western dock), in Michigan, known only from a single collection along the banks of the Escanaba River in Marquette County.

Rumex thyrsiflorus Fingerh. (Narrow-leaf sorrel), introduced Eurasian perennial of disturbed places, similar to *R. acetosa* but more densely flowered and deeply tap-rooted; Keweenaw and Mackinac counties.

1. At least some of the leaves arrowhead-shaped, the basal lobes pointing backward or outward 2
1. Leaves not arrowhead-shaped with basal lobes 3
2. Basal lobes pointing backward; valves with a conspicuous grain at base .... **R. acetosa**
2. Basal lobes pointing outward; valves without grains .. .. **R. acetosella**
3. Valves without grains .... **R. longifolius**
3. At least 1 of the valves with a prominent grain....... 4
4. Margins of mature valves entire or shallowly lobed, not toothed 5
4. Margins of mature valves with coarse or spine-tipped teeth... 9
5. Flower pedicels without a large swollen joint; base of grain distinctly above base of valve ..... **R. britannica**
5. Flower stalks with a large swollen joint below the middle or near base; base of grain even with base of valve 6
6. Fruit with 3 grains; flower pedicels 2–5 times longer than fruit .... **R. verticillatus**
6. Fruit with 1–3 grains, the grains not projecting below the valves; flower stalks 1–2 times longer than fruit .. 7
7. Leaves crisp-margined (crinkled); grains two-thirds as wide as long .... **R. crispus**
7. Leaf margins flat; grains narrower, up to half as wide as long... 8
8. Grains usually 1; leaves mostly less than 4 times longer than wide........................... **R. altissimus**
8. Grains usually 3; leaves mostly more than 4 times longer than wide .... **R. triangulivalvis**
9. Plants annual from fibrous roots; grains 3 . **R. fueginus**
9. Plants perennial from a stout taproot; grain 1......... .... **R. obtusifolius**

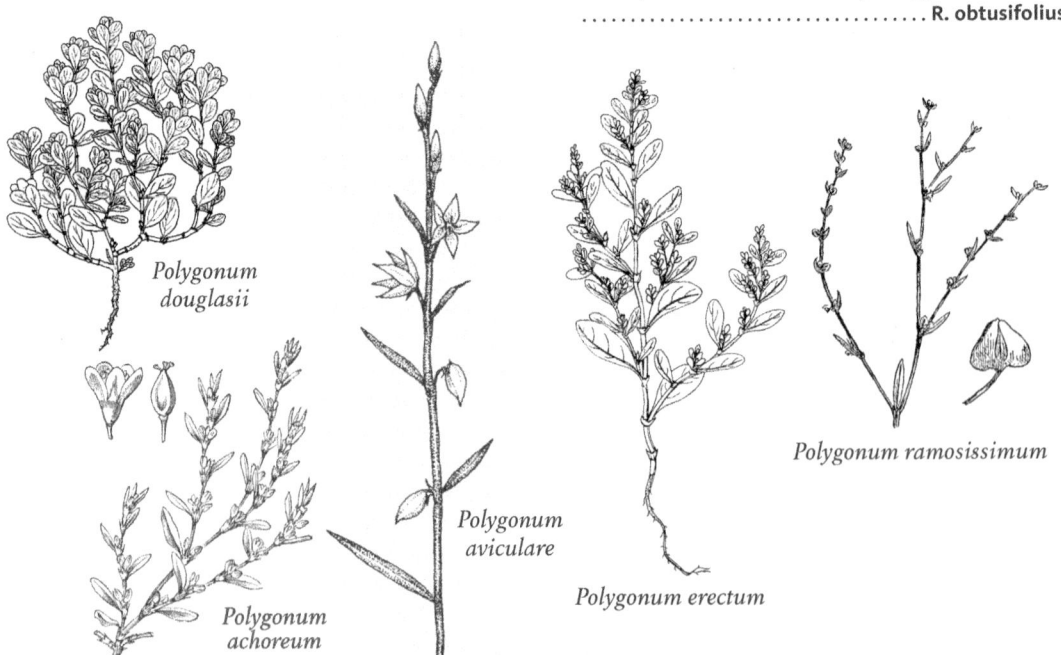

Polygonum douglasii

Polygonum achoreum

Polygonum aviculare

Polygonum erectum

Polygonum ramosissimum

Rumex acetosa L.
GREEN SORREL *introduced*
IR | WUP | **CUP** | EUP

Perennial herb. Stems stout, erect, 3–9 dm tall, usually simple to the inflorescence. Leaves oblong, all or chiefly sagittate, the lower long-petioled, the upper subsessile with triangular basal lobes directed backward. Panicle 1–2 dm long, usually leafless; sepals of the staminate flowers 2–3 mm long; outer sepals of the pistillate flowers soon reflexed, nearly 2 mm long; inner sepals in fruit 4–6 mm long and about as wide, reticulate-veined, the midrib conspicuously dilated at base into a tubercle-like appendage. Achenes dark brown, 2–2.5 mm long.

Native of Eurasia; occasionally cultivated for greens; adventive in Alger County.

Rumex acetosella L.
COMMON SHEEP SORREL *intro. (invasive)*
IR | WUP | **CUP** | EUP FACU

Perennial herb. Stems erect, simple or branched, 1–4 dm tall. Leaves variable, usually 3-lobed, the terminal lobe narrowly elliptic to oblong, the lateral much smaller, triangular, divergent; leaf base below the lobes truncate to long-cuneate. Inflorescence sometimes half as long as the plant; outer sepals lance-shaped; inner sepals in the staminate flower 1.5–2 mm long, obovate, in the pistillate flower broadly ovate. Achenes about 1.5 mm long, shining golden brown.

Naturalized from Eurasia in fields, lawns, and waste places, soils acidic; a common weed.

Rumex altissimus Wood
PALE DOCK *native*
IR | WUP | **CUP** | EUP FACW CC 2

Perennial herb, similar to willow-leaf dock (R. triangulivalvis). Stems 3–10 dm long, usually branched from base and with short branches above. Leaves all from stem, ovate to lance-shaped, 6–20 cm long and 2–6 cm wide, margins flat or slightly wavy. Flowers in panicles 1–3 dm long, the panicle branches short and more or less upright; flower stalks short, 3–5 mm long, swollen and jointed near base; valves rounded, 4–6 mm long and as wide, flattened across base, margins smooth or irregularly toothed; grains usually well developed on only 1 of the 3 valves, although sometimes present on 2–3 valves; the largest grain lance-shaped. Fruit a brown achene, 2–3 mm long. May–Aug.

Marshes, shores, streambanks, ditches; Delta County.

Rumex brittannica L.
GREAT WATER-DOCK *native*
IR | **WUP** | **CUP** | EUP OBL CC 9

Rumex orbiculatus Gray

Perennial herb. Stems stout, unbranched, 2–2.5 m long. Leaves lance-shaped or oblong lance-shaped, lower leaves 30–60 cm long, upper leaves 5–15 cm long; margins flat. Flowers in panicles to 5 dm long; valves rounded, flat at base, 5–8 mm long and as wide, smooth or with small teeth; grains 3, narrowly lance-shaped, the base distinctly above base of valve. June–Aug.

Marshes, fens, streambanks and ditches, often in shallow water.

Rumex crispus L.
CURLY DOCK *introduced*
IR | WUP | **CUP** | EUP FAC

Perennial herb, from a thick taproot. Stems stout, upright, usually single, 5–15 dm long. Basal leaves large, 10–30 cm long and 1–5 cm wide, on long petioles, often drying early in season; stem leaves smaller and with shorter petioles, oval to lance-shaped, margins strongly wavy-crisped (crinkled). Flowers in large branched panicles, the panicle branches more or less upright; flower stalks drooping at tips, 5–10 mm long, swollen-jointed near base; valves heart-shaped to broadly ovate, 4–5 mm long and as wide, margins more or less smooth; grains 3, swollen, often of unequal size, rounded at ends. Fruit a brown achene, 2–3 mm long. July–Sept.

Weed of wet meadows, shores, ditches, old fields, and other wet and disturbed areas.

Rumex fueginus Phil.
GOLDEN DOCK *native*
IR | **WUP** | **CUP** | EUP FACW CC 5

Annual herb. Stems hollow, to 8 dm long, much-branched. Leaves mostly on stems, smaller upward, lance-shaped to linear, 5–20 cm long and 0.5–4 cm wide, wedge-shaped or heart-shaped at base, margins flat to wavy-crisped. Flowers in large open panicles, the panicle branches more or less upright, leafy, the flower stalks jointed near base; valves triangular-ovate, 2–3 mm long, the margins lobed into 2–3 spine-tipped teeth on each side; grains 3. Fruit a light brown achene, 1–2 mm long. July–Aug.

Marshes, shores, streambanks and ditches, sometimes where brackish.

Rumex longifolius DC.
DOOR-YARD DOCK *introduced*
IR | WUP | **CUP** | EUP FAC

Perennial herb. Stems slender, erect, to 1 m tall. Lower leaves narrowly oblong, broadest near the middle, tapering to an acute base. Mature pedicels visibly jointed near the base; valves rotund to sub-kidney-shaped, 4–6 mm long, 5–7 mm wide, entire

272 POLYGONACEAE | Buckwheat Family

dicots

or toothed, reticulate-veined, without grains.

Native of Europe; Keweenaw (including Isle Royale) and Schoolcraft counties; waste places.

Rumex obtusifolius L.
BITTER DOCK   *introduced*
IR | WUP | CUP | EUP   FAC

Perennial herb. Stems stout, to 12 dm long, usually unbranched. Lower leaves oblong or ovate, to 30 cm long and 15 cm wide, heart-shaped or rounded at base; upper leaves smaller. Flowers in much-branched panicles, flower stalks longer than fruit, jointed near base; valves triangular-ovate, 4–5 mm long, with 2–4 spine-tipped teeth on each side; grains large and with tiny wrinkles. Fruit a shiny, red-brown achene. June–Aug.

Floodplain forests and openings, cultivated fields and disturbed areas.

Rumex triangulivalvis (Danser) Rech. f.
WILLOW-LEAF DOCK   *native*
IR | WUP | CUP | EUP   FAC CC 1

Rumex salicifolius Weinm.

Perennial taprooted herb. Stems smooth, 3–10 dm long, usually branched from base and with short branches on stem. Leaves mostly on stems, not much smaller upward, narrowly lance-shaped, tapered at both ends, pale waxy green, 5–16 cm long and 1–3 cm wide, margins mostly flat. Flowers in panicles 1–3 dm long, panicle branches few and more or less upright, with small linear leaves at base; flower stalks 2–4 mm long, swollen and jointed near base; valves thick, triangular, 3–6 mm long and wide, margins smooth or shallowly toothed; grains usually 3. Fruit a brown achene, 2 mm long. June–Aug.

Wet meadows, marshes, shores, streambanks, ditches and other low areas, sometimes where brackish.

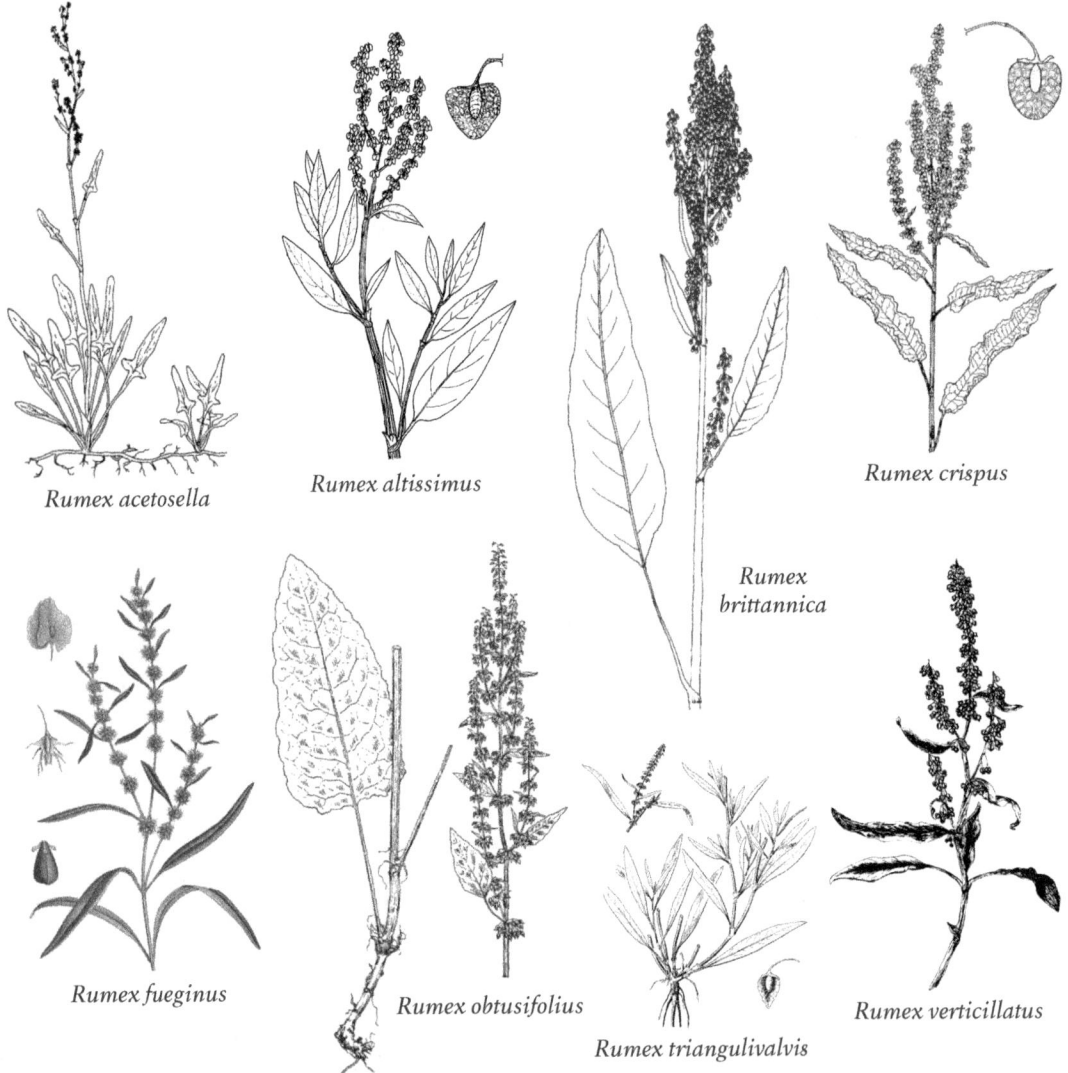

Rumex acetosella

Rumex altissimus

Rumex crispus

Rumex brittannica

Rumex fueginus

Rumex obtusifolius

Rumex triangulivalvis

Rumex verticillatus

Rumex verticillatus L.
SWAMP DOCK *native*
IR | **WUP** | CUP | EUP OBL CC 7

Perennial taprooted herb. Stems stout, 1–1.5 m long, with many short branches from leaf axils. Leaves narrowly lance-shaped, tapered to base, margins flat. Flowers in leafless panicles 2–4 dm long, the panicle branches few and more or less upright; flower stalks 10–15 mm long, jointed near base; valves triangular-ovate, 4–6 mm long and wide, thickened at center; grains 3, lance-shaped, the base blunt and projecting 0.5 mm below base of valve. June–Sept.

Marshes, swamps, wet forests, backwater areas and muddy shores, often in shallow water; Ontonagon County.

Portulacaceae
PURSLANE FAMILY

Portulaca *Purslane*

Portulaca oleracea L.
COMMON PURSLANE *introduced*
IR | **WUP** | CUP | EUP FACU

Succulent annual herb; stems prostrate, fleshy, usually purplish red, glabrous, repeatedly branched, forming large mats. Leaves succulent but flat, spatulate to obovate-cuneate, 1–3 cm long, commonly rounded at the tip; stem leaves usually alternate, occasionally opposite, the uppermost crowded and forming an involucre to the flowers. Flowers sessile, 5–10 mm wide, solitary or in small terminal glomerules, ephemeral, opening only in the sunshine; petals 4–6, commonly 5, yellow; stamens 6–10. Capsule opening near the middle, many-seeded; seeds with low blunt tubercles. All summer.

Reputedly native of w Asia, but now widely distributed as a familiar weed; sometimes cooked for greens.

Portulaca oleracea

Primulaceae
PRIMROSE FAMILY

Annual or perennial herbs. Leaves simple, opposite (sometimes whorled in *Lysimachia*), or leaves all basal. Flowers perfect (with both staminate and pistillate parts), regular, single from leaf axils, or in clusters at ends of stems; sepals 4–5, petals mostly 5 (varying from 4–9), joined, tube-shaped below and flared above, deeply cleft to shallowly lobed at tip; ovary superior, style 1; stamens 5. Fruit a 5-chambered capsule.

1 Leaves all from base of plant, inflorescence an umbel at end of naked stalk . . . . **Primula**
1 Leaves from stem; inflorescence various 2
2 Leaves in a single whorl near end of stem; flowers 7-merous . . . . **Trientalis borealis**
2 Leaves opposite or in several whorls; flowers 5–6-merous . . . . **Lysimachia**

Lysimachia *Loosestrife*

Perennial herbs, spreading by rhizomes. Stems erect. Leaves mostly opposite (sometimes appearing whorled), ovate or lance-shaped. Flowers 5-parted, single on stalks from leaf axils or in racemes or panicles; sepals green; petals bright to pale yellow. Fruit a capsule.

Genus sometimes placed in family Myrsinaceae. Purple loosestrife (*Lythrum salicaria*), an introduced weed of wetlands, is a member of the Lythraceae.

1 Plants creeping; leaves opposite, nearly round . . . . **L. nummularia**
1 Plants upright; leaves opposite or whorled, longer than wide 2
2 Flowers in terminal racemes or panicles 3
2 Flowers solitary or in clusters or spikes from the leaf axils 4
3 Plants pubescent . . . . **L. vulgaris**
3 Plants glabrous . . . . **L. terrestris**
4 Leaves rounded or heart-shaped at base; petioles 1–3 cm long, fringed with hairs . . . . **L. ciliata**
4 Leaves tapered to their base; petioles absent or short, smooth or fringed with hairs 5
5 Flowers many in racemes from leaf axils; flowers mostly 6-merous . . . . **L. thyrsiflora**
5 Flowers 1 to several from the leaf axils, 5-merous 6
6 Leaves narrowly linear, to 5 mm wide . . . **L. quadriflora**
6 Leaves lance-shaped to ovate, usually more than 8 mm wide 7
7 Flowers in clusters of several from leaf axils . . . . **L. punctata**
7 Flowers usually single from each leaf axil 8
8 Main leaves in whorls of 3 or more leaves; corolla lobes entire . . . . **L. quadrifolia**
8 Leaves opposite or whorled; corolla lobes ragged-toothed at tip . . . . **L. lanceolata**

PRIMULACEAE | Primrose Family

Lysimachia ciliata L.
FRINGED YELLOW-LOOSESTRIFE *native*
IR | WUP | CUP | EUP FACW CC 4

Perennial herb, spreading by rhizomes. Stems upright, 3–12 dm long, unbranched or with few branches above. Leaves ovate to lance-shaped, 4–15 cm long and 2–6 cm wide, rounded to heart-shaped at base, green above, slightly paler below; margins fringed with short hairs; petioles 0.5–5 cm long, fringed with hairs. Flowers yellow, single from upper leaf axils, on stalks 2–7 cm long; sepal lobes lance-shaped, often with 3–5 parallel red-brown veins; petal lobes rounded and finely ragged at tip, 4–10 mm long and 3–9 mm wide, with a short slender tip. Fruit a capsule, 4–7 mm wide. June–Aug.

Usually shaded wet areas, such as shores, streambanks, wet meadows, ditches, floodplains, wet woods and thickets.

Lysimachia lanceolata Walt.
LANCE-LEAF YELLOW-LOOSESTRIFE *native*
IR | WUP | CUP | EUP FAC CC 9

Perennial herb. Stems erect or nearly so, producing long slender stolons from the base, 3–6 dm tall, the lateral branches barely longer than the subtending leaves. Lower stem leaves ovate to obovate, petioled; principal leaves linear, to 15 cm long, paler beneath, tending to be folded along the midvein, scabrellate on the margin or ciliate near the base, gradually tapering below, with no distinction of petiole and blade. Sepals firm or thick, 5–7 mm long; petals 7–10 mm long. June–July.

Moist or wet woods or prairies; Gogebic and Menominee counties.

Lysimachia nummularia L.
CREEPING-JENNY *introduced (invasive)*
IR | WUP | CUP | EUP FACW

Perennial herb, often forming mats. Stems creeping, to 5–6 dm long. Leaves opposite, dotted with black glands, round or broadly oval, 1–2.5 cm long; petioles short. Flowers single in leaf axils, on stalks to 2.5 cm long; sepals leaflike, triangular; petals yellow, dotted with dark red, 10–15 mm long. Fruit a capsule, shorter than sepals. June–Aug.

Swamps, floodplain forests, streambanks, shores, meadows and ditches; Gogebic, Marquette and Schoolcraft counties.

Lysimachia punctata L.
LARGE YELLOW-LOOSESTRIFE *introduced*
IR | WUP | CUP | EUP OBL

Perennial herb, spreading by shallow stolons. Stems erect, rarely branched, to 1 m tall, pubescent. Leaves chiefly in whorls of 3 or 4, occasionally only opposite, lance-shaped, 5–10 cm long, pubescent. Flowers in axillary whorls, usually more numerous than the subtending leaves; uppermost whorls with smaller leaves and shorter internodes, simulating a raceme; pedicels 1–2 cm long, pubescent; sepals linear lance-shaped, pubescent; corolla yellow, 2–3 cm wide. June–July.

Native of se Europe and sw Asia; Marquette and Ontonagon counties.

Lysimachia quadriflora Sims
FOUR-FLOWER YELLOW-LOOSESTRIFE *native*
IR | WUP | CUP | EUP OBL CC 10

Perennial herb, spreading by rhizomes which form clusters of basal rosettes. Stems upright, 3–10 dm long. Leaves opposite, sometimes appearing whorled; stem leaves stalkless, often ascending, linear, 3–8 cm long and 2–7 mm wide, margins smooth or rolled under, sometimes fringed with a few hairs near base. Flowers yellow, single in clusters at ends of stems and branches, on stalks 1–4 cm long; sepal lobes lance-shaped; petal lobes oval, 7–12 mm long, entire or finely ragged at tip. Fruit a capsule, 3–5 mm wide. July–Aug.

Wet meadows, pond and marsh margins, low prairie; often where sandy and calcium-rich; Delta County.

Lysimachia quadrifolia L.
WHORLED YELLOW-LOOSESTRIFE *native*
IR | WUP | CUP | EUP FACU CC 8

Perennial herb. Stems erect, 3–9 dm tall, glabrous or sparsely pubescent, rarely branched. Leaves chiefly in whorls of 4 (3–6), lance-shaped, 5–10 cm long, widely spreading. Flowers from many of the median nodes, usually one from the axil of each leaf, on spreading pedicels 2–5 cm long; sepals oblong lance-shaped; petals yellow with dark lines, oblong or elliptic, 6–8 mm long. June–July.

Moist or dry upland soil, chiefly in open woods; Mackinac and Menominee counties.

Lysimachia terrestris (L.) B.S.P.
SWAMPCANDLES *native*
IR | WUP | CUP | EUP OBL CC 6

Perennial herb, spreading by shallow rhizomes. Stems smooth, 4–8 dm long, usually branched. Leaves opposite, dotted with glands, narrowly lance-shaped, 5–10 cm long and 2–4 cm wide, with small bulblike structures produced in leaf axils late in season; bracts awl-like, 3–8 mm long. Flowers yellow, in a single, crowded, upright raceme, 1–3 dm long; sepals lance-shaped; petal lobes oval, 5–7 mm long, with dark lines, on stalks 8–15 mm long. Fruit a capsule, 2–3 mm wide. June–Aug.

Common in marshes, fens, thickets, muddy shores, and ditches.

Lysimachia thyrsiflora L.
SWAMP LOOSESTRIFE *native*
IR | WUP | CUP | EUP   OBL CC 6

Perennial upright herb, spreading by rhizomes; plants conspicuously dotted with dark glands. Stems smooth or with patches of brown hairs, 3–7 dm long, unbranched, or branched on lower stem. Leaves opposite, linear to lance-shaped, 4–12 cm long and 0.5–4 cm wide, smooth above, smooth or sparsely hairy below; petioles absent. Flowers yellow, crowded in dense racemes from leaf axils, on spreading stalks 2–5 cm long; mostly 6-parted; sepal lobes awl-shaped; petal lobes linear, 3 mm long; stamens 2x longer than petals. Fruit a capsule, 2–4 mm wide. June–Aug.

Many types of wetlands: thickets, shores, fens and bogs, marshes, low places in conifer and deciduous swamps, often in shallow water.

Lysimachia vulgaris L.
GARDEN YELLOW-LOOSESTRIFE *introduced*
IR | WUP | **CUP** | EUP   FACW

Perennial herb. Stems erect, to 1 m tall, densely softly pubescent. Leaves either whorled or opposite, lance-shaped or ovate lance-shaped, 8–12 cm long, softly pubescent beneath. Inflorescence a terminal raceme and a series of peduncled short racemes or panicles from the upper axils, forming a terminal leafy panicle; sepals about 3 mm long; corolla yellow, about 2 cm wide, its lobes entire. July–Sept.

Native of Eurasia; escaped from cultivation, occasional on mudflats along rivers and in wet meadows; Delta County.

Primula *Primrose*
Perennial glabrous herbs. Leaves in a basal rosette. Flowers in a terminal bracted umbel of attractive nodding flowers on ascending or erect pedicels, atop a solitary, erect, leafless scape. Calyx deeply 5-parted, persistent in fruit. Corolla 5-cleft almost to the base, the lobes spreading to reflexed. Capsule ovoid to cylindric, erect, opening by 5 short terminal valves.

1 Lobes of corolla strongly reflexed .... **Primula meadia**
1 Lobes of corolla spreading or ascending ....
 .... **Primula mistassinica**

Primula meadia (L.) A.R. Mast & Reveal
EASTERN SHOOTING STAR   END *native*
IR | WUP | **CUP** | EUP   FACU CC 10
Dodecatheon meadia L.

Leaves oblong to oblong lance-shaped or rarely ovate, 6–20 cm long, usually tinged or marked with red at the base. Scapes 2–6 dm tall. Flowers few to many; corolla lobes narrow, white to lavender or lilac, 1–2.5 cm long. Capsules dark reddish brown, 7–18 mm long, thickest near the base. May–June.

Rare in a grassy prairie remnant in Menominee County.

Primula mistassinica Michx.
MISTASSINI PRIMROSE   *native*
IR | WUP | CUP | EUP   FACW CC 10

Perennial herb. Stems to 25 cm long. Leaves all at base of plant, oblong lance-shaped, 2–7 cm long, long tapered to base, smooth on upper surface, smooth or often white-yellow powdery below; margins with outward pointing teeth; bracts below flowers awl-shaped, 3–6 mm long. Flowers 1–2 cm wide, 2–10 in a cluster atop a leafless stalk; sepals joined, shorter than petals; petals joined, tubelike and flared at ends, pink and sometimes with a yellow center. Fruit an oblong, upright capsule to 1 cm long. May–June.

Moist ledges near the Great Lakes; often found with common butterwort (*Pinguicula vulgaris*).

ADDITIONAL SPECIES
Primula veris L. (Cowslip Primrose); European ornamental with yellow or purplish corolla, reported from roadside in Mackinac County.

Trientalis *Starflower*
Trientalis borealis Raf.
STARFLOWER   *native*
IR | WUP | CUP | EUP   FAC CC 5
Lysimachia borealis (Raf.) U. Manns & A. Anderb.

Low perennial herb, with slender rhizomes. Stems 1–2 dm tall, usually with a small scale-leaf near the middle and at the summit a whorl of lance-shaped acuminate leaves 4–10 cm long, from the axils of which appear 1 or several white flowers on slender pedicels 2–5 cm long. Flowers ordinarily 7-merous; calyx deeply divided into nearly separate lance-shaped sepals; corolla rotate, 8–14 mm wide, with very short tube and lance-shaped to ovate lobes; stamens inserted at base of the corolla. Fruit a 5-valved, many-seeded capsule. May–June.

Rich woods, hummocks in swamps and bogs.

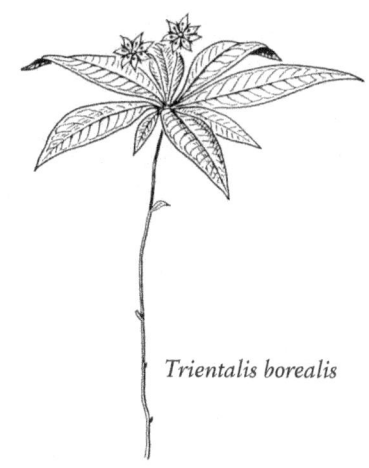

Trientalis borealis

276 RANUNCULACEAE | Buttercup Family

dicots

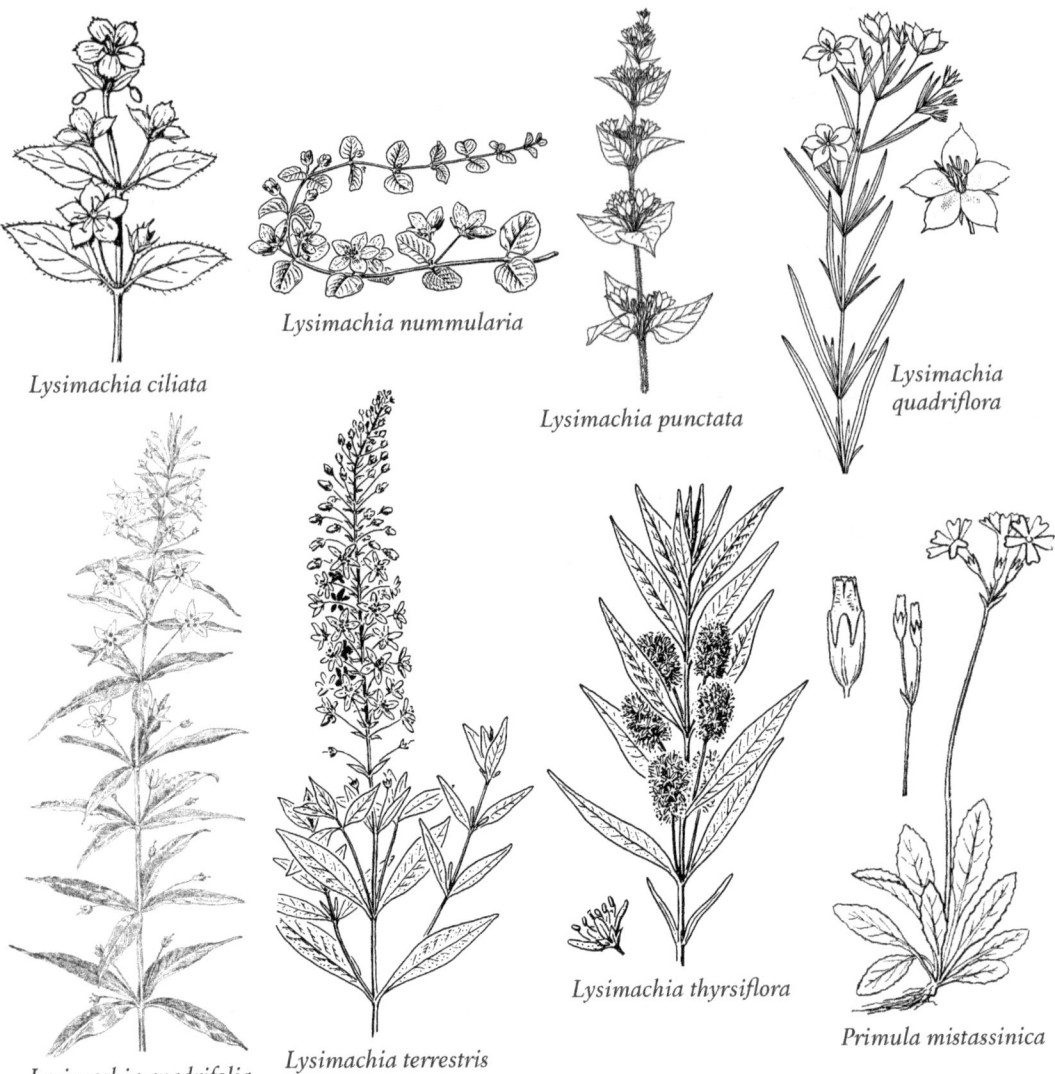

Ranunculaceae
BUTTERCUP FAMILY

Annual or perennial, aquatic or terrestrial herbs (or vines in *Clematis*). Leaves simple to compound, usually alternate, sometimes opposite or whorled, or all at base of plant. Flowers mostly white or yellow, usually with 5 (occasionally more) separate petals and sepals, or petals absent and then with petal-like sepals; flowers perfect (with both staminate and pistillate parts), stamens usually numerous; pistils several to many, ripening into beaked achenes or dry capsules (follicles).

ADDITIONAL SPECIES
Aconitum napellus L. (Garden monkshood), European garden escape known from swamps of Isle Royale's Passage Island (Keweenaw County). Very toxic if eaten.

1. Vines; leaves opposite; fruit with a long, feathery style .... **Clematis**
1. Herbs; leaves alternate or from base of plant; fruit not with a long, feathery style 2
2. Flowers spurred or strongly irregular 3
2. Flowers regular and unspurred 4
3. Sepals petal-like, unequal, the upper sepal largest and helmet-shaped, unspurred; uncommon introduction on Isle Royale..................... **Aconitum napellus***
3. Petal-like sepals equal or unequal, at least the upper sepal with a long spur .... **Aquilegia canadensis**
4. Stem leaves whorled .... **Anemone**
4. Stem leaves alternate, or all leaves from base of plant 5
5. Flowers yellow, or leaves simple and not lobed, or plants aquatic... 6

dicots ..RANUNCULACEAE | Buttercup Family 277

5   Flowers not yellow; leaves compound or 3-lobed; plants not aquatic 7
6   Leaves all alike, unlobed; sepals yellow, large and petal-like; petals absent . . . . **Caltha**
6   Leaves usually of 2 types (stem leaves different from basal leaves), or leaves deeply lobed or divided; sepals green; petals yellow or white . . . . **Ranunculus**
7   Plants with a naked scape and solitary flowers, the leaves basal 8
7   Plants with leafy stems; flowers 1 to several 9
8   Leaves glabrous, parted into 3 leaflets; sepals white . . . . **Coptis trifolia**
8   Leaves hairy, 3 lobed but not divided into leaflets; sepals white, pink, or blue . . . . **Hepatica**
9   Fruit an achene . . . . **Thalictrum**
9   Fruit berry-like, white or red . . . . **Actaea**

Actaea *Baneberry*

Perennial herbs. Leaves 2–3x 3-partedly compound, the leaflets sharply toothed. Flowers small, white, in a dense, long-peduncled, terminal raceme; the raceme at anthesis short and congested with short-pediceled flowers; the axis and pedicels elongating later, the pedicels becoming widely divergent. Sepals 3–5, obovate, petal-like. Petals 4–10, deciduous, obovate, clawed at base. Stamens numerous; filaments usually distinctly widened toward the summit. Pistil 1; stigma broad, sessile, 2-lobed. Fruit a several-seeded berry.

1   Fruit red, on slender pedicels . . . . **A. rubra**
1   Fruit white, on thicker pedicels . . . . **A. pachypoda**

Actaea pachypoda Ell.
WHITE BANEBERRY; DOLL'S EYES   native
IR | WUP | CUP | EUP   UPL CC 7

Stems 4–8 dm tall. Leaflets usually completely glabrous beneath. Petals 2.5–4 mm long, usually spatulate; stigma wider than the diameter of the ovary. Fruiting pedicels red, 1–2 mm thick; berries globose, normally white, with persistent stigma. May–June.
    Rich woods.

Actaea rubra (Ait.) Willd.
RED BANEBERRY   native
IR | WUP | CUP | EUP   FACU CC 7

Stems 4–8 dm tall. Leaflets commonly pubescent on the veins beneath. Petals 2.5–4 mm long, spatulate to obovate; stigma not so wide as the diameter of the ovary. Fruiting pedicels slender, 0.4–0.7 mm thick; berries usually red, ellipsoid. May–June.
    Rich woods.

Anemone *Thimbleweed*

Perennial herbs from a rhizome or caudex. Basal leaves few to several, deeply palmately divided; stem erect with a whorl of 3 or more involucral leaves subtending one or more elongate peduncles. Flowers white to blue or red or greenish. Sepals 4–20, petal-like; petals none; stamens numerous; pistils numerous, in a subglobose to cylindric head, pubescent. Achenes flattened, clavate, or fusiform, tipped with the persistent style.

ADDITIONAL SPECIES
Anemone blanda Schott & Kotschy (Greek thimbleweed), European ornamental perennial, with showy blue flowers (or sometimes flowers white or pink), escaped near gardens; Schoolcraft County.

1   Achenes hidden by long, cottony hairs 2
1   Achenes nearly glabrous or only short-hirsute 4
2   Involucral leaves sessile or nearly so . . . **A. caroliniana**
2   Involucral leaves with petioles 3
3   Involucral leaves 2 or 3; main leaflets thin and shallowly divided; achenes in an ovoid cluster . . . . . **A. virginiana**
3   Involucral leaves 4 or more; main leaflets thickened and deeply divided; achenes borne in an cylindrical cluster . . . . **A. cylindrica**
4   Involucral leaves with petioles; basal leaves absent or few . . . . **A. quinquefolia**
4   Involucral leaves without petioles; basal leaves usually many . . . . **A. canadensis**

Anemone canadensis L.
ROUND-LEAF THIMBLEWEED   native
IR | WUP | CUP | EUP   FACW CC 4

Perennial herb, from slender rhizomes, often forming large patches. Stems erect, 1–6 dm long, unbranched below the head. Leaves all from base of plant and with long petioles except for 2–3 stalkless leafy bracts below the head; 4–15 cm wide, deeply 3–5-lobed, round to kidney-shaped in outline, underside with long silky hairs, margins sharp-toothed. Flowers mostly single at ends of stalks, white and showy, 2–5 cm wide; sepals 5, petal-like, 1–2 cm long; petals absent; stamens and pistils many. Achenes clustered in a round, short-hairy head; achene body flat, 3–5 mm long and wide, beak 2–4 mm long. May–Aug.
    Common in wet openings, streambanks, thickets, low prairie, ditches and roadsides.

Anemone cylindrica Gray
LONG-HEAD THIMBLEWEED   native
IR | WUP | CUP | EUP   CC 6

Perennial herb. Stems stiffly erect, 3–10 dm tall. Basal and involucral leaves similar, the basal few to several, the involucral 3–10 and commonly 2x as many as the peduncles, both types petioled, broadly rounded in outline, deeply 5-parted into segments which are incised or sharply toothed only above the middle. Inflorescence usually of 2–6 erect peduncles

1–3 dm long, some of them often bearing a secondary involucre. Flowers greenish white, about 2 cm wide. Fruit a dense cylindric spike 20–35 mm long, about 8 mm thick. Achenes and style densely woolly, the style about 0.5 mm long, outwardly curved. June–Aug.

Dry open woods and prairies.

Anemone multifida Poir.
RED ANEMONE *native*
IR | WUP | CUP | EUP   CC 10

Perennial herb, from a stout (about 5 mm thick), erect or ascending, often branched caudex. Stems 1–6 dm tall, villous below the involucre. Basal leaves several, long-petioled, deeply 3-parted, the segments deeply incised or lobed into acute linear-oblong divisions; involucral leaves similar but sessile, usually near the middle, 1-flowered or in vigorous plants emitting also 1 or 2 other peduncles each bearing a small involucre and a flower; sepals usually red, rarely yellowish. May–June.

Limestone bluffs, open cliffs, hillside prairies, rocky soil.

Anemone quinquefolia L.
WOOD-ANEMONE *native*
IR | WUP | CUP | EUP   FACU CC 5

Delicate perennial herb from a slender horizontal rhizome. Stems 1–2 dm tall. Basal leaf solitary, long-petioled; leaflets 3 or apparently 5, coarsely and unevenly toothed or incised, chiefly above the middle; involucral leaves similar but smaller, the lateral leaflets commonly incised on the outer margin. Peduncle villous; sepals white or suffused with red beneath, usually 5, 10–22 mm long. Achenes fusiform, 3–4 mm long. April–June.

Moist woods.

Anemone virginiana L.
TALL THIMBLEWEED *native*
IR | WUP | CUP | EUP   FACU CC 3

Anemone riparia Fern.

Perennial herb. Leaf segments cuneate at base, the margins straight or nearly so. Flowers white, greenish white, or even red, 2–3 cm wide. Head of fruit slenderly ovoid or nearly cylindric, usually about 8 mm thick. Achenes densely woolly, the styles 1–1.5 mm long, strongly ascending, the stigma often incurved. June–Aug.

Rocky banks and open woods.

Aquilegia *Columbine*
Aquilegia canadensis L.
RED COLUMBINE *native*
IR | WUP | CUP | EUP   FACU CC 5

Perennial herb from a stout caudex-like rhizome. Stems at anthesis 3–10 dm tall, with few to several large basal leaves. Leaves compound, stem leaves gradually reduced upward, with fewer leaflets, the uppermost 3-foliolate or simple; leaflets broadly obovate, crenately toothed or lobed. Flowers nodding, 3–4 cm long; sepals 5, red; petals 5, the blade yellow, prolonged backward from the base into an elongate red spur; stamens numerous, projecting in a column; pistils usually 5, erect, each prolonged into a slender style. Fruit a several-seeded follicle. April–June.

Dry woods, rocky cliffs and ledges.

ADDITIONAL SPECIES

Aquilegia vulgaris (European columbine), resembling *A. canadensis* in foliage. Flowers nodding, blue, varying to purple, white, or pink, 3–4 cm long and wide; stamens not longer than the sepals; spurs much incurved.

Native of Eurasia; cultivated for ornament and occasionally escaped.

Caltha *Marsh-Marigold*

Succulent perennial herbs. Leaves simple, heart-shaped, mostly from base of plant, becoming smaller upward; margins entire or rounded-toothed. Flowers single at ends of stalks; sepals large and petal-like; mostly bright yellow (*C. palustris*), to pink or white (*C. natans*); petals absent; stamens many. Fruit a follicle.

1   Flowers bright yellow; stems upright; common . . . .
    . . . . **C. palustris**
1   Flowers pink or white; stems floating; rare in Baraga County . . . . **C. natans**

Caltha natans Pallas
FLOATING MARSH-MARIGOLD   THR *native*
IR | WUP | CUP | EUP   OBL CC 10

Perennial herb. Stems floating or creeping, branched, rooting at nodes. Leaves heart- or kidney-shaped, 2–5 cm wide, notched at base, upper leaves smaller. Flowers pink or white, 1 cm wide; sepals oval; petals absent; stamens 12–25. Fruit a follicle, 4–5 mm long, in dense heads of 20–40. July–Aug.

Rare in shallow water and shores of ponds and slow-moving streams; Baraga County.

Caltha palustris L.
COMMON MARSH-MARIGOLD *native*
IR | WUP | CUP | EUP   OBL CC 6

Loosely clumped perennial herb. Stems smooth, 2–6 dm long, hollow. Leaves heart-shaped to kidney-shaped, 4–10 cm wide, usually with 2 lobes at base; margins smooth or shallowly toothed; lower leaves with long petioles, stem leaves with shorter petioles. Flowers bright yellow, showy at ends of stems or in leaf axils, 2–4 cm wide; sepals 4–9, petal-like, 12–

20 mm long; petals absent; stamens many; pistils 4–15, with short styles. Fruit a follicle, 10–15 mm long. March–June.

Shallow water, swamps, wet woods, thickets, streambanks, calcareous fens, marshes, springs.

Clematis *Virgin's Bower*

Herbaceous or woody plants, erect, or climbing by the prehensile leaf-rachis. Leaves opposite, simple or compound. Flowers solitary or panicled, usually dioecious. Sepals petal-like, commonly 4. Petals none. Stamens numerous. Pistils numerous; style elongate. Fruit a flattened achene, terminated by the elongate persistent style.

1 Sepals whitish, less than 1 cm long, in a branched inflorescence .... **C. virginiana**
1 Sepals purple, 4–5 cm long, solitary ... **C. occidentalis**

Clematis occidentalis (Hornem.) DC.
PURPLE CLEMATIS   native
IR | WUP | CUP | EUP   CC 9
Clematis verticillaris DC.

Perennial, woody vine. Stems trailing or climbing, to 2 m long. Leaflets 3, long-stalked, ovate in outline; entire, crenately toothed, or lobed. Flowers chiefly axillary, solitary, on peduncles about equaling the subtending petiole; sepals 4, blue, ovate lance-shaped, 3–5 cm long, softly villous. Achenes villous, in a dense globular head; styles long-villous, 3–4 cm long. May.

Rocky woods and streambanks.

The purple flowers open early in spring with the unfolding leaves.

Clematis virginiana L.
VIRGIN'S BOWER   native
IR | WUP | CUP | EUP   FAC CC 4

Perennial, woody vine. Stems slender, to 5 m long or more, trailing on ground or over shrubs, smooth, brown to red-purple. Leaves opposite, divided into 3 leaflets, the leaflets ovate, 4–8 cm long and 2.5–5 cm wide; margins sharp-toothed or lobed; petioles 5–9 cm long. Staminate and pistillate flowers separate and on separate plants, in many-flowered, open clusters from leaf axils, on stalks 1–8 cm long, usually shorter than leaf petioles; sepals 4, creamy-white, 6–10 mm long; petals absent. Fruit a rounded head of hairy brown achenes tipped with feathery, persistent styles 2.5–4 cm long. July–Sept.

Thickets, streambanks, moist to wet woods, rocky slopes.

Coptis *Goldthread*

Coptis trifolia (L.) Salisb.
THREE-LEAF GOLDTHRAD   native
IR | WUP | CUP | EUP   FACW CC 5
Coptis groenlandica (Oeder) Fern.

Perennial herb, with slender, bright yellow rhizomes. Leaves from base of plant on long petioles, evergreen, divided into 3-leaflets, the leaflets shallowly lobed, with rounded teeth tipped by an abrupt point. Flowers single, white, 10–15 mm wide, on a stalk 5–15 cm long from base of plant; sepals 4–7, petal-like; petals absent; pistils 3–7, narrowed to a short, slender style. Fruit a beaked follicle 8–13 mm long. May–June.

Wet conifer woods and swamps, often on mossy hummocks.

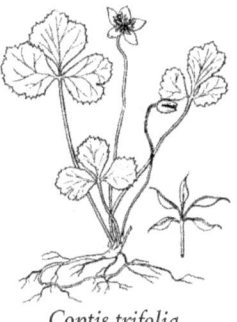

Coptis trifolia

Hepatica *Liverwort*

Perennial herbs, with several 1-flowered scapes bearing a calyx-like involucre of 3 entire bracts immediately below the flower. Leaves basal, simple, lobed, Sepals 5–12, petal-like. Petals none. Stamens numerous. Pistils numerous; ovary tapering into a short style. Achenes conic to fusiform, pubescent. The genus is very close to *Anemone* and often merged with it, differing only in the simple leaves and the position of the involucre (see species descriptions for synonyms). Leaves persist during the winter, the new leaves appearing after the very early blooming flowers.

1 Leaves lobed nearly to middle of blade, the lobes rounded .... **H. americana**
1 Leaves lobed to more than middle of blade, the lobes acute .... **H. acutiloba**

Hepatica acutiloba DC.
SHARP-LOBE HEPATICA   native
IR | WUP | CUP | EUP   CC 8
Anemone acutiloba (DC.) G. Lawson
Hepatica nobilis var. *acuta* (Pursh) Steyermark

Leaves 3-lobed or occasionally 5–7-lobed, deeply cordate at base, the lobes acute. Scapes 5–15 cm long, villous, as are also the petioles. Bracts acute, about equaling the sepals. Flowers 12–25 mm wide. March–April.

Dry or moist woods.

280 RANUNCULACEAE | Buttercup Family

Hepatica americana (DC.) Ker-Gawl.
ROUND-LOBE HEPATICA   native
IR | WUP | CUP | EUP   CC 6

Anemone americana (DC.) Hara
Hepatica nobilis var. *obtusa* (Pursh) Steyermark
Hepatica triloba Chaix

Very similar to *H. acutiloba* except in leaves and bracts. Leaves averaging smaller, 3-lobed, the lobes broadly obtuse or rounded, the terminal one often wider than long. Bracts obtuse. March–April.

Rich beech-maple forests, as for *H. acutiloba*, but more often found on drier sites with aspen, oak, hickory, pine; sometimes with spruce or cedar.

Ranunculus
Buttercup; Crowfoot; Spearwort

Aquatic, semi-aquatic, or terrestrial annual and perennial herbs. Stems erect to sprawling, sometimes floating in water. Leaves simple, or compound and finely dissected, often variable on same plant; alternate on stem or all from base of plant; petioles short to long. Flowers borne above water surface in aquatic species; sepals usually 5, green; petals usually 5, yellow or white, often fading to white, usually with a small nectary pit covered by a scale near base of petal; stamens and pistils numerous. Achenes many in a round or cylindric head; achene body thick or flattened, tipped with a straight or curved beak.

Hepatica acutiloba

Hepatica americana

Actaea rubra

Anemone canadensis

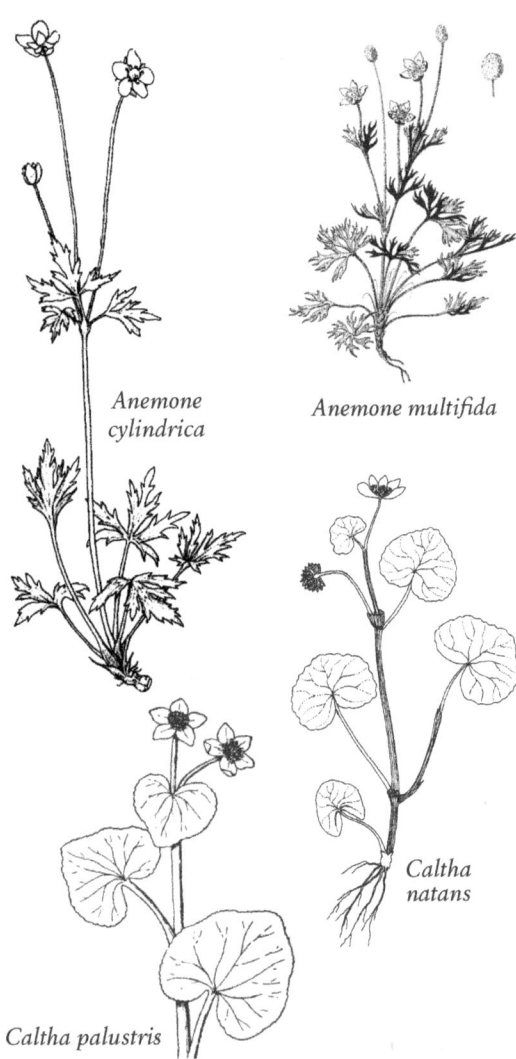

Anemone cylindrica

Anemone multifida

Caltha natans

Caltha palustris

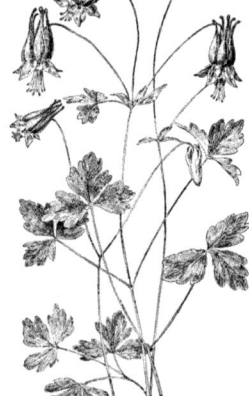

Anemone quinquefolia

Aquilegia canadensis

ADDITIONAL SPECIES

Ranunculus macounii Britton (Macoun's buttercup), native perennial; in Michigan, reported from wet, marshy places on Isle Royale.

1. Flowers white; leaves divided into linear or threadlike segments; plants typically aquatic . . . . R. aquatilis
1. Flowers yellow; leaves simple to deeply lobed or divided into narrow segments; plants aquatic or emergent . . . 2
2. Sepals 3 (rarely 4) . . . . R. lapponicus
2. Sepals 5 (or rarely more) 3
3. All leaves simple and entire, or shallowly lobed with rounded teeth 4
3. All, or at least stem leaves, deeply lobed, divided, or compound 5
4. Leaves ovate to round or kidney-shaped, shallowly lobed with rounded teeth; achenes with longitudinal ribs . . . . R. cymbalaria
4. Leaves oval to lance-shaped or linear, entire to sharp-toothed; achenes not ribbed . . . . R. flammula
5. Basal and stem leaves distinctly different in shape, the basal leaves mostly entire or with rounded teeth, the stem leaves deeply divided 6
5. Basal and stem leaves similar, all deeply lobed, divided, or compound 7
6. Flower petals 4–10 mm long, longer and wider than the sepals . . . . R. rhomboideus
6. Flower petals to 3 mm long, shorter and narrower than the sepals . . . . R. abortivus
7. Achenes swollen, without a sharp-edged margin 8
7. Achenes flattened, with a sharp or winglike margin . 10
8. Petals 2–4 mm long; achenes to 1.2 mm long, nearly beakless; plants terrestrial or in water only part of season . . . . R. sceleratus
8. Petals 4–14 mm long; achenes 1.2–2.5 mm long, beaked; plants underwater or exposed later in season 9
9. Petals more than 7 mm long; achene body more than 1.6 mm long, achene margin thickened and white-corky below the middle . . . . R. flabellaris
9. Petals less than 7 mm long; achene body less than 1.6 mm long, achene margin rounded but not thickened . . . . R. gmelinii
10. Petals 2–5 mm long 11
10. Petals 7–15 mm long 12
11. Beak of achene strongly hooked . . . . R. recurvatus
11. Beak of achene straight or only slightly curved . . . . R. pennsylvanicus
12. Style short and outcurved; introduced, weedy 13
12. Style elongate and nearly straight; native, non-weedy species 14
13. Stems creeping; terminal segment of the main leaves stalked . . . . R. repens
13. Stems upright; terminal segment of the main leaves not stalked, usually with green tissue extending to the lateral segments . . . . R. acris
14. Main leaves mostly longer than wide; mature receptacle cone-shaped . . . . R. fascicularis
14. Main leaves usually wider than long; mature receptacle often club-shaped and widest at tip . . . . R. hispidus

Ranunculus abortivus L.

KIDNEY-LEAF BUTTERCUP native
IR | WUP | CUP | EUP FAC CC 0

Biennial or perennial herb. Stems upright, 2–5 dm long, branched above, smooth or with fine hairs. Leaves at base of plant round to kidney-shaped, margins with rounded teeth, some leaves lobed; petioles long; stem leaves 3–5-divided into linear segments, margins entire or broadly toothed, petioles absent. Flowers yellow, petals 2–3 mm long, shorter than sepals. Achenes in a short, round head; achene body swollen, 1–2 mm long, with a very short, curved beak. April–June.

Wet to moist woods, floodplains, wet meadows, thickets, ditches; especially where soils disturbed or compacted.

Ranunculus acris L.

MEADOW-BUTTERCUP introduced
IR | WUP | CUP | EUP FAC

Perennial herb, with fibrous roots. Stems hairy, to 1 m long, with few branches, most leaves on lower part of stem. Leaves kidney-shaped, deeply 3–7-divided, the segments again lobed or dissected; branch leaves much smaller, 3-parted. Flowers numerous; sepals 5, half length of petals; petals 5, bright yellow, 6–15 mm long, obovate, often with a rounded notch at tip. Achenes in a round head; achene body flat, 2–3 mm long, beak 0.5 mm long. June–Aug.

Common weed of fields, thickets, ditches and shores.

Ranunculus aquatilis L.

LONG-BEAK WATER-CROWFOOT native
IR | WUP | CUP | EUP OBL CC 4

Ranunculus longirostris Godr.
Ranunculus trichophyllus Chaix

Perennial aquatic herb; plants mostly smooth. Stems underwater or floating, 3–8 dm long, with a few branches, rooting from lower nodes. Leaves round to kidney-shaped in outline, 2–3x divided into narrow threadlike segments 1–2 cm long, stiff and not collapsing when removed from water; leaf segments tipped with tiny transparent spine; petioles absent or to 4 mm long. Flowers at or below water surface, single from upper leaf axils, 1–1.5 cm wide; sepals 5, purple-green, spreading, 2–4 mm long; petals 5, white, yellow at base, 4–9 mm long. Achenes 15–25 in a round head; achene body obovate, ridged, the beak thin and straight, 1–1.5 mm long. May–Aug.

Ponds, lakes, streams, rivers and ditches.

Our plants sometimes treated as 2 species as follows:

1. Styles (at least the longest) and achene beaks 0.6–1.1 mm long, more than 1/3 the length of the achene body .... **R. longirostris**
1. Styles and achene beaks very short, less than 0.6 mm long, less than about 1/3 the length of the body....... .... **R. trichophyllus**

Ranunculus cymbalaria Pursh
ALKALI BUTTERCUP (THR) native
IR | **WUP** | CUP | EUP OBL CC 8

Cyrtorhncha cymbalaria (Pursh) Britt.
Halerpestes cymbalaria (Pursh) Greene

Perennial herb, spreading by stolons and forming dense mats. Stems 3–20 cm long, smooth. Leaves all from base of plant, ovate to kidney-shaped, 5–25 mm long and 4–30 mm wide, heart-shaped at base; margins with rounded teeth, often with 3 prominent lobes at tip; petioles sparsely hairy. Flower stalks longer than leaves, unbranched or with a few branches, with 1 to several flowers; sepals 5, green-yellow, 3–5 mm long, deciduous; petals usually 5, yellow, turning white with age, 3–5 mm long; stamens 10–30. Achenes numerous in a cylindric head to 10 mm long; achene body 1.5–2 mm long, longitudinally nerved, beak short and straight. June–Sept.

Wet meadows, streambanks, sandy or muddy shores, ditches and seeps; in the UP known from near Wakefield in Gogebic County from a single collection made in 1960.

Ranunculus fascicularis Muhl.
EARLY BUTTERCUP native
IR | **WUP** | **CUP** | **EUP** FACU CC 10

Perennial herb. Stems 1–2, or at maturity 3 dm tall, usually 2–5 from one base, strigose. Leaves mostly basal, ovate in outline, distinctly longer than wide, the terminal segment stalked, all segments deeply lobed, the lobes incised or coarsely crenately toothed; stem leaves 1–3, smaller, sessile or nearly so, less divided. Flowers long-peduncled; petals elliptic or oblong, 8–14 mm long, 3–5.5 mm wide. Achenes rotund, 2–3.3 mm long, sharply margined; beak slender, straight or nearly so, 1.4–3 mm long. April–May.

Prairies and dry woods.

Ranunculus flabellaris Raf.
GREATER YELLOW WATER BUTTERCUP native
IR | WUP | **CUP** | EUP OBL CC 10

Perennial herb; plants smooth or sometimes hairy when growing out-of-water. Stems floating, or upright from a sprawling base when exposed, branched, rooting at lower nodes, 3–7 dm long. Underwater leaves 3-parted into linear segments 1–2 mm wide, exposed leaves (when present) round to kidney-shaped in outline, 2–10 cm long and 2–12 cm wide, divided into 3 segments, the segments again 3-divided. Flowers 1 to several at ends of stems; sepals 5, green-yellow, 4–8 mm long; petals 5–8, bright yellow, 6–15 mm long. Achenes 50–75 in a round to ovate head; achene body obovate, to 2 mm long, the margin thickened and corky below middle, beak broad, flat, 1–1.5 mm long. May–July.

Shallow water or muddy shores of ponds, quiet streams, swamps, woodland pools, marshes and ditches.

Ranunculus flammula L.
CREEPING SPEARWORT native
IR | **WUP** | **CUP** | EUP FACW CC 8

Ranunculus reptans L.

Perennial herb, spreading by stolons; plants often covered with appressed hairs. Stems sprawling, rooting at nodes, unbranched or few-branched, with upright shoots 4–15 cm long. Leaves in small clusters at nodes, simple, linear or threadlike, 1–5 cm long and 1.5 mm wide, margins more or less entire; upper leaves smaller and with shorter petioles than lower. Flowers single at ends of stems; sepals 5, yellow-green, 2–4 mm long, with stiff hairs; petals 5, yellow, obovate, 3–5 mm long. Achenes 10–25 in a round head; achene body swollen, obovate, 1–1.5 mm long, smooth, the beak short, to 0.5 mm long. June–Aug.

Sandy, gravelly, or muddy shores; shallow to deep water, water usually acid.

Ranunculus gmelinii DC.
LESSER YELLOW WATER BUTTERCUP native
IR | **WUP** | CUP | EUP FACW CC 10

Perennial herb, similar to yellow water-crowfoot (*R. flabellaris*) but plants aquatic or at least partly underwater; smooth or sometimes with coarse hairs. Stems usually sprawling and rooting at nodes, 1–5 dm long, sparsely branched. Leaves all on stem or with a few basal leaves on long petioles, deeply 3-lobed or dissected, the segments again forked 2–3 times; underwater leaf segments 2–4 mm wide; exposed leaves to 2 cm long and 1.5–2.5 cm wide. Flowers usually 1 to several at ends of stems; sepals 5, green-yellow, 3–6 mm long; petals 5–8, yellow, 4–8 mm long. Achenes 50–70 in a round to ovate head; achene body obovate, 1–1.5 mm long, the margin rounded, not corky-thickened, the beak broad and thin, 0.4–0.7 mm long, somewhat curved. July–Aug.

Muddy streambanks and lakeshores, cold springs, pools in swamps and bogs.

Ranunculus hispidus Michx.
NORTHERN SWAMP BUTTERCUP native
IR | **WUP** | **CUP** | **EUP** FAC CC 5

Perennial herb; stems and leaves variable. Stems upright, 2–9 dm long, smooth to coarsely hairy. Leaves from base of plant and on stems, the basal leaves larger and with longer petioles than stem leaves; 3-lobed, heart-shaped in outline, 3–14 cm long and 4–20 cm wide, with appressed hairs on veins, upper leaves usually strongly toothed. Flowers 1 to several; sepals 5, yellow-green, 5–11 mm long, hairy; petals 5–8, yellow, fading to white, 7–15 mm long and 3–10 mm wide. Achenes 15–30 or more in a round head; achene body obovate, 2–4 mm long, smooth, winged on margin, the beak straight, 2–3 mm long. May–July.

Wet woods, floodplains and swamps, thickets, lakeshores, wet meadows and fens.

Ranunculus lapponicus L.
LAPLAND BUTTERCUP (THR) native
IR | WUP | CUP | EUP   OBL CC 10

Coptidium lapponicum (L.) Gandog.
Perennial herb, spreading by rhizomes. Stems prostrate, 1–2 dm long, sending up 1 shoot from each node, the shoots with 1–2 basal leaves, sometimes with a single smaller leaf above. Leaves kidney-shaped, deeply 3-cleft, margins with rounded teeth or shallowly lobed. Flowers single at ends of shoots; petals yellow with orange veins, 8–12 mm wide; sepals 3, curved downward. Achenes in a round head; achene body 2–3 mm long, swollen near base, flattened above, beak slender, sharply hooked. June–July.

Cedar swamps and bogs.

Plants resemble three-leaf goldthread (*Coptis trifolia*) but Lappland buttercup leaves are lobed, not compound, lighter green and deciduous.

Ranunculus pensylvanicus L. f.
BRISTLY CROWFOOT   native
IR | WUP | CUP | EUP   OBL CC 6

Annual or short-lived perennial herb. Stems erect, hollow, 3–8 dm long, branched or unbranched. Leaves at base of plant withering early, larger and with longer petioles than the few stem leaves; 4–12 cm long and 4–15 cm wide, with appressed hairs, 3-lobed and coarsely toothed, the terminal leaflet stalked. Flowers few, on short stalks; sepals 5, yellow, 4–5 mm long; petals 5, pale yellow, fading to white, shorter than the sepals, 2–4 mm long; stamens 15–20. Achenes many, in a rounded cylindric head 10–15 mm long; achene body flattened, 2–3 mm long, smooth, the beak stout, 0.5–1.5 mm long. July–Aug.

Marshes, wet meadows, ditches and streambanks, often in muck.

Ranunculus recurvatus Poir.
HOOKED CROWFOOT   native
IR | WUP | CUP | EUP   FACW CC 5

Perennial herb. Stems 2–7 dm long, usually hairy, branches few. Leaves broadly kidney-shaped or round in outline, 3-parted to below middle, covered with long, soft hairs; petioles present on all but uppermost leaves. Flowers on stalks at ends of stems; sepals curved downward, to 6 mm long; petals pale yellow, 4–6 mm long; styles strongly hooked. Achenes in a short-cylindric head; achene body flat, round, sharp-margined, to 2 mm long; beak 1 mm long, hooked or coiled. May–June.

Moist deciduous forests (especially in openings), swamps; also in drier woods.

Ranunculus repens L.
CREEPING BUTTERCUP   introduced
IR | WUP | CUP | EUP   FAC

Perennial herb. Stems normally creeping and rooting at the nodes, rarely ascending or erect; hirsute, strigose, or rarely glabrous. Leaves 3-parted, petioled, the segments broadly obovate in outline, cleft or lobed, sharply toothed. Petals 8–15 mm long, about two-thirds as wide. Achenes obovate, 2.5–3.5 mm long, sharply but narrowly margined; beak triangular, usually somewhat curved, 0.8–1.5 mm long. May–July.

Native of Europe; introduced in fields, lawns, roadsides, and wet meadows. Plants vary in size and in kind and amount of pubescence.

Ranunculus rhomboideus Goldie
LABRADOR BUTTERCUP (THR) native
IR | WUP | CUP | EUP   CC 9

Perennial herb. Stems 8–20 cm tall at anthesis. Basal leaves ovate-oblong to broadly ovate, 1–5 cm long, long-petioled, crenate mostly above the middle, tapering to rounded or acute at base; stem leaves sessile or subsessile, cleft into a few linear divisions. Flowers few to several, on villous peduncles; petals oblong-elliptic, 5–9 mm long, much longer than the villous sepals. Achenes in a globose head, obovate, 2–2.8 mm long, flattened at the base, turgid above the middle; beak very short. April–May.

Dry open woods and prairies.

Ranunculus sceleratus L.
CURSED CROWFOOT   native
IR | WUP | CUP | EUP   OBL CC 1

Weedy annual herb; plants smooth, sometimes partly submersed in shallow water. Stems upright, hollow, 1–6 dm long, branched above and with many flowers. Leaves from base of plant less deeply parted and with longer petioles than stem leaves; upper stem leaves small; leaves deeply 3-parted, the main lobes again lobed, heart-shaped at base, rounded at tip, 1–6 cm long and 3–8 cm wide. Flowers numerous at ends of stalks from upper leaf axils and branches; sepals 5, 2–3 mm long, yellow-green, tips curved downward; petals 5, light yellow, fading to white, 3–5 mm long. Achenes numerous in a short-cylindric head 4–11 mm long; achene body obovate, 1 mm long, slightly corky-thickened on margins; beak tiny, blunt. May–Sept.

Muddy shores, streambanks, wet meadows, ditches, marshes and other wet places.

Thalictrum Meadow-Rue

Perennial herbs. Leaves alternate, compound. Staminate and pistillate flowers separate, in panicles on separate plants; sepals 4–5, green or petal-like but soon deciduous; petals absent; stamens numerous, the stalks (filaments) long and slender; pistils several to many. Fruit a ribbed or nerved achene.

1   Upper stem leaves with long petioles; leaflets glabrous and not glandular; flowering in April or May before leaves fully expanded; plants less than 1m tall . . . . T. dioicum
1   Upper stem leaves sessile or nearly so (the 3 stalked leaflets appearing to arise together from the node); leaflets glabrous, or hairy, or with small, short-stalked glands; flowering in summer after leaves expanded; plants often more than 1 m tall. 2
2   Leaflets 3-lobed, each lobe tipped with 1–3 teeth . . . . T. venulosum
2   Leaflets usually 3-lobed, the lobes usually not toothed 3
3   Underside of leaflets with very short hairs (rarely smooth), not glandular; leaves odorless; widespread . . . . T. dasycarpum
3   Underside of leaflets with small beads and hairs tipped with gray or amber exudate; leaves with strong odor when crushed . . . . T. revolutum

Thalictrum dasycarpum Fisch. & Avé-Lall.
PURPLE MEADOW-RUE   native
IR | **WUP** | **CUP** | **EUP**   FACW CC 3

Perennial herb, from a short rootstock. Stems purple-tinged, 1–2 m long, branched above. Leaves divided into 3–4 groups of leaflets; leaflets 15 mm or more long, mostly tipped with 3 pointed lobes, dark green above, underside sparsely short-hairy, not waxy and without gland-tipped hairs; margins usually slightly turned under; stem leaves mostly without petioles. Flowers in panicles at ends of stems; staminate and pistillate flowers separate and on different plants (sometimes with some perfect flowers); sepals 3–5 mm long, lance-shaped; anthers linear and sharp-tipped, 2–3 mm long, filaments white; stigmas straight, 2–4 mm long. Achenes 4–6 mm long, ribbed, in a round cluster. June–July.

Common in wet to moist meadows, low prairie, swamps, thickets, streambanks.

Thalictrum dioicum L.
EARLY MEADOW-RUE   native
IR | **WUP** | **CUP** | **EUP**   FACU CC 6

Perennial herb, dioecious. Stems 3–7 dm tall at anthesis. Leaves all with long petioles, the uppermost 3–6 cm long and subtending the inflorescence; stipules of the upper leaves broadly ovate, mostly much wider than long. Filaments and anthers yellow or greenish yellow. Mature achenes sessile or subsessile, about 4 mm long, strongly ribbed, straight and essentially symmetrical.

Moist woods.

Flowering with or before the expansion of leaves on deciduous trees.

Thalictrum revolutum DC.
WAXY-LEAF MEADOW-RUE   native
IR | **WUP** | **CUP** | **EUP**   FAC CC 9

Perennial herb, from short rootstocks, with strong odor when crushed. Stems more or less smooth, often purple-tinged, 0.5–1.5 m long. Lowest leaves with petioles, middle and upper leaves stalkless; leaves divided into 3–4 groups of leaflets; leaflets variable in shape and size, usually 3-lobed, some 1–2 lobed, upper surface smooth, underside leathery and conspicuously net-veined, finely hairy with gland-tipped hairs, margins turned under. Flowers in panicles at ends of stems; staminate and pistillate flowers separate and on different plants (sometimes with some perfect flowers); anthers linear, 2–3 mm long, filaments threadlike, 2–5 mm long; pistils 6–12, stigmas 2–3 mm long. Fruit an oval or lance-shaped achene, 4–5 mm long, ridged, with tiny gland-tipped hairs. June–July.

Streambanks, thickets, moist meadows and prairies.

Thalictrum venulosum Trel.
VEINY-LEAF MEADOW-RUE   native
IR | **WUP** | **CUP** | **EUP**   FACW CC 9

Perennial herb, spreading by rhizomes; plants pale green, waxy. Stems erect, 3–10 dm long. Leaves divided into 3–4 groups of leaflets; leaflets firm, nearly circular or obovate in outline, tipped by 3–5 lobes, underside veiny, appearing wrinkled, usually sparsely covered with gland-tipped hairs; lower leaves on petioles, upper leaves stalkless. Flowers in narrow panicles at ends of stems, the panicle branches nearly erect; staminate and pistillate flowers separate and on different plants; stamens 8–20, anthers linear and pointed at tip, filaments slender. Fruit an ovate achene, 4–6 mm long, tapered to a short-beak. June–July.

Streambanks, thickets and wet, calcium-rich shores of Lake Michigan.

RANUNCULACEAE | Buttercup Family

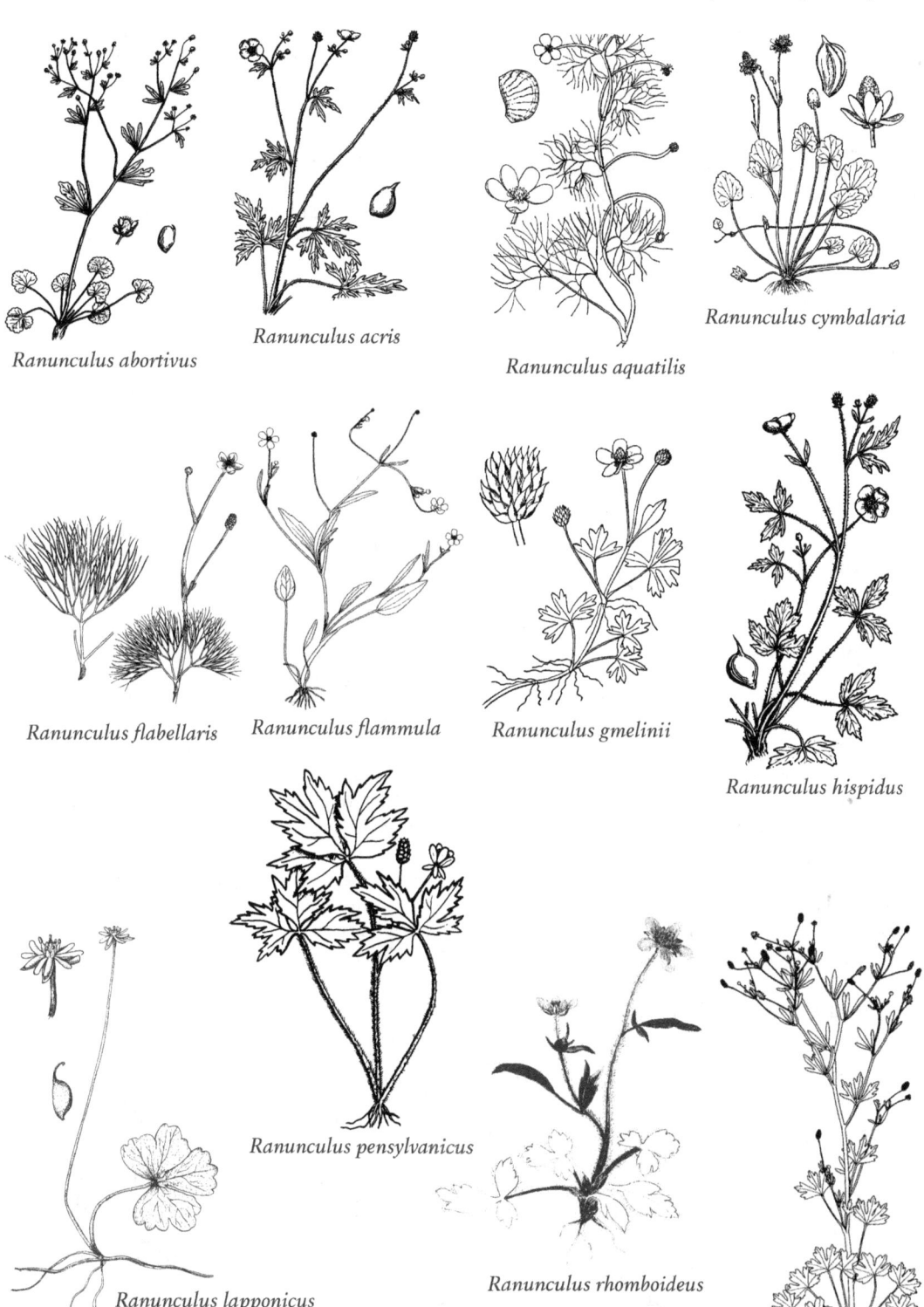

Rhamnaceae
BUCKTHORN FAMILY

Shrubs, trees, or woody vines with simple, opposite or alternate leaves, and small flowers. Flowers perfect or unisexual, regular, 4–5-merous. Petals present or lacking, small, separate. Stamens as many as and alternate with the sepals, opposite and often enfolded by the petals. Ovary 1, sessile on the disk or immersed in it; styles 2–5, united for all or part of their length. Fruit a capsule or drupe.

1. Leaves 3-veined from base of leaf; flowers white in many-flowered, stalked clusters; fruit a capsule. . . . . **Ceanothus**
1. Leaves not 3-veined from base; flowers greenish, single or few to a cluster; fruit a fleshy drupe. 2
2. Leaf margins entire or nearly so. . . . . . . **Frangula alnus**
2. Leaf margins toothed . . . . **Rhamnus**

Ceanothus *Buckbrush*

Low shrubs. Leaves alternate, 3-nerved, glandular-serrate. Flowers small, white, in sessile or short-peduncled umbels aggregated into terminal or axillary panicles. Sepals inflexed, at length deciduous above the hypanthium. Petals long-clawed. Stamens at anthesis free and exsert. Ovary 3-angled and 3-celled, immersed in the disk; style 3-lobed. Fruit a 3-lobed capsule-like drupe subtended by the persistent hypanthium. Many species highly ornamental. Characteristic are the leaves with 3 pairs of prominent parallel veins extending from the leaf base to the outer margins of the leaf tips.

ADDITIONAL SPECIES

Ceanothus sanguineus Pursh (Redstem ceanothus), native shrub of the Pacific Northwest, known in the eastern USA only from Keweenaw County along rocky roadsides and forest margins; threatened.

1. Inflorescences on peduncles arising from branches of the previous year; shrub usually over 1 m tall; Keweenaw County . . . . **C. sanguineus***
1. Inflorescences on peduncles or branchlets of the current year; shrub less than 1 m tall; widespread 2
2. Leaves elliptic, less than 2 cm wide; inflorescences at ends of current year's shoots . . . . **C. herbaceus**
2. Leaves ovate, mostly more than 2 cm wide; inflorescences from leaf axils . . . . **C. americanus**

Ceanothus americanus L.
NEW JERSEY-TEA   native
IR | **WUP** | **CUP** | EUP   CC 8

Shrub to 1 m tall, often freely branched. Leaves narrowly to broadly ovate, 3–8 cm long, usually more than half as wide, broadly cuneate to rounded or subcordate at base, the lateral nerves commonly naked for 1–3 mm at base. Inflorescences on axillary peduncles, the lower peduncles progressively longer and to 2 dm long; panicle short-cylindric to ovoid, occasionally branched, often subtended by 1–3 reduced leaves, the umbels usually separated by distinct internodes. Fruit depressed-obovoid, 5–6 mm long. June–July.

Upland woods, prairies, and barrens.

Ceanothus herbaceus Raf.
PRAIRIE REDROOT   native
IR | **WUP** | **CUP** | **EUP**   CC 9

Bushy shrub to 1 m tall. Leaves oblong to elliptic, 2–6 cm long, 1–2 cm wide, the lateral nerves never naked and often arising unevenly 1–3 mm above the base of the leaf; leaf underside usually pubescent. Panicles several to many, terminating the leafy branches of the season, on peduncles rarely to 5 cm long, hemispheric to short-ovoid, the component umbels very close together. Fruit 4–5 mm long. May–June.

Sandy or rocky soil, prairies and plains.

Thalictrum dasycarpum

Thalictrum dioicum

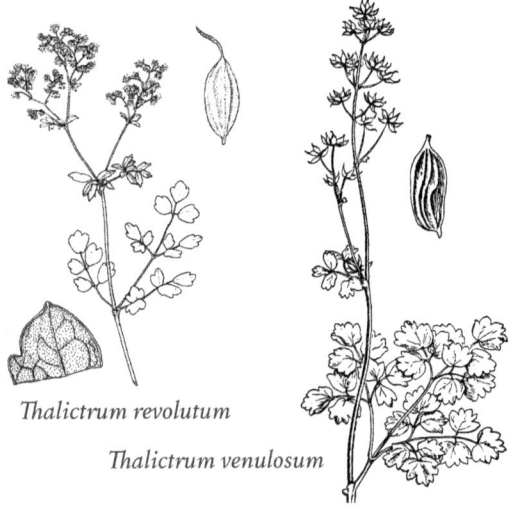

Thalictrum revolutum

Thalictrum venulosum

Frangula *False Buckthorn*

Frangula alnus P. Mill.
GLOSSY FALSE BUCKTHORN *intro. (invasive)*
IR | **WUP** | **CUP** | **EUP** FAC
Rhamnus frangula L.

Shrub to 7 m tall. Leaves usually obovate-oblong, 5–8 cm long, commonly more than half as wide, acute to abruptly short-acuminate, entire or with a few marginal glands near the tip. Umbels sessile, 2–8-flowered; pedicels usually unequal, 3–10 mm long. Flowers perfect, 5-merous; petals broadly obovate, scarcely clawed, cleft at the tip, 1–1.4 mm long; styles connate to the tip. Fruit red, ripening to nearly black, 2–3-stoned. May–June.

Native of Eurasia; escaped from cultivation, especially in wet soil.

Rhamnus *Buckthorn*

Shrubs or small trees. Leaves simple, alternate or opposite, pinnately veined, usually with stipules. Flowers perfect, or staminate or pistillate, regular, single or few from leaf axils; sepals joined, 4- or 5-parted; petals 4 or 5. Fruit a purple-black, berrylike drupe with 2–4, 1-seeded stones.

1 Leaves with 2–4 obvious pairs of lateral veins . . . .
. . . . **R. cathartica**
1 Leaves with mostly 5 or more pairs of lateral veins . . . .
. . . . **R. alnifolia**

Rhamnus alnifolia L'Hér.
ALDER-LEAF BUCKTHORN *native*
IR | **WUP** | **CUP** | **EUP** OBL CC 8

Shrub to 1 m tall, forming low thickets. Leaves alternate, oval to ovate, 6–10 cm long and 3–5 cm wide, green above, paler green below; margins with low, rounded teeth; petioles grooved, 5–12 mm long; stipules linear, to 1 cm long, deciduous before fruit mature. Flowers appearing with leaves in spring, in clusters of 1–3 flowers from leaf axils; yellow-green, usually 5-parted, 3 mm wide, on short stalks, with both stamens and pistils but one or other is non-functional, sepals 1–2 mm long, petals absent. Fruit a purple-black, berrylike drupe, 6–8 mm wide, with 1–3 nutlet-like stones. May–June.

Conifer swamps, thickets, sedge meadows, wet depressions in deciduous forests; usually where calcium-rich.

Rhamnus cathartica L.
EUROPEAN BUCKTHORN *intro. (invasive)*
IR | **WUP** | **CUP** | **EUP** FAC

Shrub or small tree to 5 m tall. Stems with pale lenticels. Leaves mostly alternate but some leaves often nearly opposite, oval or obovate, 5–8 cm long and 3–5 cm wide; margins entire or slightly wavy; petioles stout, 1–2 cm long. Flowers appearing after leaves in spring, perfect, single or in clusters of 2–8 in leaf axils, green-yellow, 5-parted, to 5 mm wide; petals 1–2 mm long. Fruit a purple-black, berrylike drupe, 7 mm wide, with 2–3 nutlike stones. May–Aug.

Conifer swamps, thickets, calcareous fens, lakeshores, moist to dry woods, especially where disturbed, heavily grazed, or cleared. Introduced from Eurasia; escaping from cultivation in ne and central North America.

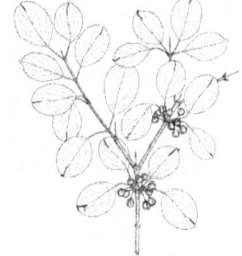
Rhamnus cathartica

Rosaceae
ROSE FAMILY

Shrubs and perennial, biennial, or annual herbs. Leaves evergreen or deciduous, mostly alternate and simple or compound. Flowers perfect (with both staminate and pistillate parts), regular, with 5 sepals and petals; stamens numerous. Fruit an achene,

Ceanothus americanus

Ceanothus herbaceus

Frangula alnus

Rhamnus alnifolia

capsule, or fleshy fruit with numerous embedded seeds (drupe), or a fleshy fruit with seeds within (pome).

ADDITIONAL SPECIES

Chamaerhodos erecta var. *nuttallii* Torr. & A. Gray (syn: *Chamaerhodos nuttallii* Rydb.), Keweenaw rockrose; a species of western North America, known in Michigan only from Brockway Mountain, Keweenaw County, where it occurs on loose conglomerate; endangered.

1. Plants trees, shrubs, or erect to trailing, thorny to bristly brambles 2
1. Plants herbs (sometimes woody at base), not thorny or bristly 14
2. Leaves mostly compound; branches or stems often thorny or bristly. 3
2. Leaves simple; branches and stems smooth or only with long stout spines 7
3. Stems biennial, prickly or bristly; leaves 3-parted or palmately compound; fruit a tight cluster of juicy drupelets; flowers usually white. . . . . **Rubus**
3. Stems perennial, smooth or thorny; leaves pinnately compound; fruit various but not a cluster of drupelets; flowers white, pink, or yellow. 4
4. Flowers pink (rarely white or yellow), 2 cm or more wide; stems thorny; fruit fleshy, red to orange . . . **Rosa**
4. Flowers white or yellow, mostly less than 2 cm wide; stems smooth; fruit various. 5
5. Flowers solitary or few in an inflorescence, the petals yellow; leaflets entire . . . . **Potentilla fruticosa**
5. Flowers many in a crowded inflorescence, the petals white; leaflets toothed. 6
6. Colony-forming shrub, occasionally escaping from cultivation; inflorescence a panicle, much longer than wide; leaflets doubly-toothed (each main tooth with several smaller teeth) . . . . **Sorbaria sorbifolia**
6. Small trees; inflorescence much wider than long; leaflets not doubly toothed . . . . **Sorbus**
7. Style and ovary 1; fruit a drupe; leaves unlobed **Prunus**
7. Styles 2 or more (1 in *Crataegus monogyna*); fruit a pome, or a cluster of drupelets or dry fruits 8
8. Ovary superior. 9
8. Ovary inferior 10
9. Leaves mostly 3–5 lobed; bark shredding into long strips . . . . **Physocarpus opulifolius**
9. Leaves not lobed; bark not shredding into long strips . . . . **Spiraea**
10. Leaves with red or black appressed glands along midrib of leaf upper surface . . . . **Aronia prunifolia**
10. Leaves without glands on midrib 11
11. Branches never thorny; flower petals white, lance-shaped and usually more than 2 times longer than wide . . . . **Amelanchier**
11. Branches sometimes with stout spines; petals less than 2 times longer than wide 12
12. Branches normally with spines; leaves toothed and often slightly lobed 13
12. Branches without spines; leaves toothed but not lobed . . . . **Malus**
13. Spines shiny; bud scales glabrous; petals white; seeds within hard nutlets . . . . **Crataegus**
13. Spines dull; bud scales hairy; petals pinkish; seeds within papery carpels . . . . **Malus**
14. Leaves 3-parted or palmately compound 15
14. Leaves pinnately compound or divided. 20
15. Styles long, jointed near middle, the lower portion persistent on the achene as a long beak. . . . . **Geum**
15. Styles short, neither jointed nor persistent on the fruit 16
16. Calyx with bractlets about as large as sepals, the calyx appearing 10-lobed 17
16. Calyx without bractlets between the sepals, the calyx 5-lobed. 19
17. Petals white; fruit fleshy and red; leaflets 3 . . . **Fragaria**
17. Petals yellow or white; fruit dry; leaflets 3, 5, or 7 . . . 18
18. Flowers yellow, leaflets 3, 5, or 7, regularly toothed, deciduous. . . . . **Potentilla**
18. Flowers white; leaflets 3, entire except for a 3 (–5)-toothed apex, evergreen . . . . **Sibbaldia**
19. Petals yellow; fruit an achene . . . . . **Geum fragarioides**
19. Petals white or pink; fruit fleshy drupelets. . **Potentilla**
20. Leaves 2–3 times compound; inflorescence a large panicle of numerous spike-like racemes. . **Aruncus dioicus**
20. Leaves once-pinnate; inflorescence various, smaller. 21
21. Calyx 5-lobed, small bractlets absent; receptacle flat or concave 22
21. Calyx 10-lobed, small bractlets alternating with sepals; receptacle hemispherical or conical 24
22. Petals pink; receptacle flat or nearly so; leaflets deeply lobed. . . . . **Filipendula**
22. Petals yellow or absent; receptacle deeply concave; leaflets not lobed. 23
23. Petals yellow; floral tube with hooked bristles at tip; inflorescence an elongate raceme. . . . . **Agrimonia**
23. Petals absent; floral tube not bristly; inflorescence short . . . . **Sanguisorba minor**
24. Styles elongating and becoming longer than achene, persistent as a beak atop the achene . . . . **Geum**
24. Styles short, deciduous 25
25. Petals deep maroon to purple; sepals red tinged; stem usually decumbent, the lower portion in water or wet ground, rooting at nodes . . . . **Comarum palustre**
25. Petals yellow or white; sepals green; stem usually erect, or with slender stolons; mostly upland 26
26. Pubescence not glandular; petals deep yellow. . . . . **Potentilla**
26. Pubescence glandular-viscid; petals white to pale yellow . . . . **Drymocallis**

Agrimonia *Agrimony; Grooveburr*

Perennial herbs from stout rhizomes. Stems erect, simple or branched above. Leaves pinnately compound, mostly below middle of stem; stipules foliaceous, usually deeply toothed or laciniate. Flowers in long, interrupted, spike-like racemes, the short peduncle subtended by a laciniate bract, the very short pedicels by a pair of 3-lobed bractlets. Hypanthium obconic to hemispheric, with hooked bristles and small resinous glands. Sepals spreading at anthesis, later incurved and forming a beak on the fruit. Petals 5, yellow, 5–8 mm wide. Stamens 5–15. Pistils 2. Fruit an achene.

1. Inflorescence rachis covered with small glands, the pubescence sparse or absent......... **A. gryposepala**
1. Inflorescence rachis without glands or nearly so; rachis covered with appressed to spreading hairs .. **A. striata**

Agrimonia gryposepala Wallr.
TALL HAIRY AGRIMONY *native*
IR | **WUP** | **CUP** | **EUP** FACU CC 2

Perennial herb, roots fibrous. Stems stout, to 15 dm tall, glandular and sparsely or densely long-hirsute throughout. Principal leaflets of the larger leaves 5–9, ovate lance-shaped to elliptic or obovate, coarsely and often bluntly serrate, glabrous or or nearly so on the surface, sparsely hirsute on the veins; stipules large and leaflike, usually 1–2 cm wide. Axis glandular, hirsute with long spreading hairs; pedicel hirsute. Hypanthium glandular only, or also with a few short stiff hairs near the base, 3–5 mm long at maturity, expanded at the summit. July–Aug.

Moist or dry open woods.

Agrimonia striata Michx.
WOODLAND AGRIMONY *native*
IR | **WUP** | **CUP** | **EUP** FACU CC 3

Perennial herb, roots fibrous. Stems stout and coarse, to 1 m tall or more, hirsute below, pubescent and glandular above. Principal leaflets of the larger leaves 7–11, the upper 5 commonly directed forwards, ovate lance-shaped, coarsely serrate, glabrous or nearly so above, sparsely pubescent beneath, especially on the veins; stipules lance-shaped, 1–2 cm long. Axis eglandular, densely pubescent with ascending hairs, commonly also with some long flexuous hairs. Flowers densely crowded; peduncle and pedicel short, the 3-cleft bractlet commonly surpassing the hypanthium; mature hypanthium reflexed, turbinate, 4–5 mm long, deeply furrowed. July–Aug.

Dry or moist woods.

Amelanchier *Serviceberry*

Trees or shrubs, without thorns. Leaves simple, alternate, serrate. Flowers in short leafy racemes (except in *A. bartramiana*) terminating the branches of the season and opening with or before the leaves. Hypanthium obconic, campanulate, or saucer-shaped. Sepals 5, spreading to recurved, persistent. Petals 5, white, oblong to oval or obovate. Stamens usually 20, shorter than the petals. Ovary 5-celled; styles 5. A confusing genus, similar to *Crataegus* in that hybridization, polyploidy, and asexual vegetative reproduction have resulted in a wide variety of forms.

1. Pedicels 1–3 in axils of leaves; petals less than twice as long as wide; leaves at least partly open and essentially glabrous (except margins and petioles) at flowering time, the blade tapering trough-like into raised petiole margins; petioles less than 8 (–15) mm long ....
.... **A. bartramiana**
1. Pedicels 4 or more (at least scars present if some have fallen with fruit), the inflorescence a raceme; petals at least twice as long as wide; leaves various (glabrous to tomentose) but the blade rounded or truncate to subcordate, not tapered at base; petioles usually longer than 8 mm; statewide.......................... 2
2. Tip of ovary glabrous; leaf blades short-acuminate, finely and closely serrate with 22–45 teeth per side .. 3
2. Tip of ovary tomentose; leaf blades variously shaped and toothed 4
3. Leaves just beginning to unfold at flowering time, densely white-tomentose beneath, otherwise green, retaining some of the pubescence on petioles and along midrib beneath into maturity............ **A. arborea**
3. Leaves mostly half-grown at flowering time, usually bronze-red, glabrous or nearly so, completely glabrous at maturity .... **A. laevis**
4. Larger leaves with 25–50 fine teeth on a side (more than twice as many teeth as lateral veins), acute, at flowering time open though not fully grown and often glabrous or soon becoming so..................... **A. interior**
4. Larger leaves with fewer than 20 (–25) teeth on a side (no more than 2x as many teeth as lateral veins), the blades at flowering time ± folded and white-tomentose beneath, when mature the tip acute to rounded 5

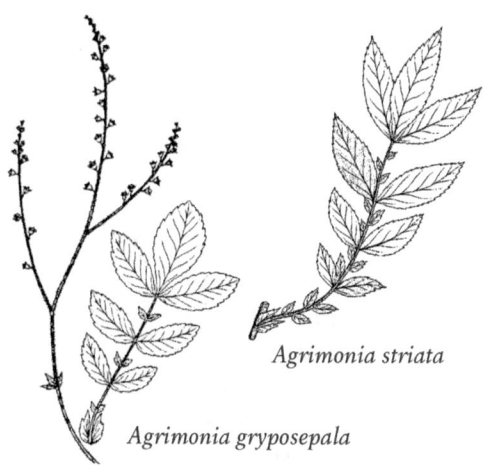

Agrimonia striata

Agrimonia gryposepala

5 Most leaves coarsely toothed (2–5 teeth per cm toward tip when mature), the veins prominent and running to tips of the teeth (or a principal fork into the teeth) at least toward tip of blade; petals 10–20 mm long; plants typically solitary or in tall clumps with many stems . . . . **A. sanguinea**

5 Most leaves finely toothed at least toward apex (5–8 teeth per cm when mature), the veins anastomosing and becoming indistinct near the margin, at most with weak veinlets ending in the teeth; petals 5–9 mm long; plants typically spreading underground and forming colonies of low shrubs. **A. spicata**

Amelanchier arborea (Michx. f.) Fern.
DOWNY SERVICEBERRY native
IR | WUP | CUP | EUP FACU CC 4

Tall shrub or small tree, rarely to 10 m tall. Leaves typically obovate, sharply and finely serrate nearly to the rounded or cordate base, densely pubescent beneath when young, at maturity nearly glabrous. Racemes drooping, many-flowered; pedicels 5–20 mm long, often covered with silky hairs; hypanthium glabrous; sepals 2–3 mm long, reflexed; petals narrowly oblong, 10–15 mm long. Pomes maroon-purple, 6–10 mm wide, insipid. April–May.

Usually in dry sandy open forests with red maple, aspen, oaks, or jack pine; sometimes in moist or swampy forests and along forest borders.

Amelanchier bartramiana (Tausch) M. Roemer
OBLONG-FRUIT SERVICEBERRY native
IR | WUP | CUP | EUP FAC CC 8

Shrub to 2 m tall, often forming clumps; twigs purplish, more or less glabrous. Leaves alternate, ovate to oval, 2–5 cm long and 1–2.5 cm wide, often tipped with a small spine, green above, paler below, often purple-tinged when unfolding; margins with small, sharp, forward-pointing teeth; petioles to 1 cm long. Flowers 1 cm or more wide, single or in groups of 2–4 at ends of branches or on pedicels 1–2 cm long from leaf axils; sepals lance-shaped; petals white, oval to oblong, 6–10 mm long. Pomes dark purple, 1–1.5 cm wide, insipid. May–Aug.

Conifer swamps, open bogs, thickets, old dune or rock ridges; borders of hardwood forests; plants may be low and sprawling on bare rock shores and ledges, otherwise a tall shrub.

Amelanchier interior Nielsen
INLAND SERVICEBERRY native
IR | WUP | CUP | EUP CC 4

Shrub or small tree. Stems 1–10 m tall, often straggling or arching; twigs glabrous at flowering. Leaves broadly ovate, 3–7 cm long and 2–5 cm wide, acute to short-acuminate, base rounded to subcordate, upper surface green, sparsely pubescent or glabrous by flowering time; margins serrate nearly to the base; petioles 1–3 cm long. Inflorescence 4–12-flowered, drooping or nodding; pedicels glabrous or nearly so; hypanthium campanulate, 3–6 mm wide; sepals recurving after flowering, 2–5 mm; petals white, obovate, 6–15 mm long; stamens 20; styles 5. Pomes purple-black, globose, 6–8 mm wide, sweet. May–June.

Sandy open savannas and dunes, shallow soil on rock outcrops and shores; sometimes at borders of hardwood forests and conifer swamps.

Amelanchier laevis Wieg.
SMOOTH SERVICEBERRY native
IR | WUP | CUP | EUP CC 4

Tall erect shrub or tree, to 10 m tall. Leaves elliptic to ovate, to 8 cm long at maturity, abruptly acute to short-acuminate, finely and sharply serrate nearly to the rounded or subcordate base, at anthesis about half grown and glabrous beneath or rarely with a few scattered hairs. Racemes many-flowered; pedicels glabrous, 1–3 cm long; hypanthium glabrous externally; sepals 3–4 mm long, reflexed; petals oblong, 10–18 mm long or rarely more; ovary glabrous at the tip. Pomes dark purple, mostly 10–15 mm wide, sweet. May.

Most often in dry sandy open forests and savannas, rocky sites, sandy bluffs and shores; also on river banks and forest and bog margins.

Amelanchier sanguinea (Pursh) DC.
NEW ENGLAND SERVICEBERRY native
IR | WUP | CUP | EUP CC 5

Amelanchier humilis Wieg.

Erect or straggling shrub or small tree, to 3 m tall, usually growing in clumps of several stems. Leaves about half grown at anthesis and then tomentose beneath, eventually glabrous, oblong to subrotund or subquadrate, to 7 cm long, finely or coarsely toothed, often only above the middle; veins often prominent and running to the teeth, especially in upper portion of blade. Inflorescences 4–10-flowered, soon arching or drooping; pedicels hairy; hypanthium saucer-shaped, 3.5–7.5 mm wide; sepals recurving or spreading after flowering, 3.5–5 mm; petals white, linear to narrowly spatulate, 11–18 mm long; stamens 20; styles 5; ovary summit rounded, densely hairy. Pomes dark purple or almost black, 5–8 mm wide, sweet. May–June.

Dry, open, sandy savannas and clearings; sandy thickets, borders of forests, gravelly shores, and low dunes.

Amelanchier spicata (Lam.) K. Koch
RUNNING SERVICEBERRY native
IR | WUP | CUP | EUP FACU CC 4

Amelanchier stolonifera Wieg.

Stoloniferous shrub 3–10 or rarely 15 dm tall, forming colonies. Leaves a quarter to half grown at anthesis and then densely tomentose beneath, at maturity glabrous and much paler beneath, ovate to oblong, or obovate-oblong, usually 2–5 cm long, finely and sharply toothed; lateral veins curved forward, branched and anastomosing near the margin; teeth almost always more than twice as many as the veins. Racemes dense; pedicels thinly pubescent, soon glabrescent, the lowest 7–15 mm long. Pomes purple-black, glaucous, 7–12 mm wide, sweet. May.

Dry, sandy plains, dunes, and savannas, usually with jack pine or oaks, often little taller than the associated shrubby species of *Comptonia* and *Vaccinium*.

Aronia *Chokeberry*

Aronia prunifolia (Marsh.) Rehder
PURPLE CHOKEBERRY native
IR | **WUP** | **CUP** | **EUP** FACW CC 5

Shrub, 1–2.5 m tall; twigs gray to purple, smooth or hairy. Leaves alternate, oval or obovate, 3–8 cm long and 1–4 cm wide, upper surface dark green and smooth (except for dark, hairlike glands along midveins), underside paler, smooth or hairy; margins with small, rounded, forward-pointing teeth, the teeth gland-tipped; petioles to 1 cm long. Flowers 5–10 mm wide, in clusters of 5–15 at ends of stems and short, leafy branches; sepals usually glandular; petals white, 4–6 mm long. Fruit a dark purple to nearly black, berrylike pome, 8–11 mm wide, not persisting into winter. May–June.

Tamarack swamps, open bogs, thickets, marshes and shores.

Aronia prunifolia

Aruncus *Goat's Beard*

Aruncus dioicus (Walt.) Fern.
GOAT'S BEARD *introduced*
IR | **WUP** | **CUP** | **EUP** FACU

Rhizomatous perennial herbs. Stems erect, 1–2 m tall. Leaves alternate, to 5 dm long; leaflets ovate lance-shaped to broadly ovate, 5–15 cm long, coarsely doubly serrate. Flowers dioecious, 5-merous, in numerous racemes aggregated into a large terminal panicle 1–3 dm long; staminate flowers with petals about 1 mm long, stamens 15 or more; pistils rudimentary, 3–5; pistillate flowers with sepals and petals as in the staminate flowers but smaller; stamens 15 or more, minute and rudimentary; pistils commonly 3, rarely 4 or 5. Follicles about 2 mm long, turgid, tipped with the persistent styles.

Rich woods.

Comarum *Marshlocks*

Comarum palustre L.
MARSH CINQUEFOIL *native*
IR | **WUP** | **CUP** | **EUP** OBL CC 7
Potentilla palustris (L.) Scop.

Perennial herb, from long, stout rhizomes. Stems 3–8 dm long, ascending to sprawling or floating in shallow water, often rooting at nodes, more or less woody at base; lower stems smooth, upper stems sparsely hairy. Leaves all from stem, pinnately divided or nearly palmate, with 3–7 leaflets; leaflets oblong to oval, 3–10 cm long and 1–3 cm wide, mostly rounded at tip, underside waxy; margins with sharp, forward-pointing teeth; lower leaves long-petioled, upper leaves nearly sessile; stipules forming wings around petioles of lower leaves, becoming shorter upward. Flowers single or paired from leaf axils, or in open clusters; sepals dark red or purple (at least on inner surface), ovate to lance-shaped, 6–20 mm long; petals 5 (sometimes 10), very dark red, 3–5 mm long, with a short slender tip; stamens about 25, dark red. Achenes red to brown, smooth, 1 mm long. June–Aug.

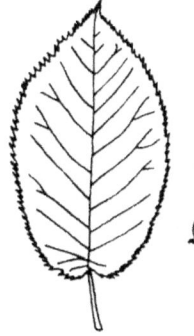
Amelanchier arborea

Amelanchier bartramiana

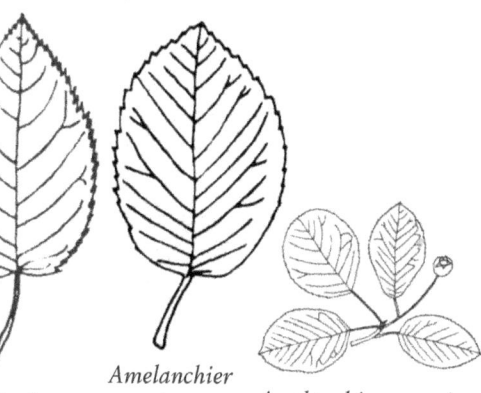
Amelanchier laevis   *Amelanchier spicata*   *Amelanchier sanguinea*

Open bogs (especially in pools and wet margins), conifer swamps, shores.

Crataegus *Hawthorn*

Small trees or shrubs with usually spiny branches. Leaves simple, deciduous, alternate, serrate or dentate and otherwise entire or variously lobed. Flowers perfect, regular, in corymbs or rarely single or 2 or 3 together. Sepals 5. Petals 5, white or rarely pink. Stamens 5–20 (the number of stamens can be counted on the flowers and also on the fruit using remnants of the filaments). Ovary of 1–5 carpels; styles 1–5, persistent. Fruit a globose pome, red or rarely yellow, blue, or black at maturity, with 1–5 bony nutlets. The leaves of sterile shoots or of the ends of branches (vegetative leaves) are often differently shaped and more deeply incised than those of the flowering branchlets.

Because of the apparent instability of many species and their tendency to hybridize, there is no generally agreed upon consensus regarding Michigan's hawthorns.

ADDITIONAL SPECIES

Crataegus dodgei Ashe (Dodge's hawthorn), native; Mackinac County.
Crataegus irrasa Sarg. (Blanchard's hawthorn), native, usually in rocky places and on lakeshores; reported from Chippewa County and western UP.
Crataegus monogyna Jacq. (English hawthorn), introduced; Houghton County; distinctive in its deeply lobed leaves, and which are smaller than in other species of hawthorn.
Crataegus submollis Sarg. (Quebec hawthorn), reported from a single Ontonagon County.

1 Leaves with some of the primary lateral veins running to (or toward, forking just before) the sinuses as well as to the points of the lobes; blades ± deltoid in general outline or small and deeply lobed; thorns under 5 cm long; stamens ca. 20 .... **C. monogyna***
1 Leaves with the primary lateral veins running only to (or toward) the points of the lobes (if any); blades, thorns, and stamens various 2
2 Nutlets with deep to shallow pits or depressions on their lateral surfaces; flowering in late May or June .. 3
2 Nutlets not pitted laterally; flowering in April–early June 4
3 Mature fruit purplish black, glaucous; thorns mostly 1.5–2.5 cm long; inflorescence glabrous or very sparsely villous; stamens 10 or fewer; nutlets 3–4 (–5), rounded at the ends .... **C. douglasii**
3 Mature fruit red or orange; thorns mostly 2.5–9.5 cm long; inflorescence densely villous to sometimes glabrous; stamens ca. 10 or 20; nutlets 2–3, or if more, then acute at the ends .... **C. succulenta**
4 Blades of at least the floral leaves (in many species also the vegetative leaves) ± acute to broadly or (more commonly) narrowly tapered or cuneate at the base 5
4 Blades of both floral and vegetative leaves mostly broadly rounded, truncate, or subcordate at the base 6
5 Blades (especially of floral leaves) mostly obovate to oblong-elliptic, broadest above or rarely at the middle, unlobed or very obscurely lobed near the apex, mostly 1.5–3 or more times as long as broad, usually thick or even stiff and leathery.................. **C. punctata**
5 Blades (at least of floral leaves) mostly elliptic to ovate, broadest at or below the middle, often ± lobed, usually 1–1.5 times as long as broad, often thin C. chrysocarpa
6 Stamens 10 or fewer; anthers pink to purple; young leaves strigose above; inflorescences completely glabrous at flowering time .... **C. macrosperma**
6 Stamens 15–20 .... **C. irrasa***

Crataegus chrysocarpa Ashe
FIREBERRY HAWTHORN *native*
IR | **WUP** | **CUP** | **EUP** CC 4

Stout, intricately branched shrub or rarely a small tree to 5–6 m tall; branchlets very thorny. Leaves elliptic, oval, or suborbicular, 4–6 cm long and 2–4 cm wide, lobed, serrate except near the base with gland-tipped teeth, roughened above with short appressed hairs while young; firm, dark yellow-green, the veins impressed above at maturity; petioles slender, sometimes slightly glandular, mostly 1/4 to 1/2 as long as the blades. Flowers 1.3–1.6 cm wide; stamens about 10, anthers white or pale yellow; sepals nearly entire or finely glandular-serrate. Fruit dull or dark red or rarely dull yellow, with thin flesh, remaining hard or dry or becoming mellow late in the season; nutlets 3–4. May; fruit ripe Sept–Oct. Sandy hillsides, stream and river banks, forest

Aruncus dioicus

Comarum palustre

borders, roadsides, fields, pastures; sometimes in wet places.

Crataegus douglasii Lindl.
BLACK HAWTHORNE native
IR | **WUP** | **CUP** | **EUP**   CC 7

Tall shrub; twigs glabrous, thornless or with scattered stout thorns 1–2.5 cm long. Leaves obovate to elliptic, mostly 2–4 long and 1.5–3 cm wide, with 2–4 pairs of small, shallow, often irregular lateral lobes (or leaves larger and more deeply cleft on vegetative shoots), firm, dark green and glossy above, essentially glabrous. Flowers 1–1.3 cm wide, in compound cymes of 5–12- flowers. Fruit to 1 cm thick, dark wine-color to black when ripe, succulent; nutlets 3compound cymes5, with a large pit on the inner surface.

Borders of forests, sometimes locally common, as at Delaware, Keweenaw County; rocky summits and openings; sand dunes and shores; more common in western USA.

Easily recognized as our only *Crataegus* with dark blue-black fruit when ripe.

Crataegus macrosperma Ashe
BIG-FRUIT HAWTHORN native
IR | **WUP** | **CUP** | EUP   FACU CC 5

Much-branched shrub to 5–6 m high. Older stems stout and armed with straight or slightly curved thorns 3–10 cm long. Leaves ovate, acuminate at tip and cuneate to truncate at base, short-pilose above; margins indented with 4–6 small lobes on each side and serrate nearly to the base; petioles 1–3 cm long, somewhat winged and grooved near the blade; leaves of the vegetative shoots more deeply lobed. Flowers 15–18 mm wide, in loose, pubescent corymbs; stamens usually 10 or less; anthers pink. Fruit oblong to nearly globose, sometimes slightly angular, 8-12 mm thick, bright red; nutlets 3–5. May–June.

Open woods, thickets, fields, and along river banks and rocky ridges.

Crataegus punctata Jacq.
DOTTED HAWTHORN native
IR | WUP | **CUP** | EUP   CC 1

Tree to 8–10 m tall with an open top of stiff spreading branches, armed with slender gray thorns 4–6 cm long, and often with compound thorns on trunk. Leaves obovate, or sometimes oblong-elliptic on shoots, dull yellowish green, mostly 2.5–6 cm long, 1.5–3 cm wide, cuneate or attenuate at base, serrate sometimes only above the middle, usually slightly lobed at least toward the apex, often deeply lobed on vegetative shoots, firm, the veins distinctly impressed on the upper surface at maturity; leaves covered with short appressed hairs above while young. Flowers in many-flowered corymbs; stamens about 20, anthers pink or pale yellow; corymbs and calyx tube gray-pubescent. Fruit subglobose or short-oblong, appearing pyriform while immature, usually 1.2–1.5 cm wide, dull red or orange-red, pale-dotted, with thick flesh becoming mellow or slightly succulent; nutlets usually 5, rounded and ridged on the back. May–June; fruit ripe Sept–Oct.

Thickets and borders of woods, often in rocky ground; Delta and Menominee counties.

Crataegus succulenta Schrad.
FLESHY HAWTHORN native
IR | **WUP** | **CUP** | **EUP**   CC 5

Tree to 7 or 8 m tall or sometimes an arborescent shrub, with slender branchlets, glabrous or rarely slightly hairy while young, often armed with long chestnut-brown thorns becoming gray and compound on the larger branches. Leaves elliptic, rhombic or rarely ovate, finely serrate and usually indented with 4–6 pairs of shallow lobes above the middle, glabrous or slightly villous along the veins beneath, firm to nearly leathery, roughened with short appressed hairs on the upper surface while young; petioles 1–2 cm long, wing-margined above, usually eglandular. Flowers 1.3–1.7 cm wide, in many-flowered, slightly villous or glabrous corymbs; stamens about 10–20, anthers white or pink; sepals glandular-serrate, reflexed after anthesis, usually deciduous from the mature fruit. Fruit subglobose, 0.7–1.2 cm wide, bright red, glabrous, lustrous; nutlets 2–3. May–June; fruit ripe Sept.

Thickets, pastures, and borders of woods, usually in dry or rocky ground.

Drymocallis   *Woodbeauty*

Drymocallis arguta (Pursh) Rydb.
TALL WOODBEAUTY native
IR | **WUP** | **CUP** | **EUP**   FACU CC 8

Potentilla arguta Pursh

Perennial from a stout rhizome, more or less viscid-pubescent throughout. Stems erect, 3–10 dm tall, simple to the inflorescence. Leaves pinnately compound, the basal leaves long-petioled; leaflets 7–11, or only 5 in the uppermost leaves. Flowers white, cream or pale yellow, 12–18 mm wide, crowded in a slender, elongate inflorescence; sepals ovate, much longer than the lance-shaped bractlets, nearly as long as the petals. Achenes obovoid, pale brown, 1 mm long, finely striate. June–July.

Dry woods and prairies.

Filipendula   *Queen-of-the-Prairie*

Filipendula rubra (Hill) B.L. Robins.
QUEEN-OF-THE-PRAIRIE (THR) native
IR | **WUP** | CUP | EUP   FACW CC 10

Perennial rhizomatous herb. Stems smooth, 1–2 m

ROSACEAE | Rose Family

Crataegus chrysocarpa   *Crataegus macrosperma*

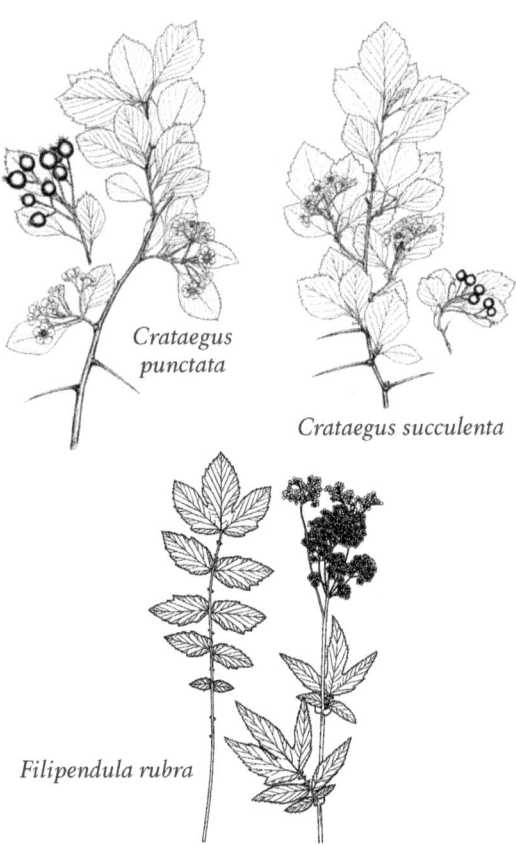

Crataegus punctata

Crataegus succulenta

Filipendula rubra

long. Leaves large, lower leaves to 8 dm long and to 2 dm wide, pinnately parted or divided into 5–9 segments, the segments opposite, stalkless, with 3–5 deep or shallow lobes; margins sharply toothed. Flowers pink-purple, fragrant, 7–10 mm wide, in a panicle 1–2 dm wide at ends of stems; petals 5, 2–4 mm long; stamens many. Fruit an erect, smooth capsule, 6–8 mm long. June–July.

Wet meadows and shores; Marquette County.

Fragaria *Strawberry*

Perennial herbs, usually spreading freely by runners and forming colonies. Leaves basal, 3-foliolate, serrate. Flowers several on peduncles. Hypanthium saucer-shaped. Sepals alternating with foliaceous bracts of nearly equal size. Petals white, obovate to subrotund. Stamens numerous, sometimes abortive. Pistils numerous. Fruit consisting of numerous minute achenes on the greatly enlarged, red, juicy receptacle, subtended by the persistent calyx and bracts.

1. Terminal center tooth of leaflets smaller than the tooth on either side of it; calyx lobes appressed to fruit .... **F. virginiana**
1. Terminal center tooth of leaflets as large or larger than the tooth on either side of it; calyx lobes spreading away from fruit .... **F. vesca**

Fragaria vesca L.
THIN-LEAVED WILD STRAWBERRY   *native*
IR | WUP | CUP | EUP   CC 2

Leaflets sessile or nearly so, ovate to obovate, more or less silky beneath; principal lateral veins diverging from the midvein at an angle of about 45 degrees; teeth sharp and divergent. Peduncles at anthesis usually shorter than the leaves, exceeding them at maturity; pedicels of unequal length, eventually forming a panicle-like inflorescence; petals commonly 5–7 mm long. April–June.

Hardwood and mixed forests, cedar and tamarack swamps, shores and forest edges.

In nearly all leaves, a line connecting the apices of the 2 uppermost lateral teeth passes across the projecting terminal tooth, which is usually more than half as wide as the lateral ones, measured from sinus to sinus.

Fragaria virginiana Duchesne
THICK-LEAVED WILD STRAWBERRY   *native*
IR | WUP | CUP | EUP   FACU CC 2

Leaflets petioled, glabrate to sericeous beneath; principal lateral veins commonly divergent from the midrib at an angle of about 30 degrees; teeth blunter and less divergent than in *F. vesca*. Inflorescence with as many as 12 flowers on pedicels of about uniform length, forming a corymbiform cluster usually shorter than the leaves; petals usually 7–10 mm long. April–June.

Fragaria vesca   *Fragaria virginiana*

Widespread in many habitats, in a diversity of deciduous, mixed, and coniferous forests, clearings, dry sandy forests, roadsides, and fields; more often in dry open sunny places than *F. vesca*. In nearly all leaves, a line connecting the tips of the 2 uppermost lateral teeth passes above the tip of the small terminal tooth, which is usually less than half as wide as the adjacent lateral ones, measured from sinus to sinus.

Geum *Avens*

Perennial herbs. Lower leaves pinnately lobed or divided, upper leaves smaller, less divided or entire. Flowers yellow, white or purple; 1 to many in clusters at ends of stems; petals 5; stamens 10 to many. Fruit an achene.

1   Leaves all 3-foliolate and basal. . . . . **G. fragarioides**
1   Leaves mostly pinnately compound or divided, if 3-foliolate, then cauline. 2
2   Calyx bell-shaped; reddish; flowers nodding; petals yellow, tinged with purple 3
2   Calyx lobes spreading, green; flowers upright; petals white or yellow 4
3   Plant of wetlands; terminal leaflet much larger than lateral leaflets; style with distinct joint near its middle, lengthening to less than 2x length of perianth **G. rivale**
3   Plant of dry habitats; terminal leaflet barely larger than lateral leaflets; style not jointed, elongating to 2x or more longer than perianth . . . . **G. triflorum**
4   Plants flowering 6
4   Plants fruiting 9
5   Petals white to pale yellow 7
5   Petals bright yellow 8
6   Petals equal to or longer than the sepals; stems glabrous or only sparsely hairy . . . . **G. canadense**
6   Petals shorter than the sepals; stems densely hairy, the hairs spreading. . . . . **G. laciniatum**
7   Terminal leaflet of basal leaves much larger than lateral segments; lower portion of style with short-stalked glands . . . . **G. macrophyllum**
7   Terminal leaflet various; lower portion of style without glands. 8
8   Native species; petals 5 mm or more long; upper portion of style long-hairy . . . . **G. allepicum**
8   Introduced weedy species; petals to 4 mm long; upper portion of style glabrous or nearly so. . . . . . **G. urbanum**
9   Receptacle glabrous or only sparsely hairy (remove a few achenes to check); plants with either the achene beak with short-stalked glands, or with the pedicels with dense long hairs over the much shorter hairs . . 10
9   Receptacle densely hairy; achene beaks neither with glands nor the pedicels with dense long hairs 11
10 Achene beak not glandular; pedicels densely long-hairy . . . . **G. laciniatum**
10 Achene beak with short-stalked glands, especially near base; pedicels finely hairy and with only scattered long hairs. . . . . **G. macrophyllum**
11 Stem leaves pinnately compound; achenes many (150 or more) in each head, the achene beak with long hairs at base . . . . **G. allepicum**
11 Stem leaves mostly 3-parted; achenes less than 100 in each head, the achene beak glabrous 12
12 Native species of natural habitats; upper segment of style long-hairy at base. . . . . **G. canadense**
12 Introduced species occasional in waste places; upper segment of style glabrous or nearly so . . . **G. urbanum**

Geum aleppicum Jacq.
YELLOW AVENS   native
IR | WUP | CUP | EUP   FAC CC 3

Perennial herb. Stems erect or ascending, to 1 m long, branched above, covered with coarse hairs. Leaves variable, basal leaves pinnately divided into 5–7 oblong leaflets, wedge-shaped at base, petioles long-hairy; stem leaves divided into 3–5 segments, stalkless or short-petioled; margins coarsely toothed. Flowers 1 to several, short-stalked, on branches at ends of stems; sepals lance-shaped; petals 5, yellow; style jointed. Achenes usually long-hairy. June–July.

Swamps, wet forests, wet meadows, marshes, calcareous fens, ditches and roadsides.

Geum canadense Jacq.
WHITE AVENS   native
IR | WUP | CUP | EUP   FAC CC 1

Perennial herb. Stems slender, 4–10 dm tall, glabrous or sparsely pubescent below, above and on the pedicels becoming densely velvety-puberulent, often with a few scattered longer hairs. Basal leaves long-petioled, commonly 3-foliolate with obovate leaflets; upper leaves short-petioled, 3-foliolate with oblong lance-shaped, sharply serrate leaflets; uppermost leaves mostly simple, lance-shaped, nearly sessile. Pedicels finely velvety hairy, with or without long scattered hairs; petals white, obovate, about as long as the sepals or distinctly exceeding them. Head of fruit obovoid, 10–15 mm long; receptacle densely bristly, the hairs protruding among the ovaries at anthesis but shorter than the mature achenes. Achenes 2.5–3.5 mm long, excluding the style. May–June.

Dry or moist woods.

Geum fragarioides (Michx.) Smedmark
BARREN STRAWBERRY   native
IR | WUP | CUP | EUP   CC 6

Waldsteinia fragarioides (Michx.) Tratt.
Perennial rhizomatous herb with the aspect of a strawberry. Leaves basal, 3-foliolate, 1–2 dm long including the petioles, ± winter-green; leaflets broadly obovate, rounded at tip, serrate with numerous

broad teeth and commonly also shallowly and irregularly lobed, the lateral leaflets unsymmetrical. Flowers in a cyme on a naked or bracted peduncle, the peduncles about equaling the leaves; hypanthium obconic; sepals triangular; petals yellow, obovate, 5–10 mm long, obtuse or rounded, much exceeding the sepals; stamens numerous, the slender filaments erect and persistent after anthesis. Fruit an achene. April–May.

Moist or dry woods, thickets, thin soil over rock outcrops.

Geum laciniatum Murr.
ROUGH AVENS native
IR | **WUP** | **CUP** | EUP   FACW CC 2
Perennial herb. Stems 4–10 dm long, covered with long, mostly downward-pointing hairs. Lower leaves pinnately divided, the segments pinnately lobed; upper leaves divided into 3 leaflets or lobes; margins coarsely toothed; petioles hairy. Flowers mostly single at ends of densely hairy stalks from ends of stems; sepals triangular, 4–10 mm long; petals 5, white, 3–5 mm long. Fruit an achene, 3–5 mm long (excluding style), grouped into round heads 1–2 cm long. May–June.

Wet woods, floodplain forests, ditches; Marquette and Ontonagon counties.

Geum macrophyllum Willd.
BIG-LEAF AVENS native
IR | **WUP** | **CUP** | **EUP**   FACW CC 5
Perennial herb. Stems to 1 m long, unbranched, or branched above, bristly-hairy. Leaves pinnately divided, basal leaves stalked, the terminal segment large, 3–7-lobed, with much smaller segments intermixed; stem leaves smaller, deeply 3-lobed or divided into 3 leaflets, short-stalked or stalkless; margins sharply toothed. Flowers 1 to several on branches at ends of stems; sepals triangular, bent backward; petals yellow, obovate, 4–7 mm long; style jointed. Achenes finely hairy. May–July.

Moist to wet forest openings, streambanks, wet meadows.

Geum rivale L.
PURPLE AVENS native
IR | **WUP** | **CUP** | **EUP**   OBL CC 7
Perennial herb. Stems erect, 3–8 dm long, mostly unbranched, hairy. Basal leaves large, 1–4 dm long, pinnately divided, the terminal 1–3 leaflets much larger than other segments; stem leaves smaller, 2–5 on stem, pinnately divided or 3-lobed; margins shallowly lobed and coarsely toothed. Flowers mostly nodding, few on pedicels at ends of stems, the pedicels with short gland-tipped hairs and longer coarse hairs; sepals 5, purple, triangular, 6–10 mm long, ascending; petals 5, yellow to pink with purple veins, tapered to a clawlike base; stamens many;

styles jointed above middle, the portion above joint deciduous, lower portion persistent and curved in fruit. Fruit a long-beaked, hairy achene, 3–4 mm long, grouped into round heads. May–July.

Conifer swamps, wet forests, bogs, fens, wet meadows; often where calcium-rich.

Geum triflorum Pursh
PRAIRIE SMOKE   (THR) native
IR | WUP | CUP | **EUP**   FACU CC 8
Perennial herb. Stems 2–4 dm tall, pubescent throughout. Basal leaves 1–2 dm long, oblong lance-shaped in outline, pinnately compound with 7–17 leaflets; lateral leaflets progressively increasing in size toward the tip, irregularly lobed; terminal leaflet similar but somewhat wider, often confluent with the upper lateral ones and scarcely larger than them; stem leaves few and small, laciniate. Peduncles eventually to 1 dm long; sepals triangular, much shorter than the linear bractlets; petals purplish, oblong lance-shaped, 8–12 mm long, about equaling the bractlets, nearly erect; styles at maturity 3–5 cm long, strongly plumose except the very tip. May–June.

Thin soil over limestone; Chippewa County (Drummond island only).

The long, plumose styles are distinctive.

Geum urbanum L.
HERB-BENNET   *introduced*
IR | WUP | **CUP** | EUP
Perennial herb. Stems 3–10 dm tall. Lower leaves long-petioled, pinnately compound, the lower leaflets small, the upper leaflets asymmetrically obovate, the terminal often lobed; upper leaves 3-foliolate with obovate leaflets. Petals bright yellow, 4–5 mm long. Achenes sparsely pubescent. May–June.

Native of Eurasia, weedy near yards and in disturbed places; Alger County.

Malus *Apple*

Trees or shrubs, sometimes thorny, with simple, alternate, toothed or lobed leaves, and large flowers in simple umbels or umbel-like clusters on dwarf lateral branches (fruit-spurs). All bloom in April or May. Hypanthium globose to obovoid. Sepals 5, spreading or ascending or recurved. Petals 5, elliptic to obovate, short-clawed. Stamens 15–50, shorter than the petals. Ovary inferior, 3–5-celled (5-celled in our species); styles as many as the cells, separate or connate at base. Fruit a fleshy pome, each cell normally with 2 seeds.

Malus pumila P. Mill.
CULTIVATED APPLE   *introduced*
IR | **WUP** | **CUP** | **EUP**
Pyrus malus L., *P. pumila* (P. Mill.) K. Koch
Widely spreading tree to 15 m tall. Leaves elliptic to

ovate, finely serrate, permanently pubescent beneath. Flowers white, tinged with pink, about 3 cm wide; hypanthium densely tomentose, open at the mouth; anthers yellow; calyx persistent on the fruit.

Native probably of w Asia; long in cultivation and occasionally escaped. Many wild apples are persistent from planted trees, but also grows from seed and appears in old fields and along fences and roads.

ADDITIONAL SPECIES

Malus baccata (L.) Borkh. (Siberian crab), Asian species with small fruit 7–11 mm wide; occasionally spreading to roadsides and fields; reported from Alger and Mackinac counties.

Physocarpus *Ninebark*

Physocarpus opulifolius (L.) Maxim.
NINEBARK   native
IR | WUP | CUP | EUP   FACW CC 4
Much-branched shrub, 2–3 m tall; twigs greenish, slightly angled or ridged, smooth or finely hairy; bark of older stems shredding in long thin strips.

Leaves alternate, ovate in outline, mostly 3-lobed, dark green above, paler and often sparsely hairy below; margins irregularly toothed; petioles 1–2 cm long, with a pair of small, deciduous stipules at base. Flowers 5-parted, white, 5–10 mm wide; many in stalked, rounded clusters at ends of branches. Fruit a red-brown pod, 5–10 mm long, in round clusters; seeds 1–2 mm long, shiny, 3–4 in each pod. June–July.

Streambanks, lakeshores, swamps, rocky shores of Lake Superior.

Potentilla *Cinquefoil*

Annual or perennial herbs, or woody in shrubby cinquefoil (*P. fruticosa*); stolons present in some species. Leaves pinnately or palmately divided, alternate or mostly from base of plant. Flowers perfect, regular; sepals 5, alternating with small bracts, the sepals and bractlets joined at base to form a saucer-shaped hypanthium; petals 5, yellow; stamens many; pistils numerous. Fruit a group of many small achenes, surrounded by the persistent hypanthium.

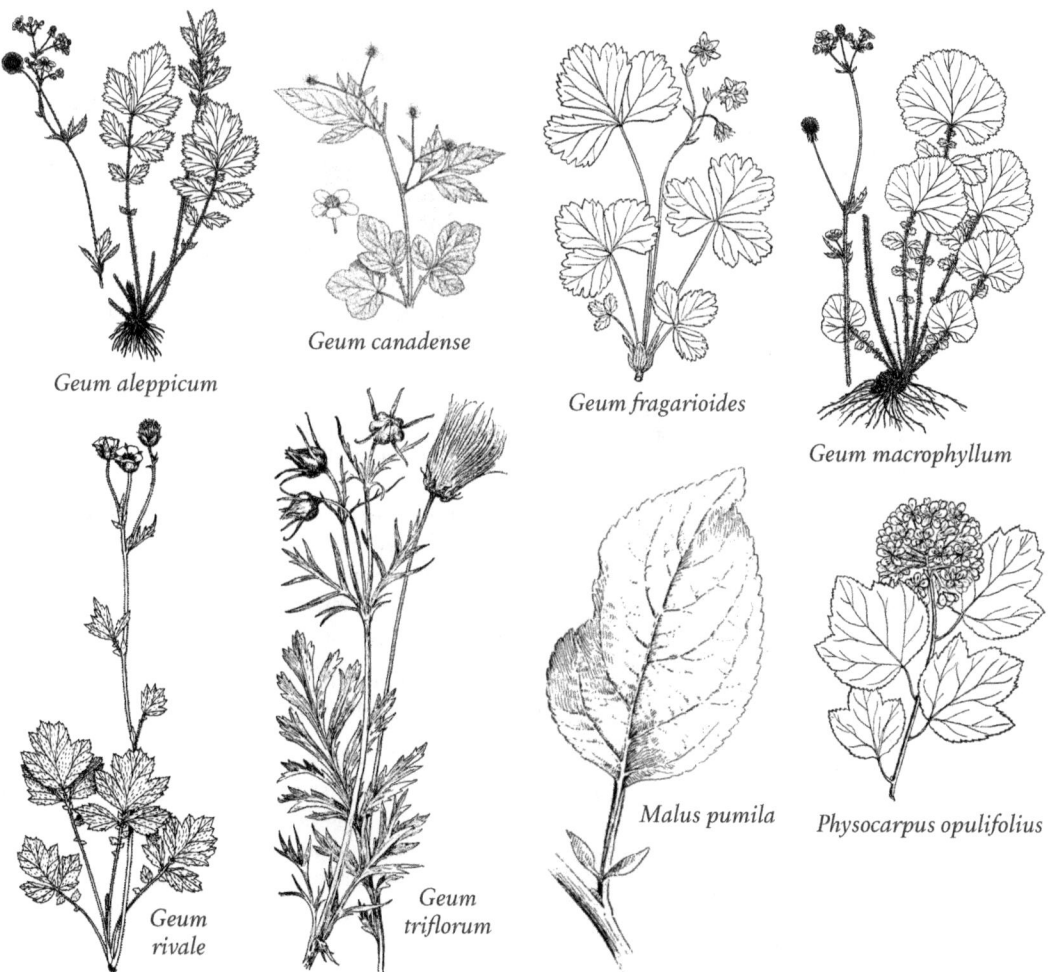

Geum aleppicum

Geum canadense

Geum fragarioides

Geum macrophyllum

Geum rivale

Geum triflorum

Malus pumila

Physocarpus opulifolius

ROSACEAE | Rose Family

ADDITIONAL SPECIES
Potentilla bipinnatifida Douglas (Tansy cinquefoil), perennial prairie species, known in Michigan from a single Schoolcraft County collection in sandy soil along a railroad.

1. Plants shrubs; leaflets 5–7, 1–2 cm long. . . . **P. fruticosa**
1. Plants herbs (or woody only at base) 2
2. Flowers solitary on naked pedicels from nodes of creeping stems 3
2. Flowers few to many in cymes 4
3. Leaves pinnately compound; leaf underside densely white-hairy . . . . **P. anserina**
3. Leaves palmately compound; leaf underside coarsely hairy . . . . **P. simplex**
4. Leaf undersides white woolly hairy 5
4. Leaf undersides with long straight hairs or glabrous, but not woolly hairy 6
5. Stem leaves reduced in size; inflorescence with few branches . . . . **P. gracilis**
5. Stem leaves well developed; inflorescence much-branched . . . . **P. argentea**
6. Main leaves below inflorescence 3-parted 7
6. Main leaves below inflorescence usually 5-parted or more 8
7. Petals and sepals about same length; stamens usually 20; achenes ridged . . . . **P. norvegica**
7. Petals much shorter than the sepals; stamens 5–10; achenes smooth . . . . **P. rivalis**
8. Leaflets essentially green beneath; larger petals ca. 3–4.3 mm; longer anthers 0.3–0.6 mm long **P. intermedia**
8. Leaflets green or with gray tomentum beneath; petals (4–) 4.5–11 mm long (if leaflets green beneath then petals (6.5–) 8–11 mm long); larger anthers 0.7–1.4 mm long 9
9. Leaflets with gray tomentum beneath in addition to straight hairs; larger petals (4–) 4.5–7 mm long, bright yellow; anthers ca. 0.7–0.9 (–1) mm long . . **P. inclinata**
9. Leaflets ± green beneath, with only ± straight hairs; petals (6.5 –) 8–11 mm long, pale yellow; anthers (at least the largest) 1–1.4 mm long . . . .**P. recta**

Potentilla anserina L.
SILVERWEED
IR | WUP | CUP | EUP   native
                                        FACW CC 5

Argentina anserina (L.) Rydb.

Perennial herb, with a stout rootstock and spreading by stolons to 1 m long. Leaves all at base of plant except for a few clustered leaves on stolons, pinnately divided into 7–25 leaflets; leaflets oblong or obovate, 1.5–5 cm long and 0.5–2 cm wide, lower leaflets much smaller; upper surface green and smooth to gray-green and silky-hairy, underside densely white-hairy; margins with sharp, forward-pointing teeth; stipules brown, membranous, at base of petiole. Flowers single from leafy axils of stolons, on stalks 5–15 cm long; sepals white silky-hairy; petals yellow, oval to obovate, 5–10 mm long; stamens 20–25. Fruit a light brown achene. May–Sept.

Wet meadows, marshes, sandy and gravelly shores and streambanks, Lake Michigan shoreline; soils often calcium-rich.

Potentilla argentea L.
SILVERY CINQUEFOIL   introduced
IR | WUP | CUP | EUP   FACU

Perennial herb, at first acaulescent, soon producing one or more long stolons which root and have small clusters of leaves at the nodes. Leaves erect, oblong lance-shaped in outline, to 3 dm long, pinnately compound with numerous leaflets often alternating with others much smaller; axis and peduncles villous; leaflets narrowly elliptic, to 4 cm long, sharply toothed, tomentose beneath and also with long appressed hairs. Flowers yellow, 15–25 mm wide, on naked peduncles, about as long as the leaves. Achenes about 2.5 mm long, deeply furrowed on the summit and back. May–Sept.

Wet sandy beaches.

Potentilla fruticosa L.
SHRUBBY CINQUEFOIL   native
IR | WUP | CUP | EUP   FACW CC 8

Dasiphora fruticosa (L.) Rydb.
Potentilla floribunda Pursh

Much-branched shrub, 0.5–1 m tall; twigs brown to red, covered with long, silky- white hairs; bark of older branches shredding. Leaves alternate, pinnately divided; leaflets 3–7 (mostly 5), the terminal 3 leaflets often joined at base, oval to oblong, 1–2 cm long and 3–7 mm wide, tapered at each end, upper surface dark green, underside paler, with silky hairs on both sides or at least on underside; margins entire, often rolled under; short-stalked. Flowers 5-parted, bright yellow, 1–2.5 cm wide, 1 to few in clusters at ends of branches; bracts much narrower than the ovate sepals; stamens 15–20. Fruit a small head of hairy achenes surrounded by the 10-parted calyx. June–Sept.

Calcareous fens, lakeshores, open bogs, conifer swamps, wet meadows.

Potentilla gracilis Dougl.
SLENDER CINQUEFOIL   introduced
IR | WUP | CUP | EUP   FAC

Potentilla flabelliformis Lehm.

Perennial from a stout caudex, erect, 4–8 dm tall. Basal leaves long-petioled. palmately compound, with 5–7 leaflets; stem leaves much smaller, often with 3 leaflets, becoming bract-like in the inflorescence; leaflets oblong lance-shaped, densely tomentose beneath, green and often glabrous above, cleft into

Potentilla inclinata Vill.
ASHY CINQUEFOIL *introduced*
IR | WUP | CUP | EUP

Potentilla canescens Besser

Perennial herb, similar to *P. argentea*, differing in having long, spreading, simple hairs on the stem, and also on the leaf veins beneath (in addition to the tomentum).

Weedy on sandy roadsides, fields, railroads, and rocks; in eastern UP, known only from Drummond Island (Chippewa County).

Potentilla norvegica L.
STRAWBERRY-WEED *native*
IR | WUP | CUP | EUP FAC CC 0

Annual herb. Stems stout and leafy, commonly branched and many-flowered, hirsute below. Leaves 3-foliolate; leaflets elliptic to broadly obovate, to 8 cm long, crenately toothed. Flowers yellow, nearly 1 cm wide; bractlets and sepals ovate lance-shaped, about equal at anthesis, the sepals expanding in fruit to 16 mm long; petals nearly as long as the sepals; stamens usually 20. Achenes pale brown, flattened, about 1 mm long, with curved longitudinal ridges. June–Aug.

Common in a wide variety of moist or dry habitats, usually where somewhat disturbed; roadsides, railroads, fields, shores, meadows, rock outcrops, gardens.

Potentilla recta L.
SULPHUR CINQUEFOIL *introduced*
IR | WUP | CUP | EUP

Perennial herb. Stems erect, simple to the inflorescence, 4–8 dm tall, pubescent. Leaves digitately compound, the basal and lower long-petioled with 5–7 leaflets, the upper short-petioled to sessile, smaller, with only 3 leaflets; leaflets radially divergent, narrowly oblong lance-shaped, deeply toothed. Inflorescence many-flowered, flattened; sepals and bractlets ovate lance-shaped, about equal; petals yellow, about 1 cm long. Achenes striate with low curved ridges. June–Aug.

Native of Europe; weedy in dry soil, roadsides, fields, railroads, gravel pits; invading dry open forests.

Sessile or short-stalked glands are usually present on the leaflet underside.

Potentilla simplex Michx.
OLDFIELD CINQUEFOIL *native*
IR | WUP | CUP | EUP FACU CC 2

Perennial herb. Stems and basal leaves from a short rhizome to 8 cm long. Stems at first erect or ascending, soon widely spreading or arching and rooting at the tips, to 1 m long, very slender, with long internodes, villous to glabrate. Leaflets 5, oblong lance-shaped to elliptic, to 7 cm long, usually less than half as wide, with numerous teeth in the upper 2/3. Peduncles slender, the lowest arising at the end of the second well developed internode. Flowers yellow, 10–15 mm wide. April–June.

Dry open sandy forests, fields, roadsides, and sandy barrens; also in moist thickets and deciduous forests, and on rocky ledges.

Our most common *Potentilla* with solitary flowers and palmately compound leaves.

Prunus *Plum; Cherry*

Trees or shrubs. Leaves alternate, simple, serrate, often with petiolar glands. Flowers umbellate or solitary from axillary buds or short lateral branches, or racemose and terminal. Hypanthium cup-shaped, obconic, or urn-shaped. Sepals spreading or reflexed, usually soon deciduous. Petals 5, white to pink or red, elliptic to obovate, spreading. Stamens about 20, with slender exserted filaments. Pistil 1. Fruit a 1-seeded drupe, the exocarp fleshy or juicy, the endocarp hard.

Many members of the genus are important for their edible fruits or attractive flowers.

ADDITIONAL SPECIES

Prunus cerasus L. (Pie cherry, Sour cherry), introduced tree, grown for its attractive flowers and edible fruit; escaping to roadsides and fencerows in Houghton and Keweenaw counties.

Prunus domestica L. (Common plum); introduced small tree, grown for its fruit and sometimes escaped to roadsides, fencerows, clearings and shores; Keweenaw and Mackinac counties.

1 Flowers 20 or more in elongate racemes; inflorescence bracts absent 2
1 Flowers 1 to several, in umbel-like clusters 3
2 Tree; leaves 2 times longer than wide, the margins with incurved teeth; fruit black . . . . **P. serotina**
2 Shrub; leaves less than 2 times longer than wide, the margins with sharp, outward pointing teeth; fruit dark red to purple . . . . **P. virginiana**
3 Plants in flower 4
3 Plants in fruit and with fully developed leaves 7
4 Sepals glabrous 6
4 Sepals hairy, at least on upper surface near base 6
5 Leaf margins mostly entire below middle; leaves widest above middle; flower pedicels mostly less than 1 cm long . . . . **P. pumila**
5 Leaf margins finely toothed for their entire length; leaves widest at or below middle; pedicels mostly more than 1 cm long . . . . **P. pensylvanica**

300 ROSACEAE | Rose Family

dicots

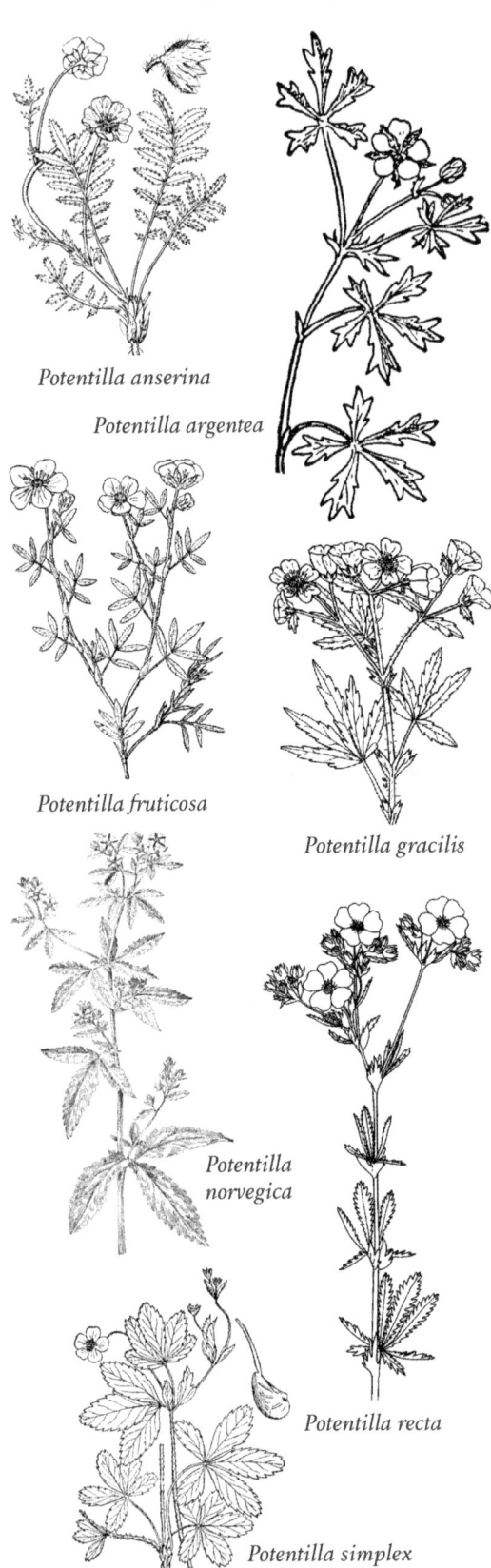

Potentilla anserina

Potentilla argentea

Potentilla fruticosa

Potentilla gracilis

Potentilla norvegica

Potentilla recta

Potentilla simplex

6 Sepals entire or with a few small teeth at tip .... **P. americana**

6 Sepal margins with gland-tipped teeth .... **P. nigra**

7 Leaves glabrous, obovate, distinctly widest above the middle; the margins entire or nearly so below the middle of the leaf........................ **P. pumila**

7 Leaves not as above 8

8 Leaves finely toothed, the teeth much less than 1 mm long; leaves more than 2x longer than wide, widest below middle...................... **P. pensylvanica**

8 Leaves coarsely toothed, the teeth 1 mm or more long; leaves less than 2x longer than wide, widest at or above the middle.. 9

9 Margin teeth sharp, often tipped with a short spine... .... **P. americana**

9 Margin teeth rounded or tapered to a tip, sometimes gland-tipped............................. **P. nigra**

Prunus americana Marsh.
WILD PLUM *native*
IR | **WUP** | **CUP** | **EUP** CC 4

Shrub or small tree to 8 m tall, spreading from the roots and forming thickets. Leaves obovate, 6–10 cm long at maturity, sharply acuminate, coarsely serrate, glandless, 1–2 mm long. Flowers 2–4 in an umbel; petals white, 10–12 mm long; sepals pubescent on the upper (inner) side, often toothed toward the end, nearly glandless. Fruit red, glaucous, about 2 cm wide; stone flattened. April–May.

Moist woods, roadsides, and fencerows.

Prunus nigra Ait.
CANADIAN PLUM *native*
IR | **WUP** | **CUP** | **EUP** FACU CC 4

Small tree, occasionally to 10 m tall. Leaves obovate, 7–12 cm long at maturity, abruptly acuminate, coarsely and often doubly serrate with irregular, gland-tipped teeth often 2 mm long, more or less pubescent beneath, at least in the vein-axils. Flowers in clusters of 3 or 4, on reddish pedicels 1–2 cm long; sepals glandular on the margin, pubescent above, glabrous on the lower (outer) side; petals white, 10–15 mm long. Fruit ellipsoid, red varying to yellow, 2–3 cm long. May.

Moist woods and thickets.

Prunus pensylvanica L. f.
PIN-CHERRY *native*
IR | **WUP** | **CUP** | **EUP** FACU CC 3

Slender shrub or small tree to 10 m tall. Leaves mostly lance-shaped, 6–12 cm long, long-acuminate, finely and irregularly serrate with rounded teeth, the gland near the sinus; petioles usually glandular at the summit. Flowers in umbel-like clusters of 2–5, on pedicels 1–1.5 cm long; sepals glabrous; petals white, about 6 mm long, villous on the back near the base. Fruit red, juicy, about 6 mm wide; stone

Prunus pumila L.
SAND-CHERRY — native
IR | **WUP** | **CUP** | **EUP** — CC 8

Low, diffusely branched, decumbent or prostrate shrub, seldom more than 1 m tall, rarely to 3 m. Leaves oblong lance-shaped, 4–10 cm long, long-tapering at base, the margin firm or cartilaginous, finely and remotely serrate with glandular teeth, glabrous, often glaucous beneath; petioles 5–12 mm long. Flowers in clusters of 2–4, on pedicels 4–12 mm long; sepals glandular-serrulate; petals white, elliptic, 4–8 mm long. Fruit nearly black, subglobose, 10–15 mm wide, edible. May.

Sand dunes and sandy soil, Great Lakes shores, dry or rocky woods.

Prunus serotina Ehrh.
WILD BLACK CHERRY — native
IR | **WUP** | **CUP** | **EUP** — FACU CC 2

Tree to 25 m tall, but often blooming when less than 5 m tall. Leaves firm, lance-shaped to oblong, 6–12 cm long, finely serrate with slender or blunt incurved teeth. Racemes terminating leafy twigs of the current season, 8–15 cm long; pedicels 3–6 mm long; sepals oblong or triangular, 1–1.5 mm long, entire or sparsely glandular-erose, persistent under the fruit; petals white, subrotund, about 4 mm long. Fruit dark purple or black, 8–10 mm wide, edible when fully ripe. May.

Formerly a forest tree, now more common as a weedy tree of roadsides, waste land, and forest margins.

Prunus virginiana L.
CHOKE-CHERRY — native
IR | **WUP** | **CUP** | **EUP** — FACU CC 2

Usually a shrub, sometimes a tree to 10 m tall. Leaves thin, oblong to obovate, 5–12 cm long, sharply serrate with slender ascending teeth. Racemes terminating leafy twigs of the season, 6–15 cm long; pedicels usually 5–8 mm long; sepals broadly triangular, 1–1.5 mm long, conspicuously glandular-erose, deciduous soon after anthesis; petals white, subrotund, about 4 mm long. Fruit dark red or crimson, 8–10 mm wide, astringent, scarcely edible. May.

In a wide variety of habitats, from rocky hills and dunes to borders of swamps.

Rosa *Rose*

Shrubs or woody vines, usually thorny. Leaves pinnately compound with 3-11 serrate leaflets, the stipules commonly large and adnate to the petiole. Hypanthium globose to urceolate with a constricted orifice. Sepals usually long-attenuate, often persistent in fruit. Petals large, spreading at anthesis, white to yellow or red. Stamens very numerous, inserted near the orifice of the hypanthium. Ovaries numerous, inserted on the bottom or also on the sides of the hypanthium; styles usually barely exsert, distinct or united. Fruit a bony achene; mature hypanthium commonly colored, pulpy or fleshy.

ADDITIONAL SPECIES

Rosa multiflora Thunb. ex Murr. (Multiflora rose); introduced white-flowered rose with spiny stems, previously planted in fencerows but escaping to roadsides and disturbed areas; Marquette and Schoolcraft counties.

Rosa rubiginosa L. (Sweetbrier; syn *Rosa eglanteria* L.), shrub with very spiny stems; introduced in several UP locations along roadsides, fields, often where sandy.

Rosa spinosissima L. (Scotch rose), introduced shrub to 1 m tall, the stems with very numerous slender straight thorns; leaflets usually 7–9, usually less than 2 cm long; flowers solitary at the end of the branches; petals white, pink, or yellow; several UP locations.

1. Flowers solitary at end of branches, the pedicel not subtended by a bract; introduced species **R. spinosissima***
1. Flowers solitary or in clusters; if solitary then the pedicel with a bract. 2
2. Young twigs densely hairy; petals 3–5 cm long; introduced species . . . . **R. rugosa**
2. Young twigs glabrous or nearly so; petals 2–3 cm long; native species except for *R. cinnamomea* 3
3. Pedicel and hypanthium with stalked glands; sepals spreading and then deciduous . . . . **R. palustris**
3. Pedicel and hypanthium glabrous; sepals persistent on fruit and typically upright 4
4. Prickles at nodes below leaf stipules present and larger than internodal prickles . . . . **R. cinnamomea**
4. Prickles at nodes below leaf stipules absent or similar to internodal prickles. 5
5. Flowers from tips of currrent year's stems and also on lateral branches on stems from previous year; leaflets mostly 9 or 11 . . . . **R. arkansana**
5. Flowers only on lateral branches of previous year's stems; leaflets mostly 5 or 7 6
6. Stems usually not prickly or bristly, or with slender prickles only on lower internodes . . . . **R. blanda**
6. Stems densely prickly on most internodes **R. acicularis**

Rosa acicularis Lindl.
BRISTLY ROSE — native
IR | **WUP** | **CUP** | **EUP** — FACU CC 4

Rosa sayi Schwein.

Stems to 1 m tall, usually densely beset with straight slender thorns, even on the flowering lateral branches.

ROSACEAE | Rose Family

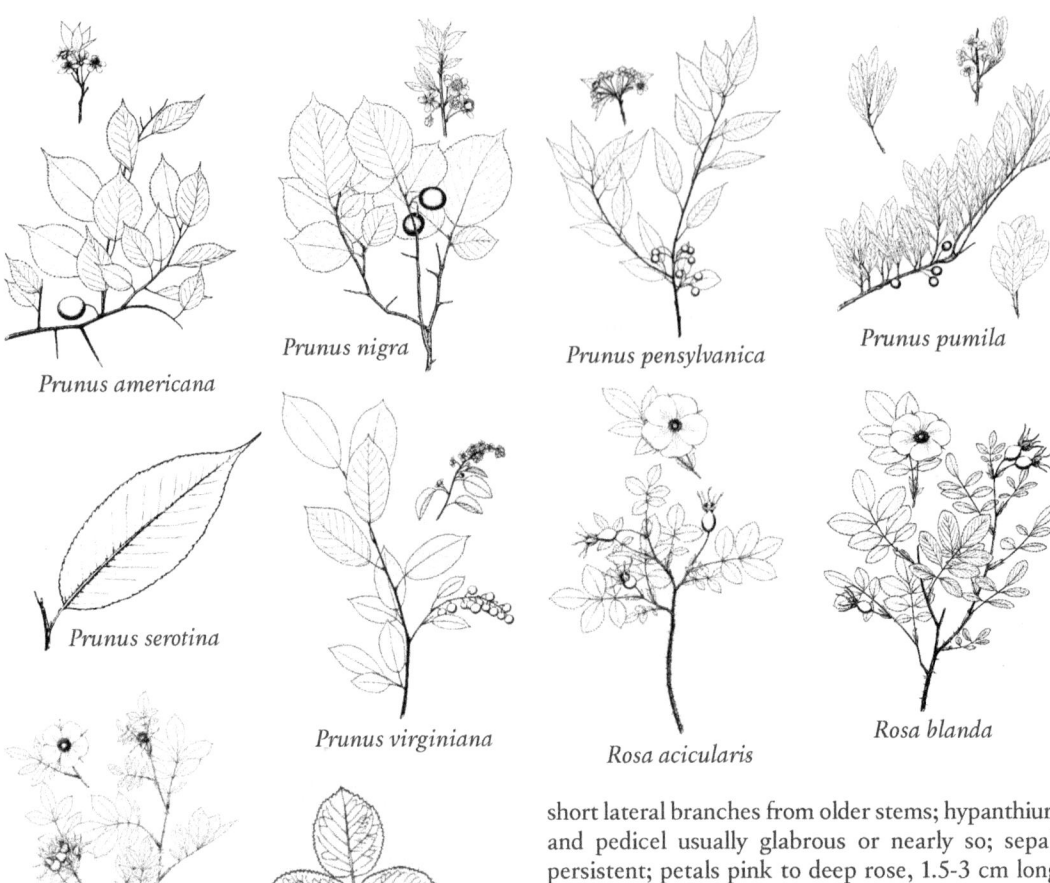

Prunus americana
Prunus nigra
Prunus pensylvanica
Prunus pumila
Prunus serotina
Prunus virginiana
Rosa acicularis
Rosa blanda
Rosa palustris
Rosa palustris

Stipules pubescent, glandular on the margin, when young densely covered with short-stipitate glands, as are also the bracts; rachis usually pubescent and glandular; leaflets 3–7, usually 5, oblong-elliptic, coarsely serrate. Flowers usually solitary; hypanthium and pedicel glabrous; petals pink, 2–3 cm long Fruit ellipsoid to pyriform or globose.

Upland woods, hills, and rocky banks; statewide.

Rosa arkansana Porter
DWARF PRAIRIE-ROSE   native
IR | **WUP** | **CUP** | **EUP**   FACU CC 4

Colonial, only half-shrubby. Stems under 1 m tall, usually densely prickly; prickles slender, straight, unequal. Stipules pubescent, usually entire, or glandular-dentate toward the tip, leaflets (7)9 or 11, 1-4 cm long, obovate, sharply serrate, very often pubescent beneath. Flowers corymbose, terminating the nearly herbaceous stems of the season and often also on short lateral branches from older stems; hypanthiurn and pedicel usually glabrous or nearly so; sepals persistent; petals pink to deep rose, 1.5-3 cm long. Fruit purplish or red, 10-15 mm wide.

Prairies, or in open or brushy sites.

Rosa blanda Ait.
SMOOTH ROSE   native
IR | **WUP** | **CUP** | **EUP**   FACU CC 3

Stems to 1.5 m tall, unarmed or with few to many slender prickles toward the base, not extending upon the flowering branches. Stipules entire to glandular-dentate; leaflets commonly 5 or 7, narrowly oblong to oval, glabrous or nearly so, coarsely toothed, especially above the middle. Flowers solitary or corymbose; pedicel and hypanthium glabrous; petals pink, 2–3 cm long.

Common in dry woods, hills, prairies, and dunes.

Thorns may be entirely absent, or may extend a variable distance up the stem.

Rosa cinnamomea L.
CINNAMON ROSE   introduced
IR | **WUP** | **CUP** | EUP

Stems 1–2 m tall, armed with curved infrastipular thorns, often also with a few internodal thorns. Leaflets 5–7, sparsely pubescent above, densely pubescent beneath, as is also the rachis. Hypanthium and pedicel glabrous, the latter short, often not surpassing the bract; petals dark rose-color.

Native of Eurasia; sometimes cultivated and rarely escaped; Delta, Keweenaw and Schoolcraft counties.

Rosa palustris Marsh.
SWAMP ROSE *native*
IR | **WUP** | **CUP** | **EUP** OBL CC 5

Stems much-branched, prickly, to 2 m tall; twigs red-brown, smooth, with a pair of broad-based, downward-curved prickles at nodes; bristles between nodes absent. Leaves alternate, pinnately divided into usually 7 leaflets; leaflets oval or obovate, 2–6 cm long and 1–2 cm wide; underside midrib often soft-hairy; margins finely toothed; stipules narrow; petioles present. Flowers single at ends of leafy branches, or in small clusters of 2–5; petals pink, 2–3 cm long; flower stalks, sepals and hypanthium with stalked glands. Fruit more or less round, red-orange, 6–10 mm wide. July–Aug.

Open bogs, conifer swamps, thickets, shores and streambanks; increasing in disturbed wetlands.

Rosa rugosa Thunb.
RUGOSA ROSE *introduced*
IR | **WUP** | **CUP** | **EUP** FACU

Stems 1–2 m tall, densely thorny, the infrastipular thorns larger, decurved; younger parts of the stem, young thorns, and thorn bases densely pubescent. Leaflets usually 7–9, rugose above. Pedicel bristly and pubescent; hypanthium smooth.

Native of e Asia; sometimes cultivated and occasionally escaped; notable for its large rose-hips (fruit), rich in vitamin C.

Rubus
Blackberry, Raspberry, Dewberry

Perennials, woody at least at base, usually with bristly stems. Stems biennial in some species, the first year's canes called primocanes, the second year's growth termed floricanes. Leaves alternate, palmately lobed or divided. Flowers 5-parted, usually perfect, white to pink or rose-purple; stamens many. Fruit a group of small, 1-seeded drupes forming a berry.

1 Stems without bristles or prickles 2
1 Stems with bristles or prickles 6
2 Leaves simple 3
2 Leaves with 3–5 leaflets 4
3 Flowers white; fruits orange to red, edible . . . .
. . . . R. parviflorus
3 Flowers rose-purple; fruits pink to red, dry and inedible
. . . . R. odoratus
4 Stems erect or arching; flowers many in long, bracted clusters; fruit black . . . . R. canadensis
4 Stems short and upright or trailing; flowers single or several in a cluster; fruit red 5
5 Flowering stems 1 or several from a short base; petals light to deep pink, 1–2 cm long . . . . R. arcticus
5 Flowering stems single from a creeping stem; petals green-white, 0.5–1 cm long . . . . R. pubescens
6 Leaves whitish or gray-hairy on underside; fruit separating easily from receptacle when ripe (raspberries) . . . 7
6 Leaves green on both sides, underside veins hairy; fruit falling with receptacle when ripe (dewberries and blackberries) 8
7 Stems erect or spreading, with stiff straight bristles; fruit red . . . . R. idaeus
7 Stems arching, often rooting at tip, with broad-based, recurved prickles; fruit black . . . . R. occidentalis
8 Plants low and trailing (less than 0.5 m tall), often rooting at nodes; flowers 1 to several in a cluster; fruit red to red-purple (dewberries) 9
8 Plants tall, to 2 m; stems erect, neither rooting at nodes nor arching and rooting at tips; flowers numerous in elongate clusters; fruit black (blackberries) 10
9 Stems with prickles, these hooked at tip and broad at base; leaves thin and deciduous; petals more than 1 cm long . . . . R. flagellaris
9 Stems with coarse hairs and slender bristles; leaves leathery and often evergreen; petals less than 1 cm long
. . . . R. hispidus
10 Stems smooth or with scattered prickles; leaves glabrous . . . . R. canadensis
10 Stems glandular-hairy, with broad-based prickles or covered with spreading bristles; leaves hairy on underside veins 11
11 Stems covered with stiff bristles, broad-based prickles absent . . . . R. setosus
11 Stems with bristles, gland-tipped hairs, and scattered broad-based prickles 12
12 Petioles and pedicels covered with gland-tipped hairs .
. . . . R. alleghaniensis
12 Petioles and pedicels without glandular hairs (or nearly so) . . . . R. pensilvanicus

Rubus alleghaniensis Porter
COMMON BLACKBERRY *native*
IR | **WUP** | **CUP** | **EUP** FACU CC 1

Stems 0.5–3 m tall, mostly erect, the young primocanes often sparsely glandular. Primocane leaves usually 5-foliolate with the intermediate pair long-petiolulate, softly pubescent beneath; terminal leaflet usually 1–2 dm long, widest near the middle, sharply serrate; lateral leaflets smaller; armature of the stem of nearly straight spines spreading at right angles, much flattened at their base; armature of the petioles, pedicels, and lower side of the midveins commonly present and consisting of spines similarly flattened but prominently hooked. Inflorescence racemose, many-flowered, the lower 1, 2, or rarely 3 flowers subtended by leaves, the others by stipules only; pedicels glandular; flowers about 2 cm wide.

Forests and forest edges, clearings, old fields, roadsides; usually on dry uplands, occasional in marshy or swampy ground. Our commonest tall blackberry.

Rubus arcticus L.
NORTHERN BLACKBERRY (END) native
IR | WUP | **CUP** | **EUP**　　　　　　　　　　CC 10

Rubus acaulis Michx.

Stems herbaceous (woody at base), 5–10 cm long, bristles or prickles absent. Leaves divided into 3 leaflets, 1–4 cm long and 0.5–3 cm wide, terminal leaflet stalked, lateral pair of leaflets nearly sessile, lateral leaflets often with a shallow lobe, upper surface smooth, underside finely hairy; margins with blunt, forward-pointing teeth; petioles long, finely hairy; stipules small, ovate. Flowers single at ends of erect stems; sepals lance-shaped, to 1 cm long; petals 5, light to dark pink, 1–2 cm long. Fruit red, nearly round, 1 cm wide, edible. June–Aug.

Conifer swamps, open bogs.

Rubus canadensis L.
SMOOTH BLACKBERRY; DEWBERRY　　　　　　native
IR | **WUP** | **CUP** | **EUP**　　　　　　　　　　CC 2

Stems erect or nearly so, to 2 m tall, spineless, or occasionally beset with stout straight spines from expanded bases. Primocane leaves commonly 5-foliolate, typically glabrous beneath, rarely softly pubescent, the 3 central leaflets long-petiolulate, the lower pair subsessile to short-petiolulate; lateral leaflets lance-shaped to broadly oblong; terminal leaflet commonly 1–2 dm long, ovate, always conspicuously acuminate, sharply serrate. Floricane leaflets much smaller, the terminal one widest above the middle, always acute to short-acuminate, sometimes entire in the basal half. Inflorescence racemose, many-flowered, the flowers on glandless pedicels 2–4 cm long, subtended by stipules 1–2 cm long and serrate.

Woods, clearings, fields, roadsides; occasionally in moist soil.

Rubus flagellaris Willd.
WHIPLASH DEWBERRY　　　　　　　　　　native
IR | **WUP** | **CUP** | EUP　　　　　　　FACU CC 1

Rubus baileyanus Britt.

Stems long-trailing, 2–4 m long, often rooting at tip, brown to red-purple, with curved, broad-based prickles; bristles absent. Primocane leaves divided into 3–5 leaflets, the terminal leaflet 2–6 cm long and 1–5 cm wide, often with small lobes above middle; floricane leaves smaller, usually divided into 3 leaflets; leaflets ovate to obovate, upper and lower surface more or less glabrous, or underside veins with appressed hairs; margins with forward-pointing teeth; petioles finely hairy, with scattered, hooked prickles. Flowers mostly 2–7 (sometimes only 1), on upright, finely hairy stalks, the stalks with scattered prickles; sepals joined, the lobes narrowed to dark tips; petals 5, white, 10–15 mm long. Fruit red, more or less round, composed of large, juicy drupelets, edible, not easily separating from receptacle. May–June.

Swamps, wetland margins; also in drier sandy woods, prairies and openings.

Rubus hispidus L.
BRISTLY DEWBERRY　　　　　　　　　　native
IR | **WUP** | **CUP** | **EUP**　　　　　　　FACW CC 4

Stems trailing or low-arching, often rooting at tip, with slender bristles or spines 2–5 mm long, these sometimes gland-tipped, not much widened at base. Leaflets 3 (rarely 5), ovate to obovate, 2–5 cm long and 1–3 cm wide, upper surface dark green and slightly glossy, slightly paler and more or less glabrous below, some leaves persisting through winter; margins with rounded teeth; petioles finely hairy and bristly; stipules linear, persistent. Flowers single in upper leaf axils or in open clusters of 2–8 at ends of short branches; sepals joined, the lobes ovate, tipped with a small dark gland; petals 5, white, 5–10 mm long. Fruit red-purple, less than 1 cm wide, sour, not easily separated from receptacle. June–Aug.

Conifer swamps, wet hardwood forests, thickets, wetland margins; usually where shaded.

Rubus idaeus L.
WILD RED RASPBERRY　　　　　　　　　native
IR | **WUP** | **CUP** | **EUP**　　　　　　　FACU CC 2

Rubus strigosus Michx.

Stems erect or spreading, to 1.5 m long, biennial; young stems bristly with slender, often gland-tipped hairs; older stems brown, smooth. Primocane leaves divided into 3 or 5 leaflets, floricane leaflets usually 3; leaflets ovate to lance-shaped, upper surface dark green and smooth or sparsely hairy, underside gray-hairy; margins with sharp, forward-pointing teeth; petioles with bristly hairs; stipules slender, soon deciduous. Flowers in clusters of 2–5 at ends of stems and 1–2 from upper leaf axils; sepals with gland-tipped hairs; petals 5, white, shorter than the sepals. Fruit red, about 1 cm wide, edible, separating from receptacle when ripe. May–Aug.

Thickets, moist to wet openings, streambanks; often where disturbed.

Our native plants sometimes considered a variety (*R. idaeus* var. *strigosus*) of the cultivated red raspberry (*R. idaeus*) from Europe.

Rubus occidentalis L.
BLACK RASPBERRY　　　　　　　　　　native
IR | WUP | **CUP** | **EUP**　　　　　　　　　　CC 1

Stems erect or ascending, or sometimes arching and

rooting at the tip, not glandular, glaucous the first year, becoming glabrous the second, sparsely beset with stout, straight or hooked spines with expanded bases, as are also the petioles and especially the pedicels. Leaflets commonly 3, occasionally 5 on the primocanes; uppermost leaves of the floricane often simple; terminal leaflet broadly ovate, rounded or subcordate at base, sharply and irregularly serrate; lower leaflets similar but smaller and narrower; all thinly gray-tomentose beneath. Flowers 3–7 in a dense umbel-like cluster; often 1 or 2 flowers also from the upper axils; petals white, shorter than the sepals, narrowly obovate, at first erect, soon deciduous. Fruit commonly black, rarely yellowish, about 1 cm wide. May–June.

Dry or moist woods, fields, and thickets; Delta and Macinac counties.

Often cultivated in many horticultural varieties.

Rubus odoratus L.
FLOWERING RASPBERRY native
IR | WUP | **CUP** | EUP CC 6

Widely branched spineless shrub 1–1.5 m tall, with shredding bark, becoming densely and coarsely glandular on the upper parts of the stem, petioles, pedicels, and calyx. Leaves subrotund to kidney-shaped in outline, 1–2 dm wide, usually 5-lobed. Flowers rose-purple, in a loose open cyme. Fruit depressed, about 1 cm across, the drupelets falling separately.

Moist shady places and margins of woods. June–Aug. Marquette County; common in New England.

Rubus parviflorus Nutt.
THIMBLEBERRY native
IR | **WUP** | **CUP** | **EUP** FACU CC 6

Stems unarmed, 1–2 in. tall, with shredding bark, the younger parts, petioles, and pedicels stipitate-glandular. Leaves rotund to kidney-shaped in outline, 1–2 dm wide, lobed to 1/3 of their width, the lobes serrate. Flowers few, white, in a long-peduncled cluster; sepals with a long caudate tip; petals elliptic-obovate, 1.5–2 cm long. Fruit red, 1.5–2 cm wide, with pubescent coherent drupelets; edible and sometimes harvested for jam-making. May–July.

Open woods and thickets.

Rubus pensilvanicus Poir.
PENNSYLVANIA BLACKBERRY native
IR | **WUP** | **CUP** | EUP CC 2

Stems usually stout, 1–3 m tall; armature of the primocanes of straight or spreading spines from expanded bases; shorter hooked spines usually occur on the petioles, often on the petiolules, and occasionally on the midveins and axis of the raceme. Primocane leaves softly pubescent beneath, 3–5-foliolate; terminal leaflet broadly ovate, 6–12 cm long at maturity, broadest well below the middle, distinctly acuminate, coarsely and irregularly serrate or doubly serrate. Floricane leaflets usually elliptic to obovate, coarsely toothed above the middle, many of them simple, ovate. Racemes usually short, few-flowered, well surpassing the leaves, sometimes loose and open.

Roadsides, fields, thickets, forests and forest borders; often in moist places such as borders of marshes and swamps.

The spines on the primocane vary from numerous to few, from stout with broad bases to very slender. The coarse serration, commonly accentuated beyond the middle of the leaflets, is characteristic, but forms with simpler serration occur.

Rubus pubescens Raf.
DWARF RASPBERRY native
IR | **WUP** | **CUP** | **EUP** FACW CC 4

Low perennial. Stems long-creeping at or near soil surface, with upright, hairy branches 1–3 dm tall; the branches herbaceous but woody at base, bristles absent; sterile branches arching to trailing, often rooting at nodes; flowering branches erect, with few leaves. Leaves alternate, divided into 3 leaflets; leaflets oval, 2–6 cm long and 1–4 cm wide, tapered to a sharp point; margins with coarse, forward-pointing teeth, often entire near base; petioles hairy. Flowers on glandular-hairy stalks, 1–3 in loose clusters at ends of erect branches, sometimes with 1–2 flowers from leaf axils; petals 5, white or pale pink, to 1 cm long. Fruit bright red, round, 5–15 mm wide, the drupelets large, juicy, edible, not separating easily from receptacle. May–July.

Common in conifer swamps, wet deciduous woods, rocky shores.

Rubus setosus Bigelow
BRISTLY BLACKBERRY native
IR | **WUP** | **CUP** | **EUP** FACW CC 3

Stems erect to spreading or arching, to 1.5 m long; branches covered with spreading bristles 1–4 mm long; older canes red-brown, ridged, not rooting at tip. Leaves alternate; primocane leaves divided into 3–5 leaflets; floricane leaves 3-divided; leaflets ovate to obovate, upper and lower surface more or less smooth but often hairy on underside veins; margins with sharp, forward-pointing teeth; petioles bristly; stipules linear, 1–2 cm long. Flowers few to many in elongate clusters at ends of stems, with small, leafy bracts throughout the head; petals 5, white, to 1 cm long. Fruit red, ripening to black, round, to 1 cm wide, dry, poor eating quality. June–Aug.

Wetland margins, shores, occasional in open bogs; also in drier sandy prairie.

ROSACEAE | Rose Family

Sanguisorba *Burnet*

Sanguisorba minor Scop.
SALAD-BURNET *introduced*
IR | WUP | **CUP** | EUP FAC

Poterium sanguisorba L.

Perennial from a caudex-like rhizome. Stems 2–7 dm tall. Basal and lower leaves numerous, 1-pinnate with usually 7–17 leaflets; upper leaves progressively reduced; leaflets ovate, 5–20 mm long, with 3–7 deep sharp teeth on each side. Heads densely flowered, several on elongate peduncles, short-ovoid to globose, 8–20 mm long; flowers 4-merous, subtended by ciliate bracts, the lower staminate, the upper pistillate or perfect; hypanthium contracted at the mouth, more or less 4-angled, not prickly; sepals green or brown, 2.5–5 mm long; petals none; stamens numerous, the long filaments drooping; pistils 2. Fruit a pair of achenes, enclosed by the indurate hypanthium. May–June.

Native of Eurasia, rarely established in roadsides, waste places, and fields; reported from Menominee County.

Sibbaldia *Fivefingers*

Sibbaldia tridentata (Aiton) Paule & Soják
SHRUBBY-FIVEFINGERS *native*
IR | WUP | CUP | EUP FACU CC 10

Potentilla tridentata Ait.
Sibbaldiopsis tridenta (Aiton) Rydb.

Stems woody at base, from a caudex, often 2 or 3 together, 1–3 dm tall, sparsely strigose. Leaves mostly near the base, digitately compound; leaflets

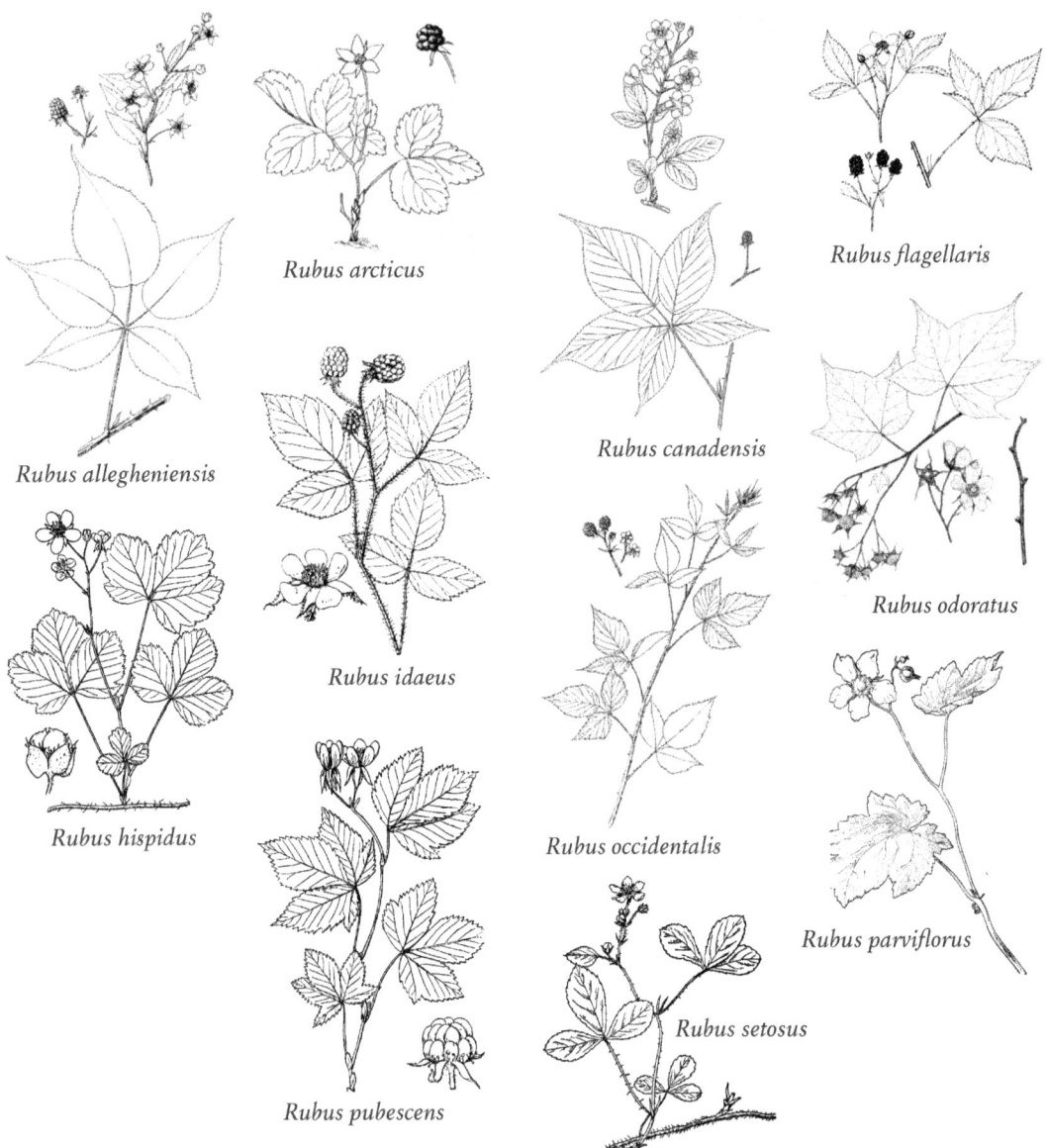

Rubus arcticus

Rubus flagellaris

Rubus alleghaniensis

Rubus canadensis

Rubus idaeus

Rubus odoratus

Rubus hispidus

Rubus occidentalis

Rubus parviflorus

Rubus setosus

Rubus pubescens

3, firm in texture, oblong lance-shaped, 15–25 mm long, entire near base, 3-toothed at the truncate tip, glabrous above, obscurely strigose beneath. Flowers several in a flattened cyme, white, about 10 mm wide; bractlets lance-shaped, somewhat shorter than the ovate-triangular sepals; ovary and achenes villous. June–Aug.

Open sandy places, dry savannas of jack pine and oak; rocky and gravelly shores, rock outcrops.

Sorbaria *False Spiraea*

Sorbaria sorbifolia (L.) A. Braun
FALSE SPIRAEA *introduced*
IR | **WUP** | CUP | EUP

Shrub 1–2 m tall, the younger parts covered with a flocculent, deciduous, stellate tomentum. Leaves 1-pinnate; stipules lance-shaped, about 1 cm long; leaflets lance-shaped, 3–7 cm long, sharply doubly serrate. Flowers 5-merous, in panicles 1–3 dm long, the branches ascending; hypanthium broadly cup-shaped; sepals to 1.5 mm long, soon reflexed, often erose; petals white, elliptic-obovate, 2.5–3 mm long; stamens numerous, to 8 mm long; pistils 5, opposite the sepals. Fruit a thin-walled follicle. July.

Native of e Asia; cultivated and occasionally escaped along roadsides and fencerows; Baraga, Luce, and Marquette counties.

Sorbus *Mountain-Ash*

Trees or shrubs. Leaves odd-pinnate with normally 11–17 serrate leaflets; flowers white, numerous, in repeatedly branched, round or flattened clusters. Hypanthium obconic. Sepals triangular, ascending. Petals 5, obovate to orbicular, rounded or cuneate at base, spreading. Stamens 15–20. Fruit a small pome, each cell with 1 or 2 elongate flattened seeds.

1 Leaflets tapered to a tip, 3–5 times longer than wide; petals obovate, to 4 mm long; fruit 5–6 mm wide . . . . **S. americana**
1 Leaflets rounded at tip or abruptly tapered to a tip, 2–3 times longer than wide; petals orbicular, 4–5 mm long; fruit 8–10 mm wide 2
2 Leaflets glabrous, pale on underside; inflorescence branches and pedicels glabrous or nearly so . **S. decora**
2 Leaflets soft-hairy on underside; inflorescence branches and pedicels with soft hairs . . . . **S. aucuparia**

Sorbus americana Marsh.
AMERICAN MOUNTAIN-ASH *native*
IR | **WUP** | CUP | EUP FAC CC 4
Pyrus americana (Marsh.) DC.

Shrub or tree to 10 m tall, the young twigs glabrous or nearly so. Winter buds glutinous, with glabrous or sparsely ciliate scales. Leaflets lance-shaped to narrowly oblong, long-acuminate, 5–9 cm long, sharply serrate, paler and usually glabrous beneath. Inflorescence 6–15 cm wide; hypanthium and sepals glabrous; petals obovate, 3–4 mm long, conspicuously longer than the stamens. Fruit bright red, 4–6 mm wide. May–June; fruit in late summer.

In moist or wet soil; swamps (both cedar and deciduous), stream banks, forest borders.

Sorbus aucuparia L.
EUROPEAN MOUNTAIN-ASH; ROWAN *introduced*
IR | **WUP** | CUP | EUP
Pyrus aucuparia (L.) Gaertn.

Tree to 10 m tall, the young twigs more or less villous. Winter buds white-villous, not glutinous. Leaflets oblong, 3–5 cm long, serrate, paler and usually long-villous beneath, at least when young. Inflorescence 10–20 cm wide; hypanthium densely white-villous; petals orbicular, 4–5 mm long, about equaling the stamens. Fruit bright red, about 10 mm wide. May–June.

Native of Europe; planted for ornament and escaped into moist woods; often mistaken for a native plant; Gogebic and Houghton counties.

Sorbus decora (Sarg.) Schneid.
NORTHERN MOUNTAIN-ASH *native*
IR | **WUP** | CUP | EUP FACU CC 4
Pyrus decora (Sarg.) Hyl.

Shrub or tree to 10 m tall, the young twigs glabrous or nearly so. Winter buds glutinous, the principal scales glabrous on the back, the inner usually conspicuously brown-ciliate. Leaflets oblong, 4–7 cm long, acute or very shortly acuminate, sharply serrate, paler and glabrous or sparsely pilose beneath. Inflorescence 6–15 cm wide; hypanthium glabrous or sparsely pilose; petals orbicular, 4–5 mm long, about equaling the stamens. Fruit bright red, 8–10 mm wide. May–June.

Moist or dry, often rocky soil. Wooded dunes and bluffs, forest margins.

Spiraea *Meadowsweet*

Shrubs with simple leaves and terminal or lateral clusters of white, pink, or purple flowers. Flowers 5-merous. Hypanthium cup-shaped or turbinate. Petals small, widely spreading. Stamens 15 to many. Pistils commonly 5, alternate with the sepals; styles terminal; ovules 2–several. Follicles firm in texture, dehiscent along the ventral suture.

Most species are attractive flowering shrubs.

ADDITIONAL SPECIES

Spiraea japonica L. f. (Japanese spiraea), introduced Asian shrub, occasionally escaping to roadsides, fields, and railways; Ontonagon and Schoolcraft counties.

ROSACEAE | Rose Family

Spiraea salicifolia L. (Willow-leaf meadowsweet), ornamental Eurasian shrub, rarely escaping to sandy shores and fields; Marquette and Schoolcraft counties.

Spiraea ×vanhouttei (Briot) Carrière (Bridal wreath), Asian ornamental shrub, rarely escaping to roadsides, railways, and lakeshores; Gogebic and Houghton counties.

1 Leaves glabrous on both sides; flowers white to pinkish . . . . **S. alba**
1 Leaf underside densely covered with light brown, woolly hairs; flowers rose-pink . . . . **S. tomentosa**

Spiraea alba Du Roi
MEADOWSWEET *native*
IR | **WUP** | **CUP** | EUP   FACW CC 4

Much-branched shrub, often forming colonies. Stems somewhat angled or ridged, 0.5–1.5 m long, smooth or short-hairy when young, becoming red-brown and smooth. Leaves alternate, often crowded on stems, oval to oblong lance-shaped, 3–7 cm long and 1–2 cm wide, smooth on both sides; margins with sharp, forward-pointing teeth; petioles 2–8 mm long; stipules absent. Flowers small, 6–8 mm wide, many in a narrow, pyramid-shaped panicle 5–25 cm long at ends of branches; sepals 5; petals 5, white. Fruit a group of 5–8 small follicles, each with several seeds; the fruiting branches often persistent over winter. June–Aug.

Common shrub of wet meadows, streambanks, lakeshores, conifer swamps; soils often sandy.

Spiraea tomentosa L.
HARDHACK *native*
IR | **WUP** | **CUP** | EUP   FACW CC 5

Sparsely branched shrub to 1 m tall. Young stems covered with brown woolly hairs, becoming smooth and red-brown. Leaves alternate, lance-shaped to ovate, 2–5 cm long and 0.5–2 cm wide; more or less smooth above, underside gray-green to tan, densely covered with feltlike hairs, the veins prominent; margins with coarse, forward-pointing teeth; petioles 1–4 mm long or absent. Flowers small, 3–4 mm wide, in spirelike panicles 5–15 cm long at ends of stems, the panicle branches covered with reddish woolly hairs; petals 5, pink or rose (rarely white). Fruit a cluster of small, hairy follicles, often persisting over winter. July–Sept.

Open bogs, conifer swamps, thickets, lakeshores, wet meadows; soils often sandy.

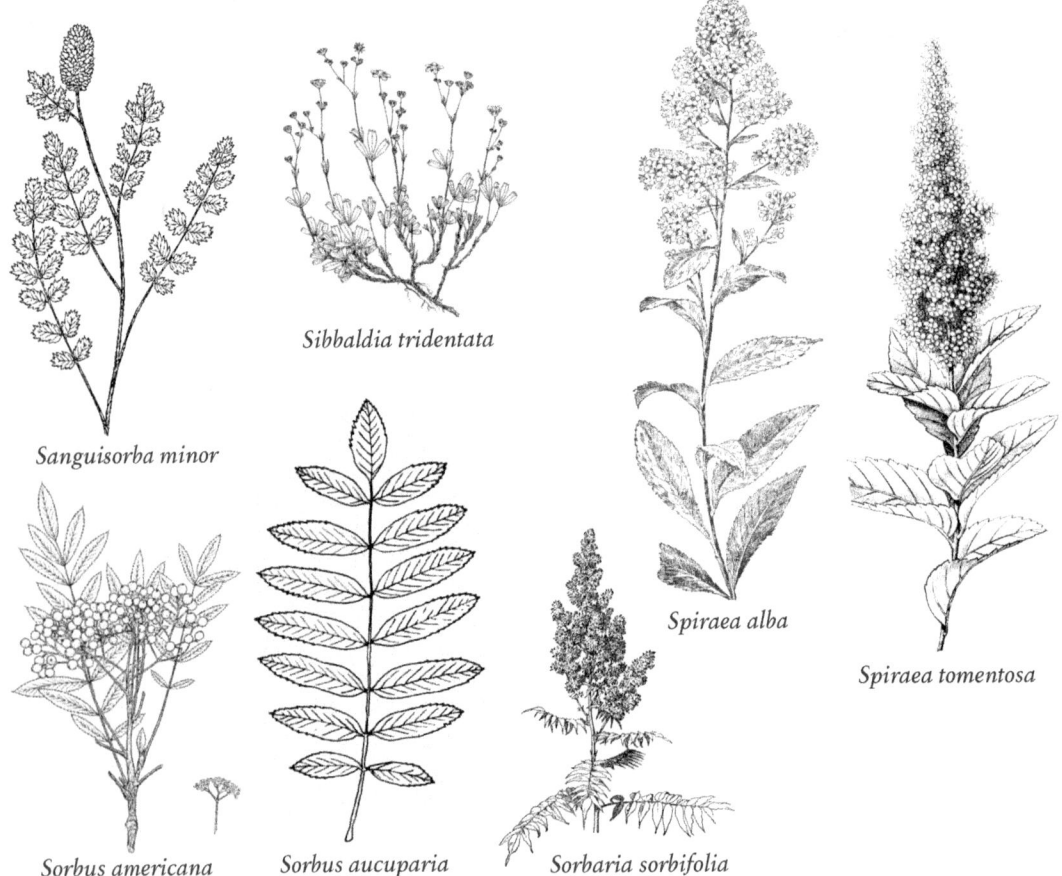

Sanguisorba minor

Sibbaldia tridentata

Spiraea alba

Spiraea tomentosa

Sorbus americana

Sorbus aucuparia

Sorbaria sorbifolia

Rubiaceae

MADDER FAMILY

Herbs (ours). Leaves simple, opposite or whorled. Flowers small, perfect (with both staminate and pistillate parts), white to green, single or in loose or round clusters; petals joined, 3–4-lobed; stamens 3–4; ovary 2-chambered. Fruit a capsule (*Galium, Houstonia*) or a berry (*Mitchella*).

1. Leaves whorled .... **Galium**
1. Leaves opposite................................... 2
2. Leaves evergreen; fruit a scarlet or white berry....... .... **Mitchella repens**
2. Leaves deciduous; fruit not a berry.................. 3
3. Stems spreading; flowers sessile; stipules with long bristles .... **Diodia teres**
3. Stems usually erect; flowers on pedicels; stipules entire or nearly so .... **Houstonia**

Galium *Bedstraw*

Annual or perennial herbs, from slender rhizomes. Stems 4-angled, ascending to reclining, smooth or bristly. Leaves entire, in whorls of 4–6. Flowers small, perfect, regular, 1 to several from leaf axils or in clusters at ends of stems; sepals absent; petals joined, 3–4-lobed, white; stamens 3–4; styles 2, ovary 2-chambered and 2-lobed, maturing as 2 dry, round fruit segments which separate when mature.

ADDITIONAL SPECIES

Galium odoratum (L.) Scop. (Sweet woodruff), European ornamental, occasionally escaping to forests and forming colonies via its rhizomes; Mackinac County (but only known from Bois Blanc, Mackinac, and Round Islands).

1. Fruit with bristly hairs 2
1. Fruit smooth or nearly so 6
2. Main leaves in whorls of 5 or more 3
2. Leaves in whorls of 4 or less..................... 4
3. Annual herb; leaves in whorls of 7 or more; flowers white, blooming completed by early summer; stems very rough-to-touch.................... **G. aparine**
3. Perennial herb; leaves in whorls of up to 6; flowers greenish, blooming beginning in early summer; stems rough or smooth..................... **G. triflorum**
4. Leaves linear to linear lance-shaped, usually less than 5 mm wide; flowers white in a large panicle.. **G. boreale**
4. Leaves broader, lance-shaped to ovate, often more than 5 mm wide; flowers greenish to purple, in few-flowered clusters................................. 5
5. Leaves lance-shaped, tapered to a tip, corolla becoming purple, glabrous .... **G. lanceolatum**
5. Leaves ovate; corolla yellowish white; rare in Chippewa County .... **G. kamtschaticum**
6. Leaves tipped with a short spine or at least sharp-pointed.. 7
6. Leaves rounded or blunt at tip 9
7. Leaves linear, the margins revolute, hairy on underside; flowers yellow in elongate terminal panicles. **G. verum**
7. Leaves narrowly lance-shaped to ovate, glabrous or rough-hairy on underside; flowers white, the inflorescence various 8
8. Leaves and stems with rough, downward-pointing hairs .... **G. asprellum**
8. Leaves and stems smooth or with short, upward-pointing hairs .... **G. mollugo**
9. Lobes of corolla 3, mostly wider than long 10
9. Lobes of corolla 4, mostly longer than wide........ 12
10. Leaves in whorls of 4; flowers and fruit on long, curved, rough-hairy pedicels .... **G. trifidum**
10. Leaves usually in whorls of 5 or more; flowers and fruit on straight glabrous pedicels...................... 11
11. Pedicels 0.5–4 mm long and often curved at maturity, solitary or in pairs in leaf axils or at ends of branches but not on a common peduncle; corolla less than 1 mm wide; mature fruit to 1 mm long; leaves mostly 2.5–7 mm long .... **G. brevipes**
11. Pedicels (at least the longest) 3–8 mm long and nearly always straight at maturity, often on a peduncle; corolla 1–1.8 mm wide; mature fruit 1–2 mm long; leaves mostly 5.5–14 (–22) mm long .... **G. tinctorium**
12. Flowers in well-branched cymes; nodes of stems glabrous .... **G. palustre**
12. Cymes only once or twice branched; nodes short-hairy 13
13. Leaves linear, bent downward, less than 2 mm wide... .... **G. labradoricum**
13. Leaves linear to oblong, spreading but not angled downward, mostly more than 2 mm wide ...... **G. obtusum**

Galium aparine L.
STICKY-WILLY; CLEAVERS *native*
IR | WUP | CUP | EUP   FACU CC 0

Annual herb. Stems weak, prostrate or reclining on bushes, 3–10 dm long, with stiff, downward-pointing hairs. Leaves on the principal stems in whorls of 8, typically oblong lance-shaped, mostly 3–8 cm long, rounded to an apiculate tip, retrorsely hispid on the margins and midvein. Peduncles axillary, exceeding the subtending leaves, divaricately branched, few-flowered. Fruit with hooked bristly hairs, 2–4 mm long. May–June.

Damp ground, usually in shade. Variable.

Galium asprellum Michx.
ROUGH BEDSTRAW *native*
IR | WUP | CUP | EUP   OBL CC 5

Perennial herb. Stems spreading or reclining on other plants, much-branched, to 2 m long, 4-angled, with rough, downward-pointing hairs on stem angles (which cling tightly to clothing). Leaves 6 in a whorl or 5-whorled on branches, narrowly oval, usually widest above middle, 1–2 cm long and 4–6 mm wide, tapered to a sharp tip; underside midvein and margins with rough hairs; petioles absent. Flowers in loose, few-flowered clusters at ends of stems and from upper leaf axils; corolla 4-lobed, white, 3 mm wide. Fruit smooth. July–Sept.

Swamps, streambanks, thickets, marshes, wet meadows, calcareous fens.

Galium boreale L.
NORTHERN BEDSTRAW native
IR | **WUP** | **CUP** | **EUP** FAC CC 3

Perennial herb. Stems erect, 2–8 dm long, 4-angled, smooth or with short hairs at leaf nodes, sometimes slightly rough-to-touch. Leaves in whorls of 4, linear to lance-shaped, 1.5–4 cm long and 3–8 mm wide, 3-nerved, tapered to a small rounded tip; margins sometimes fringed with hairs; petioles absent. Flowers many, 3–6 mm wide, in branched clusters at ends of stems; corolla lobes 4, white. Fruit with short, bristly hairs, or smooth when mature. June–Aug.

Streambanks, shores, thickets, swamps, moist meadows; also in drier woods and fields.

Galium brevipes Fern. & Weig.
LIMESTONE SWAMP BEDSTRAW native
IR | WUP | **CUP** | **EUP** OBL CC 6

Galium trifidum subsp. *brevipes* (Fernald & Wiegand) Á. & D. Löve

Perennial herb. Stems scabrous, forming sprawling, tangled mats. Leaves whorled, 4 at each node. Flowers 1 per peduncle, the peduncles very short, to only 4 mm long. Fruit smooth, lacking bristles. The very small pedicels (usually ± recurved), fruits, corollas, and leaves, if all are present, are distinctive. July–Aug.

Marshes, thickets; exposed calcareous shores, interdunal hollows, ditches.

Galium kamtschaticum Schult. & Schult. f.
BOREAL BEDSTRAW (END) native
IR | WUP | CUP | **EUP** CC 10

Low perennial herb. Flowering stems with only 3-5 whorls of leaves. Leaves in whorls of 4, ovate with 3 prominent veins, the upper leaves larger than the lower leaves. Flowers white, on short pedicels. Fruit with hooked bristles.

Low places in deciduous forests. A boreal species; rare in Michigan where known only from several locations in Chippewa County.

Galium labradoricum (Wieg.) Wieg.
NORTHERN BOG BEDSTRAW native
IR | **WUP** | **CUP** | **EUP** OBL CC 8

Perennial herb. Stems simple or branched, 1–3 dm long, 4-angled, hairy at leaf nodes, smooth on stem angles. Leaves in whorls of 4, soon curved downward, oblong lance-shaped, 1–1.5 cm long and 1–2 mm wide, blunt-tipped; underside midvein and margins with short, bristly hairs; petioles absent. Flowers single or in small groups on stalks from leaf axils; corolla lobes 4, white. Fruit smooth, dark. June–July.

Conifer swamps, sphagnum bogs, fens, sedge meadows.

Galium lanceolatum Torr.
LANCE-LEAF WILD LICORICE native
IR | **WUP** | **CUP** | **EUP** CC 4

Perennial herb. Stems slender, branched from the base, erect or ascending, 3–7 dm tall, the stems glabrous or nearly so. Leaves in whorls of 4, thin, the lower elliptic, the upper lance-shaped, 3–8 cm long, 1–2.5 cm wide, long-tapering to an acute or acuminate apex, 3–5-nerved, minutely ciliate, smooth above, finely pubescent on the midvein and sometimes on the other veins beneath. Inflorescence widely divaricate, 1–3-forked; corolla glabrous, turning purple with age, its lobes acuminate. Fruit deflexed, uncinate-hispid, 3 mm long. June–July.

Dry woods and thickets.

Galium mollugo L.
FALSE BABY'S-BREATH introduced
IR | **WUP** | **CUP** | **EUP**

Galium album Mill.

Erect perennial from a decumbent base. Stems 3–10 dm tall, smooth to finely pubescent. Leaves in whorls of 6 or 8, narrow, oblong lance-shaped, 10–20 (rarely 25) mm long, acute or apiculate, scabrous on the margin. Inflorescences several from the upper axils, forming a loose, open, elongate, divaricately branched panicle 1–3 dm long; corolla lobes white, acuminate. Fruit smooth, 1.5 mm long. May–July.

Meadows, fields, roadsides, and lawns.

Galium obtusum Bigelow
BLUNTLEAF BEDSTRAW native
IR | WUP | **CUP** | EUP FACW CC 5

Perennial herb. Stems branched, 2–6 dm long, 4-angled, hairy at leaf nodes, otherwise smooth. Leaves mostly in whorls of 4 (sometimes 5 or 6), ascending to spreading, linear to lance-shaped or oval, 1–3 cm long and 3–5 mm wide, blunt-tipped; margins with short, bristly hairs and often somewhat rolled under; petioles absent. Flowers in clusters at ends of stems; corolla lobes 4, white. Fruit smooth, dark, often with only 1 segment maturing. May–July.

Wet deciduous forests, wet meadows, streambanks, thickets, floodplains, moist prairie; Menominee Co.

Galium palustre L.
COMMON MARSH BEDSTRAW native
IR | **WUP** | **CUP** | **EUP** OBL CC 3

Perennial, simple or diffusely branched. Stems slender,

2–6 dm long, minutely and sparsely retrorse-scabrous on the angles. Leaves in whorls of 2–6, linear to narrowly oblong lance-shaped, 5–15 mm long, blunt, more or less scabrous on the margin. Inflorescences many-flowered, repeatedly forked, the short slender pedicels mostly ascending at anthesis, widely spreading or somewhat reflexed in fruit; corolla white, 4-lobed, about 4 mm wide. Fruit smooth, about 2 mm long. June–Aug.

Wet soil.

Galium tinctorium (L.) Scop.
STIFF MARSH BEDSTRAW native
IR | **WUP** | **CUP** | **EUP** OBL CC 5

Galium trifidum L. subsp. *tinctorium* (L.) Hara

Perennial herb. Stems slender, weak, 4-angled, with rough hairs on angles. Leaves in whorls of 4 or sometimes 5–6, linear to oblong lance-shaped, 1–2.5 cm long, tapered to a narrow base, dark green and dull; underside midvein and margins with rough hairs; petioles absent. Flowers in clusters of 2–3, on slender, smooth, straight stalks at ends of stems; corolla lobes 3, white. Fruit smooth. July–Sept. Plants similar to *G. trifidum* and sometimes considered a variety of that species.

Conifer swamps, open bogs, fens, thickets, wet shores and marshes.

Galium trifidum L.
NORTHERN THREE-LOBED BEDSTRAW native
IR | **WUP** | **CUP** | **EUP** FACW CC 6

Perennial herb. Stems slender, weak, 2–6 dm long, much-branched, sharply 4-angled, with rough, downward-pointing hairs on stem angles. Leaves in whorls of 4, linear to oblong lance-shaped, 5–20 mm long and 1–3 mm wide, blunt-tipped, dark green and dull on both sides; underside midvein and margins often rough-hairy; petioles absent. Flowers small, on 2–3 slender stalks from leaf axils or at ends of stems, the stalks much longer than the leaves; corolla lobes 3, white. Fruit dark, smooth. June–Sept.

Lakeshores, streambanks, swamps, marshes, bogs, springs.

Galium triflorum Michx.
SWEET-SCENTED BEDSTRAW native
IR | **WUP** | **CUP** | **EUP** FACU CC 4

Perennial herb. Stems prostrate or scrambling, 2–8 dm long, 4-angled, smooth or with rough, downward-pointing hairs on stem angles. Leaves shiny, in whorls of 6 (or 4 on smaller branches), narrowly oval to oblong lance-shaped, 2–5 cm long and to 1 cm wide, l-nerved, tipped with a short, sharp point, slightly vanilla-scented, underside midvein with rough hairs, margins with rough, forward- pointing hairs; petioles absent.Flowers 2–3 mm wide, on slender stalks from leaf axils and at ends of stems, the stalks with 3 flowers or branched into 3 short stalks, each with 1–3 flowers; corolla lobes 4, green-white. Fruit 2-lobed, covered with hooked bristles. June–Aug.

Moist to wet woods, hummocks in cedar swamps, wetland margins and shores, clearings.

Galium verum L.
YELLOW SPRING BEDSTRAW *introduced*
IR | **WUP** | **CUP** | **EUP**

Perennial herb. Stems erect from a horizontal rhizome, 3–8 dm tall, finely pubescent throughout or in the inflorescence. Leaves mostly in whorls of 8, linear, 1–3 cm long, often deflexed in age, sharply acute, usually pubescent beneath, often scabrellate above. Inflorescences numerous from the upper axils, compactly many-flowered, equaling or longer than the internodes below which they rise, forming a dense panicle; corolla yellow. Fruit smooth, about 1 mm long. June–Sept.

Fields and roadsides, usually in dry soil.

The dense, attractive, bright yellow flowers have a strong aroma. The distinctive needle-like leaves may be as many as 12 at a node.

Houstonia *Bluets*

Houstonia longifolia Gaertn.
LONG-LEAF SUMMER BLUETS native
IR | **WUP** | **CUP** | EUP CC 6

Hedyotis longifolia (Gaertn.) Hook.

Stems numerous from a perennial base, simple or branched above, 10–25 cm tall, glabrous or finely pubescent, especially at the nodes. Leaves small, opposite, sessile, broadly linear to narrowly oblong, 10–30 mm long, 2–5 mm wide, narrowed to the base, glabrous or minutely scaberulous on the margin, 1-nerved, often with a few obscure veinlets. Flowers 4-merous, short-pediceled, numerous in loose or crowded cymes; sepals linear lance-shaped, 1–2 mm long, in fruit equaling or considerably exceeding the capsule; corolla salverform or funnelform, purplish to white, 5.5–9 mm long, the lobes about half as long as the tube, pubescent within. Capsule globose, 2.3–3.1 mm long. June–Aug.

Dry to sometimes moist, sandy or gravelly soil; shallow soil over limestone; sandy fields; Delta, Gogebic, and Schoolcraft counties.

Mitchella *Partridge-Berry*

Mitchella repens L.
PARTRIDGE-BERRY native
IR | **WUP** | **CUP** | **EUP** FACU CC 5

Creeping perennial herb. Stems rooting at the nodes, 10–30 cm long, forming mats. Leaves evergreen, petioled, round-ovate, 1–2 cm long. Flowers 4-merous, dimorphic, in pairs, their hypanthia united, mostly terminal, the common peduncle shorter than the subtending leaves; corolla white, funnelform,

312 RUBIACEAE | Madder Family

10–14 mm long, with elongate tube and 4 short, spreading or recurved lobes villous on the inner face; ovary 4-celled; stigmas 4. Fruit a scarlet berry, composed of the ripened hypanthia and ovaries of the 2 flowers, 5–8 mm wide, crowned with the short sepals, edible but insipid, persistent through the winter; seeds 8. May–July.

Dry or moist woods.

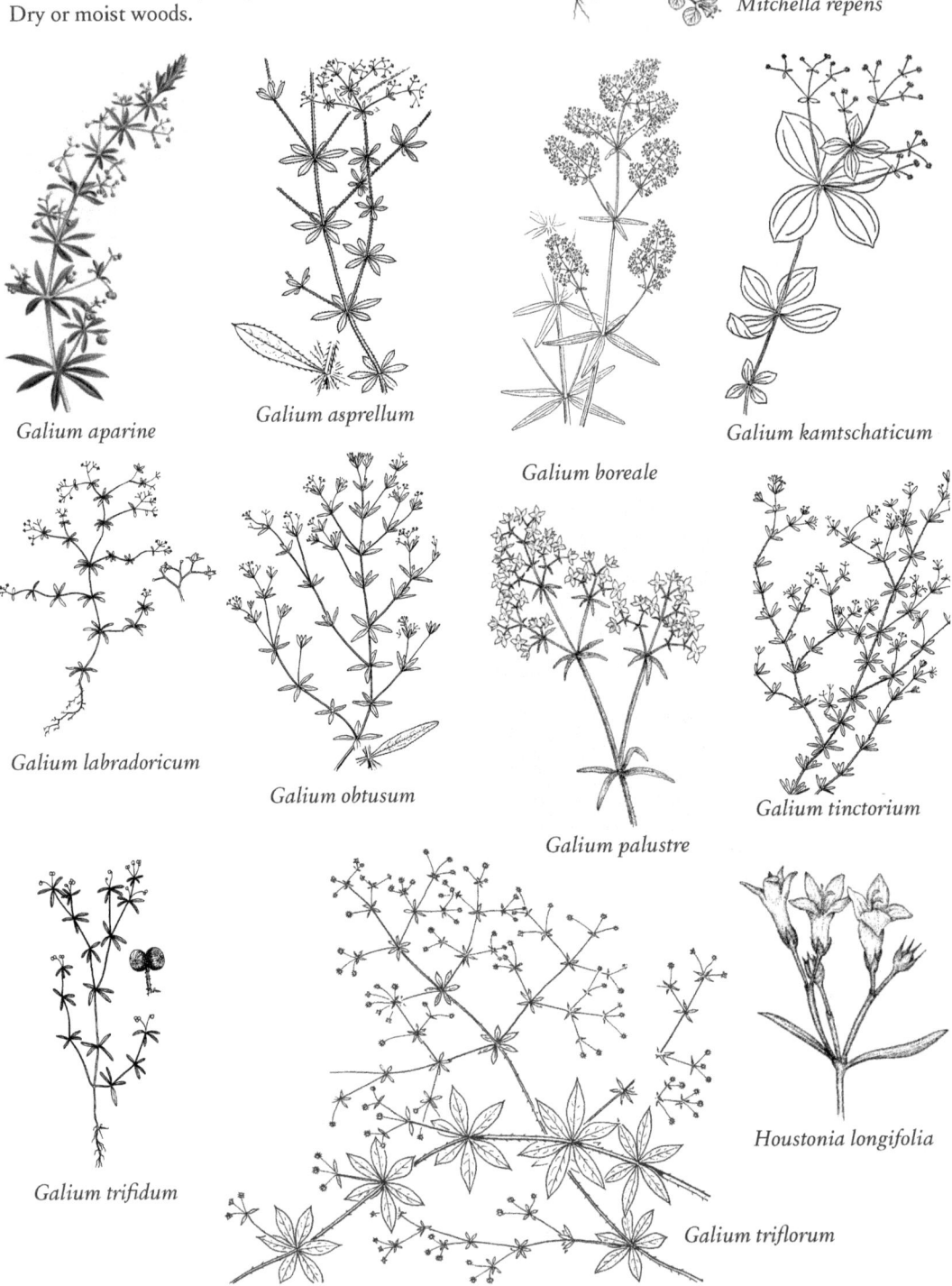

Mitchella repens

Galium aparine

Galium asprellum

Galium boreale

Galium kamtschaticum

Galium labradoricum

Galium obtusum

Galium palustre

Galium tinctorium

Galium trifidum

Houstonia longifolia

Galium triflorum

Rutaceae
RUE FAMILY

Mostly trees or shrubs with alternate, simple or compound leaves and small flowers. Flowers perfect or unisexual, usually regular. Stamens usually as many or 2x as many as the petals. Carpels commonly as many as the petals, in some genera fewer, separate, or weakly united (often by the styles only), or completely connate into a compound ovary. Fruit commonly separating into segments, in some genera a capsule, drupe, or berry. Most parts of the plant contain oil-glands; those of the leaves appear as translucent dots.

The most important economic genus is *Citrus* L., including cultivated varieties of orange, grapefruit, lemon, lime, citron, and tangerine; the two Michigan species are the northernmost members of the family.

ADDITIONAL SPECIES
Ruta graveolens L. (Common rue); introduced ornamental, rarely escaping from cultivation to old fields; in Michigan, known from Iron County.

1 Leaflets 3 . . . . **Ptelea trifoliata**
1 Leaflets 5–11 . . . . **Zanthoxylum americanum**

Ptelea *Hop-Tree*
Ptelea trifoliata L.
COMMON HOP-TREE   native
IR | WUP | **CUP** | EUP   FACU CC 4

Deciduous shrub or small tree, without spines. Leaves alternate, 3-foliolate, long-petioled; leaflets sessile, ovate, elliptic, or ovate-oblong, entire or serrulate. Flowers small, greenish white or yellowish white, with staminate, pistillate, and perfect flowers on the same plant, and produced together in terminal cymes 4–8 cm wide; sepals, petals, and stamens 4 or 5, the latter imperfect or abortive in the pistillate flowers; petals oblong, pubescent, 4–7 mm long. Fruit a thin, flat, circular samara, 15–25 mm wide, the broad wing completely surrounding the indehiscent 2-celled body, reticulately veined, with the odor of hops. May–June.

Moist or rich woods and thickets; Marquette County.

Zanthoxylum *Prickly Ash*
Zanthoxylum americanum P. Mill.
PRICKLY ASH   native
IR | WUP | **CUP** | EUP   FACU CC 3

Tall dioecious shrub or rarely a small tree to 8 m tall, foliage strongly aromatic. Stems thorny. Leaves alternate, odd-pinnately compound; leaflets 5–11, oblong to elliptic or ovate, crenate or entire, pubescent beneath, at least when young. Flowers greenish or whitish, in short-peduncled, sessile, axillary clusters on branches of the previous year; sepals none; petals 4 or 5, fringed at the tip; stamens 4 or 5, alternate with the petals; ovaries 3–5. Fruit (from each ovary) a firm-walled or somewhat fleshy follicle, about 5 mm long, the surface pitted, dehiscent across the top, with 1 or 2 seeds. April–May.

Moist woods and thickets; Delta, Dickinson, and Menominee counties.

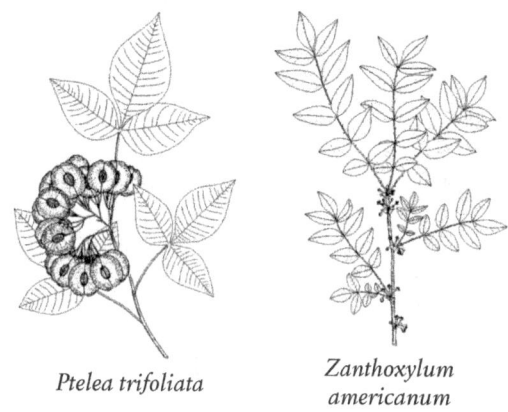

Ptelea trifoliata   *Zanthoxylum americanum*

Salicaceae
WILLOW FAMILY

Deciduous trees or shrubs. Leaves alternate, margins entire or toothed; stipules often present at base of leaf petiole, these usually soon falling. Flowers borne in catkins near ends of branches. Flowers imperfect, the staminate and pistillate flowers on separate plants, usually appearing before leaves open, or in a few species after leaves open; flowers without petals or sepals, each flower with either 1 or 2 enlarged basal glands (*Salix*) or a cup-shaped disk (*Populus*). Fruit a dry, many-seeded capsule; seeds small, covered with long, silky hairs.

1 Large trees; leaves heart-shaped to ovate, mostly less than 2 times longer than wide; buds often sticky and covered by 2 or more overlapping scales; catkins drooping, flowers subtended at base by a cup-shaped disk; stamens many, 12–80 . . . . **Populus**
1 Shrubs and trees; leaves ovate, lance-shaped or linear, 2 or more times longer than wide; buds covered by 1 scale; catkins upright or drooping, flowers subtended by 1 or 2 enlarged glands; stamens 2–8 . . . . **Salix**

Populus *Aspen; Poplar; Cottonwood*
Trees with deciduous, ovate to triangular leaves. Flowers in drooping catkins that develop and mature before and with leaves in spring; staminate and

SALICACEAE | Willow Family

pistillate flowers on separate trees; base of flower with a cup-shaped disk; stamens 10–80. Fruit a 2–4 chambered capsule with many small seeds, these covered with long, white hairs which aid in dispersal by the wind.

1. Leaf petioles round in section, leaf underside often stained brown from resin . . . . **P. balsamifera**
1. Leaf petioles strongly flattened, leaf underside not stained brown 2
2. Leaf underside and petioles densely woolly hairy . . . . **P. alba**
2. Leaf underside and petioles glabrous 3
3. Leaves strongly triangular in shape 4
3. Leaves ovate to nearly round 5
4. Leaf blades about as long or longer than wide, often with glands at tip of petiole; trees with broad crowns . . . . **P. deltoides**
4. Leaf blades wider than long, never with glands on petiole; trees narrow and spire-like . . . . **P. nigra**
5. Leaf margins coarsely wavy-toothed; leaves 7–13 cm long . . . . **P. grandidentata**
5. Leaf margins finely sharp-toothed; leaves less than 7 cm long . . . . **P. tremuloides**

Populus alba L.
WHITE POPLAR *introduced (invasive)*
IR | **WUP** | **CUP** | **EUP**

Tree with widely spreading branches and whitish gray bark; terminal bud and young twigs tomentose. Leaves white-tomentose beneath, palmately 3–7-lobed on the elongate shoots at the end of the branches, on the short lateral shoots ovate, irregularly dentate. Pistillate catkins 4–6 cm long. Capsules narrowly ovoid.

Native of Eurasia, commonly planted and spreading by root sprouts.

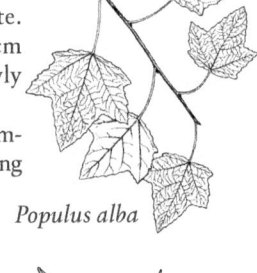
Populus alba

Populus balsamifera L.
BALSAM-POPLAR *native*
IR | **WUP** | **CUP** | **EUP** FACW CC 2

Medium to large tree to 20 m or more tall; trunk 30–60 cm wide; crown open, somewhat narrow; bark smooth when young, becoming dark gray and furrowed; twigs red-brown when young, becoming gray; leaf buds fragrant, very resinous and sticky. Leaves resinous, ovate, 8–13 cm long and 4–7 cm wide, tapered to a long tip, rounded or somewhat heart-shaped at base, dark green and somewhat shiny above, white-green or silvery and often stained with rusty brown resin below; margins with small, rounded teeth; petioles round in section, 3–4 cm long. Catkins densely flowered, drooping, appearing before leaves; scales fringed with long hairs, early deciduous; pistillate catkins 10–13 cm long; pistillate flowers with 2 spreading stigmas; stamens 20–30. Capsules ovate, 6–8 mm long, crowded on short pedicels. April–May.

Swamps, floodplain forests, shores, streambanks, forest depressions, moist dunes.

Populus deltoides Bartr.
PLAINS COTTONWOOD *native*
IR | WUP | **CUP** | **EUP** FAC CC 1

Large tree to 30 m or more tall, with a large trunk (often 1 m or more wide) and a broad, rounded crown; bark gray to nearly black, deeply furrowed; twigs olive-brown to yellow, turning gray with age; leaf buds very resinous and sticky, shiny, covered by several tan bud scales. Leaves smooth, broadly triangular, 8–14 cm long and 6–12 cm wide, short-tapered to tip, heart-shaped or truncate at base; margins with forward-pointing, incurved teeth, 2–5 large glands usually present at base of blade near petiole; petioles strongly flattened, 3–10 cm long; stipules tiny, early deciduous. Catkins loosely flowered, drooping, appearing before leaves; scales fringed, soon falling; flowers subtended by a cup-shaped disk 2–4 mm wide; pistillate catkins green, 7–12 cm long in flower, to 20 cm long in fruit; pistillate flowers with 3–4 spreading stigmas; staminate catkins dark red, soon deciduous; stamens 30–80.

Populus balsamifera

Populus deltoides

Populus grandidentata

Populus nigra

Capsules ovate, 6–12 mm long, on pedicels 3–10 mm long. April–May.

Floodplains, streambanks and bars, shores, wet meadows, ditches; Delta, Mackinac, and Menominee counties.

Populus grandidentata Michx.
BIG-TOOTH ASPEN *native*
IR | WUP | CUP | EUP FACU CC 4

Small or large tree; bark light greenish gray when young, becoming dark brown in age. Terminal buds dull brown, finely pubescent. Leaf blades broadly ovate in outline, 8–12 cm long, with 5–10 large, projecting, round-pointed teeth on each side, the lowest veins strongly ascending; petioles strongly flattened. Scales of the catkins shallowly cleft into 5–7 lance-shaped lobes; stamens 5–12; stigmas 4. Capsules slenderly conic, 3–5 mm long, on pedicels 1–2 mm long.

Dry or moist soil; common northward where it usually grows in drier soil than *P. tremuloides*.

Populus nigra L.
LOMBARDY POPLAR *introduced*
IR | WUP | CUP | EUP

Tall tree, with dull gray branches and dark furrowed bark on the older trunks; a horticultural form is the Lombardy poplar with erect branches forming a narrowly conic crown. Leaves triangular-ovate, abruptly pointed, broadly cuneate to truncate at base, finely and bluntly serrate, 5–10 cm long and usually slightly broader, glandless at the base, pubescent when young; petioles distinctly flattened. Stamens 30 or fewer; stigmas 2, broadly dilated. Capsules ovoid, 7–9 mm long, 2-valved, twice as long as their pedicels.

Native of Eurasia; often planted and occasionally escaped; Mackinac and Marquette counties.

Populus tremuloides Michx.
QUAKING ASPEN *native*
IR | WUP | CUP | EUP FACU CC 1

Slender tree with light grayish green bark, becoming dark and furrowed in age; terminal buds brown, shining, glabrous or nearly so. Leaves broadly ovate to orbicular, 3–10 cm long, abruptly pointed, broadly cuneate (rarely) to truncate or subcordate at base, finely and regularly serrate or crenate to nearly entire; lowest lateral veins strongly ascending, the venation hence apparently palmate; petioles strongly flattened. Scales of the catkins cleft to below the middle into 3–5 lance-shaped lobes; stamens, stig-

P. tremuloides

mas, and capsules as in *P. grandidentata*.

Dry or moist soil, especially in cut-over land.

Salix Willow

Shrubs and trees. Leaves variable in shape, petioles glandular in some species; stipules early deciduous or persistent, sometimes absent. Catkins (aments) stalkless or on leafy branchlets, usually shed early in season. Staminate and pistillate flowers on separate plants; staminate flowers with mostly 2–3 stamens (to 8 in some species). Fruit a 2-chambered, stalked or stalkless capsule.

1 Leaves opposite or nearly so; young branches often dark purple................................. **S. purpurea**
1 Leaves alternate; branches various colors........... 2
2 Leaf petioles with glands at or near base of blade.... 3
2 Petioles without glands............................. 9
3 Trees, usually with a single trunk; leaves narrow..... 4
3 Small trees or shrubs, usually with several to many stems; leaves broader............................ 6
4 Leaves often curved sideways (scythe-shaped), tapered to a long, slender tip; vigorous shoots with large stipules; native species...................... **S. nigra**
4 Leaves not curved sideways, tapered to a short tip; stipules small, early deciduous; introduced 5
5 Leaves glabrous; twigs easily broken at base ....
    .... **S. fragilis**
5 Leaf underside usually silky-hairy; twigs not easily broken at base .... **S. alba**
6 Leaves not waxy on underside .... **S. lucida**
6 Leaves waxy-coated on underside.................... 7
7 Leaf tips rounded or with a short point; leaf base heart-shaped or rounded; young leaves translucent; buds and leaves with a balsamlike scent........... **S. pyrifolia**
7 Leaves tapered to tip; leaf base blunt or rounded; young leaves not translucent; buds and leaves not balsam-scented 8
8 Young leaves sparsely hairy; margins with small forward-pointing teeth; flowering in early summer ......
    .... **S. amygdaloides**
8 Young leaves without hairs; margins with small, gland-tipped, forward-pointing teeth; flowering summer or fall.................................... **S. serissima**
9 Mature leaves hairy, at least on underside 10
9 Mature leaves without hairs (sometimes hairy on petiole and midvein) 19
10 Leaves linear or narrowly lance-shaped 11
10 Leaves broadly lance-shaped, oblong, or ovate 14
11 Underside of leaves with felt-like covering of white tangled hairs; young twigs white-hairy; plant of peatlands, often where calcium-rich .... **S. candida**
11 Leaves not with felt-like hairs; twigs smooth or sparsely hairy 12

SALICACEAE | Willow Family

12 Leaf margins entire and somewhat revolute; leaf underside pubescent . . . . **S. pellita**

12 Leaf margins with gland-tipped teeth; leaf underside sparsely hairy 13

13 Leaf margins with widely spaced sharp teeth; petioles 1–5 mm long; colony-forming shrub of sandy banks . . . . **S. interior**

13 Leaf margins with small teeth at least above middle of blade; petioles 3–10 mm long; stems clustered but not forming large colonies. . . . . **S. petiolaris**

14 Leaves rounded or heart-shaped at base; margins toothed; stipules present, persistent. 15

14 Leaves tapered to base; margins entire or toothed; stipules usually falling early. 16

15 Leaves oblong lance-shaped, tapered to a long tip; young leaves reddish . . . . **S. eriocephala**

15 Leaves obovate to oblong, tapered to a short tip; young leaves not reddish . . . . **S. cordata**

16 Leaves narrowly to broadly lance-shaped, more than 5 times longer than wide, underside velvety swith shiny white hairs . . . . **S. pellita**

16 Leaves obovate or elliptic, less than 5 times longer than wide, underside hairs not shiny 17

17 Small branches widely spreading; young leaves with white hairs; catkins appearing with leaves in spring; catkin bracts yellow or straw-colored; capsules on pedicels 2–5 mm long. . . . . **S. bebbiana**

17 Small branches not widely spreading; young leaves with some red or copper-colored hairs; catkins appearing before leaves in spring; catkin bracts dark brown to black; capsules on pedicels 1–3 mm long 18

18 Leaf upperside smooth or the veins slightly raised, the underside sparsely hairy; twigs often shiny. **S. discolor**

18 Leaf upperside somewhat wrinkled, the veins sunken, the leaf underside densely woolly hairy; twigs dull. . . . . **S. humilis**

19 Leaves green on both sides or slightly paler on underside, not glaucous or strongly whitened below 20

19 Leaves glaucous or whitened on underside. 22

20 Leaves ovate to oblong-ovate . . . . **S. cordata**

20 Leaves, narrower, linear to linear lance-shaped 21

21 Many-stemmed, colony-forming shrub . . . . . **S. interior**

21 Single-stemmed tree . . . . **S. nigra**

22 Leaf margins entire to shallowly lobed or with irregular teeth, sometimes revolute 23

22 Leaf margins distinctly toothed. 26

23 Leaf margins entire and somewhat revolute. 24

23 Leaf margins irregularly toothed, the teeth sharp or rounded. 25

24 Stems upright, to 1 m tall, or creeping and rooting in moss; upper surface of leaves with raised, net-like veins; catkins appearing with leaves; capsules not hairy . . . . **S. pedicellaris**

24 Stems upright, 1–4 m tall; leaf veins not net-like; catkins appearing before leaves; capsules hairy . . **S. planifolia**

25 Leaves dull green above, wrinkled below; catkins appearing with leaves; bracts of pistillate catkins green-yellow to straw-colored . . . . **S. bebbiana**

25 Leaves dark green and shiny above; catkins appearing before leaves; bracts of pistillate catkins dark brown to black. . . . . **S. discolor**

26 Leaves narrowly to broadly lance-shaped, long- or short-tapered to tip. 27

26 Leaves broadly elliptic, ovate, or obovate, rounded or somewhat abruptly short-tapered to tip 29

27 Leaves more or less equally tapered from middle of blade to tip and base . . . . **S. petiolaris**

27 Leaves unequally tapered, the tip tapered to a point; base usually rounded or heart-shaped 28

28 Young twigs glabrous; stipules small or absent; bracts of pistillate catkins pale yellow and soon deciduous . . . . **S. amygdaloides**

28 Young twigs gray-hairy; stipules large; bracts of pistillate catkins dark brown to black, persistent . . . . **S. eriocephala**

29 Leaves balsam-scented (especially when dried), underside net-veined; stipules tiny or absent; catkins appearing with leaves, on leafy or leafless branches. . . . . **S. pyrifolia**

29 Leaves neither balsam-scented nor net-veined; stipules large on vigorous shoots; catkins appearing before or with leaves, sessile or on short, leafy branches . . . . **S. myricoides**

Salix alba L.
WHITE WILLOW *introduced*
IR | **WUP** | **CUP** | EUP FACW

Tree to 20 m tall; twigs golden-yellow, often with long, silky hairs. Leaves lance-shaped, 4–10 cm long and 1–2.5 cm wide, dark green and shiny above, waxy white below, smooth to sparsely hairy on both sides, margins with small gland-tipped teeth; petioles 2–8 mm long, with silky hairs; stipules lance-shaped, 2–4 mm long, early deciduous. Catkins appearing with leaves in spring; pistillate catkins 3–6 cm long, on leafy branches 1–4 cm long; staminate catkins 3–5 cm long, stamens 2; catkin bracts pale yellow, hairy near base, early deciduous. Capsules ovate, 3–5 mm long, without hairs, stalkless or on stalks to 1 mm long. May–June.

Introduced from Europe and sometimes escaping to streambanks and other wet areas.

Salix amygdaloides Anderss.
PEACH-LEAF WILLOW *native*
IR | **WUP** | **CUP** | EUP FACW CC 3

Shrub or tree to 15 m tall, often with several trunks; twigs gray-brown to light yellow, shiny and flexible. Leaves smooth, lance-shaped, long-tapered to tip, 5–12 cm long and 1–3 cm wide, yellow-green above, waxy-white below, margins finely toothed; petioles 5–20 mm long and often twisted; stipules small and early deciduous. Catkins appearing with leaves, linear and loosely flowered; pistillate catkins 3–12 cm long, on leafy branches 1–4 cm long; catkin bracts deciduous, pale yellow, long hairy especially on inner surface; stamens 3–7 (usually 5). Capsules smooth, ovate, 3–7 mm long, on pedicels 1–3 mm long. May–June.

Floodplains, streambanks, lake and pond borders; Delta, Menominee, and Ontonagon counties.

Salix bebbiana Sarg.
BEBB'S WILLOW; BEAKED WILLOW *native*
IR | WUP | CUP | EUP FACW CC 1

Shrub or small tree to 8 m tall, stems 1 to several; twigs yellow-brown to dark brown, usually with short hairs. Leaves oval to ovate or obovate, 4–8 cm long and 1–3 cm wide, dull gray-green, hairy or sometimes smooth on upper surface, waxy-gray, hairy and wrinkled below, the veins distinctly raised on lower surface; margins entire to shallowly toothed; petioles 5–15 mm long; stipules deciduous or persistent on vigorous shoots. Catkins appearing before leaves in spring; pistillate catkins loose, 2–6 cm long, on short leafy branches to 2 cm long; catkin bracts persistent, red-tipped when young, turning brown, long hairy; stamens 2. Capsules ovate, 5–8 mm long, finely hairy, on pedicels 2–6 mm long. May–June.

Common; swamps, thickets, wet meadows, streambanks, marsh borders.

Salix candida Flueggé ex Willd.
SAGE WILLOW *native*
IR | WUP | CUP | EUP OBL CC 9

Low shrub to 1.5 m tall; twigs much-branched, covered with dense, matted white hairs. Leaves linear-oblong, tapered at tip, 4–10 cm long and 0.5–2 cm wide, dull, dark green and sparsely hairy above, veins sunken, densely white-hairy below; margins entire and rolled under; petioles 3–10 mm long; stipules persistent, 2–10 mm long, white-hairy. Catkins appearing with leaves in spring; pistillate catkins 1–5 cm long, on leafy branches 0.5–2 cm long; catkin bracts persistent, brown, hairy; stamens 2. Capsules ovate, 4–8 mm long, white-hairy, on pedicels to 1 mm long. May–June.

Fens, bogs, open swamps, streambanks, usually where calcium-rich.

Salix cordata Michx.
HEART-LEAF WILLOW *native*
IR | WUP | CUP | EUP FAC CC 10

Shrub 2–3 m tall, the vegetative parts all more or less hairy; twigs stoutish, the seasonal ones and buds densely gray-tomentose, the older less so. Leaves ovate lance-shaped to broadly ovate, 4–6 or 8 cm long, 1.5–3 or 4 cm wide, abruptly acuminate, glandular-dentate-serrate, often with stout spinulose teeth (extremely variable), rounded or cordate at base, green on both sides, more or less lanate, strongly nerved beneath; petioles stout, 4–8 mm long, somewhat clasping; stipules cordate-ovate, 6–15 mm long, dentate. Catkins appearing with leaves in spring, 5–8 cm long, on 3–5-leaved peduncles 1–2.5 cm long; catkin bracts brown, densely long-villous; stamens 2; filaments glabrous. Capsules lance-shaped, 5–8 mm long, glabrous; pedicels 0.5–1 mm long, glabrous.

Open sand dunes and sandy shores.

Salix discolor Muhl.
PUSSY-WILLOW *native*
IR | WUP | CUP | EUP FACW CC 1

Shrub or small tree to 5 m tall; twigs yellow-brown to red-brown, dull, smooth with age or with patches of fine hairs. Leaves oval and short-tapered to tip, 3–10 cm long and 1–4 cm wide, dark green and smooth above, underside red-hairy when young, becoming white-waxy, smooth and not wrinkled; margins entire or with few rounded teeth; petioles without glands; stipules deciduous, or often persistent on vigorous shoots. Catkins appearing and maturing before leaves in spring; pistillate catkins 4–8 cm long, stalkless, sometimes with 2 or 3 small, brown, bractlike leaves at the base; stamens 2. Capsules ovate with a long neck, 6–10 mm long, densely gray-hairy, on pedicels 2–3 mm long. April–May.

Swamps, fens, streambanks, floodplains, marsh borders. Common.

Salix eriocephala Michx.
MISSOURI WILLOW *native*
IR | WUP | CUP | EUP FACW CC 2

Salix cordata Muhl., *Salix rigida* Muhl.
Shrub or small tree to 6 m tall; twigs red-brown to dark brown, hairy when young. Leaves lance-shaped or oblong lance-shaped, 5–12 cm long and 1–3 cm wide, red-purple and hairy when young, upper surface becoming smooth and dark green, underside becoming pale-waxy; margins finely toothed; petioles without glands, 3–15 mm long; stipules persistent (especially on vigorous shoots), ovate or kidney-shaped, to 12 mm long, hairless, toothed. Catkins appearing with or slightly before leaves in spring; pistillate catkins 2–6 cm long, on short leafy branches to 1 cm long; catkin bracts persistent, brown to black, hairy; stamens 2. Capsules ovate with a long neck, 4–6 mm long, without hairs, on pedicels 1–2 mm long. April–May.

Shores, streambanks, floodplains, ditches and wet meadows, especially along major rivers.

Salix × fragilis L.
CRACK WILLOW *introduced*
IR | WUP | CUP | EUP FAC

Large tree to 20 m tall and 1 m diameter; twigs greenish to dark red, glabrous, very brittle at base and deciduous in strong winds. Leaves large, lance-shaped, 7–12 or 15 cm long, 2–3.5 cm wide, with 5–6 glandular serrations per cm of margin, dark green above, glaucescent to glaucous beneath, glabrous at maturity; petioles 7–15 mm long, glandular above at the outer end; stipules wanting or small, semicordate, and early deciduous. Catkins appearing with leaves in spring, lax, 4–8 cm long, on leafy peduncles 1–3 or 5 cm long, bearing 2–5 small leaves; catkin bracts greenish yellow, crisp-villous, deciduous. Capsules narrowly conic, 4–5.5 mm long, glabrous; pedicels 0.5–1 mm long. April–May. Introduced to North America from Europe in colonial times for ornament, shade, and gunpowder charcoal; common in farmyards and pastures and sometimes escaped. Considered of hybrid origin from *S. alba* and *S. euxina*.

Salix humilis Marsh.
UPLAND WILLOW   native
IR | **WUP** | **CUP** | **EUP**   FACU CC 4
Shrub 1–3 m tall; twigs yellowish to brown, pubescent to glabrate. Leaves oblong lance-shaped to narrowly obovate, 3–10 or 15 cm long, 1–2 or 3 cm wide, acute to abruptly short-acuminate, somewhat revolute, entire or sparingly undulate-crenate, dark green and often puberulent above; underside glaucous, somewhat rugose, and more or less gray-pubescent, becoming glabrate; stipules lance-shaped, acute, dentate, often deciduous. Catkins precocious, sessile or nearly so, oval-obovoid, 1.5–3 cm long, 1.5–2 cm wide; scales oblong lance-shaped, 1.5–2 mm long, blackish, long-villous; stamens 2; filaments long, free, glabrous. Capsules narrowly lance-shaped, 7–9 mm long, gray-pubescent; pedicels 1–2 mm long, pubescent. March–April.

Open woodlands, dry barrens, and prairies.

Salix interior Rowlee
SANDBAR WILLOW   native
IR | **WUP** | **CUP** | **EUP**   FACW CC 1
Salix exigua Nutt. subsp. *interior* (Rowlee) Cronq.
Shrub to 4 m tall, spreading by rhizomes and often forming dense thickets; twigs yellow-orange to brown, smooth. Leaves linear to lance-shaped, tapered at tip and base, 5–14 cm long and 5–15 mm wide, green on both sides but paler below, at first hairy but soon usually smooth; margins with widely spaced, large teeth; petioles without glands, 1–5 mm long; stipules tiny or absent. Catkins appearing with leaves in spring on short leafy branches (and plants sometimes again flowering in summer); pistillate catkins loosely flowered, 2–8 cm long; catkin bracts deciduous, yellow; stamens 2. Capsules narrowly ovate, 5–8 mm long, hairy when young, smooth when mature, on pedicels to 2 mm long. May–June.

Shores, streambanks, sand and mud bars, ditches and other wet places; often colonizing exposed banks.

Salix lucida Muhl.
SHINING WILLOW   native
IR | **WUP** | **CUP** | **EUP**   FACW CC 3
Shrub or small tree to 5 m tall; twigs yellow-brown or dark brown, smooth and shiny. Leaves lance-shaped to ovate, long-tapered and asymmetric at tip, 4–12 cm long and 1–4 cm wide, shiny green above, pale below, red-hairy when young, but soon smooth; margins with small, gland-tipped teeth; petioles with glands near base of leaf; stipules often persistent, strongly glandular. Catkins appearing with leaves in spring; pistillate catkins 2–5 cm long, on leafy branches 1–3 cm long; catkin bracts deciduous, yellow, sparsely hairy; stamens 3–6. Capsules ovate with a long neck, 4–7 mm long, not hairy, on short pedicels to 1 mm long. May.

Common; swamps, shores, wet meadows, moist sandy areas.

Salix myricoides Muhl.
BLUELEAF WILLOW   native
IR | **WUP** | **CUP** | **EUP**   FACW CC 9
Salix glaucophylloides Fern.
Shrub to 4 m tall; twigs yellow to dark brown, hairy when young. Leaves thickened, lance-shaped to ovate or oval, 4–12 cm long and 1.5–5 cm wide, dark green above, strongly waxy-white below; margins with gland-tipped teeth; petioles 5–12 mm long; stipules 5–10 mm long. Catkins appearing shortly before or with leaves; pistillate catkins 2–8 cm long, on leafy branches 5–15 mm long; catkin bracts deciduous, 1–2 mm long, brown-black and long hairy; stamens 2. Capsules lance-shaped, 5–8 mm long, not hairy, on pedicels 1–3 mm long. May.

Dune hollows and sandy shorelines, fens, mostly near Great Lakes; inland on wet, calcium-rich sites.

Salix nigra Marsh.
BLACK WILLOW   native
IR | **WUP** | **CUP** | **EUP**   OBL CC 5
Medium tree to 15 m tall, trunks 1 or several, crown rounded and open; bark dark brown, furrowed, becoming shaggy; twigs bright red-brown, often hairy when young. Leaves commonly drooping, linear lance-shaped, 6–15 cm long and 0.5–2 cm wide, long-tapered to an often curved tip, green on both sides but satiny above and paler below, lateral veins upturned at tip to form a more or less continuous vein near leaf margin; margins finely toothed; petioles 3–8 mm long, hairy, usually glandular near base of blade; stipules to 12 mm long, heart-shaped, usually deciduous. Catkins appearing with leaves in spring; pistillate catkins 3–8 cm long, on leafy branches 1–3 cm long; stamens usually 6 (varying from 3–7);

SALICACEAE | Willow Family

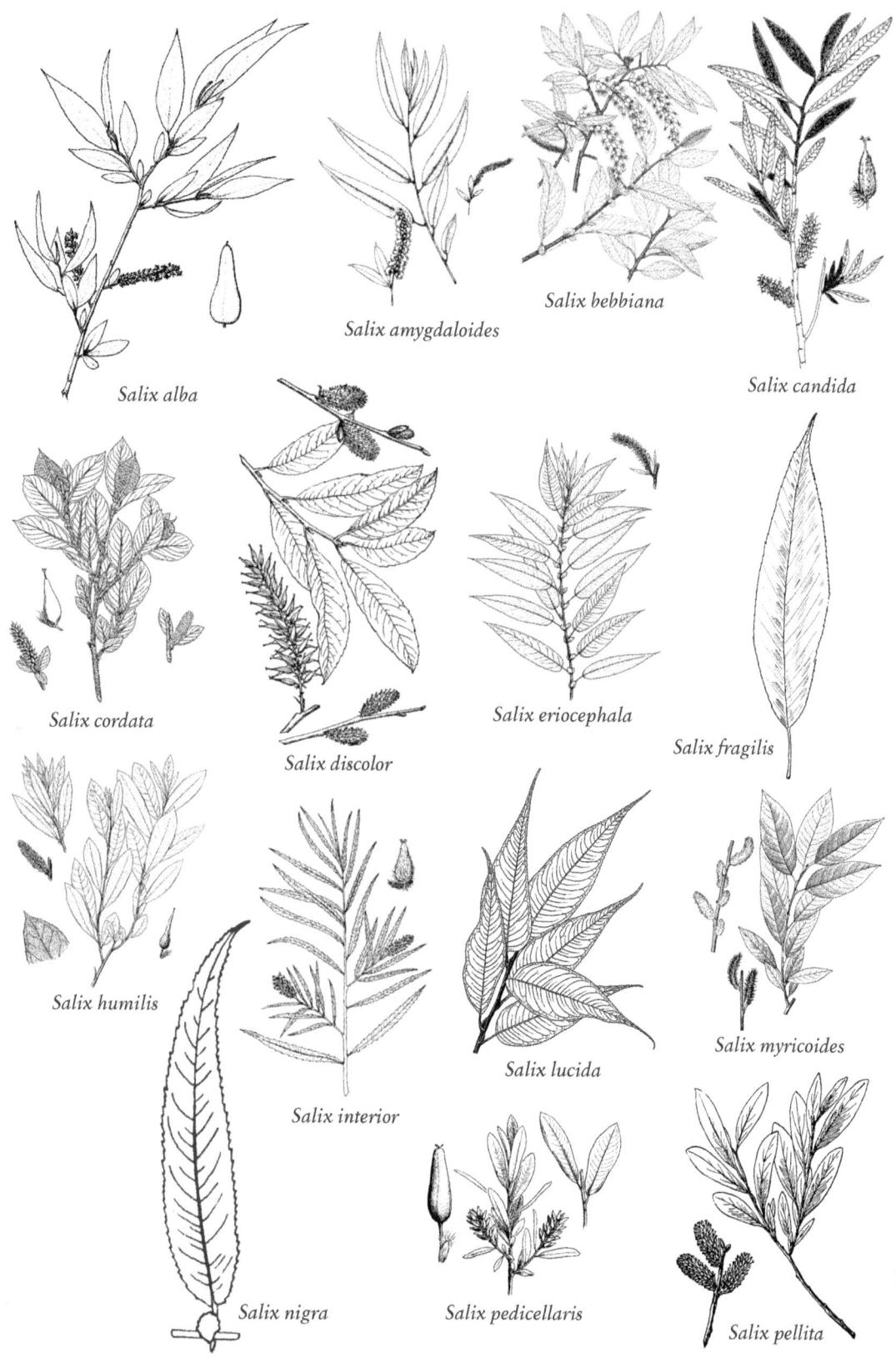

catkin bracts yellow, hairy, deciduous. Capsules ovate, 3–5 mm long, without hairs, on a short pedicel to 2 mm long. May.

Streambanks, lakeshores and wet depressions; not tolerant of shade.

Salix pedicellaris Pursh
BOG WILLOW *native*
IR | WUP | CUP | EUP   OBL CC 8

Short, sparsely branched shrub 4–15 dm tall; twigs dark brown and smooth. Leaves oblong lance-shaped to obovate, tapered to tip or blunt and often with a short point, 3–6 cm long and 0.5–2 cm wide, silky hairy when young, soon hairless, green on upper surface, white-waxy below, veins slightly raised on both sides; margins entire, often slightly rolled under; petioles without glands, 2–8 mm long; stipules absent. Catkins appearing with leaves in spring; pistillate catkins 2–4 cm long, on leafy branches 1–3 cm long; catkin bracts persistent, yellow-brown, hairy on inner surface near tip; stamens 2. Capsules lance-shaped, 4–7 mm long, without hairs, on pedicels 2–3 mm long. May–June.

Bogs, fens, sedge meadows, interdunal wetlands.

Salix pellita (Anderss.) Anderss.
SATINY WILLOW *native*
IR | WUP | CUP | EUP   CC 10

Shrub, 3–5 m tall; twigs easily broken, yellow to olive-brown or red-brown, smooth or sparsely hairy when young, becoming waxy. Leaves lance-shaped, 4–12 cm long and 1–2 cm wide, short-tapered to a tip, upper surface without hairs, veins sunken, underside waxy and satiny hairy but becoming smooth with age, with numerous, parallel lateral veins; margins rolled under, entire or with rounded teeth; petioles to 1 cm long; stipules absent. Catkins appearing and maturing before leaves in spring; pistillate catkins 2–5 cm long, stalkless or on short branches to 1 cm long; catkin bracts black, long-hairy; staminate catkins rarely seen. Capsules lance-shaped, 4–6 mm long, silky hairy, more or less stalkless. May.

Uncommon on streambanks, sandy shores and rocky shorelines.

Salix petiolaris Sm.
MEADOW WILLOW *native*
IR | WUP | CUP | EUP   FACW CC 1

Shrub to 5 m tall; twigs red-brown to dark brown, sometimes with short, matted hairs when young, smooth with age. Leaves narrowly lance-shaped, 4–10 cm long and 1–2.5 cm wide, hairy when young, becoming smooth, dark green above, white-waxy below; margins entire or with small, gland-tipped teeth; petioles without glands, 3–10 mm long; stipules absent. Catkins appearing with leaves in spring; pistillate catkins 1–4 cm long, stalkless or on short branches to 2 cm long; catkin bracts persistent, brown, with a few long, soft hairs; stamens 2. Capsules narrowly lance-shaped, 4–8 mm long, finely hairy, on pedicels 2–4 mm long. May.

Wet meadows, fens, streambanks, shores, open bogs, floating sedge mats, ditches; common.

Salix purpurea L.
BASKET WILLOW *introduced*
IR | WUP | CUP | EUP   FACW

Shrub to 2.5 m tall; twigs smooth, green-yellow to purple. Leaves more or less opposite (unique among our willows), smooth, linear to oblong lance-shaped, 4–9 cm long and 7–16 mm wide, purple-tinged, somewhat waxy below, veins raised and netlike on both sides; margins entire near base, irregularly toothed near tip; petioles short; stipules absent. Catkins appearing with and maturing before leaves in spring; pistillate catkins 2–3.5 cm long, stalkless; catkin bracts black; stamens 2 but often joined. Capsules ovate, 3–4 mm long, short-hairy, stalkless. May–June.

Introduced from Europe, occasionally escaping to lakeshores and streambanks; mostly near Lake Michigan; Delta County.

Salix pyrifolia Anderss.
BALSAM WILLOW *native*
IR | WUP | CUP | EUP   FACW CC 8

Shrub or small tree to 5 m tall; twigs smooth, yellow when young, becoming shiny red. Leaves smooth, ovate to lance-shaped, often rounded at tip, rounded to heart-shaped at base, 4–12 cm long and 2–4 cm wide, red-tinged and translucent when unfolding; green on upper surface, waxy and finely net-veined below; with balsam fragrance; margins with small gland-tipped teeth; petioles 1–2 cm long; stipules absent or small and 1–2 mm long. Catkins appearing with or after leaves in spring; pistillate catkins loosely flowered, 2–6 cm long, on leafy branches 1–3 cm long; catkin bracts red-brown, white-hairy, 2 mm long; stamens 2. Capsules lance-shaped, beaked at tip, 6–8 mm long, smooth, on pedicels 2–4 mm long. May–June.

Conifer swamps, bogs, rocky shores.

Salix serissima (Bailey) Fern.
AUTUMN WILLOW *native*
IR | WUP | CUP | EUP   OBL CC 8

Shrub to 4 m tall; twigs gray, yellow or dark brown, shiny and smooth. Leaves smooth, oval to lance-shaped, 4–10 cm long and 1–3 cm wide, red and hairless when young; green and shiny above, usually white-waxy below; margins with small gland-tipped teeth; petioles with glands near base of leaf; stipules usually absent. Catkins appearing with or after leaves in spring; pistillate catkins 2–4 cm long, on leafy branches 1–4 cm long; catkin bracts deciduous, light yellow, long hairy; stamens 3–7. Capsules nar-

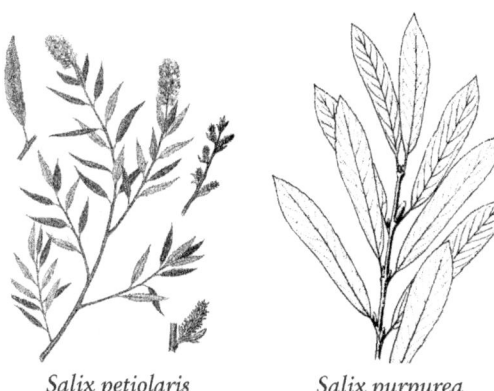
Salix petiolaris *Salix purpurea*

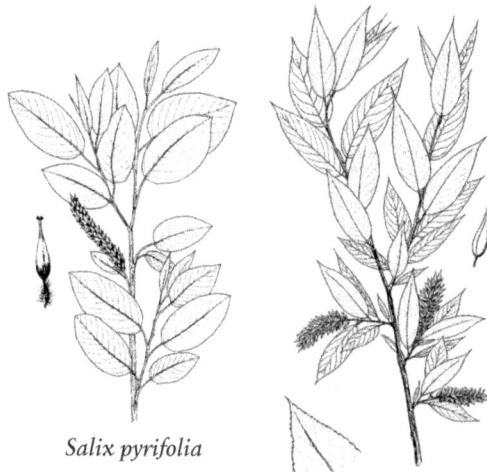
Salix pyrifolia

Salix serissima

rowly cone-shaped, 7–10 mm long, smooth, on pedicels to 2 mm long. Late May–July (our latest blooming willow).

Fens, cedar and tamarack swamps, marshes, floating sedge mats, streambanks and shores, often where calcium-rich.

Santalaceae
SANDALWOOD FAMILY

Herbs (ours), usually root-parasites. Leaves simple, alternate or opposite. Flowers perfect or unisexual (plants usually monoecious or dioecious), in terminal or axillary clusters, or solitary. Hypanthium wholly or partly enclosing the ovary, with 3–several sepals on its margin and a conspicuous disc in its center. Stamens as many as the sepals, opposite them, inserted on their base. Ovary 1-celled. Fruit a nut or drupe, indehiscent, 1-seeded. Santalum album furnishes the fragrant sandalwood of the East Indies. Both *Comandra* and *Geocaulon*, though bearing green leaves, are hemiparasitic, and are apparently always attached (by means of modified roots, or haustoria) to some other plant. Both species also serve as alternate hosts for the canker-producing Comandra blister rust fungus (*Cronartium comandrae*), which in Michigan infects trees of jack pine.

1 Plant an essentially leafless, non-green parasite on the branches of coniferous trees . **Arceuthobium pusillum**
1 Plant leafy, green, terrestrial 2
2 Flowers green-purple, 2–3 from leaf axils; fruit a juicy orange to red drupe . . . . **Geocaulon lividum**
2 Flowers white, numerous in a terminal inflorescence; fruit a dry green or yellowish drupe . . . . **Comandra umbellata**

Arceuthobium *Dwarf-Mistletoe*
Arceuthobium pusillum Peck
EASTERN DWARF-MISTLETOE   native
IR | WUP | CUP | EUP   CC 10

Woody, parasitic plants on conifers, dioecious, attached to trees by haustoria, lacking ordinary roots, but with chlorophyll. Stems short, to 2 cm long, usually only 5–10 mm long, simple or with a few short branches, greenish brown. Leaves opposite, scale-like, about 1 mm wide. Flowers perfect or unisexual, regular, resembling short lateral branches until expanded, solitary or few in the axils of the leaves; perianth simple. Fruit united with the receptacle, berry-like or drupe-like, about 2 mm long, on a short recurved pedicel about equaling the subtending leaf. June–July.

Chiefly on trees of black spruce, rarely on tamarack or white spruce; reported on white pine. On spruce it often produces witches' brooms, and may become so abundant as to endanger the host tree. Usually a single host tree supports only one sex of the mistletoe.

Arceuthobium pusillum

Comandra *Bastard Toadflax*
Comandra umbellata (L.) Nutt.
BASTARD TOADFLAX   native
IR | WUP | CUP | EUP   FACU CC 5

Stems 1–3 dm tall, from a rhizome near the surface of the soil. Leaves narrowly oblong to oval, 2–4 cm long, blunt or subacute, green on both sides, the lateral veins obscure and (except the basal) scarcely differentiated from the veinlets. Flowers white (rarely pinkish), bright green at their base; cymules terminal or subterminal, usually forming a flat-topped cluster; sepals oblong, 2–3 mm long. Fruit a dry or slightly fleshy green or yellowish drupe 4–6 mm long. May–July.

Prairies, shores, upland woods, and rock bluffs.

Geocaulon *False Toadflax*

Geocaulon lividum (Richards.) Fern.
FALSE TOADFLAX *native*
IR | **WUP** | **CUP** | **EUP** FAC CC 9

Comandra livida Richards.

Perennial herb, from a slender rhizome; at least partially parasitic on other plants. Stems smooth, 1–3 dm long. Leaves alternate, oval or ovate, 1–3 cm long and 1–1.5 cm wide, rounded at tip; margins entire; petioles short. Flowers greenish, usually 3 on slender stalks from leaf axils, the lateral 2 flowers typically staminate, the middle flower perfect; sepals 4–5, triangular, 1–2 mm long; petals absent. Fruit a round, orange or red drupe, about 6 mm wide. June–Aug.

Cedar swamps, open bogs; more commonly in sandy conifer woods and forested dune edges; in the western UP, reported only from Houghton and Keewenaw counties (including Isle Royale).

Sapindaceae
SOAPBERRY FAMILY

Soapberry Family now includes former members of Aceraceae and Hippocastanaceae.

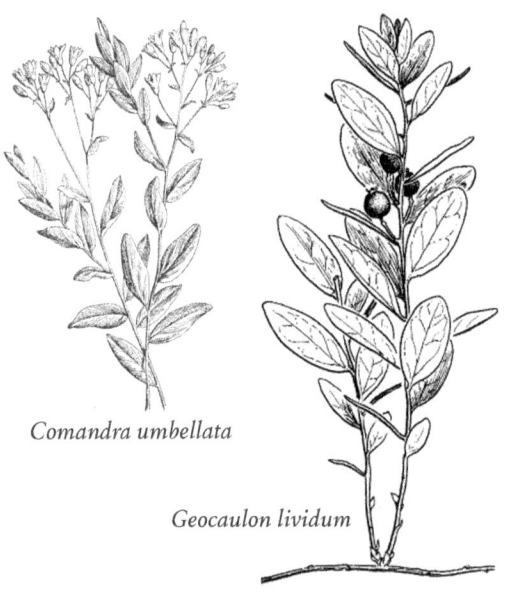

Comandra umbellata

Geocaulon lividum

1 Leaves simple and lobed or pinnately compound; fruit a 2-winged samara. . . . . **Acer**
1 Leaves palmately compound; fruit a large capsule. . . . . **Aesculus**

Acer *Maple*

Trees or shrubs. Leaves opposite, simple or compound. Staminate and pistillate flowers borne on same or separate plants. Flowers with 5 sepals and 5 petals (sometimes absent), clustered into a raceme or umbel. Fruit a samara with 2 winged achenes joined at base.

ADDITIONAL SPECIES
Acer ginnala Maxim. (Amur maple), a shrubby Asian species grown for its brilliant fall color, occasionally escaped; Gogebic and Mackinac counties.

1 Leaves pinnately compound. . . . . **A. negundo**
1 Leaves simple. 2
2 Leaf sinuses between main leaf lobes sharp at their base 3
2 Leaf sinuses rounded at their base 5
3 Leaves deeply lobed to middle of blade or below, the lobes long and narrow. . . . . **A. saccharinum**
3 Leaf lobes shorter and wider 4
4 Leaves with downy white hairs on underside, tips of twigs with appressed hairs; shrubs or small trees; fruit persistent on plants until autumn. . . . . **A. spicatum**
4 Leaves not downy-hairy on underside; twigs glabrous; medium trees; fruit shed in early summer. . **A. rubrum**
5 Leaf blades large, the margins finely doubly toothed; bark with vertical white stripes; small trees or shrubs . . . . **A. pensylvanicum**
5 Leaf blades smaller, the margins only coarsely toothed; bark not vertically white-striped; medium to large trees 6
6 Leaf petioles exuding milky juice when broken; twigs stout; samara wings widely divergent; bark becoming closely fissured, not scaly. . . . . **A. platanoides**
6 Leaf petioles not exuding milky juice when broken; twigs slender; samara wings less divergent; bark becoming deeply furrowed and plate-like . . . . . **A. saccharum**

Acer negundo L.
BOXELDER *native*
IR | **WUP** | **CUP** | **EUP** FAC CC 0

Tree to 20 m tall, the trunk soon dividing into widely spreading branches; bark brown, ridged when young, becoming deeply furrowed; twigs smooth, green and often with waxy-coated. Leaves opposite, compound, leaflets 3–7, oval to ovate, coarsely toothed or shallowly lobed, upper surface light green and smooth, underside pale green and smooth or hairy. Flowers either staminate or pistillate and on separate trees, appearing with leaves in spring; petals absent; staminate flowers in drooping, umbel-like clusters, pistillate flowers in drooping racemes. Fruit a paired

dicots   SAPINDACEAE | Soapberry Family 323

samara 3–4.5 cm long.

Floodplain forests, streambanks, shores; also fencerows, drier woods and disturbed areas.

Distinguished from the ashes (*Fraxinus*) by its paired fruit (vs. single in ash) and its green or waxy twigs.

Acer pensylvanicum L.
STRIPED MAPLE   *native*
IR | **WUP** | **CUP** | **EUP**   FACU CC 5

Shrub or slender tree to 12 m tall. Leaves 3-lobed, glabrous on both sides at maturity, finely and sharply serrate, the teeth commonly 7–12 per centimeter. Flowers mostly polygamo-monoecious, borne singly on slender pedicels along a drooping axis, forming a slender, peduncled, terminal raceme 3–10 cm long, each raceme commonly either pistillate or staminate throughout; petals bright yellow, narrowly obovate, 5–8 mm long, scarcely surpassing the oblong-oblong lance-shaped sepals; stamens 6–8. Samaras 25–30 mm long, scarcely veined over the seed, the halves diverging at 90–120 degrees.

May–June. Moist woods.

Acer platanoides L.
NORWAY MAPLE   *introduced*
IR | **WUP** | **CUP** | **EUP**

Tree with widely spreading crown. Leaves resembling those of *Acer saccharum*, with 5–7 sharply acuminate lobes and a few large teeth; juice milky (best seen at the base of a detached petiole). Flowers yellow, in erect rounded corymbs, the obovate petals 5–6 mm long, widely spreading. Samaras 35–45 mm long, scarcely distended over the seed, the halves divergent at an angle of about 180 degrees. April–May.

Native of Europe; planted as a shade tree and established as a weedy tree in vacant lots.

Acer rubrum L.
RED MAPLE   *native*
IR | **WUP** | **CUP** | **EUP**   FAC CC 1

Tree to 25 m tall; bark gray and smooth when young, becoming darker and scaly; twigs smooth, reddish with pale lenticels. Leaves opposite, 3–5-lobed (but not lobed to middle of blade), coarsely doubly toothed or with a few small lobes, upper surface green and smooth, underside pale green to white, smooth or hairy. Flowers either staminate or pistillate, usually on different trees but sometimes on same tree, in dense clusters, opening before leaves in spring; sepals oblong, 1 mm long, petals narrower and slightly longer. Fruit a paired samara, 1–2.5 cm long.

Floodplain forests, swamps; also common in drier forests.

Distinguished from silver maple (*Acer saccharinum*) by its shallowly lobed leaves vs. the deeply lobed leaves of silver maple.

Acer saccharinum L.
SILVER MAPLE   *native*
IR | **WUP** | **CUP** | **EUP**   FACW CC 2

Tree to 30 m tall; bark gray or silvery when young, becoming scaly; twigs red-brown, smooth. Leaves opposite, deeply 5-lobed to below middle of blade, sharply toothed, upper surface pale green and smooth, underside silvery white; petioles usually red-tinged. Flowers either staminate or pistillate, usually on different trees but sometimes on same tree, in dense clusters, opening before leaves in spring. Fruit a paired samara, each fruit 3–5 cm long, falling in early to mid-summer.

Floodplain forests, swamps, streambanks, shores, low areas in moist forests.

Acer saccharum Marsh.
SUGAR MAPLE   *native*
IR | **WUP** | **CUP** | **EUP**   FACU CC 5

Tree to 40 m tall, with straight central trunk when growing in a forest and a widely spreading network of branches when in the open. Leaves about as wide as long, 3–5-lobed, the lobes usually bearing a few large sharp teeth. Flowers unisexual in umbels from the terminal or uppermost lateral buds, appearing as the leaf buds open, drooping on slender pedicels up to 8 cm long; calyx gamosepalous, campanulate, 2.5–6 mm long, more or less hirsute; petals none. Samaras 2.5–4 cm, averaging 3 cm long, the seed-bearing portions diverging at right angles to the pedicel, the wings curved forward. April–May.

Rich woods, especially in calcareous soils.

Acer spicatum Lam.
MOUNTAIN MAPLE   *native*
IR | **WUP** | **CUP** | **EUP**   FACU CC 5

Shrub or small tree, occasionally 10 m tall. Leaves 3-lobed or obscurely 5-lobed, softly pubescent beneath, coarsely and irregularly serrate, the teeth 2–3 per centimeter, each tipped with a minute sharp gland. Flowers mostly polygamo-monoecious, produced in fascicles of 2–4 along an erect axis, forming a slender, terminal, long-peduncled panicle, each flower long-pediceled, the terminal one of each fascicle usually perfect, the others sterile; petals greenish, very narrowly linear-oblong lance-shaped, about 3 mm long, much exceeding the sepals; stamens usually 8. Samaras 18–25 mm long, conspicuously reticulate-veined over the seed, the halves diverging at about a right angle. June.

Moist woods.

Aesculus *Buckeye*

Aesculus glabra Willd.
OHIO-BUCKEYE   *native*
IR | **WUP** | CUP | EUP   FAC CC 5

Small to medium tree, 10–12 m tall; trunk 15–30

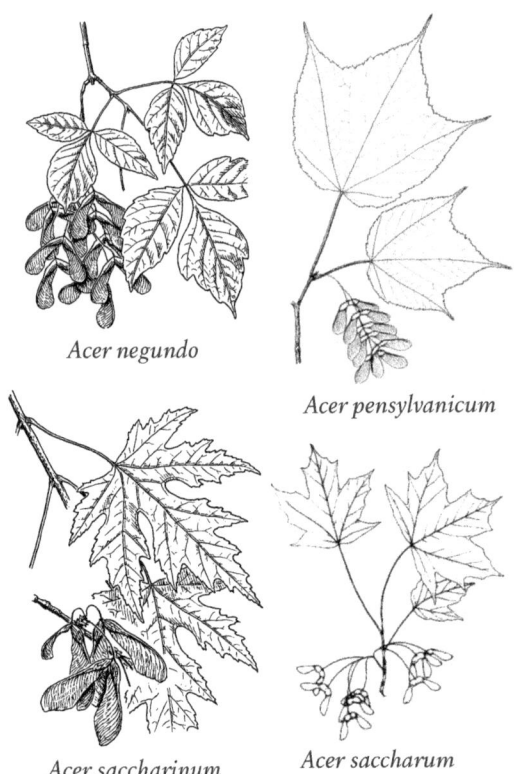
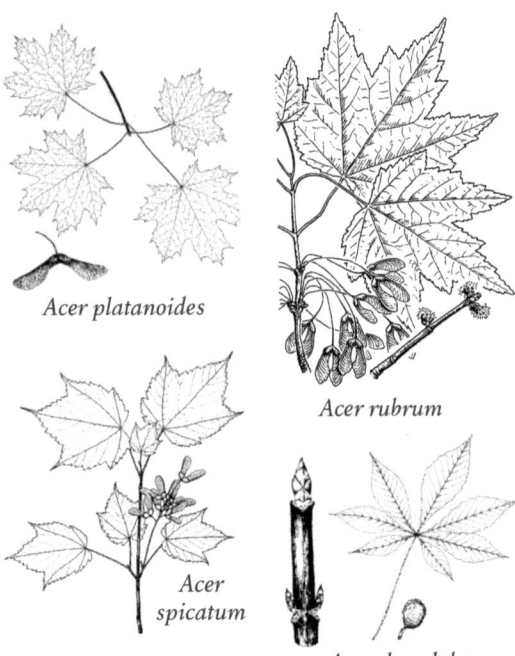

cm wide; bark thin, yellow-brown, smooth to scaly, dark brown and deeply furrowed with age, bark and leaves foul-smelling when bruised; twigs red-brown, becoming light gray. Leaves appearing early in spring, turning yellow in fall, opposite, smooth, palmately compound into usually 5 (rarely 7) leaflets; leaflets obovate, 7–15 cm long and 3–6 cm wide, tapered at both ends, margins finely toothed, petioles 10–15 cm long. Flowers perfect or either staminate or pistillate on same tree, numerous, yellow-green, appearing after leaves unfold in spring, in panicles 1–1.5 dm long and 6 cm wide at ends of branches; petals 4, pale yellow, hairy, 2 cm long, stamens 7, longer than the petals. Fruit a prickly, red-brown capsule, 2–3 cm wide, with 1 smooth, satiny brown seed. April–May.

Reported for Ontonagon County, where spread from planted trees; northern range limit for speciesis southern Michigan.

Sarraceniaceae
PITCHERPLANT FAMILY

Sarracenia *Pitcherplant*

Sarracenia purpurea L.
PITCHERPLANT   native
IR | WUP | CUP | EUP   OBL CC 10

Perennial insectivorous herb. Flower stalks leafless, 3–6 dm long. Leaves clumped, hollow and vaselike, curved and upright from base of plant, 1–2 dm long and 1–5 cm wide, green or veined with red-purple, winged, smooth on outside, upper portion of inside with downward-pointing hairs, tapered to a short petiole at base. Flowers large and nodding, 5–6 cm wide, single at ends of stalks, perfect; sepals 5; petals 5, obovate, dark red-purple, curved inward over yellow style; ovary large and round. Fruit a 5-chambered capsule; seeds small and numerous. May–July.

Sphagnum bogs, floating bog mats, occasionally in calcium-rich wetlands.

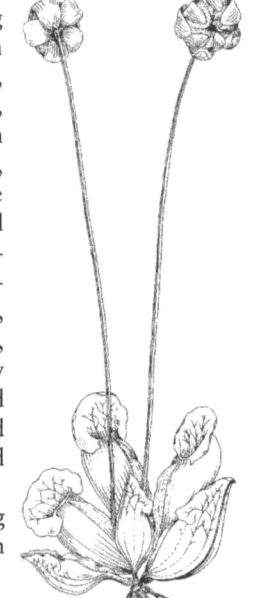

Sarracenia purpurea

Saxifragaceae
SAXIFRAGE FAMILY

Perennial herbs. Leaves alternate, opposite or basal. Flowers perfect (with both staminate and pistillate parts), regular, single on stalks or in narrow heads. Sepals 5 (4 in *Chrysosplenium*); petals 5 or absent; stamens 5 or 10, stigmas 2 or 4. Fruit mostly a 2-parted capsule.

1 Leaves all from stem; petals absent, the flowers 4-merous . . . . **Chrysosplenium americanum**
1 Leaves all (or nearly all) from base of plant; the flowers 5-merous 2
2 Stamens 5 . . . . **Heuchera richardsonii**
2 Stamens 10 3
3 Petals deeply pinnately divided; fruiting carpels widely spreading, exposing the seeds in a shallow flattish cup . . . . **Mitella**
3 Petals entire; carpels in fruit not spreading but forming an elongate capsule or pair of follicles separate nearly to the base 4
4 Leaf blades strongly cordate; carpels very unequal; inflorescence a simple raceme . . . . . **Tiarella cordifolia**
4 Leaf blades tapered at base; carpels equal; inflorescence branched . . . . **Micranthes**

Chrysosplenium *Golden-Saxifrage*

Chrysosplenium americanum Schwein.
AMERICAN GOLDEN-SAXIFRAGE *native*
IR | **WUP** | **CUP** | **EUP**   OBL CC 6

Small, perennial herb, often forming large mats. Stems creeping, branched, 5–20 cm long. Lower leaves opposite, the upper leaves often alternate, broadly ovate, 5–15 mm long and as wide, margins entire or with rounded teeth or lobes; petioles short. Flowers single and stalkless from leaf axils, 4–5 mm wide; sepals 4, green-yellow or purple-tinged; petals absent; stamens usually 8 from a red or green disk, anthers red. Fruit a 2-lobed capsule. April–June.

Springs, shallow streams, shady wet depressions; soils mucky.

Heuchera *Alumroot*

Heuchera richardsonii R. Br.
PRAIRIE ALUMROOT *native*
IR | WUP | **CUP** | EUP   FACU CC 8

Perennial herb. Stems more or less hirsute, becoming glandular in the inflorescence. Leaves broadly cordate-ovate, glabrous or nearly so above, sparsely hirsute on the veins beneath, shallowly 7–9-lobed, each lobe with 3–5 rounded or acute, mucronate lobes; petioles strongly hirsute. Flowers in relatively narrow and congested panicles; hypanthium very oblique, twice as long above as below; petals 3–4 mm long, about equaling the sepals and stamens and styles.

Prairies and dry woods; Dickinson and Menominee counties.

Micranthes *Saxifrage*

Micranthes, now separated from *Saxifraga*, includes those species with a leafless flowering stem (or at most tiny bracts), and our species with herbaceous-textured leaves.

1 Larger leaves (10–) 14–35 cm long, irregularly toothed, entire around the tip; petals 2–3 mm long; plants of wet places . . . . **M. pensylvanica**
1 Larger leaves 3–7.5 cm long or even shorter, distinctly toothed, especially around the apex; petals 3–5 mm long; plants of rocky, mostly dry places **M. virginiensis**

Micranthes pensylvanica (L.) Haw.
SWAMP SAXIFRAGE *native*
IR | **WUP** | **CUP** | **EUP**   OBL CC 10
Saxifraga pensylvanica L.

Perennial herb. Stems stout, erect, 3–10 dm long, with sticky hairs. Leaves all from base of plant, ovate to oblong ovate, 1–2 dm long and 4–8 cm wide, smooth or hairy; margins entire to slightly wavy or with irregular rounded teeth; petioles wide. Flowers small, in clusters atop stem, the head elongating with age; sepals bent backward, 1–2 mm long; petals green-white or purple-tinged, lance-

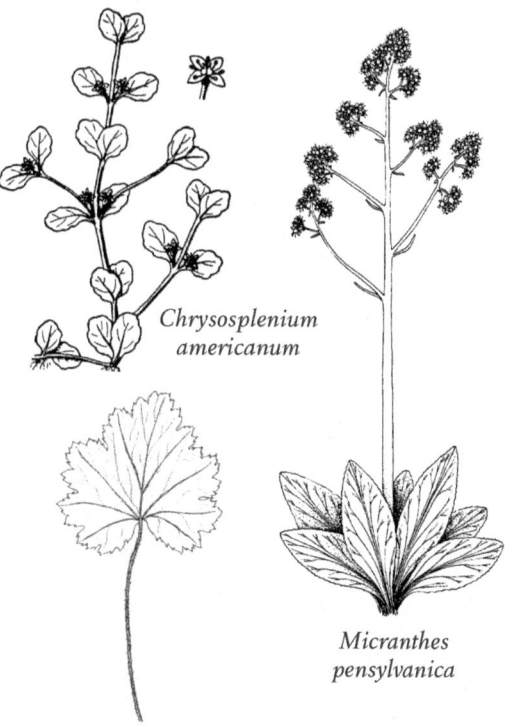

Chrysosplenium americanum

Micranthes pensylvanica

Heuchera richardsonii

shaped, 2–3 mm long; stamens 10, the filaments threadlike. Fruit a follicle. May–June.

Swamps, wet deciduous forests, marshes, moist meadows and low prairie; often where calcium-rich.

Micranthes virginiensis (Michx.) Small
EARLY SAXIFRAGE native
IR | WUP | CUP | EUP XXX   CC 10

Saxifraga virginiensis Michx.

Perennial herb. Stems to 25 cm tall, usually glandular-villous; scapes naked, rarely minutely bracted. Leaves in basal tufts, often purplish beneath; blades oblanceolate, narrowing to flattened petioles. inflorescence leafy-bracted, becoming paniculate, sometimes longer than the scape; flowers white, in congested cymules; petals oblanceolate, 5–6 mm long; tip of ovary becoming inflated and long-beaked; seed ellipsoid, not appendaged, less than 0.5 mm long.

Crevices and shallow soil on limestone (Drummond Island) and other rocky habitats, including the Porcupine and Huron Mountains.

Mitella *Mitrewort; Bishop's Cap*

Perennial rhizomatous herbs. Leaves basal or alternate from the rhizome, the flowering stems leafless or few-leaved, bearing a terminal raceme of small white, greenish, or purple flowers. Flowers perfect, regular, 5-merous. Hypanthium turbinate to saucer-shaped, adnate to the base of the ovaries, bearing the sepals, petals, and stamens at its margin. Petals narrow, deeply pinnatifid or fimbriate to entire. Stamens 10 (in our species), shorter than the sepals. Pistils 2. Carpels short, dehiscent along the ventral suture.

1 Plants small, the scape naked; the basal leaves not lobed or only slightly so; flowers green-yellow ..... **M. nuda**
1 Plants larger, with a pair of nearly sessile leaves on the scape below the inflorescence; the basal leaves clearly 3-lobed; flowers white................... **M. diphylla**

Mitella diphylla L.
TWO-LEAF MITREWORT native
IR | WUP | CUP | EUP   FACU CC 8

Flowering stems 1–4 dm tall, sparsely pubescent below, glandular-puberulent above the stem leaves. Basal leaves long-petioled, ovate-rotund, shallowly 3–5-lobed, crenate, cordate at base, pubescent. Stem leaves 2, sessile, smaller, 3-lobed, the middle lobe elongate. Raceme 5–15 cm long; pedicels 1–2 mm long. Flowers white, 5–6 mm wide; petals deeply fimbriate-pinnatifid, about 2 mm long; seeds few, black, smooth, shining, 1–1.5 mm long. May–June.

Rich woods.

The follicles at dehiscence diverge and open widely, exposing the shiny seeds.

Mitella nuda L.
NAKED MITREWORT native
IR | WUP | CUP | EUP   FACW CC 8

Small perennial herb, spreading by rhizomes or stolons. Leaves all from base of plant, or with 1 small leaf on flower stalk, rounded heart-shaped, 1–3.5 cm wide, both sides with sparse coarse hairs; margins with rounded teeth; petioles 2–8 cm long. Flowers small, green, on short stalks, in racemes of 3–12 flowers, on a glandular-hairy stalk 10–25 cm tall; calyx lobes 5, 1–2 mm long; petals green, pinnately divided into usually 4 pairs of threadlike segments, the segments 2–4 mm long; stamens 10. Fruit a capsule, splitting open to reveal the black, shiny, 1 mm long seeds. June–July.

Hummocks in swamps and alder thickets, ravines, seeps, moist mixed conifer and deciduous forests.

Saxifraga *Saxifrage*

We have two members of the genus in Michigan, both apparently confined to rocky places on Isle Royale, becoming more common northward in boreal and arctic regions.

1 Margins of basal leaves with crowded teeth and white lime-encrusted pores; Isle Royale only... **S. paniculata**
1 Margins of basal leaves entire except for (2–) 3 prominent spine-tipped apical teeth; Isle Royale only....... .... **S. tricuspidata**

Saxifraga paniculata Mill.
LIME-ENCRUSTED SAXIFRAGE (THR) native
IR | WUP | CUP | EUP   CC 10

A boreal perennial, south in North America to the northern Great Lakes, New England, and northern New York; in Michigan, rare in rock crevices along Lake Superior, Isle Royale, Keweenaw County.

Saxifraga tricuspidata Rottb.
PRICKLY SAXIFRAGE (THR) native
IR | WUP | CUP | EUP   CC 10

An arctic perennial, found in the USA only in Isle Royale National Park, where it grows on rock on the main island and on several of the offshore islands at the northeastern end of the archipelago.

Tiarella *Foam-Flower*

Tiarella cordifolia L.
FOAM-FLOWER native
IR | WUP | CUP | EUP   FACU CC 9

Perennial herb with long stolons. Flowering stems 1–3.5 dm tall, glandular-puberulent. Leaves basal, broadly cordate-ovate, shallowly 3–5-lobed, crenate, sparsely pubescent. Flowers in a raceme atop an erect, usually leafless stem; the raceme at first short and crowded, elongating to 1 dm; pedicels 5–10 mm long; hypanthium small, campanulate; sepals 2–3.5

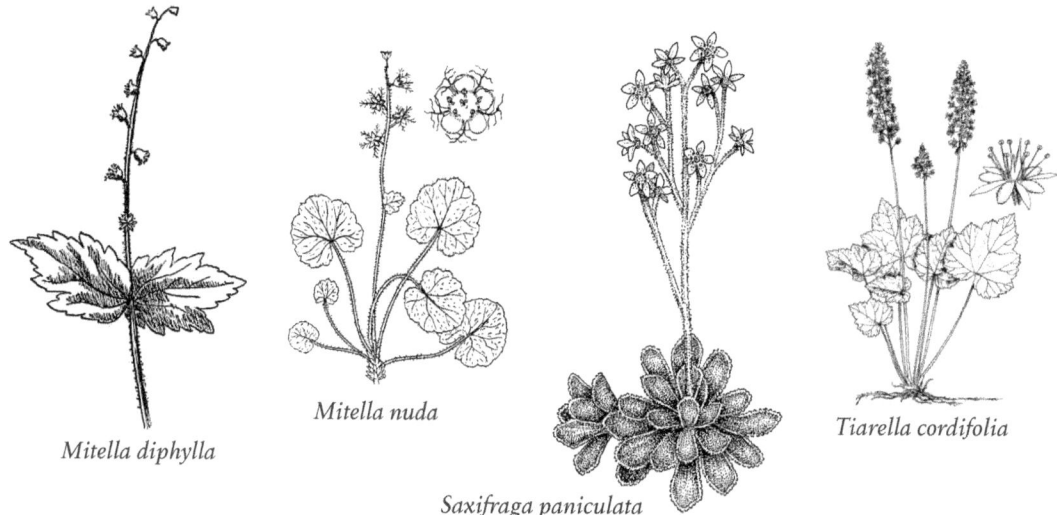

Mitella diphylla

Mitella nuda

Saxifraga paniculata

Tiarella cordifolia

mm long, blunt; petals white, elliptic, clawed, 3–5 mm long; stamens 10; pistils 2, united at base, unequal in length. Follicles 2, thin-walled, the larger about 10 mm long; seeds several, black, smooth, shining. May–early June.

Rich, mesic hardwood forests (sometimes with trees of hemlock present).

Scrophulariaceae
FIGWORT FAMILY

Annual, biennial, or perennial herbs. Leaves mostly opposite or alternate (Verbascum). Flowers single or few from leaf axils, or numerous in clusters at ends of stems or leaf axils, perfect (with both staminate and pistillate parts), usually with a distinct upper and lower lip; sepals and petals 4–5 (petals sometimes absent); stamens usually 4; pistil 2-chambered. Fruit a several- to many-seeded capsule.

Formerly a much larger family, many of our genera now segregated into other families, especially Orobanchaceae and Plantaginaceae.

ADDITIONAL SPECIES
Buddleja davidii Franch. (Butterfly-bush), introduced ornamental shrub from China; escaped from cultivation on Mackinac Island (Mackinac County); leaves opposite, nearly sessile; leaf undersides covered with stellate hairs; flowers 4-parted, ± purple with an orange center, and clustered in a dense, elongate, terminal inflorescence.

1. Stem leaves of fertile stems mostly alternate (a basal rosette may be present and the lower leaves may sometimes be opposite) . . . . **Verbascum**
1. Stem leaves of fertile stems all opposite or nearly so . . . . **Scrophularia**

Scrophularia *Figwort*
Scrophularia lanceolata Pursh
LANCE-LEAF FIGWORT   *native*
IR | WUP | CUP | EUP   FACU CC 5

Perennial herb, often forming dense colonies. Stems erect, to 2 m tall, glabrous, or minutely glandular in the inflorescence, 4-angled, the sides flat or shallowly grooved. Leaves opposite, ovate or ovate lance-shaped, 8–20 cm long, sharply serrate or incised or doubly serrate, truncate to broadly rounded at the base, glabrous beneath; petioles commonly 1.5–3 cm long, rarely as much as a third as long as the blade, narrowly margined to the base. Panicle 1–3 dm long, tending to be cylindric, rarely more than 8 cm wide; corolla ± 2-lipped, 7–11 mm long, dull reddish brown except the yellowish green lower lobe; stamens 5, 4 plus a staminodium (sterile filament) under the upper corolla lobe. Capsule dull brown, 6–10 mm long. Late May–July.

Roadsides, railroads, old roads; forests, especially in clearings; fields, fencerows, shores, swamp borders.

Verbascum *Mullein*

Biennial herbs (ours), producing a rosette of leaves the first year, from which the tall flowering stem rises the following season. Leaves alternate, entire, crenate, or rarely deeply toothed. Flowers yellow, white, or blue, in one to many spike-like racemes. Calyx regular, deeply 5-parted. Corolla rotate or saucer-shaped, nearly regular, the 3 lower lobes slightly larger than the 2 upper. Stamens 5, all fertile, more or less dimorphic. Capsule ovoid to globose, the 2 valves more or less cleft at the tip; seeds numerous, marked with longitudinal ridges.

328 SOLANACEAE | Potato Family

dicots

ADDITIONAL SPECIES

Verbascum lychnitis L. (White mullein); introduced biennial herb, recognized by its smaller flowers and much-branched inflorescence of slender stalks; found on sandy or gravelly roadsides, fields, and disturbed places; Baraga, Houghton, and Marquette counties.

1   Stems glabrous, or often glandular-hairy on upper stem and inflorescence; flowers yellow or white **V. blattaria**
1   Stems densely woolly hairy; flowers yellow. **V. thapsus**

Verbascum blattaria L.
WHITE MOTH MULLEIN *introduced*
IR | WUP | **CUP** | EUP   FACU

Stems slender, to 1 m tall, simple or branched, glandular-pubescent above; branched hairs lacking. Leaves variable, narrowly triangular to oblong or lance-shaped, sessile, not decurrent, coarsely toothed to nearly entire, glabrous, the basal larger, oblong lance-shaped, tapering to the base. Racemes elongate, loose, bearing a single flower at each node on a pedicel 8–15 mm long; calyx glandularpubescent; corolla yellow or white, about 2.5 cm wide, the filaments all about equally villous. June–Oct.

Eurasian weed of fields, roadsides, and waste places; Marquette County.

Verbascum thapsus L.
GREAT MULLEIN *introduced*
IR | WUP | **CUP** | EUP

Plants usually densely gray-tomentose throughout. Stems stout and erect, 1–2 m tall. Lower leaves oblong or oblong lance-shaped, to 3 dm long, petioled; upper leaves progressively reduced, sessile, decurrent along the stem to the next leaf below. Raceme spike-like, very dense, 2–5 dm long, about 3 cm thick, usually solitary; corolla yellow, 12–22 mm wide; upper 3 filaments short, densely white-villous; lower 2 filaments much longer, glabrous or nearly so. June–Sept.

European weed of fields, roadsides, and disturbed places.

Solanaceae
POTATO FAMILY

Herbs or shrubs, rarely climbing, or in the tropics small trees. Leaves alternate or appearing opposite. Flowers perfect, almost always 5-merous, regular (in most of our genera) or irregular. Calyx gamosepalous, persistent in fruit. Corolla rotate to funnelform or tubular. Stamens inserted on the corolla tube and alternate with its lobes, as many as the petals in most genera with regular flowers. Ovary commonly 2-celled (3–5-celled in *Nicandra*, falsely 4-celled in *Datura*). Fruit a capsule or berry. A large family, most numerous in tropical America, and with many plants, such as tomato, potato, eggplant, and peppers, of economic importance.

ADDITIONAL SPECIES

Hyoscyamus niger L. (Black henbane), plants clammy-pubescent, corolla greenish yellow with purple veins; introduced, Mackinac County.
Lycium chinense Mill (Matrimony-vine), Asian introduced shrub, rarely escaping to dry disturbed places; Houghton County.

1   Plants woody (at least at base), sprawling or climbing vines; fruit a red berry 2
1   Plants herbs; fruit various 3
2   Leaves unlobed; stems mostly woody. . . . .
    . . . . **Lycium chinense***
2   Main leaves 3–4 lobed; stems woody near base . . . .
    . . . . **Solanum dulcamara**
3   Plants in flower 4
3   Plants in fruit 8
4   Corolla with short tube and widely flared upper portion
    5
4   Corolla tube longer, the lobes joined for most of their length. 6
5   Anthers opening at tip. . . . . **Solanum**
5   Anthers opening along sides. . . . .
    . . . . **Leucophysalis grandiflora**
6   Calyx tubular, more than 3 cm long . . . . **Datura**
6   Calyx less than 3 cm long. 7
7   Corolla blue . . . . **Nicandra physalodes**

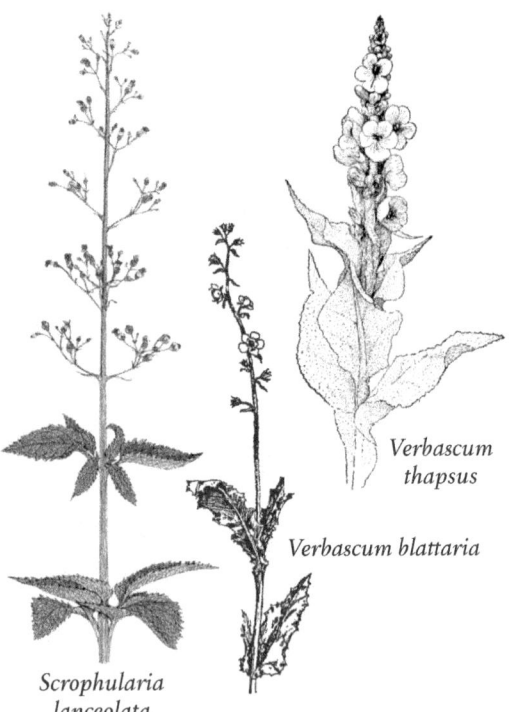

Verbascum thapsus

Verbascum blattaria

Scrophularia lanceolata

7 Corolla yellow or white.................... **Physalis**
8 Calyx or capsule spiny 9
8 Calyx and capsule not spiny 10
9 Stems and leaves spiny; calyx spiny; berry smooth .... **Solanum**
9 Plants smooth apart from spiny capsule....... **Datura**
10 Calyx inflated, covering the berry.................. 11
10 Calyx not inflated, the berry visible .... **Solanum**
11 Calyx strongly inflated, not filled by the berry .... **Physalis**
11 Calyx only slightly inflated, filled within by the berry.. .... **Leucophysalis grandiflora**

Datura *Jimsonweed*

Coarse annual herbs (ours). Leaves large, ovate, petioled. Flowers large, white to violet. Calyx elongate, tubular, unevenly 5-lobed at the tip. Corolla funnelform. Stamens barely exsert from the corolla tube. Fruit a many-seeded capsule, either 4-valved or bursting irregularly. All species are toxic.

1 Plants glabrous or nearly so; flowers less than 10 cm long .... **D. stramonium**
1 Plants covered with soft hairs; flowers more than 10 cm long .... **D. inoxia**

Datura inoxia P. Mill.
DOWNY THORN-APPLE   *introduced*
IR | WUP | CUP | EUP   FACU
  Datura wrightii Regel.
Stems to 2 m tall or rarely taller, the upper parts very finely and softly pubescent or almost velutinous, extending along the petioles and veins of the lower leaf-surface. Leaves ovate, rounded to cordate at base, mostly entire. Flowers 12–20 cm long, white. Capsule ovoid, spiny, about 5 cm in diameter. July–Oct.

Native of the southwestern states; occasionally escaped from cultivation in waste places; Gogebic County.

Datura stramonium L.
JIMSONWEED   *introduced*
IR | **WUP** | **CUP** | EUP
Stems stout, hollow, green or purple, divaricately branched, to 1.5 m tall. Leaves thin, ovate, 5–20 cm long, coarsely serrate with a few large triangular teeth, long-petioled. Flowers white or pale violet, 7–10 cm long; calyx about half as long as the corolla. Capsule ovoid, 3–5 cm

Datura stramonium

long, 4-valved at maturity, usually covered with short spines.

Widely distributed weed of fields, barnyards, and waste places.

Leucophysalis *False Ground-Cherry*

Leucophysalis grandiflora (Hook.) Rydb.
LARGE FALSE GROUND-CHERRY   *native*
IR | **WUP** | **CUP** | **EUP**   CC 5
  Chamaesaracha grandiflora (Hook.) Fernald
  Physalis grandiflora Hook.
Annual herb; thinly villous and more or less viscid. Stems to 1 m tall. Leaves ovate or ovate lance-shaped, 5–12 cm long, acute or short-acuminate, entire, at base rounded or broadly cuneate and decurrent along the petiole. Flowers commonly 2–4 from the upper nodes, on pedicels 10–15 mm long; calyx lobes narrowly triangular, acuminate; corolla white with pale yellow center, rotate, 3–4 cm wide; filaments slender; anthers about 3 mm long; fruiting calyx round-ovoid, open at the end, about 15 mm long, nearly filled by the berry; berry 8–14 mm long, ovoid, beaked, puberulent. June–Aug.

Dry sandy soil.

Nicandra *Apple-of-Peru*

Nicandra physalodes (L.) Gaertn.
APPLE-OF-PERU   *introduced*
IR | **WUP** | CUP | EUP
Glabrous annual. Stems branched, to 1.5 m tall. Leaves ovate, 1–2 dm long, coarsely and unevenly toothed, broadly cuneate to the base and decurrent partway down the long petiole. Peduncles about 1 cm long, arising from the stem at the side of a petiole, at maturity recurved at the summit; calyx deeply 5-parted; corolla blue, campanulate, shallowly 5-lobed, 2–2.5 cm long and wide; stamens included; ovary 3–5-celled; fruiting calyx 2–3 cm long, reticulately veined. Berry dry, many-seeded, enclosed by the calyx. July–Sept.

Native of Peru; cultivated for ornament and escaped on roadsides and waste places; Houghton County.

Physalis *Ground-Cherry*

Annual or perennial herbs, commonly widely branching. Leaves alternate or falsely opposite. Flowers solitary or few at the nodes, white, greenish yellow, or yellow, often with a darker center. Calyx at anthesis small, 5-lobed. Calyx tube enlarging promptly after anthesis, at maturity completely enclosing the berry or barely open at the summit, often greatly enlarged, commonly 5-angled, the calyx lobes scarcely enlarged. Corolla rotate to campanulate, shallowly lobed or entire. Stamens inserted near the base of the corolla. Berry many-seeded, pulpy.

ADDITIONAL SPECIES

Physalis grisea (Waterfall) M. Martinez (Strawberry-tomato), native; Houghton and Schoolcraft counties.

1 Colony-forming perennial herbs, spreading by rhizomes; corolla 1–2 cm long 2
1 Taprooted annual herbs; corolla less than 1 cm long (except in *P. philadelphica*)........................ 5
2 Corolla white; distinctly 5-lobed; mature calyx bright red-orange .... **P. alkekengi**
2 Corolla yellow, only slightly lobed; mature calyx green or brown 3
3 Upper stems with soft, spreading hairs **P. heterophylla**
3 Upper stems with short, stiff, appressed hairs or glabrous... 4
4 Leaves sparsely hairy on both sides; calyx tube with stiff hairs to 1.5 mm long................... **P. virginiana**
4 Leaves nearly glabrous, hairs if present mostly along main veins; calyx tube with very short, appressed hairs, the hairs less than 0.5 mm long.......... **P. longifolia**
5 Leaves grayish due to hairs, leaves also with sessile glands.................................... **P. grisea***
5 Leaves greenish, the pubescence less dense; sessile glands absent........................ **P. pubescens**

Physalis alkekengi L.
CHINESE LANTERN PLANT *introduced*
IR | **WUP** | **CUP** | EUP

Alkekengi officinarum Moench

Perennial. Stems usually erect and unbranched, 4–6 dm tall. Leaves ovate, petioled. Flowers white, about 15 mm wide. Fruiting calyx bright red, about 5 cm long.

Native of s and c Europe; often grown for its ornamental fruit and rarely appearing in waste places near gardens.

Physalis heterophylla Nees
CLAMMY GROUND-CHERRY *native*
IR | **WUP** | **CUP** | **EUP** CC 3

Perennial. Stems erect or spreading, often much branched; pubescence of the younger parts, pedicels, and calyx distinctly villous, composed of slender spreading hairs. Leaves ovate to rhombic, 3–8 cm long, shallowly and irregularly sinuate-dentate, varying to entire, at base broadly rounded or subcordate, often inequilateral, not decurrent, more or less pubescent on each side. Pedicels at anthesis about 1 cm long, to 3 cm in fruit; calyx lobes deltoid or ovate; corolla 15–20 mm long. Fruiting calyx ovoid, 3–4 cm long. June–Sept.

In dry or sandy soil, upland woods and prairies; our most abundant *Physalis*.

Physalis longifolia Nutt.
LONGLEAF GROUND-CHERRY *native*
IR | WUP | **CUP** | EUP CC 1

Perennial. Stems erect, 4–8 dm tall, usually divergently branched, the younger parts nearly glabrous to densely puberulent, the hairs always ascending or appressed, rarely more than 0.5 mm long. Leaves thin, ovate to lance-shaped, entire to sinuate-dentate, glabrous or nearly so, long-petioled. Pedicels at anthesis 1–2 cm long; calyx minutely pubescent in longitudinal strips along the nerves; calyx lobes triangular or ovate, 3–4 mm long, densely ciliate; corolla about 15 mm long. Fruiting calyx ovoid or short cylindric, 3–4 cm long. July–Aug.

Moist or dry fields, open woods, and prairies; Dickinson and Menominee counties.

Physalis pubescens L.
DOWNY GROUND-CHERRY *native*
IR | **WUP** | **CUP** | EUP CC –

Annual. In habit, flowers, and fruit very like *P. grisea*, differing chiefly in its pubescence and leaves. Pubescence of the upper part of the stem and leaves shortly villous, the hairs soft and spreading, 0.2–0.5 mm long, not dense enough to interfere with the green color. Leaves uneven in size and shape, usually broadly ovate, 3–6 cm long, commonly abruptly acuminate to an obtuse tip, entire or somewhat sinuate-dentate, but never much toothed below the middle, broadly rounded at base but not decurrent, very often inequilateral; slender-petioled. Fruiting calyx enlarges as the fruit develops, becoming inflated, ribbed, lanternlike 2–4 cm long which contains the berry. May–Sept.

Moist soil. Sometimes cultivated for its fruit, and persisting around gardens; Houghton and Schoolcraft counties.

Physalis virginiana P. Mill.
VIRGINIA GROUND-CHERRY *native*
IR | **WUP** | **CUP** | EUP CC 4

Perennial. Stems usually forked with ascending branches, 3–6 dm tall; pubescence of the younger parts, including the petioles and often the pedicels, of more or less curved hairs. Leaves lance-shaped to ovate, sinuately toothed to entire, narrowed to the base and more or less decurrent on the long petiole, sparsely to abundantly finely hirsute on both sides, margins minutely ciliolate. Pedicels at anthesis 1–2 cm long; calyx with short spreading hairs, the lobes at anthesis 2.5–5 mm long, minutely ciliate; corolla 12–18 mm long. Fruiting calyx 5-angled, notably longer than thick.

Dry or moist fields, sandy upland woods, prairies.

Solanum *Nightshade*

Herbs or vines. Corolla rotate or broadly campanulate,

regular. Stamens 5. Fruit a many-seeded berry. Our species bloom in summer, often continuing into the fall.

ADDITIONAL SPECIES
Solanum physalifolium Rusby (Ground-cherry nightshade), introduced in Menominee County.

1 Plants with spines or prickles 2
1 Plants without spines or prickles 3
2 Corolla bright yellow; leaves 1–2 times pinnately compound; calyx covered with spines . . . . **S. rostratum**
2 Corolla pale violet or white; leaves entire or toothed or lobed; calyx not spiny . . . . **S. carolinense**
3 Climbing or trailing vines, woody at base; flowers usually light blue or violet . . . . **S. dulcamara**
3 Upright herbs; flowers white 4
4 Plants sticky-hairy; calyx enlarging to cover lower half of berry . . . . **S. physalifolium***
4 Plants glabrous to pubescent, the hairs appressed and not gland-tipped; calyx covering only bottom of berry. . . . . **S. ptychanthum**

Solanum carolinense L.
HORSE-NETTLE *introduced*
IR | **WUP** | CUP | EUP   FACU

Perennial. Stems erect, branched, to 1 m tall, spiny and loosely stellate-pubescent. Leaves ovate in outline, commonly 7–12 cm long and about half as wide, with 2–5 large teeth or shallow lobes on each side, more or less spiny along the principal veins, stellate-pubescent on both sides, the hairs sessile with 4–8 branches, the central branch often elongate. Inflorescence several-flowered, elongating at maturity and forming a simple raceme-like cluster; corolla pale violet to white, about 2 cm wide. Berry yellow, 1–1.5 cm wide, subtended but not enclosed by the calyx.

Fields and waste places, especially in sandy soil; Houghton County.

Solanum dulcamara L.
CLIMBING NIGHTSHADE *introduced*
IR | **WUP** | CUP | EUP   FAC

Perennial vine, climbing 2–4 m high, the stems somewhat woody at base. Leaves either simple or deeply lobed, the simple ones and the terminal segment of the lobed ones ovate, 4–10 cm long, entire, rounded or subcordate at base; lobes, when present, 1 or 2, basal, divergent, lance-shaped or ovate, much smaller than the terminal segment. Inflorescences arising from the internodes or opposite the leaves, peduncled, loosely branched, the pedicels jointed at base; corolla pale violet or blue, varying to white, about 1 cm wide. Berry red.

Native of Eurasia; moist thickets.

Solanum ptychanthum Dunal
WEST INDIAN NIGHTSHADE *introduced*
IR | **WUP** | **CUP** | **EUP**   FACU
 Solanum nigrum L.

Annual. Stems erect, 3–6 dm tall, often widely branched. Leaves thin, long-petioled, ovate lance-shaped or ovate, cuneately narrowed to the base. Inflorescences lateral from the internodes, peduncled, umbel-like, 2–10-flowered, the pedicels not jointed at base; corolla white or very pale violet, 5–9 mm wide. Berries globose.

Both native and Eurasian forms present; a cosmopolitan weed.

Solanum rostratum Dunal
BUFFALO-BUR *introduced*
IR | **WUP** | CUP | **EUP**

Annual. Stems widely branched, to 6 dm tall, spiny and also stellate-pubescent. Leaves ovate or oblong in outline, usually spiny along the principal veins, deeply pinnately lobed, or the segments again lobed in the larger leaves, stellate-pubescent on both sides. Axis of the lateral inflorescences soon elongating and becoming raceme-like; calyx tube spiny at anthesis, later expanding, becoming 3 cm wide (including the long spines), completely enclosing the berry. Flowers yellow, 2–3 cm wide; anthers yellow, one of them much longer than the other 4.

Dry prairies.

Thymelaeaceae
MEZEREUM FAMILY

Shrubs, with showy flowers developing before the leaves in spring.

1 Flowers pink or occasionally white; stamens included in the perianth tube; fruits bright red or occasionally yellow (in white-flowered plants); buds glabrous . . . . **Daphne mezereum**
1 Flowers yellow; stamens exserted, ripe fruits yellowish-green; buds pubescent . . . . **Dirca palustris**

Daphne *Daphne*
Daphne mezereum L.
DAPHNE *introduced*
IR | **WUP** | CUP | EUP   FACU

Small shrub. Flowers fragrant, floral tube ends in 4 spreading petal-like lobes. Flowers sessile on the stems and bloom in early spring before the leaves expand.

European native, sometimes cultivated and escaping to nearby disturbed forests; Gogebic, Iron, and Ontonagon counties.

332 ULMACEAE | Elm Family

Dirca *Leatherwood*

Dirca palustris L.
EASTERN LEATHERWOOD — *native*
IR | **WUP** | **CUP** | **EUP** — FAC CC 8

Freely branched shrub 1–2 m tall, bark very tough and pliable, twigs jointed. Leaves alternate, entire, obovate, 5–8 cm long, 1/2 to 2/3 as wide, usually rounded at base, glabrous at maturity, on petioles 2–5 mm long. Flowers perfect, regular, pale yellow, 7–10 mm long, subtended by hairy bud scales in early spring before the leaves appear, in lateral clusters of 2–4; hypanthium narrowly funnelform, the limb slightly spreading; stamens 8, protruding about 3 mm from the top of the tube; sepals minute; petals none. Fruit an ellipsoid drupe, about 8 mm long. Spring.

Rich, moist deciduous woods.

Ulmaceae
ELM FAMILY

Ulmus *Elm*

Trees. Leaves alternate, simple, short-petioled or subsessile, inequilateral; margins usually doubly serrate. Flowers perfect, in short racemes or, by abbreviation of the axis, in fascicles. Calyx campanulate, 4–9-lobed. Corolla none. Stamens as many as the calyx segments, exsert. Ovary compressed, 1-celled. Fruit a flat, 1-seeded samara, usually short-stipitate and often surmounted by the persistent or enlarged styles. Several exotic elm species are sometimes planted as street trees. Hackberry (*Celtis*), a former member of this family, now included in Cannabaceae.

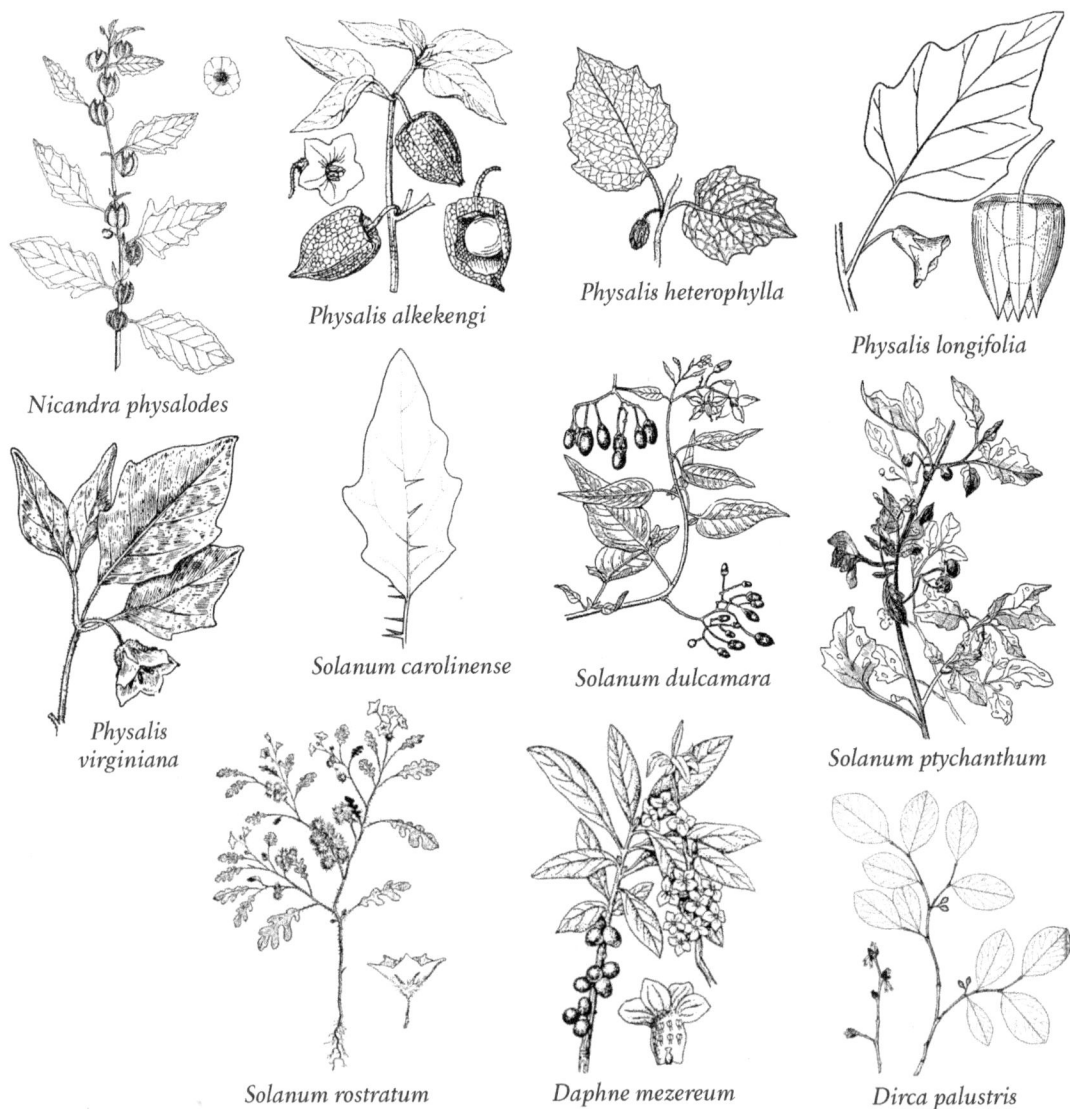

Physalis alkekengi

Physalis heterophylla

Physalis longifolia

Nicandra physalodes

Solanum carolinense

Solanum dulcamara

Physalis virginiana

Solanum ptychanthum

Solanum rostratum

Daphne mezereum

Dirca palustris

ADDITIONAL SPECIES

Ulmus glabra Huds. (Scotch elm), introduced European tree; leaves very wide, obovate and abruptly acuminate at the tip, strongly asymmetrical at base; samaras very large and completely glabrous; Chippewa County.

1   Leaf blades small, 3–7 cm long . . . . **U. pumila**
1   Leaf blades larger, 7–18 cm long 2
2   Leaves smooth on each side; branches with corky wing-like ridges, the lowermost branches short and strongly drooping; main trunk usually not dividing into several large limbs. . . . . **U. thomasii**
2   Leaves roughened on one or both sides; branches not with corky wings, the lowermost branches longer and not strongly drooping; main trunk usually dividing into several large limbs, tree vase-shaped in outline 3
3   Leaves usually rough only on upperside; bark gray, deeply fissured . . . . **U. americana**
3   Leaves usually rough on both sides; bark dark red-brown, shallowly fissured. . . . . **U. rubra**

Ulmus americana L.
AMERICAN ELM   *native*
IR | **WUP** | **CUP** | **EUP**   FACW CC 1

Tree to 25 m tall, trunk to 1 m wide, crown broadly rounded or flat-topped, smaller branches usually drooping; bark gray, furrowed, breaking into thin plates with age; twigs brown, smooth or with sparse hairs, often zigzagged; buds red-brown. Leaves to 15 cm long and 7–8 cm wide, oval, pointed at tip, base strongly asymmetrical, upper surface dark green and smooth, lower surface pale and smooth or soft-hairy; margins coarsely double-toothed; petioles short, usually yellow. Flowers small, green-red, hairy, in drooping clusters of 3–4; appearing before leaves unfold in spring. Samaras 1-seeded, oval, 1 cm wide, with a winged, hairy margin, notched at tip.

Floodplain forests, streambanks and moist, rich woods; less common now than formerly due to losses from Dutch elm disease.

Ulmus pumila L.
SIBERIAN ELM   *introduced (invasive)*
IR | WUP | CUP | **EUP**   FACU

Tree 15 to 30 m; crowns open; bark gray to brown, deeply furrowed with interlacing ridges; branches not winged; twigs gray-brown, pubescent; buds dark brown, ovoid, glabrous; scales light brown, shiny, glabrous to slightly pubescent. Leaves narrowly elliptic to lanceolate, to 6.5 cm long and 2–3.5 cm wide, base generally not oblique, margins singly serrate; upper surface with some pubescence in axils of veins, lower surface glabrous; petioles 2–4 mm long. Inflorescences tightly clustered fascicles of 6–15 sessile flowers; flowers and fruits not pendulous, sessile; calyx shallowly 4–5-lobed, glabrous; stamens 4–8; anthers brownish red; stigmas green. Samaras yellow-cream, orbiculate, 10–14 mm wide., broadly winged, glabrous, tip notched 1/3-1/2 its length.

Escaping from cultivation to waste places, roadsides, fencerows; Mackinac County.

Distinguished from our other elms by its singly serrate leaf margins.

Ulmus rubra Muhl.
SLIPPERY ELM   *native*
IR | **WUP** | CUP | EUP   FAC CC 2

Tree, to 20 m tall; twigs scabrously pubescent; winter-buds densely covered with red-brown hairs. Leaves oblong to obovate, thick and stiff, usually 10–20 cm long, very rough above. Flowers fascicled, short-pediceled to nearly sessile. Samaras nearly circular, 1.5–2 cm long, entire, the sides smooth on the wing, pubescent over the seed, scarcely reticulate.

Moist woods; Ontonagon County.

The inner bark is very mucilaginous and has some medicinal value.

Ulmus thomasii Sarg.
ROCK ELM   *native*
IR | **WUP** | **CUP** | EUP   FAC CC 4

Tree, to 30 m tall, with thinly pubescent twigs often becoming irregularly winged with 2 or more plates of cork after their second year; winter-buds thinly

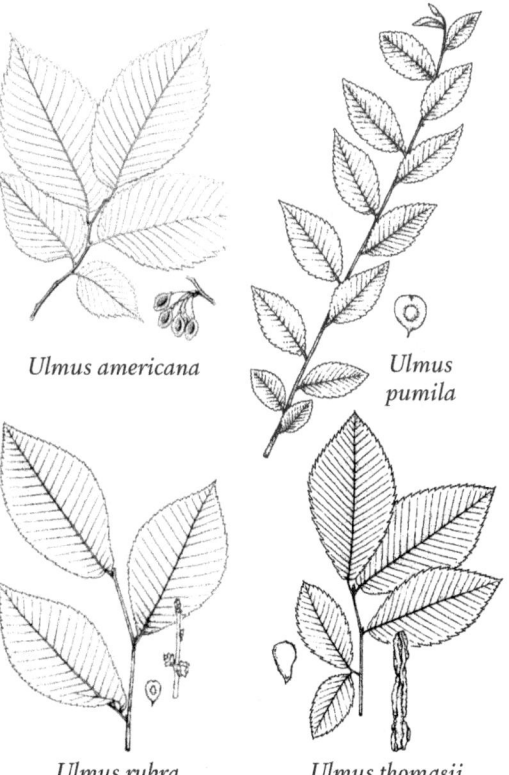

Ulmus americana

Ulmus pumila

Ulmus rubra

Ulmus thomasii

pubescent. Leaves oblong to obovate, 6–12 cm long, distinctly cordate on one side at base, glabrous above. Flowers in slender racemes to 4 cm long. Samaras elliptic, 1.5–2 cm long, ciliate, pubescent on the sides, the petiole about 1 mm long.

Rich upland woods; Iron, Menominee, and Ontonagon counties.

The thick, corky-winged branches are distinctive.

Urticaceae

NETTLE FAMILY

Annual or perennial herbs with watery juice, sometimes with stinging hairs. Leaves alternate or opposite, simple, with petioles. Flowers small, green, in simple or branched clusters from leaf axils, staminate and pistillate flowers usually separate, on same or separate plants; sepals joined, 3–5-lobed; petals absent; ovary superior, 1-chambered. Fruit an achene, often enclosed by the sepals which enlarge after flowering.

1	Leaves alternate.	2
1	Leaves opposite.	3
2	Plants large and coarse, with stiff stinging hairs, the leaves sharply toothed	**Laportea canadensis**
2	Plants smaller, without stinging hairs; leaves entire	**Parietaria pensylvanica**
3	Plants with stinging hairs; leaves lance-shaped	**Urtica dioica**
3	Plants without stinging hairs; leaves ovate	4
4	Stems translucent and fleshy; flowers in dense short clusters from leaf axils; achene equal or longer than sepals	**Pilea**
4	Stems neither translucent nor fleshy; flowers in cylindric spikes from leaf axils; achene shorter than and hidden by the sepals	**Boehmeria cylindrica**

Boehmeria *False Nettle*

Boehmeria cylindrica (L.) Sw.
SMALL-SPIKE FALSE NETTLE native
IR | WUP | **CUP** | EUP OBL CC 5

Perennial, nettle-like herb, stinging hairs absent. Stems upright, 4–10 dm long, usually unbranched. Leaves opposite, rough-textured, ovate to broadly lance-shaped, narrowed to a pointed tip, with 3 main veins; margins coarsely toothed; petioles shorter than blades. Flowers tiny, green, staminate and pistillate flowers usually on separate plants, in small clusters along unbranched stalks from upper leaf axils,

Boehmeria cylindrica

forming cylindric, interrupted spikes of staminate flowers or continuous spikes of pistillate flowers. Fruit an achene, enclosed by the enlarged bristly sepals and petals, ovate and narrowly winged. July–Aug.

Floodplain forests, swamps, marshes and bogs; Menominee County.

Laportea *Wood-Nettle*

Laportea canadensis (L.) Weddell
CANADIAN WOOD-NETTLE native
IR | **WUP** | **CUP** | **EUP** FACW CC 4

Perennial herb, spreading by rhizomes. Stems somewhat zigzagged, 5–10 dm long. Leaves alternate, 8–15 cm long, ovate and narrowed to a tip, with small stinging hairs, margins coarsely toothed. Flowers small, green, staminate and pistillate flowers separate but borne on same plant; staminate flowers in branched clusters from lower leaf axils, shorter than leaf petioles; pistillate flowers in open, spreading clusters from upper axils, usually much longer than petioles. Fruit a flattened achene, longer than the 2 persistent sepals. July–Sept.

Floodplain forests, rich moist woods, low places in hardwood forests, streambanks.

Differs from stinging nettle (*Urtica dioica*) by its shorter size, broader, alternate leaves, and the longer spike-like heads arising from upper leaf axils.

Parietaria *Pellitory*

Parietaria pensylvanica Muhl.
PENNSYLVANIA PELLITORY native
IR | **WUP** | **CUP** | EUP FACU CC 2

Annual pubescent herb; monoecious or polygamous. Stems erect, 1–4 dm tall, simple or rarely branched. Leaves alternate, entire, thin, lance-shaped, 3–8 cm long, 3-nerved from above the cuneate base, slightly scabrellate above, with a slender petiole. Flowers green, from the middle and upper axils, subtended and exceeded by narrow green bracts about 5 mm long; staminate flowers with deeply 4-parted calyx and 4 stamens; calyx of the pistillate flowers tubular at base, 4-lobed. Fruit a smooth, shining achene about 1 mm long, loosely enclosed by the expanded calyx. June–Sept.

Moist to dry forests, gravelly shores, disturbed sites; Delta and Keweenaw counties.

Pilea *Clearweed*

Annual herbs, sometimes forming colonies from seeds of previous year. Stems erect to sprawling, smooth, translucent and watery. Leaves opposite, stinging hairs absent, thin and translucent, ovate, with 3 major veins from base of leaf, margins toothed. Flowers green, staminate and pistillate flowers separate, borne on same or different plants, in clusters from leaf axils; staminate flowers with 4 sepals and 4 stamens; pistillate flowers with 3 sepals, ovary superior. Fruit a flattened, ovate achene.

1 Achenes 1–1.5 mm wide, olive-green to dark purple with a narrow pale margin, covered with low bumps . . . . **P. fontana**
1 Achenes to 1 mm wide, green to yellow, often marked with purple spots, smooth . . . . **P. pumila**

Pilea fontana (Lunell) Rydb.
LESSER CLEARWEED native
IR | WUP | **CUP** | EUP   FACW CC 5

Annual herb. Stems 1–4 dm long, often sprawling. Leaves opposite, 2–6 cm long and 1–4 cm wide; petioles 0.5–5 cm long. Flowers in clusters, staminate flowers usually innermost when mixed with pistillate flowers. Fruit a dark olive-green to purple achene, 1–1.5 mm wide, with a narrow pale margin; sepals persistent, shorter to slightly longer than achene. Aug–Sept.

Lakeshores, riverbanks, swamps, marshes and springs; Dickinson and Menominee counties.

Pilea pumila (L.) Gray
CANADIAN CLEARWEED native
IR | **WUP** | **CUP** | EUP   FACW CC 5

Annual herb, similar to *P. fontana*, but sometimes taller (to 5 dm). Leaves opposite, usually larger (to 12 cm long and 8 cm wide), thinner and more translucent than in *P. fontana*; petioles to 8 cm long. July–Sept.

Swampy woods (often on logs), wooded streambanks, floodplain forests, wet depressions, rocky hollows; usually in partial shade.

Urtica *Stinging Nettle*

Urtica dioica L.
STINGING NETTLE native
IR | **WUP** | **CUP** | **EUP**   FAC CC 1

Stout perennial herb, often forming dense patches from spreading rhizomes. Stems 8–20 dm tall, usually unbranched, with stinging hairs on stems and leaves, the hairs irritating to skin. Leaves opposite, ovate to lance-shaped, 5–15 cm long and 2–8 cm wide; margins coarsely toothed; petioles 1–6 cm long; stipules lance-shaped, to 15 mm long. Flowers small, green, staminate and pistillate flowers separate but mostly on same plants; flower clusters branched and spreading from leaf axils, the clusters usually longer than petioles, all of one sex or a mix of staminate and pistillate flowers, the pistillate clusters usually above the staminate clusters when both present on a plant. Fruit an ovate achene, 1–2 mm long, enclosed by the inner pair of sepals. July–Sept.

Moist woods, thickets, ditches, streambanks and disturbed areas.

Verbenaceae
VERBENA FAMILY

Perennial herbs with 4-angled, erect or prostrate stems. Leaves opposite, toothed. Flowers small, numerous, perfect (with both staminate and pistillate parts), in branched or unbranched spikes or heads at ends of stems or from upper leaf axils, the spikes elongating as flowers open upward from the base. Calyx 5-toothed; corolla 5-lobed, somewhat 2-lipped; stamens 4, of 2 lengths. Fruit dry, enclosed by the sepals, splitting lengthwise into 2 or 4 nutlets when mature.

Verbena *Vervain*

Annual or perennial herbs. Leaves usually opposite, simple, entire to somewhat lobed. Inflorescence of terminal spikes, usually densely many-flowered, often flat-topped, sometimes elongate with scattered flowers. Flowers solitary in the axil of a usually narrow bractlet. Calyx usually tubular, 5-angled, 5-ribbed, unequally 5-toothed, not at all or but slightly changed in fruit. Corolla funnelform, the limb weakly 2-lipped, 5-lobed. Stamens 4, inserted in the upper half of the corolla tube, included. Ovary mostly 4-lobed. Fruit mostly enclosed by the mature calyx, separating at maturity into 4 linear-oblong nutlets.

1 Plants spreading; leaves incised and often somewhat 3-lobed . . . . **V. bracteata**
1 Plants erect; leaves unlobed or lobed only near base . 2
2 Leaves narrowly lance-shaped, less than 1.5 cm wide, tapered at base to an indistinct petiole; plants glabrous or with scattered appressed hairs . . . . **V. simplex**
2 Leaves mostly ovate, 2 cm or more wide, with petioles or sessile; plants usually with at least some hairs 3
3 Plants densely gray-hairy; leaves sessile . . . . . . **V. stricta**
3 Plants not densely hairy; leaves with petioles 4
4 Flowers blue to purple, densely overlapping on spike; leaves often lobed at their base . . . . **V. hastata**
4 Flowers white, not overlapping on spike; leaves not lobed . . . . **V. urticifolia**

Verbena bracteata Lag. & Rodr.
CARPET VERVAIN native
IR | **WUP** | **CUP** | **EUP**   FACU CC -

Annual or perennial. Stems usually several from a common base, diffusely branched, decumbent or ascending, rarely erect, coarsely hirsute. Leaves 1–6.5 cm long, pinnately incised or usually 3-lobed, narrowed at base into the short margined petiole,

336 VERBENACEAE | Verbena Family
dicots

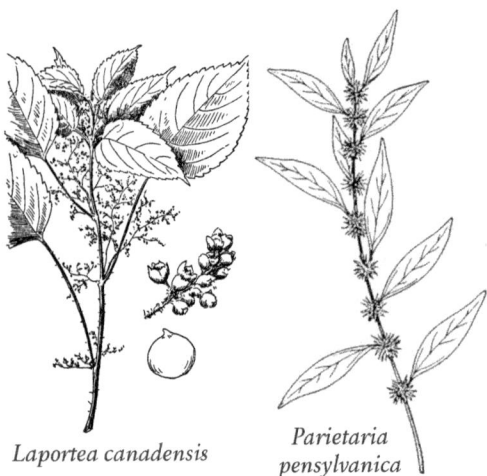
Laportea canadensis

Parietaria pensylvanica

Pilea pumila

Urtica dioica

hirsute on both surfaces, the midrib and large veins slightly prominent beneath, the lateral lobes narrow and divaricate, the middle lobe large, obovate, incised or cleft. Flowers in terminal, sessile spikes usually 10–15 mm wide, conspicuously bracteose, harshly hispid-hirsute throughout; bractlets much longer than the calyx, conspicuous, 8–15 mm long, recurved in age, coarsely hirsute, the lowermost often incised and leaf-like, the upper linear lance-shaped, entire; calyx 3–4 mm long, hirsute particularly along the nerves, its lobes very short; corolla bluish, violet, lavender, lilac, or purple, its tube protruding slightly beyond the calyx, finely pubescent outside the throat, the limb 2–3 mm wide. Nutlets linear, 2–2.5 mm long, sharply raised-reticulate above, striate below. April–Oct.

Prairies, fields, roadsides, and waste places.

Verbena hastata L.
COMMON VERVAIN native
IR | **WUP** | **CUP** | **EUP**   FACW CC 4

Perennial herb; plants with short, rough hairs. Stems stout, erect, 4–12 dm tall, 4-angled, sometimes branched above. Leaves opposite, lance-shaped to oblong lance-shaped, 4–12 cm long and 1–5 cm wide; margins with coarse, forward-pointing teeth and sometimes lobed near base; petioles short. Flowers small, numerous, slightly irregular, in long, narrow spikes 5–15 cm long at ends of stems, the spikes elongating as flowers open upward from base; calyx unequally 5-toothed, 1–3 mm long; corolla dark blue to purple, 5-lobed, trumpet-shaped, slightly 2-lipped, 2–4 mm wide. Fruit 4-angled, splitting into 4 nutlets. July–Sept.

Marshes, wet meadows, shores, streambanks, openings in swamps, ditches.

Verbena simplex Lehm.
NARROW-LEAF VERVAIN native
IR | WUP | CUP | **EUP**   CC 6

Perennial. Stems chiefly erect, 1–5 dm tall, simple or sparingly branched above, the branches ascending, usually sparsely strigose. Leaves linear, narrowly oblong, or lance-shaped, 3–10 cm long, tapering into a subsessile base, distantly or coarsely serrate, reticulately rugose above, somewhat prominently veined beneath, glabrate or sparsely strigose on both surfaces. Flowers in slender spikes, solitary at the apex of the stem and branches, usually somewhat crowded; bractlets lance-shaped, commonly shorter than the calyx, glabrous or nearly so; mature calyx 4–5 mm long, sparsely pubescent; corolla deep lavender or purple, its tube scarcely longer than the calyx, with scattered hairs at the mouth, its limb 5–6 mm wide. Nutlets linear, 2.5–3 mm long, raised-reticulate above, striate toward base. June–Aug.

Dry soil of woods, fields, rocky places, and roadsides; Chippewa County.

Verbena stricta Vent.
HOARY VERVAIN native
IR | WUP | **CUP** | EUP   CC –

Perennial. Stems 2–12 dm tall, subterete, simple or branched above, rather densely pale-pubescent or hirsute. Leaves ovate, elliptic, or suborbicular, 3–10 cm long, sessile or nearly so, thick-textured, sharply serrate, or incised, hirsute and rugose above, densely hirsute and prominently veined beneath. Flowers in 1 or several spikes, usually quite densely compact at anthesis and in fruit; bractlets lance-shaped, about as long as the calyx, hirsute, ciliate; calyx 4–5 mm long, densely hirsute; corolla deep blue or purple, or white, its tube protruding slightly beyond the calyx, pubescent outside, its limb 8–9 mm wide. Nutlets ellipsoid, about 2.5 mm long, raised-reticulate above, striate below. June–Sept.

Sandy prairies, barrens, fields, and roadsides.

Verbena urticifolia L.
WHITE VERVAIN *native*
IR | **WUP** | **CUP** | EUP FAC CC 4

Annual or perennial. Stems erect, 4–15 dm tall, solitary, simple or more often branching from near the base, finely hirsute or almost glabrous. Leaves broadly lance-shaped to obovate, petiolate, 8–20 cm long, rounded at base and decurrent into the petiole, coarsely and somewhat doubly crenate-serrate, often minutely pustulate above, hirsute on both surfaces, the hairs whitish, 1–1.3 mm long, or sometimes glabrous. Flowers in slender, usually stiffly ascending spikes, more or less sparsely flowered and remotely fruited; bractlets ovate, 1–1.5 mm long, ciliate; mature calyx 2–2.3 mm long, pubescent (especially along the nerves), the teeth short, subequal, subulate; corolla white, its tube scarcely exserted, its limb 2–3 mm wide, the lobes obtuse. Nutlets ellipsoid, about 2 mm long, corrugated or ribbed on the back. June–Oct.

Thickets, meadows, waste places.

Violaceae
VIOLET FAMILY

Viola *Violet*

Perennial herbs, with or without leafy stems. Leaves simple, all at base of plant or alternate on stems; petioles with membranous stipules. Flowers perfect, nodding, and single at ends of stems, 5-merous, with 5 unequal sepals, 2 upper petals, 2 lateral, bearded petals, and 1 lower petal prolonged into a nectar-holding spur at its base. Fruit an ovate capsule which splits to eject the numerous seeds.

ADDITIONAL SPECIES

Viola palmata L. (Wood violet), native perennial; UP plants found with *V. pedatifida* on a limestone bluff in Delta County.

1 Plants with stems; leaves and flowers borne on the upright stems 2
1 Plants without stems; leaves and flowers borne directly from rootstock 8
2 Corolla solid yellow, or white with a yellow center; stipules entire or jagged-tooth on margins 3
2 Corolla creamy-white to yellow-orange, or lavender to blue, with or without a yellow center; stipules fringed or deeply lobed 4
3 Corolla yellow; stipules ovate, widened above base before tapering to tip . . . . **V. pubescens**
3 Corolla white with yellow center; stipules long-tapered from base to tip . . . . **V. canadensis**
4 Stipules deeply lobed near base into long oblong segments 5
4 Stipules fringed with short, slender segments 6
5 Petals shorter than or about equal to the sepals; the 5 petals cream-colored on upper half, all about same length . . . . **V. arvensis**
5 Petals longer than the sepals; the upper pair of petals dark blue or purple on upper half, and longer than the other 3 petals . . . . **V. tricolor**
6 Leaves narrowly ovate to triangular, tapered to a rounded tip, often densely pubescent, the hairs tiny; margins entire or nearly so; corolla dark blue . . . . **V. adunca**

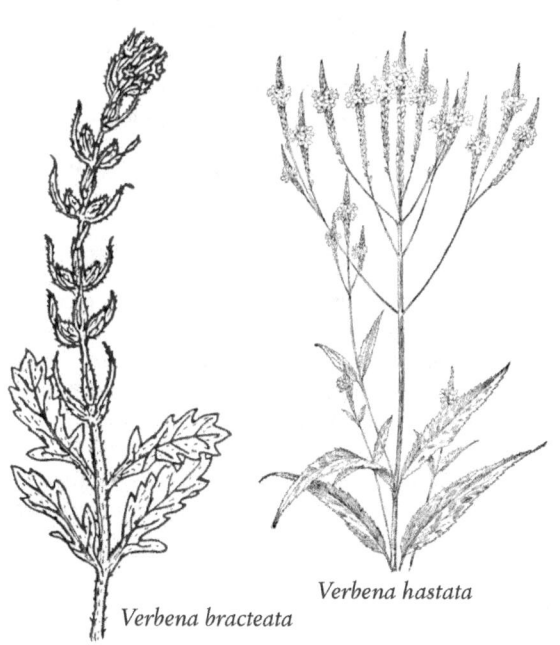

Verbena bracteata

Verbena hastata

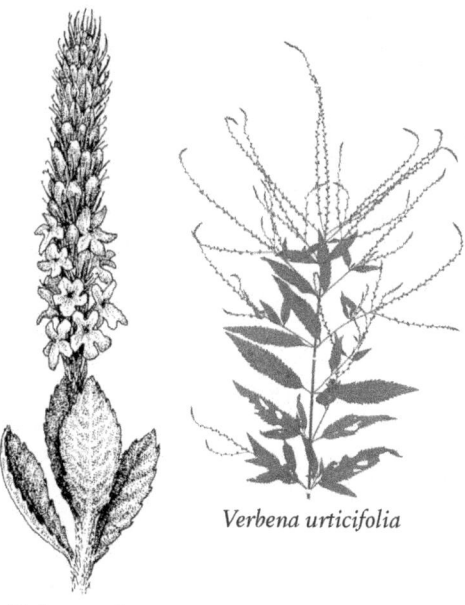

Verbena stricta

Verbena urticifolia

6 Leaves ovate to kidney-shaped, glabrous or only slightly hairy; the margins with rounded or sharp teeth; corolla creamy-white, lavender or light blue 7

7 Corolla solid light blue; lateral petals bearded on inner surface; stem leaves broadly ovate or kidney-shaped; margins with low, rounded teeth....... **V. labradorica**

7 Corolla light blue to lavender with a darker purple spot; lateral petals not bearded; stem leaves ovate to oblong-ovate; margins with scattered sharp teeth . **V. rostrata**

8 Style tipped by a slender, recurved hook; stolons green and cord-like; introduced species mostly of lawns and parks..................................... **V. odorata**

8 Style slightly expanded into a spatula-like tip, not hooked; stolons slender and pale or absent; native... 9

9 Corolla blue; petals not bearded within; spur more than 2 times longer than wide.................. **V. selkirkii**

9 Corolla white or purple; petals sometimes bearded within; spur less than 2 times longer than wide..... 10

10 Corolla white.. 11

10 Corolla purple; lateral petals bearded within; stolons absent.. 15

11 Leaf blades more than 1.5 times longer than wide ... 12

11 Leaf blades often wider than long 13

12 Leaves lance-shaped, tapered to a narrow base....... .... **V. lanceolata**

12 Leaves broader, ovate and narrowly heart-shaped at base.................................. **V. primulifolia**

13 Leaves dull, upper and lower surface without hairs, lower surface not paler than upper; margins nearly entire or with low rounded teeth; petioles often with long, soft hairs .... **V. macloskeyi**

13 Leaves shiny and smooth on upper surface, or dull and hairy on either upper or lower surface; underside paler than upper surface; margins with sharp teeth 14

14 Plants with stolons and horizontal rhizomes; upper and lower surface of leaves sparsely to densely hairy with short hairs less than 1 mm long............. **V. blanda**

14 Plants without stolons, rhizomes turned upright; leaves often shiny and smooth on upper surface, or densely hairy on upperside with hairs about 1–2 mm long and smooth on underside.................... **V. renifolia**

15 Leaf blades lobed or divided 16

15 Leaf blades toothed on margin, not lobed or divided 18

16 Leaf blades deeply divided into slender linear segments .... **V. pedatifida**

16 Leaf blades less deeply lobed or divided, the segments triangular to ovate................................ 17

17 Leaf blades much longer than wide; the spurred petal densely bearded within .... **V. sagittata**

17 Leaf blades about as long as wide; the spurred petal glabrous within or with only a few hairs..... **V. pedata**

18 Leaf blades distinctly longer than wide 19

18 Leaf blades as wide as or wider than long 22

19 Plants of wet places; leaves glabrous or nearly so; sepal margins not fringed with hairs 20

19 Plants of dry or rocky habitats; leaves sparsely to densely hairy; sepal margins usually fringed with hairs 21

20 Lateral petals with long, threadlike hairs on inner surface; spurred petal densely hairy within ..... **V. affinis**

20 Lateral petals with short, knob-tipped hairs on inner surface; spurred petal without hairs...... **V. cucullata**

21 Hairs on leaves over 1 mm long; sepals ovate, rounded at tip............................. **V. novae-angliae**

21 Hairs on leaves shorter, less than 1 mm long; sepals long-tapered to a sharp tip .... **V. sagittata**

22 Sepals long-tapered to a sharp tip; lateral petals with short, knob-tipped hairs on inner surface; spurred petal without hairs .... **V. cucullata**

22 Sepals oblong to broadly lance-shaped, rounded at tip; lateral petals with long, threadlike hairs on inner surface 23

23 Flowers held above the leaves; leaves and stems without hairs, leaves rounded at tip, margins with rounded teeth; the spurred petal densely hairy within; plants of wetlands .... **V. nephrophylla**

23 Flowers overtopped by leaves; leaves and stems usually hairy, leaves tapered to a pointed tip, margins with sharp, forward-pointing teeth; spurred petal glabrous to only slightly hairy within; plants of moist forests .... **V. sororia**

Viola adunca Sm.
HOOK-SPURRED VIOLET *native*
IR | **WUP** | **CUP** | **EUP** FACU CC 4

Perennial tufted herb. Stems several to many, 2–8 cm tall at anthesis, at first erect, later becoming prostrate and spreading and to 15 cm long. Leaves ovate to suborbicular, obtuse, crenulate, subcordate at base; stipules linear lance-shaped with fimbriate teeth. Peduncles long and slender; sepals narrowly lance-shaped; petals violet; spur 4–6 mm long, either straight and blunt or tapering to a sharp incurved point. Capsules 4–5 mm long, ellipsoid; seeds dark brown.

Dry sandy open places, often with jack pine and oaks; crevices in rock outcrops.

Viola affinis Le Conte
SAND VIOLET *native*
IR | **WUP** | **CUP** | **EUP** FACW CC 2

Perennial herb, spreading by rhizomes. Leaves all from base of plant, hairless, narrowly heart-shaped; margins with rounded teeth. Flowers violet, bearded within with long, threadlike hairs, atop stalks slightly longer than leaves. Fruit a purple-flecked capsule on horizontal or arching stalks, seeds dark. April–May.

Swamps, floodplain forests, streambanks and lakeshores, low prairie.

VIOLACEAE | Violet Family

Viola arvensis Murr.
EUROPEAN FIELD-PANSY *introduced*
IR | **WUP** | **CUP** | **EUP**

Pubescent to glabrate annual. Stems often branched from the base, to 30 cm tall, with reflexed hairs on the angles. Leaves variable, the lowest ones with orbicular to ovate blades, the upper oblong to elliptic or narrowly elliptic, blunt or acutish, all distantly crenate. Flowers 1–1.5 cm long, 0.8 cm wide; petals shorter than the broadly lance-shaped sepals, all pale yellow or occasionally with purplish-tinged tips. Capsules globose, 5–7 mm long; seeds brown.

Native of Europe; mostly in cultivated or abandoned fields.

Viola blanda Willd.
SWEET WHITE VIOLET *native*
IR | **WUP** | **CUP** | **EUP** FACW CC 5

Perennial herb, spreading by short rhizomes (and stolons later in season). Stems smooth. Leaves all from base of plant, heart-shaped, dark green and satiny, 2–5 cm wide, upper surface near base of blade usually with short, stiff white hairs; petioles usually red. Flowers white, fragrant, on stalks shorter than longer than leaves; lower 3 petals with purple veins near base, all more or less beardless; upper 2 petals narrow, twisted backward, 2 side petals forward-pointing. Fruit a purple capsule 4–6 mm long, seeds dark brown. April–May.

Hummocks in swamps and bogs, low wet areas in deciduous and conifer forests.

Viola canadensis L.
TALL WHITE VIOLET *native*
IR | **WUP** | **CUP** | **EUP** FACU CC 5

Perennial herb, arising from a short, woody rhizome; plants glabrous or minutely pubescent. Stems numerous, 2–4 dm tall, with several long-petioled basal leaves. Stem leaves numerous, the lower widely spaced, the upper crowded toward the apex, the blades cordate, 5–10 cm long; upper leaves becoming shorter-petioled and with truncate or broadly cuneate base. Stipules lance-shaped, slightly sscarious; peduncles slende. Sepals lance-shaped, ciliate; petals white inside, with yellowish eye-spot and brown-purple veins near the base, purplish-tinged outside. Capsules globose-ellipsoid, 5–7 mm long; seeds brown.

Mesic deciduous woods.

Viola cucullata Ait.
MARSH BLUE VIOLET *native*
IR | **WUP** | **CUP** | **EUP** OBL CC 5

Perennial herb, spreading by short, branched rhizomes; plants smooth. Leaves all from base of plant, ovate to kidney-shaped, to 10 cm wide, heart-shaped at base; margins coarsely toothed; blade angled from the upright petioles. Flowers light purple or white, dark at center, on slender stalks longer than leaves; the 2 side petals densely bearded with short hairs, the hairs mostly knobbed or club-tipped. Fruit a cylinder-shaped capsule, seeds dark. April–June.

Swamps, sedge meadows, shady seeps; occasionally in bogs and low areas in forests.

Viola labradorica Schrank
ALPINE VIOLET *native*
IR | **WUP** | **CUP** | **EUP** FAC CC 3

Viola adunca Sm. var. *minor* (Hook.) Fern.
Perennial herb; plants smooth. Leaves in clumps from rhizomes, at first all from base of plants, later with leafy, horizontal stems to 15 cm long; light green, ovate to kidney-shaped, 1–2.5 cm wide; margins with rounded teeth; petioles 2–6 cm long. Flowers pale blue, side petals bearded on inner surface. Fruit 4–5 mm long, seeds dark brown. April–June. Swamps, streambanks, moist hardwood forests.

Viola lanceolata L.
STRAP-LEAF VIOLET *native*
IR | **WUP** | **CUP** | **EUP** OBL CC 8

Perennial herb, spreading by rhizomes and stolons. Leaves from base of plant, narrowly lance-shaped, more than 2x longer than wide, tapered to base; margins toothed. Flowers white, all beardless; lower 3 petals purple-veined near base. Fruit a green capsule 5–8 mm long, seeds brown. April–June.

Open bogs, sedge meadows; soils sandy or mucky. The lance-shaped leaves are distinctive.

Viola macloskeyi Lloyd
WILD WHITE VIOLET *native*
IR | **WUP** | **CUP** | **EUP** OBL CC 6

Viola pallens (Banks) Brainerd
Small perennial herb (our smallest violet), spreading by rhizomes and stolons. Leaves all from base of plant, heart-shaped to kidney-shaped, 1–3 cm wide at flowering, later to 8 cm wide, underside orange-tinged; margins with rounded teeth. Flowers white, on upright stalks equal or longer than leaves, 3 lower petals purple-veined near base, 2 side petals beardless or with sparse hairs. Fruit a green capsule 4–6 mm long, seeds olive-black. April–July.

Marshes, sedge meadows, open bogs and swamps, alder thickets; sometimes in shallow water.

Viola nephrophylla Greene
NORTHERN BOG VIOLET *native*
IR | **WUP** | **CUP** | **EUP** FACW CC 8

Low perennial herb, spreading by short rhizomes. Leaves all from base of plant, smooth, heart-shaped to kidney-shaped, 1–4 cm long and 2–6 cm wide, rounded at tip; margins with rounded teeth; petioles slender, 2–16 cm long. Flowers single, nodding on

340 VIOLACEAE | Violet Family

slender stalks, the stalks longer than leaves. Flowers violet, bearded near base on inside, or upper pair of petals not bearded. Fruit a capsule 5–10 mm long. May–sometimes again flowering in Aug or Sept.

Wet meadows, calcareous fens, low areas between dunes, streambanks, rocky shores.

Viola novae-angliae House
NEW ENGLAND BLUE VIOLET (THR) *native*
IR | **WUP** | **CUP** | EUP OBL CC 10
Viola sororia var. *novae-angliae* (House) McKinney

Perennial herb. Leaves ovate, longer than wide, cordate, crenate-serrate near the base, distantly so toward the acuminate apex; petioles and lower surface of the leaves villous or pubescent. Flowers violet-purple, the three lower petals villous at the base; sepals obtuse, glabrous; cleistogamous flowers on long ascending peduncles, their capsules nearly globose, mottled with purple; seeds light brown to buff.

Uncommon on gravelly and sandy shores and in rock crevices along streams.

Viola odorata L.
SWEET BLUE VIOLET *introduced*
IR | **WUP** | CUP | EUP

Perennial herb; plants finely pubescent, from a stout or wiry, long-creeping rhizome, and spreading by numerous leafy stolons rooting at the nodes. Leaves broadly ovate to orbicular, 2–6 cm long, rounded or obtuse at the apex, evenly crenate, cordate at base. Peduncles equaling the leaves or often shorter. Flowers very fragrant, typically deep violet, varying through many paler shades to white; sepals narrowly oblong, obtuse, ciliate; lateral petals usually bearded; style hook-shaped; cleistogamous flowers on recurved peduncles. Capsules broadly ovoid, pubescent, purplish; seeds large and cream-color.

Native of Europe; commonly cultivated (both single-flowered and double-flowered forms), and sometimes escaped from gardens to waste places or woods; Houghton County.

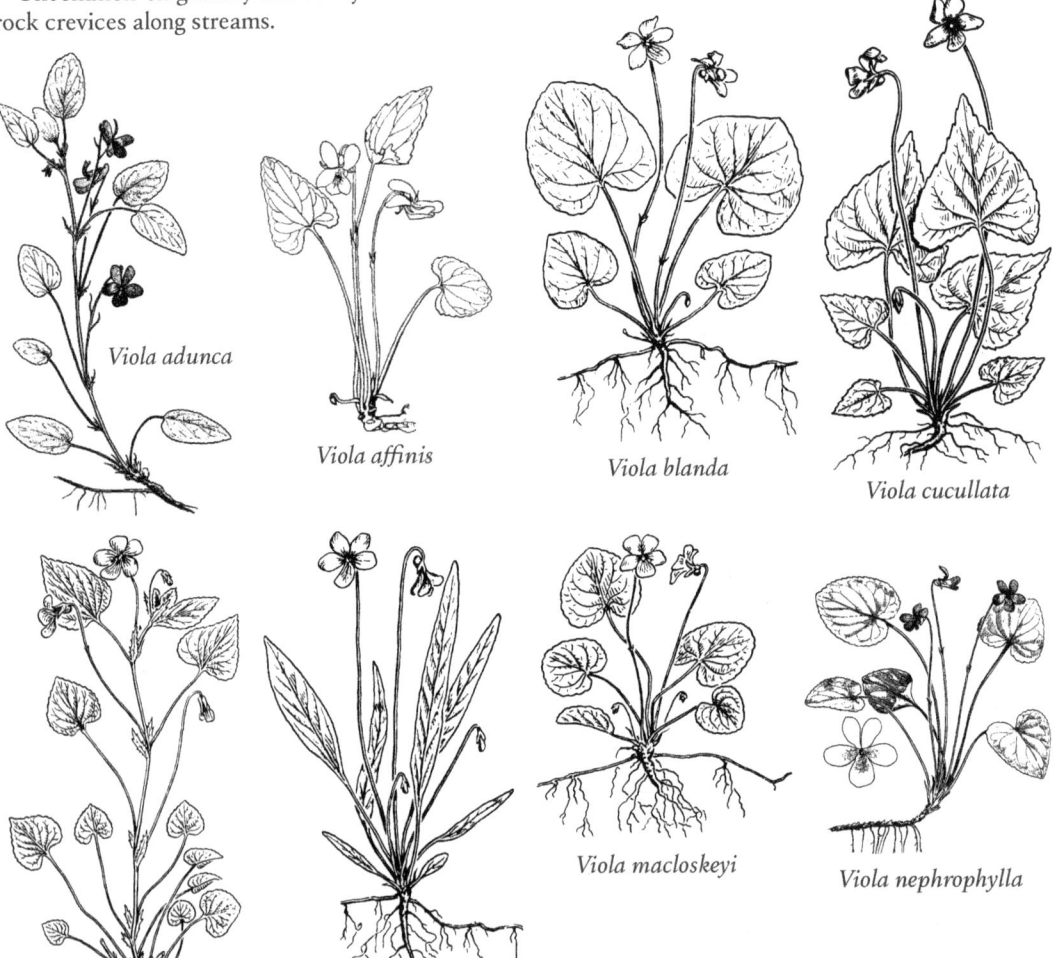

Viola adunca

Viola affinis

Viola blanda

Viola cucullata

Viola labradorica

Viola lanceolata

Viola macloskeyi

Viola nephrophylla

Viola pedata L.
BIRD-FOOT VIOLET native
IR | WUP | **CUP** | EUP CC 9

Perennial herb, plants glabrous or nearly so; rhizome erect, not stoloniferous. Leaves primarily 3-parted, the lateral divisions 3–5-parted, the segments linear to lance-shaped, often 2–4-toothed near the tip; earliest and latest leaves usually smaller and less dissected. Corolla 2–4 cm wide, the petals all beardless; tips of the large orange stamens conspicuously exsert at the center of the flower; cleistogamous flowers absent; petaliferous flowers may be produced at any time during the summer and fall. Capsules green and glabrous; seeds copper-color.

Dry fields and sandy open woods of oak and jack pine; Menominee County.

Leaves and flowers variable in shape and color.

Viola pedatifida G. Don
CROW-FOOT VIOLET (THR) native
IR | WUP | **CUP** | EUP FACU CC 10

Viola palmata var. *pedatifida* (G. Don) Cronq.
Perennial herb; plants glabrous to sparsely pilose; rhizome short, vertical. Leaves primarily 3-parted, each division again 3-parted or cleft into linear lobes, these often further cut into 2–4 lobes; earliest and latest leaves almost as deeply cut as the others. Corolla 2–4 cm wide, the three lower petals bearded, all bright violet; cleistogamous flowers on erect peduncles, their capsules yellowish when ripe; seeds light brown.

In the UP, known from a limestone bluff in Delta County with bluestem grass (*Andropogon*) and *Viola palmata*.

Viola primulifolia L.
PRIMROSE-LEAF VIOLET native
IR | **WUP** | **CUP** | EUP CC 5

Perennial herb, spreading by rhizomes and stolons. Leaves all from base of plant, oblong to ovate, rounded at tip, longer than wide; margins with small rounded teeth. Flowers white, on stalks shorter or equal to leaves, 3 lower petals purple-veined at base, 2 side petals beardless or with few hairs. Fruit a capsule 7–10 mm long; seeds red-brown to black. May.

Wet meadows and bogs, often in sphagnum moss; sandy streambanks; soil sandy or peaty, acidic.

Viola pubescens Ait.
YELLOW FOREST VIOLET FACU native
IR | **WUP** | **CUP** | **EUP** CC 4

Perennial herb; plants softly pubescent, with 1 or 2 (rarely more) stout upright stems arising from a brown woody rhizome bearing coarse fibrous roots. Leaves 2–4 near the summit, occasionally accompanied by a long-petioled kidney-shaped-cordate root leaf; stem leaves orbicular-ovate, short-pointed, crenate-dentate, cordate or truncate-decurrent at base, 4–10 cm long, and usually about 1 cm wider than long; stipules broadly ovate, the apex blunt or shallowly toothed. Flowers on downy pubescent peduncles rising little above the leaves; sepals lance-shaped, ciliate; petals clear yellow with brown-purple veins near the base, the lateral ones bearded. Capsules 10–12 mm long, woolly or glabrous; seeds pale brown.

Mesic woods.

Viola renifolia Gray
KIDNEY-LEAF WHITE VIOLET FACW native
IR | **WUP** | **CUP** | **EUP** CC 6

Perennial herb, spreading by long rhizomes. Leaves all from base of plant, mostly kidney-shaped, rounded at tip, varying from smooth and shiny above to hairy on lower surface only; margins with few rounded teeth. Flowers white, all bearded or beardless, 3 lower petals purple-veined at base. Capsules 4–5 mm long, seeds brown and dark-flecked. May–July.

Cedar swamps, sphagnum hummocks in peatlands.

Viola rostrata Pursh
LONG-SPUR VIOLET native
IR | WUP | **CUP** | EUP FACU CC 6

Perennial herb, from a branched woody rhizome. Stems glabrous, erect or spreading, 5–12 cm tall, becoming 15–25 cm tall after anthesis. Leaves ovate, 2–4 cm long, cordate at base, all but the lowermost acute or pointed at the apex; stipules lance-shaped, fimbriate-toothed to above the middle. Flowers on slender peduncles held well above the foliage; sepals narrowly lance-shaped, eciliate; petals light violet with darker veins forming a pronounced dark eye at the center of the flower; spur 10–16 mm long, slightly upcurved at the tip. Capsules ellipsoid, 5–6 mm long; seeds light yellow-brown.

Shady slopes and moist woods, usually in deep humus; Alger and Delta counties.

Viola sagittata Ait.
ARROWHEAD VIOLET native
IR | **WUP** | **CUP** | EUP FAC CC 8

Perennial herb. Leaves lance-shaped to oblong lance-shaped, hastately or sagittately incised, lobed, or toothed at the subcordate or truncate base, otherwise distantly toothed; leaves of late summer often nearly deltoid and merely crenate at the base; petioles usually longer than the blades Flowers 2–2.5 cm across, violet-purple, the lower petals usually with prominent dark veining; sepals glabrous; cleistogamous flowers on erect peduncles; seeds numerous, brown. April–June.

Open, dry pine and oak woods, usually where sandy; Keweenaw, Marquette, and Menominee counties.

342 VIOLACEAE | Violet Family

Viola selkirkii Pursh
GREAT-SPUR VIOLET — native
IR | WUP | CUP | EUP — CC 7

Perennial herb; plants rather delicate, from a slender, elongate, non-stoloniferous rhizome. Leaves all from base of plant, with minute spreading hairs on the upper surface, otherwise glabrous, at anthesis 1.5–3 cm long, later becoming larger, broadly ovate-cordate, the basal sinus narrow, the basal lobes converging or overlapping. Flowers numerous, about 1.5 cm across; sepals lance-shaped; petals pale violet, beardless, the spur large and blunt, 5–7 mm long. Capsules ellipsoid, 4–6 mm long; seeds buff-colored.

Deciduous woods and shady ravines; preferring calcareous soils.

Viola sororia Willd.
HOODED BLUE VIOLET — native
IR | WUP | CUP | EUP — FAC CC 1

Viola septentrionalis Greene

Perennial herb, spreading by short rhizomes. Leaves all from base of plant, ovate to heart-shaped, sometimes expanding to 10 cm wide in summer, with long hairs; margins with rounded teeth; blades angled from the upright petioles. Flowers blue-violet, on stalks about as high as leaves, the 2 side petals densely bearded with hairs 1 mm long and not club-tipped. Fruit a purple-flecked capsule, seeds dark brown. April–June.

Moist hardwood forests; occasionally in swamps, floodplain forests and along rocky streambanks.

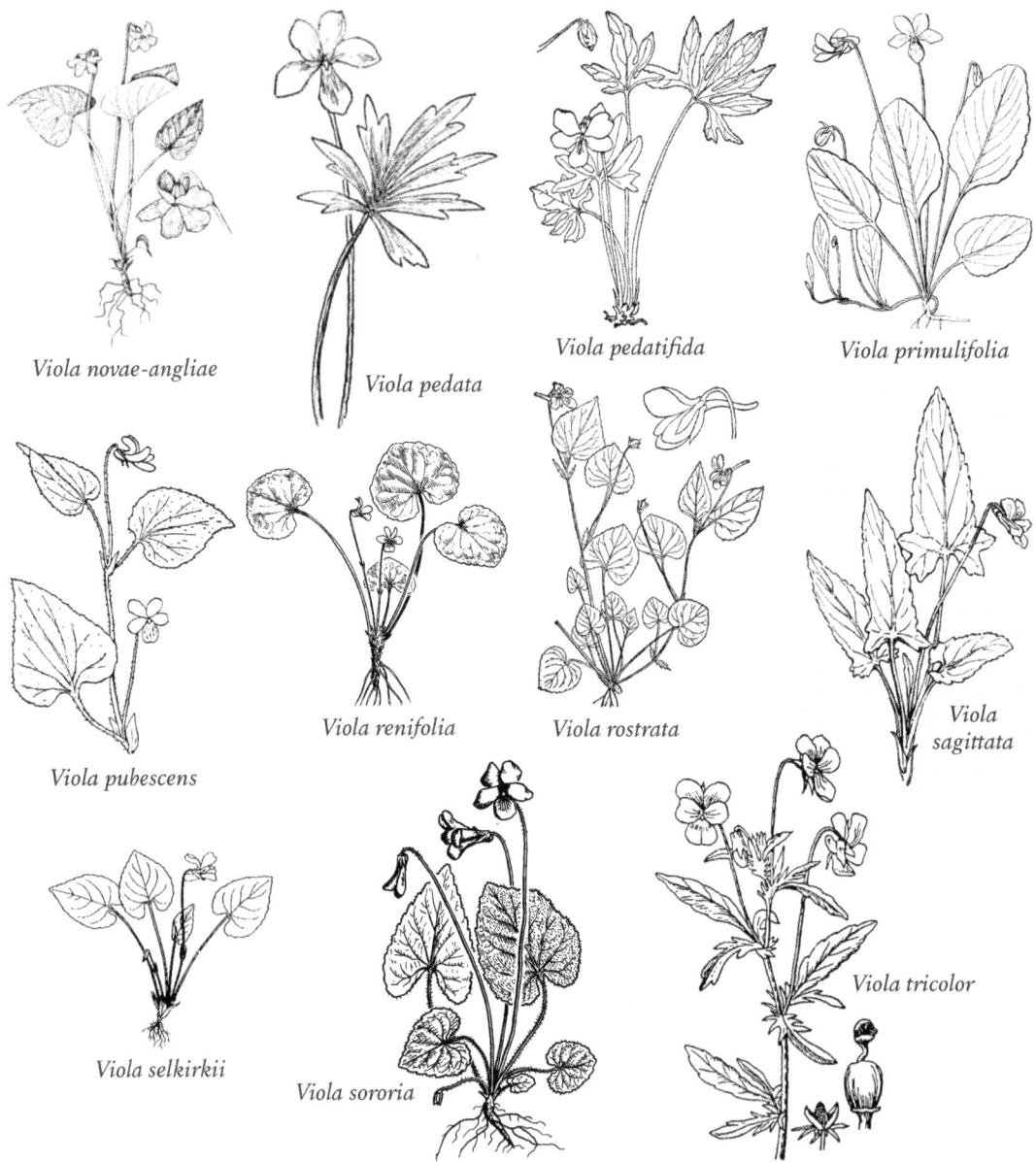

Viola novae-angliae

Viola pedata

Viola pedatifida

Viola primulifolia

Viola pubescens

Viola renifolia

Viola rostrata

Viola sagittata

Viola selkirkii

Viola sororia

Viola tricolor

Viola tricolor L.
JOHNNY-JUMP-UP *introduced*
IR | **WUP** | **CUP** | **EUP**

Annual herb, plants glabrous or pubescent. Stems to 45 cm tall, often branched from the base. Lowest leaves orbicular or cordate; upper leaves oblong to elliptic, tapering to a blunt tip; margins crenate; stipules foliaceous, laciniate to lyrate-pinnatifid, the middle lobe oblong lance-shaped. Flowers 1.5–2 cm across; sepals about 2/3 as long as the petals; petals variously marked with yellow, purple, or white, the upper petals usually darker. Capsules ellipsoid, 6–10 mm long; seeds dark brown.

Native of the Old World; escaped from gardens in waste places, only rarely persistent.

The cultivated pansy is the product of long cultivation and hybridization of *V. tricolor* with several allied European species.

Vitaceae
GRAPE FAMILY

Mostly woody vines, climbing by tendrils. Leaves alternate, simple or compound; tendrils and flower clusters produced opposite the leaves. Flowers regular, 4–5-merous, perfect or unisexual, with a hypanthium and usually with a cup-shaped disk. Calyx small or almost lacking. Petals small. Stamens as many as the petals and opposite them. Ovary 2-celled; style very short; stigma 1, slightly 2-lobed. Fruit a berry; seeds 4 or by abortion fewer.

1. Stems brown-pithy inside, the bark shredding into strips; leaves simple .... **Vitis**
1. Stems white-pithy, the bark tight, not shredding; leaves palmately compound into 5 leaflets ... **Parthenocissus**

Parthenocissus *Creeper*

Woody vines, trailing or climbing by tendrils. Leaves palmately compound with typically 5 leaflets (ours). Flowers small, in panicles borne opposite the leaves or aggregated into terminal clusters, perfect or unisexual, 5-merous. Disk none. Petals separate and spreading at anthesis. Stamens short, erect. Berries with thin flesh and 1–4 seeds.

1. Plants often climbing trees by means of adhesive disks on the tendrils; inflorescence with central axis .... **P. quinquefolia**
1. Plants not climbing, without adhesive disks on tendrils; inflorescence branched, without an evident central axis .... **P. inserta**

Parthenocissus inserta (Kerner) Fritsch
THICKET-CREEPER *native*
IR | **WUP** | **CUP** | EUP FACU CC 4

Parthenocissus vitacea (Knerr) A.S. Hitchc.

Very similar to *P. quinquefolia* in habit and foliage; tendrils few-branched, almost always without adhesive disks. Leaflets glossy green above, glabrous to thinly pubescent beneath. Inflorescence forked at the summit of the peduncle, the two branches both divergent, producing a broad rounded cluster. June.

Moist soil.

Parthenocissus quinquefolia (L.) Planch.
VIRGINIA-CREEPER *native*
IR | **WUP** | **CUP** | **EUP** FACU CC 5

High-climbing vine, adhering to its support by numerous adhesive disks at the ends of the much-branched tendrils; pubescence variable. Leaves long-petioled; leaflets dull green above, elliptic to obovate, 6–12 cm long, abruptly acuminate, sharply serrate chiefly beyond the middle, cuneate to the base, subsessile or on petioles to 15 mm long. Inflorescences terminal and from the upper axils, forming a panicle usually longer than wide, with well marked axis and divergent branches, the flowers in terminal umbel-like clusters. Berries nearly black, about 6 mm wide. June.

Moist soil, thickets, swamps.

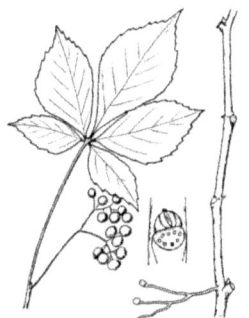

Parthenocissus quinquefolia

Vitis *Grape*

Woody vines, climbing by tendrils. Flowers actually or functionally unisexual, with a hypogynous disk, 5-merous. Calyx essentially none. Petals cohering at the tip, separating at the base, falling early. Sterile flower with 5 erect stamens and a rudimentary pistil. Fertile flower with a well developed pistil and 5 short, reflexed, functionless stamens. Fruit a juicy berry; seeds ovoid, 4, or fewer by abortion. Our species bloom in May or June.

1. Leaf underside with felt-like covering of white or rust-colored hairs .... **V. labrusca**
1. Leaf underside smooth and greenish when fully developed; often with persistent tufts of hairs in leaf axils.. .... **V. riparia**

Vitis labrusca L.
FOX GRAPE *native*
IR | WUP | **CUP** | EUP FACU CC 7

High-climbing vine; pith interrupted at the nodes by a diaphragm; tendrils or flower clusters from 3 or more successive nodes. Leaves firm, round-cordate, 1–2 dm long and wide, usually shallowly 3-lobed, occasionally unlobed or lobed nearly to the middle, shallowly serrate, the lower surface persistently covered with a dense tomentum concealing the surface, reddish or rusty in color when young and usually also at maturity, occasionally ashy-gray at maturity. Peduncles and young branches eventually glabrous. Panicles ovoid, 4–8 cm long. Berries dark red to nearly black, 1–2 cm wide.

Woods, roadsides, and thickets; Marquette County. The size of the leaf serrations varies greatly. The species has contributed to the parentage of many cultivated grape varieties.

Vitis riparia Michx.
RIVER-BANK GRAPE native
IR | WUP | CUP | EUP FAC CC 3

Vitis vulpina subsp. *riparia* (Michx.) R.T. Clausen
Woody, climbing vine to 5 m or more long; young branches green or red, hairy, becoming smooth. Leaves alternate, heart-shaped in outline, 1–2 dm long and as wide, with a triangular tip and 2 smaller lateral lobes, leaf base with a U-shaped indentation, upper surface smooth, bright green, underside paler and sparsely hairy along veins; margins with coarse, forward-pointing teeth; petioles shorter than blades. Flowers small, sweet-scented, green-white to creamy, in stalked clusters 5–10 cm long. Berries dark blue to black, 6–12 mm wide, with a waxy bloom, sour when young, becoming sweeter when ripe in fall. May–July.

Floodplain forests, moist sandy woods, streambanks, thickets, sand dunes.

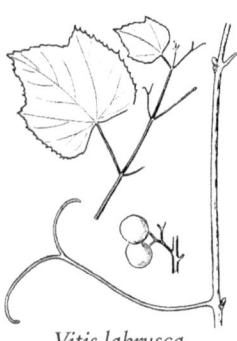
Vitis labrusca

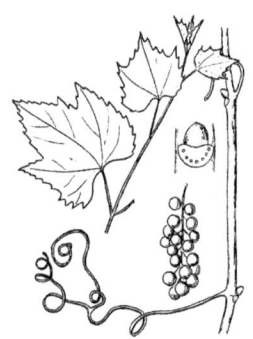
Vitis riparia

MONOCOTS

Acoraceae
CALAMUS FAMILY

Acorus *Sweetflag; Calamus*
Acorus americanus (Raf.) Raf.
AMERICAN SWEETFLAG native
IR | WUP | CUP | EUP OBL CC 6

Acorus calamus var. *americanus* (Raf.) H.D. Wulff.
Perennial herbs of wetlands; rhizomes and leaves pleasantly scented. Rhizomes branched, creeping at or near surface. Leaves linear, long and swordlike, bright green, leathery, 2-ranked, 5–15 dm long and 1–2 cm wide, translucent near base, with 1–6 prominent veins parallel along length of leaf; sweet-scented when crushed; margins entire, sharp-edged. Inflorescence a solitary yellow-green spadix, upright, 5–10 cm long and 1–2 cm wide, borne from near midway of leaf, nearly cylindric, tapering, to an rounded tip; true spathe absent. Flowers perfect, yellow or brown, composed of 6 papery tepals and 6 stamens; ovaries 1, usually 3-locular. Fruit a 1–3-seeded berry, light brown to reddish, with darker streaks, dry outside and jellylike on inside; seeds 1–6(–14), embedded in the mucilagenous jelly. June–July.

Marshes (often with cattails), bogs, streambanks.

Similar to the introduced *A. calamus* and long considered a variety of it; *A. calamus* not yet reported from the UP but is known from scattered locales in the Lower Peninsula.

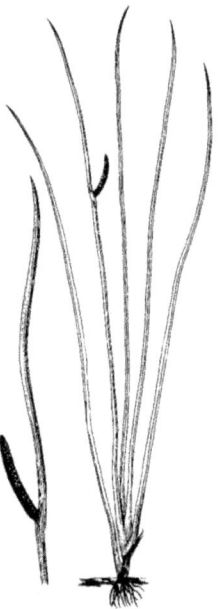
Acorus americanus

Alismataceae
WATER-PLANTAIN FAMILY

Perennial, aquatic or emergent herbs; plants swollen and tuberlike at base. Leaves all from base of plant and clasping an erect stem; underwater leaves often ribbonlike; emergent leaves broader. Flowers perfect (with both staminate and pistillate parts) or imperfect, in racemes or panicles at ends of stems, with 3 sepals and 3 petals; stamens 6 or more. Fruit a compressed achene, usually tipped by the persistent style.

1. Leaves often arrowhead-shaped; pistils or achenes in several series around a large, round receptacle, and forming a dense, round head.............. **Sagittaria**
1. Leaves never arrowhead-shaped; achenes in a single whorl on a small, flat receptacle.............. **Alisma**

Alisma *Water-Plantain*

Perennial herbs, from cormlike rootstocks. Leaves emersed or floating, ovate to lance-shaped, never arrowhead-shaped. Flowers perfect, in whorled panicles, sepals 3, green; petals 3, white or light pink; stamens 6. Fruit a flattened achene in a single whorl on a flat receptacle, style beak small or absent.

1. Flowers larger; petals about 4 mm long and to 4 mm wide.................................... **A. triviale**
1. Flowers smaller; petals to 2.5 mm long and 2 mm wide .... **A. subcordatum**

Alisma subcordatum Raf.
AMERICAN WATER-PLANTAIN   *native*
IR | WUP | **CUP** | EUP   OBL CC 1

Alisma plantago-aquatica var. *parviflorum* (Pursh) Torr.

Leaves ovate to oval, 3–15 cm long and 2–12 cm wide, rounded to nearly heart-shaped at base; petioles long. Flowers clustered on slender stalks 1–10 dm long, in whorls of 3–10; sepals 3; petals white, 3–5 mm long. Fruit an achene, 2–3 mm long, with a central groove. July–Sept.

Shallow water marshes, shores, ditches; Dickinson County.

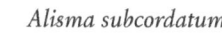
Alisma subcordatum

Alisma triviale Pursh
NORTHERN WATER-PLANTAIN   *native*
IR | **WUP** | **CUP** | EUP   OBL CC 1

Alisma brevipes Greene
Alisma plantago-aquatica L. var. *americanum* J.A. Schultes

Leaves usually long-petioled, the blade elliptic to broadly ovate, rounded to subcordate at base, 3–18 cm long. Inflorescence on a scape 1–10 dm long. Flower pedicels in whorls of 3–10; sepals obtuse, 2–3 mm long; petals white, about 4 mm long. Fruit an achene 2–3 mm long, usually with a median dorsal groove. June–Sept.

Marshes, ponds, and streams.

Sagittaria *Arrowhead*

Perennial or annual herbs, with fleshy or tuberous rootstocks. Leaves sheathing, all from base of plant, variable in shape and size. Emersed and floating leaves usually arrowhead-shaped with large lobes at base, or sometimes ovate to oval and without lobes; underwater leaves often linear in a basal rosette, normally absent by flowering time. Flowers in a raceme of mostly 3-flowered whorls; upper flowers usually staminate, lower flowers usually pistillate or sometimes perfect; sepals 3, green, persistent; petals 3, white, deciduous; stamens 7 to many; pistils crowded on a rounded receptacle. Fruit a crowded cluster of achenes in more or less round heads, the achenes flattened and winged, beaked with a persistent style.

1. Emersed leaves not arrowhead-shaped, basal lobes absent 2
1. Emersed leaves all or mostly arrowhead-shaped, with large basal lobes 3
2. Pistillate flowers and fruiting heads sessile or nearly so .... **S. rigida**
2. Pistillate flowers and fruiting heads obviously stalked. .... **S. graminea**
3. Bracts below flowers mostly less than 1 cm long; achene beak projecting horizontally from tip of achene .... **S. latifolia**
3. Bracts below flowers usually more than 1 cm long; achene beak erect .... **S. cuneata**

Sagittaria cuneata Sheldon
ARUM-LEAF ARROWHEAD   *native*
IR | WUP | **CUP** | EUP   OBL CC 6

Perennial herb, with rhizomes and large, edible tubers. Submerged leaves (if present) often awl-shaped or reduced to bladeless, expanded petioles (phyllodes); emersed leaves long-stalked, usually arrowhead-shaped, 5–20 cm long and 2–15 cm wide, the basal lobes much shorter than terminal lobe; floating leaves often heart-shaped (unlike our other species of *Sagittaria*). Flowers imperfect, the staminate flowers above the pistillate, in more or less round heads 5–12 mm wide, with 2–10 whorls of heads on a stalk 1–6 dm tall, the stalks often branched at lowest node; bracts tapered to tip, 1–4 cm long; sepals ovate, bent backward in flower and fruit; petals white, 7–15 mm long. Fruit an achene, 2–3 mm long; beak erect, small, 0.1–0.4 mm long. June–Sept.

Shallow water, lakeshores and streambanks.

Sagittaria graminea Michx.
GRASS-LEAF ARROWHEAD   *native*
IR | **WUP** | **CUP** | EUP   OBL CC 10

Perennial herb, with rhizomes. Underwater plants sometimes only a rosette of bladeless, ribbonlike petioles (phyllodes) to 1 cm wide; emergent leaves lance-shaped to oval, never arrowhead-shaped, 3–20 cm long and 0.5–3 cm wide, tapered to a blunt tip. Flowers imperfect, the staminate flowers usually above the pistillate, clustered in more or less round heads, 5–12 mm wide, the heads on spreading stalks 1–4 cm long; with 2–10 whorls of flowers along an unbranched stalk mostly shorter than leaves; bracts broadly ovate, joined in their lower portion, 2–8 mm long; sepals ovate, bent backward in fruit; petals white, equal or longer than sepals. Fruit a winged achene, 1–2 mm long, beak small or absent. June–Sept. Shallow water and shores.

Sagittaria graminea includes *S. graminea* var. *cristata* (Engelm.) Bogin, sometimes treated as a separate species, *S. cristata* Engelm., and distinguished as follows:

1 Achenes with beak 0.4–0.6 mm long; anthers clearly shorter than the filaments................ **S. cristata**
1 Achenes with beak minute, scarcely discernable, ca. 0.2 mm long; anthers as long as or longer than filaments. . ... **S. graminea**

Sagittaria latifolia Willd.
DUCK-POTATO   native
IR | **WUP** | **CUP** | **EUP**   OBL CC 4
Perennial herb, with rhizomes and edible tubers in fall. Leaves variable; emersed leaves arrowhead-shaped, mostly 8–40 cm long and 1–15 cm wide, lobes typically narrow on plants in deep water to broad on emersed plants; plants sometimes with bladeless, expanded petioles (phyllodes). Flowers staminate above and pistillate below, clustered in more or less round heads 1–2.5 cm wide, at ends of slender, spreading stalks 0.5–3 cm long, in whorls of 2–15 along a stalk 2–10 dm tall; bracts tapered to a tip or blunt, 0.5–1 cm long; sepals ovate, bent backward by fruiting time; petals white, 7–20 mm long. Fruit a winged achene, 2–4 mm long, the beak projecting horizontally, 1–2 mm long. July–Sept. Shallow water, shores, marshes and pools in bogs.

Sagittaria rigida Pursh
SESSILE-FRUIT ARROWHEAD   native
IR | **WUP** | CUP | EUP   OBL CC 6
Perennial herb, rhizomes present. Stems erect or lax. Emersed leaves lance-shaped to ovate, rarely with short, narrow basal lobes, (but not arrowhead-shaped), 4–15 cm long and to 7 cm wide; petioles sometimes bent near junction with blades; deep water plants often with only linear, bladeless, expanded petioles (phyllodes). Flowers in more or less round heads to 1.5 cm wide, the heads stalkless and bristly when mature due to achene beaks; in 2–8 whorls on a stalk 1–8 dm tall, the stalk often bent near lowest node; flowers imperfect, staminate flowers above the pistillate, staminate flowers on threadlike stalks 1–3 cm long, pistillate flowers more or less stalkless; bracts ovate, 5 mm long, joined at base; sepals ovate, 4–7 mm long, bent backward when in fruit; petals white, 1–3 cm long. Fruit a narrowly winged achene, 2–4 mm long; beak ascending, 1–1.5 mm long. June–Sept.

Shallow water, shores and streambanks.

Araceae

ARUM FAMILY

Perennial herbs with alternate, simple or compound, often fleshy leaves. Flowers small and numerous, mostly single-sexed, staminate flowers usually above pistillate, crowded in a cylindric or rounded spadix subtended by a leaflike spathe; sepals 4–6 or absent; petals absent; stamens mostly 2–6; pistils 1–3-chambered. Fruit a usually fleshy berry, containing 1 to few seeds, or the entire spadix ripening as a fruit.

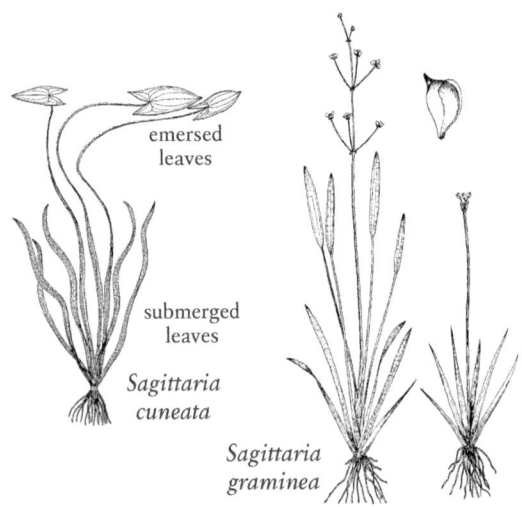

Sagittaria cuneata (emersed leaves, submerged leaves)

Sagittaria graminea

Sagittaria latifolia

Sagittaria rigida

Now included in the Araceae are tiny aquatic plants (*Lemna, Spirodela,* and *Wolffia*), genera formerly treated as their own family (Lemnaceae). These are small perennial herbs, floating at or near water surface, single or forming colonies. Plants thallus-like (not differentiated into stems and leaves), the thallus (or frond) flat or thickened; the roots, if present, unbranched, 1 or several from near center of leaf underside; reproducing vegetatively by buds from 1–2 pouches on the sides, the parent and budded plants often joined in small groups. Flowers rare, either staminate or pistillate, in tiny reproductive pouches on margins (*Lemna, Spirodela*) or upper surface (*Wolffia*) of the leaves, subtended by a small spathe within the pouch; sepals and petals absent; staminate flowers 1–2, consisting of 1 anther on a short filament; pistillate flower 1 (a single ovary), in same pouch as staminate flowers. Fruit a utricle with 1 to several seeds.

1 Plants tiny floating or submerged aquatic species less than 1 mm long, without differentiation into leaves or stems 2
1 Plants large, with clearly differentiated normal leaves and with rhizomes or tubers. 4
2 Roots absent; leaves thickened, less than 1.5 mm long . . . . **Wolffia**
2 One to several roots present on leaf underside; leaves flat, mostly more than 1.5 mm long 3
3 Each leaf with 1 root; leaf underside green or purple-tinged . . . . **Lemna**
3 Each leaf with 3 or more roots; leaf underside solid purple . . . . **Spirodela polyrhiza**
4 Leaves compound . . . . **Arisaema**
4 Leaves simple, not divided. 5
5 Leaves arrowhead-shaped, lobes equaling 1/3 or more length of blade. . . . . **Peltandra virginica**
5 Leaves heart-shaped or rounded at base 6
6 Leaves broadly heart-shaped, abruptly tapered to a tip; spathe white, long-stalked; flowering in spring . . . . **Calla palustris**
6 Leaves rounded ovate, tapered to a rounded tip; spathe green-yellow to purple-brown, short-stalked or stalkless; flowering in late winter to early spring . . . . **Symplocarpus foetidus**

Arisaema *Jack-in-the-Pulpit*
Arisaema triphyllum (L.) Schott
JACK-IN-THE-PULPIT *native*
IR | WUP | CUP | EUP FAC CC 5
Perennial herb, from bitter-tasting corms. Stems 3–12 dm long. Leaves usually longer than the flower stalk, mostly 2, divided into 3 leaflets, the terminal leaflet oval to ovate, the lateral leaflets often asymmetrical at base. Flowers staminate or pistillate and usually on separate plants, sepals and petals absent, borne near base of a cylindric, blunt-tipped spadix, subtended by a green, purple-striped spathe, rolled inward below, expanded and arched over the spadix above, abruptly tapered to a tip. Fruit a cluster of shiny red berries, each berry with 1–3 seeds, the fruit about l cm wide. April–July.

Moist forests, cedar swamps.

Calla *Water-Arum*
Calla palustris L.
WATER-ARUM *native*
IR | WUP | CUP | EUP OBL CC 10
Perennial herb, from thick rhizomes, the rhizomes creeping in mud or floating in water. Leaves broadly heart-shaped, abruptly tapered to a tip, 5–15 cm long and about as wide; petioles stout, 1–2 dm long (or longer when underwater). Flowers perfect or the uppermost staminate, on a short-cylindric spadix, 1.5–3 cm long, shorter than the spathe; the spathe white, ovate, tipped with a short, sharp point to 1 cm long; sepals and petals absent; stamens 6. Fruit a fleshy, few-seeded berry, turning red when ripe, 8–12 mm long. May–July.

Bog pools, swamps, shores and wet ditches.

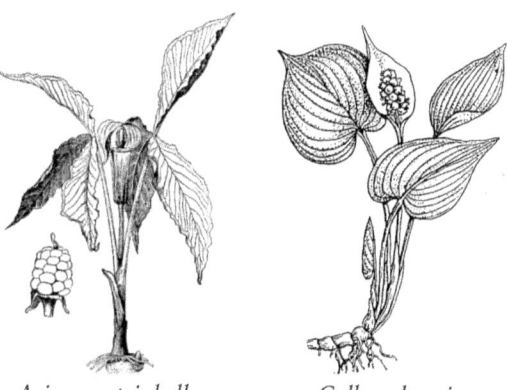

Arisaema triphyllum *Calla palustris*

Lemna *Duckweed*
Small perennial floating herbs, with 1 root per frond (or roots sometimes absent on oldest and youngest leaves). Blades single or 2 to several and joined in small colonies, floating on water surface or underwater (ivy-leaf duckweed, *L. trisulca*), varying from round, ovate, to obovate or oblong, tapered to a long point (petiole) in *L. trisulca*; green or often red-tinged; upper surface flat to slightly convex, underside flat or convex. Reproductive pouches 2, on margins of frond. Flowers uncommon, consisting of 2 stamens (staminate flowers) and a single pistil (pistillate flower) in each pouch. Fruit an utricle with 1 to several seeds. Reproduction mostly by budding of new leaves from the reproductive pouches.

ARACEAE | Arum Family

1. Fronds denticulate toward the tip, tapered to a slender stipitate base, the stipe often as long as the main body and commonly attached to the parent frond; colonies star-shaped, usually submersed .... **L. trisulca**
1. Fronds entire on the margin, nearly rounded and not obviously stipitate at the base, solitary or in tight colonies, these not star-shaped, floating on the water surface or stranded on mud 2
2. Fronds with several about equal sized, small papillae on the upper surface from the midline to the tip (often obscure), very often red-tinged on the lower surface, forming small, obovate to orbicular, rootless, dark green to brown turions under unfavorable conditions, these sinking to the bottom of the water ..... **L. turionifera**
2. Fronds lacking papillae or with one prominent papilla at the tip and another just above the node and with smaller papillae between them .... **L. minor**

Lemna minor L.
COMMON DUCKWEED native
IR | WUP | **CUP** | EUP OBL CC 5

Fronds nearly orbicular to elliptic-obovate, broadest near the middle, 2–4 mm long, symmetric or nearly so, green to yellowish green, never red-tinged or mottled on either surface (as in *L. turionifera*), obscurely 3 (5)-nerved; both surfaces flat to weakly convex, the upper surface with a low papilla at the apex and often one above the node, usually with a low median ridge or row of smaller papillae between them, the lower surface never inflated. Turions not produced. Root sheath not winged, root tip rounded. Fruit ovoid to ellipsoid, wingless, 1-seeded. July–Sept.

Quiet or stagnant water of ponds, oxbows, shores, slow-moving rivers, ditches.

Lemna trisulca L.
IVY-LEAF DUCKWEED native
IR | WUP | **CUP** | EUP OBL CC 5

Perennial floating herb, forming tangled colonies just below water surface, floating at surface only when flowering; roots single from underside of frond or absent. Fronds several to many, joined to form star-shaped colonies; oblong lance-shaped, 5–20 mm long, tapered to a slender base (petiole), flat on both sides. Fruit a 1-seeded utricle.

Ponds, streams, ditches.

Lemna turionifera Landolt
TURION DUCKWEED native
IR | WUP | **CUP** | EUP OBL CC 5

Fronds single or in groups of several, obovate, usually flat and not humped, 1–4 mm long and 1–1.5 times longer than wide, veins 3, small white dots (papillae) present on midline of upper surface (visible with naked eye but clearer with 10x hand lens); underside of frond usually red or purple and redder than upper side point of root, upper surface (especially near tip) sometimes red-spotted. Turions sometimes present, dark-green to brown, 1–1.6 mm wide, without roots, sinking to bottom and forming new plants.

Quiet water of ponds and lakes.

Peltandra *Arrow-Arum*
Peltandra virginica (L.) Schott
GREEN ARROW-ARUM native
IR | WUP | **CUP** | EUP OBL CC 6

Perennial herb, with thick, fibrous roots. Leaves all from base of plant on long petioles, bright green, oblong to triangular in outline, 1–3 dm long and 8–15 cm wide at flowering, to 8 dm long later; leaf base with a pair of lobes. Flowers in a white to orange spadix about as long as the spathe, atop a curved stalk 2–4 dm long; flowers either staminate or pistillate, the staminate flowers covering upper 3/4 of the spadix, the pistillate flowers covering lower portion; spathe green with a pale margin, 1–2 dm long, the lower portion covering the fruit. Fruit a head of green-brown berries, the berries with 1–3 seeds surrounded by a jellylike material. June–July.

Shallow water, shores, bog pools; often where shaded; Schoolcraft County.

Spirodela *Greater Duckweed*
Spirodela polyrrhiza (L.) Schleid.
GREATER DUCKWEED native
IR | **WUP** | **CUP** | EUP OBL CC 6

Perennial herb, floating on water surface; roots 5–12 per frond. Fronds usually in clusters of 2–5, flat, round to obovate, 3–6 mm long, upper surface green, underside red-purple. Flowers uncommon, comprised of 2–3 stamens (staminate flowers) and 1 pistil (pistillate flower) in each pouch. Fruit a 1–2-seeded utricle. Reproduction mainly by budding of new leaves from reproductive pouches (1 pouch on each margin of frond).

Stagnant or slow-moving water of lakes, ponds, marshes and ditches, often with *Lemna*.

Symplocarpus *Skunk-Cabbage*
Symplocarpus foetidus (L.) Salisb.
SKUNK-CABBAGE native
IR | WUP | **CUP** | EUP OBL CC 6

Perennial, foul-smelling herb, from thick rootstocks. Leaves all from base of plant, ovate to heart-shaped, 3–8 dm long and to 3 dm wide, strongly nerved; petioles short, channeled. Flowers appearing before leaves in late winter or early spring, perfect; the spathe ovate, curved over spadix, 8–15 cm long, green-purple and often mottled; sepals 4. Fruit round, 8–12 cm wide; seeds 1 cm thick. Feb–May (our earliest flowering native plant).

Floodplain forests, swamps, streambanks, calcareous fens, moist wooded slopes.

Wolffia *Watermeal*

Tiny perennial herbs, floating at or just below water surface, sometimes abundant and forming a granular scum across surface, usually mixed with other aquatic species of this family, roots absent. Leaves single or often paired, globe-shaped or ovate, flat or rounded on upper surface. Flowers uncommon, consisting of 1 stamen (staminate flower) and 1 pistil (pistillate flower) in the pouch. Fruit a round, 1-seeded utricle. Reproduction mainly by budding from the single pouch near base of frond.

Watermeal is the world's smallest flowering plant. The blades feel granular or mealy and tend to stick to the skin. One species reported from the UP.

Wolffia columbiana Karst.
COLUMBIAN WATERMEAL   native
IR | WUP | CUP | **EUP**   OBL CC 5

Fronds float low in water, only small upper surface exposed, round to broadly ovate and 1–1.5 mm long when viewed from above; nearly round when viewed from side, not raised and pointed at tip; green, not brown-dotted.

Stagnant water of ponds and marshes; Chippewa County.

Commelinaceae
SPIDERWORT FAMILY

Commelina *Day-Flower*

Commelina communis L.
COMMON DAY-FLOWER   *introduced*
IR | WUP | **CUP** | EUP   FACU

Annual herb. Stems succulent, at first erect, later diffuse and rooting from the lower nodes, to 8 dm long. Leaves alternate, ovate lance-shaped, 5–10 cm long, 1–3 cm wide. Inflorescence a small cyme, closely subtended by a folded, heart-shaped spathe from which the pedicels protrude; spathe (folded) cordate, 15–25 mm long, about half as wide, glabrous or minutely pubescent, its margins free to the base. Sepals 3, herbaceous, somewhat unequal, two usually somewhat united at base. Petals 3, the two upper blue, ovate to kidney-shaped, the lower one much smaller, usually white, sometimes absent. Fertile stamens 3; sterile stamens 3, smaller than the fertile, bearing imperfect cross-shaped anthers. Ovary 3-celled; lower two cells fertile, with 1 or 2 ovules, each with 4 seeds. Summer.

Moist or shaded ground, sometimes weedy in gardens; Schoolcraft County.

Cyperaceae
SEDGE FAMILY

Mostly perennial, grasslike, rushlike or reedlike plants. Stems 3-angled or more or less round in section, solid or pithy. Leaves 3-ranked or reduced to sheaths at base of stem; leaf blades, when present, grasslike, parallel-veined, often keeled; sheaths mostly closed around the stem. Flowers small, perfect (with

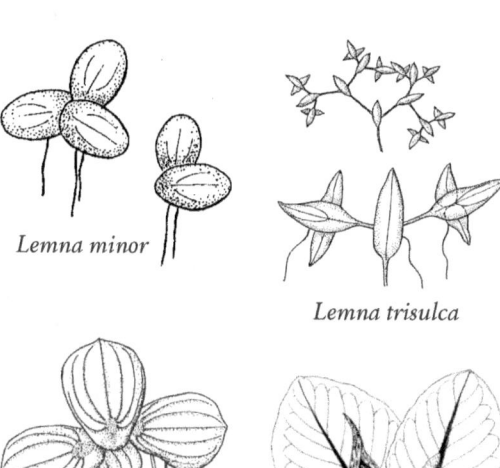

Lemna minor

Lemna trisulca

Spirodela polyrrhiza

Symplocarpus foetidus

Peltandra virginica

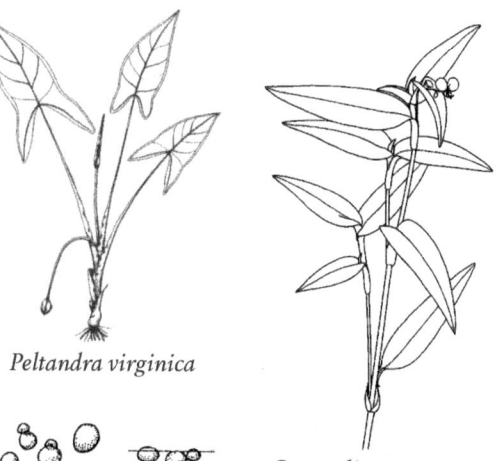

Wolffia columbiana

Commelina communis

both staminate and pistillate parts), or single-sexed, each flower subtended by a bract (scale); perianth of 1 to many (often 6) small bristles, or a single perianth scale, or absent; stamens usually 3; ovary 2–3-chambered, contained in a saclike covering (perigynium) in *Carex*, maturing into an achene, stigmas 3 or 2. Flowers arranged in spikelets (termed spikes in *Carex*), the spikelets single as a terminal or lateral spike, or several to many in various types of heads, the head often subtended by 1 to several bracts.

1 Achenes enclosed in a closed sac (perigynium) subtended by a scale, the style protruding through the apex; flowers strictly unisexual (sedges with exclusively staminate flowers should be keyed here) .... **Carex**
1 Achenes not enclosed in a closed sac, naked beside the subtending scale; at least some flowers bisexual 2
2 Scales of spikelets 2-ranked; spikelets ± flattened in cross-section and always more than one per inflorescence 3
2 Scales of spikelets spirally arranged (or if 2-ranked, the spikelet solitary); spikelets round or several-angled in cross-section, solitary or several to many per inflorescence 4
3 Stems usually ± angled, solid; inflorescences terminal; achenes without subtending bristles .... **Cyperus**
3 Stems round, hollow; inflorescences in the axils of stem leaves; achenes with subtending bristles ... **Dulichium**
4 Spikelet or cluster of spikelets borne on one side of the stem at the base of a single ± erect to somewhat angled or curved involucral bract that appears to be a continuation of the stem .... **Schoenoplectus**
4 Spikelet or spikelets terminating the stem or borne both terminally and laterally; if more than one spikelet, the inflorescence with (1–) 2 to several spreading to reflexed, leaflike involucral bracts 5
5 Spikelet solitary and terminal on the stem (very rarely a few smaller accessory spikelets occur at the base of the terminal spikelet in the bladeless genus *Eleocharis*) .. 6
5 Spikelets several to many on the stem, terminal or lateral 9
6 Sheaths totally bladeless or at most with an apical tooth up to 1 mm long; achenes usually with an apical tubercle formed by the expanded and persistent base of the style .... **Eleocharis**
6 Upper sheaths with short green blades 0.3–12 cm long; achenes blunt at apex, tubercle absent 7
7 Achenes subtended by 1–8 bristles less than twice as long as the achenes, or bristles absent .**Trichophorum**
7 Achenes subtended by conspicuous silky, white or tawny, hair-like bristles many times longer than the achenes 8
8 Bristles numerous, (12–) 15–50 or more; rhizomes erect, very short .... **Eriophorum**
8 Bristles 6; rhizomes horizontal and short-creeping .... **Trichophorum alpinum**
9 Achenes subtended by (12–) 15–50 conspicuous, silky, white or tawny, hair-like bristles many times as long as the achenes .... **Eriophorum**
9 Achenes subtended by 1–8 bristles, or bristles absent 10
10 Leaves flat or folded; with a definite, ± keeled midrib 11
10 Leaves inrolled and wiry; rounded on the back and without a definite midrib 4
11 Achenes with a conspicuous tubercle formed by the expanded, persistent style base .... **Rhynchospora**
11 Achenes blunt at apex, without a tubercle; style base, if expanded, not persistent to maturity 12
12 Widest leaves 0.5–3 mm wide; achenes lacking bristles .... **Fimbristylis**
12 Widest leaves 4–15 mm wide; achenes subtended by 1–8 bristles 13
13 Spikelets (10–) 15–36 mm long; achenes 3–5 mm long, including apiculus; anthers 4–5 mm long; stems sharply 3-angled nearly or quite to the base; colonial from rhizomes with large corm-like thickenings .... **Bolboschoenus fluviatilis**
13 Spikelets 2–10 (–12) mm long, achenes 0.9–1.2 mm long; anthers 0.5–1.3 mm long; stems terete, obtusely 3-angled, or sharply 3-angled only toward summit; tufted or with rhizomes lacking corm-like enlargements .... **Scirpus**
14 Styles 2-cleft; achenes subtended by slender bristles .. .. **Rhynchospora**
14 Styles 3-cleft; achenes lacking bristles .... **Cladium**

Bolboschoenus *Club-Rush*

Bolboschoenus fluviatilis (Torr.) Soják
RIVER CLUB-RUSH native
IR | **WUP** | **CUP** | **EUP** OBL CC 6
Schoenoplectus fluviatilis (Torr.) M.T. Strong
Scirpus fluviatilis (Torr.) Gray

Perennial, spreading by rhizomes and often forming large colonies. Stems stout, erect, 6–15 dm long, sharply 3-angled, the sides more or less flat. Leaves several on stem, smooth, 6–15 mm wide, upper leaves often longer than the head; bracts 3–5, leaflike, erect to spreading, to 3–4 dm long. Spikelets 1–3 cm long and 6–12 mm wide, clustered in an umbel of 10–20 spikelets at end of stem, several of the spikelets nearly stalkless in 1–2 clusters, others single or in groups of 2–5 at ends of spreading or drooping stalks to 8 cm long; scales gold-brown, short-hairy on back, 6–10 mm

Bolboschoenus fluviatilis

long, the midvein extended into a curved awn 1–3 mm long; bristles 6, unequal, white to copper-brown, downwardly barbed, persistent, about as long as body of achene; style yellow, 3-parted. Achenes 3-angled, dull, tan to gray-green, 3–5 mm long, with a beak to 0.5 mm long. June–Aug.

Shallow water of streams, ditches, marshes, lakes and ponds; Chippewa, Gogebic, and Schoolcraft counties.

Carex *Sedge*

Perennial grasslike plants. Stems mostly 3-angled. Leaves 3-ranked, margins often finely toothed. Flowers either staminate or pistillate, with both sexes in same spike, or in separate spikes on same plant, or the staminate and pistillate flowers on different plants. Staminate flowers with 3 or rarely 2 stamens; pistillate flowers with style divided into 2 or 3 stigmas. Achenes lens-shaped or flat on 1 side and convex on other (in species with 2 stigmas), or achenes 3-angled or nearly round (in species with 3 stigmas), enclosed in a sac called the perigynium (singular) or perigynia (plural).

Carex is the largest genus of plants in Michigan. To aid in identification, the key first identifies 41 *Carex* sections containing closely related species (section divisions largely follow those of Reznicek et al. (2011) and Hipp (2008). As evident in the keys, identification of *Carex* is often based on characteristics of the mature perigynium; a hand lens or dissecting microscope is often useful.

ADDITIONAL SPECIES

Carex scirpoidea (Bulrush sedge) is uncommon in cracks of limestone pavements (alvars) and thin soil over rock and moist, gravelly calcareous shores. Spikes usually single, but pistillate spikes sometimes with smaller lateral spikes below the terminal spike. Scales of pistillate spikes dark brown to purplish; scales of staminate spikes lighter and may be nearly white.

QUICK ENTRY KEY TO CAREX SECTIONS

Spike 1 per stem, all flowers attached to main stem in terminal spike .... **Couplet 2**

CAREX FLORAL PARTS
(*Carex chordorrhiza* shown)

Spikes 2 or more per stem, all flowers staminate .. **Couplet 9**
Spikes 2 or more per stem, at least some flowers pistillate, stigmas 2, achenes flat to biconvex in cross-section .... **Couplet 12**
Spikes 2 or more per stem, at least some flowers pistillate, stigmas usually 3, achenes usually more or less 3-angled in cross-section, body of perigynium pubescent .... **Couplet 32**
Spikes 2 or more per stem, at least some flowers pistillate, stigmas usually 3, achenes usually more or less 3-angled in cross-section, body of perigynium glabrous.. **Couplet 43**

KEY TO CAREX SECTIONS

1 Spike solitary, terminal (entirely staminate, entirely pistillate, or mixed) 2
1 Spikes 2 or more, sometimes crowded but distinguishable by the lobed appearance of inflorescence or protruding bracts or visible short segments of rachis between spikes 8
2 Styles 2-cleft; achenes 2-sided (lenticular); basal sheaths brown 3
2 Styles 3-cleft; achenes 3-sided (or nearly terete); basal sheaths brown or purple-red 4
3 Plants with slender rhizomes; perigynia obscurely or not at all serrate, plump (usually at least as convex on upper face as on the lower), the lowermost tending to be remote (as much as 1 mm apart at points of attachment); spikes without empty basal scales; anthers to 2.5 (–3) mm long .... **Carex sect. Physoglochin**
3 Plants densely tufted, not rhizomatous; perigynia minutely but strongly and regularly serrate on upper portion and beak, ± flattened, crowded; spikes usually with 1–2 empty basal scales; anthers 2–3.5 mm long .. .. **Carex sect. Stellulatae** (*C. exilis*)
4 Spikes unisexual (either staminate or pistillate); perigynia pubescent .... **Carex sect. Acrocystis**
4 Spikes containing both staminate and pistillate flowers; perigynia usually glabrous 5
5 Perigynia minutely pubescent . **Carex sect. Acrocystis**
5 Perigynia glabrous 6
6 Lower pistillate scale leaflike, at least on most spikes, much exceeding the perigynium; perigynia distinctly beaked, the body plump and filled by the mature achene .... **Carex sect. Phyllostachyae**
6 Lower pistillate scale not leaflike, scarcely if at all exceeding perigynium; perigynia essentially beakless or linear-lanceolate (tapering into an indistinct beak)... 7

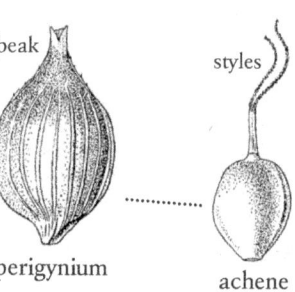

spike   scale   perigynium   achene

7   Perigynia slender, linear-lanceolate (more than 5 times as long as thick), strongly reflexed at maturity .... **Carex sect. Leucoglochin**

7   Perigynia broad, less than 5 times as long as thick, appressed-ascending ..... **Carex sect. Leptocephalae**

8   All spikes staminate 9
8   At least some spikes bisexual or pistillate 11

9   Plants with long-creeping rhizomes .... **Carex sect. Divisae** (*C. praegracilis*)
9   Plants densely tufted 10

10 Leaves flat, lax and spreading; usually in swamps and marshes ...... **Carex sect. Deweyanae** (*C. bromoides*)
10 Leaves channeled, stiff and erect; fens and other calcareous open wetlands .... **Carex sect. Stellulatae**

11 Styles 2-cleft; achenes 2-sided 12
11 Styles 3-cleft; achenes 3-sided (or nearly terete) ... 32

12 Lateral spikes peduncled, or if sessile, then elongate; terminal spike often entirely staminate 13
12 Lateral spikes sessile, short, often crowded; terminal spike at least partly pistillate (rarely staminate) 15

13 Plants slender, the stems to 3 dm tall and less than 1 mm thick (excluding leaf bases) even near the base; terminal (staminate or sometimes mixed) spike solitary, ca. 1 cm long; lowermost bract usually with a short sheath 2–7 mm long; perigynia white-pulverulent or golden-yellow at maturity 14
13 Plants coarse, the stems over (3–) 5 dm tall and usually over 1 mm thick, at least toward base; staminate spikes often 2 or more, mostly 2.5–7 cm long; lowermost bract essentially sheathless (rarely with very short sheath); perigynia neither white-pulverulent nor golden-yellow .... **Carex sect. Phacocystis**

14 Lowermost pistillate spike sessile or nearly so (except rarely one arising from near base of plant); terminal spike staminate; perigynia green or slightly glaucous, crowded .... **Carex sect. Phacocystis**
14 Lowermost pistillate spike nearly always peduncled; terminal spike often pistillate near apex, or the pistillate spikes ± loosely flowered; fresh perigynia white-pulverulent or golden-yellow ...... **Carex sect. Bicolores**

15 Stems arising mostly singly from rhizome or stolon . 16
15 Stems tufted, the tufts with or without connecting rhizomes 21

16 Perigynia plumply plano-convex to nearly terete in cross-section, not winged or sharply margined; plants of sphagnum bogs, cedar swamps, etc 17
16 Perigynia strongly flattened, with distinctly winged or sharply edged margins; plants mostly of wet or dry open habitats 18

17 Scales pale-hyaline with green midrib; perigynia apiculate or with very small beak; at least the lower few-flowered spikes ± separated; plants clumped from short, slender rhizomes .... **Carex sect. Glareosae**
17 Scales rich brown; perigynia with distinct beak ca. 0.5 mm long; spikes crowded as if in a single head; stems arising from axils of old decumbent stems (stolons) .. .. **Carex sect. Chordorrhizae**

18 Perigynia mostly over 2 mm wide; staminate flowers only at the base of some or all spikes .... **Carex sect. Ovales**

18 Perigynia mostly not over 2 mm wide; staminate flowers not restricted to base of spikes 19

19 Mature perigynia with the body ± narrowly wing-margined above and the beak bidentate (firm teeth 0.5 mm long); rhizome slender (1–1.5 mm in diameter), with brownish fibrous sheaths; spikes often dissimilar, some largely or entirely staminate or pistillate, others mixed .... **Carex sect. Ammoglochin**

19 Mature perigynia distinctly 2-edged but not winged, the beak with short weak teeth; rhizome stout (2–3 mm in diameter), with black fibrous sheaths; spikes mostly similar (each one staminate apically and pistillate basally; in section Holarrhenae the upper sometimes largely staminate) 20

20 Sheaths of upper leaves green-nerved ventrally, usually not covering the inconspicuous nodes .... **Carex sect. Holarrhenae**
20 Sheaths of upper leaves with broad white-hyaline stripe on ventral side covering the included nodes .... **Carex sect. Divisae**

21 Staminate flowers at the base of some or all spikes, not at the apex (note especially the terminal spike) 22
21 Staminate flowers at the apex of some or all spikes (even when anthers have fallen, protruding filaments are usually visible) 27

22 Perigynia with thin-winged margins, at least narrowly so along apical part of body and basal part of beak, strongly flattened and scale-like (in some species elongate), ± appressed and overlapping (or in some species spreading at the tips) 23
22 Perigynia at most with a ridge along the margin, not winged, the achene plumply filling at least the apical part of the body all the way to the margins 24

23 Bracts not resembling the leaves, narrower than 2 mm most or all their length and not over twice as long as the inflorescence; perigynia various .. **Carex sect. Ovales**
23 Bracts leaflike, the broadest 2–4 mm wide, many times exceeding the spikes (which are crowded in a dense head); perigynia very narrowly lanceolate, not over 1 mm wide .... **Carex sect. Cyperoideae**

24 Body of perigynium elliptic or nearly so (except in *C. arcta*) with at most a very short beak, and with rounded or slightly margined edges, nearly or entirely filled by the achene .... **Carex sect. Glareosae**
24 Body of perigynium ovate or lanceolate or prominently beaked, sharp-edged, only 1/2 to 2/3 filled by achene (very spongy around and below base of achene) ... 25

25 Mature perigynia loosely to strongly appressed-ascending, 4–5.7 mm long; anthers 1.3–2.6 mm long .... **Carex sect. Deweyanae**
25 Mature perigynia strongly spreading to reflexed, 2–3.6 mm long; anthers 0.8–2 mm long 26

26 Spikes 7–15, usually crowded, except sometimes the lowest, the inflorescence axis mostly concealed; beaks not bidentate .... **Carex sect. Glareosae** (*C. arcta*)

26 Spikes 3–8, not usually crowded, inflorescence axis clearly visible; beaks clearly bidentate with teeth 0.1–0.4 mm long .... **Carex sect. Stellulatae**

27 Stems stout (often 1.5 mm thick at ca. 3 cm below inflorescence) and very sharply angled (or even narrowly winged), ± soft and easily compressed (flattened in pressing); wider leaves 5–10 mm broad, with rather loose sheaths; perigynia spongy-thickened basally, on short slender stalks; anthers 1.3–2.6 mm long .... **Carex sect. Vulpinae**

27 Stems slender (not over 1.5 mm thick at ca. 3 cm below inflorescence, or rarely so in some species), firm, not wing-angled nor easily compressed (hence, not flattened in pressing); leaves, perigynia, and anthers various 28

28 Spikes 10 or fewer, usually greenish at maturity, crowded or remote in a simple inflorescence (one spike, no branches, at each node of it) 29

28 Spikes numerous (10–many), yellowish or brownish at maturity; inflorescence tending to be compound, at least its lower nodes with 2 or more spikes crowded on a lateral branch 31

29 Perigynia elliptic, essentially beakless, very plump (nearly terete) and filled by the achene; at least the lower spikes well separated, containing 1–5 perigynia . ... **Carex sect. Dispermae**

29 Perigynia ± ovate, beaked, plano-convex or lenticular; spikes various 30

30 Mature perigynia brownish; some spikes (especially terminal) entirely or mostly staminate or staminate at their bases only .... **Carex sect. Stellulatae**

30 Mature (not over-ripe) perigynia generally greenish; no spikes entirely or mostly staminate (a few may have stamens at their base in addition to their apex) .... **Carex sect. Phaestoglochin**

31 Pistillate scales terminating in a distinct rough awn; bracts, at least lower ones, very slender and exceeding spikes or branches; ventral surface of leaf sheaths usually transversely wrinkled or puckered (very rarely smooth) .... **Carex sect. Multiflorae**

31 Pistillate scales acute or minutely cuspidate; bracts mostly short, inconspicuous, or absent; leaf sheaths smooth ventrally .... **Carex sect. Heleoglochin**

32 Perigynia at least sparsely puberulent, pubescent, hispidulous, or scabrous 33

32 Perigynia glabrous (in some species, papillose or granular, but not even sparsely puberulent or scabrous) 42

33 Perigynia 12–18 mm long, in 1–2 short-oblong to spherical spikes 2–3.5 cm wide ...... **Carex sect. Lupulinae**

33 Perigynia 2–11 mm long, in 2–5 ± elongate, cylindrical spikes less than 2 cm wide 34

34 Perigynia with distinct and definite slender beak and/or the apex with 2 firm teeth 35

34 Perigynia beakless or merely apiculate ("beak" not over 0.4 mm long) and the apex not toothed 41

35 Leaves hairy 36

35 Leaves glabrous (often rough or scabrous, but not hairy) 37

36 Beak of perigynium with minute, scarcely visible teeth; body of perigynium strongly 3-angled, closely enveloping the achene, essentially nerveless, tapered to a stalk-like base; stems pubescent .... **Carex sect. Hirtifoliae**

36 Beak of perigynium with strong spreading teeth 0.8 mm or more long; body of perigynium ± rounded, loosely enveloping achene (especially at summit), strongly ribbed, ± rounded (not cuneate-tapered) at base; stems glabrous .... **Carex sect. Carex**

37 Pistillate spikes not over 10 mm long (occasionally 12 mm in *C. communis*); achenes mostly with very convex or rounded sides (the angles thus obscured), at least apically, very tightly enveloped by the perigynium, especially on the apical half; anthers 1.5–3.7 mm long; plants of dryish habitats ...... **Carex sect. Acrocystis**

37 Pistillate spikes mostly over 10 mm long; achenes with flattish to slightly concave sides (the angles thus ± evident), the summit (especially around base of style) ± loosely enveloped by the perigynium; anthers 2.5–4.7 mm long; plants of dry to wet habitats 38

38 Perigynium beak usually more than half as long as the body, the apex not or weakly and obscurely toothed; perigynia scabrous or with short stiff ascending hairs 39

38 Perigynium beak less than half as long as the body, with two firm apical teeth; perigynia ± densely short-hairy 40

39 Perigynia conspicuously 6–8 nerved; spikes densely flowered, with 20–75 perigynia; basal sheaths pale brown .... **Carex sect. Anomalae**

39 Perigynia 2-ribbed, otherwise nerveless; spikes very loosely flowered with only 3–6 (–8) perigynia; basal sheaths reddish purple .... **Carex sect. Hymenochlaenae** (*C. assiniboinensis*)

40 Perigynia 6–11 mm long, beak teeth 1.2–2.3 mm long, inner band of upper sheaths strongly purple-red tinged and thickened at apex, the thickened reddish portion opaque, smooth .... **Carex sect. Carex**

40 Perigynia 2.5–6.5 mm long, beak teeth 0.2–0.8 mm long, inner band of upper sheaths whitish to brown, brown- or purple-dotted, but not uniformly colored, not strongly opaque-thickened at apex, often scabrous ... . **Carex sect. Paludosae**

41 Leaf sheaths (and usually the blades) ± pubescent, especially toward base of plant; terminal spike pistillate toward apex, staminate toward base .... **Carex sect. Porocystis**

41 Leaf sheaths and blades glabrous; terminal spike staminate toward apex or entirely staminate .... **Carex sect. Digitatae**

42 Leaf sheaths (at least at apex) finely pubescent; blades often also pubescent or at least strongly hispidulous, especially toward base of plant 43

42 Leaf sheaths and blades completely glabrous (though sometimes scabrous) 46

43 Beak of perygynium with firm teeth ca. 1.5–3 mm long; perygynia ca. 8–10 mm long, in spikes 4–12 cm long . . . . **Carex sect. Carex** (*C. atherodes*)

43 Beak of perygynium with teeth scarcely 0.5 mm long or absent; perygynia less than 6 mm long, in spikes less than 3 cm long 44

44 Basal sheaths pale brown, leaf blades and stems glabrous or scabrous, perygynia with ca. 50 fine, impressed nerves . . . . **Carex sect. Griseae** (*C. hitchcockiana*)

44 Basal sheaths reddish purple tinged, leaf blades and stems pubescent, perygynia 5–12 nerved 45

45 Pistillate spikes laxly spreading or drooping on slender peduncles, the lowest (20–) 25–60 mm long (including portion inside sheath, if any); perygynia tapering to distinct beak . . . . **Carex sect. Hymenochlaenae**

45 Pistillate spikes erect or ascending, sessile, short-peduncled or on stiff, erect peduncles less than 20 (–25) mm long; perygynia beakless . . . . . . **Carex sect. Porocystis**

46 Perygynia ± rounded to broadly tapered at summit, beakless or essentially so (the tiny beak or apiculus less than 0.5 mm long if distinct, or up to 0.8 mm long if vaguely defined, often strongly bent or curved); beak or apiculus (if present) never toothed (or teeth scarcely 0.1 mm long) 47

46 Perygynium abruptly contracted or more gradually tapering to a definite slender beak 0.5 mm or more long, or to an indistinct tapering beak 1 mm or more long; beak in some species with short apical teeth 57

47 Leaf blades not over 0.5 mm wide, linear-filiform; perygynia dark brown or nearly black at maturity, 2 mm or less long, in few-flowered spikes, of which at least the upper ones are on peduncles usually surpassing the sessile staminate spike . . . . **Carex sect. Albae**

47 Leaf blades 0.5 mm or more wide; perygynia and spikes various (but not as above) 48

48 Bract of lowest pistillate spike sheathless (at most with a thin scarious sheath 1–3 mm long) 49

48 Bract of lowest pistillate spike with a sheath 4 mm or more long 51

49 Terminal spike partly pistillate; pistillate spikes nearly or quite sessile and erect or ascending; roots glabrous or nearly so . . . . **Carex sect. Racemosae**

49 Terminal spike normally entirely staminate; spikes and roots various 50

50 Pistillate spikes mostly drooping at maturity on slender peduncles; species of wet peatlands with roots with dense felt-like pubescence . . . . . . **Carex sect. Limosae**

50 Pistillate spikes erect or ascending, sessile or peduncled; roots glabrous **Carex sect. Vesicariae** (*C. oligosperma*)

51 Terminal spike bearing some perygynia (very rarely a few individuals with one entirely staminate); plants very strongly reddish tinged at base 52

51 Terminal spike entirely staminate; plants reddish or not at base 53

52 Staminate flowers at apex of terminal spike, pistillate flowers at base; cauline sheaths bladeless or with rudimentary blades to 2 (rarely 4) cm long; pistillate spikes short-cylindric, bearing fewer than 10 perygynia, very long-peduncled, some elongate peduncles usually arising from base of plant . . . . **Carex sect. Digitatae** (*C. pedunculata*)

52 Staminate flowers at base of terminal spike, pistillate flowers at apex; cauline sheaths with well-developed blades; pistillate spikes linear-cylindric, bearing more than 10 perygynia, on peduncles about as long as the spike or shorter, all arising from the upper part of the stem . . . . **Carex sect. Hymenochlaenae**

53 Perygynia concave- or at least cuneate-tapering toward the base, ± 3-angled and often somewhat broadly spindle-shaped 54

53 Perygynia convex-rounded toward the base, nearly or quite circular in cross-section (or very obscurely triangular), ellipsoid-cylindric to nearly spherical 55

54 Plants with elongate deep or shallow rhizomes and very slender, firm stems; leaf blades 1–4 mm wide . . . . **Carex sect. Paniceae**

54 Plants without elongate rhizomes, the stems sharply triangular, sometimes nearly wing-margined, rather weak and easily compressed, soon shriveling after maturity of the fruit; leaf blades usually more than 4 (and up to 35) mm wide . . . . **Carex sect. Laxiflorae**

55 Larger perygynia 4–5 mm long, the nerves not raised above the surface at maturity . . . . **Carex sect. Griseae**

55 Larger perygynia 2–3.5 mm long; nerves various 56

56 Perygynia with the nerves not raised above the surface, usually ± impressed; staminate spike usually long-peduncled; plants not strongly rhizomatous nor with any pistillate spikes on basal peduncles . . . . **Carex sect. Griseae** (*C. conoidea*)

56 Perygynia with the nerves slightly raised above the surface; staminate spike nearly or quite sessile or, if long-peduncled, the plants strongly rhizomatous and with basal pistillate spikes . . . . **Carex sect. Granulares**

57 Lower pistillate scales leaflike or bract-like, much exceeding the perygynia; achenes abruptly constricted to a short thick base; body of perygynium nearly terete, essentially nerveless except for 2 ribs; anthers 0.5–1.6 mm long . . . . **Carex sect. Phyllostachyae**

57 Lower pistillate scales scarcely if at all exceeding the perygynia; achenes not abruptly constricted at the base; perygynia and anthers various 58

58 Perygynia in densely crowded spherical to very short-cylindric spikes, spreading and with the lowermost usually reflexed, usually strongly few-ribbed; at least the uppermost pistillate spikes ± sessile and often crowded; the terminal spike (staminate or partly pistillate) often sessile or short-peduncled 59

58 Perygynia in elongate or long-peduncled spikes or both, all ascending, 2-ribbed or variously many-nerved; inflorescences various, but the upper spikes often not crowded and the terminal spike often long peduncled 60

59 Perigynia 2–6.2 mm long; basal sheaths brown . . . . **Carex sect. Ceratocystis**

59 Perigynia 11–18 mm long; basal sheaths reddish purple tinged . . . . **Carex sect. Lupulinae**

60 Bract of lowest pistillate spike sheathless (or pistillate spikes all crowded at base of plant in Carex tonsa); check several stems; rarely, a pistillate spike will be borne abnormally low on the stem and this spike may then have a sheath, which should be disregarded in keying 61

60 Bract of lowest pistillate spike consistently with sheath 4 mm or more long 67

61 Pistillate scales subtending at least some of the perigynia terminated by a distinct slender scabrous awn; perigynia 3–9 mm long 62

61 Pistillate scales smooth-margined and awnless or very short-awned, or at most with a scabrous margin toward an acuminate (sometimes inrolled) apex (occasionally a long rough awn in species with perigynia more than 9 mm long); perigynia (4–) 4.5–18 mm long 63

62 Scales toward apex of pistillate spikes merely acuminate or with awns shorter than their bodies (the latter easily visible, about half as long as perigynia or longer); staminate spikes 2 or more; body of perigynium rather gradually tapered to a beak 1.5 mm long, including the short (not over 0.8 mm) teeth . **Carex sect. Paludosae**

62 Scales toward apex of pistillate spikes ordinarily with awns (as on the other pistillate scales) nearly or fully as long as their bodies (the latter small and mostly hidden among the bases of the densely crowded perigynia); staminate spike solitary (or very rarely a second smaller one present); body of perigynium tapered or strongly contracted into a beak 1.2–3.5 mm long, including teeth up to 2.2 mm long . . . . **Carex sect. Vesicariae**

63 Basal sheaths pale brown; perigynia very narrowly lanceolate, 4–6.5 times as long as wide and not over 3 mm wide, many-nerved, tapering to apex (not strongly contracted into a beak); staminate spike solitary (pistillate spikes may be staminate at apex) . . . . **Carex sect. Rostrales**

63 Basal sheaths reddish purple tinged, at least on the youngest shoots; perigynia lanceolate or broader, less than 4 times as long as wide, or more than 3 mm wide, or strongly contracted into a conspicuous beak (or all of these); staminate spikes solitary or 2 or more 64

64 Perigynia strongly inflated, not tight around the achene, 2–8 mm wide 65

64 Perigynia not inflated, ± tightly enclosing achene, 1–1.6 mm wide 66

65 Perigynia 4–12 mm long, 6–12 (–15)-nerved . . . . **Carex sect. Vesicariae**

65 Perigynia 12–17 (–18) mm long, 15–20-nerved . . . . **Carex sect. Lupulinae**

66 Pistillate spikes linear-cylindric, drooping or curving on slender peduncles; perigynia (somewhat twisted) and achenes strongly angled, the latter with concave sides; tall plants (stems over 3 dm high) with scattered thin leaves . . . . . **Carex sect. Hymenochlaenae** (C. prasina)

66 Pistillate spikes short, thick, and few-flowered, often crowded at base of plant; perigynia and achenes very convex-sided; low plants (stems less than 1 dm high) with crowded, very stiff leaves . . . . **Carex sect. Acrocystis** (C. tonsa)

67 Perigynia with several to many conspicuous fine nerves on each side 68

67 Perigynia with 2 (–3) main ribs, the sides otherwise nerveless or with much less prominent nerves 71

68 Nerves of perigynia very numerous (20–65) and impressed, giving a longitudinally corrugated appearance; awns of pistillate scales rough or even ciliate . . . . **Carex sect. Griseae**

68 Nerves of perigynia several to many (5–40) and slightly raised; awns of pistillate scales absent, smooth, or rough 69

69 Awns rough and/or summit of pistillate scales minutely ciliate; lower spikes drooping on long very thin peduncles; beak slightly bidentate at maturity; plants strongly reddish at base . . . . **Carex sect. Hymenochlaenae**

69 Awns of pistillate scales usually smooth or absent; lower spikes mostly not drooping; beak not bidentate; plants pale, brown, or reddish at base 70

70 Perigynia ± sharply triangular with flattish sides, short-tapering at the base; stems bluntly trigonous, firm and not easily compressed; anthers mostly 3–4.5 mm long or lower pistillate spikes on elongate filiform spreading or drooping peduncles . . . . **Carex sect. Careyanae**

70 Perigynia ± rounded-triangular with swollen sides, long-tapering to a ± stalk-like base; stems sharply triangular to nearly wing-margined, easily compressed; anthers mostly 1.5–3 mm long and lower pistillate spikes usually on erect or ascending peduncles . . . . **Carex sect. Laxiflorae**

71 Lowermost pistillate spikes erect or ascending at maturity 72

71 Lowermost pistillate spikes drooping on long slender peduncles at maturity 73

72 Staminate spike well-peduncled; perigynia ± convex-sided toward the base; bracts with poorly developed blades; plants mat-forming from long-creeping rhizomes . . . . **Carex section Paniceae** (C. vaginata)

72 Staminate spike sessile or nearly so; perigynia tapered-cuneate toward the base; bracts with well-developed blades; plants tufted . . . . **Carex sect. Laxiflorae** (C. leptonervia)

73 Pistillate spikes not over 15 mm long . . . . **Carex sect. Chlorostachyae**

73 Pistillate spikes mostly 20 mm or more long . . . . **Carex sect. Hymenochlaenae**

Carex Section Acrocystis

First sedges to flower each year, fruits maturing in spring and soon shed. Basal leaf sheaths in most species becoming fibrous with age. Perigynium beaks bidentate, less than 0.5 mm long. Most common in dry woods, prairies, and open sandy places; less common in mesic woods or wetlands.

1. Pistillate spikes on stems of varying length, at least some of the stems short (up to 5 cm long) and partly hidden among the tufted leaf bases; anthers 1.5–2 mm long. 2
1. Pistillate spikes all on elongate stems (none borne on short basal peduncles); anthers various. 4
2. Bract of the lowest non-basal pistillate spike leaflike, equaling or exceeding the tip of the staminate spike; remnants of old leaves only slightly breaking into fibrous shreds at the base. .... **C. deflexa**
2. Bract of the lowest non-basal pistillate spike scale-like or bristle-like, not exceeding the staminate spike (or all spikes often on short basal stems, but foliage and stems stiffer and much more scabrous than in *C. deflexa*, which nearly always have some elongate stems); remnants of old leaves breaking into copious fibrous shreds at the base 3
3. Perigynia 3.2–4 mm long, the beak 1.2–1.6 (–2) mm, about half as long as the body or longer. ..... **C. tonsa**
3. Perigynia 2.5–2.9 mm long, the beak 0.4–0.9 mm, about 1/4–1/3 as long as the body. .... **C. umbellata**
4. Main body of perigynium, not including spongy-tapered base or beak, orbicular to short-obovoid, about the same diameter as length; anthers 2.1–3.7 mm long; plants either with the widest leaves 3–8 mm broad or with elongate shallow rhizomes. 5
4. Main body of perigynium ± elliptic (to slightly obovoid or oblong), definitely longer than thick; anthers 1.3–2.5 mm long; plants with mostly narrow leaves and lacking stout elongate rhizomes 8
5. Widest leaves (at least the oldest dry ones) 3–5 mm broad; stem leaves above base of plant (when present on stem) usually with the ligule longer than the width of the leaf; bract subtending the middle (and sometimes the lowest) pistillate spike(s) ± scarious-lobed at base, blade awn-like to leaflike, usually green, arising from between the lobes; staminate spike 1–2 mm thick; plants without elongate rhizomes. .... **C. communis**
5. Widest leaves 1.5–3 mm broad; stem leaves with ligule no longer than the width; bracts subtending middle pistillate spikes tapered to apex, without an elongate awn-like or leaflike blade (the lowermost bract often green but seldom lobed); staminate spike 2–4 mm thick; plants with stout, shallow elongate rhizomes with fibrous sheaths. 6
6. Larger perigynia 1.7–2.2 mm wide .... **C. inops**
6. Larger perigynia 1.2–1.7 mm wide 7
7. Beak of perigynium 1–1.6 mm, half or more as long as the body .... **C. lucorum**
7. Beak of perigynium 0.2–0.8 mm, much less than half as long as the body. .... **C. pensylvanica**
8. Widest leaves (at least the oldest dry ones) 3–5 mm broad; bract subtending the middle (and sometimes also the lowest) pistillate spike(s) ± scarious-lobed at base, the blade awn-like or leaflike, usually green, arising from between the lobes .... **C. communis**
8. Widest leaves not over 3 mm broad; bracts either scale-like or leaflike and lacking a scarious-lobed base. 9
9. Lower two pistillate spikes 7.5–22 mm distant; lowest inflorescence bracts 18–35 mm long, 3/4 as long to exceeding inflorescence; loosely mat-forming from delicate, ascending rhizomes .... **C. novae-angliae**
9. Lower two pistillate spikes mostly close together, up to 7 mm distant; lowest inflorescence bracts rarely more than 17 mm long, often less than 3/4 as long the inflorescence; ± tufted 10
10. Perigynia 2–3 mm long, minutely puberulent to short-hairy; stems very slender (seldom over 0.4 mm thick) and mostly surpassed by the leaves .... **C. deflexa**
10. Perigynia 3–4 mm long, definitely short-hairy; stems usually 1 mm or more in thickness and surpassing the leaves. .... **C. peckii**

Carex communis Bailey
FIBROUS-ROOT SEDGE   native
IR | WUP | CUP | EUP   CC 2

Plants tufted; rootstocks short, ascending, scaly, reddish purple. Stems 1.5–5 dm long, rough on angles above, purplish red at base, the old leaves conspicuous; sterile shoots numerous. Well-developed leaves several to a fertile culm, near base; blades 2–5 cm long and 2–4 mm wide, flat, flaccid, light-green, rough especially towards the tip and on margins. Terminal spike staminate; lateral spikes 2–3, pistillate; lowest bract scale-like, hyaline-margined and purplish tinged at base, the upper reduced or scale-like; scales reddish purple or -brown with hyaline margins and 3-nerved green or straw-colored center. Perigynia 3–10 to a spike, 3–3.5 mm long, ascending, light-green, puberulent, 2-keeled, the spongy base 0.75 mm long; beak 0.5 mm, flattish, bidentate. Achenes triangular with convex sides, light-brown with lighter angles, minutely pitted, truncate and bent-apiculate; stigmas 3, reddish brown.

Common, mesic forests.

Carex communis
perigynium (l)
pistillate scale (r)

Carex deflexa Hornem.
NORTHERN SEDGE   native
IR | WUP | CUP | EUP   CC 5

Plants loosely tufted. Stems 1–2 dm long, purple-tinged at base, shorter than the leaves. Leaves soft, 1–3 mm wide. Spikes either staminate or pistillate; staminate spike short, to 5 mm long; pistillate spikes on long, slender stalks near base of plant and also

2–4 spikes on stem near staminate spike; bract leaflike, to 2 cm long; pistillate scales ovate, shorter than perigynia. Perigynia green, oblong-ovate, 2–3 mm long, covered with short hairs, abruptly tapered to a small beak about 0.5 mm long. Achenes 3-angled; stigmas 3. June–Aug.

Moist woods and swamps, wetland margins, often where sandy or in sphagnum moss.

Carex inops Bailey
LONG-STOLON SEDGE native
IR | **WUP** | CUP | EUP CC 7

Carex heliophila Mackenzie
Carex pensylvanica Lam. var. *digyna* Boeckl.

Rootstocks slender; stolons long, slender, horizontal. Stems to 35 cm long, stiff, wiry, rough on angles above, reddish brown-tinged and fibrillose at base, clothed with old leaves. Well-developed leaves 5–10 to a fertile culm; blades 4–20 cm long and 1–2.5 mm wide, channeled towards base, with revolute margins, thin, stiff, dull-green, roughened, attenuate; lower sheaths breaking and filamentose. Terminal spike staminate (occasionally gynaecandrous); lateral spikes 1–2 (–3), pistillate; lowest bract scale-like, reddish brown at base, the upper reduced; scales reddish brown or tawny, with white-hyaline margins and 1–3-nerved lighter center. Perigynia 5–15 to a spike, ascending, dull-green, puberulent, 2-keeled, spongy at base; beak 0.75 mm long, serrulate, bidentate. Achenes triangular with convex sides and sharp angles, closely enveloped, minutely apiculate; stigmas 3, reddish brown.

Sandy woods and fields; rare in Keweenaw County.

Carex lucorum Willd ex Link
BLUE RIDGE SEDGE native
IR | WUP | **CUP** | **EUP** CC 4

Carex pensylvanica Lam. var. *distans* Peck

Plants tufted and stoloniferous; stolons horizontal, slender, scaly, reddish. Stems 1–3 dm long, roughened on angles above, reddish purple at base, clothed with old leaves, often fibrillose; sterile shoots lateral, long, reddish purple at base, the sheaths puberulent, becoming filamentose. Well-developed leaves 2-several to a fertile culm; blades to 3 cm long and 1.5–2.5 mm wide, flat or canaliculate, deep-green, often roughened, especially towards the tip. Terminal spike staminate; lateral spikes 2–3, pistillate; bracts scale-like, enlarged at base, hyaline-margined and reddish purble; scales reddish purple with white-hyaline margins and lighter center. Perigynia 4–10 to a spike, 3.5–4 mm long, ascending or spreading-ascending, dull- or yellowish green, puberulent, 2-keeled, the spongy base 0.5–0.75 mm long; beak 1.5–2 mm long, bidentate, hyaline and purplish tinged at mouth. Achenes triangular with convex sides and narrow angles, brown, minutely pitted, minutely apiculate; stigmas 3, brown.

Dry woods.

Carex novae-angliae Schwein.
NEW ENGLAND SEDGE (THR) native
IR | WUP | **CUP** | **EUP** FACU CC 9

Plants delicate, loosely tufted and stoloniferous; stolons slender, scaly. Stems to 40 cm long, rough above, reddish purple and fibrillose at base, the old leaves conspicuous; sterile shoots lateral, long, the leaves mostly near top. Well-developed leaves 1-several to a fertile culm, on lower 1/3; blades to 15 cm long and 1–1.5 mm wide, thin, flaccid, soft, pale-green, roughened on margins and towards tip. Terminal spike staminate; lateral spikes 2–3, pistillate; scales cuspidate, hyaline, often reddish brown-tinged, the midvein green. Perigynia 2–10 to a spike, 2.5 mm long, ascending, light-green or yellowish brown, sparsely appressed pubescent, 2-ridged, the spongy base 0.5 mm long; beak to 0.5 mm long, bidentate. Achenes dark-brown, triangular with convex sides and blunt greenish angles, minutely apiculate; stigmas 3, dark-reddish brown.

Moist deciduous woods, often on hummocks.

Carex peckii Howe
PECK'S SEDGE native
IR | **WUP** | **CUP** | **EUP** CC 3

Carex nigromarginata Schwein. var. *ellip*tica (Boott) Gleason

Plants tufted and stoloniferous; rootstocks slender, scaly. Stems to 65 cm long, roughened beneath spikes, reddish purple at base; sterile stems long, the well developed leaves towards the top. Well-developed leaves several to a fertile culm, on lower 1/4; blades 1.5–4 cm long and 1–1.5 mm wide (larger on sterile stems), flat, green, roughened on margins and towards apex. Terminal spike staminate; lateral spikes pistillate, in an inflorescence 8–20 mm long; scales reddish brown with white-hyaline margins. Perigynia 3–12 to a spike, 3.5 mm long, ascending, grayish or yellowish green, hirsute-pubescent, 2-ridged, the base spongy, 0.5 mm long; beak 0.5 mm long, obliquely cut, bidentate, hyaline at mouth. Achenes yellowish brown, triangular with convex sides and blunt green angles, minutely apiculate; stigmas 3, dark-reddish brown.

Open woods.

Carex pensylvanica Lam.
PENNSYLVANIA SEDGE native
IR | **WUP** | **CUP** | **EUP** CC 4

Plants tufted and stoloniferous; stolons horizontal, slender, scaly, fibrillose, reddish. Stems 5–40 cm long, smooth or roughened on angles above, reddish purple at base, clothed with old leaves, often fibrillose;

sterile shoots reddish purple at base, the sheaths puberulent, becoming filamentose. Well-developed leaves 2-several to a fertile culm; blades to 3 cm long and 1.5–3 mm wide, flat above, canaliculate towards base, often roughened especially towards tip. Terminal spike staminate; lateral spikes 1–4, pistillate; bracts scale-like, enlarged at base, hyaline-margined and reddish brown; pistillate scales reddish purple, with white-hyaline margins and lighter center. Perigynia 4–20 to a spike, 2.5–3 mm long, ascending, dull- or yellowish green, puberulent, 2-keeled, the spongy base to 0.75 mm long; beak 0.75 mm long, bidentate, hyaline and often purplish tinged at orifice. Achenes triangular with convex sides and narrow angles, brown, minutely pitted, minutely apiculate; stigmas 3, reddish brown.

Common in a wide range of dry to mesic woods and prairies.

Carex tonsa (Fern.) Bickn.
SHAVED SEDGE native
IR | WUP | CUP | EUP CC 5
Carex rugosperma Mackenzie
Plants tufted; rootstocks stoutish, branching; stolons short-ascending. Stems to 15 cm long, stiff, roughened, reddish brown-tinged and fibrillose at base. Leaves numerous; blades 5–25 cm long and 2–4.5 mm wide, channeled with revolute margins, thick, stiff, deep-green, rough towards the tip. Terminal spike staminate; pistillate spike occasionally present near terminal spike; basal pistillate spikes long-peduncled; bract of upper spike setaceous, reddish at base; pistillate scales conspicuous, whitish or straw-colored, with 3-nerved greenish or straw- colored center. Perigynia 3–20 to a spike, 3.5–4.5 mm long, appressed-ascending, compressed-orbicular, somewhat leathery, light- green, sparsely pubescent, 2-keeled, the base 0.75 mm long; beak to 2.5 mm long, 2-edged, serrulate, bidentate. Achenes triangular with convex sides and sharp angles, brownish, shiny, truncate and minutely apiculate; stigmas 3.

Dry sandy fields and open woods.

Carex umbellata Schkuhr
PARASOL SEDGE native
IR | WUP | CUP | EUP CC 5
Plants densely tufted; rootstocks short, stout. Stems to 15 cm long, stiff, rough on angles, reddish brown-tinged and fibrillose at base. Leaves numerous; blades to 3 dm long and 1.5–2.5 mm wide, channeled towards base, flat and rough above, with revolute margins, firm, light-green. Terminal spike staminate; lateral spikes 3–4, pistillate or androgynous; bract of upper spike scale-like, reddish tinged at base; pistillate scales hyaline with several-nerved green center, the upper reddish brown-tinged. Perigynia 4–20 to a spike, 2–3 mm long, ascending, triangular-orbicular, dull-green, pubescent above, 2-keeled, the base 0.5 mm long; beak to 1 mm long, 2-edged, bidentate, hyaline-tipped. Achenes triangular with convex sides and sharp angles, filling perigynia, brownish black, shining, minutely pitted, minutely apiculate; stigmas 3.

Dry, often calcareous fields and prairies.

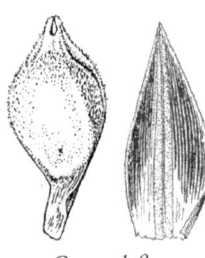

Carex deflexa
perigynium (l)
pistillate scale (r)

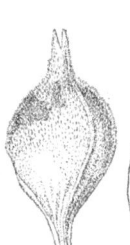

Carex inops
perigynium (l)
pistillate scale (r)

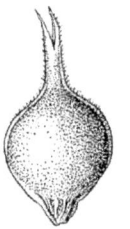

Carex lucorum
perigynium (l)
pistillate scale (r)

Carex novae-angliae
perigynium (l)
pistillate scale (r)

Carex peckii
perigynium (l)
pistillate scale (r)

Carex pensylvanica
perigynium (l)
pistillate scale (r)

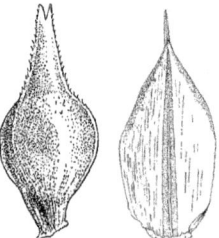

Carex tonsa
perigynium (l)
pistillate scale (r)

Carex umbellata
perigynium (l)
pistillate scale (r)

CYPERACEAE | Sedge Family

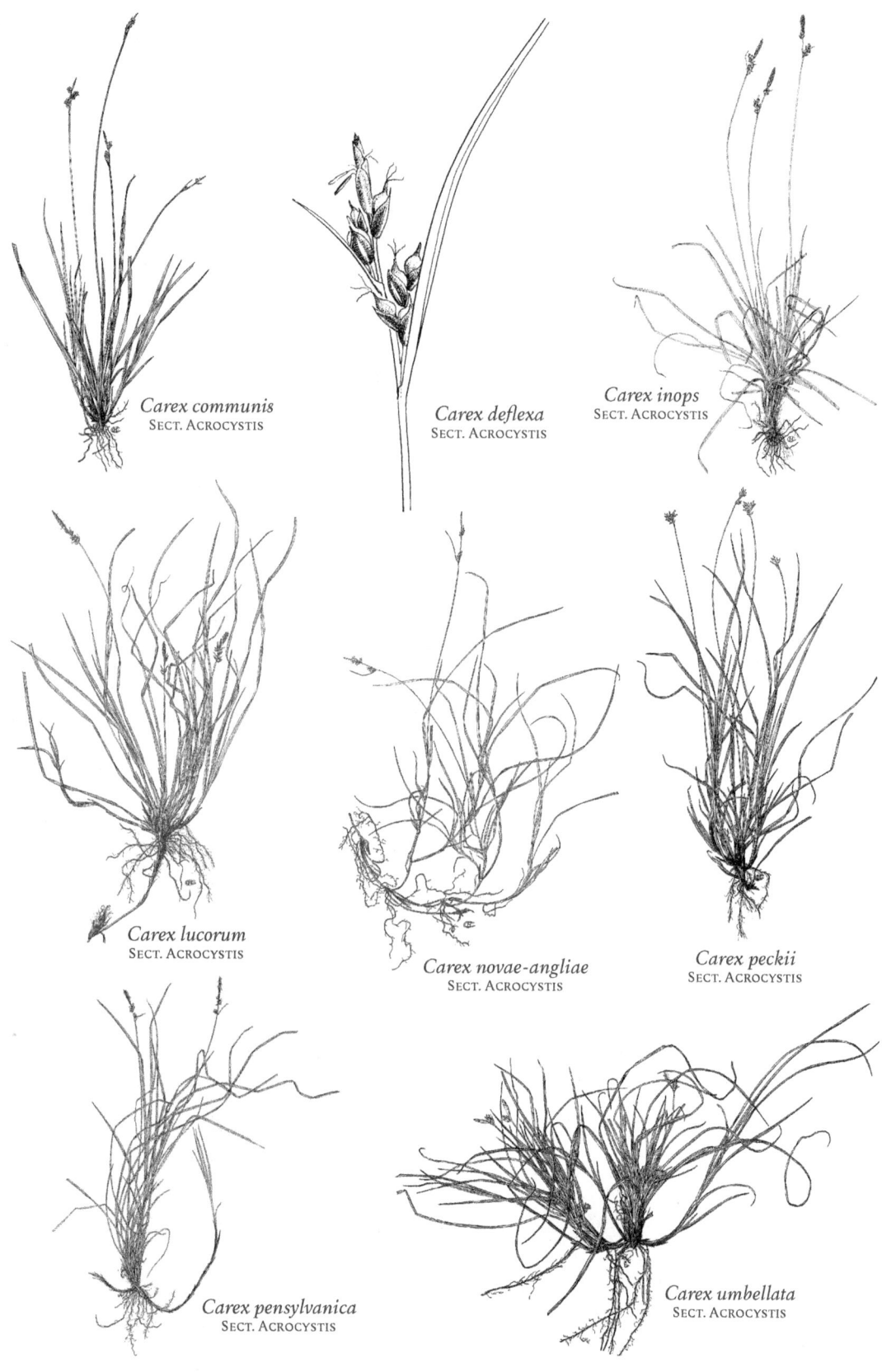

Carex Section Albae

One member of the section in the Upper Peninsula. Rhizomes elongate, the plants forming mats. Leaf blades involute, wiry. Perigynia becoming dark in age, beaks short, white-tipped.

Carex eburnea Boott
BRISTLE-LEAF SEDGE   *native*
IR | WUP | CUP | EUP   FACU CC 7

Plants tufted; rootstocks long, slender, brownish. Stems 1–3.5 dm, obtusely triangular, brownish tinged at base. Well-developed leaves 3–6 to a fertile culm, near base; blades 5–25 cm long and 0.5 mm wide, often recurved-spreading, involute, firm, green, roughened. Terminal spike staminate; lateral spikes 2–4, pistillate, on peduncles 1–2.5 cm long; bracts bladeless, tubular, greenish or greenish yellow with white margins; pistillate scales whitish with green midrib, often yellowish brown-tinged. Perigynia 2–6 to a spike, 2 mm long, triangular, light-green or brownish, shining, puncticulate, 2-ribbed, finely nerved; beak short, cylindric, obliquely cut, hyaline at orifice. Achenes triangular with concave sides and thickened angles, closely enveloped, brown, granular, apiculate; stigmas 3, brownish.

Dry sand prairies, and rarely in fens.

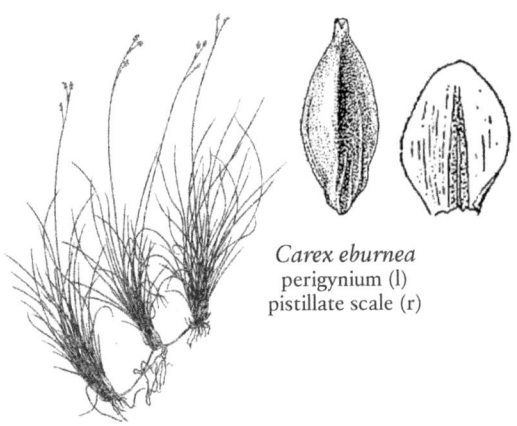

Carex eburnea
perigynium (l)
pistillate scale (r)

Carex Section Ammoglochin

One member of the section in the Upper Peninsula.

Carex siccata Dewey
DRY-SPIKE SEDGE   *native*
IR | WUP | CUP | EUP   CC 5

Plants tufted; rootstocks short, black, fibrillose. Stems 4–10 dm long, often nodding, roughened on angles beneath head, brownish at base, clothed with old leaves. Well-developed leaves 3–5 to a fertile culm, on lower third; blades 1–4 dm long and 2–4.5 mm wide, flat, green, roughened towards tip and on margins; sheaths green-and-white-mottled dorsally. Spikes 4–15, gynaecandrous, in a flexuous linear inflorescence 2–6 cm long; bracts scale-like; scales silvery-green, often brownish tinged, with 3-nerved green center; staminate flowers few except in terminal spike. Perigynia 6–20 to a spike, 3–4.5 mm long, appressed-ascending, nearly concealed by scales, green or silvery-green, winged, strongly nerved, serrulate; beak 1–1.5 mm long, flat, serrulate obliquely cut, bidentate, hyaline-tipped, the orifice white-margined. Achenes lenticular, dull-yellowish brown, apiculate; stigmas 2, dark-reddish brown.

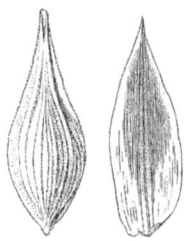

Carex siccata
perigynium (l)
pistillate scale (r)

Dry sandy prairies and woods.

Carex Section Anomalae

One member of the section in the Upper Peninsula. Upper surface of leaf blade and perigynia are scabrous.

Carex scabrata Schwein.
EASTERN ROUGH SEDGE   *native*
IR | WUP | CUP | EUP   OBL CC 4

Plants colony-forming, rough-to-touch. Stems loosely clustered, 4–9 dm long. Leaves 4–14 mm wide, lowest leaves not reduced to scales. Spikes either staminate or pistillate; staminate spike single, 2–4 cm long, short-stalked; pistillate spikes 3–6, cylindric, 2–4 cm long, upright, the lower on long stalks, the upper stalkless or short-stalked; bracts leaflike; pistillate scales lance-shaped, about as long as the perigynia, tapered to a tip. Perigynia obovate, 3-angled, 2-ribbed, 3–5 mm long, finely coarse-hairy, few-nerved, abruptly tapered to a slightly curved, notched beak. Achenes 3-angled; stigmas 3. May–Aug.

Carex scabrata
perigynium (l)
pistillate scale (r)

Low shaded areas in forests, streambanks, seeps.

Carex Section Bicolores

Plants short, colonial, loosely tufted, shoots arising singly or few in a clump; rhizomes elongate; bases brown. Terminal spike staminate or gynecandrous, hidden by the crowded lateral spikes. Perigynia plump, golden to whitish, weakly veined; margins and apex rounded, beakless to short-beaked. Stigmas 2. Calcium-rich sites where somewhat disturbed.

1 Mature perigynia golden-orange when fresh (drying dark brown or, especially if immature, ± white); terminal

Carex siccata
Sect. Ammoglochin

Carex scabrata
Sect. Anomalae

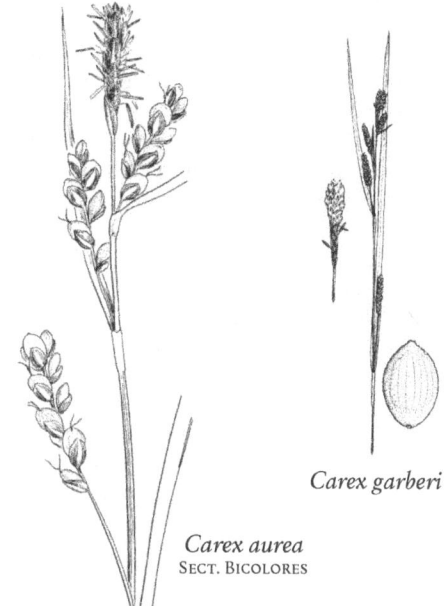

Carex garberi

Carex aurea
Sect. Bicolores

spikes mostly all staminate (occasionally with a very few perigynia); pistillate scales ± loosely spreading, distinctly shorter than the mature perigynia (usually averaging 3/4 or less as long), most of them acute to cuspidate.................................. **C. aurea**

1 Mature perigynia white-pulverulent when fresh; terminal spikes usually staminate at base only, with several to numerous perigynia apically; pistillate scales ± appressed, nearly (averaging about 3/4) to quite as long as the perigynia, most of them blunt to acute ....
.... **C. garberi**

Carex aurea Nutt.
GOLDEN-FRUIT SEDGE   native
IR | WUP | CUP | EUP   FACW CC 3

Plants small, loosely tufted. Stems upright, 3-angled, 5–30 cm long. Leaves 1–4 mm wide. Spikes 2–5 per stem, the lower spikes stalked; spikes at ends of stems staminate, 3–18 mm long; lateral spikes pistillate, 8–20 mm long, the spikes clustered to widely spaced; bract of lowest spike longer than the head; pistillate scales white-tinged to yellow-brown, with a green midvein, tipped with a short, sharp point, shorter than the perigynia. Perigynia with short white hairs when young, becoming a distinctive gold-orange when mature (drying paler), round to obovate, beakless or with a very short beak, several-ribbed, 2–3 mm long. Achenes dark brown to black, lens-shaped; stigmas 2. May–July.

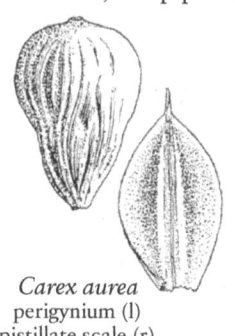

Carex aurea
perigynium (l)
pistillate scale (r)

Moist to wet meadows, low prairie, swales, wet woods and along sandy or gravelly shores; often where calcium-rich.

Carex garberi Fern.
ELK SEDGE   native
IR | WUP | CUP | EUP   FACW CC 8

Similar to *Carex aurea*; one distinction between the 2 species is terminal spike of *C. garberi* is tipped with pistillate flowers (with staminate flowers below); in *Carex aurea*, terminal spike is of staminate flowers only. Also, in *C. garberi*, the perigynia are more granular, more crowded, and more overlapping than in *C. aurea*.

Wet sandy, gravelly, or marly shores, limestone pavements, interdunal flats, and edges of cedar thickets.

Carex Section Carex

Plants typically colonial; rhizomes elongate. Vegetative stems prominent. Perigynia long-beaked with prominent beak teeth.

1 Perigynia covered with hairs....................... 2
1 Perigynia smooth and hairless 3
2 Inner band of the uppermost leaf sheaths red to purple and thickened at the summit, glabrous; native, Ontonagon County .... **C. trichocarpa**
2 Inner band of leaf sheaths not colored, pubescent; introduced in Mackinac County .... **C. hirta**
3 Inner band of the uppermost leaf sheaths red to purple and thickened at the summit, glabrous. **C. trichocarpa**
3 Inner band of leaf sheaths pale or brown, not thickened at the summit, glabrous or pubescent ... **C. atherodes**

Carex atherodes Spreng.
SLOUGH SEDGE native
IR | **WUP** | **CUP** | EUP OBL CC 5

Plants loosely tufted, from long scale-covered rhizomes. Stems 3-angled, 5–12 dm long. Leaves 3–12 mm wide; sheaths hairy on back, brown to purple-tinged at the mouth, the lower sheaths shredding into narrow strands. Spikes either staminate or pistillate; staminate spikes 2–6 at ends of stems; pistillate spikes 2–4, widely spaced, cylindrical, 2–11 cm long; bracts leaflike, longer than the stems; pistillate scales thin, translucent or pale brown, shorter than the perigynia, tipped with a slender awn. Perigynia ovate, 6–11 mm long, long-tapered to a smooth beak, with many distinct nerves, the beak with spreading teeth 1.5–3 mm long. Achenes 3-angled; stigmas 3. June–Aug.

Marshes, wet meadows, prairie swales, stream and pond margins, usually in shallow water where may form dense colonies.

Carex hirta L.
HAMMER SEDGE introduced
IR | WUP | CUP | **EUP**

Plants loosely tufted and stoloniferous; stolons long, stout, horizontal, scaly. Stems 2–10 dm long, obtusely triangular, brownish or purplish at base, the basal sheaths filamentose; sterile shoots long, the leaves clustered at top. Well-developed leaves 2–5 to a fertile culm, 5–25 cm long and 2–6 mm wide, flat, thin, light-green, soft-hairy, roughened towards the tip; sheaths white-pilose at mouth. Upper 1–3 spikes staminate; lower 2–3 spikes pistillate, the lowest nearly basal; bracts leaflike, the lowest strongly sheathing; scales ciliate and white-hairy, purplish brown with hyaline margins and 3-nerved green center. Perigynia 10–35 to a spike, 5–9 mm long, ascending, suborbicular, inflated, greenish straw-colored or light-brownish, white-pubescent, 15–20-ribbed; beak 1.5–2.5 mm, bidentate, the teeth hispidulous. Achenes triangular with obtuse angles, loosely enveloped, yellowish, apiculate; stigmas 3, blackish.

Introduced from Europe; reported from Mackinac County.

Carex trichocarpa Muhl.
HAIRY-FRUIT SEDGE native
IR | **WUP** | CUP | EUP OBL CC 8

Plants loosely tufted, with short rhizomes. Stems stout, 6–12 dm long, smooth below, rough-to-touch above. Leaves 2–6 mm wide, rough-to-touch on margins, upper leaves and bracts often longer than stems. Spikes either all staminate or pistillate, the upper 2–6 spikes staminate, long-stalked; pistillate spikes 2–4, cylindric, 4–10 cm long, the upper spikes more or less stalkless, the lower spikes on slender stalks; pistillate scales ovate, with white translucent margins, about half as long as perigynia. Perigynia ovate, usually covered with short white hairs, prominently ribbed, gradually tapered to a 2-toothed beak. Achenes 3-angled; stigmas 3. May–Aug.

Riverbanks and old river channels, marshes, wet meadows, low prairie; Ontonagon County.

Similar to slough sedge (*Carex atherodes*) but sheaths strongly purple-tinged at tip, the leaf blades not hairy on underside, and the perigynia with short white hairs (vs. smooth in *C. atherodes*).

Carex Section Careyanae
Resembling section Laxiflorae in appearance, but stems generally firm. Perigynia acutely angled, tightly enclosing the achene; veins more than 40, impressed in fresh plants, raised when dried.

Carex plantaginea Lam.
PLANTAIN-LEAF SEDGE native
IR | **WUP** | **CUP** | **EUP** CC 8

Plants tufted; rootstocks short. Stems 2.5–6 dm long, purple-tinged at base, the lower bladeless sheaths conspicuous. Basal leaves and those of sterile stems 15–35 cm long and 10–25 mm wide, flat, the mid-nerve prominent below, 2 lateral nerves prominent above, roughened on margins; fertile stem leaves bladeless or nearly so; sheaths purple-tinged. Terminal spike staminate; lateral spikes about 3, pistillate, the lower on slender peduncles; bracts bladeless, purple-tinged; pistillate scales white-hyaline with green midrib, purplish tinged. Perigynia 4–12

Carex atherodes
perigynium (l)
pistillate scale (r)

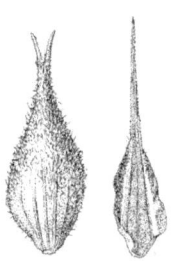
Carex hirta
perigynium (l)
pistillate scale (r)

Carex trichocarpa
perigynium (l)
pistillate scale (r)

Carex plantaginea
perigynium (l)
pistillate scale (r)

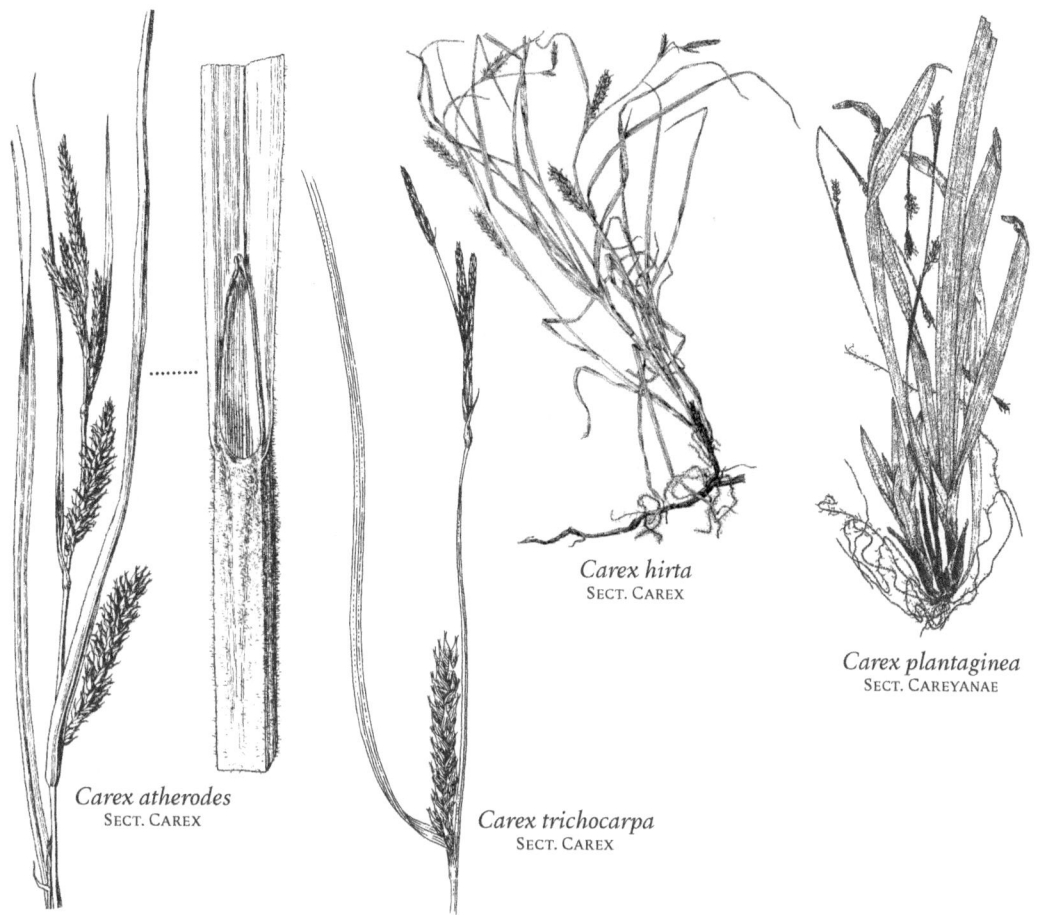

Carex atherodes
Sect. Carex

Carex trichocarpa
Sect. Carex

Carex hirta
Sect. Carex

Carex plantaginea
Sect. Careyanae

to a spike, 4–5 mm long, triangular; beak 1 mm, erect or curved, the orifice entire, hyaline. Achenes triangular with deeply concave sides, filling perigynia, brownish, apiculate; stigmas 3.

Mesic forests.

Carex Section Ceratocystis

Plants tufted; rhizomes short; bases brown. Terminal spike staminate, occasionally androgynous. Lateral spikes pistillate, densely flowered, globose to oblong. Perigynia strongly veined, abruptly beaked; beak toothed, generally reflexed. Stigmas 3. Usually where wet and calcareous.

1. Larger perigynia ca. 2–3 mm long, horizontally spreading, the beak about 1/4 to nearly 1/2 as long as the body .... **C. viridula**
1. Larger perigynia (3–) 3.5–6.2 mm long, at least the beaks becoming conspicuously reflexed on lower half of spike, the beak nearly or fully half as long as the body 2
2. Pistillate scales at maturity strongly flushed with shiny brown or reddish color, hence conspicuous in the spike; widest leaves 3–5 mm wide .... **C. flava**
2. Pistillate scales greenish or yellowish, the same color as the perigynia and essentially invisible in the spikes; widest leaves 1.5–4 mm wide .... **C. cryptolepis**

Carex cryptolepis Mackenzie
NORTHEASTERN SEDGE *native*
IR | WUP | CUP | EUP   OBL CC 8

Plants tufted. Stems 2–6 dm long and longer than leaves. Leaves 2–4 mm wide. Spikes staminate or pistillate; staminate spikes short-stalked or stalkless, the stalk shorter than the pistillate spikes; pistillate spikes 3–4, the upper 2 spikes grouped, the third separate, the fourth spike lower on stem, short-cylindric, 1–2 cm long, stalkless; bracts leaflike and spreading; pistillate scales narrowly ovate, same color as perigynia and as long as perigynia body. Perigynia yellow-brown when mature, lower ones curved outward and downward, body obovate, 3–5 mm long, 2-ribbed and several nerved, contracted into a smooth beak 1–1.5 mm long. Achenes 3-angled; stigmas 3. June–Aug.

Wet meadows and marshy areas, peatlands, swamp margins; often where calcium-rich. Similar to *C. flava*.

Carex flava L.
YELLOW-GREEN SEDGE native
IR | WUP | CUP | EUP OBL CC 4

Plants densely tufted, from short rootstocks. Stems stiff, 1–7 dm long, usually longer than the leaves. Leaves 4–8 to a stem, mostly near base, 3–5 mm wide. Terminal spike staminate (or rarely partly pistillate), stalkless or short-stalked; pistillate spikes 2–5, sometimes with staminate flowers at tip, the uppermost spikes nearly stalkless, the lower stalked; bracts conspicuous, leaflike, spreading outward, much longer than the head; pistillate scales ovate, narrower and much shorter than the perigynia, red-tinged except for the pale, three-nerved middle and the narrow translucent margins. Perigynia 15–35, crowded in several to many rows, 4–6 mm long, obovate, yellow-green becoming yellow with age, conspicuously ribbed, tapered to a slender, finely toothed beak about as long as the body, the tip notched. Achenes obovate, 3-angled, yellow-brown; stigmas 3. May–Aug.

Wet, peaty meadows, often where calcium-rich.

Carex viridula Michx.
LITTLE GREEN SEDGE native
IR | WUP | CUP | EUP OBL CC 4

Plants tufted. Stems stiff, slightly 3-angled, 0.5–4 dm long, longer than leaves. Leaves 1–3 mm wide; sheaths white-translucent. Spikes either staminate or pistillate (or sometimes mixed), the terminal spike staminate or with a few pistillate flowers at tip or middle, 3–15 mm long, short-stalked or stalkless, longer than the pistillate spikes or clustered with them; lateral spikes pistillate, 2–6, ovate to short-cylindric, 5–10 mm long, clustered and stalkless above, the lower spikes often separate and on short stalks; bracts leaflike, usually upright, much longer than the heads; pistillate scales brown on sides, rounded or with a short, sharp point, about equal to perigynia. Perigynia yellow-green to brown, rounded 3-angled, obovate, 2–4 mm long, 2-ribbed, tapered to a slightly notched beak 0.5–1 mm long. Achenes 3-angled; stigmas 3. May–Aug.

Wet meadows, sandy lake margins, fens and seeps; often where calcium-rich.

Carex Section Chlorostachyae
One member of the section in the Upper Peninsula. Plants small, densely tufted, with fibrous basal leaf sheaths and small beadlike perigynia borne in slender spikes on threadlike stalks.

Carex capillaris L.
HAIR-LIKE SEDGE native
IR | WUP | CUP | EUP FACW CC 9

Plants small, densely tufted. Stems slender, 3-angled, 1.5–4 dm long. Leaves mostly at base of plant and much shorter than stems, 1–3 mm wide; sheaths tight. Spikes either staminate or pistillate; terminal spike staminate, 4–8 mm long; lateral spikes 1–4, separated on stem, loosely flowered, short-cylindric, on threadlike, spreading to drooping stalks 5–15 mm long; pistillate scales white, translucent on outer edges, green or light brown in middle, blunt or acute at tip, shorter but usually wider than perigynia, deciduous. Perigynia shiny brown to olive-green, ovate, round in section, 2–4 mm long, 2-ribbed, otherwise without nerves, tapered to a translucent-tipped beak 0.5 mm or more long. Achenes 3-angled with concave sides; stigmas 3. June-July.

Alder thickets, wetland margins, usually in shade.

Carex Section Chordorrhizae
One member of the section in the Upper Peninsula. Plants stoloniferous, the stolons arching and rooting.

Carex chordorrhiza Ehrh.
ROPE-ROOT SEDGE native
IR | WUP | CUP | EUP OBL CC 10

Plants from long, creeping stems. Flowering stems upright, rounded 3-angled in section, 1–3 dm tall, single or several together, arising from axils of dried leaves on older, reclining sterile stems. Leaves several on stem, the lower ones often bladeless, 1–2 mm wide; sheaths translucent. Spikes 3–8, with both staminate and pistillate flowers, staminate flowers borne above pistillate, crowded in an ovate head 5–15 mm long; bracts absent; pistillate scales dark brown, ovate, about equaling the perigynia. Perigynia brown, compressed, ovate, 2–3.5 mm long, leathery,

Carex cryptolepis
perigynium (l)
pistillate scale (r)

Carex flava
perigynium (l)
pistillate scale (r)

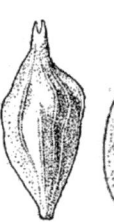

Carex viridula
perigynium (l)
pistillate scale (r)

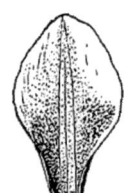

Carex capillaris
perigynium (l)
pistillate scale (r)

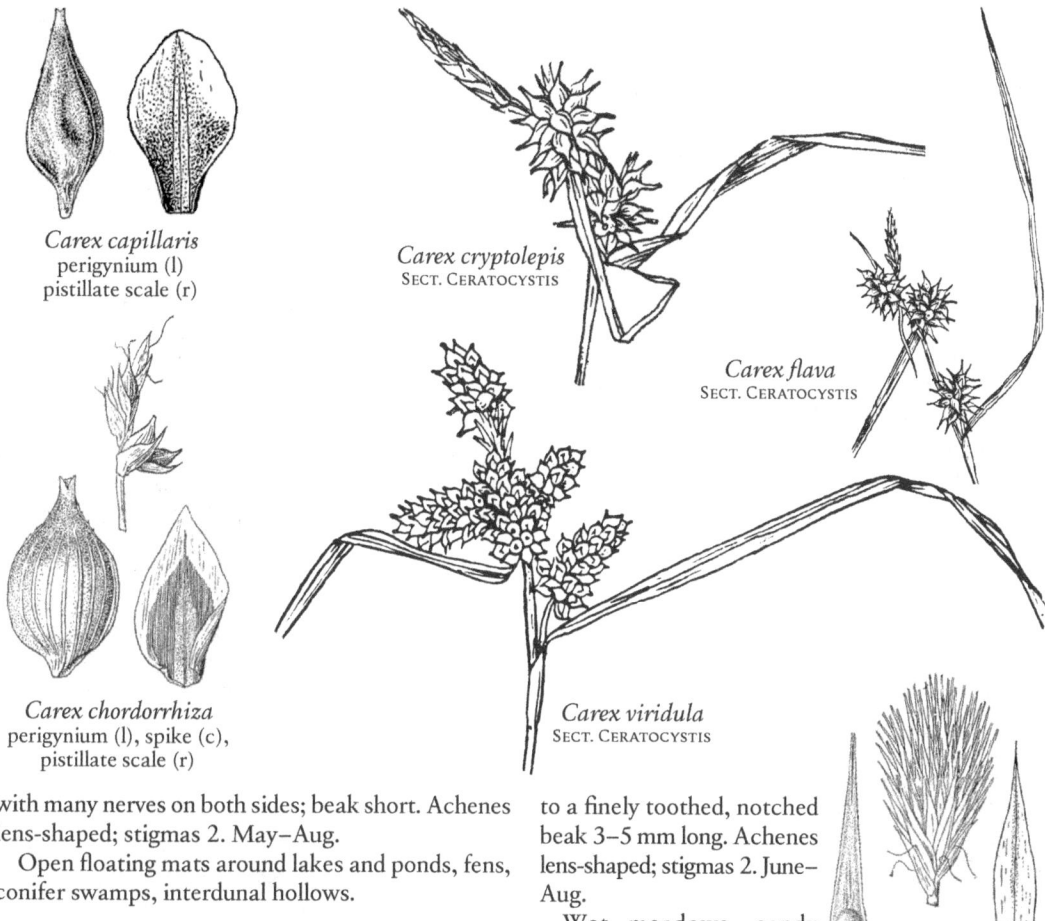

Carex capillaris
perigynium (l)
pistillate scale (r)

Carex cryptolepis
Sect. Ceratocystis

Carex flava
Sect. Ceratocystis

Carex chordorrhiza
perigynium (l), spike (c),
pistillate scale (r)

Carex viridula
Sect. Ceratocystis

Carex sychnocephala
perigynium (l), spike (c),
pistillate scale (r)

with many nerves on both sides; beak short. Achenes lens-shaped; stigmas 2. May–Aug.

Open floating mats around lakes and ponds, fens, conifer swamps, interdunal hollows.

Carex Section Cyperoideae

One member of the section in the Upper Peninsula. Similar to the Ovales and sometimes placed within that section; distinguished by leafy bracts more than 3 times as long as the inflorescence and very long-tapering perigynia, the beak often twice as long as the body.

Carex sychnocephala Carey
MANY-HEAD SEDGE   native
IR | WUP | **CUP** | EUP   FACW CC 6

Plants tufted, from fibrous roots. Stems many and crowded, rounded 3-angled, 0.5–6 dm long. Leaves 1.5–4 mm wide; sheaths tight, white-translucent. Spikes with both staminate and pistillate flowers, pistillate flowers borne above staminate, densely clustered in ovate heads 1.5–3 cm long; bracts leaflike, 2–4 per head, the longest bracts much longer than the heads; pistillate scales thin and translucent with a green midvein, 2/3 length of perigynia, tapered to a tip or with a short sharp point. Perigynia green to straw-colored, flat, lance-shaped, 5–7 mm long and to 1 mm wide, narrowly wing-margined, spongy at base when mature, tapered to a finely toothed, notched beak 3–5 mm long. Achenes lens-shaped; stigmas 2. June–Aug.

Wet meadows, sandy lakeshores, marshes.

Carex Section Deweyanae

Two members of the section in the Upper Peninsula. Plants tufted; rhizomes mostly short; bases brown. Inflorescence slender, open, at least the lowest spike(s) distinct; bracts setaceous. Spikes mostly gynecandrous, lateral spikes sometimes pistillate, mixed, or (rarely) staminate. Perigynia appressed to ascending, ovate to lanceolate, plano-convex, slender; base spongy; beak distinct, margins serrate, tip bidentate. Achenes mostly filling the perigynium body. Usually in moist to wet shaded places.

1 Perigynia 0.8–1.2 mm wide and 4–5 times as long as wide, conspicuously nerved on dorsal face, weakly to strongly nerved on ventral face .... **C. bromoides**
1 Perigynia 1.3–1.6 mm wide and usually 3–3.5 times as long as wide, faintly nerved or nerveless on both faces .... **C. deweyana**

Carex bromoides Schkuhr
BROME-LIKE SEDGE native
IR | **WUP** | **CUP** | EUP FACW CC 6

Plants densely tufted. Stems very slender, 3–8 dm long. Leaves 1–2 mm wide. Spikes 3–7, narrowly oblong, 1–2 cm long, terminal spike with both staminate and pistillate flowers, the staminate below pistillate; lateral spikes all pistillate or with a few staminate flowers at base, the spikes clustered or overlapping; pistillate scales obovate, about as long as perigynia body, pale brown or orange-tinged with translucent margins, tapered to tip or short-awned. Perigynia lance-shaped, flat on 1 side and convex on other, light green, 4–6 mm long, nerved on both sides, gradually tapered to a finely sharp-toothed beak, the beak 1/2–2/3 as long as body. Achenes lens-shaped, in upper part of perigynium body; stigmas 2. April–July.

Floodplain forests, old river channels, swamps.

Carex deweyana Schwein.
DEWEY'S SEDGE native
IR | **WUP** | **CUP** | EUP FACU CC 3

Plants loosely tufted, from short rhizomes. Stems weak and spreading, 2–12 dm long, rough-to-touch below the head. Leaves shorter than stems, yellow-green to waxy blue-green, soft, flat, 2–5 mm wide; sheath tight. Spikes 2–6, the lower separate, the upper grouped, forming a head 2–6 cm long and often drooping near tip; terminal spike with staminate flowers at base, lateral spikes usually pistillate, the perigynia upright; pistillate scales ovate, blunt to short-awned at tip, thin and translucent with green center, slightly shorter than perigynia. Perigynia flat on 1 side and convex on other, 4–6 mm long, pale-green, very spongy at base, the beak 2–3 mm long, finely toothed and weakly notched. Achenes lens-shaped, nearly round, yellow-brown; stigmas 2. May–Aug.

Common; moist to dry woods, thickets, swamps.

Carex Section Digitatae
Basal sheaths not fibrous. Bracts reduced to bladeless sheaths. Perigynium beaks untoothed, mostly less than 0.5 mm long. Similar to section Acrocystis but basal sheaths not fibrous.

1   Terminal spike pistillate at base; basal spikes usually present, on long very thin peduncles; pistillate scales abruptly truncate and awned; anthers 2–3 mm long . . .
. . . . **C. pedunculata**

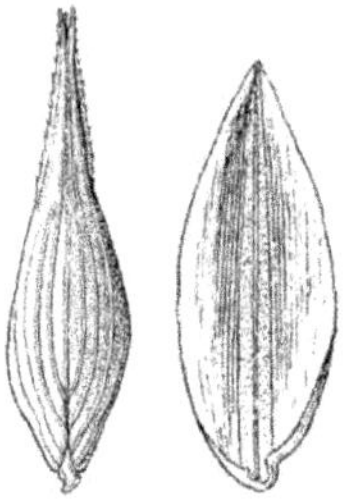

Carex bromoides
perigynium (l)
pistillate scale (r)

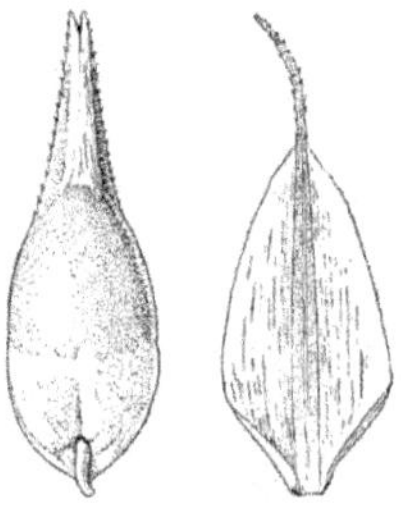

Carex deweyana
perigynium (l)
pistillate scale (r)

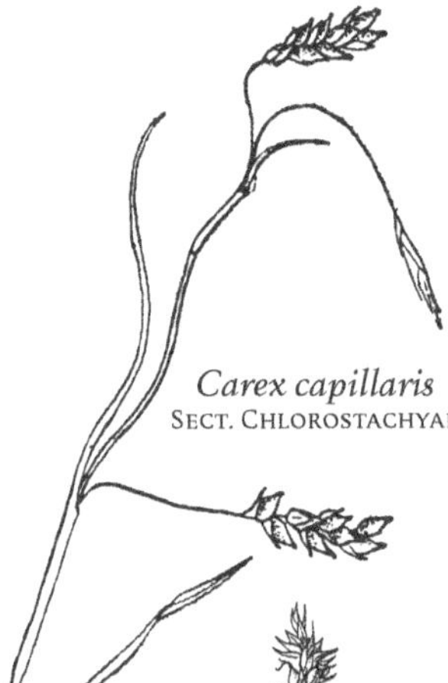

Carex capillaris
SECT. CHLOROSTACHYAE

Carex sychnocephala
SECT. CYPEROIDEAE

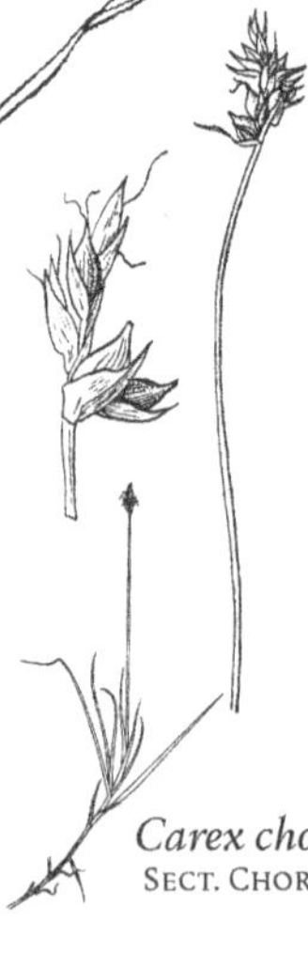

Carex chordorrhiza
SECT. CHORDORRHIZAE

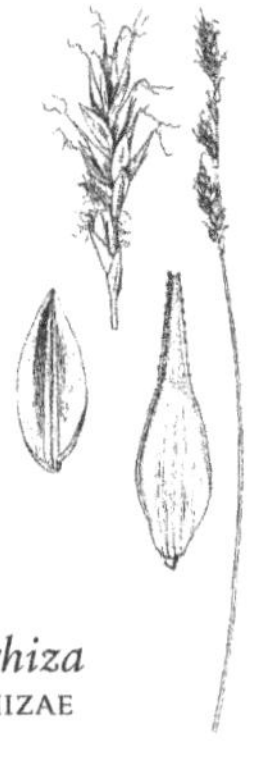

Carex bromoides
SECT. DEWEYANAE

Carex deweyana
SECT. DEWEYANAE

1. Terminal spike usually entirely staminate; basal spikes not present; pistillate scales not awned; anthers various 2
2. Staminate spike 4–6 (–8) mm long; pistillate spikes less than 10 mm long; pistillate scales obtuse, minutely ciliate, distinctly shorter than the perigynia; anthers 1–1.5 mm long. .... **C. concinna**
2. Staminate spike 10–22 mm long; pistillate spikes (often staminate at their tips) (8–) 10 mm long; pistillate scales mostly acute to acuminate, glabrous, and equaling or exceeding the perigynia; anthers 2–3.5 mm long .... **C. richardsonii**

Carex concinna R. Br.
LOW NORTHERN SEDGE *native*
IR | **WUP** | **CUP** | **EUP** FACU CC 10

Plants loosely tufted; rootstocks slender, often long, brownish black, scaly, ascending. Fertile stems 5–20 cm long, erect or incurved, roughened on angles above, dark-brownish tinged and fibrillose at base. Well-developed leaves 5–9 to a fertile culm, near base, the upper reduced; blades 5–10 cm long and 2–2.5 mm wide, flat with recurved margins, involute at base, thick, light-green. Terminal spike staminate; lateral spikes 2–3, pistillate, the lower rarely long-peduncled; bracts reduced to sheaths 7 mm long or less, light-reddish brown-tipped; pistillate scales dark-reddish brown with white-hyaline margins and obsolete or straw-colored midrib, hairy. Perigynia 5–12 to a spike, 3–3.5 mm long, ascending, obtusely triangular, whitish or greenish, hirsute, 2-ribbed and several-nerved; beak chestnut-brown, the orifice hyaline. Achenes triangular with sides convex above, closely enveloped; stigmas 3, short, blackish.

Open, moist, sandy places, usually underlain by dolomitic limestone.

Carex pedunculata Muhl.
LONG-STALK SEDGE *native*
IR | **WUP** | **CUP** | **EUP** FACU CC 5

Rootstocks stout, woody, branching. Fertile stems 2–3 dm long, decumbent, roughened on angles, purple-tinged at base; sterile shoots purple-tinged. Upper leaves of fertile stems bladeless, the sheaths loose, reddened at base; basal leaves 3–5; blades 15–35 cm long and 2–3 mm wide, flat, thickish, pale-green, glaucous, roughened, especially on margins and nerves above. Terminal spike staminate or usually androgynous; lateral spikes 3–4, pistillate or usually androgynous, the lowest basal, long-peduncled; bracts sheathing, reddish purple-tinged at base; pistillate scales ciliate, purple with 3-nerved green center. Perigynia 1–8 to a spike, 3.5–4.5 mm long, appressed, triangular, deep-green, minutely puberulent, 2-ridged, spongy at base; beak minute, usually bent, the orifice entire. Achenes triangular with concave sides, closely enveloped in upper part of perigynia, short bent-apiculate; stigmas 3, deciduous.

Rich, mesic forests.

Carex richardsonii R. Br.
RICHARDSON'S SEDGE *native*
IR | **WUP** | **CUP** | **EUP** CC 9

Plants loosely tufted; rootstocks long, ascending, slender, brownish black, scaly. Stems 15–35 cm long, roughened on angles above, dark-brownish at base. Well-developed leaves 6–10 to a fertile culm, near base; blades 1–2.5 dm long and 2–2.5 mm wide, thick, light-green, rough on margins, especially towards the tip; upper leaves bladeless, the sheaths reddish purple with hyaline margins. Terminal spike staminate; lateral spikes usually 2, pistillate; bracts bladeless, purple-tinged and white-hyaline-margined; pistillate scales dark-purplish with hyaline margins and lighter midvein. Perigynia 10–25 to a spike, 2.5–3 mm long, ascending, obscurely triangular, straw-colored or light-brownish above, appressed-pubescent, 2-keeled; beak 0.5 mm, the orifice obliquely cut. Achenes triangular with sides convex above, closely enveloped, brownish, shining, conic-apiculate; stigmas 3, blackish.

Uncommon in dry sandy prairies and barrens; rarely in fens.

Carex Section Dispermae

One member of the section in the Upper Peninsula. Plants slender, shoots arising singly or in small bunches from pale, slender rhizomes; spikes few-flowered, androgynous; perigynia spreading, darkening at maturity, plump.

Carex disperma Dewey
SOFT-LEAF SEDGE *native*
IR | **WUP** | **CUP** | **EUP** OBL CC 10

Plants small, loosely tufted, from slender rhizomes. Stems slender, weak, 3-angled, 1–4 dm long, shorter to longer than leaves. Leaves soft and spreading, 1–2 mm wide; sheaths tight, translucent. Spikes with both staminate and pistillate flowers, staminate flowers borne above pistillate, 2–5, few flowered

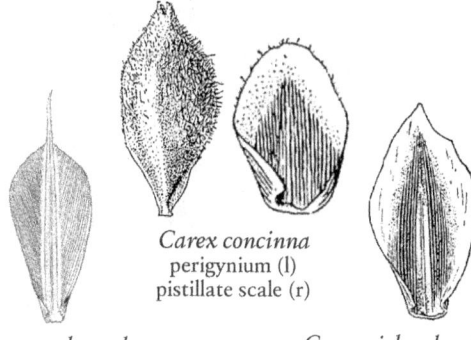

Carex concinna
perigynium (l)
pistillate scale (r)

Carex pedunculata
pistillate scale

Carex richardsonii
pistillate scale

and small, with 1–6 perigynia and 1–2 staminate flowers, to 5 mm long, stalkless, separate or upper spikes grouped in interrupted heads 1.5–2.5 cm long; bracts sheathlike and resembling the pistillate scales, or threadlike and to 2 cm long; pistillate scales white, translucent except for the darker midrib, tapered to tip or short-awned, 1–2 mm long. Perigynia convex on both sides to nearly round in section, oval, 2–3 mm long, strongly nerved and rounded on the margins, beak tiny. Achenes lens-shaped, oval; stigmas 2. May–July.

Hummocks in conifer swamps and alder thickets, wetland margins; usually where shaded.

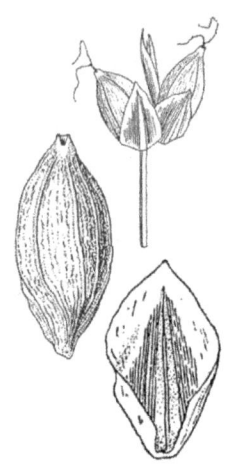

Carex disperma
perigynium (l), spike (c), pistillate scale (r)

Spikes with both staminate and pistillate flowers, staminate flowers above pistillate, or spikes nearly all staminate or pistillate, 4–8 mm long, upper spikes crowded, lower spikes separated, in narrowly ovate heads 1–4 cm long; bracts absent; pistillate scales brown, shiny, shorter or equal to perigynia. Perigynia green-brown, turning dark brown, flat on 1 side and convex on other, ovate to lance-shaped, 3–4 mm long and 1 mm wide, sharp-edged, spongy at base, tapered to a finely toothed beak 2 mm long, unequally notched. Achenes lens-shaped, 1–2 mm long; stigmas 2. May–June.

Wet to moist meadows, shores, streambanks and ditches; most common along salted highways. Native of western USA, considered adventive in the UP.

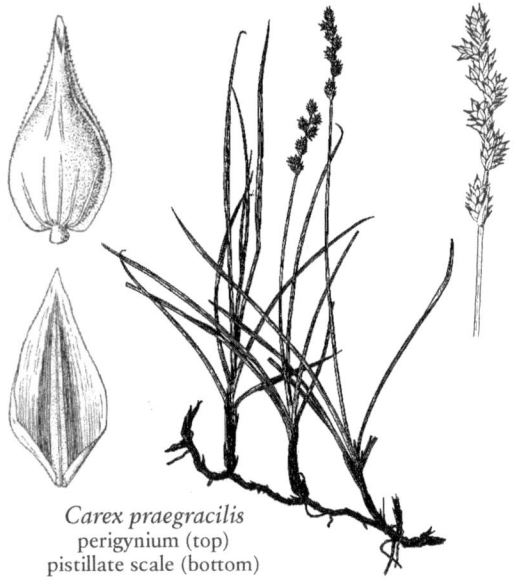

Carex Section Divisae
One member of the section in the Upper Peninsula. Plants strongly rhizomatous, unisexual. Not native in most of e North America but spreading, especially along expressways, where tolerant of road salt.

Carex praegracilis W. Boott
CLUSTERED FIELD SEDGE *introduced*
IR | **WUP** | **CUP** | EUP   FACW

Plants colony-forming, from long black rhizomes. Stems single or few together, 3-angled, 1–7 dm long, longer than the leaves. Leaves on lower part of stems, 2–3 mm wide; sheaths white-translucent.

Carex praegracilis
perigynium (top)
pistillate scale (bottom)

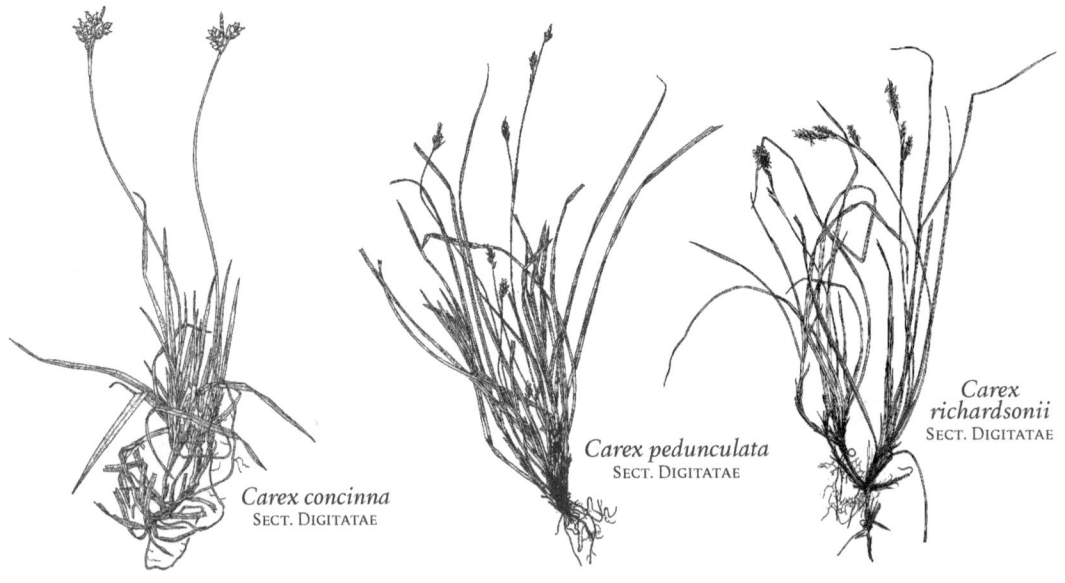

Carex concinna
Sect. Digitatae

Carex pedunculata
Sect. Digitatae

Carex richardsonii
Sect. Digitatae

Carex Section Glareosae

Tufted sedges of wetlands, soils often peaty. Spikes distinct, mostly nonoverlapping (except *Carex arcta* which has spikes overlapping, the upper not separated), mostly or all gynecandrous, lateral spikes sometimes pistillate. Perigynia ascending to spreading; margins rounded in most species, smooth or finely serrate, often finely papillose.

1. Lowest bract bristle-like, several times as long as its spike; perigynia mostly 2.8–3.8 (–4) mm long, including very short smooth beak; spikes widely separated, containing 1–5 perigynia each .... **C. trisperma**
1. Lowest bract absent or at most about twice as long as its spike (if rarely prolonged, the perigynia smaller and often with serrulate beak); perigynia and spikes various 2

2. Perigynia broadest near the base of the body, with a conspicuous beak 0.7–1.1 mm long; spikes mostly 7–15, usually ± overlapping or crowded into an ovoid to narrowly pyramidal head 2–4.5 cm long .... **C. arcta**
2. Perigynia broadest at or near the middle of the body; beak essentially absent or less than 0.6 mm long; spikes 2–8, at least the lower spikes well separated or, if crowded, the inflorescence only 0.6–2 cm long 3

3. Spikes 2–4, crowded into a short inflorescence 0.6–2 cm long; perigynia 2.5–3.5 mm long, beak often smooth-margined .... **C. tenuiflora**
3. Spikes 4–8 (–10), remote or ± crowded, but total inflorescence over 2 cm long; perigynia 1.7–2.6 mm long, beak serrulate usually minutely or scabrous......... 4

4. Perigynia 3–9 per spike (occasionally one or two spikes on a plant, especially terminal one, with as many as 15), loosely spreading, becoming rich brown in age; largest leaves 1–2 mm wide; foliage and perigynia green when fresh .... **C. brunnescens**
4. Perigynia mostly 10–many per spike, appressed-ascending, greenish or dull brown in age; largest leaves 2–2.7 (–3.7) mm wide; foliage and perigynia glaucous or gray-green at least when fresh .... **C. canescens**

Carex arcta Boott
NORTHERN CLUSTER SEDGE native
IR | **WUP** | **CUP** | EUP   OBL CC 8

Plants loosely to densely tufted, from very short thick rhizomes. Stems 2–8 dm long, soft, sharply triangular, very rough-to-touch above. Leaves clustered near base, light-green, flat, 2–4 mm wide, very rough; sheaths loose, purple-dotted. Spikes 5–15, each with both staminate and pistillate flowers, the staminate small and below the pistillate; flowers crowded in oblong heads, 1.5–3 cm long, upper spikes densely packed, lower spikes slightly separate; pistillate scales ovate, acute, translucent with a brown-tinged center, shorter than the perigynia. Perigynia flat on 1 side and convex on other, ovate, 2–3 mm long, green to straw-colored or brown when mature, covered with white dots, widest near the broad base, tapered to a sharp-toothed, notched beak 0.5–1.5 mm long. Achenes lens-shaped, brown; stigmas 2. June–Aug.

Floodplain forests, old river channels, swamps and wetland margins.

Carex brunnescens (Pers.) Poir.
BROWNISH SEDGE native
IR | **WUP** | **CUP** | **EUP**   FACW CC 5

Plants densely tufted, from a short fibrous rootstock. Stems sharply 3-angled, to 5 dm long, smooth or slightly rough-to-touch below the head. Leaves 1–3 mm wide; sheaths tight, thin and translucent. Spikes 5–10 in a head 2–5 cm long, all with pistillate flowers borne above staminate, each spike with 5–15 perigynia, lower spikes separated; lowermost bract bristlelike, shorter or longer than lowermost spike; pistillate scales ovate, rounded or acute at tip, shorter than the perigynia. Perigynia 3-angled, not winged or sharp-edged, 2–3 mm long, faintly nerved on both sides, not spongy-thickened at base, tapered at tip to a short, minutely notched beak, the beak and upper body finely toothed and white-dotted. Achenes lens-shaped; stigmas 2. June–Aug.

Common in wet forests and swamps, peatland margins.

Carex canescens L.
HOARY SEDGE native
IR | **WUP** | **CUP** | **EUP**   OBL CC 8

Plants tufted. Stems 2–6 dm long. Leaves waxy blue- or gray-green, 2–4 mm wide, mostly near base of plant and shorter than stems. Spikes 4–8, silvery green or grayish, with both staminate and pistillate flowers, the staminate below the pistillate, ovate to cylindric, 5–10 mm long, the lower spikes more or less separate, each spike with 10–30 perigynia. Perigynia flat on one side and convex on other, 2–3 mm long and 1–2 mm wide, with a beak to 0.5 mm long, not noticeably finely toothed on the margins; pistillate scales shorter than perigynia. Achenes lens-shaped; stigmas 2. May–July.

Peatlands (including hummocks in patterned fens), tamarack swamps, floating mats, swamps, alder thickets, wet forest depressions.

Similar to *C. brunnescens* but leaves waxy blue-green rather than green and spikes somewhat larger and silver-green vs. brown.

Carex tenuiflora Wahlenb.
SPARSE-FLOWER SEDGE native
IR | **WUP** | **CUP** | **EUP**   OBL CC 10

Plants delicate, loosely tufted; spreading from long, slender rhizomes. Stems very slender, 2–6 dm long. Leaves 1–2 mm wide. Spikes 2–4, with both staminate and pistillate flowers, the staminate below the pistillate, stalkless, clustered into a head 8–15 mm long; pistillate scales white-translucent with green center, covering most of the perigynium. Perigynia 3–15, oval, flat on 1 side and convex on other, 3–4 mm long, dotted with small white depressions, sharp-edged, beakless. Achenes lens-shaped, nearly filling the perigynia; stigmas 2. June–Aug.

Hummocks in peatlands, floating mats, conifer swamps.

Carex trisperma Dewey
THREE-SEED SEDGE
IR | WUP | CUP | EUP OBL CC 9 native

Loosely tufted perennial, with short, slender rhizomes. Stems very slender and weak, 2–7 dm long. Leaves 1–2 mm wide. Spikes 1–3 (usually 2), stalkless, 1–4 cm apart in a slender, often zigzagged head, each spike with 2–5 perigynia and a few staminate flowers at the base; lowest spike subtended by a bristlelike bract 2–4 cm long; pistillate scales ovate, translucent with a green center, shorter or equal to the perigynia. Perigynia flat on 1 side and convex on other, oval, 3–4 mm long, finely many-nerved, tapered near tip to a short, smooth beak 0.5 mm long. Achenes oval-oblong, filling the perigynia; stigmas 2. May–Aug.

Forested wetlands and conifer swamps, alder thickets.

Carex Section Granulares
Plants tufted or shoots arising singly from elongate rhizomes. Pistillate spikes oblong to narrowly oblong, densely packed with perigynia. Pistillate scales and perigynia dotted or finely streaked with red. Perigynia more than 25 per pistillate spike; veins 25–40, raised.

1. Staminate spike long-peduncled, elevated above summit of uppermost pistillate spikes; lowest pistillate spike usually on a separate basal peduncle; stems mostly solitary from elongate rhizomes; widest leaves 1.5–4 mm wide .... **C. crawei**
1. Staminate spike sessile or nearly so; lowest pistillate spike not on a basal peduncle; stems clumped, without elongate rhizomes; widest leaves 4.5–10 mm wide .... **C. granularis**

Carex crawei Dewey
CRAWE'S SEDGE
IR | WUP | CUP | EUP FACW CC 10 native

Plants from long-creeping rhizomes. Stems single or several together, faintly 3-angled, 0.5–4 dm long. Leaves 1–4 mm wide. Spikes either staminate or pistillate, cylindric, densely flowered, 1–3 cm long, terminal spike staminate; lateral spikes pistillate, 2–5, separate, the lowest spike near base of plant; bract leaflike, the blade shorter than the terminal spike; pistillate scales red-brown with a pale or green midrib, shorter and narrower than the perigynia. Perigynia green to brown, ovate, 2–3.5 mm long, many-nerved; beak absent or very short, entire

Carex crawei
perigynium (l)
pistillate scale (r)

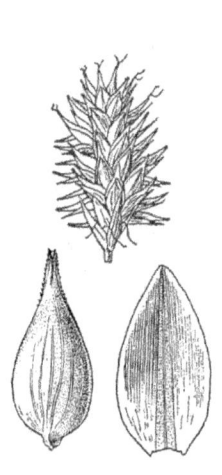

Carex arcta
perigynium (l), spike (c),
pistillate scale (r)

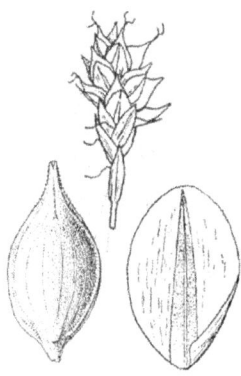

Carex brunnescens
perigynium (l), spike (c),
pistillate scale (r)

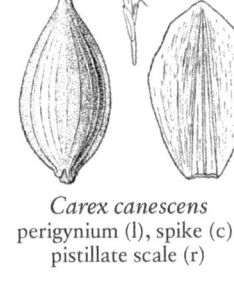

Carex canescens
perigynium (l), spike (c),
pistillate scale (r)

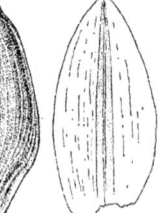

Carex tenuiflora
perigynium (l)
pistillate scale (r)

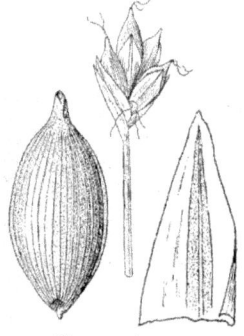

Carex trisperma
perigynium (l), spike (c),
pistillate scale (r)

to notched. Achenes 3-angled; stigmas 3. May–July.

Wet to moist meadows and prairies, marly lakeshores, ditches, especially where calcium-rich.

Carex granularis Muhl.
LIMESTONE-MEADOW SEDGE   native
IR | **WUP** | **CUP** | **EUP**   FACW CC 2

Plants tufted, from short rhizomes. Stems rounded 3-angled, 1–5 dm long. Leaves often longer than stems, 3–13 mm wide; sheaths membranous on front, divided-with small swollen joints on back. Spikes either all staminate or pistillate, the terminal spike staminate, stalkless; the lateral spikes pistillate, clustered around the staminate spike; bracts longer than the head; pistillate scales brown, tapered to tip or with a short, sharp point, half as long as perigynia. Perigynia crowded in several rows, green or olive to brown, oval to obovate, 2–3 mm long, 2-ribbed, strongly nerved; beak tiny or absent, entire to slightly notched. Achenes 3-angled; stigmas 3. May–July.

Wet to moist meadows and swales, streambanks and pond margins, especially where calcium-rich.

Carex granularis
perigynium (l)
pistillate scale (r)

Carex Section Griseae

Perigynia round or obtusely angled in cross-section, many-veined; veins impressed on both fresh and dried plants. Pistillate scales awned.

1 Perigynia contracted to a distinct beak 0.5–1.3 mm long
 . . . . **C. hitchcockiana**
1 Perigynia essentially beakless 2
2 Peduncles of lateral spikes finely scabrous; staminate spike long-peduncled; perigynia 2.5–3.6 (–4) mm long,

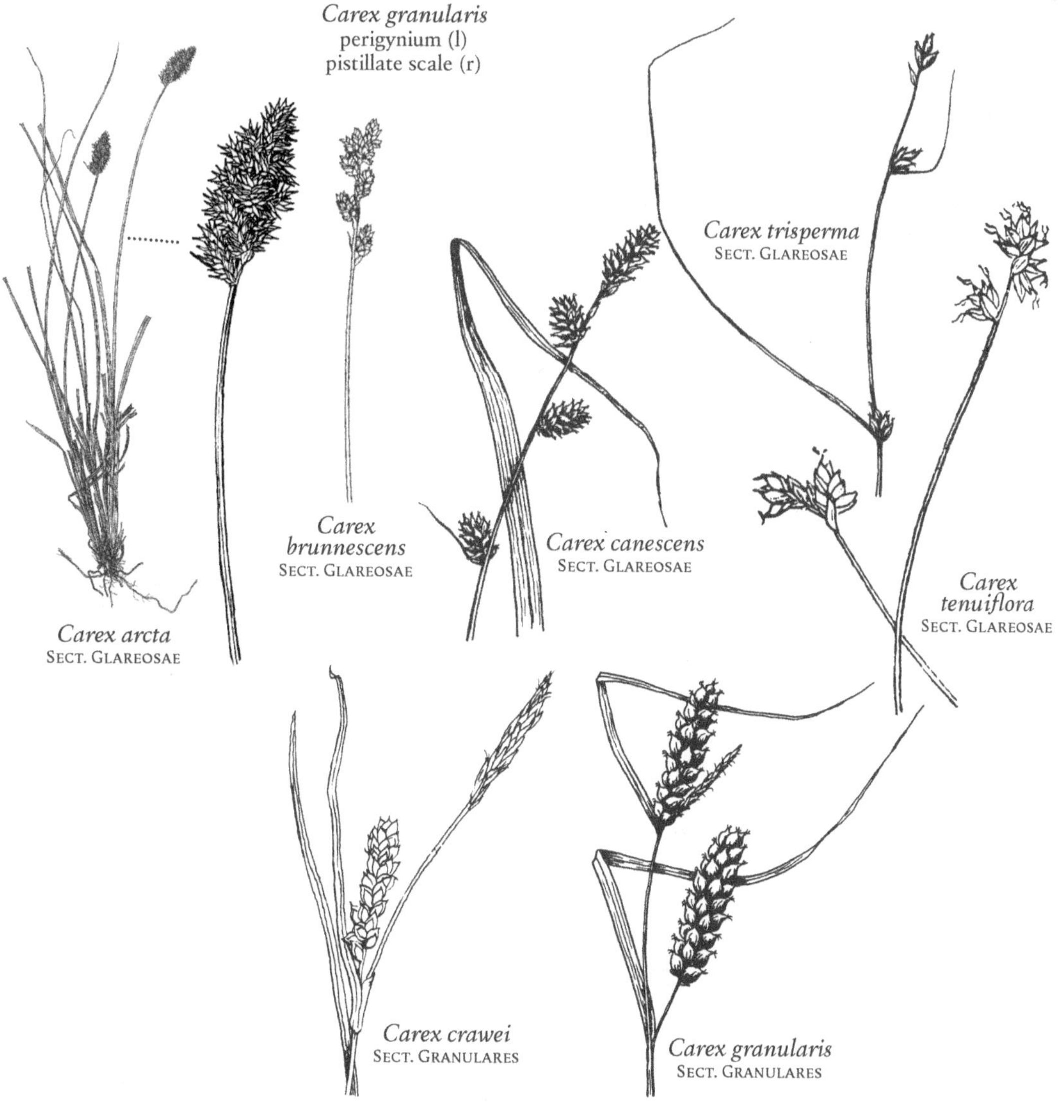

Carex arcta
SECT. GLAREOSAE

Carex brunnescens
SECT. GLAREOSAE

Carex canescens
SECT. GLAREOSAE

Carex trisperma
SECT. GLAREOSAE

Carex tenuiflora
SECT. GLAREOSAE

Carex crawei
SECT. GRANULARES

Carex granularis
SECT. GRANULARES

usually more than 20 per spike .... **C. conoidea**

2 Peduncles of lateral spikes smooth; staminate spike sessile or nearly so; perigynia 4–5 mm long, usually fewer than 15 per spike .... **C. grisea**

Carex conoidea Schkuhr
OPEN-FIELD SEDGE *native*
IR | **WUP** | CUP | **EUP**   FACW CC 9

Plants tufted. Stems 1–7 dm long, much longer than leaves. Leaves 2–4 mm wide. Spikes either staminate or pistillate; staminate spike on a long stalk and overtopping pistillate spikes, linear, 1–2 cm long; pistillate spikes 2–4, widely spaced or upper 2 grouped, short cylindric, 1–2 cm long, on short, rough stalks; bract leaflike with a rough sheath; pistillate scales ovate and much shorter than perigynia, with a green midvein prolonged into an awn. Perigynia oval, 3–4 mm long and 1–2 mm wide. Achenes 3-angled; stigmas 3. May–July.

Wet calcareous prairies, sedge meadows; also in drier old fields; Keweenaw and Schoolcraft counties.

Carex grisea Wahlenb.
INFLATED NARROW-LEAF SEDGE *native*
IR | WUP | **CUP** | EUP   FAC CC 3

Carex amphibola Steud. var. *turgida* Fern.
Plants tufted; rootstocks short. Stems 2–6 dm long, purple-tinged at base. Leaves 1–3 dm long and 2–4 mm wide, flat, thin, deep-green, the midvein prominent below, roughened towards the tip; sheaths red-dotted. Terminal spike staminate, rough-peduncled; lateral spikes 3–5, pistillate, the lowest nearly basal; bracts sheathing; pistillate scales awned, white-hyaline with green midvein, yellowish brown-tinged and red-dotted. Perigynia 4–12 to a spike, 3.5–4.5 mm long, erect, exceeding scales, suborbicular, somewhat leathery, light-green or yellowish brown, minutely puncticulate, the orifice hyaline, entire. Achenes triangular with concave sides, loosely enveloped, yellowish brown, granular, apiculate; stigmas 3, reddish brown.

Mesic to wet deciduous forests, roadside ditches; Menominee County.

Carex hitchcockiana Dewey
HITCHCOCK'S SEDGE *native*
IR | WUP | **CUP** | **EUP**   CC 5

Plants tufted; rootstocks short. Stems 1.5–7 dm long, roughened above, brownish tinged at base. Leaves 3–4 to a culm, 1–2.5 dm long and 3–7 mm wide, flat, thin, light-green, the midvein conspicuous below, roughened on margins and towards tip on veins; sheaths conspicuously prolonged, cinnamon brown-tinged; ligule ciliate. Terminal spike staminate, the peduncle rough; lateral spikes 3–4, pistillate, the rachis zigzag, the lowest separate; bracts leaflike, reduced upwards, the sheaths rough-hairy; pistillate scales rough-awned, serrulate, keeled, all white-hyaline with 3-nerved green center. Perigynia 1–9 to a spike, 4.5–5 mm long, ascending, obtusely triangular, somewhat leathery, yellowish or grayish green, puncticulate, spongy at base; beak 1 mm long, straight or bent, the orifice hyaline, entire. Achenes triangular, filling perigynia, yellowish brown, granular, bent-apiculate; stigmas 3, red-brown.

Mesic woods; Delta and Mackinac counties.

Carex Section Heleoglochin

Plants densely tufted; bases brown. Stems narrowing toward the tip, typically arching at maturity. Inner band of the leaf sheaths smooth, pigmented toward the summit. Leaf blades less than 3 mm wide (ours). Spikes androgynous, the lower branched. Perigynia plano-convex to biconvex, darkening at maturity, mostly less than 3 mm long; beak short-triangular, scabrous on the margin, bidentate. Wetlands, primarily in peaty soils.

1 Leaf sheaths whitish or pale ventrally except for purplish dots; inflorescence ± crowded, the lowermost spike (or branch) usually at least slightly overlapping the next above it (occasionally separated by a distance no more than its total length); perigynia tending to spread at maturity, therefore not concealed by the scales .... **C. diandra**

1 Leaf sheaths strongly tinged with copper color toward their summits ventrally; inflorescence ± interrupted, the

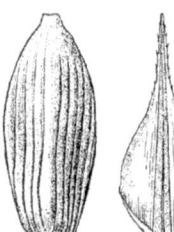

Carex conoidea
perigynium (l)
pistillate scale (r)

 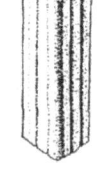

Carex grisea
perigynium (l), stem-section (c),
pistillate scale (r)

Carex hitchcockiana
perigynium (l), stem-section (c),
pistillate scale (r)

lowermost spikes (or branches) often well separated or even peduncled; perigynia ± appressed at maturity, nearly or completely concealed by the large scales....
.... **C. prairea**

Carex diandra Schrank
LESSER TUSSOCK SEDGE native
IR | **WUP** | **CUP** | **EUP** OBL CC 8
Plants densely tufted. Stems sharply 3-angled, 3–8 dm long, usually longer than leaves. Leaves 1–3 mm wide; sheaths white with fine pale lines, translucent on front or slightly copper-colored at mouth. Spikes with both staminate and pistillate flowers, staminate flowers borne above pistillate, clustered in ovate heads 1–4 cm long; bracts small and inconspicuous, shorter than the spikes; pistillate scales brown, tapered to tip or with a short sharp point, about equaling the perigynia. Perigynia brown, shiny, unequally convex on both sides, broadly ovate, 2–3 mm long, beak finely toothed, entire to notched, 1–2 mm long. Achenes lens-shaped; stigmas 2. May–July.

Wet meadows, ditches, peatlands (especially calcareous fens), floating mats.

Carex prairea Dewey
PRAIRIE SEDGE native
IR | WUP | **CUP** | **EUP** FACW CC 10
Plants densely tufted, from short rootstocks. Stems sharply 3-angled, 5–10 dm long. Leaves 2–3 mm wide; sheaths translucent, yellow-brown or bronze-colored. Spikes with both staminate and pistillate flowers, staminate flowers borne above pistillate, ovate, 4–7 mm long, lower spikes usually separate, in linear-oblong heads 3–8 cm long; bracts small; pistillate scales red-brown, tapered to tip, as long as and covering most of perigynia. Perigynia dull brown, flat on 1 side and convex on other, lance-shaped to ovate, 2–3 mm long, tapered to a finely toothed, unequally notched beak 1–2 mm long. Achenes lens-shaped; stigmas 2. May–July.

Wet meadows, calcareous fens, marshes, tamarack swamps and peaty lakeshores.

Carex Section Hirtifoliae

One member of the section in the Upper Peninsula; recognized by the soft pubescence covering the entire plant, including the distinctly beaked, 2-ribbed perigynia.

Carex hirtifolia Mackenzie
PUBESCENT SEDGE native
IR | **WUP** | **CUP** | **EUP** CC 5
Plants loosely tufted; rootstocks slender, branched. Stems 3–6 dm long, pubescent, roughened above, brownish red at base. Well-developed leaves 3–4 to a fertile culm, more on sterile stems; blades to 35 cm long and 3–7 mm wide, flat, flaccid, hirsute, the nerves prominent above; sheaths cinnamon-brown ventrally. Terminal spike staminate; lateral spikes 2–4, pistillate; lowest bract 1.5–7 cm long, the

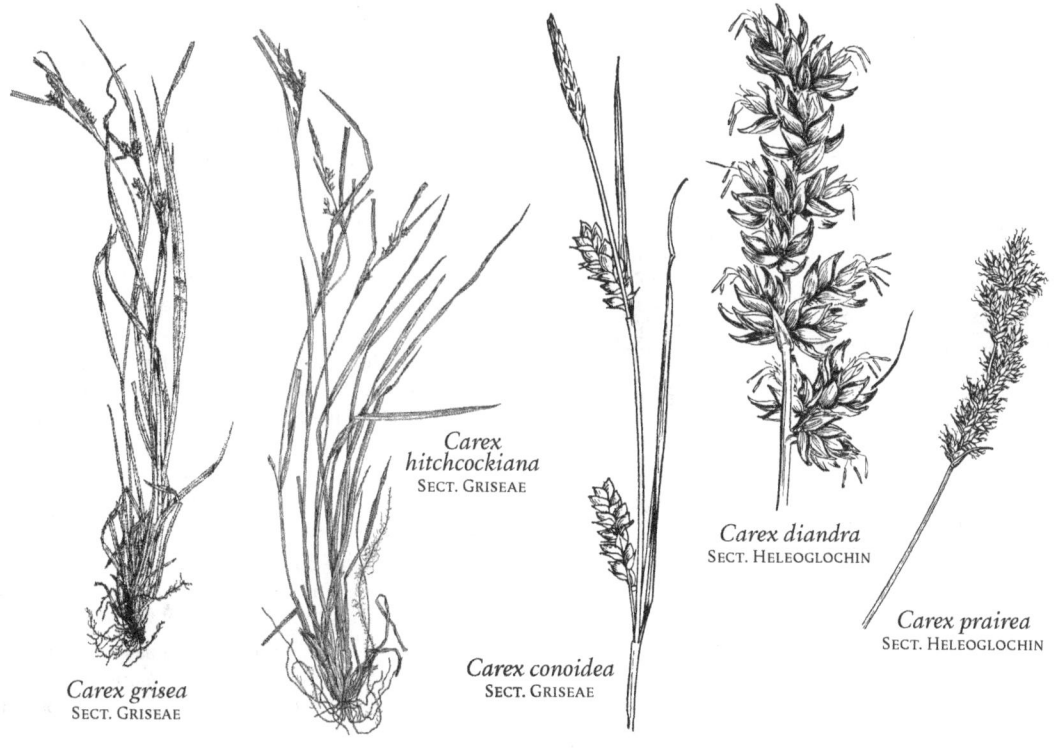

Carex grisea
SECT. GRISEAE

Carex hitchcockiana
SECT. GRISEAE

Carex conoidea
SECT. GRISEAE

Carex diandra
SECT. HELEOGLOCHIN

Carex prairea
SECT. HELEOGLOCHIN

upper shorter; pistillate scales whitish, ciliate, with green excurrent midrib. Perigynia 10–25 to a spike, 3.5–5 mm long, triangular, green, pubescent, nerveless; beak ca. 1 mm long, obliquely cut, 2-toothed. Achenes sharply triangular with concave sides, short-apiculate; stigmas 3, reddish brown.

Rich mesic woods.

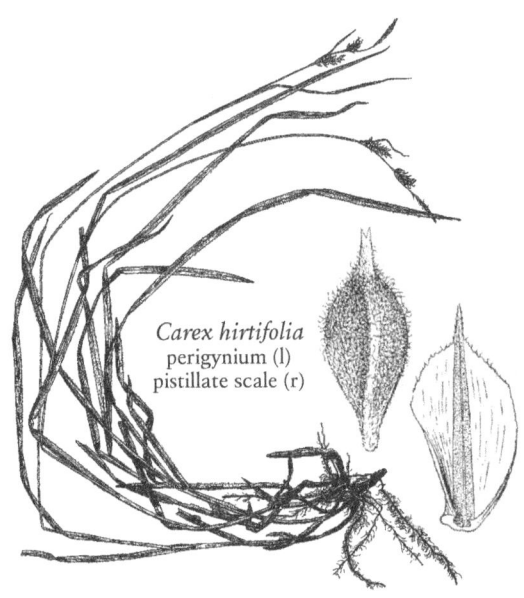

Carex hirtifolia
perigynium (l)
pistillate scale (r)

Carex Section Holarrhenae

One member of the section in the Upper Peninsula; resembling those of sections Divisae and Ammoglochin but distinguished by the green-veined inner band of its leaf sheaths.

Carex sartwellii Dewey
SARTWELL'S SEDGE   native
IR | WUP | **CUP** | **EUP**   OBL CC 5

Plants colony-forming, from long black rhizomes. Stems single or few together, stiff, sharply 3-angled, 3–8 dm long. Leaves 2–4 mm wide, few per stem, the lowest leaves small and without blades; sheaths with green lines on front, and a translucent ligule around stem. Spikes with both staminate and pistillate flowers, staminate flowers above pistillate, or upper spikes staminate; clustered or lower spikes separate, 5–10 mm long, in cone-shaped

Carex sartwellii
perigynium (l)
pistillate scale (r)

heads, 3–6 cm long; bracts small, the lower bracts sometimes bristlelike and longer than the spike; pistillate scales brown with a prominent green midvein, about equal to perigynia. Perigynia tan to brown, flat on 1 side and convex on other, ovate, 2.5–3.5 mm long, finely nerved on both sides, sharp-edged, tapered to a short, finely toothed beak. Achenes lens-shaped; stigmas 2. May–July.

Wet to moist meadows, marshes, fens and shores, often where calcium-rich; Chippewa and Delta counties.

Carex Section Hymenochlaenae

Includes nearly all of the forest understory sedges with long, nodding pistillate spikes. Superficially similar to section Gracillimae but plants more delicate. Terminal spike wholly staminate. Perigynia 8–45 per spike (fewer in *Carex assiniboinensis*), narrow and long-tapering to the beak. Woodlands and wetlands.

1   Terminal spike gynecandrous; sheaths ± softly pubescent or perigynia essentially beakless (except *C. prasina*) 2
1   Terminal spike staminate; sheaths glabrous (except *C. castanea*) and perigynia conspicuously beaked....... 3
2   Perigynia to 1.6 mm wide, beakless; sheaths and blades glabrous............................ **C. gracillima**
2   Perigynia 1.7–2.5 mm wide, abruptly contracted to a short beak; sheaths and leaf blades ± softly pubescent, at least below (sometimes very sparsely so).......... .... **C. formosa**
3   Perigynia pubescent............. **C. assiniboinensis**
3   Perigynia glabrous 4
4   Leaf sheaths and blades (at least toward the base) ± hairy; pistillate spikes 1–2.5 cm long ...... **C. castanea**
4   Leaf sheaths and blades glabrous (at most the lowermost bladeless sheaths minutely hispidulous); pistillate spikes mostly 2.5–6.5 cm long..................... 5
5   Basal sheaths reddish purple for at least several cm above the base; perigynia clearly nerved between the 2 ribs 6
5   Basal sheaths brown, lacking any trace of reddish purple color (at most a small trace on the smaller sheaths in *C. prasina*); perigynia 2-ribbed, but otherwise nerveless or faintly nerved................................... 7
6   Perigynia short-stalked, the achene within sessile or nearly so; broadest leaves 6–12 mm wide; pistillate scales mostly awned or cuspidate.......... **C. arctata**
6   Perigynia sessile but the achene within on a definite short stalk 0.5–1 mm long; broadest leaves 2.5–4.5 (–5.5) mm wide; pistillate scales mostly not awned **C. debilis**
7   Perigynia (somewhat twisted) gradually tapering to a poorly defined conical beak, the cylindrical apical portion only 0.2–0.5 mm long .... **C. prasina**

7 Perigynia (symmetrical) tapering to abruptly contracted into a well developed beak ca. 1.2–4.5 mm long . . . . **C. sprengelii**

Carex arctata Boott.
DROOPING WOODLAND SEDGE *native*
IR | WUP | CUP | EUP   CC 3

Plants tufted. Stems 3–9 dm long, the basal sheaths purple-tinged at base. Leaves 2–3 per fertile culm; blades 2–3 dm long and 5–10 mm wide long, ca. 3 mm wide, flat, thin, soft, deep-green, roughened on margins and towards the tip; sheaths minutely roughened, yellowish brown-tinged and red-dotted ventrally. Terminal spike staminate; lateral spikes 3–5, pistillate, nodding on slender rough peduncles; lowest bract leaflike, the upper reduced; pistillate scales awned, greenish white with green center, ciliate, thin. Perigynia 15–45 to a spike, 3–5 mm long, ascending, deep-green, puncticulate, 2-ribbed; beak 0.75 mm long, bidentate, hyaline above, ciliate between teeth. Achenes triangular with sides concave below, closely enveloped, yellowish brown, granular, apiculate; stigmas 3, short, blackish.

Common in mesic deciduous forests.

Similar to *C. gracillima* but perigynia of *C. arctata* taper to the beak and are constricted at base to form a short stipe, and its terminal spikes are staminate.

Carex assiniboinensis W. Boott
ASSINIBOIA SEDGE   (THR) *native*
IR | WUP | CUP | EUP   CC 9

Plants tufted; rootstocks short, stout, with unique, long-arching vegetative shoots produced in summer and tipped with new plantlets; these rooting to form new plants. Stems 35–75 cm long, compressed-triangular, weak, smooth or roughened on angles above, purple-tinged at base, basal sheaths filamentose. Well-developed leaves 3–5 to a culm, on lower third; blades 1–2 dm long and 1.5–3 mm wide, flat, thin, green, roughened towards the tip. Terminal spike staminate, a narrow bract from its base usually subtending one perigynium; lateral spikes about 3, pistillate, on slender rough peduncles 2–6 cm long; lower bracts leaflike, sheathing, the upper reduced; pistillate scales greenish straw-colored with white-hyaline margins and sharp green midrib. Perigynia 1–8 to a spike, 6 mm long, erect, leathery, pale-green or straw-colored, tuberculate-hispid, 2-ridged; beak 2.5 mm long, oblique and white-tipped. Achenes triangular with sides concave, closely enveloped, apiculate; stigmas 3, long, reddish brown.

Uncommon in rich mesic woods.

Carex castanea Wahlenb.
CHESTNUT-COLOR SEDGE *native*
IR | WUP | CUP | EUP   FACW CC 6

Plants tufted. Stems 3–10 dm long, purple-tinged at base. Leaves 3–6 mm wide, softly hairy. Spikes either staminate or pistillate; the terminal spike staminate, upright atop a long stalk; lateral spikes pistillate, usually 3, on slender, drooping stalks, short cylindric; pistillate scales ovate, brown-tinged, about as long as perigynia. Perigynia lance-shaped, 4–6 mm long, somewhat 3-angled, strongly 2-ribbed with several faint nerves, tapered to a notched beak up to half length of body. Achenes 3-angled; stigmas 3. June–July.

Swamps, moist openings, wetland margins and ditches.

Carex debilis Michx.
WHITE-EDGE SEDGE *native*
IR | **WUP** | **CUP** | EUP   FACW CC 6

Plants tufted. Stems 6–10 dm long, purple-tinged at base. Leaves 2–4 mm wide. Staminate spike linear, sometimes with a few pistillate flowers near tip; pistillate spikes 2–4, separate along stem, spreading or nodding, flowers loose in spikes; pistillate scales oblong, half the length of perigynia with translucent or brown margins and a green midrib. Perigynia lance-shaped, somewhat 3-angled, 2-ribbed, 5–8 mm long, narrowed to a beak. Achenes 3-angled; stigmas 3. May–Aug.

Wet woods (usually under conifers), swamp margins, wet sandy ditches.

Carex formosa Dewey
HANDSOME SEDGE *native*
IR | WUP | **CUP** | EUP   FAC CC 10

Plants densely tufted. Stems 3–8 dm long, dark maroon at base; flowering stems 0.5–1 mm thick,

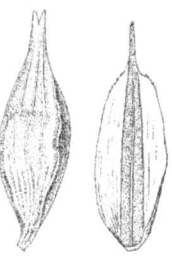

Carex arctata
perigynium (l)
pistillate scale (r)

Carex assiniboinensis
perigynium (l)
pistillate scale (r)

Carex castanea
perigynium (l)
pistillate scale (r)

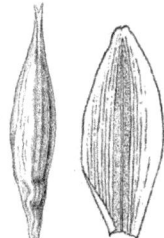

Carex debilis
perigynium (l)
pistillate scale (r)

scabrous on angles within inflorescence. Leaves flat, 3–6 mm wide, glabrous on undersurface, pilose on uppersurface and margins; sheaths pubescent; lowest bracts shorter or equal to tip of the inflorescence. Spikes usually all gynecandrous, pubescent; lateral spikes 2–4, well separated, on slender peduncles to 10 cm long and usually much longer than spikes, drooping at maturity, pistillate except for 1 or 2 basal staminate flowers in each; pistillate scales hyaline tinged with chestnut brown, the broad green midrib red-dotted, shorter than the mature perigynia. Perigynia intermediate in size between those of *C. davisii* and *C. gracillima*, copiously red dotted, 2-ribbed, loosely enveloping achene, tip narrowed to abrupt beak; beak minutely bidentate, less than 0.5 mm long; stigmas 3.

Rich mesic forests, usually where soils calcareous; Menominee County.

Carex gracillima Schwein.
GRACEFUL SEDGE   *native*
IR | WUP | CUP | EUP   FACU CC 4

Plants tufted; rootstocks short, slender. Stems 2–9 dm long, purple-tinged at base. Leaves 3–4 to a culm, on lower half; blades 1–3 dm long, flat, flaccid, deep-green, roughened on margins and towards tip; sheaths yellowish brown-tinged and reddish dotted. Terminal spike gynaecandrous or staminate; lateral spikes 3–4, pistillate, separate on slender nodding roughish peduncles; lowest bract leaflike, sheathing, the upper shorter; scales whitish or yellowish brown with green midrib. Perigynia 10–45 to a spike, 2.5–3.5 mm long, ascending, puncticulate, few-nerved, beakless. Achenes triangular with concave sides and thick angles; stigmas 3, short, blackish.

Common in mesic to wet forests, sometimes in drier oak woods.

Carex sprengelii Dewey
LONG-BEAK SEDGE   *native*
IR | WUP | CUP | EUP   FAC CC 5

Plants tufted; rootstocks long, stout, matted. Stems 3–9 dm long, erect or decumbent, rough on angles above, brownish tinged and long-fibrillose at base; Leaves 1–4 dm long and 2.5–4 mm wide, flat, roughened below and on margins towards the tip. Upper 1–3 spikes staminate (occasionally androgynous), the upper peduncled; pistillate spikes pendulous on rough capillary peduncles; pistillate scales straw-colored or greenish white, with 3-nerved light green center. Perigynia 10–40 to a spike, 5–6 mmlong, spreading-ascending, globose, greenish straw-colored, shining, laterally 2-ribbed; beak equaling or exceeding body, white above, obliquely cut at orifice, bidentate, teeth scarious. Achenes obtusely triangular with sides concave below, closely enveloped, yellowish, bent-apiculate; stigmas 3, brown.

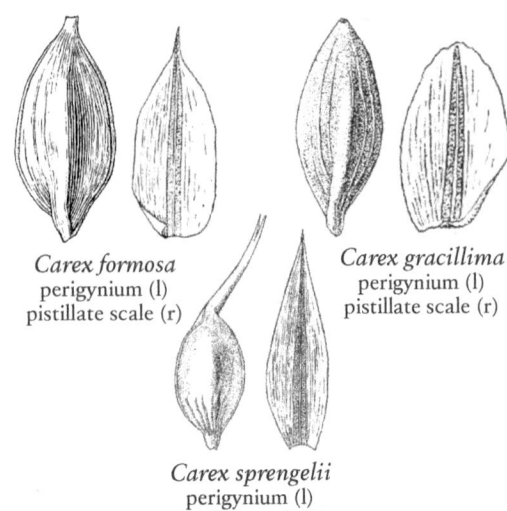

Carex formosa
perigynium (l)
pistillate scale (r)

Carex gracillima
perigynium (l)
pistillate scale (r)

Carex sprengelii
perigynium (l)
pistillate scale (r)

Mesic and floodplain forests, sometimes where disturbed, not tolerant of heavy shade.

Carex Section Laxiflorae

Plants tufted; bases pale to brown or occasionally reddish. Stems weak, ascending to decumbent, sharply triangular in cross-section, angles sometimes winged. Perigynia triangular in cross-section with rounded edges, 25–40-veined (except *Carex leptonervia*); beak (in our species) abrupt, short, often bent. Woodland species.

1   Sides of perigynia with at most 1 main nerve, otherwise nerveless or each with up to 6 obscure nerves; perigynium with a straightish or slightly bent short beak ....
    .... **C. leptonervia**
1   Sides of perigynia each with 7 or more conspicuous nerves; perigynium with straightish or strongly bent beak 2
2   Angles of bract sheaths smooth or nearly so (granular-papillose in *C. ormostachya*); beak of perigynium usually straight or slightly bent 3
2   Angles of bract sheaths minutely ciliate-serrulate; beak of perigynium strongly bent....................... 4
3   Perigynia mostly more than twice as long as wide, tapered to the straightish beak........... **C. laxiflora**
3   Perigynia mostly twice as long as wide, or shorter, abruptly contracted to a very short bent beak ....
    .... **C. ormostachya**
4   Widest leaves 8 mm or more broad; pistillate scales broadly obtuse or truncate, at most scarcely toothed at apex; staminate spike sessile or nearly so. **C. albursina**
4   Widest leaves often less than 8 mm broad; pistillate scales acuminate, awned, or cuspidate; staminate spike sessile or nearly so .... **C. blanda**

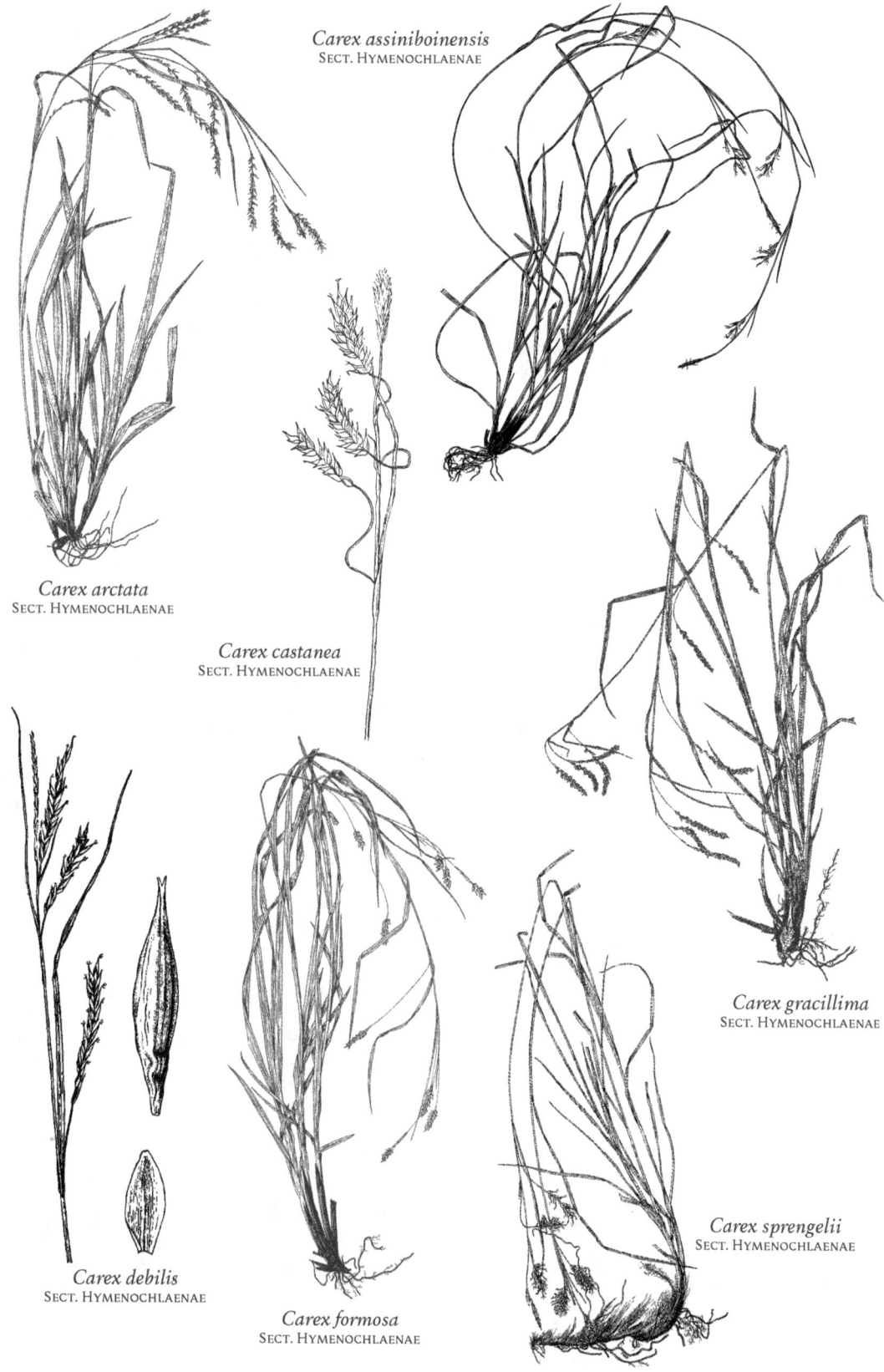

Carex albursina Sheldon
WHITE BEAR SEDGE *native*
IR | WUP | CUP | **EUP** CC 5

Plants loosely tufted; rootstocks short. Stems 1–6 dm, winged, roughened on angles, dark-brown at base. Well-developed leaves 2–5 to a fertile culm; blades 6–25 cm long and 7–15 mm wide, thin, flaccid, light-green, the midvein prominent below, the midlaterals above, roughened towards tip and on margins; sheaths loose. Terminal spike staminate; lateral spikes 3–4, pistillate, the lower separate; bracts sheathing, rough-edged, the lower leaflike; scales white-hyaline with 3-nerved green center. Perigynia 3–18 to a spike, 3–4 mm long, erect, obtusely triangular, yellowish green, the base spongy; beak 0.5 mm long, bent, the orifice entire, hyaline. Achenes triangular with concave sides, closely enveloped, yellowish brown; stigmas 3, reddish brown. Mesic forests; Mackinac County.

Carex blanda Dewey
EASTERN WOODLAND SEDGE *native*
IR | WUP | **CUP** | EUP FAC CC 1

Plants tufted; rootstocks short. Stems 1–6 dm long, slightly winged, 2-edged and flattened in drying, minutely serrulate above, brownish at base; sterile stem leaf blades 1–3.5 dm long and 4–15 mm wide, flat, thin, flaccid, light-green, the midvein prominent below, roughened on margins; blades of fertile stems smaller. Terminal spike staminate or gynaecandrous; lateral spikes 2–5, pistillate, the lower separate on slender, 2-edged peduncles; bracts leaflike; pistillate scales awned, greenish white with 3-nerved green center. Perigynia 8–25 to a spike, 3–4 mm long, ascending, obtusely triangular, yellowish green, strongly nerved, the base spongy; beak 0.5 mm long, bent, the orifice entire, hyaline. Achenes triangular, yellowish brown, granular; stigmas 3, short, reddish brown.

Mesic to wet deciduous forests, sometimes in moist open places, tolerant of disturbance; Dickinson and Menominee counties.

Carex laxiflora Lam.
BROAD LOOSE-FLOWER SEDGE *native*
IR | WUP | **CUP** | **EUP** CC 8

Plants tufted; rootstocks short. Stems 1.5–4 dm long, narrowly winged, brownish at base; sterile shoots reduced to tufts of large leaves, the blades to 2.5 dm long and 7–20 mm wide, flat, thin, flaccid, the midvein prominent below, roughened on margins. Well-developed leaves about 2 per fertile stem; blades 7–15 cm long and 5–8 mm wide; sheaths long, enlarged upward; ligule long. Terminal spike staminate; lateral spikes 3–4, pistillate, the lower long-peduncled; bracts sheathing, the lower leaflike; pistillate scales white-hyaline with 3-nerved green center. Perigynia 5–15 to a spike, 3–5 mm long, appressed-ascending, obtusely triangular, light-green, puncticulate, the base spongy; beak 0.5 mm long, straight or curved, the orifice white-hyaline, entire. Achenes triangular with concave sides and blunt angles, filling perigynia, yellowish brown, granular; stigmas 3, short, reddish brown.

Mostly in beech woods; Chippewa and Menominee counties.

Carex leptonervia (Fern.) Fern.
NERVELESS WOODLAND SEDGE *native*
IR | **WUP** | **CUP** | **EUP** FAC CC 3

Plants tufted; rootstocks slender. Stems 1.5–7 dm long, weakly erect or decumbent, the angles minutely serrulate, brownish at base. Leaves to 3.5 dm long and 3–10 mm wide, flat, flaccid, deep-green, the midvein prominent below, roughened on margins towards tip; sheaths enlarged upward; ligule long. Terminal spike staminate; lateral spikes 2–4, pistillate, the lower 1–2 separate, rough-peduncled; bracts with sheath-margins serrulate; pistillate scales white-hyaline, brownish tinged; anthers 1.5 mm. Perigynia 10–20 to a spike, 3.5–4.5 mm long, erect-ascending, obtusely triangular, light-green, glandular-puncticulate, 2-ribbed, the base spongy; beak 0.5 mm long, the orifice entire, oblique. Achenes triangular with concave sides, filling perigynia, brownish, granular; stigmas 3, reddish brown.

Common in woods.

Carex ormostachya Wieg.
NECKLACE SPIKE SEDGE *native*
IR | WUP | CUP | EUP CC 5

Plants tufted; rootstocks short. Stems 2–6 dm long, minutely granular, crenulate on angles, purplish at

Carex albursina
perigynium (l)
pistillate scale (r)

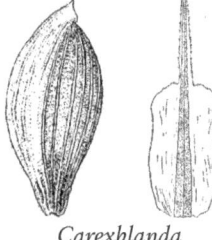

Carexblanda
perigynium (l)
pistillate scale (r)

Carex laxiflora
perigynium (l)
pistillate scale (r)

Carex leptonervia
perigynium (l)
pistillate scale (r)

monocots   **CYPERACEAE** | Sedge Family   379

base. Leaves 5–20 cm long and 2–8 mm wide, flat, flaccid, light-green, the midvein prominent below, roughened on margins towards the sharp tip; sheaths tight. Terminal spike staminate; lateral spikes 3–5, pistillate, loosely flowered below, the lower separate, the peduncles minutely serrulate; pistillate scales mucronate to awned, hyaline with 3-nerved greenish center, sometimes reddish brown-tinged. Perigynia 10–20 to a spike, 2.5–3.5 mm long, appressed-ascending, obtusely triangular, dull-brownish, strongly nerved, the base spongy; beak minute, straight or bent, the orifice hyaline, truncate or oblique. Achenes triangular with concave sides and blunt angles, yellowish; stigmas 3, short, reddish brown.

Woodlands.

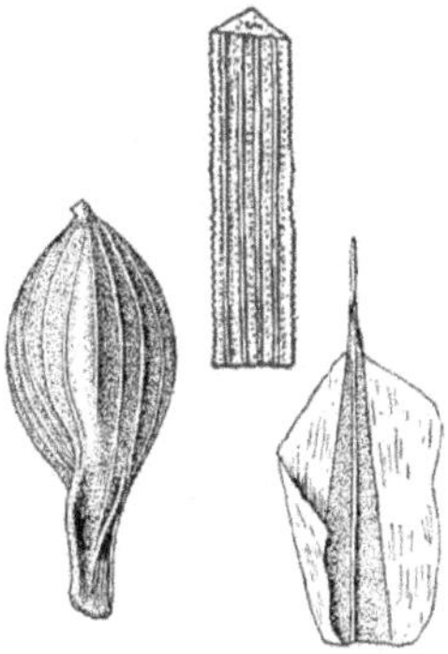

Carex ormostachya
perigynium (l),
stem-section (c),
pistillate scale (r)

Carex albursina
Sect. Laxiflorae

Carex ormostachya
Sect. Laxiflorae

Carex blanda
Sect. Laxiflorae

Carex leptoneria
Sect. Laxiflorae

Carex laxiflora
Sect. Laxiflorae

Carex Section Leptocephalae

One member of the section in the Upper Peninsula. Plants soft, very slender, rhizomatous; spike androgynous (with staminate flowers at tip, pistillate flowers below); perigynia few.

Carex leptalea Wahlenb.
BRISTLY STALK SEDGE — *native*
IR | WUP | CUP | EUP — OBL CC 5

Densely tufted perennial. Stems slender, rounded 3-angled, 1–7 dm long, equal or longer than leaves. Leaves narrow, 0.5–1.5 mm wide; sheaths tight, white, translucent on front. Spikes single on the stems, few-flowered, 5–15 mm long, with both staminate and pistillate flowers, the staminate flowers borne above pistillate; bracts absent; pistillate scales rounded or with a short sharp point, shorter than the perigynia (or the tip of lowest scale sometimes longer than the perigynium). Perigynia yellow-green, nearly round in section to slightly flattened, oblong to oval, 3–5 mm long, finely many-nerved, beakless or with a short beak. Achenes 3-angled; stigmas 3. May–July.

Swamps, alder thickets, open bogs, calcareous fens; usually in partial shade.

Carex Section Leucoglochin

One member of the section in the Upper Peninsula. Spike solitary; perigynia several. Sphagnum moss wetlands.

Carex pauciflora Lightf.
FEW-FLOWER SEDGE — *native*
IR | WUP | CUP | EUP — OBL CC 10

Perennial, from long slender rhizomes. Stems single or several together, 1–4 dm long, longer than leaves. Leaves 1–2 mm wide, lower stem leaves reduced to scales; bract absent. Spike single, to l cm long, with both staminate and pistillate flowers, the staminate above the pistillate; staminate scales infolded to form a slender terminal cone; pistillate scales lance-shaped, 4–6 mm long, pale brown, soon deciduous. Perigynia 1–6, soon turned downward, slender, spongy at base, nearly round in section, straw-colored or pale brown, deciduous when mature, 6–8 mm long. Achenes 3-angled, not filling the perigynium; stigmas 3. June–July.

Open peatlands and floating mats in sphagnum moss, true bogs.

Carex Section Limosae

Plants loosely tufted or stems arising singly, strongly rhizomatous; bases reddish. Roots covered in a dense yellow felt-like tomentum. Vegetative shoots becoming decumbent, behaving like stolons, producing shoots at the nodes. Pistillate spikes pendulous on slender stalks. Perigynia pale, short-beaked, papillose. Stigmas 3. Common in northern bogs and fens.

1. Pistillate scales nearly or quite as broad as the perigynia and often only slightly if at all longer; staminate spike (12–) 15–30 (–50) mm long; plants strongly stoloniferous .... **C. limosa**
1. Pistillate scales distinctly narrower than perigynia, generally with narrowly acuminate tips much exceeding them; staminate spike 5–12 (–15) mm long; plants loosely clumped .... **C. magellanica**

Carex limosa L.
MUD SEDGE — *native*
IR | WUP | CUP | EUP — OBL CC 10

Plants loosely tufted, from long, scaly, yellow-felted rhizomes. Stems sharply 3-angled, 3–5 dm long, longer than leaves, usually rough-to-touch above. Leaves involute, 1–3 mm wide; sheaths translucent, shredding into threadlike fibers near base. Spikes either all staminate or pistillate, the terminal spike staminate; the lower 1–3 spikes pistillate, drooping on lax, threadlike stalks 1–3 cm long; pistillate scales brown, rounded or with a short, sharp point, about same size as perigynia. Perigynia waxy blue-green, ovate, flattened except where filled by achene, 2.5–4 mm long, strongly 2-ribbed with a few faint nerves on each side; beak tiny. Achenes 3-angled; stigmas 3. May–July.

Open bogs and floating mats.

Poor sedge (*C. magellanica*) similar but has scales much narrower than perigynia; *C. buxbaumii* also similar but lacks yellow roots.

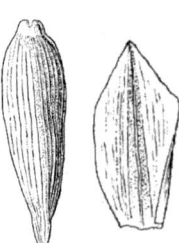

Carex leptalea
perigynium (l)
pistillate scale (r)

Carex pauciflora
perigynium (l)
pistillate scale (r)

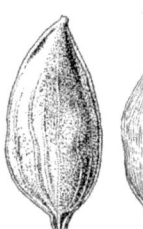

Carex limosa
perigynium (l)
pistillate scale (r)

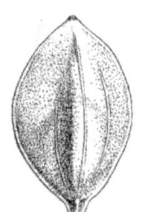

Carex magellanica
perigynium (l)
pistillate scale (r)

Carex magellanica Lam.
POOR SEDGE native
IR | **WUP** | **CUP** | EUP OBL CC 8
Carex paupercula Michx.

Plants loosely tufted, from slender, branching, yellow-felted rhizomes. Stems slender, 1–8 dm long, longer than the leaves, red-brown at base. Leaves 3–12 on lower half of stem, flat but with slightly rolled under margins, 2–4 mm wide, the dried leaves of previous year conspicuous; sheaths red-dotted. Terminal spike staminate (or sometimes with a few pistillate flowers at tip), on a long stalk, usually upright; pistillate spikes 1–4 (rarely with several staminate flowers at base), clustered, usually drooping on slender stalks; lowest bract leaflike, equal or longer than the head; pistillate scales lance-shaped to ovate, narrower but usually longer than the perigynia, brown or green in center, margins brown. Perigynia broadly ovate or oval, 2–3 mm long, flattened and 2-ribbed, with several evident nerves, pale or somewhat waxy blue-green, covered with many small bumps, the tip rounded and barely beaked. Achenes 3-angled, obovate, 2 mm long; stigmas 3. July–Aug.

Open bogs, partly shaded peatlands, floating mats, cedar swamps and thickets, usually in sphagnum moss.

Carex Section Lupulinae
Distinctive sedges of wet forests; recognized by the strongly inflated, ribbed perigynia, 1–2 cm long.

1 Pistillate spikes cylindrical or short-oblong, usually definitely longer than broad; sheath of uppermost leaf usually 1.7 cm or longer; style strongly bent and contorted immediately above the body of the achene; beak of perigynium nearly or quite as long as the body . . . . **C. lupulina**

1 Pistillate spikes spherical or nearly so, scarcely if at all longer than wide; sheath of uppermost leaf absent or less than 1.5 (–2.5) cm; style straight or sinuous or contorted (especially in *C. intumescens*) just below or at the middle; beak of perigynium much shorter than the body 2

2 Perigynia (7–) 10–31 per spike, radiating in all directions, narrowed at the base to a ± broad cuneate stalk, sometimes hispidulous basally; pistillate spikes 1–2 (–3) . . . . **C. grayi**

2 Perigynia 2–8 (–12) per spike, mostly spreading-ascending, rounded at the base, glabrous (and often very shiny); pistillate spikes (1–) 2–5 . . . . **C. intumescens**

Carex grayi Carey
GRAY'S SEDGE native
IR | **WUP** | **CUP** | EUP FACW CC 6

Rhizomes absent. Stems single or forming small clumps, 3–9 dm long, rough on upper stem angles, sheaths at base of stem persistent, red-purple. Leaves 5–12 mm wide. Spikes either staminate or pistillate; terminal spike staminate, stalked; pistillate spikes 1–2, rounded, stalked; bracts leaflike; pistillate scales ovate, body shorter than perigynia but sometimes tipped with an awn to 7 mm long. Perigynia 10–30 per spike, spreading in all directions, not shiny, 10–20 mm long, strongly nerved, tapered from widest point to a notched beak 2–3 mm long. Achenes

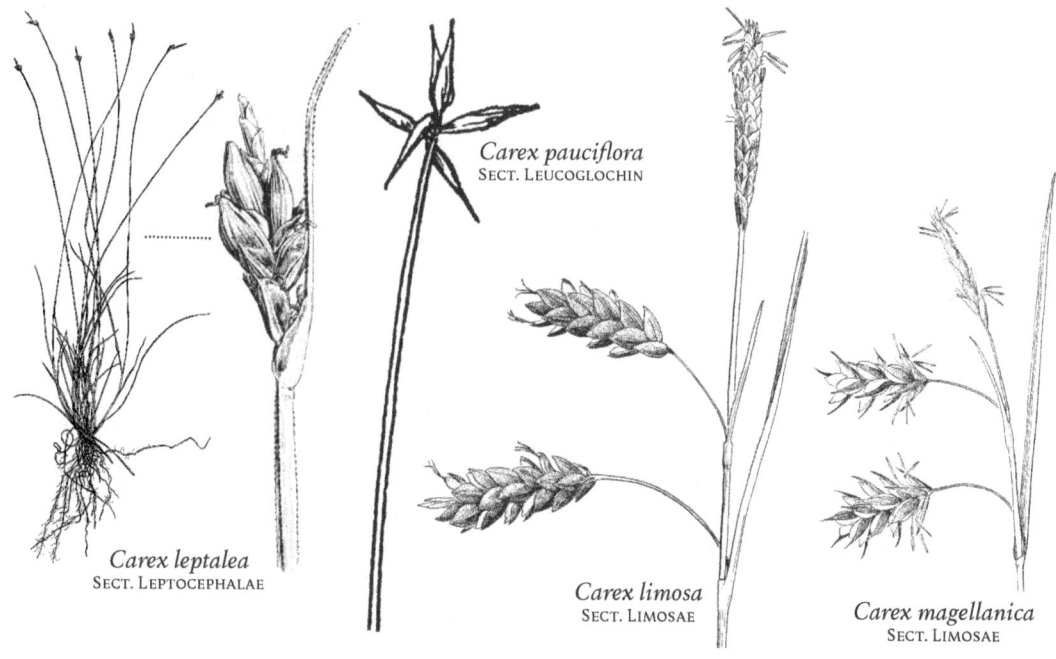

Carex leptalea
SECT. LEPTOCEPHALAE

Carex pauciflora
SECT. LEUCOGLOCHIN

Carex limosa
SECT. LIMOSAE

Carex magellanica
SECT. LIMOSAE

with a persistent, withered style; stigmas 3. June–Sept.

Floodplain forests and backwater areas; Iron and Menominee counties.

Carex intumescens Rudge
GREATER BLADDER SEDGE   native
IR | WUP | CUP | EUP   FACW CC 3

Rhizomes absent. Stems single or in small clumps, 3–9 dm long, rough on upper stem angles; sheaths at base of stem persistent, red-purple. Leaves 4–12 mm wide, bracts leaflike. Spikes either staminate or pistillate, or sometimes staminate spikes with a few pistillate flowers; terminal spike staminate, stalked; pistillate spikes 1–4, rounded, on stalks to 1.5 cm long; pistillate scales narrowly ovate, shorter and narrower than perigynia. Perigynia 1–12 per spike, spreading in all directions, satiny (not dull), 10–17 mm long, tapered to a beak 2–4 mm long. Achenes flattened; stigmas 3. May–Aug.

Common in mixed and deciduous moist forests, kettle wetlands in woods, swamps and alder thickets.

Carex lupulina Muhl.
HOP SEDGE   native
IR | WUP | CUP | EUP   OBL CC 4

Plants loosely tufted, from rhizomes. Stems stout, 3–12 dm long. Leaves much longer than head, 4–15 mm wide; upper sheaths white and translucent, the lower sheaths brown. Spikes either all staminate or pistillate, the upper spike staminate, short-stalked, 2–5 cm long; pistillate spikes 2–6, clustered or overlapping, the lowermost sometimes separate; bracts leaflike and spreading, much longer than head; pistillate scales narrowly ovate, tapered to tip or with a short awn, much shorter than the perigynia. Perigynia many, upright, dull green-brown, lance-shaped, inflated, 10–20 mm long, many-nerved, tapered to a finely toothed bidentate beak 5–10 mm long. Achenes 3-angled; stigmas 3. June–Aug.

Wet woods, swamps, wet meadows and marshes, ditches and shores.

Shining bur sedge (*C. intumescens*) is similar but differs from hop sedge by having fewer, uncrowded perigynia which are olive-green and glossy.

Carex Section Multiflorae

Plants tufted; bases fibrous, brown or pale. Inner band of the leaf sheaths hyaline, corrugated. Inflorescence compound, cylindrical, densely flowered, stiff. Bracts setaceous. Spikes androgynous, at least the lowest branched. Perigynia plano-convex, weakly or inconspicuously spongy at the base. Primarily in wetlands. Characterized by the corrugated inner band of the leaf sheaths; firm, narrow stems; and densely flowered, straight, compound inflorescence.

Carex vulpinoidea Michx.
COMMON FOX SEDGE   native
IR | WUP | CUP | EUP   OBL CC 1

Plants densely tufted, from short rootstocks. Stems stiff, sharply 3-angled, 3–9 dm long. Leaves 2–4 mm wide; sheaths tight, cross-wrinkled and translucent on front, mottled green and white on back. Spikes with both staminate and pistillate flowers, staminate flowers borne above pistillate; heads oblong to cylindric, 3–9 cm long, with several spikes per branch at lower nodes; bracts small and bristlelike, longer than the spikes; pistillate scales awn-tipped, the awns equal or longer than the perigynia. Perigynia yellow-green, becoming straw-colored or brown when mature, flat on 1 side and convex on other, ovate to nearly round, 2–3 mm long, abruptly contracted to a notched, finely toothed beak 1 mm long. Achenes lens-shaped, 1–2 mm long; stigmas 2. May–Aug.

Common in wet to moist meadows, marshes, lakeshores, streambanks, roadside ditches.

Carex Section Ovales

In general, Ovales are characterized by a tufted habit, brownish basal sheaths, and sterile shoots with both nodes and internodes; this is in contrast to the sterile shoots of most species of *Carex*, where the stem-like portion is formed only of overlapping leaf sheaths, and nodes and internodes are absent. Mature perigynia (and often a dissecting microscope) are often needed for accurately identifying species in this large group. Considering the preferred moisture regime of Michigan Ovales may help narrow the list of possible species:

 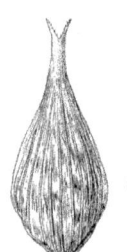 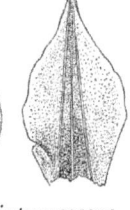 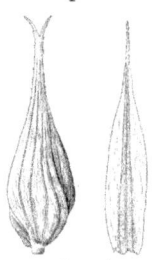 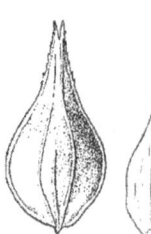

Carex grayi
perigynium (l)
pistillate scale (r)

Carex intumescens
perigynium (l)
pistillate scale (r)

Carex lupulina
perigynium (l)
pistillate scale (r)

Carex vulpinoidea
perigynium (l)
pistillate scale (r)

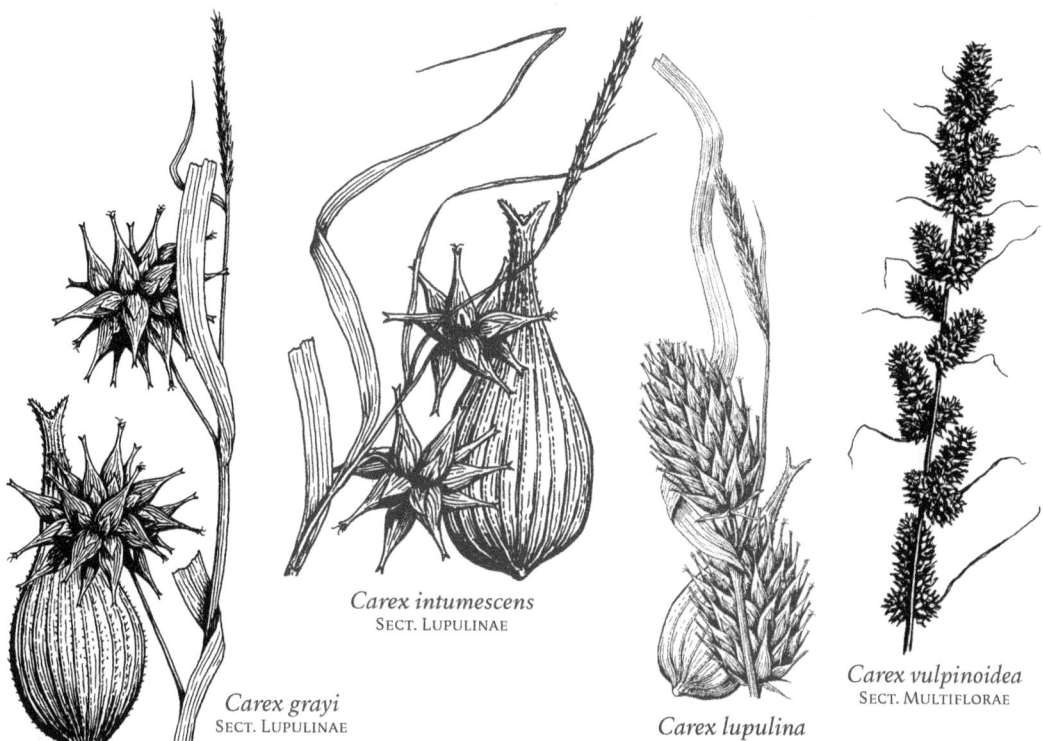

Carex intumescens
Sect. Lupulinae

Carex grayi
Sect. Lupulinae

Carex lupulina
Sect. Lupulinae

Carex vulpinoidea
Sect. Multiflorae

Wetlands: *C. bebbii, C. crawfordii, C. cristatella, C. normalis, C. projecta, C. scoparia, C. tenera, C. tribuloides.*

Non-wetlands: *C. adusta, C. brevior, C. cumulata, C. foenea, C. merritt-fernaldii, C. tincta.*

1   Pistillate scales about or fully as long as the perigynia and nearly the same width as the beaked portion (not necessarily the body), so that the tip of each perigynium is largely concealed; anthers 1.5–3 mm long 2
1   Pistillate scales (or most of them) both shorter and narrower than beaks of perigynia, so the mature perigynia are largely exposed at the tip; anthers various 3
2   Inflorescence stiff, the spikes close together, mostly overlapping; pistillate scales nearly as wide as the bodies of the perigynia, almost concealing them **C. adusta**
2   Inflorescence ± lax or flexuous, the lowermost spikes usually remote; pistillate scales distinctly narrower than bodies of perigynia (the wings of which clearly protrude at maturity)........................**C. foenea**
3   Pistillate scales in the middle or lower portions of the spikes acuminate with a subulate tip or awned 4
3   Pistillate scales obtuse, acute or acuminate, sometimes inconspicuous in the spikes........................5
4   Perigynia 0.9–1.2 mm wide; achenes 0.6–0.8 mm wide; inflorescences dense, lowest inflorescence internodes 2–3 (–5) mm long ....**C. crawfordii**
4   Perigynia 1.2–2.0 mm wide; achenes 0.7–1.1 mm wide; inflorescences dense to open or flexuous, lowest internodes 2–17 mm long......................**C. scoparia**
5   Mature perigynia more than 2 mm wide at widest part 6
5   Mature perigynia less than 2 mm wide 11
6   Perigynium bodies obovate, widest above the middle; leaf sheaths green-nerved ventrally nearly to the summit with at most a narrow V-shaped hyaline area 7
6   Perigynium bodies lanceolate, ovate, elliptic, or orbicular, widest at or below the middle; leaf sheaths various, some with prominent hyaline band near the apex ventrally..8
7   Perigynium beaks spreading, slender and abruptly contracted from the body, the distance from beak tip to top of achene 1–2 mm; styles with strong lateral sinuosity at the base ....**C. albolutescens**
7   Perigynium beaks appressed-ascending, triangular and gradually tapered from the distance from beak tip to top of achene for 2 mm or more; styles straight or less often sinuous near middle ....**C. cumulata**
8   Perigynium bodies narrowly to broadly ovate, greenish, gradually tapered to the beak; pistillate scales with a green midstripe and hyaline or pale margins, rarely brown tinged; leaves 2.5–6.5 mm wide, the sheaths green-mottled, the mouth of sheaths truncate and prolonged up to 2 mm above the base of leaf blades ......
    ....**C. normalis**
8   Perigynium bodies broadly ovate, broadly elliptic, or orbicular, yellowish to tan brown, often abruptly con-

tracted to the beak; pistillate scales greenish to dark brown; leaves 1.5–4 (–5) mm wide, the sheaths evenly colored, the mouth of sheaths concave (prolonged above base of leaf blades in *C. merritt-fernaldii*)...... 9

9 Leaf sheaths smooth .... **C. brevior**
9 Leaf sheaths finely papillose at high magnification (30–40×), especially near leaf base 10

10 Pistillate scales dark rust or brown; leaves of fertile shoots 2–4, the leaf sheaths with ventral hyaline area sometimes puckered or cross-corrugated .... **C. tincta**
10 Pistillate scales greenish to yellowish; leaves of fertile shoots 3–6, the leaf sheaths not puckered............ .... **C. festucacea**

11 Perigynia thin, ± scale-like, often not winged to the base; leaf sheaths somewhat expanded towards apex and bearing narrow wings continuous with midrib and edges of leaf blade, blades 3–7 mm wide; vegetative shoots tall, conspicuous, and with numerous leaves spaced along upper 1/2 of stem 12
11 Perigynia thicker, plano-convex, winged to the base; leaf sheaths with ± rounded edges, not distinctly expanded towards apex, blades 1–4.5 mm wide (except in *C. normalis*); vegetative shoots usually inconspicuous, with leaves relatively few and clustered at apex 14

12 Perigynia stiffly spreading or recurved; spikes ± spherical; pistillate scales hidden, 1.6–2.3 mm long.. **C. cristatella**
12 Perigynia loosely spreading or appressed ascending; spikes nearly spherical to ovate-oblong; pistillate scales evident, 2–3 mm long 13

13 Inflorescences stiff, spikes overlapping; perigynia usually more than 40, beaks appressed- ascending; leaf sheaths firm at summit................ **C. tribuloides**
13 Inflorescences flexuous, the lower spikes usually separated; perigynia usually 15–40, the beaks spreading; leaf sheaths firm or friable at summit .... **C. projecta**

14 Perigynia 2.6–4 times longer than wide, the bodies lanceolate, the distance from beak tip to top of achene 2.2–5.0 mm (as little as 1.8 mm long in *C. crawfordii* with perigynia less than 1.2 mm wide) 15
14 Perigynia less than 2.5 times longer than wide, the bodies obovate, orbicular, or ovate, the distance from beak tip to top of achene 0.8–2.2 mm 16

15 Perigynia 0.9–1.2 mm wide; achenes 0.6–0.8 mm wide; inflorescences dense, lowest inflorescence internodes 2–3 (–5) mm long .... **C. crawfordii**
15 Perigynia 1.2–2 mm wide; achenes 0.7–1.1 mm wide; inflorescences dense to open or even flexuose, lowest internodes 2–17 mm long .... **C. scoparia**

16 Inflorescences on tallest stems compact, 1.5–3 times as long as wide, erect, the spikes overlapping; lowest inflorescence internodes 1–6 (–7.5) mm long, 1/12–1/5 (–1/4) the total length of the inflorescence............... 17
16 Inflorescences on tallest stems elongate, ± open proximally, (2.5–) 3–5.1 times as long as wide, often arching or nodding; lowest inflorescence internodes (5–) 7–19 mm long, mostly 1/5–1/3 (–1/2) the total length of the inflorescence 19

17 Achenes 0.6–0.9 mm wide; perigynia nerveless or with 1–3 faint or basal nerves on the ventral face; inflorescences less than 3 cm long .... **C. bebbii**
17 Achenes 0.9–1.3 mm wide; perigynia often with 3 or more well-defined ventral nerves; inflorescences 1–6 cm long.. 18

18 Sheaths smooth, whitish mottled, the inner band not corrugated; perigynia greenish at maturity **C. normalis**
18 Sheaths finely papillose (under magnification, most easily seen near the leaf base) not whitish mottled, the inner band sometimes corrugated; perigynia pale brown at maturity .... **C. tincta**

19 At least some sheaths papillose near the collar (30–40×), not prominently whitish mottled; perigynium beaks appressed or ascending in the spikes, exceeding pistillate scales by 0–0.8 mm; beaks and shoulders of perigynia stramineous to reddish brown at maturity .. .. **C. tenera**
19 Sheaths totally smooth, often whitish mottled; perigynium beaks spreading, mostly exceeding pistillate scales by 0.7–1.6 mm; beaks and shoulders of perigynia greenish to yellowish or greenish brown at maturity 20

20 Inflorescences erect to somewhat bent, the lowest internodes mostly 6–10long, the rachis stiff; leaves 2.2–6.5 mm wide; larger perigynia mostly 3.1–3.8 mm long and 1.8–2.2 times longer than wide; plants forming small, ± erect clumps often with fewer than 20 stems . ... **C. normalis**
20 Inflorescences arching or nodding, the lowest internodes (6–) 10–21 mm long, the rachis usually thin and wiry; leaves 1.5–3.5 mm wide; larger perigynia mostly 3.5–4.6 mm long, 2.1–3 times longer than wide; plants often forming large, spreading clumps of more than 30 stems .... **C. echinodes**

Carex adusta Boott
LESSER BROWN SEDGE *native*
IR | **WUP** | **CUP** | **EUP** CC 4

Plants tufted; rootstocks short, blackish. Stems 2–8 dm long, stiff, obtusely triangular, brownish tinged at base, clothed with old leaves. Well-developed leaves 4–7 to a fertile culm, on lower fourth; blades 5–20 cm long and 2–4 mm wide, flat or canaliculate, stiff, yellowish green, roughened towards tip and on margins above; sheaths tight, striate dorsally, white-hyaline ventrally. Spikes 4–15, gynaecandrous; lowest bract dilated, the upper scale-like; scales light-reddish brown with white-hyaline margins and 3-nerved lighter center; staminate flowers inconspicuous except in terminal spike. Perigynia 4–5 mm long, appressed-ascending or looser, concealed by scales, plano-convex, leathery, olive-green or blackish, shining, strongly nerved dorsally, narrowly wing-margined, serrulate above; beak ca. 1.5 mm, flat, serrulate, obliquely cut, bidentate, yellowish brown-tinged. Achenes lenticular, brown, shining; stigmas 2, reddish brown.

Dry soil.

Carex bebbii Olney
BEBB'S SEDGE native
IR | WUP | CUP | EUP OBL CC 4

Plants tufted. Stems sharply 3-angled, 2–8 dm long. Leaves 2–5 mm wide; sheaths white, thin and translucent. Spikes 5–10, with both staminate and pistillate flowers, pistillate flowers above staminate, 5–8 mm long, clustered in an ovate head 1.5–3 cm long; pistillate scales tapered to tip, narrower and slightly shorter than the perigynia. Perigynia green to brown, flat on 1 side and convex on other, ovate, 2.5–3.5 mm long, finely nerved on back, nerveless on front, wing-margined, with a finely toothed beak 1/3–1/2 the length of the body, shallowly notched at tip. Achenes lens-shaped; stigmas 2. June–Aug.

Wet to moist meadows, marshes, streambanks, ditches and other wet places.

Carex brevior (Dewey) Mackenzie
SHORT-BEAK SEDGE native
IR | **WUP** | **CUP** | EUP FAC CC 5

Plants tufted; rootstocks short, somewhat woody, black. Stems 3–10 dm long, clothed with old leaves. Well-developed leaves 3–6 to a culm, on lower third; blades 1–2 dm long and 1.5–4 mm wide, thickish, light-green, roughened towards tip especially on margins; sheaths tight, white-hyaline ventrally; sterile shoots conspicuous, the leaves at top. Spikes 3–10, gynaecandrous, in a narrow head; lowest bracts lowest often 1–4 cm long, the upper acuminate or awned; scales yellowish brown with hyaline margins and 3-nerved green center. Perigynia 8–20 to a spike, 4–5.5 mm long, ascending-spreading, thick, leathery, green above, greenish white beneath, strongly nerved dorsally, winged, serrulate; beak 1 mm long, flat, serrulate, obliquely cut, bidentate, reddish brown-tipped. Achenes yellowish brown; stigmas 2, long, reddish brown.

Open places.

Carex crawfordii Fern.
CRAWFORD'S SEDGE native
IR | WUP | CUP | EUP FACW CC 4

Plants densely tufted. Stems 1–8 dm long, stiff. Leaves 3–4 on each stem, 1–4 mm wide. Spikes 3–15, with both staminate and pistillate flowers, the staminate below the pistillate, grouped into a narrowly oblong, sometimes drooping head; pistillate scales light brown with green center, shorter and about as wide as perigynia. Perigynia flattened except where enlarged by the achenes, lance-shaped, 3–4 mm long, brown, narrowly winged nearly to the base, finely toothed above the middle, tapered to a long, slender, toothed, notched beak. Achenes brown, lens-shaped; stigmas 2. July–Sept.

Moist openings and wetland margins, sandy shorelines.

Carex cristatella Britt.
CRESTED SEDGE native
IR | WUP | CUP | EUP FACW CC 3

Plants tufted, from short rhizomes. Stems sharply 3-angled, 3–10 dm long. Leaves 3–7 mm wide; sheaths loose, with fine green lines. Spikes with both staminate and pistillate flowers, pistillate flowers borne above staminate; spikes 5–12, crowded in an ovate to oblong head; bracts much reduced; pistillate scales tapered to tip, shorter than the perigynia. Perigynia widely spreading when mature, green to pale brown, flat on 1 side and convex on other, 2.5–4 mm long, faintly nerved on both sides, strongly winged above the middle, tapered to a finely toothed, notched beak 1–2 mm long. Achenes lens-shaped; stigmas 2. June–Aug.

Carex cristatella
perigynium (l)
pistillate scale (r)

Wet meadows, ditches, floodplains, marshy shores and streambanks.

Carex cumulata (Bailey) Fern.
CLUSTERED SEDGE native
IR | WUP | **CUP** | **EUP** FACU CC 8

Plants densely tufted. Stems 3–9 dm, roughened beneath head, brownish at base, clothed with old leaves. Well-developed leaves 2–4 to a stem, on lower third, not bunched; blades 7–25 cm long and 3–5 mm wide, flat, thickish, light-green, roughened towards tip; sheaths loose, green-striate ventrally. Spikes 5–30, the upper gynaecandrous, the lower

Carex bebbii
perigynium (l)
pistillate scale (r)

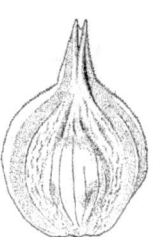
Carex brevior
perigynium (l)
pistillate scale (r)

Carex crawfordii
perigynium (l)
pistillate scale (r)

pistillate, greenish or brownish, in a head 2–4 cm long; scales white-hyaline with 1–3-nerved green center, yellowish brown-tinged at tip. Perigynia numerous, 3–4 mm long, ascending, thin plano-convex, greenish or brownish, lightly nerved dorsally, wing-margined, serrulate above; beak 0.6–1 mm long, flat, serrulate, obliquely cut. Achenes lenticular, yellowish brown, bent-apiculate; stigmas 2, short, light-reddish.

Rocky woods and sandy places; Chippewa and Schoolcraft counties.

Carex foenea Willd.
BRONZE-HEAD OVAL SEDGE native
IR | WUP | CUP | EUP CC 3

Carex aenea Fernald

Plants densely tufted. Stems 2–12 dm long. Leaves 3–6 per fertile culm, green, 8–30 cm long and 2–4 mm wide. Inflorescences open, usually with widely spaced spikes, brown or greenish brown; bracts scalelike, sometimes bristlelike. Spikes usually 3–7, nodding, copper-colored; pistillate scales usually reddish brown, with 3-veined green or brown midstripe, equaling, and more or less covering the perigynia. Perigynia erect-ascending, green or brown, conspicuously 4–9-veined 3–5 mm long, margin flat, including small wing 0.2–0.4 mm wide; beak white or brown, white margined at tip, flat, serrulate, distance from beak tip to achene 1.7–2.5 mm. Achenes dark brown at maturity, ovoid-orbicular; stigmas 2.

Dry open sandy places, ro

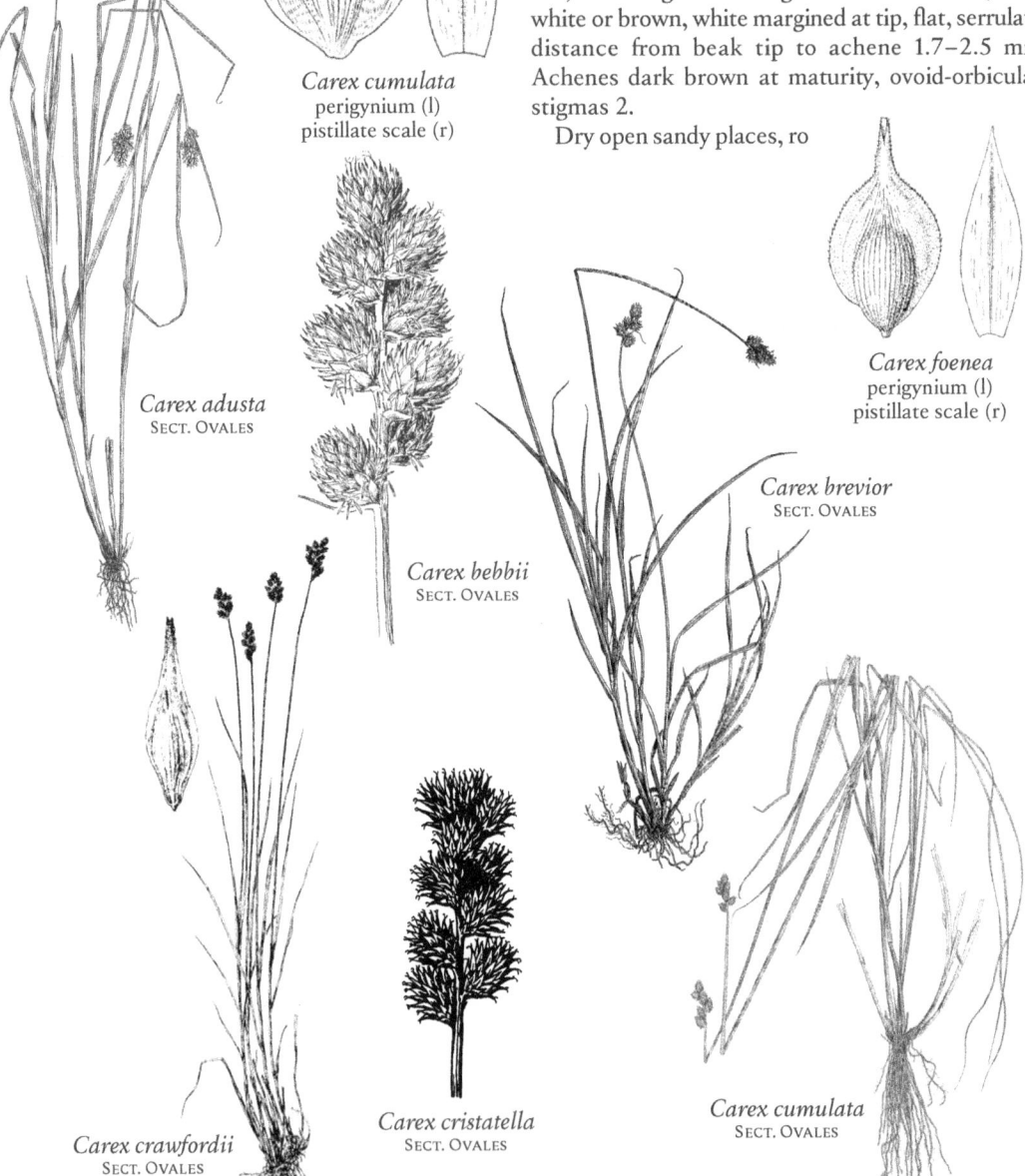

Carex merritt-fernaldii Mackenzie
MERRITT FERNALD'S SEDGE native
IR | **WUP** | **CUP** | EUP CC 4

Plants tufted; rootstocks short, somewhat woody, black. Stems 3–10 dm, stiff, clothed with old leaves. Well-developed leaves 3–6 to a culm, on lower third; blades 1–2 dm long and 1.5–3 mm wide, yellowish green, roughened towards tip especially on margins; sheaths tight, white-hyaline ventrally, papillate dorsally. Spikes 4–10, gynaecandrous, in a head 1.5–5 cm long; bracts scale-like, the lowest often 1–4 cm, the upper acuminate or awned; scales yellowish brown with hyaline margins and 3-nerved green center. Perigynia 15–30 to a spike, 4–5 mm long, appressed-ascending, yellowish green or straw-colored, strongly nerved dorsally, winged, serrulate; beak 1 mm, flat, serrulate, obliquely cut, bidentate, reddish brown-tipped. Achenes brown, shining; stigmas 2, long, reddish brown.

Dry woodlands.

Carex normalis Mackenzie
GREATER STRAW SEDGE native
IR | **WUP** | **CUP** | EUP FACW CC 5

Plants tufted. Stems 3–8 dm long. Leaves 2–6 mm wide, lower stem leaves reduced to scales. Spikes 5–10, with both staminate and pistillate flowers, the staminate below the pistillate, round in outline, stalkless, loosely grouped in heads; pistillate scales translucent, lightly brown-tinged, with green midvein, shorter than the perigynia. Perigynia upright, flat on 1 side and convex on other, green or pale green-brown, 3–4 mm long, finely nerved, tapered to a finely toothed beak. Achenes lens-shaped; stigmas 2. June–Aug.

Moist to wet deciduous woods, floodplain forests, alder thickets, marshes, pond margins.

Carex projecta Mackenzie
NECKLACE SEDGE native
IR | **WUP** | **CUP** | **EUP** FACW CC 3

Plants tufted, from short rhizomes. Stems slender and weak, 3-angled, 4–10 dm long, upper stems rough. Leaves stiff, 3–7 mm wide; sheaths loose. Spikes 7–15, with both staminate and pistillate flowers, pistillate flowers above staminate in each spike, obovate to nearly round, straw-colored, distinct and more or less separated (at least the lower spikes) in a somewhat lax and zigzagged inflorescence; bracts inconspicuous; pistillate scales straw-colored, narrower and shorter than the perigynia. Perigynia ascending to spreading when mature, 3–5 mm long, dull brown, flattened except where filled by the achene, winged on margin, the wing gradually narrowing from middle to base, tapered to a notched, finely toothed beak 1–2 mm long. Achenes lens-shaped; stigmas 2. June–Aug.

Common in floodplain forests, swamps, thickets, wet openings, shaded slopes.

Similar to *Carex tribuloides* but the perigynia tips spreading rather than erect as in *C. tribuloides*.

Carex scoparia Schkuhr
POINTED BROOM SEDGE native
IR | **WUP** | **CUP** | **EUP** FACW CC 4

Plants densely tufted, sometimes spreading by surface runners. Stems 2–10 dm long, sharply 3-angled. Leaves 1–3 mm wide; sheaths tight, white-translucent. Spikes 4–10, with both staminate and pistillate flowers, pistillate flowers borne above staminate, ovate to broadest at middle, clustered or separate, in a narrowly ovate head; bracts small, the lowest often bristlelike; pistillate scales slightly shorter than perigynia. Perigynia greenish white, flat, 3–7 mm long, margins narrowly winged, tapered to a finely toothed, slightly notched beak 1–2 mm long. Achenes lens-shaped; stigmas 2. May–July.

Wet meadows and openings, low prairie, swamps and sandy lakeshores.

Carex tenera Dewey
QUILL SEDGE native
IR | **WUP** | **CUP** | EUP FAC CC 4

Plants tufted, from short rhizomes. Stems slender, sharply 3-angled, 3–8 cm long, rough-to-touch above. Leaves 0.5–3 mm wide; sheaths white-translucent on front, mottled green and white on back. Spikes 4–8, with both staminate and pistillate flowers, pistillate flowers borne above staminate, ovate to round, loose in nodding heads; bracts small, sometimes

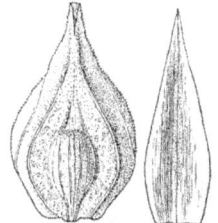

Carex merritt-fernaldii
perigynium (l)
pistillate scale (r)

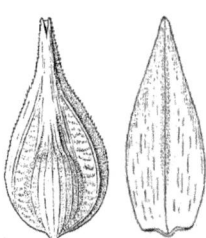

Carex normalis
perigynium (l)
pistillate scale (r)

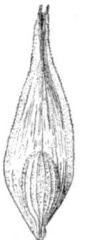

Carex projecta
perigynium (l)
pistillate scale (r)

Carex scoparia
perigynium (l)
pistillate scale (r)

bristlelike, longer than the spike; pistillate scales slightly shorter than perigynia. Perigynia ovate, flat on 1 side and convex on other, straw-colored when mature, 2.5–4 mm long, wing-margined, tapered to a notched, finely toothed beak 1–2 mm long. Achenes lens-shaped; stigmas 2. June–Aug.

Wet to moist meadows, streambanks, floodplains and moist woods.

Carex tincta (Fern.) Fern.
TINGED SEDGE (THR) native
IR | WUP | CUP | EUP   CC 4

Plants tufted; rootstocks short, black. Stems 4–8 dm long, roughened beneath head, brownish black at base, clothed with old leaves. Well-developed leaves 3–4 to a fertile culm, on lower fourth; blades 1–3 dm long and 2–4 mm wide, flat, firm, light-green, roughened towards the tip; sheaths septate-nodulose dorsally, hyaline and cross-rugulose ventrally. Spikes 5–10, gynaecandrous; bracts scale-like; scales acute, light-reddish brown with hyaline margins and 3-nerved green center. Perigynia 12–30 to a spike, 3.5–4 mm wide, appressed-ascending, plano-convex, flat, thickish, firm, membranous, green or straw-colored, strongly nerved, winged, serrulate above; beak 1.25 mm, broad, flat, serrulate, obliquely cut dorsally bidentate, reddish brown. Achenes lenticular; stigmas 2, reddish brown.

Rare in woods; Dickinson County, disjunct from main range of northeastern North America.

Carex tribuloides Wahlenb.
BLUNT BROOM SEDGE   native
IR | WUP | CUP | EUP   FACW CC 3

Plants tufted, from short rhizomes. Stems sharply 3-angled, 3–9 dm long, longer than leaves. Leaves stiff, 3–7 mm wide; sheaths loose, with green lines. Spikes 5–15, with both staminate and pistillate flowers, pistillate flowers borne above staminate, obovate, densely to loosely clustered into an ovate or oblong head; bracts inconspicuous; pistillate scales tapered to tip, shorter than the perigynia. Perigynia light green to pale brown, flattened except where filled by the achenes, lance-shaped, 3–6 mm long, broadly winged near middle, tapered to a notched, finely toothed beak 1–2 mm long. Achenes lens-shaped; stigmas 2. June–July.

Local in floodplain forests, shady low areas in woods, pond and lake margins, marshes, low prairie.

Carex Section Paludosae

Mostly slender, long-rhizomatous plants, with red basal leaf sheaths (and ladder-fibrillose in all but *C. houghtoniana*), and pubescent perigynia. *Carex lacustris* is somewhat different, having glabrous perigynia.

1   Perigynia glabrous . . . . **C. lacustris**
1   Perigynia pubescent 2
2   Perigynia 4.5–6.5 mm long, sparsely hairy, the strong nerves of perigynium and even cellular detail of body therefore evident; plants usually of dry and sandy habitats . . . . **C. houghtoniana**
2   Perigynia 3–4.5 (–5.2) mm long, densely pubescent, nerving of perigynium and cellular detail therefore obscured; plants usually of wetlands 3
3   Leaf blades involute to triangular-channeled, 0.7–2 mm wide, those of vegetative shoots especially long-prolonged into a curled, filiform tip; leaves and lowermost bracts with the midvein low, rounded, and forming an inconspicuous keel (at least proximally) . **C. lasiocarpa**
3   Leaf blades flat or folded into an M-shape except at the base and near the tip, 2–5 mm wide, not prolonged into a long filiform tip; leaves and lowest bract with the midvein forming a prominent and sharply pointed keel for much of the length. . . . . **C. pellita**

Carex houghtoniana Torr. ex Dewey
HOUGHTON'S SEDGE   native
IR | WUP | CUP | EUP   CC 5

Plants loosely tufted and stoloniferous; stolons long, slender, horizontal, scaly. Stems 1.5–6.5 dm long, stiff, rough above, purplish at base, the basal sheaths filamentose. Well-developed leaves 2–4 to a fertile culm, on lower third; blades 8–20 cm long and 2.5–4 mm wide, flat with revolute margins, septate-nodulose, deep-green, roughened especially on margins and towards tip; sheaths tight, thin and yellowish brown-tinged ventrally. Terminal spike staminate, rough-peduncled, often with a shorter spike near base; lateral spikes 1–3, pistillate; lowest bract leaflike; pistillate scales reddish brown with hyaline margins and 3-nerved green center. Perigynia 15–30 to a spike, 5–6 mm long, spreading or ascending, obscurely triangular, olive or brownish green, short-hirsute, 15–20-ribbed; beak 2 mm long, bidentate, purple between teeth, scabrous within. Achenes triangular with concave sides, closely enveloped, yellowish brown; stigmas 3, short, blackish.

Open sandy or rocky soil.

Carex lacustris Willd.
LAKEBANK SEDGE   native
IR | WUP | CUP | EUP   OBL CC 6

Plants large, tufted, from scaly rhizomes. Stems erect, 3-angled, 6–13 dm long, rough-to-touch. Leaves 6–15 mm wide; sheaths often red-tinged, the lower ones disintegrating into a network of fibers. Spikes either staminate or pistillate, the upper 2–4 staminate, stalkless; the lower 2–4 spikes pistillate, erect, usually separate, stalkless or short-stalked, cylindric; bracts leaflike, some or all longer than the head; pistillate scales awned or tapered to tip, the body shorter than the perigynia, the sides

CYPERACEAE | Sedge Family

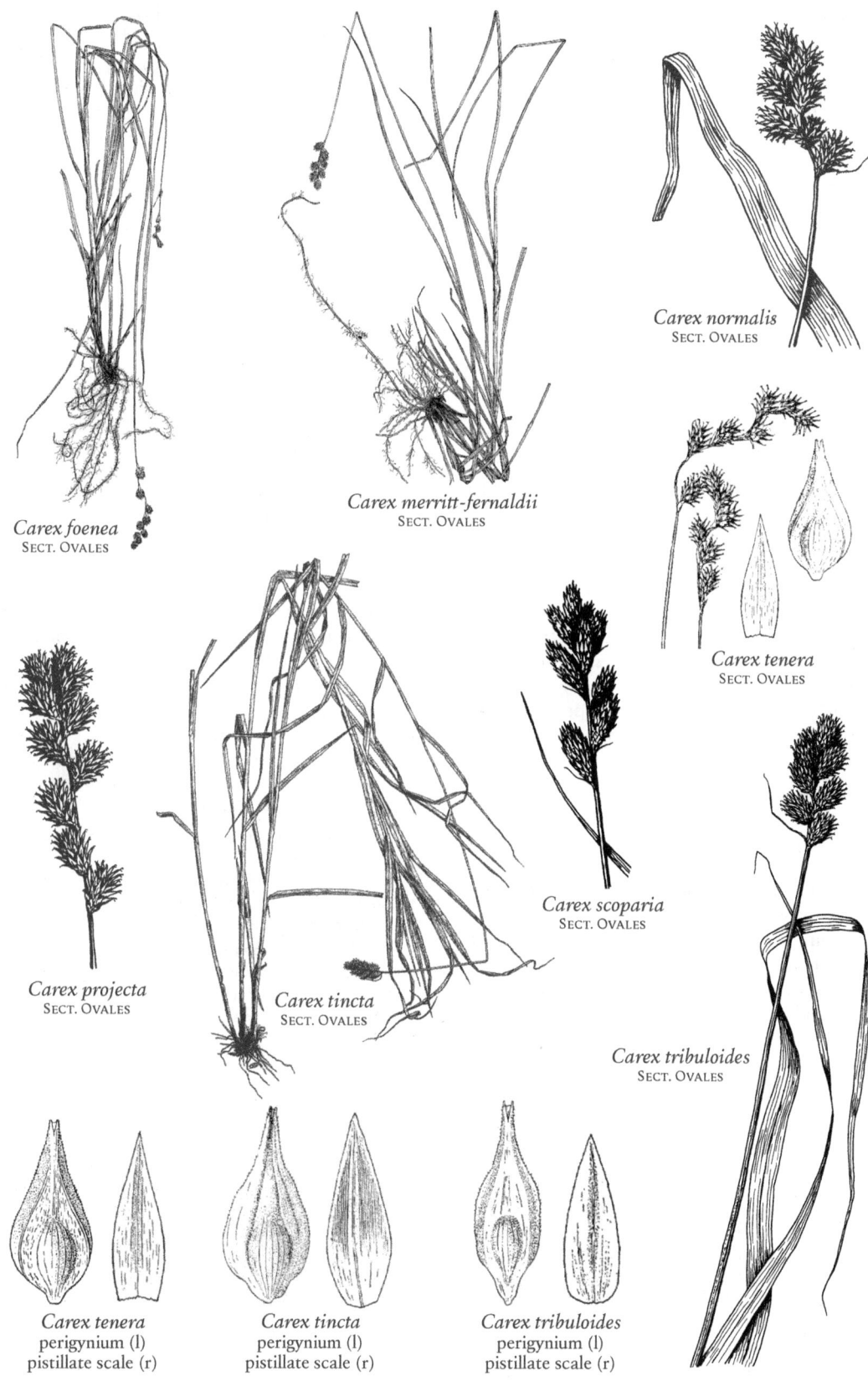

thin and translucent to pale brown. Perigynia olive, flattened to nearly round in section, narrowly ovate, 5–7 mm long, with more than 10 raised nerves, tapered to a smooth beak about 1 mm long. Achenes 3-angled; stigmas 3. May–Aug.

Swamps, marshes, kettle wetlands, wetland margins, usually in shallow water; low areas in tamarack swamps.

Carex lasiocarpa Ehrh.
SLENDER SEDGE native
IR | WUP | CUP | EUP OBL CC 8

Carex lanuginosa Michx. var. *americana* (Fern.) Boivin

Colony-forming perennial, from long, scaly rhizomes. Stems loosely tufted, 3-angled, 3–10 dm long. Leaves elongate and inrolled, 1–2 mm wide; sheaths tinged with yellow-brown. Spikes either all staminate or pistillate, usually the upper 2 staminate; the staminate spikes slender, on a long stalk; the lower 1–3 spikes pistillate, widely separate, more or less stalkless, cylindric; bracts leaflike, the lowest usually longer than the stem; pistillate scales purple-brown with a green center, narrowly ovate. Perigynia dull brown green, obovate, nearly round in section, 3–5 mm long, densely soft hairy, contracted to a beak about 1 mm long, the beak teeth erect. Achenes yellow-brown, 3-angled with concave sides; stigmas 3. June–Aug.

Peatlands and wet peaty soils, open bogs, pond margins (where a pioneer mat-former).

Carex pellita Muhl
WOOLLY SEDGE native
IR | WUP | CUP | EUP OBL CC 2

Carex lanuginosa auct. non Michx.
Carex lasiocarpa Ehrh. var. *latifolia* (Boeckl.) Gilly

Plants tufted and stoloniferous; stolons long, horizontal, scaly. Stems 3–10 dm long, stiff, rough above, dark-purplish red at base, the lower sheaths filamentose;. Well-developed leaves 2–5 to a fertile culm; blades 2–6 dm long and 1.5–5 mm wide, flat with revolute margins, septate-nodulose, rough especially towards the tip; sheaths purplish tinged. Usually 2 upper spikes staminate, long-peduncled; lower 2–3 spikes pistillate; bracts sheathless or nearly so; pistillate scales acuminate, mucronate or awned, ciliate, reddish brown with hyaline margins and 3-nerved green center. Perigynia 25–75 to a spike, 2.5–3.5 mm long, ascending, suborbicular, inflated, leathery, dull-brownish green, densely hairy, many-ribbed; beak 1 mm, bidentate. Achenes triangular with concave sides and blunt angles, loosely enveloped, yellowish brown, punctate; stigmas 3, blackish.

Swamps.

Carex Section Paniceae

Plants colonial, shoots arising singly or few together; rhizomes elongate; bases brown to maroon. Leaf blades typically stiff. Terminal spike staminate, typically raised above the uppermost pistillate spike. Lateral spikes generally cylindrical, ascending (except *Carex vaginata*). Perigynia several-veined, mostly short-beaked, papillose (except *C. vaginata*). Calciphiles, growing mostly in wet soils (but *C. meadii* common in dry calcareous prairies). The section is fairly distinctive and easy to recognize, apart from *C. vaginata*, which is morphologically distinct.

1 Perigynium with a beak 1 mm long. . . . . **C. vaginata**
1 Perigynium beakless, indistinctly beaked, or contracted to beak less than 0.5 mm 2
2 Perigynia strongly ascending, beakless or tapering to an erect, very short straight beak; leaves stiff, thick, channeled, strongly glaucous. . . . . **C. livida**
2 Perigynia ascending to spreading, tapering to a bent apex; leaves relatively thin and flexible, flat or folded, green to somewhat glaucous 3
3 Bladeless basal sheaths and proximal leaf sheaths strongly tinged with reddish purple; plants forming loose clumps to extensive closed colonies of vegetative shoots from superficial rhizomes; perigynia ± 2-ranked; plants of rich forests . . . . **C. woodii**
3 Bladeless basal sheaths and proximal leaf sheaths brownish, green, or faintly, irregularly tinged with reddish purple; plants usually with vegetative shoots widely scattered and inconspicuous from deep rhizomes; perigynia 3–6-ranked; plants of moist, usually sunny habitats. . . . . **C. tetanica**

Carex houghtoniana
perigynium (l)
pistillate scale (r)

Carex lacustris
perigynium (l)
pistillate scale (r)

Carex lasiocarpa
perigynium (l)
pistillate scale (r)

Carex pellita
perigynium (l)
pistillate scale (r)

Carex houghtoniana
SECT. PALUDOSAE

Carex lacustris
SECT. PALUDOSAE

Carex lasiocarpa
SECT. PALUDOSAE

Carex pellita
SECT. PALUDOSAE

Carex livida (Wahlenb.) Willd.
LIVID SEDGE *native*
IR | WUP | CUP | EUP   OBL CC 10

Plants forming small clumps, from long slender rhizomes. Stems erect, to 6 dm long, light brown at base. Leaves 6–12 on lower third of stem, strongly waxy blue-green, channeled, 0.5–4 mm wide, dried leaves of the previous year conspicuous; sheaths thin. Terminal spike staminate (or rarely with both staminate and pistillate flowers, the staminate below the pistillate), linear; pistillate spikes 1–3, the lowest more or less separate, sometimes long-stalked, the upper grouped, stalkless or short-stalked, with 5–15 upright perigynia; bracts leaflike, sometimes longer than the head; pistillate scales shorter than the perigynia, light purple with broad green center and white translucent margins. Perigynia slightly flattened and rounded 3-angled, 2–5 mm long, strongly waxy blue-green, with small dots, two-ribbed and with fine nerves, tapered to a beakless tip. Achenes 3-angled with prominent ribs, brown-black; stigmas 3. July–Aug.

Wet meadows and fens, especially where calcium-rich.

Carex tetanica Schkuhr
RIGID SEDGE *native*
IR | WUP | CUP | EUP   FACW CC 9

Tufted perennial from slender rhizomes. Stems 3-angled, 1–6 dm long, rough-to-touch above. Leaves 1–5 mm wide; sheaths tight, white or yellow and translucent. Spikes either all staminate or pistillate, terminal spike staminate; lateral spikes pistillate, usually widely separated, the lower spikes short-cylindric, stalked, loosely flowered with perigynia in 3 rows; bracts shorter than the head; pistillate scales purple-brown on margins, as wide as but shorter than the perigynia. Perigynia green, faintly 3-angled, obovate, 2–4 mm long, 2-ribbed; beak tiny, bent. Achenes 3-angled with concave sides; stigmas 3. May–July.

Wet meadows and openings, low prairies, marshy areas; Delta, Keweenaw (including Isle Royale), and Menominee counties.

Carex vaginata Tausch
SHEATHED SEDGE *native*
IR | WUP | CUP | EUP   OBL CC 10
Carex saltuensis Bailey

Perennial, from long rhizomes. Stems 2–6 dm long, several together. Leaves 2–5 mm wide, not scale-like at base of stem. Terminal spike staminate, 1–2 cm long; pistillate spikes 1–3, sometimes staminate at tip, loosely spreading, widely separated, the lower stalks long, the upper shorter; bracts with loose sheaths and blades shorter than the spikes; pistillate scales purple-brown, sometimes with a narrow green center. Perigynia usually in 2 rows, the lower separate, the upper overlapping, 3–5 mm long, narrowly obovate, with a curved beak 1 mm long. Achenes 3-angled, nearly filling the perigynia; stigmas 3. June–Aug.

Swamps and thickets, especially where calcium-rich.

Carex woodii Dewey
PRETTY SEDGE   native
IR | **WUP** | **CUP** | EUP   FACU CC 8

Carex tetanica Schkuhr var. *woodii* (Dewey) Wood
Plants loosely tufted and stoloniferous; stolons slender, purple, scaly. Stems 3–7 dm long, roughened above; sterile shoots numerous, long. Well-developed leaves 2–4 to a fertile culm, near base; blades 5–20 cm long and 2.5–4 mm wide, flat with revolute margins, flaccid, light-green, white-lined below; sheaths loose, overlapping, white or yellowish hyaline ventrally. Terminal spike staminate, the peduncle roughish; lateral spikes 2–3, pistillate, on slender roughish peduncles; bracts long-sheathing, the sheaths tight; pistillate scales purplish or reddish brown with hyaline margins and 3-nerved green center. Perigynia 6–15 to a spike, 3.5–4 mm long, ascending, yellowish green, puncticulate, 2-keeled, lightly nerved; beak 0.5 mm long, excurved, the orifice oblique, entire, hyaline. Achenes triangular with concave sides and blunt angles, closely enveloped, yellowish brown; stigmas 3, reddish brown.

Dry woods.

Carex Section Phacocystis

Plants often cespitose; rhizomes short or long. Lower leaf sheaths brown to red, fibrous in some species. Terminal spike typically staminate, ascending. Lateral spikes pistillate or androgynous, ascending to nodding or drooping, elongate. Perigynia biconvex with distinct marginal veins. Stigmas 2. Mostly common species of Michigan wetlands, ranging from floodplains and wet forests xxxcheck (*Carex crinita, C. gynandra, C. emoryi*), to sedge meadows (*C. stricta*), wet prairies (*C. haydenii*), bogs and marshes (*C. aquatilis*), and wet roadsides and ditches.

ADDITIONAL SPECIES
Carex nigra (L.) Reichard, reported from sandy wetlands in Schoolcraft County; distinguished by dark-spotted perigynia 2–3.5 mm long and black pistillate scales; endangered.

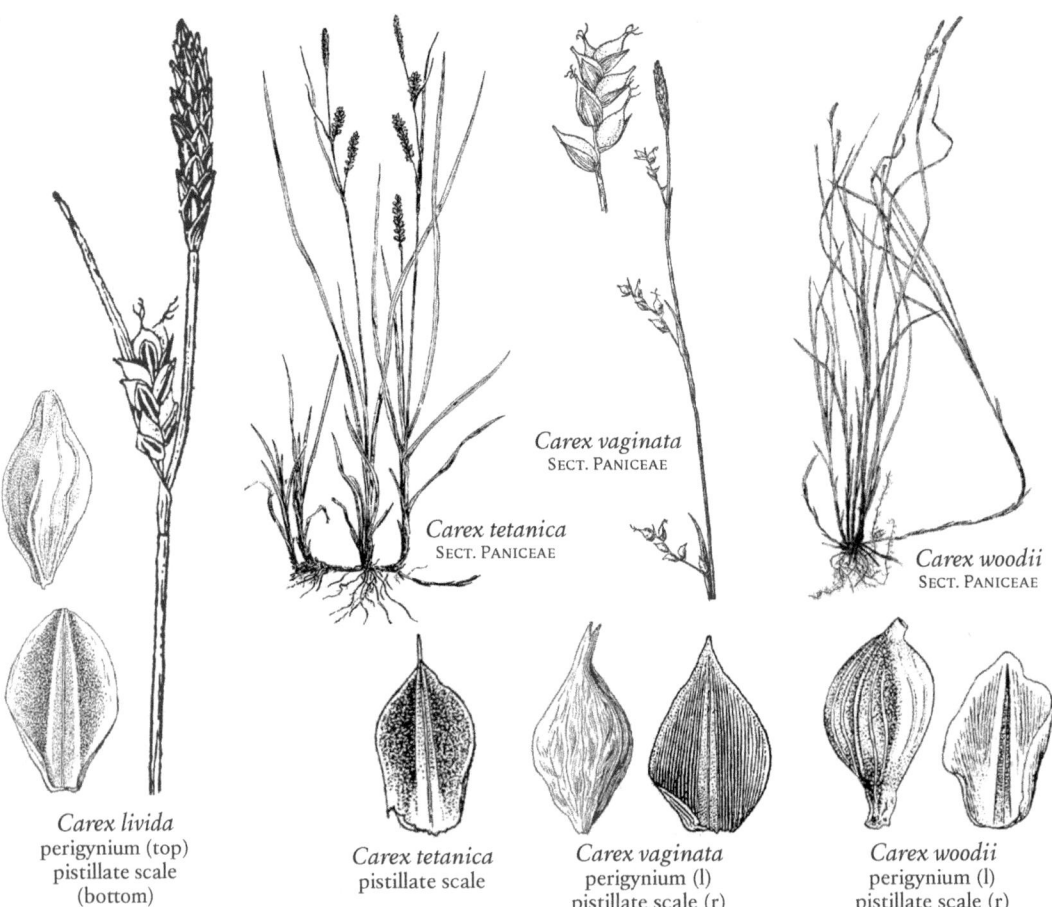

Carex vaginata
SECT. PANICEAE

Carex tetanica
SECT. PANICEAE

Carex woodii
SECT. PANICEAE

Carex livida
perigynium (top)
pistillate scale
(bottom)

Carex tetanica
pistillate scale

Carex vaginata
perigynium (l)
pistillate scale (r)

Carex woodii
perigynium (l)
pistillate scale (r)

1. Pistillate spikes on ± lax peduncles, at length drooping, the scales prominently awned; body of achene with an irregular notch, constriction, or wrinkle on one side. . 2
1. Pistillate spikes erect or strongly ascending, often sessile, the scales acute or acuminate, not awned; body of achene smooth and ± regular 3
2. Sheaths smooth; bodies of most if not all pistillate scales shallowly lobed at summit (on each side of base of the awn) . . . . C. crinita
2. Sheaths scabrous-hispidulous; bodies of most or all pistillate scales on lower part of spike truncate or tapered at summit . . . . C. gynandra
3. Fertile stems of current year with conspicuous bladeless sheaths at base, not surrounded by dried-up bases of the previous year's leaves but arising laterally; lowest bract usually shorter than to approximately equaling the inflorescence 4
3. Fertile stems of current year mostly lacking bladeless sheaths at base, arising centrally from tufts of dried-up bases of previous years leaves; lowest bract usually conspicuously longer than the inflorescence 5
4. Ligule longer than width of leaf blade (deeply inverted V-shaped); ventral surface of lower leaf sheaths tearing to form a ladder-like arrangement of fibers and usually minutely scabrous and red-dotted, especially near the tip . . . . C. stricta
4. Ligule shorter than width of leaf blade (often nearly horizontal); ventral surface of lower leaf sheaths not tearing to form a ladder like arrangement of fibers, smooth and whitish . . . . C. emoryi
5. Perigynia essentially nerveless, except sometimes at the base only; staminate spikes usually 2 or more . . . . C. aquatilis
5. Perigynia conspicuously few-ribbed on both sides; staminate spike usually 1 6
6. Plants densely tufted, without long rhizomes; scales with a broad central green portion about as wide as the darker margins; leaves mostly overtopping spikes . . . . C. lenticularis
6. Plants colonial from elongated rhizomes; scales with very narrow green portion much narrower than the broad, dark margins, scarcely if at all broader than the midrib; leaves mostly shorter than stems . . . . . C. nigra

Carex aquatilis Wahlenb.
WATER SEDGE *native*
IR | **WUP** | **CUP** | **EUP** OBL CC 7

Plants large, tufted or forming turfs; spreading by many slender rhizomes. Stems 3–12 dm long, 3-angled, usually rough-to-touch below the spikes. Leaves waxy blue-green, 2–7 mm wide; sheaths white or purple-dotted. Spikes 3–5, the upper spikes staminate, the middle and lower spikes pistillate or

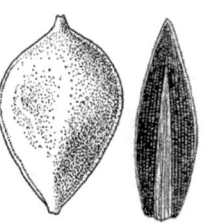

Carex aquatilis
perigynium (l)
pistillate scale (r)

often with staminate flowers borne above pistillate; pistillate scales tapered to tip. Perigynia pale green to yellow-brown or red-brown, broadest near tip, not inflated, 2–3 mm long; beak tiny. Achenes lens-shaped; stigmas 2. May–Aug.

Wet meadows, marshes, shores, streambanks, kettle lakes, ditches and fens.

Carex crinita Lam.
FRINGED SEDGE *native*
IR | **WUP** | **CUP** | **EUP** OBL CC 4

Plants large, densely tufted. Stems 5–15 dm long. Leaves 7–13 mm wide, lowest stem leaves reduced to scales; sheaths smooth. Spikes staminate or pistillate, drooping on slender stalks; staminate spikes 1–3, above pistillate spikes; pistillate spikes 2–5, narrowly cylindric; bract leaflike, without a sheath; pistillate scales rounded and notched at tip with pale midvein prolonged into a toothed awn to 10 mm long, scale edges coppery-brown. Perigynia green, 2-ribbed, nerves faint or absent, round in cross-section, abruptly tapered to a tiny beak. Achenes lens-shaped; stigmas 2. May–July.

Swamps and alder thickets, wet openings, ditches and potholes.

Similar to *C. gynandra* but with smooth sheaths, lower pistillate scales rounded at tip, and perigynia round in section and inflated.

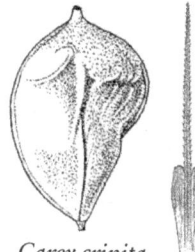

Carex crinita
perigynium (l)
pistillate scale (r)

Carex emoryi Dewey
EMORY'S SEDGE *native*
IR | **WUP** | **CUP** | EUP OBL CC 7

Plants loosely tufted, from scaly rhizomes. Stems 3-angled, 4–12 dm long. Leaves 3–7 mm wide, lowest leaves reduced to red-brown sheaths; upper sheaths white or yellow-tinged and translucent, lower sheaths red-brown. Spikes 3–7, the terminal 1 or 2 all staminate, the lateral spikes all pistillate or with staminate flowers above the pistillate; lowest bract leaflike; pistillate scales narrower than the perigynia. Perigynia light green, becoming straw-colored at maturity, convex on both sides, 1.5–3 mm long; stigmas 2. May–July.

Shores, streambanks, wet meadows and floodplain forests, sometimes forming pure stands, especially along rivers.

Carex gynandra Schwein.
NODDING SEDGE *native*
IR | **WUP** | **CUP** | **EUP** OBL CC 3

Carex crinita var. *gynandra* (Schwein.) Schwein. & Torr.
Plants large, tufted. Stems 5–15 dm long, longer than leaves. Leaves 7–14 mm wide, lowest leaves

reduced to scales; sheaths finely hairy; bracts leaflike, lowest bract 1–3.5 dm long. Spikes either staminate or pistillate, spreading or drooping and often curved, stalked; staminate spikes 1–3, above pistillate; pistillate spikes 2–5, long-cylindric; lower pistillate scales 5–6 mm long, with a pale midrib, tapered to an awned tip about 5 mm long. Perigynia green, ovate to oval, somewhat flattened, not inflated, 3–4 mm long. Achenes lens-shaped, stigmas 2. June–July.

Similar to *C. crinita*, but with finely hairy sheaths, lower pistillate scales tapered to an awned tip, and perigynia somewhat flattened and not inflated.

Carex lenticularis Michx.
LAKESHORE SEDGE *native*
IR | WUP | CUP | EUP OBL CC 10
Plants densely tufted. Stems 1–6 dm long, upright, slender, brown at base. Leaves clustered on lower third of stem, upright, long-tapered to tip, 1–2 mm wide; sheaths dotted with yellow-brown on front. Staminate spike single, sometimes with a few pistillate flowers, stalked, linear; pistillate spikes 3–5, upright, the upper stalkless, the lower stalked, the upper grouped, the lower separate, linear; lowest bract leaflike, erect, much longer than the head, the upper bracts shorter; pistillate scales red or red-brown, with a 3-veined, green center, the margins translucent near tip, narrower and usually shorter than the perigynia. Perigynia upright, soon deciduous, flattened, convex on both sides and sharply two-edged, 2–3 mm long, waxy blue-green, with a few yellow glandular dots or bumps, tapered at the abruptly pointed tip; the beak small, to 0.2 mm long. Achenes lens-shaped, brown; stigmas 2. June–Sept.

Rocky and sandy lakeshores, rock pools along Lake Superior, shallow ponds, sedge mats.

Carex stricta Lam.
TUSSOCK SEDGE *native*
IR | WUP | CUP | EUP OBL CC 4
Plants densely tufted, from long scaly rhizomes, forming large raised hummocks to 1 m tall. Stems 3-angled, 3–10 dm long, rough-to-touch. Leaves 2–6 mm wide, the lower leaves reduced to sheaths around the base of stem; sheaths white to red-brown on front, green on back, the lower sheaths breaking into ladderlike thin strands. Spikes mostly all staminate or pistillate (sometimes mixed), the upper 1–3 spikes staminate, the terminal spike 1.5–5 cm long, the lower 2–5 spikes pistillate or some with staminate flowers borne above pistillate; lowest bract leaflike; pistillate scales equal or longer than the perigynia but narrower. Perigynia green at tip and margins, golden to yellow-brown in middle, with white or brown bumps, convex on both sides to nearly flat, 2–3 mm long, 2-ribbed with a few faint nerves on both sides; beak short, to only 0.3 mm long. Achenes lens-shaped; stigmas 2. May–July.

Often dominant sedge of wet meadows, marshes, fens, shores, streambanks, ditches.

Carex Section Phaestoglochin

Plants tufted; rhizomes short or inconspicuous; bases pale to brown, occasionally reddish. Inner band of the leaf sheaths hyaline, corrugated or smooth. Spikes all or mostly androgynous, simple in most taxa, the lower branched in some species. Perigynia mostly plano-convex, beaks typically bidentate. Mostly upland species of forests and open, sometimes disturbed habitats, including several of non-native species.

1 Leaf sheaths loose, white with green veins or mottled green and white on back; wider blades 5–10 mm broad 2
1 Leaf sheaths ± tight and slender and uniform green or whitish on back; wider blades 0.9–4.3 mm broad 3
2 Spikes close together, the lower not separated more than their length, usually ± overlapping; perigynia 3.6–4.5 mm long, 2–3 times longer than wide, the bodies not wing-margined; widest leaf blades 5–7 (–8) mm wide .. .. **C. cephaloidea**
2 Spikes well separated below, the lower ones ± remote; perigynia 3–4.1 mm long, 1.3–2 times as long as wide, the bodies ± narrowly thin-winged; widest leaf blades 5.5–10 mm wide .... **C. sparganioides**
3 Perigynia mostly widely spreading at maturity, conspicuously spongy-thickened at their bases and there puckered in drying, the wire-like margin above the base tending to turn inward 4
3 Perigynia mostly ascending and not widely spreading, at most with thin spongy area at base not conspicuously puckered in drying (unless immature), the margin above flat or slightly incurved 5
4 Wider leaf blades mostly 0.9–1.8 (very rarely 2.5) mm wide; stigmas reddish to dark brown, slender and elongate (when intact), often protruding 1–1.5 mm or more, often reflexed but otherwise straight or slightly sinuous .... **C. radiata**
4 Wider leaf blades mostly (1.5–2.7 mm wide; stigmas very dark reddish brown, comparatively short and stout, strongly curled .... **C. rosea**
5 Inflorescence crowded to oblong and interrupted (the lower spikes overlapping but distinct); leaf blades densely papillose above (at 20×–30×; bodies of scales more (often much more) than half as long as bodies of the perigynia they subtend; larger perigynia in spike 3–4.1 mm long, 2–2.6 mm wide .... **C. muehlenbergii**
5 Inflorescence densely crowded, ± ovoid, the spikes in a close head and nearly indistinguishable except by the slightly protruding setaceous bracts; leaf blades smooth above the collar or the cellular outlines conspicuous, but only rarely some leaves papillose; bodies of scales

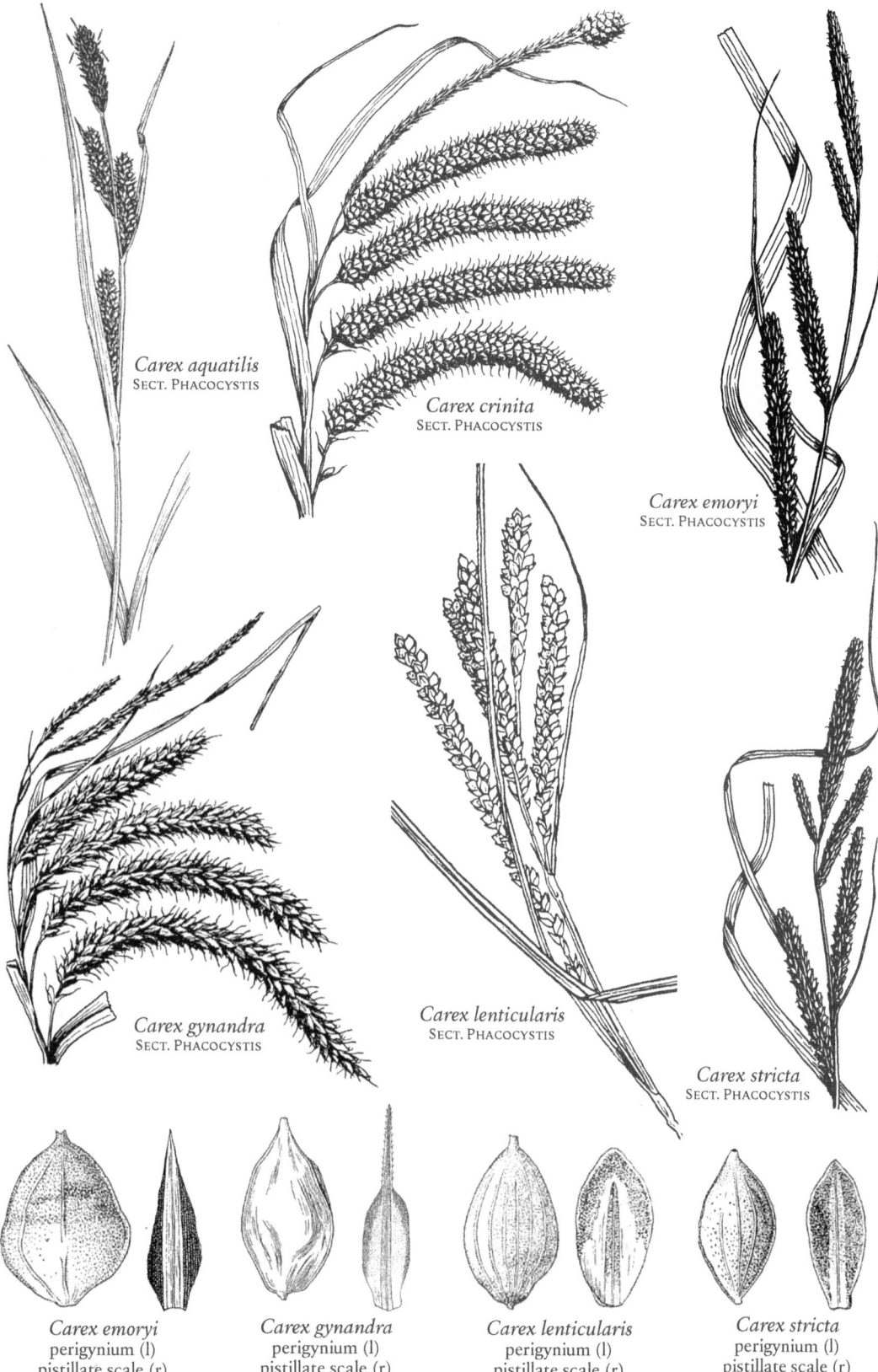

usually about or only slightly more than half as long as bodies of the perigynia; perigynia 2.5–3.2 mm long, 1.5–2 mm wide . . . . **C. cephalophora**

Carex cephaloidea (Dewey) Dewey
THIN-LEAF SEDGE native
IR | **WUP** | **CUP** | EUP FACU CC 5

Carex sparganioides Muhl. var. *cephaloidea* (Dewey) Carey

Rootstocks short, stout, brownish black, fibrillose. Stems 3–5 dm long, weak, serrulate on angles above, light-brownish tinged at base. Leaves 25–35 cm long and 3–7 mm wide (lower leaves shorter), flat, thin, weak, roughened on margins and towards tip; sheaths loose, overlapping, green-and-white-mottled dorsally, the lowest transversely rugulose, ; ligule dark-margined. Spikes 5–10, androgynous; bracts scale-like or setiform; scales thin, greenish hyaline with green midrib; staminate flowers few. Perigynia 3.5–4 mm long, exceeding scales, spreading, plano-convex, deep-green, sharp-edged; beak 0.5 mm, serrulate, bidentate, white-hyaline within. Achenes lenticular; stigmas 2, short, reddish brown.

Woods.

Carex cephalophora Muhl.
OVAL-LEAF SEDGE native
IR | **WUP** | CUP | EUP FACU CC 3

Plants tufted; rootstocks short, black, fibrillose. Stems 2–5 dm long, stiff, roughened beneath head, light-brownish tinged at base. Well-developed leaves 3–5 to a stem, on lower fourth; blades 1–4 dm long and 2–4.5 mm wide, flat, flaccid, pale-green, roughened on margins and towards tip; sheaths tight. Spikes 3–8, androgynous; bracts setiform, 1–5 cm long; scales greenish hyaline with green midrib; staminate flowers few. Perigynia 2.5 mm long, ascending or spreading, plano-convex, light-green or yellowish, nerveless or 2–3-nerved dorsally, the margins raised, sharp; beak 0.5–0.75 mm long, serrulate, bidentate, the teeth triangular, white-hyaline within. Achenes lenticular, filling perigynia; stigmas 2, short, reddish brown.

Dry woods; Ontonagon County.

Carex muehlenbergii Schkuhr ex Willd.
MUEHLENBERG'S SEDGE native
IR | WUP | **CUP** | **EUP** CC 7

Plants tufted; rootstocks short, somewhat woody, dark, fibrillose. Stems 2–9 dm long, stiff, rough above, light-brownish at base, the old leaves conspicuous. Well-developed leaves 5–10 to a stem, on lower fifth; blades 1–3 dm long and 2–4 mm wide, flat or channeled, thick, light-green, roughened on margins and towards the tip; sheaths tight, yellowish brown-tinged at mouth. Spikes 3–10, androgynous; bracts setiform, short; staminate flowers few; scales greenish hyaline with 3-nerved green center. Perigynia 8–20 to a spike, 3–3.5 mm long, ascending or spreading, plano-convex, somewhat leathery, pale-green, many-ribbed with sharp slightly raised margins, serrulate above; beak 1 mm long, bidentate, the teeth hyaline within. Achenes filling perigynia, lenticular; stigmas 2, long, reddish brown.

Sand hills and dry places; Dickinson and Mackinac counties.

Carex radiata (Wahlenb.) Small
EASTERN STAR SEDGE native
IR | **WUP** | **CUP** | **EUP** CC 2

Plants tufted. Stems slender, 2.5–5 dm long, weak, roughened above, light-brown to blackish tinged and fibrillose at base. Well-developed leaves 4–6 to a fertile stem, on lower fourth; blades 1–2 mm wide, flat, light-green; sheaths tight. Spikes 4, androgynous; staminate flowers few; lowest bract setaceous, the upper smaller; scales thin, white-hyaline with green midvein. Perigynia 2–6 to a spike, 2–3 mm long, exceeding scales, deep-green, erect, nerveless or nearly so, serrulate above, spongy at base; beak to 1 mm long, bidentate, white-hyaline between teeth. Achenes lenticular, filling perigynia; stigmas 2, short, twisted, dark-brownish red.

Dry woods.

Carex rosea Schkuhr
ROSY SEDGE native
IR | **WUP** | **CUP** | **EUP** CC 2

Carex convoluta Mackenzie

Plants tufted; rootstocks short, dark, fibrillose. Stems slender 2–5 dm long, smooth or serrulate above,

Carex cephaloidea
perigynium (l)
pistillate scale (r)

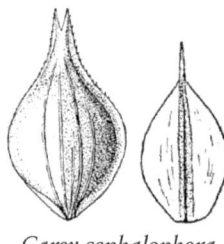

Carex cephalophora
perigynium (l)
pistillate scale (r)

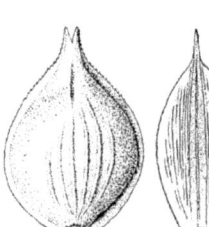

Carex muehlenbergii
perigynium (l)
pistillate scale (r)

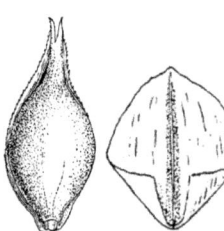

Carex radiata
perigynium (l)
pistillate scale (r)

light-brownish tinged and fibrillose at base. Well-developed leaves 3–6 to a stem, on lower third; blades 3 dm long and 1–2 mm wide, flat, light-green, serrulate on margins and on veins towards tip; sheaths tight. Spikes 4–8, androgynous; bracts to 10 cm long, the upper reduced; scales thin, greenish hyaline with green midrib; staminate flowers inconspicuous. Perigynia 4–12 to a spike, 3–3.5 mm long, exceeding scales, ascending or widely radiating, plano-convex, light-green, nerveless or nearly so, serrulate above, spongy at base; beak ca. 0.5 mm long, bidentate. Achenes lenticular, filling perigynia; stigmas 3, long, light-reddish brown.

Dry woodlands.

Carex sparganioides Muhl.
BUR-REED SEDGE native
IR | **WUP** | **CUP** | **EUP** FACU CC 5

Plants tufted; rootstocks short, dark, somewhat woody. Stems 3–7.5 dm long, ascending or erect, narrowly margined, serrulate above, brownish yellow-tinged at base. Well-developed leaves 3–6 to a stem; blades 2–4 dm long and 5–10 mm wide, flat, weak, serrulate on margins, roughened on veins; sheaths loose, overlapping, green-and-white-mottled dorsally, the lower sheaths transversely rugulose. Spikes 6–12, androgynous; bracts short, often rudimentary; scales thin, greenish hyaline; staminate flowers inconspicuous. Perigynia 5–50 to a spike, 3

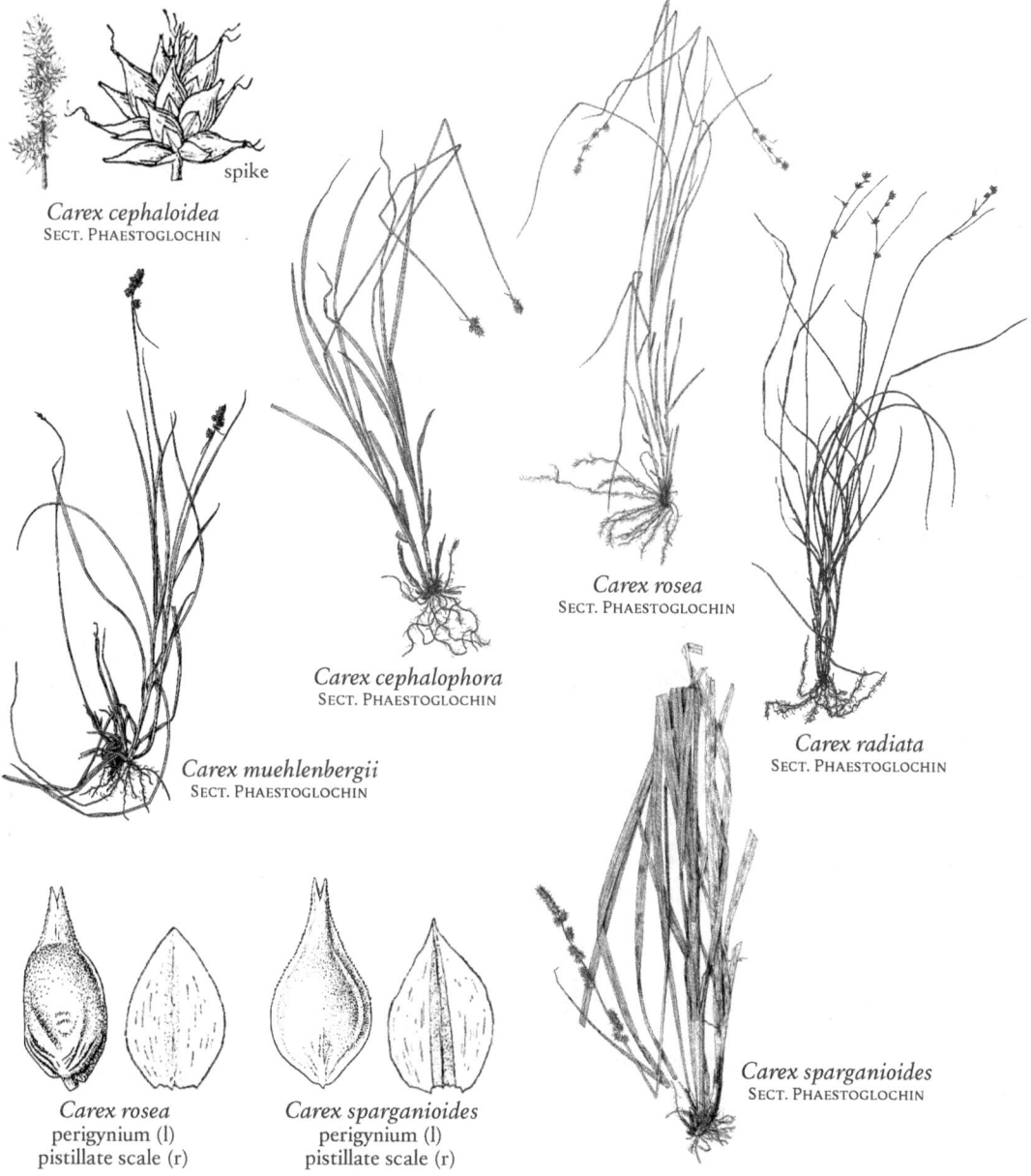

mm long, ascending or spreading, plano-convex, nerveless or nearly so, with sharp and somewhat raised margins, serrulate above; beak serrulate, bidentate, white-hyaline within. Achenes lenticular; stigmas 2, short, reddish brown.

Dry woods.

Carex Section Phyllostachyae

One member of the section in the Upper Peninsula. Plants tufted, bases brown. Bracts lacking. Lateral spikes absent or basal, pistillate or androgynous. Terminal spike androgynous. Lowest pistillate scale foliose, suggesting the lowest bract of the inflorescence in most other sections, exceeding the tip of the spike. Perigynia 2-ribbed, beak untoothed.

Carex backii Boott
BACK'S SEDGE   native
IR | **WUP** | **CUP** | **EUP**   CC 8

Plants tufted; rootstocks short, dark-brown. Stems to 25 cm long, weak, narrowly winged, serrulate on angles, enlarged upward. Well-developed leaves 2–6 to a stem, near base; blades 1–3 dm long and 2.5–6 mm wide, flat, erect or curved, thickish, deep-green papillate, roughened especially on margins and towards tip; sheaths thin and hyaline ventrally, yellowish brown-tinged, oblique at mouth. Spikes 1–3, androgynous, the lower long-peduncled; pistillate scales bract-like, 3–4 cm long and 5 mm wide, nerved, tapering; staminate flowers few. Perigynia 2–5 to a spike, 5–6 mm long, erect on a zigzag winged rachis, concealed by scales, light-green, many-nerved, 2-keeled, spongy at base; beak 2 mm, 2-edged the orifice entire, truncate, hyaline, tawny-tinged below. Achenes triangular-globose with convex sides, closely enveloped, yellowish green or blackish, triangular; stigmas 3, short, dark.

Dry woods.

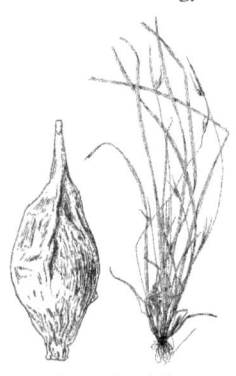

Carex backii perigynium (l)

Carex Section Physoglochin

One member of the section in the Upper Peninsula; sphagnum moss peatlands.

Carex gynocrates Wormsk.
NORTHERN BOG SEDGE   native
IR | **WUP** | **CUP** | **EUP**   OBL CC 10

Carex dioica L.

Small perennial, from long, slender rhizomes. Stems single or few together, 0.3–3 dm long, smooth, brown at base. Leaves clustered near base of plant, blades inrolled and threadlike, to 1 mm wide. Spikes only 1 per stem, all staminate or all pistillate, or with both staminate and pistillate flowers and with the staminate flowers borne above the pistillate; the staminate spike or portion of spike narrowly cylindric, the pistillate spike or portion short-cylindric; bract absent; pistillate scales brown or red-brown, tapered to tip, shorter but wider than perigynia. Perigynia 4–10, widely spreading, yellow to dark brown, shiny, plump, obovate, 2–4 mm long, spongy at base, abruptly contracted to the beak; beak nearly entire to unequally notched, 0.5 mm long. Achenes lens-shaped; stigmas 2. June–July.

Conifer swamps and open peatlands, usually in sphagnum and wet, peaty soils; in the western UP, known only from Keweenaw County (including Isle Royale).

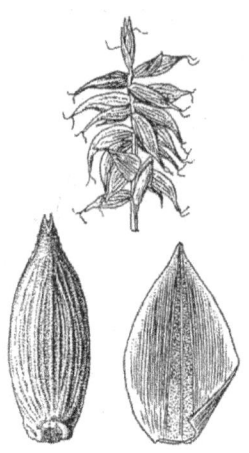

Carex gynocrates perigynium (l), spike (c), pistillate scale (r)

Carex Section Porocystis

Plants tufted. Leaves and stems usually pubescent, at least sparsely. Pistillate spikes erect to spreading, ovoid to oblong-cylindrical. Perigynia beakless or very short-beaked, glabrous or pubescent. Stigmas 3. The cylindrical pistillate spikes of *C. pallescens* and *C. swanii* resemble the spikes of *C. granularis* and relatives.

1 Perigynia pubescent; terminal spike pistillate at apex, staminate at base . . . . **C. swanii**
1 Perigynia glabrous; terminal spike entirely staminate . . . . **C. pallescens**

Carex pallescens L.
PALE SEDGE   native
IR | **WUP** | **CUP** | **EUP**   FAC CC 5

Plants tufted; rootstocks short; stolons short-ascending. Stems 2–6 dm long, triangular with concave sides, rough above, pubescent, brownish red-tinged at base. Well-developed leaves 2–3 to a stem, on lower third; blades 8–35 cm long and 2–3 mm wide, flat with revolute margins, deep-green, soft-pubescent below; sheaths tight, soft-pubescent, cinnamon-brown-tinged. Terminal spike staminate; lateral spikes 2–3, pistillate, on capillary peduncles to 15 mm long; lowest bract leaflike, sheathless or nearly so, the upper reduced; pistillate scales yellowish brown or greenish white with 3-nerved green center. Perigynia 15–40 to a spike, 2.5–3 mm long, erect to spreading, greenish or yellowish green, minutely puncticulate, finely nerved, beakless. Achenes triangular with concave sides, loosely enveloped, short-apiculate; stigmas 3, short.

Dry banks and meadows.

Carex swanii (Fern.) Mackenzie
SWAN'S SEDGE native
IR | WUP | **CUP** | EUP FACU CC 4

Plants tufted; rootstocks short. Stems mostly 1.5–6 dm long, roughened above, sparsely hairy, reddish purple-tinged at base, the basal sheaths filamentose. Well-developed leaves 3–6 to a stem; blades 1.5–3 dm long and 1.5–3 mm wide, flat, flaccid, dull-green, short-pilose; sheaths long, tight, short-pilose, the lowest yellowish brown-tinged ventrally. Spikes 2–5, the terminal gynaecandrous, the lateral pistillate; bracts sheathless, the lowest setaceous, 3–6 cm long and 0.5 mm wide, the upper smaller; scales awned, hyaline with green midrib. Perigynia 10–30 to a spike, ca. 2 mm long, erect-appressed, compressed-triangular, green, white-hirsute, nerved dorsally, beakless, the orifice entire. Achenes triangular with concave sides, yellowish brown, bent-apiculate; stigmas 3, short, brownish.

Dry woods; Alger and Schoolcraft counties.

Carex Section Racemosae

Plants loosely to densely tufted; rhizomes variable in length; bases dark red, generally fibrous; roots not clothed with yellow felt. Terminal spike gynecandrous (in our species). Pistillate scales dark, often black. Perigynia pale, often greenish, very short-beaked to beakless, smooth or papillose, 2-ribbed, inconspicuously veined (in our species). Stigmas 3.

1 Pistillate scales mostly awned or narrowly acuminate, exceeding the perigynia; ventral surface of lower leaf sheaths tearing into fibers . . . . **C. buxbaumii**
1 Pistillate scales obtuse or acute, equaling or shorter than the perigynia; ventral surface of lower sheaths not tearing to form fibers. . . . . **C. media**

Carex buxbaumii Wahlenb.
BROWN BOG SEDGE native
IR | WUP | CUP | EUP OBL CC 10

Loosely tufted perennial, from long rhizomes. Stems single or few together, 3-angled, 3–10 dm long, rough-to-touch above, red-tinged near base. Leaves 1–3 mm wide, the lowest leaves without blades; lower sheaths shredding into thin strands, the upper sheaths membranous and purple-dotted. Spikes 2–5, terminal spike with pistillate flowers above staminate and larger than the lateral spikes, lateral spikes pistillate, short-cylindric, stalkless or nearly so; bracts leaflike, the lowest shorter than the head; pistillate scales dark brown, tapered to an awn at tip. Perigynia light green, golden brown near base, oval, 2.5–3.5 mm long, 2-ribbed, with 6–8 faint nerves on each side; beak tiny, notched. Achenes 3-angled; stigmas 3. May–Aug.

Wet meadows and fens, shallow marshes, low prairie, hollows in patterned peatlands.

Carex media R. Br.
MONTANA SEDGE THR native
IR | **WUP** | CUP | EUP FACW CC 10
Carex norvegica Retz.

Plants loosely tufted, from short rhizomes. Stems slender, not stiff, 2–8 dm long, smooth or slightly rough-to-touch above, sharply triangular above, much longer than the leaves, red-tinged at base. Leaves 7–15 and mostly near base of stem, pale-green, flat or margins slightly rolled under, 2–3 mm wide, rough-to-touch on margins, the dried leaves of previous year conspicuous; sheaths translucent. Spikes usually 3, densely flowered, the terminal with both staminate and pistillate flowers, the staminate below the pistillate, clustered, upright, stalkless; the lateral spikes pistillate, on short stalks; lowest bract usually shorter than the head; pistillate scales 2–3 mm long, purple-black, margins white-translucent, nearly as wide as perigynia but much shorter. Perigynia 2–4 mm long, rounded 3-angled, slightly inflated, yellow-green to brown, two-ribbed, otherwise

Carex pallescens
perigynium (l)
pistillate scale (r)

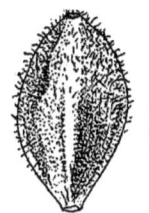

Carex swanii
perigynium (l)
pistillate scale (r)

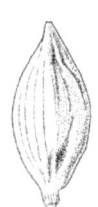

Carex buxbaumii
perigynium (l)
pistillate scale (r)

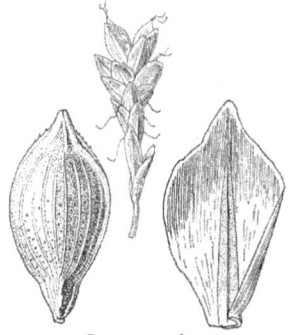

Carex media
perigynium (l), spike (c),
pistillate scale (r)

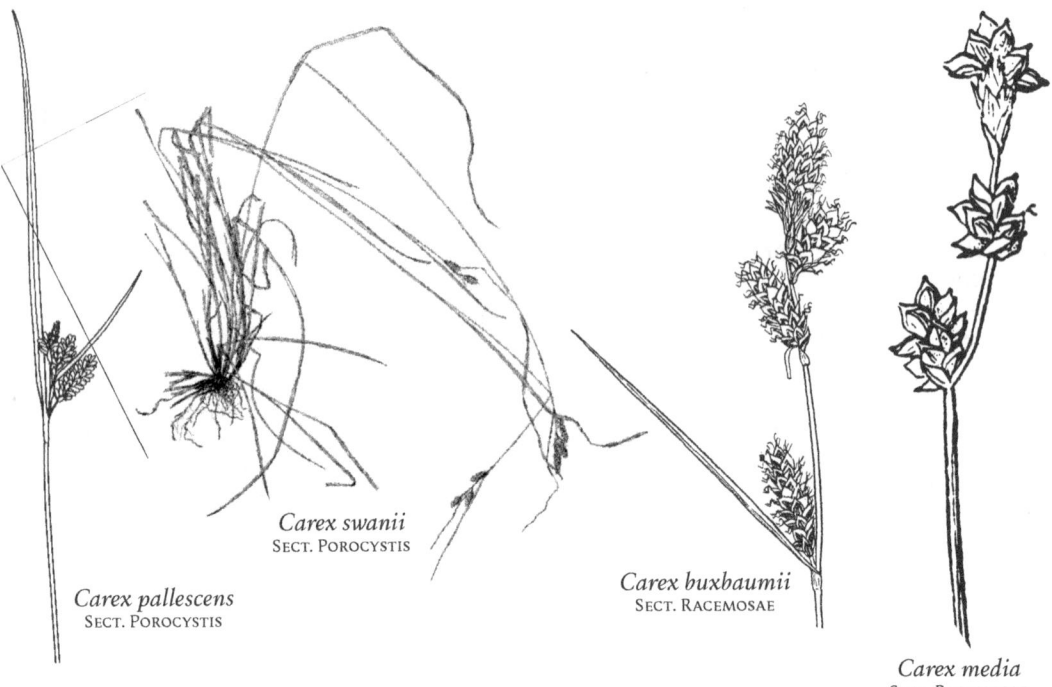

Carex swanii Sect. Porocystis

Carex pallescens Sect. Porocystis

Carex buxbaumii Sect. Racemosae

Carex media Sect. Racemosae

without nerves, tip rounded and abruptly beaked; beak short, to 0.5 mm long, red-tinged, with a small notch. Achenes obovate, 3-angled, yellow-brown; stigmas 3. June–July.

Rare in cracks of rocky Lake Superior shoreline; open rocky woods; Keweenaw County (including Isle Royale).

Carex Section Rostrales

One member of the section in the Upper Peninsula. Plants tufted; rhizomes elongate; bases brown, not reddish or purplish. Staminate spike solitary, the base lower than or roughly equaling the apex of the uppermost pistillate spike. Pistillate spikes 2–6, approximately as long as thick. Perigynia ≤20 per spike, divergent or the lowermost reflexed, somewhat inflated, narrow, tapering continuously to the apex, generally 4–7x as long as wide, beakless, subtly bidentate.

Species in this section superficially resemble the more common *Carex intumescens*, but differ in their narrower perigynia.

Carex folliculata L.
NORTHERN LONG SEDGE   native
IR | WUP | CUP | EUP   OBL CC 10

Plants large, tufted. Stems to 1 m long. Leaves 5–15 mm wide. Staminate spike single, long-stalked; pistillate spikes 2–5, widely separated, upright, stalked; bracts leaflike, longer than stems; pistillate scales translucent or brown-tinged, green in center, much shorter than perigynia. Perigynia lance-shaped, many-nerved, 10–15 mm long, tapered to a long, finely toothed beak, the teeth upright. Achenes 3-angled; stigmas 3. June–Aug.

Wet woods and cedar swamps.

Carex Section Stellulatae

Plants tufted; rhizomes short; bases brown, not fibrous. Inflorescence mostly open, spikes readily distinguished from each other, the lowest in our more common species not overlapping; bracts inconspicuous or lacking. Spikes 2–10 (solitary in *Carex exilis*), gynecandrous (unisexual in *C. sterilis*). Perigynia spreading to reflexed, typically plano-convex, widest at base, generally chestnut brown to dark brown or blackish at maturity; margins acute; base spongy; beak generally bidentate, margins finely serrate. Achenes much smaller than the perigynia. Wetlands. The distinctions between species in this section are subtle. When examining perigynia, view the lowest 2–3 perigynia in the spike; the upper perigynia are very similar in this section.

ADDITIONAL SPECIES

Carex weigandii Mack. is uncommon in acidic, peaty or sandy wetlands in the c and e UP.

1   Spikes solitary; leaves involute; anthers 2–3.6 mm long
    .... **C. exilis**
1   Spikes 2–8; leaves flat or plicate; anthers 0.6–2.2 mm long...2

2 Terminal spikes entirely staminate .... **C. sterilis**
2 Terminal spikes partly or wholly pistillate 3
3 Terminal spikes without a distinct clavate base of staminate scales, staminate portion less than 1 mm long.... .... **C. sterilis**
3 Terminal spikes with a distinct clavate base of staminate scales mostly 1–8 mm long 4
4 Lower perigynia mostly 3–3.6 mm long, 1.8–3.6 times longer than wide; beaks 0.9–2 mm long, mostly 0.5–0.8 times longer than the body .... **C. echinata**
4 Lower perigynia mostly 1.9–3 mm long, 1–2 times longer than wide; beaks 0.4–0.9 mm long, mostly 0.2–0.5 times longer than the body .... **C. interior**

Carex echinata Murr.
STAR SEDGE   *native*
IR | WUP | CUP | EUP   OBL CC 6
Carex angustior Mackenzie

Plants tufted. Stems 1–6 dm long, rough above. Leaves scale-like at base of stem; leaves with blades 3–6 on lower stem, 1–3 mm wide. Spikes 3–7, stalkless, few-flowered; terminal spike with a slender staminate portion near its base; lateral spikes usually all pistillate; bract small; pistillate scales shorter than perigynia, yellow-tinged with green midvein. Perigynia 5–15 and crowded in each spike, spreading or curved downward, green or light brown, flat on 1 side and convex on other, spongy-thickened at base, 3–4 mm long, tapered to a toothed, notched beak 1–2 mm long. Achenes lens-shaped; stigmas 2. July–Sept.

Swamp margins, wet sandy lakeshores, hummocks in peatlands.

Carex exilis Dewey
COASTAL SEDGE   *native*
IR | WUP | CUP | EUP   OBL CC 10

Plants densely tufted. Stems stiff, 2–7 dm long and longer than the leaves. Leaves narrow and rolled inward. Spike usually 1, either staminate or pistillate, or with both staminate and pistillate flowers, the staminate below the pistillate, 1–3 cm long; lateral spikes (if present) 1 or 2 and much smaller than terminal spike; lower 2 scales empty and upright; pistillate scales red-brown with translucent margins, about as long as perigynia. Perigynia spreading or drooping, flat on 1 side and convex on other, 3–5 mm long, spongy-thickened at base, tapered to a toothed beak to 2 mm long. Achenes lens-shaped; stigmas 2. June–Aug.

Sphagnum peatlands, interdunal wetlands near Great Lakes; coastal disjunct.

Carex interior Bailey
INLAND SEDGE   *native*
IR | WUP | CUP | EUP   OBL CC 3

Plants densely tufted. Stems slender, sharply 3-angled, 1–6 dm long, equal or longer than the leaves. Leaves 1–2 mm wide; sheaths tight, thin and translucent. Spikes 2–4, the terminal spike with pistillate flowers borne above staminate (or rarely all staminate), the lateral spikes pistillate (or rarely with pistillate flowers borne above staminate), more or less overlapping; bracts small or absent; pistillate scales much shorter than the perigynia. Perigynia green-brown to brown, filled to margins by the achenes, sharp-edged but not wing-margined, 2–3 mm long, the base spongy so that achene fills upper perigynium body, tapered to a finely toothed beak to 1 mm long; the beak teeth small, not longer than 0.3 mm. Achenes lens-shaped; stigmas 2. May–Aug.

Swamps, tamarack bogs, alder thickets, wet meadows and wetland margins.

Carex sterilis Willd.
DIOECIOUS SEDGE   *native*
IR | WUP | CUP | EUP   OBL CC 10
Carex muricata L. var. *sterilis* (Willd.) Gleason

Plants tufted. Stems stiff, 1–7 dm long, longer than the leaves, rough-to-touch on the upper stem angles. Leaves 3–5 from lower part of stem, 1–4 mm wide, rough, lower stem leaves reduced to scales. Spikes 3–8, stalkless, clustered or the lower separate; staminate and pistillate flowers mostly on separate plants; pistillate

Carex sterilis
perigynium (l)
pistillate scale (r)

Carex folliculata
perigynium (l)
pistillate scale (r)

Carex echinata
perigynium (l)
pistillate scale (r)

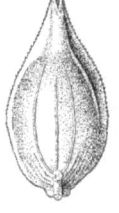
Carex exilis
perigynium (l)
pistillate scale (r)

Carex interior
perigynium (l)
pistillate scale (r)

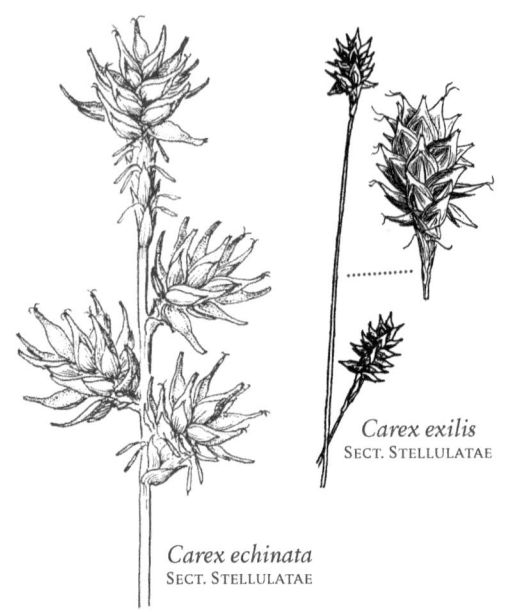

Carex echinata
SECT. STELLULATAE

Carex exilis
SECT. STELLULATAE

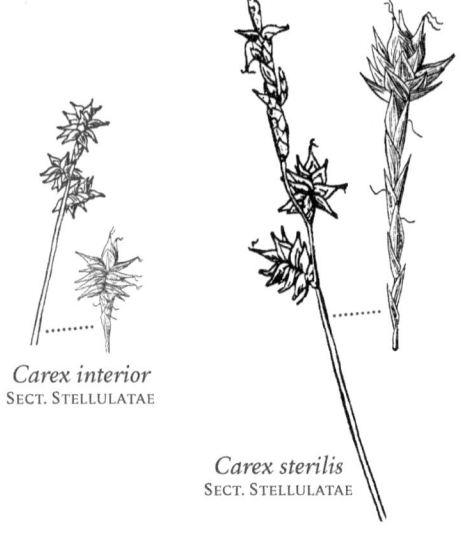

Carex interior
SECT. STELLULATAE

Carex sterilis
SECT. STELLULATAE

scales red-brown with green midvein and translucent marginperigynium (l), pistillate scale (r)s, about as long as body of perigynia. Perygynia 5–25, the lower spreading, 2–4 mm long, red-brown, flat on 1 side and convex on other, spongy-thickened at base, tapered to a finely toothed, notched beak 0.5–1.5 mm long, the beak teeth sharp, to 0.5 mm long. Achenes lens-shaped; stigmas 2. April–June.

Spring-fed calcareous fens, calcium-rich wet meadows. Similar to inland sedge (*C. interior*).

Carex Section Vesicariae

Includes typical bottlebrush sedges of former section Pseudocypereae, with pistillate spikes tightly packed with perigynia, and pistillate scales with scabrous awns conspicuous between the perigynia; and also former section Vesicariae, with pistillate spikes often narrower, longer, less densely packed with perigynia in some species, and pistillate scales mostly not awned, hidden by the perigynia.

1 Pistillate scales with a prominent, scabrous awn; often the body also ciliate 2
1 Pistillate scales smooth-margined, obtuse to acuminate, awnless (rarely the lowermost awned in *C. rostrata* and *C. utriculata*) 5
2 Perigynia ± reflexed at maturity, hard-walled, uninflated, flattened-triangular in cross-section, strongly and closely nerved with most nerves separated by less than three times their width; longest beak teeth 0.7–2.2 mm long 3
2 Perigynia spreading to ascending, thin-textured, ± inflated, ± round in cross-section; many nerves separated by more than three times their width; longest beak teeth 0.3–0.7 mm long 4
3 Spikes 12–18 mm thick; beak teeth strongly outcurved, the longest 1.3–2.1 mm long . . . . **C. comosa**
3 Spikes 9–12 mm thick; beak teeth straight or slightly outcurved, the longest 0.7–1.2 mm . **C. pseudocyperus**
4 Staminate scales (except sometimes the lowermost) acute to acuminate, essentially smooth-margined except at the very tip; plants extensively colonial from elongate, creeping rhizomes; perigynia 7–11-nerved . . . . **C. schweinitzii**
4 Staminate scales (at least some) with a distinct, scabrous awn and sometimes also ciliate-margined; plants densely to loosely tufted, rhizomes connecting individual stems in a clump not more than 10 cm long; perigynia 15-20-nerved . . . . **C. hystericina**
5 Leaf blades and bracts involute-filiform, wiry, 1–3 mm wide; stems round or obtusely 3-angled in cross-section, smooth; pistillate spikes 3–15-flowered, nearly spherical or short-oblong (not over 2 cm long); staminate spike usually solitary . . . . **C. oligosperma**
5 Leaf blades and bracts flat, U-, V-, or W-shaped in cross-section, 1.5–12 mm wide; stems round to 3-angled, often scabrous-angled; pistillate spikes usually more than 15-flowered, oblong to long-cylindric; staminate spikes normally 2 or more (often 1 in *C. retrorsa*) 6
6 Perigynia 4–7 mm thick; achenes with a deep notch or constriction on one angle . . . . **C. tuckermanii**
6 Perigynia 2.5–3.5 mm thick; achenes symmetrical, not notched on one angle 7
7 Lowest pistillate bract 3–9 times as long as the entire inflorescence; mature perigynia 7–12 mm long, at least the lower reflexed or widely spreading; staminate spike often 1, its base (or base of lowest staminate spike if more than one) slightly if at all elevated above summit of the crowded pistillate spikes (rarely lower spike remote) . . . . **C. retrorsa**

7 Lowest pistillate bract less than 3 times as long as inflorescence; perigynia 4–7.5 mm long, ascending or spreading; staminate spikes mostly 2–4, generally well elevated above the pistillate spikes 8

8 Leaves strongly papillose on upper surface, U-shaped in cross-section, glaucous, widest leaves 1.5–4.5 (–7.5) mm wide; stems round or very obtusely triangular, smooth below inflorescence .... **C. rostrata**

8 Leaves smooth or scabrous on upper surface, flat or folded, pale to dark green, widest leaves 3–12mm wide; stems triangular, often scabrous below the inflorescence 9

9 Colonial from long-creeping rhizomes; widest leaves 5–12 mm wide; ligules about as long as wide; basal sheaths usually spongy-thickened with little or no red tingeing; perigynia (at least those on lower portion of fully mature spike) ± widely spreading; stems bluntly triangular and sparsely and irregularly scabrous below the inflorescence .... **C. utriculata**

9 Tufted; widest leaves 3–6 mm wide; ligules longer than wide; basal sheaths not spongy-thickened and often tinged with reddish purple; perigynia ascending; stems sharply triangular and scabrous-angled below the inflorescence............................. **C. vesicaria**

Carex comosa Boott
BEARDED SEDGE *native*
IR | WUP | CUP | EUP   OBL CC 5

Plants large, often forming large clumps. Stems stout, sharply 3-angled, 5–15 dm long. Leaves 5–12 mm wide; sheaths translucent on front, with small swollen joints on back. Spikes either staminate or pistillate; terminal spike staminate; lateral spikes pistillate, 3–5, cylindric, the lower spikes longer stalked and drooping when mature; bracts leaflike, much longer than the head; pistillate scales with translucent margins, tapered into a long, rough awn. Perigynia numerous, spreading outward when ripe, flattened 3-angled, lance-shaped, 5–8 mm long, shiny, strongly nerved, gradually tapered to the 2–3 mm long beak, the beak with curved teeth 1–2 mm long. Achenes 3-angled; stigmas 3. June–Aug.

Marshes, wetland margins, floating mats, ditches.

Carex hystericina Muhl.
PORCUPINE SEDGE *native*
IR | WUP | CUP | EUP   OBL CC 2

Plants from short rhizomes, often forming large clumps. Stems upright or leaning, 3-angled, 2–10 dm long, usually longer than the leaves. Leaves yellow-green, 3–8 mm wide; sheaths white, thin and translucent on front, green to yellow or red on back, the lower sheaths breaking into threadlike fibers. Spikes either all staminate or pistillate, the terminal spike staminate, usually short-stalked and often with a bract; lateral spikes pistillate or occasionally with staminate flowers above pistillate, 1–4, short-cylindric, separate or clustered, the lower spikes usually nodding on slender stalks, the upper spikes short-stalked and upright; pistillate scales small, narrow and much shorter than the perigynia, tipped with a rough awn. Perigynia spreading or upright, green to straw-colored, ovate, round in section when mature, 5–8 mm long, strongly nerved, abruptly tapered to a slender, toothed beak 3–4 mm long; the beak teeth to 1 mm long. Achenes 3-angled with concave sides; stigmas 3. May–July.

Swamps, alder thickets, wet meadows and ditches.

Carex oligosperma Michx.
FEW-SEED SEDGE *native*
IR | WUP | CUP | EUP   OBL CC 10

Plants forming colonies from creeping rhizomes. Stems slender, 4–10 dm long, purple-tinged at base. Leaves stiff, rolled inward, 1–3 mm wide. Spikes either all staminate or pistillate; staminate spike usually single; pistillate spikes 1 (or 2–3 and widely separated), stalkless or nearly so, ovate to short-cylindric, lowest bract leaflike. Perigynia 3–15, ovate, somewhat inflated, compressed, 4–7 mm long, strongly several-nerved, abruptly tapered to a beak 1–2 mm long. Achenes 3-angled, 2–3 mm long; stigmas 3. June–Aug.

Open bogs and swamps, floating mats, pioneer mat-former along pond margins. Sometimes a dominant sedge in poor fens in the UP.

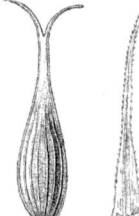
Carex comosa
perigynium (l)
pistillate scale (r)

Carex hystericina
perigynium (l)
pistillate scale (r)

Carex oligosperma
perigynium (l)
pistillate scale (r)

Carex pseudocyperus
perigynium (l)
pistillate scale (r)

Carex pseudocyperus L.
CYPRESS-LIKE SEDGE native
IR | **WUP** | **CUP** | **EUP** OBL CC 5

Plants large, tufted. Stems stout, 3–10 dm long, 3-angled, rough-to-touch. Leaves 5–15 mm wide; sheaths translucent, yellow-tinged on back. Spikes either all staminate or pistillate, the terminal spike staminate; lateral spikes pistillate, 2–6, cylindric, lower spikes drooping on slender stalks; bracts much longer than the head; pistillate scales tipped by an awn, the awn shorter or longer than the perigynia. Perigynia spreading, 3-angled, 4–6 mm long, shiny, strongly nerved, tapered to a toothed beak, the beak teeth 0.5–1 mm long. Achenes 3-angled; stigmas 3. June–Aug.

Marshy lake margins, swamps, fens, wet ditches; in Minn, an indicator of calcium-rich fens in the Red Lake peatland. Similar to *Carex comosa*.

Carex retrorsa Schwein.
RETRORSE SEDGE native
IR | **WUP** | **CUP** | **EUP** OBL CC 3

Plants densely tufted. Stems 4–10 dm long. Leaves 3–4 dm long and 4–10 mm wide, flat and soft; sheaths dotted with small bumps. Spikes either all staminate or pistillate, or the terminal 1–2 spikes with both staminate and pistillate flowers, the staminate above the pistillate, stalkless or lowest spike on a slender stalk; lower spikes 3–8, pistillate; pistillate scales conspicuous, shorter and narrower than the perigynia. Perigynia crowded in rows, spreading or the lowest perigynia angled downward, smooth and shiny, 6–13-nerved, 7–10 mm long, somewhat inflated, tapered to a long, smooth beak 2–4 mm long, the beak teeth short, to 1 mm long. Achenes dark brown, 3-angled, loose in the lower part of the perigynium; stigmas 3. June–Aug.

Floodplain forests, swamps, thickets and marshes.

Carex rostrata Stokes
SWOLLEN BEAKED SEDGE native
IR | WUP | **CUP** | **EUP** OBL CC 10

Plants with short to long-creeping rhizomes. Stems round or bluntly 3-angled, 3–10 dm long, smooth below inflorescence. Leaves waxy blue, with many fine bumps on upper surface, to 4 mm wide, inrolled or channeled in section. Spikes either staminate or pistillate, the upper 2–5 staminate; lower 2–5 spikes pistillate or sometimes 1 or 2 with staminate flowers above the pistillate, cylindric. Perigynia upright when young, becoming widely spreading when mature, yellow-green to brown, shiny, ovate, nearly round in section, inflated, 2–6 mm long, narrowed to a beak about 1 mm long. Achenes 3-angled; stigmas 3. July–Sept.

Peat mats or shallow water.

Similar to *Carex utriculata*, but much less common, and with the leaves waxy blue and dotted with fine bumps on upper surface, v-shaped in section or inrolled, and only 2–4 mm wide.

Carex schweinitzii Dewey
SCHWEINITZ'S SEDGE native
IR | WUP | **CUP** | EUP OBL CC 10

Plants loosely tufted. Stems 3–7 dm long, single or few together from rhizomes, sharply 3-angled. Leaves 4–10 mm wide, rough-to-touch near tip, lower stem leaves reduced to scales. Spikes either all staminate or pistillate; terminal spike staminate, on a slender stalk and usually with a bract; pistillate spikes 2–5, grouped or the lowest separate, cylindric, spikes ascending or spreading; pistillate scales translucent or brown-tinged, the midvein prolonged into a finely toothed awn often longer than the perigynium. Perigynia spreading or upright, inflated, round in section, 5–7 mm long, with 7–9 nerves, abruptly tapered to a beak, the beak teeth upright or spreading. Achenes 3-angled, loosely enclosed in the perigynia; stigmas 3. May–July.

Shaded streambanks; Delta and Marquette counties.

Carex tuckermanii Dewey
TUCKERMAN'S SEDGE native
IR | **WUP** | **CUP** | **EUP** OBL CC 8

Plants tufted, from short rhizomes. Stems 4–8 dm long. Leaves 2–4 dm long and 3–6 mm wide, soft and flat. Spikes either staminate or pistillate; staminate spikes usually 2, separated, raised above pistillate spikes; pistillate spikes 2–4, separated, cylindric. Perigynia overlapping and ascending in 6 rows, 7–

Carex retrorsa
perigynium (l)
pistillate scale (r)

Carex rostrata
perigynium

Carex schweinitzii
perigynium (l)
pistillate scale (r)

Carex tuckermanii
perigynium (l)
pistillate scale (r)

10 mm long and 4–7 mm wide, inflated, tapered to a notched beak 2 mm long. Achenes 3-angled, obovate, with a deep indentation near the middle of 1 angle; stigmas 3. June–Aug.

Swamps, alder thickets, low areas in forests, pond margins.

Carex utriculata Boott
NORTHWEST TERRITORY SEDGE native
IR | WUP | CUP | EUP OBL CC 5

Carex rostrata Stokes var. *utriculata* (Boott) Bailey
Plants large, densely tufted, from short rootstocks, also forming turfs from long rhizomes. Stems bluntly 3-angled, 3–12 dm long, spongy at base. Leaves strongly divided with swollen joints 4–12 mm wide; sheaths white-translucent on front, divided with swollen joints on back. Spikes either staminate or pistillate, the upper 2–5 staminate, held well above the pistillate spikes; lower 2–5 spikes pistillate or sometimes 1 or 2 with staminate flowers above the pistillate, usually separate, cylindric, the upper spikes stalkless or short-stalked, lower spikes stalked, upright; bracts shorter to slightly longer than the head; pistillate scales acute to awn-tipped, body of scale shorter than perigynia. Perigynia upright at first to widely spreading when mature, in many rows, yellow-green to brown, shiny, nearly round in section, inflated, 3–8 mm long, strongly 7–9-nerved, contracted to a toothed beak 1–2 mm long, the teeth mostly straight, 0.5 mm long. Achenes 3-angled; stigmas 3. June–Aug.

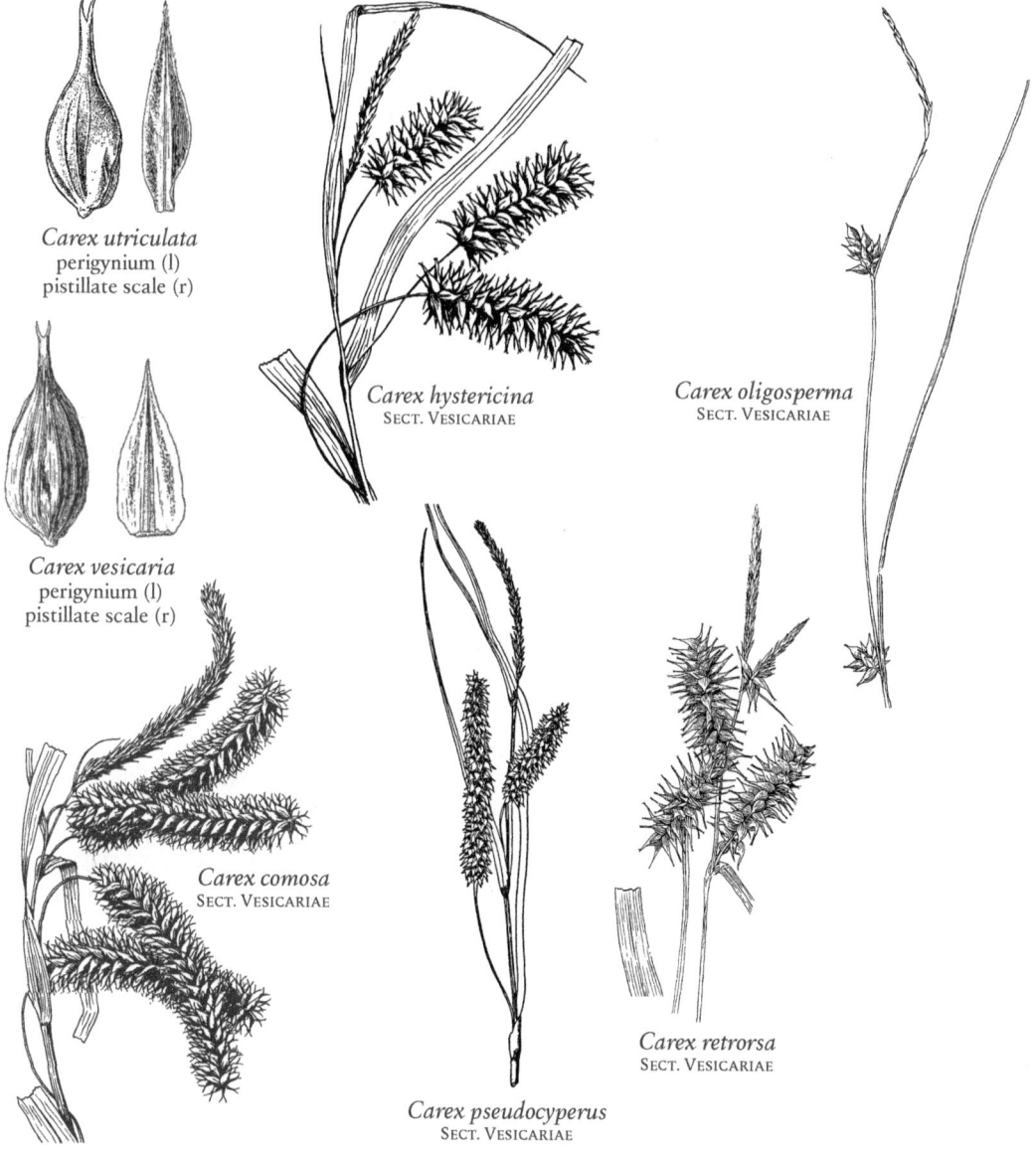

Carex utriculata
perigynium (l)
pistillate scale (r)

Carex vesicaria
perigynium (l)
pistillate scale (r)

Carex hystericina
SECT. VESICARIAE

Carex oligosperma
SECT. VESICARIAE

Carex comosa
SECT. VESICARIAE

Carex retrorsa
SECT. VESICARIAE

Carex pseudocyperus
SECT. VESICARIAE

Wet meadows, marshes, fens, swamps and lakeshores.

Long confused with *Carex rostrata*, a boreal species with waxy blue leaves to only 4 mm wide and which has numerous small bumps on upper leaf surface.

Carex vesicaria L.
LESSER BLADDER SEDGE native
IR | **WUP** | **CUP** | EUP OBL CC 7
Plants tufted, from stout, short rhizomes. Stems 3–10 dm long, sharply 3-angled and rough-to-touch below the head, not spongy at base (as in *C. utriculata*). Leaves 2–7 mm wide; sheaths white-translucent on front, not conspicuously divided-with small swollen joints on back, the lowest sheaths often shredding into ladderlike fibers. Spikes either all staminate or pistillate, the upper 2–4 staminate, held well above the pistillate; lower 1–3 spikes pistillate, separate, cylindric, stalkless or short-stalked, erect; lowest bract usually longer than the head; pistillate scales acute to awn-tipped, shorter to as long as perigynia. Perigynia upright and overlapping in rows, dull yellow-green to brown, inflated, 3–8 mm long, strongly nerved, abruptly tapered to a toothed beak 1–2 mm long, the teeth 0.5–1 mm long. Achenes 3-angled; stigmas 3. June–Aug.

Wet meadows, marshes, forest depressions and shores.

Carex Section Vulpinae

Plants tufted; bases generally pale. Inner band of the leaf sheaths hyaline, in other regards various: corrugated or smooth, thickened or fragile at the summit, sparsely purple-dotted or lacking pigmentation, and combinations of the above. Stems thick, spongy, weak, the angles narrowly winged, scabrous. Inflorescence longer than wide (ours), ovate to cylindrical. Bracts setaceous. Spikes densely flowered, the lower branched, mostly or all androgynous (the terminal always androgynous). Perigynia plano-convex, bases spongy (not spongy in *Carex alopecoidea*). Wetlands. The thick, spongy stems, branched lower spikes, and spongy perigynium bases (except in *C. alopecoidea*) are characteristic.

1 Perigynia contracted into a beak no longer than the body, 3–4.5 mm long, essentially nerveless ventrally; ventral surface of leaf sheaths sparsely to strongly dotted with purplish, especially toward the tip .... **C. alopecoidea**
1 Perigynia somewhat contracted or ± cuneately tapered into the beak (this then difficult to define, but about equaling or slightly exceeding the body, if the latter is measured from the base of perigynium to tip of achene), 4–6.2 mm long, with at least a few nerves ventrally; ventral surface of leaf sheaths not dotted with purplish .... **C. stipata**

Carex alopecoidea Tuckerman
FOX-TAIL SEDGE native
IR | **WUP** | **CUP** | EUP FACW CC 3
Plants tufted. Stems soft, 4–10 dm long, 3-angled and sharply winged. Leaves 3–8 mm wide; sheaths purple-dotted, not cross-wrinkled. Spikes with both staminate and pistillate flowers, staminate flowers above pistillate, in heads 1.5–5 cm long; pistillate scales tapered to tip or with a short sharp tip. Perigynia yellow-brown when mature, ovate, flat on 1 side and convex on other, 3–5 mm long, spongy-thickened at base, narrowed to a beak half to as long as the body. Achenes lens-shaped, 1–2 mm long; stigmas 2. May–July.

Swamps and floodplain forests, streambanks, swales and moist fields; Menominee and Ontonagon counties.

Carex stipata Muhl.
STALK-GRAIN SEDGE native
IR | **WUP** | **CUP** | EUP OBL CC 1
Plants densely tufted. Stems 3-angled and slightly winged, 2–12 dm long. Leaves 4–8 mm wide; sheaths cross-wrinkled on front, divided with small swollen joints on back. Spikes with both staminate and pistillate flowers, staminate flowers borne above pistillate, clustered or the lowest spikes often separate; bracts small and sometimes bristle-like, longer than the spike; pistillate scales tapered to a tip or with a short, sharp point, half to 3/4 as long as the perigynia. Perigynia yellow-green to dull brown, flat on 1 side and convex on other, 3–5 mm long, strongly several-nerved on both sides, tapered to a finely toothed, notched beak 1–3 mm long. Achenes lens-shaped; stigmas 2. May–July.

Common in floodplain forests and swamps, thickets, wet meadows, wetland margins and ditches; usually not in sphagnum bogs.

Cladium *Saw-Grass*

Cladium mariscoides (Muhl.) Torr.
SMOOTH SAW-GRASS native
IR | **WUP** | **CUP** | EUP OBL CC 10
Grasslike perennial, spreading by rhizomes and forming colonies. Stems single or in small groups, stiff, slender, smooth, 0.3–1 m tall. Leaves 1–3 mm wide, upper portion round in section, middle portion flattened. Flowers in lance-shaped spikelets, 3–5 mm long, in branched clusters (umbels) at end of stem and also with 1–2 clusters on slender stalks from leaf axils; uppermost flower perfect, the style 3-parted; middle flowers staminate; lowest scale of each spikelet empty; scales overlapping, ovate, brown; bristles absent. Achenes dull brown, 2–3 mm long, pointed at tip; tubercle absent. June–Aug.

Shallow water, sandy or mucky shores, floating

bog mats, calcium-rich wet meadows, seeps, fens and low prairie.

Cyperus *Flat Sedge*

Small to medium, annual or perennial, grasslike plants. Stems often clumped, unbranched, sharply 3-angled. Leaves mostly from base of plants, with 1 or more leaflike bracts near top of stems, the blades

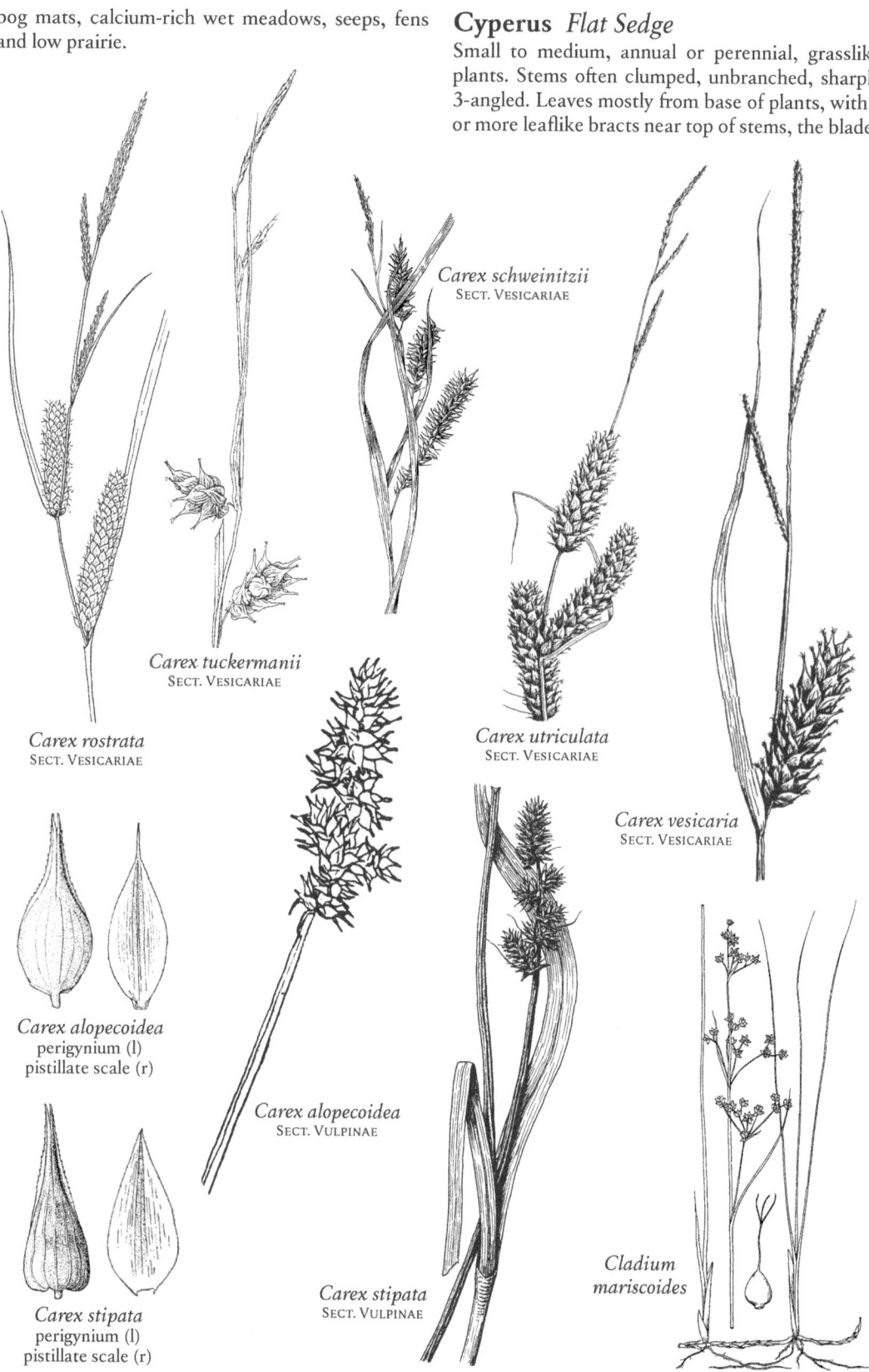

Carex schweinitzii
SECT. VESICARIAE

Carex tuckermanii
SECT. VESICARIAE

Carex rostrata
SECT. VESICARIAE

Carex utriculata
SECT. VESICARIAE

Carex vesicaria
SECT. VESICARIAE

Carex alopecoidea
perigynium (l)
pistillate scale (r)

Carex alopecoidea
SECT. VULPINAE

Carex stipata
perigynium (l)
pistillate scale (r)

Carex stipata
SECT. VULPINAE

Cladium mariscoides

flat or folded along midvein. Flower heads in umbels at ends of stems; the spikelets many, grouped in 1 to several rounded or cylindric spikes. Flowers perfect; bristlelike sepals and petals absent; stamens 1–3; styles 2–3-parted. Achenes lens-shaped or 3-angled, beakless.

1. Achenes lens-shaped; stigmas 2 . . . . **C. bipartitus**
1. Achenes 3-angled; stigmas 3 2
2. Rachilla of spikelets continuous, scales gradually deciduous, falling from base of rachilla to apex . . . . **C. esculentus**
2. Rachilla of spikelets articulated; scales persistent and then falling all at once from the rachilla 3
3. Rachilla articulating at the base of each scale . . . . **C. odoratus**
3. Rachilla not separating into joints (short segments) . 4
4. Scales with strongly recurved acuminate tips . . . . **C. squarrosus**
4. Scales with incurved or straight blunt tips 5
5. Rachilla of spikelet with wings to 1.2 mm wide . . . . **C. strigosus**
5. Rachilla of spikelet essentially wingless 6
6. Culms scabrous, rarely smooth; mucro of the scales 0.3–1.5 mm long; achenes 2.2–2.6 mm long. **C. schweinitzii**
6. Culm smooth; mucro of the scale very short, to 0.3 mm long; achenes 1.4–2.2 mm long 7
7. Involucral leaves strongly ascending, slightly scabrous or smoothish toward base; stem leaves smooth or nearly so . . . . **C. houghtonii**
7. Involucral leaves mostly widely spreading or recurved at maturity, the margins of these and of the stem leaves scabrous . . . . **C. lupulinus**

Cyperus bipartitus Torr.
SHINING FLAT SEDGE   native
IR | **WUP** | **CUP** | EUP   FACW CC 3

Cyperus niger Ruiz & Pavón var. *rivularis* (Kunth) V. Grant, *Cyperus rivularis* Kunth, *Pycreus rivularis* (Kunth) Palla

Tufted grasslike annual. Stems 3-angled, 1–3 dm tall. Leaves usually shorter than stems; leaves and bracts 0.5–2 mm wide, the bracts usually 3, longer than the spikes. Spikelets linear, 10–15 mm long and 2–3 mm wide, in clusters (spikes) of 3–10, the spikes stalkless or on stalks to 10 cm long; scales overlapping, ovate, shiny, purple-brown on margins; stamens 2 or 3; style 2-parted, the lower 1/3 not divided. Achenes lens-shaped, 1–2 mm long, hidden by the scales. July–Sept.

Wet, sandy, gravelly or muddy shores, streambanks, wet meadows, ditches.

Very similar to umbrella flatsedge (*C. diandrus*), found in lower Mich and Wisc, but the scales shiny and the styles not as deeply divided (vs. dull scales and the styles cleft nearly to base in *C. diandrus*).

Cyperus esculentus L.
CHUFA   native
IR | **WUP** | **CUP** | EUP   FACW CC 1

Grasslike perennial, with rhizomes ending in small tubers. Stems single, 3-angled, erect, 2–7 dm long. Leaves light green, mostly from base of plant, about as long as stems, 3–10 mm wide, with a prominent midvein; the bracts 3–6, usually much longer than the spikes. Spikelets linear, 3–12 cm long and 1–2 mm wide; pinnately arranged on a stalk, forming loose cylindrical spikes, the spikes to 5 cm long and 1–2 mm wide; scales straw-colored, 2–3 mm long, overlapping; stamens 3; style 3-parted. Achenes pale brown, 3-angled, 1–2 mm long. July–Sept.

Sandy or muddy shores, streambanks, marshes, ditches and other wet places; weedy in wet or moist cultivated fields; Houghton and Schoolcraft counties.

Cyperus houghtonii Torr.
HOUGHTON'S FLAT SEDGE   native
IR | WUP | **CUP** | EUP   CC 5

Grasslike perennial. Stems smooth, 2–6 dm tall. Leaves 2–4 mm wide; bracts 3–6, mostly ascending, at least one exceeding the inflorescence. Inflorescence of 1 or 2 sessile spikes in an irregular head, usually with 2–5 rays to 10 cm long, each with a similar but smaller head. Spikelets usually 5–15, mostly ascending, flattened, usually 5–12-flowered; scales rotund, 2–2.5 mm long, in half-view from the side more than half as wide as long, many-nerved, obtuse or the uppermost minutely mucronate. Achenes 1.5–2 mm long.

Dry, especially sandy soil.

Cyperus lupulinus (Spreng.) Marcks
GREAT PLAINS FLAT SEDGE   native
IR | WUP | **CUP** | EUP   FACU CC 2

Grasslike perennial. Stems 1–5 dm tall. Leaves shorter than the culm, 2–4 mm wide; bracts 3 or 4, usually widely spreading or even decurved. Inflorescence usually a single subglobose or hemispheric sessile spike, or occasionally with a few rays to 7 cm long, each bearing a similar but smaller spike. Spikelets very crowded, radiating from the axis, flattened, usually 6–12-flowered but often fewer; scales oblong-elliptic, 2.5–3.5 mm long, in half-view from the side about a third as wide as long, many-nerved, obtuse or minutely mucronulate. Achenes 1.5–2 mm long.

Dry woods and fields; Dickinson and Menominee counties.

Cyperus odoratus L.
RUSTY FLAT SEDGE   native
IR | WUP | **CUP** | EUP   OBL CC 3

Cyperus engelmannii Steud., *Cyperus ferruginescens* Boeckl., *Cyperus speciosus* Vahl

Stout, grasslike, fibrous-rooted annual. Stems tufted

or single, 3-angled, 2–7 dm long. Leaves mostly from base of plant, shorter to longer than flowering stems, the blades 2–8 mm wide; the involucral bracts much longer than the spikes. Spikelets linear, 1–2 cm long, pinnately arranged along a stalk, forming several to many cylindrical spikes, the spikes stalkless or stalked; scales red-brown, 2–3 mm long, overlapping; stamens 3; style 3-parted. Achenes brown, 3-angled, 1–2 mm long. July–Sept.

Sandy or muddy shores, floating mats, ditches; Delta and Schoolcraft counties.

Cyperus schweinitzii Torr.
SAND FLAT SEDGE native
IR | WUP | **CUP** | **EUP** FACU CC 5
Cyperus × mesochoreus Geise
Mariscus schweinitzii (Torr.) T. Koyama
Grasslike perennial. Stems rough, 1–8 dm tall. Leaves 2–4 mm wide; bracts 3–6, usually much longer than the inflorescence. Sessile spike obconic to oblong; rays 1–6, rarely more than 10 cm long; spikelets 5–15 in the sessile spike, fewer in the peduncled ones, all crowded, ascending, flattened, usually 8–12-flowered. Scales broadly ovate-elliptic to rotund, the body 3–3.7 mm long, many-nerved, the uppermost with a conspicuous mucro to 1 mm long. Achenes oblong, 2–3 mm long.

Dry or moist sandy soil.

Cyperus squarrosus L.
AWNED FLAT SEDGE native
IR | WUP | **CUP** | EUP OBL CC 5
Cyperus aristatus Rottb., *Cyperus inflexus* Muhl.
Small, tufted, sweet-scented, grasslike annual. Stems very slender, 3-angled, 3–15 cm long. Leaves few, all at base of plant, 1–2 mm wide; bracts 2–3, longer than the spikes. Spikelets linear, flattened, 3–10 mm long, in 1–4 dense, rounded spikes, 1 spike stalkless, the other spikes on stalks to 3 cm long; scales 1–2 mm long, tipped by an awn to 1 mm long, pale brown; stamens 1; style 3-parted. Achenes brown, 3-angled, 0.5–1 mm long. July–Sept.

Wet, sandy or muddy lakeshores, streambanks, mud and gravel bars, wet meadows; Dickinson and Menominee counties.

Cyperus strigosus L.
STRAW-COLOR FLAT SEDGE native
IR | WUP | **CUP** | EUP FACW CC 3
Cyperus hansenii Britt., *Cyperus stenolepis* Torr.
Grasslike perennial, from tuberlike corms. Stems single or few, slender, sharply 3-angled, 1–8 dm long. Leaves mostly at base of plants, the blades 2–12 mm wide, margins rough-to-touch; the bracts mostly longer than the spikes. Spikelets flat, linear, 6–20 mm long and 1–2 mm wide, golden-brown, pinnately arranged and spreading, in several to many cylindric spikes, the spikes often bent downward, on stalks 1–12 cm long, the stalks sometimes branched; scales straw-colored, 3–5 mm long; stamens 3; style 3-parted. Achenes brown, 3-angled, 1–2 mm long. July–Sept.

Wet, sandy or muddy shores, streambanks, marshes, wet meadows, ditches; Alger and Menominee counties.

Dulichium *Three-Way Sedge*

Dulichium arundinaceum (L.) Britt.
THREE-WAY SEDGE native
IR | **WUP** | **CUP** | **EUP** OBL CC 8
Grasslike perennial, spreading by rhizomes and often forming large colonies. Stems stout, erect, 3–10 dm long, jointed, hollow, rounded in section. Leaves 3-ranked, flat, short, 4–15 cm long and 3–8 mm wide; lower leaves reduced to sheaths. Flower heads from leaf axils, in linear clusters of 5–10 spikelets, the clusters 1–2.5 cm long; scales lance-shaped, green to brown, 5–8 mm long. Flowers perfect; sepals and petals reduced to 6–9 downwardly barbed bristles; stamens 3; style 2-parted. Achenes light brown, oblong, 2–4 mm long, beaked by the persistent, slender style. July–Sept.

Shallow marshes, wet meadows, shores, bog margins.

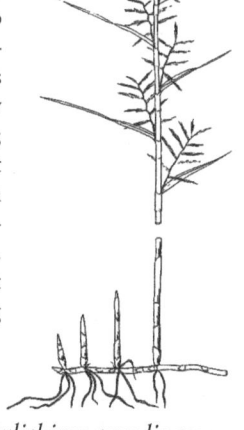

Dulichium arundinaceum

Eleocharis *Spike-Rush*

Small to medium rushlike plants, perennial from rhizomes, or annual, often forming large, matlike colonies. Stems round, flattened, or angled in section. Leaves reduced to sheaths at base of stems. Flower head a single spikelet at tip of stem; scales of the spikelets spirally arranged and overlapping. Flowers perfect; sepals and petals bristlelike or absent, the bristles usually 6 if present; stamens 3; styles 2–3-parted, the base of style swollen and persistent as a projection (tubercle) atop the achene, or sometimes joined with the achene body. Achenes rounded on both sides or 3-angled.

1   Mature spikelet scarcely if at all thicker than main portion of stem; scales persistent; stems triangular . . . . . .
    . . . . **E. robbinsii**

CYPERACEAE | Sedge Family

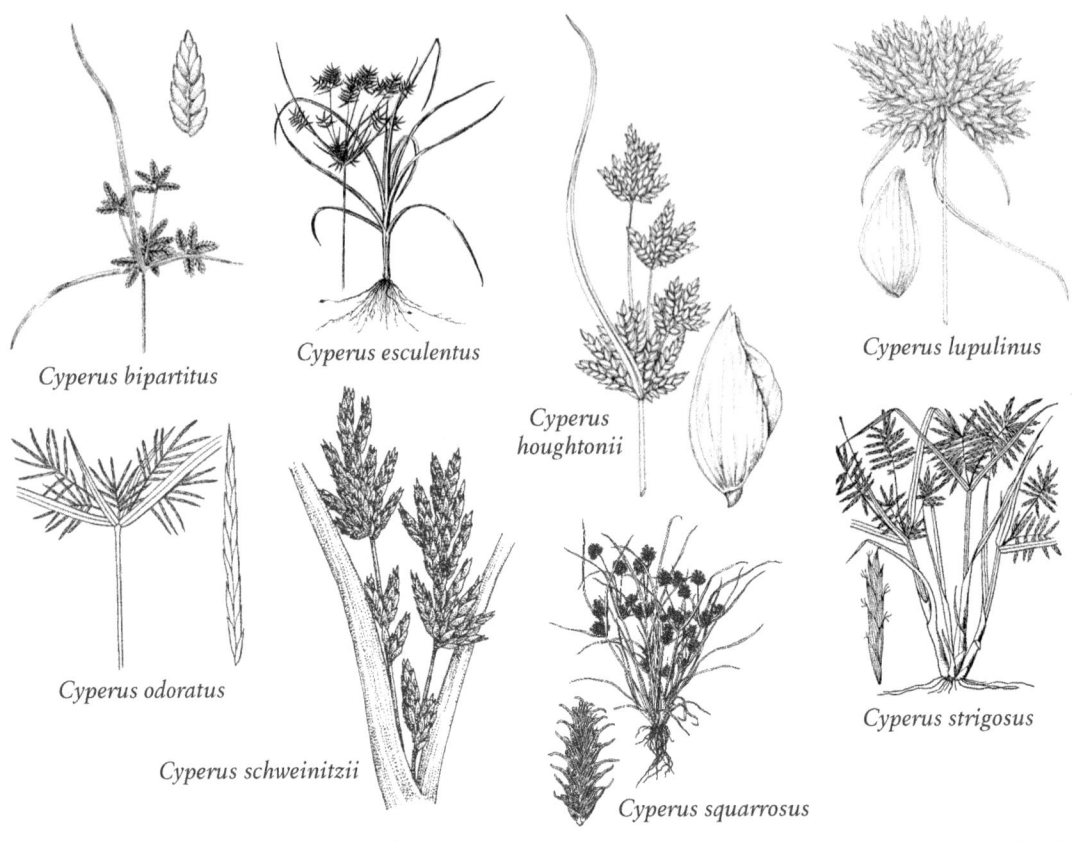

Cyperus bipartitus
Cyperus esculentus
Cyperus houghtonii
Cyperus lupulinus
Cyperus odoratus
Cyperus schweinitzii
Cyperus squarrosus
Cyperus strigosus

1. Mature spikelet decidedly thicker than stem; scales usually deciduous; stems terete (or sometimes flattened or many-ridged), not cross-partitioned 2
2. Tubercle a slender or tiny conical continuation of the body of the achene, slightly differentiated in texture or color, not separated by a constriction or shaped as a distinct apical cap; stigmas 3; tip of leaf sheath without a prominent tooth 3
2. Tubercle differentiated in shape as well as texture, and usually separated from body of achene by a narrow constriction to form a distinct apical cap; stigmas 2 or 3; leaf sheaths sometimes with a prominent tooth at tip. 4
3. Fertile stems 20–70 cm tall, flattened, stout; vegetative stems often as long or longer and rooting at their tips; spikelets 9–17 mm long. . . . . **E. rostellata**
3. Fertile stems to 35 cm tall, but often all tufted and less than 20 cm tall, very slender; stems not rooting at tips; spikelets 4–7 mm long. . . . . **E. quinqueflora**
4. Achenes 3-sided (the angles sharp, or obscure and the achene plumply rounded); styles 3-cleft; surface of achene normally ridged, reticulate, roughened, or in a few species only minutely punctate 5
4. Achenes 2-sided (lenticular or biconvex); styles 2- or 3-cleft; surface of achene smooth, usually ± shiny 9
5. Achenes white or pearly, with prominent longitudinal ridges connected by numerous tiny cross-bars; basal scales of spikelet fertile. . . . . **E. acicularis**
5. Achenes greenish, yellow, golden, brown, black (rarely whitish), and reticulate, smooth, or roughened; basal scales of spikelets sterile 6
6. Plants tufted, without rhizomes; achenes whitish, greenish, olive, or black, smooth to finely reticulate . . . . **E. intermedia**
6. Plants with very stout rhizomes; achenes yellow, golden, or brown, the surface strongly papillate-roughened or honeycombed 7
7. Stems very strongly flattened and often ± twisted, with obscure ridges; scales at middle of spikelet reddish brown with narrow, deeply bifid scarious whitish tips to 1 mm long. . . . . **E. compressa**
7. Stems slightly or not at all flattened, prominently ridged; scales at middle of spikelet deep reddish brown to nearly black, with short, entire, lacerate, or bifid tips mostly less than 0.6 mm long. 8
8. Stems usually 10–50 cm tall, 0.4–0.8 mm wide, scales 2–3.4 mm long . . . . **E. elliptica**
8. Stems ca. 3–10 (–15) cm tall, 0.2–0.3 mm wide; scales 1–1.4 mm long. . . . . **E. nitida**
9. Top of leaf sheaths thin and membranous, cleft on one side, usually whitish; achene olive green to brown, ca. 1–1.5 mm long, including the green tubercle; anthers to 1 mm long . . . . **E. flavescens**
9. Top of leaf sheaths thin to firm, truncate, not split (sometimes with a tooth); achenes and anthers various 10

ELEOCHARIS ACHENES

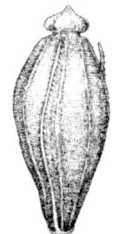
Eleocharis acicularis

Eleocharis compressa

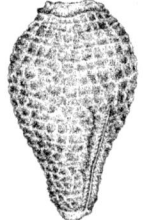
Eleocharis elliptica

Eleocharis flavescens

Eleocharis intermedia

Eleocharis nitida

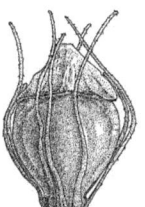
Eleocharis obtusa

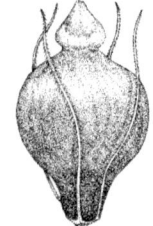
Eleocharis palustris

Eleocharis quinqueflora

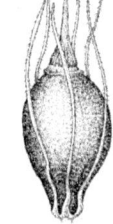
Eleocharis robbinsii

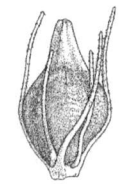
Eleocharis rostellata

10 Plants perennial, with stiff stems and rhizomes; scales acute to acuminate at tip (or somewhat obtuse); achenes 1.5–2.8 mm long, including tubercle; anthers ca. 1–3 mm long . . . . **E. palustris**

10 Plants annual, with soft, easily compressed, densely tufted stems; scales broadly rounded at tip; achenes 1.1–1.5 mm long, including the strongly flattened tubercle; anthers to 0.7 mm long 14 . . . . **E. obtusa**

Eleocharis acicularis (L.) Roemer & J.A. Schultes
NEEDLE SPIKE-RUSH   native
IR | WUP | CUP | EUP   OBL CC 7

Small, tufted, mat-forming perennial, from slender rhizomes. Stems threadlike, 3–15 cm long and to 0.5 mm wide, somewhat 4-angled and grooved; sheaths membranous, usually red at base. Spikelets narrowly ovate, 3–6 mm long and 1–1.5 mm wide; scales with a green midvein and chaffy margins; sepals and petals reduced to 3–4 bristles or absent; style 3-parted. Achenes gray, rounded 3-angled, ridged, to 1 mm long; tubercle cone-shaped, constricted at base. May–Sept.

Shallow water, exposed muddy or sandy shores, marshes and streambanks.

Eleocharis compressa Sullivant
FLAT-STEM SPIKE-RUSH   (THR) native
IR | WUP | CUP | **EUP**   FACW CC 9
Eleocharis elliptica var. *compressa* (Sullivant) Drapalik & Mohlenbrock
Eleocharis tenuis var. *atrata* (Svens.) Boivin

Tufted perennial, from stout black rhizomes. Stems flattened and often twisted, 1.5–4 dm long and 0.5–1 mm wide, shallowly grooved; sheaths red or purple at base. Spikelets ovate, 4–10 mm long and 3–4 mm wide; lowest scale sterile and encircling the stem; fertile scales with a green midvein, purple-brown on sides, and white translucent margins; sepals and petals absent or reduced to 1–5 bristles; style 3-parted. Achenes yellow-brown, covered with small bumps, somewhat 3-angled, 1–1.5 mm long; tubercle small, constricted at base. May–Aug.

Crevices in dolomite on Drummond Island (Chippewa County), calcareous meadows and shores; Mackinac County.

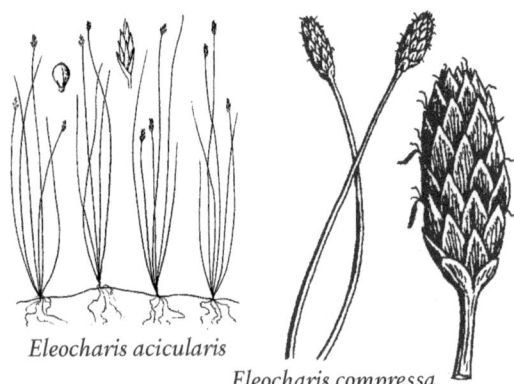

Eleocharis acicularis

Eleocharis compressa

Eleocharis elliptica Kunth
ELLIPTIC SPIKE-RUSH *native*
IR | WUP | CUP | EUP OBL CC 6

Mat-forming perennial, with long rhizomes. Stems subterete to compressed, often with 5–10 ridges, to 90 cm long and 0.3–0.8 mm wide, spongy; sheaths persistent, not splitting, dark red at base, usually redbrown at tip, tooth to 0.5 mm long. Spikelets ovoid, 3–8 cm long and 2–3 mm wide, scales spreading in fruit, 10–30, brown, midrib region often paler, entire or shallowly notched; perianth bristles absent or rarely 1–3, pale brown, to 1/2 of achene length, sparsely retrorsely spinulose; style 3-parted. Achenes yellow, orange, or medium brown, obpyriform, angles evident to prominent, ca. 1 mm long, rugulose (visible at 10x), with 12–20 horizontal ridges in a vertical series; tubercle brown to whitish, depressed, apiculate.

Wet places, often where marly, including fens, rocky beaches, sandy swales and ditches; sometimes in shallow water.

Eleocharis flavescens (Poir.) Urban
YELLOW SPIKE-RUSH *native*
IR | WUP | CUP | EUP OBL CC 7

Eleocharis flaccida var. *olivacea* (Torr.) Fern. & Grisc., *Eleocharis olivacea* Torr.

Small, tufted, mat-forming perennial, spreading by slender rhizomes. Stems bright green, flattened, 3–15 cm long. Spikelets ovate, 2–7 mm long and much wider than stem; scales ovate, red-brown, with a green midvein; sepals and petals reduced to 6–8 barbed bristles; style 2-parted (rarely 3-parted). Achenes lens-shaped, brown, 1 mm long; tubercle pale, cone-shaped, constricted at base.

Shallow water, sandy or muddy shores, mud flats; sometimes where calcium-rich.

Eleocharis intermedia J.A. Schultes
INTERMEDIATE SPIKE-RUSH *native*
IR | WUP | CUP | EUP OBL CC 7

Small, densely tufted annual. Stems threadlike, grooved, of unequal lengths, 5–20 cm long; sheaths toothed on 1 side. Spikelets long-ovate, wider than stem; scales oblong lance-shaped, purple-brown, with a green midvein and white, translucent margins; sepals and petals reduced to barbed bristles or sometimes absent; style 3-parted. Achenes light brown to olive, 3-angled, 1 mm long; tubercle cone-shaped, constricted at base. June–Sept.

Wet, sandy or mucky shores, streambanks, mud flats.

Eleocharis nitida Fern.
QUILL SPIKE-RUSH (END) *native*
IR | WUP | CUP | **EUP** OBL CC 10

Small, mat forming perennial, from matted or creeping purplish rhizomes. Stems round in section to somewhat 4- angled, to 15 cm long, thin and delicate, to only 0.3 mm wide. Spikelets ovoid, small, to 4 mm long and 2 mm wide; fertile scales 1–1.3 mm long, brown to dark brown, midrib usually pale or greenish, the scales often early-deciduous; bristles absent; style 3-parted. Achenes persistent after scales fall, dark yellow-orange or brown, 3-angled (the angles evident), covered with small bumps (under magnification); tubercle brown, flattened and saucer-like, with a tiny central tip. May–June.

Moist sandy depressions in jack pine woods; in Michigan, known only from a single Schoolcraft County location.

Eleocharis obtusa (Willd.) J.A. Schultes
OVOID SPIKE-RUSH *native*
IR | WUP | CUP | EUP OBL CC 3

Eleocharis ovata (Roth) Roem. & Schult.

Tufted, fibrous-rooted annual. Stems slender, round in section, ribbed, 0.5–5 dm long and 0.5–2 mm wide; sheaths green, with a small tooth. Spikelets ovate to cylindric, 4–15 mm long and 2–4 mm wide; scales orange-brown, with a green midvein and pale margins; sepals and petals reduced to 6–7 brown bristles, or absent; styles 2- or 3-parted. Achenes lens-shaped, light to dark brown or olive, shiny, ca. 1 mm long; tubercle flattened-triangular, 2/3 to nearly as wide as the broad top of achene. June–Sept.

May form large colonies, especially on exposed mud flats and drying shores of receding lakes.

Eleocharis palustris (L.) Roemer & J.A. Schultes
COMMON SPIKE-RUSH *native*
IR | WUP | CUP | EUP OBL CC 5

Eleocharis calva var. *australis* (Nees) St. John
Eleocharis erythropoda Steud.
Eleocharis mamillata (H.Lindb.) H.Lindb.
Eleocharis smallii Britt.

Perennial, spreading by rhizomes. Stems single or in small clusters, slender to stout, round in section, 1–8 dm long and 1–3 mm wide; sheaths red or purple at base. Spikelets long-ovate, 5–30 mm long and 2–4 mm wide, wider than stems; lowest scale sterile, encircling the stem; fertile scales lance-shaped to ovate, 2–5 mm long, brown or red-brown, with a green or pale midvein; sepals and petals reduced to usually 4, pale brown, barbed bristles; style 2-parted. Achenes lens-shaped, yellow to brown, 1–2 mm long; tubercle flattened-triangular, constricted at base. May–Aug.

Shallow water of marshes, wet meadows, muddy shores, bogs, ditches, streambanks and swamps.

A variable and common species known by a number of synonyms.

Eleocharis quinqueflora (F.X. Hartmann) Schwarz
FEW-FLOWER SPIKE-RUSH *native*
IR | **WUP** | **CUP** | **EUP** OBL CC 10

Eleocharis bernardina Munz & Johnston
Scirpus pauciflorus Lightf.
Scirpus quinqueflorus F.X. Hartmann

Small, tufted perennial, spreading by rhizomes. Stems threadlike, grooved, 1–3 dm long and less than 1 mm wide. Spikelets ovate, 4–8 mm long and 2–3 mm wide; scales ovate, brown, chaffy on margins, 2–5 mm long; sepals and petals reduced to bristles or absent; style 3-parted. Achenes gray-brown or brown, 3-angled, 1–3 mm long; tubercle slender, joined to the achene and beaklike. June–Aug.

Wet, sandy or gravelly shores and flats marshes and fens; often where calcium-rich.

Eleocharis robbinsii Oakes
ROBBINS' SPIKE-RUSH *native*
IR | **WUP** | **CUP** | **EUP** OBL CC 8

Tufted perennial, spreading by rhizomes. Stems slender, 3-angled, 2–6 dm long and 1–2 mm wide; when underwater, plants often with numerous sterile stems from base; sheaths brown. Spikelets lance-shaped, 1–2 cm long and 2–3 mm wide, barely wider than stems; scales narrowly ovate, margins chaffy; sepals

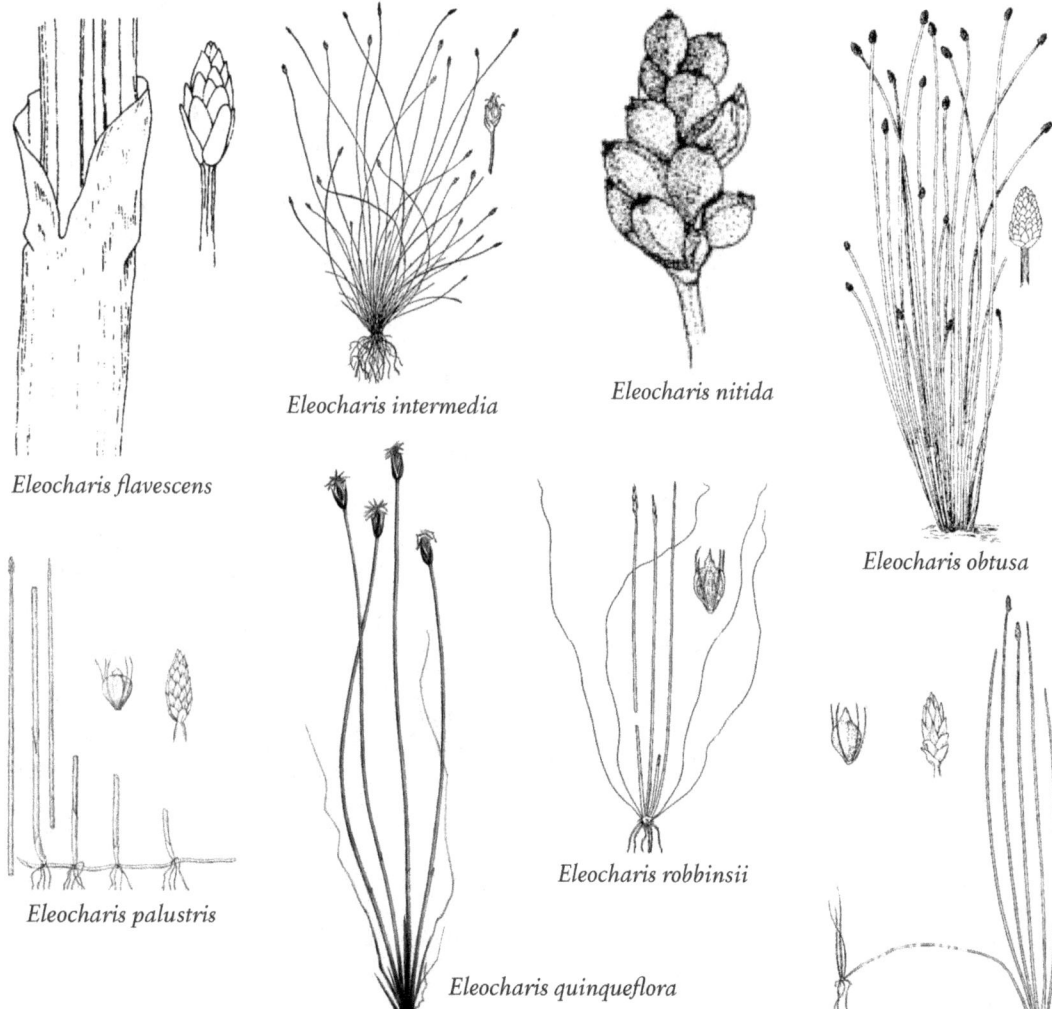

Eleocharis flavescens

Eleocharis intermedia

Eleocharis nitida

Eleocharis obtusa

Eleocharis palustris

Eleocharis quinqueflora

Eleocharis robbinsii

Eleocharis rostellata

and petals reduced to 6 barbed bristles; style 3-parted. Achenes rounded on both sides, light brown, 2–3 mm long; tubercle flattened and cone-shaped, with a raised ring at base. July–Aug.

Local on wet, sandy or mucky lake and pond shores, marshes, exposed flats.

Eleocharis rostellata (Torr.) Torr.
BEAKED SPIKE-RUSH native
IR | WUP | **CUP** | **EUP**   OBL CC 10

Scirpus rostellatus Torr.

Tufted perennial, without creeping rhizomes. Stems flattened, wiry, 3–10 dm long and 1–2 mm wide; the fertile stems upright, the sterile stems often arching and rooting at tip; sheaths brown. Spikelets oblong, tapered at both ends, 5–15 mm long and 2–5 mm wide, wider than the stem; scales ovate, 3–5 mm long, green to brown with a darker midvein and translucent margins; sepals and petals reduced to 4–8 barbed bristles; style 3-parted. Achenes olive to brown, rounded 3-angled, 2–3 mm long; tubercle cone-shaped, joined with the achene body and beaklike. July–Sept.

Shores, wet meadows, calcareous fens and mud flats; typically where calcium-rich and often associated with mineral springs.

Eriophorum Cotton-Grass

Grasslike perennials. Stems clumped or single, round to rounded 3-angled in section. Leaves mostly at base of plant, the blades flat, folded or inrolled; upper leaves often reduced to bladeless sheaths. Flower heads at ends of stems, with 1 or several spikelets; spikelets resemble cottonballs when mature; scales many, spirally arranged, chaffy on margins; involucral bracts leaflike in species with several spikelets in the head, or reduced to scales in species with 1 spikelet at end of stems (*E. chamissonis, E. vaginatum*). Flowers perfect; sepals and petals numerous, reduced to long, cottony, persistent, white to tawny brown bristles; stamens 3; styles 3-parted. Achenes brown, more or less 3-angled, sometimes with a short beak formed by the persistent style.

1 Head a single spikelet at end of stem; leaflike bracts absent .... **E. vaginatum**
1 Head of 2 or more spikelets; leaflike bracts present .. 2
2 Leaves 1–2 mm wide; leaflike bract 1, erect, the head appearing lateral from side of stem 3
2 Leaves 3 mm or more wide; leaflike bracts 2 or 3, the head appearing terminal 4
3 Blade of uppermost stem leaf much shorter than its sheath .... **E. gracile**
3 Blade as long or longer than its sheath .... **E. tenellum**
4 Scales 3–7-nerved, copper-brown on sides ....
   .... **E. virginicum**
4 Scales with 1 nerve, sides olive-green to nearly black . 5
5 Midvein of scale slender, fading before reaching tip of scale .... **E. angustifolium**
5 Midvein of scale widening toward tip of scale and reaching scale tip .... **E. viridicarinatum**

Eriophorum angustifolium Honckeny
THIN-SCALE COTTON-GRASS native
IR | WUP | **CUP** | **EUP**   OBL CC 10

Eriophorum polystachion L.

Grasslike perennial, spreading by rhizomes and forming colonies. Stems mostly single, 2–8 dm long and 2–3 mm wide, more or less round in section, becoming 3-angled below the head. Leaves few, flat or folded along midrib, 3–8 mm wide, often dying back from the tips; sheaths sometimes red, dark-banded at tip. Spikelets 3–10, clustered in heads 1–3 cm wide when mature, the heads drooping on weak stalks; involucral bracts leaflike, often black at base, the main bract upright and usually longer than the head; scales lance-shaped, brown or purple-green, 4–6 mm long, the midvein not extending to tip of scale; bristles bright white, 2–3 cm long. Achenes brown to nearly black, 2–3 mm long. May–July.

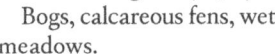
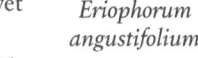

Eriophorum angustifolium

Bogs, calcareous fens, wet meadows.

Similar to *E. viridicarinatum*, which see.

Eriophorum gracile W.D.J. Koch
SLENDER COTTON-GRASS native
IR | WUP | **CUP** | **EUP**   OBL CC 10

Grasslike perennial, spreading from rhizomes. Stems single, spreading or reclining, slender, more or less round in section, 2–6 dm long and 1–2 mm wide. Leaves few, channeled on upper side, 1–2 mm wide, the basal leaves often withered by flowering time, blades of uppermost leaves small. Spikelets in clusters of 2–5 at ends of stems, on spreading to nodding stalks 2–3 cm long; involucral bract leaflike and erect, shorter than spikelet cluster; scales ovate, pale to black-brown with a prominent midvein; bristles bright white. Achenes light brown, 3–4 mm long. May–July.

Fens and bogs.

Eriophorum tenellum Nutt.
FEW-NERVE COTTON-GRASS native
IR | WUP | **CUP** | **EUP**   OBL CC 10

Grasslike perennial, with rhizomes and forming colonies. Stems single, slender, erect, 3–8 dm long, rounded 3-angled, rough-to-touch on upper angles.

Leaves linear, 1–2 mm wide, channeled, not reduced and bladeless on upper stem. Spikelets 3–6, in short-stalked clusters at ends of stems, or with 1–2 rough, drooping stalks to 5 cm long; involucral bract leaflike, stiff and erect, usually shorter than the spikelet cluster; scales ovate, straw-colored to red-brown; bristles white. Achenes brown, 2–3 mm long.

Bogs and conifer swamps.

Eriophorum vaginatum L.
TUSSOCK COTTON-GRASS *native*
IR | WUP | CUP | EUP OBL CC 10
Eriophorum spissum Fern.
Densely tufted, grasslike perennial, forming large hummocks. Stems stiff, rounded 3-angled, 2–7 dm long. Leaves at base of stems, mostly shorter than stems, only 1 mm wide, with 1–3 inflated, bladeless sheaths on stem. Spikelets clustered in a single head at end of stems; involucral bracts absent; scales narrowly ovate, purple-brown to black, with white margins, spreading when mature; bristles usually white (rarely red-brown). Achenes obovate, 3–4 mm long. June.

Sphagnum bogs and tamarack swamps.

Eriophorum virginicum L.
TAWNY COTTON-GRASS *native*
IR | WUP | CUP | EUP OBL CC 8
Large grasslike perennial, with slender rhizomes. Stems single or in small groups, stiff, erect, to 1 m long, leafy, mostly smooth. Leaves flat, 2–4 mm wide, the uppermost often longer than the head. Spikelets in dense clusters of several to many at ends of stems, on short stalks of more or less equal lengths, the clusters wider than long; involucral bracts 2–3, leaflike, spreading or bent downward, unequal, much longer than the head; scales ovate, thick, copper-brown with a green center; bristles tawny or copper-brown. Achenes light brown, 3–4 mm long. July–Aug.

Sphagnum moss peatlands.

Eriophorum viridicarinatum (Engelm.) Fern.
DARK-SCALE COTTON-GRASS *native*
IR | WUP | CUP | EUP OBL CC 8
Grasslike perennial, forming colonies from spreading rhizomes. Stems mostly single, more or less round in section, 3–7 dm long. Leaves flat except at tip, the uppermost leaves 10–15 cm long; sheaths green. Spikelets usually 20–30, clustered in heads at ends of stems, on short to long, finely hairy stalks; involucral bracts 2–4, not black at base, longer or equal to head; scales narrowly ovate, black-green, the midvein pale, extending to tip of scale; bristles white. Achenes brown, 3–4 mm long. May–July.

Bogs and open conifer swamps.

Similar to *E. angustifolium*, but usually with more spikelets, the scale midvein extending to tip of scale, and the leaf sheaths not dark-banded at tip.

Fimbristylis *Fimbry*

Fimbristylis autumnalis (L.) Roemer & J.A. Schultes
SLENDER FIMBRY *native*
IR | WUP | **CUP** | **EUP** FACW CC 6
Tufted, grasslike annual, with shallow fibrous roots. Stems flattened, slender, sharp-edged, 0.5–3 dm long. Leaves shorter than the stems, flat, 1–2 mm wide. Spikelets usually many in an open umbel-like cluster, the spikelets lance-shaped, 3–8 mm long, single or several at ends of threadlike, spreading stalks; involucral bracts 2–3, leaflike, usually shorter than the head; scales spirally arranged and overlapping, ovate, golden-brown with a prominent green midvein, 1–2 mm long; style 3-parted. Flowers perfect, sepals

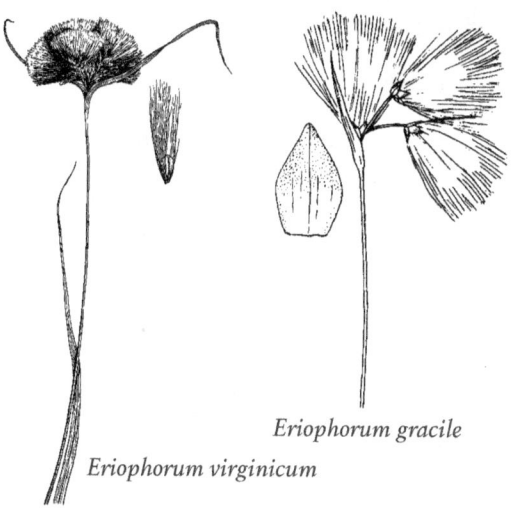

Eriophorum gracile

Eriophorum virginicum

Eriophorum tenellum

Eriophorum vaginatum

and petals absent, stamens 1–3; styles 2–3-parted, swollen at base, deciduous when mature. Achenes ivory to tan, 3-angled and ribbed on the angles, to 0.5 mm long. July–Sept.

Sandy or mucky shores (especially where seasonally flooded and then later exposed), streambanks, wet meadows, ditches.

Fimbristylis autumnalis

Rhynchospora *Beak Sedge*

Grasslike perennials (ours), clumped or spreading by rhizomes. Stems erect, leafy, usually 3-angled or sometimes round. Leaves flat or rolled inward. Spikelets clustered in dense heads, the heads open to crowded; scales overlapping in a spiral. Flowers perfect, or sometimes upper flowers staminate only; sepals and petals reduced to usually 6 (1–20) bristles or sometimes absent; stamens usually 3; styles 2-parted, swollen at base and persistent on the achene as a tubercle. Achenes lens-shaped.

1 Spikelets white to tan; bristles 8 or more ...... **R. alba**
1 Spikelets brown, dark olive-green or nearly black; bristles 5–6 2
2 Scales dark olive-green to black; bristles with upward-pointing barbs, at least some of the bristles longer than the tubercle .... **R. fusca**
2 Scales brown; bristles with downward pointing barbs (rarely smooth), the bristles shorter to as long as the tubercle 3
3 Stems narrow and threadlike; achene margins not translucent, achene body less than half as wide as long .... **R. capillacea**
3 Stems stout; achene with translucent margins, body more than half as wide as long .... **R. capitellata**

Rhynchospora alba (L.) Vahl
WHITE BEAK SEDGE native
IR | **WUP** | **CUP** | **EUP** OBL CC 6

Tufted, grasslike perennial. Stems slender, erect, 1–6 dm long. Leaves bristlelike, 0.5–3 mm wide, shorter than the stems. Spikelets in 1–3 rounded heads, 5–20 mm wide, at or near ends of stems, the lateral heads usually long-stalked; the spikelets oblong, narrowed at each end, 4–5 mm long, white, becoming pale brown; bristles 8–15, downwardly barbed, about equaling the tubercle. Achenes lens-shaped, brown-green, 1–2 mm long; tubercle triangular, about half as long as achene. June–Sept.

Bogs, open conifer swamps of black spruce and tamarack, fens.

Rhynchospora capillacea Torr.
NEEDLE BEAK SEDGE native
IR | **WUP** | **CUP** | **EUP** OBL CC 10

Small, tufted, grasslike perennial. Stems slender, 0.5–4 dm long. Leaves threadlike, rolled inward, to only 0.5 mm wide, much shorter than the stem. Spikelets in 1–2 small, separated clusters, each cluster subtended by 1 to several short, bristlelike bracts; the spikelets ovate, 3–7 mm long; scales overlapping, ovate, brown with a paler, sharp-tipped midvein; bristles 6, downwardly barbed, longer than the achenes; style 2-parted. Achenes lens-shaped, satiny yellow-brown, 2 mm long; tubercle dull brown, narrowly triangular, about 1 mm long. June–Aug.

Calcareous fens, interdunal flats, wet sandy or gravelly shores, seeps; usually where calcium-rich.

Rhynchospora capitellata (Michx.) Vahl
BROWNISH BEAK SEDGE native
IR | **WUP** | **CUP** | **EUP** OBL CC 6

Rhynchospora glomerata var. *capitellata* (Michx.) Kük.

Tufted, grasslike perennial. Stems erect, 3-angled, 3–8 dm long. Leaves flat, 2–4 mm wide, rough on margins, shorter than stems. Spikelets 3–5 mm long, several to many in 2–7 rounded, more or less loose clusters 1–1.5 cm wide, the lateral clusters often in pairs on slender stalks; bristles 6, 2–3 mm long, usually downwardly barbed, about equaling the achene; style 2-parted. Achenes lens-shaped, dark brown, 1–2 mm long; tubercle triangular, about as long as achene. June–Sept.

Wet sandy or mucky shores and flats, wet meadows, bogs, calcareous fens, ditches.

Rhynchospora fusca (L.) Ait. f.
BROWN BEAK SEDGE native
IR | **WUP** | **CUP** | **EUP** OBL CC 7

Tufted, grasslike perennial, spreading by short rhizomes and forming colonies. Stems slender, 3-angled, 1–3 dm long. Leaves very slender, rolled inward, mostly shorter than the stems. Spikelets spindle-shaped, dark brown, 4–6 mm long, in 1–4 loose clusters, the lower clusters on long stalks, each cluster subtended by an erect, leafy bract, the bract longer than the cluster; bristles 6, upwardly barbed; style 2-parted. Achenes light brown, 1–1.5 mm long; tubercle flattened-triangular, nearly as long as achene.

Wet sandy shores, interdunal wetlands, sedge meadows, bog mats.

Schoenoplectus *Club-Rush*

Perennial or annual, tufted or rhizomatous herbs. Stems cylindric to strongly 3-angled, smooth, spongy with internal air cavities. Leaves basal, rarely 1(–2) on stem; sheaths tubular; ligules membranous; blades well-developed to rudimentary. Inflorescences ter-

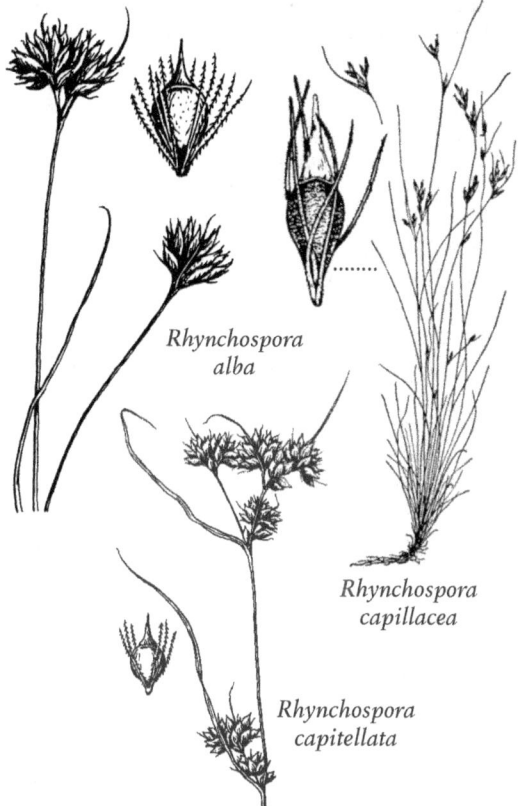

Rhynchospora alba

Rhynchospora capillacea

Rhynchospora capitellata

minal, head-like to openly paniculate; spikelets 1–100 or more; involucral bracts 1–5, leaflike, proximal bract erect to spreading. Spikelets terete; scales deciduous, spirally arranged, each subtending a flower, or proximal scale empty, midrib usually prolonged into short awn, margins ciliate. Flowers bisexual; perianth of 0–6(–8) spinulose bristles shorter than to somewhat longer than the achene; stamens 3. Achenes biconvex to trigonous, with apical beak, rugose or with transverse wavy ridges.

Our 2 annual species, S. purshianus and S. smithii, are sometimes placed in genus *Schoenoplectiella*.

1 Spikelets (at least several of them) distinctly pediceled (sometimes congested in S. acutus); stems terete, often over 1 m tall 2
1 Spikelets 1-few, crowded, sessile or nearly so (rarely one on a short pedicel); stems 3-angled or terete (if terete, then slender, soft, and not over 1 m tall) 3
2 Stems firm and dark olive-green when fresh; spikelets ovoid to cylindrical (often 2.5 or more times as long as wide), usually in a stiffer, sometimes condensed, inflorescence; scales dull, pale or whitish brown, the midrib not strongly contrasting, the margins often more copiously ciliate than in S. tabernaemontani, and the backs copiously flecked with shiny red dots, often puberulent; mature achenes ca. 2.2–2.7 mm long, including apiculus, completely hidden by the scales............ **S. acutus**
2 Stems rather soft and easily compressed, pale blue-green when fresh; spikelets ovoid (about twice as long as wide, or shorter), in an open, lax inflorescence; scales ± shiny, rich orange-brown, often with prominent greenish midrib, the margins ciliate but the backs essentially glabrous (puberulence and swollen red flecks, if any, limited to region of midrib); mature (dark gray or lead-colored) achenes ca. 1.6–2.1(–2.4) mm long, including apiculus, barely covered by the scale............
.... **S. tabernaemontani**
3 Spikelet 1, strongly ascending, the involucral bract surpassing its tip by not more than 15(–20) mm; leaves normally many, hair-like, submersed; stem seldom over 1 mm thick; anthers (2.2–3.5 mm long; achenes 3-sided, the body ca. 2.5–3 mm long ....
.... **S. subterminalis**
3 Spikelets usually more than 1 and the involucral bract surpassing them by more than 15 mm (except in smallest plants of some populations); leaves stiff and stems thicker; anthers and achenes various 4
4 Plants annual, with soft, terete or obscurely 3-angled, tufted stems; anthers 0.3–0.7 mm long 5
4 Plants perennial, with elongate rhizomes; stems sharply 3-angled, at least distally; anthers 1–3 mm long...... 6
5 Taller shoots with stems (base of plant to inflorescence) more than 3/4 as long as height of the plant, including the involucral bract; achenes thickly and asymmetrically biconvex (inner face slightly but clearly convex, outer faces forming a clear angle)............ **S. purshianus**
5 Taller shoots with stems to 3/4 as long as height of the plant; achenes flattened-plano-convex (inner face essentially flat, the outer faces gently rounded) ......
.... **S. smithii**
6 Midrib of scale ± greenish, excurrent as a short (not over 0.5 mm) tip extending beyond the tapered (sometimes very slightly notched) apex of the scale; bristles slightly exceeding body of achene; rhizome soft; achene with apiculus 0.5 mm or more in length; styles 3-cleft and achenes 3-sided; leaves more than half as tall as the stems .... **S. torreyi**
6 Midrib of scale brown, excurrent as a 0.5–1 mm longtip equaling or exceeding lobes; bristles shorter than body of achene; rhizome firm and hard; achene with apiculus shorter than 0.5 mm; styles usually 2-cleft and achenes biconvex to plano-convex (occasionally some styles 3-cleft and achenes 3-sided in a spikelet); leaves less than half as tall as the stems.................. **S. pungens**

Schoenoplectus acutus (Muhl.) A. & D. Löve
HARDSTEM CLUB-RUSH   native
IR | WUP | CUP | EUP   OBL CC 5
Scirpus acutus Muhl.

Perennial, from stout rhizomes and often forming large colonies. Stems round in section, 1–3 m long. Leaves reduced to 3–5 sheaths near base of stem, blades absent, or upper leaves with blades to 25 cm long; main bract erect, appearing as a continuation of stem, 2–10 cm long, eventually turning brown.

Spikelets 5–15 mm long and 3–5 mm wide, in clusters of mostly 3–7, the clusters grouped into a branched head of up to 60 spikelets, the head appearing lateral from side of stem, the branches stiff and spreading; scales chaffy, mostly translucent, 3–4 mm long, often with red-brown spots, usually tipped with an awn to 1 mm long; bristles 6, unequal, usually shorter than achene; style 2-parted (rarely 3-parted). Achenes light green to dull brown, flat on 1 side and convex on other, 2–3 mm long, the style beak small, to 0.5 mm long. May–Aug.

Usually emergent in shallow to deep water (1–2 m deep) of marshes, ditches, ponds and lakes; sometimes where brackish.

Schoenoplectus pungens (Vahl) Palla
COMMON THREESQUARE native
IR | WUP | CUP | EUP OBL CC 5

Scirpus americanus var. *pungens* (Vahl) Barros & Osten, *Scirpus pungens* Vahl

Perennial, from slender rhizomes and forming colonies. Stems erect to somewhat curved, 2–12 dm long, 3-angled, the sides concave to slightly convex. Leaves mostly 1–3 near base of stem, usually folded, or channeled near tip, reaching to about middle of stem and 1–3 mm wide; main bract erect, sharp-tipped, resembling a continuation of the stem, 2–15 cm long. Spikelets 5–20 mm long and 3–5 mm wide, clustered in heads of 1–6 stalkless spikelets, the head appearing lateral; scales brown and translucent, 3–5 mm long, notched at tip, with a midvein extended into a short awn 1–2 mm long; bristles 4–6, unequal, shorter than achene; style 2–3-parted. Achenes light green or tan to dark brown, 3-angled or flat on 1 side and convex on other, 2–3 mm long, the beak to 0.5 mm long. May–Sept.

Shallow water (to about 1 m deep), wet sandy, gravelly or mucky shores, streambanks, wet meadows, ditches, seeps and other wet places.

Schoenoplectus purshianus (Fern.) M.T. Strong
WEAK-STALK CLUB-RUSH native
IR | WUP | CUP | EUP OBL CC 8

Schoenoplectiella purshiana (Fern.) Lye

Annual. Stems often arching (to decumbent), cylindric, to 1 m long and to 2 mm wide. Leaves 1, to as long as the stem; blade absent, or if present, C-shaped in cross-section, 0.5–1 mm wide; bract erect or often divergent, to 15 cm long. Spikelets 1–12; scales straw-colored to orange-brown, midrib often greenish, broadly obovate, 2.5–3 mm long, margins ciliolate at tip and with a small sharp point; perianth members 6, brown, bristle-like, equaling to slightly exceeding achene, densely retrorsely spinulose. Achenes biconvex, brown, turning blackish, 1.6–2.2 mm long, rounded at base to a distinct stipelike constriction; beak 0.1–0.3 mm long.

Sandy to mucky, sometimes marly, shores, especially where water levels have receded.

Schoenoplectus smithii (Gray) Soják
SMITH'S CLUB-RUSH native
IR | WUP | CUP | EUP OBL CC 8

Schoenoplectiella smithii (Gray) Hayas.
Scirpus smithii Gray

Tufted annual. Stems slender, smooth, round or rounded 3-angled, to 6 dm long. Leaves reduced to sheaths, or some with short blades; bract narrow, upright, 2–10 cm long, appearing to be a continuation of stem. Spikelets ovate, 5–10 mm long, in a single cluster of 1–12 spikelets; scales yellow-brown with a green midvein; bristles 4–6, barbed or smooth, longer than achene, sometimes smaller or absent; style 2-parted. Achenes lens-shaped or flat on 1 side and convex on other, glossy brown to black, 1–2 mm long. July–Aug.

Sandy, gravelly or mucky shores, floating mats, bogs; Baraga and Menominee counties.

Schoenoplectus subterminalis (Torr.) Soják
SWAYING CLUB-RUSH native
IR | WUP | CUP | EUP OBL CC 8

Scirpus subterminalis Torr.

Aquatic perennial, spreading by rhizomes. Stems slender, weak, round in section, to 1 m or more long, floating or slightly emergent from water surface near tip. Leaves many, threadlike, channeled, from near base of stem and extending to just below water surface; bract 1–6 cm long, appearing to be a continuation of stem. Spikelets single at ends of stems, with several flowers, light brown, narrowly ovate, tapered at each end, 7–12 mm long; scales thin, 4–6 mm long, light brown with a green midvein; bristles shorter to about as long as achene, downwardly barbed; style 3-parted. Achenes 3-angled, brown, 2–4 mm long, tipped with a slender beak to 0.5 mm long. July–Aug.

In water to about 1 m deep of lakes, ponds and bog margins.

Schoenoplectus tabernaemontani (K. C. Gmel.) Palla
SOFT-STEM CLUB-RUSH native
IR | WUP | CUP | EUP OBL CC 4

Scirpus tabernaemontani K.C. Gmel.
Scirpus validus Vahl

Perennial, spreading by rhizomes and sometimes forming large colonies. Stems stout, smooth, erect, 1–3 m long, round in section. Leaves reduced to 4–5 sheaths at base of stem, or upper leaves with a blade to 7 cm long; main bract erect, 1–10 cm long, shorter than the head. Spikelets red-brown, 4–12 mm long and 3–4 mm wide, single or in clusters of 2–5 at ends of stalks, the stalks spreading or drooping, the clusters in paniclelike heads; scales ovate, light to dark brown, 2–3 mm long, the midvein usually extended into a short awn to 0.5 mm long; bristles 4–6, downwardly barbed, equal or longer than achene; style 2-parted. Achenes flat on 1 side and convex on other, brown to black, about 2 mm long, tapered to a very small beak to 0.2 mm long. June–Aug.

Shallow water and shores of lakes, ponds, marshes, streams, and ditches. Similar to hardstem club-rush (*S. acutus*) but the stems easily crushed between the fingers, plants generally smaller and more slender, and the head more open.

Schoenoplectus torreyi (Olney) Palla
TORREY'S CLUB-RUSH native
IR | **WUP** | **CUP** | **EUP** OBL CC 10
Scirpus torreyi Olney

Perennial, spreading by rhizomes and often forming colonies. Stems erect, sharply 3-angled, 5–10 dm long. Leaves several, narrow, often longer than the stem; bract erect, 5–15 cm long, appearing to be a continuation of stem. Spikelets ovate, light brown, 8–15 mm long, in a single head of 1–4 spikelets, the head appearing lateral from side of stem; scales ovate, shiny brown, with a greenish midvein sometimes extended as a short awn to 0.5 mm long; bristles about 6, downwardly barbed, longer than achene; style 3-parted. Achenes compressed 3-angled, shiny, light brown, 3–4 mm long, tipped by a slender beak to 0.5 mm long. June–Aug.

Uncommon in shallow water, wet sandy or mucky shores.

Scirpus *Bulrush*

Stout, rushlike perennials, mostly spreading by rhizomes. Stems unbranched, 3-angled or round in section, solid or pithy. Leaves broad and flat, to narrow and often folded near tip, or reduced to sheaths at base of stems; involucral bracts several and leaflike, or single and appearing like a continuation of the stem. Spikelets single, or in panicle-like or umbel-like clusters at ends of stems, or appearing lateral from the stem; the spikelets stalked or stalkless; scales overlapping in a spiral. Flowers perfect; sepals and petals reduced to 1–6 smooth or downwardly barbed bristles, or sometimes absent; stamens 2 or 3; styles 2–3-parted. Achenes lens-shaped, flat on 1 side and convex on other, or 3-angled, usually tipped with a beak.

1 Lower sheaths red-tinged . . . . **S. microcarpus**
1 Sheaths green or brown. 2
2 Spikelets many in dense, more or less round heads; bristles about as long as achene or shorter . . **S. atrovirens**
2 Spikelets few in open clusters; bristles much longer than achene 3
3 Mature bristles equal or only slightly longer than scales, spikelets not woolly . . . . **S. pendulus**
3 Mature bristles longer than scales, giving spikelets woolly appearance . . **S. cyperinus complex** (see desc.)

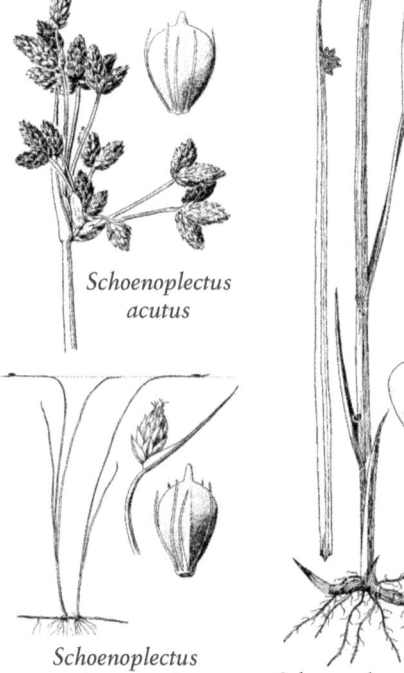

Schoenoplectus acutus

Schoenoplectus subterminalis

Schoenoplectus pungens

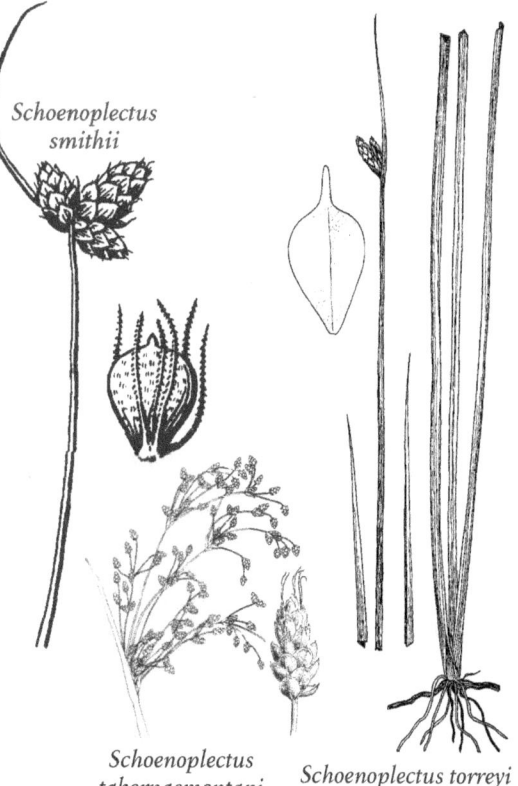

Schoenoplectus smithii

Schoenoplectus tabernaemontani

Schoenoplectus torreyi

Scirpus atrovirens Willd.
DARK-GREEN BULRUSH native
IR | **WUP** | **CUP** | **EUP** OBL CC 3

Loosely tufted perennial, with short rhizomes. Stems 3-angled, leafy, 0.5–1.5 m long. Leaves mostly on lower half of stem, blades ascending, usually shorter than the head, 6–18 mm wide; bracts 3–4, leaflike, to 15 cm long, mostly longer than the head. Spikelets many, 2–8 mm long and 1–3 mm wide, crowded in rounded heads at end of stems, the heads on stalks to 12 cm long; scales brown–black, translucent except for the broad green midvein, 1–2 mm long, tipped by an awn to 0.5 mm long; bristles 6, white or tan, shorter or equal to the achene; style 3-parted. Achenes tan to nearly white, compressed 3-angled, about 1 mm long, with a short beak 0.2 mm long. June–Aug.

Wet meadows, shores, ditches, streambanks, swamps, springs and other wet places.

ADDITIONAL SPECIES

Scirpus hattorius Makino, formerly included in *S. atrovirens*, and known from similar habitats, now sometimes considered separate species; distinguished from *S. atrovirens* in the following key. Plants of *S. hattorius* tend to be more slender, and the scales of its spikelets are usually black, rather than brown as in *S. atrovirens*.

1 Bristles ± straight or stiffly curved to follow the achene outline, at least the longer clearly slightly exceeding the achene; lower leaf sheaths and blades distinctly septate-nodulose.................... **S. atrovirens**
1 Bristles weak and contorted, all shorter than the achene; lower leaf sheaths and blades weakly septate-nodulose .... **S. hattorianus**

Scirpus cyperinus (L.) Kunth
WOOL-GRASS native
IR | **WUP** | **CUP** | **EUP** OBL CC 5

Coarse, densely tufted perennial, rhizomes short. Stems leafy, to 2 m tall, rounded 3-angled to nearly round in section. Leaves flat, 3–10 mm wide, rough-to-touch on margins; sheaths brown; bracts 2–4, leaflike, spreading, usually drooping at tip, often red-brown at base. Spikelets numerous, ovate, 3–8 mm long and 2–3 mm wide, appearing woolly due to the long bristles, in clusters of 1 to several spikelets; the spikelet clusters grouped into large, spreading, branched heads at ends of stems; scales ovate, 1–2 mm long; bristles 6, smooth, brown, much longer than achene and scale; styles 3-parted. Achenes white to tan, flattened 3-angled, 0.5–1 mm long, with a short beak. July–Sept.

Common in wet meadows, marshes, swamps, ditches, bog margins, thickets; where wet or in very shallow standing water.

The *Scirpus cyperinus* complex, including this species, *S. atrocinctus* Fern., and *S. pedicellatus* Fern., is often regarded as one highly variable species. Alternately, the 3 taxa can be separated as follows:

1 Spikelets all or mostly all sessile in clusters of (2–) 3–7 or more........................... **S. cyperinus**
1 Spikelets mostly pediceled, the ultimate branches of the inflorescence typically bearing 1 central, sessile spikelet with 2–3 pediceled ones 2
2 Scales and bases of bracts dark blackish green; plants slender with leaves 2–5 mm wide ...... **S. atrocinctus**
2 Scales and bases of bracts brown or gray-brown; plants more robust with leaves 3–10 mm wide **S. pedicellatus**

Scirpus atrocinctus flowers and fruits earlier than the other two species, often with inflorescences fully developed by late June, and achenes ripe by late July. *S. atrocinctus* readily hybridizes with *S. cyperinus* to form hybrid swarms. Scales of *S. atrocinctus* are usually distinctly blackened, at least near the tip, while those of *S. pedicellatus* have no, or only very slight, black pigment. *S. pedicellatus* is paler and larger than *S. atrocinctus*, the spikelets greenish to pale brown.

Scirpus microcarpus J. & K. Presl
RED-TINGE BULRUSH native
IR | **WUP** | **CUP** | **EUP** OBL CC 5

Perennial, from stout rhizomes. Stems single or few together, 5–15 dm long, weakly 3-angled. Leaves several along stem, flat, ascending, 7–15 mm wide, the upper leaves longer than the head, margins rough-to-touch; sheaths often red-tinged; bracts 3–4, leaflike, to 2–3 dm long. Spikelets numerous, 3–6 mm long and 1–2 mm wide; in a loose, spreading, umbel-like head, the head formed of clusters of 4–20 or more spikelets on stalks to 15 cm long; scales 1–2 mm long, brown and translucent except for green midvein; bristles 4–6, white to tan, downwardly barbed, longer than achene; style 2-parted. Achenes lens-shaped, pale tan to nearly white, about 1 mm long, the beak tiny. June–July.

Streambanks, wet meadows, marshes, wet shores, thickets, swamps, springs; not in dense shade.

Scirpus pendulus Muhl.
RUFOUS BULRUSH native
IR | **WUP** | **CUP** | **EUP** OBL CC 3

Loosely tufted perennial, from short, thick rhizomes. Stems upright, rounded 3-angled, to 1.5 m long, lower stem covered by old leaf bases. Leaves several on stem, flat, 4–10 mm wide, shorter than head; bracts leaflike, 3 or more, shorter than the head, pale brown at base. Spikelets many, cylindric, 4–10 mm long and 2–4 mm wide; in an open, umbel-like head at end of stem, the spikelets drooping and clustered in groups of 1 stalkless and several stalked spikelets; scales about 2 mm long, red-brown with a green midvein; bristles 6, brown, smooth, longer than

Scirpus atrovirens

Scirpus cyperinus

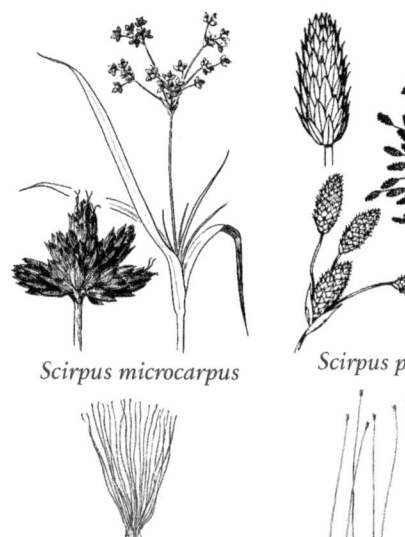
Scirpus microcarpus   *Scirpus pendulus*

Trichophorum alpinum   *Trichophorum caespitosum*

achene and about as long as scale; style 3-parted. Achenes compressed 3-angled, light brown, about 1 mm long, with a short, slender beak. June–Aug.

Marshes, wet meadows, streambanks, swamp openings and ditches; in the western UP, known only from Keweenaw County.

Trichophorum   Leafless-Bulrush

Tufted perennials. Stems 3-angled or terete. Leaves basal or nearly so; sheaths bladeless or with very short blades less than 1 cm long and to 1 mm wide. Inflorescences terminal; spikelets 1; involucral bracts 1, suberect, scale-like, tip mucronate or awned. Spikelets with 3–9 spirally arranged scales, each subtending a flower. Flowers bisexual; perianth of 0–6 bristles, straight, shorter than to about 20 times as long as the achene, smooth or scabrous; stamens 3. Achenes 3-angled or plano-convex.

1 Stems more or less round in section, smooth . . . .
   . . . . T. caespitosum
1 Stems 3-angled, rough on angles 2
2 Perianth bristles ciliate, slightly or not at all exceeding the blunt achene; scales of spikelet not more than 7; achenes ca. 1.6–1.8 mm long; not in wetlands . . . .
   . . . . T. clintonii
2 Perianth bristles smooth, several times as long as the apiculate achene at maturity; scales of spikelets slightly more than 7; body of achene less than 1.5 mm long; wetland species. . . . . T. alpinum

Trichophorum alpinum (L.) Pers.
ALPINE LEAFLESS-BULRUSH   native
IR | WUP | CUP | EUP   OBL CC 10
Eriophorum alpinum L.
Scirpus hudsonianus (Michx.) Fern.

Perennial, from short rhizomes. Stems single to clustered, slender, 1–4 dm long, sharply 3-angled, rough-to-touch on the angles. Leaves reduced to scales at base of stem, with 1–2 leaves upward on stem, these with short narrow blades 5–15 mm long. Spikelets single at ends of stems, brown, 5–7 mm long, with 10–20 flowers, involucral bract awl-shaped, shorter than spikelet, sometimes absent; scales ovate, blunt-tipped, yellow-brown; bristles 6, white, flattened, longer than the scales, when mature forming a white tuft 1–2 cm longer than the spikelet. Achenes 3-angled, dull brown, 1–4 mm long.

Open bogs, conifer swamps, wet meadows, wet sandy shores; sometimes where calcium-rich.

Trichophorum caespitosum (L.) Hartman
TUFTED LEAFLESS-BULRUSH   native
IR | WUP | CUP | EUP   OBL CC 10
Scirpus caespitosus L.

Densely tufted perennial, rhizomes short. Stems slender, smooth, more or less round in section, 1–4 dm long. Leaves light brown and scalelike at base of stems, and also usually 1 leaf upward on stem, the blade narrow, short, to 6 mm long. Spikelets 1 at end of stems, brown, 4–6 mm long, several-flowered; scales yellow-brown, deciduous, the lowest scale about as long as spikelet; bristles 6, usually slightly longer than achene; style 3-parted. Achenes brown, 3-angled, 1.5 mm long.

Open bogs, cedar swamps, calcareous fens, wet swales between dunes; also Lake Superior rocky shores.

Trichophorum clintonii (Gray) S.G. Sm.
CLINTON'S LEAFLESS-BULRUSH   native
IR | WUP | CUP | EUP   FACU CC 10
Scirpus clintonii Gray

Tufted perennial from short rhizomes. Stems slender,

erect, 1–4 dm tall, trigonous, scabrous on the angles. Lower leaves reduced to bladeless sheaths or with short rudimentary blades; uppermost leaves usually prolonged into a blade shorter than the culm and to 1 mm wide. Spikelet solitary, terminal, ovoid, 4–5 mm long, 4–7-flowered; bract erect, ovate, prolonged into a stout mucro shorter than the spike; scales ovate, the midvein often not reaching the apex. Achenes pale brown, trigonous, obovoid, 1.4–2 mm long, obtuse; bristles 3–6, equaling or exceeding the achene.

Dry sandy hillsides, openings in oak and pine woods; rare in Marquette and Schoolcraft counties.

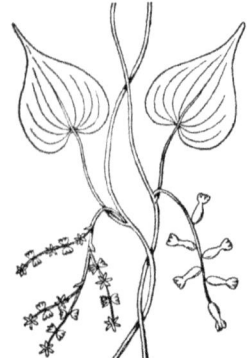

Dioscorea villosa

Eriocaulon aquaticum

Dioscoreaceae
YAM FAMILY

Dioscorea *Yam*
Dioscorea villosa L.
WILD YAM; COLIC-ROOT native
IR | WUP | **CUP** | EUP   FAC CC 4
  Dioscorea hirticaulis Bartlett
  Dioscorea quaternata J.F. Gmel.

Perennial dioecious herb. Stems twining, to 5 m long. Leaves alternate, cordate-ovate, 5–10 cm long, abruptly acuminate, 7–11-nerved; petioles glabrous or nearly so. Flowers regular; unisexual; small, white to greenish yellow; perianth 6-parted, the sepals and petals similar; staminate panicle widely branched, 3–10 cm long; pistillate spikes 5–10 cm long, bearing 5–10 solitary flowers; stamens 6 or 3; ovary 3-celled; styles 3. Fruit a 3-winged capsule 16–26 mm long; seeds very flat, broadly winged, 8–18 mm long. June, July.

Moist to dry woods, thickets, pond and marsh borders, river bottoms, roadsides and railroads. Menominee County; more common in southern Michigan.

Eriocaulaceae
PIPEWORT FAMILY

Eriocaulon *Pipewort*
Eriocaulon aquaticum (Hill) Druce
SEVEN-ANGLE PIPEWORT native
IR | WUP | **CUP** | EUP   OBL CC 9
  Eriocaulon septangulare Withering

Perennial, spongy at base, with fleshy roots. Stems usually single, leafless, slightly twisted, 5–7-ridged, 3–20 cm long (or reaching 2–3 m long when in deep water). Leaves grasslike, in a rosette at base of plant, thin and often translucent, 2–10 cm long and 2–5 mm wide, 3–9-nerved with conspicuous cross-veins. Flowers either staminate or pistillate, grouped together in a single, more or less round head at end of stem, the heads white-woolly, 4–6 mm wide. Fruit a 2–3-seeded capsule. July–Sept.

Shallow water, sandy or peaty shores.

Hydrocharitaceae
TAPE-GRASS FAMILY

Aquatic herbs. Stems leafy, the leaves opposite (*Najas*), whorled (*Elodea*), or plants stemless with clusters of long, linear, ribbonlike leaves (*Vallisneria*). Flowers usually either staminate or pistillate and borne on separate plants, small and stalkless, or in a spathe at end of a stalk; sepals 3; petals 3 or absent; staminate flowers with 3 or more stamens; stigmas 3. Fruit several-seeded, maturing underwater.

ADDITIONAL SPECIES
Hydrocharis morsus-ranae L. (European Frogbit) introduced and invasive in shallow open water of marshes, ditches, and slow moving streams. A serious invasive pest spreading in the eastern Great Lakes region, and able to quickly cover large areas of shallow open water. First collected in Michigan in 1996 in Wayne County and now known from additional southeastern Michigan locations; in the UP, reported from Chippewa County.

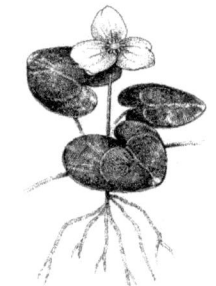

H. morus-ranae

1   Leaves floating, blades orbicular with a cordate base . . . . **Hydrocharis morsus-ranae***
1   Leaves submerged, linear 2
2   Leaves very long and ribbon-like (mostly 3–11 mm wide), in a basal rosette . . . . **Vallisneria americana**

2   Leaves to 6 (–12) cm long, opposite or whorled 3
3   Leaves whorled, entire . . . . **Elodea**
3   Leaves opposite, minutely denticulate to visibly toothed
    . . . . **Najas**

Elodea *Waterweed*

Aquatic perennial herbs, rooting from lower nodes or free-floating. Stems slender, leafy, branched. Leaves crowded near tip of stem, mostly in whorls of 3–4, or opposite, stalkless; margins finely sharp-toothed. Flowers either staminate or pistillate and on separate plants, tiny, single in upper leaf axils, subtended by a 2-parted spathe, usually extended to the water surface by a long, threadlike hypanthium, or stalkless and breaking free to float to water surface in staminate flowers of *E. nuttallii*; sepals 3; petals 3 or absent, white or purple; staminate flowers with 9 stamens; pistillate flowers with 3 stigmas, the stigmas entire or 2-parted. Fruit a capsule, ripening underwater. *Anacharis* or *Philotria* in older floras.

1   Leaves mostly 2 mm or more wide; staminate flowers long-stalked in a spathe, the spathe more than 7 mm long, extended to water surface by a long, threadlike hypanthium. . . . . **E. canadensis**
1   Leaves to 1.5 mm wide; staminate flowers stalkless in a spathe, the spathe 2–4 mm long, breaking free to float to water surface at flowering time. . . . . **E. nuttallii**

Elodea canadensis Michx.
CANADIAN WATERWEED native
IR | **WUP** | **CUP** | **EUP** OBL CC 1

Anacharis canadensis (Michx.) Planch.

Submerged perennial herb. Stems round in section, usually branched, 2–10 dm long. Leaves bright green, firm; lower leaves opposite, reduced in size, ovate or lance-shaped; upper leaves in whorls of 3, the uppermost crowded and overlapping, lance-shaped, 5–15 mm long and about 2 mm wide, rounded at tip. Flowers at ends of threadlike stalks, 2–30 cm long; staminate flowers in spathes from upper leaf axils, the spathes about 10 mm long and to 4 mm wide; sepals green, 3–5 mm long; petals white, 5 mm long; stamens 9; pistillate flowers in spathes from upper leaf axils, the spathes 10–20 mm long, extended to water surface by a threadlike hypanthium; sepals 2–3 mm long; petals white, 2–3 mm long. Fruit a capsule, 5–6 mm long, tapered to a beak 4–5 mm long. June–Aug.

Shallow to deep water of lakes, streams and ditches.

Elodea nuttallii (Planch.) St. John
WESTERN WATERWEED native
IR | **WUP** | **CUP** | **EUP** OBL CC 5

Anacharis nuttallii Planch.

Submerged perennial herb. Stems slender, round in section, usually branched, 3–10 dm long. Lower leaves opposite, reduced in size, ovate to lance-shaped; upper leaves in whorls of 3 (or sometimes 4), not densely overlapping at tip, linear to lance-shaped, 6–13 mm long and 0.5–1.5 mm wide, tapered to a pointed tip. Staminate flowers in stalkless spathes from middle leaf axils, the spathes ovate, 2–3 mm long, the flowers single and stalkless in the spathe, breaking free and floating to water surface and then opening; sepals green or sometimes red, 2 mm long; petals absent or very short (to 0.5 mm long); stamens 9; pistillate flowers in cylindric spathes from upper leaf axils, the spathes 1–2.5 cm long, extended to water surface by a threadlike stalk to 10 cm long; sepals green, about 1 mm long; petals white, longer than sepals. Fruit a capsule, 5–7 mm long. June–Aug.

Shallow to deep water of lakes, streams and ditches.

Similar to *E. canadensis* but less common, smaller and more delicate overall, the leaves narrower, paler green, and not closely overlapping at stem tips, and the staminate flowers not elevated on a long slender stalk.

Najas *Waternymph*

Aquatic annual herbs, roots fibrous, rhizomes absent. Stems wavy, with slender branches. Leaves simple, opposite or in crowded whorls, stalkless, abruptly widened at base to sheath the stem; margins toothed

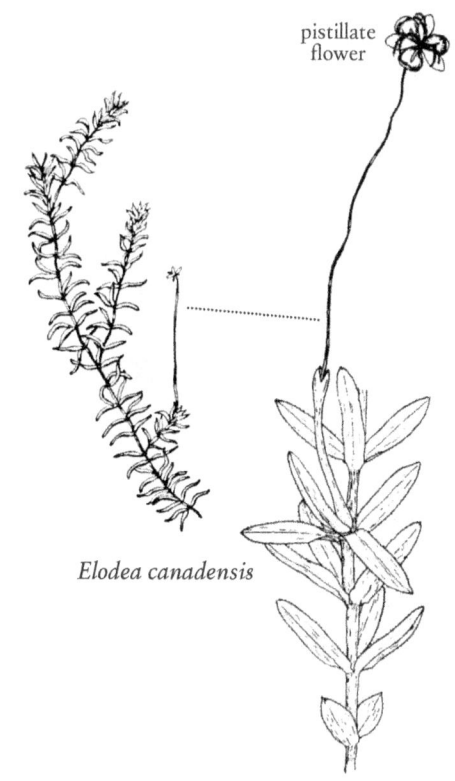

Elodea canadensis

to nearly entire, the teeth sometimes spine-tipped. Flowers either staminate or pistillate, separate on same plant or on different plants, tiny, single and stalkless in leaf axils, enclosed by the sheathing leaf bases; staminate flowers a single anther within a membranous envelope (spathe), this surrounded by perianth scales, the scales sometimes joined into a tube; pistillate flowers surrounded by 1–2 spathes, pistils 1, stigmas 2–4, style usually persistent. Fruit a 1-seeded achene.

1. Leaves with auriculate, scarious basal lobes (these toothed at tip), blade to 0.5 mm wide ... **N. gracillima**
1. Leaves expanded at the base, but expanded portion tapering to the slender blade (not auriculate or broadly truncate), blade to 2 mm wide 2
2. Fruit 2–4 mm long, smooth and glossy, with very fine reticulate pattern; leaves very slender (to 1 mm wide at the middle, sometimes to 3 cm long), tapering to tip .. .. **N. flexilis**
2. Fruit ca. 2 mm long, the fruit surface reticulate and shallowly pitted; leaves to 2.2 mm wide at middle, up to 1.8 cm long, and appearing more abruptly acute at tip.... .... **N. guadalupensis**

Najas flexilis (Willd.) Rostk. & Schmidt
WAVY WATERNYMPH native
IR | **WUP** | **CUP** | **EUP** OBL CC 5

Stems branched, 5–40 cm long. Leaves densely clustered at tips of stems, linear, tapered to a long slender point, spreading or ascending, 1–4 cm long and to 0.5 mm wide; margins with tiny sharp teeth. Flowers either staminate and pistillate, separate on same plant. Achenes oval, olive-green to red, the beak 1 mm or more long; seeds straw-colored, shiny, 2–4 mm long. July–Sept.

Ponds, lakes, streams.

Najas gracillima (A. Braun) Magnus
SLENDER WATERNYMPH native
IR | **WUP** | **CUP** | EUP OBL CC 8

Plants light green. Stems very slender, branched, 0.5–5 dm long. Leaves opposite or in groups of 3 or more, bristlelike, 0.5–3 cm long and to 0.5 mm wide, spreading or ascending; margins with very small teeth. Flowers either staminate or pistillate and on the same plant. Achenes cylindric, narrowed at ends; seeds light brown, 2–3 mm long.

Shallow water of lakes, usually in muck; intolerant of polluted water; Houghton and Marquette counties.

Najas guadalupensis (Spreng.) Magnus
SOUTHERN WATERNYMPH native
IR | WUP | **CUP** | EUP OBL CC 7

Stems much branched, 1–6 dm long. Leaves numerous, linear, spreading and often curved downward at tip, 1–3 cm long and 0.5–2 mm wide; groups of smaller leaves also present in leaf axils; margins with very small teeth. Flowers either staminate or pistillate, separate on same plant. Achenes cylindric, the beak to 0.5 mm long; seeds brown or purple, 1–3 mm long. July–Sept.

Shallow to deep water of lakes, ponds and sometimes rivers; Schoolcraft County.

Vallisneria *Eel-Grass*
Vallisneria americana Michx.
AMERICAN EEL-GRASS native
IR | **WUP** | **CUP** | **EUP** OBL CC 7

Submerged perennial herb, fibrous rooted, spreading by stolons and often forming large colonies. Stems absent. Leaves long and ribbonlike, in tufts from a small crown, to 1 m or more long and 3–10 mm wide, rounded at tip, margins smooth. Flowers either staminate or pistillate and on separate plants; staminate flowers small, about 1 mm wide, in a many-flowered head, the head within a stalked spathe from base of plant, the stalk 3–15 cm long; sepals 3, petals 1, stamens 2; the staminate flowers released singly from the spathe and floating to water surface where they open; pistillate flowers single in a spathe, on long slender stalks that extend to water surface, the stalk contracting and coiling after flowering to draw the fruit underwater; sepals 3, petals small, 3; stigmas 3. Fruit a cylindric, curved capsule, 4–10 cm long. July–Sept.

Shallow to sometimes deep water of lakes and streams.

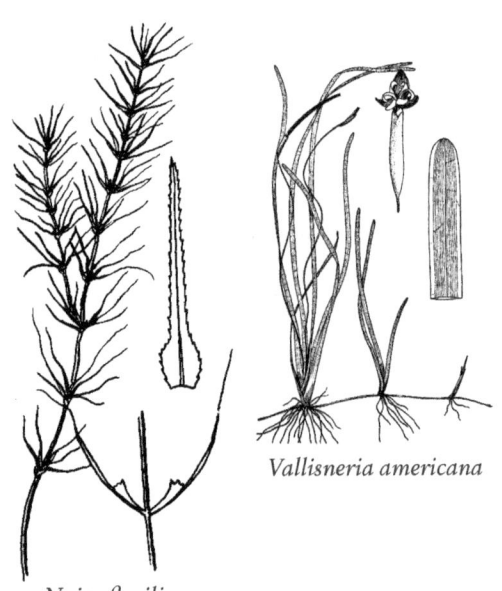

Vallisneria americana

Najas flexilis

Hypoxidaceae
LILIID MONOCOT FAMILY

Hypoxis *Star-Grass*

Hypoxis hirsuta (L.) Coville
EASTERN YELLOW STAR-GRASS   *native*
IR | WUP | **CUP** | EUP   FAC CC 10

Low perennial herb, from a small, shallow corm. Stems leafless, lax, 1 to several, silky-hairy in upper part, shorter than leaves when flowering, to 4 dm long when mature. Leaves from base of plant, linear, hairy, to 6 dm long and 2–10 mm wide. Flowers 1–6 (usually 2), yellow, 1–2.5 cm wide, in racemes at ends of stems, tepals hairy on outside, 5–12 mm long, spreading in flower, closing and turning green after flowering, persistent. Capsule oval, 3–6 mm long; seeds black. May–July.

Wet meadows, shores, moist prairie; often where calcium-rich; Menominee County.

Iridaceae
IRIS FAMILY

Perennial herbs with rhizomes, bulbs, or fibrous roots. Leaves parallel-veined, narrow, 2-ranked, the margins joined to form an edge facing the stem (equitant). Flowers perfect, with 6 petal-like segments, single or in clusters at ends of stem, stamens 3, style 3-parted. Fruit a 3-chambered capsule.

1. Flowers more than 2 cm wide; stems not winged; leaves more than 6 mm wide.........................**Iris**
1. Flowers to 2 cm wide; stems winged; leaves to 6 mm wide..............................**Sisyrinchium**

Iris *Iris; Flag*

Perennial herbs, spreading by thick rhizomes. Stems erect. Leaves swordlike, erect or upright, the margins joined to form an edge facing the stem. Flowers 1 or several at ends of stems; yellow or blue-violet; sepals 3, spreading or bent downward, longer and wider than the petals; petals 3, erect or arching; stamens 3; styles 3-parted, the divisions petal-like and arching over the stamens. Fruit an oblong capsule.

ADDITIONAL SPECIES
Iris sibirica L. (Siberian Iris), cultivated Eurasian iris, escaped to moist roadsides and meadows near Cooks, Schoolcraft County.

1. Plants dwarf, the flowering stems less than 15 (usually less than 10) cm tall....................**I. lacustris**
1. Plants more than 15 cm tall.........................2
2. Flowers yellow, mature capsules spreading or pendant ....**I. pseudacorus**
2. Flowers blue, capsules erect........................3
3. Base of expanded portion of sepal with a bright yellow spot, finely pubescent with hairs as long as the thickness of the sepal; outer spathe bracts of uniform texture and color; seeds round to D-shaped, irregularly (but shallowly) pitted ....**I. virginica**
3. Base of expanded portion of sepal at most with a greenish yellow spot, with papillae shorter than thickness of the sepal; outer spathe bracts with the margins generally darker and more shiny than the rest of the dull surface; seeds D-shaped, with a ± regularly pebbled surface................................**I. versicolor**

Iris lacustris Nutt.
DWARF LAKE IRIS   (THR) *native*
IR | WUP | **CUP** | EUP   FAC CC 9

Perennial herb, mostly less than 15 cm tall. Leaves broadly linear, curved-arching, 4–6 cm long at anthesis, later to 18 cm long and 5–10 mm wide. Petals blue or rarely, white; perianth tube dull yellow, 1–2 cm long, dilated upwards, shorter than the sepals and petals; perianth limb 4–5 cm across. Fruit a capsule; seeds dark brown. May.

Gravelly shores in calcareous soil, sandy beach ridges and stabilized dunes; near shores of Lakes Huron and Michigan.

Iris pseudacorus L.
PALE-YELLOW IRIS   *introduced*
IR | WUP | CUP | **EUP**   OBL

Perennial herb, from thick rhizomes. Stems 0.5–1 m long, shorter or equal to the leaves. Leaves sword-shaped, stiff and erect, waxy, 1–2 cm wide. Flowers several at end of stems, yellow, 7–9 cm wide, sepals spreading, upper portion marked with brown; petals erect, narrowed in middle, 1–2.5 cm long. Capsules oblong, 6-angled, 5–9 cm long. May–June.

Lakeshores, streambanks, marshes, ditches; Chippewa and Houghton counties.

Iris versicolor L.
NORTHERN BLUEFLAG   *native*
IR | WUP | **CUP** | EUP   OBL CC 5

Perennial herb, from thick, fleshy rhizomes and forming colonies. Stems more or less round in section, often branched above, 4–9 dm long. Leaves sword-shaped, erect or arching, somewhat waxy, 2–3 cm wide, usually shorter than stem. Flowers several on short stalks at ends of stems, blue-violet, 6–8 cm wide; sepals spreading, unspotted, or with a green-yellow spot near base, surrounded by white streaks and purple veins; petals erect, about half as long as sepals. Capsules oblong, 3–6 cm long. June–July.

Marshes, shores, wet meadows, open bogs, swamps, thickets, forest depressions; often in shallow water.

Iris virginica L.
SOUTHERN BLUEFLAG native
IR | WUP | **CUP** | EUP OBL CC 5

Perennial herb, from thick rhizomes, often forming large colonies. Stems more or less round in section, to 1 m long. Leaves sword-shaped, erect or arching, 2–3 cm wide, usually longer than stems. Flowers several on short-stalks at ends of stems, blue-violet, often with darker veins, 6–8 cm wide; sepals spreading, curved backward at tip, with a hairy, bright yellow spot near base; petals shorter than sepals. Capsules ovate to oval capsule, 4–7 cm long. May–July.

Swamps, thickets, shores, streambanks, marshes, ditches; Delta and Menominee counties.

Sisyrinchium *Blue-Eyed-Grass*

Tufted perennial herbs, from fibrous roots. Stems slender, leafless, flattened or winged. Flowers in an umbel at end of stem, above a pair of erect green bracts (spathe), blue-violet or rarely white, with 6 spreading segments, the segments joined only at base, the tips rounded but with an small bristle. Leaves narrow and linear, from base of plant, the margins joined and turned to form an edge facing the stem. Fruit a rounded capsule; seeds round, black.

1. Spathes on peduncles arising from a leaflike bract, usually more than one, the upper portion of the stem thus appearing branched 2
1. Spathes sessile or nearly so at the end of a simple stem 3
2. Stems 1.5–2.5 mm wide; spathe 16–22 mm long, the bracts subequal; pedicels (at least some of them) wing-margined basally more than half their length; capsule pale whitish tan (sometimes purplish at tip). . . . . **S. strictum**
2. Stems 2.5–3.5 mm wide; pedicels merely 2-edged or slightly winged basally; capsules dark. **S. angustifolium**
3. Plant very slender, the stem usually 1 mm or less wide and barely margined or narrowly winged; largest leaves to 1.5 mm wide; capsule 2.5–4 mm long. . . . . . **S. mucronatum**
3. Plants stout, the stem usually 2–2.5 mm wide, winged; largest leaves 2–3 mm wide; mature capsule 5–7 mm long. . . . . **S. montanum**

Sisyrinchium angustifolium P. Mill.
NARROW-LEAF BLUE-EYED-GRASS native
IR | **WUP** | **CUP** | EUP FAC CC 4

Plants bright green, drying dark. Stems 1–5 dm tall, usually erect. Scapes narrowly winged but stout, 3–4 mm wide. Spathes often purplish tinged, the inner bracts 1.5–3 cm long; perianth bright violet, 8–12 mm long. Capsules obovoid, brown, 4–6 mm long.

Local in meadows, fields, and open woods.

Sisyrinchium montanum Greene
STRICT BLUE-EYED-GRASS native
IR | **WUP** | **CUP** | **EUP** FAC CC 4

Plants pale-green and waxy. Stems stiff and erect, leafless, flattened and winged, 1–5 dm tall and 2–4 mm wide. Leaves mostly from base of plant, narrow and grasslike, about half as long as stem, 1–3 mm wide. Flowers in head of 1 to several flowers at end of stem, subtended by a spathe, the spathe of 2 bracts, the outer bract 3–7 cm long, the inner bract about half as long; the flower segments (tepals) blue-violet with a yellow center, 5–15 mm long, with a short, slender tip. Capsules more or less round, 4–7 mm wide, pale brown, on an erect stalk shorter than the inner bract. May–July.

Wet meadows, shores, thickets, ditches, swales; also in drier woods and fields.

Iris versicolor

Hypoxis hirsuta

Iris lacustris

Sisyrinchium montanum

Sisyrinchium mucronatum

Sisyrinchium mucronatum Michx.
NEEDLE-TIP BLUE-EYED-GRASS *native*
IR | WUP | **CUP** | **EUP** FAC CC 10

Plants dark green. Stems very slender, to 1 mm wide, leafless, margins not or barely winged. Leaves from near base of plant, narrow and linear, to 1.5 mm wide. Flowers in a single head at end of stem, subtended by a spathe, the spathe of 2 bracts, the bracts often purple-tinged, the outer bract 2–3 cm long, the inner bract shorter, 1–2 cm long; the segments (tepals) deep violet-blue, 8–10 mm long, tipped with a sharp point. Capsules more or less round, 2–4 mm long, pale brown, on spreading stalks. May–June. Wet meadows, calcareous fens.

Sisyrinchium strictum E.P. Bicknell
BLUE-EYED-GRASS *native*
IR | **WUP** | CUP | EUP FAC CC 10

Plants generally light green to yellowish green when dry, not glaucous. Stems to 5 dm tall, branched. Spathes usually with purplish tinge on margins, wider than supporting branch, glabrous; perianth pale to deep bluish violet, bases yellow; ovary similar in color to foliage. Capsules subglobose, tan to nearly white, sometimes with purplish tip, 3.5–5 mm long; pedicel erect or ascending; seeds globose to obconic, lacking obvious depression, rugulose or granular. June–July.

Openings with gravelly and sandy soil; rare in Baraga and Gogebic counties.

Juncaceae
RUSH FAMILY

Distinguished from grasses and sedges by the presence of a true perianth of 6 tepals and a 3–many-seeded capsule rather than a 1-seeded grain (grasses) or achene (sedges). No ligule (as in the grasses) is present at junction of leaf blade and sheath; however an auricle (an ear-like appendage) may occur at top of leaf sheath.

1 Foliage completely glabrous; capsules usually many-seeded . . . . **Juncus**
1 Foliage ± hairy, at least toward summit of sheaths; capsules 3-seeded . . . . **Luzula**

Juncus *Rush*

Clumped or rhizomatous rushes, mostly perennial (annual in *Juncus bufonius*). Stems erect and unbranched. Leaves from base of plant or along stem, alternate, round in section, flat to involute, or reduced to sheaths at base of stem. Flowers perfect, regular, in compact to open clusters of few to many flowers, subtended by 1 or several leaflike involucral bracts; sepals and petals of 6 chaffy, scalelike, green to brown tepals (perianth); stamens 6 or 3; stigmas 3, ovary 1 or 3-chambered. Capsule many-seeded; seeds with a short slender tip or with a tail-like appendage at each end.

ADDITIONAL SPECIES
Juncus inflexus L., European native, established in roadside ditches in Baraga and Houghton counties.

1 Head from side of stem, the involucral bract erect, round in section and appearing to be a continuation of stem; basal and stem leaves absent 2
1 Head at end of stem; basal leaves present, stem leaves present or absent 4
2 Involucral bract more than half height of plant . . . . **J. filiformis**
2 Involucral bract less than half height of plant 3
3 Stems densely clumped; stamens 3 . . . . **J. effusus**
3 Stems single from rhizomes, the stems often in rows; stamens 6 . . . . **J. balticus**
4 Leaves flat or somewhat channeled 5
4 Leaves round, inrolled, or narrowly channeled 12
5 Flowers in dense heads; leaves 2 mm or more wide . . 6
5 Flowers single on branches of head; leaves usually less than 1.5 mm wide 7
6 Leaves folded and joined near tip, with 1 edge facing the stem (as in *Iris*) . . . . **J. ensifolius**
6 Leaves not folded or joined, the leaf surface facing the stem . . . . **J. marginatus**
7 Plants annual, to 10 cm tall . . . . **J. bufonius**
7 Plants perennial, more than 10 cm tall 8
8 Larger stem leaves more than 3 mm wide, with 5 main veins . . . . **J. marginatus**
8 Stem leaves to 3 mm wide, with less than 5 main veins 9
9 Leaves flat or inrolled but open for entire length; seeds less than 0.5 mm long 10
9 Leaves round in section, closed for more or less entire length; seeds more than 0.7 mm long 11
10 Auricles flaplike, prolonged into a membranaceous or scarious projection 3–5 mm long . . . . **J. tenuis**
10 Auricles shorter, not flaplike, prolonged up to 2 mm beyond the sheath, cartilaginous, very rigid **J. dudleyi**
11 Ends of seeds with white "tails" about half as long as the slender body; sepals ca. (3.5–) 4 mm long; longest involucral bract 1–6 cm long (often less than 3 cm) . . . . **J. vaseyi**
11 Ends of seeds without "tails" or these at most half the width of the plump body; sepals 3–4 mm long; longest involucral bract to 21 cm long . . . . **J. greenei**
12 Leaf blades without cross-partitions at regular intervals . . . . **J. stygius**
12 Leaf blades with cross-partitions at regular intervals 13
13 Heads with 1–2 flowers; the head less than 1/4 total height of plant . . . . **J. pelocarpus**

13 Heads with 3–13 flowers; the head more than 1/4 total height of plant 14
14 Leaf sheath distinctly ribbed; seeds 1–2 mm long, with short to long white tails 15
14 Leaf sheath smooth or veined; seeds less than 1 mm long, without white tails or with very short dark tails 17
15 Petals short-tapered to a rounded tip; seeds with a short tail (about 1/10 of body length) .... J. brachycephalus
15 Petals short-tapered to a pointed tip; seeds with long or short tails. 16
16 Head narrowly cylindric in outline, the branches erect; heads with 2–5 flowers .... J. brevicaudatus
16 Head ovate in outline, the branches ascending or spreading; heads with 5–40 flowers ..... J. canadensis
17 Flowers in dense round heads, the flowers radiating in all directions 18
17 Heads various but the flowers not radiating in all directions. 20
18 Heads to 10 mm wide; most capsules longer than sepals .... J. nodosus
18 Heads more than 10 mm wide; capsules shorter to nearly equaling the sepals. 19
19 Stamens 6; capsules lance-shaped, about as long as sepals. .... J. torreyi
19 Stamens 3; capsules ovate, shorter to about as long as sepals .... J. acuminatus
20 Branches of head ascending; petals shorter than sepals .... J. alpinoarticulatus
20 Branches of head spreading; petals as long or slightly longer than sepals. .... J. articulatus

Juncus acuminatus Michx.
KNOTTY-LEAF RUSH *native*
IR | **WUP** | CUP | EUP OBL CC 8

Tufted perennial rush. Stems erect, slender, 2–8 dm tall, with 1–2 leaves. Leaves from stem and at base of plant, round to compressed in section, 5–40 cm long and 1–3 mm wide; auricles rounded, 1–2 mm long; bract erect, round, 1–4 cm long, shorter than the head. Flowers in an open, pyramid-shaped inflorescence, 5–12 cm long and less than half as wide, composed of 5–50 rounded heads 6–10 mm wide, each head with 5–30 flowers, the branches spreading, 1–10 cm long; tepals lance-shaped, green or straw-colored, 3–4 mm long; stamens 3, shorter than the tepals. Capsule oval, straw-colored to light brown, 3–4 mm long, about as long as the tepals, tipped with a short, blunt point. June–Aug.

Wet sandy shores, streambanks and ditches; not in open bogs; Houghton County.

Juncus alpinoarticulatus Chaix
NORTHERN GREEN RUSH *native*
IR | **WUP** | CUP | EUP OBL CC 5

Juncus alpinus Vill.

Perennial rush, spreading by rhizomes. Stems in small clumps, 1.5–4 dm long. Leaves mostly from base of plant and with 1–2 stem leaves, round in section, hollow, with small swollen joints, 2–12 cm long and 0.5–1 mm wide; sheaths green to red, auricles rounded, 0.5–1 mm long; bract round in section, 2–6 cm long and shorter than the head. Flowers in an open panicle of 5–25 heads, 2–15 cm long and 1–5 cm wide, the heads oblong pyramid-shaped, 2–6 mm wide, mostly 2–5-flowered, the branches upright, 1–7 cm long; tepals green to brown, 2–3 mm long, the inner tepals shorter, the margins chaffy; stamens 6. Capsule oblong, 3-angled, straw-colored to chestnut brown, satiny, 2–3 mm long, slightly longer than the tepals, tapered to a rounded tip. June–Sept.

Sandy or gravelly shores, streambanks, fens; often where calcium-rich.

Juncus articulatus L.
JOINT-LEAF RUSH *native*
IR | **WUP** | **CUP** | **EUP** OBL CC 3

Perennial rush, with coarse white rhizomes. Stems usually tufted, 2–6 dm long. Leaves from stem and at base of plant, more or less round in section, hollow, with small swollen joints, 4–12 cm long and 1–3 mm wide; sheaths green or sometimes red, auricles rounded, about 1 mm long; bract erect, round in section, 1–4 cm long, shorter than the head. Flowers in open panicles, 4–10 cm long and 3–6 cm wide, composed of 3–30 heads, the heads rounded, 6–8 mm wide, 3–10-flowered, panicle branches erect to widely spreading, 1–4 cm long; tepals green to dark brown, 2–3 mm long; stamens 6. Capsule oval, dark brown, shiny, 3–4 mm long, longer than the tepals, tapered to a tip. July–Sept.

Sandy, gravelly or mucky shores, streambanks and springs.

Juncus balticus Willd.
SMALL-HEAD RUSH *native*
IR | **WUP** | **CUP** | **EUP** OBL CC 4

Juncus arcticus Willd.

Perennial rush, spreading by stout, brown to black rhizomes. Stems slender and tough, dark green, 3–9 dm long, in rows from the rhizomes. Leaves reduced to red-brown sheaths at base of stem; bract erect, round in section, 1–2 dm long, longer than the head and resembling a continuation of stem. Flowers single on stalks, in dense to spreading heads, the heads appearing lateral, extending outward from stem 1–7 cm; tepals lance-shaped, dark brown, 3–5 mm long, margins chaffy; stamens 6. Capsule ovate, somewhat 3-angled, red-brown, 3–4 mm long, shorter to slightly longer than the tepals, tapered to a sharp point. May–Aug.

Wet sandy or gravelly shores, interdunal wetlands near Lake Michigan, meadows, ditches, marshes, seeps.

Juncus brachycephalus (Engelm.) Buch.
SMALL-HEAD RUSH native
IR | WUP | CUP | EUP   OBL CC 7

Densely tufted perennial rush. Stems erect, round in section, 3–7 dm long. Leaves from stem and base of plant, round in section, 2–20 cm long and 1–2 mm wide, often spreading; auricles rounded, to 1 mm long; bract erect, round in section, 1–5 cm long, shorter than the head. Flowers in an open raceme or panicle of 10–80 heads, 5–25 cm long and 2–12 cm wide, the heads oval, 2–5 mm wide, 2–6-flowered, branches upright to spreading, 1–5 cm long; tepals lance-shaped, green to light brown, 3-nerved, 2–3 mm long, margins chaffy; stamens 3 or sometimes 6. Capsule ovate, more or less 3-angled, light brown, 3–4 mm long, longer than the tepals, abruptly narrowed to a short beak. June–Sept.

Sandy or gravelly shores, streambanks, open bogs, calcium-rich springs.

Juncus brevicaudatus (Engelm.) Fern.
NARROW-PANICLE RUSH native
IR | WUP | CUP | EUP   OBL CC 8

Densely tufted perennial rush. Stems erect, round in section, 1.5–5 dm long. Leaves from stem and base of plant, round in section, hollow, with small swollen joints, 3–20 cm long and 1–2 mm wide; sheaths green or sometimes red, auricles rounded, 1–2 mm long; bract erect, round in section, 2–7 cm long, shorter to longer than the head. Flowers in a raceme or panicle of 3–35 heads, 3–12 cm long and 1–4 cm wide, the heads oval, 2–6 mm wide, 2–7-flowered, branches upright, 0.5–3.5 cm long; tepals green to light brown, often red-tinged near tip, 3-nerved, 3–4 mm long, margins chaffy; stamens 3. Capsule oval, 3-angled, dark brown, 3–5 mm long, longer than the tepals, tapered to a sharp point. Aug–Sept.

Wet meadows, marshes, fens, sandy lakeshores, rocks along Lake Superior.

Juncus bufonius L.
TOAD RUSH native
IR | WUP | CUP | EUP   FACW CC 2

Small annual rush. Stems tufted, erect to spreading, 5–20 cm long. Leaves from stem and at base of plant, flat or channeled, 1–7 cm long and to 1 mm wide, usually shorter than stem; sheaths green to red or brown, auricles absent; bract erect, 1–10 cm long, shorter than the head. Flowers single, mostly stalkless, with 1–7 flowers along each branch of the inflorescence, the inflorescence comprising half or more of the entire length of plant; tepals lance-shaped, green to straw-colored, 4–6 mm long, margins chaffy; stamens 6. Capsule ovate, brown or green, 3–4 mm long, rounded at tip, shorter than the tepals. June–Aug.

Sandy or silty shores, mud flats, streambanks, wet compacted soil of trails and wheel ruts.

Juncus canadensis J. Gay
CANADIAN RUSH native
IR | WUP | CUP | EUP   OBL CC 6

Tufted perennial rush. Stems erect, rigid, round in section, 3–9 dm long. Leaves from stem and at base of plant, round in section, hollow, with small swollen joints, 3–20 cm long and 1–3 mm wide; sheaths green to red, auricles rounded, 1–2 mm long; bract erect, round in section, 3–7 cm long and shorter than the head. Flowers in an open or crowded raceme or panicle of few to many heads, the heads more or less round, 3–8 mm wide, with 5–40 or more flowers, the branches upright, 1–10 cm long; tepals narrowly lance-shaped, green to brown, 3–5 mm long; stamens 3. Capsule ovate, 3-angled, light to dark brown, 3–5 mm long, equal or longer than the tepals, rounded to a short tip. July–Sept.

Sandy, muddy or mucky shores, marshes, streambanks, thickets, ditches.

Juncus dudleyi Wieg.
DUDLEY'S RUSH native
IR | WUP | CUP | EUP   FACW CC 1

Juncus tenuis var. *dudleyi* (Wiegand) F.J. Herm.
Tufted perennial rush, from branching rhizomes. Stems 2–10 dm long. Leaves basal, 2–3; auricles carilaginous, yellowish, to 0.4 mm long; blade flat, 5–30 cm long; bract usually longer than the head. Inflorescences compact to loose and lax; primary bract usually exceeding inflorescence. Capsules tan, ca. 3 mm long.

Damp to drier open places, including lakeshores, marsh margins, ditches.

Juncus effusus L.
LAMP RUSH native
IR | WUP | CUP | EUP   OBL CC 3

Densely tufted perennial rush. Stems erect, round in section, to about 1 m long. Leaves reduced to bladeless sheaths at base of stem, the sheaths to 2 dm long, mostly red-brown; bract round in section, 10–30 cm long, appearing like a continuation of stem, longer than the head. Flowers in a many-flowered inflorescence, with 2–4 flowers along each branch of the inflorescence, the inflorescence appearing lateral, the branches upright to spreading or bent downward; tepals lance-shaped, green to straw-colored, 2–3 mm long; stamens 3. Capsule broadly ovate, olive-green to brown, 2–3 mm long, about as long as tepals, sometimes tipped with a short point. June–July.

Marshes, shores, thickets, streambanks, bog margins, wet meadows.

ADDITIONAL SPECIES

Juncus pylaei Laharpe, sometimes treated as a species or as *J. effusus* var. *pylaei* (Laharpe) Fern. & Weig., and reported from Chippewa and Houghton counties. *J. effusus* has soft, broad stems rather smooth because the ridges are tiny and numerous and readily flattened by pressing; *J. pylaei* has fewer, prominent ridges; and stems that are narrower, sturdier, and generally not flattened by pressing.

Juncus ensifolius Wikstr.
DAGGER-LEAF RUSH *introduced*
IR | **WUP** | **CUP** | EUP   FACW

Perennial rush, spreading by rhizomes. Stems single or in loose clumps, erect, flattened and narrowly winged, 2–6 dm long. Leaves from stem and at base of plant, the blade folded along midrib, the margins joined, with one edge turned toward the stem, 5–12 cm long and 2–6 mm wide; sheaths green or red, with broad chaffy margins, auricles absent; bract erect, 1–4 cm long, shorter than the head. Flowers in an open panicle of 3–11 heads, the heads more or less round, 8–10 mm wide, with 15 or more flowers, the branches upright, 1–10 cm long; tepals lance-shaped, straw-colored to red-brown, 3–4 mm long; stamens 3 (or 6). Capsule oval, dark brown, 3–4 mm long, tapered to a short beak, shorter to longer than the tepals. July–Sept.

Margins of streams, ponds and springs.

Disjunct from main range of western USA in the Upper Peninsula (Gogebic, Ontonagon, Schoolcraft counties) and in Ashland County, Wisconsin.

Juncus filiformis L.
THREAD RUSH   *native*
IR | **WUP** | **CUP** | **EUP**   FACW CC 10

Perennial rush, with short or long rhizomes. Stems tufted or in rows from the rhizomes, erect, round in section, 1–5 dm long. Leaves reduced to bladeless sheaths at base of stem, the sheaths pale brown, to 6 cm long; bract erect, round in section, 6–20 cm long, appearing to be a continuation of stem, longer than the head. Flowers in an branched inflorescence, 1–3 cm long, with 1–3 flowers along each branch of the inflorescence, the inflorescence appearing lateral, the branches erect to spreading, to 1 cm long; tepals lance-shaped, green to straw-colored, 2–3 mm long, margins chaffy; stamens 6. Capsule broadly ovate, light brown, 2–3 mm long, slightly longer than the tepals, tipped by a short beak.

Sandy, mucky, or gravelly shores, streambanks, thickets.

Juncus greenei Oakes & Tuckerman
GREENE'S RUSH   *native*
IR | **WUP** | **CUP** | **EUP**   FAC CC 7

Perennial rush, from a short rhizome. Stems tufted, 2–8 dm long. Basal leaves filiform, nearly terete, 5–20 cm long; involucral leaves similar, length varies from 2–15 cm long. Inflorescence small and compact, obpyramidal, 2–5 cm high. Perianth segments oblong lance-shaped, appressed, acute or aristulate; sepals 2.3–3.5 mm long; petals 1.9–3.4 mm long. Capsule ovoid-cylindric, 3–4 mm long, truncate, at maturity conspicuously exceeding the perianth.

Moist to dry sandy open places: shores, swales, fields, clearings, dunes.

Juncus marginatus Rostk.
GRASS-LEAF RUSH   *native*
IR | **WUP** | CUP | EUP   FACW CC 8
Juncus biflorus Ell.

Perennial rush, spreading by rhizomes. Stems single or in small clumps, erect, compressed, 2–5 dm long, bulblike at base. Leaves from base of plant and on stem, flat, grasslike, 2–30 cm long and 1–3 mm wide; sheaths green, membranous on margins, auricles rounded, to 0.5 mm long; bract erect to spreading, flat, 1–8 cm long, shorter to slightly longer than the head. Flowers in an open panicle, 2–8 cm long and 1–6 cm wide, composed of 5–15 heads, the heads rounded, 3–6 mm wide, 6–20-flowered, branches upright, 0.5–2.5 cm long; tepals lance-shaped, green with red spots, 2–3 mm long, margins chaffy; stamens 3. Capsule more or less round, brown with red spots, 2–3 mm long, slightly longer than the tepals, rounded at tip. June–Aug.

Sandy shores and streambanks, wet meadows, marshes; Houghton County.

Juncus nodosus L.
INLAND RUSH   *native*
IR | **WUP** | **CUP** | **EUP**   OBL CC 5

Perennial rush, spreading by rhizomes. Stems erect, slender, round in section, 1.5–6 dm long. Leaves on stem and one at base of plant, round in section, hollow, with small swollen joints, 3–30 cm long and 1–2 mm wide, upper leaves usually longer than the head; sheaths green, their margins green, becoming yellow and membranous toward tip, auricles rounded, yellow, 0.5–1 mm long; bract erect to spreading, round in section, 2–12 cm long, usually much longer than head. Flowers in a raceme or panicle of several heads, 1–6 cm long and 1–3 cm wide, the heads more or less round, 6–10 mm wide, 6–20-flowered, the branches erect to spreading, 0.5–3 cm long; tepals narrowly lance-shaped, green to light brown, 3–4 mm long; the margins narrowly translucent; stamens 6. Capsule awl-shaped, brown, 4–5 mm long, longer than the tepals, tapered to a sometimes curved beak. July–Sept.

Sandy, gravelly or clayey shores and streambanks, wet meadows, fens, ditches, springs; often where calcium-rich.

JUNCACEAE | Rush Family

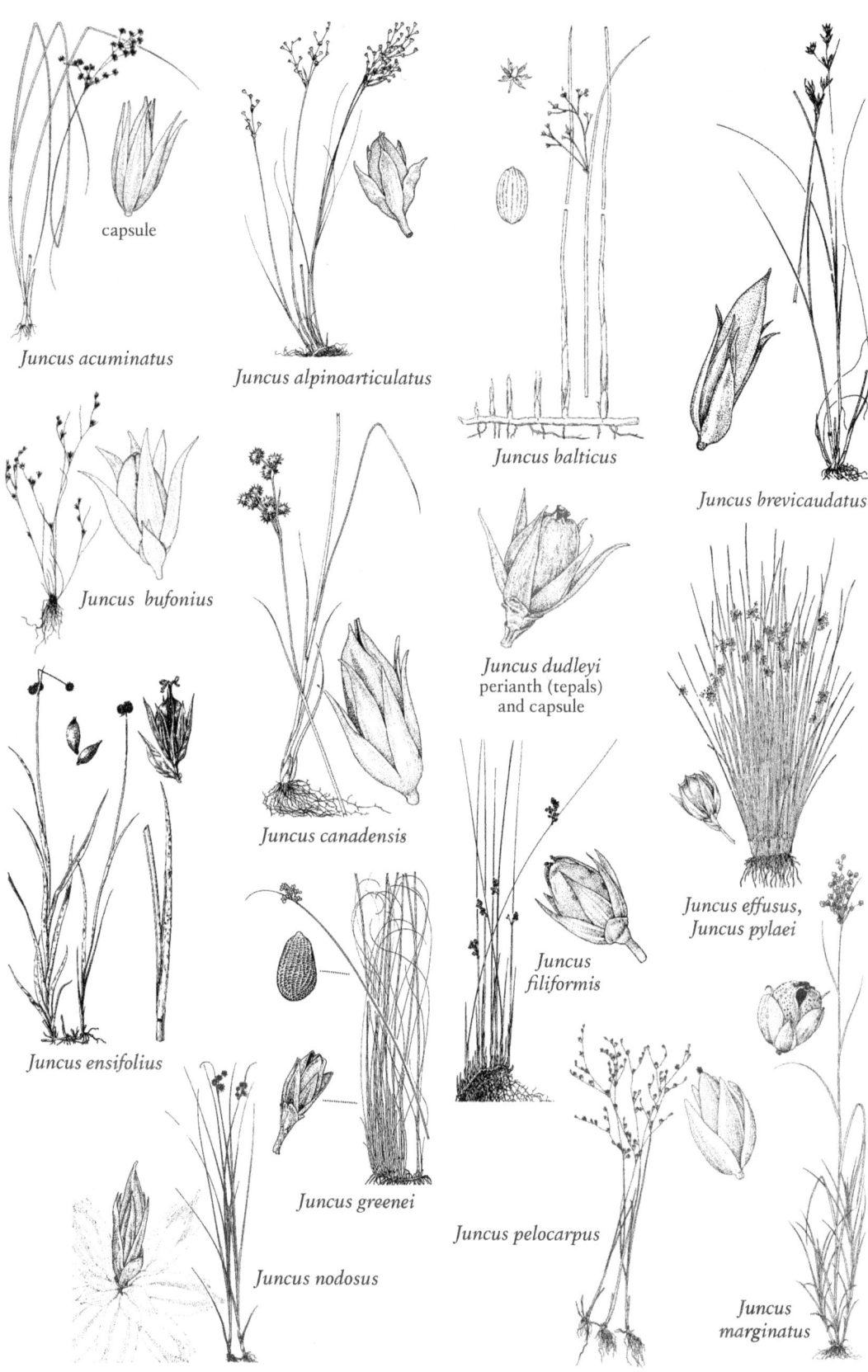

JUNCACEAE | Rush Family

Juncus pelocarpus E. Mey.
BROWN-FRUIT RUSH — native
IR | WUP | CUP | EUP OBL CC 8

Perennial rush, spreading by rhizomes and forming colonies. Stems erect, round in section, 1–4 dm long. Leaves from stem and at base of plant, round in section, very slender, 2–10 cm long and about 1 mm wide; auricles absent or short and straw-colored; bract erect, round in section, 2–4 cm long, shorter than the head. Flowers single or paired in a much-branched inflorescence, 5–15 cm long and 4–10 cm wide, the flowers on mostly 1 side of each branch, the branches upright to widely spreading, 1–4 cm long, with at least some of the flowers usually replaced by clusters of awl-shaped leaves; tepals ovate, dark brown, about 2 mm long, margins chaffy; stamens 6. Capsule narrowly ovate, dark brown, satiny, 2–3 mm long, equal or slightly longer than tepals, tapered to a slender beak. July–Aug.

Shallow water, sandy or mucky shores, bog margins.

Juncus stygius L.
MOOR RUSH — THR native
IR | WUP | CUP | EUP OBL CC 10

Perennial rush, from slender rhizomes. Stems single or few together, erect, round in section, 1–4 dm long. Leaves 1–3 from near base of plant, with 1 leaf above middle of stem, round in section or somewhat flattened, 3–15 cm long and 0.5–2 mm wide; auricles short and rounded or absent; bract erect, round in section, 1–2 cm long, shorter than the head. Flowers in an inflorescence of 1–3 heads, the heads obovate, 5–10 mm wide, 1–4-flowered, branches erect, to 1 cm long; tepals lance-shaped, straw-colored to red-brown, 4–5 mm long, margins chaffy; stamens 6, nearly as long as the tepals. Capsule oval, 3-angled, green-brown, 6–8 mm long, longer than the tepals, tipped with a distinct point.

Rare in open bogs, marshes, and shallow water.

Juncus tenuis Willd.
POVERTY RUSH — native
IR | WUP | CUP | EUP FAC CC 1

Tufted, perennial rush. Stems erect, round in section to slightly flattened, 1–6 dm long. Leaves near base of stem, flat to broadly channeled, 10–15 cm long and to 1 mm wide; sheaths green, the margins yellow and glossy, auricles triangular, 1–3 mm long; bracts 1–3 (usually 2), the lowest erect, flat, 6–10 cm long, longer than the head. Flowers stalkless or on short stalks to 3 mm long, on branches with 1–7 flowers, in a crowded to spreading head 2–5 cm long; tepals lance-shaped, green to straw-colored or light brown, 3–5 mm long, margins narrowly translucent; stamens 6. Capsule ovate, green to straw-colored, 2–5 mm long, shorter or equaling the tepals, rounded at tip. June–July.

Wet meadows, shores, streambanks, springs, common in disturbed places (often where soils compacted) such as trails, roadsides, ditches; also in drier woods and meadows.

Juncus torreyi Coville
TORREY'S RUSH — native
IR | WUP | CUP | EUP FACW CC 4

Perennial rush, from tuber-bearing rhizomes. Stems single, erect, round in section, 4–8 dm long. Leaves from stem and base of plant, round in section, hollow, with small swollen joints, 15–30 cm long and 1–2 mm wide, the upper leaves often longer than the head; sheaths green, the margins white and translucent, auricles 1–3 mm long; bract erect or spreading, round in section, 4–12 cm long, longer than the head. Flowers a crowded, rounded raceme or panicle of 3–23 heads, 2–5 cm long and as wide, the heads round, 10–15 mm wide, with 25 to many flowers, branches erect to spreading 1–4 cm long; tepals narrowly lance-shaped, green to brown, 3–5 mm long, margins narrowly translucent; stamens 6.

Juncus stygius

Juncus tenuis

Juncus torreyi

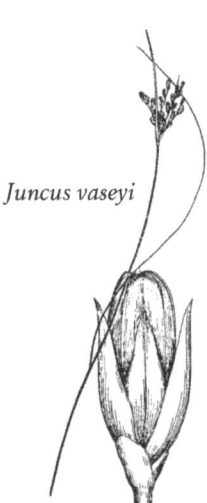

Juncus vaseyi

Capsule awl-shaped, brown, 4–6 mm long, equal or longer than the tepals, tapered to a short beak. June–Sept.

Sandy shores, streambanks, wet meadows, marsh borders, springs, ditches; Chippewa and Menominee counties.

Juncus vaseyi Engelm.
VASEY'S RUSH (THR) *native*
IR | WUP | **CUP** | **EUP** FACW CC 10
Juncus greenei Oakes & Tuckerman var. *vaseyi* (Engelm.) Boivin

Tufted perennial rush. Stems erect, 2–6 dm long. Leaves all at base of plant, round in section, solid, narrowly channeled on upper surface, to 3 dm long and to 1 mm wide, usually shorter than stem; sheaths green or red, the margins membranous, auricles short or absent; bract upright, usually shorter than the head. Flowers single, stalkless or on short stalks, in a crowded inflorescence, 1–4 cm long; tepals lance-shaped, green to light brown, 4–6 mm long, margins narrowly translucent; stamens 6. Capsule cylindric, 1–2 mm long, equal or slightly longer than the tepals. July–Aug.

Rare in wet meadows, sandy shores.

Luzula *Wood-Rush*

Perennial grasslike herbs, with narrow, flat, more or less pubescent leaves often involute toward the tip, and an umbel-like or spike-like inflorescence, the flowers solitary, rarely paired, or glomerulate. Perianth as in *Juncus*, white, green, or brown, often scarious. Stamens 6. Ovary and capsule 1-celled; ovules 3. Seeds plump, ellipsoid, tipped by a short appendage (caruncle).

Distinguished from *Juncus* by the presence of few to many long hairs on the leaves, especially toward the base of the blades.

ADDITIONAL SPECIES
Luzula pallidula Kirschner (Eurasian Wood-rush), Eurasian perennial, in Michigan known only from a single moist grassy location near Manistique, Schoolcraft County.
Luzula parviflora (Ehrh.) Desv. (Small-Flowered Woodrush), native perennial, known from moist open woods and shores in Alger and Keweenaw counties (including Isle Royale); threatened.

1 Flowers grouped in heads or short dense spikes; capsule usually no longer than the perianth. . . . . . **L. multiflora**
1 FFlowers single at ends of inflorescence branches; capsule at maturity slightly longer than perianth 2
2 Inflorescence umbelliform, bases of the primary branches close together; branches unforked or at most with 1 or 2 flowers in addition to terminal flower; mature perianth 2.5–4 mm long; outer end of seed with conspicuous appendage . . . . **L. acuminata**
2 Inflorescence cymose, primary branches arising from bases distinctly spaced along a short axis; branches forked at least once or twice; mature perianth 2–2.5 mm long; seeds without appendage . . . . **L. parviflora***

Luzula acuminata Raf.
HAIRY WOOD-RUSH *native*
IR | WUP | **CUP** | **EUP** FACU CC 5

Stems tufted, often stoloniferous 1–4 dm tall. Basal leaves elongate, to 3 dm long and 10 mm wide; stem leaves 2–4, shorter and somewhat narrower, all sparsely pilose, with a blunt callous apex. Inflorescence 3–6 cm high, the loosely spreading, almost filiform pedicels mostly simple with a single terminal flower, or a few with a lateral flower, or a few branched and bearing 3 or 4 flowers; perianth segments lance-shaped, usually chestnut-brown in the center, with scarious margins, 2.6–4.3 mm long, the petals slightly exceeding the sepals, Capsule ovoid, 3.2–4.5 mm long, mucronate; seeds purple-brown, the body about 1 mm long, with a pale appendage nearly as long. May–June.

Moist or dry woods and forest openings, meadows, streambanks, hillsides.

Luzula multiflora (Ehrh.) Lej.
COMMON WOOD-RUSH *native*
IR | WUP | **CUP** | EUP FACU CC 5
Luzula campestris var. *multiflora* (Ehrh.) Celak.

Stems loosely tufted, 2–4 or rarely 5 dm tall. Basal leaves several, stem leaves 2 or 3, flat, except toward the callous-pointed tip, 2–6 mm wide. Inflorescence usually with a few slender peduncles and 1 or more sessile glomerules; bracts usually scarious toward the acute tip, not elongate; perianth segments lance-shaped, 2–3.7 mm long. Capsule obovoid, from somewhat shorter to somewhat longer than the perianth; seeds ellipsoid, 1–1.3 mm long, with an ovate-triangular appendage 0.4–0.5 mm long.

Forests, sometimes with *L. acuminata*; also in swamps and moist grassy areas.

Luzula acuminata

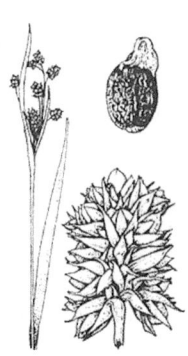

Luzula multiflora

Juncaginaceae
ARROW-GRASS FAMILY

Triglochin *Arrow-Grass*

Grasslike perennial herbs, clumped from creeping rhizomes, often in brackish habitats. Stems slender, leafless. Leaves all from base of plant, slender, linear, round or somewhat flattened in section, sheathing at base. Flowers perfect, regular, on short stalks in a spike-like raceme at end of stem; flower segments (tepals) 6; stigmas 3 or 6, styles short or absent; stamens 6, anthers stalkless, nearly as large as tepals. Fruit of 3 or 6 carpels, these splitting when mature into 1-seeded segments.

1 Plants generally small and slender; stigmas 3; fruit linear, clublike toward tip .... **T. palustris**
1 Plants larger, usually 3 dm or more tall; stigmas 6; fruit short-cylindric .... **T. maritima**

Triglochin maritima L.
SEASIDE ARROW-GRASS *native*
IR | WUP | CUP | EUP OBL CC 8

Tufted perennial herb, from a thick crown and spreading by rhizomes. Stems more or less round in section, leafless, 2–8 dm long. Leaves upright to spreading, somewhat flattened, to 5 dm long and 1–3 mm wide. Flowers 2–3 mm wide, in densely flowered, spike-like racemes 1–4 dm long; the flowers on upright stalks 4–6 mm long, the stalks extending downward on the stem as a wing; tepals 6, 1–2 mm long; stigmas 6; stamens 6. Fruit of 6 ovate carpels, 2–5 mm long and 1–3 mm wide, the carpel tips curved backward. June–Aug.

Sandy, gravelly, or marly lakeshores and streambanks; marshes, brackish wetlands. Plants larger than marsh arrow-grass (*T. palustris*) and the fruit ovate rather than linear.

Triglochin palustris L.
MARSH ARROW-GRASS *native*
IR | WUP | CUP | EUP OBL CC 8

Small tufted perennial herb. Stems slender, leafless, 2–4 dm long. Leaves erect, round in section, to 3 dm long and 1–2 mm wide. Flowers small, 1–2 mm wide, in loosely flowered racemes, 10–25 cm long; the flowers on erect stalks, 2–5 mm long; tepals 6, 1–2 mm long; stigmas 3; stamens 6. Fruit of 3 narrow, clublike carpels, 5–8 mm long and 1 mm wide, splitting upward from base into 3 segments. June–Sept.

Sandy, gravelly, or marly lakeshores and streambanks, calcareous fens, marshes, interdunal swales; often where calcium-rich; in the western UP, known only from Keweenaw County.

Liliaceae
LILY FAMILY

Perennial herbs, from corms, bulbs or rhizomes. Stems leafy or leafless. Leaves linear to ovate, usually from base of plant, sometimes along stem, alternate to opposite or whorled. Flowers perfect (with both staminate and pistillate parts), regular; sepals and petals of 6 petal-like tepals in 2 series of 3; stamens 6; ovary superior or inferior, 3-chambered. Fruit a capsule or round berry.

Under the Angiosperm Phylogeny Group III system (APG III), genera in the Liliaceae have been placed into various new familes (see page 3). However, the family designations are still in a state of flux, and may change in the future. As a convenience, our genera are retained within the traditional Lily Family grouping, with the proposed new family name noted in parentheses after each genus name.

1 Sepals and petals of quite different color and/or texture, the former green or brownish .... **Trillium**
1 Sepals and petals both colored and petal-like, usually similar in shape (tepals) or the sepals (in Iris) of different size and shape 2
2 Ovary inferior (flowers bisexual) 3
2 Ovary superior (or flowers unisexual) 4
3 Ovary clearly inferior; uncommon garden escape .... **Narcissus**
3 Ovary half-inferior, part of it adnate to the perianth, glabrous (at most granular-roughened) ...... **Anticlea**
4 Flowers or inflorescences lateral, arising from the axils of alternate stem leaves or scales 5
4 Flowers or inflorescences terminal on scapes or leafy (simple or branched) stems........................ 6

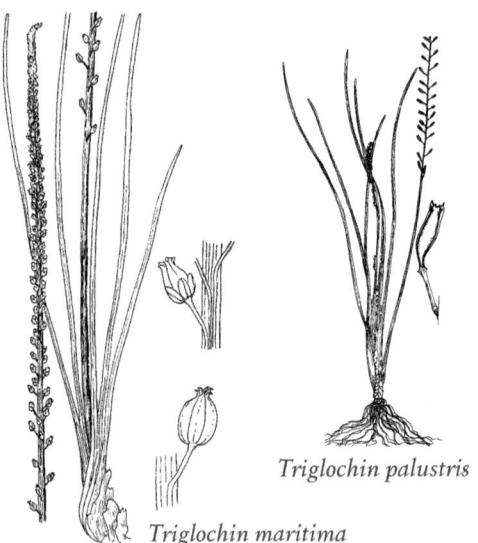

Triglochin palustris

Triglochin maritima

LILIACEAE | Lily Family

5. Leaves scale-like, mostly brownish or yellowish, those on the much-branched upper portion of the plant subtending short green filiform branches (often mistaken for leaves) . . . . **Asparagus**
5. Leaves broad, flat, green (scale-like leaves or bracts may be present in addition to normal leaves) . . . . (go to 24)
6. Leaves all withering before plant flowers . . . . **Allium tricoccum**
6. Leaves present at flowering time 7
7. Leaves all in one or two whorls on the stem . . **Medeola**
7. Leaves alternate or basal, or if in whorls these more than 2 or some alternate leaves also present 8
8. Flowers more than 3.5 cm long 9
8. Flowers less than 3.5 cm long 13
9. Leaves perfoliate . . . . **Uvularia grandiflora**
9. Leaves not perfoliate 10
10. Principal leaves cauline, not crowded toward base of plant . . . . **Lilium**
10. Principal leaves basal or nearly so; stem leafless above or with very small bracts 11
11. Flowers 2–many . . . . **Hemerocallis**
11. Flowers solitary 12
12. Leaves basal . . . . **Erythronium**
12. Leaves cauline . . . . **Lilium**
13. Flowers in a many (7–60 or more) flowered umbel on an unbranched stem or scape; plants with odor of onion or garlic . . . . **Allium**
13. Flowers solitary or in a raceme, panicle, or corymb, on a simple or branched stem (if in a few-flowered umbel, then either the stem branched or forked or flowers less than 7) 14
14. Plants with principal leaves clearly along the stem; basal leaves (at least of current season) absent or at most apparently one 15
14. Plants with the principal leaves all basal (at ground level) or nearly so; stem leaves absent, reduced to bracts, or much smaller or fewer than basal leaves . . 18
15. Stem forked or branched; perianth over 12 mm long . . . . (go to 24)
15. Stem unbranched (above the ground and below the inflorescence); perianth usually less than 12 mm long 16
16. Ovary with 1 style; fruit a berry; inflorescence a raceme (tepals up to 7 mm long) or if a panicle, the tepals less than 3 mm long; leaves ovate . . . . (go to 24)
16. Ovary with 3 styles (one on each lobe); fruit a capsule; inflorescence a panicle; tepals 5–13 mm long; leaves very elongate 17
17. Plants with at least some principal leaves clearly along the stem; inflorescences paniculate (except in depauperate individuals) . . . . **Anticlea**
17. Plants with the principal leaves all basal (at ground level) or nearly so, stem leaves absent, reduced to bracts, or much smaller or fewer than basal leaves; inflorescences racemose . . . . **Triantha**
18. Flower solitary . . . . **Erythronium**
18. Flowers 2 or more in an inflorescence 19
19. Tepals united for half or more of their length 20
19. Tepals completely separate or united at base only . . . 21
20. Perianth blue, less than 6 mm long; leaves linear; plants bulbous . . . . **Muscari**
20. Perianth white, 5–10 mm long when mature; leaves lanceolate to elliptic; plants not bulbous . . **Convallaria**
21. Ovary with 3 styles (1 on each lobe) in bisexual or pistillate flowers (or flowers all staminate) 22
21. Ovary with a single style; flowers bisexual 23
22. Plants with at least some principal leaves clearly cauline; inflorescences paniculate . . . . **Anticlea**
22. Plants with the principal leaves all basal or nearly so, stem leaves absent, reduced to bracts, or at most much smaller or fewer than basal leaves; inflorescences racemose . . . . **Triantha**
23. Leaves broad (over 3 cm wide); perianth yellow; fruit a blue berry . . . . **Clintonia**
23. Leaves long and narrow (less than 1.5 cm wide); perianth white; fruit a capsule . . . . **Ornithogalum**

Lead 24. Fruit red, blue, or black berries, except in *Uvularia*, which has a capsule.

24. Leaves all in 1 or 2 whorls on the stem . . . . **Medeola**
24. Leaves alternate or basal 25
25. Plants with the leaves all basal or nearly so; stem leaves absent, reduced to bracts, or at most much smaller or fewer than the basal leaves 26
25. Plants with leaves clearly along the stem 27
26. Flowers yellow, in an umbel; fruit blue . . . . . . **Clintonia**
26. Flowers white, in a raceme; fruit (rarely produced) orange or red . . . . **Convallaria**
27. Plant unbranched 28
27. Plant branched (above the ground) 29
28. Flowers in a terminal raceme or panicle, white, tepals separate; ripe fruit red, with dark stripes in one species . . . . **Maianthemum**
28. Flowers in the axils of leaves, greenish, greenish white, or yellowish, tepals united most of their length; ripe fruit blue to black . . . . **Polygonatum**
29. Perianth pale to deep yellow; fruit a glabrous capsule; stem and pedicels glabrous . . . . **Uvularia**
29. Perianth greenish, rose-purple, white, or creamy, fruit a pubescent to glabrate or tuberculate red berry; stem (at least when young) and pedicels often pubescent . . . . **Streptopus**

Allium *Onion; Leek; Garlic*
[ALLIACEAE]

Biennial or perennial herbs from a coated bulb, with a strong odor of onion or garlic, the leaves usually narrow, basal, or on the lower part of the stem, the scape-like stem erect, terminated by an umbel sub-

LILIACEAE | Lily Family

tended by 1–3 bracts, the flowers white to pink or purple, in some species wholly or partly replaced by sessile bulblets. Flowers perfect, in umbels. Perianth segments 6, uniform in color, but the inner circle often somewhat different in shape or size, withering and persistent below the capsule. Stamens 6. Ovary 3-celled. Capsule short, ovoid, globose, or obovoid, 3-lobed, loculicidal; seeds black, 1 or 2 in each cell. *Allium* includes the cultivated onion, garlic, chives, and leek.

1. Leaves usually over 2 cm wide, flat, petiolate, withering before the plant flowers . . . . **A. tricoccum**
1. Leaves linear, flat or terete, less than 2 cm wide (usually less than 1 cm), not petiolate, present at flowering time 2
2. Umbel nodding, on bent or reflexed tip of scape; leaves flat. . . . . **A. cernuum**
2. Umbel erect on straight tip of scape; leaves flat or terete 3
3. Leaf blades terete, hollow, at least most of their length (else flattened where pressed in drying, but the base of blade, just above summit of sheath, will not show 2 distinct surfaces). . . . . **A. schoenoprasum**
3. Leaf blades flat (sometimes keeled). . . . . . **A. canadense**

Allium canadense L.
MEADOW GARLIC native
IR | WUP | **CUP** | EUP FACU CC 4

Bulb ovoid-conic, 1–3 cm long, its coats fibrous-reticulate. Stems erect, stout, 2–6 dm tall, leafy in the lower third. Leaves elongate, flat, commonly 2–4 mm, occasionally to 7 mm wide; bracts 2 or 3, broadly ovate, acuminate. Umbels bearing bulblets only, or with 2–5 flowers also; bulblets ovoid to fusiform, to 1 cm long; pedicels slender, 15–30 mm long; perianth segments pink or white, oblong lance-shaped, 6–9 mm long, acute. Capsule rarely developed, subglobose, not crested. May, June.

Moist or dry open woods and meadows; Mackinac and Menominee counties.

Allium cernuum Roth
NODDING ONION native
IR | WUP | **CUP** | EUP FACU CC 5

Bulb slenderly conic, very gradually tapering into the stem. Leaves several, arising near together at the surface of the soil, shorter than the stem, commonly 2–4 mm, occasionally to 8 mm wide. Scape 3–6 dm tall, abruptly declined or decurved near the summit. Umbel nodding, many-flowered, without bulblets; pedicels 12–25 mm long, becoming rigid in fruit; perianth segments white to rose, ovate or elliptic, 4–6 mm long, obtuse or subacute; stamens exsert; filaments barely widened toward the base. Capsule obovoid, 3-lobed, about 4 mm long, each valve (and each lobe of the ovary) bearing 2 erect triangular processes near the summit. July, Aug.

Dry woods and prairies; Menominee County.

Allium schoenoprasum L.
WILD CHIVES (THR) native-introduced
IR | **WUP** | **CUP** | EUP FACU CC 10

Bulb slender, often scarcely thicker than the stem. Stems 2–5 dm tall. Leaves erect, terete, hollow, the longest nearly equaling the stem. Umbel compact, hemispheric, subtended by 2 ovate bracts; pedicels 3–7 mm long. Flowers numerous; perianth segments bright rose–color, ovate to lance-shaped, 10–14 mm long, acuminate, prominently 1-nerved. Capsule ovoid, 3-lobed, about half as long as the perianth. June–July.

Rocky Lake Superior shore, rocky riverbanks.

Subsp. *schoenoprasum* is the cultivated chives, common in gardens, and not easily distinguished from the uncommon native var. *sibiricum* (L.) Hartm.

Allium tricoccum Ait.
WILD LEEK native
IR | **WUP** | **CUP** | EUP FACU CC 5

Bulb ovoid, 2–6 cm long, its coats finely fibrous-reticulate. Leaves 2–3 dm long, including the slender petiole, the blades flat, lance-elliptic, 1–2 dm long, 2–6 cm wide. Scape 1.5–6 dm tall. Umbel erect, subtended by 2 ovate deciduous bracts; pedicels 1–2 cm long; perianth segments ovate to oblong-obovate, white, 5–7 mm long obtuse, about equaling the stamens; filaments greatly widened toward the base. Capsule depressed, deeply 3-lobed, each valve often gibbous on the back below the middle.

Rich woods, often in large colonies.

The leaves develop in early spring and disappear before the flowers appear in June and July.

Anticlea *Death-Camas*
[MELANTHIACEAE]

Anticlea elegans (Pursh) Rydb.
MOUNTAIN DEATH-CAMAS native
IR | WUP | **CUP** | EUP FACW CC 10

Zigadenus elegans Pursh

Perennial herb, from an ovate bulb; plants waxy, especially when young. Stems erect, 2–6 dm long. Leaves mostly from base of plant, linear, 2–4 dm long and 4–12 mm wide; stem leaves much smaller. Flowers green-yellow or white, in a raceme or panicle, 1–3 dm long, the branches upright, subtended by large, lance-shaped, green or purplish bracts; tepals 6, obovate, 7–12 mm long, usually purple-tinged near base; stamens 6. Fruit an ovate capsule, 10–15 mm long; seeds 3 mm long. July–Aug.

Sandy or rocky shores of Great Lakes, open bogs, calcareous fens. Highly toxic.

monocots LILIACEAE | Lily Family 437

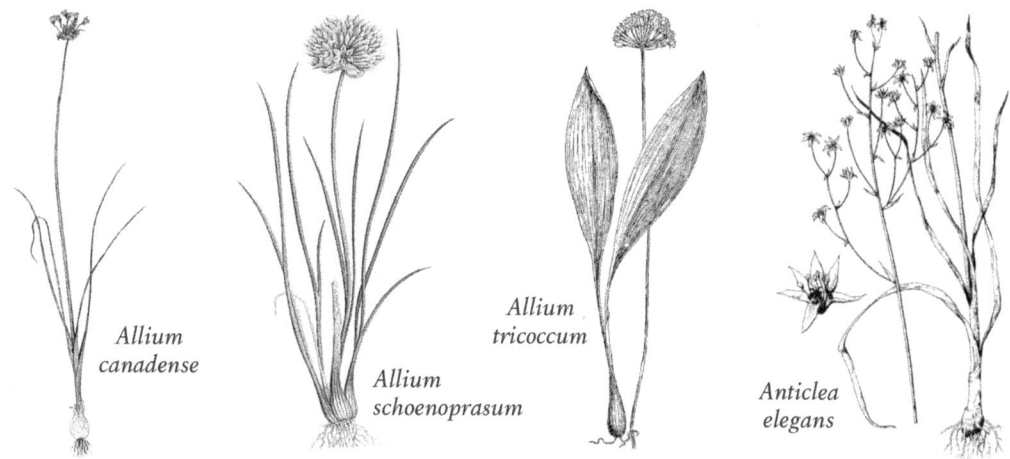

Allium canadense

Allium schoenoprasum

Allium tricoccum

Anticlea elegans

Asparagus *Asparagus*
[ASPARAGACEAE]

Asparagus officinalis L.
ASPARAGUS *introduced*
IR | **WUP** | **CUP** | **EUP** FACU

Perennial herb of various forms, the leaves reduced to small scales and replaced functionally by branches sometimes leaflike in appearance. Stems perennial from a rhizome, freely branched, to 2 m tall; ultimate branches filiform, 8–15 mm long; pedicels solitary or paired, lateral, 5–10 mm long, jointed in the middle. Flowers perfect or unisexual, greenish white, campanulate, 3–5 mm long; stamens 6; stigmas 3; ovary 3-celled. Fruit a red, spherical berry, about 8 mm wide, with a few large rounded seeds. May–June.

Native of Europe; commonly cultivated in home and truck gardens and escaped in waste places.

Clintonia *Bluebead-Lily*
[LILIACEAE]

Clintonia borealis (Ait.) Raf.
YELLOW BLUEBEAD-LILY *native*
IR | **WUP** | **CUP** | **EUP** FAC CC 5

Perennial herbs from a rhizome, bearing 2–4 ample basal leaves, the bases sheathing a leafless erect scape bearing a few-flowered umbel of conspicuous flowers. Leaves 2–5, dark glossy green, oblong to elliptic or obovate, eventually to 3 dm long, abruptly acuminate, finely ciliate. Scape 1.5–4 dm tall, usually pubescent at the summit, or glabrous at maturity. Umbel 3–8-flowered; pedicels 1–3 cm, long, erect

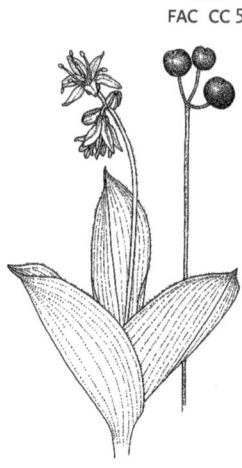

Clintonia borealis

in fruit, softly pubescent. Flowers perfect, nodding; perianth segments distinct, narrow, greenish yellow, 15–18 mm long; stamens 6, inserted on the very base of the perianth; ovary 3-celled; stigma obscurely 3-lobed. Fruit a blue berry, rarely varying to white, spherical, about 8 mm wide, containing a few to several seeds. May–June.

Rich moist woods and swamps.

Convallaria *Lily-of-the-Valley*
[ASPARAGACEAE]

Convallaria majalis L.
EUROPEAN LILY-OF-THE-VALLEY *introduced*
IR | **WUP** | **CUP** | **EUP**

Perennial herb from a rhizome, the short stem bearing a few leafless sheaths and 2 or 3 broad leaves, the scape terminating in a bracted raceme. Leaves narrowly elliptic, to 2 dm long, acuminate. Scape 1–2 dm tall. Raceme loosely flowered, one-sided; bracts small, lance-shaped; pedicels drooping. Flowers perfect, white, fragrant, 6–9 mm long; perianth globose-campanulate, with 6 short recurved lobes; stamens 6, inserted on the perianth near its base; style straight, included; ovary 3-celled. Fruit a spherical, many-seeded red berry, about 1 cm wide. May.

Widely distributed in n Eurasia, commonly cultivated, and occasionally escaped near gardens.

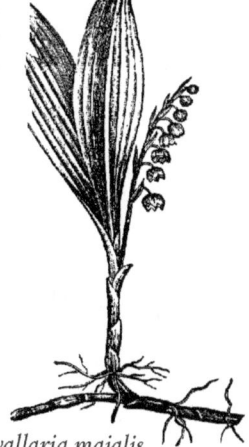

Convallaria majalis

Erythronium
Trout-Lily; Fawn-Lily [LILIACEAE]

Perennial from a deep solid corm, the slender stem about half underground; leaves borne near middle

LILIACEAE | Lily Family

of the stem and therefore appearing basal, usually mottled with brown, lance-shaped to oblance-shaped or elliptic; scape bearing a single nodding flower. Our species grow in colonies, producing numerous 1-leaved sterile plants and a few 2-leaved fertile ones; reproduction often vegetatively by bulbs at the tip of slender lateral offshoots from the corm or stem. Perianth segments (tepals) separate to the base but connivent, lance-shaped, at anthesis spreading and usually eventually recurved. Stamens 6. Ovary 3-celled; style slender below, thickened above to the 3 short stigmas. Fruit an obovoid to oblong capsule.

1 Flowers white............................ **E. albidum**
1 Flowers yellow...................... **E. americanum**

Erythronium albidum Nutt.
SMALL WHITE FAWN-LILY *native*
IR | **WUP** | CUP | EUP FACU CC 7

Scape stout, 1–2 dm tall. Perianth segments (tepals) normally bluish white, varying to light pink, often suffused with green or blue externally, yellow at base within, without marginal glands; stigmas stout, separate, divergent from the linear-clavate style. April–May.

Moist woods; Houghton and Ontonagon counties.

Erythronium americanum Ker-Gawl.
TROUT-LILY *native*
IR | **WUP** | **CUP** | **EUP** CC 5

Scape stout, 1–2 dm tall. Perianth segments (tepals) normally yellow, often spotted toward the base within or darker colored without, the petals bearing a glandular spot on each margin near the base; stigmas very short, scarcely separate, terminating the club-shaped style. April–May.

Moist woods.

Hemerocallis *Day-Lily*
[XANTHORRHOEACEAE]

Tall perennial herbs, with numerous, elongate, linear, basal leaves and leafless scapes bearing a terminal cluster of large flowers, each lasting a single day. Perianth funnelform, its segments spreading or recurved, joined below into a short tube. Stamens inserted at the summit of the tube. Ovary 3-celled with numerous ovules; style slender, declined; stigma capitate.

1 Flowers orange........................... **H. fulva**
1 Flowers yellow.................... **H. lilioasphodelus**

Hemerocallis fulva (L.) L.
ORANGE DAY-LILY *introduced (invasive)*
IR | **WUP** | CUP | **EUP**

Scapes about 1 m tall. Flowers tawny-orange, about 12 cm wide. June–July.

Long in cultivation and freely escaped.

Hemerocallis lilioasphodelus L.
YELLOW DAY-LILY *introduced*
IR | **WUP** | **CUP** | EUP

Hemerocallis flava (L.) L.

Similar to *H. fulva*. Flowers lemon-yellow, about 10 cm wide. May–June.

Commonly cultivated and occasionally escaped along roadsides.

Lilium *Lily* [LILIACEAE]

Tall perennial herbs from a scaly bulb, in our species the erect stem bearing numerous narrow leaves, either alternate or whorled, and at the summit 1 to many, large, erect or nodding, yellow to red flowers. Perianth campanulate or funnelform, its 6 segments clawed or sessile, erect or spreading or recurved, in many species connivent at base, in ours spotted with purple toward the base. Stamens 6. Ovary 3-celled with numerous ovules; style 1; stigmas 3-lobed. Fruit a more or less 3-angled capsule with numerous closely packed, flat seeds.

1 Flowers erect; tepals narrowed at the base to a slender claw; leaves to 8 (–14) mm wide.... **L. philadelphicum**
1 Flowers nodding (fruit becoming erect); tepals narrowed gradually toward base, not clawed; widest leaves 8–35 mm wide..................... **L. michiganense**

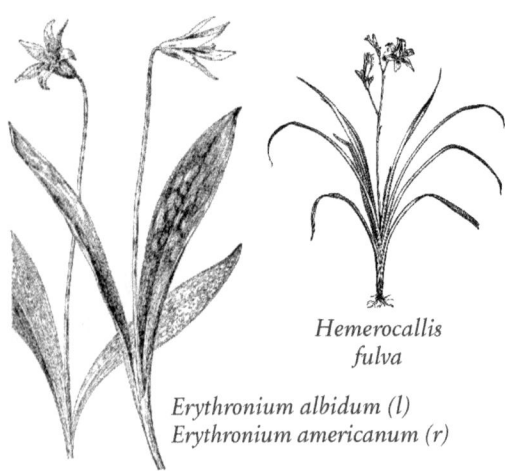

Hemerocallis fulva

Erythronium albidum (l)
Erythronium americanum (r)

Lilium michiganense Farw.
MICHIGAN LILY native
IR | **WUP** | **CUP** | EUP FACW CC 5

Stems stout and erect, to 2.5 m tall. Principal leaves whorled, the upper stem leaves and those of the inflorescence alternate; blades lance-shaped, tapering to both ends, smooth, the larger commonly 8–12 cm long and to 2 cm wide. Flowers occasionally solitary, usually several or many, partly in an umbel from the uppermost leaf-whorl and partly in a terminal raceme, nodding from long, erect or ascending pedicels; perianth segments strongly recurved, lance-shaped, 6–9 cm long, orange or orange-red, spotted with purple, bright green at the base within. July, Aug.

Wet meadows and low ground.

Lilium philadelphicum L.
WOOD-LILY native
IR | **WUP** | **CUP** | **EUP** FAC CC 7

Stems erect, 3–8 dm long. Leaves all from stem, narrowly lance-shaped, 4–10 cm long and 3–9 mm wide, parallel-veined; lower leaves alternate, upper leaves opposite or whorled; petioles absent. Flowers 1–5, erect, large and showy, on stalks 1–8 cm long at ends of stem; perianth segments orange-red, yellow and dark-spotted toward base, lance-shaped, 4–8 cm long and 0.8–2.8 cm wide, stamens and pistil about as long as tepals; stigma 3-parted; ovary superior. Capsule oblong, 2.5–4 cm long; seeds flat. June–July.

Wet meadows, low prairie, fens and open bogs, seeps, ditches; also in drier meadows, prairies and woods.

Maianthemum *False Solomon's-Seal*
[ASPARAGACEAE]

Perennial herbs from a slender creeping rhizome, the erect or ascending stems bearing few to many alternate, sessile or nearly sessile leaves, and a terminal raceme or panicle of small white flowers. Perianth regular, spreading, the segments equal and distinct, 6 (4 in *M. canadense*). Stamens 6 (4 in *M. canadense*). Ovary globose, 2- or 3-celled; style very short; stigma obscurely 2- or 3-lobed. Fruit a globose berry, usually with only 1 or 2 seeds. Includes species formerly included in *Smilacina*.

1. Perianth of 4 parts; leaves 3 or fewer (very rarely 4, usually 2), sometimes pubescent beneath.. **M. canadense**
1. Perianth of 6 parts; leaves often more than 3 (1–4 in one species, where completely glabrous) 2
2. Inflorescence a panicle; perianth 1–2.5 mm long, the stamens up to 3 mm long .... **M. racemosum**
2. Inflorescence a raceme; perianth 2.5–9 mm long, exceeding the stamens........................... 3
3. Stem leaves more than 6, finely pubescent beneath (rarely almost glabrous); uppermost leaves surpassing the top of the inflorescence .... **M. stellatum**
3. Stem leaves 1–4, completely glabrous; inflorescence almost always overtopping leaves .... **M. trifolium**

Maianthemum canadense Desf.
FALSE LILY-OF-THE-VALLEY native
IR | **WUP** | **CUP** | **EUP** FACU CC 4

Stems erect, 5–20 cm long, spreading by rhizomes. Leaves usually 2 along stem, ovate, heart-shaped at base, 3–10 cm long; petioles short or absent. Flowers small, white, 4–6 mm wide, stalked, in a short raceme at end of stem, the raceme 3–6 cm long; tepals 4, spreading; stamens 4; style 2-lobed. Fruit a pale red berry, 3–4 mm wide; seeds 1–2. May–July.

Common in moist to dry woods; also on hummocks in swamps, open bogs and thickets.

Maianthemum racemosum (L.) Link
FALSE SOLOMON'S-SEAL native
IR | **WUP** | **CUP** | **EUP** FACU CC

Smilacina racemosa (L.) Desf.

Stems usually curved-ascending, 4–8 dm tall, finely pubescent. Leaves spreading horizontally in two ranks, elliptic, 7–15 cm long, 2–7 cm wide, obtuse or rounded at base, short-acuminate, finely pubescent beneath. Panicle peduncled or rarely sessile, ovoid to cylindric, 3–15 cm long. Flowers very numerous, short-pediceled, 3–5 mm wide. Berry red, dotted with purple. May–June.

Rich woods.

Maianthemum stellatum (L.) Link
STARRY FALSE SOLOMON'S-SEAL native
IR | **WUP** | **CUP** | **EUP** FAC CC 5

Smilacina stellata (L.) Desf.

Stems ascending or usually erect, 2–6 dm tall, finely pubescent or glabrous. Leaves spreading or oftener strongly ascending, usually folded along the midvein, sessile and somewhat clasping, mostly lance-shaped, 6–15 cm long and 2–5 cm wide, gradually tapering to the acute tip, finely pubescent beneath. Raceme short-peduncled or nearly sessile, 2–5 cm long, with few to several flowers 8–10 mm wide. Fruit black or green with black stripes, 6–10 mm wide. May–June. Moist, especially sandy soil of woods, shores, and prairies.

Maianthemum trifolium (L.) Sloboda
THREE-LEAF FALSE SOLOMON'S-SEAL native
IR | **WUP** | **CUP** | **EUP** OBL CC 10

Smilacina trifolia (L.) Desf.

Stems erect, 1–5 dm long at flowering time, from long rhizomes. Leaves alternate, smooth, usually 3 (2–4), oval or oblong lance-shaped, 6–12 cm long and 1–4 cm wide; petioles absent. Flowers small, white, 8 mm wide, stalked, 3–8 in a raceme; tepals

LILIACEAE | Lily Family

Lilium philadelphicum

Maianthemum canadense

Maianthemum stellatum (l)
Maianthemum trifolium (r)

6, spreading; stamens 6. Fruit a dark red berry, 3–5 mm wide; seeds 1–2. May–June.

Open bogs, conifer swamps, thickets.

Medeola *Cucumber-Root*
[LILIACEAE]

Medeola virginiana L.
INDIAN CUCUMBER-ROOT *native*
IR | WUP | **CUP** | EUP CC 10

Perennial herb from a thick, tuber-like, horizontal rhizome, the slender stem bearing 2 whorls of leaves and a sessile, few-flowered, terminal umbel of greenish yellow flowers. Stems erect, 3–7 dm tall; when young, more or less covered with flocculent wool, which persists about the base of the leaves. Lower leaves 5–11 in a whorl, oblong-oblance-shaped, 6–12 cm long and a fourth to a third as wide, acuminate at both ends; upper leaves normally 3 in a whorl, ovate, 3–6 cm long, half to two-thirds as wide, rounded at base, acuminate. Umbel 3–9-flowered; pedicels spreading or deflexed, 15–25 mm long; perianth segments essentially similar, separate, recurved, about 8 mm long; ovary ovoid, 3-celled. Fruit a globose, few-seeded, dark purple berry. May–July.

Rich woods.

Muscari *Grape-Hyacinth*
[ASPARAGACEAE]

Muscari botryoides (L.) P. Mill.
COMMON GRAPE-HYACINTH *introduced*
IR | **WUP** | **CUP** | EUP

Perennial herb from a bulb, the linear leaves basal, the short erect scape bearing a dense raceme of flowers. Leaves flat, oblong lance-shaped, at maturity to 2.5 dm long and 1 cm wide. Scape 1–2 dm tall at anthesis, to 4 dm in fruit. Raceme compact, ovoid-cylindric, 2–4 cm long at anthesis, elongating in fruit. Flowers blue, all fertile except a few at the tip, nodding, exceeding their slender pedicels; perianth globular, 4–5 mm long; stamens 6, inserted on the perianth tube, included. Fruit a 3-celled capsule, distinctly 3-angled or almost winged, with 2 angular seeds in each cell. Spring.

Commonly cultivated and sometimes escaped; Baraga and Schoolcraft counties.

ADDITIONAL SPECIES

Muscari neglectum Guss. (Starch grape-hyacinth), introduced, reported to occur in Houghton and Marquette counties as a weed in waste places. In contrast to *M. botryoides*, its leaves are only 1–2 mm wide and nearly terete rather than flat.

Ornithogalum *Star-of-Bethlehem*
[ASPARAGACEAE]

Ornithogalum umbellatum L.
STAR-OF-BETHLEHEM *introduced*
IR | **WUP** | CUP | EUP FACU

Perennial herb from a tunicate bulb, with linear basal leaves, an erect or ascending scape, and a short bracted raceme of white flowers. Leaves elongate, 2–4 mm wide. Scape 1–3 dm tall. Raceme 3–7-flowered, the ascending pedicels longer than the internodes. Flowers perfect, erect; perianth segments 6, separate, widely spreading, oblong lance-shaped, 15–20 mm long, white above, with a broad green median stripe beneath; stamens 6, free from the perianth. Fruit a capsule, obtusely 3-angled, with a few seeds in each cell. May–June.

Native of Europe; uncommon escape from cultivation in gardens, roadsides, and occasionally fields and woods; Houghton County.

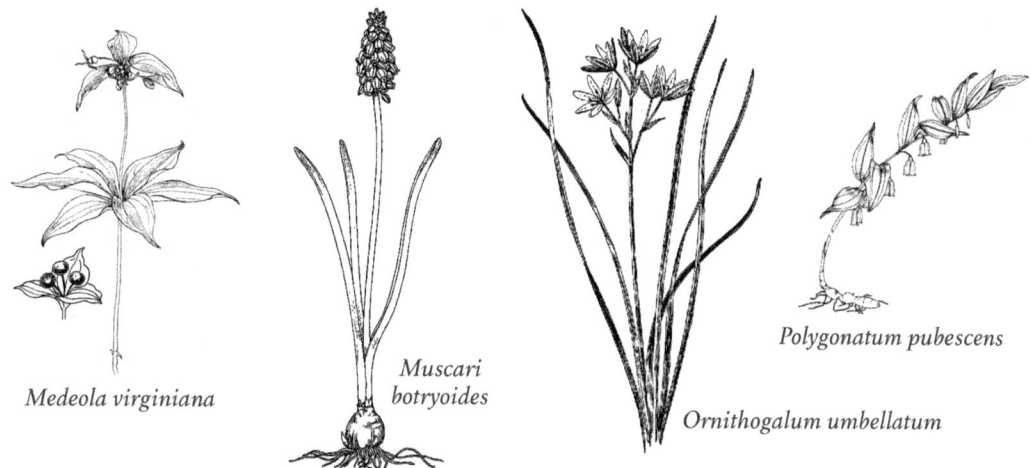

Medeola virginiana
Muscari botryoides
Ornithogalum umbellatum
Polygonatum pubescens

Polygonatum *Solomon's-Seal*
[ASPARAGACEAE]

Polygonatum pubescens (Willd.) Pursh
HAIRY SOLOMON'S-SEAL   *native*
IR | WUP | CUP | EUP   CC 5

Perennial herb from a horizontal knotty rhizome; stems slender, 5–9 dm tall, mostly erect, bearing in the upper portion numerous alternate leaves in two ranks and short, axillary, 1–several-flowered peduncles with pendent, white to greenish or yellow flowers. Leaves narrowly elliptic to broadly oval, 4–12 cm long, 1–6 cm wide, narrowed below to a short petiole, glabrous above, glaucous and pubescent on the veins beneath, with 3–9 prominent nerves. Peduncles slender, usually 1–2 (rarely 3–4)-flowered; pedicels usually shorter than the peduncle. Flowers yellowish green, 10–13 mm long; perianth regular, tubular, shortly 6-lobed; stamens included, inserted on the perianth tube. Fruit a dark blue or black, several-seeded berry. May–July.

Moist woods and thickets.

The similar *P. biflorum* (Walt.) Ell. is known from northern Wisconsin and may be expected to occur in the western UP. The leaves of *P. biflorum* are completely glabrous (vs. leaves finely hairy on the underside veins in *P. pubescens*).

Streptopus *Twisted Stalk*
[LILIACEAE]

Perennial herbs from a rhizome, stems often branched, with alternate sessile or clasping leaves; and small, greenish white to purple, solitary or paired, axillary flowers. Perianth campanulate to rotate, its segments separate to the base, essentially alike, the outer whorl usually slightly wider. Stamens 6, adnate to the base of the perianth; filaments widened at base; anthers oblong to linear, apiculate or aristate. Ovary 3-celled with several ovules; style slender (in our species), 3-cleft, 3-lobed, or entire. Fruit a red, ellipsoid to subglobose, many-seeded berry.

1 Leaves entire or minutely denticulate, strongly clasping at the base, glaucous beneath; nodes and upper internodes glabrous (lower internodes sometimes hispid); tepals spreading or curving from near the middle; flowers whitish green .... **S. amplexifolius**

1 Leaves prominently ciliate on the margins, the cilia usually visible to the naked eye, sessile or slightly clasping (the larger ones subtending branches more strongly clasping), sometimes paler but not glaucous beneath; nodes and upper internodes ± pubescent or sparsely hispidulous; tepals spreading or recurved only at the tips; flowers usually pinkish (or even maroon) ....
.... **S. lanceolatus**

Streptopus amplexifolius (L.) DC.
CLASPING TWISTED STALK   *native*
IR | WUP | CUP | EUP   FAC   CC 8

Stems 4–10 dm tall, glabrous. Leaves ovate, varying to ovate or ovate-lance-shaped, cordate and clasping at base, entire or very minutely toothed, the principal leaves 6–12 cm long and 2–5.5 cm wide. Free portion of the peduncle and the pedicel together 3–5 cm long, jointed at about 2/3 of its length, above the joint 1-flowered or sometimes 2-flowered and abruptly deflexed or twisted. Perianth segments greenish white, about 1 cm long, spreading from near the middle; anthers 1-pointed; stigma entire or barely 3-lobed. Berry red, usually ellipsoid, about 15 mm long. June, July.

Rich moist woods.

Streptopus lanceolatus (Ait.) Reveal
LANCE-LEAF TWISTED STALK native
IR | WUP | CUP | EUP FACU CC 5

Streptopus roseus Michx.

Stems simple or in larger plants commonly branched, 3–8 dm tall, sparsely and finely pubescent, especially at the nodes. Leaves ovate lance-shaped, broadly rounded to a sessile base, finely ciliate, the principal ones 5–9 cm long, 2–3.5 cm wide. Peduncle and pedicel combined 1–3 cm long, jointed at or below the middle, always 1-flowered. Perianth segments rose-color, about 1 cm long, spreading only near the tip; anthers each double-pointed; lobes of the style nearly 1 mm long. Berry red, globose, about 1 cm long. May–July.

Rich woods.

Streptopus lanceolatus

Streptopus amplexifolius

Triantha *False Asphodel*
[TOFIELDIACEAE]

Triantha glutinosa (Michx.) Baker
STICKY FALSE ASPHODEL native
IR | WUP | CUP | EUP OBL CC 10

Tofieldia glutinosa (Michx.) Pers.

Perennial herb, from a bulb. Stems erect, nearly leafless, 2–5 dm long, covered with sticky hairs. Leaves 2–4 from base of plant, linear, hairy, 8–20 cm long and to 8 mm wide, sometimes with 1 bractlike leaf near middle of stem. Flowers white, on sticky-hairy stalks 3–6 mm long, in a raceme 2–5 cm long when in flower, becoming longer when fruiting, 2–3 at each node of the raceme, upper flowers opening first; tepals 6, oblong lance-shaped, 4 mm long; stamens 6. Fruit an oblong capsule, 5–6 mm long; seeds about 1 mm long, with a slender tail at each end. June–July.

Sandy or gravelly shores, interdunal wetlands, calcareous fens, rocky shores of Lake Superior.

Trillium *Trillium; Wake-Robin*
[MELANTHIACEAE]

Perennial herbs from a stout short rhizome, the erect stem bearing a single whorl of 3 ample leaves and a single, large, terminal, sessile or peduncled flower. Perianth segments distinct to the base, the sepals green, the petals white or colored, often of a different shape. Ovary 3-lobed or 3–6-angled or winged, 3-celled with numerous ovules; style short or none; stigmas 3. Fruit a many-seeded berry. A few species are cultivated, especially *T. grandiflorum*.

1. Petals white to pink, 3.5–9 cm long, distinctly longer than the sepals, ± obtuse (occasional small plants with shorter petals—though still longer than sepals—may be recognized by the straight styles and broad obovate petals); stigmatic styles straight (though sometimes spreading) or slightly curved at very tip, uniform in diameter; peduncles held above the leaves . . . . **T. grandiflorum**
1. Petals white, usually less than 3.5 cm long, seldom much longer than sepals; stigmatic styles spreading, thick at base, tapering, and recurved; peduncles in white-flowered plants (and often also in maroon ones) usually reflexed and held below the leaves **T. cernuum**

Trillium cernuum L.
WHIP-POOR-WILL-FLOWER native
IR | WUP | CUP | EUP FAC CC 5

Stems slender, 2–4 dm tall. Leaves broadly rhombic-obovate, commonly 6–10 cm long at anthesis, acuminate, narrowed from near the middle to an acute base and obscurely petioled. Peduncle 1–4 cm long, reflexed or recurved below the leaves. Sepals lance-shaped, acuminate, about equaling the petals; petals normally white, 1.5–2.5 cm long; anthers 3–7 mm long, to 1/3 longer than the filaments; ovary white or pinkish. May–June.

Conifer swamps, bog margins, moist or wet mixed forests, often with paper birch; thickets along streams; less often in rich hardwood forests.

Trillium grandiflorum (Michx.) Salisb.
LARGE-FLOWER WAKEROBIN native
IR | WUP | CUP | EUP CC 5

Stems 2–4 dm tall at anthesis. Leaves ovate to rhombic: or subrotund, at anthesis commonly 8–12 cm long, short-acuminate, narrowed from below the middle to an acute base. Peduncle erect or declined, usually 5–8 cm long. Sepals lance-shaped, spreading, 3–5 cm long; petals normally white, ascending from the base, spreading above, obovate, 4–6 cm long; filaments nearly as long as the anthers and scarcely wider, the whole stamen 15–25 mm long.

Moist to rather dry deciduous forests, forming large colonies in beech-maple forests; less common in oak-hickory woods, swamps, mixed conifer-hardwoods.

Uvularia Bellwort
[COLCHICACEAE]

Perennial herbs from a slender rhizome, the erect stem forked above the middle, the lower portion bearing a few bladeless sheaths and up to 4 leaves; leaves sessile or perfoliate, reaching full size after anthesis; flowers yellow or greenish yellow, terminal, but appearing axillary by prolongation of the branches, nodding. Flowers perfect. Perianth segments 6, elongate, distinct. Stamens 6; filaments short. Ovary shallowly 3-lobed, with several ovules in each cell; styles separate to the base or united to beyond the middle. Capsule obovoid and 3-lobed or 3-winged. Seeds subglobose, few in each cell.

1   Leaves perfoliate, finely puberulent (rarely almost glabrous) and usually light or dark green but not glaucous beneath; rhizome short with many crowded roots; mature capsule less than usually 8–10 mm long. . . . . **U. grandiflora**
1   Leaves sessile, glaucous but glabrous beneath; rhizome elongate, bearing scattered small roots; mature capsule over 15 mm long . . . . **U. sessilifolia**

Uvularia grandiflora Sm.
LARGE-FLOWER BELLWORT   native
IR | **WUP** | **CUP** | EUP   CC 5

Stem at anthesis 2–5 dm tall, at maturity to 1 m tall, forking above, bearing 0–2 leaves below the fork, 4–8 on the sterile branch, and several leaves and 1–4 flowers on the fertile branch. Leaves perfoliate, broadly oval to oblong, to 12 cm long at maturity, minutely pubescent beneath. Flowers yellow, nodding; perianth segments 25–50 mm long, acute or acuminate, smooth within. April–May.

Rich woods, preferring calcareous soil.

Uvularia sessilifolia L.
SESSILE-LEAF BELLWORT   native
IR | **WUP** | CUP | EUP   FACU CC 5

Stem at anthesis 1–3 dm tall, smooth, bearing 0–2 leaves below the fork and 1 or 2 flowers. Leaves at anthesis lance-oblong, acute at both ends, glaucous beneath, at maturity elliptic, to 8 cm long and 3 cm wide, obtuse or rounded at base, nearly smooth on the margins. Flowers pale straw-color, perianth segments 12–25 mm long; styles joined for about 3/4 of their length, about equaling the perianth, much exceeding the anthers. Capsule 3-angled, distinctly stipitate, commonly 15–20 mm long. April–May.

Dry or moist woods; Gogebic and Ontonagon counties.

Orchidaceae
ORCHID FAMILY

Perennial herbs, from fleshy or tuberous roots, corms, or bulbs. Leaves simple, along the stem and alternate, or mostly at base of plant, stalkless and usually sheathing the stem, parallel-veined, often somewhat fleshy. Flowers perfect (with both staminate and pistillate parts), irregular, showy in some species, in heads of 1 or 2 flowers at ends of stems, or with several to many flowers in a spike, raceme or panicle, each flower usually subtended by a bract; sepals 3, green or colored, sometimes resembling the lateral petals, the lateral sepals free, or joined to form an appendage below the lip, or joined with the lateral petals to form a hood over the lip (*Spiranthes*); petals 3 white or colored, the 2 lateral petals alike, the lowest petal different and called the lip; stamens 1–2, attached to the style and forming a stout column; ovary inferior. Fruit a many-seeded capsule, opening by 3 or sometimes 6 longitudinal slits, but remaining closed at tip and base; seeds very small.

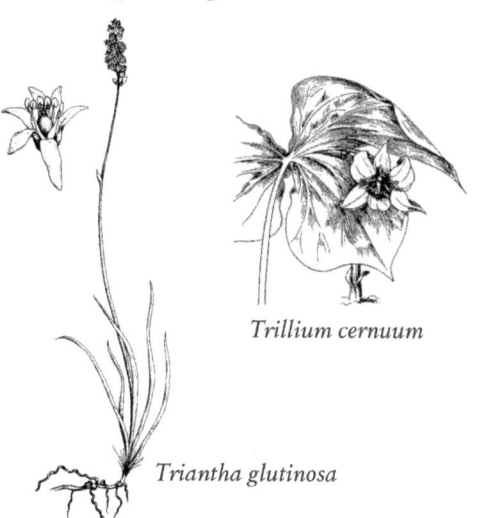

Triantha glutinosa

Trillium cernuum

Trillium grandiflorum

Uvularia grandiflora (l)
Uvularia sessilifolia (r)

ORCHIDACEAE | Orchid Family

One of the world's largest families of vascular plants, with over 900 genera and an estimated 25,000–30,000 species, most of which occur in the tropics.

1. Lip a showy inflated pouch 1–5 cm long. 2
1. Lip showy or inconspicuous, but not an inflated pouch with a small opening, usually ± flat with or without a slender basal spur (or if somewhat saccate, hardly showy and less than 1 cm long) 4
2. Plants with leafy stems; lip a closed pouch (i.e., open only at base above). **Cypripedium**
2. Plants with leaves basal; lip split down middle above or open at base about half its length 3
3. Basal leaf single, petiolate, the blade less than 7 cm long, produced in late summer and withering after the plant blooms the following spring; lip ca. 1.5–2 cm long; plants less than 20 cm tall . . . . **Calypso bulbosa**
3. Basal leaves 2, longer, tapered to sheathing bases and not distinctly petiolate, present throughout the summer (but not winter); lip ca. 4–5 cm long; plants more than 20 cm tall . . . . **Cypripedium acaule**
4. Flower solitary (rarely plants with 2 flowers in a population of 1-flowered ones) 5
4. Flowers 2 or more on one plant 6
5. Leaf linear, at most up to 7 (very rarely 10) mm wide, often poorly developed at flowering time, ± folded or plicate longitudinally, sheathing stem at base; plant from a small bulbous corm. . . . . **Arethusa bulbosa**
5. Leaf ± elliptic or lanceolate, usually over 7 mm wide and well developed at flowering time, flat, arising near middle of stem, sessile but not sheathing at base; plant from slender roots and rhizome . . **Pogonia ophioglossoides**
6. Lip prolonged into a distinct (usually slender and elongate) spur at base 2–40 mm long (pouch-like and only 2–3 mm in *Coeloglossum*) 7
6. Lip at most somewhat swollen or saccate (but not with a spur 2 mm or more long) 11
7. Leaves cauline 8
7. Leaves all basal or nearly so, or absent at flowering time (bracts subtending flowers may be leaflike) 9
8. Spur a thick pouch 2–3 mm long, much shorter than the lip . . . . **Coeloglossum viride**
8. Spur slender, sometimes ± clavate, 7–40 mm long, ± equaling (at most slightly shorter than) to much longer than the lip. . . . . **Platanthera**
9. Flowers entirely white and/or green, the lip lanceolate to narrowly linear, entire; lateral petals free Platanthera
9. Flowers with white lip (spotted or not) broadly ovate to oblong, often crenate or lobed; lateral petals connivent or fused with dorsal sepal to form a pink to purple hood 10
10. Leaves 1; lip less than 1 cm long, spotted, notched at apex and with a lateral lobe on each side. . . . . **Amerorchis rotundifolia**
10. Leaves normally 2; lip over 1 cm long, unspotted, not lobed. . . . . **Galearis spectabilis**
11. Plants lacking green color (except sometimes in fruit), leafless with red, yellow, brown, or purplish stems arising from a coralloid rhizome . . . . **Corallorhiza**
11. Plants with green color, bearing leaves at some time in the year (if leaves absent at flowering time or plants apparently lacking green, arising from tubers, corms, or short rhizomes, not a coralloid mass) 12
12. Leaves a single opposite pair, definitely cauline, not at all sheathing the stem . . . . **Neottia**
12. Leaves solitary, alternate, absent, or basal (or almost basal, with sheathing bases) 13
13. Stem leafy, with 4 or more conspicuous broadly ovate-lanceolate to elliptic leaves; perianth ca. 7–10 mm long; flowers greenish, at least the petals suffused with pink; upper part of stem and axis of inflorescence finely pubescent. . . . . **Epipactis helleborine**
13. Stem with the leaves fewer than 4, narrow, and/or basal (or absent); perianth various, but if pinkish then 10 mm or more long and the vegetative parts completely glabrous. 14
14. Perianth 10–12 mm long, white or creamy; inflorescence dense, spike-like . . . . **Spiranthes**
14. Perianth longer or shorter, or not whitish and the inflorescence not spike-like 15
15. Perianth 10 mm or more long, at least in part usually with some shade of pink or purple (yellowish in a form of *Aplectrum*) 16
15. Perianth less than 10 mm long, greenish, white, or yellowish, with no trace of pink or purple 18
16. Flowers 2–3 cm or more wide, the lip uppermost, bearded with a tuft of yellow-tipped hairs; leaf solitary (rarely 2), several times longer than wide . . **Calopogon**
16. Flowers less than 1.5 cm broad, the lip lowermost and not bearded; blade of leaf not over 3.5 times longer than wide. 17
17. Leaf solitary, petioled, developing in fall and overwintering, usually withered before plant flowers . . . . **Aplectrum hyemale**
17. Leaves 2, sheathing at base, developing in current season and present at flowering . . . . **Liparis**
18. Leaves 1 or 2, sheathing at the base, the scape naked to the inflorescence; flowers on short pedicels, the raceme glabrous and not 1-sided nor noticeably twisted 19
18. Leaves 3 or more (or withering at flowering time), the stem above them bearing small bracts or scales; flowers sessile or almost so in a narrow spike-like inflorescence, which is 1-sided or spirally twisted, or pubescent (or both) 20
19. Leaves 1 (very rarely 2); perianth less than 4 mm long . . . . **Malaxis**
19. Leaves 2; perianth over 4 mm long. . . . . **Liparis**
20. Leaves ovate to elliptic, basal or nearly so, present and firm at flowering time, the midvein and/or other veins margined in white or pale green (not always visible in

dry plants); lip pouched or saccate at the base........ .... **Goodyera**
20 Leaves ovate-elliptic to linear and grass-like, sometimes cauline, often withering at flowering time (in wider-leaved species), not marked with whitish; lip not pouched .... **Spiranthes**

Amerorchis *Round-Leaf Orchid*

Amerorchis rotundifolia (Banks) Hultén
ROUND-LEAF ORCHID   (END) *native*
IR | **WUP** | CUP | EUP   OBL CC 10

Galearis rotundifolia (Pursh) R. M. Bateman, *Orchis rotundifolia* Banks, *Platanthera rotundifolia* (Banks ex Pursh) Lindl.

Perennial herb, roots few from a slender rhizome. Stems leafless, smooth, 15–30 cm long. Leaves single from near base of plant, oval, 4–15 cm long and 2–8 cm wide; usually with 1–2 bladeless sheaths below. Flowers 4 or more, in a raceme 3–8 cm long; sepals white to pale pink; petals white to pink or purple-tinged, the 2 lateral petals joined with the upper sepal to form somewhat of a hood over the column; lip white, with purple spots, 6–10 mm long and 4–7 mm wide, 3-lobed, the terminal lobe largest and notched at tip; spur about 5 mm long, shorter than lip. June–July.

Rare in conifer swamps (on moss under cedar, tamarack, or black spruce); southward in cold conifer swamps of balsam fir, black spruce and cedar; usually found over underlying limestone and where sphagnum mosses not predominant.

Aplectrum *Adam-and-Eve*

Aplectrum hyemale (Muhl.) Torr.
ADAM-AND-EVE; PUTTY-ROOT   *native*
IR | **WUP** | CUP | EUP   FAC CC 10

Perennial from globose corms, which produce a single leaf in the late summer and a bracted scape the following spring; the globose corms are connected by a slender rhizome. Leaf single, basal; leaf blade elliptic 10–15 cm long. Scape 3–6 dm tall, with a few linear-oblong sheathing bracts. Flowers 7–15 in a loose terminal raceme; sepals and petals similar, 10–15 mm long, purplish toward the base, brown toward the summit, the sepals spreading, the petals projecting forward over the column; lip white, marked with violet, 10–15 mm long, broadly obovate, with 3 low parallel ridges near the center, obliquely 3-cleft, the lateral lobes curved upward, the terminal dilated and curved upward at the margin. May–June.

Rich woods; Keweenaw County.

Arethusa *Dragon's-Mouth*

Arethusa bulbosa L.
DRAGON'S-MOUTH   *native*
IR | **WUP** | CUP | EUP   OBL CC 10

Perennial herb; roots few, fibrous, from a corm. Stems leafless, smooth, 1–4 dm long. Leaves 1, linear, small and bractlike at flowering time, later expanding to 2 dm long and 3–8 mm wide; lower stem with 2–4 bladeless sheaths. Flowers single at ends of stems, sepals rose-purple, oblong, 2.5–5 cm long; petals joined and more or less hoodlike over the column; lip pink, streaked with rose-purple, 2.5–4 cm long, curved downward near middle. June–July.

Open bogs and conifer swamps (in sphagnum moss), floating mats around bog lakes, calcareous fens; often with grass-pink (*Calopogon tuberosus*) and rose pogonia (*Pogonia ophioglossoides*).

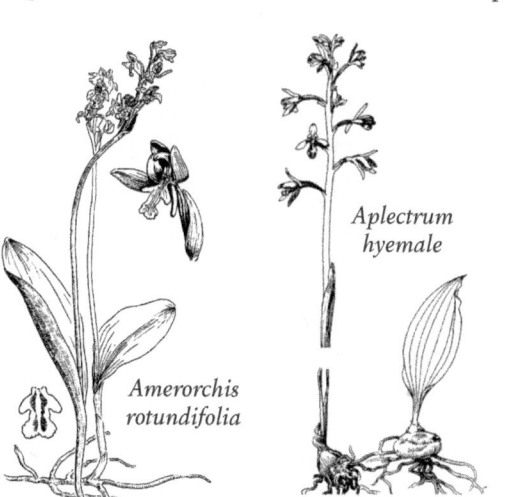

Aplectrum hyemale

Amerorchis rotundifolia

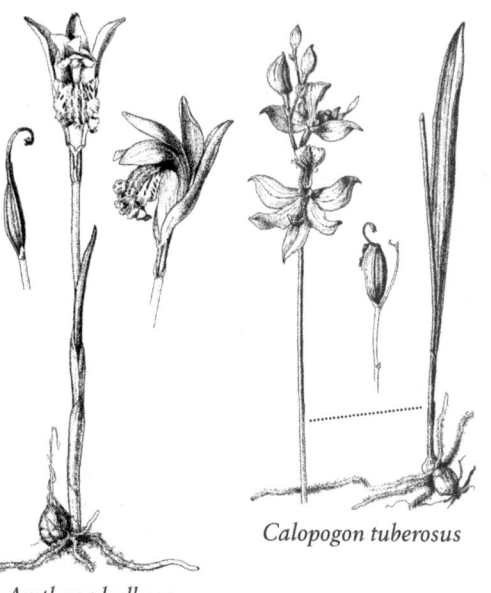

Calopogon tuberosus

Arethusa bulbosa

Calopogon Grass-Pink

Calopogon tuberosus (L.) B.S.P.
GRASS-PINK   native
IR | WUP | CUP | EUP   OBL CC 9

Perennial herb, from a corm. Stems leafless, smooth, 2–7 dm long. Leaves 1 near base of plant, linear, 1–4 dm long and 2–15 mm wide. Flowers pink to purple, 2–15 in a loose raceme, 3–12 cm long; sepals ovate, 1–2.5 cm long; petals oblong, 1–2.5 cm long, the lip located above the lateral petals, 1–2 cm long, bearded on inside with yellow-tipped bristles.

Open bogs and floating mats, openings in conifer swamps, calcareous fens near Great Lakes shoreline.

Distinguished from swamp-pink (*Arethusa bulbosa*) and rose pogonia (*Pogonia ophioglossoides*) by having a raceme of several flowers vs. single flowers in *Arethusa* and *Pogonia*.

Calypso Fairy-Slipper Orchid

Calypso bulbosa (L.) Oakes
CALYPSO; FAIRY-SLIPPER   (THR) native
IR | WUP | CUP | EUP   FACW CC 10

Perennial herb, from a corm. Stems 0.5–2 dm long, with 2–3 bladeless sheaths on lower portion. Leaves single from the corm, ovate, 3–5 cm long and 2–3 cm wide, petioles 1–5 cm long. Flowers 1, nodding at end of stem; sepals and lateral petals similar, pale purple to pink, lance-shaped, 1–2 cm long and 3–5 mm wide, lip white to pink, streaked with purple, 1.5–2 cm long and 5–10 mm wide, the lip extended to form a white "apron" with several rows of yellow bristles. May–June.

Mature conifer forests or mixed forests of conifers and deciduous trees (such as balsam fir, hemlock, and paper birch), usually in shade; soils rich in woody humus.

The single leaf of calypso appears in late August or September, persists through the winter, and withers after flowering in spring. Between fruiting in June and July and the emergence of the new leaf in late summer of fall, no aboveground portions of the plant may be visible.

Coeloglossum Bracted Orchid

Coeloglossum viride (L.) Hartman
BRACTED ORCHID   native
IR | WUP | CUP | EUP   FAC CC 8

Habenaria viridis (L.) R. Br.

Stems slender or stout, 2–5 dm tall. Lowest 1 or 2 leaves reduced to bladeless sheaths; principal foliage leaves obovate, 5–12 cm long, to 5 cm wide, the upper progressively narrower and shorter and passing gradually into the bracts. Inflorescence loose or compact, 5–20 cm long; bracts foliaceous, lance-shaped, exceeding the flowers, the lowest to 5 cm long. Flowers greenish, often tinged with purple; lip oblong, 6–10 mm long, slightly widened distally, terminating in 3 teeth, the central one the shortest; petals lance-shaped, nearly concealed by the incurved sepals; spur pouch-like, 2–3 mm long. June–Aug.

Moist woods.

Corallorhiza Coral-Root

Yellow, brown, or purplish saprophytic herbs, lacking in chlorophyll, and parasitic on fungi inhabiting their characteristic coral-like rhizomes. Stems with a few sheathing scales toward the base and a terminal raceme of small, usually bicolored flowers. Sepals and lateral petals narrow, similar and nearly equal, spreading or projecting over the column; lateral sepals united with the base of the column. Lip deflexed, oblong to rotund, often with two lateral lobes, the margins usually upturned. Fruit a pendent capsule.

1 Lip with a small lobe or elongate tooth on each side near the base (sometimes difficult to see in dried specimens) 2
1 Lip entire, or merely denticulate or erose 3
2 Sepals and petals 3-nerved; summit of ovary with a low protuberance (like a rudimentary spur) usually visible below the base of the lip; lip 4.5–7 mm long .... **C. maculata**
2 Sepals and petals 1-nerved (or the latter rarely weakly 3-nerved); summit of ovary without visible protuberance; lip 2.5–4.5 mm long....................... **C. trifida**
3 Sepals and petals 3–5-nerved, 8–15 mm long, conspicuously striped with purple, the lip solid purplish apically .... **C. striata**
3 Sepals and petals 1-nerved (or faintly 3-nerved), less than 6 mm long, not conspicuously striped .. **C. trifida**

Corallorhiza maculata (Raf.) Raf.
SPOTTED CORAL-ROOT   native
IR | WUP | CUP | EUP   FACU CC 5

Perennial saprophytic herb. Stems pinkish purple, 2–5 dm tall. Raceme 5–15 cm long, 10–40-flowered. Sepals and lateral petals more or less spotted or suffused with purple, narrowly oblong to oblance-shaped, 6–8 mm long, 3-nerved, the lateral sepals somewhat divergent; spur a prominent, sometimes divergent swelling near the summit of the ovary; lip white, irregularly spotted with purple, 6–8 mm long, bearing 2 conspicuous lateral lobes below the middle and 2 short parallel ridges on the face, the terminal lobe rounded and deflexed, 3–4 mm wide. July–Sept.

Woods.

ORCHIDACEAE | Orchid Family

Corallorhiza striata Lindl.
STRIPED CORAL-ROOT   native
IR | WUP | CUP | EUP   FACU CC 6
Perennial saprophytic herb. Stems purple or magenta, rather stout, 2–4 dm tall, with 2–4 acute or cuspidate scales. Raceme 5–12 cm long, usually 10–20-flowered. Sepals and lateral petals arching forward, oblong lance-shaped, 10–14 mm long, yellowish white with 3 conspicuous longitudinal stripes of purple, usually also purple on the margin; lip 8–12 mm long, white, heavily striped with purple, or purple throughout. May–July.

Moist or dry woods.

Corallorhiza trifida Chatelain
YELLOW CORALROOT   native
IR | WUP | CUP | EUP   FACW CC 6
Perennial saprophytic herb, roots absent. Stems yellow-green, smooth, 1–3 dm long, single or in clusters from the coral-like rhizome. Leaves reduced to 2–3 overlapping sheaths on lower stem. Flowers yellow-green, 5–15 in a raceme 3–8 cm long; sepals and lateral petals yellow-green, linear, 3–5 mm long, lip white, sometimes with purple-spots, obovate, 3–5 mm long and 2–3 mm wide. Capsules drooping, 1–1.5 cm long and 3–7 mm wide. May–June.

Moist to wet, mostly conifer woods, swamps (often under white cedar); usually where shaded.

Cypripedium *Lady's-Slipper*

Erect perennial herbs, from coarse, fibrous roots. Stems unbranched, often clumped, hairy. Leaves 2 or more at base of plant or along stem, broad. Flowers 1 or 2, large and mostly showy at ends of stems, white, pink or yellow; lateral sepals similar to lateral petals, the sepals joined to form a single appendage below the lip; lateral petals free and spreading, lip inflated and pouchlike, projecting forward; stamens 2, 1 on each side of column. Fruit a many-seeded capsule.

1 Lip pouch pink to purple; leaves 2 at base of stem..... .... **C. acaule**
1 Lip pouch yellow or white; leaves 3 or more on stem . 2
2 Pouch yellow, sometimes brown- or purple-dotted... 3
2 Pouch white to pink, or pink with white patches..... 4
3 Sepals and petals red-brown; lateral petals strongly twisted, brown-purple; pouch less than 4 cm long .... **C. parviflorum var. makasin**
3 Sepals and petals yellow to brown-green; lateral petals wavy, green with red-brown streaks; pouch more than 4 cm long............. **C. parviflorum var. pubescens**
4 Pouch projected downward into a cone-shaped spur .. .. **C. arietinum**
4 Pouch not spurred........................ **C. reginae**

Cypripedium acaule Ait.
PINK LADY'S-SLIPPER   native
IR | WUP | CUP | EUP   FACW CC 5
Perennial herb, from coarse rhizomes; roots long and cordlike. Stems leafless, 2–4 dm long, glandular-hairy. Leaves 2 at base of plant, opposite, oval to obovate, 1–2 dm long and 3–10 cm wide, thinly hairy, stalkless. Flowers 1, nodding at end of stem; sepals and lateral petals yellow-green to green-brown, the 2 lower sepals joined to form a single sepal below the lip; lip drooping, pink with red veins, 3–5 cm long, cleft along the upper side and hiding the opening. May–June.

Forests, typically where shaded, acidic, and nutrient-poor; sometimes on hummocks in conifer swamps.

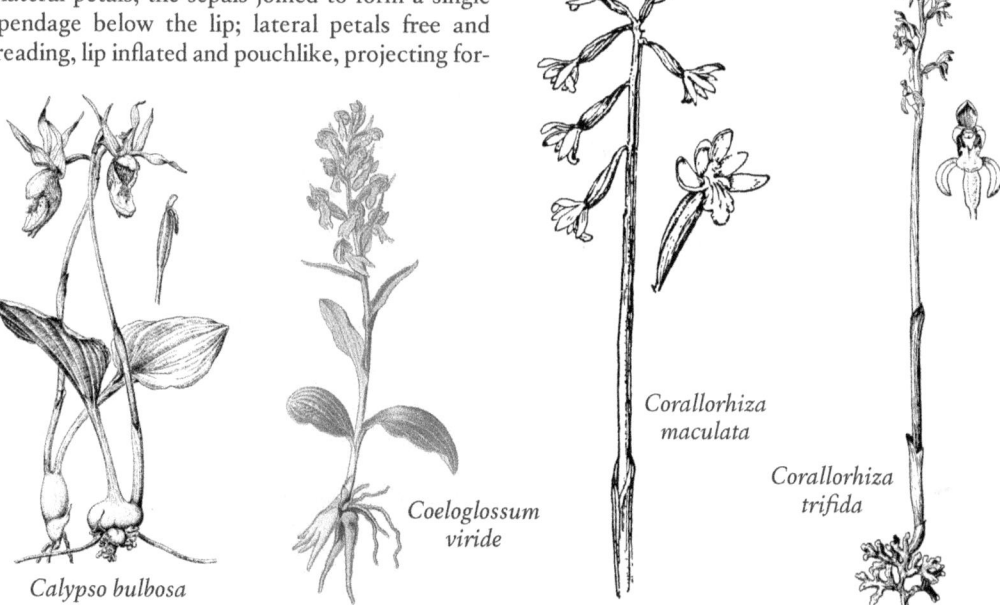

Calypso bulbosa

Coeloglossum viride

Corallorhiza maculata

Corallorhiza trifida

Cypripedium arietinum Ait. f.
RAM'S-HEAD LADY'S-SLIPPER native
IR | WUP | CUP | EUP FACW CC 10

Perennial herb, from a coarse rhizome, roots long and cordlike. Stems slender, 1–4 dm long, thinly hairy. Leaves 3–5, above middle of stem, stalkless, oval, often folded, 5–10 cm long and 1.5–3 cm wide, finely hairy. Flowers 1 or sometimes 2 at ends of stems; sepals and lateral petals similar, green-brown; lip an inflated pouch, 1.5–2.5 cm long, white or pink-tinged, with prominent red-veins, extended downward to form a conical pouch. Late May–June.

Conifer swamps, wet forest openings (often with white cedar); also in drier, sandy, conifer and mixed conifer-deciduous forests, and on low dunes under conifers near shores of Great Lakes (usually with tamarack near Lake Michigan). Our smallest and rarest lady's-slipper.

Cypripedium parviflorum Salisb.
var. **makasin** (Farw.) Sheviak
YELLOW LADY'S-SLIPPER native
IR | WUP | CUP | **EUP** FAC CC 5

Cypripedium calceolus var. *parviflorum* (Salisb.) Fernald

Perennial herb, from rhizomes, roots long and numerous. Stems 1.5–6 dm long, glandular-hairy. Leaves 2–5, alternate along stem, ascending, oval, 5–18 cm long and 2–7 cm wide, sparsely hairy, stalkless. Flowers 1 (rarely 2) at ends of stems; sepals purple-brown, the lateral sepals joined below the lip, notched at tip; lateral petals linear, purple-brown, spirally twisted, 2–5 cm long; lip an inflated pouch, 1.5–3 cm long, yellow, often with purple veins and spots near opening. May–July. Conifer swamps, wet meadows, fens, and moist forests (often under cedar); sphagnum mosses are usually sparse; sites are shaded or sunny, with organic or mineral, often calcium-rich soil; reported for Chippewa (drummond Island) and Mackinac counties, more common in Lower Michigan.

Our two varieties may be distinguished by the size of the pouch (lip) and the color of the sepals and petals: in var. *makasin*, the lip is mostly 2–3 cm long, and the sepals and petals are dark red; in var. *pubescens*, the lip is mostly 3–6 cm long and the sepals and petals are yellow-green; however, intermediate forms may occur.

Cypripedium parviflorum Salisb.
var. **pubescens** (Willd.) Knight
YELLOW LADY'S-SLIPPER native
IR | **WUP** | **CUP** | **EUP** FAC CC 5

Cypripedium calceolus var. *pubescens* (Willd.) Correll

Perennial herb, from a rhizome, roots long and numerous. Stems 1.5–6 dm long, glandular-hairy. Leaves 3–6, alternate along stem, ascending, ovate to oval, 8–20 cm long and 3–8 cm wide, sparsely hairy. Flowers 1 (rarely 2) at ends of stems; sepals yellow-green, the lateral sepals joined below the lip, notched at tip; lateral petals linear, yellow-green, often streaked with red-brown, usually spirally twisted, 4–8 cm long; lip an inflated pouch, 3–6 cm long, yellow, often with purple veins near opening. May–July.

Conifer swamps, bogs, fens, prairies, especially where soils derived from limestone; also in wetter hardwood forests.

Cypripedium reginae Walt.
SHOWY LADY'S-SLIPPER native
IR | WUP | CUP | EUP FACW CC 9

Perennial herb, from a coarse rhizome, roots many, long and cordlike. Stems 4–10 dm long, strongly glandular-hairy. Leaves 4–12, alternate along stem, spreading or ascending, broadly oval, 10–25 cm long, 4–12 cm wide, abruptly tapered to tip, nearly smooth to hairy, stalkless; reduced to sheaths at base. Flowers 1 or often 2 at ends of stems, the subtending bract leaflike, 6–12 cm long; sepals and lateral petals white, the lateral sepals joined to form an appendage under the lip, rounded at tip; lip an inflated pouch, 3–5 cm long, white, streaked and spotted with pink or purple. June–July.

Conifer and hardwood swamps (especially balsam fir-cedar-tamarack swamps), bogs, calcareous fens, sedge meadows, floating mats, wet openings, wet clayey slopes, ditches; especially where open and sunny; most abundant in openings in wet forests and swamps not dominated by sphagnum mosses.

Showy lady's-slipper is our largest lady's-slipper. Avoid touching plants as the hairs can be irritating.

Epipactis *Helleborine*

Epipactis helleborine (L.) Crantz
HELLEBORINE *introduced (invasive)*
IR | **WUP** | **CUP** | **EUP** UPL

Perennial herb. Stems erect, to 8 dm tall. Leaves alternate, sessile and clasping, ovate to lance-shaped, the lower to 10 cm long, the upper progressively shorter and narrower. Flowers in a terminal, many-flowered raceme 1–3 dm long; bracts linear or narrowly lance-shaped, the lower surpassing the flowers; sepals and lateral petals ovate-lance-shaped, 10–14 mm long, acute, dull green, strongly veined with purple; lip greenish and purple, strongly saccate in the basal half, the terminal lobe broadly ovate, crested at its base with two elevated swellings. July–Aug.

Native of Europe; established and spreading in deciduous and mixed woods.

Galearis Showy Orchid

Galearis spectabilis (L.) Raf.
SHOWY ORCHID   (THR) *native*
IR | **WUP** | CUP | EUP   CC 10
Orchis spectabilis L.

Perennial herb. Leaves 2, rather fleshy, narrowly obovate to broadly elliptic, 8–15 cm long. Scape 1–2 dm tall, stout; bracts foliaceous, oblong lance-shaped, 15–50 mm long. Sepals and lateral petals pink to pale purple, 13–18 mm long, all connivent; lip white, 15–20 mm long, rhombic-obovate; spur stout, about equaling the lip. May–June.

Rare in rich moist woods and on streambanks; Baraga and Ontonagon counties.

Goodyera Rattlesnake-Plantain

Perennial herbs from a short rhizome, plants glandular-pubescent on the scape, bracts, ovary, and sepals. Leaves in a basal cluster, commonly reticulated with white, narrowed to a broad, petiole-like base. Flowers in a spike-like raceme of white or greenish flowers, atop an erect scape with several scale-like bract. Upper sepal and lateral petals coherent by their margins, forming a concave galea extending forward over the lip. Lateral sepals free, scarcely spreading, except at the tip. Lip shorter than the galea, conspicuously pouch-like at base, prolonged upwards into a horizontal or deflexed beak; lateral petals white.

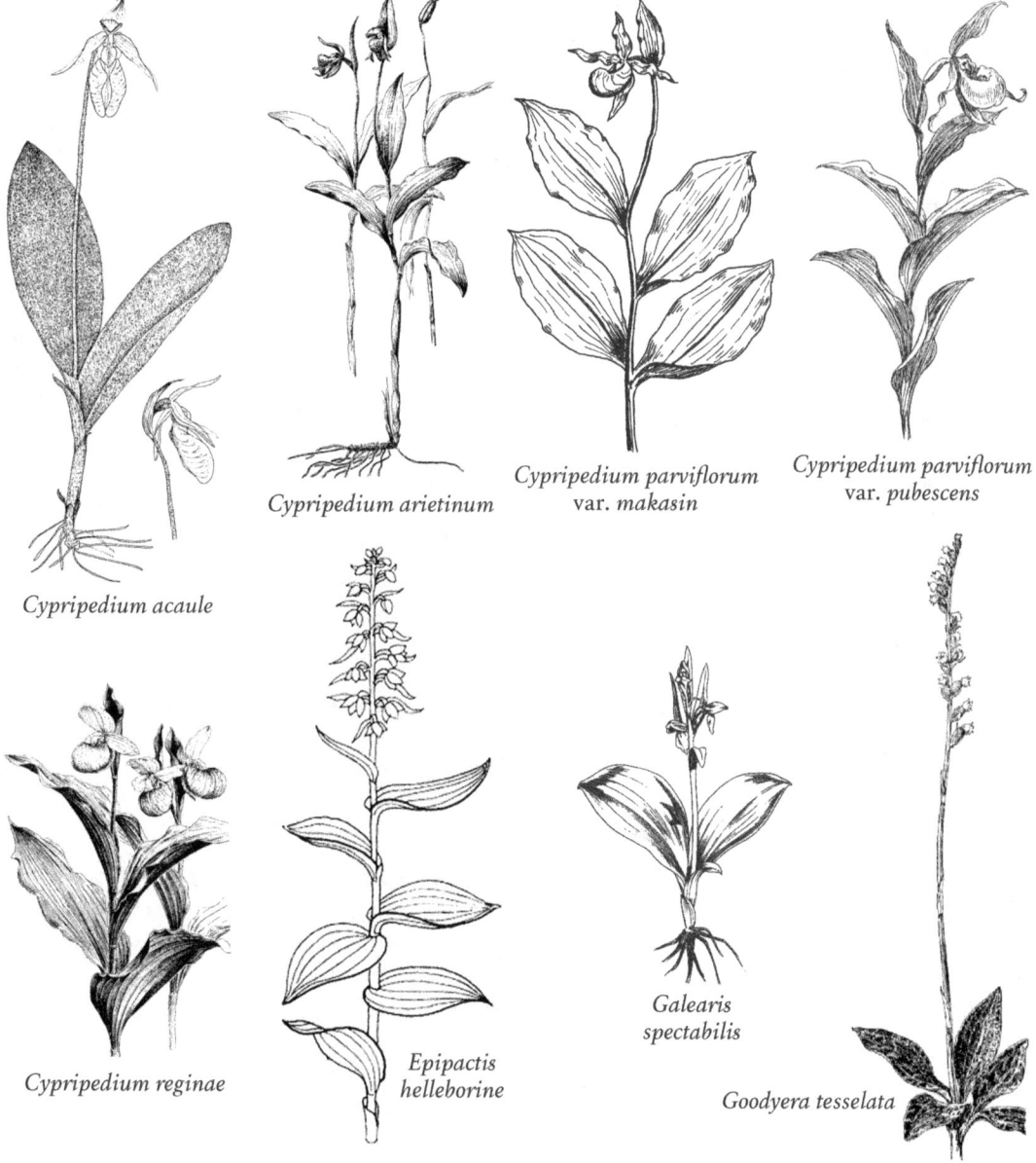

Cypripedium arietinum

Cypripedium parviflorum var. *makasin*

Cypripedium parviflorum var. *pubescens*

Cypripedium acaule

Cypripedium reginae

Epipactis helleborine

Galearis spectabilis

Goodyera tesselata

ORCHIDACEAE | Orchid Family

1. Leaf blades with only the midvein outlined above in white or pale green, the largest blades usually 4–6 cm long; plants 20–50 cm tall; perianth 6–9 mm long. . . . . **G. oblongifolia**
1. Leaf blades with white or pale green reticulation ± throughout (sometimes not on the midvein), the largest blades often less than 4 cm long; plants 5–30 cm tall; perianth 2.5–5.5 (rarely to 6.5) mm long. 2
2. Stem with usually 7–10 cauline bracts (undeveloped leaves); beak of lip (beyond the large pouch) less than 1 mm long, about 1/4 the total length of the lip or usually less; inflorescence ± densely flowered on all sides . . . . **G. pubescens**
2. Stem with 2–5 cauline bracts; beak of lip 1–2 mm long, about 1/2 the total length of the lip; pouch shallow or deep; inflorescence strongly one-sided or ± loosely flowered on all sides. 3
3. Lip deeply pouched, the pouch about as deep as long, the beak often strongly turned downward at maturity; plants mostly 10–20 cm tall; largest leaf blades mostly 1–2 cm long; cauline bracts 2–4 (usually 3) . . **G. repens**
3. Lip shallowly pouched, the pouch longer than deep, the beak horizontal or slightly recurved; plants usually 17–25 cm tall; largest leaf blades mostly 2–4 cm long; cauline bracts usually 4–5. . . . . **G. tesselata**

Goodyera oblongifolia Raf.
WESTERN RATTLESNAKE-PLANTAIN native
IR | WUP | CUP | EUP FACU CC 8

Stem stout, 3–4.5 dm tall. Leaf blades ovate-lance-shaped to narrowly elliptic, to 8 cm long, about a third as wide. Raceme 1-sided, densely flowered, 6–12 cm long. Galea 8–10 mm long; lateral sepals lance-shaped, 6–7 mm long, acute; lip ovate in general outline, 6–7 mm long, its beak about two-thirds as long as the semi-globose body, curved-deflected, broad, with upturned sides. July–Aug.

Dry woods.

Goodyera pubescens (Willd.) R. Br.
DOWNY RATTLESNAKE-PLANTAIN native
IR | WUP | CUP | EUP FACU CC 7

Stem rather stout, 2–4 dm tall. Leaf blades ovate or ovate-lance-shaped, 3–6 cm long, with 5 or 7 white veins and numerous white reticulate veinlets. Raceme dense, many-flowered, 4–10 cm long. Galea broadly elliptic, very convex, 4–4.5 mm long, upturned at the summit; lateral sepals broadly ovate to obovate, 3.5–4 mm long, abruptly short-acuminate; lip subglobose, 3.5–4 mm long, its straight beak less than 1 mm long and scarcely projecting beyond the ventricose body. July–Aug.

Dry woods.

Goodyera repens (L.) R. Br.
DWARF RATTLESNAKE-PLANTAIN native
IR | WUP | CUP | EUP FACU CC 9

Stem slender, 1–3 dm tall, glandular-pubescent. Leaf blades ovate to oblong, 1.5–3 cm long. Raceme loosely flowered, one-sided, 3–6 cm long. Galea 3.5–5 mm long; lateral sepals slightly shorter, broadly ovate; lip deeply pouchlike at base, its beak triangular, acute, abruptly deflected. Anther blunt. July–Aug.

Dry woods.

Goodyera tesselata Lodd.
CHECKERED RATTLESNAKE-PLANTAIN native
IR | WUP | CUP | EUP FACU CC 8

Stem 1.5–4 dm tall. Leaf blades ovate or ovate-lance-shaped, 3–5 cm long. Raceme loosely flowered, often spiral, 4–10 cm long. Galea strongly convex, 4–5 mm long; lateral sepals ovate, about 4 mm long; lip about 3.5 mm long, strongly saccate at base, narrowed distally to a beak about as long as the body, with upturned sides and blunt or rounded at the tip; anther abruptly narrowed to a conspicuous sharp beak. July–Aug.

Dry woods.

Liparis *Wide-Lip Orchid*

Liparis loeselii (L.) L.C. Rich.
FEN-ORCHID native
IR | WUP | CUP | EUP FACW CC 5

Small, smooth perennial herb, from a bulblike base. Stems erect, 1–2.5 dm long, upper stem somewhat angled in section. Leaves 2 from base of plant, ascending, sheathing at base, shiny, lance-shaped to oval, 4–15 cm long and 1–4 cm wide. Flowers 2–15, yellow-green, small, upright, atop a naked scape in an open raceme 2–10 cm long and 1–2 cm wide; sepals narrowly lance-shaped, 4–6 mm long and 1–2 mm wide; lateral petals linear, 3–5 mm long, often twisted and bent forward under the lip and appearing thread-like; lip yellow-green, obovate, 4–5 mm long, tipped with a short point. Capsules persistent, short-cylindric, 8–12 mm long. June–Aug.

Conifer swamps, fens, floating mats, streambanks, sandy shores, ditches; soils peaty to mineral, acid to calcium-rich.

Malaxis *Adder's-Mouth Orchid*

Small perennial herbs. Leaves 1–5 from base of plant or single along stem. Flowers green-white, spaced or crowded in slender or cylindric racemes at ends of stems.

1. Flowers evenly spaced in a raceme 5–11 cm long . . . . **M. monophyllos**
1. Flowers crowded near top of raceme, the raceme 2–5 cm long. . . . . **M. unifolia**

Malaxis monophyllos (L.) Sw.
WHITE ADDER'S-MOUTH ORCHID *native*
IR | WUP | CUP | EUP FACW CC 10
Malaxis brachypoda (Gray) Fern.
Perennial herb, from a bulblike base; roots few, fibrous. Stems smooth, 1–2 dm long. Leaves single, appearing to be attached well above base of stem, the leaf base clasping stem, ovate to oval, 3–7 cm long and 1.5–4 cm wide. Flowers small, green-white, 14–30 or more, in a long, slender, spike-like raceme 4–11 cm long and to 1 cm wide; on stalks 1–2 mm long, the flowers evenly spaced in the raceme; lip heart-shaped, bent downward, 2–3 mm long and 1–2 mm wide, narrowed at middle to form a long, lance-shaped tip, with a pair of lobes at base. June–Aug. Ours are var. *brachypoda* (Gray) F. Morris & Eames.

Conifer swamps (white cedar, balsam fir, black spruce) on sphagnum moss hummocks, and in wet depressions and often where soils are marly; wet hardwood forests.

Malaxis unifolia Michx.
GREEN ADDER'S-MOUTH ORCHID *native*
IR | WUP | CUP | EUP FAC CC 8
Small perennial herb, from a bulblike base; roots few, fibrous. Stems smooth, 1–3 dm long. Leaves single, attached near middle of stem, ovate, 2–7 cm long and 1–4 cm wide. Flowers small, green, numerous in a cylindric raceme 1.5–6 cm long and 1–2 cm wide, the upper flowers crowded, the lower flowers more widely spaced; lip very small, 1–2 mm long, with 3 teeth at tip. June–Aug.

Sphagnum moss hummocks in swamps, sedge meadows, thickets; also in drier forests including pine plantations.

Neottia *Twayblade*
Perennial herbs. Stems with a pair of opposite leaves near middle, stems smooth below leaves, hairy above. Leaves broad, stalkless. Flowers small, green to purple, in a raceme at end of stem, the lip 2-lobed or deeply parted. Formerly considered part of genus *Listera*.

1 Lip 3–5 mm long, divided to about middle into 2 narrow segments................................ **N. cordata**
1 Lip 7–12 mm long, shallowly notched or divided 1/3 of length, the segments broad 2
2 Lip wide at base, with a pair of auricles.. **N. auriculata**
2 Lip narrowed to base, auricles absent................ .... **N. convallarioides**

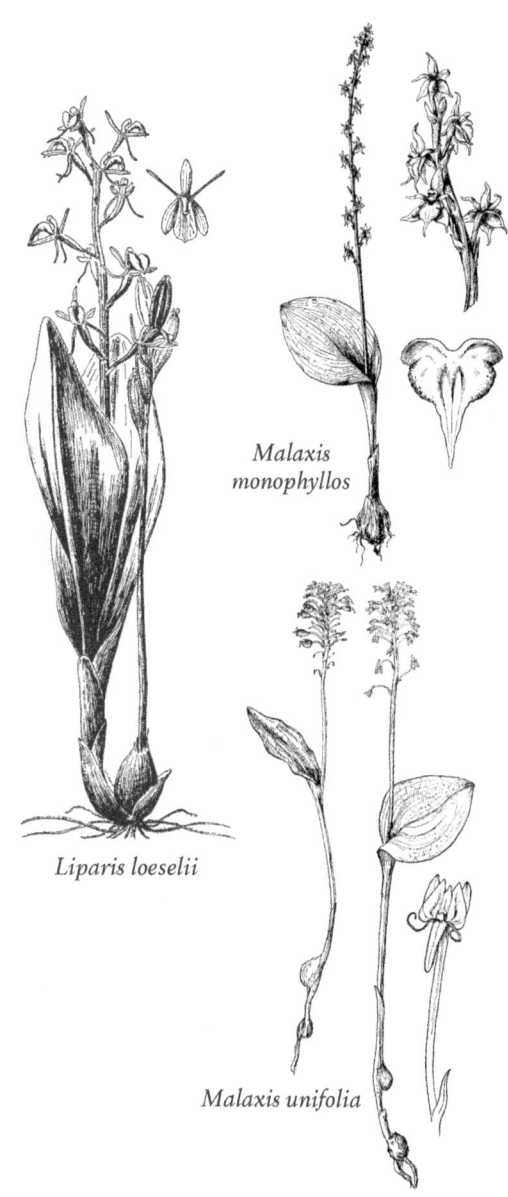

Malaxis monophyllos

Liparis loeselii

Malaxis unifolia

Neottia auriculata (Wiegand) Szlach.
AURICLED TWAYBLADE *native*
IR | WUP | CUP | EUP FACW CC 9
Listera auriculata Wieg.
Perennial herb, roots fibrous. Stems 1–2 dm long, smooth below leaves, hairy above. Leaves 2 near middle of stem, opposite, ovate, 2–5 cm long and 2–4 cm wide. Flowers pale green, 8–15 in a raceme 4–8 cm long and 2–3 cm wide, on stalks 2–5 mm long; lip oblong, 6–10 mm long and 2–5 mm wide, the base with a pair of small clasping auricles, the tip cleft for about 1/4–1/3 of its length. June–Aug.

Sand along rivers and Lake Superior, often under alders, occasionally in moist conifer or mixed conifer and deciduous forests; usually where shaded.

Neottia convallarioides (Sw.) Rich.
BROAD-LIP TWAYBLADE *native*
IR | WUP | CUP | EUP FACW CC 10
Listera convallarioides (Sw.) Nutt.
Perennial herb, roots fibrous. Stems 1–3 dm long,

glandular-hairy above leaves, smooth below. Leaves 2, opposite near middle of stem, broadly ovate, 3–6 cm long and 2–5 cm wide, stalkless. Flowers yellow-green, 6–20 in a raceme 4–10 cm long and 2–3 cm wide; lip wedge-shaped, 9–11 mm long and to 6 mm wide at tip, usually with a small tooth on each side near the base, the tip shallowly 2-lobed. July–Aug.

Seeps in forests, cedar swamps, wet, mixed conifer-deciduous woods, streambanks.

Neottia cordata (L.) Rich.
HEART-LEAF TWAYBLADE   native
IR | WUP | CUP | EUP   FACW CC 10
Listera cordata (L.) R. Br.
Perennial herb, roots fibrous. Stems 1–3 dm long, glandular-hairy above the leaves, smooth below. Leaves 2, opposite near middle of stem, 1–4 cm long and 1–3 cm wide, stalkless. Flowers green to red-purple, 6–20 in a raceme 3–12 cm long and 1–2 cm wide; lip slender, 3–5 mm long, with 2 teeth on side near base, the tip cleft halfway or more into spreading linear lobes. June–July.

Bogs and conifer swamps, where usually on sphagnum moss hummocks; hemlock groves.

Platanthera *Rein-Orchid*

Perennial herbs, from a cluster of fleshy roots. Stems erect, smooth. Leaves mostly along the stem, upright, reduced to sheaths at base and upward on stem; leaves basal in P. orbiculata. Flowers white or green, several to many in a spike or raceme; upper sepal joined with petals to form a hood over the column; lateral sepals spreading; lip linear to ovate or 3-lobed, entire, toothed or fringed, extended backward into a spur, the spur commonly curved; stamens 1, the anther attached to the top of the short column. Fruit a many-seeded capsule.

Previously included in *Habenaria*, that genus now considered tropical, and *Platanthera* occurring in temperate regions.

ADDITIONAL SPECIES AND HYBRIDS
Platanthera unalascensis (Spreng.) Kurtz. [Alaska orchid, syn: *Piperia unalascensis* (Spreng.) Rydb.]; open woods of aspen and birch, clearings on thin soil over dolomite; native western species, disjunct in northern Great Lakes region in eastern UP and Bruce Peninsula of Ontario; Chippewa and Mackinac counties.
Platanthera × andrewsii (M. White) Luer [syn: *Platanthera lacera* var. *terrae-novae* (Fern.) Luer], is a hybrid between *P. lacera* and *P. psycodes*; reported from Keweenaw County.

1 Lip prominently ciliate or fringed 2
1 Lip entire or toothed, but not fringed. 3
2 Flowers pink-purple; divisions of the lip broadly fan-shaped, copiously lacerate-fringed, but the fringe usually cut less than half the distance to the base of the division of the lip . . . . **P. psycodes**
2 Flowers cream or greenish; at least the lateral divisions of the lip more narrowly cuneate, mostly cut into a long fringe more than half their length . . . . **P. lacera**
3 Leaves all basal, the stem at most with reduced bracts
    4
3 Leaves cauline (along the stem). 8
4 Leaves about twice as long as wide, or longer; spur less than 12 mm long. 5
4 Leaves less than twice as long as broad, orbicular or almost so; spur 16–40 mm long 7
5 Lip with truncate 3-toothed or crenate apex; spur 7–11 mm long, much exceeding the lip . . . . **P. clavellata**
5 Lip tapered to a pointed or rounded untoothed tip ; spur about equaling lip or at most ca. 2 mm longer . . 6
6 Leaves 1 (–2), present through anthesis; ovary short-pediceled (evident on older flowers or fruit), the inflorescence usually a raceme shorter than 10 cm; lateral sepals 3.5–6 mm long . . . . **P. obtusata**
6 Leaves usually 2–3, withering during or before anthesis; ovary sessile, the inflorescence a spike usually 10 cm or more long; sepals 1.5–3 mm long . . . . . **P. unalascensis***
7 Scape naked (rarely with a bract); spurs (16–24 mm long) tapered ± evenly to rounded tip; lip yellowish green, tending to turn upward near the end **P. hookeri**
7 Scape with 1–6 bracts between leaves and inflorescence; spurs parallel-sided or even somewhat club-shaped toward tip; lip whitish green, tending to turn downward
    . . . . **P. orbiculata**
8 Lip truncate and 2–3-toothed or -lobed at tip. . . . .
    . . . . **P. clavellata**
8 Lip tapered, rounded (or almost truncate and obscurely crenulate) but not 2–3-toothed at tip. 9
9 Flowers pure white, lip strongly expanded basally . . . .
    . . . . **P. dilatata**
9 Flowers green, greenish yellow or greenish white, lip cuneate to strap-shaped, not or only slightly widened at base 10
10 Anther sacs essentially in contact above the rounded stigma (separated at tip by less than 0.3 mm); lips 2.5–5 mm long. . . . . **P. aquilonis**
10 Anther sacs separated at tip by ca. 0.4 mm or more, stigma pointed; lips 4–8 mm long . . . . . . **P. huronensis**

Platanthera aquilonis Sheviak
BOG ORCHID   native
IR | WUP | CUP | EUP   CC 5
Perennial herb. Stems 60 cm long. Leaves several, ascending to spreading, gradually reduced to bracts upwards; blade linear-lanceolate, 3–23 cm long and to 4 cm wide. Spikes lax to very dense. Flowers not showy, yellowish green with dull yellowish lip; lateral sepals spreading to reflexed; petals ovate, margins

entire; lip descending, 2.5–6 cm long and to 1.5 mm wide, projecting, not thickened at base, margins entire; spur clavate or sometimes rather cylindric, 2–5 mm long, apex usually broadly obtuse. May–Aug.

Moist to wet including moist forests, cedar swamps, riverbanks, wet meadows, fens, ditches and borrow pits.

Platanthera clavellata (Michx.) Luer
GREEN WOODLAND ORCHID   native
IR | WUP | CUP | EUP   FACW CC 6
Habenaria clavellata (Michx.) Spreng.
Perennial herb. Stems slender, 1–4 dm long. Foliage leaf 1, linear-oblong to oblong lance-shaped, commonly 7–16 cm long, to 3 cm wide, blunt; upper leaves 1 or few, much reduced, the uppermost linear, scale-like. Inflorescence open, 5–15-flowered, 2–6 cm long; bracts narrowly lance-shaped, shorter than the flowers. Flowers divergent from the axis, white or tinged with green or yellow, twisted to one side so that the spur is lateral; lip broadly cuneate, 3–5 mm long, shallowly 3-lobed at the summit; petals and sepals broadly ovate, about equal; spur strongly curved, dilated at the tip, 8–12 mm long. July–Aug.

Acid bogs and wet soils, especially in sphagnum.

Platanthera dilatata (Pursh) Lindl.
WHITE BOG-ORCHID   native
IR | WUP | CUP | EUP   FACW CC 10
Habenaria dilatata (Pursh) Hook.
Piperia dilatata (Pursh) Szlach. & Rutk.
Perennial herb, strongly clove-scented, roots fleshy. Stems stout or slender, to 1 m long. Leaves 3–6, alternate along stem, upright, lance-shaped, to 10–20 cm long and 1–3 cm wide, with 1–2 small, bractlike leaves above and 1 bladeless sheath at base of stem. Flowers 10–60, bright white, upright, in a raceme 1–2.5 dm long; lateral sepals lance-shaped, 4–9 mm long and 1–3 mm wide; lateral petals similar but joined with upper sepal to form somewhat of a hood over the column; lip lance-shaped, widened at base, 6–8 mm long; spur slender, 4–8 mm long. June–July.

Wet, open bogs and floating mats, conifer swamps, streambanks, shores and seeps; often where sandy or calcium-rich (as in calcareous fens), not in deep sphagnum moss.

Similar to northern bog-orchid (*P. huronensis*) but with white rather than green-tinged flowers as in *P. huronensis*.

Platanthera hookeri (Torr.) Lindl.
HOOKER'S ORCHID   native
IR | WUP | CUP | EUP   FAC CC 8
Habenaria hookeri Torr.
Perennial herb; scape 2–4 dm tall, bractless, rising from a few fleshy roots. Leaves 2, basal, commonly broadly elliptic to rotund, 6–12 cm long, usually 2/3 to fully as wide, blunt or rounded, abruptly narrowed to the base. Flowers sessile, yellowish green, ascending; lip triangular-lance-shaped, directed outward, upcurved, 8–12 mm long; lateral petals lance-shaped, incurved and more or less adjacent under the upper sepal; spur 13–24 mm long, directed, downward, tapering to the tip. June–July.

Coniferous or mixed forests, wooded dunes; soils often sandy.

Platanthera huronensis (Nutt.) Lindl.
NORTHERN BOG-ORCHID   native
IR | WUP | CUP | EUP   FACW CC 5
Habenaria hyperborea (L.) R. Br.
Perennial herb, roots fleshy. Stems 2–8 dm long. Leaves 2–7, alternate on stem, linear to oblong, 5–30 cm long and 2–5 cm wide, with 1–3 smaller leaves above. Flowers small, green, erect, many in a raceme 4–25 cm long; lateral sepals ovate and spreading; lateral petals lance-shaped, curved upward and joined with upper sepal to form a loose hood over column; lip lance-shaped, 3–7 mm long, not abruptly widened at base; spur curved forward under the lip, about as long as lip, 3–7 mm long. June–Aug.

Moist to wet forests and swamps, thickets, streambanks, wet meadows, wet sand along Great Lakes shoreline, ditches.

The *Platanthera hyperborea* complex, including *P. dilatata*, *P. aquilonis*, and *P. huronensis*, are often difficult to separate; living rather than dried plants are easiest to identify.

Platanthera lacera (Michx.) G. Don
GREEN FRINGED ORCHID   native
IR | WUP | CUP | EUP   FACW CC 6
Habenaria lacera (Michx.) R. Br.
Perennial herb, roots fleshy. Stems 3–8 dm long. Leaves 3–7, alternate on stem, lance-shaped to oval, to 5–15 cm long and 1–4 cm wide; upper leaves much smaller. Flowers white or green-white, in a usually compact, many-flowered raceme, 5–20 cm long and 2–5 cm wide; sepals broadly oval, 4–7 mm long, the lateral ones deflexed behind the lip; lateral petals linear, entire; lip 10–16 mm long and 5–20 mm wide, deeply 3-lobed, each lobe fringed with a few long segments; spur curved, 1–2 cm long. June–Aug.

Hummocks in open sphagnum bogs, conifer bogs, swamps, wet meadows, sandy prairie, thickets, ditches.

Platanthera obtusata (Banks) Lindl.
BLUNT-LEAF ORCHID   native
IR | WUP | CUP | EUP   FACW CC 10
Habenaria obtusata (Banks) Richards.
Perennial herb, roots fleshy. Stems leafless, slender, 1–3 dm long. Leaves 1 at base of stem, ascending,

persistent through flowering, obovate, 5–15 cm long and 1–4 cm wide, blunt-tipped, long-tapered to base. Flowers 4–20, green-white, in a raceme 3–12 cm long and 1–2 cm wide; lateral sepals ovate, spreading; petals ascending, widened below middle; lip lance-shaped, widened at base, 4–6 mm long; spur curved, tapered to a thin tip, 5–8 mm long. June–Aug.

Shaded hummocks in conifer swamps (especially under cedar, black spruce or balsam fir), wet mixed conifer-deciduous forests, alder thickets.

Platanthera orbiculata (Pursh) Lindl.
ROUND-LEAF ORCHID native
IR | WUP | CUP | EUP FAC CC 10

Habenaria orbiculata (Pursh) Torr.

Perennial herb, roots fleshy. Stems 2–6 dm long, leafless apart from 1–6 small bracts. Leaves 2, opposite at base of plant, spreading or lying flat on ground, more or less round, shiny, 6–15 cm long and 4–15 cm wide. Flowers green-white, several in a raceme 5–20 cm long and 3–6 cm wide; sepals ovate, to 1 cm long; petals ovate, 6–7 mm long; lip entire, rounded at tip, 10–15 mm long and 2 mm wide; spur 2–3 cm long, somewhat widened at tip. Late June–Aug.

Shaded conifer swamps of white cedar, balsam fir, and black spruce, especially where underlain by marl; also in drier pine forests.

ADDITIONAL SPECIES

Platanthera macrophylla (Goldie) P. M. Br., an uncommon species of rich deciduous or mixed woods, formerly included in *P. orbiculata*; Alger and Marquette counties. The flowers of the two species differ as follows:

1   Spurs 16–27 mm long .... **P. orbiculata**
1   Spurs 29–43 mm long .... **P. macrophylla**

Platanthera psycodes (L.) Lindl.
LESSER PURPLE FRINGED ORCHID native
IR | WUP | CUP | EUP FACW CC 7

Habenaria psycodes (L.) Spreng.

Perennial herb, roots thick and fleshy. Stems stout, 3–10 dm long. Leaves 4–12, alternate on stem, lance-shaped or oval, the upper much smaller and narrow. Flowers rose-purple, in a densely flowered, cylindric raceme 4–20 cm long and 3–5 cm wide; sepals oval to obovate, 4–6 mm long; petals spatula-shaped, finely toothed on margins; lip broad, 8–14 mm wide, deeply 3-lobed, the lobes fan-shaped, fringed to less than half way to base; spur curved, about 2 cm long. July–Aug.

Wetland margins, shores, wet forests, wet meadows, low prairie, roadside ditches; typically not on sphagnum moss.

Pogonia *Snake-Mouth*

Pogonia ophioglossoides (L.) Ker-Gawl.
ROSE POGONIA native
IR | WUP | CUP | EUP OBL CC 10

Perennial herb, spreading by surface runners (stolons) which send up a stem every 10 cm or more apart. Stems slender, smooth, 1.5–4 dm long. Leaves single, attached about halfway up stem, narrowly oval, 3–10 cm long and 1–2.5 cm wide, stalkless. Flowers pink to purple, usually 1 at end of stems; sepals widely spreading, petals oval, hovering over the column; lip pink with purple veins, 1.5–2 cm long and 5–10 mm wide, fringed at tip, bearded with yellow bristles. June–July.

Conifer swamps and open bogs in sphagnum moss, floating sedge mats, sedge meadows, sandy interdunal wetlands.

Spiranthes *Ladies'-Tresses*

Perennial herbs, from a cluster of tuberous roots. Stems slender, erect. Leaves largest at base of plant, becoming smaller upward on stem, the stem leaves erect and sheathing. Flowers small, white or creamy, spirally twisted in a densely flowered, spike-like raceme; sepals and lateral petals similar, the lateral petals joined with all 3 sepals or with only the upper sepal to form a hood over lip and column; lip folded upward near middle so that margins embrace the column, curved downward beyond the middle, with a pair of bumps or thickenings at base; anthers 1, from back of the short column.

1   Leaves widely spreading or lying flat in a basal rosette, short-petioled, sometimes withered at flowering time, their blades less than 4.5 cm long, about 2/5 as wide as long or wider; perianth 2.5–5.5 mm long ..... **S. lacera**
1   Leaves ascending, not distinctly petioled, usually present at flowering time, their blades (non-sheathing portion) over 4.5 cm long and less than 2/5 as wide as long; flowers usually in 2 or more rows in a ± crowded spike (sometimes one-sided); perianth larger than 5.5 mm long. 2
2   Plants flowering in June and early July; largest leaves (non-sheathing portion) about 5–10x longer than wide; lip bright yellow or yellowish orange; leaves all basal (rarely 1 on the stem), 4.5–10 cm long, the stem with 1–2 bracts (including reduced leaf, if present) .. **S. lucida**
2   Plants flowering in mid- to late summer and fall; largest leaves commonly over 10x longer than wide, or leaves absent; lip white or creamy, sometimes the central portion pale yellow; leaves often present on lower portion of stem, the stem bracts and leaves numbering 3–6. .. 3
3   Lower flowers with perianth 3–7.5 mm long (most easily measured using the dorsal sepal); flowers in one row, usually in a loose spiral because of the twisted rachis . ... **S. casei**

3 Lower flowers with perianth 7–11 mm long (most easily measured using the dorsal sepal); flowers in 2 or more rows, often tightly spiraled........................ 4

4 Lip fiddle-shaped, strongly constricted behind expanded tip; lateral sepals united for at least half their length with dorsal sepal and lateral petals, forming a hood...
.... **S. romanzoffiana**

4 Lip oblong, often erose-margined but not strongly constricted; at least the lateral sepals free (or easily separated if connivent when young)........ **S. cernua**

Spiranthes casei Catling & Cruise
CASE'S LADIES'-TRESSES *native*
IR | **WUP** | **CUP** | **EUP** CC 8

Perennial herb. Stems 2–4 dm tall. Basal leaves lance-shaped or broader, 5–15 cm long and 8–15

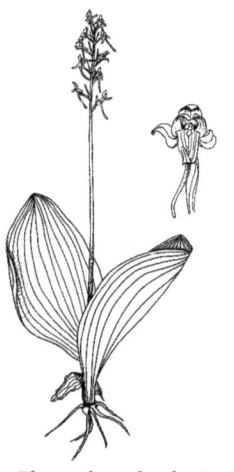

Platanthera hookeri

Platanthera huronensis

Platanthera dilatata

Platanthera clavellata

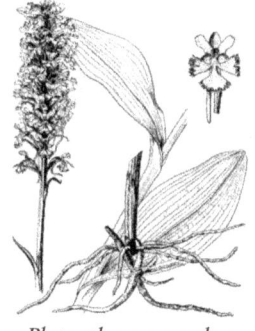

Platanthera psycodes

Pogonia ophioglossoides

Platanthera lacera

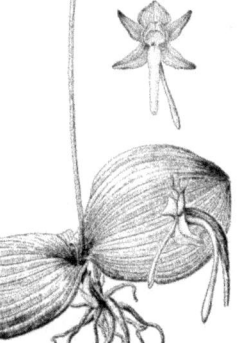

Platanthera orbiculata

mm wide; stem sheaths 3 or 4, the lower with a short blade. Inflorescence 2–16 cm long, loosely to fairly densely flowered. Flowers whitish or greenish yellow, more greenish near base; lip fleshy, obovate, papillate below, the basal bumps about 1 mm long. Aug–Sept.

Sandy acidic soil, often with bracken fern (*Pteridium*).

Spiranthes cernua (L.) L.C. Rich.
WHITE NODDING LADIES'-TRESSES   *native*
IR | WUP | CUP | EUP   FACW CC 4
Perennial herb, roots fleshy. Stems 1–5 dm long, upper stem short-hairy, lower stem smooth. Leaves mostly at base of plant, usually present at flowering time, linear to oblong lance-shaped, 6–25 cm long and 5–15 mm wide; upper stem leaves 3–5, much smaller and bractlike. Flowers white, in a spike-like raceme 3–15 cm long, with 2–4 vertical rows of flowers, the rows spirally twisted; sepals and petals hairy on outside; lateral petals joined with upper sepal to form a hood; lip white, yellow-green at center, 6–10 mm long and 3–6 mm wide, slightly narrowed at middle, curved downward, the tip curved inward toward stem, the tip wavy-margined or with small rounded teeth, the base of lip with a pair of backward-pointing bumps. Aug–Oct.

Open, usually sandy wetlands such as wet meadows, lakeshores, moist prairies, ditches and roadsides.

Spiranthes lacera (Raf.) Raf.
NORTHERN SLENDER LADIES'-TRESSES   *native*
IR | WUP | CUP | EUP   FAC CC 8
Perennial herb. Stems arising from a cluster of thickened roots, very slender, 1–6 dm tall. Basal leaves often present at anthesis, oval to oblong, 2–5 cm long, a to a 1/3 as wide; stem leaves reduced to small scales. Flowers white, in a very slender twisted raceme 4–10 cm long, glabrous or nearly so; lip oblong, about 4 mm long, white with green median area, its tip abruptly deflexed, crisped on the margin. Aug–Sept.

Dry sandy soil, often with blueberry and bracken fern in open woods of jack pine, red pine, and oak; moist aspen groves, conifer thickets along shores and on dunes.

Spiranthes lucida (H. H. Eat.) Ames
SHINING LADIES'-TRESSES   *native*
IR | WUP | CUP | EUP   FACW CC 7
Small perennial herb, roots fleshy. Stems slender, smooth to finely hairy above, 1–3 dm long. Leaves mostly at base of plant, oblong lance-shaped, shiny, 5–10 cm long and 5–15 mm wide; stem leaves usually 2, small and bractlike. Flowers white, nodding, in a spike-like raceme 2–7 cm long, the flowers in 1–2 vertical rows, the rows spirally twisted; upper sepal and petals forming a hood over the column; lip oblong, 5–6 mm long, the outer half bright yellow or yellow-orange, with white margins, bumps at base of lip small, less than 1 mm long. June–July.

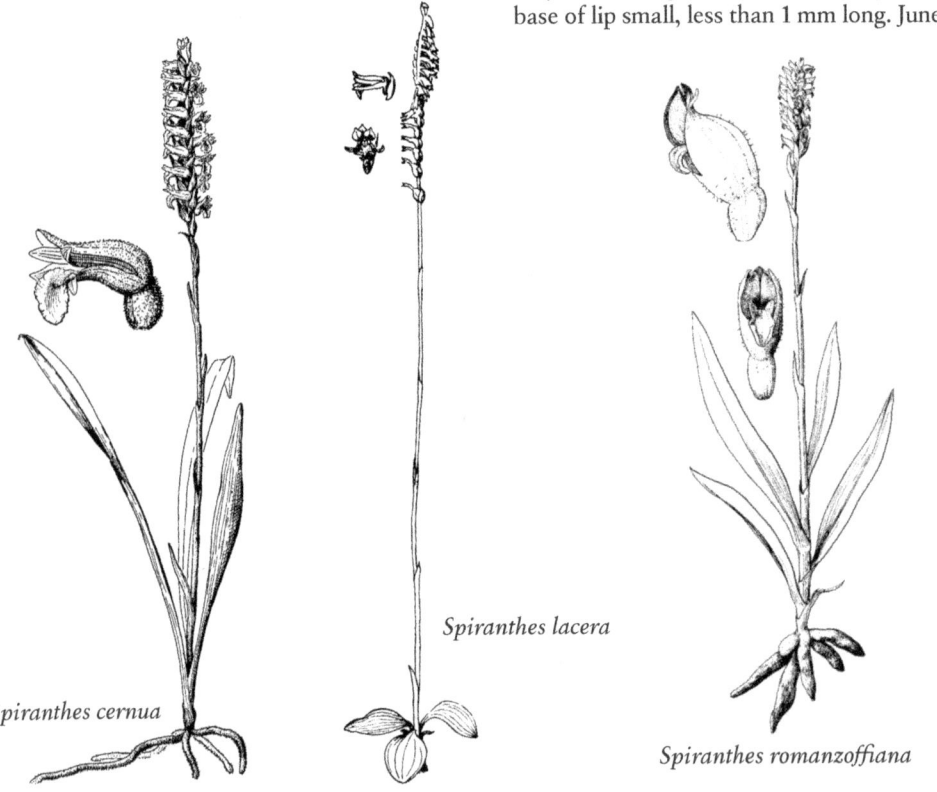

Spiranthes cernua

Spiranthes lacera

Spiranthes romanzoffiana

Streambanks, lakeshores, wet meadows, ditches; especially on calcium-rich soils and limestone gravels, often where somewhat disturbed; Menominee and Ontonagon counties.

Spiranthes romanzoffiana Cham.
HOODED LADIES'-TRESSES *native*
IR | **WUP** | **CUP** | **EUP** OBL CC 8

Perennial herb, roots thick and fleshy. Stems 1–4 dm long, upper stem finely hairy. Leaves mostly from base of plant, present at flowering time, upright, linear to narrowly lance-shaped, 5–20 cm long and 3–9 mm wide, the stem leaves becoming smaller and bractlike. Flowers white or cream-colored, in a spike-like raceme 3–10 cm long, with 1–3 vertical rows of flowers, the rows spirally twisted; sepals and lateral petals joined to form a hood over the lip; lip ovate, strongly constricted near middle (violin-shaped), curved downward, the tip ragged and bent inward toward stem, the bumps at base very small. July–Sept.

Open wetlands including wet meadows, fens, lakeshores, open swamps, ditches, seeps; usually in neutral or calcium-rich habitats.

Poaceae

GRASS FAMILY

Perennial or annual herbs, clumped or spreading by rhizomes. Stems (culms) usually hollow, with swollen, solid nodes. Leaves linear, parallel-veined, alternate in 2 ranks or rows, sheathing the stem, the sheaths usually split vertically, sometimes joined and tubular as in brome (*Bromus*) and mannagrass (*Glyceria*); with a membranous or hairy ring (ligule) at top of sheath between blade and stem, or the ligule sometimes absent; a pair of projecting lobes (auricles) sometimes present at base of blade.

Flowers (florets) small, usually perfect (with both staminate and pistillate parts), or sometimes either staminate or pistillate, the staminate and pistillate flowers separate on the same or different plants. Florets grouped into spikelets, each spikelet with 1 to many florets, the florets stalkless and alternate along a small stem or axis (rachilla), with a pair of small bracts (glumes) at base of each spikelet (the glumes rarely absent); the glumes usually of different lengths, the lowermost (or first) glume usually smaller, the upper (or second) glume usually longer. Within the spikelet, each floret subtended by 2 bracts, the larger one (lemma) containing the flower, the smaller one (palea) covering the flower; the lemma and palea often enclosing the ripe fruit (grain or caryopsis); stamens usually 3 or sometimes 6, usually exserted when flowering; ovary superior, never enclosed in a sac (as in sedges); styles 2–3-parted, the stigmas often feathery.

Spikelets grouped in a variety of heads, most commonly in branching heads (panicles), or stalked along an unbranched stem (rachis) in a raceme, or the spikelets stalkless along an unbranched stem in a spike; spikelets breaking (disarticulating) either above or below the glumes when mature, the glumes remaining in the head if falling above the glumes, or the glumes falling with the florets if disarticulation is below the glumes.

ADDITIONAL SPECIES
Holcus mollis L. (Creeping Velvet-Grass), introduced perennial grass; in UP, known only from Alger and Houghton counties.

KEY TO POACEAE GROUPS

1. Tip of plant with a large "tassel" or spike-like raceme bearing staminate florets in pairs, the pistillate florets either sunken in hardened joints of rachis below the staminate portion or in separate "ears" lower on the plant .... **KEY 1**
1. Tip of plant not as above; upper portion of inflorescence bearing pistillate or bisexual florets, or rarely staminate florets in short one-sided spikes 2
2. Spikelets concealed within ± globular, hard, bur-like structures .... **Cenchrus**
2. Spikelets exposed, not concealed within globular, bur-like structures 3
3. Spikelets all unisexual, segregated into different and dissimilar parts of the inflorescence .... **Zizania**
3. Spikelets perfect, or if unisexual, then scattered among bisexual spikelets 4
4. Spikelets forming a simple spike or spikes, directly sessile or subsessile on main axis of inflorescence or at most on secondary branches 5
4. Spikelets not forming simple spikes as above; pedicelled and/or on tertiary or further branches of the inflorescence; in some species congested and hence spike-like, but not directly sessile or subsessile (reduced panicle branches usually visible upon removal of some spikelets) 7
5. Spike solitary, terminal (its rachis a continuation of the culm), the spikelets on opposite sides of rachis key 2
5. Spikes several, one-sided (spikelets in two rows on one side of rachis) 6
6. Glumes keeled and ± equal (or the smaller half or more as long as the larger) .... **KEY 3**
6. Glumes rounded on the back (not keeled), very unequal .... **KEY 4**
7. Spikelets with only 1–2 florets 8
7. Spikelets with 3 or more florets, including any sterile ones 14
8. Spikelets with an involucre consisting of long subtending bristles .... **KEY 4**
8. Spikelets without an involucre of bristles (although glumes or lemmas may be awned) 9

9   Glumes or lemmas (or both) ± laterally compressed or keeled (lateral nerves, if present, less prominent than midnerve) go to couplet 14
9   Glumes and lemmas rounded on back, not keeled (nerves, if present, about equally prominent). 10
10 Glumes very unequal in length, one of them minute, or absent, or at most about half as long as the spikelet. (Note: In Key 4 species, a sterile lemma is present that closely resembles the large second glume opposite it and might easily be misinterpreted as a glume; the true first glume is a small, sometimes minute and membranous, even deciduous, scale at the very base of the spikelet; a reduced palea, often associated with the sterile lemma, will also help to identify the latter as part of a sterile floret and not a glume) 11
10 Glumes ± equal in length, neither of them much reduced nor absent 12
11 Spikelets disarticulating below the glumes (except in *Setaria italica*), ± elliptic (less than 3 times as long as wide); a sterile lemma resembling the larger glume present .... **KEY 4**
11 Spikelets disarticulating above the glumes, ± lanceolate (3–10 times as long as wide); no sterile lemma present .... **KEY 5**
12 Spikelets paniculate, all (or mostly) 1-flowered, bisexual, the florets all alike (no sterile lemmas or separate sterile pedicels present); spikelets less than 4 mm long, except in *Stipa* with awns over 5 cm long .... **KEY 5**
12 Spikelets paniculate or racemose, basically 2-flowered (the lower floret staminate or sterile with often suppressed palea); spikelets 3 mm or more long 13
13 Spikelets all alike, not paired with a pedicel bearing a rudimentary, staminate, or no floret **Panicum virgatum**
13 Spikelets in pairs, usually of two kinds: one sessile and with a bisexual floret, the other a hairy pedicel with or without a staminate or rudimentary floret (rarely 2 stalked florets with 1 sessile one) .... **KEY 1**
14 Spikelets all or mostly containing 1 bisexual floret and no sterile or vestigial ones below it 1 5
14 Spikelets all or mostly containing 2–several florets, the lower ones sometimes staminate or rudimentary (scale-like or reduced to tiny hairy appendages) 16
15 Glumes both completely absent; spikelets strongly flattened, appressed and ± overlapping, the lemmas scabrous or hispid-ciliate .... **Leersia**
15 Glumes (one or both) usually present; spikelets various but not as above .... **KEY 5**
16 Glumes shorter than the lowest floret (excluding awns if present); awn of lemma none, terminal, arising from between terminal teeth and not twisted, or at most subterminal .... **KEY 7**
16 Glumes (at least one of them, not necessarily the first glume) longer than lowest floret; awn of lemma none or arising from between terminal teeth and strongly twisted below, or (the usual condition) inserted on the middle or lower part of the lemma 17
17 Spikelets containing one bisexual awnless floret (the lemma sometimes membranaceous) with two additional, often dissimilar (sometimes awned) staminate, sterile, or vestigial lemmas below it .... **KEY 8**
17 Spikelets usually containing 2 or more bisexual florets (staminate or sterile florets, if present, above the fertile one and/or fertile lemma awned) .... **KEY 9**

KEY 1

Tribe Sacchareae

Subtribe Andropogoninae (*Andropogon, Schizachyrium*)
Subtribe Sorghinae (*Sorghastrum*)

1   Inflorescence an open to contracted panicle .... **Sorghastrum**
1   Inflorescence of 1 or several narrow or spike-like simple racemes (some spikelets sessile and some pediceled) 2
2   Spike-like simple racemes solitary at the ends of the branches .... **Schizachyrium**
2   Spike-like simple racemes 2–ca. 20 at the ends of the branches .... **Andropogon**

KEY 2

Tribe Hordeeae

Subtribe Hordeinae (*Elymus, Hystrix, Hordeum, Leymus, Pascopyrum, Secale*)

Subtribe Triticinae (*Triticum*)

Tribe Poeae

Subtribe Loliinae (*Lolium*)

1   Lemmas smooth and glabrous, except for a spiny-ciliate keel and exposed margin, tapering into a long awn ... . **Secale**
1   Lemmas smooth to scabrous or pubescent, but not simply with spiny-ciliate keel and margin, awned or awnless 2
2   Larger glumes 3.3–6.5 mm broad with at least 3 prominent nerves, the keel or midnerve not centered 3
2   Larger glumes less than 2.5 mm broad, variously nerved (or glumes absent) 4
3   Glumes glabrous, or pubescent toward the base on nerves and margins (rarely pubescent throughout), the larger ones 4–6.5 mm wide, less than 3 times as long (excluding awns if present); lemmas awned or awnless .... **Triticum**
3   Glumes softly hairy or glabrous throughout, 3.3–4.2 mm wide, ca. 6–10 times as long; lemmas awnless . **Leymus**
4   Spikelets mostly 2–3 at each node of the rachis; glumes usually 4–6 (the spikelet arrangement may be obscured by reduction or asymmetric positions of some spikelets, but the basic structure is revealed by the presence of usually 4–6 glumes subtending the entire group of spikelets); glumes usually vestigial or absent in *Elymus hystrix* with mostly 2 easily recognized narrow spikelets at each node 5
4   Spikelets 1 at each node (or most nodes) of the rachis; glumes variously arranged (or absent), but not more than 2 8

5 Spikelets 2 at each node of the rachis (or at some nodes, only 1, rarely 3, but total number of glumes [awn-like or broader] developed at a node not more than 4) **Elymus**

5 Spikelets basically 3 at each node of the rachis (the lateral 2 in commonest species reduced to bristles), this arrangement most easily recognized by the presence of 6 awn-like or narrowly lanceolate and awn-tipped glumes at a node 6

6 Body of larger lemmas ca. 3.5–6 mm long; rachis of spike readily disintegrating at maturity .... **Hordeum**

6 Body of larger lemmas ca. 8–12 mm long; rachis not disintegrating 7

7 Awn of lemmas much stouter than awn of glumes, ± straight .... **Hordeum vulgare**

7 Awn of lemmas as slender as awn of glumes, spreading to recurved at maturity .... **Elymus**

8 Glumes 1 (except terminal spikelet with 2), the narrow edge of the spikelet against the rachis and lacking a glume .... **Lolium**

8 Glumes 2 on all spikelets............................ 9

9 Lemmas with awns strongly divergent or recurved when mature .... **Elymus trachycaulus**

9 Lemmas with awns ± straight or absent 10

10 Lemmas densely hairy .... **Elymus lanceolatus**

10 Lemmas glabrous (rarely slightly pubescent), smooth or scabrous 11

11 Leaf blades mostly broad and flat, slightly or not at all involute when dry, no more deeply grooved above than below between the numerous fine nerves (not strongly scabrous, usually with scattered long hairs above); glumes completely lacking cilia toward the base; cartilaginous belt at upper nodes nearly or fully as long as its diameter .... **Elymus**

11 Leaf blades strongly involute when dry, deeply grooved above between the prominent raised nerves (and usually strongly scabrous above); glumes mostly with margin minutely ciliate toward base; cartilaginous belt (sharply defined, usually darker, non-green zone) at upper nodes of culm usually less than half as long as its diameter .... **Pascopyrum smithii**

KEY 3
Tribe Chlorideae
  Subtribe Bouteloinae (*Bouteloua*)
  Subtribe Eleusininae (*Eleusine*)
Tribe Poeae
  Subtribe Poinae (*Beckmannia*)
Tribe Zoysieae
  Subtribe Sporobolinae (*Spartina*)

1 Spikelets concealed within ± globular, hard, bur-like structures .... **Buchloe*** (female plants)

1 Spikelets exposed, not concealed within bur-like structures 2

2 Spikes radiating from summit of culm (i.e., umbellate or nearly so) or at least the lower ones whorled (solitary in depauperate individuals); anthers not over 1.5 mm long .... **Eleusine**

2 Spikes all racemose or panicled; anthers various 3

3 Glumes equal, ca. 2–3 mm long, deeply pouch-like and largely covering the floret, the spikelet strongly flattened and about as wide as long; ligule membranous, not ciliate; anthers 0.5–1 mm long ...... **Beckmannia**

3 Glumes unequal, the longer ones ca. 2.5–11 mm long, not pouch-like, equaling or shorter than florets, the spikelet not strongly flattened, much narrower than long; ligules and anthers various 4

4 Spikes short peduncled, peduncles glabrous to hispid, 0.5–20 mm long .... **Spartina**

4 Spikes ± sessile or on finely hispidulous or short-pubescent peduncles to 3 mm long .... **Bouteloua**

KEY 4
Tribe Paniceae
  Subtribe Anthephorinae (*Digitaria*)
  Subtribe Boivinellinae (*Echinochloa*)
  Subtribe Cenchrinae (*Cenchrus, Setaria*)
  Subtribe Panicinae (*Dicanthelium, Panicum*)

1 Spikelets with an involucre consisting of a spiny bur or of long subtending bristles 2

1 Spikelets without an involucre (although glumes or lemmas may be awned) 3

2 Involucre a spiny bur enclosing much or all of the spikelets, the whole readily disarticulating **Cenchrus**

2 Involucre of long slender bristles subtending but not concealing the spikelet, remaining attached to pedicels when spikelets disarticulate .... **Setaria**

3 Spikelets ± spiny-hispid and usually also awned; ligule none .... **Echinochloa**

3 Spikelets glabrous or pubescent but not coarsely hispid and not awned; ligule present, distinct or nearly absent (of hairs or membranous) 4

4 Inflorescence composed of 1-sided spikes or spike-like racemes, the rachis of each winged or at least flat on the side opposite the spikelets .... **Digitaria**

4 Inflorescence an open panicle, not spike-like or distinctly one-sided 5

5 Ligule a membranous collar 1–1.5 mm high, without hairs; base of leaf blade and very summit of sheath without special zone of short pubescence or long hairs or cilia; first glume minute or vestigial; fertile lemma leathery in texture, with thin flat translucent margins .... **Digitaria cognata**

5 Ligule usually partly or entirely of short or long hairs; (if ligule membranous, plants not otherwise as above: ligule ca. 0.5 mm long or virtually absent; or summit of sheath or basal margin of blade pubescent or ciliate; and/or first glume more than 0.5 mm long); fertile lemma hard and shiny 6

6 Spikelets at least sparsely pubescent towards their margins .... **Dichanthelium**

6 Spikelets glabrous 7

7 Terminal panicle 8–40 cm long, (smaller in occasional depauperate individuals of annual species); annuals or perennials, but without clear remnants of overwintering basal rosette leaves; flowering and fruiting summer-fall .... **Panicum**
7 Terminal panicle 2.5–8 (–12) cm long; tufted perennials with clear remnants of old, dead leaves from the previous year present at the base, these sometimes formed into a clear overwintering rosette; flowering spring and fruiting in late spring–early summer .. **Dichanthelium**

KEY 5
Tribe Aristideae (*Aristida*)
Tribe Brachyelytreae (*Brachyelytrum*)
Tribe Poeae
  Subtribe Agrostidinae (*Agrostis, Ammophila, Calamagrostis*)
  Subtribe Miliinae (*Milium*)
  Subtribe Phleinae (*Phleum*)
  Subtribe Poinae (*Alopecurus, Apera, Beckmannia, Cinna*)
Tribe Stipeae
  Subtribe Stipinae (*Oryzopsis, Piptatherum, Hesperotipa*)
Tribe Zoysieae
  Subtribe Sporobolinae (*Calamovilfa, Sporobolus*)

1 Lemma with awn (or awns) strictly terminal 2
1 Lemma with awn absent or dorsal or subterminal ... 7
2 Awns of lemma 3 (lateral ones sometimes very short) . ... **Aristida**
2 Awn of lemma solitary 3
3 Body of lemma 8–23 mm long 4
3 Body of lemma less than 7 mm long 5
4 Glumes 9.5–45 mm long .... **Hesperostipa**
4 Glumes rudimentary or one of them up to 5 mm long . ... **Brachyelytrum**
5 Glumes acuminate, not over 1 mm wide, keeled, the spikelets somewhat compressed; lemma ± keeled, membranous or thin .... **Muhlenbergia**
5 Glumes acute to obtuse, more than 1 mm wide, scarcely if at all keeled, the spikelets nearly terete; lemma rounded on the back, ± firm and hardened 6
6 Principal leaf blades basically flat (just the margins involute), basal or nearly so, densely and very finely rough-puberulent, with strong closely spaced veins .. .. **Oryzopsis**
6 Principal leaf blades involute or, if flat, all cauline and glabrous .... **Piptatherum**
7 Spikelets 10–15 mm long; anthers (4–) 5–8 mm long; panicle crowded and ± spike-like, (10–) 12–20 (–28) mm across at the middle .... **Ammophila**
7 Spikelets less than 8 mm long (excluding awns); anthers to 4.5 mm long (usually much shorter); panicle various 8
8 Spikelets sessile or nearly so, crowded in a very dense spike-like panicle (branches of panicle suppressed, scarcely if at all visible without dissection of panicle) 9
8 Spikelets in ± open or contracted (but not densely spike-like) inflorescences, with evident pedicels and/or panicle branches 11
9 Glumes awnless .... **Alopecurus**
9 Glumes awned 10
10 Plants with scaly rhizomes; glumes gradually tapered into awn; ligule to ca. 1 mm long ...... **Muhlenbergia**
10 Plants without scaly rhizomes; glumes abruptly rounded or truncate, the awn distinct; ligule over 1 mm long ... . **Phleum**
11 Spikelets rounded on back, not keeled (neither glumes nor lemma with a midvein more prominent than other nerves), at least 2.5 mm long; lemma ± shiny, distinctly firmer in texture than the glumes 12
11 Spikelets keeled (glumes and/or lemmas with midvein more prominent than other nerves) or less than 2.5 mm long; lemma no firmer in texture than the glumes... 13
12 Lemmas with appressed pubescence; leaves with blades usually involute; upper ligules not over 3 mm long ... . **Piptatherum pungens**
12 Lemmas glabrous; leaves with blades broad and flat; upper ligules mostly 4–6 (–8) mm long .... **Milium**
13 Ligule a fringe of short hairs; lemmas awnless 14
13 Ligule membranous (at most minutely ciliate at summit of membrane); lemmas awned or awnless 15
14 Lemma (4.7–) 5–6.7 mm long, surrounded with a tuft of long hairs (more than half its length) at its base .... **Calamovilfa**
14 Lemma 1.5–5.5 mm long, without long hairs at its base .... **Sporobolus**
15 Lemma with long hairs at base (on or near callus)... 16
15 Lemma without long hairs at base (at most with hairs on callus less than 0.5 mm long) 17
16 Long hairs at least in part arising from lower portion of lemma; glumes (excluding awn-tips if present) shorter than lemma .... **Muhlenbergia**
16 Long hairs restricted to callus at base of floret; glumes slightly exceeding lemma .... **Calamagrostis**
17 Glumes both distinctly shorter than lemma; lemma awnless .... **Muhlenbergia**
17 Glumes (one or both of them) equaling or exceeding the lemma and/or the lemma awned 18
18 Floret raised above base of glumes on a minute stalk; spikelet articulated below the glumes; lemma with a small subterminal awn; stamen 1 .... **Cinna**
18 Floret not stalked; spikelets articulated above the glumes; lemma awnless or with long subterminal awn or with dorsal awn; stamens 3 19
19 Lemma with a long subterminal awn, much exceeding the body in length; rachilla prolonged (scarcely 0.5 mm) behind the palea .... **Apera***
19 Lemma awnless or with mid-dorsal awn; rachilla not prolonged .... **Agrostis**

KEY 6

Tribe Oryzeae (*Leersia, Zizania*)

1. Spikelets bisexual; plant variously scabrous-pubescent, at least the nodes retrorsely bearded; stamens 3. . . . . **Leersia**
1. Spikelets all unisexual, segregated into different parts of the inflorescence, upper panicle branches bearing awned pistillate spikelets, lower branches bearing staminate spikelets; plant glabrous; stamens 6 . . . **Zizania**

KEY 7

Tribe Arundineae (*Phragmites*)
Tribe Bromeae (*Bromus*)
Tribe Eragrostideae
  Subtribe Eragrostidinae (*Eragrostis*)
Tribe Meliceae (*Glyceria, Melica, Schizachne*)
Tribe Poeae
  Subtribe Aveninae (*Graphephorum, Trisetum*)
  Subtribe Coleanthinae (*Puccinellia, Sclerochloa*)
  Subtribe Cynosurinae (*Cynosurus*)
  Subtribe Dactylidinae (*Dactylis*)
  Subtribe Holcinae (*Deschampsia*)
  Subtribe Loliinae (*Festuca, Schedonorus*)
  Subtribe Torreyochloinae (*Torreyochloa*)

1. Plants tall and stout (usually over 1.5 m tall) with larger leaf blades 1–3.5 cm wide; spikelets ca. 11–17 mm long; ligule a densely ciliate brown band; rachilla (above the lowest floret) with silky beard about equaling or exceeding the lemmas . . . . **Phragmites australis**
1. Plants generally less than 1.5 m tall with narrow leaves less than 1 cm wide; spikelets and ligule various; rachilla with beard shorter or absent 2
2. Spikelets sessile or at most very short-pediceled, crowded into dense clusters, these either at the ends of elongate panicle branches or in a single congested, rather spike-like inflorescence 3
2. Spikelets short- to long-pediceled in a ± open panicle 6
3. Lemma with a prominent, somewhat twisted or spreading dorsal awn; rachilla villous . . . . **Trisetum**
3. Lemma with awn absent or short and strictly terminal; rachilla not villous 4
4. Clusters of spikelets at the ends of elongate naked branches of the panicle; sheaths closed much of their length; ligule 2–8 mm long . . . . **Dactylis**
4. Clusters of spikelets all crowded into a congested, rather spike-like inflorescence; sheaths open their entire length; ligule 1 mm or less long 5
5. Spikelets of two kinds in a cluster: normal fertile and special sterile fan-like ones; fertile lemmas mostly short- or long-awned . . . . **Cynosurus**
5. Spikelets all similar, fertile; lemmas not awned . . **KEY 3**
6. Callus at base of floret with dense beard of straight hairs 0.5 mm or more long 7
6. Callus glabrous, minutely puberulent, or cobwebby (not bearded with straight hairs) 11
7. Awn of lemma arising near base . . . . **Deschampsia**
7. Awn of lemma absent, terminal, subterminal, or arising between terminal teeth 9
8. Lemmas awnless, weakly 5-nerved; sheaths open . . . . **Graphephorum**
8. Lemmas with short or long awn, either 3-nerved or plants with closed sheaths 10
9. Sheaths open; lemmas 3-nerved, ± truncate (ragged or lobed) at tip, the nerves hairy . . . . **Tridens**
9. Sheaths closed; lemmas 5–7-nerved, tapering to an apparently 2-lobed or sharply bifid tip, glabrous or hairy 10
10. Callus with distinct beard, the lemma glabrous or nearly so; grain glabrous . . . . **Schizachne**
10. Callus lacking a distinct beard, the pubescence like that of the lemma; grain pubescent at the summit . **Bromus**
11. Glumes (at least one of them) and usually also lemmas strongly keeled (lemmas in a few species rounded on the back); awns absent or not over 2 mm 12
11. Glumes and lemmas rounded on back, not keeled (or obscurely so toward tip); awns absent or present . . . 15
12. Larger glumes ca. 4.5–7 mm long . . . . **Dactylis**
12. Larger glumes not over 4.4 mm long 13
13. Larger glumes ± obovate, broadest above the middle . . . . **KEY 3**
13. Larger glumes broadest at or below the middle 14
14. Ligule a fringe of hairs; lemmas with 3 prominent nerves, glabrous; spikelets 2–30-flowered . . **Eragrostis**
14. Ligule a membranous scale, the cilia, if any, shorter than the scale; lemmas with 3–5 nerves, glabrous, hairy, and/or cobwebby at base; spikelets various . . . . **Poa**
15. Lemmas usually 2-toothed or minutely 2-lobed at the tip and usually with at least a short awn arising from just below or between the teeth (if teeth apparently united, as in some species of *Bromus,* the awn thus subterminal); sheaths closed nearly to their summit . . . 16
15. Lemmas not 2-lobed or 2-toothed at tip, awnless or with strictly terminal awn; sheaths open (or closed in *Glyceria,* with prominently nerved and awnless lemmas, and in the youngest shoots of *Festuca rubra*) 17
16. Spikelets narrowly linear-lanceolate on much shorter, densely hispid pedicels; lemmas awned, minutely strigose or scabrous, at least on the nerves; sheaths retrorsely scabrous; ligules 3–6 mm long; grain glabrous . . . . **Melica**
16. Spikelets broadly linear to oblong, usually on ± elongate pedicels; lemmas various; sheaths glabrous or pubescent but not scabrous; ligules less than 2.5 (–4) mm long; grain pubescent at the summit . . . . **Bromus**
17. Lemmas acute at the tip, awned 18
17. Lemmas acutish or obtuse, awnless 19
18. Blades of leaves flat (or merely once-folded), at least the larger ones (2.5–) 3–8 mm wide . . . . **Schedonorus**

18 Blades of leaves ± strongly involute, less (usually much less) than 3 mm broad . . . . **Festuca**

19 Nerves of lemma prominent, straight and becoming parallel at the tip; sheaths closed or open 20

19 Nerves of lemma very weak (or if visible, then converging, not parallel, at the tip); sheaths open 21

20 Sheaths closed much of their length (but easily splitting); second (larger) glume with one distinct nerve; plants rhizomatous . . . . **Glyceria**

20 Sheaths completely open; second glume with 3 (–5) nerves distinct at its base; plants without rhizomes (though culms may be decumbent or prostrate) . . . . **Torreyochloa**

21 Lemmas ca. 2 mm long . . . . **Puccinellia**

21 Lemmas 2.5–8 mm long 22

22 Blades of leaves strongly involute, usually much less than 3 mm wide . . . . **Festuca**

22 Blades of leaves flat (or merely once-folded), at least the larger ones 3–8 mm wide 23

23 Larger lemmas 2.5–4.5 mm long; anthers 0.8–1.4 mm long; spikelets mostly containing 2–4 (–5) florets and borne beyond the middle of the primary panicle branches . . . . **Festuca subverticillata**

23 Larger lemmas 5.5–8 mm long; anthers 2.2–3.5 (–3.8) mm long; spikelets often containing 5 or more florets, borne below as well as above the middle of the primary panicle branches . . . . **Schedonorus**

KEY 8

Tribe Poeae

Subtribe Phalaridinae (*Anthoxanthum, Hierochloe, Phalaris*)

1 Panicle open, pyramidal, the branches spreading or drooping; glumes nearly equal in length, with lateral nerves obscure or prominent only on basal half; lower florets staminate, at least as large as bisexual floret, awnless . . . . **Hierochloe**

1 Panicle contracted, the branches ascending or suppressed; glumes equal or not, with lateral nerves (at least on larger glumes) prominent beyond the middle; lower florets sterile, either vestigial or large and awned 2

2 Glumes very unequal; lower lemmas with prominent dorsal awns, concealing the awnless bisexual floret . . . . **Anthoxanthum**

2 Glumes nearly or quite equal; lower lemmas awnless, small and inconspicuous, only the awnless bisexual floret evident . . . . **Phalaris**

KEY 9

Tribe Danthonieae (*Danthonia*)

Tribe Poeae

Subtribe Aveninae (*Arrhenatherum, Avena, Graphephorum, Koeleria, Sphenopholis, Trisetum*)

Subtribe Holcinae (*Deschampsia, Holcus*)

1 Lemmas all awnless; larger glumes ± obovate (broadest above the middle), generally shorter than the lowest floret 2

1 Lemmas with distinct twisted, jointed, or curved awn (sometimes largely hidden by the glumes or absent on some florets of a spikelet); glumes mostly ovate to lanceolate, at least one of them longer than the lowest floret 4

2 Rachilla and callus prominently bearded with long straight hairs. **Graphephorum**

2 Rachilla and callus glabrous or at most with short hairs (under 0.5 mm long) 3

3 Axis and branches of inflorescence glabrous, at most scabrous; larger glumes not over 3 (–3.2) mm long . . . . **Sphenopholis**

3 Axis and branches of inflorescence densely short-pubescent; larger glumes 3–4.5 mm long . . . . **Koeleria**

4 Larger glumes 6–27 mm long 5

4 Larger glume less than 6 mm long 7

5 Ligule a fringe of short hairs with a long tuft at each side; lemma with awn between terminal teeth. . . . . **Danthonia**

5 Ligule membranous, hairless; lemma with awn arising dorsally 6

6 Spikelets less than 10 mm long (excluding awns), the lower floret staminate with strong awn and the upper floret bisexual with (usually) weak awn . . . . **Arrhenatherum**

6 Spikelets ca. 20–27 mm long, the florets all bisexual or the upper rudimentary; awns various . . . . **Avena**

7 Bisexual (lowermost) floret awnless; awn subterminal on a reduced staminate floret; nodes pubescent . . . . **Holcus mollis***

7 Bisexual florets awned; awn twisted or spreading; nodes glabrous or (sometimes in *Trisetum*) pubescent. 8

8 Awn arising above middle of lemma; panicle ± crowded and spike-like . . . . **Trisetum**

8 Awn arising well below middle of lemma; panicle at maturity very open and diffuse. . . . . **Deschampsia**

Agrostis Bent-Grass; Bent

Perennial grasses, clumped or spreading by rhizomes or sometimes by stolons. Leaves soft, auricles absent, ligules membranous, sheaths open, usually smooth and glabrous. Head an open panicle. Spikelets small, 1-flowered, breaking above glumes; glumes more or less equal length, 1-veined; floret shorter than glumes; lemma awnless or with a short straight awn; palea small or absent; stamens usually 3.

ADDITIONAL SPECIES

Agrostis capillaris L. (Colonial Bent), European perennial grass, often planted for lawns in the northeastern USA; in the UP, rarely escaping to fields and dry woods; Baraga, Keweenaw, and Marquette counties.

GRASS TERMS

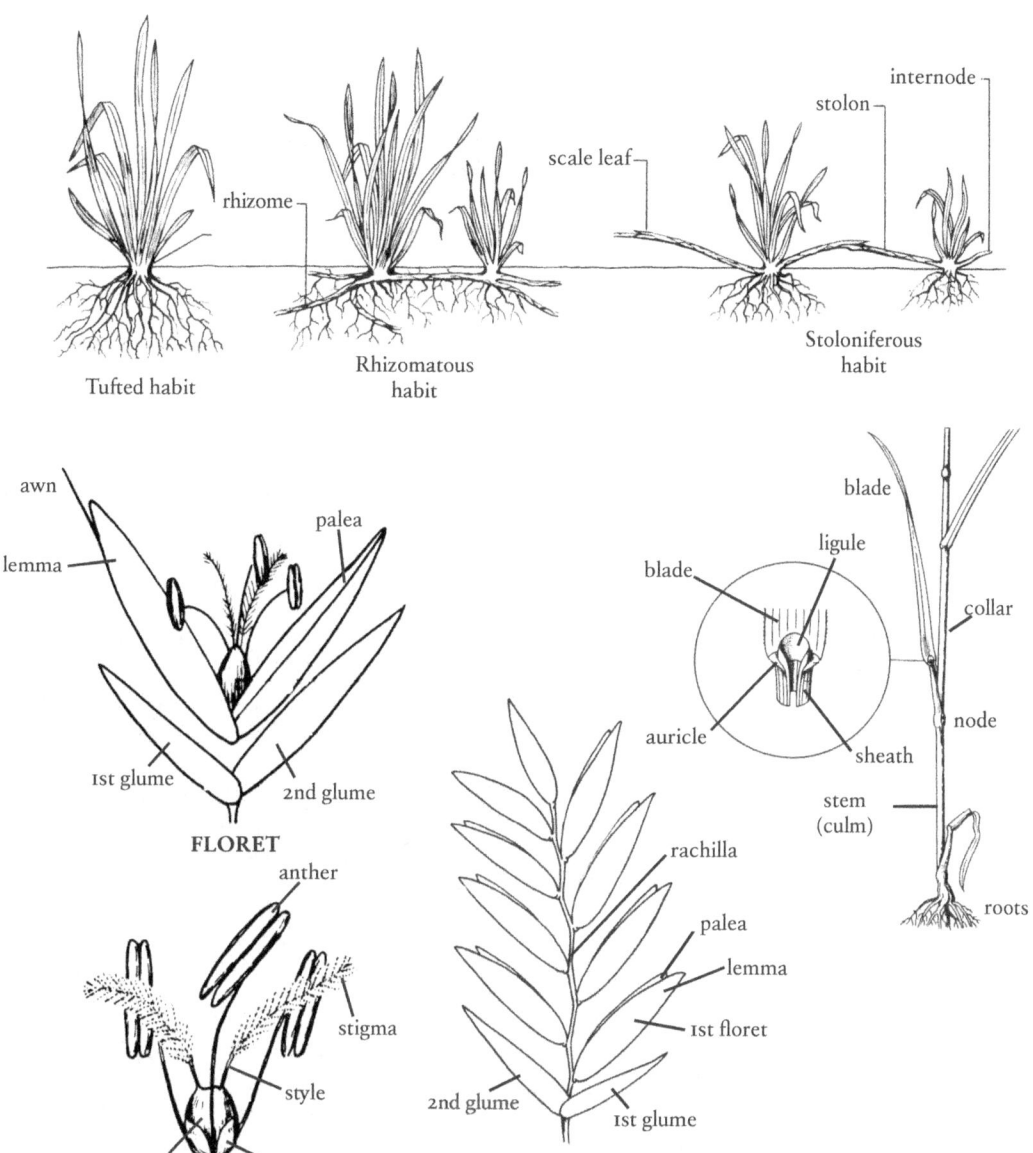

1. Palea present, about half as long as the lemma or longer; anthers ca. 0.8–1.5 mm long 2
1. Palea absent or vestigial; anthers ca. 0.6 mm long or shorter 3
2. Plants rhizomatous but not stoloniferous, i.e., stems arising from underground rhizomes, straight or curved at the very base, otherwise erect and nearly or quite straight; larger leaf blades mostly 3–7 (–10) mm wide; spikelets usually flushed with red or purplish; bases of middle panicle branches mostly meeting the axis of the panicle at an angle of 30–45° (except when very immature) .... **A. gigantea**
2. Plants stoloniferous but not rhizomatous, i.e., stems usually decumbent at their bases, the lower nodes often strongly bent and/or rooting, but underground rhizomes absent; larger leaf blades 1.7–3 (–4) mm wide; spikelets pale, greenish; bases of middle panicle branches usually strongly ascending or appressed to axis of panicle, at most diverging about 15° (but panicle branches often spreading distally) ..... **A. stolonifera**
3. Longest panicle branches less than 6 (–12) cm long and the uppermost leaf blade more than 5 cm long; leaf blades flat, the wider ones 1.5–3.5 mm wide; panicle branches forked about or below the middle, often smooth or only sparingly hispidulous-scabrous; panicle pale, greenish (very rarely red-tinged) .. **A. perennans**

3   Longest panicle branches more than 6 cm long or uppermost leaf blade less than 5 cm long (or both conditions); leaf blades usually ± involute, the widest to 1.5 (rarely 3) mm wide; panicle branches often not forked until beyond the middle, copiously hispidulous-scabrous; panicle ± flushed with red . . . . **A. scabra**

Agrostis gigantea Roth
BLACK BENT *introduced*
IR | **WUP** | **CUP** | **EUP**   FACW

Perennial grass, rhizomatous and sod-forming, not stoloniferous. Stems to 10 dm long or sometimes more; auricles absent; ligules membranous, larger (upper) ligules mostly 2.5–6 mm long, higher than wide; leaf blades 3–8 mm wide. Panicle 10–20 cm long, notably suffused with purplish red, at anthesis triangular-ovoid, with widely spreading unequal branches, sometimes later more contracted; at least some of the panicle branches floriferous to the base; panicle branches and often the pedicels scabrous. Spikelets rather crowded, 2–3.5 mm long; glumes scabrous along the keel; lemma 2/3 as long as the glumes, scabrous near tip, usually unawned; callus minutely bearded.

Native of Europe, cultivated and escaped into moist meadows, shores, coastal marshes, and other moist places.

Agrostis perennans (Walt.) Tuckerman
UPLAND BENT *native*
IR | **WUP** | **CUP** | **EUP**   FACU CC 5

Tufted perennial grass. Stems 5–10 dm tall; auricles absent; ligules membranous, leaf blades flat, 2–6 mm wide, elongate, the uppermost blade more than 5 cm long. Panicle mostly pale greenish, 10–25 cm long, notably longer than wide, the smooth or sparsely scabrous branches forking near or below the middle, soon divaricate, the longest ones often less than 6 cm long. Spikelets 1.8–2.8 mm long; glumes subequal, scabrous on the midvein; lemma 1.3–2 mm, awnless or rarely with a very short, slender awn near the tip; palea obsolete.

Various habitats, usually in dry soil.

Agrostis scabra Willd.
ROUGH BENT *native*
IR | **WUP** | **CUP** | **EUP**   FAC CC 4

Tufted perennial grass. Stems slender, 3–9 dm long; auricles absent; ligules membranous; leaves mostly basal or below the middle of the stem, usually erect, the blade flat or more often involute, 1–2 mm wide. Panicle ovoid or pyramidal, 1–3 dm long, sometimes half as long as the whole plant, very diffuse, more or less reddish, the scabrous filiform branches divaricate, mostly forking well above the middle. Spikelets 1.2–3.2 mm long; glumes scabrous on the midvein; lemma awnless or with a short straight awn; callus short-bearded; palea obsolete or to 0.3 mm long.

Abundant, widely distributed in many habitats, and variable.

Agrostis stolonifera L.
REDTOP; SPREADING BENT *introduced*
IR | **WUP** | **CUP** | **EUP**   FACW

Agrostis alba var. *palustris* (Huds.) Pers.
Agrostis palustris Huds.

Perennial grass, spreading by rhizomes and also sometimes by stolons. Stems erect or more or less horizontal at base, 3–10 dm or more long; auricles absent; ligules membranous, usually splitting at tip, 2–5 mm long; leaf blades ascending, 2–8 mm wide, rough-to-touch. Panicle open, 3–20 cm long, the branches spreading, branched and with spikelets along their entire length. Spikelets 1-flowered, usually purple, 2–4 mm long; glumes lance-shaped, 1.5–2.5 mm long; lemma 2/3 length of glumes, 1–2 mm long; palea present, about half as long as lemma. July–Sept.

Wet meadows, ditches, streambanks and shores; disturbed areas.

Alopecurus Meadow-Foxtail

Annual or perennial grasses. Stems erect or more or less horizontal at base. Leaves mostly from lower 1/2 of the stems; sheaths open; auricles absent; ligules membranous, entire to lacerate. Heads densely flowered, cylindric, spike-like panicles. Spikelets 1-flowered, flattened, breaking below the glumes; glumes equal length, 3-nerved, often silky hairy on back, awnless; lemma about as long as glumes or shorter, awned from the back, the awn shorter to longer than the glume tips; palea absent. The narrow panicles resemble those of timothy (*Phleum*).

1   Spikelets (excluding awns) ca. 4–6.5 mm long; awns mostly exserted 3.5–6 mm beyond tips of glumes . . . . **A. pratensis**
1   Spikelets not over 3 mm long; awns at most exserted ca. 1 mm . . . . **A. aequalis**

Alopecurus aequalis Sobol.
SHORT-AWN FOXTAIL *native*
IR | **WUP** | **CUP** | **EUP**   OBL CC 4

Annual or short-lived perennial grass. Stems single or in small clumps, slender, erect to more or less horizontal, 2–6 dm long, often rooting at the nodes; auricles absent; ligules 2–6.5 mm long, obtuse; leaf blades 1–5 mm wide, finely rough-to-touch above; ligule membranous, rounded to elongate, 2–7 mm long. Panicle erect, spike-like, 2–7 cm long and 3–5 mm wide. Spikelets 1-flowered; glumes 2–3 mm long, blunt-tipped, hairy on the keel and veins; lemma about equaling the glumes, awned from back,

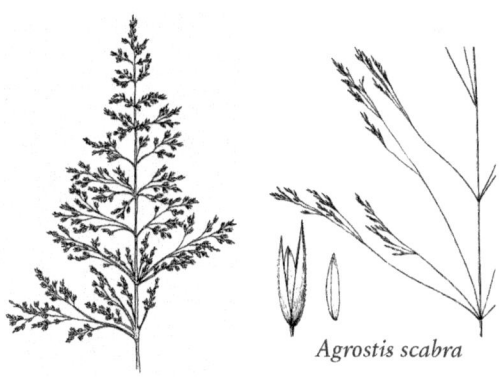

Agrostis gigantea

Agrostis scabra

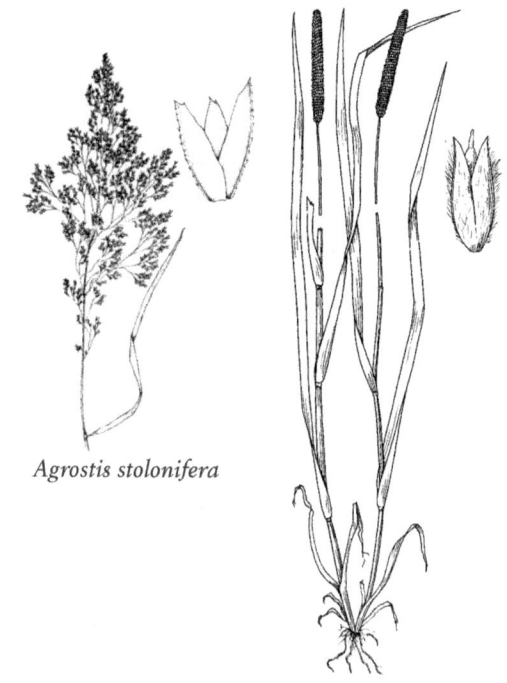

Agrostis stolonifera

Alopecurus aequalis

the awn straight, to 1.5 mm longer than glume tips. June–Aug.

Shallow water or mud of wet meadows, marshes, ditches, springs, open bogs, fens, shores and streambanks; sometimes where calcium-rich.

Alopecurus pratensis L.
FIELD MEADOW-FOXTAIL *introduced*
IR | WUP | CUP | EUP FAC

Perennial grass, shortly rhizomatous. Stems erect or decumbent at base, 4–8 dm long; auricles absent; ligules 1.5-3 mm long, obtuse to truncate; upper sheaths not or scarcely inflated. Panicle spike-like, 2–8 cm long, 5–10 mm wide, scarcely tapering; glumes 4–5.5 mm long, the keel narrowly winged, conspicuously ciliate, especially above the middle, with hairs 1–1.5 mm long; awn inserted about halfway between the base and middle of the lemma, exserted 2–6 mm.

Native of Eurasia; naturalized in moist meadows, fields, and waste places.

Ammophila *Beach-Grass*

Ammophila breviligulata Fern.
AMERICAN BEACH-GRASS *native*
IR | WUP | CUP | EUP CC 10

Coarse, stiff, perennial grass, from long running rhizomes. Stems stout, erect, glabrous, 5–10 dm tall; sheaths glabrous; ligule membranous, ovate or truncate, 1–3 mm long; leaf blades flat at base, involute above, 4–8 mm wide when unrolled, scabrous above, glabrous beneath. Panicle dense, 1–4 dm long, 1–2.5 cm thick, its base often enclosed in the upper sheath. Spikelets 1-flowered, strongly flattened, articulated above the glumes; glumes about equal, 10–15 mm long, linear lance-shaped, keeled, the first 1-nerved, the second 3-nerved, scabrous on the keel; lemmas shorter than the glumes, scaberulous; obscurely 3–5-nerved, awnless, subtended by a tuft of short hairs 1–3 mm long from the callus.

Dunes and dry sandy shores along the Great Lakes; useful as sand-binders in dune control.

Andropogon *Bluestem*

Andropogon gerardii Vitman
BIG BLUESTEM; TURKEY-FOOT *native*
IR | WUP | CUP | EUP FACU CC 5

Perennial grass. Stems stout, 1–3 m tall, forming large bunches or extensive sod; ligules membranous; leaf blades usually 5–10 mm wide, the lower ones and the sheaths sometimes villous. Racemes 2–6, subdigitate, on a long-exserted peduncle, 5–10 cm long; joints of the rachis and pedicels equal, sparsely or usually densely ciliate, densely bearded at the summit. Spikelets of two kinds, in pairs at the joints of the rachis, one sessile and perfect, the other pediceled and staminate, sterile, or abortive; glumes of the fertile spikelet equal or nearly so, leathery, flat to concave on the back, lacking a midnerve, often ciliate; fertile lemma shorter than the glumes, narrow, hyaline, usually ending in a long awn awn 8–15 mm long, twisted below and more or less bent.

Moist or dry soil of prairies, roadsides, railroads; in dry open woods, old fields, rarely in fens and sedge meadows.

Anthoxanthum *Sweet Vernal Grass*

Anthoxanthum odoratum L.
LARGE SWEET VERNAL GRASS *introduced*
IR | WUP | CUP | **EUP** FACU

Sweetly scented perennial grass. Stems tufted, 3–7 dm tall; internodes hollow; sheaths open; auricles 0.5–1 mm long, pilose-ciliate, sometimes absent; ligules membranous, 2–7 mm long, truncate; leaves mostly near the base, blades flat, 3–8 mm wide, the upper much shorter. Panicle spike-like, 3–6 cm long. Spikelets 1-flowered, articulated above the glumes; glumes unequal, much exceeding the lemmas, scabrous on the keel to villous throughout, the first ovate, about 4 mm long 1-nerved; the second lance-shaped, 7–9 mm long 3-nerved; lower sterile lemmas about equal, 3–3.5 mm long, golden-silky, the first awned on the back near the tip, the second near the base, its awn twisted below, geniculate at about the summit of the lemma, the awn of the second about equaling the second glume; fertile lemma smaller than and enclosed by the sterile, at first hyaline, at maturity brown and shining.

Native of Europe; shores, meadows, roadsides, and waste places. *Hierochloe* sometimes placed in this genus.

Aristida *Three-Awn*

Aristida basiramea Engelm.
FORKED THREE-AWN *native*
IR | WUP | **CUP** | EUP CC 3

Annual grass. Stems tufted, erect, 3–6 dm tall, usually branched from some or all of the nodes; ligules membranous, about 0.3 mm long and long-ciliate; leaf blades very narrow, usually about 1 mm wide, often involute. Flowers in a terminal, slender, rather loose panicle, the terminal 5–10 cm long, the lateral shorter and more slender, scarcely surpassing the subtending sheath. Spikelets 1-flowered, articulated above the glumes; glumes membranous, 1-nerved, distinctly unequal, the first 6–12 mm, the second 9.5–15 mm long; lemma usually about equaling the first glume, 7.5–10.5 mm long; awns elongate, normally 3; central awn divergent, 11–19 mm long, coiled at base when dry into 1–3 turns; lateral awns erect to curved-divergent but not coiled, 7.5–13 mm long.

Dry sterile or sandy soil; Delta and Schoolcraft counties.

Aristida basiramea

Arrhenatherum *Oatgrass*

Arrhenatherum elatius (L.) Beauv.
TALL OATGRASS *introduced*
IR | **WUP** | **CUP** | **EUP** FACU

Tall perennial grass; loosely tufted, sometimes rhizomatous. Stems erect, to 2 m tall, smooth, or minutely pubescent at the nodes; sheaths smooth, open, not overlapping; auricles absent; ligules membranous, 1–3 mm long, obtuse to truncate, usually ciliate; leaf blades flat, 4–8 mm wide, sometimes scabrous. Panicle shining, slender, 1–3 dm long, the short branches in fascicles. Spikelets 2-flowered, disarticulating above the glumes and between the lemmas; glumes unequal, hyaline in age, the first shorter, 4.5–8 mm long, 1-nerved, the second 6.6–10 mm long, equaling the lemmas, 3-nerved; lemmas thin, rounded on the back, short-bearded at base, 5–7-nerved, the lower enclosing a staminate flower and bearing below the middle a long awn 10–20 mm long, geniculate near the middle, the upper one enclosing a perfect flower, awnless or bearing just below the tip a much shorter straight awn to 6 mm long.

Meadows, roadsides, and waste ground, usually in moist soil.

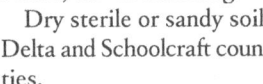

Andropogon gerardii

Ammophila breviligulata

Anthoxanthum odoratum

Avena Oat

Mostly annual grasses with broad flat blades and ample panicles of large spikelets. Sheaths open, auricles absent, ligules membranous. Spikelets 2–3-flowered, articulated above the glumes and usually between the lemmas; rachilla often hirsute, at least at the base of the lemmas. Glumes nearly equal, exceeding the lemmas, 1–11-nerved. Lemmas indurate, often scarious toward the tip, rounded on the back, obscurely 5–9-nerved, or prominently nerved toward the apex. Awn arising about the middle of the lemma, bent near the middle, or straight, or lacking.

1 Lemmas with a stout, strongly twisted awn and often with stiff hairs on the back; florets falling from the spikelet by a distinct oval disarticulation surface . . . . **A. fatua**
1 Lemmas with the awn usually straight, weak, or absent, and the back glabrous; florets falling by fracture of rachilla at base of spikelet. . . . . **A. sativa**

Avena fatua L.
WILD OAT *introduced*
IR | **WUP** | CUP | **EUP**

Annual grass. Stems stout, smooth, usually 5–8 dm tall; sheaths of the basal leaves with scattered hairs, distal sheaths glabrous; ligules 4-6 mm long, acute; leaf blades 5–15 mm wide, scaberulous. Panicle lax, the fascicled branches usually spreading horizontally. Spikelets 3-flowered; rachilla hirsute, readily disarticulating, the lemmas falling separately; glumes 7–11-nerved, about 2 cm long; lemmas sparsely to densely hirsute on the back, or rarely glabrous; awn 3–4 cm long.

European native; disturbed sites such as railroads, roadsides, and beaches; Chippewa and Houghton counties.

Avena sativa L.
OAT *introduced*
IR | **WUP** | **CUP** | EUP

Annual grass. Stems branching from the base, stout, 3–10 dm tall; sheaths smooth or finely scabrous; ligules 3-8 mm long, acute; leaf blades 5–15 mm wide, finely scabrous. Panicle lax, many-flowered, with rough, slender, fascicled branches. Spikelets 2-flowered; glumes 7–11-nerved, about 25 mm long; lemmas long remaining attached to the rachilla, glabrous, 15–20 mm long, awnless or with a short straight awn.

An important cultivated species apparently derived from *A. fatua* (with which it readily hybridizes); often adventive along roads and railways, probably not persistent.

Beckmannia Slough Grass

Beckmannia syzigachne (Steud.) Fern.
AMERICAN SLOUGH GRASS (THR) *native*
IR | WUP | CUP | **EUP** OBL CC 4

Stout annual grass. Stems single or in small clumps, 4–12 dm long; sheaths overlapping, smooth, the upper sheath often loosely enclosing lower part of panicle; auricles absent; ligule membranous, acute, 3–6 mm long; leaf blades flat, 3–10 mm wide, rough-to-touch. Head of many 1-sided spikes in a narrow panicle 10–30 cm long, the panicle branches erect, overlapping, 1–5 cm long; each spike 1–2 cm long, with several to many spikelets in 2 rows on the rachis. Spikelets with 1 floret, a second undeveloped (or sometimes well-developed) floret occasionally present; overlapping, nearly round, 2–4 mm long, straw-colored when mature, breaking below the glumes; glumes equal, broad, inflated along midvein, with a short, slender tip; lemma about as long as glumes but narrower; palea nearly as long as lemma. June–Sept.

Wet meadows, marshes, ditches, shores and streambanks; more common in Great Plains region; Isle Royale and Mackinac County.

Arrhenatherum elatius

Avena fatua

Beckmannia syzigachne

Bouteloua *Grama-Grass*

Bouteloua curtipendula (Michx.) Torr.
SIDE-OATS GRAMA   (THR) *native*
IR | WUP | **CUP** | EUP   CC 10

Perennial grass from slender rhizomes. Stems erect, 3–10 dm tall. Leaves mostly basal, sheaths open, smooth or nearly so; ligules to 0.5 mm long, membranous, ciliate; leaf blades elongate, 2–5 mm wide, scabrous on the margins. Spikes 10–50, racemose on a common axis, spreading or nodding, 8–15 mm long, secund along an axis 1–3 dm long, falling entire. Spikelets usually 3–6, each with 1 perfect flower and 1 sterile rudiment, articulated above the glumes, inserted on one side of a narrow flat rachis; glumes unequal; first glume linear-subulate, 3–4 mm long; second glume lance-shaped, 4–7 mm long; fertile lemma usually somewhat exceeding the glumes, acuminate, its lateral nerves prolonged into awns about 1 mm long; rudiment with a long central awn and 2 shorter lateral ones arising below the middle, or greatly reduced, or lacking.

Dry open places and prairies; often included in prairie seed mixtures; Schoolcraft County.

Brachyelytrum aristosum

Bouteloua curtipendula

Brachyelytrum *Shorthusk*

Perennial forest understory grasses from knotty rhizomes. Leaves broad, mostly along the stem; sheaths open; auricles absent; ligules short, membranous; lower leaf blades absent or reduced; upper leaf blades flat. Panicles narrow, few-flowered. Spikelets readily deciduous, 1-flowered, articulated above the glumes. Glumes minute, subulate to triangular, 1-nerved. Lemma linear-subulate, rounded on the back, sharply 5-nerved, gradually tapering to an elongate awn. Palea 2-keeled; rachilla prolonged into an elongate bristle appressed to the furrow of the palea.

Brachyelytrum aristosum (Michx.) Beauv. ex Branner & Coville
BEARDED SHORTHUSK   *native*
IR | **WUP** | **CUP** | **EUP**   CC 7

Perennial grass, previously included within *Brachyelytrum erectum* (Shreb. ex Spreng.) Beauv., (found in the Lower Peninsula), and distinguished from that species by pubescence of lemma, having 3–5-nerved lemmas, florets ca. 8–10 mm long (excluding awn) and 0.7–1.3 mm wide, anthers not over 4 mm long, and more than 15 cilia per 5 mm of leaf margin. *B. erectum* has more strongly 7–9-nerved lemmas, at least the larger florets are 10–12 mm long and 1–1.6 mm wide, anthers are more than 5 mm long, and with fewer than 10 cilia per 5 mm of leaf margin.

Moist to dry deciduous forests, lowland forests, moist thickets, sandy pine forests, and coniferous swamps.

Bromus *Brome; Chess; Cheat-Grass*

Perennial grasses. Leaves generally flat; sheaths closed to near top, usually pubescent; auricles usually absent; ligules membranous, usually erose or lacerate. Head a panicle of drooping spikelets. Spikelets with several to many flowers, breaking above the glumes; glumes shorter than lemmas; lemmas awned or unawned; stamens usually 3.

ADDITIONAL SPECIES

Bromus nottowayanus Fernald (Satin Brome), previously included in *B. pubescens*; moist, rich hardwood forests; occasionally in drier forests on banks and slopes; Menominee and Ontonagon counties. This species is distinguished from *B. pubescens* in the field by the satiny sheen to the underside of the leaf (which, however, may appear uppermost due to the 180 degree twist of the leaves). Note that *B. pubescens* is found in the central and southern Lower Peninsula and not known from the UP.

1. First glume with one distinct nerve; second glume with 3 (–5) nerves.................................... 2
1. First glume with 3 (–5) distinct nerves; second glume with 5–7 nerves 5
2. Awns 10–30 mm long, as long as or longer than their lemmas; apex of lemma beyond insertion of awn 1.5–2.7 mm long; annual weed.................. **B. tectorum**
2. Awns absent or up to 7 (–9) mm long, shorter than their lemmas; apex of lemma less than 1.5 mm long; perennials, mostly native (*B. inermis* introduced) 3
3. Plants with elongate rhizomes; lemmas (at least when fresh) usually ± flushed with purplish, especially toward

the margins, the awns absent or less than 4 (–5.5) mm long . . . . **B. inermis**

3 Plants without elongated rhizomes; lemmas (when fresh) green (very rarely flushed with purple), the larger awns 3–7 (–9) mm long 4

4 Nodes and also number of leaves usually 8–15; leaf sheaths longer than the internodes, thus overlapping and covering all the nodes, the summit of the sheath with a band of dense pubescence and (when intact) with a pair of prominent tooth-like auricles; anthers 1.5–2.2 mm long . . . . **B. latiglumis**

4 Nodes and leaves usually not more than 6 (–8 in *B. ciliatus*); leaf sheaths shorter than at least the upper internodes, exposing one or more of them, the summit of the sheath glabrous or pubescent but lacking auricles; anthers various . . . . **B. ciliatus**

5 Lemmas pubescent all across the back, at least near tip; glumes pubescent; awns straight . . . . **B. kalmii**

5 Lemmas glabrous or scabrous on the back; glumes glabrous; awns usually divaricate or undulate (or absent) 6

6 Lemma equaling or slightly shorter than the tip of the mature palea; sheaths glabrous (or occasionally the lowermost with some short hairs); margins of ripe lemmas strongly inrolled, exposing the rachilla; lemmas ca. 7–8.2 mm long, the awns ± undulate, sometimes as long as lemma but usually much shorter, rudimentary, or occasionally absent . . . . **B. secalinus**

6 Lemma at least slightly exceeding tip of palea; sheaths of at least middle and lower leaves ± densely (though sometimes finely) hairy; lemmas and awns various . . . 7

7 Longest awns in a spikelet longer than their lemmas and more than twice as long as awn on lowest lemma of the spikelet; branches of inflorescence lax and flexuous; hairs of sheath very fine and delicate, and tending to be ± crooked or tangled toward their tips (though basically ± retrorse); tip of palea 1–2.5 mm shorter than tip of lemma; anthers 0.5–1.5 mm long . . . . **B. arvensis**

7 Longest awns in a spikelet about as long as their lemmas or shorter, and usually less than twice as long as awn on lowest lemma of the spikelet; branches of inflorescence rather stiff (whether spreading or ascending); hairs of sheath usually fine but stiffish and straight (spreading to retrorse); tip of palea less than 2 mm shorter than tip of its lemma; anthers 1–2 mm long . . . . **B. racemosus**

Bromus arvensis L.
FIELD BROME *introduced*
IR | **WUP** | CUP | EUP FACU

Bromus japonicus Thunb.

Weedy annual grass. Stems glabrous, 3–9 dm tall; lower sheaths softly and densely appressed-hairy; ligules 1–1.5 mm long, hairy, erose; leaf blades 10–20 cm long and 2–6 mm wide, coarsely pilose on both surfaces. Panicle 1–2 dm long, loose and open, the slender branches spreading or drooping, much longer than the spikelets. Spikelets mostly 6–10-flowered, often purple-tinged, glabrous or nearly so; first glume 3-nerved, 4–6 mm long; second glume 5-nerved, 5–7.5 mm long; lemmas 7-nerved, 6–8 mm long, including the triangular teeth 2–3 mm long; palea 1.5–2 mm, about equaling or shorter than the lemma; awns more or less twisted and divergent when dry, the lowest 2.5–5 mm, the uppermost 8–12 mm long; anthers 2.5–5 mm long.

Eurasian native; introduced as a weed in waste places; Gogebic County.

B. japonicus sometimes treated as a separate species, distinguished by generally smaller size, spikelets not purple-tinged, and smaller anthers (to 1.5 mm long).

Bromus ciliatus L.
FRINGED BROME *native*
IR | WUP | CUP | EUP FACW CC 6

Perennial grass, rhizomes absent. Stems single or few together, smooth or hairy at nodes, 5–12 dm long; sheaths usually with long hairs; ligule membranous, short, to 2 mm long, ragged across tip; leaf blades flat, 4–10 mm wide, usually with long, soft hairs mainly on upper surface. Panicle loose, open, 1–3 dm long, the branches usually drooping. Spikelets large, 4–10-flowered, 1.5–3 cm long and 5–10 mm wide; glumes usually more or less smooth, lance-shaped, the first glume 4–9 mm long, the second glume 6–10 mm long, often tipped with a short awn; lemma 10–15 mm long, more or less smooth on back, usually long-hairy along lower margins, tipped with an awn 2–6 mm long; palea about as long as body of lemma. July–Sept.

Streambanks, shores, thickets, sedge meadows, fens, marshes; also in moist woods.

Bromus inermis Leyss.
SMOOTH BROME *introduced (invasive)*
IR | WUP | CUP | EUP UPL

Perennial grass, with short to long-creeping rhizomes. Stems 5–10 dm tall; sheaths glabrous; auricles sometimes present; ligules to 3 mm long, glabrous, truncate, erose; leaf blades glabrous, 8–15 mm wide. Panicle spreading at anthesis, later contracted, 1–2 dm long, often with 4–10 branches from a node. Spikelets 15–30 mm long, about 3 mm wide, 7–11-flowered; first glume 1-nerved, 4–8 mm long; second glume 3-nerved, 7–11 mm long; lemmas 10–12 mm long, 3–5-nerved, the outer pair of nerves often inconspicuous, obtuse or retuse, glabrous or scaberulous, awnless or with an awn to 2 mm long.

Native of Europe, cultivated for forage and often escaped.

Bromus kalmii Gray
KALM'S BROME   *native*
IR | **WUP** | **CUP** | **EUP**   FAC CC 8

Perennial grass, not rhizomatous. Stems slender, loosely tufted or solitary, 5–10 dm tall, mostly glabrous, often pubescent at the nodes; sheaths usually villous, varying to glabrous; auricles absent; ligules 0.5–1 mm long, glabrous, truncate, erose; leaf blades 1–2 dm long and 5–10 mm wide, glabrous or pubescent on both sides. Panicle nodding, 5–10 cm long or rarely longer, the relatively few spikelets drooping on slender flexuous pedicels. Spikelets 15–25 mm long, 6–11-flowered, softly villous; first glume 3-nerved, 6–7 mm long; second glume 5-nerved, 7–9 mm long, lemmas 7-nerved, 8–10 mm long, obtuse, the awn 2–3 mm long; the teeth about a third as wide as long.

Dry woods, rocky banks, and sandy or gravelly soil.

Bromus latiglumis (Shear) A.S. Hitchc.
EARLY-LEAF BROME   *native*
IR | **WUP** | **CUP** | EUP   FACW CC 6
Bromus altissimus Pursh

Perennial grass, not rhizomatous. Stems single or in small clumps, more or less smooth, 6–15 dm long; sheaths overlapping, retrorsely pilose or glabrous, with a dense ring of hairs at top; auricles 1–2.5 mm long; ligules 0.8–1.4 mm long, hirsute, ciliate, truncate, erose; leaf blades flat, 8–20 along stem; 10–15 mm wide. Panicle 1–2 dm long, the branches spreading or drooping. Spikelets several-flowered, 2–3 cm long, first glume 5–8 mm long, awl-shaped, second glume wider, 6–10 mm long; lemmas 10–12 mm long, hairy, awned, the awn 2–7 mm long.

Floodplain forests, thickets and streambanks, sometimes in rocky woods.

Bromus racemosus L.
BALD BROME   *introduced*
IR | WUP | **CUP** | EUP

Annual grass. Stems 4–9 dm tall, glabrous or nearly so; sheaths finely villous with retrorse hairs; ligules 1–2 mm long, glabrous or hairy, erose; leaf blades 3–6 mm wide, pubescent on both sides. Panicle normally 1–2 dm long, at first erect, nodding at maturity, at least some of its branches elongate and exceeding the spikelets. Spikelets 5–10-flowered, glabrous or minutely scaberulous; first glume narrowly lance-shaped, 3-nerved, 4.5–6.5 mm long; second glume broader, 5-nerved, 5.6–9 mm long; lemmas elliptic or somewhat obovate, 6.5–10 mm long, 7–9-nerved, the awns 3–10 mm long, the upper often twice as long as the lower; anthers about 3 times as long as wide.

Native of Europe; introduced in fields, roadsides, and waste places; Delta County.

Bromus secalinus L.
RYE BROME; CHESS   *introduced*
IR | **WUP** | CUP | **EUP**

Annual grass. Stems glabrous, 3–8 dm tall; middle and upper sheaths glabrous; ligules 2–3 mm long, glabrous, obtuse; leaf blades glabrous or pubescent, 3–8 mm wide. Panicle loose and open, 7–15 cm long, its branches several from a node, simple or again branched. Spikelets usually drooping, 6–11-flowered, 1–2 cm long; first glume oblong, 3–5-nerved, 4–6 mm long; second glume similar, 5–7-nerved, 5–7 mm long; lemmas elliptic, obtuse, obscurely 7-nerved, 6–9 mm long, glabrous or minutely scaberulous, awnless or with an awn to 5 mm long between the broad teeth; margins of the lemmas soon involute, causing the florets to diverge and expose the flexuous rachilla.

Native of Europe; introduced in grainfields, roadsides, and waste places; Houghton, Keweenaw, and Mackinac counties.

Bromus tectorum L.
CHEAT GRASS   *introduced*
IR | **WUP** | **CUP** | **EUP**

Annual grass. Stems tufted, 3–7 dm tall; sheaths and blades softly pubescent, the latter 2–4 mm wide; ligules 2–3 mm long, glabrous, obtuse, lacerate. Panicle 1–2 dm long, repeatedly branched, bearing rather crowded, drooping spikelets to 3 cm long, on slender pedicels; first glume subulate, 1-nerved, 5.5–7 mm long; second glume subulate, 3-nerved, 8–10 mm long; lemmas narrowly lance-shaped, 5–7-nerved, 10–12 mm long, pubescent throughout, usually hirsute toward the tip, acuminate into slender scarious teeth; awn 12–17 mm long; palea conspicuously ciliate.

Native of s Europe, widely established as a weed in waste ground and on roadsides; now a prominent feature of many sagebrush ecosystems in the western states.

Calamagrostis *Reed-Grass*

Perennial grasses, spreading by rhizomes. Stems single or in clumps. Leaves flat or inrolled, green or waxy blue-green, smooth or rough-to-touch; sheaths smooth; ligule large, membranous, usually with an irregular, ragged margin. Head a loose and open or dense and contracted panicle. Spikelets 1-flowered, breaking above glumes; glumes nearly equal, lance-shaped; lemma shorter than glumes, lance-shaped, awned from back, the awn about as long as lemma, the base of lemma (callus) bearded with a tuft of hairs, these shorter to as long as lemma; palea shorter than lemma; stamens 3.

POACEAE | Grass Family

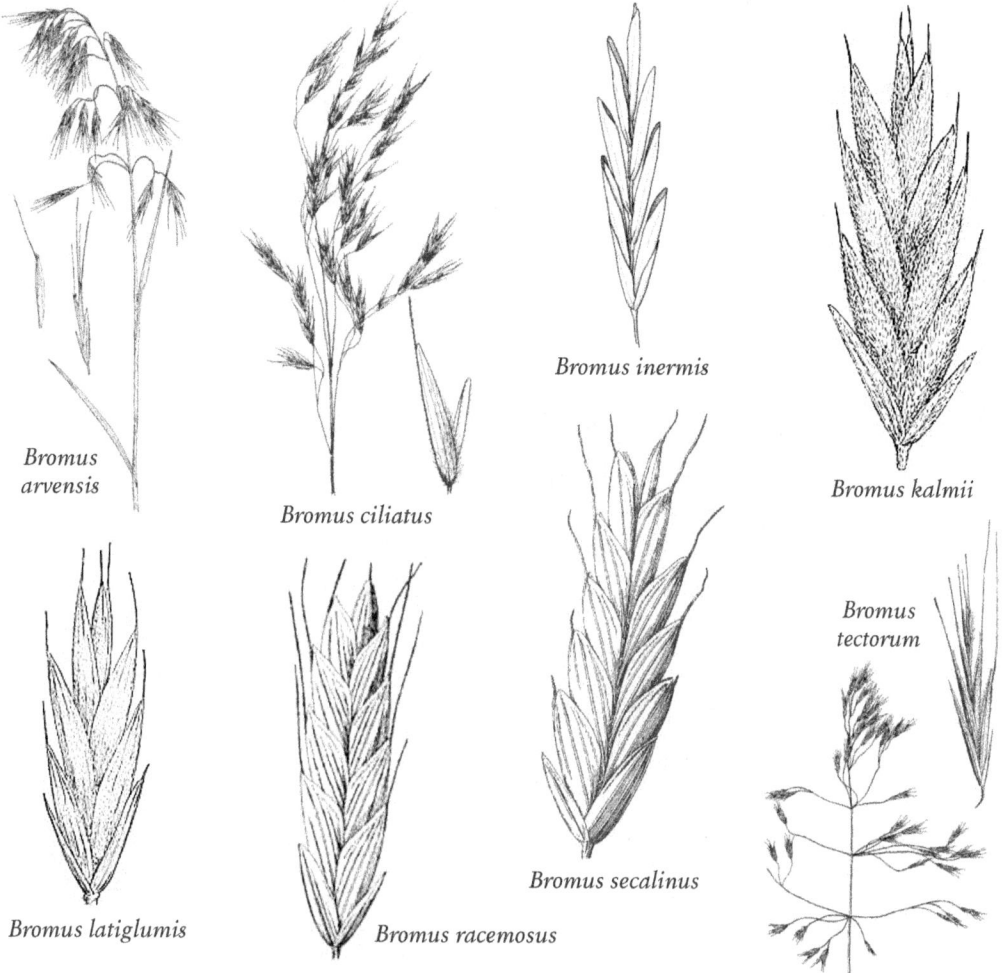

Bromus arvensis
Bromus ciliatus
Bromus inermis
Bromus kalmii
Bromus tectorum
Bromus latiglumis
Bromus racemosus
Bromus secalinus

1 Callus hairs and glumes at least 1.5 times as long as the lemma .... **C. epigeios**
1 Callus hairs and glumes barely if at all exceeding lemma 2
2 Leaf blades rather lax, to 10 mm wide; panicle mostly open with rather loosely ascending to spreading branches at flowering time; lemma nearly or quite smooth, membranous and translucent for at least the apical half; awn nearly or quite smooth, at least on basal half; callus hairs about as long as lemma (occasionally shorter), ± uniform in length and distribution; palea not over 2 mm long .... **C. canadensis**
2 Leaf blades stiff, to 4 mm wide; panicle mostly narrow and contracted with strongly ascending branches at flowering time; lemma usually firm and prominently scabrous, colorless and translucent only toward the tip; awn distinctly but minutely antrorsely scabrous its entire length (at 20×); callus hairs generally shorter than lemma, ± unequal in length or distribution (those immediately below the middle of the lemma shorter than those at the side, or absent; do not confuse the hairy prolongation of the rachilla behind the palea); palea often longer than 2 mm .... **C. stricta**

Calamagrostis canadensis (Michx.) Beauv.
BLUEJOINT *native*
IR | **WUP** | **CUP** | EUP   OBL CC 3
Perennial grass, from creeping rhizomes. Stems erect, in small clumps, 6–15 dm long, often rooting from lower nodes when partly underwater; leaf blades flat, green to waxy blue-green, 3–8 mm wide, rough-to-touch on both sides; sheaths smooth; ligules 3–7 mm long. Panicle more or less open, 8–20 cm long, the branches upright or spreading. Spikelets 1-flowered, 2–6 mm long; glumes more or less equal, 2–4 mm long, smooth or finely rough-hairy on back; lemma more or less smooth, awned from middle of back, the awn straight, base with dense callus hairs about as long as lemma. June–Aug.

Wet meadows, shallow marshes, calcareous fens, streambanks, thickets. Common.

Calamagrostis epigeios (L.) Roth
FEATHERTOP *introduced*
IR | **WUP** | **CUP** | EUP
Perennial grass. Stems 10–15 dm long, spreading by rhizomes and forming dense patches. Panicle dense,

elongate, at maturity silky from the protruding callus-hairs; glumes lance-subulate, 4–5.2 mm long; lemma membranous, 2–2.5 mm long, the slender awn inserted variously from near the base to somewhat above the middle; callus hairs copious, much exceeding the lemma and nearly equaling the glumes.

Native of Eurasia; disturbed places, Gogebic and Marquette counties.

Calamagrostis stricta (Timm) Koel.
SLIM-STEM REED-GRASS native
IR | WUP | CUP | EUP FACW CC 10
Calamagrostis inexpansa Gray

Perennial grass, spreading by rhizomes; plants waxy blue-green. Stems erect, 3–12 dm long; leaf blades stiff, often inrolled, 1–4 mm wide when flattened. Panicle narrow, 5–15 cm long, the branches short, upright to erect. Spikelets 1-flowered; glumes 3–6 mm long, smooth or rough-hairy on back; lemma rough-hairy, 2–4 mm long, awned, the awn straight, from near middle of back, base with many callus hairs, half to as long as lemma. June–Sept.

Wet meadows, shallow marshes, shores, streambanks; rocky shore of Lake Superior.

C. stricta subsp. *inexpansa* is uncommon in rock crevices and rocky openings on igneous rocks, mostly on or near the shores of Lake Superior (Houghton, Iron, and Marquette counties plus Isle Royale; threatened), and differ from the more common *C. stricta* subsp. *inexpansa* by having the awn twisted at base, mostly bent at middle (the tip therefore protruding from the sides of some of the spikelets), inserted on lower third of lemma; palea nearly or quite as long as lemma; and callus lacking hairs immediately below the middle of the lemma.

Calamovilfa *Sand-Reed*

Calamovilfa longifolia (Hook.) Scribn.
SAND-REED native
IR | **WUP** | **CUP** | EUP CC 10

Perennial grass, from creeping rhizomes, the rhizomes covered with shiny, scale-like leaves. Stems stout, stiffly erect, to 2 m tall; sheaths much overlapping at base, glabrous except usually more or less villous at the throat; ligule a ring of short hairs 1–2 mm long; leaf blades flat and 3–8 mm wide at base, involute above, tapering to a fine point. Panicle open, 1–4 dm long, with ascending branches. Spikelets 1-flowered, articulated above the glumes; glumes 1-nerved; first glume ovate, 3.5–6 mm long, second glume 4.5–7.5 mm long; lemma glabrous on the back, equaling or slightly shorter than the second glume, awnless, subtended by a conspicuous tuft of hairs from the callus, the hairs about half as long as the lemma.

Dry sandy openings and dunes.

Cenchrus *Sandbur*

Cenchrus longispinus (Hack.) Fern.
COMMON SANDBUR native
IR | WUP | **CUP** | EUP CC 0

Annual grass. Stems ascending or spreading, 2–6 dm long, often with many branches arising from the base; sheaths strongly compressed-keeled, villous-ciliate toward the summit; ligules 0.6-1.8 mm long, villous; leaf blades usually 5–12 cm long. Inflorescences terminal, spikelike panicles of highly reduced branches termed fascicles ("burs"); fascicles consisting of 1-2 series of many, stiff, sharp bristles surrounding 1-4 spikelets. Spikelets with 1 perfect flower, narrowly ovoid, acuminate, permanently enclosed by a spiny bur composed of several to many concrescent, flattened bristles. Burs subglobose, pubescent, usually 4–5 mm wide, excluding the spines, the latter 3–5 mm long; tips of the spikelets conspicuously exsert,

Calamagrostis stricta

Calamagrostis canadensis

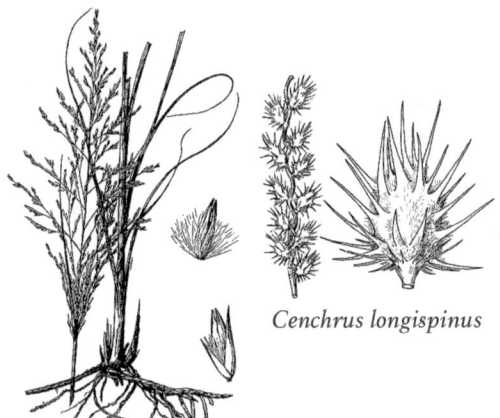

Cenchrus longispinus

Calamovilfa longifolia

their bodies visible to the middle through the lateral cleft of the bur; glumes hyaline, the first 1-nerved, the second 3–5-nerved.

Disturbed places, especially where dry and sandy; Delta and Menominee counties.

The spines are painfully sharp if walking barefoot.

Cinna *Wood-Reed*

Tall, perennial grasses, rhizomes weak or absent. Leaves wide, flat and lax; auricles absent; ligule brown, membranous, with an irregular, jagged margin. Head a large, closed to open panicle, the branches upright to spreading or drooping. Spikelets small, 1-flowered, laterally compressed, breaking below the glumes; glumes nearly equal, lance-shaped, keeled; lemma similar to glumes, with a short awn from just below the tip; palea shorter than lemma; stamens 1.

1 Panicle more or less crowded and narrow, the branches upright; second glume 4–6 mm long .. **C. arundinacea**
1 Panicle open, the branches spreading to drooping; second glume 2–4 mm long. . . . . **C. latifolia**

Cinna arundinacea L.
SWEET WOOD-REED   *native*
IR | **WUP** | CUP | EUP   FACW CC 7

Perennial grass, rhizomes weak or absent. Stems 1 or few together, erect, 6–15 dm long, often swollen at base; sheaths smooth; ligules red-brown, 3–10 mm long; leaf blades 4–12 mm wide, margins rough-to-touch. Panicle narrow, dull gray-green, 1–3 dm long, the branches upright. Spikelets 1-flowered; glumes narrowly lance-shaped, 3–5 mm long, the first glume 1-veined, the second glume 3-veined, usually rough-hairy; lemma 3–5 mm long, rough-hairy on back, usually with an awn to 0.5 mm long, attached just below tip and mostly shorter than lemma tip. Aug–Sept.

Swamps, floodplain forests, streambanks, pond margins, moist woods; Menominee and Ontonagon counties.

Distinguished from *C. latifolia* by its 3-veined upper glumes and larger spikelets.

Cinna latifolia (Trev.) Griseb.
DROOPING WOOD-REED   *native*
IR | **WUP** | CUP | EUP   FACW CC 5

Perennial grass, with weak rhizomes. Stems single or in small groups, erect, 5–13 dm long, not swollen at base; sheaths smooth to finely roughened; ligules pale, 2–7 mm long; leaf blades 5–15 mm wide, usually rough-to-touch. Panicle loose, open, pale green, satiny, 1–3.5 dm long, the branches spreading to drooping. Spikelets 1-flowered; glumes narrowly lance-shaped, 1-veined, 2–4 mm long; lemma 2–4 mm long, finely rough-hairy on back, usually with an awn to 1.5 mm long from just below the tip, the awn usually longer than the tip. July–Aug.

Common in wet woods, swamps, and near springs.

Cynosurus *Dog's-Tail Grass*

Cynosurus cristatus L.
CRESTED DOG'S-TAIL GRASS   *introduced*
IR | **WUP** | CUP | EUP   FAC

Perennial, densely tufted grass. Stems 3–8 dm tall; sheaths glabrous, open to the base; auricles absent; ligules truncate, entire, erose or ciliolate, to 2 mm long; leaf blades flat few, 1–3 mm wide. Panicle slender, long-exsert, 3–10 cm long. Spikelets densely crowded, dimorphic, nearly sessile in short-peduncled sterile and fertile pairs; sterile spikelets of 2 glumes and several narrow, scabrous, acuminate lemmas on a continuous rachilla, borne in front of the fertile spikelets and nearly covering them; fertile spikelets 2-4-flowered, the rachilla disarticulating; glumes

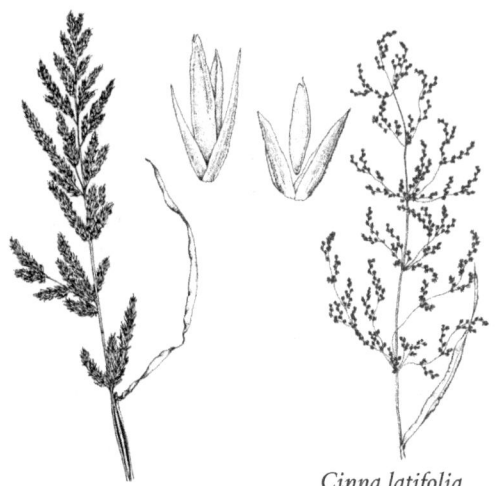

Dactylis glomerata

Cinna arundinacea   *Cinna latifolia*   *Cynosurus cristatus*

narrow, unequal; lemmas broader, 3–3.5 mm long, rounded on the back, scabrous, awn–tipped.

Introduced from Europe in fields, roadsides, and waste places; Isle Royale and Gogebic County.

Dactylis *Orchard-Grass*

Dactylis glomerata L.
ORCHARD-GRASS   *introduced*
IR | WUP | CUP | EUP   FACU

Perennial, densely tufted grass. Stems 5–12 dm tall; sheaths closed for at least 1/2 their length compressed or keeled, scaberulous; auricles absent; ligules membranous, 3–11 mm long, truncate to acuminate; leaf blades flat, elongate, 3–8 mm wide, with a conspicuous midrib and white, scabrous margins. Panicles open, 1–2 dm long, the lower branches naked at base, erect or divergent. Spikelets few-flowered, flat, disarticulating above the glumes and between the lemmas, nearly sessile in dense one-sided clusters, 3–6-flowered; glumes unequal, nearly as long as the lemmas, lance-shaped, 1–3-nerved, keeled, usually ciliate on the keel; lemmas 5–8 mm long, usually ciliate on the keel, awnless or with an awn to 2 mm long.

Introduced from Europe, cultivated for hay or pasture; occasional escape to moist fields, meadows, lawns, and roadsides.

Danthonia *Wild Oatgrass*

Tufted perennial grasses with narrow, often involute leaves, the ligule reduced to a tuft of hairs, auricles absent, and small panicles of large spikelets. Spikelets several-flowered (usually 4–6), articulated above the glumes and between the lemmas; rachilla not prolonged beyond the upper palea. Glumes extending beyond the lemmas, 3-nerved, obscurely keeled. Lemmas closely imbricate, rounded on the back, obscurely several-nerved, more or less pilose, ending in 2 triangular teeth. Awn arising between the teeth, the flat brown base tightly twisted when dry, the upper portion straight, usually divergent from the spikelet. Palea extending about to the base of the awn.

ADDITIONAL SPECIES

Danthonia intermedia Vasey (Oatgrass); plants have purplish spikelets and lemmas glabrous on the back; uncommon in central and western UP.

1 Longest pedicels on lowest panicle branches 1.1–1.8 times longer than the spikelet (including awns); lower panicle branches spreading to reflexed at maturity ... . **D. compressa**
1 Longest pedicels on lowest panicle branches 0.3–0.9 times longer than the spikelet; lower panicle branches ascending to erect at maturity 2
2 Spikelets strongly purple or bronze; lemmas pilose only on margins and at base (on callus), usually 5 mm or more in length (including teeth but not awn) .... **D. intermedia***
2 Spikelets ± greenish or at most purplish at tips; lemmas at least sparsely pilose across the back, less than 6 mm long .... **D. spicata**

Danthonia compressa Austin
FLAT OATGRASS   *native*
IR | WUP | CUP | EUP   FACU CC 7

Perennial grass. Similar to *D. spicata*; see key for differences, also, the leaves of *D. compressa* do not curl tightly upon drying like those of *D. spicata*.

Northern hardwood forests, especially on margins and in openings.

Danthonia spicata (L.) Beauv.
POVERTY WILD OATGRASS   *native*
IR | WUP | CUP | EUP   CC 4

Perennial grass. Stems densely tufted, erect, 2–6 dm tall; sheaths glabrous or sparsely pilose; leaves mostly at or near the base, blades usually involute, 1–2 mm wide, seldom more than 10 cm long, glabrous or sparsely pilose; uppermost stem blades erect to ascending. Panicle contracted, racemiform, 2–5 cm long, the short branches rarely bearing more than 1 spikelet; glumes 8.5–13 mm long; lemmas broadly ovate, sparsely pilose on the back, 3.4–5.2 mm long, including the triangular teeth which are 0.8–1.8 mm long; awn 4.5–7 mm long.

Common in dry woods in sandy or stony soil, especially on jack pine plains, where it may form extensive colonies following disturbance; occasionally found in marshy or boggy places.

The basal leaves tend to curl and form distinctive tufts.

Deschampsia *Hairgrass*

Tufted perennial grasses. Leaves usually mainly basal, narrow, flat or involute; sheaths open; auricles absent; ligules membranous. Flowers in panicles; spikelets yellowish or purple, 2-flowered, disarticulating above the glumes and between the lemmas; rachilla hairy, prolonged beyond the base of the upper lemma. Glumes membranous, usually shining, equaling or longer than the lemmas. Lemmas membranous, obtuse or truncate and erose-toothed, rounded on the back, obscurely 5-nerved, the midnerve diverging at or below the middle into a short awn, the callus bearded.

1 Leaf blades involute, 1–2 mm wide; ligule 1–2.5 mm long; lemmas minutely scabrous-pubescent, bearing a conspicuously bent awn 1–3 mm longer than the lemmas; palea not bifid at tip..................... **D. flexuosa**
1 Leaf blades flat or conduplicate, 1–5 mm wide; ligule

usually 3–12 mm long; lemmas glabrous, bearing a ± straight awn shorter than to slightly exceeding the lemmas; palea bifid at the tip.............. **D. cespitosa**

Deschampsia caespitosa (L.) Beauv.
TUFTED HAIRGRASS native
IR | WUP | CUP | EUP FACW CC 9
Stems densely tufted stiff, erect, 3–10 dm long. Leaves mostly from base of plant, usually shorter than head, 5–30 cm long, usually at least some flat and 1–4 mm wide, the remainder folded or rolled and to 1 mm wide; sheaths glabrous; ligules 2–13 mm long, white, translucent. Panicles narrow to open, 1–4 dm long, the panicle branches threadlike, upright to spreading, the lower branches in groups of 2–5, flowers mostly near branch tips. Spikelets 2-flowered, purple-tinged, fading to silver with age, 2–5 mm long, breaking above the glumes; glumes shiny, 2–5 mm long, the first glume slightly shorter than second glume; lemma smooth, 2–4-toothed across the flat tip, awned from near base on back, the awn shorter to about as long as lemma. June–July.

Wet meadows, streambanks, shores, calcium-rich seeps, rocky shores of Great Lakes.

Deschampsia flexuosa (L.) Trin.
HAIRGRASS native
IR | WUP | CUP | EUP CC 6
Avenella flexuosa (L.) Drej.
Stems densely tufted, 3–10 dm long. Leaves mostly at or near the base; blades involute, 1–2 mm wide;

ligules 1.5–3.5 mm long. Panicle loose and open, somewhat nodding, to 15 cm long, the lowest branches in fascicles of 2–5. Spikelets 4.3–6 mm long; first glume 3–4.5 mm long; second glume acuminate, 3.6–5.3 mm long; lemmas minutely scabrous; awn twisted below the middle, the distal half somewhat divergent, surpassing the lemma by 1–3 mm.

Dry woods, fields, and sand hills.

Dichanthelium *Panic-grass*

Perennial grasses, tufted or sometimes rhizomatous, sometimes with hard, corm-like bases. Stems hollow, usually erect or ascending, sometimes decumbent in the fall, usually branching from the lower stem nodes in summer and fall, terminating in small panicles that are usually partly included in the sheaths. Basal rosettes of winter leaves sometimes present. Stem leaves usually markedly longer and narrower than the rosette blades; ligules of hairs, membranous, or membranous and ciliate, sometimes absent. Flowers in terminal panicles (vernal) developing late spring to early summer, and sometimes lateral panicles (autumnal) in late-summer or fall; disarticulation below the glumes.

Dichanthelium is often included in genus *Panicum*, the two genera being similar in form. However, molecular data reinforce the separation of *Dichanthelium* as a distinct genus.

ADDITIONAL SPECIES

Dichanthelium clandestinum (L.) Gould (Deertongue grass), known from along a railway in Gogebic County; more common in southern Lower Peninsula.

Dichanthelium spretum (Schult.) Freckmann; wet sandy shores, Luce and Schoolcraft counties; sometimes included in *D. acuminatum* as subsp. *spretum* (Schult.) Freckmann & Lelong.

1 Basal leaf blades similar in shape to the lower stem leaves, usually erect to ascending, clustered at the base, sometimes vestigial; stems branching from near the base in the fall, with 2-4 leaves, only the upper 2-4 internodes elongated.............................2

1 Basal leaf blades usually well-differentiated from the stem blades, spreading, forming a rosette, or basal blades absent; stems usually branching from the mid-culm nodes in the fall, with 3-14 leaves, usually all internodes elongated............................3

2 Upper glumes and lower lemmas forming a beak extending 0.2-1 mm beyond the upper florets; spikelets 3.2-4.3 mm long; primary panicles with 7-25 spikelets .
.................................**D. depauperatum**

2 Upper glumes and lower lemmas equaling or exceeding the upper florets by no more than 0.3 mm, not forming a beak; spikelets 2-3.4 mm long; primary panicles with 12-70 spikelets..................... **D. linearifolium**

Deschampsia caespitosa

Danthonia spicata

Deschampsia flexuosa

3. Lower glumes thinner and more weakly veined than the upper glumes, attached about 0.2 mm below the upper glumes, the bases clasping the pedicels; spikelets attenuate basally . . . . **D. portoricense**

3. Lower glumes similar in texture and vein prominence to the upper glumes, attached immediately below the upper glumes, the bases not clasping the pedicels; spikelets usually not attenuate basally 4

4. Ligules with a membranous base, ciliate distally; stems usually arising from slender rhizomes; lower florets often staminate; stem blades 5-40 mm wide, often with a cordate base 5

4. Ligules of hairs; stems arising from caudices; lower florets sterile; stem blades 1-18 mm wide, bases usually tapered, rounded, or truncate at the base, sometimes cordate 6

5. Spikelets ellipsoid, not turgid, with pointed apices; stem blades 4-6, cordate at the base; sheaths without papillose-based hairs . . . . **D. latifolium**

5. Spikelets obovoid, turgid, with rounded apices; stem blades 3-4, tapered, rounded or truncate to cordate at the base; sheaths with papillose-based hairs . . . .
 . . . . **D. xanthophysum**

6. Spikelets 2.5-4.3 mm long, usually obovoid, turgid; upper glumes usually with an orange or purple spot at the base, the veins prominent. . . . . **D. oligosanthes**

6. Spikelets 0.8-3 mm long, ellipsoid or obovoid, not turgid; upper glumes lacking an orange or purple spot at the base and the veins not prominent 7

7. Ligules absent or to 1.8 mm long, without adjacent pseudoligules; stems and at least the upper sheaths glabrous or sparsely pubescent with hairs of 1 length only; spikelets glabrous or pubescent. . . . . . **D. boreale**

7. Ligules 1-5 mm long, or the stems and sheaths with long hairs and also puberulent; spikelets variously pubescent to subglabrous 8

8. Spikelets 1.1-2.1 mm long; sheaths glabrous or pubescent with hairs no more than 3 mm long . . . **D. acuminatum**

8. Spikelets 1.8-3 mm long; sheaths with hairs to 4 mm long . . . . **D. ovale**

Dichanthelium acuminatum (Sw.) Gould & C. A. Clark
HAIRY PANIC-GRASS native
IR | **WUP** | **CUP** | **EUP** FACU CC -
Dichanthelium implicatum (Scribn.) Kerguélen
Panicum acuminatum Sw.

Stems densely tufted, erect to prostrate, usually straight and radiating from the base, glabrous to pilose or villous; sheaths softly pubescent, papillose-pilose, or glabrous; ligule hairs 2–5 mm long; leaf blades 4–12 cm long, 5–12 mm wide, glabrous or pubescent on either or both sides. Primary panicle ovoid, with divergent, often flexuous branches, the axis glabrate to villous. Spikelets plumpellipsoid to obovoid, 1–2 mm long, finely pubescent; first glume broadly angular-rotund, usually less than a third as long than the second glume. Autumnal phase spreading or prostrate, copiously branched chiefly from the middle nodes; blades about half as large as the vernal ones; panicles few-flowered, mostly surpassed by the leaves.

Moist or dry situations, open woods, dunes, shores, and prairies.

Dichanthelium boreale (Nash) Freckmann
NORTHERN PANIC-GRASS native
IR | **WUP** | **CUP** | **EUP** FAC CC 7
Panicum boreale Nash

Stems upright, 2–6 dm long; sheaths hairy; ligule a fringe of short hairs to 0.5 mm long; leaf blades upright to spreading, 5–20 cm long and 1–2 cm wide, smooth or sometimes hairy on underside, base of blade often fringed with hairs. Panicle open, 5–12 cm long, the branches spreading or upright. Spikelets oval in outline, finely hairy, about 2 mm long, on long stalks with 1 fertile flower; first glume 0.5-1 mm long, triangular-ovate; second glume and lemma purple-tinged, about equal, and as long as fruit. Autumnal phase with decumbent stems, branches arising from the lower and mid-stem nodes, and rebranching 2–3x. June–Aug.

Moist to wet sandy or rocky places; sometimes in drier aspen or oak woods.

Dichanthelium depauperatum (Muhl.) Gould
STARVED PANIC-GRASS native
IR | **WUP** | **CUP** | **EUP** CC 4
Panicum depauperatum Muhl.

Stems tufted, erect or nearly so, 1–4 dm long, very slender, glabrous to puberulent; sheaths and leaves glabrous to sometimes long-pilose; leaf blades erect, 8–15 cm long, 2–5 mm wide; ligules about 0.5 mm long. Primary panicle eventually exsert, 3–6 cm long, not much exceeding the leaves. Spikelets ellipsoid, 2.7–4.1 mm long, averaging 3.2 mm, glabrous or minutely pubescent; first glume membranous, ovate or triangular, about a third as long; second glume and sterile lemma sharply nerved, pointed, projecting 0.5–1.5 mm beyond the fertile lemma. Autumnal phase similar, the panicles much reduced and usually concealed among a dense mass of erect leaves.

Dry or sandy soil, usually in open woods.

Dichanthelium latifolium (L.) Gould & C. A. Clark
BROADLEAVED PANIC-GRASS native
IR | **WUP** | **CUP** | **EUP** FACU CC 5
Panicum latifolium L.

Stems tufted, slender, erect, 4–10 dm long, usually glabrous, rarely sparsely puberulent; sheaths pubescent; ligules to 0.7 mm long, membranous, ciliate, the cilia longer than the membranous portion; leaf blades lance-shaped, spreading, glabrous. or nearly so on both sides, ciliate at the cordate base, the larger usually 10–16 cm long, 15–30 mm wide. Primary panicle tardily exsert, ovoid with ascending branches, 6–12 cm long. Spikelets oblong-obovoid, 2.9–3.7 mm long, averaging 3.3 mm, softly villosulous; first glume about half as long, acute; second glume and sterile lemma shorter than the fruit. Autumnal phase sparsely branched from the middle nodes, the leaf blades not much reduced or greatly crowded; panicles small, included at base.

Moist or dry woods and thickets.

Dichanthelium linearifolium (Scribn.) Gould
LINEAR-LEAVED PANIC-GRASS   native
IR | WUP | CUP | EUP   CC 4
Panicum linearifolium Scribn.

Stems densely tufted, 2–6 dm long, glabrous or nearly so; sheaths glabrous or pilose with dense, fine, papillose-based hairs; ligules about 0.5 mm long; leaf blades erect, usually 10–20 cm long, 2–5 mm wide, glabrous to sparsely pilose. Primary panicle 3–8 cm long, usually much surpassing the blades. Spikelets ellipsoid to oblong-obovoid, 1.7–3.1 mm long, glabrous to pilose; first glume about a third as long, ovate to triangular; second glume and sterile lemma blunt, equaling the fruit. Autumnal phase similar, the greatly reduced panicles concealed among the leaves.

Dry or stony soil, open woods and banks.

Dichanthelium oligosanthes (J.A. Schultes) Gould
FEW-FLOWERED PANIC-GRASS   native
IR | WUP | **CUP** | EUP   FACU CC 5
Panicum oligosanthes J.A. Schultes

Stems few to several, loosely tufted, erect or ascending, 2–7 dm long, often purplish; sheaths not overlapping, glabrous or puberulent, margins ciliate, collars loose, puberulent; ligules 1-3 mm long, of hairs; leaf blades spreading, lance-shaped, glabrous or rarely sparsely papillose-pilose above; glabrous, softly pubescent, or sparsely papillose-pilose beneath, usually papillose-ciliate and densely long-hairy at base, the larger 6–12 cm long and 7–12 mm wide. Primary panicle short-exsert (0–5 cm), becoming long-exsert in age, ovoid, 5–10 cm long. Spikelets ellipsoid to obovoid, 2.7–4 mm long, glabrous or minutely villous; first glume about 2/5 as long, broadly ovate; second glume and sterile lemma about equal, barely equaling the fruit. Autumnal phase sparsely branched, chiefly from the middle and upper nodes, forming loose bunches; leaf blades not greatly reduced, surpassing the few-flowered panicles.

Dry or moist, often sandy soil, open woods and prairies.

Dichanthelium ovale (Elliott) Gould & C.A. Clark
STIFF-LEAVED PANIC-GRASS   native
IR | WUP | **CUP** | EUP   FACU CC -

Stems tufted, 20–60 cm long, usually more than 1 mm thick, not delicate, ascending or spreading and often decumbent; sheaths shorter than the internodes, pilose, hairs to 4 mm, occasionally with shorter, spreading hairs underneath; ligules 1-5 mm long, of hairs; leaf blades 4–10 cm long and 3–10 mm wide, relatively firm, mostly ascending or spreading, sparsely to densely pubescent with appressed or erect hairs to 5 mm long, margins often whitish. Primary panicle 3-10 cm long, nearly as wide when fully expanded; branches often stiffly ascending or spreading, usually

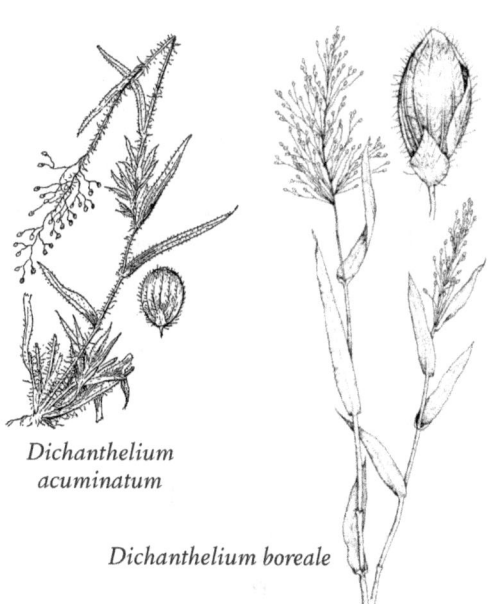

Dichanthelium acuminatum

Dichanthelium boreale

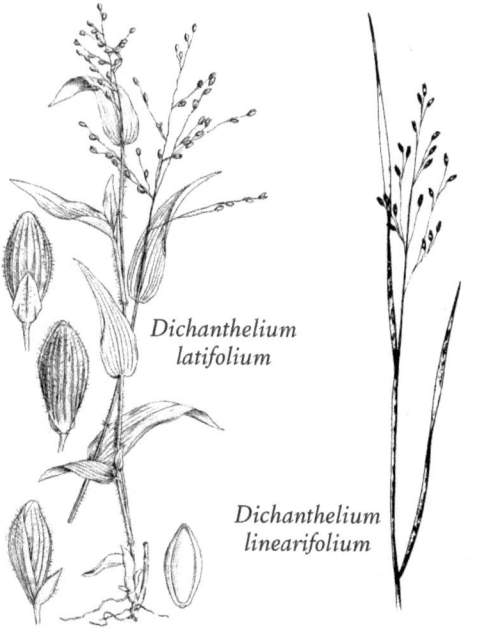

Dichanthelium latifolium

Dichanthelium linearifolium

pilose at base. Spikelets 2–3 mm long, ellipsoid or obovoid, densely to sparsely pilose; first glume to 1/2 as long as the spikelets, often triangular, not strongly veined; second glume usually slightly shorter than the lower lemmas and upper florets, not strongly veined. Autumnal phase with decumbent to prostrate stems, with erect, slightly reduced blades and greatly reduced secondary panicles.

Dry, sandy openings and meadows; Menominee County.

Dichanthelium portoricense (Desv. ex Ham.) Hansen & Wunderlin
BLUNT-GLUMED PANIC-GRASS *native*
IR | **WUP** | **CUP** | EUP CC 5
Dichanthelium columbianum (Scribn.) Freckmann
Panicum columbianum Scribn.

Stems densely tufted, erect or ascending, 2–5 dm tall, often purplish, densely short-pubescent with minute hairs 0.1–0.4 mm long, or toward the tips of the lower internodes sometimes 1 mm long; sheaths similarly pubescent; ligule hairs 0.5–1.5 mm long; leaf blades 3–7 cm long and 3–7 mm wide, glabrous above or with a few widely scattered hairs, minutely puberulent beneath. Primary panicle ovoid, 2–6 cm long, its axis puberulent. Spikelets obovoid, obtuse, 1.4–1.9 mm long, finely pubescent; first glume averaging 2/5 as long, triangular-ovate. Autumnal phase spreading or decumbent, branched early from most of the nodes; blades scarcely reduced; panicles smaller, surpassed by the leaves.
Moist or dry, especially sandy soil.

Dichanthelium xanthophysum (Gray) Freckmann
PALE PANIC-GRASS *native*
IR | **WUP** | **CUP** | **EUP** CC 6
Panicum xanthophysum A. Gray

Stems few or several in loose tufts, erect or ascending, 2–5 dm long, glabrous; sheaths loose, often exceeding the internodes, glabrous to pilose or papillose-pilose; ligules to 0.5 mm long, membranous, ciliate, the cilia longer than the membranous bases; leaf blades yellowish green, erect or nearly so, the larger 10–15 cm long and 10–20 mm wide, glabrous on both sides, slightly narrowed to the rounded, papillose-ciliate base. Primary panicle 5–10 cm long, very narrow, with erect branches. Spikelets obovoid, 3.3–3.8 mm long, minutely puberulent; first glume about half as long, triangular-ovate; second glume slightly shorter than the sterile lemma and fruit. Autumnal phase with 1 or 2 erect branches, bearing scarcely reduced leaf blades equaling or exceeding the shorter panicles.

Dry sandy soil of open woodlands.

Digitaria *Crabgrass*

Annual or perennial grasses. Stems spreading, branched from the base. Leaves wide, flat, prostrate, with the tips ascending; sheaths open, ligules membranous. Flowers in several terminal, digitate, spike-like racemes. Spikelets 1-flowered, single or in clusters of 2 or 3 on unequal pedicels on one side of an elongate rachis. First glume minute or lacking; second glume a third to fully as long as the spikelet, conspicuously 5–7-nerved. Fertile lemma cartilaginous with hyaline margins, acute, often shining.

1 Spikelets ca. 2–2.3 mm long, the fertile lemma dark brown; second glume nearly or fully as long as the floret; sheaths and blades usually nearly or quite glabrous (except around summit of sheath) ..... **D. ischaemum**
1 Spikelets ca. 2.5–3 mm long, the fertile lemma light or dark grayish; second glume only about half as long as the floret; sheaths and usually blades ± pilose, at least toward base of plant .... **D. sanguinalis**

Digitaria ischaemum (Schreb.) Muhl.
SMOOTH CRABGRASS *introduced*
IR | **WUP** | **CUP** | **EUP** FACU

Annual grass (or sometimes longer-lived). Stems branched and spreading from the decumbent base, rooting at the nodes, 2–5 dm long; sheaths glabrous or sparsely pubescent; ligules 0.6–2.5 mm long; leaf blades to 9 cm long and 3–5 mm wide, glabrous. Racemes 2–5, or rarely to 8, 4–10 cm long; rachis broadly winged, about 1 mm wide. Spikelets elliptic or somewhat obovate, 1.7–2.1 mm long, often purple; first glume lacking or minute and hyaline; second glume and sterile lemma equal and about as long as

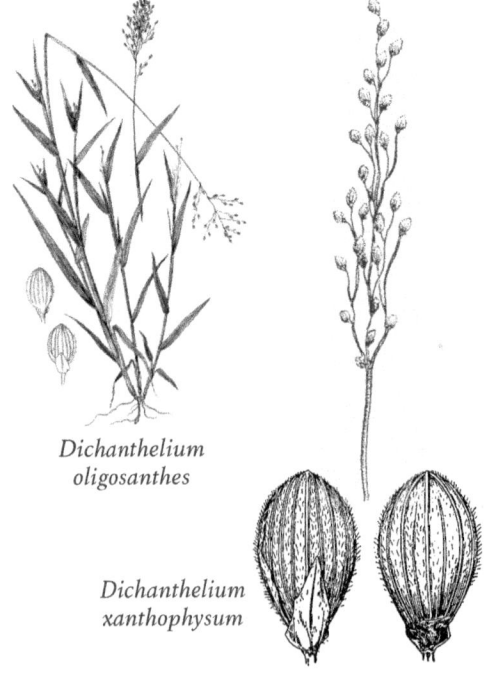

Dichanthelium oligosanthes

Dichanthelium xanthophysum

the spikelet, both more or less pubescent or subtomentose with capitellate hairs, especially in stripes between the nerves; fertile lemma purple-black.

Eurasian weed of lawns, waste places.

Digitaria sanguinalis (L.) Scop.
HAIRY CRABGRASS *introduced*
IR | WUP | CUP | EUP   FACU

Annual grass. Stems decumbent or prostrate, much branched, rooting at the nodes, usually 3–6 dm long. leaf blades 4–10 cm long, 5–10 mm wide, pilose. Racemes 3–6 in each of 1–3 whorls, 5–15 cm long; rachis broadly winged, 1 mm wide, scabrous on the margins; pedicels triquetrous, scabrous. Spikelets 2.4–3 mm long; first glume minute, often deciduous, the second half as long as the spikelet; sterile lemma usually scabrous on the 5 strong nerves; fertile lemma greenish brown.

Eurasian weed of fields, gardens, lawns, and waste ground.

Echinochloa crus-galli (L.) Beauv.
LARGE BARNYARD-GRASS *introduced*
IR | WUP | CUP | EUP   FAC

Weedy annual grass. Stems 1 m or more long; sheaths glabrous; ligules absent; leaf blades 7–30 mm wide. Panicle erect, green to purple, 1–2.5 dm long; panicle branches spreading to erect, long-hairy, some of the hairs as long or longer than spikelets (excluding spikelet awns). Spikelets 3–5 mm long (excluding awns); glumes awnless; sterile lemma awnless or with an awn to 4 cm or more long; tip of fertile lemma firm, shiny, rounded or broadly tapered to a point, the beak usually green and withered, the lemma body and beak separated by a line of tiny hairs. July–Sept.

Shores, wet meadows, ditches, streambanks, mud flats, moist disturbed areas. Introduced and naturalized throughout most of USA. *E. muricata* is similar in form and habitat, but distinguished by features of the lemma (see key).

Echinochloa muricata (Beauv.) Fern.
BARNYARD-GRASS *native*
IR | WUP | CUP | EUP   OBL CC 1

Annual grass. Stems 1 m or more long; sheaths glabrous; ligules absent; leaf blades 5–30 mm wide. Panicle green to purple, sometimes strongly purple, 1–3 dm long, panicle branches spreading, hairs on branches absent or to 3 mm long and shorter than spikelets. Spikelets 2–4 mm long (excluding awns); glumes awnless; sterile lemma awnless or with an awn 5–10 mm long; tip of fertile lemma firm, shiny, gradually tapered to the stiff beak, the lemma body and beak not separated by a line of tiny hairs (the beak itself often short-hairy). July–Sept.

Shores, streambanks and ditches, where sometimes in shallow water.

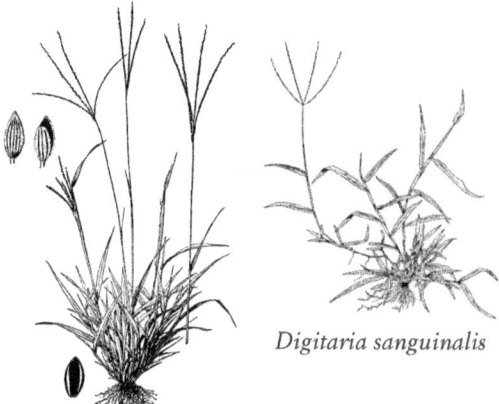

Digitaria sanguinalis

Digitaria ischaemum

Echinochloa *Barnyard-Grass*

Large, weedy, annual grasses. Stems single or several together, erect to more or less horizontal, to 1 m or more long. Leaves flat, wide and smooth; sheaths smooth or hairy; ligules usually absent. Head a dense panicle, the branches crowded with spikelets forming racemes or spikes. Spikelets with 1 terminal fertile floret and 1 sterile floret, breaking below the glumes, nearly stalkless; glumes unequal, the first glume 3-veined, to half the length of second glume, the second glume 5-veined; sterile lemma similar to second glume, awned or awnless; fertile lemma smooth and shiny.

1 Fertile lemma rounded or broadly tapered to a thin, membranous, withered beak .... **E. crus-galli**
1 Fertile lemma tapered to a stiff, persistent beak .... **E. muricata**

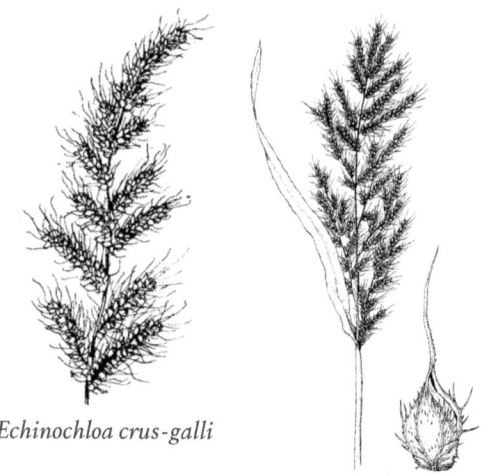

Echinochloa crus-galli

Echinochloa muricata

Eleusine *Goose Grass*

Eleusine indica (L.) Gaertn.
INDIAN GOOSE GRASS *introduced*
IR | WUP | **CUP** | EUP FACU

Annual grass. Stems mostly 3–6 dm long, compressed, branched from the base, spreading or ascending; sheaths compressed and keeled; ligules membranous, ciliate; leaf blades flat, soft, with prominent white midveins; lower margins and/or undersurface often with papillose-based hairs. Spikes digitate, usually 3–8, 4–10 cm, long and about 5 mm wide, spreading or ascending. Spikelets articulated above the glumes, crowded, 3–6-flowered; glumes unequal, the first shorter, 1-nerved, the second 2–3 mm long, strongly 3–5-nerved, shorter than the lemmas but resembling them; lemmas 2.5–4 mm long, compressed, strongly 3–5-nerved.

Native of the Old World; weedy in lawns, gardens, and waste places; Marquette County. Often popularly confused with hairy crabgrass (*Digitaria sanguinalis*).

Eleusine indica

Elymus *Wild Rye*

Tufted perennial grasses. Leaves flat, sheaths open for most of their length, auricles often present, ligules short. Head a densely flowered spike. Spikelets usually 2 at each node of spike, breaking above or below glumes; glumes narrow and awnlike; lemmas tipped with a long awn; stamens 3.

1 Spikelets 1 at each node (or most nodes) of the rachis; glumes not more than 2......................... 2
1 Spikelets mostly 2–3 at each node of the rachis...... 5
2 Lemmas densely hairy; leaves with narrow (rarely as much as 4.5 mm wide) often involute blades, the whole plant usually strongly glaucous......... **E. lanceolatus**
2 Lemmas glabrous (rarely slightly pubescent), smooth or scabrous; leaves various........................ 3
3 Stems tufted, rhizomes absent; anthers 1–2.2 (–2.4) mm long; rachilla readily disarticulating between the florets when mature (on dry specimens, the florets very easily dislodged and empty glumes often remaining on older plants)............................. **E. trachycaulus**
3 Stems from elongate rhizomes; anthers 3–6 mm long; rachilla often not readily disarticulating (florets not easily dislodged on dry specimens except over-ripe ones, empty glumes seldom if ever present)............... 4
4 Spikelets 15-30 mm long, with 6-16 florets; leaf blades stiff, deeply grooved on the upper surface; cartilaginous band of upper nodes of stem shorter than thick .... **E. smithii**
4 Spikelets 10-18 mm long, with 3-6 florets; leaf blades lax, not deeply grooved; cartilaginous band of upper nodes of stem as long as thick .... **E. repens**
5 Glumes absent or vestigial, or, if present, slenderly awn-like their entire length and at least one much shorter than the others at a node; spikelets horizontally spreading at maturity (± ascending when young), well separated, clearly revealing the entire rachis **E. hystrix**
5 Glumes present, awn-like to lanceolate, of about equal length; spikelets ascending at maturity, usually concealing much of the rachis 6
6 Larger paleas (lowest in each spikelet) 8.6–13 mm long; awns of lemmas usually widely spreading at maturity 7
6 Larger paleas 5.5–8.5 mm long; awns of lemmas mostly straight 9
7 Body of glume about twice as long as its awn, or longer; awns of lemmas usually straight at maturity; spike curved to erect...................... (go to lead 10)
7 Body of glume about equaling its awn, or shorter; awns of lemmas ± curved at maturity (straight when young); spike curved to strongly nodding 8
8 Leaves 5–8 on a stem, the broadest blades rarely as much as 15 mm wide, glabrous above ... **E. canadensis**
8 Leaves 10–12 on a stem, the broadest blades 15–19 mm wide, finely hairy above .... **E. wiegandii**
9 Glumes, at least the broadest, 1–2 mm wide, clearly expanded and flattened above the base........... 10
9 Glumes less than 1 mm wide, scarcely if at all widened above the base 11
10 Base of glumes not conspicuously bowed out, but flattened, hardened for less than 1 mm; glumes not thickened above the base on inner face, with very narrow, thin, translucent margins, often slightly overlapping; stem leaves 5–6.............. **E. glaucus**
10 Base of glumes ± bowed out, terete and hard for 1 mm or more; glumes also thickened, pale, and hardened on inner face for about the basal half or more, with firm margins, not at all overlapping; stem leaves 6–10 .... **E. virginicus**
11 Palea of lowest floret in spikelet 7–8.5 mm long; leaves 8–10, glabrous........................... **E. riparius**
11 Palea 5.5–7 mm long; leaves 6–7, the sheaths and upper surface of blades finely villous............. **E. villosus**

Elymus canadensis L.
NODDING WILD RYE *native*
IR | **WUP** | **CUP** | **EUP** FACU CC 5

Stems loosely tufted, stout, 1 m or more tall; sheaths often reddish brown; auricles 1.5–4 mm long, brown or purplish black; ligules to 2 mm long, truncate, ciliolate; leaf blades flat, or involute when dry, usually 8–20 mm wide, glabrous to sparsely pilose. Spike 10–15 cm long, usually nodding, often interrupted at base by the elongation of the lower internodes, 1–2 cm thick, excluding the awns. Spikelets 3–7-flowered; glumes, including the awns, usually 15–30 mm long, 3–5-nerved, glabrous, scabrous, or pubescent; lemmas, including the outwardly curved awns, 3–5 cm long, glabrous, scabrous, or pubescent. Variable.

Dry or moist soil, often where sandy or gravelly, usually in full sun.

Elymus glaucus Buckl.
BLUE WILD RYE   *native*
IR | **WUP** | **CUP** | **EUP**   FACU CC 8

Tufted perennial lacking rhizomes or with short rhizomes. Stems hollow, to 12 dm tall; sheaths glabrous or sometimes those at the base hirsute; ligules to 1 mm long, entire or ciliate; auricles well developed, usually clasping the stem leaf blades mostly flat, usually glaucous, 5–9 mm wide, pilose on the veins above, scabrid or glabrous below. Spikes erect to arching, exserted, mostly 6–12 cm long; spikelets 1 or 2 per node, each with 3–6 florets; glume bodies linear-lanceolate, 7–14 mm long and 1–2 mm wide, shallowly keeled or flat, margins hyaline, apex sharp pointed or with an awn to 5 mm long; lemma bodies 8.5–13 mm long, awn usually straight, 10–25 mm long.

Dry, sandy or rocky woods and rock shores, mostly near Lake Superior.

Elymus hystrix L.
BOTTLEBRUSH-GRASS   *native*
IR | **WUP** | **CUP** | **EUP**   FACU CC 5

Hystrix patula Moench.

Plants occasionally glaucous, particularly the spikes; stems usually solitary or loosely tufted, not rhizomatous, 6–10 dm tall; sheaths usually glabrous, sometimes pilose, often purplish; auricles usually present, 0.5–3 mm long, brown to black; ligules 1–3 mm long; leaf blades 8–13 mm wide. Spikes 5–12 cm long, the internodes of the flexuous 2-edged rachis 4–10 mm long. Spikelets usually in pairs; glumes varying, even on the same plant, from none to setaceous and to 16 mm long; lemmas 8–11 mm long, tipped with a rough awn 1–4 cm long.

Moist deciduous woods, especially in wet or slightly disturbed areas. Spikelets soon horizontally divergent, the lemmas easily detached.

Elymus lanceolatus (Scribn. & J.G. Sm.) Gould
STREAMSIDE WILD RYE   *native*
IR | **WUP** | **CUP** | **EUP**   FACU CC 10

Agropyron dasystachyum (Hook.) Scribn. & J. Sm.

Plants strongly rhizomatous, often glaucous; stems slender, 5–8 dm tall from long rhizomes; leaves often mostly basal; sheaths glabrous or pubescent; auricles usually present on the lower leaves; ligules to 0.5 mm long, erose, sometimes ciliolate; leaf blades involute when dry, 1–3 mm wide. Spikes 7–15 cm long. Spikelets 4–8-flowered, 12–20 mm long; glumes narrowly lance-shaped, villous to glabrous, 3–5nerved, the first 7–10 mm long, the second somewhat longer; lemmas 9–12 mm long, usually awnless; in the typical variety, moderately hairy, the hairs stiff, shorter than 1 mm.

Sandy shores and dunes along Lake Michigan.

Elymus repens (L.) Gould
QUACK-GRASS   *introduced (invasive)*
IR | **WUP** | **CUP** | **EUP**   FACU

Agropyron repens (L.) Beauv.
Elytrigia repens (L.) Nevski

Plants strongly rhizomatous, sometimes glaucous. Stems erect, usually 5–10 dm tall; sheaths pilose or glabrous near base; auricles to 1 mm long; ligules to 1.5 mm long; leaf blades flat, soft, 5–10 mm wide, with numerous slender nerves about 0.2 mm apart. Spikes 6–17 cm long, with numerous ascending, overlapping spikelets; rachis joints usually flat on one side, rounded on the other. Spikelets 10–18 mm long, 4–8-flowered; glumes narrowly oblong to lance-shaped, 8–14 mm long, sharply nerved, acuminate or short-awned; lemmas similar in size and shape, less sharply nerved, acuminate or with an awn to 10 mm long.

Eurasian native, abundant and often a noxious weed in meadows, fields, roadsides, and waste places.

Highly variable in color from green to glaucous, in pubescence, and in presence and length of awns.

Elymus canadensis   *Elymus hystrix*

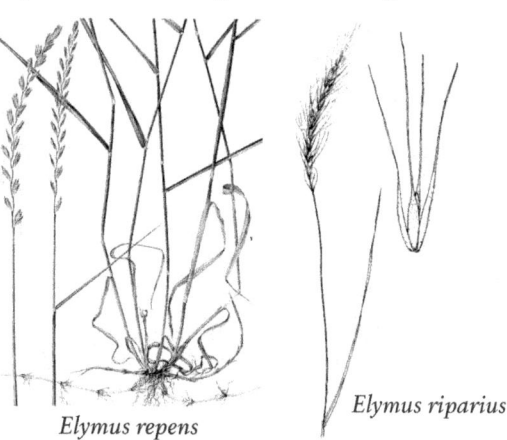

Elymus repens   *Elymus riparius*

Elymus riparius Wieg.
RIVERBANK WILD RYE native
IR | WUP | CUP | EUP   FACW CC 8

Plants tufted, not rhizomatous, often somewhat glaucous. Stems 1 m or more tall; sheaths usually glabrous, often reddish brown; auricles absent or to 2 mm long, brown; ligules shorter than 1 mm; leaf blades 5–15 mm wide, flat, lax, dull green, drying to grayish, upper surface smooth to rough. Spikes 6–20 cm long, somewhat nodding. Spikelets mostly 2 at each node, 2–4-flowered, finely hairy, breaking above glumes; glumes narrow, to 1 mm wide at middle, not bowed-out at base; lemma finely hairy to smooth, tipped with a straight awn 2–3 cm long.

Streambanks, floodplain forests.

Similar to nodding wild rye (*E. canadensis*), a species of drier, sandy places, but awns straight rather than bent and curved.

Elymus trachycaulus (Link) Gould ex Shinners
SLENDER WILD RYE native
IR | WUP | CUP | EUP   FACU CC 8

Agropyron caninum (L.) Beauv.
Agropyron trachycaulum (Link) Steud.

Stems loosely tufted, erect, 4–10 dm tall; sheaths usually glabrous, sometimes hirsute or villous; auricles absent or to 1 mm long; ligules 0.2–0.8 mm long, truncate; leaf blades 4–10 mm wide, flat to involute, usually straight and ascending, with numerous fine sharp nerves. Spikes 6–20 cm long. Spikelets erect or ascending, few-flowered, in ours mostly not imbricate, the tip of one not reaching to the base of the next one above on the same side; glumes 5–7-nerved, acuminate or short-awned; lemmas awnless or with straight awns to 2 cm long (rarely more); rachilla readily disarticulating between the lemmas, leaving the persistent glumes attached.

Dry, open, rocky woods, sandy shores and barrens; rarely in fens and tamarack swamps.

Variable but distinguished by the short anthers (when young), and by the readily disintegrating spikelets (when mature); the rachilla is also nearly always villous.

Elymus villosus Muhl.
HAIRY WILD RYE native
IR | WUP | **CUP** | EUP   FACU CC 5

Plants often persistently deep green. Stems tufted, slender, 5–10 dm tall; sheaths glabrous to pilose; auricles 1–3 mm long, brownish; ligules less than 1 mm long, entire or erose; leaf blades 4–12 mm wide, lax, dark glossy green, softly villous on the upper side. Spikes slightly or strongly nodding, dense, 5–12 cm long; glumes setaceous, not widened above the base, 0.4–1 mm wide, strongly 1–3-nerved, 15–30 mm long, including the awn; lemmas 2–4 cm long, including the straight ascending awn.

Swampy forests and riverbanks; also in drier woods.

Glumes and lemmas usually conspicuously hirsute.

Elymus virginicus L.
VIRGINIA WILD RYE native
IR | WUP | CUP | EUP   FACW CC 4

Plants sometimes glaucous, especially in the spikes. Stems tufted, 6–12 dm long; sheaths usually glabrous, rarely hirsute, occasionally reddish or purplish; auricles absent or to 1.8 mm long, pale brown; ligules less than 1 mm long; leaf blades flat, lax, 5–15 mm wide, rough-to-touch on both sides. Spikes erect, 5–15 cm long, the base of spike often covered by top of upper sheath. Spikelets usually 2 at each node, 2–4-flowered, breaking below glumes; glumes firm, 1–2 mm wide, yellowish, bowed-out at base, tapered to a straight awn about 1 cm long; lemmas 6–9 mm long, smooth to hairy, usually with a straight awn to 3 cm long. July–Aug.

Common in floodplain forests, thickets, and on streambanks.

Elymus wiegandii Fern.
WIEGAND'S WILD RYE native
IR | WUP | CUP | EUP   FAC CC 8

Similar to *E. canadensis* but plants taller, the inflorescence is drooping, not merely arching, and the leaves are broader (in *E. canadensis* the leaves are stiff and often involute, especially toward the tip); larger glumes in *E. wiegandii* are 0.4–0.7 mm wide, glumes of *E. canadensis* are 0.7–1.6 mm wide.

Moist forests, especially along streams.

Eragrostis *Lovegrass*

Annual grasses (ours), perfect-flowered or with staminate and pistillate flowers on different plants. Stems clumped, or spreading and rooting at lower nodes and with creeping stolons. Leaves with short, flat to folded blades; sheaths open, short-hairy near top; ligule a ring of short hairs. Heads usually many, in an open or narrow panicle. Spikelets few- to many-flowered, breaking above glumes, laterally compressed, the florets overlapping; glumes unequal; lemmas 3-veined; palea shorter than lemma, 2-veined.

1 Plants prostrate basally, rooting at lower nodes; nodes of stem bearded (very rarely glabrous) .. **E. hypnoides**
1 Plants ± erect or spreading from the base, not rooting at nodes; nodes of stem glabrous 2
2 Margins (often inrolled) of leaves and also (usually at least sparsely) pedicels and keels of lemmas and glumes ± glandular-warty 3
2 Margins of leaves, pedicels, and keels of lemmas and glumes not glandular-warty 4

3. Well-developed spikelets 2.5–3.5 mm wide; larger glume 1.7–2.5 mm long; sheaths essentially glabrous except at summit . . . . **E. cilianensis**
3. Well-developed spikelets 1.5–2 mm wide; larger glume 1–1.5 mm long; sheaths sparsely pilose . . . . **E. minor**
4. Spikelets reddish to purplish; plants perennial, with hard knotty base; lowest panicle branches usually with a long-pilose white to yellowish or red pubescence in the axil . . . . **E. spectabilis**
4. Spikelets greenish gray to dark lead-colored (occasionally with purplish flush besides); plants annual, with relatively soft base; lowest panicle branches glabrous to sparsely pilose 5
5. Larger spikelets mostly 6–11 (–15)-flowered, usually on ± appressed pedicels (though panicle branches may be widely spreading); lowest lemma ca. 1.4–2 mm long; lateral nerves of lemma distinct . . . . **E. pectinacea**
5. Larger spikelets mostly 2–4 (–6)-flowered, on spreading pedicels; lowest lemma ca. 1.2–1.6 mm long; lateral nerves of lemma obscure . . . . **E. frankii**

Eragrostis cilianensis (All.) Vign.
STINK-GRASS *introduced*
IR | **WUP** | **CUP** | EUP   FACU

Stems densely tufted, spreading or ascending from a decumbent base, rarely erect, 1–4 dm long; sheaths glabrous, occasionally glandular, tips hairy, the hairs to 5 mm long; ligules to 0.8 mm long, ciliate; leaf blades 5–20 cm long, 2–6 mm wide. Panicle ovoid to subcylindric, 5–15 cm long, the branches spreading, the pedicels usually 1–2 mm long. Spikelets broadly linear, 2.5–3 mm wide, 10–40-flowered; first glume 1.3–1.9 mm long; second glume 1.5–2 mm long; lemmas broadly elliptic-ovate, closely imbricate, 2.1–2.6 mm long, glandular on the keel; grain 0.7 mm long, dull brown.

Native of Europe; a weed of moist ground.

Eragrostis frankii C.A. Mey.
SANDBAR LOVEGRASS *native*
IR | WUP | **CUP** | EUP   FACW CC 4

Stems densely tufted, branched, 1–5 dm long; sheaths mostly glabrous but long-hairy at tip, the hairs to 4 mm long, often also with glandular pits; ligules to 0.5 mm long, ciliate; leaf blades 1–4 mm wide, flat to involute. Panicle open, 5–20 cm long, the branches mostly ascending. Spikelets 3–6-flowered, 2–3 mm long and 1–2 mm wide. Aug–Sept.

Wet, muddy areas, streambanks, sandbars, roadside ditches, cultivated fields; Menominee County.

Eragrostis hypnoides (Lam.) B.S.P.
TEAL LOVEGRASS *native*
IR | **WUP** | **CUP** | EUP   OBL CC 8

Stems mostly spreading and rooting at lower nodes, 5–15 cm long, smooth but short-hairy at nodes, stoloniferous and forming mats; sheaths pilose on the margins, collars and tips, the hairs to 0.6 mm long; ligule of short hairs about to 0.6 mm long; leaf blades to 5 cm long, 1–3 mm wide, flat to involute, upper surface hairy. Panicle loose, 2–6 cm long. Spikelets 10–35-flowered, linear, 3–10 mm long; glumes 1-veined, 0.5–1.5 mm long; lemma smooth and shiny, 1–2 mm long. July–Sept.

Wet, sandy or muddy shores and streambanks, sand bars, mud flats.

Eragrostis minor Host
LITTLE LOVEGRASS *introduced*
IR | **WUP** | **CUP** | **EUP**

Stems tufted, slender, ascending or decumbent, 1–4 dm long; sheaths sometimes glandular on the midveins, hairy at the tips, hairs to 4 mm; ligules to 0.5 mm long, ciliate; leaf blades 1–3 mm wide, flat, glabrous or sparsely white-hairy. Panicle ovoid or oblong, 3–10 cm long, with spreading branches. Spikelets pediceled, 10–20-flowered, 1.5–2 mm wide; first glume 1.2–1.5 mm long; second glume 1.4–1.7 mm long; lemmas broadly elliptic-ovate, 1.7–1.9 mm long, glandular on the keel, closely imbricate; grain 0.6–0.8 mm long, bright brown.

Introduced from Europe in moist soil, waste places, gardens, railways, and roadsides.

Eragrostis pectinacea (Michx.) Nees
CAROLINA LOVEGRASS *native*
IR | **WUP** | **CUP** | EUP   FAC CC 0

Stems densely tufted, erect or ascending, often repeatedly branched, 1–5 dm tall; sheaths hirsute at the tips, the hairs to 4 mm long; ligules to 0.5 mm long; leaf blades 2–20 cm long 1–4 mm wide, flat to involute. Panicle diffusely branched, often half as long as the entire plant, the spikelets tending to be appressed along the branches. Spikelets 5–11-flowered, linear, 1–1.5 mm wide; first glume 0.8–1.2 mm long; second glume 1.1–1.6 mm long; lowest lemma 1.5–1.8 mm long; grain to 1 mm long.

Moist ground, especially as a weed in gardens, roadsides, railways, and waste places.

Eragrostis spectabilis (Pursh) Steud.
PURPLE LOVEGRASS *native*
IR | WUP | **CUP** | EUP   CC 3

Stems tufted, erect or ascending, 3–6 dm tall; sheaths hairy on the margins and at tips, the hairs to 7 mm long; ligules to 0.2 mm long; leaf blades 3–7 mm wide, flat to involute, both surfaces usually pilose. Panicle ovoid, about 2/3 as long as the entire plant, its base usually included in the upper sheath, its scabrous branches rigid, divaricate, pilose in the axils, the lateral spikelets pediceled and more or less spreading. Spikelets purple, 5–10-flowered; first glume 1–2 mm long; lemmas 1.6–2.1 mm long, scabrous on the keel, the lateral nerves evident; palea conspicuously short-ciliate on the keels.

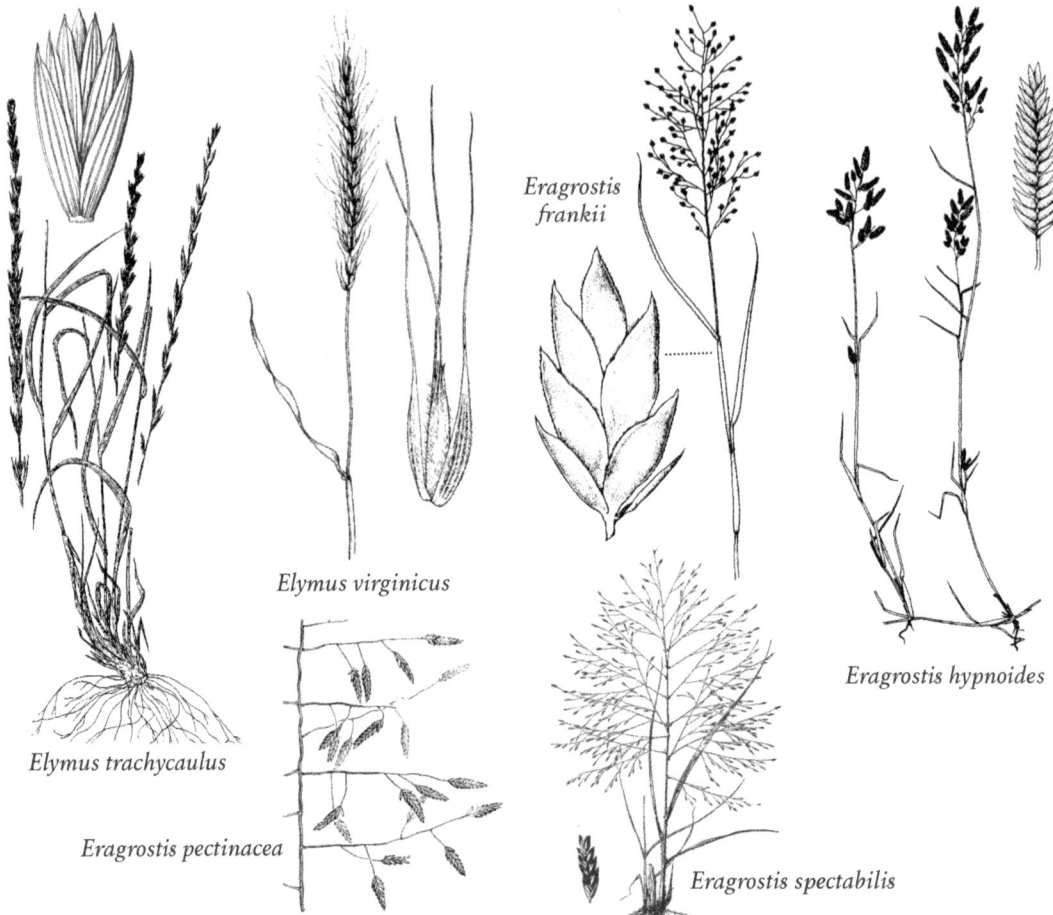

Dry soil, fields and open woods; Menominee County.

Whole panicle eventually detached and behaving as a tumbleweed. Available commercially for planting as an ornamental grass.

Festuca *Fescue*

Annual or perennial grasses, often densely tufted. Leaves flat to involute, auricles absent; ligules membranous, usually truncate, usually ciliate. Flowers in open or contracted panicles. Spikelets 3–11-flowered, the rachilla disarticulating above the glumes and between the lemmas. Glumes narrow, unequal, 1–3-nerved, usually shorter than the lemmas. Lemmas rounded on back, obscurely 5-nerved, usually awned from the apex. Paleas about equaling the lemmas. Stamens 1 or 3.

1 Blades of leaves flat (or merely once-folded), at least the larger ones 3–8 mm wide; lemmas awnless or rarely with awn less than 1 mm long . . . . **F. subverticillata**
1 Blades of leaves strongly involute, usually much less than 3 mm wide; lemmas awned or awnless 2
2 Margins of lemmas thin and membranous; tip of ovary bristly-pubescent; awns mostly more than 3 mm long, nearly equaling or longer than the bodies of their lemmas; mature panicle open and lax . . . . . **F. occidentalis**
2 Margins of lemmas at most very narrowly membranous-bordered, the lemmas firm and thick throughout; summit of ovary glabrous; awns all less than 3 mm long, shorter than the bodies of their lemmas; mature panicle rather narrow, crowded, and compact, the branches strongly ascending or, if spreading, very short 3
3 Sheaths closed in young leaves, the old ones ± dark reddish brown basally, becoming fibrous by splitting between the prominent pale veins; basal shoots usually arising laterally, the stems thus tending to be strongly curved or bent at the base; anthers mostly 2–3.5 mm long . . . . **F. rubra**
3 Sheaths open most of their length even in young leaves (margins ± overlapping), the old ones mostly pale or drab brown, not becoming fibrous; basal shoots erect, the stems thus nearly or quite straight from the base upwards; anthers various 4
4 Lower panicle branches often spreading; anthers 2–3 mm long . . . . **F. trachyphylla**
4 Lower panicle branches strongly ascending; anthers less than 2 mm long . . . . **F. saximontana**

Festuca occidentalis Hook.
WESTERN FESCUE *native*
IR | WUP | CUP | EUP CC 6

Stems slender, tufted, 4–8 dm tall, glabrous, shining; sheaths closed for much less than 1/2 their length, glabrous, somewhat persistent or slowly shredding into fibers; collars glabrous; ligules to 0.4 mm long, usually longer at the sides; leaf blades 0.3–0.7 mm wide, conduplicate, upper surface smooth or finely roughened. Panicle narrow, flexuous, more or less secund, 5–20 cm long. Spikelets on slender pedicels, 6–10 mm long, 3–5-flowered; first glume 2.7–3.6 mm long; second glume 3.5–4.5 mm long; lemmas green or suffused with purple, soft and membranous, the body 4.5–6.5 mm long, the awn two-thirds to fully as long.

Cobble beaches and stabilized dunes along Great Lakes, dry woods.

Festuca rubra L.
RED FESCUE *introduced*
IR | WUP | CUP | EUP FACU

Stems glabrous, 3–10 dm tall, usually loosely tufted, often decumbent at base, frequently rhizomatous; sheaths closed for about 3/4 their length when young, soon disintegrating into loose fibers, usually pubescent, reddish; collars glabrous; ligules 0.1–0.5 mm long; leaf blades usually conduplicate, to 2.5 mm wide, sometimes flat and 1.5–7 mm wide. Panicle 5–20 cm long, narrow with ascending branches, or in some forms loosely spreading. Spikelets 4–7-flowered; first glume subulate, 2.6–4.5 mm long; second glume broader, 3.5–5.5 mm long; lemmas 4.8–6.1 mm long; longest awns 1–3 mm long.

Variable, and many horticultural varieties exist. Widely distributed in n Europe and North America; Michigan plants considered adventive.

Festuca saximontana Rydb.
ROCKY MOUNTAIN FESCUE *native*
IR | WUP | CUP | EUP CC 6
Festuca brachyphylla J.A. Schultes

Stems very slender, densely tufted, not stoloniferous, glabrous; sheaths closed for about 1/2 their length, usually persistent, rarely slowly shredding into fibers, mostly pale or drab brown; collars glabrous; ligules to 0.5 mm long; leaf blades 0.5–1.2 mm wide, conduplicate, upper surface glabrous or sparsely puberulent, undersurface scabrous or puberulent. Panicle 1–10 cm, narrow and spiciform, or somewhat open at anthesis, the first pedical of the lowermost branches usually no more than 5 mm, from the base. Spikelets 2–4-flowered, the first glume 2–3 mm, 1-nerved, the second 2.5–4.5 mm, 3-nerved; lemmas mostly 3.5–6) mm, with a short awn 1–3 mm.

Dry forests, shores, dunes, and disturbed places; rock crevices near Lake Superior. Similar to *F. trachyphylla* in general appearance.

Festuca subverticillata (Pers.) Alexeev
NODDING FESCUE *native*
IR | WUP | CUP | EUP FACU CC 5

Stems few in a tuft, 6–12 dm tall, smooth; sheaths closed for less than 1/3 their length, glabrous or sparsely pilose, shredding into fibers; ligules mostly 0.5–1 mm long; leaf blades 4–10 mm wide, flat or loosely convolute, glabrous or sparsely pilose. Panicle long-exsert, 15–30 cm long; branches slender, elongate, racemiform, eventually widely spreading, bearing spikelets only above the middle. Spikelets relatively

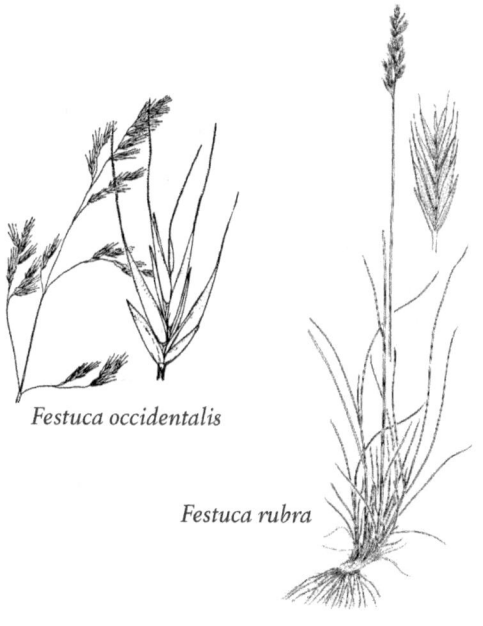

Festuca occidentalis

Festuca rubra

Festuca subverticillata

Festuca trachyphylla

remote, the tip of one barely reaching the base of the next, usually 3-flowered, occasionally 4–5-flowered, 4–6 mm long; first glume subulate, averaging 2.8 mm long; second glume ovate, averaging 3.4 mm long; lemmas acute, averaging 3.7 mm long, appressed till maturity.

Moist forests of beech-maple or oak-hickory; occasionally in wet conifer woods.

Festuca trachyphylla (Hack.) Krajina
HARD FESCUE; SHEEP FESCUE *introduced*
IR | WUP | CUP | EUP UPL

Festuca ovina auct. p.p. non L.

Stems densely tufted, without rhizomes, 20–80 cm tall, glabrous or with sparse hairs; sheaths closed for less than 1/3 their length, usually glabrous, rarely pubescent, persistent; collars glabrous; ligules to 0.5 mm long; leaf blades about 1 mm in diameter, usually conduplicate, rarely flat. Panicle 3–15 cm long, contracted, with 1–2 branches per node; branches erect or stiffly spreading, lower branches with 2 or more spikelets. Spikelets 5–9 mm long, with 3–7 florets; glumes exceeded by the upper florets, mostly glabrous; lower glumes 2–4 mm long; upper glumes 3–5 mm long; lemmas lance-shaped, usually smooth on the lower portion and scabrous or pubescent upwards, especially on the margins, awns 0.5–2.5 mm long, usually less than 1/2 as long as the lemma body.

Native of Europe, introduced as a turf grass and sometimes weedy.

Glyceria *Manna Grass*

Perennial grasses, loosely clumped or spreading by rhizomes. Stems upright, or reclining at base and often rooting at lower nodes. Leaves flat or folded; sheaths closed for most of their length; ligules scarious, erose to lacerate. Head an open panicle. Spikelets 3-flowered, ovate to linear, round in section or somewhat flattened, breaking above the glumes; glumes unequal, shorter than lemmas, 1-veined; lemmas unawned, usually 7-veined; palea about as long as lemma; stamens 3 or 2.

Glyceria, *Puccinellia*, and *Torreyochloa* are often confused because of similarities in form and their occurrence in wetlands; only *Glyceria* has closed leaf sheaths and 1-veined upper glumes (the other two genera have open leaf sheaths and 3-veined upper glumes). Only *Puccinellia* has inconspicuous veins on the lemmas (the other two genera generally have conspicuous veins on the lemmas).

1 Spikelets linear-cylindric, 10 mm long or longer. . . . . **G. borealis**
1 Spikelets ovate, 2–7 mm long. 2
2 Spikelets 3–4 mm wide; veins of lemma not raised . . . . **G. canadensis**
2 Spikelets 2–2.5 mm wide; veins of lemma raised 3
3 Spikelets 4–7 mm long . . . . **G. grandis**
3 Spikelets 2–4 mm long . . . . **G. striata**

Glyceria borealis (Nash) Batchelder
NORTHERN MANNA GRASS *native*
IR | WUP | CUP | EUP OBL CC 6

Stems erect or reclining at base, often rooting from lower nodes, 6–12 dm long. Leaves flat or folded, 2–5 mm wide, smooth; sheaths smooth; ligule 3–10 mm long. Panicle 2–4 dm long, with stiff, erect to ascending, branches to 8–12 cm long, each with several spikelets. Spikelets linear, mostly 6–12-flowered, 1–1.5 cm long; glumes rounded at tip, 2–3 mm long; lemmas 3–4 mm long, 7-veined. June–Aug.

Marshes, ponds, stream, ditches, often in shallow water or mud.

Glyceria canadensis (Michx.) Trin.
RATTLESNAKE MANNA GRASS *native*
IR | WUP | CUP | EUP OBL CC 8

Stems single or few together, erect, 6–15 dm long. Leaves 3–7 mm wide, upper surface rough; ligules 2–5 mm long. Panicle open, 1–3 dm long, the branches drooping, with spikelets mostly near tips. Spikelets ovate, 5–10-flowered, 5–7 mm long, the florets spreading; glumes 2–3 mm long, the first glume lance-shaped, the second glume ovate; lemma veins not raised.

Marshes, swamps, thickets, open bogs, fens.

Glyceria grandis S. Wats.
AMERICAN MANNA GRASS *native*
IR | WUP | CUP | EUP OBL CC 6

Stems loosely tufted, erect, stout, 1–1.5 m long and 4–6 mm wide. Leaves flat, smooth, 6–12 mm wide; sheaths smooth; ligules translucent, 3–6 mm long. Panicle large, open, much-branched, 2–4 dm long, usually nodding at tip, branches lax and drooping when mature. Spikelets ovate, purple, slightly flattened, 5–9-flowered, 4–7 mm long; glumes pale or white, 1–3 mm long; lemmas purple, 2–3 mm long. June–Sept.

Marshes, ditches, streams, lakes and ponds, open bogs, fens; usually in shallow water or mud.

Glyceria striata (Lam.) A.S. Hitchc.
FOWL MANNA GRASS *native*
IR | WUP | CUP | EUP OBL CC 4

Plants pale green. Stems loosely tufted erect, slender, 3–10 dm long. Leaves flat or folded, smooth, 2–6 mm wide; sheaths smooth; ligules 1–3 mm long. Panicle open, loose, 1–2 dm long, the branches lax, drooping. Spikelets ovate, often purple, 3–7-flowered, 3–4 mm long; glumes 0.5–1.5 mm long; lemma 2 mm long, strongly 7-veined. June–Aug.

Common in swamps, thickets, low areas in forests, wet meadows, springs, streambanks.

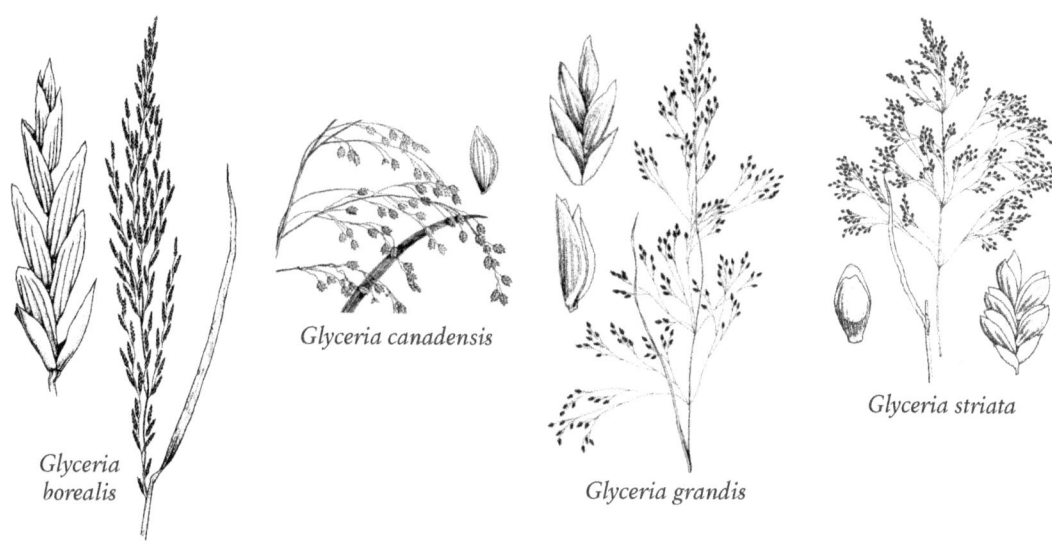

Glyceria borealis
Glyceria canadensis
Glyceria grandis
Glyceria striata

Graphephorum *False Oat*

Graphephorum melicoides (Michx.) Desv.
PURPLE FALSE OAT native
IR | **WUP** | **CUP** | **EUP** FACW CC 10
Trisetum melicoides (Michx.) Vasey

Tufted perennial grass. Stems smooth or finely hairy, 4–9 dm long. Leaves flat, 3–8 mm wide, sparsely long-hairy; sheaths glabrous or hairy; ligules membranous, 1.5–3.5 mm long, rounded or truncate-ragged at tip. Head a slender, nodding panicle, 10–20 cm long, the branches upright to drooping, to 6 cm long, the spikelets mostly above middle of branch. Spikelets 2-flowered, 6–7 mm long, finely hairy; glumes somewhat unequal, 4–7 mm long, the first glume 1-veined, the second glume 3-veined; lemma unawned; stalk within spikelet (rachilla) and base of lemma white-hairy.

Mixed forests and ridge and swale ecosystems near Lake Michigan, shoreline dolomitic sites, and seepage areas on shoreline bluffs.

Sometimes retained in genus *Trisetum*.

Hesperostipa *Needlegrass*

Hesperostipa spartea (Trin.) Barkworth
PORCUPINE GRASS native
IR | WUP | **CUP** | **EUP** CC 7
Stipa spartea Trin.

Tufted perennial grass; stems in small tufts, 6–12 dm tall; sheaths glabrous; ligules membranous, of the upper leaves 4–6 mm long, the lower much shorter; auricles absent; leaf blades 2–5 mm wide, glabrous beneath, scabrous and usually also pubescent above, the lower elongate, tapering to a fine point. Panicle narrow, more or less nodding, 1–2 dm long, the few branches each bearing 1–few spikelets. Spikelets 1-flowered, articulated above the glumes; glumes about equal, lance-shaped, papery, with broad scarious margins, the first 3-nerved, the second 5-nerved, nerves parallel, only the middle one extending to the tip; 28–42 mm long, tapering to a very slender point; mature lemma 18–21 mm long, brown, pubescent at base, less so above; awn terminal, greatly elongate, articulated with the lemma but persistent, stiff, 12–20 cm long, twice geniculate near the middle, the central segment usually 1.5–3 cm long. The awn is hygroscopic, imparting a twisting motion to the fruit as it winds or unwinds, pushing the sharp basal callus into the soil, and serving to bury the grain.

Sandy, often calcareous places; dune ridges, oak savanna, dry prairies, railways; Mackinac and Menominee counties.

Hierochloe *Sweetgrass*

Hierochloe odorata (L.) Beauv.
SWEETGRASS native
IR | **WUP** | **CUP** | **EUP** CC 9
Anthoxanthum hirtum (Schrank) Y.Schouten & Veldkamp, *Hierochloe hirta* (Schrank) Borbás

Perennial grass, from creeping rhizomes; plants nicely sweet-scented, especially when dried. Stems erect, 2–6 dm tall, smooth; sheaths brownish or reddish; ligules membranous, 2–5 mm long; leaf blades of basal and stem leaves 2.5–5.5 mm wide, upper surface glabrous and shiny, undersurface pilose; fertile stem leaves short, 1–4 cm long, leaves on sterile shoots much longer. Head a pyramid-shaped panicle, 5–10 cm long, the branches spreading to drooping. Spikelets 3-flowered, the lower 2 florets staminate, the terminal spikelet perfect, golden brown, or green or purple at base and golden near tips, 5 mm long, breaking above the glumes; glumes ovate, shiny, 4–6 mm long; lemmas 3–4 mm long, the staminate lemma hairy. May–July.

Wet meadows, shores, low prairie; often where sandy.

The fragrance emitted when fresh plants are crushed or burned is from coumarin, an anti-coagulant agent. *Hierchloe* sometimes merged into genus *Anthoxanthum*.

Hordeum *Barley*

Annual or perennial grasses with flat blades, scarious truncate ligules, and dense bristly spikes which disarticulate at each joint. Spikelets 1-flowered or rarely 2-flowered, not disarticulating, aggregated in groups of three at each joint of the rachis, the lateral spikelets often pediceled and sterile, the central sessile and fertile. Glumes elongate, awned or awnlike, setaccous throughout or widened at base, the 6 in each triad of spikelets forming a false involucre to the florets. Lemma of the lateral spikelets often reduced in size or abortive; that of the central spikelet indurate, obscurely nerved, its rounded back turned away from the rachis, long-awned; rachilla prolonged behind the palea as a short bristle.

1. Body of larger lemmas ca. 8–11 mm long; leaves glabrous, with prominent auricles at base of blade; awns of lemmas much stouter than those of glumes; rachis of spike not disintegrating . . . . **H. vulgare**
1. Body of larger lemmas ca. 3.5–6 mm long; leaves (at least lower sheaths) ± pubescent, without auricles; awns of lemmas as slender as those of glumes; rachis of spike readily disintegrating as it matures. . **H. jubatum**

Hordeum jubatum L.
FOXTAIL-BARLEY *introduced*
IR | **WUP** | **CUP** | **EUP** FAC

Tufted perennial grass; plants smooth to densely hairy. Stems erect or reclining at base, 2–7 dm long; sheaths glabrous or pubescent; auricles absent; ligules less than 1 mm long; leaf blades usually flat, 2–5 mm wide. Head a terminal spike, erect to nodding, 3–10 cm long, appearing bristly due to the long, spreading awns from glumes and lemmas. Spikelets 1-flowered, 3 at each node, the center spikelet fertile, stalkless, the 2 lateral spikelets sterile, short-stalked, reduced to 1–3 spreading awns; the 3 spikelets at each node falling as a unit; glumes of fertile spikelet awnlike; lemma lance-shaped, tipped by a long awn; the glume and lemma awns 2–7 cm long. June–Sept.

Wet meadows, ditches, shores, shallow marshes, disturbed areas; often where brackish; native and common in the western USA, considered adventive in Michigan.

Hordeum vulgare L.
COMMON BARLEY *introduced*
IR | **WUP** | CUP | EUP

Summer or winter annual grass, loosely tufted. Stems erect, in wild plants 3–12 dm tall; lower sheaths pilose; upper sheaths glabrous; auricles to 6 mm long; leaf blades 3–15 mm wide. Spike erect or nearly so, dense, 3–10 cm long, excluding the awns; glumes flat, narrowly linear, pubescent, often tapering into a slender soft awn; lemmas about 1 cm long, tapering into a stout, flat, erect, 1-nerved awn 6–15 cm long or with a terminal 3-lobed appendage.

Cultivated grain and an occasional waif along roads and railways; Houghton and Keweenaw counties.

Graphephorum melicoides

Hesperostipa spartea

Hierochloe odorata

Hordeum jubatum

Hordeum vulgare

Koeleria *Junegrass*

Koeleria macrantha (Ledeb.) J.A. Schultes
PRAIRIE JUNEGRASS *native*
IR | WUP | **CUP** | EUP CC 9
  Koeleria pyramidata (Lam.) Beauv.

Perennial tufted grass. Stems erect, 3–6 dm tall, pubescent below the panicle. Leaves mostly basal; sheaths glabrous or pubescent; ligules membranous, 0.5-2 mm long, erose; blades 1–3 mm wide, flat, or involute when dry, glabrous or pubescent. Panicle spike-like, shining, silvery-green, 5–12 cm long, 1–2 cm thick. Spikelets normally 2-flowered, disarticulating above the glumes and between the lemmas, subsessile, overlapping, more or less scabrous; glumes unequal, obscurely keeled, scariously margined, the first 1-nerved, the second broadest above the middle, 3–5-nerved ; lemma about as long as the glumes, rounded on the back, acute, scarious at margin and tip, obscurely 5-nerved awnless; palea hyaline, nearly as long as the lemma. Variable, especially in pubescence.

Dry soil, prairies, sand hills, open woods; Marquette and Menominee counties. An important grass of the western USA, Michigan at eastern edge of species' range.

Leersia *Cut-Grass*

Perennial grasses, spreading by long rhizomes. Stems slender, somewhat weak. Leaves flat, smooth to hairy or rough-to-touch; sheaths open; auricles absent; ligules membranous, short. Head an open panicle. Spikelets 1-flowered, laterally compressed, falling as a unit from the stalk; glumes absent; lemmas smooth to hairy, 5-veined; palea narrow, about as long as lemma; stamens 2–3 (ours).

1 Stems round in section; leaves very rough-to-touch; spikelets 4–6 mm long . . . . **L. oryzoides**
1 Stems flattened in section; leaves smooth or finely roughened; spikelets to 3.5 mm long . . . . . . **L. virginica**

Leersia oryzoides (L.) Sw.
RICE CUT-GRASS *native*
IR | WUP | **CUP** | EUP OBL CC 3

Loosely tufted perennial grass, from creeping rhizomes. Stems weak and sprawling, rooting at nodes, 1–1.5 m long. Leaves flat, 2–3 dm long and 5–10 mm wide, rough-to-touch, the margins fringed with short spines; sheaths rough-hairy; ligules 0.5–1 mm long, flat-topped. Panicle open at end of stem and from leaf axils (these often partly enclosed by leaf sheaths), 1–2 dm long, the branches ascending to spreading. Spikelets 1-flowered, oval, 5 mm long and 1–2 mm wide, compressed, pale green, turning brown with age; glumes absent; lemma covered with bristly hairs. July–Sept.

Muddy or sandy streambanks, shores, swales and marshes; sometimes forming large patches.

Leersia virginica Willd.
WHITE GRASS *native*
IR | **WUP** | **CUP** | EUP FACW CC 5

Perennial grass, spreading by rhizomes. Stems slender and weak, often more or less horizontal at base and rooting at nodes, 5–12 dm long. Leaves rough-hairy, especially along margins, 5–20 cm long and 5–15 mm wide; sheaths smooth or finely hairy; ligules 1–3 mm long, flat-topped. Panicle open, 1–2 dm long, the branches separated along the rachis, stiffly spreading, the spikelets from middle to tip of branches. Spikelets oblong, barely overlapping one another, 3 mm long and 1 mm wide, sparsely hairy; glumes absent; lemma 3–4 mm long, the keel and margins sparsely hairy. July–Sept.

Swamps, floodplain forests, shaded forest depressions, streambanks; Menominee and Ontonagon counties.

Leymus *Lyme Grass*

Leymus mollis (Trin.) Pilger
AMERICAN LYME GRASS *native*
IR | WUP | **CUP** | **EUP** CC 10
  Elymus mollis Trin.

Stout perennial grass, green and somewhat glaucous, with long rhizomes (not tufted); stems 5–15 dm tall, glabrous apart from fine hairs for 1–4 cm below the spike; ligules to 2.5 mm long; auricles short, to 0.7 mm long; leaf blades to 9 dm long, 3–15 mm wide, lower surfaces rough-to-touch. Spikelets in dense spikes to 30 cm long and 1–2 cm wide, usually with 2 spikelets per node; disarticulation above the glumes, beneath the florets. Glumes usually 2, equal or nearly so, lanceolate, pubescent; lemmas hairy like the glumes (or more so), tapering to a slender, awnless tip.

Sandy shores and dunes along Lake Superior. Sometimes mistaken for *Ammophila breviligulata*, which appears similar and grows in the same shoreline habitat, but spikelets in that species have only a single floret.

Lolium *Rye Grass*

Lolium perenne L.
ENGLISH RYEGRASS *introduced*
IR | WUP | **CUP** | EUP FACU

Perennial tufted grass. Stems slender, 3–7 dm tall, glabrous throughout; sheaths open; ligules membranous, to 4 mm long; leaf blades flat, glossy, 2–4 mm wide. Spike slender, 1–2 dm long, smooth on the back of each joint opposite the spikelet, minutely scabrous on the sharp margin. Spikelets solitary at

POACEAE | Grass Family

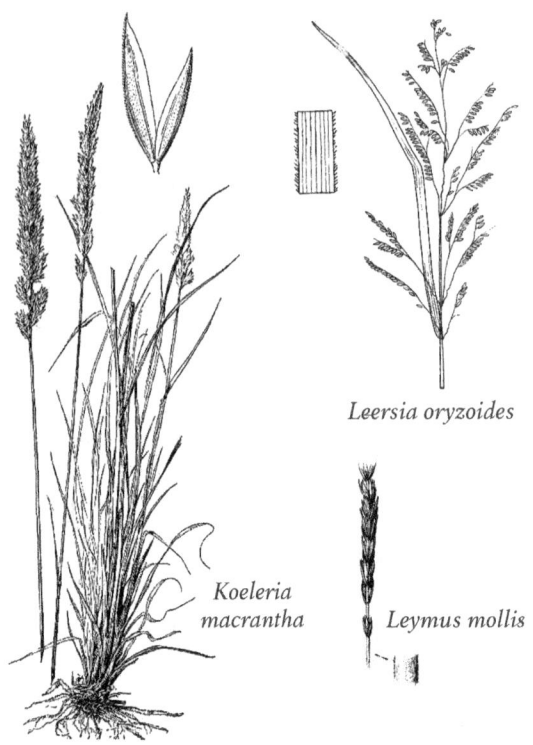

Koeleria macrantha

Leersia oryzoides

Leymus mollis

Leersia virginica

Lolium perenne

each node, usually 5–10-flowered, placed edgewise to the rachis, the edge fitting into a concavity in the axis, disarticulating above the glumes and between the lemmas; first glume absent except in the terminal spikelet; second glume (on the side of the spikelet away from the rachis) strongly 3–5-nerved; lemmas awnless, the lowest 5.5–8 mm long, the upper progressively reduced.

Native of Europe; cultivated in meadows and lawns and often included in commercial seed mixes; and escaped and established on roadsides and in waste places.

Melica *Melic Grass*

Melica smithii (Porter) Vasey
SMITH'S MELIC GRASS native
IR | WUP | CUP | EUP CC 7

Perennial, loosely tufted grass, not rhizomatous. Stems 5–10 dm tall, thickened and corm-like at the base; sheaths glabrous, or often rough-hairy, particularly at the throat, veins often prominent; ligules thinly membranous, erose to lacerate, 2–4 mm long; leaf blades flat, soft, 5–12 mm wide, both surfaces usually finely roughened. Panicle few-flowered, with several spreading, reflexed branches bearing several spikelets near the branch tips. Spikelets linear, with 3–5 bisexual florets, green or purplish, 12–20 mm long; lemmas and glumes narrowly lance-shaped, sharply but finely 3–5-nerved; lemmas bifid at the tip, with an erect awn 2–5 mm long; sterile lemma similar but much smaller; disarticulation above the glumes.

Moist hardwood forests, wooded dunes.

Milium *Millet Grass*

Milium effusum L.
AMERICAN MILLET GRASS native
IR | WUP | CUP | EUP CC 8

Perennial rhizomatous grass. Stems erect from a bent base, 6–12 dm tall, glabrous; sheaths open; ligule membranous, 3–9 mm long, obtuse-erose; leaf blades flat, broad, 8–17 mm wide, glabrous, or scaberulous on the margin. Panicle 1–3 dm long, ovoid or pyramidal, the branches in fascicles of 2 or 3, widely spreading and bearing drooping spikelets beyond their middle. Spikelets 1-flowered, disarticulation above the glumes; glumes equal, ovate or elliptic, rounded on the back, scaberulous, about 3 mm long, 3-nerved; lemma about as long as the glumes, awnless, nerveless, obtuse, rounded on the back, at first thin, at maturity firm, white, and shining, its margins partly covering a palea of similar texture.

Rich, usually moist woods.

Muhlenbergia *Muhly*

Perennial grasses, clumped or with creeping rhizomes. Stems erect or reclining at base, often branching from base. Leaves smooth to hairy, ligules membranous. Head a panicle, usually narrow and spike-like,

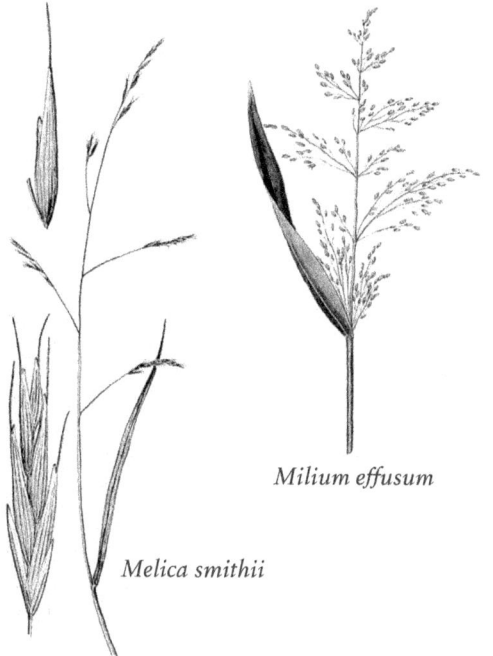

Milium effusum

Melica smithii

sometimes open and spreading, at ends of stems and sometimes also from leaf axils. Spikelets 1-flowered, breaking above glumes; glumes usually nearly equal in length, 1-veined, the tip often awned; lemma lance-shaped, 3-veined, sometimes awned, some species with long, soft hairs at lemma base; palea about as long as lemma.

1. Lemmas not pilose at the base, glabrous (or with minute, even pubescence on back), awnless; stems loosely or densely tufted or matted, rhizomes, if present, thin and wiry, not densely clothed with overlapping scales 2
1. Lemmas pilose at base, glabrous or short-pubescent on back, awned or awnless; stems arising from conspicuous, elongate scaly rhizomes 3
2. Spikelets less than 2 mm long, mostly on pedicels more than twice as long, in an open panicle . . . . . **M. uniflora**
2. Spikelets ca. 2.4–3.5 mm long, mostly on pedicels less than twice as long, in a slender contracted panicle . . . . **M. richardsonis**
3. Glumes (including prominent awn-tip) 3.5–7.5 mm long, mostly distinctly longer than the body of the lemma; lemma at most short-awned; anthers 0.5–1.3 mm long 4
3. Glumes generally less than 3.6 mm long (rarely, especially on lower spikelets of panicle, up to 4 mm), mostly about equaling or shorter than the body of the lemma; lemma awnless to long-awned; anthers not over 0.5 mm long 5
4. Internodes minutely puberulent or roughened over much of their surface (rarely nearly glabrous); ligule (excluding cilia) 0.5–0.7 mm long or shorter; anthers 0.8–1.3 mm long . . . . **M. glomerata**
4. Internodes of stem smooth and glabrous over most of their surface; ligule 0.7–1 mm long; anthers ca. 0.5–0.8 mm long . . . . **M. racemosa**
5. Larger glumes 0.6–1 mm wide, less than 4 times as long, hence ovate and usually ± abruptly tapered at the tip; stems puberulent below the nodes . . . . . . **M. sylvatica**
5. Larger glumes not over 0.6 mm wide, more than 4 times as long, hence narrowly lanceolate and usually ± attenuate at the tip; stems puberulent or glabrous below the nodes 6
6. Stem smooth and glabrous throughout, sometimes decumbent at base and rooting at nodes, generally much branched and bushy above; inflorescences (except terminal one) with base often enclosed in upper leaf sheath . . . . **M. frondosa**
6. Stem puberulent below the nodes, ± erect, simple or branched; inflorescences all generally exserted 7
7. Ligules ca. 1 mm long or shorter; some spikelets in panicle sessile or subsessile . . . . **M. mexicana**
7. Ligules, at least the longest, 1.3–2 mm long; spikelets all on distinct (though sometimes rather short) pedicels . . . . **M. sylvatica**

Muhlenbergia frondosa (Poir.) Fern.
WIRESTEM MUHLY *native*
IR | **WUP** | **CUP** | EUP   FACW CC 3

Perennial grass, from stout, scaly rhizomes. Stems 4–10 dm long, unbranched and erect when young, becoming branched and sprawling with age, smooth and shiny between nodes; sheaths glabrous, margins hyaline; ligules membranous, 1–2 mm long, fringed; leaf blades lax, smooth or finely roughened, 2–6 mm wide. Panicle narrow, to 10 cm long, from ends of stems and leaf axils (where partly enclosed by sheaths), the branches erect to spreading, with spikelets from near base to tip. Spikelets 1-flowered; glumes 2–3 mm long, tipped with a short awn; lemma 3–4 mm long, usually with an awn to 1 cm long, short-hairy at base. Aug–Sept.

Floodplain forests, streambanks, thickets, shores; also somewhat weedy in disturbed areas such as along railroads; Houghton, Menominee, and Ontonagon counties.

Muhlenbergia glomerata (Willd.) Trin.
MARSH MUHLY *native*
IR | **WUP** | **CUP** | **EUP**   OBL CC 10

Perennial grass, spreading from rhizomes. Stems upright, 3–9 dm long, sometimes with a few branches from base, dull and finely hairy between nodes; sheaths finely roughened, slightly keeled; ligules membranous, to 0.6 mm long, truncate, fringed; leaf blades flat, lax, 2–6 mm wide, usually scabrous. Panicle narrow, crowded, cylindric, 2–10 cm long and 5–10 mm wide, the lower clusters of spikelets often separate from one another. Spikelets 1-flowered, often purple-tinged, 5–6 mm long; glumes nearly equal, longer than the floret, tipped with an awn 1–5 mm long; lemma lance-shaped, 2–3 mm long, with long, soft hairs at base. Aug–Sept.

Swamps, wet meadows, marshes, springs, open bogs, fens, calcareous shores.

Muhlenbergia mexicana (L.) Trin.
MEXICAN MUHLY   native
IR | **WUP** | **CUP** | **EUP**   FACW CC 3
Perennial grass, from scaly rhizomes. Stems upright, 2–8 dm long, sometimes branched from base; dull and finely hairy between nodes; sheaths smooth or finely roughened, somewhat keeled; ligules membranous, to 1 mm long, truncate, fringed; leaf blades flat, lax, 2–6 mm wide, scabrous or smooth. Panicle narrow, densely flowered, 5–15 cm long and 2–10 mm wide, from ends of stems and leafy branches. Spikelets 1-flowered, green or purple, 2–3 mm long; glumes nearly equal, lance-shaped, 3–4 mm long, about as long as floret, tipped with a short awn about 1 mm long; lemma lance-shaped, 2–3 mm long, unawned or with an awn to 7 mm long. Aug–Sept.

Swamps, floodplain forests, thickets, wet meadows, marshes, springs, fens and streambanks.

Muhlenbergia racemosa (Michx.) B.S.P.
GREEN MUHLY   introduced
IR | **WUP** | **CUP** | EUP   FACU
Perennial grass, from scaly rhizomes. Stems erect or declined, 5–12 dm tall, rather stout, simple or sparsely branched; sheaths finely roughened, slightly keeled; ligules membranous, 0.6–1.5 mm long, truncate, fringed; leaf blades ascending or appressed, flat, 2–5 mm wide, usually scabrous. Panicle narrow, usually compact and dense, often interrupted toward the base, 5–10 cm long, about 1 cm thick. Spikelets on short pedicels, crowded and much overlapping; glumes subulate, tapering into an awn, 1.5–2× as long as the lemmas, the first 4.4–6.1, the second 4.5–7.5 mm long, including their awns; lemma scabrous, pilose at base, 2.6–4 mm long, acuminate to a slender point, awnless.

Moist or wet soil in open places; Delta and Menominee counties; native west of Michigan, considered adventive here.

Muhlenbergia richardsonis (Trin.) Rydb.
MATTED MUHLY   (THR) native
IR | **WUP** | **CUP** | **EUP**   FACW CC 10
Loosely tufted perennial grass, rooting from lower nodes and forming mats. Stems very slender, erect or more or less horizontal at base, 2–6 dm long; sheaths shorter or longer than the internodes, glabrous; ligules membranous, 1–3 mm long, acute to truncate, erose; leaf blades upright, flat or involute, 0.5–4 mm wide. Panicle narrow and spike-like, 2–12 cm long. Spikelets 1-flowered, uncrowded, green or gray-green, 2–3 mm long; glumes nearly equal, ovate, to half as long as floret; lemma lance-shaped, smooth, 2–3 mm long tipped with a short point. July–Sept.

Low prairie, wet meadows, marshes and seeps; Delta, Keweenaw, and Mackinac counties.

Muhlenbergia sylvatica Torr.
WOODLAND MUHLY   native
IR | WUP | **CUP** | EUP   FACW CC 8
Perennial grass, spreading by rhizomes. Stems erect, or sprawling when old, 4–10 dm long, coarse-hairy between nodes; sheaths smooth for most of their length, roughened near tip, margins hyaline; ligules membranous, 1–2.5 mm long, truncate, lacerate-ciliolate; leaf blades flat, 3–7 mm wide, scabrous. Panicle slender, often nodding, 5–20 cm long and 2–7 mm wide. Spikelets 1-flowered, 2–4 mm long, at ends of stalks about 3 mm long; glumes nearly equal, sharp-tipped, shorter than lemma; lemma 2–4 mm long, short hairy at base, tipped with an awn 5–15 mm long. Aug–Sept.

Streambanks, shaded wet areas; Delta County.

Muhlenbergia uniflora (Muhl.) Fern.
BOG MUHLY   native
IR | **WUP** | **CUP** | **EUP**   OBL CC 8
Perennial grass, loosely matted. Stems very slender, 2–4 dm long, often more or less horizontal and rooting at base; leaves crowded near base of plant; sheaths longer than the internodes, keeled; ligules membranous, 0.5–1.5 mm long, truncate, erose, without lateral lobes; leaf blades 1–2 mm wide, usually flat, upper surface smooth or finely roughened abaxially, midveins thickened and whitish near base of blade. Panicle loose, open, 7–20 cm long and 2–4 cm wide, the branches threadlike. Spikelets 1-flowered (rarely 2-flowered), oval, purple-tinged, 1–2 mm long; glumes about equal, ovate, to half the length of spikelet; lemma 1–2 mm long, unawned.

Wetland margins, exposed sandy shores.

Oryzopsis *Mountain Ricegrass*

Oryzopsis asperifolia Michx.
WHITE-GRAIN MOUNTAIN RICEGRASS   native
IR | **WUP** | **CUP** | **EUP**   CC 6
Loosely tufted perennial grass. Stems 3–7 dm tall, often widely spreading. Leaves mostly basal; sheaths open, glabrous; auricles absent; basal ligules to 0.7 mm long, sometimes longest at the sides, ciliate; blades of basal leaves 30–90 cm long and 4–9 mm wide, upper surface glaucous; upper stem leaves with greatly reduced blades 3 cm long or less or lacking. Raceme slender, 2–6 cm long, the paired branches each with a single spikelet. Spikelets 1-

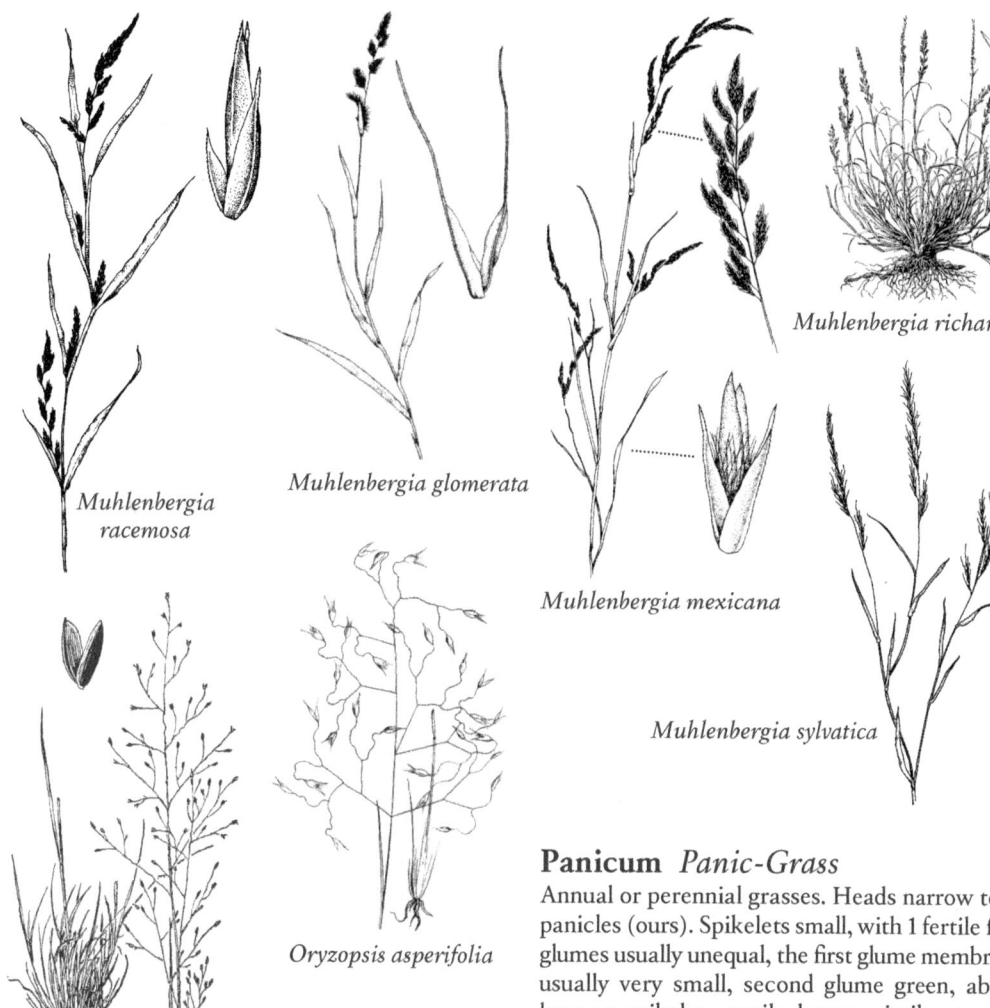

Muhlenbergia racemosa

Muhlenbergia glomerata

Muhlenbergia richardsonis

Muhlenbergia mexicana

Muhlenbergia sylvatica

Oryzopsis asperifolia

Muhlenbergia uniflora

Panicum Panic-Grass

Annual or perennial grasses. Heads narrow to open panicles (ours). Spikelets small, with 1 fertile flower; glumes usually unequal, the first glume membranous, usually very small, second glume green, about as long as spikelet; sterile lemma similar to second glume, enclosing the palea and sometimes a staminate flower; fertile lemma whitish, smooth.

1 Spikelets all or mostly 3 mm or more in length, strongly nerved 2
1 Spikelets less than 3 mm long, strongly nerved or not 4
2 First glume more than half as long as second glume; plants over 5 dm tall, essentially glabrous (except for margin and throat of leaf sheath), from strong scaly rhizome, with panicle terminal and over 15 cm tall .... **P. virgatum**
2 First glume not over half as long as second glume (except in *P. miliaceum*); plants shorter, usually ± pubescent, not rhizomatous, with panicles usually shorter or several 3
3 Leaves (blades and sheaths) essentially glabrous .... **P. dichotomiflorum**
3 Leaves pubescent (at least on sheaths) .. **P. miliaceum**
4 Sheaths of middle and upper leaves glabrous on back (may be ciliate on margins) ...... **P. dichotomiflorum**
4 Sheaths sparsely to heavily pilose on back 5

flowered, articulated above the glumes; glumes equal or nearly so, broad, 7–8.5 mm long, abruptly acute or mucronate; lemma about equaling the glumes, becoming indurate at maturity, pale green or yellowish; awn 6–14 mm long, straight or twisted, articulated with the lemma, readily detached.

Moist or dry open woods, forested dunes. New leaves start to develop in mid-summer, the blades at first erect, then bending downward and remaining green through winter. Sheaths below the level of the duff are usually bright purple.

Our other species formerly in this genus are now placed in *Piptatherum* (or *Piptateropsis* by some).

5 Panicle 2–3 times as long as wide, with clearly ascending branches .... **P. flexile**
5 Panicle less than 2 times as long as wide, branches ± spreading at maturity............................ 6
6 Peduncles only slightly exserted from sheaths, exserted portion less than half as long as panicle ... **P. capillare**
6 Peduncles long-exserted from sheaths, exserted portion half as long as panicle or longer................... 7
7 Spikelets lanceolate in outline, acuminate at tip, the longest 2.4–3.3 mm long................. **P. capillare**
7 Spikelets ovate or narrowly ovate in outline, acute at tip, the longest 1.9–2.3 mm long ... **P. philadelphicum**

Panicum capillare L.
COMMON PANIC-GRASS *native*
IR | **WUP** | **CUP** | **EUP** FAC CC 0

Annual grass, hirsute or hispid, hairs often purplish. Stems branched from base, erect, ascending, or decumbent, to 7 dm long; sheaths rounded; ligules membranous, ciliate, the cilia 0.5–1.5 mm long; leaf blades spreading, 3–18 mm wide. Panicles diffusely branched, sometimes 2/3 as long as the entire plant. Spikelets all or mostly on long pedicels, the first glume to half as long as the sterile lemma; margins of the lemma distinctly inrolled.

Dry or moist soil, often a weed in fields in gardens, widely distributed and variable.

Panicum dichotomiflorum Michx.
FALL PANIC-GRASS *native*
IR | **WUP** | **CUP** | **EUP** FACW CC 0

Annual or short-lived perennial grass. Stems erect to decumbent or diffuse, often 1 m long or more, rooting at the lower nodes when in water; sheaths compressed, inflated, sparsely pubescent near the base, elsewhere mostly glabrous; ligules 0.5–2 mm long; leaf blades 3–25 mm wide, glabrous or sparsely pilose, often scabrous near the margins; midribs whitish. Panicle in large plants to 4 dm long, widely branched. Spikelets green or tinged with purple, ellipsoid to oblong; first glume broad, obtuse or rounded, 0.5–1.1 mm long; second glume and sterile lemma acute, 7-nerved.

Moist soil and shores, sometimes in shallow water, often a weed in cultivated land.

Panicum flexile (Gattinger) Scribn.
WIRY PANIC-GRASS *native*
IR | WUP | **CUP** | **EUP** FACW CC 8

Delicate annual grass, green or yellow-green. Stems erect, 2–7 dm long, branched from base, hairy at nodes; sheaths longer than the internodes, green to purplish, hispid, margins ciliate; ligules 0.5–1.5 mm long; leaf blades ascending to erect, 1–7 mm wide, flat or the margins involute, surfaces sparsely hirsute. Panicle narrow panicle, 10–20 cm long and about a third as wide, the branches threadlike, upright to spreading. Spikelets lance-shaped, 3–4 mm long, with 1 fertile flower; first glume about half as long as second glume and sterile lemma. Aug–Sept.

Sandy and gravelly shores, fens, marshes; often where calcium-rich; Delta, Mackinac, and Menominee counties.

Panicum miliaceum L.
BROOMCORN MILLET *introduced*
IR | **WUP** | **CUP** | **EUP**

Annual grass. Stems stout, 2–6 (rarely 10) dm tall; sheaths overlapping, terete, densely pilose; ligules membranous, ciliate, the cilia 1–3 mm long; leaf blades 7–25 mm wide. Panicle included at base, pyramidal to cylindric, 8–20 cm long, often nodding at maturity. Spikelets 4.5–5.5 mm long; first glume half as long, 5-nerved; second glume and sterile lemma equal, distinctly 7–9-nerved. Grain straw-colored to brown, 3–3.5 mm long.

Asian native; occasionally grown for forage or for bird seed; adventive on roadsides and in waste places.

Panicum philadelphicum Bernh.
PHILADELPHIA PANIC-GRASS (THR) *native*
IR | **WUP** | **CUP** | **EUP** FAC CC 8

Annual grass. Stems slender, erect or rarely decumbent, 1–5 dm tall, branched from the base; leaves often crowded at base; sheaths usually longer than the internodes, hispid; ligules 0.5–1.5 mm long; leaf blades linear, ascending to erect, flat, 2–12 mm wide, hirsute to sparsely pilose, greenish or purplish. Panicle ovoid, usually about 1/3 as long as the entire plant; peduncle well exsert. Spikelets tending to be paired at the ends of the capillary branches; ovate to elliptic, 1.6–2.4 mm long; first glume obtuse or acute, about 2/5 as long as the abruptly acuminate second glume and sterile lemma, margins of the lemma barely inrolled. Grain plump, becoming blackish.

Dry soil and sandy fields.

Panicum virgatum L.
SWITCHGRASS *native*
IR | **WUP** | **CUP** | EUP FAC CC 4

Perennial grass, from hard, scaly rhizomes, often forming large bunches. Stems stout, erect, to 3 m tall; sheaths longer than the lower internodes, shorter than those above, glabrous or pilose, especially on the throat, margins ciliate; ligules a dense zone of silky hairs 2–6 mm long; leaf blades flat, erect to spreading, 2–5 dm long, 2–15 mm wide, undersurface sometimes densely pubescent, margins scabrous. Panicle open, freely branched, pyramidal, usually 2–4 dm long. Spikelets ovoid, soon widened distally by spreading of the glumes and sterile lemma, 2.2–5.6 mm long; first glume half or more as long as the spikelet; second glume and sterile lemma about

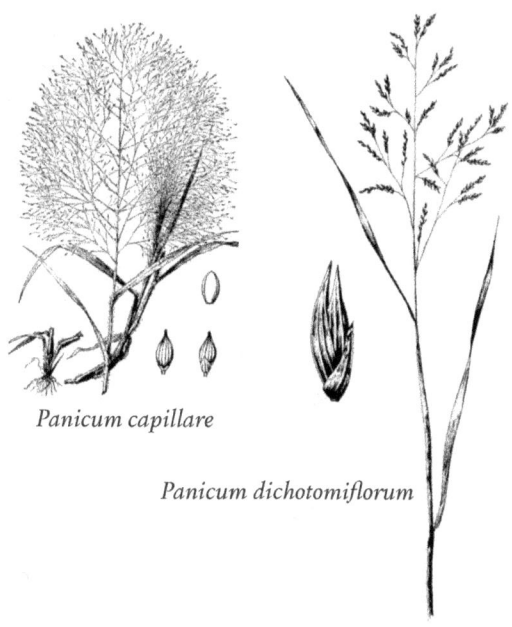

Panicum capillare

Panicum dichotomiflorum

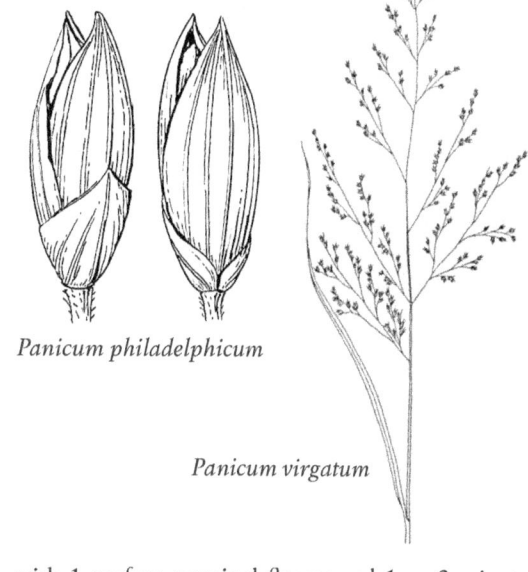

Panicum philadelphicum

Panicum virgatum

equal, conspicuously nerved, acute to long-acuminate.

Open woods, prairies, dunes, and shores, and brackish marshes.

Variable in length and shape of the glumes, especially the first glume.

Pascopyrum *Western Wheatgrass*

Pascopyrum smithii (Rydb.) Barkworth & D.R. Dewey
WESTERN WHEATGRASS   *introduced*
IR | **WUP** | **CUP** | **EUP**   FACU

Agropyron smithii Rydb., *Elytrigia smithii* (Rydb.) Nevski, *Elymus smithii* (Rydb.) Gould

Perennial grass, usually glaucous. Stems from long rhizomes, stout, usually 4–6 dm tall; leaves mostly at base; sheaths striate when dry, usually glabrous, rarely pilose; auricles 0.2–1 mm long, often purple; ligules membranous, about 0.1 mm long; leaf blades 2–5 mm wide, involute when dry. Spike usually 7–15 cm long, dense. Spikelets 12–22 mm long, 6–12-flowered; glumes narrowly lance-shaped, 9–14 mm long, strongly nerved, attenuate from below the middle with nearly straight sides; lemmas similar in shape, 10–14 mm long, acurninate to a stout stiff point or with an awn to 1 mm long.

Roadsides, railways; dry or sandy soil. An important grass of grasslands and high deserts in the western USA, considered adventive in Michigan.

Phalaris *Canary-Grass*

Annual or perennial grasses. Leaves glabrous, auricles absent; ligule large, membranous. Flowers in dense or spike-like panicles of medium-sized or large spikelets. Spikelets articulated above the glumes, with 1 perfect terminal flower and 1 or 2 minute sterile lemmas below it. Glumes about equal, compressed and keeled, usually winged along the midnerve; lateral nerves usually stronger than the midnerve. Lemmas awnless, shorter than the glumes, the sterile linear, resembling tufts of hairs at the base of a solitary functional floret; the fertile lemma firm or leathery, often shining.

1. Rhizomatous perennial, with elongate lobed panicle (or the lower branches spreading at anthesis); glumes mostly 4–5.7 mm long, the keel not winged .... **P. arundinacea**
1. Annual, with very dense, compact, ovoid panicle; glumes mostly 6–8 mm long, the keel prominently winged .... **P. canariensis**

Phalaris arundinacea L.
REED CANARY-GRASS   *introduced (see desc.)*
IR | **WUP** | **CUP** | **EUP**   FACW CC 0

Tall perennial grass, spreading by scaly rhizomes and typically forming large, dense colonies. Stems stout, smooth, 5–20 dm long; sheaths smooth; ligules membranous, 4–9 mm long, truncate, lacerate; leaf blades flat, 5–20 mm wide, surfaces scabrous, margins serrate. Panicle narrow, densely flowered, 5–25 cm long, often purple-tinged, the branches short and upright. Spikelets 4–6 mm long, breaking above glumes, with 1 fertile flower and 2 small sterile lemmas below; glumes nearly equal, longer than fertile floret, lance-shaped, tapered to tip or short-awned, becoming straw-colored with age, 3-veined; fertile lemma ovate, 3 mm long, shiny; palea as long as lemma. June–July.

Wet meadows, shallow marshes, ditches, shores and streambanks.

Reed canary-grass is an aggressive, highly competitive wetland species, now widely naturalized, often to the detriment of our native flora. Our populations are likely a mix of native and Eurasian strains, including cultivars developed for forage.

Phalaris canariensis L.
COMMON CANARY-GRASS *introduced*
IR | **WUP** | CUP | EUP FACU

Annual grass. Stems erect, 3–9 dm tall; ligules membranous, 3–6 mm long, rounded to obtuse, lacerate; leaf blades 2–10 mm wide. Panicle dense, ovoid, usually about 3 cm long; glumes broad, 7–8 mm long, broadly winged on the keel, the midnerve marked by a broad green stripe; fertile lemma about 5 mm long; sterile lemmas 2, linear, about 2.5 mm long.

Native of Europe; introduced and adventive; sometimes grown for bird seed; Gogebic and Houghton counties.

Distinguished by the exposed, nearly semi-circular ends of the glumes.

Phleum *Timothy*

Phleum pratense L.
COMMON TIMOTHY *introduced*
IR | **WUP** | CUP | EUP FACU

Tufted perennial grass. Stems mostly 5–10 dm tall; sheaths open; auricles absent or inconspicuous; ligules membranous, 2–4 mm long, not ciliate; leaf blades flat, typically 5–8 mm wide, rough-margined. Panicle spike-like and cylindric, usually 5–10 cm long, 6–8 mm thick. Spikelets 1-flowered, strongly flattened, articulated above the glumes; glumes equal, compressed and keeled, hispid-ciliate on the keel, 3-nerved, 2.6–3.2 mm long, rounded to the tip; lemma much shorter than the glumes, thin and delicate, 3–5-nerved, awnless; palea narrow, somewhat shorter than to nearly equaling the lemma.

Introduced from Eurasia as a forage grass, commonly cultivated for hay and pasture; escaped to fields, roadsides, and disturbed places.

Phragmites *Reed*

Phragmites australis (Cav.) Trin. ex Steud.
COMMON REED *native (see desc.)*
IR | **WUP** | CUP | EUP FACW CC 5
Phragmites communis Trin.

Tall, stout perennial reed, from deep, scaly rhizomes, or the rhizomes sometimes exposed and creeping over the soil; often forming large colonies. Stems erect, hollow, 1–4 m long and 5–15 mm wide near base, the internodes often purple; sheaths open, mostly overlapping; ligule membranous, white, 1 mm long, ciliate. leaf blades flat, long, 1–3 cm wide. Head a large, plumelike panicle, purple when young, turning yellow-brown with age, 15–40 cm long, much-branched, the branches angled or curved upward. Spikelets 3–7-flowered, linear, 10–15 mm long, breaking above the glumes; the stem within the spikelet (rachilla) covered with long silky hairs, these longer than the florets and becoming exposed as the lemmas spread after flowering; glumes unequal, the first glume half the length of second glume. Grain seldom maturing. Aug–Sept.

Fresh to brackish marshes, shores, streams, ditches, occasional in tamarack swamps; sometimes in shallow water.

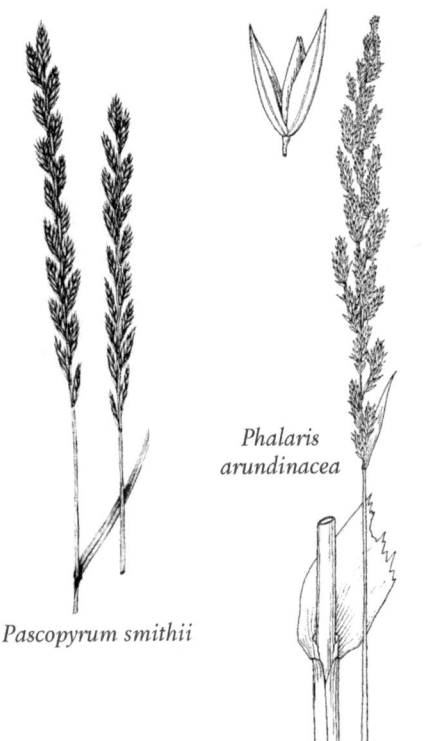

Pascopyrum smithii

Phalaris arundinacea

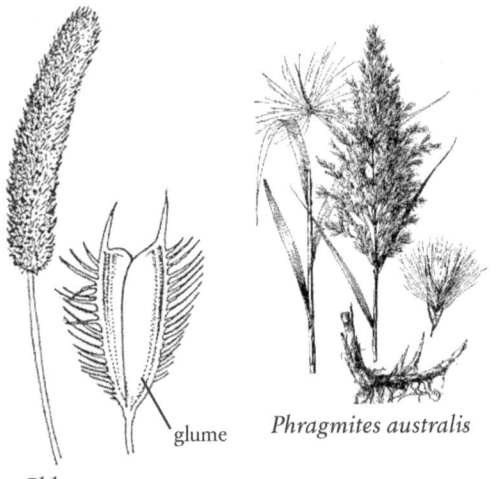

Phleum pratense

Phragmites australis

Two subspecies in Michigan, one native (subsp. *americanus*, whose distribution is poorly understood) and one introduced (subsp. *australis*):

1. Plants rarely forming a monoculture; ligules 1–1.7 mm long; lower glumes 3–6.5 mm long; upper glumes 5.5–11 mm long; lemmas 8–13.5 mm long; leaf sheaths deciduous, exposing stems in winter . . . . . **subsp. americanus**
1. Plants invasive and often forming a monoculture; ligules 0.4–0.9 mm long; lower glumes 2.5–5 mm long; upper glumes 4.5–7.5 mm long; lemmas 7.5–12 mm long; leaf sheaths not deciduous, stems not exposed in winter . . . . **subsp. australis**

Piptatherum *Ricegrass*

Tufted perennial grasses, sometimes rhizomatous. Leaf blades with flat or involute, auricles absent, ligules membranous to hyaline. Spikelets often large, in contracted or open panicles. Spikelets 1-flowered, articulated above the glumes. Glumes equal or nearly so, broad. Lemma about equaling the glumes, becoming indurate at maturity, with a terminal, readily detached awn.

1. Blades flat, mostly 5–18 mm wide; body of lemma 5.5–7 mm long; ligules absent or to 0.5 mm long . . . . **P. racemosum**
1. Blades involute, less than 2 mm wide; body of lemma 2.5–4 mm long; ligules of upper leaves 1.5–3 mm long 2
2. Awn 6–9 mm long, ± twisted; glumes completely smooth . . . . **P. canadense**
2. Awn absent or less than 2 (–3) mm long, nearly straight; glumes very minutely scabrous toward tip (20×). . . . . **P. pungens**

Piptatherum canadense (Poir) Dorn
CANADIAN MTN. RICEGRASS　　　(THR) native
IR | **WUP** | **CUP** | **EUP**　　　　　　　　　　CC 9

Oryzopsis canadensis (Poir.) Torr. ex A. Gray.
Piptatheropsis canadensis (Poir) Romasch., P.M. Peterson & Soreng

Stems loosely tufted, not rhizomatous, 3–8 dm tall; sheaths smooth or finely roughened; ligules 1–4 mm long, hyaline; basal leaf blades 4–15 cm long, 1–1.5 mm wide when flat, less than 1 mm wide when folded. Panicle lax and open, ovoid, 8–15 cm long, with flexuous, capillary, widely spreading branches. glumes elliptic-obovate, 3.5–4.8 mm long, very thin, the lateral nerves inconspicuous; mature lemma dull brown; awn persistent, 7–11 mm long, crooked, twisted and often somewhat coiled in the basal half.

Dry, sandy or rocky woods, often with jack pine and white spruce. The persistent, longer awns distinguish *P. canadense* from *P. pungens*.

Piptatherum pungens (Torr. ex Spreng.) Dorn
SHORT-AWN MOUNTAIN RICEGRASS　　　native
IR | **WUP** | **CUP** | **EUP**　　　　　　　　　　CC 9

Oryzopsis pungens (Torr.) Hitchc.
Piptatheropsis pungens (Torr. ex Spreng.) Romasch., P.M. Peterson & Soreng

Stems densely tufted, 2–5 dm tall; leaves mostly basal; sheaths smooth or somewhat scabrous; ligules 0.5–2.5 mm long, truncate to acute; blades 0.5–1.8 mm wide, flat to involute (at least when dry), scaberulous. Panicle 3–8 cm long, usually slender with appressed or strongly ascending branches, or ovoid and open at anthesis; glumes elliptic to obovate, 3.5–4 mm long, very thin, the lateral nerves inconspicuous; lemma gray or pale green; awn 1–2 mm long, straight or slightly bent.

Sandy dry woods, usually with aspen, oak, jack pine, and red pine; dunes and rocky places.

The fragile awn is readily broken off and is lacking in most herbarium specimens.

Piptatherum racemosum (Sm) Barkworth
BLACK-SEED MOUNTAIN RICEGRASS　　　native
IR | **WUP** | **CUP** | **EUP**　　　　　　　　　　CC 8

Oryzopsis racemosa (Sm.) Ricker ex Hitchc.
Patis racemosa (Sm.) Romasch., P.M. Peteron & Soreng

Stems 4–10 dm tall, loosely tufted from a knotty rhizome; sheaths usually glabrous; basal leaves more or less absent; upper stem leaves distinctly longer than the lower, the blades usually 1–2 dm long and 8–15 mm wide, scaberulous above, pubescent beneath;

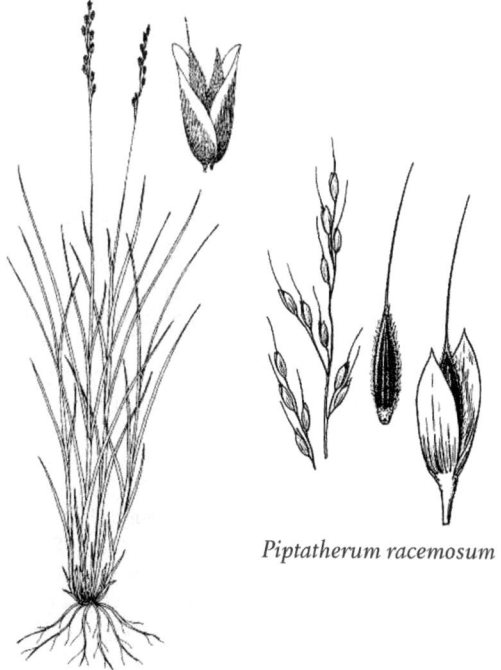

Piptatherum racemosum

Piptatherum pungens

ligules 0.3–0.7 mm long, truncate. Panicle sparsely branched, 1–2 dm long, the few straight branches spreading or ascending, bearing the few appressed spikelets toward the ends; glumes herbaceous, narrowly elliptic, 7–9 mm long, 7-nerved; lemma dark brown and shining, somewhat shorter than the glumes; awn 12–22 mm long.

Moist, rich deciduous forests and wooded dunes, sometimes in disturbed places; rarely in dry woods of jack pine or oak.

Poa *Bluegrass*

Annual or perennial grasses, with or without rhizomes or stolons, densely to loosely tufted or the culms solitary. Leaves mostly near base, flat to folded, midrib 2-grooved, the tip keeled similar to the bow of a boat; sheaths partly closed, auricles absent, ligules membranous. Head an open panicle. Spikelets small, with 2 to several flowers breaking above the glumes; glumes nearly equal, the first glume usually 1-veined, the second glume 3-veined; lemmas often with a tuft of distinctive cobwebby hairs at base; palea nearly as long as lemma.

ADDITIONAL SPECIES

Poa alpina L. (Alpine Bluegrass), native perennial of mostly arctic regions; in Michigan, found in rock crevices along Lake Superior shore in Keweenaw County and on Drummond Island, Chippewa County; threatened.

Poa secunda J. Presl (Canby's Bluegrass), native perennial, disjunct in Michigan from the western USA in rock crevices on Isle Royale; endangered.

POA GROUP KEY

Some species appear more than once in the keys; 'flag leaf' refers to the uppermost leaf which is often angled outward from the stem.

1. Stems with bulbous bases; spikelets often bulbiferous. .... **P. bulbosa**
1. Stems with non-bulbous bases. 2
2. Plants annual or perennial; anthers to 1 mm long in all florets and well developed, or only the upper 1–2 florets with rudimentary anthers .... **Subkey 1**
2. Plants perennial; some anthers 1.3–4 mm long, or the florets pistillate and all anthers vestigial and to 0.2 mm long, or longer and poorly developed 3
3. Plants rhizomatous or stoloniferous, rhizomes or stolons usually longer than 5 mm; basal leaves of the erect shoots with well-developed blades; plants densely to loosely tufted or the stems solitary ...... **Subkey 2**
3. Plants neither rhizomatous nor stoloniferous; basal leaves of the erect shoots sometimes without blades; plants densely tufted .... **Subkey 3**

POA SUBKEY 1

Plants annual or perennial. Stems not bulbous at base. Basal leaf sheaths not swollen at the base. Spikelets not bulbiferous, florets developing normally. Anthers 0.1–1 mm long.

1. Plants annual, sometimes surviving for a second season, introduced, weedy species; calluses glabrous; lemmas usually softly puberulent to long-villous on the keel and marginal veins, often also on the lateral veins, glabrous between the veins, non-alpine plants rarely glabrous throughout; palea keels smooth, usually short- to long-villous near the apices, rarely glabrous; panicle branches and glume keels smooth .... **P. annua**
1. Plants perennial, native, sometimes growing in disturbed habitats; calluses webbed or glabrous, if glabrous, the lmma pubescence not as above or the palea keels at least slightly scabrous near the apices; panicle branches and glume keels smooth of scabrous 2
2. Calluses webbed; lemma keels glabrous throughout or, if hairy on the proximal 1/2, the marginal veins glabrous 3
2. Calluses webbed or glabrous, if webbed, the lemmas hairy on the keel and marginal veins 4
3. Lemmas hairy only on the keels; branches in whorls of (2)3–5(7) .... **P. alsodes**
3. Lemmas usually glabrous, marginal veins rarely sparsely hairy at the base, hairs to 0.15 mm long; branches 1–3 per node. .... **P. saltuensis**
4. Sheaths closed for 1/5–7/8 their length; stems with or without bladeless leaves; anthers 0.2–1.2 mm long, well developed. .... **P. sylvestris**
4. Sheaths closed for 1/10–1/5 their length; lower 1–3 leaves along the stems; anthers 0.8–1.2 mm long, sometimes poorly developed 5
5. Flag leaf nodes at or above mid-stem length. .... **P. nemoralis**
5. Flag leaf nodes usually in the basal 1/3 of the stem. ... . **P. glauca**

POA SUBKEY 2

Plants with rhizomes or stolons, densely to loosely tufted or the stems solitary.

1. Stems and nodes strongly compressed; stems usually geniculate; lower stem nodes usually exserted; panicle branches angled, scabrous on the angles; sheaths closed for 1/10–1/5 their length .... **P. compressa**
1. Stems terete to somewhat compressed, nodes not or only weakly compressed; stems geniculate or not; lower stem nodes exserted or not; panicle branches angled or terete, smooth or scabrous; sheath closure varied... 2
2. Lemma keels softly puberulent for 3/5 their length, hairs usually sparse, marginal veins glabrous or puberulent to 1/4 their length, intercostal regions smooth and glabrous; lateral veins prominent; calluses webbed; palea keels smooth, muriculate, tuberculate, or scabridulous; lower glumes 1-veined, usually arched to

sickle-shaped; ligules 3–10 mm long, acute to acuminate; panicle branches angled, angles densely scabrous; plants usually weakly stoloniferous . . . . **P. trivialis**

2 Lemmas glabrous or variously pubescent, if as above, the lateral veins faint or moderately prominent or the calluses glabrous or the palea keels distinctly scabrous or hairy or the lower glumes 3-veined; calluses glabrous or hairy; palea keels scabrous at least near the apices; lower glumes 1–3-veined, not arched, not sickle-shaped; ligules 0.5–18 mm long, truncate to acuminate; panicle branches terete or angled, smooth or scabrous; plants stoloniferous or not 3

3 Sheaths closed for 1/10–1/5 their length; spikelets 3–5 mm long; lemmas glabrous between the keels and marginal veins; panicle branches angled, angles densely scabrous; plants sometimes stoloniferous, sometimes branching above the stem bases; florets bisexual . . . . **P. palustris**

3 Sheaths closed for 1/5–9/10 their length; spikelets 3.5–12 mm long; lemmas glabrous or hairy between the keels and marginal veins; panicle branches terete or angled, smooth or scabrous; plants rarely stoloniferous, usually rhizomatous, never branching above the stem bases; florets bisexual or unisexual . . . . **P. pratensis**

POA SUBKEY 3

Plants perennial, loosely to densely tufted, rhizomes and stolons absent. Stems not bulbous at base. Basal sheaths not swollen. Spikelets not bulbiferous, florets developing normally.

1 Calluses usually dorsally webbed . . . . **P. trivialis**
1 Calluses glabrous or with a crown of hairs 2

2 Lemma lateral veins pronounced, keels pubescent, marginal veins glabrous or softly puberulent at the base, lemmas glabrous elsewhere; lower glumes 1-veined, subulate to narrowly lanceolate, usually arched to sickle-shaped; callus web well-developed. . . **P. trivialis**

2 Lemma lateral veins obscure to pronounced, keels glabrous throughout or, if pubescent, the marginal veins distinctly pubescent for more than 1/4 their length, lemma lateral veins and intercostal regions glabrous or pubescent, or, if pubescent as in P. trivialis, then the callus web short, scant, poorly developed and the lower glumes 3-veined and lanceolate or broader. 3

3 Panicles open, conical, with whorls of 3–10, spreading to eventually reflexed, scabrous-angled branches at the lower nodes; lemmas hairy on the keel and veins; callus webs well developed . . . . **P. sylvestris**

3 Panicles contracted to open, if open then not conical and without whorls of (2)3–10, eventually reflexed, scabrous-angled branches at the lower nodes; branches smooth or scabrous-angled; lemmas glabrous or hairy; calluses glabrous, with diffuse hairs, or with a scanty or well-developed web 4

4 Sheaths closed for 1/3–3/4 their length . . **P. saltuensis**
4 Sheaths closed for up to 1/4 their length 5

5 Flag leaf nodes usually in the lower 1/10–1/3 of the stems; flag leaf blades usually distinctly shorter than their sheaths; lemmas sometimes softly puberulent between the veins, lateral veins usually with at least a few minute hairs; ligules 1–4 mm long . . . . **P. glauca**

5 Flag leaf nodes usually in the upper 2/3 of the stems; flag leaf blades shorter or longer than their sheaths; lemmas glabrous between the veins, lateral veins usually glabrous, rarely with 1 to several minute hairs; ligules 0.2–6 mm long 6

6 Spikelets lanceolate; glumes subulate to narrowly lanceolate, gradually tapering to narrowly acuminate tips; ligules to 0.5 mm long, truncate; flag leaf nodes at or above the middle of the stems; flag leaf blades usually longer than their sheaths; rachillas usually hairy, hairs to 0.15 mm long; webs usually short, scanty . . . . **P. nemoralis**

6 Spikelets and glumes not as above or, if so, the ligules 1.5–6 mm long, truncate to acute, and the rachillas glabrous; flag leaf nodes at or above the lower 1/3 of the stem; flag leaf blades longer or shorter than their sheaths; webs short or long, scanty or not 7

7 Panicles 10–30 cm long, branches 4–15 cm long; stems closely spaced to isolated at the base; lower glumes tapering to the apices; lemma keels abruptly inwardly arched beneath the scarious tips; lemma margins distinctly inrolled; rachillas usually muriculate, rarely sparsely hispidulous; web hairs usually longer than 2/3 the length of the lemmas . . . . **P. palustris**

7 Panicles 3–15 cm long, branches 0.4–8 cm long; stems closely spaced at the base; lower glumes abruptly narrowing to the apices, lengths 4.5–6.3 times the widths; lemma keels not abruptly inwardly arched beneath the scarious apices; lemma margins not or slightly inrolled; rachillas usually softly puberulent; web hairs shorter than 1/2(2/3) the length of the lemmas. . . . . . **P. interior**

Poa alsodes Gray
GROVE BLUEGRASS *native*
IR | WUP | CUP | EUP FAC CC 9

Loosely tufted perennial grass, rhizomes absent. Stems slender, 3–8 dm long; sheaths closed for 1/2–7/8 their length; ligules to 2 mm long, smooth or sparsely scabrous, truncate to obtuse; leaf blades 1–4 mm wide, flat, lax. Panicle open, lax, 10–20 cm long, the branches becoming widely spreading, mostly in groups of 4–5, with 1 to few spikelets near tip of branch; base of panicle sometimes remaining enclosed by sheath. Spikelets ovate, 2–3-flowered, 3–5 mm long; glumes nearly equal, 2–4 mm long; lemmas 2–4 mm long, with cobwebby hairs at base. May–July.

Alder thickets, swamp hummocks, most common in moist deciduous or mixed conifer-deciduous forests.

Poa annua L.
ANNUAL BLUEGRASS *introduced*
IR | WUP | CUP | EUP FACU

Annual grass, densely tufted. Stems to 3 dm long, prostrate to ascending; sheaths closed for about 1/3 their length, terete or weakly compressed, smooth; ligules 0.5–3 mm long, glabrous, decurrent, obtuse to truncate; leaf blades 1–10 cm long, 1–4 mm wide, flat or weakly folded, thin, soft, smooth, margins usually slightly scabrous, broadly prow-shaped at tip. Panicle ovoid, 2–8 cm long, with few ascending branches bearing rather crowded spikelets above the middle. Spikelets green, 3–6-flowered, 3–5 mm long; glumes broadly lance-shaped, acute, scarious-margined, indistinctly nerved, the first 1.5–2.4 mm, the second 1.8–2.8 mm long; lemmas thin, elliptic, 5-nerved, obtuse, pubescent on the nerves, not webbed at base.

Native of Eurasia and a widely distributed weedy species of roadsides, lawns, forest trails, clearings, shores, and disturbed places.

Poa bulbosa L.
BULBOUS BLUEGRASS *introduced*
IR | WUP | **CUP** | **EUP**

Densely tufted perennial grass; rhizomes and stolons absent. Stems erect from bulbous-thickened bases, purplish below, 2–5 dm tall; sheaths closed for about 1/4 their length, terete, lowest sheaths with swollen bases; ligules 1–3 mm long, smooth or scabrous; leaf blades 1–2.5 mm wide, flat, thin, lax, soon withering. Panicles compact and crowded, ovoid, 4–8 cm long; florets mostly converted into turgid purple bulblets, the bracts prolonged into linear tips 5–15 mm long.

Native of Europe; introduced in fields, lawns, and roadsides.

Poa compressa L.
CANADA BLUEGRASS *introduced (invasive)*
IR | WUP | **CUP** | **EUP** FACU

Perennial grass; the shoots usually solitary, sometimes loosely tufted, extensively rhizomatous. Stems erect, 2–7 dm tall, strongly flattened, especially above; sheaths closed for 1/10–1/5 their length, distinctly compressed; ligules 1–3 mm long, scabrous, ciliolate; leaf blades 1.5–4 mm wide, flat. Panicle usually compact and narrow, bluish or grayish green, 2–8 cm long, the branches usually in pairs, bearing spikelets nearly to the base; pedicels of the lateral spikelets 0.5 mm long. Spikelets 3–6-flowered, 4–6 mm long; first glume 1.7–2.4 mm long; second glume 1.8–2.6 mm long; lemmas firm, obscurely nerved, 2–2.8 mm long, slightly pubescent on the nerves below, somewhat webbed at base.

Native of Europe; open, usually dry places, especially in acidic soil.

Along with *P. pratensis*, a very common grass in Michigan; *P. compressa* differs from *P. pratensis* in its flattened, less tufted stems, lemmas with sparse or even absent web at the base, and a more slender panicle with fewer branches at each node.

Poa glauca Vahl
WHITE BLUEGRASS *native*
IR | **WUP** | **CUP** | **EUP** CC 10

Perennial grass, usually glaucous; densely tufted, rhizomes and stolons absent. Stems tufted, erect, 2–5 dm tall; sheaths closed for 1/10–1/5 their length, terete; ligules 1–4 mm long, sparsely to densely scabrous, obtuse to acute; leaf blades 1–2.5 mm wide, flat or folded, thin, soft, narrowly prow-shaped at tip. Panicle long-exsert, narrow, rather dense, 6–12 cm long, the branches in fascicles of 2–5, at first ascending, later spreading, each bearing a few spikelets. Spikelets 2–4-flowered; glumes lance-shaped, nearly equal, 2.3–3.8 mm long, less than half as wide; lemmas obscurely nerved, 2.5–3.5 mm long, densely sericeous on the lower half of the keel and marginal nerves, not webbed at base.

Open, sandy forests; rock crevices and rocky shores, sometimes where underlain by limestone.

Poa interior Rydb.
INTERIOR BLUEGRASS *native*
IR | **WUP** | **CUP** | **EUP** FAC CC 10

Densely tufted perennial grass, green or less often glaucous, rhizomes and stolons absent. Stems to 80 cm long, erect or ascending; sheaths closed for up to 1/5 their length, terete; ligules 0.5–1.5 mm long, scabrous, truncate to obtuse, ciliolate; leaf blades mostly flat, thin, soft, 1–3 mm wide, narrowly prow-shaped at tip. Panicles to 15 cm long; branches to 8 cm long, ascending to widely spreading, angled, the angles scabrous. Spikelets mostly 2–3-flowered, narrowly ovate, laterally compressed 3–6 mm long, usually not glaucous; glumes lance-shaped, distinctly keeled, keels smooth or sparsely scabrous; calluses usually webbed, webs usually scant, less than 1/2 the lemma length, frequently tiny; lemmas 2.4–4 mm long, lance-shaped, distinctly keeled, straight or gradually arched, keels and marginal veins short-villous.

Shallow rocky or sandy soil of outcrops and talus slopes; Chippewa County records are from Drummond Island only.

Distinguished from *P. nemoralis* by its longer ligules and wider glumes and lemmas; differs from *P. palustris* in having a densely tufted habit, scantly webbed calluses, and lemmas with wider hyaline margins.

Poa nemoralis L.
WOODLAND BLUEGRASS *introduced*
IR | **WUP** | **CUP** | **EUP** FACU

Densely tufted perennial grass, green or glaucous, rhizomes and stolons absent. Stems slender, 4–8 dm tall; sheaths closed for 1/10–1/5 their length, terete; ligules 0.2–0.8 mm long, sparsely to densely scabrous, truncate; leaf blades 1–3 mm wide, mostly flat, narrowly prow-shaped at tip. Panicle narrowly ovoid, 1–2 dm long, eventually loose and open, the slender branches in fascicles of about 5, bearing spikelets above the middle. Spikelets 2–4-flowered; glumes narrowly lance-shaped, long-acuminate, the first 2.2–3 mm long, conspicuously narrower than the first lemma, the second 2.3–3.3 mm long; lemmas 3-nerved, 2.1–3.1 mm long.

Dry, sandy or rocky soil, forest borders and clearings, old farmsteads.

Poa palustris L.
FOWL BLUEGRASS *native*
IR | WUP | CUP | EUP FACW CC 3
Loosely tufted perennial grass, often with stolons. Stems smooth, 4–12 dm long, reclining at base and rooting from lower nodes, lower portion often purple-tinged; sheaths closed for 1/10–1/5 their length, slightly compressed, glabrous or sparsely retrorsely scabrous; ligules 1.5–6 mm long, smooth or sparsely scabrous, tips obtuse to acute, frequently lacerate; leaf blades 1.5–8 mm wide, flat, narrowly prow-shaped at tip. Panicle loosely spreading (narrow when emerging from sheath), 1–3 dm long, the branches in mostly widely separated groups along panicle stem (rachis). Spikelets 2–4-flowered, 2–5 mm long and 1–2 mm wide; glumes nearly equal, lance-shaped, 2–3 mm long, often purple; lemma 2–3 mm long, often purple on sides, with cobwebby hairs at base. June–Sept.

Wet meadows, marshes, shores, streambanks, ditches and low prairie; also moist woods. Native to boreal regions of North America and n Eurasia.

Poa pratensis L.
KENTUCKY BLUEGRASS *introduced (invasive)*
IR | WUP | CUP | EUP FACU
Perennial grass, sometimes glaucous, densely to loosely tufted or the shoots solitary; extensively rhizomatous and sod-forming. Stems 3–10 dm tall; sheaths closed for 1/4–1/2 their length, terete to slightly compressed, glabrous; collars glabrous; ligules mostly 1–2 mm long, ciliolate or glabrous; leaf blades 0.4–4.5 mm wide, flat, folded, or involute, soft and lax to moderately firm, tips usually broadly prow-shaped. Panicle ovoid, rather dense, its branches spreading or ascending, the lower chiefly in fascicles of 4 or 5. Spikelets 3–5-flowered, with very short rachilla-joints; first glume 1.8–2.9 mm long; second glume 2.3–3 mm long; lemmas distinctly 5–nerved, thinly to densely pubescent on the nerves, webbed at base, glabrous between the nerves, the lowest 2.5–3.5 mm long; anthers 1–1.4 mm long.

Moist or dry soil, disturbed places, woods, fields, avoiding acidic soils and heavy shade, often cultivated in lawns and meadows. Introduced from Europe and naturalized in much of North America.

Poa saltuensis Fern. & Wieg.
OLD-PASTURE BLUEGRASS *native*
IR | WUP | CUP | EUP CC 5
Loosely tufted perennial grass; rhizomes and stolons absent. Stems slender, usually weak, 3–10 dm tall; sheaths closed for 1/3–2/3 their length; ligules 0.2–3 mm long, smooth or sparsely scabrous, truncate to obtuse; leaf blades 1–4 mm wide, flat, thin, lax, veins prominent. Panicle loose, more or less nodding, 5–10 cm long, the slender branches bearing a few spikelets beyond the middle, the lower branches usually in pairs, rarely solitary or in 3's. Spikelets ovate, 2–4-flowered, 3–4 mm long; glumes acute, the first lance-shaped to ovate, 1.7–2.6 mm. long, the second ovate, 2–3 mm long; lemmas firm, obscurely nerved, oblong, 2.4–3.2 mm long, glabrous except the webbed base.

Dry or rocky deciduous and mixed woods.

Poa sylvestris Gray
WOODLAND BLUEGRASS *native*
IR | WUP | **CUP** | EUP FAC CC 8
Loosely tufted perennial grass; rhizomes and stolons absent, or sometimes evidently shortly rhizomatous Stems erect, usually 4–8 dm tall; sheaths closed for 1/2–7/8 their length, terete, throats often ciliate near their junction; ligules 0.5–2.7 mm long, smooth or sparsely scabrous, truncate to obtuse; leaf blades 0.7–5 mm wide, flat, thin, lax. Panicle rather narrow, oblong, 1–2 dm long, its slender flexuous branches in fascicles of 4–8, soon divaricate or reflexed, bearing a few spikelets much above the middle. Spikelets 2–5-flowered; glumes scarious-margined, acute, the first lance-shaped, 1.5–2.7 mm long, the second oblong, 1.9–3.4 mm long; lemmas distinctly 5-nerved, 2.1–3.5 mm long, villous on the marginal nerves, at least toward the base, and nearly or quite to the tip of the keel, webbed at base.

Rich deciduous woods; Marquette County.

Poa trivialis L.
ROUGH-STALK BLUEGRASS *introduced*
IR | **WUP** | **CUP** | EUP FACW
Short-lived perennial grass, tufted and forming mats from aboveground stolons. Stems slender to stout, erect from a decumbent base, 5–10 dm tall, scabrous below the panicle; sheaths closed for about 1/3–1/2 their length, compressed, usually densely scabrous; ligules 3–10 mm long, scabrous, acute to acuminate; leaf blades 1–5 mm wide, flat, lax, soft, sparsely scabrous over the veins, margins scabrous, tips narrowly prow-shaped. Panicle soon long-exsert, ovoid, the ascending branches in fascicles of 5–8 with numerous crowded spikelets; pedicels scabrous. Spikelets ovate or elliptic, 2- or 3-flowered; glumes lance-shaped, incurved, the first 1.7–2.9 mm., the second 2–3.3 mm long; lemmas thin, narrowly ovate, sharply

502 POACEAE | Grass Family

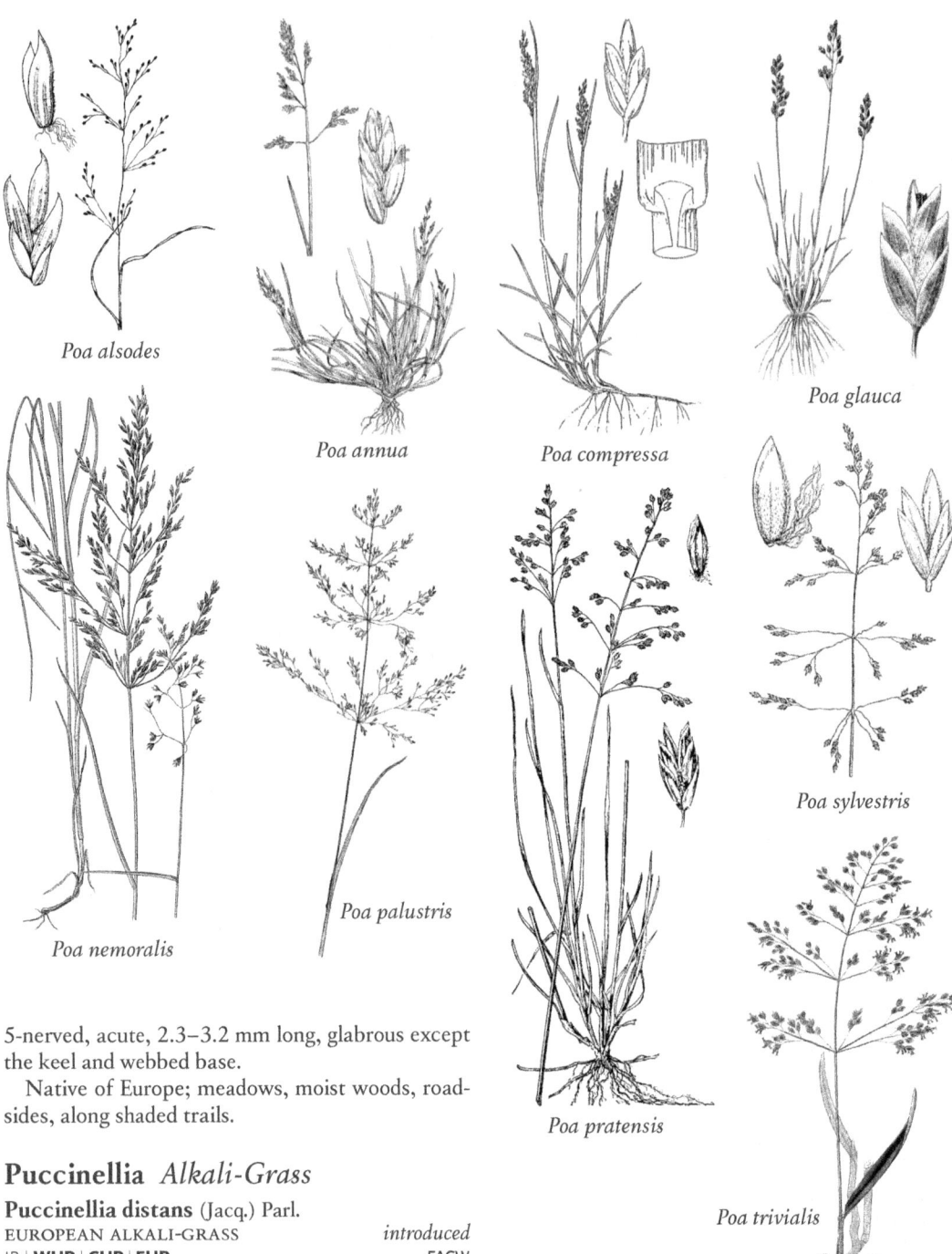

5-nerved, acute, 2.3–3.2 mm long, glabrous except the keel and webbed base.

Native of Europe; meadows, moist woods, roadsides, along shaded trails.

Puccinellia Alkali-Grass

Puccinellia distans (Jacq.) Parl.
EUROPEAN ALKALI-GRASS *introduced*
IR | **WUP** | **CUP** | **EUP** FACW

Tufted perennial grasses, glabrous, usually in brackish habitats. Stems erect or reclining at base, 1–5 dm long; sheaths open to the base or nearly so; auricles absent; ligules membranous. Leaves mostly from base of plants, flat to slightly inrolled, 1–6 mm wide; ligules about 1 mm long, obtuse to truncate, usually entire. Panicle loose, pyramid-shaped, 5–15 cm long, the branches in groups, the lower branches angled downward. Spikelets 3–7-flowered, 4–6 mm long; glumes ovate, 1–2 mm long; lemmas about 2 mm long, smooth or short-hairy at base. May–Aug.

Waste areas and ditches along salted highways.

Schedonorus *Tall Fescue*

Previously in *Festuca*, *Schedonorus* includes the large, broad- and flat-leaved species with awned or at least sharply pointed lemmas. It is very closely related to *Lolium*, with which it hybridizes.

1. Auricles at top of leaf sheath ciliate, having at least 1 or 2 hairs along the margins; panicle branches at the lowest node usually paired, the shorter with 1–13 spikelets, the longer with 3–19 spikelets; lemmas 5.5–7 mm long, usually scabrous at least distally, unawned or with an awn up to 4 mm long .... **S. arundinaceus**
1. Auricles glabrous; panicle branches at the lowest node 1 or 2, if paired the shorter with 1–2(3) spikelets, the longer with 2–6(9) spikelets; lemmas 7–8.5 mm long, usually smooth, sometimes slightly scabrous distally, unawned or with a mucro to 0.2 mm long. **S. pratensis**

Schedonorus arundinaceus (Schreb.) Dumort.
TALL RYE GRASS *introduced*
IR | **WUP** | **CUP** | **EUP** FACU

Festuca arundinacea Schreb.
Lolium arundinaceum (Schreb.) S.J. Darbyshire

Tufted perennial grass, sometimes with rhizomes. Stems erect above a geniculate base, to 1.5 m tall, glabrous; sheaths smooth; auricles ciliate, having at least 1 or 2 hairs along the margins; ligules ca. 1 mm long; leaf blades glabrous or scaberulous, 4–8 mm wide, dilated at base into conspicuous auricles. Panicle erect or nodding at the tip, 1–2 dm long, contracted at least after flowering, the internodes of the branches less than 2x as long as the spikelets. Spikelets 4–11-flowered, usually 7–8-flowered; first glume subulate, 2.5–4.5 mm long; second glume lance-shaped, sharply nerved, 3.5–7 mm long; lemmas 5.5–8 mm long, scarious at the acute tip, occasionally with a short awn.

Native of Europe; cultivated for forage and as a turfgrass; established in fields and meadows.

Schedonorus pratensis (Huds.) Beauv.
MEADOW RYE GRASS *introduced*
IR | **WUP** | **CUP** | **EUP** FACU

Festuca elatior L. p.p.
Festuca pratensis Huds.
Lolium pratense (Huds.) S.J. Darbyshire

Tufted perennial grass. Stems often basally decumbent, to 1.3 m tall, glabrous; old sheaths brown, decaying to fibers; ligules glabrous to 0.5 mm long; leaf blades lax, 3–5 mm wide, dilated at base into conspicuous auricles. Panicle 1–2.5 dm, erect or nodding at the tip, contracted at least after anthesis, the internodes of the branches less than 2x as long as the spikelets. Spikelets 10–15 mm long, 4–10-flowered; first glume subulate, 2.5–4 mm, 1-veined, the second lance-shaped, 3.5–5 mm, 3–5-veined, with hyaline margins; lemmas 5.5–7 mm long, usually glabrous, 5-veined, the tip hyaline, acute, rarely with a short awn to 2 mm long.

Native of Europe, cultivated for forage and established in fields, meadows, and moist soil.

Schizachne *False Melic Grass*

Schizachne purpurascens (Torr.) Swallen
FALSE MELIC GRASS *native*
IR | **WUP** | **CUP** | **EUP** FACU CC 5

Loosely tufted perennial grass. Stems erect from a short-decumbent base, 3–10 dm tall; sheaths closed almost to the top; ligules membranous, 0.5–1.5 mm long; leaf blades mostly erect, elongate, 1–5 mm wide. Panicle with few drooping branches each bearing 1–3 slender spikelets about 2 cm long. Spikelets 3–5-flowered, usually purplish, disarticulating above the glumes and between the lemmas; glumes purple at base, unequal, 5–8 mm long, 3–5-nerved; fertile lemmas 8–10 mm long, strongly nerved, densely short-bearded at base, bifid for a fourth of their length, with an awn 8–15 mm long between the teeth, awns at length divergent.

Drier, sandy or rocky woods and openings; deciduous forests.

Schedonorus arundinaceus

Schedonorus pratensis

Schizachne purpurascens

Schizachyrium scoparium

Schizachyrium *Little Bluestem*

Schizachyrium scoparium (Michx.) Nash
LITTLE BLUESTEM *native*
IR | **WUP** | **CUP** | **EUP** FACU CC 5

Andropogon scoparius Michx.

Perennial grass; loosely or densely tufted or with rhizomes, green to purplish, sometimes glaucous. Stems 5–12 dm tall, often freely branched above; sheaths rounded or keeled, glabrous or pubescent; auricles absent; ligules membranous, 0.5–2 mm long; leaf blades 3–7 mm wide. Racemes solitary, usually long-exsert, bearing 5–20 pairs of spikelets on a straight or flexuous, white-ciliate rachis. Sessile spikelets with 2 florets; glumes exceeding the florets; lower florets reduced to hyaline lemmas; upper florets bisexual, lemmas hyaline, bilobed or bifid to 7/8 of their length, awned from the sinuses. Pedicellate spikelets usually shorter than the sessile spikelets, sterile or staminate, with 1 floret. Late summer and early fall.

An important prairie species; in Michigan often in drier sandy woods and openings, old fields, sand dunes and shores.

Secale *Rye*

Secale cereale L.
RYE *introduced*
IR | **WUP** | **CUP** | **EUP**

Annual or biennial grass. Stems branched from the base 5–10 dm tall; sheaths open; ligules membranous, truncate, often lacerate; leaf blades 4–12 mm wide, usually glabrous. Spikes densely flowered, 8–15 cm long, often distinctly nodding when mature. Spikelets usually 2-flowered, solitary at each joint of the rachis; glumes linear-subulate, 1-nerved, shorter than the lemmas; lemmas lance-subulate, 5-nerved, with their sides toward the axis, tapering into a long awn 3–8 cm long; disarticulation in the rachis, at the nodes, tardy or the spikes not disarticulating.

An important Eurasian cereal grass, also widely used for soil stabilization and, especially in Canada, for whisky. Mostly along roadsides, where planted for erosion control following construction; also on shores, dunes, along railroads, and in old fields; not long-persisting.

Setaria *Bristle Grass*

Tufted annual grasses (ours). Ligules membranous and ciliate or of hairs. Spikelets all alike, with 1 perfect flower, turgid or plano-convex, subtended by an involucre of 1 to many slender bristles, articulated and eventually deciduous above the bristles, aggregated into cylindric, spike-like, terminal panicles. First glume triangular to ovate, 3–5-nerved, half as long as the spikelet or less. Second glume longer, sometimes equaling the spikelet. Sterile lemma equaling the spikelet, several-nerved. Fertile lemma indurate, smooth or transversely rugose.

1. Bristles, summit of stem, and axis of panicle scabrous with retrorse barbs; panicle branches tending to appear whorled, the panicle ± interrupted toward its base. . . . . **S. verticillata**
1. Bristles, summit of stem, and axis of panicle scabrous or pubescent with antrorse barbs or hairs; panicle very compact throughout. 2
2. Fertile lemmas mostly ca. (2.7–) 3 (–3.4) mm long, rugose with distinctly transverse ridges, the upper half exposed at maturity; bristles 5 or more per spikelet, becoming orange or golden-brown; sheaths glabrous . . . . **S. pumila**
2. Fertile lemmas less than 3 mm long, evenly and finely rugose or reticulate or smooth (without transverse ridges), the upper half largely or entirely concealed at maturity; bristles fewer than 5 per spikelet, pale greenish or purple (rarely yellow) at maturity; sheaths ciliate with long hairs on the margins. 3
3. Spikelet articulated above the glumes and sterile lemma; fertile lemma distinctly yellow or darker at maturity; panicle very dense, often ± lobed in appearance . . . . **S. italica**
3. Spikelet articulated below the glumes; fertile lemma pale green or brown; panicle not lobed 4
4. Panicle strongly nodding, bent below the middle; spikelets mostly over 2.5 mm long, the fertile lemma ± tapering to a distinctly exposed tip; leaf blades ± hairy above . . . . **S. faberi**
4. Panicle straight and erect or rarely slightly nodding; spikelets not over 2.5 mm long, the blunt fertile lemma nearly or quite concealed by the second glume; leaf blades glabrous above . . . . **S. viridis**

Setaria faberi Herrm.
JAPANESE BRISTLE GRASS *introduced*
IR | **WUP** | **CUP** | **EUP** FACU

Annual grass; much like *S. viridis*, but more robust. Stems 5–20 dm tall; sheaths glabrous, fringed with white hairs; ligules about 2 mm long; leaf blades 15–30 cm long and 1–2 cm wide, with long, soft papillose-based hairs on upper surface. Panicle spike-like, 6–20 cm long, drooping from near the base. Spikelets 2.5–3 mm long, subtended by 1–6 (usually 3) bristles, the second glume 2/3 to 3/4 as long as the more strongly rugose fertile lemma; sterile lemma with a palea, two-thirds as long.

Unintentionally introduced into North America from China in the 1920s, now a serious weed in corn and soybean fields of the midwest USA.

Setaria italica (L.) Beauv.
FOXTAIL-MILLET *introduced*
IR | **WUP** | CUP | EUP FACU

Annual grass. Stems to ca. 1 m tall; sheaths mostly

glabrous, margins sparsely ciliate; ligules 1–2 mm long; leaf blades flat, to 3 cm wide, scabrous. Panicle to 25 cm long and 3 cm thick, purple or tawny; bristles 1–3 below each spikelet, upwardly barbed, 3–10 mm long. Spikelets 2–3 mm long, articulated above the sterile lemma, the grain readily detached from the persistent glumes; sterile lemma usually somewhat shorter than the fertile.

Native of the Old World; sometimes cultivated and escaped to ditches, fields and disturbed places; Keweenaw County.

Setaria pumila (Poir.) Roem. & Schult.
PEARL-MILLET *introduced*
IR | **WUP** | **CUP** | **EUP** FAC
Setaria glauca (L.) Beauv.

Annual grass. Stems usually erect, solitary or tufted, 4–8 dm tall; sheaths glabrous; ligules ciliate; leaf blades 4–10 mm wide, loosely twisted, upper surface with papillose-based hairs near base. Panicles spike-like, usually 5–10 cm long, or sometimes longer, the axis pubescent; bristles 3–10 mm long, yellow or tawny at maturity. Spikelets thick, 3–3.5 mm long; first glume 5-nerved, half as long as the spikelet; second glume 5-nerved or usually 7-nerved, 2/3 as long as the spikelet; fertile lemma transversely rugose.

Native of Europe, introduced in lawns, roadsides, railroads, cultivated fields, and disturbed places.

Setaria verticillata (L.) Beauv.
ROUGH BRISTLE GRASS *introduced*
IR | WUP | CUP | EUP FACU

Annual grass. Stems often branched at base, usually 6–10 dm tall; sheaths glabrous, margins ciliate upwards; ligules to 1 mm long, densely ciliate; leaf blades flat, 5–15 mm wide, undersurface scabrous. Panicles often tapering upward, 5–15 cm long, usually more or less lobed or interrupted near the base where the short branches tend to be whorled; bristle one below each spikelet, retrorsely barbed, purplish or tawny, 2–8 mm long. Spikelets about 2 mm long; sterile lemma equaling the finely rugose fertile one.

European native, weedy in cultivated or waste ground; Menominee County.

Setaria viridis (L.) Beauv.
GREEN FOXTAIL-GRASS *introduced*
IR | **WUP** | **CUP** | **EUP**

Annual grass. Stems usually branched and often geniculate at base; sheaths mostly glabrous, margins ciliate upwards; ligules 1–2 mm long, ciliate; leaf blades flat, 4–25 mm wide, scabrous or smooth. Panicle 1–7 cm long, the short branches uniformly spaced on the rachis; bristles 1–3 below each spikelet, upwardly barbed, green, purple, or tawny, usually 2–10 mm long. Spikelets 2–2.5 mm long; sterile lemma usually as long as the finely rugose fertile one.

Native of Eurasia; weedy in gardens, cultivated fields, and disturbed places.

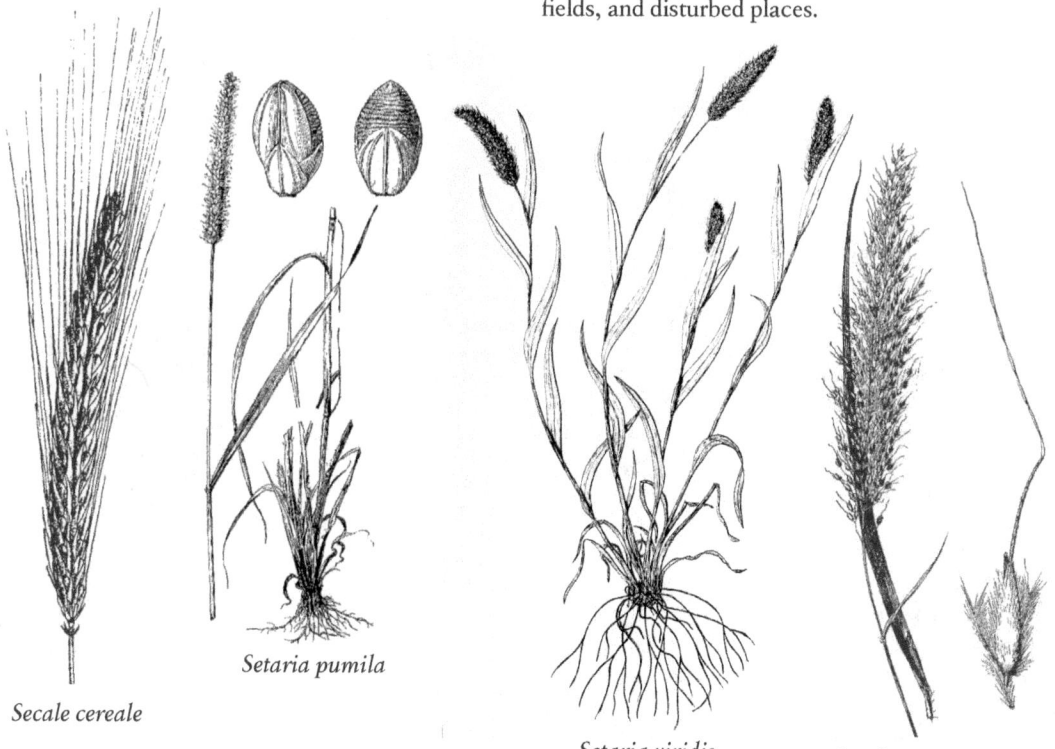

Secale cereale

Setaria pumila

Setaria viridis

Sorghastrum nutans

Sorghastrum *Indian Grass*

Sorghastrum nutans (L.) Nash
YELLOW INDIAN GRASS   native
IR | **WUP** | **CUP** | EUP   FACU CC 6

Large perennial grass from short scaly rhizomes. Stems in loose tufts, 1–2 m tall, the nodes densely pubescent; sheaths glabrous to hirsute; ligules membranous, 2–6 mm long, usually with thick, pointed auricles; leaf blades 10–70 cm long and to 1 cm wide, usually glabrous. Panicle narrow, 10–25 cm long, the ultimate branches of the panicle bearing short racemes of 1–5 spikelets; terminal spikelet usually with 2 sterile pedicels adjacent the nodes. Spikelets lance-shaped, 6–8 mm long; sessile spikelet perfect, subterete; pediceled spikelet absent, represented by its pedicel only; first glume pale brown, villous, its edges enclosing the margin of the glabrous or ciliate second glume; lemmas hyaline, bifid, awned from the sinuses; the awn 9–15 mm long, twisted below, bent at about a third of its length; sterile pedicel densely villous, 4–5 mm long.

Moist or dry prairies, open woods, fields, shores, and rarely, in marshes; sometimes spreading in disturbed places as along roadsides and railroads; Delta and Menominee counties; an important species of tallgrass prairies.

Spartina *Cord-Grass*

Spartina pectinata Bosc
FRESHWATER CORD-GRASS   native
IR | **WUP** | **CUP** | EUP   FACW CC 5

Stout perennial grass, strongly rhizomatous, the rhizomes scaly, purplish-brown or light brown (drying white). Stems tough, 1–2 m long; sheaths open, mostly glabrous, throats often pilose; ligules membranous, 1–3 mm long, ciliate; leaf blades flat to inrolled, 3–10 mm wide, margins strongly scabrous. Head a spike-like raceme of mostly 10–30, 1-sided spikes, the spikes upright to sometimes appressed, 3–10 cm long. Spikelets 1-flowered, flattened, 8–11 mm long, overlapping in 2 rows on one side of the rachis, breaking below the glumes; first glume nearly as long as floret, hispid on the keels, tapered to tip or with an awn 1–5 mm long, second glume longer than floret, tipped with an awn 2–8 mm long; lemma 7–9 mm long, shorter than second glume. July–Sept.

Shallow marshes, wet meadows.

Sphenopholis *Wedgescale*

Perennial grasses. Leaf blades flat, sheaths open, auricles absent; ligules membranous, erose. Panicles slender or spike-like, shining. Spikelets 2-flowered (rarely with a rudimentary third flower), disarticulating below the glumes and below the upper lemma; rachilla prolonged behind the second palea. Glumes unequal, keeled, the first linear, 1-nerved, the second obovate, scarious-margined, 3–5-nerved. Lemmas slender, firm in texture, rounded on the back or keeled toward the summit, awnless. Palea hyaline, about equaling the lemma.

1 Larger (second) glume distinctly swollen or distended, abruptly truncate and usually shallowly 2-lobed at tip .... **S. obtusata**
1 Larger glume not swollen or distended, obtuse to acute at tip .... **S. intermedia**

Sphenopholis intermedia (Rydb.) Rydb.
SLENDER WEDGESCALE   native
IR | **WUP** | **CUP** | EUP   CC 4

Sphenopholis obtusata var. *major* (Torr.) K.S. Erdman
Perennial grass. Stems tufted, 3–12 dm tall; sheaths smooth or finely roughened, sometimes pubescent; ligules 1.5–2.5 mm, erose-ciliate, often lacerate; leaf blades flat, soft, 2–5 mm wide, Panicle slender, 8–15 cm long, 1–3 cm wide, more or less lobed or irregular in contour, not spike-like except when very young; first glume subulate, 1.5–2.4 mm long, less than 0.5 mm wide; second glume obovate, broadly acute or apiculate, thin in texture, obscurely nerved, 2–2.7 mm long; lower lemma 2.3–2.9 mm long.

Moist to wet gravelly shores, tamarack swamps, marsh borders, thickets, forest depressions; sometimes in moist woods.

Sphenopholis intermedia

Spartina pectinata

Similar in appearance to *Koeleria macrantha*, but differs in its more open panicle, the very narrow first glume, and the essentially glabrous foliage and panicle.

Sphenopholis obtusata (Michx.) Scribn.
PRAIRIE WEDGESCALE native
IR | **WUP** | CUP | EUP   FAC CC 8

Tufted perennial (sometimes annual) grass; plants smooth to rough-hairy. Stems slender, 2–10 dm long; sheaths glabrous or hairy; ligules 1.5–3 mm long, more or less lacerate; leaf blades usually flat, 2–8 mm wide, scabrous or pubescent. Panicle spike-like, dense, shiny, 5–20 cm long, the spikes often (in part) separate from one another. Spikelets 2-flowered, 3–4 mm long, unawned, breaking below the glumes; glumes 2–3 mm long, the first glume linear, 1-veined, the second glume broader, 3–5-veined; lemma 2–3 mm long, 1-veined; palea linear, about as long as lemma. June–Aug.

Dry forests, moist to wet meadows, gravelly shores; Houghton and Keweenaw counties.

The inflorescence is more dense and contracted compared to the more open and lax panicle of *S. intermedia*.

Sporobolus *Dropseed*

Annual or perennial grasses. Leaf blades narrow, often involute; sheaths open, usually glabrous, often ciliate at the top; ligules of short hairs. Panicles open or contracted. Spikelets 1-flowered, articulated above the glumes. Glumes lance-shaped to ovate, 1-nerved, from much shorter than to somewhat longer than the lemma. Lemma rounded on the back, nerveless or 1-nerved, awnless. Palea about as long as the lemma or longer. Fruit differs from a true grain in that the pericarp is free from the seed coat, and often slipping away, at least when moist.

1	Plants annuals or short-lived perennials flowering in the first year 2
1	Plants perennial 3
2	Lemmas strigose; spikelets 2.3-6 mm long; mature fruits 1.8-2.7 mm long .... **S. vaginiflorus**
2	Lemmas glabrous; spikelets 1.6-3 mm long; mature fruits 1.2-1.8 mm long .... **S. neglectus**
3	Spikelets 1-2.5 mm long .... **S. cryptandrus**
3	Spikelets 2.5-10 mm long 4
4	Mature panicles to 30 cm wide, pyramidal; panicle branches appressed or spreading ...... **S. heterolepis**
4	Mature panicles to 4 cm wide, spikelike; panicle branches appressed 5
5	Lemmas minutely pubescent or scabridulous, chartaceous and opaque; pericarps loose but neither gelatinous nor slipping off the seeds when wet; fruits 2.0-3.5 mm long .... **S. clandestinus**
5	Lemmas usually glabrous and smooth, membranous to chartaceous and hyaline; pericarps gelatinous, slipping off the seeds when wet; fruits 1-2 mm long .... **S. compositus**

Sporobolus compositus (Poir.) Merr.
TALL DROPSEED native
IR | WUP | **CUP** | EUP CC –*Sporobolus asper* (Beauv.) Kunth

Tufted perennial grass, sometimes with rhizomes. Stems stout, erect, to 12 dm tall; sheaths sparsely hairy at tips, the hairs to 3 mm long; ligules to 0.5 mm long; leaf blades long, 2–5 mm wide, tapering to a filiform point, flat or involute at least when dry. Panicle pale or purplish, 5–15 cm long, about 1 cm thick, long enclosed in the upper sheath; glumes and lemma boat-shaped and somewhat carinate, obtuse to acute; first glume 2–3.5 mm long; second glume 2.5–4.6 mm long; lemma glabrous, 3.5–6 mm long, about equaling the palea.

Dry or sandy soil of prairies, roadsides and along railways; Menominee County.

Sporobolus cryptandrus (Torr.) Gray
SAND-DROPSEED native
IR | **WUP** | **CUP** | **EUP**   FACU CC 3

Tufted perennial grass, rhizomes absent, bases not hard and knotty. Stems solitary or in small tufts, 3–10 dm tall, the lower portion usually covered by sheaths; sheath tips with conspicuous tufts of hairs, the hairs to 4 mm long; ligules to 1 mm long; leaf blades 2–6 mm wide, flat or drying involute, 2–6 mm wide, tapering to a long point. Panicle ovoid or pyramidal, 1–2 dm long, at base usually partly included in the upper sheath; branches alternate, soon widely divergent, the branchlets more or less appressed and forming a dense narrow cluster. Spikelets 2–3 mm long; glumes acute, the first half or less as long as the second; lemma about equaling the second glume.

Dry, especially sandy soil, cedar glades, barrens, fields and dunes; often in sandy disturbed areas such as roadsides and railways.

Sporobolus heterolepis (Gray) Gray
PRAIRIE-DROPSEED native
IR | WUP | **CUP** | **EUP**   FACU CC 10

Tufted perennial grass, rhizomes absent. Stems erect, 4–10 dm tall; sheaths dull and fibrous at base, glabrous or sparsely pilose below, the hairs contorted, to 4 mm long; ligules to 0.3 mm long; leaf blades very long, narrow, flat or folded, 1–2.5 mm wide, margins scabrous. Panicle cylindric to narrowly ovoid, with ascending branches, 1–2 dm long; branches mostly raceme-like, they and the pedicels irregularly interrupted by paler, slightly widened segments. Spikelets 3–6 mm long; first glume about half as long as the second, subulate above a broader base; second glume acuminate into an involute tip, usually slightly longer than the lemma; palea usually slightly exceeding the lemma; mature grain spherical, spreading the parts of the spikelet and splitting the palea.

Moist to dry prairies, sometimes in fens and in shallow soil on dolomite pavement; Chippewa, Delta, and Mackinac counties.

Plants have a distinctive musky smell especially noticeable in hot weather.

Sporobolus neglectus Nash
SMALL DROPSEED   native
IR | **WUP** | **CUP** | **EUP**   FACU CC 2

Tufted annual grass; plants delicate, slender, very similar to S. vaginiflorus. Stems 10–45 cm tall, wiry, erect to decumbent; sheaths inflated, mostly glabrous but the tips with small tufts of hairs to 3 mm long; ligules to 0.3 mm long; leaf blades to 2 mm wide, flat to loosely involute. Panicle rarely exsert, usually permanently exceeded by the uppermost blade. Spikelets smaller, the glumes and lemma less acuminate and proportionately wider; first glume 1.5–2.4 mm long; second glume 1.7–2.7 mm long; lemma 2–3 mm long, glabrous, about equaling the wide palea.

Dry sterile or sandy soil of roadsides and fields; also along shores and on mudflats.

Sporobolus vaginiflorus (Torr.) Wood
POVERTY-GRASS   native
IR | **WUP** | **CUP** | **EUP**   CC 2

Annual grass. Stems tufted, erect to spreading, 15–60 cm tall, very thin and wiry, seldom more than 1 mm in diameter; sheaths often inflated, sometimes with sparse hairs at base, glabrous or the tips with small tufts of hairs to 3 mm long; ligules to 0.3 mm long; leaf blades to 2 mm wide, the lower elongate, the upper progressively shorter to only 1–2 cm long. Panicle slender, 2–5 cm long, eventually exsert; axillary panicles also developed and mostly included in the lower sheaths. Spikelets crowded; glumes and lemma lance-shaped, straight; first glume 2.8–4.1 mm long; second glume 2.9–4.6 mm long; lemma 3–5 mm long, minutely villous; palea equaling or somewhat exceeding the lemma.

Dry sandy or sterile soil as along roadsides (especially where gravelly) and in fields.

Very similar to *S. neglectus* and impossible to distinguish without spikelets; *S. vaginiflorus* differs in having strigose lemmas, sheaths that are sparsely hairy towards the base and, usually, longer spikelets. Both differ from our other species in their annual habit, and by having nearly equal glumes.

Torreyochloa *False Mannagrass*

Torreyochloa pallida (Torr.) Church
FALSE MANNAGRASS   native
IR | **WUP** | **CUP** | **EUP**   OBL CC 7

Puccinellia pallida (Torr.) Clausen
Torreochloa fernaldii (A.S. Hitchc.) Church

Perennial rhizomatous grass. Stems slender and flaccid, 3–10 dm tall, usually more or less decumbent and creeping at base; sheaths open to the base; auricles absent; ligules membranous, 2–9 mm long, truncate or acute; leaf blades flat, soft, 2-15 mm wide. Panicle with relatively few branches, eventually diffuse, 5–15 cm long. Spikelets narrowly ovate, 4–7 mm, long, 4–6-flowered; glumes broadly rounded at the scarious tip; lemmas ovate, sharply nerved, finely pubescent or scaberulous, erose at the rounded tip; palea 4–5 times longer than wide.

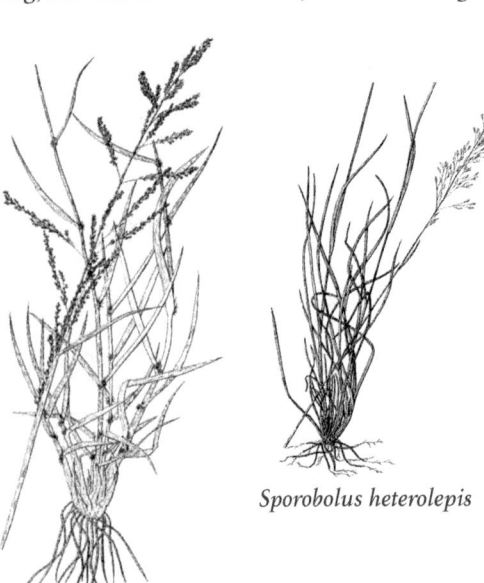

Sporobolus heterolepis

Sporobolus cryptandrus

Sporobolus neglectus

Sporobolus vaginiflorus

Cat-tail marshes, bogs, shorelines, wet forest depressions, often in shallow water.

Trisetum *False Oat*

Trisetum spicatum (L.) Richter
NARROW FALSE OAT *native*
IR | WUP | CUP | EUP   FAC CC 10

Tufted perennial grass, with both fertile and sterile shoots; rhizomes absent. Stems erect, 1–5 dm tall, glabrous or pubescent below, pubescent below the panicle; sheaths open, pubescent or glabrous; auricles absent; ligules membranous, 0.5–4 mm long, truncate or rounded; leaf blades flat or involute, glabrous to pubescent, 1–4 mm wide. Panicle from spikelike to open, often interrupted basally, green, purplish, or tawny, usually silvery-shiny; dense, or interrupted at base, 3–10 cm long. Spikelets usually 2-flowered; disarticulation usually above the glumes and between the florets; glumes thin or membranous, with broad hyaline margins, about as long as the spikelet; first glume 1-nerved, 3–5 mm long; second glume, 3-nerved, 4–6 mm long; lemmas slightly surpassing the glumes, bifid, teeth usually less than 1 mm long, awned; the awns 3-8 mm long, from the upper 1/3 of the lemma, geniculate, twisted near the base. Highly variable.

Exposed or partly shaded rocks and crevices, including dolomite on Drummond Island.

Triticum *Wheat*

Triticum aestivum L.
COMMON WHEAT *introduced*
IR | WUP | CUP | EUP

Annual grass. Stems erect, single or branched at base, 5–12 dm tall; internodes usually hollow, even immediately below the spikes; sheaths open; auricles present, often deciduous at maturity; ligules membranous; leaf blades flat, 6–15 mm wide, glabrous or pubescent. Spikes densely flowerd, to 18 cm long. Spikelets 2–5-flowered, single and sessile at each joint of the rachis; glumes broadly ovate, the broader side truncate or notched at the tip; lemma sides turned toward the axis, the midnerve prolonged into a short point or awn to 8 cm long.

Commonly cultivated, sometimes appearing on roadsides from spilled grain; probably never persisting in our flora.

Zizania *Wild Rice*

Large annual grasses (ours) of marshes and shallow water, with tall stems, wide flat blades, and fleshy yellow roots. Sheaths open, not inflated; ligules membranous or scarious. Spikelets 1-flowered, articulated at the base, readily deciduous, nearly terete, unisexual, the staminate on the lower, the pistillate on the upper branches of the large panicle. Glumes absent. Lemma of the staminate spikelet thin, herbaceous, linear, acuminate or short-awned, 5-nerved. Pistillate spikelets inserted in a cup-shaped excavation at the summit of the pedicels; lemma firm at maturity, prominently 3-ribbed, awned.

Zizania palustris L.
NORTHERN WILD RICE *native*
IR | WUP | CUP | EUP   OBL CC 8

Stems to 3 m tall, usually at least partly immersed in water; sheaths glabrous or with scattered hairs; ligules 3–15 mm long; leaf blades 0.5–1.7 cm wide. Panicles 25–60 cm long; staminate and pistillate flowers separate on same plant, the staminate flowers on lower panicle branches, pistillate flowers on upper branches; staminate branches ascending or

Torreyochloa pallida

Trisetum spicatum

Triticum aestivum

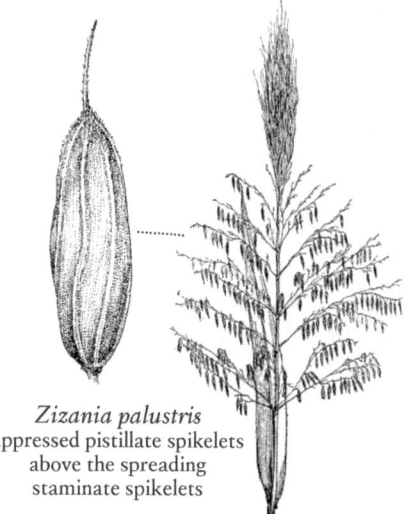

Zizania palustris appressed pistillate spikelets above the spreading staminate spikelets

divergent; pistillate branches mostly appressed or ascending, a few sometimes divergent. Staminate spikelets 6–17 mm long, lanceolate, acuminate or awned, the awns to 2 mm long. Pistillate spikelets 8–33 mm long, lanceolate or oblong, leathery or indurate, lustrous, glabrous or with lines of short hairs, tips usually hirsute and abruptly narrowed, awned, the awns to 10 cm long; lemmas and paleas remaining clasped at maturity. Grain 6–30 mm long.

Ours are var. *palustris*. Rangewide, *Z. palustris* grows mostly to the north of the similar *Z. aquatica* L. (not known from the UP), but their ranges overlap in portions of the Great Lakes region (e.g., Michigan Lower Peninsula, Minnesota, Wisconsin).

Z. palustris is the source of commercial wild rice (California is the nation's largest producer); in the Great Lakes region, harvesting is most common in Minnesota, especially by Native Americans where large areas of lakes and shallow marshes may be dominated by this plant. The grain is also an excellent food for waterfowl. Many of our populations are intentional introductions.

Pontederiaceae

PICKERELWEED FAMILY

Mostly perennial, aquatic or emergent herbs. Leaves alternate, stalkless and straplike, or with a petiole and broad blade. Flowers perfect (with both staminate and pistillate parts), regular or irregular, single from leaf axils or in spikes or panicles, subtended by leaflike bracts (spathes), light yellow, white or blue-purple, perianth of 6 petal-like lobes, usually joined near base to form a tube; stamens 3–6, the filaments attached to throat of perianth tube; ovary superior, 3-chambered, style 1. Fruit a many-seeded capsule inside the spathe, or a 1-seeded, achene-like utricle.

1   Flowers 2-lipped, each lip 3-lobed, the 3 lower lobes spreading; stamens 6, 3 longer than petals, 3 shorter; fruit 1-seeded................... **Pontederia cordata**
1   Flowers regular, the lobes more or less equal; stamens 3, all longer than petals; fruit a many-seeded capsule . ... **Heteranthera dubia**

Heteranthera *Mud-Plantain*

Heteranthera dubia (Jacq.) MacM.
GRASS-LEAF MUD-PLANTAIN   native
IR | **WUP** | **CUP** | **EUP**   OBL CC 6
   Zosterella dubia (Jacq.) Small

Aquatic perennial herb, with lax stems and leaves, or plants sometimes exposed and forming small, leafy rosettes. Stems slender, forked, often rooting at lower nodes, to 1 m long. Leaves alternate, linear, flat, translucent, rounded at tip or tapered to a point, 2–12 cm long and 2–6 mm wide, the midrib and veins inconspicuous; petioles absent. Flowers 1, opening on water surface, light yellow, enclosed in a spathe from upper leaf axils, the spathe membranous, 2–5 cm long, surrounding much of the slender perianth tube; perianth tube often curved, 2–8 cm long, the 6 perianth segments linear, 4–6 mm long; stamens 3, all alike. Fruit a many-seeded capsule about 1 cm long. July–Sept.

Shallow water, muddy shores of ponds, lakes, streams and marshes.

Distinguished from the pondweeds (*Potamogeton*) by lack of a leaf midrib.

Pontederia *Pickerelweed*

Pontederia cordata L.
PICKERELWEED   native
IR | **WUP** | **CUP** | **EUP**   OBL CC 8

Perennial emergent herb, spreading from rhizomes and forming colonies. Stems stout, upright, to 12 dm long, with 1 leaf. Leaves lance-shaped to ovate, 5–20 cm long and 2–15 cm wide, heart-shaped at base; petioles 3–7 cm long, sheathing on stem. Flowers blue-purple (rarely white), many in a spike 5–15 cm long, subtended by a bractlike spathe 3–6 cm long; perianth funnel-like, the tube 6 mm long, 2-lipped above, upper lip with 3 ovate lobes, lower lip with 3 slender, spreading lobes, the lobes 7–10 mm long. Fruit a 1-seeded utricle, 5–10 mm long. June–Sept.

Shallow water (to 1 m deep) of lakes, ponds, rivers and swamps.

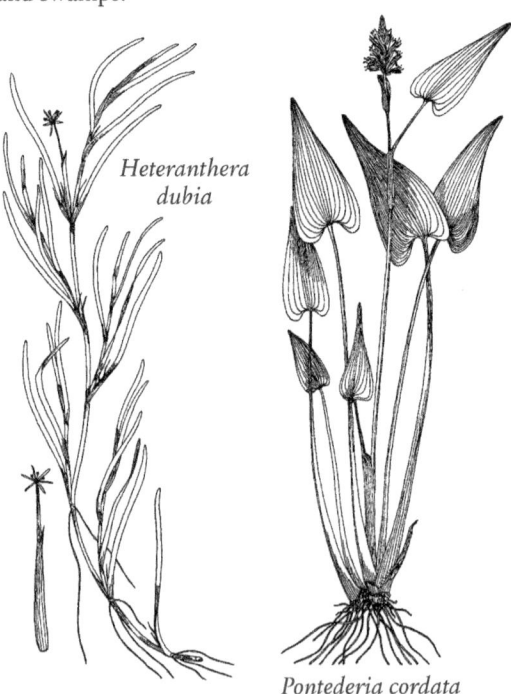

Heteranthera dubia

Pontederia cordata

Potamogetonaceae

PONDWEED FAMILY

This treatment includes two genera, *Ruppia* and *Zannichellia*, previously included in separate families.

1 Submersed leaves opposite or whorled, floating leaves absent................................ **Zannichellia**
1 Submersed leaves alternate, floating leaves (sometimes present) alternate or opposite..................... 2
2 Flowers 2, at first enclosed in sheathing leaf base, the peduncle elongating and often spiraled or coiled at its base; fruit long-stalked; stipular sheath lacking free ligule at summit (the stipule wholly adnate to the leaf blade and merely rounded at the summit); leaf blade terete...................................... **Ruppia**
2 Flowers several to many in a peduncled head or spike; perianth of 4 tepals; fruit ± sessile; stipular sheath absent (stipules entirely free from leaf) or with a short ligule-like extension if stipules fused to the leaf blade 3
3 Stipules adnate to the leaves for 10–30 mm or more (at least on the larger leaves), adnate for ca. 2/3 of the length of the stipule; leaves all submersed, filiform to narrowly linear (up to 2.5 mm wide)........ **Stuckenia**
3 Stipules free from the leaves or adnate for less than half the length of the stipule (adnate for 5 mm or less except in *P. robbinsii*); leaves submersed or floating, filiform to ovate, oblong, or elliptic .... **Potamogeton**

Potamogeton *Pondweed*

Aquatic perennial herbs, with only underwater leaves or with both underwater and floating leaves, from rhizomes or tubers, sometimes reproducing and over-wintering by free-floating winter buds. Stems long, wavy, anchored to bottom by roots and rhizomes. Leaves alternate, or becoming opposite upward in some species, simple, with an open or closed sheath at base. Underwater leaves usually linear and threadlike, sometimes broader, margins often wavy, usually stalkless. Floating leaves, if present, oval or ovate, stalked, with a waxy upper surface. Flowers perfect, regular, green to red, in stalked spikes at ends of stems or from leaf axils, usually raised above water surface, the spikes with few to many small flowers; perianth of 4 sepal-like bracts; stamens 4. Fruit a 4-parted, beaked achene. The narrow-leaved pondweeds (leads 8–16 in Group 1 key), although important as a group as waterfowl food, are often difficult to positively identify in the field, the distinguishing features being somewhat hard to see.

Key to Potamogeton Groups

1 Plants with underwater leaves only, these all alike ....
 .... **Group 1**
1 Plants with 2 kinds of leaves: broad floating leaves and broad or narrow underwater leaves.......... **Group 2**

Potamogeton Group 1

Plants with underwater leaves only, these all alike.

1 Leaves broad, lance-shaped to oval or ovate, never linear 2
1 Leaves linear... 7
2 Leaf margins wavy-crisped, finely toothed .. **P. crispus**
2 Leaf margins flat or sometimes wavy, entire (or rarely finely toothed at tip)............................. 3
3 Base of leaf blade tapered, not clasping stem 4
3 Base of leaf blade clasping stem 5
4 Plants green; upper leaves stalked; leaf margins finely toothed near tip...................... **P. illinoensis**
4 Plants red-tinged; upper leaves more or less stalkless; leaf margins entire .... **P. alpinus**
5 Stems whitish; leaves 10–30 cm long; fruit 4–5 mm long
 .... **P. praelongus**
5 Stems green; leaves 1–12 cm long; fruit 2–4 mm long . 6
6 Leaves ovate, mostly 1–5 cm long, margins flat; stipules small or absent; plants drying olive-green ....
 .... **P. perfoliatus**
6 Leaves lance-shaped, mostly more than 5 cm long; margins wavy-crisped; stipules conspicuous, persisting as shreds; plants drying light green ...... **P. richardsonii**
7 Stipules joined with lower part of leaf to form a sheath at least 1 cm long .... **P. robbinsii**
7 Stipules free from leaf, or rarely joined to leaf base for only 1–2 mm 8
8 Plants with slender creeping rhizomes............. 9
8 Plants with short rhizomes or rhizomes absent (plants often rooting at lower nodes of stem) 10
9 Flower clusters on stalks at ends of stems, the stalks mostly 5–25 cm long; leaves threadlike, narrower than stems .... **P. confervoides**
9 Flower clusters on stalks from leaf axils, the stalks less than 3 cm long; leaves linear, wider than stems ....
 .... **P. foliosus**
10 Leaves 9- to many-veined (with 1–2 main veins and many finer ones) .... **P. zosteriformis**
10 Leaves 1–7-veined................................. 11
11 Leaves without glands at base............. **P. foliosus**
11 At least some of leaves with pair of glands at base .. 12
12 Leaves with 5–7 nerves .... **P. friesii**
12 Leaves with 3 (rarely 1 or 5) nerves................. 13
13 Leaves gradually tapered to a bristlelike tip 14
13 Leaves rounded at tip or tapered to a point, not bristle-tipped... 15
14 Leaf margins rolled under; widespread . **P. strictifolius**
14 Leaf margins flat, not rolled under .... **P. hillii**
15 Leaves 1–4 mm wide, rounded at tip; body of achene 2.5–4 mm long...................... **P. obtusifolius**
15 Leaves to 2.5 mm wide, usually tapered to a sharp tip; body of achene to 2 mm long.............. **P. pusillus**

POTAMOGETONACEAE | Pondweed Family

Potamogeton Group 2
Plants with 2 kinds of leaves: broad floating leaves and broad or narrow underwater leaves.

1. Underwater leaves broad, never linear.............. 2
1. Underwater leaves linear......................... 6
2. Floating leaves with 30–55 nerves; underwater leaves with 30–40 nerves................. **P. amplifolius**
2. Floating leaves with fewer than 30 nerves; underwater leaves with less than 30 nerves.................... 3
3. Underwater leaves with more than 7 nerves, all leaves stalked............................... **P. nodosus**
3. Underwater leaves mostly with 7 nerves, at least the lower leaves stalkless.......................... 4
4. Margins of underwater leaves finely toothed near tip. .... **P. illinoensis**
4. Margins of underwater leaves entire 5
5. Plants red-tinged; underwater leaves 5–20 cm long and at least as wide as floating leaves, mostly on main stem .... **P. alpinus**
5. Plants green; underwater leaves 3–8 cm long and narrower than floating leaves, often numerous on short branches from leaf axils................ **P. gramineus**
6. Spikes of 2 kinds: those in axils of lower underwater leaves round, on short stalks; those in axils of upper or floating leaves cylindric, often emersed on long stalks; fruit flattened; stipules of leaves (or at least some of lower leaves) joined with leaf base .... **P. spirillus**
6. Spikes of 1 kind only; fruit not (or only slightly) compressed; stipules not joined with leaf base 7
7. Floating leaves less than 1 cm wide and less than 2 cm long .... **P. vaseyi**
7. Floating leaves more than 1 cm wide and more than 2 cm long 8
8. Underwater leaves flat and tapelike, 2–10 mm wide... .... **P. epihydrus**
8. Underwater leaves round in cross-section, often reduced to a petiole, mostly less than 1.5 mm wide... 9
9. Blade of floating leaves oval, tapered to base; fruit 3-keeled............................... **P. nodosus**
9. Blade of floating leaves ovate to nearly heart-shaped at base; fruit barely keeled........................ 10
10. Floating leaves mostly 3–10 cm long; spikes 3–6 cm long .... **P. natans**
10. Floating leaves 2–5 cm long; spikes 1–3 cm long....... .... **P. oakesianus**

Potamogeton alpinus Balbis
REDDISH PONDWEED native
IR | **WUP** | **CUP** | **EUP** OBL CC 10

Plants red-tinged. Stems round in section, unbranched or sometimes branched above, to 1 m long and 1–2 mm wide. Underwater leaves linear lance-shaped, 4–20 cm long and 5–15 mm wide, 7–9-veined, usually rounded at tip, narrowed to a stalkless base. Floating leaves often absent, if present, thin, obovate, 4–6 cm long and 1–2 cm wide, 7- to many-veined, rounded at tip, tapered to a narrow base; stipules not joined to leaf base, membranous, 1–3 cm long and to 1.5 cm wide. Flowers in cylindric spikes, 1–3 cm long, with 5–9 whorls of flowers, on stalks 6–15 cm long and about as thick as stem. Achenes yellow-brown to olive, flattened, 3 mm long, the beak short, curved backward. July–Sept.

Shallow to deep water of lakes and streams.

Potamogeton amplifolius Tuckerman
LARGE-LEAF PONDWEED native
IR | **WUP** | **CUP** | **EUP** OBL CC 6

Stems round in section, usually unbranched, to 1 m or more long and 2–4 mm wide. Upper underwater leaves ovate, folded and sickle-shaped, 8–20 cm long and 2–7 cm wide, many-veined; lower underwater leaves lance-shaped, to 2 cm wide, often not folded, usually decayed by fruiting time, many-veined; petioles 1–5 cm long. Floating leaves usually present at flowering time, ovate 5–10 cm long and 3–6 cm wide, many-veined, rounded at tip or abruptly tapered to a sharp tip, rounded at base; petioles 5–15 cm long; stipules open and free of the petioles, 5–12 cm long, long-tapered to a sharp tip. Flowers in dense cylindric spikes, 3–6 cm long in fruit; stalks 6–20 cm long, widening near tip. Achenes green-brown to brown, 4–5 mm long, beak to 1 mm long. July–Aug.

Shallow to deep water of lakes and rivers.

Potamogeton confervoides Reichenb.
TUCKERMAN'S PONDWEED native
IR | **WUP** | **CUP** | **EUP** OBL CC 10

Stems slender, to 8 dm long, branched, the branches forking; from a long rhizome. Leaves many, all underwater, delicate, flat, bright green, 2–5 cm long and about 0.3 mm wide, tapered to a hairlike tip, l-veined; stipules short-lived, 1–5 cm long. Flowers in a short spike 5–10 mm long, at end of an erect stalk 5–20 cm long. Achenes 2–3 mm long, with a sharp keel. June–Aug.

Uncommon in shallow water of lakes, kettle hole ponds and peatlands; in western UP, reported from Keweenaw County only.

Unique among our pondweeds in its much-branched stems with linear leaves and the flower spike atop an elongate, leafless stalk.

Potamogeton crispus L.
CURLY PONDWEED *introduced (invasive)*
IR | **WUP** | **CUP** | **EUP** OBL

Stems compressed, with few branches, to 8 dm long and 1–2 mm wide. Leaves all underwater, oblong, 3–9 cm long and 5–10 mm wide, rounded at tip, slightly clasping at base, stalkless, 3–5-veined, margins wavy-crisped, finely toothed; stipules 4–10 mm long, slightly joined at base, early shredding. Flowers in dense cylindric spikes, 1–2 cm long, appearing bristly in fruit from long achene beaks; on stalks 2–6 cm long. Achenes brown, 2–3 mm long, with a beak 2–3 mm long. April–June.

Shallow to deep water of lakes and rivers; pollution-tolerant.

Potamogeton epihydrus Raf.
RIBBON-LEAF PONDWEED   native
IR | **WUP** | **CUP** | **EUP**   OBL CC 8

Stems slender, compressed, sparingly branched, to 2 m long and 1–2 mm wide. Underwater leaves linear, ribbonlike, 10–20 cm long and 3–8 mm wide, with a translucent strip on each side of midvein forming a band 1–3 mm wide, 5–13-veined, stalkless; stipules 1–3 cm long, not joined to leaf. Floating leaves usually present and numerous, opposite, oval to obovate, 3–8 cm long and 1–2 cm wide, mostly obtuse to bluntly abruptly short-awned at the tip, 11–25-veined, tapered to flattened petioles; stipules free, 1–3 cm long. Flowers in dense, cylindric spikes 2–3 cm long, on stalks 2–6 cm long and about as thick as stem. Achenes olive to brown, 2–3 mm long; beak tiny. July–Sept.

Water to 2 m deep in lakes, ponds and rivers.

Potamogeton foliosus Raf.
LEAFY PONDWEED   native
IR | **WUP** | **CUP** | **EUP**   OBL CC 4

Stems compressed, much-branched, to 8 dm long and 1 mm wide. Leaves all underwater, linear, 1–8 cm long and 1–2 mm wide, 1–3-veined, stalkless; stipules free, 0.5–2 cm long, glands usually absent at base of stipules. Flowers in rounded to short-cylindric spikes, 2–7 mm long, with 1–2 whorls of flowers, on stalks 5–12 mm long, widened at tip. Achenes green-brown, 1.5–3 mm long, winged, the beak to 0.5 mm long. June–Aug.

Shallow to deep water of lakes, ponds, rivers and streams.

Potamogeton friesii Rupr.
FLAT-STALK PONDWEED   native
IR | **WUP** | **CUP** | **EUP**   OBL CC 6

Stems compressed, branched, 1–1.5 m long and to 1 mm wide. Leaves all underwater, linear, 3–7 cm long and 1.5–3 mm wide, tip rounded with a short slender point, tapered to the base, 5–7-veined, stalkless, margins flat or becoming rolled under; stipules free, 5–20 mm long, fibrous, often shredding above, 2 glands present at base of stipule. Flowers in cylindric spikes, 8–16 mm long, with 2–5 whorls of flowers, on stalks 1.5–6 cm long. Achenes olive-green to brown, 2–3 mm long, beak flat, short. June–Aug.

Shallow to deep water of lakes, ponds, rivers and streams.

Potamogeton gramineus L.
GRASSY PONDWEED   native
IR | **WUP** | **CUP** | **EUP**   OBL CC 5

Stems slender, slightly compressed, much-branched, to 8 dm long and 1 mm wide. Underwater leaves variable, linear to lance-shaped or oblong lance-shaped, 3–9 cm long and 3–12 mm wide, 3–7-veined, tapered to a stalkless base. Floating leaves usually present, oval, 2–6 cm long and 1–3 cm wide, 11–19-veined, rounded at base; petioles 2–10 cm long, shorter to longer than blade; stipules free, persistent, 1–4 cm long. Flowers in dense, cylindric

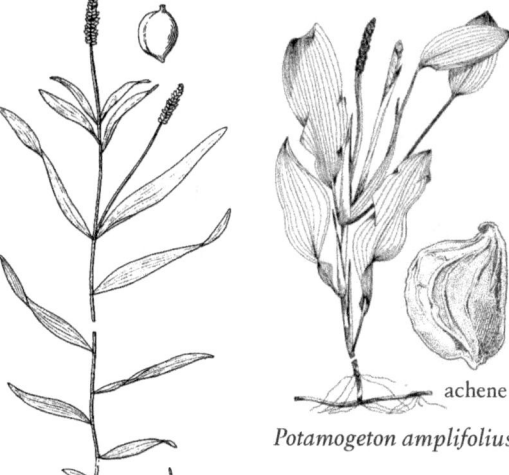

Potamogeton alpinus

Potamogeton amplifolius

Potamogeton crispus

Potamogeton foliosus

spikes, 1.5–4 cm long, the stalks thicker than stem, 2–10 cm long. Achenes dull green, 2–3 mm long. June–Aug.

Shallow to deep water of lakes and ponds.

Potamogeton hillii Morong
HILL'S PONDWEED   (THR) *native*
IR | WUP | CUP | **EUP**   OBL CC 9

Stems slender, slightly compressed, much-branched, to 1 m long; rhizomes ± absent. Leaves all underwater, linear, 3–7 cm long and 1–2 mm wide, 3-veined, the lateral veins nearer margins than midvein; stipules white or cream-colored, free, 1–2 cm long, becoming fibrous. Flowers in rounded spikes, 4–8 mm long, with 1 (sometimes 2) whorls of flowers, on stalks 5–15 mm long. Achenes flattened, 2–4 mm long, the beak 0.5 mm long.

Shallow water of ponds and streams, often where calcium-rich; Mackinac County.

Potamogeton illinoensis Morong
ILLINOIS PONDWEED   *native*
IR | **WUP** | **CUP** | **EUP**   OBL CC 5

Stems nearly round in section, usually branched, to 2 m long and 2–5 mm wide. Underwater leaves lance-shaped to obovate, 6–20 cm long, 2–4 cm wide, 9–17-veined, tapered to a broad, flat petiole, 2–4 cm long; stipules free, persistent, 3–8 cm long. Floating leaves sometimes absent, opposite, lance-shaped to oval, 5–14 cm long and 2–6 cm wide, 13- to many-veined, often short-awned from the rounded tip, rounded to wedge-shaped at base; petioles 3–10 cm long, shorter than blades. Flowers in dense cylin-dric spikes, 2–6 cm long, on stalks 4–20 cm long, usually wider than stem. Achenes olive-green, 3–4 mm long, the beak short, blunt. July–Sept.

Shallow to deep water of lakes and rivers.

Potamogeton natans L.
FLOATING PONDWEED   *native*
IR | **WUP** | **CUP** | **EUP**   OBL CC 5

Stems slightly compressed, usually unbranched, 0.5–2 m long and 1–2 mm wide. Underwater leaves reduced to linear, bladeless, expanded petioles (phyllodes), these often absent by flowering time, 10–30 cm long and 1–2 mm wide. Floating leaves ovate to oval, 4–10 cm long and 2–5 cm wide, usually tipped with a short point, rounded to heart-shaped at base, many-veined; petioles usually much longer than blades, the blade often angled at juncture with petiole; stipules free, 4–10 cm long, persistent or shredding with age. Flowers in dense cylindric spikes, 2–5 cm long, stalks thicker than the stem, 6–14 cm long. Achenes green-brown to brown, 3–5 mm long, with a loose, shiny covering, the beak short. June–Aug.

Usually shallow water (to 2 m deep) of ponds, lakes, rivers and peatlands.

Potamogeton nodosus Poir.
LONG-LEAF PONDWEED   *native*
IR | **WUP** | CUP | EUP   OBL CC 6

Stems round in section, branched, to 2 m long and 1–2 mm wide. Underwater leaves commonly decayed by fruiting time, lance-shaped to linear, translucent, 10–30 cm long and 1–3 cm wide, 7–15-veined, gradually tapered to a petiole 4–10 cm long. Floating leaves oval, thin, 5–12 cm long and 1–5 cm wide, ta-

Potamogeton illinoensis

Potamogeton gramineus

Potamogeton natans

Potamogeton nodosus

pered at both ends, many-veined; petioles somewhat winged, 5–20 cm long and 2–3 mm wide, usually longer than blades; stipules free, those of underwater leaves often absent by flowering time, those of floating leaves persistent, 3–10 cm long. Flowers in dense cylindric spikes, 2–6 cm long, on stalks 3–15 cm long and thicker than stem. Achenes red-brown to brown, 3–4 mm long, the beak short. July–Aug.

Shallow water to 2 m deep of rivers and lakes; Houghton County.

Potamogeton oakesianus J.W. Robbins
OAKES' PONDWEED native
IR | **WUP** | **CUP** | **EUP** OBL CC 10

Stems slender, often much-branched, to 1 m long. Underwater leaves bladeless, petiolelike, 0.5–1 mm wide, often persistent. Floating leaves oval, 3–6 cm long and 1–2 cm wide, rounded at base, 12- to many-veined; petioles 5–15 cm long; stipules free, 2.5–4 cm long. Flowers in cylindric spikes, 1.5–3 cm long, on stalks 3–8 cm long and wider than stem. Achenes 2–4 mm long, with a tight, dull covering, the beak flat.

Ponds and streams, peatland pools.

Similar to floating pondweed (*P. natans*) but plants smaller and the fruit more or less smooth on sides (vs. depressed in *P. natans*).

Potamogeton obtusifolius Mert. & Koch
BLUNT-LEAF PONDWEED native
IR | **WUP** | **CUP** | **EUP** OBL CC 10

Stems slender, compressed, much-branched, to 1 m long; rhizomes more or less absent. Leaves all underwater, linear, stalkless, often red-tinged, 3–10 cm long and 1–4 mm wide, rounded at tip, the midvein broad, base usually with pair of translucent glands; stipules free, white, 1–2 cm long. Flowers in thick cylindric spikes, 8–14 mm long, on slender, upright stalks 1–3 cm long. Achenes 2–3 mm long, the beak rounded, 0.5 mm long.

Lakes, ponds, streams, peatland pools.

Potamogeton perfoliatus L.
REDHEAD-GRASS native
IR | **WUP** | **CUP** | **EUP** OBL CC 6

Stems slender, to 2.5 m long, often much-branched. Leaves all underwater, ovate to nearly round or sometimes lance-shaped, 1–7 cm long and 5–30 mm wide, tip often very finely toothed, base cordate and clasping stem, stalkless; stipules free, soon decaying. Flowers on underwater cylindric spikes, 1–5 cm long, on upright stalks 1–7 cm long and about as wide as stem. Achenes 2–3 mm long, the beak short, curved.

Lakes and streams; Houghton, Luce, and Marquette counties.

Potamogeton praelongus Wulfen
WHITE-STEM PONDWEED native
IR | **WUP** | **CUP** | **EUP** OBL CC 8

Stems white-tinged, compressed, branched, to 2–3 m long and 2–4 mm wide, the shorter internodes often zigzagged. Leaves all underwater, lance-shaped, 10–30 cm long and 1–4 cm wide, with 3–5 main veins, rounded and hoodlike at tip, base more or less heart-shaped and clasping stem, stalkless, margins entire and gently wavy; stipules free, white, 1–3 cm long, fibrous at tip. Flowers in dense, cylindric spikes 2–5 cm long; stalks erect, 1–4 dm long, as wide as stem. Achenes green-brown, swollen, 4–5 mm long, the beak rounded, 0.5 mm long. June–Aug.

Shallow to deep water of lakes (including Great Lakes), streams.

Potamogeton pusillus L.
SLENDER PONDWEED native
IR | **WUP** | **CUP** | **EUP** OBL CC 4

Stems very slender, round in section, usually freely branched, 2–10 dm long and about 0.5 mm wide; rhizomes more or less absent. Leaves all underwater,

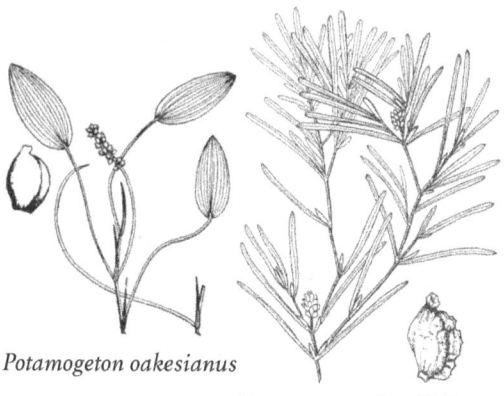

Potamogeton oakesianus

Potamogeton obtusifolius

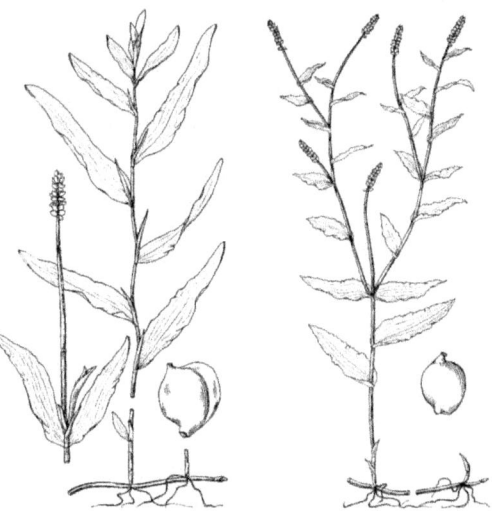

Potamogeton praelongus   *Potamogeton richardsonii*

516 POTAMOGETONACEAE | Pondweed Family   *monocots*

linear, 1–7 cm long and 0.5–2 mm wide, tapered to a stalkless base, the midvein broad; stipules free, boat-shaped, brown-green, 4–10 mm long and 2x width of leaf base, soon decaying, glands sometimes present at stipule base. Flowers in short-cylindric spikes 2–10 mm long, the flowers in 1–3 whorls, on slender, upright stalks 1–5 cm long. Achenes green to brown, 1–2 mm long, the beak flat. June–Aug.

Shallow water (to 2 m deep) of lakes and ponds, less often in streams.

Includes plants treated as *Potamogeton pusillus* subsp. *tenuissimus*, which are sometimes separated as *Potamogeton berchtoldii* Fieber.

Potamogeton richardsonii (Benn.) Rydb.
RED-HEAD PONDWEED   native
IR | WUP | CUP | EUP   OBL CC 5

Stems brown to yellow-green, round in section, sparingly to freely branched, mostly 3–10 dm long and 1–2.5 mm wide, the shorter internodes rarely zigzagged. Leaves all underwater, lance-shaped, 5–12 cm long and 1–2.5 cm wide, with 13 or more prominent veins, base heart-shaped and clasping stem, stalkless, margins entire and gently wavy; stipules free, 1–2 cm long, soon shredding into white fibers. Flowers in dense cylindric spikes 1.5–4 cm long, on stalks 2–20 cm long, the stalks strongly curved when in fruit. Achenes green to brown, 2–4 mm long, the beak short. July–Aug.

Shallow to deep water of lakes (including Great Lakes), streams.

Similar to *P. perfoliatus* but that species uncommon in the UP, its leaves narrower and often longer, and the stipules persisting as fibers, vs. soon decayed in *P. perfoliatus*.

Potamogeton robbinsii Oakes
FERN PONDWEED   native
IR | WUP | CUP | EUP   OBL CC 10

Stems few-branched below, much-branched above, to 1 m long; rhizomes not tuberous. Leaves all underwater, crowded in 2 ranks, linear, 4–10 cm long and 3–7 mm wide, tapered to a pointed tip, abruptly narrowed at base, with rounded auricles where joined with stipule, midvein pronounced, margins pale; stipules joined to leaf for 5–15 mm, soon decaying into fibers. Flowers on underwater, cylindric spikes 1–2 cm long, with 3–5 separated whorls of flowers, the inflorescence often branched into 5–20 stalks, 2–5 cm long, at ends of stems. Achenes rarely produced, 3–5 mm long, the beak thick, somewhat curved; reproduction most commonly by stem fragments which root from the nodes. July–Aug.

Shallow to deep water of lakes, ponds and streams.

Potamogeton spirillus Tuckerman
SPIRAL PONDWEED   native
IR | WUP | CUP | EUP   OBL CC 8

Stems compressed, to 1 m long, branched, the branches short and often curved. Underwater leaves 1–8 cm long and 0.5–2 mm wide, rounded at tip, stalkless; stipules joined for most of length. Floating leaves, if present, 1–4 cm long and 5–12 mm wide, 5–13-veined, the veins sunken on underside of blade, petioles 2–4 cm long; stipules free. Flowers in 2 types of spikes, the underwater spikes round, with 1–8 fruit, more or less stalkless in the leaf axils; emersed spikes longer, cylindric, to 8–12 mm long, on stalks from leaf axils. Achenes 1–2.5 mm long, flattened, winged, spiraled on surface, the beak absent.

Shallow water of lakes and ponds.

Potamogeton strictifolius Benn.
STRAIGHT-LEAF PONDWEED   native
IR | WUP | CUP | EUP   OBL CC 6

Stems slender, slightly compressed, unbranched or branched above, to 1 m long and 0.5 mm wide. Leaves all underwater, linear, upright, 1–6 cm long and 0.5–2 mm wide, 3–5-veined, the veins prominent on underside, tapered to stalkless base, margins often rolled under; stipules free, white, shredding at tip, 5–20 mm long; 2 glands present at base of stipules. Flowers in cylindric spikes 6–15 mm long, with 3–5 whorls of flowers, on stalks 1–5 cm long. Achenes green-brown, 2 mm long, the beak broad, rounded. June–Aug.

Shallow to deep water of lakes and rivers.

Potamogeton vaseyi J. W. Robbins.
VASEY'S PONDWEED   (THR) native
IR | WUP | CUP | EUP   OBL CC 10

Stems threadlike, 2–10 dm long, much-branched, the upper branches short. Underwater leaves transparent, linear, 2–6 cm long and to 1 mm wide, tapered to a sharp tip, 1-veined or rarely with 2 weak lateral nerves, stalkless; stipules free, linear, white, 1–2 cm long, sometimes with 2 glands at base. Floating leaves on flowering plants only, opposite, obovate, leathery, 8–15 mm long and 4–7 mm wide, 5–9-veined, the veins sunken on underside, petiole about as long as blade. Flowers in cylindric spikes 3–8 mm long, with 1–4 whorls of flowers, on stems 1–3 cm long. Achenes 2–3 mm long, the beak short.

Shallow to deep water of ponds and small lakes.

Potamogeton zosteriformis Fern.
FLAT-STEM PONDWEED   native
IR | WUP | CUP | EUP   OBL CC 5

Stems strongly flattened, sometimes winged, freely branched, to 1 m long and 1–3 mm wide; rhizomes more or less absent. Leaves all underwater, linear, 5–20 cm long and 3–5 mm wide, 15- to many-veined, tapered to a tip, or sometimes with a short, sharp point, slightly narrowed to the stalkless base; stipules free, white, shredding with age, 1–4 cm long. Flowers in cylindric spikes, 1–2.5 cm long, with 7–11 whorls of flowers, on curved stalks 2–6 cm long. Achenes dark green to brown, 4–5 mm long, the beak short and blunt. July–Aug.

Shallow to deep water of lakes (including Great Lakes) and streams.

The flat stem and many-nerved parallel-sided leaves mostly 3–5 mm wide and acute to mucronate at the tip, are distinctive.

Ruppia *Ditch-Grass*

Ruppia cirrhosa (Petag.) Grande
SPIRAL DITCH-GRASS (THR) native
IR | WUP | CUP | **EUP** OBL CC 10

Aquatic perennial herb. Stems slender, round in section, white-tinged, wavy, to 6 dm long, branching at base and with short branches above, the internodes often zigzagged. Leaves simple, alternate or opposite, stalkless, threadlike, mostly 5–25 cm long and 0.5 mm wide, 1-veined, with a sheathing stipule at base. Flowers very small, perfect, in small, 2-flowered spikes from leaf axils, the spikes enclosed by the leaf sheath at flowering time, the flower stalks elongating and usually coiling as fruit mature; sepals and petals absent; stamens 2; pistils typically 4 (varying from 2–8), raised on a slender stalk in fruit and becoming umbel-like. Fruit an olive-green to black, ovate drupelet, 1.5–3 mm long. July–Aug.

In Michigan, known only from Manistique Lake, straddling Luce and Mackinac counties, where collected in water to more than 2 m deep, and often common on the shoreline following storms.

Stuckenia *False Pondweed*

Stuckenia is a small genus of perennial aquatic herbs, now segregated from *Potamogeton*. In Stuckenia, the stipules are joined to the blade for 2/3 to nearly the entire length of the stipule; in *Potamogeton*, the stipules in most species are free, or if adnate, joined for well less than half the length of the stipule. Also, submersed leaves of *Potamogeton* are translucent, flat, and without grooves or channels; in *Stuckenia*, submersed leaves are opaque, channeled, and turgid.

1 Leaves gradually tapered to tip; rhizomes tuber-bearing; stigmas raised on a tiny style............ **S. pectinata**
1 Leaves rounded, blunt-tipped or tipped with a short, sharp point, stigmas inconspicuous, broad and not raised .... **S. filiformis**

Stuckenia filiformis (Pers) Boerner
THREADLEAF FALSE PONDWEED native
IR | WUP | CUP | **EUP** OBL CC 7
Potamogeton filiformis Pers.

Stems more or less round in section, branched from

Potamogeton robbinsii

Potamogeton strictifolius

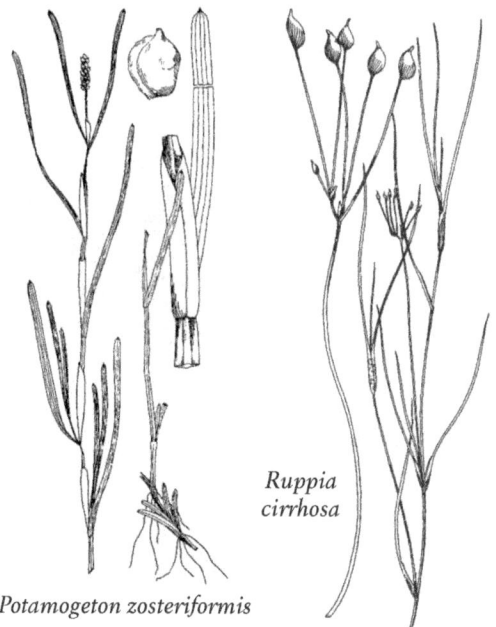

Potamogeton zosteriformis

Ruppia cirrhosa

518 SCHEUCHZERIACEAE | Scheuchzeria Family

base, mostly unbranched above, 1–5 dm or more long and 1 mm wide, from a long, tuber-bearing rhizome. Leaves all underwater, narrowly linear, 5–10 cm long and 0.2–2 mm wide, 1-veined; stipules 1–3 cm long, joined to base of leaf blade, forming a tight sheath around stem. Flowers in underwater spikes, 1–5 cm long, with 2–5 separated whorls of flowers, on slender stalks 2–12 cm long. Achenes olive-green, 2–3 mm long, the beak flat, tiny. July–Aug.

Mostly shallow water (to 1 m) in lakes and rivers.

Stuckenia pectinata (L.) Boerner
SAGO FALSE PONDWEED   *native*
IR | **WUP** | **CUP** | **EUP**   OBL CC 3
Potamogeton pectinatus L.
Stems slender, round in section, 3–10 dm long and 1–2 mm wide much-branched and forking above, fewer branched near base, from rhizomes tipped with a white tuber. Leaves all underwater, threadlike to narrowly linear, 3–12 cm long and 0.5–1.5 mm wide, stalkless; stipules joined to base of blade for 1–3 cm, forming a sheath around stem. Flowers on underwater, cylindric spikes 1–5 cm long, with 2–5 whorls of flowers, on lax, threadlike stalks to 15 cm long. Achenes yellow-brown, 3–4 mm long, the beak to 0.5 mm long; the large fruit an important waterfowl food. June–Sept.

Shallow to deep water of lakes, ponds and streams; tolerant of brackish water.

Zannichellia Horned-Pondweed
Zannichellia palustris L.
HORNED-PONDWEED   *native*
IR | WUP | **CUP** | **EUP**   OBL CC 6
Perennial aquatic herb, with creeping rhizomes, and often forming extensive underwater mats. Stems slender and delicate, wavy, 0.5–5 dm long, branched from base. Leaves simple, opposite (or upper leaves appearing whorled), threadlike, 2–8 cm long and 0.5 mm wide, stalkless; stipules membranous and soon deciduous. Flowers small, produced underwater, either staminate or pistillate, separate on plant but from same leaf axil, with 1 staminate flower and usually 4 (varying from 1–5) pistillate flowers at each node, surrounded by a membranous, spathelike bract; petals and sepals absent; staminate flower a single anther. Fruit a brown to red-brown, crescent-shaped nutlet, gently wavy on margins, 2–3 mm long, tipped by a beak 1–2 mm long; the fruit mostly 2–6 per node. June–Aug.

Submerged in fresh or brackish water of streams, muddy lake and pond bottoms, marshes and ditches; Mackinac and Menominee counties.

Scheuchzeriaceae
SCHEUCHZERIA FAMILY

Scheuchzeria Pod-Grass
Scheuchzeria palustris L.
POD-GRASS   *native*
IR | **WUP** | **CUP** | **EUP**   OBL CC 10
Perennial rushlike herb, from creeping rhizomes. Stems 1 to several, 1–4 dm long, remains of old leaves often persistent at base of plant. Leaves alternate, several from base and 1–3 along stem, 1–3 dm long and 1–3 mm wide, the stem leaves smaller; lower part of blade half-round in section, with an

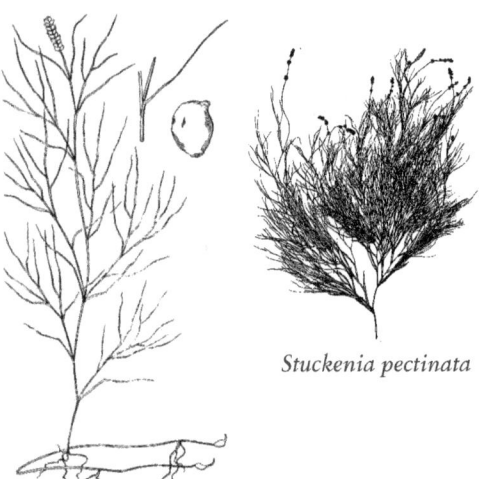

Stuckenia pectinata

Zannichellia palustris

Scheuchzeria palustris

expanded sheath at base, upper portion of blade flat, with a small pore at leaf tip. Flowers perfect, regular, green-white, in a several-flowered raceme 3–10 cm long, the flowers on stalks 1–2.5 cm long; tepals 6, in 2 series, ovate, 2–3 mm long; stamens 6. Fruit a group of 3 (rarely to 6) spreading follicles, 5–10 mm long, each with 1–2 seeds; seeds brown-black, 4–5 mm long. May–June.

Wet sphagnum peatlands.

Smilacaceae
GREENBRIER FAMILY

Smilax *Greenbrier*

Perennial herbs (with annual stems), or vining shrubs, climbing by tendrils terminating the stipules, with wide, longitudinally nerved, net-veined, alternate leaves and axillary peduncled umbels of small yellow or greenish yellow flowers. Flowers dioecious, the staminate often the larger. Perianth segments alike, spreading. Stamens in the staminate flower 6; filaments slender or flattened; anthers oblong. Stamens of the pistillate flower reduced to 6 filiform staminodes. Ovary 3-celled, with 1 or 2 ovules in each cell; style none or very short; stigmas solitary or 3, oblong, recurved. Fruit a 1–6-seeded berry. Leaves of all species vary greatly in size and shape.

1. Stems woody and prickly; leaves glabrous beneath (sometimes roughened on main veins) ..... **S. hispida**
1. Stems herbaceous, never prickly; leaves finely puberulent, at least on the veins beneath. 2
2. Stem of mature plants more than 1 m long, the main stem or elongate branches climbing (or resting on other objects for support); plant almost always branched, with total of more than 25 leaves; tendrils conspicuously curled, present at most nodes, including those from which peduncles arise; peduncles longer than petioles (sometimes several times as long), all or most arising from axils of foliage leaves; flowers (at least on main stem) more than 25 in an umbel (but not all develop into fruit) .... **S. lasioneura**
2. Stem less than 1 m tall, stiffly erect much of its length; plant unbranched, with fewer than 25 leaves (in S. illinoensis rarely more); tendrils absent or at most poorly developed and limited to uppermost nodes (never at the lower nodes from which peduncles arise); peduncles longer or, more often, shorter than petioles, at least the lowest ones usually arising from scale-like bracts on the stem below the foliage leaves; flowers more or fewer than 25 in an umbel. 3
3. Pistillate (and usually also staminate) flowers fewer than 25 in an umbel; leaves fewer than 20 (usually 7–9) on a plant; stems under 50 cm tall; peduncles usually shorter than the petioles or slightly longer; tendrils completely absent (rarely on upper 2–3 nodes) .... **S. ecirrata**
3. Pistillate and staminate flowers usually more than 25 in an umbel and plants with one or more other exceptions to the above (i.e., leaves more than 20, stems over 50 cm tall, peduncles more than 2 cm longer than petioles, tendrils present on several upper nodes). **S. illinoensis**

Smilax ecirrata (Engelm.) S. Wats.
UPRIGHT CARRION-FLOWER native
IR | **WUP** | **CUP** | EUP CC 6

Perennial herb. Stems annual, erect, usually without tendrils, or producing a few tendrils from the upper leaves only, to 8 dm tall. Leaves narrowly to broadly ovate, truncate to cordate at base, convexly narrowed to a short cusp, pubescent beneath. Umbels 1–3, to 25-flowered; peduncles arising from the axils of lance-linear bracts along the lower leafless portion of the stem, or rarely also from the axil of the lowest leaf. Perianth green, the tepals 4–6 mm long. Berries purplish black, globose, ca. 10 mm wide, not glaucous. May–June.

Rich deciduous woods, floodplain forests.

Smilax hispida Muhl. ex Torr.
CHINAROOT native
IR | **WUP** | **CUP** | EUP FAC CC 5

Smilax tamnoides L.

Vine, from a short knotty rhizome. Stems often climbing high (to 7 m or more), usually conspicuously thorny and often densely so; branches nearly terete. Leaves thin, ovate to rotund; commonly 8–12 cm long and 6–10 cm wide at maturity; acute to rounded or cuspidate; at base rounded, truncate, or cordate; not thickened at the margin; minutely serrulate (at 10x), at least near base; 5–7-nerved, the reticulate veinlets not prominently raised. Umbels many, axillary to leaves, to 25-flowered; peduncle often drooping, 1.5–6.5 cm long; perianth green to bronze. Berries black, globose, 6–10 mm wide, not glaucous. May–June.

Moist woods and thickets.

Smilax illinoensis Mangaly
ILLINOIS GREENBRIER native
IR | **WUP** | **CUP** | EUP CC 4

Perennial herb. Stems annual, erect, unbranched, to 1 m long; prickles absent. Leaves narrowly ovate, pubescent and not glaucous on underside, base rounded to truncate, margins convex, tip acute to acuminate; petiole thin, equaling or longer than blade; tendrils few, short. Umbels 3–10, axillary to leaves and bracts, 10–50-flowered; tepals 3.5–4.5 mm long. Berries blue to black, globose. May–June.

Woods, thickets.

Smilax lasioneura Hook.
BLUE RIDGE CARRION-FLOWER native
IR | **WUP** | CUP | EUP CC 5

Smilax herbacea var. *lasioneura* (Hook.) A. DC. Perennial herb. Stems annual, climbing, often to 2 m tall and freely branched. Leaves ovate to rotund, at base cordate to rounded; acuminate to cuspidate or broadly rounded at tip; underside often somewhat glaucous; lateral margins always convex. Umbels many, axillary to leaves, to 35-flowered; peduncles arising from the axils of foliage leaves, flattened; perianth greenish. Berries bluish black to black, subglobose, 8–10 mm wide, glaucous. May–June.

Moist soil of open woods, roadsides, and thickets.

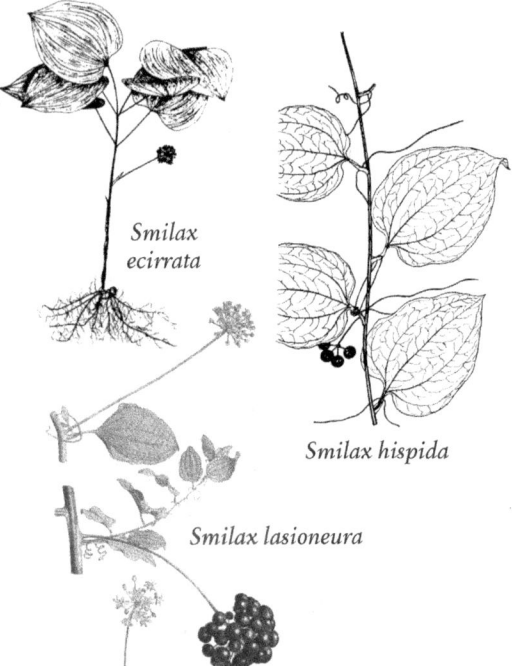

Typhaceae
CAT-TAIL FAMILY

Family now includes genus *Sparganium* from former family Sparganiaceae (discontinued under APG III).

1 Pistillate flowers in one to several spherical heads; perianth of greenish sepals; leaves strongly keeled (3-angled in cross-section) . . . . **Sparganium**
1 Pistillate flowers in an elongate densely flowered spike; perianth of white hairs; leaves flat-elliptic in cross-section . . . . **Typha**

Sparganium *Bur-reed*

Perennial sedgelike herbs, floating or emergent in shallow water, from rhizomes and forming colonies. Stems stout, usually erect, unbranched, round in section. Leaves long, broadly linear, sheathing stem at base. Flowers crowded in round heads, the heads with either staminate or pistillate flowers; staminate heads few to many, borne above pistillate heads in a unbranched or sparsely branched inflorescence; the pistillate heads 1 to several, from leaf axils or borne above axils on upper stem; sepals and petals reduced to chaffy, spatula-shaped scales, these appressed to the achenes in the mature pistillate heads; staminate flowers with mostly 3–5 stamens; pistillate flowers with a 1–2-chambered pistil, stigmas 1 or 2. Fruit a beaked, nutletlike achene, stalkless or short-stalked.

1 Plants large, about 1 m tall; leaves usually erect; stigmas 2; achenes broadly oblong pyramid-shaped . . . . **S. eurycarpum**
1 Plants smaller, leaves erect or floating; stigmas 1; achenes slender 2
2 Fruiting heads about 1 cm wide; staminate head 1 (often absent by fruiting time); achene beaks less than 1 mm long . . . . **S. natans**
2 Fruiting heads 1.5 cm or more wide; staminate heads 2 or more; achene beaks 2 mm or more long 3
3 Fruiting heads 1.5–2 cm wide; anthers and stigma less than 1 mm long; leaves mostly flat 4
3 Fruiting heads larger mostly 2–3 cm wide; anthers and stigma 1–4 mm long; leaves oftenkeeled 5
4 Staminate heads several, separate from the pistillate heads; achene not shiny. . . . . **S. fluctuans**
4 Staminate heads usually 1 (sometimes 2) and near upper pistillate head; achene shiny . . . . **S. glomeratum**
5 Fruiting heads or branches all from leaf axils . . . . **S. americanum**
5 At least some fruiting heads or branches borne above leaf axils 6
6 Leaves floating; achene beak 1–3 mm long . . . . **S. angustifolium**
6 Leaves usually stiffly erect and emersed; achene beak 3–5 mm long . . . . **S. emersum**

Sparganium americanum Nutt.
AMERICAN BUR-REED native
IR | **WUP** | CUP | EUP OBL CC 6

Perennial herb. Stems stout, erect, mostly unbranched, 3–10 dm long. Leaves linear, flat to somewhat keeled, to 1 m long and 4–12 mm wide; leaflike bracts on upper stem shorter than leaves, widened at base. Inflorescence usually unbranched, or with a few, straight branches; pistillate heads sessile, 2–4 on main stem, sometimes with 1–3 on branches, 2 cm wide when mature; scales widest at tip; staminate heads 3–10 on main stem, sometimes with 1–5 on branches. Achenes widest at middle, tapered to both ends, dull brown, 3–5 mm long, the beak straight, 2–4 mm long. July–Aug.

Marshes, shallow water, streambanks.

Sparganium angustifolium Michx.
NARROW-LEAF BUR-REED *native*
IR | WUP | CUP | EUP OBL CC 10

Sparganium acaule (Beeby) Rydb., *Sparganium chlorocarpum* var. *acaule* (Beeby) Fern., *Sparganium emersum* var. *angustifolium* (Michx.) Taylor & MacBryde, *Sparganium multipedunculatum* (Morong) Rydb.

Perennial herb. Stems long and usually floating. Leaves floating, mostly 2–3 mm wide, often wider at base. Inflorescence unbranched; pistillate heads 1–3, shiny, about 2 cm wide, the lowest stalked, the upper pistillate heads sessile; scales spatula-shaped, ragged at tip; staminate heads 2–6, close together above pistillate heads. Achenes spindle-shaped, 5–7 mm long, dull brown except at red-brown base, abruptly contracted to a beak 1 mm long. July–Aug.

Lakes, ponds and shores.

Sparganium emersum Rehmann
NARROW-LEAF BUR-REED *native*
IR | WUP | CUP | EUP OBL CC 6

Sparganium chlorocarpum Rydb.
Sparganium simplex Huds.

Perennial herb. Stems usually erect, sometimes lax and trailing in water, 2–6 dm long. Leaves linear, yellow-green, flat to keeled, 3–7 dm long and 3–6 mm wide, usually longer than stems; bracts leaflike, erect, barely widened at base. Inflorescence unbranched, 1–2 dm long; pistillate heads 1–4, sessile or lowest head often stalked, at least 1 head on stem above leaf axils, 1.5–2.5 cm wide when mature; scales spatula-shaped, widest at tip; staminate heads usually 2–5, 1.5–2 cm wide at flowering time. Achenes widest at middle, tapered to both ends, 4–5 mm long, shiny olive-green, the beak 3–5 mm long. June–Aug.

Shallow water or mud of marshes, streams, ditches, open bogs, ponds.

Sparganium eurycarpum Engelm.
BROAD-FRUIT BUR-REED *native*
IR | WUP | CUP | EUP OBL CC 5

Sparganium californicum Greene
Sparganium greenei Morong

Perennial herb. Stems stout, branched, 4–10 dm long. Leaves linear, bright green, keeled, 8–10 dm long and 5–12 mm wide; bracts leaflike, slightly widened at base. Inflorescence 1–3 dm long, branched from the bract axils; lower branches with 1 pistillate head and several staminate heads, main stem and upper branches with 6–10 staminate heads; pistillate heads 2–6, 1.5–2.5 cm wide in fruit, scales spatula-shaped; staminate heads numerous, 1–2 cm wide. Achenes oblong pyramid-shaped, 6–8 mm long, the

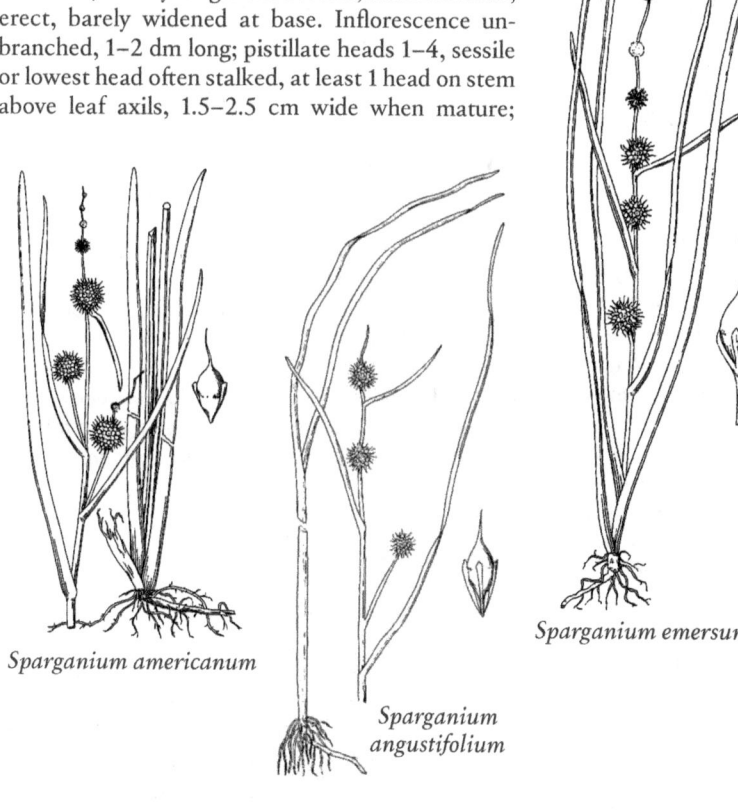

Sparganium americanum

Sparganium angustifolium

Sparganium emersum

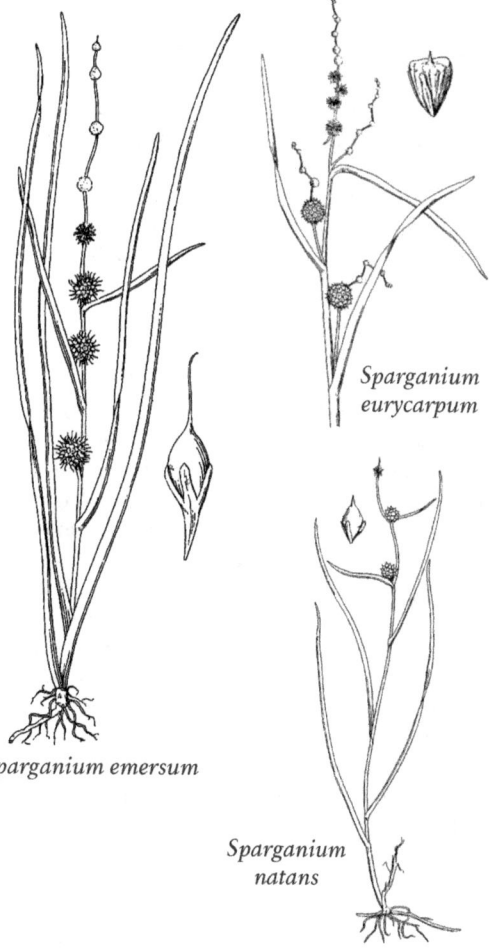

Sparganium eurycarpum

Sparganium natans

top flattened, 4–7 mm wide, brown to golden-brown, the beak 2–4 mm long. June–Aug.

Usually in shallow water of marshes, streams, ditches, ponds and lakes, often with cat-tails (*Typha*); our most common bur-reed.

Sparganium fluctuans (Morong) B.L. Robins.
FLOATING BUR-REED   native
IR | WUP | CUP | EUP   OBL CC 10

Perennial herb. Stems slender, floating, to 15 dm long. Leaves floating, linear, flat, translucent, 3–10 mm wide, underside with netlike veins; bracts leaflike, short, widened at base. Inflorescence usually branched, the main stem with 2–4 staminate heads, the branches with 1 pistillate head near base and 2–3 staminate heads above; pistillate heads 2–4, 1.5–2 cm wide when mature, scales oblong; staminate heads to 1 cm wide. Achenes obovate, 3–4 mm long, sometimes narrowed near middle, brown, the beak curved, 2–3 mm long.

In shallow water of ponds and lakes.

Sparganium natans L.
ARCTIC BUR-REED   native
IR | WUP | CUP | EUP   OBL CC 8

Sparganium minimum (Hartman) Wallr.
Perennial herb. Stems usually long and floating, sometimes shorter and upright, 1–3 dm or more long. Leaves linear, dark green, thin, flat, 2–6 mm wide; bracts leaflike, short, somewhat widened at base. Inflorescence unbranched; pistillate heads 2–3, from bract axils, sessile or the lowest sometimes short-stalked, 1 cm wide when mature; scales spatula-shaped, widest at tip; staminate heads usually 1 (rarely 2). Achenes broadly oval, 3–4 mm long, dull green-brown, the beak 1–2 mm long.

Shallow water, pond margins.

Typha *Cat-tail*

Large, familiar, reedlike perennials, from fleshy rhizomes and forming colonies. Stems erect, unbranched, round in section, sheathed for most of length by overlapping leaf sheaths. Leaves mostly near base of plant, alternate in 2 ranks, erect, linear, spongy. Flowers tiny, either staminate or pistillate, separate on same plant; petals and sepals reduced to bristles. Staminate flowers usually of 3–5 stamens, bristles absent or 1–3 or more. Pistillate flowers intermixed with some sterile flowers; pistil 1, raised on a short stalk (gynophore), with numerous bristles near base, the bristles longer than pistil; small bracts (bractlets) also sometimes present, these intermixed with the bristles, slender but with a widened brown tip. Heads with staminate flowers above pistillate in a single, dense, cylindric spike, the staminate and pistillate portions of the spike unalike, contiguous in broad-leaf cat-tail (*T. latifolia*) or separated in narrow-leaf cat-tail (*T. angustifolia*); the mature spike brown and fuzzy in appearance due to the crowded stigmas and gynophore bristles. Fruit a yellow-brown achene, 1–2 mm long, the style persistent, long and slender with an expanded stigma.

A hybrid between *T. angustifolia* and *T. latifolia* is termed *Typha* × *glauca* Godr. Usually larger than either parent, staminate and pistillate portions of hybrid plants are usually separated by a space to 4 cm long. The staminate portion of the spike is light brown, 0.5–2 dm long and about 1 cm wide at flowering time; the pistillate portion is dark brown, 10–20 cm long and 1–2 cm wide. Since *Typha* × *glauca* is sterile, reproduction is vegetative by rhizomes. The hybrid can occur wherever populations of *T. angustifolia* and *T. latifolia* overlap, but to date, is only reported from Schoolcraft County.

1   Staminate and pistillate portions of spike usually separated; leaves to 1 cm wide; stigmas long and slender, pale brown . . . . **T. angustifolia**
1   Staminate and pistillate portions of spike usually contiguous, not separated; leaves mostly 1–2 mm wide; stigmas broad and flattened, dark brown . . . **T. latifolia**

Typha angustifolia L.
NARROW-LEAF CAT-TAIL   introduced
IR | WUP | CUP | EUP   OBL

Perennial emergent herb. Stems erect, 1–2 m long. Leaves upright, flat, 4–10 mm wide. Flowers either staminate or pistillate, on separate portions of the spike, separated by an interval of 2–10 cm; staminate portion 7–20 cm long and 7–15 mm wide, staminate bractlets brown; pistillate portion of spike dark brown, 10–20 cm long and 1–2 cm wide; each flower with 1 bristlelike bractlet, these flat and brown at the widened tip, gynophore hairs brown-tinged at tips; stigmas pale brown, linear, 1 mm long. Fruit 5–7 mm long, subtended by many fine hairs, the hairs slightly widened and brown at tip. June.

Marshes, lakeshores, streambanks, roadside ditches, pond margins, usually in shallow water; more tolerant of brackish conditions than *Typha latifolia*.

Typha latifolia L.
BROAD-LEAF CAT-TAIL   native
IR | WUP | CUP | EUP   OBL CC 1

Perennial emergent herb. Stems erect, 1–2.5 m long. Leaves upright, mostly 1–2 cm wide. Flowers either staminate or pistillate, the staminate and pistillate portions of spike normally contiguous, rarely separated by 3–4 mm; staminate portion 5–15 cm long and 1.5–2 cm wide at flowering time, staminate bractlets white; pistillate portion of spike dark brown, 10–15 cm long and 2–3 cm wide when mature, pistillate bractlets absent, gynophore hairs white; stigma lance-shaped, becoming dark brown, less than 1 mm long. Fruit 1 cm long, with many white, linear hairs from base. June.

Marshes, lakeshores, streambanks, ditches, pond margins, usually in shallow water; less tolerant of brackish conditions than *Typha angustifolia*.

Xyridaceae
YELLOW-EYED-GRASS FAMILY

Xyris Yellow-Eyed-Grass
Perennial rushlike herbs. Stems erect, leafless, straight or sometimes ridged. Leaves all from base of plant, upright to spreading, linear, often twisted, usually dark green. Flowers small, perfect, yellow, from base of tightly overlapping bracts or scales, in rounded or cylindric heads at ends of stems; sepals 3, petals 3; stamens 3; style 3-parted. Fruit an oblong, 3-chambered capsule.

1	Plants swollen and hard at base . . . . **X. torta**
1	Plants flattened and soft at base 2
2	Leaves 5 mm or more wide; upper flower scales with a green spot, 2–3 mm long, near center . . . . **X. difformis**
2	Leaves to 2 mm wide; flower scales without central green spot . . . . **X. montana**

Xyris difformis Chapman
BOG YELLOW-EYED-GRASS   *native*
IR | WUP | **CUP** | **EUP**   OBL CC 8

Perennial herb. Stems leafless, 1.5–6 dm long, lower stem round in section and twisted, upper stem compressed and straighter, with 2 prominent ridges. Leaves linear, not twisted, 1–5 dm long and 5–15 mm wide, widened to a soft base. Flowers yellow, in round to ovate spikes 0.5–1 cm long; scales ovate, entire; lateral sepals shorter than scales, the margins finely fringed from middle to tip; petals obovate, 4 mm long. Seeds 0.5 mm long. July–Aug.

Sandy or peaty lakeshores, sphagnum peatlands, floating sedge mats; Alger and Luce counties.

Xyris montana Ries
NORTHERN YELLOW-EYED-GRASS   *native*
IR | WUP | **CUP** | **EUP**   OBL CC 10

Densely tufted perennial herb. Stems leafless, 0.5–3 dm long, round in section, straight or lower part of stem slightly twisted. Leaves narrowly linear, flat or only slightly twisted, 5–20 cm long and 1–2 mm wide, rough, dark green, red-purple at base. Flowers yellow, in ovate spikes less than 1 cm long; scales obovate, finely fringed at tip; lateral sepals about as long as scales, linear, margins entire or finely hairy near tip. Seeds 1 mm long.

Wet sandy shores, pools in sphagnum peatlands.

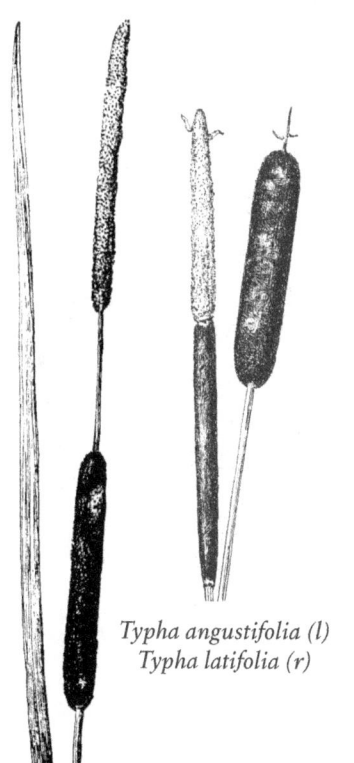

Typha angustifolia (l)
Typha latifolia (r)

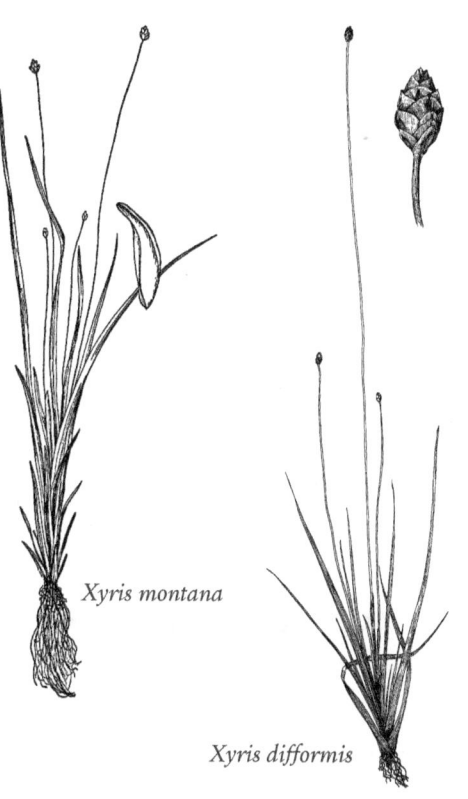

Xyris montana

Xyris difformis

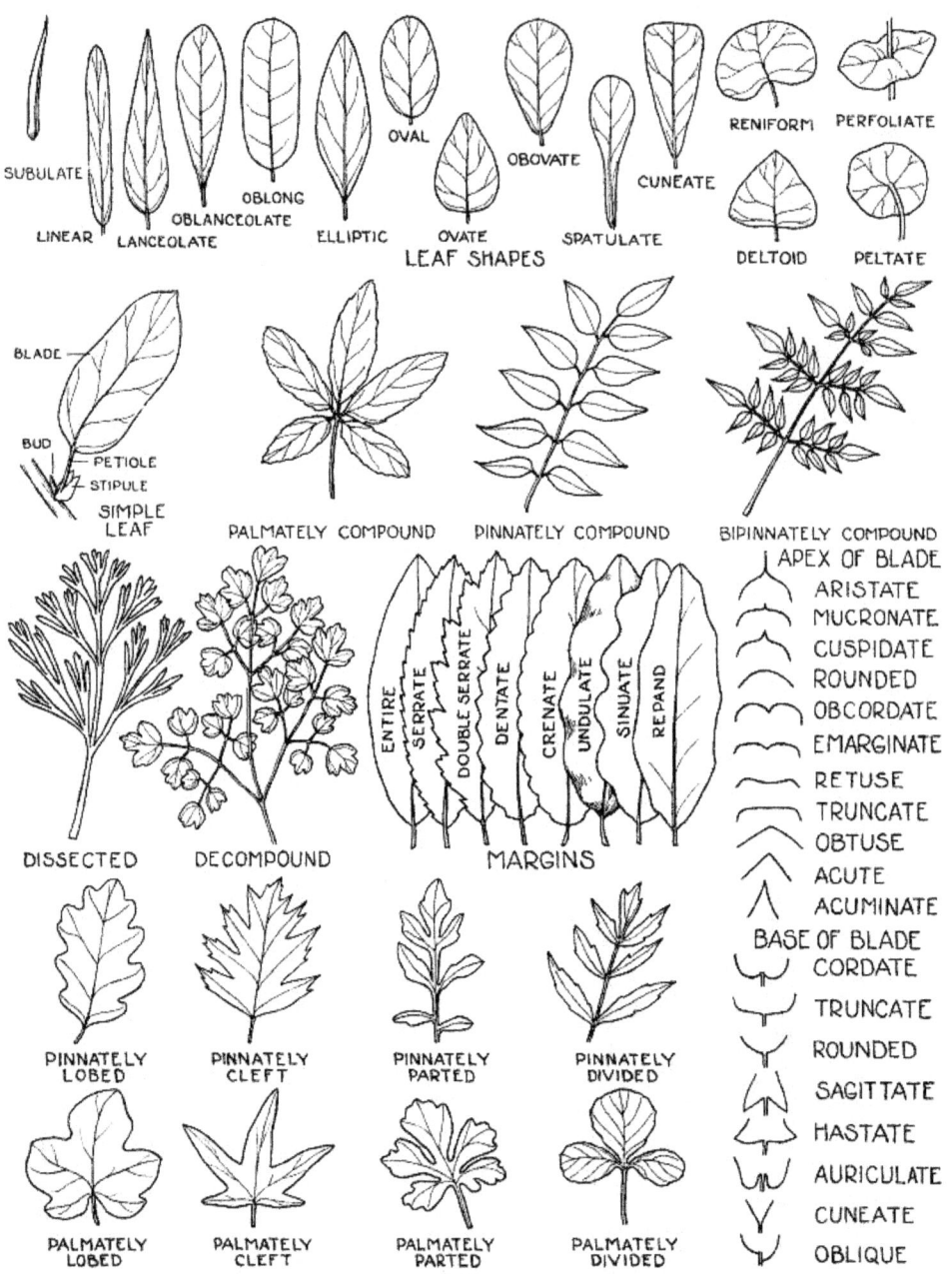

Leaf characteristics. (From H. L. Mason, *A Flora of the Marshes of California*. University of California Press, 1957.)

FAMILY GROUP KEYS: Summary

Ferns and Fern Relatives

- Aquatic ferns; plants floating, or rooted and submersed .... **Isoetes**

- Fern relatives (horsetails, quillworts, clubmosses); leaves unlobed, awl-shaped, scalelike, or grasslike, and not "fern-like" .... **GROUP 1**

- Small ferns, growing on rock or on rock in thin soil .... **GROUP 2**

- Small ferns, growing in soil, not associated with rock outcrops .... **GROUP 3**

- Medium to large ferns, growing on rock, or over rock in thin soil .... **GROUP 4**

- Medium to large ferns, growing in soil (not on rock outcrops) .... **GROUP 5**

Seed Plants

- Aquatic plants, with all leaves underwater or floating on water surface .... **GROUP 6**

- Woody plants (trees, shrubs, woody vines, and small evergreen creeping plants) .... **GROUP 7**

- Herbaceous plants lacking both green color and developed leaves at flowering time .... **GROUP 8**

- Inflorescence apparently converted to bulblets, tufts of leaves, etc .... **GROUP 9**

- Monocots; leaves with parallel veins ...... **GROUP 10**

- Flowers in an involucrate head (i.e., the flowers clustered in a head above a whorl of bracts) **GROUP 11**

- Herbaceous plants with single-sex flowers . **GROUP 12**

- Herbaceous dicots, flowers bisexual
    Perianth of 1 series (calyx or corolla present but not both)
        Ovary inferior .... **GROUP 13**
        Ovary superior .... **GROUP 14**
    Perianth of 2 series (both calyx and corolla present)
        Ovaries 2 or more in each flower ... **GROUP 15**
        Ovary inferior .... **GROUP 16**
        Ovary superior
            Stamens more numerous than the petals ..
            .... **GROUP 17**
            Stamens the same number as the petals or fewer, and petals separate ...... **GROUP 18**
            Corolla regular and stamens the same number as its lobes, and petals united .....
            .... **GROUP 19**
            Corolla either bilaterally symmetrical or stamens fewer than its lobes, or both and petals united .... **GROUP 20**

FERNS AND FERN RELATIVES

Fern Groups

Includes true ferns and fern relatives (quillworts, clubmosses, spike-mosses).

1 Plants aquatic, either floating and unattached, or rooting and often completely submersed; Leaves linear, from a swollen, corm-like base (*Isoetes*) . **ISOETACEAE**
1 Plants of various wetland, upland, and rock habitats; leaves not as above 2

2 Leaves not "fern-like;" unlobed, variously awl-shaped, scale-like, or grass-like (the fern relatives will key here plus one species of *Asplenium*) .... **GROUP 1**
2 Leaves "fern-like," variously lobed or divided (true ferns will key here; the following groups are based on size of frond and habitat, either soil or rock) 3

3 Plants small, leaf blades (not including the stipe) small, less than 30 cm long or wide (some species will key both here and in the next lead no. 3) 4
3 Plants larger, leaf blades medium to large, more than 30 cm long or wide 5

4 Plants growing on rock, rock walls, or over rock in thin soil .... **GROUP 2**
4 Plants terrestrial, growing in soil, not associated with rock outcrops .... **GROUP 3**

5 Plants growing on rock, or over rock in thin soil mats or pockets of soil .... **GROUP 4**
5 Plants growing in soil, not associated with rock outcrops .... **GROUP 5**

GROUP 1

Fern relatives; leaves unlobed, awl-shaped, scalelike, or grasslike, and not "fern-like."

1 Stems obviously jointed; leaves small and scalelike, in a whorl from the joints or nodes, or sometimes absent; spores borne in a terminal conelike strobilus covered with peltate scales (i.e., scales more or less round and attached at middle like an umbrella) . **EQUISETACEAE**
1 Stems not jointed; leaves scalelike or larger, but if scalelike not in whorls from the nodes; spores borne variously, but if in a terminal strobilus the scales not peltate 2

2 Leaves linear, grasslike, 1-50 cm long (*Isoetes*) .... **ISOETACEAE**
2 Leaves various (scalelike, awl-like, moss-like, or flat), but not linear and grasslike 3

3 Leaves very numerous and overlapping along creeping, ascending, or erect stems; the leaves usually scalelike or awl-like, 0.5-2 (-3) mm wide, typically sharp- or hair-tipped; sporangia borne in strobili 4
3 Leaves not as above 8

4 Sporangia borne in flattened or 4-sided strobili sessile at tips of leafy branches; spores and sporangia of two sizes, the megasporangia larger and borne at base of strobili (*Selaginella*)............. **SELAGINELLACEAE**
4 Sporangia borne either in axils of normal foliage leaves, or in strobili sessile at tips of leafy branches, or stalked on specialized branches with fewer and smaller leaves; spores and sporangia of one size 5

5 Leafy stems erect, simple or dichotomously branched, the ultimate branches upright; sporophylls like the sterile leaves or only slightly smaller, in annual bands along the stem; vegetative reproduction by leafy gemmae near tip of stem (*Huperzia*) **LYCOPODIACEAE**
5 Leafy stems prostrate or erect, if erect then generally branched, the ultimate branches spreading (horizontal) or ascending; sporophylls differing from sterile leaves, either broader and shorter, or more spreading, grouped into terminal cones; lacking vegetative reproduction by gemmae 6

6 Plants of wetlands, mostly on moist or wet sand or peat; leaves herbaceous, pale or yellow-green, dull, deciduous; leafy stems creeping; rhizome dying back annually to an underground vegetative tuber at tip (*Lycopodiella*) .... **LYCOPODIACEAE**
6 Plants of uplands, mostly in moist to dry soils; leaves stiff, bright to dark green, shiny, evergreen; leafy stems mainly erect, treelike, fanlike, or creeping (if creeping, then the leaves with a hairlike tip); rhizome trailing, perennial 7

7 Shoots flat-branched, 1-5 mm wide (including the leaves); leaves scalelike, dimorphic, overlapping and appressed to stem, in 4 ranks; strobili on long, branched stalks (*Diphasiastrum*) .... **LYCOPODIACEAE**
7 Shoots round-branched, usually 5-8 mm wide (including the leaves), leaves awl-shaped, monomorphic (though sometimes differing in size), separate, spreading or ascending, in 6 ranks; strobili sessile at stem tips (*Lycopodium*)................... **LYCOPODIACEAE**

8 Plants with 1 (-several) leaves, the sterile leaf blade ovate, entire-margined, obtuse, the longer fertile portion with 2 rows of sporangia somewhat embedded in it (*Ophioglossum*) .... **OPHIOGLOSSACEAE**
8 Plant with many leaves, generally 5 or more, not divided into separate sterile and fertile segments, the leaves lance-shaped with a long-tapering tip (often with a plantlet which can root to form new plants) (*Asplenium rhizophyllum*) .... **ASPLENIACEAE**

GROUP 2
Small ferns, growing on rock or on rock in thin soil.

1 Fronds pinnatifid or 2-pinnatifid, most of the pinnae not fully divided from one another (the rachis winged by leaf tissue for most or all its length)..................... 2
1 Fronds pinnate, pinnate-pinnatifid, 2-pinnate, or even more divided (rachis naked for most of its length, but often winged in upper portion) 4

2 Fronds 2-pinnatifid, at least the lowermost pinnae deeply lobed (*Phegopteris*) ..... **THELYPTERIDACEAE**
2 Fronds 1-pinnatifid.................................. 3

3 Blades with long, narrow tapering tip, upper portion of blade unlobed or only slightly lobed; sori elongate (*Asplenium*)....................... **ASPLENIACEAE**
3 Fronds without a long, narrow tapering tip; blade lobed for most of its length; sori round (*Polypodium*)........ .... **POLYPODIACEAE**

4 Fronds 1-pinnate or 1-pinnate-pinnatifid 5
4 Fronds 2-pinnate or more divided 7

5 Sori on the undersurface of the leaf, away from the margins (*Asplenium*)................. **ASPLENIACEAE**
5 Sori on the undersurface of the leaf, along margins and more-or-less hidden beneath either the unmodified inrolled leaf margin or under a modified, reflexed false indusium 6

6 Stipes green to straw-colored for at least the upper 1/3, rachis green; fronds dimorphic, the fertile longer than the sterile and with narrower segments (*Cryptogramma*) .... **PTERIDACEAE**
6 Stipes and rachis dark brown to almost black throughout; fronds similar or somewhat different (*Pellaea*) .... **PTERIDACEAE**

7 Blade broadly triangular in outline (*Gymnocarpium*)... .... **CYSTOPTERIDACEAE**
7 Blade elongate, mostly lance-shaped, generally 4x or more as long as wide, not notably triangular in outline 8

8 Sori on margins, usually more-or-less hidden under the inrolled margin of the pinnule 9
8 Sori not on margins, either naked, or slightly to strongly hidden by indusia 10

9 Sori round or oblong, distinct and separate along the pinnule margins; fronds bright-green, smooth, herbaceous, delicate, and flexible (*Adiantum pedatum*) .... **PTERIDACEAE**
9 Sori continuous along the pinnule margins; fronds mostly dark-green, often hairy, leathery, tough, and stiff (*Pellaea*) .... **PTERIDACEAE**

10 Blades 3-12 cm long; sori elongate, covered by a flap-like, entire indusium (*Asplenium*)......... **ASPLENIACEAE**
10 Blades 4-30 (-50) cm long; sori round, surrounded or covered by an entire, fringed, or divided indusium... 11

11 Veins reaching margin; indusium attached under one side of sorus, hoodlike or pocketlike, arching over sorus; stipes smooth or sparsely covered with scales, stipe bases not persistent (*Cystopteris*) .... **CYSTOPTERIDACEAE**
11 Veins ending short of margin; indusium attached under sorus, cuplike (divided into 3-6 lobes which surround the sorus from below) or of numerous tiny hairs, which extend out from under sorus on all sides; stipes often densely covered with scales, stipe bases persistent (*Woodsia*) .... **WOODSIACEAE**

GROUP 3

Small ferns, terrestrial, growing in soil, not associated with rock outcrops.

1. Stipe branched once dichotomously, each branch with 3-7 pinnae in one direction only, the outline of the blade fan-shaped, often wider than long (*Adiantum pedatum*) .... **PTERIDACEAE**
1. Stipe not branched dichotomously, the outline of the blade either longer than wide or triangular and about as wide as long 2
2. Fronds pinnatifid or 2-pinnatifid, most of the pinnae not fully divided from one another (the rachis winged by leaf tissue for most or all of its length) 3
2. Fronds 1-pinnate, 1-pinnate-pinnatifid, 2-pinnate, or even more divided (the rachis naked for most of its length, or often winged in the upper portion) 5
3. Sporangia borne on an erect stalk that arises at or above ground level from stipe of sterile leaf blade (joining stipe of sterile leaf above the rhizome) (*Botrychium*, *Botrypus*) .... **OPHIOGLOSSACEAE**
3. Sporangia either borne on normal leaf blades or on specialized (fertile) fronds. 4
4. Fronds all alike, sori on normal leaf blades (*Phegopteris*) .... **THELYPTERIDACEAE**
4. Fronds of two types; sori on fronds significantly different than normal fronds (*Onoclea sensibilis*) .... **ONOCLEACEAE**
5. Fronds broadly triangular in outline, about as broad as long; sporangia borne on an erect stalk that arises at or above ground level from the stipe of the sterile leaf blade (joining the stipe of the sterile leaf above the rhizome) 6
5. Fronds lance-shaped in outline, much longer than broad; sporangia either borne on normal leaf blades, on slightly dimorphic blades, or on an erect stalk that arises at or above ground level from the stipe of the sterile leaf blade (joining the stipe of the sterile leaf above the rhizome) 7
6. Sporangia borne on normal leaf blades (*Gymnocarpium*) .... **CYSTOPTERIDACEAE**
6. Sporangia borne on an erect stalk that arises at or above ground level from the stipe of the sterile leaf blade (joining the stipe of the sterile leaf above the rhizome) (*Sceptridium*) .... **OPHIOGLOSSACEAE**
7. Blades 1-8 cm long; sporangia borne on an erect stalk that arises at or above ground level from the stipe of the sterile leaf blade (joining the stipe of the sterile leaf above the rhizome) (*Botrychium*) **OPHIOGLOSSACEAE**
7. Blades 10-30 (-100) cm long; sporangia either on normal leaf blades or on slightly modified blades 8
8. Fronds evergreen, dark green, somewhat leathery (*Polystichum*)................ **DRYOPTERIDACEAE**
8. Fronds light to medium green, herbaceous, deciduous to semi-evergreen................................. 9
9. Sori elongate; leaf blades somewhat dimorphic, the fertile larger and erect, the sterile smaller and prostrate, the larger leaf blades 2-4 (-6.5) cm wide (*Asplenium platyneuron*) .... **ASPLENIACEAE**
9. Sori round; leaf blades monomorphic; the larger leaf blades 5-15 cm wide (*Thelypteris*) .... **THELYPTERIDACEAE**

GROUP 4

Medium to large ferns, growing on rock, or over rock in thin soil.

1. Fronds 1-pinnate-pinnatifid or less divided, the pinnae entire, toothed, lobed or pinnatifid 2
1. Fronds 2-pinnate or more divided, the pinnae divided to their midribs 6
2. Sori elongate, the indusium flap-like, attached along the side; leaf blades (if more than 30 cm long) less than 7 cm wide (*Asplenium platyneuron*) .... **ASPLENIACEAE**
2. Sori circular or globular, the indusium peltate, kidney-shaped, or cuplike; leaf blades (if more than 30 cm long) more than 5 cm wide 3
3. Fronds 1-pinnate, the pinnae toothed and each with a slight to prominent lobe near the base on the side towards the leaf tip, dark green, somewhat leathery; indusia peltate (*Polystichum*).... **DRYOPTERIDACEAE**
3. Fronds 1-pinnate-pinnatifid, the pinnae pinnatifid, generally lacking a prominent basal lobe, light green to dark green, herbaceous to slightly leathery; indusium either kidney-shaped or cuplike 4
4. Vascular bundles in the stipe 3-7 (*Dryopteris*)......... .... **DRYOPTERIDACEAE**
4. Vascular bundles in the stipe 2, uniting above 5
5. Indusium kidney-shaped, arching over the sorus (*Thelypteris*) .... **THELYPTERIDACEAE**
5. Indusium cuplike, attached beneath sorus and consisting of 3-6 lanceshaped to ovate segments (*Woodsia*) .... **WOODSIACEAE**
6. Sori marginal and borne on underside of the false indusium; stipes and rachis shiny black or reddish-black, glabrous except at the very base of the stipe; pinnules fanshaped or obliquely elongate (*Adiantum*) .... **PTERIDACEAE**
6. Sori not marginal, borne on undersurface of leaf blade (or if marginal, as in *Pteridium*, borne on undersurface of the leaf); stipes darkened only near base (if at all), rachis green, tan, or reddish; pinnules not notably fan-shaped or obliquely elongate 7
7. Blades broadly triangular in outline, about as long as wide (*Pteridium aquilinum*). .... **DENNSTAEDIACEAE**
7. Blades elongate, mostly lanceolate, generally 4x or more as long as wide.................................. 8
8. Outline of leaf blade narrowed to base, the widest point more than 7 pinna pairs above the base, the lowermost pinnae 1/4 or less as long as the longest pinnae; rhizomes long-creeping, the fronds scattered, forming clonal patches (*Parathelypteris*) . **THELYPTERIDACEAE**

8   Outline of the leaf blade slightly if at all narrowed to the base, the widest point less than 5 pinna pairs from the base, the lowermost pinnae more than 1/2 as long as the longest pinnae; rhizomes short-creeping, the fronds clustered, not forming clonal patches 9

9   Vascular bundles (3-) 5 (-7) in the stipe; mostly larger woodland ferns (*Dryopteris*). . . . . . **DRYOPTERIDACEAE**
9   Vascular bundles 2 in the stipe (or joining near the leaf blade into 1); ferns of woodlands and rocky places . . 10

10 Fronds 1-pinnate-pinnatifid; indusium cuplike, attached beneath the sorus and consisting of 3-6 lanceolate to ovate segments; mostly smaller ferns on rock (*Woodsia*) . . . . **WOODSIACEAE**
10 Fronds 2-pinnate-pinnatifid; indusium flaplike or pocketlike, attached at one side of the sorus and arching over it. 12

11 Fronds 10-30 cm wide, the tip acute to acuminate; indusium flaplike (*Athyrium filix-femina*) . . . . **ATHYRIACEAE**
11 Fronds 4-9 cm wide, the tip long-attenuate; indusium pocketlike or hoodlike (*Cystopteris bulbifera*) . . . . **CYSTOPTERIDACEAE**

GROUP 5

Medium to large ferns, growing in soil (not on rock outcrops).

1   Blades broadly (about equilaterally) triangular, pentagonal, or flabellate in outline, 0.7-1.3x as long as wide 2
1   Fronds elongate in outline, mostly ovate, lanceolate, oblanceolate, or narrowly triangular, 1.5-10x or more as long as wide 4

2   Blades fan-shaped in outline, the stipe branched once dichotomously, each branch bearing 3-7 pinnae (*Adiantum pedatum*) . . . . **PTERIDACEAE**
2   Blades broadly triangular in outline, the stipe not branched dichotomously 3

3   Sporangia in a stalked, specialized, fertile portion of the blade; texture of mature blades somewhat fleshy; plants solitary from a short underground rhizome with thick, mycorrhizal roots; plants of moist woods (*Botrypus virginianus*) . . . . **OPHIOGLOSSACEAE**
3   Sporangia in marginal, linear sori, indusium absent, protected by the revolute leaf margin and a minute false indusium; texture of mature leaf blades hard and stiff; plants colonial from deep rhizomes; plants of moist to dry woodlands and openings (*Pteridium aquilinum*) . . . . **DENNSTAEDIACEAE**

4   Fronds 2-pinnate or more divided, the pinnae divided to their midribs. 5
4   Fronds 1-pinnate-pinnatifid or less divided; the pinnae entire, toothed, lobed or pinnatifid 9

5   Blade divided into sterile and fertile portions; sterile pinnae located below terminal fertile pinnae, the sterile pinnules 30-70 mm long and 8-23 mm wide, finely toothed, tip rounded to somewhat pointed; fertile pinnae greatly reduced in size, the fertile pinnules 7-11 mm long and 2-3 mm wide (*Osmunda regalis*) . . . . **OSMUNDACEAE**
5   Blade not divided into sterile and fertile portions, the pinnules bearing sporangia only slightly if at all reduced in size, both fertile and sterile pinnules mostly 4-20 mm long and 2-10 mm wide 6

6   Vascular bundles (3-) 5 (-7) in the stipe (*Dryopteris*) . . . . **DRYOPTERIDACEAE**
6   Vascular bundles 2 in the stipe (or joining upwards near leaf blade into 1 7

7   Fronds more than 10 cm wide, the tip acute to acuminate; indusium flaplike; pealike bulblets absent (*Athyrium filix-femina*) . . . . **ATHYRIACEAE**
7   Fronds 4-9 cm wide, the tip long-tapering; indusium pocketlike or hoodlike; bulblets often present on upper portion of blade (*Cystopteris bulbifera*) . . . . **CYSTOPTERIDACEAE**

8   Fronds 1-pinnate or 1-pinnate-pinnatifid, the pinnae fully divided from one another (rachis naked for most of its length, but often winged in upper portion); fronds dimorphic or not (*Onoclea sensibilis*) . **ONOCLEACEAE**
8   Fronds 1-pinnatifid, most of the pinnae not fully divided from one another (rachis winged by leaf tissue for most or all of its length); fronds dimorphic, the fertile much modified, stiff and/or woody 9

9   Rhizomes long-creeping, fronds scattered, forming patches (*Thelypteris*). . . . . **THELYPTERIDACEAE**
9   Rhizomes short-creeping, the fronds clustered, not forming patches (or rhizomes of both long and short, but fronds borne only in clusters on the short erect rhizomes in *Matteucia*). 10

10 Plants medium to large, fronds typically 60-300 cm tall; fronds either strongly dimorphic, the fertile fronds very unlike the sterile, brown at maturity (*Matteucia* and *Osmunda cinnamomea*) or fertile pinnae very unlike the sterile, brown at maturity, borne as an interruption in the blade, with normal green pinnae above and below (*Osmunda claytoniana*); rachises scaleless, stipes scaleless (except at the base in *Matteucia*). 11
10 Plants mostly smaller, the fronds 30-100 cm tall (except *Dryopteris goldiana* to 15 dm); fronds not at all or only slightly dimorphic, the fertile differing in various ways, such as having narrower pinnae (as in *Diplazium* and *Thelypteris palustris*) or fertile fronds taller and more deciduous (as in *Asplenium platyneuron* and *Dryopteris cristata*), but not as described in the first lead; rachises and stipes variously scaly or scaleless, but at least the stipe and often also the rachis scaly if the plants over 1 m tall. 12

11 Fronds strongly tapering to the base from the broadest point (well beyond the midpoint of the blade), lowermost pinnae much less than 1/2 as long as the largest pinnae (*Matteucia struthiopteris*) . . . . **ONOCLEACEAE**
11 Fronds slightly if at all tapering to the base, about equally broad through much of their length, lowermost

pinnae much more than 1/2 as long as the largest pinnae (*Osmunda*) .... **OSMUNDACEAE**

12 Sori elongate, the indusium elongate, attached along one side as a flap 13
12 Sori roundish; the indusium kidney-shaped or nearly round, attached by a central stalk, or sometimes absent 14
13 Stipe and rachis lustrous brownish black; fertile fronds 2-8 (-12) cm wide (*Asplenium platyneuron*) .... **ASPLENIACEAE**
13 Stipe and rachis green; fertile fronds 10-20 (-30) cm wide (*Deparia acrostichoides*) .... **ATHYRIACEAE**
14 Vascular bundles in the stipe 4-7 (*Dryopteris*) .... **DRYOPTERIDACEAE**
14 Vascular bundles in the stipe 2, uniting upwards (*Thelypteris*) .... **THELYPTERIDACEAE**

SEED PLANTS

SEED PLANT GROUPS include leads to 15 Group Keys and to several families with specialized features.

1 Plants aquatic, the leaves or plant body entirely submersed or floating on the surface of the water (at most, the inflorescence and bracts, not leaves, held above the surface .... **GROUP 6**
1 Plants with at least some leaves or stems above the water, or plants terrestrial 2
2 Plants woody (trees, shrubs, and woody vines), with erect, trailing, or viny aboveground stems living through the winter and continuing to grow the next season (leaves may be evergreen or deciduous) .... **GROUP 7**
2 Plants herbaceous, the perennial parts, if any, below or on the surface of the ground (to which the stems die back each year), not producing woody stems which survive the winter well above ground [hence, without aerial evergreen leaves (although there may be basal winter-green leaves)] 3
3 Plant lacking green color (often wholly parasitic or saprophytic) and the leaves none at flowering time or reduced to tiny scales) .... **GROUP 8**
3 Plant with green color and the leaves usually developed (occasionally the stems photosynthetic, as in cacti) .. 4
4 Inflorescences producing only small bulblets or tufts of little leaves (or modified floral parts), but no flowers or fruit .... **GROUP 9**
4 Inflorescences normal 5
5 Perianth parts (2), 3, (4), or 6 (never 5) and leaves (or other green photosynthetic parts when leaves are absent or reduced) parallel-veined (the 3 or more main veins running from base of blade to apex and ± parallel, with or without minute cross-veins), entire, simple ... . **GROUP 10**
5 Perianth parts various (often 5) but leaves netted-veined (or with only the midvein conspicuous), entire or toothed, simple or compound (the main veins, if more than 1, branching and ± reticulate) 6
6 Inflorescence a dense "head" (either a true head or a spadix), of a few to many small sessile flowers on a common receptacle (not merely an elongate spike), subtended by 1 or more small or large bracts **GROUP 11**
6 Inflorescence not an involucrate head, or if head-like the individual flowers short-pediceled and/or the "head" not immediately subtended by 1 or more bracts 7
7 Plants leafless but with thick, fleshy, green stem segments often bearing strong spines (*Opuntia*) .... **CACTACEAE**
7 Plants not both leafless and with spiny fleshy stems . 8
8 Inflorescence of "false flowers" consisting of small cup-like structures (uniform in texture and not composed of separate parts like bracts or scales) each bearing 1-5 glands on its rim (sometimes with additional petaloid appendages) and including 2 or more stamens and 1 central stalked 3-lobed pistil (which ripens into an exserted, 3-lobed capsule); sap milky (*Euphorbia*) .... **EUPHORBIACEAE**
8 Inflorescence various, but not composed of such structures; pistil only rarely stalked (and if so, not 3-lobed); sap various 9
9 Anthers and stigma fused into a central structure obscuring the individual reproductive parts; ovaries 2, ripening into follicles, the seeds each with a tuft of hairs (except *Vinca*); sap milky .... **APOCYNACEAE**
9 Anthers and stigmas not fused to each other, of diverse but recognizable structure; ovaries, seeds, and sap various but not combined as above................. 10
10 Flowers unisexual, containing one or more stamens or pistils, but not both .... **GROUP 12**
10 Flowers all or mostly bisexual, containing both stamen(s) and pistil(s) (although these may not all be equally mature at the same time).................. 11
11 Perianth none....................................... 12
11 Perianth present (but not always conspicuous) 13
12 Leaves deeply lobed or compound **RANUNCULACEAE**
12 Leaves unlobed, entire (*Callitriche, Hippuris*) .... **PLANTAGINACEAE**
13 Perianth of only one series (calyx or corolla) 14
13 Perianth of two series (both calyx and corolla)...... 15
14 Ovary inferior........................... **GROUP 13**
14 Ovary (or ovaries) superior .... **GROUP 14**
15 Ovaries 2 or more in each flower .... **GROUP 15**
15 Ovary 1 in each flower (styles or stigmas may be separate) 16
16 Ovary inferior .... **GROUP 16**
16 Ovary superior 17
17 Stamens more numerous than the petals... **GROUP 17**
17 Stamens the same number as the petals or lobes (not lips) of the corolla, or fewer 18

18 Petals separate .... **GROUP 18**
18 Petals connate at least at the base 19

19 Corolla regular and the stamens the same number as its lobes................................... **GROUP 19**
19 Corolla either bilaterally symmetrical or the stamens fewer than its lobes (not lips), or both conditions present .... **GROUP 20**

GROUP 6
Aquatic plants, with all leaves underwater or floating on water surface.

1 Plants without distinct stem and leaves, free-floating at or below surface of water (except where stranded by drop in water level), the segments (internodes) small (to 15 mm long, but in most species much smaller), often remaining attached where budded from parent plant 2
1 Plants with distinct stem and/or leaves, usually anchored in substrate, mostly larger................ 3
2 Plant body once to several times equally 2-lobed or 2-forked ... **RICCIACEAE** (a family of aquatic liverworts)
2 Plant body not consistently 2-lobed or 2-forked .... **ARACEAE**
3 Plants with floating leaves present (blades, or at least their terminalportions, floating on the surface of the water, usually ± smooth and firm in texture, especially compared with submersed leaves, or submersed leaves none) 4
3 Plants without any floating leaves, entirely submersed (except sometimes for inflorescences and associated bracts) 13
4 Blades of some or all floating leaves on a plant sagittate or deeply lobed at base, or compound, or peltate (± circular, with the stalk attached on the underside).... 5
4 Blades of floating leaves all unlobed (at most subcordate at base), simple, the petiole small or absent in ribbon-like leaves 7
5 Floating blades (at least some of them) sagittate, the tip and lobes acute (Note that plants with sagittate leaves extending above the water surface will not key here) (*Sagittaria*)....................... **ALISMATACEAE**
5 Floating (and any other) blades circular to ± elliptic in outline, peltate or rounded at tip with deep sinus at base 6
6 Leaves rounded at tip with deep sinus at base .... **NYMPHAEACEAE**
6 Leaves peltate (*Brasenia*) .... **CABOMBACEAE**
7 Floating leaves small (less than 1 cm long), crowded in a terminal rosette; submersed leaves distinctly opposite; flowers solitary, axillary (*Callitriche*) .... **PLANTAGINACEAE**
7 Floating leaves larger, not in a rosette; submersed leaves alternate, basal, or absent; flowers mostly in a terminal inflorescence 8

8 Leaves narrow and ribbon-like, the blades many times longer than wide, without distinct petiole (though in some species a sheath surrounds the stem) 9
8 Leaves (at least floating ones) with ± elliptic blades and distinct petioles................................. 10
9 Leaves ± rounded at tip (even if tapered), the floating portion smooth and shiny, somewhat yellow-green to bright green when fresh, occasionally keeled but midvein scarcely if at all more prominent than others; leaf not differentiated into blade and sheath, the submersed portion similar to the floating but more evidently with a fine closely checkered pattern; flowers and fruit in spherical heads (*Sparganium*) **TYPHACEAE**
9 Leaves sharply acute at tip, the floating portion rather dull, ± blue-green when fresh, with midrib; leaf including a sheath around stem and a membranous ligule at junction of sheath and blade; flowers and fruit in paniculate spikelets .... **POACEAE**
10 Leaves all basal; petals 3, white ..... **ALISMATACEAE**
10 Leaves cauline (along the stem), alternate or opposite; petals 4-6, pink or dull and inconspicuous (white in the rare *Caltha natans*)................................. 11
11 Flowers individually peduncled in a few-flowered, open inflorescence; rare (*Caltha natans*) **RANUNCULACEAE**
11 Flowers sessile in spikes, pink or dull and inconspicuous; mostly common species........................... 14
12 Veins netted; flowers bright pink, in dense ovoid to cylindrical spikes (*Persicaria amphibia*).............. .... **POLYGONACEAE**
12 Venation parallel; flowers dull, in narrow cylindrical spikes.................... **POTAMOGETONACEAE**
13 Leaves (or leaf-like structures) all basal and simple . 14
13 Leaves cauline, simple or compound (basal and dissected in one species) 31
14 Leaves flat, widest about the middle or parallel-sided 15
14 Leaves (or similar vegetative stems) filiform or terete or only slightly flattened (especially basally), elongate and limp to short and quill-like, less than twice as broad as thick... 20
15 Leaf blades not over twice as long as wide juvenile ... . **NYMPHAEACEAE**
15 Leaf blades more than twice as long as broad 16
16 Leaves stiff and erect or somewhat outcurved, less than 20 cm long.. 17
16 Leaves limp, more than 20 cm long, ribbon-like 18
17 Base of leaf somewhat sheathing, with a membranous ligule (as in a grass) at base of spreading blade (*Pontederia*) .... **PONTEDERIACEAE**
17 Base of leaf not sheathing and with no ligule (*Sagittaria*) .... **ALISMATACEAE**
18 Midvein not evident, all veins of essentially equal prominence, with the tiny cross-veins giving a checkered appearance to the leaf, which is thus uniformly marked with minute rectangular cells ca. 1-2 mm long or smaller (*Sparganium*) .... **TYPHACEAE**

18 Midvein (and usually some additional longitudinal veins) evident, the veins not all of equal prominence, not dividing the leaf into minute rectangular cells .. 19

19 Leaves with the central third (or more) of distinctly different pattern (more densely reticulate) than the two marginal zones; plants dioecious, the staminate flowers eventually liberated from a dense inflorescence submersed at base of plant, the pistillate solitary on a long ± spiraled stalk which reaches the surface of the water; plants without milky juice (*Vallisneria*)......... .... **HYDROCHARITACEAE**

19 Leaves ± uniform in venation, not 3-zoned; plants monoecious, with emergent inflorescence of white-petaled flowers (but these scarce on plants with submersed tape-like leaves); plants often with milky juice (*Sagittaria*)................... **ALISMATACEAE**

20 Major erect structures solitary, spaced along a simple or branched delicate rhizome, consisting either of rather yellowish stems bearing minute alternate bumps as leaves or of filiform leaves mostly buried in the substrate and with a few minute bladder-like organs 21

20 Major erect structures solitary to densely tufted, consisting of filiform or quill-like leaves or stems, with neither alternate bumps or bladders 22

21 Leaves merely minute alternate bumps on stem; bladders not present; flowers sessile, inconspicuous, regular (*Myriophyllum tenellum*).... **HALORAGACEAE**

21 Leaves filiform, mostly buried in substrate (only the green tips, incurled when young, protruding); bladders (minute) usually present on the delicate branching rhizomes and buried leaf bases; flowers short-pediceled, showy (yellow or purple), bilaterally symmetrical (*Utricularia*)....... **LENTIBULARIACEAE**

22 Leaves very limp (retaining no stiffness when removed from water and hence irregularly sinuate, bent, or matted on herbarium specimens) though a stiffer straight stem may also be present, mostly more than 20 cm long, ca. 0.2-1 mm wide...................... 23

22 Leaves usually firm (retaining stiffness when removed from water and hence straight or with an even curve in herbarium specimens), less (in most species much less) than 20 cm long, of various widths................ 25

23 Leaves slightly expanded basally for ca. (0.7-) 2-10 cm, sheathing the next inner leaf at least dorsally (usually the sheath continued ventrally as an almost invisible membrane), with tiny ligule or pair of auricles at the summit; rhizome various; inflorescence a lateral spikelet or terminal cyme (*Schoenoplectus subterminalis*) .... **CYPERACEAE**

23 Leaves (actually vegetative stems) terete their entire length, not expanded basally nor sheathing each other, but each separate and closely surrounded at base for ca. (0.6-) 1 cm or more by a very delicate membranous tubular sheath (this sometimes requiring careful dissection to distinguish); rhizome less than 2 mm in diameter; inflorescence (rare on plants otherwise entirely submersed) a single strictly terminal spikelet 24

24 Rhizome reddish, at least on older portions; leaves (vegetative culms) mostly over 20 cm long, very limp; fertile culm triangular in cross-section on emersed portion, much larger in diameter than the hair-like vegetative culms, but spikelet no thicker than culm (*Eleocharis robbinsii*) .... **CYPERACEAE**

24 Rhizome whitish throughout; leaves often shorter, usually stiffer; fertile culms terete, no larger than the vegetative culms, but spikelet distinctly thicker than culm (*Eleocharis acicularis*) .... **CYPERACEAE**

25 Leaves filiform throughout, not broader basally nor sheathing each other, solitary (rarely) or in small tufts along a filiform whitish rhizome, each leaf (actually a vegetative stem) closely surrounded at its base for ca. 6 mm or more by a very delicate membranous tubular sheath (this sometimes requiring careful dissection to distinguish); inflorescence (rare on completely submersed plants) a single terminal spikelet (*Eleocharis acicularis*)........................... **CYPERACEAE**

25 Leaves linear or tapered from base to apex, or if otherwise uniformly filiform then expanded at base or sheathing each other, without individual tubular sheaths as described above; inflorescence various . 26

26 Leaf in cross-section appearing composed of 2 hollow tubes, linear (± parallel-sided), broadly rounded at tip; flowers bilaterally symmetrical, in a few-flowered raceme (*Lobelia dortmanna*)...... **CAMPANULACEAE**

26 Leaf not (or rarely) of 2 hollow tubes, tapered and ± acute (or filiform); flowers regular and racemose, or solitary, or in a dense head or spike, or plant producing spores at base ....

27 Roots with prominent cross-septate appearance (checkered with fine transverse lines); inflorescence a small whitish or gray head (flowering in shallow water and on wet shores) .... **ERIOCAULACEAE**

27 Roots not distinctly septate or cross-lined; inflorescence not as above.................................... 28

28 Leaves rather abruptly expanded at base to enclose sporangia, often dark green, composed of 4 hollow tubes (in cross-section), surrounding a hard corm-like stem; plant submersed (unless stranded), non-flowering (*Isoetes*, a fern relative) .... **ISOETACEAE**

28 Leaves gradually and slightly expanded or grooved on one side at a somewhat sheathing base but not composed of 4 tubes nor enclosing sporangia and no corm-like stem present; plants (except *Subularia*) not flowering when submersed but only on wet shores . 29

29 Leaves somewhat flattened at least basally, widest at the base, gradually tapered to sharp apex; plants with buried rhizome (*Juncus pelocarpus*) ...... **JUNCACEAE**

29 Leaves ± terete, scarcely or no wider at base than at middle, of ± uniform width at least to the middle (or even slightly thicker there before tapering to apex); plants with rhizomes or stolons at, near, or above surface of substrate 30

30 Plants with green stolons strongly arching above substrate; leaves filiform, ± uniform in diameter, 0.5-1 mm thick, truncate at tip (*Ranunculus flammula*)...... .... **RANUNCULACEAE**
30 Plants producing delicate horizontal white to green stolons at or near (above or below) surface of substrate (in addition to stouter short rhizome); leaves ca. 0.7-3 mm thick at middle, whence tapered to apex (*Littorella*) .... **PLANTAGINACEAE**
31 Leaves compound, dissected, forked, or deeply lobed 32
31 Leaves simple, unlobed, usually entire (toothed in a few species) 41
32 Leaves apparently in a basal rosette, few (*Sium suave*). .... **APIACEAE**
32 Leaves definitely cauline: opposite, whorled, or alternate 33
33 Leaves all or mostly opposite or whorled 34
33 Leaves definitely all alternate 38
34 Leaves (or whorled branches) rolled inward at tip when young, bearing tiny stalked bladders; flowers emersed, bilaterally symmetrical, purple or yellow (*Utricularia*) . ... **LENTIBULARIACEAE**
34 Leaves not inrolled at tip, without bladders; flowers various but not as above 35
35 Petiole evident (5-15 mm long on well developed leaves), the blade fan-shaped and much dissected; flowers emergent, white (*Cabomba*) .... **CABOMBACEAE**
35 Petiole absent or nearly so, the blade pectinate (with straight central axis following midrib, once-pinnatifid or comb-like on both sides) or much dissected or soon forking once or twice; flowers inconspicuous or yellow 36
36 Leaves once or twice dichotomously forked, the segments usually sparsely toothed along one edge; flowers inconspicuous, axillary, submersed............ .... **CERATOPHYLLACEAE**
36 Leaves not dichotomously forked, the segments entire; flowers emersed or (rarely) submersed 37
37 Leaves pectinate; flowers inconspicuous, in all but the rarest species emersed in terminal spike (*Myriophyllum*) .... **HALORAGACEAE**
37 Leaves with no definite central axis, much dissected; flowers emersed in a showy yellow head (usually with at least one pair of merely serrate opposite leaves below it) (*Bidens beckii*) .... **ASTERACEAE**
38 Leaves with a definite central axis (following midvein); flowers various................................. 39
38 Leaves with no definite central axis (except sometimes after initially forking at the stem); flowers emersed, with conspicuous corolla........................... 40
39 Leaves pectinate (the lateral segments not again branched); flowers inconspicuous, axillary; fruit a nutlet (*Proserpinaca*).................... **HALORAGACEAE**
39 Leaves with lateral segments further narrowly divided; flowers with white corollas, in emersed raceme; fruit a silique (*Rorippa aquatica*)............ **BRASSICACEAE**
40 Petiole present (sometimes very short), ± adnate to a stipular sheath; plants without bladders; flowers regular, white or yellow, with numerous separate carpels forming achenes (*Ranunculus*).... **RANUNCULACEAE**
40 Petioles and stipular sheaths absent; plants with small stalked bladders on leaves or on separate branches; flowers bilaterally symmetrical, yellow or purplish, with a single pistil producing a capsule (*Utricularia*) .... **LENTIBULARIACEAE**
41 Leaves much reduced, ± scale-like, not over 7 mm long, never distinctly opposite or whorled 42
41 Leaves much longer or distinctly opposite or whorled (or both conditions) 43
42 Leaves minute, yellowish, merely widely spaced bumps or scales on stem (*Myriophyllum tenellum*) .... **HALORAGACEAE**
42 Leaves to 7 mm long, green or brownish, loosely overlapping .... liverworts, aquatic mosses
43 Leaves alternate, with ligule-like stipules (these wholly adnate to leaves in *Ruppia*)....................... 44
43 Leaves opposite or whorled, without stipules 46
44 Leaf blades ± filiform, terete or at least half as thick as broad, and the stipule adnate to leaf base for 10-30 mm or more, forming a sheath around the stem .... **POTAMOGETONACEAE**
44 Leaf blades definitely flattened and several times as broad as thick (even if narrow), or stipule little if at all adnate to blade (or both conditions) 45
45 Blades flattened, ribbon-like (up to 5 or even 7.5 mm wide), with no definite midrib (no central vein more prominent than others except rarely toward base); flowers solitary, rare, cleistogamous in axils of submersed leaves or (these almost never on submersed plants) with 6 bright yellow tepals (*Heteranthera*)..... .... **PONTEDERIACEAE**
45 Blades flattened with a definite midrib or filiform; flowers in spherical or cylindrical spikes, neither cleistogamous nor with showy yellow perianth .... **POTAMOGETONACEAE**
46 Leaves nearly filiform, not over 0.5 mm wide, very gradually tapered from base to apex but not abruptly expanded basally, perfectly smooth; plants perennial by slender rhizomes; flowers axillary, 1 staminate flower (a single stamen) and (1) 2-several carpels at a node; fruit slightly curved and minutely toothed on convex side (*Zannichellia*) .... **POTAMOGETONACEAE**
46 Leaves broader; or if filiform then abruptly expanded basally and with apiculate or toothed margins, the plants annual, and the fruit solitary and ellipsoid ... 47
47 Leaves definitely whorled 48
47 Leaves opposite (in some species, with bushy axillary tufts of leaves which may give a falsely whorled appearance).. 49

48 Whorled structures ("branches") cylindrical, elongate, usually stiff with calcium deposits; plants with distinctive musky odor (*Chara*, a macro-algae). . . . . CHARACEAE
48 Whorled structures (true leaves) flattened, short (not over 20 mm long) or elongate and very limp; plants without odor 51

49 Leaves 6-12 (usually 9) in a whorl, not over 2.5 mm wide, 12-25 times as long as wide; flowers bisexual, apetalous, sessile in axils of emersed leaves or bracts (*Hippuris*). . . . . PLANTAGINACEAE
49 Leaves mostly 3-4 (rarely 6) in a whorl, 0.8-5 mm wide, at most 10-13 times as long; flowers bisexual or unisexual, but with petals 50

50 Leaves mostly 3 (rarely 6) in a whorl, very thin (2 cell layers) and delicate; stem round (not angled), smooth; flowers unisexual, with 3 often pink petals, at least the pistillate long-stalked from entirely submersed stem (*Elodea*) . . . . HYDROCHARITACEAE
50 Leaves mostly 4 in a whorl, stiff and firm; stem 4-sided, often with minutely retrorse-scabrous angles; flowers bisexual, with 3-4 white petals (usually not developed on wholly submersed plants) (*Galium*) . . . RUBIACEAE

51 Largest leaves at least 1-4 cm long, with distinct petiole and expanded, entire blade 52
51 Largest leaves smaller, or sessile, or toothed (or all of these) 53

52 Leaf blades ± orbicular, with orange to black glandular dots on underside; flowers 5-merous with showy yellow petals and superior ovary (*Lysimachia nummularia*) . . . . PRIMULACEAE
52 Leaf blades ± diamond-shaped, without glandular dots; flowers 4-merous, inconspicuous, with inferior ovary (*Ludwigia palustris*) . . . . ONAGRACEAE

53 Leaves large, 3-13 cm long, 5-20 mm wide 54
53 Leaves small (shorter or narrower than the above, or usually both) 55

54 Leaves sessile and clasping, limp, at most obscurely and remotely toothed; flowers (rarely present on plants with all foliage submersed) in axillary racemes (*Veronica anagallis-aquatica*) . . . . PLANTAGINACEAE
54 Leaves sessile, clasping, tapered, or petioled, stiff, often regularly crenate or toothed; flowers various; includes submersed plants of normally terrestrial or emergent plants, often members of LAMIACEAE. see description

55 Leaves linear and bidentate at apex when well submersed, often becoming obovate, ± weakly 3-nerved, and not necessarily bidentate toward summit of stem (or in floating rosettes); fruit solitary in axils, somewhat heart-shaped, of two 2-seeded segments (*Callitriche*) . . . . PLANTAGINACEAE
55 Leaves filiform to orbicular or tapered from base to apex, but essentially uniform on a plant and if linear not bidentate at apex; fruit various 56

56 Leaves at least 3 times as long as wide, broader at base than at middle; fruit absent or solitary in axils of leaves and ± ellipsoid 57

56 Leaves less than 3 times as long as wide, often nearly round 58

57 Leaves (especially lower ones) ± evenly tapered from broad base to minutely but bluntly bidentate apex, 3-10 times as long as wide, strictly entire, not subtending axillary tufts of leaves or fruit; plant often with a few scattered pale glandular dots on surface toward upper portion (*Gratiola aurea*) . . . . PLANTAGINACEAE
57 Leaves filiform to linear-lanceolate, ± expanded at very base, acute or apiculate at apex, at least 6 times as long as wide, minutely apiculate to conspicuously toothed on margins, usually subtending axillary tufts of leaves and/or flowers or ellipsoid fruit; plant without glands on surface (*Najas*) . . . . HYDROCHARITACEAE

58 Stems forming moss-like mats but the erect or ascending tips (above rooted nodes) less than 3 cm long; leaves with at most 1 weak nerve; stipules minute but usually evident with some leaves; flowers axillary, inconspicuous . . . . ELATINACEAE
58 Stems greatly elongate (generally 10-30 cm); leaves more evidently veined; stipules none; flowers terminal, yellow (but usually absent on plants with all leaves submersed) 59

59 Stems stiffly erect; leaves weakly pinnately veined (with evident midvein), with reddish to blackish shiny dots (these often also on stem) (*Lysimachia terrestris*) . . . . PRIMULACEAE
59 Stems ± lax; leaves 3-nerved, without dark dots or flecks (though emersed leaves have translucent dots) (*Hypericum boreale*) . . . . HYPERICACEAE

GROUP 7

Woody plants (trees, shrubs, woody vines, and small evergreen creeping plants such as *Vinca*, *Linnaea*, and *Mitchella*).

1 Leaves scalelike (ca. 4 mm or less long and often appressed/imbricate) or needle-like, evergreen (except in *Larix* in Pinaceae) 2
1 Leaves with expanded (or dissected) blades, neither scale-like nor needle-like, if linear, then herbaceous, not stiff; deciduous or occasionally evergreen; occasionally absent at flowering time 9

2 Plant with leaves scale-like (or less than 3 mm long) . 3
2 Plant with leaves needle-like or narrowly linear (over 3 mm long) 6

3 Leaves alternate (*Hudsonia*) . . . . CISTACEAE
3 Leaves opposite or whorled 4

4 Plants fragrant when crushed, producing small dry or berry-like female cones but never flowers or true fruit . . . . CUPRESSACEAE
4 Plants not fragrant, producing flowers and fruit in season 5

5 Plant a parasite, less than 1.5 cm high, on branches of conifers, blooming in very early spring without showy perianth (*Arceuthobium*) . . . . SANTALACEAE

5. Plant a small terrestrial shrub, blooming in late summer with showy pink flowers; reported from Mackinac County (*Calluna*) . . . . **ERICACEAE**

6. Leaves opposite or whorled . . . . **CUPRESSACEAE**
6. Leaves alternate or in clusters 7

7. Seed solitary in a red, fleshy, cup-like aril; leaves flattened, with strongly decurrent base, persistent, appearing 2-ranked, all green on both sides (may be yellowish beneath) . . . . **TAXACEAE**
7. Seeds borne on scales of a dry woody cone; leaves flattened or not (but if so, not strongly decurrent, readily falling when dry, not 2-ranked, and/or with white lines beneath) 8

8. Leaves evergreen (except *Larix* with leaves spirally arranged), arranged in clusters, spiraled around the stem, or in flattened 2-ranked sprays; cones slightly to very much longer than wide, the cone scales flattened . . . . **PINACEAE**
8. Leaves deciduous, arranged in flattened 2-ranked sprays; cones globular, with peltate cone scales . . . . **CUPRESSACEAE**

9. Leaves opposite or whorled or nearly so (evident from scars if leaves not expanded at anthesis) 10
9. Leaves alternate 47

10. Flowers appearing before leaves are expanded 11
10. Flowers appearing after the leaves have expanded (i.e., leaves present) 20

11. Perianth of both calyx and corolla 12
11. Perianth of only one cycle of parts, or none 14

12. Ovary superior; petals separate; flowers often unisexual (*Acer*) . . . . **SAPINDACEAE**
12. Ovary inferior; petals united; flowers bisexual 13

13. Flowers numerous in terminal cymes (*Sambucus*) . . . . **ADOXACEAE**
13. Flowers in pairs on axillary peduncles (*Lonicera*) . . . . **CAPRIFOLIACEAE**

14. Inflorescence an ament (catkin); bud scale 1 (*Salix purpurea*) . . . . **SALICACEAE**
14. Inflorescence otherwise, of clustered or pediceled flowers but not an elongate ament; bud scales more than 1 15

15. Flowers staminate or bisexual 16
15. Flowers pistillate 18

16. Stamens 2 (-4) (*Fraxinus*) . . . . **OLEACEAE**
16. Stamens 5 or more 17

17. Calyx lobes 4; stamens 8; buds scurfy-pubescent (*Shepherdia*) . . . . **ELAEAGNACEAE**
17. Calyx lobes 5; stamens 5-10; buds not scurfy-pubescent (*Acer*) . . . . **SAPINDACEAE**

18. Ovary with 2 divergent lobes (*Acer*) . . . **SAPINDACEAE**
18. Ovary unlobed 19

19. Floral tube with a prominent disk at its summit; buds scurfy-pubescent; young fruit rotund (*Shepherdia*) . . . . **ELAEAGNACEAE**
19. Floral tube without a prominent disk; buds not scurfy; young fruit strongly flattened (*Fraxinus*) . . . **OLEACEAE**

20. Leaves compound 21
20. Leaves simple 25

21. Plant a climbing or trailing vine (*Clematis*) . . . . **RANUNCULACEAE**
21. Plant erect, not a vine 22

22. Petals none; fruit a samara (winged) 23
22. Petals well developed and conspicuous; fruit various but not a samara 24

23. Ovary 2-lobed; fruit united in pairs; stamens ca. 5-10; leaflets usually 3-5 (*Acer negundo*) . . . . **SAPINDACEAE**
23. Ovary not lobed; fruits not paired; stamens 2 (-4); leaflets 5-11 (*Fraxinus*) . . . . **OLEACEAE**

24. Petals united; leaves pinnately compound with 5 or more leaflets; fruit fleshy (*Sambucus*) . . . **ADOXACEAE**
24. Petals separate; leaves palmately or pinnately compound; fruit dryish (*Aesculus*) . . . . **SAPINDACEAE**

25. Stamens more numerous than the petals or lobes of the corolla (or of the calyx if corolla is absent), or flowers strictly pistillate 26
25. Stamens the same number as the lobes or petals of the corolla or fewer 36

26. Petals united . . . . **ERICACEAE**
26. Petals separate or none 27

27. Stamens usually more than 10; corolla yellow or white 28
27. Stamens 10 or fewer, or flowers strictly pistillate; corolla pink, green, greenish-yellow, or white 33

28. Leaves scale-like; style 1; ± prostrate shrub less than 2 dm tall (*Hudsonia*) . . . . **CISTACEAE**
28. Leaves well developed; styles 3-5 (sometimes ± coherent); shrubs to 1 m tall (*Hypericum kalmianum*) . . . . **HYPERICACEAE**

29. Leaves palmately lobed, toothed; fruit a samara, united in pairs (*Acer*) . . . . **SAPINDACEAE**
29. Leaves unlobed, entire or toothed; fruit a berry or capsule, not paired 34

30. Plant a bushy shrub, with scurfy or stellate pubescence (*Shepherdia*) . . . . **ELAEAGNACEAE**
30. Plant barely woody at base, glabrous to somewhat tomentose but not scurfy or stellate pubescent; flowers bisexual with showy pink (to white) petals; fruit a capsule 35

31. Leaves evergreen, very shiny, toothed; stigma nearly sessile (*Chimaphila*) . . . . **ERICACEAE**
31. Leaves deciduous, dull, entire; stigma on an elongate style . . . . **LYTHRACEAE**

32. Petals separate 37
32. Petals united 39

33 Flowers in terminal inflorescences ..... **CORNACEAE**
33 Flowers axillary.................... **RHAMNACEAE**
34 Ovary inferior 40
34 Ovary superior 44
35 Flowers and fruits in dense spherical peduncled heads or paired at the ends of trailing branches; leaves entire, with broad stipules between the petiole bases .... **RUBIACEAE**
35 Flowers and fruits pediceled in small clusters or ± branched inflorescences; leaves entire or toothed, with stipules none or slender and partly adnate to petioles 41
36 Leaves of flowering shoots or flowering portions of shoots entire or somewhat undulate or sinuous, not sharply or regularly toothed ...... **CAPRIFOLIACEAE**
36 Leaves with margins lobed, ± regularly toothed, crenate, or finely crenulate, or at least with regular minute gland-like teeth 42
37 Calyx lobes up to 1.5 mm long and broadly triangular to broadly rounded or virtually absent; corolla rotate (flat with very short tube); style very short or essentially absent; fruit fleshy with one pit (*Viburnum*) .... **ADOXACEAE**
37 Calyx lobes (1.6-) 2-6.5 (-7.5) mm long, linear or narrowly lanceolate; corolla tubular; style elongate, conspicuous; fruit dry .... **CAPRIFOLIACEAE**
38 Corolla bilaterally symmetrical......... **LAMIACEAE**
38 Corolla regular 46
39 Ovaries 2 (but styles and stigmas united); plants evergreen creeper with blue flowers solitary in the leaf axils (*Vinca*) .... **APOCYNACEAE**
39 Ovary 1; plant erect, with flowers in inflorescences ... . **OLEACEAE**
40 Leaves deeply dissected into linear-filiform segments, aromatic (*Artemisia*).................. **ASTERACEAE**
40 Leaves simple, compound (then leaflets broader than linear-filiform), or absent at anthesis, aromatic or not 41
41 Plants dioecious 42
41 Plants not dioecious, the flowers either bisexual or unisexual (if the latter, then both sexes on the same individual) 62
42 Plant a climbing vine (or trailing in absence of support) 43
42 Plant ± erect, not climbing 48
43 Stems with tendrils............................... 44
43 Stems without tendrils (aerial roots may be present along stem).................................... 45
44 Leaves entire; stems prickly (at least below); perianth of 6 tepals .... **SMILACACEAE**
44 Leaves toothed; stems unarmed; perianth of 5 petals and 5 (sometimes vestigial) sepals......... **VITACEAE**
45 Leaves trifoliolate; plants climbing by adventitious roots (*Toxicodendron*).................. **ANACARDIACEAE**
45 Leaves simple or with more than 3 leaflets; plants climbing by twining stems 46

46 Leaves pinnately veined, simple (*Celastrus*).......... .... **CELASTRACEAE**
46 Leaves palmately veined or compound 47
47 Sepals and petals each 6; leaves ± peltate (petiole attached in from margin of the blade), at most somewhat lobed but not toothed **MENISPERMACEAE**
47 Sepals (often vestigial) and petals each 5; leaves with marginal petiole, toothed.................. **VITACEAE**
48 Flowers (at least the male) in cylindrical to nearly spherical aments (catkins) 49
48 Flowers not in aments............................ 51
49 Twigs and leaves with milky sap; calyx minute; leaves palmately or pinnately veined .... **MORACEAE**
49 Twigs and leaves with watery sap; calyx none; leaves pinnately veined 50
50 Crushed foliage pungently aromatic; twigs resin-dotted; fruit an achene or dry drupe .... **MYRICACEAE**
50 Crushed foliage in most species not at all aromatic; twigs without resinous dots (may be generally shiny); fruit a capsule .... **SALICACEAE**
51 Leaves compound, present at anthesis............. 52
51 Leaves simple, or unexpanded at anthesis.......... 53
52 Leaves punctate with translucent oil glands; stems sometimes prickly .... **RUTACEAE**
52 Leaves without translucent oil dots; stems unarmed .. .. **ANACARDIACEAE**
53 Flowers pistillate 54
53 Flowers staminate............................... 59
54 Calyx and corolla both present (the former sometimes very small and inconspicuous) 55
54 Calyx and corolla not differentiated, or absent...... 57
55 Inflorescences terminal, ± crowded or many-flowered .... **ANACARDIACEAE**
55 Inflorescences axillary 56
56 Stigma nearly or quite sessile ...... **AQUIFOLIACEAE**
56 Stigma clearly on an elongate style ... **RHAMNACEAE**
57 Style and stigma 1; leaves entire ... **AQUIFOLIACEAE**
57 Style divided above, stigmas 2-4; leaves toothed (at least on apical half) 58
58 Leaf blades asymmetrical at base (*Celtis*)............ .... **CANNABACEAE**
58 Leaf blades ± symmetrical at base .... **RHAMNACEAE**
59 Inflorescences terminal, ± crowded or many-flowered. .... **ANACARDIACEAE**
59 Inflorescences axillary........................... 60
60 Stamens alternate with the sepals (opposite the petals if any) .... **RHAMNACEAE**
60 Stamens opposite the sepals (alternate with the petals if any).. 61
61 Leaf blades ± symmetrical at base, entire or toothed.. .... **AQUIFOLIACEAE**
61 Leaf blades asymmetrical at base, toothed (at least on apical half) (*Celtis*) .... **CANNABACEAE**

62 Flowers (at least the staminate) in aments or dense spherical heads (always unisexual and individually inconspicuous) 63
62 Flowers not in aments or heads (often bisexual and/or conspicuous) 67
63 Staminate flowers in dense spherical heads (*Fagus*) ... . **FAGACEAE**
63 Staminate flowers in cylindrical to ellipsoid aments 64
64 Pistillate flowers solitary or in small clusters; styles 3 or leaves compound............................... 65
64 Pistillate flowers in aments, heads, or cone-like structures (in *Corylus*, the red styles protruding from an ament resembling a leaf bud); styles 2; leaves simple.. 66
65 Leaves pinnately compound; styles 2 **JUGLANDACEAE**
65 Leaves simple (may be deeply lobed); styles usually 3 . ... **FAGACEAE**
66 Twigs and leaves with milky sap; calyx minute, 4-parted; leaves palmately veined (or fruit in a large spherical fleshy structure) .... **MORACEAE**
66 Twigs and leaves with watery sap; calyx usually none or 2-parted; leaves pinnately veined (fruit in aments or small clusters) .... **BETULACEAE**
67 Perianth none or apparently of a single series of parts 68
67 Perianth clearly differentiated into calyx and corolla 76
68 Stamens more numerous than the segments or lobes (if any) of the perianth (or perianth none) 69
68 Stamens the same number as the lobes or segments of the perianth 70
69 Stamens 8; perianth lobes 4 (or essentially none).... .... **THYMELAEACEAE**
69 Stamens 5-7 or 9; perianth lobes or segments 5 or 6 (*Rhododendron*) .... **ERICACEAE**
70 Styles 2, 3, or 5 71
70 Style 1 (may be branched above) 72
71 Leaves with prominent, straight ± parallel lateral veins running into the principal teeth; flowers bisexual, the perianth shallowly lobed; ovary flattened and winged; fruit a samara........................... **ULMACEAE**
71 Leaves with lateral veins curved and ascending, weaker and the branches anastomosing near the margins; flowers usually unisexual, the perianth lobed nearly or quite to the base; ovary not flattened, fruit a drupe (*Celtis*)........................... **CANNABACEAE**
72 Plant a vine, climbing or trailing by tendrils **VITACEAE**
72 Plant an erect shrub or tree 73
73 Inflorescences terminal .... **CORNACEAE**
73 Inflorescences lateral 74
74 Leaves beneath and branchlets silvery-scurfy; stamens 4 .... **ELAEAGNACEAE**
74 Leaves and branchlets glabrous or nearly so, not scurfy; stamens 4-6 75
75 Stamens alternating with the sepals .. **RHAMNACEAE**
75 Stamens opposite the sepals....... **AQUIFOLIACEAE**
76 Ovaries at least 3, distinct .... **ROSACEAE**
76 Ovary 1 77
77 Corolla bilaterally symmetrical (or petal only 1); stamens 10 (usually with some of the filaments connate) .... **FABACEAE**
77 Corolla essentially regular; stamens various........ 78
78 Petals united 79
78 Petals separate 83
79 Stamens more numerous than the corolla lobes .... **ERICACEAE**
79 Stamens the same number as the corolla lobes 80
80 Stamens adnate to the corolla (and falling with it if the corolla is deciduous); plants vining to shrubby; fruit a red berry or drupe 81
80 Stamens free from the corolla; plant an erect or trailing shrub (not climbing); fruit various................. 82
81 Flowers white, on short (< 5 mm) pedicels; leaves toothed....................... **AQUIFOLIACEAE**
81 Flowers purple (except in rare albinos), pedicels > 7 mm; leaves entire-margined (though sometimes lobed) ... . **SOLANACEAE**
82 Stigma on a well developed style; fruit a capsule .... **ERICACEAE**
82 Stigma nearly sessile; fruit a red drupe .... **AQUIFOLIACEAE**
83 Ovary at least partly inferior 84
83 Ovary entirely superior 91
84 Stamens more than the number of petals 85
84 Stamens the same number as the petals 87
85 Style 1... 86
85 Styles 2-5 .... **ROSACEAE**
86 Petals united; unarmed shrubs (*Vaccinium*)**ERICACEAE**
86 Petals free; thorny tree or large shrub (*Crataegus monogyna*).......................... **ROSACEAE**
87 Petals 4 88
87 Petals 5 89
88 Flowers white, in terminal cymes, blooming in early to mid-summer; fruit fleshy; leaves entire.. **CORNACEAE**
88 Flowers yellow, in small axillary clusters, blooming in late fall; fruit a capsule; leaves with rounded teeth (*Hamamelis virginiana*) .... **HAMAMELIDACEAE**
89 Flowers in umbels, umbels either solitary or arranged in larger inflorescences................... **ARALIACEAE**
89 Flowers in racemes, small axillary clusters, or domes or flat-topped corymbs 90
90 Flowers in racemes or small axillary clusters; small to medium shrubs less than 2 m tall .... **GROSSULARIACEAE**
90 Flowers in terminal, domed or flat-topped corymbs; large shrubs or small trees more than 2 m tall (*Crataegus*) .... **ROSACEAE**

91 Stamens more than twice as many as the petals ...92
91 Stamens twice as many as the petals or fewer......94
92 Corolla yellow; fruit a capsule .... **CISTACEAE**
92 Corolla white to pink; fruit indehiscent 93
93 Inflorescence apparently borne at the middle of a tongue-shaped bract; leaves palmately veined (*Tilia*) .. .. **MALVACEAE**
93 Inflorescence borne normally; leaves pinnately veined .... **ROSACEAE**
94 Leaves compound 95
94 Leaves simple 100
95 Leaves even-pinnate or even-bipinnate; fruit a large woody legume (pod splitting on 2 sutures) (*Robinia*) .. .. **FABACEAE**
95 Leaves odd-pinnate, trifoliolate, or palmate; fruit a samara, drupe, berry, or 4-5-lobed capsule 96
96 Inflorescences terminal 97
96 Inflorescences lateral or axillary 98
97 Fruit a samara, in loose open cymes; leaves trifoliolate, punctate with translucent oil glands (*Ptelea*) .... **RUTACEAE**
97 Fruit a glandular-pubescent drupe, in dense panicles; leaves pinnately compound, without translucent glands .... **ANACARDIACEAE**
98 Leaflets strongly spiny-toothed; flowers yellow (*Berberis*) .... **BERBERIDACEAE**
98 Leaflets without spines; flowers greenish yellow ...99
99 Leaves palmately compound with mostly 5-7 leaflets, if trifoliolate then leaflets sharply toothed or pinnately lobed; plant a vine with tendrils; stamens opposite the petals (i.e., alternate with the sepals) (*Parthenocissus*) . ... **VITACEAE**
99 Leaves trifoliolate or pinnately compound with entire or nearly entire leaflets; plant a shrub or vine with adventitious roots (not tendrils); stamens alternating with the petals (i.e., opposite the sepals) (*Toxicodendron*) .... **ANACARDIACEAE**
100 Styles 2, separate to the base; petals 4, yellow, linear . ... **HAMAMELIDACEAE**
100 Style 1 or 3 (may be lobed or cleft at summit); petals various 101
101 Stems spiny; flowers yellow, 6-merous (*Berberis*)...... .... **BERBERIDACEAE**
101 Stems unarmed; flowers white, pink, or greenish, 4-5-merous...102
102 Stamens more numerous than the petals; inflorescence an umbel or raceme; plant a low evergreen subshrub . ... **ERICACEAE**
102 Stamens the same number as the petals; inflorescence various; plant a bushy shrub, deciduous except in *Rhododendron*...................................103
103 Leaves evergreen, densely white- or brown-tomentose beneath, revolute (*Rhododendron*) .... **ERICACEAE**
103 Leaves deciduous, glabrous or nearly so, with flat margins 104
104 Stamens alternating with the sepals (i.e., opposite the petals); style 3-lobed .... **RHAMNACEAE**
104 Stamens opposite the sepals (i.e., alternating with the petals); style nearly or quite absent **AQUIFOLIACEAE**

GROUP 8
Herbaceous plants lacking both green color and developed leaves at flowering time.

1 Plants not anchored in the ground, solely parasitic on and attached to stems of other plants at maturity ...2
1 Plants clearly anchored in the ground, not attached to other above-ground plants 3
2 Stem up to 15 mm long, with minute opposite leaves (scale-like); flowers in May, unisexual (plants dioecious), the staminate with stamens adnate to calyx lobes, the pistillate with inferior ovary; parasites on conifers (*Arceuthobium*) .... **SANTALACEAE**
2 Stem elongate, with minute alternate leaves; flowers in late summer, bisexual, the stamens partly adnate to corolla and the ovary superior; parasites on flowering plants (*Cuscuta*) .... **CONVOLVULACEAE**
3 Stem buried in ground; flowers in late winter or earliest spring, crowded in a spadix with a nearly or partly buried hood-like brownish or mottled spathe (green leaves from rhizome appearing after flowering); stamens 4; plant with skunk-like odor (*Symplocarpus*) .. **ARACEAE**
3 Stem or flower stalk above ground; flowers later, solitary or in a few- to many-flowered raceme, umbel, or head; stamens various; plant with odor, if any, not skunk-like 4
4 Flowers completely 3-merous and regular, in an umbel on a naked peduncle arising from an underground, onion-smelling bulb .... **LILIACEAE**
4 Flowers not completely 3-merous, regular or bilaterally symmetrical, not in an umbel, on aerial stems 5
5 Scale-like leaves (and branches if any) opposite; flowers less than 5 mm long 6
5 Scale-like leaves alternate (or apparently none); flowers of various size....................................7
6 Stem thick and fleshy, appearing jointed, the flowers deeply embedded in it (*Salicornia*) .... **AMARANTHACEAE**
6 Stem normal, slender and wiry, the flowers not at all embedded in it (*Bartonia*) .... **GENTIANACEAE**
7 Inflorescence a single dense, short spike with spirally arranged scales; flowers lacking petals and sepals; stem with tubular sheaths at base (*Eleocharis*) **CYPERACEAE**
7 Inflorescence of normal flowers not aggregated into a single dense spike, flowers with at least tiny petals, often showy; stem without tubular sheaths at base (except some Orchidaceae) 8

8 Petals 5, mostly united in a tube, the flower slightly to distinctly bilaterally symmetrical, not spurred; stamens 4 .... **OROBANCHACEAE**
8 Petals 3-5 but not united in a tubular corolla, the flower regular or strongly bilaterally symmetrical (sometimes spurred); stamens various........................... 9

9 Perianth strongly bilaterally symmetrical; stamens 1-2. 10
9 Perianth regular; stamens 4-10 11

10 Sepals and petals 3, the lower petal a definite lip, the others little modified; ovary inferior; plants of various habitat but not aquatic; perianth of various colors .... **ORCHIDACEAE**
10 Sepals apparently 2 and petals 5, but corolla basically 2-lipped; ovary superior; plants of wet shores, ponds, and bog pools, with perianth yellow or purple (*Utricularia*) .... **LENTIBULARIACEAE**

11 Corolla at least 5 mm long; stamens 8-10 . **ERICACEAE**
11 Corolla less than 5.5 mm long; stamens 4 12

12 Flowers sessile; plant of wet lake shores, nearly or quite aquatic; stigmas 4, conspicuously exposed (corolla barely 2 mm long); fruit an indehiscent nutlet (*Myriophyllum tenellum*)........... **HALORAGACEAE**
12 Flowers long-pediceled; plant of peaty habitats but not aquatic; stigma inconspicuous (corolla longer); fruit a capsule (*Bartonia*) .... **GENTIANACEAE**

GROUP 9

Inflorescence apparently converted to bulblets, tufts of leaves, etc.

1 Leaves with flat, net-veined (or dissected) blades.... 2
1 Leaves terete or slender and parallel-veined 3

2 Leaves with narrow, sparsely toothed leaflets or further dissected; stem hollow; bulblets produced in the axils of broad-based acuminate bracts or leaves, not transversely segmented (*Cicuta bulbifera*).. **APIACEAE**
2 Leaves simple and entire; stem solid; bulblets otherwise (*Lysimachia terrestris*) .... **PRIMULACEAE**

3 Bulblets in a ± spherical head or umbel; plants with odor of onion or garlic (*Allium*) .... **LILIACEAE**
3 Bulblets not in a distinct umbel or spherical shead; plants without strong odor 4

4 Leaves terete, septate (with hard cross-partitions, easily seen on dry specimens or felt by gently pinching a leaf and drawing it between the fingers) (*Juncus*) .... **JUNCACEAE**
4 Leaves flat, neither terete nor septate 5

5 Stem ± triangular and solid (*Scirpus*).... **CYPERACEAE**
5 Stem terete, with hollow internodes .... **POACEAE**

GROUP 10

Monocots; leaves with parallel veins.

1 Plant a climbing or twining vine, in most species with tendrils; flowers unisexual; leaves net-veined 2
1 Plant not a vine and without tendrils; flowers bisexual or unisexual; leaves parallel- or net-veined 3

2 Inflorescence an umbel; plants with tendrils; ovary superior; fruit a berry..................... **LILIACEAE**
2 Inflorescence spicate to paniculate; plant without tendrils; ovary inferior; fruit a capsule .... **DIOSCOREACEAE**

3 Inflorescence a spadix, subtended by a spathe which may be broad and hood-like or elongate; leaves in some species compound or net-veined 4
3 Inflorescence not a spadix (if flowers in a head, this with neither an elongate fleshy axis nor a conspicuous subtending spathe); leaves simple, rarely net-veined (in *Smilax*, *Trillium*, and some Alismataceae)............ 5

4 Leaves narrow, sword-like, with ± parallel sides; spathe appearing like a continuation of the leaf-like peduncle (the spadix thus apparently lateral) ..... **ACORACEAE**
4 Leaves expanded; spathe clearly differentiated from peduncle .... **ARACEAE**

5 Perianth much reduced: absent, or composed solely of bristles (these small and stiff or elongate and cottony), or of chaffy or scale-like parts, never conspicuously petaloid 6
5 Perianth at least in part of ± conspicuous white or colored petals................................... 14

6 Individual flowers subtended by 1 or 2 scales; leaves ± elongate, grass-like, usually with a sheath at the base surrounding the stem; fruit a 1-seeded grain or nutlet (achene) 7
6 Individual flowers subtended by no scales or only by bristles, or with a regular perianth of chaffy scales (or tepals); leaves and fruit various..................... 8

7 Each fertile flower subtended by a single scale (others may be at base of spikelet); sheaths of leaves closed (margins connate); stems frequently triangular (but 4-several-angled or terete in many species), usually solid; leaves usually 3-ranked (especially in a species with terete hollow stem); stamens with filament attached to end of anther; fruit a definitely 2- or 3-sided (rarely nearly terete) nutlet .... **CYPERACEAE**
7 Each flower subtended by 2 scales (almost opposite each other, one rarely absent); sheaths often open; stems ± terete (sometimes flattened), never triangular; leaves not clearly 3-ranked (basically 2-ranked); stamens with filament attached near middle of anther (or apparently so because of sagittate anthers); fruit usually a grain neither flattened (2-sided) nor triangular .... **POACEAE**

8 Inflorescence a single, very compact, almost spherical head (terminating an erect scape), less than 12 mm across.. 9

8 Inflorescence not a single terminal head and/or exceeding 12 mm 10
9 Surface of head (tips of receptacular bracts) white-woolly; flowers chaffy, not concealed by involucral bracts; roots with abundant conspicuous transverse markings .... **ERIOCAULACEAE**
9 Surface of head (bracts) glabrous; flowers yellow or largely concealed by bracts; roots without transverse markings .... **XYRIDACEAE**
10 Inflorescence composed of separate staminate and pistillate portions, the former consisting of conspicuous stamens, sooner or later withering, leaving only the pistillate portion conspicuous .... **TYPHACEAE**
10 Inflorescence composed of bisexual flowers, without conspicuously separate staminate and pistillate portions 11
11 Flowers in a branched or umbellate inflorescence, solitary or, more often, clustered into small heads of 2 or more; fruit a 3- to many-seeded capsule .... **JUNCACEAE**
11 Flowers in a single elongate spike or zigzag raceme; fruit indehiscent or a 1-2-seeded follicle 12
12 Spike (truly a spadix) apparently lateral; fruit of each flower indehiscent .... **ACORACEAE**
12 Spike or raceme terminal; fruit of each flower consisting of 3 or 6 1-2-seeded follicles 13
13 Pedicels bractless; carpels 3 or 6, erect and ± adherent to a central axis at maturity; leaves all basal or nearly so, without a terminal pore........... **JUNCAGINACEAE**
13 Pedicels bracted; carpels 3, widely divergent in fruit; leaves mostly cauline, each with a terminal pore (*Scheuchzeria*) .... **SCHEUCHZERIACEAE**
14 Flowers bilaterally symmetrical 15
14 Flowers regular (radially symmetrical) 17
15 Ovary inferior; fertile stamens 1 or 2, united with the pistil; flowers not blue (almost any other color)....... .... **ORCHIDACEAE**
15 Ovary superior; fertile stamens 3 or 6, free; flowers blue (except albinos), at least in part 16
16 Sepals colored like the petals; stamens 6, all fertile; flowers in a dense elongate inflorescence (*Pontederia*) .... **PONTEDERIACEAE**
16 Sepals greenish, unlike the petals; stamens 6, 3 with imperfect anthers; flowers few (*Commelina*).......... .... **COMMELINACEAE**
17 Sepals and petals of quite different color and/or texture, the former green or brownish 18
17 Sepals and petals both colored and petaloid, usually similar in shape (tepals) or the sepals (in Iris) of different size and shape 21
18 Leaves in a single whorl of 3 on the stem .. **LILIACEAE**
18 Leaves all basal or, if cauline, not in a single whorl of 3 19
19 Petals yellow; flowers in a single compact head less than 12 mm across........................ **XYRIDACEAE**

19 Petals blue, purple, white, or pink; flowers in a more open or larger inflorescence 20
20 Pistils several in each flower, each developing into an achene; stamens 6-many; flowers unisexual or bisexual; petals white or pinkish; leaves often broadly elliptic or sagittate, usually ± net-veined, all basal .... **ALISMATACEAE**
20 Pistil 1 in each flower, developing into a capsule; stamens 6; flowers bisexual; petals blue, purple, or deep pink (except in occasional albinos); leaves elongate, clearly parallel-veined, basal and cauline............. .... **COMMELINACEAE**
21 Ovary inferior (flowers bisexual).................... 22
21 Ovary superior (or flowers unisexual) 25
22 Stamens 3; leaves equitant .... **IRIDACEAE**
22 Stamens 6; leaves not equitant 23
23 Ovary only half-inferior, part of it adnate to the perianth, glabrous (at most granular-roughened) (*Anticlea*).............................. **LILIACEAE**
23 Ovary clearly inferior 24
24 Foliage and ovary hairy (*Hypoxis*) ... **HYPOXIDACEAE**
24 Foliage and ovary glabrous .... **LILIACEAE**
25 Stamens 3; tepals 6, yellow; plants creeping on wet shores (*Heteranthera*) .... **PONTEDERIACEAE**
25 Stamens and tepals 4 or 6, the latter yellow or not; plants erect, of various habitats .... **LILIACEAE**

GROUP 11

Flowers in an involucrate head (i.e., the flowers clustered in a head above a whorl of bracts).

1 Flowers on a thick fleshy axis (inflorescence a spadix) subtended by a single large overtopping or enveloping bract (spathe); perianth none or of 4 tepals 2
1 Flowers not in a spadix overtopped by a spathe; perianth various 3
2 Leaves narrow, sword-like, with ± parallel sides; spathe appearing like a continuation of the leaf-like peduncle (the spadix thus apparently lateral) ..... **ACORACEAE**
2 Leaves expanded; spathe clearly differentiated from peduncle and terminal .... **ARACEAE**
3 Leaves parallel-veined, all basal, and less than 5 mm broad 4
3 Leaves net-veined or if parallel-veined then cauline and more than 5 mm broad........................... 5
4 Flowers yellow, mostly concealed by bracts; roots without transverse markings; surface of head (bracts) glabrous............................. **XYRIDACEAE**
4 Flowers chaffy, not concealed by involucral bracts; roots with abundant conspicuous transverse markings; surface of head white-woolly (tips of receptacular bracts) .... **ERIOCAULACEAE**
5 Ovary inferior 6
5 Ovary superior 7

6 Leaves opposite (very rarely whorled), toothed or pinnatifid; corolla 4-lobed, lilac-purple (sometimes pale); stamens 4, separate .... **DIPSACACEAE**
6 Leaves and corolla not combined as above, e.g., leaves alternate and/or entire or corolla 5-lobed and/or not lilac-purple .... **ASTERACEAE**

7 Leaves alternate, compound; involucral bract 3-foliolate; flowers strongly bilaterally symmetrical, papilionaceous (as in other legumes) (*Trifolium*) .... **FABACEAE**
7 Leaves opposite, simple; flowers often nearly or quite regular 8

8 Plants with minty odor; ovary deeply 4-lobed, with 1 style; petals united .... **LAMIACEAE**
8 Plants without minty odor; ovary not lobed, with 2 styles; petals separate (*Petrorhagia, Dianthus*) .... **CARYOPHYLLACEAE**

GROUP 12

Herbaceous plants with single-sex flowers.

1 Leaves compound 2
1 Leaves simple 9

2 Leaves palmately compound (or 3-foliolate) 3
2 Leaves pinnately compound or more than once compound 7

3 Flowers in umbels 4
3 Flowers in spikes or panicles 5

4 Leaves cauline, in a single whorl (*Panax*) **ARALIACEAE**
4 Leaves alternate and basal (*Sanicula*) ...... **APIACEAE**

5 Margins of leaflets entire; flowers at the base of a prolonged fleshy spadix subtended by a single large bract (spathe) (*Arisaema*) .... **ARACEAE**
5 Margins of leaflets toothed; flowers on normal herbaceous (but not fleshy) pedicels or axes 6

6 Leaves all opposite; plant a vine; perianth showy (*Clematis*) .... **RANUNCULACEAE**
6 Leaves alternate on upper part of stem; plant erect; perianth minute and inconspicuous (*Cannabis*) .... **CANNABACEAE**

7 Flowers in panicles (*Thalictrum*) .. **RANUNCULACEAE**
7 Flowers in tight ovoid heads or umbels 8

8 Leaves once pinnately compound; flowers in tight heads; perianth 4-merous (*Sanguisorba*)... **ROSACEAE**
8 Leaves 2-3 times compound; flowers pediceled, in umbels; perianth 5-merous (*Aralia*) ..... **ARALIACEAE**

9 Plant with leaves all basal 10
9 Plant with leaves all or mostly cauline............. 11

10 Flowers in dense spikes (or 1-3 at base in *Littorella*) ... . **PLANTAGINACEAE**
10 Flowers pediceled in panicles (*Rumex*)............... .... **POLYGONACEAE**

11 Leaves peltate or pubescent with forked/stellate hairs .... **EUPHORBIACEAE**
11 Leaves neither peltate nor with forked/stellate hairs 12

12 Leaves opposite or whorled 13
12 Leaves alternate (at least at upper nodes).......... 18

13 Flowers solitary in axils of leaves; perianth none; stamen 1 .... **PLANTAGINACEAE**
13 Flowers in axillary or terminal inflorescences 14

14 Leaves hastate, otherwise unlobed but entire to coarsely or irregularly toothed; pistillate flowers and fruit mostly concealed by a pair of bracts with margins ± united at base (*Atriplex*) .... **AMARANTHACEAE**
14 Leaves not hastate, in some species deeply lobed, in some closely toothed; pistillate flowers without 2 basal bracts 15

15 Inflorescence terminal; corolla white or colored 16
15 Inflorescence axillary; corolla none or of reduced scales 17

16 Stem leaves deeply pinnately lobed; style 1; stamens 3-4 (*Valeriana*).................... **CAPRIFOLIACEAE**
16 Stem leaves unlobed; styles 3-7; stamens 10 (*Silene*)... .... **CARYOPHYLLACEAE**

17 Plant a vine; leaves deeply 3-7-lobed (*Humulus*)....... .... **CANNABACEAE**
17 Plant erect, not a vine; leaves unlobed . **URTICACEAE**

18 Flowers with 6 petaloid tepals and 6 stamens or 3 carpels (dioecious); inflorescences on long peduncles from the nodes (not terminal); leaves with several prominent longitudinal veins (including midrib) (*Dioscorea*) .... **DIOSCOREACEAE**
18 Flowers either with other numbers of tepals, stamens, and carpels or the inflorescence terminal (on main stem or branches); leaves various but without several prominent long veins 19

19 Perianth with both calyx and corolla (sometimes very inconspicuous); plants climbing or trailing, with tendrils .... **CUCURBITACEAE**
19 Perianth absent or of 1 series of parts (tepals); plants erect or prostrate, without tendrils 20

20 Flowers very small, in axillary clusters [plants monoecious; look for pistillate flowers for keying] .. 21
20 Flowers small or not, in chiefly terminal inflorescences (spikes, panicles, or racemes on main stem and/or branches) 23

21 Style 1; stamens 4 or 5.................. **URTICACEAE**
21 Styles (or sessile stigmas) 2-3; stamens various..... 22

22 Styles 3, branched; bracts in inflorescence well developed and at least 5-10-lobed (*Acalypha*)......... .... **EUPHORBIACEAE**
22 Styles 2-3, unbranched; bracts in inflorescence unlobed (may be toothed) (*Amaranthus, Atriplex*) .... **AMARANTHACEAE**

23 Flowers consistently 3-merous (tepals 6, stamens 6, carpels 3); stipules united into a sheath (ocrea) surrounding the stem above each node (*Rumex*) .... **POLYGONACEAE**

23 Flowers not consistently 3-merous (tepals 5 or fewer, stamens usually 5, styles often 2); stipules none . . . . **AMARANTHACEAE**

GROUP 13
Herbaceous dicots with bisexual flowers, perianth in 1 series, ovary inferior.

1 Stamens more numerous than the 1-4 perianth lobes or parts 2
1 Stamens the same number as or fewer than the perianth lobes or parts, or perianth 5-merous (or both conditions) 3

2 Perianth with 1-3 (rarely 4) lobes; stamens 6 or 12 . . . . **ARISTOLOCHIACEAE**
2 Perianth 4-parted; stamens 8 (*Chrysosplenium*) . . . . **SAXIFRAGACEAE**

3 Leaves all or mostly opposite or whorled 5
3 Leaves alternate or basal 12

4 Inflorescence a dense terminal cluster of flowers (sessile or nearly so) 6
4 Inflorescence of solitary, axillary, or clearly pediceled flowers 8

5 Leaves apparently whorled; bracts below the inflorescence large and white (*Cornus canadensis*) . . . . **CORNACEAE**
5 Leaves clearly opposite; bracts below the inflorescence greenish or inconspicuous 7

6 Heads subtended by several involucral bracts below a receptacle with sessile flowers . . . . **DIPSACACEAE**
6 Heads not subtended by a distinct involucre, with visible branching structure; flowers sessile but not on a common receptacle (*Valeriana*) . . . . **CAPRIFOLIACEAE**

7 Leaves compound, in a single whorl (*Panax*) . . . . **ARALIACEAE**
7 Leaves simple or deeply lobed, opposite or in several whorls (rarely the lower alternate in *Valeriana*) 8

8 Leaves in whorls (*Galium*) . . . . **RUBIACEAE**
8 Leaves opposite 9

9 Plants low and densely matted, with linear leaves; perianth 5-merous (*Scleranthus*) **CARYOPHYLLACEAE**
9 Plants prostrate or erect, but with broader leaves; perianth 5- or 4-merous 10

10 Flowers in rather dense terminal inflorescences (at ends of stem and branches); stamens 3 (occasionally 4) (*Valeriana*) . . . . **CAPRIFOLIACEAE**
10 Flowers 1-few in axils or solitary at ends of branches; stamens various 11

11 Styles 2; flowers solitary at ends of branches; plant flowering in May (*Chrysosplenium*) . **SAXIFRAGACEAE**
11 Style 1; flowers sessile, axillary; plant flowering in summer (*Ludwigia palustris*) . . . . **ONAGRACEAE**

12 Leaves entire, simple and unlobed; flowers in cymes or few-flowered cymules; style 1 . . . . **SANTALACEAE**

12 Leaves (at least the cauline ones) toothed or crenulate, often deeply lobed or compound; flowers in umbels, axillary, or ovoid to cylindric heads; styles 2, 3, or 5 (1 in *Sanguisorba*) 13

13 Tepals and stamens each 5 14
13 Tepals and stamens each 3 or 4 16

14 Styles 5; fruit fleshy, berry-like (*Aralia*) . . **ARALIACEAE**
14 Styles 2; fruit dry, splitting into 2 achene-like indehiscent parts (mericarps) 15

15 Leaves simple, with crenate margins (*Hydrocotyle*) . . . . **ARALIACEAE**
15 Leaves, at least the cauline, compound, dissected, or deeply lobed . . . . **APIACEAE**

16 Stamens and tepals 3 (*Proserpinaca*) **HALORAGACEAE**
16 Stamens and tepals 4 18

17 Leaves with conspicuous stipules, pinnately compound and strongly toothed (*Sanguisorba minor*) . **ROSACEAE**
17 Leaves without stipules . . . . go to couplet 10

GROUP 14
Herbaceous dicots with bisexual flowers, perianth in 1 series, superior ovary.

1 Ovaries more than 1 in each flower, the carpels separate at least above the middle of the ovaries 2
1 Ovary 1 in each flower (bearing 1 or more styles), the carpels united at least below the styles 4

2 Stipules conspicuous; leaves pinnately compound (*Sanguisorba minor*) . . . . **ROSACEAE**
2 Stipules none or leaves simple 3

3 Ovaries united for most of lower half; leaves simple, unlobed (*Penthorum*) . . . . **PENTHORACEAE**
3 Ovaries distinct; leaves of most species lobed or compound . . . . **RANUNCULACEAE**

4 Leaves bipinnately compound, fruit a legume . . . . **FABACEAE**
4 Leaves simple or compound (but not bipinnate); fruit not a legume 5

5 Plants with a solitary large (ca. 3-5 cm wide) white flower between a single usually opposite or subopposite pair of long-petioled cauline eccentrically peltate and deeply lobed leaves (*Podophyllum*) . . **BERBERIDACEAE**
5 Plants with more flowers per stem or, if only one, then leaves not as above 6

6 Stamens more than twice as many as the perianth lobes or parts 7
6 Stamens only twice as many as the perianth lobes or parts, or fewer 10

7 Leaves tubular, open at apex and pitcher-like . . . . **SARRACENIACEAE**
7 Leaves flat, of normal structure, simple or compound but not hollow 8

8 Perianth small and inconspicuous (stamens more showy); leaves compound with definite flat broad leaflets .... **RANUNCULACEAE**
8 Perianth well developed, showy; leaves simple or dissected into very narrowly linear segments 9
9 Leaf blades entire, unlobed except for deeply cordate base; plants aquatic (*Nuphar*)........ **NYMPHAEACEAE**
9 Leaf blades deeply lobed or dissected; plants terrestrial .... **PAPAVERACEAE**
10 Style 1 or none (stigmas may be 2 or more) 11
10 Styles 2 or more 18
11 Stamens more numerous than the perianth divisions 12
11 Stamens the same number as or fewer than the perianth lobes.................................. 13
12 Flowers bilaterally symmetrical; perianth colorful (white, yellow, or pink)............. **PAPAVERACEAE**
12 Flowers regular; perianth dull, greenish .... **BRASSICACEAE**
13 Leaves alternate or basal 14
13 Leaves opposite 16
14 Perianth parts (and stamens) 6, 8, or 9............. .... **BERBERIDACEAE**
14 Perianth parts (and usually stamens) 4............. 15
15 Leaves simple, entire (*Parietaria*) .... **URTICACEAE**
15 Leaves pinnately compound with toothed leaflets (*Sanguisorba minor*).................... **ROSACEAE**
16 Flowers solitary or few in axils of leaves, sessile or nearly so .... **LYTHRACEAE**
16 Flowers in terminal inflorescences (on stems and branches).. 17
17 Perianth showy, pink to purple; inflorescences each subtended by a conspicuous petaloid or papery 5-lobed involucre which enlarges as fruit matures .... **NYCTAGINACEAE**
17 Perianth reduced, inconspicuous, whitish or scarious; inflorescences subtended at most by very small bracts (*Froelichia*).................... **AMARANTHACEAE**
18 Flowers embedded in a succulent segmented stem; leaves reduced to tiny opposite scales (*Salicornia*) .... **AMARANTHACEAE**
18 Flowers not embedded in a succulent stem; leaves not scalelike.. 19
19 Leaves opposite or whorled 20
19 Leaves alternate 22
20 Margins of leaves crenate; stamens normally 8 (*Chrysosplenium*)................ **SAXIFRAGACEAE**
20 Margins of leaves entire; stamens various, but usually not 8.. 21
21 Leaves opposite................ **CARYOPHYLLACEAE**
21 Leaves whorled .... **MOLLUGINACEAE**
22 Plant with a ± membranous stipular sheath (ocrea) surrounding the stem above each node .... **POLYGONACEAE**
22 Plant lacking stipules of any kind **AMARANTHACEAE**

GROUP 15

Herbaceous dicots with bisexual flowers, perianth of 2 series, ovaries 2 or more in each flower.

1 Style and/or stigmas united (i.e., 1 in each flower, but style may be branched) 2
1 Style and stigmas separate (1 on each ovary, or scarcely developed) 5
2 Ovaries 2; corolla regular, of united petals; stamens 5; sap in most species milky .... **APOCYNACEAE**
2 Ovaries 4 or more; corolla regular or bilaterally symmetrical, of united or separate petals; stamens 2, 4, 5, or numerous; sap not milky 3
3 Petals separate; ovaries apparently 5 or more; stamens numerous, their filaments connate, at least for much of their length, into a tube around the style; leaves palmately veined (may be deeply lobed) **MALVACEAE**
3 Petals united; ovaries apparently 4; stamens 2, 4, or 5; their filaments not connate (but ± adnate to corolla); leaves mostly pinnately veined 4
4 Leaves alternate; stamens 5; corolla regular (bilaterally symmetrical only in the very bristly *Echium*); stems not angled (rarely winged) and foliage not aromatic .... **BORAGINACEAE**
4 Leaves opposite; stamens 2 or 4; corolla bilaterally symmetrical or in a few genera essentially regular; stems usually 4-angled ("square") and foliage often aromatic when bruised (minty or citrus-like) .... **LAMIACEAE**
5 Perianth bilaterally symmetrical 6
5 Perianth regular................................... 7
6 Leaves deeply cleft; stamens numerous; perianth with spurs; fruit a follicle (3 per flower) (*Aconitum*) .... **RANUNCULACEAE**
6 Leaves shallowly lobed; stamens 5; perianth without spurs; fruit a capsule (*Heuchera*) ... **SAXIFRAGACEAE**
7 Sepals (or sepal-like bracts) 3 8
7 Sepals 4 or more 10
8 Plant aquatic, with peltate (often floating) round or shield-shaped alternate floating leaf blades or palmately dissected opposite submersed leaves **CABOMBACEAE**
8 Plant terrestrial, with leaves neither peltate nor palmately dissected 9
9 Petals and usually carpels 3; stamens 3 or 6; leaves cauline, deeply and narrowly pinnate-lobed .... **LIMNANTHACEAE**
9 Petals 5 or more, carpels and stamens numerous; leaves basal, with 3 (-7) broad lobes (*Hepatica*).............. .... **RANUNCULACEAE**
10 Sepals separate to the base; stamens and petals individually falling from the receptacle after anthesis . ... **RANUNCULACEAE**

10 Sepals, petals, and stamens united to form a saucer- or cup-like floral tube ("hypanthium") at the margin of which the stamens and petals are borne. 11
11 Carpels as many as, or more than, the petals 12
11 Carpels fewer than the petals 13
12 Leaves succulent, simple, entire, estipulate . . . . **CRASSULACEAE**
12 Leaves not succulent, deeply lobed or compound, toothed, stipulate . . . . **ROSACEAE**
13 Leaves simple, at most shallowly lobed . . . . **SAXIFRAGACEAE**
13 Leaves clearly compound . . . . **ROSACEAE**

GROUP 16

Herbaceous dicots with bisexual flowers, perianth of 2 series, ovary inferior.

1 Stamens twice as many as the petals (or nearly so) . . 2
1 Stamens the same number as the petals or corolla lobes, or fewer 5
2 Style 1 (sometimes very short) 3
2 Styles 2 or more. 4
3 Petals spreading; herbaceous plants; fruit a capsule or dry and indehiscent . . . . **ONAGRACEAE**
3 Petals strongly reflexed; creeping evergreen wetland subshrubs; fruit a berry (*Vaccinium oxycoccos* and *V. macrocarpon*) . . . . **ERICACEAE**
4 Sepals 2; leaves succulent; styles 3. **PORTULACACEAE**
4 Sepals (4-) 5; leaves not succulent; styles 2 . . . . **SAXIFRAGACEAE**
5 Petals united 6
5 Petals separate 14
6 Stem leaves alternate. 7
6 Stem leaves opposite or whorled (rarely the lower alternate in *Valeriana* with pinnate leaves) 8
7 Corolla bilaterally symmetrical (*Lobelia*) . . . . **CAMPANULACEAE**
7 Corolla regular . . . . **CAMPANULACEAE**
8 Leaves whorled . . . . **RUBIACEAE**
8 Leaves opposite 9
9 Stipules present (connate around the stem) . . . . **RUBIACEAE**
9 Stipules absent. 10
10 Flowers sessile in terminal heads 11
10 Flowers visibly pediceled (even if crowded) or axillary. 12
11 Heads subtended by several involucral bracts below a receptacle with sessile flowers . . . . **DIPSACACEAE**
11 Heads not subtended by a distinct involucre, with visible branching structure; flowers sessile but not on a common receptacle (*Valeriana*). . . . **CAPRIFOLIACEAE**
12 Flowers numerous, in rather dense terminal inflorescences (at ends of stem and branches) (*Valeriana*) . . . . **CAPRIFOLIACEAE**
12 Flowers axillary or on paired pedicels on a peduncle. 13
13 Leaves strictly entire, stems erect (*Triosteum*) . . . . **CAPRIFOLIACEAE**
13 Leaves shallowly toothed on apical half; stems trailing; flowers on paired pedicels on a peduncle (*Linnaea*). . . . . **CAPRIFOLIACEAE**
14 Stamens and petals each 2 (*Circaea*) . . **ONAGRACEAE**
14 Stamens (fertile) and petals each 4 or 5 (stamens sometimes alternating with staminodia, which may have gland-tipped divisions) 15
15 Petals 4 16
15 Petals 5 17
16 Principal leaves apparently whorled; flowers in a dense head like terminal cluster subtended by 4 large white bracts (*Cornus canadensis*) . . . . **CORNACEAE**
16 Principal leaves alternate; flowers neither in a head-like terminal cluster nor subtended by 4 large white bracts (*Ludwigia*) . . . . **ONAGRACEAE**
17 Leaves simple; styles 2 or stigmas 4 and sessile; inflorescence various. 18
17 Leaves compound; inflorescence an umbel 19
18 Flowers in panicles, in cymes, or solitary. . . . . **SAXIFRAGACEAE**
18 Flowers in umbels . . . . **APIACEAE**
19 Styles 5; fruit berry-like (*Aralia*) . . . . **ARALIACEAE**
19 Styles 2-3; fruit various 20
20 Leaves alternate or basal; fruit dry, splitting into 2 achene-like indehiscent parts (mericarps) . **APIACEAE**
20 Leaves in a single whorl; fruit berry-like (*Panax*) . . . . **ARALIACEAE**

GROUP 17

Herbaceous dicots with bisexual flowers, perianth of 2 series, ovary 1 and superior, stamens more numerous than the petals.

1 Corolla bilaterally symmetrical 2
1 Corolla regular (radially symmetrical) 10
2 Sepals all or partly petal-like in appearance or prolonged into a spur. 3
2 Sepals not petal-like in form or appearance, usually green 4
3 Spur none; stamens 6, 7, or 8; leaves entire. . . . . **POLYGALACEAE**
3 Spur present on one of the sepals . **BALSAMINACEAE**
4 Sepals 2, separate, usually deciduous early in anthesis; leaves dissected or twice-compound **PAPAVERACEAE**
4 Sepals 4 or more, usually ± connate 5
5 Lower 2 petals forming a laterally compressed "keel" that encloses the stamens; leaves once-compound . . . . **FABACEAE**

5	Lower petals not forming a keel nor enclosing the stamens. 6
6	Flowers completely 5-merous (sepals and petals 5, stamens 5 or usually 10); pistil long-beaked; corolla pink or purple; leaves deeply lobed or cleft or compound (the main stem leaves opposite and toothed or cleft). . . . . **GERANIACEAE**
6	Flowers with at least the carpels fewer than 5; corolla and leaves various 7
7	Leaves simple, deeply lobed to entire. 8
7	Leaves compound 9
8	Sepals and petals each 4; stamens 6. . **BRASSICACEAE**
8	Sepals and petals each (4) 5-7; stamens 2x as many . . . . **LYTHRACEAE**
9	Petals and sepals each 4; fruit a capsule (*Polanisia*). . . . . **CLEOMACEAE**
9	Petals and sepals each 5; fruit a legume . . . **FABACEAE**
10	Leaves tubular, open at apex and pitcher-like; style greatly expanded, large and umbrella-shaped (*Sarracenia*). . . . . **SARRACENIACEAE**
10	Leaves flat or at most succulent, of usual shapes; style not unusually expanded 11
11	Plants with a solitary large (ca 3-5 cm wide) white flower between a single, usually opposite, long-petioled pair of deeply lobed leaves (*Podophyllum*). . **BERBERIDACEAE**
11	Plants with more flowers per stem or, if only one, then leaves not as above 12
12	Sepals 2 13
12	Sepals 3 or more 14
13	Leaves lobed, compound, or coarsely toothed, not succulent; sap in most species colored (yellow to orange) . . . . **PAPAVERACEAE**
13	Leaves unlobed, entire, succulent; sap watery . . . . **PORTULACACEAE**
14	Stamens more than 2x as many as the petals 15
14	Stamens twice as many as the petals or fewer 21
15	Leaves compound 16
15	Leaves simple 17
16	Plant clammy-pubescent; leaves palmately compound with 3 entire leaflets (*Polanisia*) . . . . **CLEOMACEAE**
16	Plant glabrous or with a little non-glandular pubescence; leaves twice-compound with numerous sharply toothed leaflets . . . . **RANUNCULACEAE**
17	Plant truly aquatic, with all leaves basal, the petioles all arising from a rhizome buried under water (except when stranded) 18
17	Plant terrestrial, with at least some leaves on the stem 19
18	Leaves rounded at tip with deep sinus at base . . . . **NYMPHAEACEAE**
18	Leaves peltate (*Brasenia*) . . . . **CABOMBACEAE**
19	Style 1 (or none, with 3 sessile stigmas) . . . **CISTACEAE**
19	Styles 2 or more, evident 20
20	Leaves opposite, with translucent dots; petals yellow (*Hypericum*) . . . . **HYPERICACEAE**
20	Leaves alternate, without translucent dots; petals of various colors . . . . **MALVACEAE**
21	Stamens fewer than twice as many as the petals . . . 22
21	Stamens exactly twice as many as the petals 26
22	Styles 2-5; leaves opposite or whorled, simple and entire 23
22	Style 1 or none; leaves usually alternate, simple or compound, entire or toothed 25
23	Petals yellow (*Hypericum*) . . . . **HYPERICACEAE**
23	Petals white, pink, or red 24
24	Stamens 9, in 3 distinct groups of 3 each, with 3 conspicuous glands alternating with the groups (*Triadenum*) . . . . **HYPERICACEAE**
24	Stamens various but neither 9 nor in groups . . . . **CARYOPHYLLACEAE**
25	Sepals 5 (of which the 2 outer ones may be much reduced); petals 3, minute (shorter than the calyx), reddish (*Lechea*) . . . . **CISTACEAE**
25	Sepals and petals each 4; petals usually ± showy, colors various . . . . **BRASSICACEAE**
26	Petals 3 27
26	Petals 4 or more 28
27	Leaves alternate, compound or deeply pinnately lobed (*Floerkea*) . . . . **LIMNANTHACEAE**
27	Leaves in a single whorl, simple and unlobed (*Trillium*) . . . . **LILIACEAE**
28	Sepals and petals each 6 or more (*Lythrum*) . . . . **LYTHRACEAE**
28	Sepals and petals each 4 or 5 29
29	Leaves compound or deeply divided nearly to base of blade 30
29	Leaves simple and entire, toothed, or shallowly lobed 33
30	Leaves opposite 31
30	Leaves alternate 32
31	Leaves uniformly trifoliolate . . . . **OXALIDACEAE**
31	Leaves palmately compound or lobed (*Geranium*) . . . . **GERANIACEAE**
32	Styles 5; leaves with 3 obcordate leaflets . . . . **OXALIDACEAE**
32	Style 1; leaves pinnately compound . . . . **FABACEAE**
33	Style 1. 34
33	Styles 2 or more. 37
34	Floral tube ("hypanthium") present, with petals and sepals borne at its margin; petals white to (usually) pink-purple 35
34	Floral tube none, all parts arising directly from the receptacle; petal color various 36
35	Anthers opening by terminal pores, very showy (curved, yellow, appearing set at 90° on the filament) and

stamens becoming skewed toward one side of the flower .... **MELASTOMATACEAE**
35 Anthers opening by longitudinal slits, not especially showy; stamens not skewed.......... **LYTHRACEAE**
36 Sepals of 2 sizes, the 2 outer ones very much narrower and often shorter than the 3 inner ones (appearing as mere appendages on them); petals yellow **CISTACEAE**
36 Sepals all of nearly the same size and shape; petals white, greenish, or pink................ **ERICACEAE**
37 Ovary lobed, with a style on each lobe 38
37 Ovary unlobed, the styles all arising together 39
38 Leaves not succulent, all or mostly basal, (stem leaves, if any, few and small or a single pair); lobes of ovary 2 . ... **SAXIFRAGACEAE**
38 Leaves succulent, all or mostly cauline; lobes of ovary 4 or 5 .... **CRASSULACEAE**
39 Petals yellow; leaves with translucent dots (*Hypericum*) .... **HYPERICACEAE**
39 Petals white to pink or red (never yellow); leaves without translucent dots....... **CARYOPHYLLACEAE**

GROUP 18

Herbaceous dicots with bisexual flowers, perianth of 2 series, ovary 1 and superior, stamens the same number as the petals or fewer, and petals separate.

1 Leaves compound or dissected 2
1 Leaves entire or toothed to deeply lobed 4
2 Flowers solitary on leafless peduncles arising from the ground .... **VIOLACEAE**
2 Flowers on leafy stems 3
3 Petals and stamens each 6; leaves 2-3-times compound, with flat, broad leaflets (*Caulophyllum*) .... **BERBERIDACEAE**
3 Petals and stamens each 5; leaves dissected .... **GERANIACEAE**
4 Leaves opposite or whorled 5
4 Leaves alternate or basal 13
5 Sepals 2 or 3; petals 2-6........................... 6
5 Sepals and petals each 4-6 (or more) 8
6 Plant aquatic or stranded on wet shores; sepals and petals each 2 or 3 .... **ELATINACEAE**
6 Plant terrestrial; sepals 2; petals 5 or 6.............. 7
7 Stem leaves 2; flowers pedunculate ... **MONTIACEAE**
7 Stem leaves numerous; flowers essentially sessile .... **PORTULACACEAE**
8 Leaves deeply palmately lobed ...... **GERANIACEAE**
8 Leaves entire or merely toothed 9
9 Style 1, sometimes very short or the stigma ± sessile 10
9 Styles 2-5...................................... 11
10 Floral tube or disk well developed, with sepals and petals borne at its margin .... **LYTHRACEAE**

10 Floral tube or disk none............ **GENTIANACEAE**
11 Flowers completely 5-merous, including 5 styles; stamens with filaments connate at the base around the ovary; ovary 5- (or 10-) locular .... **LINACEAE**
11 Flowers with styles usually fewer than 5 (and petals sometimes 4); stamens not connate; ovary with 1 locule 12
12 Petals yellow; leaves with translucent dots (*Hypericum*) .... **HYPERICACEAE**
12 Petals white to pink or red; leaves without translucent dots .... **CARYOPHYLLACEAE**
13 Leaves shallowly to deeply palmately lobed 14
13 Leaves unlobed or pinnately lobed 15
14 Corolla bilaterally symmetrical, spurred; style 1 .... **VIOLACEAE**
14 Corolla regular or nearly so, not spurred; styles 2 (*Heuchera*)...................... **SAXIFRAGACEAE**
15 Styles 2 or more 16
15 Style 1 or none 17
16 Leaves essentially all basal; flowers white............ .... **DROSERACEAE**
16 Leaves cauline; flowers yellow .... **LINACEAE**
17 Floral tube well developed and prolonged, with sepals and petals borne at its margin .... **LYTHRACEAE**
17 Floral tube none or very little developed........... 18
18 Corolla bilaterally symmetrical, saccate or spurred at the base .... **VIOLACEAE**
18 Corolla regular, without a spur 19
19 Petals and sepals each 4 .... **BRASSICACEAE**
19 Petals and sepals each 5 20
20 Leaves pinnately lobed or dissected (*Erodium*)........ .... **GERANIACEAE**
20 Leaves entire or merely toothed 21
21 Flowers solitary, terminal; styles essentially none (stigmas 4, nearly sessile); stamens alternating with cleft, gland-tipped staminodia (*Parnassia*)............ .... **CELASTRACEAE**
21 Flowers in a terminal umbel or raceme; style present; staminodia none 22
22 Principal leaves all basal; inflorescence a stalked umbel .... **PRIMULACEAE**
22 Principal leaves all or partly cauline; inflorescence a terminal raceme (*Lysimachia*) .... **PRIMULACEAE**

GROUP 19

Herbaceous dicots with bisexual flowers, perianth of 2 series, ovary 1 and superior, corolla regular and stamens the same number as its lobes, and petals united.

1 Leaves all basal 2
1 Leaves all or mostly cauline 4

2. Leaves covered with conspicuous stalked glands. . . . . **DROSERACEAE**
2. Leaves without stalked glands 3
3. Perianth 4-merous; flowers in spikes or heads; corolla scarious . . . . **PLANTAGINACEAE**
3. Perianth 5-merous; flowers in umbels; corolla petaloid . . . . **PRIMULACEAE**
4. Ovary deeply 4-lobed, appearing like 4 separate ovaries [and also keyed as such] but with one style arising deep in the midst of the lobes 5
4. Ovary not conspicuously lobed (may be slightly 4- or 2-lobed or notched at apex, where style arises) 6
5. Leaves opposite; stamens 2 or 4; stems 4-angled ("square") and foliage aromatic (minty or citrus-like) . . . . **LAMIACEAE**
5. Leaves alternate; stamens 5; stem not angled (rarely winged) and foliage not aromatic . . **BORAGINACEAE**
6. Leaves opposite (or whorled), at least below the inflorescence 7
6. Leaves alternate, at least below the inflorescence. . . 12
7. Flowers in dense heads or short spikes; corolla 4-lobed (*Plantago*) . . . . **PLANTAGINACEAE**
7. Flowers in crowded or more open racemes or other inflorescences; corolla lobes 4-7 8
8. Stamens opposite the corolla lobes (i.e., each stamen arising and oriented above the middle of a lobe) and readily visible . . . . **PRIMULACEAE**
8. Stamens alternating with the corolla lobes (sometimes hidden in a corolla tube or closed corolla). 9
9. Lobes of corolla 4 . . . . **GENTIANACEAE**
9. Lobes of corolla 5. 10
10. Stigmas 3; ovary with 3 locules. . . . . **POLEMONIACEAE**
10. Stigma 1 (may be 2-lobed); ovary with 2 (or 4 or 5) locules. 11
11. Leaves glabrous; ovary 1-locular; fruit a 2-valved capsule . . . . **GENTIANACEAE**
11. Leaves strongly clammy-pubescent; ovary 2-locular; fruit a berry or a 2-valved capsule (*Leucophysalis*, in part) . . . . **SOLANACEAE**
12. Blades of leaves deeply lobed, dissected, or compound 13
12. Blades of leaves entire, toothed, or at most shallowly lobed (or merely cordate) 18
13. Plant a twining or trailing vine 14
13. Plant, whether erect or prostrate, not a vine 15
14. Corolla deeply funnel-shaped (or even trumpet-shaped) . . . . **CONVOLVULACEAE**
14. Corolla ± flat (rotate) (*Solanum*). . . . . . . **SOLANACEAE**
15. Anthers forming a cone around the pistil . . . . **SOLANACEAE**
15. Anthers clearly separate 16
16. Leaves 3-foliolate . . . . **MENYANTHACEAE**
16. Leaves otherwise lobed, compound, or dissected . . . 17

17. Leaves pinnately compound or pinnately dissected into entire filiform lobes; ovary 3-locular; stigmas or style branches 3; capsule 3-valved . . . . . **POLEMONIACEAE**
17. Leaves not compound: pinnatifid or bipinnatifid, the segments not both entire and filiform; ovary 1-locular; stigmas or style branches 2; capsule 2-valved . . . . **BORAGINACEAE**
18. Leaves reduced to small scales; flowers 4-merous (*Bartonia*) . . . . **GENTIANACEAE**
18. Leaves developed; flowers mostly 5-merous 19
19. Flowers or inflorescences axillary 20
19. Flowers or inflorescences terminal 21
20. Fruit a 4-seeded capsule; corolla large, funnel-shaped; stigmas clearly 2, separate . . . . . . **CONVOLVULACEAE**
20. Fruit a many-seeded berry or capsule; corolla large and funnel shaped (*Datura*) or ± flat (rotate) or bell-shaped; stigma 1 . . . . **SOLANACEAE**
21. Flowers solitary; corolla over 7 cm long (*Datura*). . . . . **SOLANACEAE**
21. Flowers in clusters; corolla smaller 22
22. Inflorescence branched (panicle or cyme) (*Collomia*) . . . . **POLEMONIACEAE**
22. Inflorescence simple (spike, raceme, or umbel) 23
23. Corolla bell-shaped, yellow with purple stripes; adventive in Mackinac County (*Hyoscyamus*) . . . . **SOLANACEAE**
23. Corolla flat or saucer-shaped; mostly widespread in the UP. 24
24. Anthers separate, at least some of them on hairy filaments; fruit a capsule (*Verbascum*) . . . . **SCROPHULARIACEAE**
24. Anthers forming a cone around the pistil, on glabrous filaments; fruit a berry (*Solanum*) . . . . . . **SOLANACEAE**

GROUP 20

Herbaceous dicots with bisexual flowers, perianth of 2 series, ovary 1 and superior, corolla either bilaterally symmetrical or stamens fewer than its lobes, or both and petals united.

1. Fertile (anther-bearing) stamens 5 2
1. Fertile stamens 2 or 4. 4
2. Ovary deeply 4-lobed; plant strongly bristly-hairy; fruit (1-) 4 nutlets (*Echium*) . . . . **BORAGINACEAE**
2. Ovary not lobed; plants glabrous or with dense clammy pubescence; fruit a capsule 3
3. Corolla flat or saucer-shaped, white or yellow; common (*Verbascum*) . . . . **SCROPHULARIACEAE**
3. Corolla bell-shaped, greenish yellow with purple veins; reported from Mackinac County (*Hyoscyamus*) . . . . **SOLANACEAE**
4. Corolla with a spur or sac at the base 5
4. Corolla not prolonged into a spur or sac at the base. . 7
5. Calyx 2-parted (*Utricularia*) . . . . . **LENTIBULARIACEAE**

5 Calyx 5-parted 6
6 Leaves all basal, glandular-sticky above; flowers solitary on scapes (*Pinguicula*) .... **LENTIBULARIACEAE**
6 Leaves all or mostly cauline and not sticky (glandular in *Chaenorrhinum*); flowers not solitary .... **PLANTAGINACEAE**
7 Stem leaves all alternate 8
7 Stem leaves all or mostly opposite or whorled 11
8 Corolla nearly regular, the lobes equaling or exceeding the tube (*Veronica*) .... **PLANTAGINACEAE**
8 Corolla bilaterally symmetrical, ± 2-lipped, the lobes (not lips) distinctly shorter than the tube 9
9 Bracts of inflorescence contrasting with the leaves, cream, yellow, or red at least apically (*Castilleja*) .... **OROBANCHACEAE**
9 Bracts of inflorescence the same color as the leaves, green or purplish-green 10
10 Stem leaves deeply pinnately lobed (*Pedicularis*) .... **OROBANCHACEAE**
10 Stem leaves unlobed (at most shallowly toothed) .... **PLANTAGINACEAE**
11 Ovary deeply 4-lobed, appearing like 4 separate ovaries around the base of the single style [and also keyed as such], the fruit (1-) 4 nutlets; plants usually with a 4-angled ("square") stem and often a minty or citrus-like aroma when bruised .... **LAMIACEAE**
11 Ovary not 4-lobed (at most, somewhat 2-lobed), the fruit a capsule; stem in only a few species 4-angled or with aroma when bruised 12
12 Fertile stamens 2 13
12 Fertile stamens 4 16
13 Flowers in axillary racemes or spikes .... **SCROPHULARIACEAE**
13 Flowers in terminal racemes or spikes, or solitary or paired in the axils of the leaves 14
14 Corolla almost regular, with a 4-5-lobed limb .... **PLANTAGINACEAE**
14 Corolla clearly two lipped 15
15 Pedicels minutely glandular-pubescent, with a pair of sepal-like bractlets at their summit, subtending the calyx; capsules ovoid or spherical (*Gratiola*) .... **PLANTAGINACEAE**
15 Pedicels smooth and glabrous, without bractlets; leaves not gland-dotted; corolla whitish to purple; capsule distinctly ellipsoid .... **LINDERNIACEAE**
16 Corolla nearly regular, the lobes about equal 17
16 Corolla strongly bilaterally symmetrical 20
17 Corolla salverform (trumpet-shaped, with a slender tube of almost uniform diameter) (*Verbena*) **VERBENACEAE**
17 Corolla funnel-shaped or bell-shaped, with a tube broad toward its summit 18
18 Corolla pink to purple (white in albinos); calyx and other parts glabrous (at most scabrous) or with hairs of distinctly different lengths (*Agalinis*) .... **OROBANCHACEAE**
18 Corolla bright yellow; calyx tube, pedicels, and/or stems with hairs of uniform or mixed lengths (not of 2 distinct lengths and only rarely completely glabrous) 19
19 Capsule (like the calyx tube, pedicels, and stem) ± densely pubescent with short uniform eglandular hairs (*Aureolaria*) .... **OROBANCHACEAE**
19 Capsule glabrous; calyx tube, pedicels, and/or stems with viscid or minute gland-tipped hairs (*Mimulus*).... .... **PHRYMACEAE**
20 Mature flowers and fruit strongly reflexed, nearly sessile and in remote pairs on opposite sides of a spike-like terminal raceme; calyx with 3 upper teeth bristle-like and 2 lower teeth broadly triangular (*Phryma*)........ .... **PHRYMACEAE**
20 Mature flowers and fruit not strongly reflexed, and otherwise not as above (long-pediceled, alternate, and/or crowded); calyx with teeth equal or subequal (never bristle-like)............................. 21
21 Upper lip of corolla apparently absent (the corolla split lengthwise above) or much shorter than the lower lip and 4-lobed............................ **LAMIACEAE**
21 Upper lip of corolla well developed, of 2 lobes (or these ± fused into one), often nearly or quite as long as the lower lip.. 22
22 Inflorescence terminal and branched (± paniculate); stamens 4 fertile plus 1 staminodium 23
22 Inflorescence a spike or raceme (no branched stalks), or flowers all axillary; stamens 4, fertile, in most genera with no staminodium (or only a very rudimentary one) 24
23 Leaves below the inflorescence distinctly petioled; corolla brownish, less than 12 mm long; staminodium broad (ca. 1-2 mm) and flat at the free apex (mostly adnate to the upper lip), glabrous (*Scrophularia*) .... **SCROPHULARIACEAE**
23 Leaves below the inflorescence sessile; corolla white to purple-violet, ca. 15-30 (-45) mm long; staminodium slender, elongate (of similar diameter and length as the style), close to lower lip of corolla, bearded at the apex (*Penstemon*).................... **PLANTAGINACEAE**
24 Leaves (especially middle and lower ones) deeply pinnately toothed or lobed ca. one-third or more the distance to the midrib .... **OROBANCHACEAE**
24 Leaves of main stem toothed or entire but not so deeply pinnately toothed or lobed (uppermost leaves or bracts may have small basal lobes) 25
25 Sepals separate nearly or quite to the base; flowers in a compact terminal inflorescence (*Chelone*)............ .... **PLANTAGINACEAE**
25 Sepals (at least at anthesis) fused ca. 1/3 or more the length of the calyx; flowers solitary in the leaf axils or in a loose terminal inflorescence.................. 26

26 Flowers (all or many of them, especially lower ones) in the axils of alternate bracts in a distinct terminal or racemose inflorescence .... **OROBANCHACEAE**

26 Flowers all solitary in the axils of opposite (or whorled) leaves or bracts 27

27 Lobes less than 1/3 the total length of the calyx, or corolla bright yellow (or both conditions) (*Mimulus*) .. .. **PHRYMACEAE**

27 Lobes ca. 1/2 or more the total length of the calyx; corolla blue, purple, white, or cream (rarely yellow). 28

28 Flowers and fruit on pedicels about equaling or exceeding the calyx; corolla with at least the lower lip deep blue (*Collinsia*)............. **PLANTAGINACEAE**

28 Flowers and fruit sessile or on pedicels distinctly shorter than the calyx; corolla with whitish to pink or magenta ground color (plus dark spots and/or yellow markings) .... **OROBANCHACEAE**

Inflorescence types. (From H. L. Mason, *A Flora of the Marshes of California.* University of California Press, 1957.)

REFERENCES

Listed below are some of the more popular and readily available field guides and other resources to the plants found in Michigan and the Midwest region. A list of additional technical references is available at: *http://www.michiganflora.net*

Barnes, B., and W. Wagner. 1981. *Michigan Trees.* The University of Michigan Press. Ann Arbor, MI. 383 p.

Billington, C. 1952. *Ferns of Michigan.* Cranbrook Institute of Science Bulletin No. 32. Bloomfield Hills, MI 240 p.

Black, M., and E. Judziewicz. 2009. *Wildflowers of Wisconsin and the Great Lakes Region: A Comprehensive Field Guide.* The University of Wisconsin Press. Madison, WI. 320 p.

Case, F., Jr. 1987. *Orchids of the Western Great Lakes Region.* Cranbrook Institute of Science Bulletin No. 48. Bloomfield Hills, MI. 240 p.

Chadde, S. 2012. *A Great Lakes Wetland Flora. A Complete, Illustrated Guide to the Aquatic and Wetland Plants of the Upper Midwest* (4th ed.). 683 p.

Chadde, S. 2012. *Wetland Plants of Michigan.* 2nd ed. 684 p.

Chadde, S. 2013. *Midwest Ferns.* 450 p.

Chadde, S. 2013. *Wisconsin Flora.* 818 p.

Chase M. W. and J. L. Reveal. 2009. *A phylogenetic classification of the land plants to accompany APG III.* Botanical Journal of the Linnean Society 161: 122 127.

Christenhusz, M. J. M., X.-C. Zhang, and H. Schneider. 2011. *A linear sequence of extant families and genera of lycophytes and ferns.* Phytotaxa 19: 7 54.

Cody, W., and D. Britton. 1989. *Ferns and Fern Allies of Canada.* Publication 1829/E. Research Branch, Agriculture Canada. Ottawa, Canada. 430 p.

Crow, G., and C. Hellquist. 2000. *Aquatic and Wetland Plants of Northeastern North America* (2 vols.). University of Wisconsin Press. Madison, WI.

Crum, H. 1988. *A Focus on Peatlands and Peat Mosses.* The University of Michigan Press. Ann Arbor, MI. 306 p.

Curtis, J. 1971. *The Vegetation of Wisconsin.* The University of Wisconsin Press. Madison, WI. 657 p.

Eastman, J. 1995. The *Book of Swamp and Bog: Trees, Shrubs, and Wildflowers of Eastern Freshwater Wetlands.* Stackpole Books. Mechanicsburg, PA. 237 p.

Eggers, S., and D. Reed. 1997. *Wetland Plants and Plant Communities of Minnesota and Wisconsin* (2nd ed.). U.S. Army Corps of Engineers, St. Paul District. 264 p.

Fassett, N.C. 1951. *Grasses of Wisconsin.* Madison: University of Wisconsin Press.

Fassett, N. 1957. *A Manual of Aquatic Plants.* The University of Wisconsin Press. Madison, WI. 405 p.

Fassett, N. C. 1976. *Spring Flora of Wisconsin.* 4th ed. Madison: University of Wisconsin Press.

Flora of North America Editorial Committee. 1993. *Flora of North America North of Mexico.* Set, partially published. Oxford University Press. New York, NY.

Gleason, H., and A. Cronquist. 1991. Manual of Vascular Plants of Northeastern United States and Adjacent Canada (2nd Ed.). The New York Botanical Garden. Bronx, NY. 910 p.

Hipp, A. 2008. *Field Guide to Wisconsin Sedges: An Introduction to the Genus* Carex *(Cyperaceae).* The University of Wisconsin Press. Madison, WI. 280 p.

Holmgren, N. (editor). 1998. *Illustrated Companion to Gleason and Cronquist's Manual.* New York Botanical Garden. Bronx, NY. 937 p.

Kartesz, J.T. 2010. *Floristic Synthesis of North America*, Version 1.0. Biota of NorthAmerica Program (BONAP). Chapel Hill, NC.

Kost, M., D. Albert, J. Cohen, B. Slaughter, R. Schillo, C. Weber, and K. Chapman. 2007. *Natural Communites of Michigan: Classification and Description.* Michigan Natural Features Inventory, Lansing, MI.

Lichvar, R.W. 2013. *The National Wetland Plant List: 2013 wetland ratings.* Phytoneuron 2013-49: 1-241.

Lichvar, R.W, and J. T. Kartesz. 2012. *North American Digital Flora: National Wetland Plant List*, v.2.4.0 (https:// wetland_plants. usace.army.mil). U.S. Army Corps of Engineers, Engineer Research and Development Center, Cold Regions Research and Engineering Laboratory, Hanover, NH, and BONAP, Chapel Hill, NC.

Reznicek, A. A., E. G. Voss, and B. S. Walters. 2011. *Michigan Flora Online.* University of Michigan. http://www.michiganflora.net.

Smith, A.R., K.M. Pryer, E. Schuettpelz, P. Korall, H. Schneider, and P. G. Wolf. 2006. *A classification for extant ferns.* Taxon 55: 705–731.

Smith, W. 1993. *Orchids of Minnesota.* The Univ. of Minnesota Press. Minneapolis, MN. 172 p.

Smith, W. 2008. *Trees and Shrubs of Minnesota*. The Univ. of Minnesota Press. Minneapolis, MN.

Soper, J., and M. Heimburger. 1982. *Shrubs of Ontario*. The Royal Ontario Museum. Toronto, Ontario. 495 p.

Swink, F., and G. Wilhelm. 1994. *Plants of the Chicago Region* (4th Ed.). Indiana Academy of Science. Indianapolis, IN. 921 p.

Tryon, R. 1980. *Ferns of Minnesota* (2nd Ed.). The Univ. of Minnesota Press. Minneapolis, MN. 165 p.

Tryon, R., N. Fassett, D. Dunlop, and M. Diemer. 1953. *The Ferns and Fern Allies of Wisconsin*. The University of Wisconsin Press. Madison, WI. 158 p.

Voss, E.G. 1972. *Michigan Flora, Part I. Gymnosperms and Monocots*. Cranbrook Institute of Science Bulletin 55 and University of Michigan Herbarium. 488 p.

Voss, E.G. 1985. *Michigan Flora, Part II. Dicots (Saururaceae - Cornaceae)*. Cranbrook Institute of Science Bulletin 59 and University of Michigan Herbarium. 724 p.

Voss, E.G. 1996. *Michigan Flora, Part III. Dicots (Pyrolaceae - Compositae)*. Cranbrook Institute of Science Bulletin 61 and University of Michigan Herbarium. 622 p.

Voss, E.G. and A. A. Reznicek. 2011. *Field Manual of Michigan Flora*. University of Michigan Press. 1008 p.

Wagner, W.H. Jr., and F.S. Wagner. 1990. *Moonworts (*Botrychium *subg.* Botrychium*) of the upper Great Lakes region, U.S and Canada, with descriptions of two new species*. Contr. Univ. Mich. Herb. 17:313 325.

Online

A wealth of information about Michigan's flora is available online; especially useful is the website of the University of Michigan Herbarium:

http://www.michiganflora.net/

The ongoing Flora of North America project is located at:

http://www.efloras.org/

Distribution maps were obtained from the Biota of North America Program (bonap):

http://www.bonap.org

For species of conservation concern (endangered, threatened, special concern), information was obtained from the Michigan Natural Features Inventory:

http://mnfi.anr.msu.edu/

GLOSSARY

abaxial On the side away from the axis, usually refers to the underside of a leaf (compare with adaxial).

acaulescent Without an upright, leafy stem.

achene A one-seeded, dry, indehiscent fruit with the seed coat not attached to the mature wall of the ovary.

acid Having more hydrogen ions than hydroxyl (OH) ions; a pH less than 7.

acuminate Tapering to a narrow point, more tapering than acute, less than attenuate.

acute Gradually tapered to a tip.

adaxial On the side toward the axis, usually refers to the top side of a leaf (compare with abaxial).

adnate Fused with a structure different from itself, as when stamens are adnate to petals (compare with connate).

adventive Not native to and not fully established in a new habitat.

alkaline Having more hydroxyl ions than hydrogen ions; a pH greater than 7.

alluvial Deposits of rivers and streams.

alternate Borne singly at each node, as in leaves on a stem.

ament Spikelike inflorescence of same-sexed flowers (either male or female); same as catkin.

androgynous Spike with both staminate and pistillate flowers, the pistillate located at the base, below the staminate (compare with gynaecandrous).

angiosperm A plant producing flowers and bearing seeds in an ovary.

annual A plant that completes its life cycle in one growing season, then dies.

anther Pollen-bearing part of stamen, usually at the end of a stalk called a filament.

anthesis The period during which a flower is fully open and functional.

anthocyanic Pigmented with anthocyanins, this usually manifested as a tinging or suffusion of pink, red, or purple.

aphyllopodic Having basal sheaths without blades; with new shoots arising laterally from parent shoot (compare with phyllopodic).

apiculate Having an apiculus.

apiculus An abrupt, very small, projected tip.

appressed Lying flat to or parallel to a surface.

aquatic Living in water.

areole In leaves, the spaces between small veins.

aril A specialized appendage on a seed, often brightly colored, derived from the seed coat.

aristate Tipped with a slender bristle.

armed Bearing a sharp projection such as a prickle, spine, or thorn.

aromatic Strongly scented.

ascending Angled upward.

asymmetrical Not symmetrical.

attenuate Tapering gradually to a prolonged tip.

auricle An ear-shaped appendage to a leaf or stipule.

awl-shaped Tapering gradually from a broad base to a sharp point.

awn A bristle-like organ.

axil Angle between a stem and the attached leaf.

barb Sharp, thorn-like projection.

basal From base of plant.

basic A pH greater than 7.

beak A slender, terminal appendage on a 3-dimensional organ.

beard Covering of long or stiff hairs.

berry Fruit with the seeds surrounded by fleshy material.

biennial A plant that completes its life cycle in two growing season, typically flowering and fruiting in the second year, then dying.

bifid Cleft into two more or less equal parts.

blade Expanded, usually flat part of a leaf or petiole.

bloom A whitish powdery or waxy coating that can be rubbed away.

bog A wet, acidic, nutrient-poor peatland characterized by sphagnum and other mosses, shrubs and sedges. Technically, a type of peatland raised above its surroundings by peat accumulation and receiving nutrients only from precipitation.

boreal Far northern latitudes.

brackish Salty.

bract An accessory structure at the base of some flowers, usually appearing leaflike.

bractlet A secondary bract (*Typha*).

branchlets A small branch.

bristle A stiff hair.

bud An undeveloped shoot, inflorescence, or flower, in woody plants often covered by scales and serving as the overwintering stage.

bulb A group of modified leaves serving as a food-storage organ, borne on a short, vertical, underground stem (compare with corm).

bulbil A bulb-like structure borne in the leaf axils or in place of flowers.

bulblet Small bulb borne above ground, as in a leaf axil.

ca. About, approximately (Latin *circa*).

caducous Falling off early, as stipules that leave behind a scar.

callosity A hardened thickening.

callus A firm, thickened portion of an organ; the firm base of the lemma in the Poaceae.

calcareous fen An uncommon wetland type associated with seepage areas, and which receive groundwater enriched with primarily calcium and magnesium bicarbonates.

calcium-rich Refers to wetlands underlain by limestone or receiving water enriched by calcium compounds.

calyx All the sepals of a flower.

campanulate Bell-shaped.

capillary Very fine, hair-like, not-flattened.

capitate Abruptly expanded at the apex, thereby forming a knob-like tip.

capsule A dry, dehiscent fruit splitting into 3 or more parts.

carpel Fertile leaf of an angiosperm, bearing the ovules. A pistil is made up of one or more carpels.

caruncle An appendage at or near the hilum of some seeds.

caryopsis The dry, indehiscent seed of grasses.

catkin Spikelike inflorescence of same-sexed flowers (either male or female); same as ament.

caudex Firm, hardened, summit of a root mass that functions as a perennating organ.

cauline Of or pertaining to the aboveground portion of the stem.

cespitose Growing in a compact cluster with closely spaced stems; tufted, clumped.

chaff Thin, dry scales; in the Asteraceae, sometimes found as chaffy bracts on the receptacle.

cilia Hairs found at the margin of an organ.

ciliate Provided with cilia.

circumboreal Refers to a species distribution pattern which circles the earth's boreal regions.

clasping Leaves that partially encircle the stem at the base.

clavate Widened in the distal portion, like a baseball bat.

claw The narrow, basal portion of perianth parts.

cleistogamous Type of flower that remains closed and is self-pollinated.

clumped Having the stems grouped closely together; tufted.

colony-forming A group of plants of the same species, produced either vegetatively or by seed.

column The joined style and filaments in the Orchidaceae.

coma A tuft of fine hairs, especially at the tip of a seed.

composite An inflorescence that is made up of many tiny florets crowded together on a receptacle; members of the Aster Family (Asteraceae).

compound leaf A leaf with two or more leaflets.

concave Curved inward.

conduplicate Folded lengthwise into nearly equal parts.

cone The dry fruit of conifers composed of overlapping scales.

conifer Cone-bearing woody plants.

connate Two like parts that are fused (compare with adnate).

connivent Converging and touching but not actually fused, applies to like organs.

convex Curved outward.

convolute Arranged such that one edge is covered and the other is exposed, usually referring to petals in bud.

cordate With a rounded lobe on each side of a central sinus; heart-shaped.

coriaceous With a firm, leathery texture.

corm A short, vertical, enlarged, underground stem that serves as a food storage organ (compare with bulb).

corolla Collectively, all the petals of a flower.

corymb An indeterminate inflorescence, somewhat similar to a raceme, that has elongate lower branches that create a more or less flat-topped inflorescence.

costa (plural costae) A prominent midvein or midrib of a leaflet.

crenate With rounded teeth.

crenulate Finely crenate.

crisped An irregularly crinkled or curled leaf margin.

crown Persistent base of a plant, especially a grasses.

culm The stem of a grass or grasslike plant, especially a stem with the inflorescence.

cuneate Tapering to the base with relatively straight, non-parallel margins; wedge-shaped.

cyme A type of inflorescence in which the central flowers open first.

deciduous Not persistent.

decumbent A stem that is prostrate at the base and curves upward to have an erect or ascending, apical portion.

decurrent Possessing an adnate line or wing that extends down the axis below the node, usually referring to leaves on a stem.

dehiscent Splitting open at maturity.

deltate Triangle-shaped.

dentate Provided with outward oriented teeth.

depauperate Poorly developed due to unfavorable conditions.

dicots One of two main divisions of the Angiosperms (the other being the Monocots); plants having 2 seed leaves (cotyledons), net-venation, and flower parts in 4s or 5s (or multiples of these numbers).

dioecious Bearing only male or female flowers on a single plant.

dimorphic Having two forms.

disarticulation Spikelets breaking either above or below the glumes when mature, the glumes remaining in the head if disarticulation above the glumes, or the glumes falling with the florets if disarticulation is below the glumes.

discoid In composite flowers (Asteraceae), a head with only disk (tubular) flowers, the ray flowers absent.

disjunct A population of plants widely separated from its main range.

disk In the Asteraceae, the central part of the head, composed of tubular flowers.

dissected Leaves divided into many smaller segments.

disturbed Natural communities altered by human influences.

divided Leaves which are lobed nearly to the midrib.

dolomite A type of limestone consisting of calcium magnesium carbonate.

dorsal Underside, or back of an organ.

driftless area Portions of sw Wisconsin, ne Iowa, and se Minnesota that are not covered by glacial drift.

drupe A fleshy fruit with a single large seed such as a cherry.

echinate With spines.

eglandular Without glands.

elliptic Broadest at the middle, gradually tapering to both ends.

emergent Growing out of and above the water surface.

emersed leaf Growing above the water surface or out of water.

endangered A species in danger of extinction throughout all or most of its range if current trends continue.

endemic A species restricted to a particular region.

entire With a smooth margin.

erect Stiffly upright.

erose With a ragged edge.

escape A cultivated plant which establishes itself outside of cultivation.

evergreen Plant retaining its leaves throughout the year.

excurrent With the central rib or axis continuing or projecting beyond the organ.

exserted Extending beyond the mouth of a structure such as stamens extending out from the mouth of the corolla.

falcate Sickle-shaped

false indusium A modified tooth or reflexed margin of a fern leaf that covers the sorus.

fen An open wetland usually dominated by herbaceous plants, and fed by in-flowing, often calcium- and/or magnesium-rich water; soils vary from peat to clays and silts.

fern Perennial plants with spore-bearing leaves similar to the vegetative leaves and bearing sporangia on their underside, or the spore-bearing leaves much modified.

fibrous A cluster of slender roots, all with the same diameter.

filament The stalk of a stamen which supports the anther.

filiform Thread-like.

flexuous An elongate axis that arches or bends in alternating directions in a zig-zag fashion.

floating mat A feature of some ponds where plant roots form a carpet over some or all of the water surface.

floodplain That part of a river valley that is occasionally covered by flood waters.

floret A small flower in a dense cluster of flowers; in grasses the flower with its attached lemma and palea.

follicle A dry, dehiscent fruit that splits along one side when mature.

floricane the second-year flowering stem of *Rubus* (compare with primocane).

genus The first part of the scientific name for a plant or animal (plural genera).

glabrate Nearly glabrous or becoming so.

glabrous Lacking hairs.

gland An appendage or depression which produces a sticky or greasy substance.

glandular Bearing glands.

glaucous Having a bluish appearance.

glumes A pair of small bracts at base of each spikelet the lowermost (or first) glume usually smaller the upper (or second) glume usually longer.

grain The fruit of a grass; the swollen seedlike protuberance on the fruit of some *Rumex*.

gymnosperm Plants in which the seeds are not produced in an ovary, but usually in a cone.

gynaecandrous Having both staminate and pistillate flowers on the same spike, the staminate located at the base, below the pistillate (compare with androgynous).

gynophore The central stalk of some flowers, especially in cat-tails (*Typha*).

halophyte A plant adapted to growing in a salty substrate.

hastate More or less triangular in outline with outward-oriented basal lobes.

haustorium A specialized, root-like connection to a host plant that a parasite uses to extract nourishment.

hardwoods Loosely used to contrast most deciduous trees from conifers.

herb A herbaceous, non-woody plant.

herbaceous Like an herb; also, leaflike in appearance.

hilum The scar at the point of attachment of a seed.

hirsute Pubescent with coarse, somewhat stiff, usually curving hairs, coarser than villous but softer than hispid.

hispid Pubescent with coarse, stiff hairs that may be uncomfortable to the touch, coarser than hirsute but softer than bristly.

hummock A small, raised mound formed by certain species of sphagnum moss.

humus Dark, well-decayed organic matter in soil.

hybrid A cross-breed between two species.

hydric Wet (compare with mesic, xeric).

hypanthium A ring, cup, or tube around the ovary; the sepals, petals and stamens are attached to the rim of the hypanthium.

imbricate Overlapping, as shingles on a roof.

indehiscent Not splitting open at maturity.

indusium In ferns, a membranous covering over the sorus (plural indusia).

inferior The position of the ovary when it is below the point of attachment of the sepals and petals.

inflorescence A cluster of flowers.

insectivorous Refers to the insect trapping and digestion habit of some plants as a nutrition supplement.

interdunal swale Low-lying areas between sand dune ridges.

internode Portion of a stem between two nodes.

introduced A non-native species.

invasive Non-native species causing significant ecological or economic problems.

involucral bract A single member of the involucre; sometimes called phyllary in composite flowers (Asteraceae).

involucre A whorl of bracts, subtending a flower or inflorescence.

irregular flower Not radially symmetric; with similar parts unequal.

joint A node or section of a stem where the branch and leaf meet.

keel A central rib like the keel of a boat.

lance-shaped Broadest near the base, gradually tapering to a narrower tip.

lateral Borne on the sides of a stem or branch.

lax Loose or drooping.

leaf axil The point of the angle between a stem and a leaf.

leaflet One of the leaflike segments of a compound leaf.

lemma In grasses, the lower bract enclosing the flower (the upper, smaller bract is the palea).

lens-shaped Biconvex in shape (like a lentil).

lenticel Blisterlike openings in the epidermis of woody stems, admitting gases to and from the plant, and often appearing as small oval dots on bark.

ligulate Having a ligule; in the Asteraceae, the strap-shaped corolla of a ray floret.

ligule In grasses and grasslike plants, the membranous or hairy ring at top of sheath between the blade and stem.

linear Narrow and flat with parallel sides.

lip Upper or lower part of a 2-lipped corolla; also the lower petal in most orchid flowers.

lobed With lobes; in leaves divisions usually not over halfway to the midrib.

local Occurring sporadically in an area.

low prairie Wet and moist herbaceous plant community, typically dominated by grasses.

margin The outer edge of a leaf.

marl A calcium-rich clay.

marsh Wetland dominated by herbaceous plants, with standing water for part or all the growing season, then often drying at the surface.

megaspore Large, female spores.

mesic Moist, neither dry nor wet (compare with hydric, xeric).

microspore Small, male spores.

midrib The prominent vein along the main axis of a leaf.

mixed forest A type of forest composed of both deciduous and conifer trees.

moat The open water area ringing the outer edge of a peatland or floating mat.

monecious Having male and female reproductive parts in separate flowers on the same plant.

monocots One of two main divisions of the Angiosperms (the other being the Dicots); plants with a single seed leaf (cotyledon); typically having narrow leaves with parallel veins, and flower parts in 3s or multiples of 3.

muck An organic soil where the plant remains are decomposed to the point where the type of plants forming the soil cannot be determined.

mucro A sharp point at termination of an organ or other structure.

naked Without a covering; a stalk or stem without leaves.

native An indigenous species.

naturalized An introduced species that is established and persistent in an ecosystem.

needle A slender leaf, as in the Pinaceae.

nerve A leaf vein.

neutral A pH of 7.

node The spot on a stem or branch where leaves originate.

nutlet A small dry fruit that does not split open along a seam.

oblanceolate Reverse lance-shaped; broadest at the apex, gradually tapering to the narrower base.

oblique Emerging or joining at an angle other than parallel or perpendicular.

oblong Broadest at the middle, and tapering to both ends, but broader than elliptic.

obovate Broadly rounded at the apex, becoming narrowed below.

ocrea A tube-shaped stipule or pair of stipules around the stem; characteristic of the Smartweed Family (Polygonaceae).

opposite Leaves or branches which are paired opposite one another on the stem.

organic Soils composed of decaying plant remains.

oval Elliptical.

ovary The lower part of the pistil that produces the seeds.

ovate Broadly rounded at the base, becoming narrowed above; broader than lanceolate.

palea The uppermost of the two inner bracts subtending a grass flower (the lower bract is the lemma).

palmate Divided in a radial fashion, like the fingers of a hand.

panicle An arrangement of flowers consisting of several racemes.

papilla (plural: papillae) A short, rounded or cylindrical projections.

pappus The modified sepals of a composite flower which persist atop the ovary as bristles, scales or awns.

parallel-veined With several veins running from base of leaf to leaf tip, characteristic of most monocots.

peat An organic soil formed of partially decomposed plant remains.

peatland A wetland whose soil is composed primarily of organic matter (mosses, sedges, etc.); a general term for bogs and fens.

peltate More or less circular, with the stalk attached at a point on the underside.

pepo A fleshy, many-seeded fruit with a tough rind, as a melon.

perennial Living for 3 or more years.

perfect A flower having both male (stamens) and female (pistils) parts.

perianth Collectively, all the sepals and petals of a flower.

perigynium A sac-like structure enclosing the pistil in *Carex* (plural perigynia).

petal An individual part of the corolla, often white or colored.

petiole The stalk of a leaf.

phyllary An involucral bract subtending the flower head in composite flowers (Asteraceae).

phyllode An expanded petiole.

phyllopodic Having the basal sheaths blade-bearing; with new shoots arising from the center of parent shoot (compare with aphyllopodic).

pinna The primary or first division in a fern frond or leaf (plural pinnae).

pinnate Divided once along an elongated axis into distinct segments.

pinnule The pinnate segment of a pinna.

pistil The seed-producing part of the flower, consisting of an ovary and one or more styles and stigmas.

pith A spongy central part of stems and branches.

pollen The male spores in an anther.

prairie An open plant community dominated by herbaceous species, especially grasses.

primocane The first-year, vegetative stem in *Rubus* (compare with floricane).

pro sp. When a taxon is transferred from the non-hybrid category to the hybrid category, the author citation remains unchanged, but may be followed by an indication in parentheses of the original category.

prostrate Lying flat on the ground.

raceme A grouping of flowers along an elongated axis where each flower has its own stalk.

rachilla A small stem or axis.

rachis The central axis or stem of a leaf or inflorescence.

radiate heads In composite flowers, heads with both ray and disk flowers (Asteraceae).

ray flower A ligulate or strap-shaped flower in the Asteraceae, where often the outermost series of flowers in the head.

receptacle In the Asteraceae, the enlarged summit of the flower stalk to which the sepals, petals, stamens, and pistils are usually attached.

recurved Curved backward.

regular Flowers with all the similar parts of the same form; radially symmetric.

rhizome An underground, horizontal stem.

rib A pronounced vein or nerve.

rootstock Similar to rhizome but referring to any underground part that spreads the plant.

rosette A crowded, circular clump of leaves.

samara A dry, indehiscent fruit with a well-developed wing.

saprophyte A plant that lives off of dead organic matter.

scale A tiny, leaflike structure; the structure that subtends each flower in a sedge (Cyperaceae).

scape A naked stem (without leaves) bearing the flowers.

section Cross-section.

secund Flowers mostly on 1 side of a stalk or branch.

sedge meadow A community dominated by sedges (Cyperaceae) and occurring on wet, saturated soils.

seep A spot where water oozes from the ground.

sepal A segment of the calyx; usually green in color.

sheath Tube-shaped membrane around a stem, especially for part of the leaf in grasses and sedges.

shrub A woody plant with multiple stems.

silicle Short fruit of the Mustard Family (Brassicaceae), normally less than 2x longer as wide.

silique Dry, dehiscent, 2-chambered fruit of the Mustard Family (Brassicaceae), longer than a silicle.

simple An undivided leaf.

sinus The depression between two lobes.

smooth Without teeth or hairs.

sorus Clusters of spore containers (plural sori).

spadix A fleshy axis in which flowers are embedded.

spathe A large bract subtending or enclosing a cluster of flowers.

spatula-shaped Broadest at tip and tapering to the base.

sphagnum moss A type of moss common in peatlands and sometimes forming a continuous carpet across the surface; sometimes forming layers several meters thick; also loosely called peat moss.

spike A group of unstalked flowers along an unbranched stalk.

spikelet A small spike; the flower cluster (inflorescence) of grasses (Poaceae) and sedges (Cyperaceae).

sporangium The spore-producing structure (plural sporangia).

spore a one-celled reproductive structure that gives rise to the gamete-bearing plant.

sporophyll A modified, spore-bearing leaf.

spreading Widely angled outward.
spring A place where water flows naturally from the ground.
spur A hollow, pointed projection of a flower.
stamen The male or pollen-producing organ of a flower.
staminode An infertile stamen.
stem The main axis of a plant.
stigma The terminal part of a pistil which receives pollen.
stipe A stalk.
stipule A leaflike outgrowth at the base of a leaf stalk.
stolon A horizontal stem lying on the soil surface.
style The stalklike part of the pistil between the ovary and the stigma.
subspecies A subdivision of the species forming a group with shared traits which differ from other members of the species (subsp.).
subtend Attached below and extending upward.
succulent Thick, fleshy and juicy.
superior Referring to the position of the ovary when it is above the point of attachment of sepals, petals, stamens, and pistils.
swale A slight depression.
swamp Wooded wetland dominated by trees or shrubs; soils are typically wet for much of year or sometimes inundated.

talus Fallen rock at the base of a slope or cliff.
taproot A main, downward-pointing root.
tendril A threadlike appendage from a stem or leaf that coils around other objects for support (as in *Vitis*).
tepal Sepals or petals not differentiated from one another.
terete Circular in cross-section.
terminal Located at the end of a stem or stalk.
thallus A small, flattened plant structure, without distinct stem or leaves.
thicket A dense growth of woody plants.
threatened A species likely to become endangered throughout all or most of its range if current trends continue.
translucent Nearly transparent.
tree A large, single-stemmed woody plant.
tuber An enlarged portion of a root or rhizome.
truncate Abruptly cut-off.
tubercle Base of style persistent as a swelling atop the achene different in color and texture from achene body.

tundra Treeless plain in arctic regions, having permanently frozen subsoil.
turion A specialized type of shoot or bud that overwinters and resumes growth the following year.

umbel A cluster of flowers in which the flower stalks arise from the same level.
umbelet A small, secondary umbel in an umbel, as in the Apiaceae.
upright Erect or nearly so.
urceolate Constricted at a point just before an opening; urn-shaped.
utricle A small, one-seeded fruit with a dry, papery outer covering.

valve A segment of a dehiscent fruit; the wing of the fruit in *Rumex*.
variety Taxon below subspecies and differing from other varieties within the same subspecies (var.).
vein A vascular bundle, as in a leaf.
velum The membranous flap that partially covers the sporangium in *Isoetes*.
venation The pattern of veins on an organ.
ventral Front side.
ventricose Inflated or distended.
verrucose Covered with small, wart-like projections.
verticil One whorled cycle of organs.
verticillate Arranged in whorls.
villous Pubescent with long, soft, bent hairs, the hairs not crimped or tangled.
vine A trailing or climbing plant, dependent on other objects for support.
viscid Sticky, glutinous.

whorl A group of 3 or more parts from one point on a stem.
wing A thin tissue bordering or surrounding an organ.
woody Xylem tissue (the vascular tissue which conducts water and nutrients).

xeric Dry (compare with hydric, mesic).

558 GLOSSARY

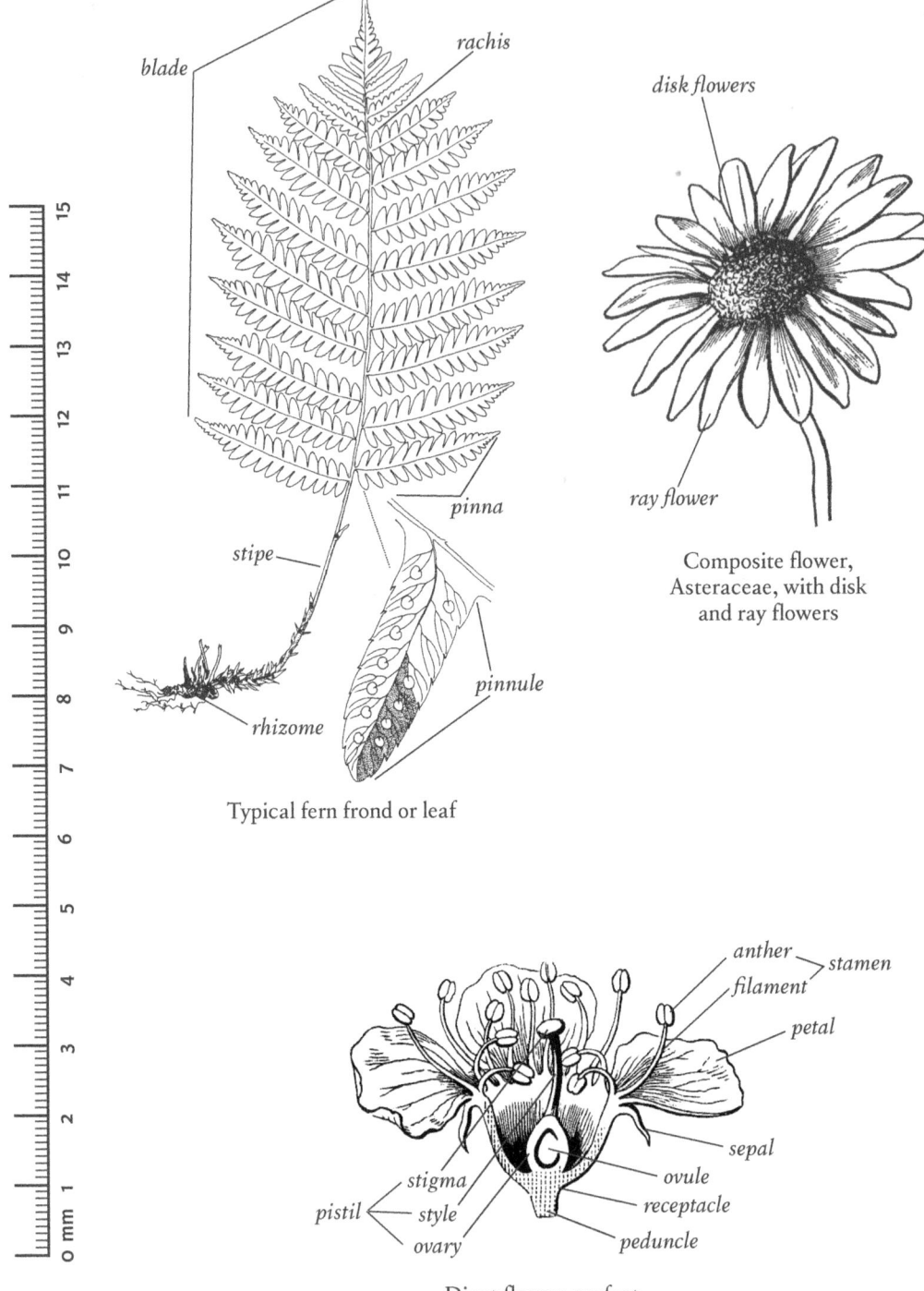

Typical fern frond or leaf

Composite flower, Asteraceae, with disk and ray flowers

Dicot flower, perfect, with male and female parts

Synonyms are listed in *italics*.

A

Abies 34
  balsamea 34
Abutilon 227
  theophrasti 227
Acalypha 184
  rhomboidea 184
  virginica 184
Acer 322
  ginnala 322
  negundo 322
  pensylvanicum 323
  platanoides 323
  rubrum 323
  saccharinum 323
  saccharum 323
  spicatumm 323
Achillea 64
  millefolium 64
  ptarmica 64
Acinos arvensis 213
Aconitum napellus 276
Acoraceae 344
Acorus 344
  americanus 344
  calamus 344
Actaea 277
  pachypoda 277
  rubra 277
Adam-and-Eve 445
Adder's-Mouth Orchid 450
Adder's-Tongue 24
Adder's-Tongue Family 22
Adenocaulon bicolor 64
Adiantum pedatum 28
Adlumia fungosa 247
Adoxaceae 37
Aegopodium podagraria 47
Aesculus glabra 323
Agalinis 241
  besseyana 241
  paupercula 241
  purpurea 241
  tenuifolia 241
Agastache foeniculum 212
Ageratina altissima 65
Agrimonia 289
  gryposepala 289
  striata 289
Agrimony 289
Agropyron caninum 482
Agropyron dasystachyum 481
Agropyron repens 481
Agropyron smithii 495
Agropyron trachycaulum 482
Agrostemma githago 153
Agrostis 462
  alba 464
  capillaris 462

  gigantea 464
  palustris 464
  perennans 464
  scabra 464
  stolonifera 464
Ajuga 212
  genevensis 212
  reptans 212
Alcea rosea 228
Alder 117
alder-leaf buckthorn 287
Alexanders 52
alfalfa 192
Alisma 345
  brevipes 345
  plantago-aquatica 345
  subcordatum 345
  triviale 345
Alismataceae 344
alkali buttercup 282
Alkali-Grass 502
Alkekengi officinarum 330
Allegheny monkey-flower 250
Allegheny-vine 247
allheal 152
Alliaria 129
  alliaria 129
  officinalis 129
  petiolata 129
Allium 435
  canadense 436
  cernuum 436
  schoenoprasum 436
  tricoccum 436
Alnus 117
  crispa 118
  incana 117
  rugosa 117
  viridis 118
Alopecurus 464
  aequalis 464
  pratensis 465
alpine bistort 264
alpine enchanter's nightshade 235
alpine leafless-bulrush 421
alpine violet 339
alpine woodsia 32
alsike clover 194
alternate-flower water-milfoil 205
alternate-leaf dogwood 169
Althaea rosea 228
Alumroot 325
Alyssum 129
  alyssoides 129
  murale 129
Amaranth 40
Amaranth Family 39
Amaranthaceae 39
Amaranthus 40
  albus 40
  blitoides 40

  powellii 40
  retroflexus 41
Ambrosia 65
  artemisiifolia 65
  psilostachya 65
  trifida 65
Amelanchier 289
  arborea 290
  bartramiana 290
  humilis 290
  interior 290
  sanguinea 290
  spicata 290
  stolonifera 290
  laevis 290
American beach-grass 465
American beech 197
American bittersweet 164
American bur-reed 520
American cow-parsnip 50
American cow-wheat 244
American dragonhead 213
American eel-grass 424
American elm 333
American germander 221
American ginseng 58
American golden-saxifrage 325
American hazelnut 119
American hog-peanut 188
American lopseed 250
American lyme grass 489
American manna grass 486
American marsh-pennywort 58
American millet grass 490
American mountain-ash 307
American plantain 257
American shoreweed 254
American slough grass 467
American squawroot 242
American sweetflag 344
American trailplant 64
American vetch 195
American water-awlwort 142
American water-plantain 345
American wild mint 217
American willowherb 236
American witch-hazel 207
American yellow-rocket 130
American-brooklime 258
Amerorchis rotundifolia 445
Ammophila breviligulata 465
Amorpha fruticosa 188
Amphicarpaea bracteata 188
Anacardiaceae 44
Anacharis canadensis 423
Anacharis nuttallii 423
Anaphalis margaritacea 66
Anchusa azurea 120
Andromeda 176
  glaucophylla 176
  polifolia 176

Andropogon 465
   gerardii 465
   scoparius 504
Anemone 277
   acutiloba 279
   americana 280
   blanda 277
   canadensis 277
   cylindrica 277
   multifida 278
   quinquefolia 278
   riparia 278
   virginiana 278
Anethum graveolens 48
Angelica atropurpurea 48
aniseroot 51
annual bluegrass 499
annual false foxglove 242
annual knawel 158
annual wallrocket 136
Antennaria 66
   howellii 66
   neglecta 66
   neodioica 66
   parlinii 67
   plantaginifolia 67
Antenoron virginianum 267
Anthemis 67
   arvensis 67
   cotula 67
   tinctoria 67
Anthoxanthum 466
   hirtum 487
   odoratum 466
Anthriscus sylvestris 48
Anthyllis vulneraria 188
Anticlea elegans 436
Apiaceae 46
Apios americana 189
Aplectrum hyemale 445
Apocynaceae 53
Apocynum 53
   androsaemifolium 53
   cannabinum 53
apple 296
apple-of-Peru 329
Aquifoliaceae 56
Aquilegia 278
   canadensis 278
   vulgaris 278
Arabidopsis 129
   thaliana 130
   lyrata 129
Arabis 130
   caucasica 130
   divaricarpa 131
   drummondii 132
   glabra 143
   hirsuta 130
   holboelii 131
   pycnocarpa 130

   lyrata 129
Araceae 346
Aralia 57
   hispida 57
   nudicaulis 57
   racemosa 57
Araliaceae 56
Arbor-Vitae 33
Arceuthobium pusillum 321
Arctic bur-reed 522
Arctic grass-of-parnassus 164
Arctium minus 68
Arctostaphylos uva-ursi 176
Arenaria 153
   macrophylla 157
   michauxii 157
   serpyllifolia 153
Arethusa bulbosa 445
Argentina anserina 298
Arisaema triphyllum 347
Aristida basiramea 466
Aristolochiaceae 58
Armoracia 130
   aquatica 141
   rusticana 130
   lacustris 141
Arnica 68
   cordifolia 68
   lonchophylla 68
Arnoglossum plantagineum 69
Aronia prunifolia 291
Arrhenatherum elatius 466
Arrow-Arum 348
Arrow-Grass 434
Arrow-Grass Family 434
arrow-leaf s colt's-foot 97
arrow-leaf tearthumb 267
arrow-leaved aster 112
Arrow-Wood 37
Arrowhead 345
arrowhead violet 341
Artemisia 69
   abrotanum 69
   absinthium 69
   biennis 69
   campestris 70
   pontica 70
   stelleriana 70
   vulgaris 70
   ludoviciana 70
Arum Family 346
arum-leaf arrowhead 345
Aruncus dioicus 291
Asarum canadense 58
Asclepias 54
   exaltata 54
   incarnata 54
   ovalifolia 54
   speciosa 55
   syriaca 55
   tuberosa 55

   verticillata 55
Ash 234
ashy cinquefoil 299
ashy sunflower 87
asparagus 437
Asparagus officinalis 437
Aspen 313
Aspleniaceae 5
Asplenium 5
   platyneuron 5
   rhizophyllum 5
   ruta-muraria 5
   scolopendrium 5
   trichomanes 6
   trichomanes-ramosum 6
   viride 6
Assiniboia sedge 375
Aster Family 58
Aster borealis 108
Aster brachyactis 108
Aster ciliolatus 108
Aster dumosus 109
Aster ericoides 109
Aster firmus 111
Aster hesperius 110
Aster hirsuticaulis 110
Aster inngifolius 112
Aster interior Wieg 110
Aster junciformis 108
Aster laevis 109
Aster lanceolatus 110
Aster lateriflorus 110
Aster lindleyanus 108
Aster macrophyllus 81
Aster nemoralis 95
Aster novae-angliae 110
Aster ontarionis 111
Aster oolentangiensis 111
Aster pilosus 111
Aster ptarmicoides 104
Aster pubentior 79
Aster puniceus 79
Aster sagittifolius 112
Aster sericeus 112
Aster umbellatus 79
Asteraceae 58
Astragalus 189
   canadensis 189
   neglectus 189
Athyriaceae 6
Athyrium 6
   angustum 6
   filix-femina 6
   thelypterioides 6
Atocion armeria 159
Atriplex patula 41
Aureolaria pedicularia 242
auricled twayblade 451
autumn olive 174
autumn water-starwort 252
autumn willow 320

INDEX

Avena 467
   fatua 467
   sativa 467
Avenella flexuosa 475
Avens 295
awned flat sedge 409

B

Baby's-Breath 156
Back's sedge 398
bald brome 470
Balkan catchfly 160
balsam fir 34
balsam willow 320
balsam-poplar 314
Balsaminaceae 115
Balsamita major 113
Baneberry 277
Barbarea 130
   orthoceras 130
   vulgaris
Barberry 116
Barberry Family 116
Barley 488
barnyard-grass 479
barren strawberry 295
Bartonia 198
   paniculata 198
   virginica 198
basal-leaved rosinweed 101
basil-thyme   213
basket willow 320
Bassia scoparia 41
basswood 228
bastard toadflax 321
Bayberry 231
Bayberry Family 231
Beach-Grass 465
beach-pea 190
Beak Sedge 416
beaked hazelnut 119
beaked spike-rush 414
beaked willow 317
Bearberry 176
bearded sedge 403
bearded shorthusk 468
Beardtongue 255
Bebb's sedge 385
Bebb's willow 317
beck's water-marigold 71
Beckmannia syzigachne 467
Bedstraw 309
Beebalm 217
Beech 197
Beech Family 196
Beech-Fern 30
beechdrops 242
beggar's-lice 122, 123
Beggarticks 71
Bellflower 144
Bellflower Family 144

Bellis perennis 70
Bellwort 443
Bent 462
Bent-Grass 462
Berberidaceae 116
Berberis 116
   aquifolium 116
   thunbergii 116
   vulgaris 116
Berteroa incana 131
Betula 118
   alleghaniensis 118
   glandulosa 118
   lutea 118
   papyrifera 118
   pumila 118
Betulaceae 117
Bidens 71
   acuta 72
   beckii 71
   cernua 71
   comosa 72
   comosa 72
   connata 72
   discoidea 72
   frondosa 72
   puberula 72
   trichosperma 72
   tripartita 72
   vulgata 72
biennial wormwood 69
big bluestem 465
big-fruit hawthorn 293
big-leaf avens 296
big-tooth aspen 315
bigflower tickseed 78
Bindweed 168
Birch 118
Birch Family 117
bird-eye pearlwort 158
bird-foot violet 341
bird-vetch 195
bird's-foot-trefoil 192
birdseye speedwell 260
Birthwort Family 58
Bishop's Cap 326
Bishop's goutweed 47
Bistort 264
Bistorta vivipara 264
bitter dock 272
Bittercress 134
bitternut hickory 210
Bittersweet 164
Bittersweet Family 164
black ash 234
black bent 464
black bindweed 265
black crowberry 176
black hawthorne 293
black huckleberry 178
black medick 192

black mustard 132
black raspberry 304
black spruce 35
black twinberry 150
black willow 318
Black-Bindweed 264
black-eyed susan 99
black-seed mountain ricegrass 497
Black-Snakeroot 51
Blackberry 303
black locust 193
bladder fern 7
Bladder Fern 7
Bladder Fern Family 7
bladder-campion 160
Bladderwort 223
Bladderwort Family 221
Blanket-Flower 83
Blazing Star 93
Blechnaceae 7
Bleedinghearts 247
bloodroot 248
blue bugle 212
blue cohosh 116
blue giant-hyssop 212
Blue Ridge carrion-flower 520
Blue Ridge sedge 357
blue scorpion-grass 125
blue skullcap 220
blue water speedwell 258
blue wild rye 481
Blue-Eyed Mary 253
blue-eyed-grass 426, 427
blue-pod lupine 192
Bluebead-Lily 437
bluebell-of-Scotland 145
Bluebells 124
Blueberry 181
bluebuttons 172
Bluegrass 498
bluejoint 471
blueleaf willow 318
Bluestem 465
Bluets 311
blunt broom sedge 388
blunt-glumed panicgrass 478
blunt-leaf orchid 453
blunt-leaf pondweed 515
blunt-lobe grape fern 26
blunt-lobed woodsia 32
bluntleaf bedstraw 310
Boechera 131
   grahamii 131
   retrofracta 131
   stricta 132
Boehmeria cylindrica 334
bog birch 118
Bog Clubmoss 19
bog muhly 492
bog orchid 452
bog valerian 152

bog willow 320
bog willowherb 237
bog yellow-eyed-grass 523
bog-laurel 178
bog-rosemary 176
Bolboschoenus fluviatilis 350
boneset 81
Borage Family 119
Boraginaceae 119
Borago officinalis 120
boreal bedstraw 310
Botrychium 22
    acuminatum 22
    campestre 22
    dissectum 26
    hesperium 22
    lanceolatum 23
    lunaria 23
    matricariifolium 23
    minganense 23
    mormo 23
    multifidum 26
    neolunaria 23
    obliquum 26
    oneidense 26
    pallidum 23
    rugulosum 26
    simplex 24
    spathulatum 24
    virginianum 24
Botrypus virginianus 24
bottle-gentian 199
bottlebrush-grass 481
bouncing-bet 158
Bouteloua curtipendula 468
boxelder 322
Brachyelytrum aristosum 468
bracken fern 9
Bracken Fern Family 9
bracted orchid 446
branched centaury 199
Brasenia schreberi 143
Brassica 132
    alba 142
    arvensis 142
    campestris 132
    hirta 142
    juncea 132
    kaber 142
    napus 132
    nigra 132
    rapa 132
Brassicaceae 126
Braun's holly fern 12
Braya humilis 126
Bristle Grass 504
bristle-leaf sedge 360
bristly black gooseberry 203
bristly blackberry 305
bristly crowfoot 283
bristly dewberry 304

bristly lady's-thumb 267
bristly locust 193
bristly rose 301
bristly sarsaparilla 57
bristly stalk sedge 380
brittle bladder fern 7
brittle-stem hemp-nettle 213
broad beech fern 30
broad-fruit bur-reed 521
broad-leaf cat-tail 522
broad-leaf toothwort 134
broad-lip twayblade 451
broadleaved panicgrass 476
broad loose-flower sedge 378
Brome 468
brome-like sedge 366
Bromus 468
Bromus altissimus 470
    arvensis 469
    ciliatus 469
    inermis 469
    japonicus 469
    kalmii 470
    latiglumis 470
    nottowayanus 468
    racemosus 470
    secalinus 470
    tectorum 470
bronze-head oval sedge 386
brook lobelia 146
brooklime 259
Broom-Rape 244
Broom-Rape Family 240
broomcorn millet 494
brown beak sedge 416
brown bog sedge 399
brown-fruit rush 432
brown-ray knapweed 74
brownish beak sedge 416
brownish sedge 369
buckbean 230
Buckbean Family 230
Buckbrush 286
Buckeye 323
Buckthorn 287
Buckthorn Family 286
Buckwheat Family 264
Buddleja davidii 327
Buffalo-Berry 174
buffalo-bur 331
buffalo-currant 204
Bugle 212
bugseed 43
bulblet fern 7
bulblet-bearing water-hemlock 49
bulbous bluegrass 500
bull thistle 77
Bulrush 419
bulrush sedge 351
bunchberry 169
bur-oak 197

Bur-reed 520
bur-reed sedge 397
Burdock 68
Burnet 306
burnet saxifrage 51
Bush-Clover 191
Bush-Honeysuckle 148
bushy aster 109
butter-and-eggs 254
Buttercup 280
Buttercup Family 276
butterfly weed 55
butternut 210
Butterwort 223

C
Cabombaceae 143
Cacalia plantaginea 69
Cactaceae 143
Cactus Family 143
Cakile 132
    edentula 132
    lacustris 132
Calamagrostis 470
    canadensis 471
    epigeios 471
    inexpansa 472
    stricta 472
Calamovilfa longifolia 472
Calamus 344
Calamus Family 344
Calendula officinalis 59
Calla palustris 347
Callitriche 252
    hermaphroditica 252
    palustris 252
Calluna vulgaris 175
Calopogon tuberosus 446
Caltha 278
    natans 278
    palustris 278
calypso 446
Calypso bulbosa 446
Calystegia 168
    sepium 168
    spithamaea 168
Camelina 133
    microcarpa 133
    sativa 133
Campanula 144
    aparinoides 144
    rapunculoides 145
    rotundifolia 145
    trachelium 145
Campanulaceae 144
Campion 159
Camptosorus rhizophyllus 5
Canada bluegrass 500
Canada yew 36
Canadanthus modestus 59
Canadian clearweed 335

Canadian honewort 49
Canadian milk-vetch 189
Canadian moonseed 230
Canadian mtn ricegrass 497
Canadian plum 300
Canadian rush 429
Canadian thistle 76
Canadian waterweed 423
Canadian wild ginger 58
Canadian wood-nettle 334
Canary-Grass 495
Cannabaceae 147
Cannabis sativa 147
Capnoides sempervirens 247
Caprifoliaceae 148
Capsella bursa-pastoris 133
Caragana arborescens 188
caraway 48
Cardamine 134
   arenicola 134
   concatenata 134
   diphylla 134
   flexuousa 134
   maxima 134
   parviflora 134
   pensylvanica 134
   pratensis 134
Cardaria pubescens 139
cardinal-flower 145
Carduus 73
   acanthoides 73
   crsipus 73
   nutans 73
carelessweed 78
Carex 351
   Section Acrocystis 355
   Section Albae 360
   Section Ammoglochin 360
   Section Anomalae 360
   Section Bicolores 360
   Section Carex 361
   Section Careyanae 362
   Section Ceratocystis 363
   Section Chlorostachyae 364
   Section Chordorrhizae 364
   Section Cyperoideae 365
   Section Deweyanae 365
   Section Digitatae 366
   Section Dispermae 367
   Section Divisae 368
   Section Glareosae 369
   Section Granulares 370
   Section Griseae 371
   Section Heleoglochin 372
   Section Hirtifoliae 373
   Section Holarrhenae 374
   Section Hymenochlaenae 374
   Section Laxiflorae 376
   Section Limosae 380
   Section Leptocephalae 380
   Section Leucoglochin 380

   Section Lupulinae 381
   Section Multiflorae 382
   Section Ovales 382
   Section Paludosae 388
   Section Paniceae 390
   Section Phacocystis 392
   Section Phaestoglochin 394
   Section Phyllostachyae 398
   Section Physoglochin 398
   Section Porocystis 398
   Section Racemosae 399
   Section Rostrales 400
   Section Stellulatae 400
   Section Vesicariae 402
   Section Vulpinae 406
   adusta 384
   aenea 386
   albursina 378
   alopecoidea 406
   amphibola 372
   angustior 401
   aquatilis 393
   arcta 369
   arctata 375
   assiniboinensis 375
   atherodes 362
   aurea 361
   backii 398
   bebbii 385
   blanda 378
   brevior 385
   bromoides 366
   brunnescens 369
   buxbaumii 399
   canescens 369
   capillaris 364
   castanea 375
   cephaloidea 396
   cephalophora Muhl 396
   chordorrhiza 364
   communis 356
   comosa 403
   concinna 367
   conoidea 372
   convoluta 396
   crawei 370
   crawfordii 385
   crinita 393
   crinitam 393
   cristatella 385
   cryptolepis 363
   cumulata 385
   debilis 375
   deflexa 356
   deweyana 366
   diandra 373
   dioica 398
   disperma 367
   eburnea 360
   echinata 401
   emoryi 393

   exilis 401
   flava 364
   foenea 386
   folliculata 400
   formosa 375
   garberi 361
   gracillima 376
   granularis 371
   grayi 381
   grisea 372
   gynandra 393
   gynocrates 398
   heliophila Mackenzie 357
   hirta 362
   hirtifolia 373
   hitchcockiana 372
   houghtoniana 388
   hystericina 403
   inops 357
   interior 401
   intumescens 382
   lacustris 388
   lanuginosa 390
   lasiocarpa 390
   lasiocarpa 390
   laxiflora 378
   lenticularis 394
   leptalea 380
   leptonervia 378
   limosa 380
   livida 391
   lucorum 357
   lupulina 382
   magellanica 381
   media 399
   merritt-fernaldii 387
   muehlenbergii 396
   muricata 401
   nigra 392
   nigromarginata 357
   normalis 387
   norvegica 399
   novae-angliae 357
   oligosperma 403
   ormostachya 378
   pallescens 398
   pauciflora 380
   paupercula 381
   peckii 357
   pedunculata 367
   pellita 390
   pensylvanica 357
   pensylvanica 357
   plantagineam 362
   praegracilis 368
   prairea 373
   projecta 387
   pseudocyperus 404
   radiata 396
   retrorsa 404
   richardsonii 367

rosea 396
rostrata 404, 405
saltuensis 391
sartwellii 374
scabrata 360
schweinitzii 404
scirpoidea 351
scoparia 387
siccata 360
sparganioides 396
sparganioides 397
sprengelii 376
sterilis 401
stipata 406
strictam 394
swanii 399
sychnocephala 365
tenera 387
tenuiflora 369
tetanica 392
tetanica 391
tincta 388
tonsa 358
tribuloides 388
trichocarpa 362
trisperma 370
tuckermanii 404
umbellata 358
utriculata 405
vaginata 391
vesicaria 406
viridula 364
vulpinoidea 382
weigandii 400
woodii 392
Carey's smartweed 266
Carolina crane's-bill 201
Carolina lovegrass 483
Carolina springbeauty 231
carpet vervain 335
Carpetweed 230
Carpetweed Family 230
Carpinus caroliniana 119
Carrot 49
Carrot Family 46
Carum carvi 48
Carya 210
  cordiformis 210
  ovata 210
Caryophyllaceae 152
Case's ladies'-tresses 455
Castilleja septentrionalis 242
Cat-tail 522
Cat-Tail Family 520
Cat's-Ear 90
catberry 56
Catchfly 159
catnip 218
Caulophyllum thalictroides 116
Ceanothus 286
  americanus 286

herbaceus 286
sanguineus 286
celandine 247
Celastraceae 164
Celastrus scandens 164
Celtis occidentalis 147
Cenchrus longispinus 472
Centaurea 74
  cyanus 74
  diffusam 74
  jacea 74
  macrocephala 74
  maculosam 75
  montana 74
  nigra 75
  nigrescens 74
  stoebe 75
Centaurium 199
  erythraea 199
  pulchellum 199
Centaury 199
Cerastium 154
  arvense 154
  brachypodum 154
  brachypodum 155
  fontanum 154
  nutans 155
  semidecandrum 155
  tomentosum 154
  vulgatum 154
Ceratophyllaceae 165
Ceratophyllum 165
  demersum 165
  echinatum 165
  muricatum 165
Chaenorhinum minus 252
Chain Fern 7
Chain Fern Family 7
Chamaedaphne calyculata 176
Chamaenerion angustifolium 235
Chamaerhodos erecta 288
Chamaesaracha grandiflora 329
Chamaesyce geyeri 185
Chamaesyce glyptosperma 186
Chamaesyce maculata 186
Chamaesyce nutans 186
Chamaesyce polygonifolia 186
Chamaesyce serpyllifolia 187
Chamaesyce vermiculata 187
Chamomile 67
Chamomilla inodora 95
cheat grass 468, 470
checkered rattlesnake-plantain 450
Chelidonium majus 247
Chelone glabra 253
Chenopodium 41
  album 42
  ambrosioides 44
  bonus-henricus 42
  botrys 44
  capitatum 42

glaucum 42
pratericola 42
simplex 42
standleyanum 43
Cherry 299
Chervil 48
chess 468, 470
chestnut-color sedge 375
Chickweed 162
chicory 75
Chimaphila umbellata 176
chinaroot 519
Chinese lantern plant 330
Chinese mustard 132
choke-cherry 301
Chokeberry 291
Chorispora tenella 127
Chrysanthemum balsamita 113
Chrysanthemum leucanthemum 93
Chrysanthemum parthenium 113
Chrysopsis villosa 89
Chrysosplenium americanum 325
chufa 408
Cichorium intybus 75
Cicuta 48
  bulbifera 49
  maculata 49
Cinna 473
  arundinacea 473
  latifolia 473
cinnamon rose 302
cinnamon-fern 26
Cinquefoil 297
Circaea 235
  alpina 235
  canadensis 236
  lutetiana 236
Cirsium 76
  arvense 76
  discolor 76
  hillii 76
  muticum 76
  palustre 76
  pitcheri 77
  vulgare 77
Cistaceae 166
Citrullus natus 171
Cladium mariscoides 406
clammy ground-cherry 330
clammy hedge-hyssop 254
clammy rabbit-tobacco 98
Clammyweed 167
clasping twisted stalk 441
clasping-leaf venus'-looking-glass 147
Claytonia 230
  caroliniana 231
  sibirica 231
  virginica 231
Clearweed 334
cleavers 309

INDEX 565

Clematis 279
    occidentalis 279
    verticillaris 279
    virginiana 279
Cleomaceae 167
Cleome Family 167
Cliff Fern 31
Cliff Fern Family 31
Cliffbrake 28
climbing nighshade 331
Clinopodium 213
    arkansanum 213
    glabrum 213
    vulgare 213
Clinton's leafless-bulrush 421
Clintonia borealis 437
Clover 194
Club-Rush 350, 416
Clubmoss Family 16
clustered black-snakeroot 51
clustered field sedge 368
clustered sedge 385
coastal jointweed 268
coastal sedge 401
Cocklebur 115
Coeloglossum viride 446
colic-root 422
Collinsia parviflora 253
Collomia linearis 260
Columbian watermeal 349
Columbine 278
Comandra 321
    livida 322
    umbellata 321
Comarum palustre 291
Comfrey 126
Commelina communis 349
Commelinaceae 349
common barley 488
common blackberry 303
common blanket-flower 83
common butterwort 223
common canary-grass 496
common chickweed 163
common cocklebur 115
common comfrey 126
common corncockle 153
common dandelion 114
common day-flower 349
common dodder 168
common dog-mustard 137
common duckweed 348
common elder 37
common enchanter's nightshade 236
common evening-primrose 239
common false foxglove 241
common flat-topped goldenrod 82
common fox sedge 382
common foxglove 253
common goldenrod 103
common grape-hyacinth 440

common hackberry 147
common hop-tree 313
common hops 148
common horsetail 13
common juniper 33
common lilac 234
common mallow 228
common mare's-tail 254
common marsh bedstraw 310
common marsh-marigold 278
common mermaid-weed 206
common milkweed 55
common moonwort 23
common nipplewort 92
common panic-grass 494
common plantain 257
common purslane 273
common ragweed 65
common reed 496
common sandbur 472
common scouring-rush 14
common sheep sorrel 271
common sneezeweed 84
common sow-thistle 107
common speedwell 259
common spike-rush 412
common St John's-wort 209
common sunflower 86
common tansy 113
common threesquare 418
common timothy 496
common vervain 336
common vetch 195
common viper's-bugloss 121
common water-hemlock 49
common water-milfoil 206
common wheat 509
common wood-rush 433
common wormwood 69
common yarrow 64
common yellow wood-sorrel 246
common yellowcress 141
compass-plant 101
Comptonia peregrina 231
Coneflower 79, 99
Conium maculatum 49
Conopholis americana 242
Conringia orientalis 135
Convallaria majalis 437
Convolvulaceae 167
Convolvulus arvensis 168
Conyza canadensis 77
coon's-tail 165
Cooper's milk-vetch 189
Copperleaf 184
Coptidium lapponicum 283
Coptis 279
    groenlandica 279
    trifolia 279
Coral-Root 446
coralberry 151

Corallorhiza 446
    maculata 446
    striata 447
    trifida 447
Cord-Grass 506
Coreopsis 78
    grandiflora 78
    lanceolata 78
Corispermum 43
    americanum 43
    orientale 43
    pallasii 43
    villosum 43
corn chamomile 67
corn poppy 248
corn speedwell 259
corn spurry 161
corn-mustard 142
Cornaceae 168
Corncockle 153
Cornus 168
    alba 169
    alternifolia 169
    amomum 169
    canadensis 169
    foemina 170
    obliqua 169
    racemosa 170
    rugosa 170
    sericea 169
    stolonifera 169
Coronilla varia 194
Corydalis 247
    aurea 247
    sempervirens 247
Corylus 119
    americana 119
    cornuta 119
Cosmos bipinnatus 59
costmary 113
Cota tinctoria 67
Cotton-Grass 414
Cottonweed 44
Cottonwood 313
Cow-Parsnip 50
Cow-Wheat 244
cowcockle 163
Crabgrass 478
crack willow 317
Crane's-Bill 201
Crassulaceae 170
Crataegus 292
    chrysocarpa 292
    dodgei 292
    douglasii 293
    irrasa 292
    macrosperma 293
    monogyna 292
    punctata 293
    submollis 292
    succulenta 293

Crawe's sedge 370
Crawford's sedge 385
cream vetchling 191
Creeper 343
creeping bellflower 145
creeping bladderwort 223
creeping buttercup 283
creeping juniper 33
creeping snowberry 177
creeping spearwort 282
creeping yellowcress 141
creeping-jenny 274
Crepis tectorum 78
crested dog's-tail grass 473
crested sedge 385
crested wood-fern 10
Crocanthemum 166
    bicknellii 166
    canadense 166
crow-foot violet 341
Crowberry 176
Crowfoot 280
Crown-Vetch 194
crowned beggarticks 72
Cryptogramma 28
    acrostichoides 28
    stelleri 28
Cryptotaenia canadensis 49
cuckoo-flower 134
Cucumber Family 171
Cucumber-Root 440
Cucurbita 171
    maxima 171
    pepo 171
Cucurbitaceae 171
Cudweed 84
cultivated flax 225
Culver's-root 260
Cupressaceae 33
curly dock 271
curly pondweed 512
curly-top gumweed 84
Currant 202
Currant Family 202
cursed crowfoot 283
Cuscuta gronovii 168
Cut-Grass 489
cut-leaf grape fern 26
cut-leaf teasel 172
cut-leaf toothwort 134
cut-leaf water-horehound 216
cutleaf coneflower 99
Cyclachaena xanthiifolia 78
Cycloloma atriplicifolium 43
Cymbalaria muralis 253
Cynoglossum 121
    boreale 121
    officinale 121
    virginianum 121
Cynosurus cristatus 473
Cyperaceae 349

Cyperus 407
    aristatus 409
    bipartitus 408
    engelmannii 408
    esculentus 408
    ferruginescens 408
    hansenii 409
    houghtonii 408
    inflexus 409
    lupulinus 408
    × *mesochoreus 409*
    niger 408
    odoratus 408
    rivularis 408
    schweinitzii 409
    speciosus 408
    squarrosus 409
    stenolepis 409
    strigosus 409
Cypress Family 33
cypress spurge 185
cypress-like sedge 404
Cypripedium 447
    acaule 447
    arietinum 448
    calceolus 448
    parviflorum 448
    reginae 448
Cyrtorhncha cymbalaria 282
Cystopteridaceae 7
Cystopteris 7
    bulbifera 7
    fragilis 7
    protrusa 8
    tenuis 8
    laurentiana 8

D

Dactylis glomerata 474
dagger-leaf rush 430
Daisy 80
daisy-leaf moonwort 23
dalmatian toadflax 254
dame's rocket 138
Dandelion 114
Danthonia 474
    compressa 474
    intermedia 474
    spicata 474
daphne 331
Daphne mezereum 331
dark-green bulrush 420
dark-scale cotton-grass 415
Dasiphora fruticosa 298
Datura 329
    inoxia 329
    stramonium 329
    wrightii 329
Daucus carota 49
David's spurge 185
Day-Flower 349

Day-Lily 438
Dead Nettle 214
Death-Camas 436
Decodon verticillatus 226
deep-root ground-pine 17
Dendrolycopodium dendroideum 20
Dendrolycopodium hickeyi 20
Dendrolycopodium obscurum 21
Dennstaedtiaceae 9
Dentaria diphylla 134
Dentaria maxima 134
Dentariaciniata 134
Deparia acrostichioides 6
Deptford pink 155
Deschampsia 474
    caespitosa 475
    flexuosa 475
Descurainia 135
    pinnata 136
    sophia 136
desert goosefoot 42
Desmodium 189
    canadense 189
    glutinosum 189
devil's-bite 94
devil's-paintbrush 89
devil's-pitchfork 72
dewberry 304
dewey's sedge 366
Dianthus 155
    armeria 155
    barbatus 155
    carthusianorum 155
    deltoides 155
    plumarius 155
    sylvestris 155
Dicentra 247
    canadensis 247
    cucullaria 248
    eximia 247
Dichanthelium 475
    acuminatum 476
    boreale 476
    clandestinum 475
    columbianum 478
    depauperatum 476
    implicatum 476
    latifolium 476
    linearifolium 477
    oligosanthes 477
    ovale 477
    portoricense 478
    spretum 475
    xanthophysum 478
Diervilla lonicera 148
Digitalis 253
    grandiflora 253
    purpurea 253
Digitaria 478
    ischaemum 478
    sanguinalis 479

dill 48
dioecious sedge 401
Dioscorea 422
   hirticaulis 422
   quaternata 422
   villosa 422
Dioscoreaceae 422
Diphasiastrum 16
   complanatum 17
   digitatum 17
   × habereri 16
   sabinifolium 17
   tristachyum 17
   × zeilleri 17
Diplotaxis muralis 136
Dipsacaceae 171
Dipsacus 172
   fullonum 172
   laciniatus 172
   sylvestris 172
Dirca palustris 332
Ditch-Grass 516
ditch-stonecrop 248
dock-leaf smartweed 266
Dodder 168
Dodecatheon meadia 275
Doellingeria umbellata 79
Dog-Mustard 137
Dog's-Tail Grass 473
Dogbane 53
Dogbane Family 53
Dogwood 168
Dogwood Family 168
doll's eyes 277
door-yard dock 271
dotted hawthorn 293
dotted smartweed 267
Douglas' knotweed 269
downy arrow-wood 39
downy ground-cherry 330
downy rattlesnake-plantain 450
downy serviceberry 290
downy thorn-apple 329
downy willowherb 237
Draba 136
   arabisans 136
   cana 137
   glabella 137
   incana 136
   nemorosa 137
   verna 137
Dracocephalum parviflorum 213
dragon's-mouth 445
Dragonhead 213
drooping wood-reed 473
drooping woodland sedge 375
Dropseed 507
Drosera 172
   anglica 172
   intermedia 173
   linearis 173

   rotundifolia 173
Droseraceae 172
drug eyebright 243
Drummond's rockcress 132
dry-spike sedge 360
Drymocallis arguta 293
Dryopteridaceae 10
Dryopteris 10
   assimilis 11
   austriaca 10
   × boottii 10
   carthusiana 10
   cristata 10
   expansa 11
   filix-mas 11
   fragrans 11
   goldiana 11
   intermedia 11
   marginalis 11
   × montgomeryi 10
   phegopteris 30
   × slossoniae 10
   spinulosa 10, 11
   × triploidea 10
duck-potato 346
Duckweed 347
Dudley's rush 429
Dulichium arundinaceum 409
dune spurge 185
dune thistle 77
dusty miller 70
Dutchman's-breeches 248
dwarf bilberry 182
dwarf cornel 169
dwarf ginseng 58
dwarf lake iris 425
dwarf mallow 228
dwarf milkweed 54
dwarf prairie-rose 302
dwarf raspberry 305
dwarf rattlesnake-plantain 450
dwarf scouring-rush 15
Dwarf-Dandelion 91
Dwarf-Mistletoe 321
dwarf-snapdragon 252
Dysphania 44
   ambrosioides 44
   botrys 44

E
early buttercup 282
early goldenrod 104
early lowbush blueberry 183
early meadow-rue 284
early saxifrage 326
early-leaf brome 470
earth-nut vetchling 191
eastern daisy fleabane 80
eastern dwarf-mistletoe 321
eastern hemlock 36
eastern leatherwood 332

eastern lined aster 110
eastern prickly gooseberry 203
eastern rough sedge 360
eastern shooting star 275
eastern star sedge 396
eastern tansy 113
eastern waterleaf 123
eastern white pine 35
eastern woodland sedge 378
eastern yellow star-grass 425
ebony-spleenwort 5
Echinacea pallida 79
Echinochloa 479
   crus-galli 479
   muricata 479
Echinocystis lobata 171
Echium vulgare 121
Eel-Grass 424
Elaeagnaceae 173
Elaeagnus umbellata 174
Elatinaceae 174
Elatine minima 174
Elder 37
elecampane 91
Eleocharis 409
   acicularis 411
   bernardina 413
   calva 412
   compressa 411
   elliptica 411
   elliptica 412
   erythropoda 412
   flaccida 412
   flavescens 412
   intermedia 412
   mamillata 412
   nitida 412
   obtusa 412
   olivacea 412
   ovata 412
   palustris 412
   quinqueflora 413
   robbinsii 413
   rostellata 414
   smallii 412
   tenuis 411
Eleusine indica 480
elk sedge 361
elliptic shinleaf 180
elliptic spike-rush 412
Elm 332
Elm Family 332
Elodea 423
   canadensis 423
   nuttallii 423
Elymus 480
   canadensis 480
   glaucus 481
   hystrix 481
   lanceolatus 481
   mollis 489

repens 481
riparius 482
smithii 495
trachycaulus 482
villosus 482
virginicus 482
wiegandii 482
Elytrigia repens 481
Elytrigia smithii 495
Emory's sedge 393
Empetrum nigrum 176
Enchanter's Nightshade 235
English daisy 70
English plantain 257
English ryegrass 489
English sundew 172
Epifagus virginiana 242
Epigaea repens 177
Epilobium 236
  angustifolium 235
  brachycarpum 236
  ciliatum 236
  coloratum 236
  leptophyllum 237
  glandulosum 236
  hirsutum 237
  palustre 237
  parviflorum 236
  strictum 237
Epipactis helleborine 448
Equisetaceae 12
Equisetum 12
  affine 14
  arvense 13
  × ferrissii 13
  fluviatile 14
  hyemale 14
  laevigatum 14
  × mackaii 13
  palustre 14
  pratense 14
  scirpoides 15
  sylvaticum 15
  telmateia 13
  variegatum 15
Eragrostis 482
  cilianensis 483
  frankii 483
  hypnoides 483
  minor 483
  pectinacea 483
  spectabilis 483
Erechtites hieraciifolia 80
erect knotweed 269
Ericaceae 175
Erigeron 80
  acris 80
  annuus 80
  canadensis 77
  hyssopifolius 80
  philadelphicus 80

pulchellus 80
ramosus 81
strigosus 81
Eriocaulaceae 422
Eriocaulon 422
  aquaticum 422
  septangulare 422
Eriophorum 414
  alpinum 421
  angustifolium 414
  gracile 414
  polystachion 414
  spissum 415
  tenellum 414
  vaginatum 415
  virginicum 415
  viridicarinatum 415
Erodium cicutarium 200
Erucastrum gallicum 137
Erysimum 138
  asperum 138
  capitatum 138
  cheiranthoides 138
  hieraciifolium 138
  inconspicuum 138
Erythronium 437
  albidum 438
  americanum 438
Eupatorium 81
  maculatum 82
  perfoliatum 81
  rugosum 65
Euphorbia 184
  corollata 185
  cyparissias 185
  davidii 185
  esula 187
  geyeri 185
  glyptosperma 186
  helioscopia 186
  maculata 186
  marginata 186
  nutansg 186
  peplus 186
  polygonifolia 186
  serpyllifolia 187
  vermiculata 187
  virgata 187
Euphorbiaceae 183
Euphrasia 243
  hudsoniana 243
  nemorosa 243
  officinalis 243
  stricta 243
Eurasian water-milfoil 206
European alkali-grass 502
European barberry 116
European buckthorn 287
European field-pansy 339
European gromwell 124
European lily-of-the-valley 437

European mountain-ash 307
European swamp thistle 76
European wallflower 138
Eurybia macrophylla 81
Euthamia 82
  caroliniana 82
  graminifolia 82
  remota 82
Eutrochium maculatum 82
Evening-Primrose 238
Evening-Primrose Family 235
everlasting-pea 190
eyebane 186
eyebright 243, 244

F
Fabaceae 188
Fagaceae 196
Fagopyrum esculentum 264
Fagus grandifolia 197
fairy-slipper 446
fall panic-grass 494
fall phlox 262
Fallopia 264
  cilinodis 265
  convolvulus 265
  japonica 265
  sachalinensis 265
False Asphodel 442
false baby's-breath 310
False Buckthorn 287
False Dragonhead 219
False Flax 133
False Foxglove 241, 242
False Ground-Cherry 329
false indigo-bush 188
false lily-of-the-valley 439
false mannagrass 508
false melic grass 503
false mermaidweed 224
False Nettle 334
False Oat 487, 509
false pennyroyal 214, 221
False Pimpernel 226
False Pondweed 517
false Solomon's-seal 439
false spiraea 307
False Spiraea 307
false toadflax 322
False Toadflax 322
fan ground-pine 17
fancy wood-fern 11
farewell-summer 110
farwell's water-milfoil 205
Fawn-Lily 437
feathered pink 155
feathertop 471
fen grass-of-parnassus 164
fen-orchid 450
fern pondweed 516
Fescue 484

INDEX

Festuca 484
   arundinacea 503
   brachyphylla 485
   elatior 503
   occidentalis 485
   ovina 486
   pratensis 503
   rubra 485
   saximontana 485
   subverticillata 485
   trachyphylla 486
feverfew 96, 113
Feverwort 152
few-flower sedge 380
few-flower spike-rush 413
few-flowered panicgrass 477
few-nerve cotton-grass 414
few-seed sedge 403
fibrous-root sedge 356
field bindweed 168
field brome 469
field chickweed 154
field horsetail 13
field meadow-foxtail 465
field mustard 132
field pennycress 142
field pussytoes 66
field sagewort 70
field thistle 76
field-cress 139
Figwort 327
Figwort Family 327
Filaree 200
Filipendula rubra 293
Fimbristylis autumnalis 415
Fimbry 415
Fir 34
Fir-Moss 18
fireberry hawthorn 292
fireweed 80, 235, 236
five-stamen mouse-ear chickweed 155
Fivefingers 306
Flag 425
flat oatgrass 474
Flat Sedge 407
flat-leaf bladderwort 224
flat-stalk pondweed 513
flat-stem pondweed 516
flat-stem spike-rush 411
Flat-Topped Goldenrod 82
Flat-Topped White Aster 79
Flax 225
Flax Family 225
Fleabane 80
fleshy hawthorn 293
fleshy stitchwort 162
floating bur-reed 521
floating marsh-marigold 278
floating pondweed 514
Floerkea proserpinacoides 224
flowering raspberry 305

flowering spurge 185
fly-honeysuckle 149
foam-flower 326
forest phlox 262
Forget-Me-Not 124
forked catchfly 160
forked three-awn 466
four-flower yellow-loosestrife 274
Four-O'clock 232
Four-O'clock Family 232
fowl bluegrass 501
fowl manna grass 486
fox grape 343
fox-tail sedge 406
Foxglove 253
foxglove beardtongue 256
foxtail-barley 488
foxtail-millet 504
Fragaria 294
   vesca 294
   virginiana 294
fragile rockbrake 28
fragrant rabbit-tobacco 98
fragrant wood fern 11
Frangula alnus 287
Fraser's marsh-St. John's-wort 210
Fraxinus 234
   americana 234
   nigra 234
   pennsylvanica 234
freshwater cord-grass 506
fringed black bindweed 265
fringed brome 469
fringed polygala 263
fringed sedge 393
fringed yellow-loosestrife 274
Fringed-Gentian 199
Froelichia gracilis 44
Frostweed 166
Fuller's teasel 172
Fumewort 247

G

Gaillardia aristata 83
Gale palustris 231
Galearis 449
   rotundifolia 445
   spectabilis 449
Galeopsis tetrahit 213
Galinsoga 83
   parviflora 83
   quadriradiata 83
Galium 309
   album 310
   aparine 309
   asprellum 309
   boreale 310
   brevipes 310
   kamtschaticum 310
   labradoricum 310
   lanceolatum 310

mollugo 310
obtusum 310
odoratum 309
palustre 310
tinctorium 311
trifidum 310, 311
triflorum 311
verum 311
gallant-soldier 83
garden baby's-breath 156
garden black currant 203
garden cornflower 74
garden forget-me-not 126
garden red currant 205
garden yellow-loosestrife 275
garden yellow-rocket 130
Garlic 435
garlic-mustard 129
Gaultheria 177
   hispidula 177
   procumbens 178
Gay Feather 93
Gaylussacia baccata 178
gaywings 263
Gentian 199
Gentian Family 198
Gentiana 199
   andrewsii 199
   linearis 199
   rubricaulis 199
Gentianaceae 198
Gentianopsis 199
   procera 199
   virgata 199
Geocaulon lividum 322
Geraniaceae 200
Geranium 201
   bicknellii 201
   carolinianum 201
   maculatum 201
   pusillum 201
   robertianum 201
   sanguineum 201
Geranium Family 200
Gerardia purpurea 241
Gerardia tenuifolia 241
Germander 221
germander speedwell 259
Geum 295
   aleppicum 295
   canadense 295
   fragarioides 295
   laciniatum 296
   macrophyllum 296
   rivale 296
   triflorum 296
   urbanum 296
giant chickweed 162
giant knotweed 265
giant ragweed 65
giant sunflower 86

Giant-Hyssop 212
Ginseng 58
Ginseng Family 56
Glaucium flavum 246
glaucous white lettuce 98
Glechoma hederacea 214
globular coneflower 99
glossy false buckthorn 287
Glyceria 486
    borealis 486
    canadensis 486
    grandis 486
    striata 486
Gnaphalium 84
    macounii 98
    obtusifolium 98
    sylvaticum 84
    uliginosum 84
goat's beard 291
Goat's-Beard 114
goblet-aster 110
golden alexanders 53
Golden Aster 89
golden corydalis 247
golden dock 271
golden hedge-hyssop 254
golden-chamomile 67
golden-fruit sedge 361
Golden-Heather 166
Golden-Saxifrage 325
Goldenrod 101
Goldie's wood-fern 11
Goldthread 279
Goodyera 449
    oblongifolia 450
    pubescens 450
    repens 450
    tesselata 450
Goose Grass 480
Gooseberry 202
Goosefoot 41
Goutweed 47
graceful sedge 376
Grama-Grass 468
Grape 343
Grape Family 343
Grape Fern 22, 24
grape honeysuckle 150
Grape-Hyacinth 440
Graphephorum melicoides 487
Grass Family 457
grass-leaf arrowhead 345
grass-leaf mud-plantain 510
grass-leaf rush 430
grass-leaf speedwell 260
grass-leaf stitchwort 162
Grass-of-Parnassus 164
grass-pink 446
grassy pondweed 513
Gratiola 253
    aurea 254
    lutea 254

    neglecta 254
gray dogwood 170
gray goldenrod 104
Gray's sedge 381
great blue lobelia 146
Great Lakes gentian 199
great mullein 328
Great Plains flat sedge 408
great St. John's-wort 208
great water-dock 271
great waterleaf 123
great-spur violet 342
greater bladder sedge 382
greater bladderwort 224
greater Canadian St. John's-wort 209
greater duckweed 348
greater hop clover 194
greater straw sedge 387
greater yellow water buttercup 282
green adder's-mouth orchid 451
green alder 118
green amaranth 40
green arrow-arum 348
green ash 234
green carpetweed 230
green foxtail-grass 505
green fringed orchid 453
green muhly 492
green sorrel 271
green spleenwort 6
green woodland orchid 453
green-flower wintergreen 180
Greenbrier 519
Greenbrier Family 519
Greene's rush 430
Grindelia 84
    ciliata 84
    squarrosa 84
Gromwell 123
groove-stem Indian-plantain 69
Grooveburr 289
grooved yellow flax 225
Grossulariaceae 202
ground hemlock 36
Ground-Cherry 329
ground-ivy 214
Ground-Pine 16, 19
groundnut 189
Groundsel 95, 100
grove bluegrass 499
Grove-Sandwort 157
Gumweed 84
Gymnocarpium 8
    dryopteris 9
    jessoense 9
    robertianum 9
Gypsophila 156
    elegans 156
    muralis 156
    paniculata 156
    scorzonerifolia 156

H

Habenaria clavellata 453
Habenaria dilatata 453
Habenaria hookeri 453
Habenaria hyperborea 453
Habenaria lacera 453
Habenaria obtusata 453
Habenaria orbiculata 454
Habenaria psycodes 454
Habenaria viridis 446
Hackberry 147
Hackelia 122
    deflexa 122
    virginiana 122
hair-like sedge 364
hairgrass 474, 475
hairy beardtongue 256
hairy cat's-ear 90
hairy crabgrass 479
hairy evening-primrose 240
hairy four-o'clock 232
hairy golden aster 89
hairy goldenrod 103
hairy honeysuckle 150
hairy panicgrass 476
hairy rock cress 130
hairy Solomon's-seal 441
hairy sunflower 86
hairy sweet-cicely 50
hairy vetch 196
hairy wild rye 482
hairy willowherb 237
hairy wood-rush 433
hairy-fruit sedge 362
hairy-stem gooseberry 203
halberd-leaf orache 41
Halenia deflexa 200
Halerpestes cymbalaria 282
Haloragaceae 205
Hamamelidaceae 207
Hamamelis virginiana 207
hammer sedge 362
handsome sedge 375
hard fescue 486
hardhack 308
hardstem club-rush 417
hare's-ear-mustard 135
Harebell 144
Hart's-tongue fern 5
Hawk's-Beard 78
Hawkbit 92
Hawkweed 89
Hawthorn 292
Hazelnut 119
heart-leaf alexanders 53
heart-leaf arnica 68
heart-leaf four-o'clock 232
heart-leaf twayblade 452
heart-leaf willow 317
heart-leaved groundsel 95
Heath Family 175

Hedeoma hispida 214
hedge-bindweed 168
Hedge-Hyssop 253
hedge-mustard 142
hedge-nettle 221
Hedge-Nettle 220
Hedyotis longifolia 311
Hedysarum alpinum 188
Helenium 84
    autumnale 84
    flexuosum 85
Helianthemum 166
    bicknellii 166
    canadense 166
Helianthus 85
    annuus 86
    divaricatus 86
    giganteus 86
    hirsutus 86
    × laetiflorus 85
    maximiliani 87
    mollis 87
    occidentalis 87
    pauciflorus 87
    petiolaris 88
    strumosus 88
    tuberosus 88
Heliopsis helianthoides 88
helleborine 448
Hemerocallis 438
    flava 438
    fulva 438
    lilioasphodelus 438
Hemlock 36
hemlock water-parsnip 52
hemp 147
Hemp Family 147
Hemp-Nettle 213
henbit 215
Hepatica 279
    acutiloba 279
    americana 280
    nobilis 279, 280
    triloba 280
Heracleum 50
    lanatum 50
    mantegazzianum 50
    maximum 50
    sphondylium 50
herb-bennet 296
herb-robert 201
herb-sophia 136
Herniaria hirsuta 153
Hesperis matronalis 138
Hesperostipa spartea 487
Heteranthera dubia 510
Heterotheca villosa 89
Heuchera richardsonii 325
Hickory 210
hidden spikemoss 29
hidden-fruit bladderwort 223

Hieracium 89
    aurantiacum 89
    caespitosum 90
    canadense 90
    florentinum 90
    kalmii 90
    lachenalii 89
    maculatum 89
    murorum 89
    piloselloides 90
    pratense 90
    scabriusculum 90
    scabrum 90
    umbellatum 90
    venosum 89
Hierochloe 487
    hirta 487
    odorata 487
high mallow 228
high-bush cranberry 38
Hill's pondweed 514
Hill's thistle 76
Hippuris vulgaris 254
Hitchcock's sedge 372
hoary alyssum 131
hoary frostweed 166
hoary puccoon 124
hoary sedge 369
hoary vervain 336
hoary whitlow-grass 137
Hog-Peanut 188
Holcus mollis 457
Holly 56
Holly Family 56
holly fern 12
hollyhock 228
Honewort 49
Honeysuckle 149
Honeysuckle Family 148
hooded blue violet 342
hooded ladies'-tresses 457
hooded skullcap 220
hook-spurred violet 338
hooked crowfoot 283
Hooker's orchid 453
hop sedge 382
hop-hornbeam 119
Hop-Tree 313
Hops 148
Hordeum 488
    jubatum 488
    vulgare 488
hornbeam 119
horned bladderwort 223
horned-pondweed 518
Hornwort 165
Hornwort Family 165
horse-gentian 152
horse-mint 218
horse-nettle 331
horse-radish 130

Horsetail 12
Horsetail Family 12
horseweed 77
Houghton's goldenrod 103
Houghton's flat sedge 408
Houghton's sedge 388
hound's-tongue 121
Houstonia langifolia 311
Huckleberry 178
Hudson Bay currant 203
Hudson Bay eyebright 243
Hudsonia tomentosa 166
Humulus lupulus 148
Huperzia 18
    appalachiana 18
    appressa 18
    lucidula 18
    porophila 18
    selago 19
Hydrocharis morsus-ranae 422
Hydrocharitaceae 422
Hydrocotyle americana 58
Hydrophyllum 122
    appendiculatum 123
    virginianum 123
Hylodesmum glutinosum 189
Hylotelephium telephium 171
Hyoscyamus niger 328
Hypericaceae 207
Hypericum 208
    ascyron 208
    boreale 208
    canadense 208
    ellipticum 208
    kalmianum 208
    majus 209
    perforatum 209
    punctatumm 209
    pyramidatum 208
    virginicum 210
Hypochaeris radicata 90
Hypopitys americana 179
Hypopitys monotropa 179
Hypoxidaceae 425
Hypoxis hirsuta 425
hyssop hedge-nettle 221
Hystrix patula 481

I
Iberis 127
    sempervirens 127
    umbellata 127
Ilex 56
    mucronata 56
    verticillata 56
Illinois greenbrier 519
Illinois pondweed 514
Impatiens 115
    biflora 115
    capensis 115
Indian cucumber-root 440

Indian goose grass 480
Indian Grass 506
Indian Plantain 69
Indian-hemp 53
Indian-Paintbrush 242
Indian-Pipe 178
Indian-tobacco 146
Indigo-Bush 188
inflated narrow-leaf sedge 372
inland rush 430
inland sedge 401
inland serviceberry 290
interior bluegrass 500
intermediate spike-rush 412
interrupted fern 27
Inula helenium 91
Iowa moonwort 22
Iridaceae 425
Iris 425
 lacustris 425
 pseudacorus 425
 sibirica 425
 versicolor 425
 virginica 426
Iris Family 425
ironwood 119
Isanthus brachiatus 221
Isoetaceae 16
Isoetes 16
 braunii 16
 echinospora 16
 hieroglyphica 16
 lacustris 16
 macrospora 16
 muricata 16
Iva xanthifolia 78
ivy-leaf duckweed 348

J
jack pine 35
jack-go-to-bed-at-noon 115
jack-in-the-pulpit 347
Jacobea vulgaris 59
Japanese barberry 116
Japanese bristle grass 504
Japanese knotweed 265
Jerusalem-artichoke 88
Jerusalem-oak 44
jewelweed 115
jimsonweed 329
Joe-Pye-Weed 81, 82
Johnny-jump-up 343
joint-leaf rush 428
Jointweed 268
Juglandaceae 210
Juglans cinerea 210
jumpseed 267
Juncaceae 427
Juncaginaceae 434
Juncus 427
 acuminatus 428

 alpinoarticulatus 428
 alpinus 428
 arcticus 428
 articulatus 428
 balticus 428
 biflorus 430
 brachycephalus 429
 brevicaudatus 429
 bufonius 429
 canadensis 429
 dudleyi 429
 effusus 429
 ensifolius 430
 filiformis 430
 greenei 433
 greenei 430
 inflexus 427
 marginatus 430
 nodosus 430
 pelocarpus 432
 pylaei 430
 stygius 432
 tenuis 429
 tenuis 432
 torreyi 432
 vaseyi 433
Junegrass 489
Juniper 33
Juniperus 33
 communis 33
 horizontalis 33

K
Kalm's brome 470
Kalm's St. John's-wort 208
Kalmia 178
 angustifolia 178
 polifolia 178
Kenilworth-ivy 253
Kentucky bluegrass 501
kidney-leaf buttercup 281
kidney-leaf white violet facw 341
Knapweed 74
Knautia arvensis 172
Knawel 158
knotty pearlwort 158
knotty-leaf rush 428
Knotweed 269
Kochia scoparia 41
Koeleria 489
 macrantha 489
 pyramidata 489
Krigia biflora 91

L
Labrador buttercup 283
Labrador Indian-paintbrush 242
Lactuca 91
 biennis 92
 canadensis 92
 hirsuta 92

 scariola 92
 serriola 92
Ladies'-Tresses 454
lady fern 6
Lady Fern Family 6
Lady's-Slipper 447
lady's-thumb 265, 267
lake quillwort 16
lakebank sedge 388
lakecress 141
lakeshore sedge 394
lamb's quarters 42
Lamiaceae 211
Lamium 214
 amplexicaule 215
 maculatum 215
 purpureum 215
lamp rush 429
lance-leaf figwort 327
lance-leaf tickseed 78
lance-leaf twisted stalk 442
lance-leaf wild licorice 310
lance-leaf yellow-loosestrife 274
Lapland buttercup 283
Laportea canadensis 334
Lappula squarrosa 123
Lapsana communis 92
larch 34
large barnyard-grass 479
large clammyweed 167
large cranberry 182
large false ground-cherry 329
large sweet vernal grass 466
large yellow-loosestrife 274
large-flower bellwort 443
large-flower wakerobin 442
large-fruit black-snakeroot 52
large-leaf grove-sandwort 157
large-leaf pondweed 512
large-leaf wood-aster 81
large-seed false flax 133
Larix 34
 decidua 34
 laricina 34
Lathyrus 190
 japonicus 190
 latifolius 190
 maritimus 190
 ochroleucus 191
 palustris 191
 pratensis 190
 sylvestris 190
 tuberosus 191
 venosus 191
Laurel 178
lavender bladderwort 224
lawn daisy 70
Leafless-Bulrush 421
leafy pondweed 513
leafy spurge 187
least moonwort 24

leatherleaf 176
leatherwood 332
leathery grape fern 26
leathery knotweed 269
Lechea 166
    intermedia 167
    pulchella 167
ledge spike-moss 29
Ledum groenlandicum 180
Leek 435
Leersia 489
    oryzoides 489
    virginica 489
Lemna 347
    minor 348
    trisulca 348
    turionifera 348
Lentibulariaceae 221
lentil vetch 196
Leontodon 92
    saxatilism 92
    taraxacoides 92
Leonurus cardiaca 215
Lepidium 139
    appelianum 139
    campestre 139
    densiflorum 139
    virginicum 140
Leptilon canadense 77
Lespedeza capitata 191
lesser bladder sedge 406
lesser bladderwort 224
lesser brown sedge 384
lesser burdock 68
lesser Canadian St. John's-wort 208
lesser clearweed 335
lesser fringed-gentian 199
lesser hop clover 194
lesser knapweed 75
lesser periwinkle 55
lesser purple fringed orchid 454
lesser tussock sedge 373
lesser yellow water buttercup 282
Lettuce 91
Leucanthemella serotina 59
Leucanthemum vulgare 93
Leucophysalis grandiflora 329
Levisticum officinale 46
Leymus mollis 489
Liatris 93
    aspera 94
    pycnostachya 94
    scariosa 94
Liliaceae 434
Liliid Monocot Family 425
Lilium 438
    michiganense 439
    philadelphicum 439
Lily Family 434
Lily-of-the-Valley 437
limber honeysuckle 149

lime-encrusted saxifrage 326
limestone oak fern 9
limestone swamp bedstraw 310
limestone wild basil 213
limestone-meadow sedge 371
Limnanthaceae 224
Linaceae 225
Linaria 254
    canadensis 255
    dalmatica 254
    spartea 254
    vulgaris 254
linden 228
Lindernia dubia 226
Lindernia Family 226
Linderniaceae 226
linear-leaved panicgrass 477
Linnaea borealis 148
Linum 225
    catharticum 225
    sulcatum 225
    usitatissimum 225
Liparis loeselii 450
Listera auriculata 451
Listera convallarioides 451
Listera cordata 452
Lithospermum 123
    canescens 124
    caroliniense 124
    officinale 124
little bluestem 504
little goblin moonwort 23
little green sedge 364
little hawkbit 92
little lovegrass 483
little prickly pear 143
little shinleaf 180
little skullcap 220
little sundrops 239
little yellow rattle 245
little-pod false flax 133
Littorella 254
    americana 254
    uniflora 254
live forever 171
Liverwort 279
livid sedge 391
Lobelia 145
    cardinalis 145
    dortmanna 146
    inflata 146
    kalmii 146
    siphilitica 146
    spicatam 147
Lobularia maritima 127
Locust 193
Logfia arvensis 59
Lolium 489
    arundinaceum 503
    perenne 489
    pratense 503

Lombardy poplar 315
long-beak sedge 376
long-beak water-crowfoot 281
long-branch frostweed 166
long-head thimbleweed 277
long-leaf pondweed 514
long-leaf speedwell 259
long-leaf summer bluets 311
long-leaved arnica 68
long-leaved aster 112
long-leaved stitchwort 162
long-spur violet 341
long-stalk sedge 367
long-stalk starwort 162
long-stolon sedge 357
longleaf ground-cherry 330
Lonicera 149
    × bella 149
    caerulea 150
    canadensis 149
    dioica 149
    hirsuta 150
    involucrata 150
    morrowii 150
    oblongifolia 150
    prolifera 150
    reticulata Raf 150
    tatarica 150
    villosa 150
Loosestrife 226, 273
Loosestrife Family 226
Lopseed 250
Lopseed Family 248
Lotus corniculatus 192
Lousewort 244
Lovegrass 482
low baby's-breath 156
low bindweed 168
low northern sedge 367
lowbush blueberry 182
lowland bladder fern 8
Ludwigia 237
    palustris 237
    polycarpa 238
Lunaria annua 127
Lupine 192
Lupinus polyphyllus 192
Luzula 433
    acuminata 433
    campestris 433
    multiflora 433
    pallidula 433
    parviflora 433
Lychnis alba 160
Lychnis chalcedonica 159
Lychnis coronaria 159
Lycium chinense 328
Lycopodiaceae 16
Lycopodiella 19
    inundata 19
    subappressa 19

Lycopodium 19
    annotinum 19
    canadense 20
    clavatum 20, 21
    dendroideum 20
    digitatum 17
    hickeyi 20
    inundatum 19
    lagopus 21
    lucidulum 18
    obscurum 21
    porophilum 18
    rupestre 29
    sabinifolium 17
    selago 18, 19
    tristachyum 17
Lycopus 216
    americanus 216
    uniflorus 216
    virginicus 216
Lyme Grass 489
lyre-leaf rockcress 129
Lysimachia 273
    borealis 275
    ciliata 274
    lanceolata 274
    nummularia 274
    punctata 274
    quadriflora 274
    quadrifolia 274
    terrestris 274
    thyrsiflora 275
    vulgaris 275
Lythraceae 226
Lythrum 226
    alatum 226
    salicaria 227

M

Madder Family 309
Madwort 129
Maianthemum 439
    canadense 439
    racemosum 439
    stellatum 439
    trifolium 439
maiden pink 155
Maidenhair Fern 28
Maidenhair Fern Family 28
maidenhair-spleenwort 6
Malaxis 450
    brachypoda 451
    monophyllos 451
    unifolia 451
male fern 11
Mallow 228
Mallow Family 227
Maltese cross 159
Malus 296
    baccata 297
    pumila 296

Malva 228
    alcea 228
    moschata 228
    neglecta 228
    pusilla 228
    rotundifolia 228
    sylvestris 228
Malvaceae 227
Manna Grass 486
many-fruit primrose-willow 238
many-head sedge 365
Maple 322
maple-leaf arrow-wood 38
maple-leaf goosefoot 42
Mare's-Tail 254
marginal wood fern 11
marijuana 147
Mariscus schweinitzii 409
marsh arrow-grass 434
marsh bellflower 144
marsh blue violet 339
marsh cinquefoil 291
marsh cudweed 84
Marsh Fern Family 30
marsh muhly 491
marsh primrose-willow 237
Marsh St. John's-Wort 210
marsh vetchling 191
marsh willowherb 237
Marsh-Elder 78
marsh-fern 31
marsh-horsetail 14
Marsh-Marigold 278
Marshlocks 291
Maryland black-snakeroot 51
mat amaranth 40
Matricaria 94
    chamomilla 95
    discoidea 95
    maritima 95
    matricarioides 95
    parthenium 113
    recutita 95
matted muhly 492
Matteuccia struthiopteris 21
maximilian sunflower 87
may-apple 116
Mayweed 94
meadow garlic 436
meadow goat's-beard 114
meadow rye grass 503
meadow willow 320
meadow-buttercup 281
Meadow-Foxtail 464
meadow-horsetail 14
Meadow-Pitchers 229
Meadow-Rue 284
Meadowfoam Family 224
meadowsweet 307, 308
Medeola virginiana 440
Medicago 192

    lupulina 192
    sativa 192
Medick 192
Megalodonta beckii 71
Melampyrum lineare 244
Melastomataceae 229
Melastome Family 229
Melic Grass 490
Melica smithii 490
Melilotus 192
    albus 193
    officinalis 193
Menispermaceae 230
Menispermum canadense 230
Mentha 216
    aquatica 216
    arvensis 217
    canadensis 217
    × gracilis 216
    × piperita 217
    spicata 217
    × villosa 216
Menyanthaceae 230
Menyanthes trifoliata 230
Mermaid-Weed 206
Merritt Fernald's sedge 387
Mertensia 124
    paniculata 124
    virginica 124
Mexican muhly 492
Mexican tea 44
Mexican-fireweed 41
Mezereum Family 331
Michaux's stitchwort 157
Michigan lily 439
Michigan monkey-flower 250
Micranthes 325
    pensylvanica 325
    virginiensis 326
Midwestern sundrops 240
mild water-pepper 266
Milium effusum 490
Milk-Vetch 189
Milkweed 54
Milkwort 262
Milkwort Family 262
Millet Grass 490
Mimulus 249
    glabratus 249
    guttatus 250
    michiganensis 250
    moschatus 250
    ringens 250
Mingan moonwort 23
Mint 216
Mint Family 211
Minuartia 156
    dawsonensis 157
    michauxii 157
Mirabilis 232
    albida 232

hirsuta 232
nyctaginea 232
Missouri willow 317
Mistassini primrose 275
Mitchella repens 311
Mitella 326
   diphylla 326
   nuda 326
Mitrewort 326
Moehringia macrophylla 157
Molluginaceae 230
Mollugo verticillata 230
Monarda 217
   didyma 217
   fistulosa 217
   punctata 218
Moneses uniflora 178
Monkey-Flower 249
Monotropa 178
   hypopithys 179
   uniflora 179
Montana sedge 399
Montia Family 230
Montiaceae 230
Moonseed 230
Moonseed Family 230
Moonwort 22
moor rush 432
Moraceae 231
Morning-Glory Family 167
Morrow's honeysuckle 150
Morus alba 231
moss-pink 262
mossy stonecrop 171
motherwort 215
Mount Albert goldenrod 105
mountain cranberry 183
mountain death-camas 436
mountain fir-moss 18
mountain holly 56
mountain maple 323
Mountain Ricegrass 492
mountain sweet-cicely 50
Mountain-Ash 307
Mountain-Mint 219
Mountain-Trumpet 260
mouse-ear chickweed 154
mouse-ear cress 130
mud sedge 380
Mud-Plantain 510
Muehlenberg's sedge 396
mugwort 70
Muhlenbergia 490
   frondosa 491
   glomerata 491
   mexicana 492
   racemosa 492
   richardsonis 492
   sylvatica 492
   uniflora 492
Muhly 490

Mulberry 231
Mulberry Family 231
Mullein 327
Muscari 440
   botryoides 440
   neglectum 440
musk mallow 228
musk thistle 73
Muskroot Family 37
musky monkey-flower 250
Mustard 132
Mustard Family 126
Mycelis muralis 59
Myosotis 124
   arvensis 125
   laxa 125
   palustris 125
   scorpioides 125
   stricta 125
   sylvatica 126
   verna 126
   virginica 126
Myosoton aquaticum 162
Myrica gale 231
Myricaceae 231
Myriophyllum 205
   alterniflorum 205
   exalbescens 206
   farwellii 205
   heterophyllum 206
   sibiricum 206
   spicatum 206
   tenellum 206
   verticillatum 206
Myrrhis odorata 46

N

Nabalus albus 98
Nabalus racemosus 98
Nahanni oak fern 9
Najas 423
   flexilis 424
   gracillima 424
   guadalupensis 424
naked broom-rape 244
naked mitrewort 326
naked-stemmed sunflower 87
nanny-berry 38
narrow false oat 509
narrow-leaf blue-eyed-grass 426
narrow-leaf bur-reed 521
narrow-leaf cat-tail 522
narrow-leaf gentian 199
narrow-leaf hawk's-beard 78
narrow-leaf hawkweed 90
narrow-leaf mountain-mint 219
narrow-leaf mtn-trumpet 260
narrow-leaf vervain 336
narrow-panicle rush 429
Nasturtium 140
   microphyllum 140

officinale 140
necklace sedge 387
necklace spike sedge 378
needle beak sedge 416
needle spike-rush 411
needle-tip blue-eyed-grass 427
Needlegrass 487
Nemopanthus mucronatus 56
Neobeckia aquatica 141
Neottia 451
   auriculata 451
   convallarioides 451
   cordata 452
Nepeta cataria 218
nerveless woodland sedge 378
Nettle Family 334
New England aster 110
New England blue violet 340
New England sedge 357
New England serviceberry 290
New Jersey-tea 286
New York fern 30
Nicandra physalodes 329
night-flowering catchfly 160
Nightshade 330
ninebark 297
Nipplewort 92
nodding bur-marigold 71
nodding fescue 485
nodding mouse-ear chickweed 155
nodding onion 436
nodding sedge 393
nodding stickseed 122
nodding wild rye 480
none-so-pretty 159
northeastern sedge 363
northern adder's-tongue 24
northern bedstraw 310
northern beech-fern 30
northern blackberry 304
northern bluebells 124
northern blueflag 425
northern bog bedstraw 310
northern bog clubmoss 19
northern bog sedge 398
northern bog violet 339
northern bog-aster 108
northern bog-goldenrod 105
northern bog-orchid 453
northern bush-honeysuckle 148
northern club-moss 20
northern cluster sedge 369
northern crane's-bill 201
northern fir-moss 19
northern gooseberry 204
northern green rush 428
northern heart-leaved aster 108
northern maidenhair fern 28
northern manna grass 486
northern mountain-ash 307
northern oak fern 9

northern panicgrass 476
northern pin oak 197
northern red oak 198
northern running-pine 17
northern sedge 356
northern slender ladies'-tresses 456
northern spikemoss 30
northern St. John's-wort 208
northern stitchwort 162
northern swamp buttercup 282
northern sweet colt's-foot 97
northern three-lobed bedstraw 311
northern water-horehound 216
northern water-plantain 345
northern white cedar 33
northern wild rice 509
northern wood-sorrel 246
northern yellow-eyed-grass 523
northern long sedge 400
Northwest Territory sedge 405
Norway maple 323
Norway spruce 34
Nuphar 233
   lutea 233
   microphylla 233
   pumila 233
   × *rubrodisca 233*
   variegata 233
Nuttallanthus canadensis 255
Nyctaginaceae 232
Nymphaea 233
   leibergii 233
   odorata 233
   tuberosa 233
Nymphaeaceae 232

O

Oak 197
Oak fern 8
oak-leaf goosefoot 42
Oakes' evening-primrose 239
Oakes' pondweed 515
oat 467
Oatgrass 466
obedience 219
oblong-fruit serviceberry 290
Oclemena nemoralis 95
Odontites 244
   serotinus 244
   vulgaris 244
Oenothera 238
   biennis 239
   fruticosa 239
   nuttallii 239
   oakesiana 239
   parviflora 239
   perennis 239
   pilosella 240
   villosa 240
Ohio flat-topped goldenrod 104
Ohio-buckeye 323

old-man-in-the-spring 100
old-pasture bluegrass 501
oldfield cinquefoil 299
oldfield-toadflax 255
Oleaceae 234
Oleaster Family 173
Oligoneuron album 104
Oligoneuron houghtonii 103
Oligoneuron ohioense 104
Oligoneuron rigidum 104
Olive Family 234
Omalotheca sylvatica 84
Onagraceae 235
one-cone ground-pine 21
one-flower Indian-pipe 179
one-flowered shinleaf 178
one-sided shinleaf 179
Onion 435
Onoclea sensibilis 21
Onocleaceae 21
Ontario aster 111
open-field sedge 372
Ophioglossaceae 22
Ophioglossum pusillum 24
opium poppy 248
Oplopanax horridus 57
Opuntia fragilis 143
Orache 41
orange day-lily 438
orange dwarf-dandelion 91
orchard-grass 474
Orchid Family 443
Orchidaceae 443
Orchis rotundifolia 445
Orchis spectabilis 449
Oregon woodsia 33
Origanum vulgare 211
Ornithogalum umbellatum 440
Orobanchaceae 240
Orobanche uniflora 244
Orthilia secunda 179
Oryzopsis 492
   asperifolia 492
   canadensis 497
   pungens 497
   racemosa 497
Osmorhiza 50
   berteroi 50
   chilensis 50
   claytonii 50
   depauperata 50
   longistylis 51
Osmunda 26
   cinnamomea 26
   claytoniana 27
   regalis 27
Osmundaceae 26
Osmundastrum cinnamomeum 26
ostrich fern 21
Ostrya virginiana 119
Oswego tea 217

oval-leaf blueberry 182
oval-leaf sedge 396
ovoid spike-rush 412
ox-eye daisy 93
Oxalidaceae 245
Oxalis 245
   acetosella 246
   dillenii 246
   fontana 246
   montana 246
   stricta 246

P

Packera 95
   aurea 95
   indecora 96
   insulae-regalis 95
   paupercula 96
pagoda dogwood 169
pale dock 271
pale madwort 129
pale moonwort 23
pale panicgrass 478
pale sedge 398
pale St. John's-wort 208
pale-leaf woodland sunflower 88
pale-yellow iris 425
Panax 58
   quinquefolius 58
   trifolius 58
Panic-Grass 475, 493
Panicum 493
   acuminatum 476
   boreale 476
   capillare 494
   columbianum 478
   depauperatum 476
   dichotomiflorum 494
   flexile 494
   latifolium 476
   linearifolium 477
   miliaceum 494
   oligosanthes 477
   philadelphicum 494
   virgatum 494
   xanthophysum 478
Papaver 248
   rhoeas 248
   somniferum 248
Papaveraceae 246
paper birch 118
parasol sedge 358
Parathelypteris noveboracensis 30
Parietaria pensylvanica 334
Parlin's pussytoes 67
Parnassia 164
   glauca 164
   palustris 164
   parviflora 164
Parsnip 51
Parthenium integrifolium 96

INDEX 577

Parthenocissus 343
  inserta 343
  quinquefolia 343
  vitacea 343
partridge-berry 311
Pascopyrum smithii 495
Pastinaca sativa 51
Patis racemosa 497
Pea Family 188
peach-leaf willow 316
pearl-millet 505
Pearlwort 157
pearly-everlasting 66
Peck's sedge 357
Pedicularis 244
  canadensis 244
  lanceolata 245
Pellaea 28
  atropurpurea 28
  glabella 28
Pellitory 334
Peltandra virginica 348
Pennsylvania bittercress 134
Pennsylvania blackberry 305
Pennsylvania ground-pine 20
Pennsylvania pellitory 334
Pennsylvania sedge 357
Pennycress 142
Pennywort 58
Penstemon 255
  calycosus 256
  digitalis 256
  gracilis 256
  hirsutus 256
Penthoraceae 248
Penthorum Family 248
Penthorum sedoides 248
peppermint 217
Pepperwort 139
perennial knawel 159
perennial ragweed 65
perennial sow-thistle 107
Periwinkle 55
Persicaria 265
  amphibia 266
  careyi 266
  hydropiper 266
  hydropiperoides 266
  lapathifolia 266
  longiseta 267
  maculosa 267
  pensylvanica 267
  punctata 267
  sagittata 267
  virginiana 267
Petasites 96
  frigidus 97
  hybridus 96
  palmatus 97
  sagittatus 97
Petrorhagia saxifraga 157

Petroselinum crispum 46
petty spurge 186
Phacelia franklinii 120
Phalaris 495
  arundinacea 495
  canariensis 496
Phedimus spurius 170
Phegopteris 30
  connectilis 30
  hexagonoptera 30
Philadelphia daisy 80
Philadelphia panic-grass 494
Phleum pratense 496
Phlox 262
  divaricata 262
  paniculata 262
  subulata 262
Phlox Family 260
Phragmites 496
  australis 496
  communis 496
Phrymaceae 248
Phryma leptostachya 250
Phyllitis scolopendrium 5
Physalis 329
  alkekengi 330
  grandiflora 329
  grisea 330
  heterophylla 330
  longifolia 330
  pubescens 330
  virginiana 330
Physocarpus opulifolius 297
Physostegia virginiana 219
Picea 34
  abies 34
  glauca 35
  mariana 35
pickerelweed 510
Pickerelweed Family 510
pigweed 42
Pilea 334
  fontana 335
  pumilaGray 335
Pimpernel 52
Pimpinella saxifraga 51
pin-cherry 300
Pinaceae 34
Pine 35
Pine Family 34
pineapple-weed 95
Pinedrops 179
pinesap 179
Pinguicula vulgaris 223
Pink 155
Pink Family 152
pink lady's-slipper 447
pink shinleaf 180
pinkweed 267
Pinus 35
  banksiana 35

  resinosa 35
  strobus 35
  sylvestris 36
Pinweed 166
Piperia dilatata 453
Pipewort 422
Pipewort Family 422
pipsissewa 176
Piptatheropsis canadensis 497
Piptatheropsis pungens 497
Piptatherum 497
  canadense 497
  pungens 497
  racemosum 497
pitcherplant 324
Pitcherplant Family 324
plains cottonwood 314
plains puccoon 124
plains sunflower 88
Plantaginaceae 250
Plantago 256
  arenaria 257
  lanceolata 257
  major 257
  patagonica 257
  rugelii 257
Plantain 256
Plantain Family 250
plantain-leaf sedge 362
Platanthera 452
  × andrewsii 452
  aquilonis 452
  clavellata 453
  dilatata 453
  hookeri 453
  huronensis 453
  lacera 453
  macrophylla 454
  obtusata 453
  orbiculata 454
  psycodes 454
  rotundifolia 445
  unalascensis 452
Plum 299
Poa 498
  alpina 498
  alsodes 499
  annua 499
  bulbosa 500
  compressa 500
  glauca 500
  interior 500
  nemoralis 500
  palustris 501
  pratensis 501
  saltuensis 501
  secunda 498
  sylvestris 501
  trivialis 501
Poaceae 457
pod-grass 518

Podophyllum peltatum 116
Pogonia ophioglossoides 454
pointed broom sedge 387
pointed-leaf tick-clover 189
poison-hemlock 49
Poison-Ivy 45
poke milkweed 54
Polanisia dodecandra 167
Polemoniaceae 260
Polygala 262
    paucifolia 263
    polygama 263
    sanguinea 263
    senega 263
    vulgaris 263
Polygalaceae 262
Polygaloides paucifolia 263
Polygonaceae 264
Polygonatum pubescens 441
Polygonella articulata 268
Polygonum 269
    achoreum 269
    amphibium 266
    articulatum 268
    aviculare 269
    caespitosum 267
    careyi 266
    cilinode 265
    convolvulus 265
    cuspidatum 265
    douglasii 269
    erectum 269
    hydropiper 266
    hydropiperoides 266
    lapathifolium 266
    pensylvanicum 267
    persicaria 267
    punctatum 267
    ramosissimum 269
    sachalinense 265
    sagittatum 267
    virginianum 267
    viviparum 264
Polypodiaceae 27
Polypodium 27
    virginianum 27
    vulgare 27
Polypody 27
Polypody Fern Family 27
Polystichum 12
    braunii 12
    lonchitis 12
Pondweed 511
Pondweed Family 511
Pontederia cordata 510
Pontederiaceae 510
poor sedge 381
poor-man's pepper 140
Poplar 313
Poppy 248
Poppy Family 246

Populus 313
    alba 314
    balsamifera 314
    deltoides 314
    grandidentata 315
    nigra 315
    tremuloides 315
porcupine grass 487
porcupine sedge 403
Portulaca oleracea 273
Portulacaceae 273
Potamogeton 511
    alpinus 512
    amplifolius 512
    confervoides 512
    crispus 512
    epihydrus 513
    filiformis 517
    foliosus 513
    friesii 513
    gramineus 513
    hillii 514
    illinoensis 514
    natans 514
    nodosus 514
    oakesianus 515
    obtusifolius 515
    pectinatus 517
    perfoliatus 515
    praelongus 515
    pusillus 515
    richardsonii 516
    robbinsii 516
    spirillus 516
    strictifolius 516
    vaseyi 516
    zosteriformis 516
Potamogetonaceae 511
Potato Family 328
Potentilla 297
    anserina 298
    argentea 298
    arguta 293
    bipinnatifida 298
    canescens 299
    flabelliformis 298
    floribunda 298
    fruticosa 298
    gracilis 298
    inclinata 299
    norvegica 299
    palustris 291
    recta 299
    simplex 299
    tridentata 306
Poterium sanguisorba 306
poverty rush 432
poverty wild oatgrass 474
poverty-grass 508
prairie alumroot 325
prairie coneflower 79

prairie flat-topped goldenrod 104
prairie heart-leaved aster 111
prairie junegrass 489
prairie pepperwort 139
prairie redroot 286
prairie sedge 373
prairie smoke 296
prairie wedgescale 507
prairie-dropseed 507
prairie-meadow sage 220
Prenanthes 97
    alba 98
    racemosa 98
pretty sedge 392
prickly ash 313
prickly lettuce 92
Prickly Pear 143
prickly Russian-thistle 44
prickly saxifrage 326
Primrose 275
Primrose Family 273
primrose-leaf violet 341
Primrose-Willow 237
Primula 275
    meadia 275
    mistassinica 275
    veris 275
Primulaceae 273
prince's pine 176
princess-pine 21
Proserpinaca 206
Proserpinaca palustris 206
Prunella vulgaris 219
Prunus 299
    americana 300
    cerasus 299
    domestica 299
    nigra 300
    pensylvanica 300
    pumila 301
    serotina 301
    virginiana 301
Pseudognaphalium 98
    helleri 98
    macounii 98
    micradenium 98
    obtusifolium 98
Ptelea trifoliata 313
Pteridaceae 28
Pteridium aquilinum 9
Pterospora andromedea 179
pubescent sedge 373
Puccinellia 502
    distans 502
    pallida 508
Puccoon 123
purple avens 296
purple bladderwort 224
purple chokeberry 291
purple clematis 279
purple crown-vetch 194

purple false foxglove 241
purple false oat 487
purple loosestrife 227
purple lovegrass 483
purple meadow-rue 284
purple milkwort 263
purple-head sneezeweed 85
purple-leaf willowherb 236
purple-stem angelica 48
purple-stem aster 111
purple-stem cliffbrake 28
Purslane 273
Purslane Family 273
purslane speedwell 259
pussy-willow 317
Pussytoes 66
putty-root 445
Pycnanthemum 219
    tenuifolium 219
    virginianum 219
Pycreus rivularis 408
Pyrola 180
    americana 180
    asarifolia 180
    chlorantha 180
    elliptica 180
    minor 180
    rotundifolia 180
    secunda 179
    uniflora 178
Pyrus americana 307
Pyrus aucuparia 307
Pyrus decora 307
Pyrus malus 296
Pyrus pumila 296

Q

quack-grass 481
quaking aspen 315
queen anne's-lace 49
queen-of-the-prairie 293
Quercus 197
    alba 197
    ellipsoidalis 197
    macrocarpa 197
    rubra 198
    × schuettei 197
Quickweed 83
quill sedge 387
quill spike-rush 412
Quillwort 16
Quillwort Family 16

R

rabbit-foot clover 194
Rabbit-Tobacco 98
racemed milkwort 263
Radish 140
Ragweed 65
Ragwort 100
ram's-head lady's-slipper 448

Ranunculaceae 276
Ranunculus 280
    abortivus 281
    acris 281
    aquatilis 281
    cymbalaria 282
    fascicularis 282
    flabellaris 282
    flammula 282
    gmelinii 282
    hispidus 282
    lapponicus 283
    longirostris 281
    macounii 281
    pensylvanicus 283
    recurvatus 283
    repens 283
    reptans 282
    rhomboideus 283
    sceleratus 283
    trichophyllus 281
Raphanus 140
    raphanistrum 140
    sativus 140
Raspberry 303
Ratibida pinnata 99
rattlesnake fern 24
rattlesnake manna grass 486
Rattlesnake-Plantain 449
Rattlesnake-Root 97
rayless alpine groundsel 96
rayless mountain groundsel 96
red anemone 278
red baneberry 277
red bearberry 176
red clover 195
red columbine 278
red fescue 485
red maple 323
red osier-dogwood 169
red pine 35
red-berried elder 37
red-head pondweed 516
red-root amaranth 41
red-seed dandelion 114
red-tinge bulrush 420
reddish pondweed 512
redhead-grass 515
redstem-filaree 200
redtop 464
Reed 496
reed canary-grass 495
Reed-Grass 470
Rein-Orchid 452
retrorse sedge 404
Reynoutria japonica 265
Reynoutria sachalinensis 265
Rhamnaceae 286
Rhamnus 287
    alnifolia 287
    cathartica 287

    frangula 287
Rheum x hybridum 264
Rhexia virginica 229
Rhinanthus 245
    crista-galli 245
    minor 245
Rhododendron groenlandicum 180
rhombic copperleaf 184
Rhus 44
    × borealis 45
    aromatica 45
    glabra 45
    hirta 45
    radicans 45
    typhina 45
Rhynchospora 416
    alba 416
    capillacea 416
    capitellata 416
    fusca 416
    glomerata 416
rib-seed sandmat 186
ribbon-leaf pondweed 513
Ribes 202
    americanum 202
    aureum 204
    cynosbati 203
    glandulosum 203
    hirtellum 203
    hudsonianum 203
    lacustre 203
    nigrum 203
    odoratum 204
    oxyacanthoides 204
    rubrum 205
    sativum 205
    triste 205
rice cut-grass 489
Ricegrass 497
Richardson's sedge 367
rigid sedge 391
river club-rush 350
river-bank grape 344
river-bank wild rye 482
roadside sandspurry 161
Robbins' spike-rush 413
robin's plantain 80
Robinia 193
    hispida 193
    pseudoacacia 193
rock clubmoss 18
rock cress 131
rock elm 333
rock polypody 27
rock stitchwort 157
rock whitlow-grass 136
rock-harlequin 247
Rock-Rose Family 166
Rockbrake 28
Rockcress 130, 131
Rocky Mountain fescue 485

580 INDEX

Roman wormwood 70
rope-root sedge 364
Rorippa 140
  aquatica 141
  nasturtium-aquaticum 140
  palustris 141
  sylvestris 141
Rosa 301
  acicularis 301
  arkansana 302
  blanda 302
  cinnamomea 302
  multiflora 301
  palustris 303
  rubiginosa 301
  rugosa 303
  sayi 301
  spinosissima 301
Rosaceae 287
Rose 301
rose campion 159
Rose Family 287
rose pogonia 454
Rosinweed 100
rosy sedge 396
rough avens 296
rough bedstraw 309
rough bent 464
rough bristle grass 505
rough false pennyroyal 214
rough fleabane 81
rough forget-me-not 125
rough hawkweed 90
rough-leaved goldenrod 104
rough-stalk bluegrass 501
round-head bush-clover 191
round-leaf dogwood 170
round-leaf monkey-flower 249
round-leaf orchid 445, 454
round-leaf sundew 173
round-leaf thimbleweed 277
round-lobe hepatica 280
rowan 307
royal fern 26, 27
Royal Fern Family 26
Rubiaceae 309
Rubus 303
  acaulis 304
  allegheniensis 303
  arcticus 304
  baileyanus 304
  canadensis 304
  flagellaris 304
  hispidus 304
  idaeus 304
  occidentalis 304
  odoratus 305
  parviflorus 305
  pensilvanicus 305
  pubescens 305
  setosus 305

strigosus 304
Rudbeckia 99
  hirta 99
  laciniata 99
  triloba 100
Rue Family 313
rufous bulrush 420
rugosa rose 303
Rumex 270
  acetosa 271
  acetosella 271
  altissimus 271
  brittannica 271
  crispus 271
  fueginus 271
  longifolius 271
  obtusifolius 272
  occidentalis 270
  orbiculatus 271
  salicifolius 272
  thyrsiflorus 270
  triangulivalvis 272
  verticillatus 273
running ground-pine 20
running serviceberry 290
Ruppia cirrhosa 517
Rush 427
Rush Family 427
russet buffalo-berry 174
Russian-Olive 174
Russian-Thistle 44
rusty cliff fern 32
rusty flat sedge 408
rusty Labrador-tea 180
Ruta graveolens 313
Rutaceae 313
rye 504
rye brome 470
Rye Grass 489

S

Sage 69, 220
sage willow 317
Sagina 157
  nodosa 158
  procumbens 158
Sagittaria 345
  cuneata 345
  graminea 345
  latifolia 346
  rigida 346
sago false pondweed 517
salad-burnet 306
Salicaceae 313
Salix 315
  alba 316
  amygdaloides 316
  bebbiana 317
  candida 317
  cordata 317
  cordata 317

discolor 317
eriocephala 317
exigua 318
× fragilis 317
glaucophylloides 318
humilis 318
interior 318
lucida 318
myricoides 318
nigra 318
pedicellaris 320
pellita 320
petiolaris 320
purpurea 320
pyrifolia 320
rigida 317
serissima 320
salsify 115
Salsola 44
  kali 44
  tragus 44
Salvia pratensis 220
Sambucus 37
  canadensis 37
  nigra 37
  pubens 37
  racemosa 37
sand flat sedge 409
sand golden-heather 166
sand plantain 257
sand violet 338
sand-cherry 301
sand-dropseed 507
sand-reed 472
Sandalwood Family 321
sandbar lovegrass 483
sandbar willow 318
Sandbur 472
Sandspurry 161
Sandwort 153
Sanguinaria canadensis 248
Sanguisorba minor 306
Sanicula 51
  gregaria 51
  marilandica 51
  odorata 51
  trifoliata 52
Santalaceae 321
Sapindaceae 322
Saponaria officinalis 158
Sarracenia purpurea 324
Sarraceniaceae 324
Sarsaparilla 57
Sartwell's sedge 374
satiny willow 320
Satureja acinos 213
Satureja arkansana 213
Satureja glabella 213
Satureja hortensis 211
Satureja vulgaris 213
savanna pinweed 167

INDEX 581

savin-leaf creeping-cedar 17
Saw-Grass 406
Saxifraga 326
    paniculata 326
    pensylvanica 325
    tricuspidata 326
    virginiensis 326
Saxifragaceae 325
Saxifrage 325, 326
Saxifrage Family 325
saxifrage-pink 157
scarlet beebalm 217
scarlet Indian-paintbrush 242
scentless chamomile 95
Sceptridium 24
    dissectum 26
    multifidum 26
    oneidense 26
    rugulosum 26
Schedonorus 503
    arundinaceus 503
    pratensis 503
Scheuchzeria Family 518
Scheuchzeria palustris 518
Scheuchzeriaceae 518
Schizachne purpurascens 503
Schizachyrium scoparium 504
Schoenoplectiella purshiana 418
Schoenoplectiella smithii 418
Schoenoplectus 416
    acutus 417
    fluviatilis 350
    pungens 418
    purshianus 418
    smithii 418
    subterminalis 418
    tabernaemontani 418
    torreyi 419
Schweinitz's sedge 404
Scirpus 419
    acutus 417
    americanus 418
    atrovirens 420
    caespitosus 421
    clintonii 421
    cyperinus 420
    fluviatilis 350
    hattorius 420
    hudsonianus 421
    microcarpus 420
    pauciflorus 413
    pendulus 420
    pungens 418
    quinqueflorus 413
    rostellatus 414
    smithii 418
    subterminalis 418
    tabernaemontani 418
    torreyi 419
    validus 418
Scleranthus 158

annuus 158
perennis 159
Scorpion Grass 124
Scorzoneroides autumnalis 59
Scotch pine 36
Scouring-Rush 12
Screwstem 198
Scrophulariaceae 327
Scrophularia lanceolata 327
Scutellaria 220
    x churchilliana 220
    epilobiifolia 220
    galericulata 220
    lateriflora 220
    parvula 220
sea-rocket 132
seaside arrow-grass 434
seaside sandmat 186
Secale cereale 504
Securigera varia 194
Sedge 351
Sedge Family 349
Sedum 170
    acre 171
    purpureum 171
    telephium 171
Selaginella 29
    apoda 29
    eclipes 29
    rupestris 29
    selaginoides 30
Selaginellaceae 29
Selaginella Family 29
self-heal 219
Seneca-snakeroot 263
Senecio 100
    aureus 95
    hieraciifolius 80
    indecorus 96
    pauperculus 96
    sylvaticus 100
    vulgaris 100
sensitive fern 21
Sensitive Fern Family 21
Serviceberry 289
sessile-fruit arrowhead 346
sessile-leaf bellwort 443
Setaria 504
    faberi 504
    glauca 505
    italica 504
    pumila 505
    verticillata 505
    viridis 505
seven-angle pipewort 422
shagbark hickory 210
shaggy-soldier 83
sharp-lobe hepatica 279
shaved sedge 358
sheathed sedge 391
sheep fescue 486

sheep-laurel 178
shepherd's-purse 133
Shepherdia 174
    argentea 174
    canadensis 174
shining fir-moss 18
shining flat sedge 408
shining ladies'-tresses 456
shining willow 318
Shinleaf 180
Shoreweed 254
short-awn foxtail 464
short-awn mountain ricegrass 497
short-beak sedge 385
Shorthusk 468
showy lady's-slipper 448
showy goldenrod 105
showy milkweed 55
showy orchid 449
showy tick-trefoil 189
shrubby cinquefoil 298
shrubby-fivefingers 306
shy wallflower 138
Sibbaldia tridentata 306
Sibbaldiopsis tridenta 306
Siberian elm 333
side-oats grama 468
Sidebells 179
Silene 159
    antirrhina 159
    armeria 159
    chalcedonica 159
    coronaria 159
    csereii 160
    dichotoma 160
    latifolia 160
    nivea 160
    noctiflora 160
    vulgaris 160
silky dogwood 169
Silphium 100
    laciniatum 101
    rumicifolium 101
    terebinthinaceum 101
silver buffalo-berry 174
silver maple 323
Silver-Berry 174
silverweed 298
silvery cinquefoil 298
Silvery Glade Fern 6
silvery spleenwort 6
Sinapis 141
    alba 142
    arvensis 142
single-delight 178
Sisymbrium 142
    altissimum 142
    officinale 142
Sisyrinchium 426
    angustifolium 426
    montanum 426

mucronatum 427
strictum 427
Sium suave 52
Skullcap 220
skunk currant 203
skunk-cabbage 348
sleepy catchfly 159
slender beardtongue 256
slender cinquefoil 298
slender cotton-grass 414
slender cottonweed 44
slender fimbry 415
slender goldentop 82
slender pondweed 515
slender sedge 390
slender water-milfoil 206
slender waternymph 424
slender wedgescale 506
slender wild rye 482
slender-leaf sundew 173
slim-stem reed-grass 472
slippery elm 333
Slough Grass 467
slough sedge 362
small beggarticks 72
small cranberry 183
small dropseed 508
small pussytoes 66
small waterwort 174
small white fawn-lily 438
small-flower blue-eyed mary 253
small-flower crane's-bill 201
small-flower evening-primrose 239
small-flower false foxglove 241
small-flowered bittercress 134
small-flowered grass-of-parnassus 164
small-head rush 428, 429
small-spike false nettle 334
smaller forget-me-not 125
Smartweed 265, 269
Smilacaceae 519
Smilacina racemosa 439
Smilacina stellata 439
Smilacina trifolia 440
Smilax 519
   ecirrata 519
   herbacea 520
   hispida 519
   illinoensis 519
   lasioneura 520
   tamnoides 519
Smith's club-rush 418
Smith's melic grass 490
smooth blackberry 304
smooth blue aster 109
smooth brome 469
smooth cliffbrake 28
smooth crabgrass 478
smooth goldenrod 103
smooth hedge-nettle 221
smooth rose 302

smooth saw-grass 406
smooth scouring-rush 14
smooth serviceberry 290
smooth sumac 45
smooth whitlow-grass 137
Smotherweed 41
Snake-Mouth 454
Snakeroot 65
sneezeweed 64, 84
snow-on-the-mountain 186
snowberry 151
snowy catchfly 160
Soapberry Family 322
Soapwort 158
soft-leaf sedge 367
soft-stem club-rush 418
Solanaceae 328
Solanum 330
   carolinense 331
   dulcamara 331
   nigrum 331
   physalifolium 331
   ptychanthum 331
   rostratum 331
Solidago 101
   altissima 102
   caesia 102
   canadensis 103
   flexicaulis 103
   gigantea 103
   graminifolia 82
   hispida 103
   houghtonii 103
   juncea 104
   nemoralis 104
   ohioensis 104
   patula 104
   ptarmicoides 104
   rigida 104
   rugosa 105
   serotina 103
   simplex 105
   spathulata 105
   speciosa 105
   uliginosa 105
Solomon's-Seal 441
Sonchus 105
   arvensis 107
   asper 107
   oleraceus 107
Sorbaria sorbifolia 307
Sorbus 307
   americana 307
   aucuparia 307
   decora 307
Sorghastrum nutans 506
Sorrel 270
southern blueflag 426
southern waternymph 424
Sow-Thistle 105
Sparganium 520

   acaule 521
   americanum 520
   angustifolium 521
   californicum 521
   chlorocarpum 521
   emersum 520
   emersum 521
   eurycarpum 521
   fluctuans 522
   greenei 521
   minimum 522
   multipedunculatum 520
   natans 522
   simplex 521
sparse-flower sedge 369
Spartina pectinata 506
spearmint 217
Spearscale 41
Spearwort 280
speckled alder 117
Speedwell 257
Spergula arvensis 161
Spergularia rubra 161
Sphenopholis 506
   intermedia 506
   obtusata 506, 507
Spiderwort Family 349
Spike-Rush 409
spiked lobelia 147
Spikemoss 29
spikenard 57
spineless hornwort 165
spinulose wood-fern 10
Spinulum annotinum 19
Spinulum canadense 20
spiny plumeless-thistle 73
spiny-leaf sow-thistle 107
spiny-spored quillwort 16
Spiraea 307
   alba 308
   japonica 307
   salicifolia 308
   tomentosa 308
   × vanhouttei 308
spiral ditch-grass 516
spiral pondweed 516
Spiranthes 454
   casei 455
   cernua 456
   lacera 456
   lucida 456
   romanzoffiana 457
Spirodela polyrrhiza 348
Spleenwort 5
Spleenwort Family 5
spoon-leaf moonwort 24
spoon-leaf sundew 173
Sporobolus 507
   compositus 507
   cryptandrus 507
   heterolepis 507

neglectus 508
vaginiflorus 508
spotted coral-root 446
spotted crane's-bill 201
spotted dead nettle 215
spotted joe-pye-weed 82
spotted knapweed 75
spotted sandmat 186
spotted St. John's-wort 209
spreading bent 464
spreading dogbane 53
spreading rockcress 131
spreading wood fern 11
spring forget-me-not 126
spring speedwell 260
spring whitlow-grass 137
Springbeauty 230
Spruce 34
Spurge 184
Spurge Family 183
spurred gentian 200
Spurry 161
squashberry 37, 38
squaw-bush 45
Squawroot 242
squirrel-corn 247
St. John's-Wort 208
St. John's-Wort Family 207
St. Lawrence bladder fern 8
Stachys 220
    hyssopifolia 221
    pilosa 221
    tenuifolia 221
staghorn sumac 45
stalk-grain sedge 406
star sedge 401
Star-Grass 425
Star-of-Bethlehem 440
Star-Thistle 74
starflower 275
starry false Solomon's-seal 439
starved panicgrass 476
Stellaria 162
    aquatica 162
    borealis 162
    crassifolia 162
    graminea 162
    longifolia 162
    longipes 162
    media 163
Stickseed 122, 123
sticky false asphodel 442
sticky-willy 309
stiff goldenrod 104
stiff ground-pine 19
stiff marsh bedstraw 311
stiff sunflower 87
stiff-leaved panicgrass 477
stinging nettle 335
stink-grass 483
stinking chamomile 67

Stipa spartea 487
Stitchwort 156
Stonecrop 170
Stonecrop Family 170
Stoneseed 123
Stork's Bill 200
straight-leaf pondweed 516
strap-leaf violet 339
straw-color flat sedge 409
Strawberry 294
strawberry-blite 42
strawberry-weed 299
streamside wild rye 481
Streptopus 441
    amplexifolius 441
    lanceolatus 442
    roseus 442
strict blue-eyed-grass 426
striped coral-root 447
striped maple 323
Stuckenia 517
    filiformis 517
    pectinata 518
Subularia aquatica 142
sugar maple 323
sulphur cinquefoil 299
Sumac 44
Sumac Family 44
sun spurge 186
Sundew 172
Sundew Family 172
sundrops 239
Sunflower 85
sunflower-everlasting 88
swamp dock 273
swamp fly-honeysuckle 150
swamp loosestrife 275
swamp lousewort 245
swamp milkweed 54
swamp red currant 205
swamp rose 303
swamp saxifrage 325
swamp smartweed 266
swamp thistle 76
swamp-loosestrife 226
swampcandles 274
Swan's sedge 399
swaying club-rush 418
sweet American wintergreen 180
sweet blue violet   340
Sweet Colt's-Foot 96
sweet gale 231
Sweet Vernal Grass 466
sweet white violet 339
sweet wood-reed 473
Sweet-Cicely 50
Sweet-Clover 192
sweet-fern 231
sweet-scented bedstraw 311
Sweetflag 344
sweetgrass 487

sweetwilliam 155
Swida alternifolia 169
switchgrass 494
swollen beaked sedge 404
Symphoricarpos 151
    albus 151
    occidentalis 151
    orbiculatus 151
Symphyotrichum 107
    boreale 108
    ciliatum 108
    dumosum 109
    ericoides 109
    laeve 109
    lanceolatum 110
    lateriflorum 110
    novae-angliae 110
    ontarionis 111
    oolentangiense 111
    pilosum 111
    puniceum 111
    robynsianum 112
    sericeum 112
    urophyllum 112
Symphytum officinale 126
Symplocarpus foetidus 348
Syringa 234
    reticulata 235
    vulgaris 234

T

Taenidia integerrima 52
tag alder 117
tall annual willowherb 236
tall baby's-breath 156
tall beggarticks 72
tall bilberry 182
tall blue lettuce 92
tall dropseed 507
Tall Fescue 503
tall flat-topped white aster 79
tall gayfeather 94
tall goldenrod 102
tall hairy agrimony 289
tall hawkweed 90
tall lettuce 92
tall oatgrass 466
tall rye grass 503
tall thimbleweed 278
tall white violet 339
tall woodbeauty 293
tamarack 34
Tanacetum 112
    balsamita 113
    bipinnatum 113
    coccineum 112
    huronense 113
    parthenium 113
    vulgare 113
Tansy 112
tansy-mustard 135, 136

Tape-Grass Family 422
Taraxacum 114
   erythrospermum 114
   laevigatum 114
   officinale 114
   palustre 114
Tartarian honeysuckle 150
tawny cotton-grass 415
Taxaceae 36
Taxus canadensis 36
Teaberry 177
teal lovegrass 483
Tearthumb 269
Teasel 172
Teasel Family 171
ternate grape fern 26
Tetraneuris herbacea 59
Teucrium canadense 221
Thalecress 129
Thalictrum 284
   dasycarpum 284
   dioicum 284
   revolutum 284
   venulosum 284
Thelypteridaceae 30
Thelypteris 31
   hexagonoptera 30
   noveboracensis 30
   palustris 31
   phegopteris 30
thick-leaved wild strawberry 294
thick-spike blazing star 94
thicket-creeper 343
thimbleberry 305
Thimbleweed 277
thin-leaf sedge 396
thin-leaved wild strawberry 294
thin-scale cotton-grass 414
Thistle 73, 76
Thlaspi arvense 142
thread rush 430
threadleaf false pondweed 517
Three-Awn 466
three-leaf false Solomon's-seal 439
three-leaf goldthrad 279
three-leaf toothwort 134
three-lobe beggarticks 72
three-lobed coneflower 100
three-seed sedge 370
three-way sedge 409
throatwort 145
Thuja occidentalis 33
thyme-leaf sandwort 153
thyme-leaf speedwell 260
thyme-leaved spurge 187
Thymelaeaceae 331
Thymus pulegioides 211
Tiarella cordifolia 326
Tick-Clover 189
Tick-Trefoil 189
Tickseed 78

Tilia americana 228
Timothy 496
tinged sedge 388
toad rush 429
Toadflax 254
Tofieldia glutinosa 442
Toothwort 134
Torreochloa fernaldii 508
Torrey's club-rush 419
Torrey's rush 432
Torreyochloa pallida 508
Touch-Me-Not 115
Touch-Me-Not Family 115
tower-mustard 143
Toxicodendron rydbergii 45
Tragopogon 114
   dubius 114
   major 114
   porrifolius 115
   pratensis 115
trailing arbutus 177
Trailplant 64
tree ground-pine 20
Trefoil 192
Triadenum fraseri 210
triangle moonwort 23
Triantha glutinosa 442
Trichophorum 421
   alpinum 421
   caespitosum 421
   clintonii 421
Trichostema brachiatum 221
Trientalis borealis 275
Trifolium 194
   arvense 194
   aureum 194
   campestre 194
   hybridum 194
   incarnatum 194
   pratense 195
   repens 195
Triglochin 434
   maritima 434
   palustris 434
Trillium 442
   cernuum 442
   grandiflorum 442
Triodanis perfoliata 147
Triosteum aurantiacum 152
Tripleurospermum inodorum 95
Tripleurospermum perforata 95
Trisetum 509
   melicoides 487
   spicatum 509
Triticum aestivum 509
trout-lily 437, 438
true forget-me-not 125
Tsuga canadensis 36
Tuckerman's pondweed 512
Tuckerman's sedge 404
tufted hairgrass 475

tufted leafless-bulrush 421
tumbleweed 40
tumbling mustard 142
turion duckweed 348
turkey-foot 465
turnip 132
Turritis glabra 143
Turtlehead 253
tussock cotton-grass 415
tussock sedge 394
Twayblade 451
twinflower 148
twining screwstem 198
Twisted Stalk 441
two-leaf mitrewort 326
two-leaf water-milfoil 206
two-row stickseed 123
Typha 522
   angustifolia 522
   latifolia 522
Typhaceae 520

U
Ulmaceae 332
Ulmus 332
   americana 333
   glabra 333
   pumila 333
   rubra 333
   thomasii 333
upland bent 464
upland brittle bladder fern 8
upland willow 318
upright carrion-flower 519
Urtica dioica 335
Urticaceae 334
Utricularia 223
   cornuta 223
   geminiscapa 223
   gibba 223
   intermedia 224
   macrorhiza 224
   minor 224
   purpurea 224
   resupinata 224
   vulgaris 224
Uvularia 443
   grandiflora 443
   sessilifolia 443

V
Vaccaria hispanica 163
Vaccinium 181
   angustifolium 182
   caespitosum 182
   macrocarpon 182
   membranaceum 182
   myrtilloides 182
   ovalifolium 182
   oxycoccos 183
   pallidum 183

uliginosum 181
vitis-idaea 183
Valerian 152
Valeriana 152
   officinalis 152
   sitchensis 152
   uliginosa 152
Vallisneria americana 424
variegated scouring-rush 15
Vasey's pondweed 516
Vasey's rush 433
veiny vetchling 191
veiny-leaf meadow-rue 284
velvet-leaf blueberry 182
velvetleaf 227
Venus'-Looking-Glass 147
Verbascum 327
   blattaria 328
   thapsus 328
   lychnitis 328
Verbena 335
   bracteata 335
   hastata 336
   simplex 336
   stricta 336
   urticifolia 337
Verbenaceae 335
Verbena Family 335
vernal water-starwort 252
Veronica 257
   americana 258
   anagallis-aquatica 258
   arvensis 259
   beccabunga 258, 259
   chamaedrys 259
   filiformis 257
   longifolia 259
   officinalis 259
   peregrina 259
   persica 260
   scutellata 260
   serpyllifolia 260
   verna 260
Veronicastrum virginicum 260
Vervain 335
Vesiculina purpurea 224
Vetch 195
Vetchling 190
Viburnum 37
   acerifolium 38
   affine 39
   cassinoides 38
   edule 38
   lantana 38
   lentago 38
   nudum 38
   opulus 38
   pauciflorum 38
   rafinesquianum 39
   recognitum 37
   trilobum 38

Vicia 195
   americana 195
   cracca 195
   sativa 195
   sepium 195
   tetrasperma 196
   villosa 196
Vinca minor 55
Viola 337
   adunca 338, 339
   affinis 338
   arvensis 339
   blanda 339
   canadensis 339
   cucullata 339
   labradorica 339
   lanceolata 339
   macloskeyi 339
   nephrophylla 339
   novae-angliae 340
   odorata 340
   pallens 339
   palmata 337, 341
   pedata 341
   pedatifida 341
   primulifolia 341
   pubescens 341
   renifolia 341
   rostrata 341
   sagittata 341
   selkirkii 342
   septentrionalis 342
   sororia 340, 342
   tricolor 343
Violaceae 337
Violet 337
Violet Family 337
Viper's-Bugloss 121
virgin's bower 279
Virginia bluebells 124
Virginia chain fern 7
Virginia ground-cherry 330
Virginia mountain-mint 219
Virginia springbeauty 231
Virginia water-horehound 216
Virginia wild rye 482
Virginia-creeper 343
Vitaceae 343
Vitis 343
   labrusca 343
   riparia 344
   vulpina 344

W

Wake-Robin 442
Waldsteinia fragarioides 295
walking fern 5
Wallflower 138
Wallrocket 136
Walnut 210
Walnut Family 210

water lobelia 146
water sedge 393
water smartweed 266
water-arum 347
Water-Awlwort 142
Water-Hemlock 48
Water-Horehound 216
water-horsetail 14
Water-Lily 233
Water-Lily Family 232
Water-Milfoil 205
Water-Milfoil Family 205
Water-Parsnip 52
Water-Plantain 345
Water-Plantain Family 344
Water-Starwort 252
Water-Willow 226
waterberry 150
watercress 140
Watermeal 349
Waternymph 423
watershield 143
Watershield Family 143
Waterweed 423
Waterwort 174
Waterwort Family 174
wavy waternymph 424
waxy-leaf meadow-rue 284
wayfaring-tree 38
weak-stalk club-rush 418
Wedgescale 506
West Indian nightshade 331
western annual aster 108
western fescue 485
western monkey-flower 250
western poison-ivy 45
western rattlesnake-plantain 450
western silvery aster 112
western snowberry 151
western wallflower 138
western waterweed 423
western wheatgrass 495
Wheat 509
whip-poor-will-flower 442
whiplash dewberry 304
white adder's-mouth orchid 451
white ash 234
white avens 295
white baneberry 277
white beak sedge 416
white bear sedge 378
white birch 118
white bluegrass 500
white bog-orchid 453
white campion 160
white clover 195
white grass 489
white heath aster 109
white knapweed 74
white moth mullein 328
white mulberry 231

white nodding ladies'-tresses 456
white oak 197
white oldfield aster 111
white poplar 314
white rattlesnake-root 98
white sage 70
white snakeroot 65
white spruce 35
white turtlehead 253
white vervain 337
white walnut 210
white water-lily 233
white willow 316
white-edge sedge 375
white-grain mountain ricegrass 492
white-mustard 141, 142
white-stem evening-primrose 239
white-stem pondweed 515
Whitlow-Grass 136
whorled milkweed 55
whorled water-milfoil 206
whorled yellow-loosestrife 274
Wide-Lip Orchid 450
Wiegand's wild rye 482
Wild Aster 107
wild basil 213
wild bergamot 217
wild black cherry 301
wild black currant 202
wild chamomile 95
wild chervil 48
wild chives 436
wild comfrey 121
wild cucumber 171
Wild Geranium 201
Wild Ginger 58
wild leek 436
wild oat 467
Wild Oatgrass 474
wild parsnip 51
Wild Pea 190
wild plum 300
wild quinine 96
wild radish 140
wild raisin 38
wild red raspberry 304
Wild Rice 509
Wild Rye 480
wild sarsaparilla 57
wild white violet 339
wild yam 422
Willow 315
Willow Family 313

willow-leaf dock 272
Willowherb 236
wing-stem meadow-pitchers 229
winged loosestrife 226
winged-pigweed 43
winterberry 56
wintergreen 178, 180
wirestem muhly 491
wiry panic-grass 494
Witch-Hazel 207
Witch-Hazel Family 207
withe-rod 38
Wolffia columbiana 349
wood betony 244
Wood Fern Family 10
wood-anemone 278
Wood-Aster 81
Wood-Fern 10
wood-lily 439
Wood-Nettle 334
Wood-Reed 473
Wood-Rush 433
Wood-Sorrel 245
Wood-Sorrel Family 245
Woodbeauty 293
woodland agrimony 289
woodland bluegrass 500, 501
woodland cudweed 84
woodland goosefoot 43
woodland muhly 492
woodland pinedrops 179
woodland ragwort 100
woodland sunflower 86
woodland whitlow-grass 137
woodland-horsetail 15
Woodsia 31
 × abbeae 31
 alpina 32
 ilvensis 32
 obtusa 32
 oregana 33
Woodsiaceae 31
Woodwardia virginica 7
wool-grass 420
woolly plantain 257
woolly sedge 390
worm-seed sandmat 187
worm-seed wallflower 138
wormseed 44
Wormwood 69
wreath goldenrod 102
wrinkle-leaved goldenrod 105

X
Xanthium strumarium 115
Xyridaceae 523
Xyris 523
 difformis 523
 montana 523

Y
Yam 422
Yam Family 422
yard knotweed 269
Yarrow 64
yellow avens 295
yellow birch 118
yellow bluebead-lily 437
yellow coralroot 447
yellow day-lily 438
yellow forest violet facu 341
yellow foxglove 253
yellow Indian grass 506
yellow king-devil 90
yellow lady's-slipper 448
yellow pond-lily 233
Yellow Rattle 245
yellow screwstem 198
yellow spike-rush 412
yellow spring bedstraw 311
yellow sweet-clover 193
Yellow Water-Lily 233
Yellow-Eyed-Grass 523
Yellow-Eyed-Grass Family 523
yellow-flower knotweed 269
yellow-green sedge 364
yellow-pimpernel 52
Yellow-Rocket 130
yellow-seed false pimpernel 226
Yellowcress 140
Yew 36
Yew Family 36

Z
Zannichellia palustris 518
Zanthoxylum americanum 313
Zigadenus elegans 436
zigzag goldenrod 103
Zizania palustris 509
Zizia 52
 aptera 53
 aurea 53
Zosterella dubia 510

www.ingramcontent.com/pod-product-compliance
Lightning Source LLC
Chambersburg PA
CBHW081152020426
42333CB00020B/2480